T0190364

Lecture Notes in Computer Science   10211

Commenced Publication in 1973
Founding and Former Series Editors:
Gerhard Goos, Juris Hartmanis, and Jan van Leeuwen

More information about this series at http://www.springer.com/series/7410

Jean-Sébastien Coron · Jesper Buus Nielsen (Eds.)

Advances in Cryptology – EUROCRYPT 2017

36th Annual International Conference on the Theory
and Applications of Cryptographic Techniques
Paris, France, April 30 – May 4, 2017
Proceedings, Part II

 Springer

Editors
Jean-Sébastien Coron
University of Luxembourg
Luxembourg
Luxembourg

Jesper Buus Nielsen
Aarhus University
Aarhus
Denmark

ISSN 0302-9743   ISSN 1611-3349   (electronic)
Lecture Notes in Computer Science
ISBN 978-3-319-56613-9   ISBN 978-3-319-56614-6   (eBook)
DOI 10.1007/978-3-319-56614-6

Library of Congress Control Number: 2017936355

LNCS Sublibrary: SL4 – Security and Cryptology

Printed on acid-free paper

This Springer imprint is published by Springer Nature
The registered company is Springer International Publishing AG
The registered company address is: Gewerbestrasse 11, 6330 Cham, Switzerland

Preface

Eurocrypt 2017, the 36th annual International Conference on the Theory and Applications of Cryptographic Techniques, was held in Paris, France, from April 30 to May 4, 2017. The conference was sponsored by the International Association for Cryptologic Research (IACR). Michel Abdalla (ENS, France) was responsible for the local organization. He was supported by a local organizing team consisting of David Pointcheval (ENS, France), Emmanuel Prouff (Morpho, France), Fabrice Benhamouda (ENS, France), Pierre-Alain Dupoint (ENS, France), and Tancrède Lepoint (SRI International). We are indebted to them for their support and smooth collaboration.

The conference program followed the now established parallel track system where the works of the authors were presented in two concurrently running tracks. Only the invited talks spanned over both tracks.

We received a total of 264 submissions. Each submission was anonymized for the reviewing process and was assigned to at least three of the 56 Program Committee members. Submissions co-authored by committee members were assigned to at least four members. Committee members were allowed to submit at most one paper, or two if both were co-authored. The reviewing process included a first-round notification followed by a rebuttal for papers that made it to the second round. After extensive deliberations the Program Committee accepted 67 papers. The revised versions of these papers are included in these three-volume proceedings, organized topically within their respective track.

The committee decided to give the Best Paper Award to the paper "Scrypt Is Maximally Memory-Hard" by Joël Alwen, Binyi Chen, Krzysztof Pietrzak, Leonid Reyzin, and Stefano Tessaro. The two runners-up to the award, "Computation of a 768-bit Prime Field Discrete Logarithm," by Thorsten Kleinjung, Claus Diem, Arjen K. Lenstra, Christine Priplata, and Colin Stahlke, and "Short Stickelberger Class Relations and Application to Ideal-SVP," by Ronald Cramer, Léo Ducas, and Benjamin Wesolowski, received honorable mentions. All three papers received invitations for the *Journal of Cryptology*.

The program also included invited talks by Gilles Barthe, titled "Automated Proof for Cryptography," and by Nigel Smart, titled "Living Between the Ideal and Real Worlds."

We would like to thank all the authors who submitted papers. We know that the Program Committee's decisions, especially rejections of very good papers that did not find a slot in the sparse number of accepted papers, can be very disappointing. We sincerely hope that your works eventually get the attention they deserve.

We are also indebted to the Program Committee members and all external reviewers for their voluntary work, especially since the newly established and unified page limits and the increasing number of submissions induce quite a workload. It has been an honor to work with everyone. The committee's work was tremendously simplified by Shai Halevi's submission software and his support, including running the service on IACR servers.

Finally, we thank everyone else —speakers, session chairs, and rump session chairs — for their contribution to the program of Eurocrypt 2017. We would also like to thank Thales, NXP, Huawei, Microsoft Research, Rambus, ANSSI, IBM, Orange, Safran, Oberthur Technologies, CryptoExperts, and CEA Tech for their generous support.

May 2017   Jean-Sébastien Coron
                                                             Jesper Buus Nielsen

Eurocrypt 2017

The 36th Annual International Conference on the Theory and Applications of Cryptographic Techniques

Sponsored by *the International Association for Cryptologic Research*

30 April – 4 May 2017
Paris, France

General Chair

Michel Abdalla   ENS, France

Program Co-chairs

Jean-Sébastien Coron   University of Luxembourg
Jesper Buus Nielsen   Aarhus University, Denmark

Program Committee

Gilad Asharov   Cornell Tech, USA
Nuttapong Attrapadung   AIST, Japan
Fabrice Benhamouda   ENS, France and IBM, USA
Nir Bitansky   MIT, USA
Andrey Bogdanov   Technical University of Denmark
Alexandra Boldyreva   Georgia Institute of Technology, USA
Chris Brzuska   Technische Universität Hamburg, Germany
Melissa Chase   Microsoft, USA
Itai Dinur   Ben-Gurion University, Israel
Léo Ducas   CWI, Amsterdam, The Netherlands
Stefan Dziembowski   University of Warsaw, Poland
Nicolas Gama   Inpher, Switzerland and University of Versailles, France
Pierrick Gaudry   CNRS, France
Peter Gaži   IST Austria, Austria
Niv Gilboa   Ben-Gurion University, Israel
Robert Granger   EPFL, Switzerland
Nathan Keller   Bar Ilan University, Israel
Aggelos Kiayias   University of Edinburgh, UK
Eike Kiltz   Ruhr-Universität Bochum, Germany

Foteini Baldimtsi
Marshall Ball
Valentina Banciu
Subhadeep Banik
Razvan Barbulescu
Guy Barwell
Carsten Baum
Anja Becker
Christof Beierle
Amos Beimel
Sonia Belaïd
Shalev Ben-David
Iddo Bentov
Jean-François Biasse
Begul Bilgin
Olivier Blazy
Xavier Bonnetain
Joppe Bos
Christina Boura
Florian Bourse
Luis Brandao
Dan Brownstein
Chris Campbell
Ran Canetti
Anne Canteaut
Angelo De Caro
Ignacio Cascudo
David Cash
Wouter Castryck
Hubert Chan
Nishanth Chandran
Jie Chen
Yilei Chen
Nathan Chenette
Mahdi Cheraghchi
Alessandro Chiesa
Ilaria Chillotti
Sherman S.M. Chow
Kai-Min Chung
Michele Ciampi
Ran Cohen
Craig Costello
Alain Couvreur
Claude Crépeau
Edouard Cuvelier
Guillaume Dabosville

Ivan Damgård
Jean Paul Degabriele
Akshay Degwekar
David Derler
Apoorvaa Deshpande
Julien Devigne
Christoph Dobraunig
Frédéric Dupuis
Nico Döttling
Maria Eichlseder
Keita Emura
Xiong Fan
Pooya Farshim
Sebastian Faust
Omar Fawzi
Dario Fiore
Ben Fisch
Benjamin A. Fisch
Nils Fleischhacker
Georg Fuchsbauer
Eiichiro Fujisaki
Steven Galbraith
Chaya Ganesh
Juan Garay
Sumegha Garg
Romain Gay
Ran Gelles
Mariya Georgieva
Benedikt Gierlichs
Oliver W. Gnilke
Faruk Göloğlu
Sergey Gorbunov
Dov Gordon
Rishab Goyal
Hannes Gross
Vincent Grosso
Jens Groth
Daniel Gruss
Jian Guo
Siyao Guo
Qian Guo
Benoît Gérard
Felix Günther
Britta Hale
Carmit Hazay
Felix Heuer

Shoichi Hirose
Viet Tung Hoang
Justin Holmgren
Fumitaka Hoshino
Pavel Hubáček
Ilia Iliashenko
Laurent Imbert
Takanori Isobe
Tetsu Iwata
Malika Izabachene
Kimmo Jarvinen
Eliane Jaulmes
Dimitar Jetchev
Daniel Jost
Marc Joye
Herve Kalachi
Seny Kamara
Chethan Kamath
Angshuman Karmakar
Pierre Karpman
Nikolaos Karvelas
Marcel Keller
Elena Kirshanova
Fuyuki Kitagawa
Susumu Kiyoshima
Thorsten Kleinjung
Lars Knudsen
Konrad Kohbrok
Markulf Kohlweiss
Ilan Komargodski
Venkata Koppula
Thomas Korak
Lucas Kowalczyk
Thorsten Kranz
Fabien Laguillaumie
Kim Laine
Virginie Lallemand
Adeline Langlois
Hyung Tae Lee
Jooyoung Lee
Kwangsu Lee
Troy Lee
Kevin Lewi
Huijia (Rachel) Lin
Jiao Lin
Wei-Kai Lin

Feng-Hao Liu
Atul Luykx
Vadim Lyubashevsky
Xiongfeng Ma
Houssem Maghrebi
Mohammad Mahmoody
Daniel Malinowski
Alex Malozemoff
Antonio Marcedone
Daniel P. Martin
Daniel Masny
Takahiro Matsuda
Christian Matt
Alexander May
Sogol Mazaheri
Peihan Miao
Kazuhiko Minematsu
Ameer Mohammed
Tal Moran
Fabrice Mouhartem
Pratyay Mukherjee
Elke De Mulder
Pierrick Méaux
Michael Naehrig
Yusuke Naito
Kashif Nawaz
Kartik Nayak
Khoa Nguyen
Ryo Nishimaki
Olya Ohrimenko
Elisabeth Oswald
Ayoub Otmani
Giorgos Panagiotakos
Alain Passelègue
Kenneth G. Paterson
Serdar Pehlivanoglou
Alice Pellet–Mary
Pino Persiano
Cécile Pierrot
Rafaël Del Pino
Bertram Poettering
David Pointcheval
Antigoni Polychroniadou

Romain Poussier
Thomas Prest
Erick Purwanto
Carla Rafols
Ananth Raghunathan
Srinivasan Raghuraman
Sebastian Ramacher
Somindu Ramanna
Francesco Regazzoni
Ling Ren
Oscar Reparaz
Silas Richelson
Thomas Ricosset
Thomas Ristenpart
Florentin Rochet
Mike Rosulek
Yannis Rouselakis
Sujoy Sinha Roy
Michal Rybár
Carla Ràfols
Robert Schilling
Jacob Schuldt
Nicolas Sendrier
Yannick Seurin
Ido Shahaf
Sina Shiehian
Siang Meng Sim
Dave Singelee
Luisa Siniscalchi
Daniel Slamanig
Benjamin Smith
Akshayaram Srinivasan
François-Xavier Standaert
Ron Steinfeld
Noah
  Stephens-Davidowitz
Katerina Stouka
Koutarou Suzuki
Alan Szepieniec
Björn Tackmann
Stefano Tessaro
Adrian Thillard
Emmanuel Thomé

Mehdi Tibouchi
Elmar Tischhauser
Yosuke Todo
Ni Trieu
Roberto Trifiletti
Yiannis Tselekounis
Furkan Turan
Thomas Unterluggauer
Margarita Vald
Prashant Vasudevan
Philip Vejre
Srinivas Vivek Venkatesh
Daniele Venturi
Frederik Vercauteren
Ivan Visconti
Vanessa Vitse
Damian Vizár
Petros Wallden
Michael Walter
Lei Wang
Huaxiong Wang
Mor Weiss
Weiqiang Wen
Mario Werner
Benjamin Wesolowski
Carolyn Whitnall
Friedrich Wiemer
David Wu
Keita Xagawa
Sophia Yakoubov
Shota Yamada
Takashi Yamakawa
Avishay Yanay
Kan Yasuda
Eylon Yogev
Kazuki Yoneyama
Henry Yuen
Thomas Zacharias
Karol Zebrowski
Rina Zeitoun
Bingsheng Zhang
Ryan Zhou
Dionysis Zindros

Contents – Part II

Blockchain

Contents – Part III

Contents – Part I

Functional Encryption II

On Removing Graded Encodings
from Functional Encryption

Nir Bitansky[1], Huijia Lin[2], and Omer Paneth[1]

[1] MIT, Cambridge, USA
omerpa@gmail.com
[2] UCSB, Santa Barbara, USA
rachel.lin@cs.ucsb.edu

Abstract. Functional encryption (FE) has emerged as an outstanding concept. By now, we know that beyond the immediate application to computation over encrypted data, variants with *succinct ciphertexts* are so powerful that they yield the full might of indistinguishability obfuscation (IO). Understanding how, and under which assumptions, such succinct schemes can be constructed has become a grand challenge of current research in cryptography. Whereas the first schemes were based themselves on IO, recent progress has produced constructions based on *constant-degree graded encodings*. Still, our comprehension of such graded encodings remains limited, as the instantiations given so far have exhibited different vulnerabilities.

Our main result is that, assuming LWE, *black-box constructions* of *sufficiently succinct* FE schemes from constant-degree graded encodings can be transformed to rely on a much better-understood object — *bilinear groups*. In particular, under an *über assumption* on bilinear groups, such constructions imply IO in the plain model. The result demonstrates that the exact level of ciphertext succinctness of FE schemes is of major importance. In particular, we draw a fine line between known FE constructions from constant-degree graded encodings, which just fall short of the required succinctness, and the holy grail of basing IO on better-understood assumptions.

In the heart of our result, are new techniques for removing ideal graded encoding oracles from FE constructions. Complementing the result, for weaker ideal models, namely the generic group model and the random oracle model, we show a transformation from *collusion-resistant* FE in either of the two models directly to FE (and IO) in the plain model, without assuming bilinear groups.

N. Bitansky—Supported by NSF Grants CNS-1350619 and CNS-1414119, and the Defense Advanced Research Projects Agency (DARPA) and the U.S. Army Research Office under contracts W911NF-15-C-0226. Part of this research was done while visiting Tel Aviv University and supported by the Leona M. & Harry B. Helmsley Charitable Trust and Check Point Institute for Information Security.

H. Lin—Partially supported by NSF grants CNS-1528178 and CNS-1514526.

J.-S. Coron and J.B. Nielsen (Eds.): EUROCRYPT 2017, Part II, LNCS 10211, pp. 3–29, 2017.
DOI: 10.1007/978-3-319-56614-6_1

1   Introduction

Functional Encryption (FE) is a fascinating object. It enables fine-grained control of encrypted data, by allowing users to learn only specific functions of the data. This ability is captured trough the notion of *function keys*. A function key SK_f, associated with a function f, allows to partially decrypt a ciphertext CT_x encrypting an input x in a way that reveals $f(x)$ and nothing else.

A salient aspect of FE schemes is their *ciphertext succinctness*. Focusing on the setting of (indistinguishability-based) *single-key* FE where only one function key SK_f is supported, we say that an FE scheme is *weakly succinct* if the ciphertext size scales *sub-linearly* in the size of the circuit f; namely,[1]

$$|\mathsf{CT}_x| \leq |f|^\gamma \cdot \mathrm{poly}(|x|), \qquad \text{for some constant } compression\,factor\,\gamma < 1.$$

While non-succinct single-key FE schemes (where we allow the size of ciphertexts to grow polynomially with $|f|$) are equivalent to public-key encryption (or just one-way functions, in the secret-key setting) [28,41], weakly succinct schemes are already known to be extremely strong. In particular, subexponentially-secure weakly-succinct FE for functions in $\mathbf{NC}^1$ implies indistinguishability obfuscation (IO) [1,6,11], and has far reaching implications in cryptography and beyond (*e.g.*, [7,10,18,23,24,42]).[2]

Thus, understanding how, and under which assumptions, weakly-succinct FE can be constructed has become a central question in cryptographic research. While schemes for Boolean functions in $\mathbf{NC}^1$ have been constructed from LWE [27], the existence of such FE scheme for non-Boolean functions (which is required for the above strong implications) is still not well-founded, and has been the subject of a substantial body of work. The first construction of general purpose FE that achieves the required succinctness relied itself on IO [24]. Subsequent constructions were based on the algebraic framework of *multilinear graded encodings* [22]. Roughly speaking, this framework extends the traditional concept of *encoding in the exponent* in groups. It allows *encoding* values in a field (or ring), evaluating polynomials of a certain bounded *degree d* over the encoded values, and testing whether the result is zero.

Based on graded encodings of polynomial degree Garg, Gentry, Halevi, and Zhandry [25] constructed *unbounded-collusion* FE, which in turn is known to lead to weakly succinct FE [2,11]. Starting from the work of Lin [31], several works [3,32,35] have shown that assuming also pseudorandom generators with constant locality, weakly-succinct FE can be constructed based on *constant-degree* graded encodings under simple assumptions like asymmetric DDH. However, these constructions require constant degree $d \geq 5$.

---

[1] Here *weak* succinctness is in contrast to *full* succintness, where the ciphertext size does not depend at all on the function size.

[2] Formally, [1,11] require that not only the ciphertext is succinct, but also the encryption circuit itself. This difference can be bridged assuming LWE [34], and for simplicity is ignored in this introduction. Our results will anyhow rely on LWE.

Despite extensive efforts, our understanding of graded encodings of any degree larger than two is quite limited. Known instantiations are all based on little-understood lattice problems, and have exhibited different vulnerabilities [15, 20–22, 38]. In contrast, bilinear group encodings [14, 29], akin to degree-2 graded encodings, have essentially different instantiations based on elliptic curve groups, which are by now quite well understood and considered standard. Bridging the gap between degree 2 and degree $d > 2$ is a great challenge.

Our Main Result in a Nutshell: Size Matters. We show that the exact level of succinctness in FE schemes has a major impact on the latter challenge. Roughly speaking, we prove that *black-box constructions* [40] of weakly-succinct FE from degree-d graded encodings, with compression factor $\gamma < \frac{1}{d}$, can be transformed to rely *only on bilinear groups*. Specifically, assuming LWE [3] and for any constant ε, starting from $\frac{1}{d+\varepsilon}$-succinct FE in the *ideal degree-d graded encoding model*, we construct weakly-succinct FE in the *ideal bilinear model*.

The ideal graded encoding model generalizes the classical generic-group model [43]. In this model, the construction as well as the adversary perform all graded encoding operations through an *ideal oracle*, without access to an explicit representation of encoded elements. Having this ideal model as a starting point allows capturing a large class of constructions and assumptions, as it models *perfectly secure* graded encodings. Indeed, the FE schemes in [3, 31, 32, 35] can be constructed and proven secure in this model.

The resulting construction from ideal bilinear encodings can further be instantiated in the plain model using existing bilinear groups, and proven secure under an *über assumption* on bilinear groups [13, 16]. In particular, assuming also subexponential-security, it implies IO in the plain model.

How Close are We to IO from Bilinear Maps? Existing weakly-succinct FE schemes in the ideal constant-degree model [3, 31, 32, 35] have a compression factor $\gamma = C/d$, for some absolute constant $C > 1$. Thus, our result draws a fine line that separates known FE constructions based on constant-degree graded encodings and constructions that would already take us to the promised land of IO based on much better-understood mathematical objects. Crossing this line may very well require a new set of techniques. Indeed, one may also interpret our result as a negative one, which puts a barrier on black-box constructions of FE from graded encodings.

Discussion: Black-Box vs Non-Black-Box Constructions. For IO schemes (rather than FE), a combination of recent works [12, 39] demonstrates that black-box constructions from constant-degree graded encodings are already very powerful. They show that any IO construction relative to a constant-degree oracle can be converted to a construction in the plain model (under standard assumptions, like DDH). Since weakly-succinct FE schemes imply IO, we may be lead to think that weakly-succinct black-box constructions of FE from constant-degree

---

[3] More precisely, we need to assume the hardness of LWE with subexponential modulus-to-noise ratio. For simplicity, we ignore the parameters of LWE in this introduction; see Sect. 4 for more details.

graded encodings would already imply IO in the plain model from standard assumptions. Interestingly, this is not the case.

The crucial point is that the known transformations from FE to IO [1,11] are *non-black-box*, they use the code of the underlying FE scheme, and thus do not *relativize* with respect to graded encoding oracle. That is, we do not know how to move from an FE scheme based on graded encodings to an IO scheme that uses graded encodings in a black-box way. Indeed, if there existed such a black-box transformation between FE and IO, then combining [12,31,35,36,39], IO in the plain model could be constructed from standard assumptions.

Instead, we show how to directly remove constant-degree oracles from FE. Our transformation relies on new techniques that are rather different than those used in the above works for removing such oracles from IO.

1.1   Our Results in More Detail

We now describe our results in further detail. We start by describing the ideal graded encoding model and the ideal bilinear encoding model more precisely.

The Ideal Graded Encoding Model. A graded encoding [22] is an encoding scheme for elements of some field.[4] The encoding supports a restricted set of homomorphic operations that allow one to evaluate certain polynomials over the encoded field elements and test whether these polynomials evaluate to zero or not. Every field element is encoded with respect to a *label* (sometimes called the *level* of the encoding). For a given sequence of encodings, their labels control which polynomials are valid and can be evaluated over the encodings. The *degree* of the graded encoding is the maximal degree of a polynomial that is valid with respect to any sequence of labels.

In the ideal graded encoding model, explicit encodings are replaced by access to an oracle that records the encoded field elements and provides an interface to perform operations over the elements. Different formalizations of such ideal graded encoding oracles exist in the literature (*e.g.* [4,5,17,39]) and differ in details. In this work, we follow the model of Pass and shelat in [39].

The ideal graded encoding oracle $\mathcal{M}$ is specified by a field $\mathbb{F}$ and a validity predicate V operating on a polynomial and labels taken from a set $\mathbb{L}$. The oracle $\mathcal{M} = (\mathbb{F}, V)$ provides two functions — encoding and zero-testing.

Encoding: Given a field element $\xi \in \mathbb{F}$ and a label $\ell \in \mathbb{L}$ the oracle $\mathcal{M}$ samples a sufficiently long random string r to create a *handle* $h = (r, \ell)$. It records the pair (h, ξ) associating the handle with the encoded field element.

Zero-testing: a query to $\mathcal{M}$ consists of a polynomial p and a sequence of handles $h_1, \cdots, h_m$ where h_i encodes the field elements ξ_i relative to label ℓ_i. $\mathcal{M}$ tests if the polynomial and the labels satisfy the validity predicate and whether the polynomial vanishes on the corresponding field elements. That is, $\mathcal{M}$ returns **true** if and only if $V(p, \ell_1, \cdots \ell_m) = \textbf{true}$ and $p(\xi_1, \cdots \xi_m) = 0$.

---

[4] For ease of exposition, we consider graded encodings over fields. Our results can also be obtained with any commutative ring in which it is computationally hard to find non-unit elements.

Like in [39], we restrict attention to well-formed validity predicates. For such predicates, a polynomial p is valid with respect to labels $\ell_1, \cdots, \ell_m$, if and only if every monomial Φ in p is valid with respect to the labels of the handles that Φ acts on. Indeed, existing graded encodings all consider validity predicates that are *well-formed*.[5]

The Ideal Bilinear Encoding Model. The ideal bilinear encoding model corresponds to the ideal graded encoding model where valid polynomials are of degree at most two. We note that in the ideal graded encoding model described above, encoding is a randomized operation. In particular, encoding the same element and label (ξ, ℓ) twice gives back two different handles. In contrast, traditional instantiations of the ideal bilinear encoding model are based on bilinear pairing groups (such as elliptic curve groups) where the encoding is a deterministic function. We can naturally capture such instantiations, by augmenting the ideal bilinear encoding model to use a *unique handle* for every pair of field element and label (as done for instance in [4, 35, 44]).

The Main Result. Our main result concerns FE schemes in the ideal graded encoding model. In such FE schemes, all algorithms (setup, key derivation, encryption, and decryption), as well as all adversaries against the scheme, have access to a graded encoding oracle $\mathcal{M}$. We show:

Theorem 1 (Informal). *Assume the hardness of LWE. For any constants $d \in \mathbb{N}$ and $\gamma \leq \frac{1}{d}$, any γ-succinct secret-key FE scheme for $P/poly$, in the ideal degree-d graded encoding model, can be transformed into a weakly-succinct public-key FE scheme for $P/\textbf{poly}$ in the ideal bilinear encoding model.*

IO in the Plain Model under an Über Assumption. Our main transformation results in a weakly-succinct public-key FE scheme in the ideal bilinear encoding model. By instantiating the ideal bilinear encoding oracle with concrete bilinear pairing groups, we get a corresponding FE scheme in the plain model. For security to hold, we make an über assumption [13] on the bilinear groups. An über assumption essentially says that two encoded sequences of elements in the plain model can be distinguished only if they are also distinguishable in the ideal model. There are no known attacks on the über security of existing instantiations of bilinear pairing groups.

Since weakly-succinct public-key FE with subexponential security in the plain model implies IO we deduce the following corollary

Corollary 1 (Informal). *Assume subexponential hardness of LWE and bilinear groups with subexponential über security. For any constants $d \in \mathbb{N}, \gamma < \frac{1}{d}$, any subexponentially-secure, γ-succinct, secret-key FE for $P/\textbf{poly}$ in the ideal degree-d graded encoding model, can be transformed into an IO scheme for $P/\textbf{poly}$ in the plain model.*

---

[5] In the body, we make another structural requirement on validity predicates called *decomposability*. This requirement is somewhat more technical, but is also satisfied by all known formulations of graded encodings. For the simplified technical exposition in this introduction it can be ignored. See further details in Definition 4.

FE in Weaker Ideal Models. We also consider FE schemes in ideal models that are weaker than the ideal bilinear encoding model. Specifically, we consider the generic-group model (that corresponds to the ideal degree-1 graded encoding model) and the random-oracle model. We give transformations from FE in these models directly to FE in the plain model *without relying on bilinear encodings*.

In the transformation given by Theorem 1, from the ideal constant-degree graded encoding model to the ideal bilinear encoding model, we considered the notion of single-key weakly succinct FE. In contrast, our transformations from the generic-group model and the random-oracle model to the plain model require that we start with a stronger notion of *collusion-resistant FE*. Collusion-resistance requires security in the presence of an *unbounded* number of functional keys. Crucially, ciphertexts are required not to grow with the number of keys (but are allowed to grow polynomially in the size of the evaluated functions).

Collusion-resistant FE is known to imply weakly-succinct FE through a black-box transformation [2,11]. In the converse direction, only a non-black-box transformation is known [26,30], and therefore we cannot apply it to ideal model constructions of FE.

Theorem 2 (Informal). *Assume the hardness of LWE. Any collusion-resistant secret-key FE scheme in the generic-group model, or in the random-oracle model, can be transformed into a collusion-resistant public-key FE scheme in the plain model.*

1.2   Our Techniques

We next give an overview of the main ideas behind our degree-reduction transformation given by Theorem 1.

Can We Adopt Techniques from IO? As already mentioned, we do not know how to transform FE schemes into IO schemes in a black-box way. Thus, we cannot rely directly on existing results that remove ideal oracles from IO [39]. Furthermore, trying to import ideas from these results in the IO setting to the setting of FE encounters some inherent difficulties, which we now explain.

Roughly speaking, removing ideal oracles from IO is done as follows. Starting with a scheme in an ideal oracle model, we let the obfuscator emulate the oracle by itself and publish, together with the obfuscated circuit, some *partial view* of the self-emulated oracle. This partial view is on one hand, sufficient to preserve the functionality of the obfuscated circuit on most inputs, and on the other hand, does not compromise security. The partial view is obtained by evaluating the obfuscation on many random inputs (consistently with the self-emulated oracle), observing how evaluation interacts with the oracle, and performing a certain *learning process*. Arguing that the published partial view does not compromise security crucially relies on the fact that *evaluating the obfuscated program is a public procedure that does not share any secret state with the obfuscator*.

The setting of FE, however, is somewhat more complicated. Here rather than an evaluator we have a decryptor that given a function key SK_f and ciphertext

CT encrypting x, should be able to compute $f(x)$. In contrast to the evaluator in obfuscation, the state of the decryptor is *not publicly samplable*. Indeed, generating function keys SK_f for different functions requires knowing a master secret key. Accordingly, it is not clear how to follow the same approach as before.

XIO instead of IO. Nevertheless, we observe that there is a way to reduce the problem to a setting much more similar to IO. Specifically, *there exists* [9] *a black-box transformation from FE to a weaker version of IO called XIO.* XIO [33], which stands for *exponentially-efficient IO*, allows the obfuscation and evaluation algorithms to run in exponential time $2^{O(n)}$ in the input size n, and only requires that the *size* of an obfuscation $\widetilde{C}$ of a circuit C is slightly subexponenetial in n:

$$|\widetilde{C}| \leq 2^{\gamma n} \cdot \mathrm{poly}(|C|) \qquad \text{for some constant } compression \, factor \, \gamma < 1.$$

Despite this inherent inefficiency, [33] show that XIO for *logarithmic-size* inputs implies IO assuming subexponential hardness of LWE. A natural direction is thus to try and apply the techniques used to remove oracles from IO to remove the same oracles also from XIO; indeed, if this can be done, such oracles can also be removed from FE, due to the black-box transformation between the two.

This, again, does not work as is. The issue is that the transformations removing degree-d graded encoding oracle from IO may blow up the size of the original obfuscation from $|\widetilde{C}|$ in the oracle model to roughly $|\widetilde{C}|^{2d}$ in the plain model. However, the known black-box construction of XIO from FE [9] is not sufficiently compressing to account for this blowup. Even starting from FE with great compression, say $\gamma_{\mathrm{FE}} < d^{-10}$, the resulting XIO has a much worst compression factor $\gamma_{\mathrm{XIO}} > 1/2$. In particular, composing the two would result in a useless plain model obfuscation of exponential size $2^{n \cdot d}$.

Motivating our Solution. To understand our solution, let us first describe an over-simplified candidate transformation for reducing XIO with constant-degree graded encoding oracles to XIO with degree-1 oracles (akin to the generic-group model). This transformation will suffer from the same size blowup of the transformations mentioned above.

For simplicity of exposition, we first restrict attention to XIO schemes with the following simple structure:

– Any obfuscated circuit $\widetilde{C}$ consists of a set of handles $h_1, \ldots, h_m$ corresponding to field elements $\xi_1, \ldots, \xi_m$ encoded during obfuscation, under certain labels $\ell_1, \ldots, \ell_m$.
– Evaluation on any given input x consists of performing valid zero-tests over the above handles, which are given by degree-d polynomials $p_1, \ldots, p_k$.

A simple idea to reduce the degree-d oracle to a linear oracle is to change the obfuscation algorithm so that it computes ahead of time the field elements ξ_Φ corresponding to all valid degree-d monomials $\Phi(\xi_1, \ldots, \xi_m) = \prod_{i \in [d]} \xi_{j_i}$. Then, rather than using the degree-d oracle, it uses the linear oracle to encode the field elements ξ_Φ, and publishes the corresponding handles $\{h_\Phi\}_\Phi$. Evaluation is done in a straight forward manner by writing any zero-test polynomial p of degree d as a linear function in the corresponding monomials

$$p(\xi_1, \cdots, \xi_m) = \sum_{\Phi} \alpha_{\Phi} \Phi(\xi_1, \ldots, \xi_m),$$

and making the corresponding zero-test query $L_p(\{h_{\Phi}\}) := \sum_{\Phi} \alpha_{\Phi} h_{\Phi}$ to the linear oracle.

Indeed, the transformation blows up the size of the obfuscated circuit from roughly m, the number of encodings in the original obfuscation, to m^d, the number of all possible monomials. While such a polynomial blowup is acceptable in the context of IO, for XIO with compression $d^{-1} \leq \gamma < 1$, it is devastating.

Key Idea: XIO in Decomposable Form. To overcome the above difficulty, we observe that the known black-box construction of XIO from FE [9] has certain structural properties that we can exploit. At a very high level, it can be decomposed into smaller pieces, so that instead of computing *all* monomials over *all* the encodings created during obfuscation, we only need to consider a much smaller subset of monomials. In this subset, each monomial only depends on a few small pieces, and thus only on few encodings.

To be more concrete, we next give a high-level account of this construction. To convey the idea in a simple setting of parameters, let us assume that we have at our disposal an FE scheme that support an unbounded number of keys, rather than a single key scheme, with the guarantee that the size of ciphertexts does not grow with the number of keys. In this case, the XIO scheme in [9] works as follows:

- To obfuscate a circuit C with n input bits, the scheme publishes a collection of function keys $\{\mathsf{SK}_{D_\tau}\}_\tau$ for circuits D_τ, indexed by prefixes $\tau \in \{0,1\}^{n/2}$ (will be specified shortly), and a collection of ciphertexts $\{\mathsf{CT}_{\rho\|C}\}_\rho$, each encrypting the circuit C and a suffix $\rho \in \{0,1\}^{n/2}$.
- Decrypting a ciphertext $\mathsf{CT}_{\rho\|C}$ with key SK_{D_τ} reveals $D_\tau(\rho\|C) := C(\tau, \rho)$.

The obfuscated circuit indeed has slightly subexponential size. It contains:

- $2^{n/2}$ function keys SK_{D_τ}, each of size $\mathrm{poly}(|C|)$,
- $2^{n/2}$ ciphertexts $\mathsf{CT}_{\rho\|C}$, each of size $\mathrm{poly}(|C|)$.

Going back to the ideal graded-encoding model, the FE key generation and encryption algorithms use the ideal oracle to encode elements. Therefore, generating the obfuscation involves generating a set of k encodings $\boldsymbol{h}_\tau = \{h_{\tau,i}\}_{i \in [k]}$ for each secret key SK_{D_τ} and a set of k encodings $\boldsymbol{h}_\rho = \{h_{\rho,i}\}_{i \in [k]}$ for each ciphertext $\mathsf{CT}_{\rho\|C}$, for some $k = \mathrm{poly}(|C|)$. The crucial point is that now, evaluating the obfuscation on a given input (τ, ρ) only involves the two small sets of encodings $\boldsymbol{h}_\tau, \boldsymbol{h}_\rho$. In particular, any zero-test made by the decryption algorithm is a polynomial defined only over the underlying field elements $\boldsymbol{\xi}_\tau = \{\xi_{\tau,i}\}_{i \in [k]}$ and $\boldsymbol{\xi}_\rho = \{\xi_{\rho,i}\}_{i \in [k]}$.

This gives rise to the following degree reduction strategy. In the obfuscation, rather then precomputing all monomials in all encodings as before, we precompute only the monomials corresponding to the different pieces $\left\{\Phi(\boldsymbol{\xi}_\rho)\right\}_{\rho,\Phi}, \left\{\Phi(\boldsymbol{\xi}_\tau)\right\}_{\tau,\Phi}$. Now, rather than representing zero-tests made by the decryption algorithm as linear polynomials in these monomials, they can be represented as quadratic polynomials

$$p(\boldsymbol{\xi}_\tau, \boldsymbol{\xi}_\rho) = Q_p \left(\{\Phi(\boldsymbol{\xi}_\tau)\}_\Phi, \{\Phi(\boldsymbol{\xi}_\rho)\}_\Phi \right).$$

To support such quadratic zero tests, we resort to bilinear groups. We use the bilinear encoding oracle to encode the values $\left\{\Phi(\boldsymbol{\xi}_\rho)\right\}_{\rho,\Phi}, \left\{\Phi(\boldsymbol{\xi}_\tau)\right\}_{\tau,\Phi}$, and publish the corresponding handles $\{h_{\tau,\Phi}\}_{\tau,\Phi}, \{h_{\rho,\Phi}\}_{\rho,\Phi}$. Evaluation is done in a straight forward manner by testing the quadratic polynomial Q_p.

The key gain of this construction is that now the blowup is tolerable. Now, each set of k encodings, blows up to k^d, which is acceptable since $k = \mathrm{poly}(|C|)$ is small (and not proportional to the size of the entire obfuscation as before, which is exponential in n). In the body, we formulate a general *product form* property for XIO schemes, which can be used as the starting point of the above-described transformation; we further show that single-key FE schemes with $\frac{1}{d+\varepsilon}$-succinctness implies such XIO schemes.

A Closer Look. The above exposition is oversimplified. To actually fulfill our strategy, we need to overcome two main challenges.

Challenge 1: Explicit Handles. The core idea described above assumes that the obfuscation is simply given as an explicit list of handles, which may not be the case starting from an arbitrary FE scheme. In particular, the obfuscator may use the oracle $\mathcal{M}$ to produce a set of encodings, but not output them explicitly; indeed, it can output an arbitrary string. In this case, we can no longer apply the degree reduction technique, since we do not know which encodings are actually contained in the obfuscation. Naïvely publishing all monomials in all field elements ever encoded by the obfuscator may be insecure — some of these encodings, which are never explicitly included in the obfuscation, may leak information.

To handle XIO schemes constructed from general FE schemes, we need a way to make any "implicit" handles explicit, without compromising security. Our idea is to *learn* the *significant handles* that would later suffice for evaluation on most inputs, and publish them explicitly. This idea is inspired by [19,36,39] and their observation (already mentioned above) that in obfuscation, the evaluator's view, including all the handles it sees, is publicly and efficiently samplable.

Roughly speaking, the learning process involves evaluating the obfuscated circuit on many random inputs and making explicit all handles involved in these evaluations. When doing this naïvely, the number of such test evaluations required to guarantee reasonable correctness is proportional to the number of elements encoded by the obfuscator. This would result in a quadratic overhead in the size of the obfuscation, which would again completely foil XIO compression.

Avoiding the blowup requires a somewhat more sophisticated learning process that once again exploits the local structure of the construction in [9].

The scheme resulting from the above learning process is only approximately correct — the obfuscation with explicit handles errs on say 10% of the inputs. We show that even such *approximate XIO* is sufficient for obtaining FE and IO in the plain model (this step is described later in this overview).

Challenge 2: Invalid Monomials. Another main challenge is that it may be insecure to publish encodings of all the monomials $\{\Phi(\boldsymbol{\xi}_\rho)\}_{\rho,\Phi}$, $\{\Phi(\boldsymbol{\xi}_\tau)\}_{\tau,\Phi}$. The problem is that some products $\Phi(\boldsymbol{\xi}_\rho) \cdot \Phi'(\boldsymbol{\xi}_\tau)$ may result in monomials that would have been invalid in the degree-d ideal model. For example, $\Phi(\boldsymbol{\xi}_\rho)$ could correspond to a degree-$(d-2)$ monomial Φ. In the degree-d ideal model, it would only be possible to multiply such a monomial by degree-2 monomials $\Phi'(\boldsymbol{\xi}_\tau)$, and zero test. In the the described new scheme, however, it can multiply monomials $\Phi'(\boldsymbol{\xi}_\tau)$ of degree 3, or even d, which might compromise security.

Our solution proceeds in two steps. First, we show how to properly preserve validity by going to a more structured model of bilinear encodings that generalizes *asymmetric* bilinear groups. In this model, every encoding contains one of many labels and only pairs of encodings with valid labels can be multiplied. We then encode the monomials $\{\Phi(\boldsymbol{\xi}_\rho)\}_{\rho,\Phi}$, $\{\Phi(\boldsymbol{\xi}_\tau)\}_{\tau,\Phi}$ with appropriate labels that preserve the information regarding the original set of labels. This guarantees that the set of monomials that can be zero-tested in this model corresponds exactly to the set of valid monomials in the constant-degree graded encoding model we started from.

Second, we show how to transform any construction in this (more structured) ideal model into one in the standard ideal bilinear encoding model (corresponding to *symmetric* bilinear maps). At a very high-level, we develop a "secret-key transformation" from asymmetric bilinear groups to symmetric bilinear groups. The transformation allows anyone in the possession of a secret key to translate encodings in the asymmetric setting to new encodings in the symmetric setting in a manner that enforces the asymmetric structure.

From Approximately-Correct XIO back to FE. After applying all the above steps, we obtain an approximately-correct XIO scheme in the ideal bilinear encoding model. The only remaining step is going from such an XIO scheme back to FE. The work of [34] showed how to construct FE from XIO with *perfect* correctness, assuming in addition LWE. We modify their transformation to construct FE starting directly from approximately-correct XIO. This is done using appropriate Error Correcting Codes to accommodate for the correctness errors from XIO.[6] The transformation uses XIO as a black-box, and can thus be performed in the ideal bilinear model.

Putting It All Together. Putting all pieces together, we finally obtain our transformation from $\frac{1}{d+\varepsilon}$-succinct FE in the constant-degree graded encoding

---

[6] We note that existing transformations for removing errors from IO [12] do not work for XIO. See Sect. 4 for details.

model to weakly-succinct FE in the bilinear encoding model. To recap the structure of the transformation:

1. Start with a $\frac{1}{d+\varepsilon}$-succinct (single-key) FE in the ideal constant-degree graded encoding model.
2. Transform it into an XIO scheme in the ideal constant-degree graded encoding model satisfying an appropriate decomposition property (which we call product form).
3. Transform it into an approximate XIO scheme in the ideal bilinear encoding model.
4. Use the resulting approximate XIO scheme and LWE to get a weakly-succinct FE (still, in the ideal bilinear encoding model).

Instantiating the oracle in bilinear groups with über security gives a corresponding construction in the plain model.

Organization. In Sect. 2, we define (oracle-aided) XIO, and introduce the constant-degree oracles considered in this work. In Sect. 3, we show how to transform XIO, in a certain product form, relative to constant-degree oracles to approximate XIO relative to symmetric bilinear oracles. In Sect. 4, we explain how to move from approximate XIO and LWE, to IO. Due to the space limit, some of the details and proofs are omitted. These can be found in the full version of this paper [8], where we additionally describe how to remove generic-group oracles and random oracles from unbounded collusion FE schemes.

2 Preliminaries

2.1 XIO

We next formally define the notion of exponentially-efficient indistinguishability obfuscation (XIO) for any collection of circuit classes $\mathcal{C} \subseteq \mathbf{P}^{\log}/\mathbf{poly}$, where $\mathbf{P}^{\log}/\mathbf{poly}$ is the collection of all classes of polynomial-size circuits with logarithmic size input. The definition extends the one in [33] by considering also approximate correctness.

Definition 1 ($\mathbf{P}^{\log}/\mathbf{poly}$). *The collection $\mathbf{P}^{\log}/\mathbf{poly}$ includes all classes $\mathcal{C} = \{\mathcal{C}_\lambda\}$ for which there exists a constant $c = c(\mathcal{C})$, such that the input of any circuit $C \in \mathcal{C}_\lambda$ is bounded by $c \log \lambda$ and the size of C is bounded by λ^c.*

Definition 2 (XIO [33]). *A pair of algorithms $\mathsf{xiO} = (\mathsf{xiO.Obf}, \mathsf{xiO.Eval})$ is an exponentially-efficient indistinguishability obfuscator (XIO) for a collection of circuit classes $\mathcal{C} = \{\mathcal{C} = \{\mathcal{C}_\lambda\}\} \subseteq \mathbf{P}^{\log}/\mathbf{poly}$ if it satisfies:*

– ***Functionality:*** *for any $\mathcal{C} \in \mathcal{C}$, security parameter $\lambda \in \mathbb{N}$, and $C \in \mathcal{C}_\lambda$ with input size n,*

$$\Pr_{\substack{\mathsf{xiO} \\ x \leftarrow \{0,1\}^n}} [\mathsf{xiO.Eval}(\widetilde{C}, x) = C(x) : \widetilde{C} \leftarrow \mathsf{xiO.Obf}(C, 1^\lambda)] \geq 1 - \alpha(\lambda).$$

We say that $\mathsf{xiO.Obf}$ is correct if $\alpha(\lambda) \leq \mathsf{negl}(\lambda)$ and approximately-correct if $\alpha(\lambda) \leq 1/100$.

- **Non-trivial Efficiency:** *there exists a constant $\gamma < 1$ and a fixed polynomial* poly($\cdot$), *depending on the collection $\boldsymbol{C}$ (but not on any specific class $\mathcal{C} \in \boldsymbol{C}$), such that for any class $\mathcal{C} \in \boldsymbol{C}$ security parameter $\lambda \in \mathbb{N}$, circuit $C \in \mathcal{C}_\lambda$ with input length n, and input $x \in \{0,1\}^n$ the running time of both* $\mathsf{xiO.Obf}(C, 1^\lambda)$ *and* $\mathsf{xiO.Eval}(\widetilde{C}, x)$ *is at most* poly($2^n, \lambda, |C|$) *and the size of the obfuscated circuit $\widetilde{C}$ is at most $2^{n\gamma} \cdot$ poly($|C|, \lambda$). We call γ the compression factor, and say that the scheme is γ-compressing.*
- **Indistinguishability:** *for any $\mathcal{C} = \{\mathcal{C}_\lambda\} \in \boldsymbol{C}$ and polynomial-size distinguisher $\mathcal{D}$, there exists a negligible function $\mu(\cdot)$ such that the following holds: for all security parameters $\lambda \in \mathbb{N}$, for any pair of circuits $C_0, C_1 \in \mathcal{C}_\lambda$ of the same size and such that $C_0(x) = C_1(x)$ for all inputs x,*

$$\left| \Pr\left[\mathcal{D}(\mathsf{xiO.Obf}(C_0, 1^\lambda)) = 1 \right] - \Pr\left[\mathcal{D}(\mathsf{xiO.Obf}(C_1, 1^\lambda)) = 1 \right] \right| \le \mu(\lambda).$$

We further say that $\mathsf{xiO.Obf}$ *is δ-secure, for some concrete negligible function $\delta(\cdot)$, if for all polynomial-size distinguishers the above indistinguishability gap $\mu(\lambda)$ is smaller than $\delta(\lambda)^{\Omega(1)}$.*

Remark 1 (Logarithmic Input). Indeed, for XIO to be useful, we must restrict attention to circuit collections $\boldsymbol{C} \subseteq \mathbf{P}^{\log}/\mathbf{poly}$. This ensures that obfuscation and evaluation are computable in time $2^{O(n)} = \text{poly}(\lambda)$.

Remark 2 (Probabilistic $\mathsf{xiO.Eval}$). Above, we allow the evaluation algorithm $\mathsf{xiO.Eval}$ to be probabilistic. Throughout most of the paper, we restrict attention to *deterministic* evaluation algorithms. This typically will simplify exposition and is without loss of generality.

XIO with an Oracle. We say that an XIO scheme $\mathsf{xiO} = (\mathsf{xiO.Obf}, \mathsf{xiO.Eval})$ is constructed relative to an oracle $\mathcal{O}$ if the corresponding algorithms, as well as the adversary, may access the oracle $\mathcal{O}$. Namely, the obfuscation algorithm $\mathsf{xiO.Obf}^{\mathcal{O}}(C, 1^\lambda)$ and the evaluation algorithm $\mathsf{xiO.Eval}^{\mathcal{O}}(\widetilde{C}, x)$ are given oracle access to $\mathcal{O}$. In the security definition, the adversarial distinguisher $\mathcal{D}^{\mathcal{O}}$ also gets access to the oracle.

2.2 The Ideal Graded Encoding Model

The ideal graded-encoding model we consider is inspired by previous generic group and ideal graded-encoding models [5,17,37,43] and is closest to the model of Pass and Shelat [39]. As in [39], we consider well-formed predicates that are determined by the validity of monomials.

Definition 3 (Well-Formed Validity Predicate). *V is well-formed if for any $d \in \mathbb{N}$ and degree-d polynomial $V(p, \ell_1, \ldots, \ell_m) = \bigwedge_{i \le d, j_1, \ldots, j_i \in [m], \rho_{j_1, \ldots, j_i} \ne 0}$ $V(\{\ell_{j_1}, \ldots, \ell_{j_i}\})$; namely, p is valid relative to the labels $\ell_1, \ldots, \ell_m$ if every monomial of p is valid relative to the corresponding multi-set of labels $\{\ell_{j_1}, \ldots, \ell_{j_i}\}$.*

We additionally consider the following decomposability requirement.

Definition 4 (Decomposable Validity Predicate). *V is decomposable if it is well-formed and there exist a projection function Π and a two-input predicate V_Π satisfying: For every two multisets $A = \{\ell_{1,1}, \ldots, \ell_{1,k_1}\}$ and $B = \{\ell_{2,1}, \ldots, \ell_{2,k_2}\}$ of labels, the validity of their union is given by[7]*

$$V(A \uplus B) = V_\Pi(\Pi(A), \Pi(B)).$$

The arity *of a decomposable predicate V is*

$$\mathsf{Arity}(V) := \max_A |\{\Pi(B) \; : \; V_\Pi(\Pi(A), \Pi(B)) = 1 \}|;$$

namely, it is the maximum number of projections $\Pi(B)$ that satisfy the validity predicate together with any given projection $\Pi(A)$, where A and B are multisets of labels.

Intuitively, a decomposable validity predicate has the property that any two different pairs of multi-sets $(A, B) \neq (A', B')$ share the same validity decision if they have the same projection $(\Pi(A), \Pi(B)) = (\Pi(A'), \Pi(B'))$. In other words, any information about the multi-sets beyond their projection does not matter. In the literature, all known ideal graded encoding models consider decomposable validity predicates with arity bounded by the degree (or even less). For instance, in set-based graded encodings, the labels correspond to subsets of some fixed universe $\mathbb{U}$, and a set of labels $\{S_1, \ldots, S_k \mid S_i \subseteq \mathbb{U}\}$ is valid if the sets are disjoint and $\uplus S_i = \mathbb{U}$. Therefore, we can define the projection of any $A = \{S_1, \ldots, S_i\}$ to be $\Pi(A) = \uplus S_i$ (or $\perp$ if the sets are not disjoint), in which case the arity is exactly one (indeed, for any $\Pi(A)$ only $\mathbb{U} \setminus \Pi(A)$ may satisfy the induced validity predicate).

We now formally define the ideal graded encoding model.

Definition 5 (Ideal Graded Encoding Oracle). *The oracle $\mathcal{M}_{\mathbb{F},V}$ is a stateful oracle, parameterized by a field $\mathbb{F}$ and a validity predicate V. The oracle answers queries of two forms:*

1. **Encoding Queries:** *Given a field element $\xi \in \mathbb{F}$ and label ℓ, the oracle samples a uniformly random string $r \leftarrow \{0,1\}^{\log |\mathbb{F}|}$, returns the handle $h = (r, \ell)$, and stores (h, ξ).*
2. **Zero-Test Queries:** *Given a polynomial $p \in \mathbb{F}[v_1, \ldots, v_m]$, and handles $h_1, \ldots, h_m$, the oracle does the following:*
   - *For each $i \in [m]$, obtains a tuple (h_i, ξ_i) from the stored list. If no such tuple exists, stops and returns* false.
   - *From each $h_i = (r_i, \ell_i)$, obtains ℓ_i, and checks that $V(p, \ell_1, \ldots, \ell_m) = $* true *to verify the query is valid and if not, returns* false.
   - *Performs a zero test, returning* true *if $p(\xi_1, \ldots, \xi_m) = 0$ and* false *otherwise.*

---

[7] For two multisets $A = \{a_1, \ldots, a_n\}, B = \{b_1, \ldots, b_m\}$, their union $A \uplus B = \{a_1, \ldots, a_n, b_1, \ldots, b_m\}$ counts multiplicity; e.g., $\{1,1\} \uplus \{1,2\} = \{1,1,1,2\}$.

An ideal graded encoding oracle $\mathcal{M} = \{\mathcal{M}_{\mathbb{F}_\lambda, V_\lambda}\}$ is a collection of oracles $\mathcal{M}_{\mathbb{F}_\lambda, V_\lambda}$, one for each $\lambda \in \mathbb{N}$, where $|\mathbb{F}| = 2^{\Theta(\lambda)}$.

The oracle $\mathcal{M}$ is said to be degree-d, if for every polynomial p of degree $\deg(p) > d$, and any label vector ℓ, $V(p, \ell) = \texttt{false}$. We say that an oracle $\mathcal{M}$ is decomposable if it has a decomposable validity predicate with bounded polynomial arity $\text{poly}(\lambda)$.

Remark 3. In some previous models (e.g., [39]), the ability to make encoding queries is further restricted. The above definition does not enforce any such restrictions. The results in this paper are presented in a public encoding model, which allows anyone to encode at any time. Our results on removing generic group oracle and random oracle from FE schemes can be extended to the model of private encodings, and the same holds for our results on reducing the degree of graded encoding oracles (Sect. 3), under certain mild assumptions. See the full version [8] for more details.

3   Reducing Constant-Degree Oracles to Bilinear Oracles

We show that any XIO scheme with a constant-degree decomposable ideal oracle can be transformed into an approximately-correct one with an ideal symmetric bilinear oracle (analogous to symmetric bilinear groups), provided that the XIO scheme is in a certain *product form*. We start by defining formally the notion of XIO in product form and of a symmetric bilinear oracle.

Definition 6 (Product Collection). $\mathcal{X} = \{\mathcal{X}_n\}_{n \in \mathbb{N}}, \mathcal{Y} = \{\mathcal{Y}_n\}_{n \in \mathbb{N}}$ *are said to be a product collection if:*

1. **Equal-Size Partition:** *For any* $X, X' \in \mathcal{X}_n$ *and* $Y, Y' \in \mathcal{Y}_n$:

$$|X| = |X'|, X \cap X' = \emptyset \qquad |Y| = |Y'|, Y \cap Y' = \emptyset,$$

2. **Product Form:** *let* $\boldsymbol{X}_n = \biguplus_{X \in \mathcal{X}_n} X, \boldsymbol{Y}_n = \biguplus_{Y \in \mathcal{Y}_n} Y$ *then the input space* $\{0,1\}^n$ *factors:*

$$\{0,1\}^n \cong \boldsymbol{X}_n \times \boldsymbol{Y}_n.$$

Definition 7 (XIO in Product Form). *We say that an XIO scheme* $\textsf{xiO} = (\textsf{xiO.Obf}^\mathcal{O}, \textsf{xiO.Eval}^\mathcal{O})$, *relative to oracle* $\mathcal{O}$, *for a collection of circuit classes* $\mathcal{C}$, *is in* $(\mathcal{X}, \mathcal{Y})$-*product form for a product collection* $(\mathcal{X}, \mathcal{Y})$ *if:*

- *The obfuscation algorithm* $\textsf{xiO.Obf}^\mathcal{O}$ *factors into two algorithms* $(\textsf{xiO.Obf}_\mathcal{X}^\mathcal{O}, \textsf{xiO.Obf}_\mathcal{Y}^\mathcal{O})$, *such that for any circuit* $C \in \mathcal{C}$, $\textsf{xiO.Obf}^\mathcal{O}(C, 1^\lambda; r)$, *outputs*

$$\left(\left\{ \widetilde{C}_X \leftarrow \textsf{xiO.Obf}_\mathcal{X}^\mathcal{O}(C, X, 1^\lambda; r) \right\}_{X \in \mathcal{X}_n}, \left\{ \widetilde{C}_Y \leftarrow \textsf{xiO.Obf}_\mathcal{Y}^\mathcal{O}(C, Y, 1^\lambda; r) \right\}_{Y \in \mathcal{Y}_n} \right),$$

and all executions may use joint randomness r.

- *There is an evaluation algorithm* $\mathsf{xiO.Eval}^{\mathcal{O}}_{\mathcal{X},\mathcal{Y}}$ *such that for any* $(X, Y) \in \mathcal{X}_n \times \mathcal{Y}_n$,

$$\mathsf{xiO.Eval}^{\mathcal{O}}_{\mathcal{X},\mathcal{Y}}(\widetilde{C}_X, \widetilde{C}_Y) = \left(\mathsf{xiO.Eval}^{\mathcal{O}}(\widetilde{C}, (x, y))\right)_{(x,y) \in X \times Y}.$$

Corresponding notation:

- *We denote by* $q^{\mathcal{X}}_o = q^{\mathcal{X}}_o(C, \lambda)$ *the maximal total size* $\sum_{Q \in \mathbf{Q}^X_o} |Q|$ *of all oracle queries* $\mathbf{Q}^X_o = \{Q\}$ *made by* $\mathsf{xiO.Obf}^{\mathcal{O}}_{\mathcal{X}}(C, X, 1^\lambda)$ *when obfuscating an n-bit input circuit* $C \in \mathcal{C}$ *for any* $X \in \mathcal{X}_n$. *Symmetrically, we denote by* $q^{\mathcal{Y}}_o = q^{\mathcal{Y}}_o(C, \lambda)$ *the bound on the total size* $\sum_{Q \in \mathbf{Q}^Y_o} |Q|$ *of oracle queries* $\mathbf{Q}^Y_o = \{Q\}$ *made by* $\mathsf{xiO.Obf}^{\mathcal{O}}_{\mathcal{Y}}(C, Y, 1^\lambda)$ *for any* $Y \in \mathcal{Y}_n$.

Definition 8 (Symmetric Bilinear Oracle). *The symmetric Bilinear Oracle* $\mathcal{B}^2 = \{\mathcal{B}^2_{\mathbb{F}_\lambda, V}\}$ *is a special case of the ideal graded encoding oracle, where the validity predicate* V *is of degree two and is defined over a single label* $\ell_\mathcal{B}$. *That is,* $V(L) = \mathtt{true}$ *for a multiset of labels* L, *if and only if* $L \subseteq \{\ell_\mathcal{B}, \ell_\mathcal{B}\}$.

We now state the main theorem of this section.

Theorem 3. *Let* $\mathsf{xiO} = (\mathsf{xiO.Obf}^{\mathcal{M}}, \mathsf{xiO.Eval}^{\mathcal{M}})$ *be an* $\mathsf{xiO.Obf}$ *scheme, relative to a degree-d decomposable ideal graded encoding oracle* $\mathcal{M}$, *for a collection of circuit classes* $\mathcal{C}$ *that is in* $(\mathcal{X}, \mathcal{Y})$-*product form, for some product collection* $(\mathcal{X}, \mathcal{Y})$. *Further assume that for some constant* $\gamma < 1$,

$$|\mathcal{X}_n| \cdot \left(q^{\mathcal{X}}_o \cdot \min\left(q^{\mathcal{X}}_o, |\mathcal{Y}_n| \cdot \log q^{\mathcal{X}}_o\right)\right)^d + |\mathcal{Y}_n| \cdot \left(q^{\mathcal{Y}}_o \cdot \min\left(q^{\mathcal{Y}}_o, |\mathcal{X}_n| \cdot \log q^{\mathcal{Y}}_o\right)\right)^d$$
$$\leq 2^{\gamma n} \cdot \mathrm{poly}(|C|, \lambda).$$

Then xiO *can be converted into an approximately-correct scheme* $\mathsf{xiO}^\star$ *relative to the symmetric bilinear oracle* $\mathcal{B}^2$.

Remark 4. A slightly easier to parse version of the above condition, with some loss in parameters, is that $|\mathcal{X}_n| \cdot \left(q^{\mathcal{X}}_o\right)^{2d} + |\mathcal{Y}_n| \cdot \left(q^{\mathcal{Y}}_o\right)^{2d} \leq 2^{\gamma n} \cdot \mathrm{poly}(|C|, \lambda)$.

Remark 5. Our ideal symmetric bilinear oracle captures symmetric bilinear pairing groups, but with two small gaps: Our oracle generates randomized encodings (following the Pass-shelat model) whereas bilinear pairing groups have unique encodings (of the form g^a), and our oracle does not support homomorphic opeartions whereas bilinear paring groups do. These differences are not consequential. In the full version of this paper [8], we show how to instantiate the transformed XIO schemes produced by the above theorem using concrete bilinear pairing groups.

Without Loss of Generality. Throughout this section, we make the following assumptions without loss of generality.

- **Obfuscator only encodes:** The XIO obfuscation algorithm only performs encoding queries and does not perform any zero tests. This is without loss of generality, as the obfuscator knows the field elements and labels underlying any generated handle (it encoded them itself), so zero-tests can be internally simulated.
- **Evaluator and adversary only zero-test:** The XIO evaluation algorithm as well as the adversary only perform zero tests and do not encode any elements themselves. Indeed, encoding of any (ξ, ℓ) can be internally simulated by sampling a corresponding handle $\tilde{h}$. Then, whenever a zero-test $(p, h_1, \ldots, h_m, \tilde{h}_1, \ldots, \tilde{h}_{\tilde{m}})$ includes such self-simulated handles $\tilde{h}_i$, it is translated to a new zero test that does not include such handles, by hardwiring the required field elements into the polynomial p.

3.1   Step 1: Explicit Handles

In this section, we show how to transform any XIO in product form relative to an ideal degree-d oracle (not necessarily decomposable) into one where all handles required for evaluation are given explicitly (also in product form). We start by defining the notion of explicit handles in product form and then state and describe the transformation.

Definition 9 (Explicit Handles in Product Form). *An XIO scheme* xiO $=$ (xiO.Obf$^{\mathcal{M}}$, xiO.Eval$^{\mathcal{M}}$), *relative to an ideal graded encoding oracle, for a collection of circuit classes $\mathcal{C}$, is said to have explicit handles in $(\mathcal{X}, \mathcal{Y})$-product form, for a product collection $(\mathcal{X}, \mathcal{Y})$, if the obfuscation and evaluation algorithms satisfy the following structural requirement:*

- *The algorithm* xiO.Obf$^{\mathcal{M}}(C, 1^\lambda)$ *outputs* $\tilde{C} = \left(\tilde{Z}, \{\tilde{H}_X\}_{X \in \mathcal{X}_n}, \{\tilde{H}_Y\}_{Y \in \mathcal{Y}_n} \right)$, *where each $\tilde{H}_X$ and $\tilde{H}_Y$ are sets of handles generated by the oracle $\mathcal{M}$ during obfuscation, and $\tilde{Z}$ is arbitrary auxiliary information.*
- *All* true *zero-test queries $(p, h_1, \ldots, h_m)$ — that is, zero-test queries that evaluate to* true *— made by the evaluation algorithm* xiO.Eval$^{\mathcal{M}} \left(\tilde{C}, (x, y) \right)$ *are such that for all $j \in [m]$, $h_j \in \tilde{H}_X \cup \tilde{H}_Y$, where $(X, Y) \in \mathcal{X}_n \times \mathcal{Y}_n$ are the (unique) sets such that $(x, y) \in X \times Y$.*

Corresponding notation:

- *We denote by $q_h^{\mathcal{X}} = q_h^{\mathcal{X}}(C, \lambda)$ the bound $\max_{X \in \mathcal{X}_n} |\tilde{H}_X|$ on the maximum size of the set of explicit handles corresponding to any $X \in \mathcal{X}_n$. We denote by $q_h^{\mathcal{Y}} = q_h^{\mathcal{Y}}(C, \lambda)$ the bound on $\max_{Y \in \mathcal{Y}_n} |\tilde{H}_Y|$.*

We show that any xiO.Obf scheme relative to an ideal graded encoding oracle that is in product form can be turned into one that has explicit handles in product form, but is approximately correct.

Lemma 1. *Let* xiO $=$ (xiO.Obf$^{\mathcal{M}}$, xiO.Eval$^{\mathcal{M}}$) *be an* xiO.Obf *scheme, relative to an ideal graded encoding oracle* $\mathcal{M}$, *for a collection of circuit classes* $\mathcal{C}$, *that is in* $(\mathcal{X}, \mathcal{Y})$-*product form, for some product collection* $(\mathcal{X}, \mathcal{Y})$. *Then* xiO *can be converted into a new approximately-correct scheme* xiO* *with explicit handles in* $(\mathcal{X}, \mathcal{Y})$-*product form (relative to the same oracle* $\mathcal{M}$).

Furthermore, the size of the explicit handle sets are bounded as follows

$$q_h^{\mathcal{X}} \le O\left(q_o^{\mathcal{X}}\right) \cdot \min\left(q_o^{\mathcal{X}}, |\mathcal{Y}_n| \cdot \log q_o^{\mathcal{X}}\right), \ \ q_h^{\mathcal{Y}} \le O\left(q_o^{\mathcal{Y}}\right) \cdot \min\left(q_o^{\mathcal{Y}}, |\mathcal{X}_n| \cdot \log q_o^{\mathcal{Y}}\right).$$

Our New XiO Scheme with Explicit Handles. We now describe the new obfuscator xiO. We assume w.l.o.g that $q_o^{\mathcal{X}} \ge q_o^{\mathcal{Y}}$ (otherwise, the obfuscator reverses the roles of $\mathcal{X}, \mathcal{Y}$).

The Obfuscator xiO*.Obf: Given a circuit $C \in \mathcal{C}$ with input size n, and security parameter 1^{λ}, xiO*.Obf$^{\mathcal{M}}(C, 1^{\lambda})$ does the following:

– **Obfuscate:** Emulate the obfuscator xiO.Obf$^{\mathcal{M}}(C, 1^{\lambda})$ to obtain

$$\left(\{\widetilde{C}_X \leftarrow \text{xiO.Obf}_{\mathcal{X}}^{\mathcal{M}}(C, X, 1^{\lambda})\}_{X \in \mathcal{X}_n}, \{\widetilde{C}_Y \leftarrow \text{xiO.Obf}_{\mathcal{Y}}^{\mathcal{M}}(C, Y, 1^{\lambda})\}_{Y \in \mathcal{Y}_n} \right).$$

  For each $X \in \mathcal{X}_n$ store a list L_X of all tuples (h, ξ) such that xiO.Obf$_{\mathcal{X}}^{\mathcal{M}}(C, X, 1^{\lambda})$ requested the oracle $\mathcal{M}$ to encode (ξ, ℓ) and obtained back a handle $h = (r, \ell)$. Store a similar list L_Y for each execution xiO.Obf$_{\mathcal{Y}}^{\mathcal{M}}(C, Y, 1^{\lambda})$.
– **Learn Heavy Handles for** $\mathcal{X}_n$: for each $X \in \mathcal{X}_n$, let $\widetilde{H}_X = \emptyset$.
  For $i \in \left\{1, \ldots, K_{\mathcal{X}} = \min\left(400q_o^{\mathcal{X}}, |\mathcal{Y}_n| \cdot \log\left(400q_o^{\mathcal{X}}\right)\right)\right\}$ do:
  • Sample a random $Y_i \leftarrow \mathcal{Y}_n$.
  • Emulate xiO.Eval$_{\mathcal{X}, \mathcal{Y}}^{(\cdot)}(\widetilde{C}_X, \widetilde{C}_{Y_i})$. To answer zero-test queries, emulate $\mathcal{M}$ using the lists (L_X, L_{Y_i}) constructed during the obfuscation phase.
  • In the process, for every zero-test query $(p, h_1, \ldots, h_m)$, if $\mathcal{M}(p, h_1, \ldots, h_m) = \texttt{true}$, namely it is a valid zero test and the answer is indeed zero, add $h_1, \ldots, h_m$ to $\widetilde{H}_X$.
  Store the resulting $\widetilde{H}_X$.
– **Learn Remaining Handles for** $\mathcal{Y}_n$: for each $Y \in \mathcal{Y}_n$, let $\widetilde{H}_Y = \emptyset$.
  For $i \in \left\{1, \ldots, K_{\mathcal{Y}} = \min\left(200q_o^{\mathcal{Y}}, |\mathcal{X}_n| \cdot \log\left(200q_o^{\mathcal{Y}}\right)\right)\right\}$ do the following:
  • Sample a random $X_i \leftarrow \mathcal{X}_n$, and let $\widetilde{H}_{X_i, Y} = \emptyset$.
  • Emulate xiO.Eval$_{\mathcal{X}, \mathcal{Y}}^{(\cdot)}(\widetilde{C}_{X_i}, \widetilde{C}_Y)$. To answer zero-test queries, emulate $\mathcal{M}$ using the lists (L_{X_i}, L_Y) constructed during the obfuscation phase.
  • In the process, for every zero-test query $(p, h_1, \ldots, h_m)$, if $\mathcal{M}(p, h_1, \ldots, h_m) = \texttt{true}$, namely it is a valid zero test and the answer is indeed zero, add $h_1, \ldots, h_m$ to $\widetilde{H}_{X_i, Y}$.
  • Remove from $\widetilde{H}_{X_i, Y}$ all handles in $\widetilde{H}_{X_i}$.
  • If $|\widetilde{H}_{X_i, Y}| \le q_o^{\mathcal{Y}}(C, \lambda)$, add $\widetilde{H}_{X_i, Y}$ to $\widetilde{H}_Y$. Otherwise discard $\widetilde{H}_{X_i, Y}$.
  Store the resulting $\widetilde{H}_Y$.
– **Output:**

$$\widetilde{C}^{\star} = (\widetilde{Z}, \{\widetilde{H}_X\}_{X \in \mathcal{X}_n}, \{\widetilde{H}_Y\}_{Y \in \mathcal{Y}_n}), \text{ where } \widetilde{Z} = (\{\widetilde{C}_X\}_{X \in \mathcal{X}_n}, \{\widetilde{C}_Y\}_{Y \in \mathcal{Y}_n}).$$

The Evaluator xiO*.Eval: Given an obfuscation $\widetilde{C}^\star = (\widetilde{C}, \{\widetilde{H}_X\}_{X \in \mathcal{X}_n}, \{\widetilde{H}_Y\}_{Y \in \mathcal{Y}_n})$, $(x, y) \in \boldsymbol{X}_n \times \boldsymbol{Y}_n$, xiO*.Eval$^\mathcal{M}(\widetilde{C}^\star, (x, y))$ does the following:

- Let $(X, Y) \in \mathcal{X}_n \times \mathcal{Y}_n$ be the (unique) sets such that $(x, y) \in X \times Y$.
- Emulate xiO.Eval$^{(\cdot)}_{\mathcal{X}, \mathcal{Y}}(\widetilde{C}_X, \widetilde{C}_Y)$.
- Whenever xiO.Eval makes a zero-test query $(p, h_1, \ldots, h_m)$:
  - If for some i, $h_i \notin \widetilde{H}_X \cup \widetilde{H}_Y$, answer `false`.
  - Forward any other zero-test to the oracle $\mathcal{M}$ and return its answer.

In the full version [8], we show that the new obfuscator is approximately correct, secure, and efficient as stated in Lemma 1.

3.2 Step 2: From Constant-Degree to Degree Two

We show that any XIO scheme with explicit handles in product form, relative to a degree-d decomposable ideal oracle (for arbitrary $d = O(1)$), can be transformed into one relative to a degree-2 decomposable ideal oracle. The resulting degree-2 oracle is defined with respect to a validity predicate V^2 related to the validity predicate V^d of the degree-d oracle we start with.

Intuitively, this model can be seen as an extension of the standard asymmetric bilinear maps, where instead of two base groups we may have more. That is, instead of two asymmetric base-groups G_1, G_2 where $(g_1^a, g_2^b) \in G_1 \times G_2$ can be mapped to $e(g_1, g_2)^{ab}$ in the target group G_T, we possibly have a larger number of groups $G_1, \ldots, G_n$ and a collection of *valid mappings* $\{e_k : G_{i_k} \times G_{j_k} \to G_T\}$, which may be a strict subset of all possible bilinear maps.

Lemma 2. *Let* xiO $= (\mathsf{xiO.Obf}^{(\cdot)}, \mathsf{xiO.Eval}^{(\cdot)})$ *be an XIO scheme, for a collection of circuit classes* $\mathcal{C}$, *defined relative to a degree-d decomposable ideal oracle* $\mathcal{M}^d = \{\mathcal{M}^d_{\mathbb{F}_\lambda, V_\lambda}\}$, *with explicit handles in* $(\mathcal{X}, \mathcal{Y})$-*product form, for some product collection* $(\mathcal{X}, \mathcal{Y})$. *Assume further that for some constant* $\gamma < 1$,

$$|\mathcal{X}_n| \cdot \left(q_h^\mathcal{X}\right)^d + |\mathcal{Y}_n| \cdot \left(q_h^\mathcal{Y}\right)^d \leq 2^{\gamma n} \cdot \mathrm{poly}(|C|, \lambda).$$

Then xiO *can be converted to a new scheme* xiO*, *also with explicit handles in* $(\mathcal{X}, \mathcal{Y})$-*product form, relative to a degree-2 decomposable oracle* $\mathcal{M}^2$.

We now present our new XiO scheme relative to a degree-2 decomposable oracle; see the full version for its analysis.

The New XiO Scheme Relative to a Degree-2 Oracle $\mathcal{M}^2$. In what follows, let xiO $= (\mathsf{xiO.Obf}^{(\cdot)}, \mathsf{xiO.Eval}^{(\cdot)})$ be an XIO scheme with explicit handles in product form, defined relative to a degree-d decomposable ideal oracle $\mathcal{M}^d = \{\mathcal{M}^d_{\mathbb{F}_\lambda, V_\lambda}\}$. We describe a new scheme xiO* $= (\mathsf{xiO^\star.Obf}^{(\cdot)}, \mathsf{xiO^\star.Eval}^{(\cdot)})$ (also, with explicit handles in product form) defined relative to a degree-2 decomposable ideal oracle $\mathcal{M}^2 = \{\mathcal{M}^2_{\mathbb{F}_\lambda, V_\lambda^\star}\}$.

The Obfuscator xiO*.Obf: Given a circuit $C \in \mathcal{C}$ with input size n, and security parameter 1^λ, and oracle access to $\mathcal{M}^2$, xiO*.Obf$^{\mathcal{M}^2}(C, 1^\lambda)$ does as follows:

- **Emulate Obfuscation:**
  - Emulate $\mathsf{xiO.Obf}^{\mathcal{M}^d}(C, 1^\lambda)$.
  - Throughout the emulation, emulate the oracle $\mathcal{M}^d$, storing a list $L = \{(h, \xi)\}$ of encoded element-label pairs (ξ, ℓ) and corresponding handles $h = (r, \ell)$.
  - Obtain the obfuscation $(\widetilde{Z}, \{\widetilde{H}_X\}_{X \in \mathcal{X}_n}, \{\widetilde{H}_X\}_{Y \in \mathcal{Y}_n})$.
- **Encode Monomials:**
  - For each $X \in \mathcal{X}_n$:
    1. Retrieve $\widetilde{H}_X = (h_1, \ldots, h_m)$ and the corresponding field elements and labels $(\xi_1, \ell_1), \ldots, (\xi_m, \ell_m)$ from the stored list L.
    2. For every formal monomial $\Phi(v_1, \ldots, v_m) = v_{i_1} \ldots v_{i_j}$, where $j \leq d$ and $i_1, \ldots, i_j \in [m]$, compute

    $$\Phi(\boldsymbol{\xi}) := \xi_{i_1} \cdots \xi_{i_j}, \quad \Phi(\boldsymbol{\ell}) := \{\ell_{i_1}, \ldots, \ell_{i_j}\}, \quad \Phi(\boldsymbol{h}) := \{h_{i_1}, \ldots, h_{i_j}\}.$$

    (For simplicity of notation, we overload Φ to describe different functions when acting on field elements, labels, and handles.) Then, request $\mathcal{M}^2$ to encode the field element and label $(\xi^\star_{X,\Phi}, \ell^\star_{X,\Phi}) := (\Phi(\boldsymbol{\xi}), \Phi(\boldsymbol{\ell}))$, and obtain a handle $h^\star_{X,\Phi}$.
    3. Store $\widetilde{H}^\star_X = \{(h^\star_{X,\Phi}, \Phi(\boldsymbol{h}))\}_\Phi$
  - For each $Y \in \mathcal{Y}_n$:
    1. Symmetrically perform the above two steps with respect to $\widetilde{H}_Y$ (instead of $\widetilde{H}_X$).
    2. Store $\widetilde{H}^\star_Y = \{(h^\star_{Y,\Phi}, \Phi(\boldsymbol{h}))\}_\Phi$.
- **Output:**

  $$\widetilde{C}^\star = (\widetilde{C}, \{\widetilde{H}^\star_X\}_{X \in \mathcal{X}_n}, \{\widetilde{H}^\star_Y\}_{Y \in \mathcal{Y}_n}), \text{ where } \widetilde{C} := (\widetilde{Z}, \{\widetilde{H}_X\}_X, \{\widetilde{H}_Y\}_Y).$$

The Evaluator $\mathsf{xiO}^\star.\mathsf{Eval}$: Given an obfuscation $\widetilde{C}^\star = (\widetilde{C}, \{\widetilde{H}^\star_X\}_{X \in \mathcal{X}_n}, \{\widetilde{H}^\star_Y\}_{Y \in \mathcal{Y}_n})$, input $(x, y) \in \boldsymbol{X}_n \times \boldsymbol{Y}_n$, and oracle $\mathcal{M}^2$, $\mathsf{xiO}^\star.\mathsf{Eval}^{\mathcal{M}^2}(\widetilde{C}^\star, (x, y))$ does the following:

- Emulate $\mathsf{xiO.Eval}^{\mathcal{M}^d}(\widetilde{C}, (x, y))$.
- Emulate any zero-test query $(p, h_1, \ldots, h_m)$ it makes to $\mathcal{M}^d$ as follows:
  1. Parse $\widetilde{C} = (\widetilde{Z}, \{\widetilde{H}_X\}_{X \in \mathcal{X}_n}, \{\widetilde{H}_Y\}_{Y \in \mathcal{Y}_n})$.
  2. Let $(X, Y) \in \mathcal{X}_n \times \mathcal{Y}_n$ be the (unique) sets such that $(x, y) \in X \times Y$. Retrieve $\widetilde{H}_X, \widetilde{H}_Y$.
  3. Split $\boldsymbol{h} = (h_1, \ldots, h_m)$ into two verctors of handles $\boldsymbol{h}_X \subseteq \widetilde{H}_X$ and $\boldsymbol{h}_Y \subseteq \widetilde{H}_Y$. (Such a partition always exists, by the guarantee of explicit handles in product form.)
  4. Viewing $p(\boldsymbol{h})$ as a formal polynomial in variables $\boldsymbol{h}$, factor it as

  $$p(\boldsymbol{h}) = \sum_i \gamma_i \Phi_i(\boldsymbol{h}) = \sum_i \gamma_i \Phi_{X,i}(\boldsymbol{h}_X) \Phi_{Y,i}(\boldsymbol{h}_Y),$$

  where $\gamma_i \in \mathbb{F} \backslash \{0\}$ are the coefficients, and each monomial $\Phi_i(\boldsymbol{h})$ is factored into $\Phi_{X,i}(\boldsymbol{h}_X) \cdot \Phi_{Y,i}(\boldsymbol{h}_Y)$.

5. Translate $\{\Phi_{X,i}(h_X), \Phi_{Y,i}(h_Y)\}_i$ into handles $\{h^\star_{X,i}, h^\star_{Y,i}\}_i$ by locating $(h^\star_{X,i}, \Phi_{X,i}(h_X)) \in \widetilde{H}^\star_X$ and $(h^\star_{Y,i}, \Phi_{Y,i}(h_Y)) \in \widetilde{H}^\star_Y$.

6. Consider the degree-2 formal polynomial:

$$p^\star(h^\star) = \sum_i \gamma_i h^\star_{X,i} h^\star_{Y,i}.$$

7. Make the zero-test $(p^\star, h^\star)$ to the oracle $\mathcal{M}^2$ and return the result.

Labels and Validity Predicate V^2 of Oracle $\mathcal{M}^2$. Note that labels with respect to $\mathcal{M}^2$ are subsets of the label set of $\mathcal{M}$. Let V^d be the decomposable validity predicate associated with $\mathcal{M}^d$. We define a new validity predicate of degree 2, which is also decomposable. For this purpose, we need to define V^2 for labels corresponding to bilinear monomials given by a multi-set $\{\ell^\star_1, \ell^\star_2\}$. For all other multi-sets L (with cardinality larger than 2), $V^2(L) = \texttt{false}$, capturing that this is a degree 2-predicate.

The validity predicate $V^2(\{\ell^\star_1, \ell^\star_2\})$ is computed as follows:

- Parse $\ell^\star_1$ and $\ell^\star_2$ as as two multi-sets $\{\ell_{1,1}, \ldots, \ell_{1,k_1}\}, \{\ell_{2,1}, \ldots, \ell_{2,k_2}\}$.
- Apply the original predicate to the disjoint union multi-set:

$$V^2(\{\ell^\star_1, \ell^\star_2\}) := V^d(\ell^\star_1 \uplus \ell^\star_2) = V^d(\{\ell_{1,1}, \ldots, \ell_{1,k_1}\} \uplus \{\ell_{2,1}, \ldots, \ell_{2,k_2}\}).$$

Recall that the fact that V^d is decomposable means that there exist a projection function Π^d and predicate V^d_Π, such that, for every two multi-sets $\ell^\star_1, \ell^\star_2$, $V^d(\ell^\star_1 \uplus \ell^\star_2) = V^d_\Pi(\Pi^d(\ell^\star_1), \Pi^d(\ell^\star_2))$. We show that V^2 is also decomposable, by defining its corresponding projection function Π^2 and predicate V^2_Π, and showing that on input two multisets $A = \{\ell^\star_i\}_i$ and $B = \{\ell^\star_j\}_j$, $V^2(A \uplus B) = V^2_\Pi(\Pi^2(A), \Pi^2(B))$. The projection function Π^2 on input a multiset A computes: $\Pi^2(A) = (|A|, \Pi^d(\uplus_{\ell^\star \in A} \ell^\star))$. The predicate V^2_Π on input two multisets A, B outputs $\texttt{false}$ if $|A| + |B| > 2$. Otherwise, if A, B contain exactly two labels $\ell^\star_1, \ell^\star_2$, the predicate computes:

$$V^2_\Pi(\Pi^2(A), \Pi^2(B)) = V^d(\Pi^d(\uplus_{\ell^\star \in A} \ell^\star), \Pi^d(\uplus_{\ell^\star \in B} \ell^\star))$$
$$= V^d((\uplus_{\ell^\star \in A} \ell^\star) \uplus (\uplus_{\ell^\star \in B} \ell^\star)) = V^d(\ell^\star_1 \uplus \ell^\star_2) = V^2(A \uplus B)$$

Therefore V^2 is decomposable. Moreover, it is easy to see that the arity of V^2 is exactly that of V^d, which is bounded by a fixed polynomial.

3.3 Step 3: Asymmetric Oracles to Symmetric Oracles

We show that any XIO scheme with explicit handles relative to the oracle $\mathcal{M}^2$ can be converted to a scheme relative to a symmetric bilinear oracle $\mathcal{B}^2$ (also with explicit handles). This model is analogous to the symmetric bilinear pairing groups where there is a single base group G with a bilinear map $e: G \times G \to G_T$ (Definition 8). The transformation will incur a certain blowup depending on the arity of the oracle $\mathcal{M}^2$, which is a bounded polynomial.

Lemma 3. *Let* $\mathsf{xiO} = (\mathsf{xiO.Obf}^{(\cdot)}, \mathsf{xiO.Eval}^{(\cdot)})$ *be an XIO scheme, for a collection of circuit classes* $\mathcal{C}$, *defined relative to the (asymmetric) decomposable oracle* $\mathcal{M}^2$, *with explicit handles in* $(\mathcal{X}, \mathcal{Y})$-*product form, for some product collection* $(\mathcal{X}, \mathcal{Y})$. *Then* xiO *can be converted to a new scheme* $\mathsf{xiO}^\star$ *relative to the (symmetric) oracle* $\mathcal{B}^2$, *also with explicit handles in* $(\mathcal{X}, \mathcal{Y})$-*product form.*

Towards the lemma, we show a transformation that reduces the oracle $\mathcal{M}^2$ to a symmetric bilinear oracle $\mathcal{B}^2$. In the full version [8], we use this transformation to convert any XiO scheme relative to $\mathcal{M}^2$ to one relative to $\mathcal{B}^2$.

Reducing Oracle $\mathcal{M}^2$ to Oracle $\mathcal{B}^2$. The transformation consists of a recoding process $\mathcal{E}$ that takes a secret key K, and an arbitrary encoding query of the form (ξ, ℓ) to $\mathcal{M}^2$, and transforms it into a set of new encoding queries $(\xi_1^\star, \ell_\mathcal{B}), \ldots, (\xi_k^\star, \ell_\mathcal{B})$ which it gives $\mathcal{B}^2$ (all with respect to the unique label $\ell_\mathcal{B}$). $\mathcal{E}$ then outputs a handle $\boldsymbol{h}$ representing (ξ, ℓ) consisting of a list of handles $\boldsymbol{h} = (h_1^\star, \ldots, h_k^\star)$ generated by $\mathcal{B}^2$ for $\xi_1^\star, \ldots, \xi_k^\star$.

The encoder $\mathcal{E}$ is associated with a (public) decoder $\mathcal{D}$. The decoder $\mathcal{D}$ is given as input a zero-test query $(p, \boldsymbol{h}_1, \ldots, \boldsymbol{h}_m)$ for $\mathcal{M}^2$ to be evaluated over underlying field elements $\boldsymbol{\xi} = (\xi_1, \ldots, \xi_m)$, and now represented by $\boldsymbol{\xi}^\star = (\xi_{1,1}^\star, \ldots, \xi_{1,k}^\star, \ldots, \xi_{m,1}^\star, \ldots, \xi_{m,k}^\star)$ encoded in $\mathcal{B}^2$ with handles $\boldsymbol{h}^\star = (h_{1,1}^\star, \ldots, h_{1,k}^\star, \ldots, h_{m,1}^\star, \ldots, h_{m,k}^\star)$. The decoder then translates it into a new zero-test query $(p^\star, \boldsymbol{h}^\star)$ and submits it to $\mathcal{B}^2$, with the guarantee that if the zero test is valid with respect to the validity predicate V associated with $\mathcal{M}^d$, then $p(\boldsymbol{\xi}) = p^\star(\boldsymbol{\xi}^\star)$, and otherwise, $p^\star(\boldsymbol{\xi}^\star)$ evaluates to non-zero with overwhelming probability.

We next turn to a more formal description of the transformation. In what follows, let V be an arbitrary degree-2 decomposable validity predicate, defined over pairs of labels $(\ell, \ell') \in \mathbb{L} \times \mathbb{L}$ from a label set $\mathbb{L}$, and associated with projection function Π and predicate V_Π with bounded arity $\mathsf{Arity}(V_\Pi) \le \mathrm{poly}(\lambda)$.

Secret Encoding Key. The secret key K consists of random invertible field elements $\eta_\ell, \varphi_\ell \leftarrow \mathbb{F} \setminus \{0\}$ for each label $\ell \in \mathbb{L}$, and random invertible field elements $\alpha_\pi, \beta_\pi, \gamma_\pi, \delta_\pi \leftarrow \mathbb{F} \setminus \{0\}$ for every π in the corresponding set of projections $\Gamma = \{\Pi(\{\ell\}) : \ell \in \mathbb{L}\}$.

Remark 6 (Lazy Secret-Key Sampling). Note that the total number of labels and their projection could be superpolynomial, making the secret key superpolynomial in length. To deal with such cases, the recoder uses lazy sampling to sample the above random invertible elements only when needed and keeps a record of all sampled elements. As we argue below, the total number of random invertible elements to be sampled is polynomial in the number of tuples (ξ, ℓ) to be recoded. For simplicity of exposition, we describe the procedure with respect to a key consisting of all possible random invertible elements.

Recoding. Given the secret key K and $(\xi, \ell) \in \mathbb{F} \times \mathbb{L}$, the encoder $\mathcal{E}^{\mathcal{B}^2}((\xi, \ell), K)$ does the following:

- Samples two secret shares ξ_L, ξ_R at random from $\mathbb{F}$ subject to $\xi_L + \xi_R = \xi$.
- Let $\pi = \Pi(\{\ell\})$ be the projection of $\{\ell\}$. Generates the field elements:

$$\boldsymbol{\xi}_\circ^\star := \left(\xi_{\circ,\alpha,L}^\star = \alpha_\pi \cdot \xi_L,\ \xi_{\circ,\beta,R}^\star = \beta_\pi \cdot \xi_R,\ \xi_{\circ,\gamma,L}^\star = \gamma_\pi \cdot \xi_L,\ \xi_{\circ,\delta,R}^\star = \delta_\pi \cdot \xi_R \right).$$

- Let $\mathsf{match}(\pi) = \{\pi' : \mathsf{V}_\Pi(\pi,\pi') = \mathsf{true}\}$ be the set of projections that evaluates to true with π. (For every $\pi' \in \mathsf{match}(\pi)$, and every ℓ', such that, $\pi' = \Pi(\{\ell'\})$, it holds that $V(\{\ell,\ell'\}) = \mathsf{true}$.)
  For each $\pi' \in \mathsf{match}(\pi)$, generates the field elements:

$$\boldsymbol{\xi}_{\pi'}^\star := \left(\xi_{\pi',\frac{1}{\alpha},L}^\star = \frac{1}{\alpha_\pi} \cdot \xi_L,\qquad \xi_{\pi',\frac{1}{\beta},L}^\star = \frac{1}{\beta_\pi} \cdot \xi_L, \right.$$
$$\left. \xi_{\pi',\frac{1}{\gamma},R}^\star = \frac{1}{\gamma_\pi} \cdot \xi_R,\qquad \xi_{\pi',\frac{1}{\delta},R}^\star = \frac{1}{\delta_\pi} \cdot \xi_R \right).$$

- If $V(\{\ell\}) = \mathsf{true}$, generates field elements

$$\boldsymbol{\xi}_\Delta^\star := \left(\xi_{\Delta,\eta,L}^\star = \eta_\ell \cdot \xi_L,\qquad \xi_{\Delta,\frac{1}{\eta}}^\star = \frac{1}{\eta_\ell},\qquad \xi_{\Delta,\varphi,R}^\star = \varphi_\ell \cdot \xi_R,\qquad \xi_{\Delta,\frac{1}{\varphi}}^\star = \frac{1}{\varphi_\ell} \right),$$

- Asks $\mathcal{B}^2$ to encode (with respect to the unique label $\ell_\mathcal{B}$) the field elements $\boldsymbol{\xi}_\circ^\star, (\boldsymbol{\xi}_{\pi'}^\star)_{\pi' \in \mathsf{match}(\pi)}, \boldsymbol{\xi}_\Delta^\star$ generated above, obtaining corressponding handles

$$\boldsymbol{h}^\star = \left(\boldsymbol{h}_\circ^\star, (\boldsymbol{h}_{\pi'}^\star)_{\pi' \in \mathsf{match}(\pi)}, \boldsymbol{h}_\Delta^\star \right).$$

- Outputs handles $\boldsymbol{h}^\star$.

We argue that when V has bounded $\mathsf{poly}(\lambda)$ arity, the size of the new encoding $\boldsymbol{h}^\star$ is bounded by $\mathsf{poly}(\lambda)$. This is because, $\boldsymbol{h}_\circ^\star$ and $\boldsymbol{h}_\Delta^\star$ each consists of 4 encodings, while $(\boldsymbol{h}_{\pi'}^\star)_{\pi' \in \mathsf{match}(\pi)}$ consists of $O(|\mathsf{match}(\pi)|) = \mathsf{Arity}(V_\Pi) \leq \mathsf{poly}(\lambda)$.

Decoding. Given a degree-2 polynomial p and handles $(\boldsymbol{h}_1^\star, \ldots, \boldsymbol{h}_m^\star)$, where $\boldsymbol{h}_i^\star = \boldsymbol{h}_{i,\circ}^\star, \left(\boldsymbol{h}_{i,\pi'}^\star \right)_{\pi' \in \mathsf{match}(\pi)}, \boldsymbol{h}_{i,\Delta}^\star$ the decoder $\mathcal{D}^{\mathcal{B}^2}(p, \boldsymbol{h}_1^\star, \ldots, \boldsymbol{h}_m^\star)$:

- Writes p as a formal polynomial

$$p(\boldsymbol{h}_1^\star, \ldots, \boldsymbol{h}_m^\star) = \sigma + \sum_k \rho_k \boldsymbol{h}_k^\star + \sum_{i \leq j} \rho_{i,j} \boldsymbol{h}_i^\star \boldsymbol{h}_j^\star.$$

- If for any monomial $\boldsymbol{h}_k^\star$ in p, $V(\{\ell_k\}) = \mathsf{false}$, or for any monomial $\boldsymbol{h}_i^\star \boldsymbol{h}_j^\star$, $V(\{\ell_i, \ell_j\}) = \mathsf{false}$, return false. Otherwise, continue.
- Generates a new degree-2 formal polynomial

$$p^\star(\boldsymbol{h}^\star) = \sigma + \sum_k \rho_k \cdot \left(h_{k,\Delta,\eta,L}^\star h_{k,\Delta,\frac{1}{\eta}}^\star + h_{k,\Delta,\varphi,R}^\star h_{k,\Delta,\frac{1}{\varphi}}^\star \right) +$$
$$\sum_{i \leq j} \rho_{i,j} \cdot \left(h_{i,\circ,\alpha,L}^\star h_{j,\pi_i,\frac{1}{\alpha},L}^\star + h_{i,\circ,\gamma,L}^\star h_{j,\pi_i,\frac{1}{\gamma},R}^\star + h_{i,\circ,\beta,R}^\star h_{j,\pi_i,\frac{1}{\beta},L}^\star + h_{i,\circ,\delta,R}^\star h_{j,\pi_i,\frac{1}{\delta},R}^\star \right).$$

- It submits to $\mathcal{B}^2$ the zero test $(p^\star, \boldsymbol{h}^\star)$ and returns the result.

3.4 Putting It All Together

We conclude the proof of Theorem 3.

Proof (of Theorem 3). To obtain $\mathsf{xiO}^{\star}$, we apply to xiO Lemmas 1, 2, 3.

- Lemma 1 turns xiO into an approximately-correct XIO scheme xiO_1 with explicit handles, relative to the same degree-d decomposable oracle $\mathcal{M}^d$ that xiO uses.
- Lemma 2 turns xiO_1 into an approximately-correct XiO scheme xiO_2 with explicit handles, relative to an asymmetric bilinear oracle $\mathcal{M}^2$ that is also decomposable.
- Lemma 3 turns xiO_2 into an approximately-correct XiO scheme xiO_3 with explicit handles, relative to a symmetric bilinear oracle $\mathcal{B}^2$.

The final XiO scheme xiO_3 is exactly the new XiO scheme $\mathsf{xiO}^{\star}$. By composing the three lemmas, we have that $\mathsf{xiO}^{\star}$ is approximately correct and secure. The only thing to argue that $\mathsf{xiO}^{\star}$ is also weakly succinct. Note that the obfuscated circuits of $\mathsf{xiO}^{\star}$ have the form

$$\widetilde{C} = \left(\widetilde{Z}, \{\widetilde{H}_X\}, \{\widetilde{H}_Y\}, \ \{\widetilde{H}_X^{\star}\}, \{\widetilde{H}_Y^{\star}\}, \ \{\widetilde{H}_X'\}, \{\widetilde{H}_Y'\} \right)$$

where $\widetilde{Z}$ is an obufscated circuit of the original scheme xiO, $\widetilde{H}_X$ and $\widetilde{H}_Y$ are the sets of explicit handles of $\mathcal{M}^d$ added by Lemma 1, $\widetilde{H}_X^{\star}$ and $\widetilde{H}_Y^{\star}$ are the encodings of monomials of $\mathcal{M}^2$ added by Lemma 2, $\widetilde{H}_X'$ and $\widetilde{H}_Y'$ are the re-encodings of $\mathcal{B}^2$ added by Lemma 3. By the three lemmas and the fact that the original scheme xiO is $\gamma^{\star}$-compressing and satisfies the efficiency requirement stated in Theorem 3, we have,

$$|\widetilde{C}| \leq |\widetilde{Z}| + O\left(\left| \{\widetilde{H}_X'\}, \{\widetilde{H}_Y'\} \right| \right)$$

$$\leq 2^{\gamma^{\star} n} \mathrm{poly}(\lambda, |C|) + \left(|\mathcal{X}_n| \cdot \left(q_o^{\mathcal{X}} \cdot \min\left(q_o^{\mathcal{X}}, |\mathcal{Y}_n| \cdot \log q_o^{\mathcal{X}} \right) \right)^d \right.$$

$$\left. + |\mathcal{Y}_n| \cdot \left(q_o^{\mathcal{Y}} \cdot \min\left(q_o^{\mathcal{Y}}, |\mathcal{X}_n| \cdot \log q_o^{\mathcal{Y}} \right) \right)^d \right) \cdot \mathrm{poly}(\lambda)$$

$$\leq \left(2^{\gamma^{\star} n} + 2^{\gamma n} \right) \cdot \mathrm{poly}(\lambda, |C|) \leq 2^{\gamma' n} \cdot \mathrm{poly}(\lambda, |C|),$$

for some $\gamma' < 1$. Thus,the new XIO scheme is weakly succinct.

4 From (Approximate) XIO and LWE to FE

We describe at a high-level how to use approximate XIO to construct 1-key weakly succinct FE for $\mathbf{P/poly}$, assuming LWE. The formal transformation can be found in the full version of this paper [8].

Theorem 4. *Assuming LWE with subexponential modulus-to-noise ratio and the existence of an approximate XIO scheme for $\mathbf{P}^{\log}/\mathbf{poly}$, there exists a single-key weakly-succinct FE scheme* FE *for $\mathbf{P/poly}$.*

A Failed Attempt. Lin, Pass, Seth and Telang [33] showed a transformation from correct XIO for $\mathbf{P}^{\log}/\mathbf{poly}$ to IO for $\mathbf{P}/\mathbf{poly}$, assuming LWE.[8] Previously, Bitansky and Vaikuntanathan [12] showed how to make any approximately correct IO correct (assuming, say, LWE). Thus, to prove the above theorem, a natural idea is to amplify the correctness of approximate XIO to obtain correct XIO by [12], and then invoke the transformation of [33]. This approach turns out to completely fail. Indeed, the [12] transformation only works for classes of circuits that are expressive enough; in particular, it relies on the ability of circuits in the class to process encrypted inputs, which must inherently be of super-logarithmic length in the security parameter. However, XIO for such circuit classes, which lie outside of $\mathbf{P}^{\log}/\mathbf{poly}$, is inefficient (see Remark 1).

Instead, we show how to modify the transformation of [33], based on error-correcting codes, so that, it works directly with approximate XIO. Below, we briefly review the [33] transformation and describe our key ideas.

Review of the [33] Transformation. Goldwassar et al. [27] constructed, from LWE with subexponential modulus-to-noise ratio, a fully succinct, public-key, single-key, FE scheme for *Boolean* $\mathbf{NC}^1$ circuits; namely, the encryption circuit of their scheme has size $\mathrm{poly}(n, \lambda)$, where n is the message length.

Starting from such an FE scheme bFE for Boolean circuits, the first observation in [33] is as follows: To construct an FE scheme, FE for any (possibly non-Boolean) circuit C, one can use bFE to issue a key for the corresponding Boolean circuit B that produces *one output bit at a time*, that is, $B(m, i) = (C(m))_i$. Then to enable evaluating the circuit C, it suffices to publish a list of bFE ciphertexts encrypting all pairs (m, i). This, however, leads to a scheme with encryption time linear in the length of the output (as it needs to produce a ciphertext for every output bit), and is not weakly succinct. The key idea in [33] is using XIO to generate the list of encrypted pairs (m, i). Namely, obfuscate a circuit that given as input i, outputs the encryption of (m, i), where randomness is derived with a pseudorandom function. Since XIO achieves "sublinear compression", the resulting FE scheme is now weakly succinct for all of $\mathbf{NC}^1$, including circuits with non-Boolean output.

Our Approach. The basic idea behind replacing XIO with approximate XIO is to use good error-correcting codes to allow recovering the output of a given function even if some of the encryptions (m, i) are faulty. Specifically, we make the following modification to the transformation of [33]. Instead of deriving a key for the Boolean function $B(m, i) = (C(m))_i$, which computes the i-th bit of the circuit's output, we consider the function $B^\star(m, i) = (\mathsf{ECC}(C(m)))_i$ that outputs the i-th bit of an error-corrected version of this output. As before, we use XIO to to generate the list of encryptions (m, i), only that now, with approximate XIO, some of these encryptions may be faulty. Nevertheless, we can still recover $(\mathsf{ECC}(C(m)))_i$ for a large enough fraction of indices i, and can thus correct, and obtain $C(m)$. By using codes with constant rate, and a linear-size constant-depth encoding circuit, we can show that this transformation achieves the required compression.

---

[8] The LWE assumption was later weakened to the existence of public key encryption by [9], but only for sufficiently-compressing XIO.

Acknowledgements. We thank V. Vaikuntanathan for enlightening discussions.

References

1. Ananth, P., Jain, A.: Indistinguishability obfuscation from compact functional encryption. In: Gennaro, R., Robshaw, M. (eds.) CRYPTO 2015. LNCS, vol. 9215, pp. 308–326. Springer, Heidelberg (2015). doi:10.1007/978-3-662-47989-6_15
2. Ananth, P., Jain, A., Sahai, A.: Achieving compactness generically: indistinguishability obfuscation from non-compact functional encryption. IACR Cryptology ePrint Archive 2015, 730 (2015)
3. Ananth, P., Sahai, A.: Projective arithmetic functional encryption and indistinguishability obfuscation from degree-5 multilinear maps. IACR Cryptology ePrint Archive 2016, 1097 (2016)
4. Applebaum, B., Brakerski, Z.: Obfuscating circuits via composite-order graded encoding. In: Dodis, Y., Nielsen, J.B. (eds.) TCC 2015. LNCS, vol. 9015, pp. 528–556. Springer, Heidelberg (2015). doi:10.1007/978-3-662-46497-7_21
5. Barak, B., Garg, S., Kalai, Y.T., Paneth, O., Sahai, A.: Protecting obfuscation against algebraic attacks. In: Nguyen, P.Q., Oswald, E. (eds.) EUROCRYPT 2014. LNCS, vol. 8441, pp. 221–238. Springer, Heidelberg (2014). doi:10.1007/978-3-642-55220-5_13
6. Barak, B., Goldreich, O., Impagliazzo, R., Rudich, S., Sahai, A., Vadhan, S.P., Yang, K.: On the (im)possibility of obfuscating programs. J. ACM **59**(2), 6 (2012)
7. Bitansky, N., Goldwasser, S., Jain, A., Paneth, O., Vaikuntanathan, V., Waters, B.: Time-lock puzzles from randomized encodings. In Sudan, M. (ed.) ITCS 2016: 7th Innovations in Theoretical Computer Science, Cambridge, MA, USA, pp. 345–356. Association for Computing Machinery, 14–16 January 2016
8. Bitansky, N., Lin, H., Paneth, O.: On removing graded encodings from functional encryption. IACR Cryptology ePrint Archive 2016, 962 (2016)
9. Bitansky, N., Nishimaki, R., Passelègue, A., Wichs, D.: From cryptomania to obfustopia through secret-key functional encryption. In: Hirt, M., Smith, A. (eds.) TCC 2016. LNCS, vol. 9986, pp. 391–418. Springer, Heidelberg (2016). doi:10.1007/978-3-662-53644-5_15
10. Bitansky, N., Paneth, O., Rosen, A.: On the cryptographic hardness of finding a Nash equilibrium. In: Guruswami, V. (ed.) 56th Annual Symposium on Foundations of Computer Science, Berkeley, CA, USA, pp. 1480–1498. IEEE Computer Society Press, 17–20 October 2015
11. Bitansky, N., Vaikuntanathan, V.: Indistinguishability obfuscation from functional encryption. In: IEEE 56th Annual Symposium on Foundations of Computer Science, FOCS 2015, Berkeley, CA, USA, pp. 171–190, 17–20 October 2015
12. Bitansky, N., Vaikuntanathan, V.: Indistinguishability obfuscation: from approximate to exact. In: Kushilevitz, E., Malkin, T. (eds.) TCC 2016. LNCS, vol. 9562, pp. 67–95. Springer, Heidelberg (2016). doi:10.1007/978-3-662-49096-9_4
13. Boneh, D., Boyen, X., Goh, E.-J.: Hierarchical identity based encryption with constant size ciphertext. In: Cramer, R. (ed.) EUROCRYPT 2005. LNCS, vol. 3494, pp. 440–456. Springer, Heidelberg (2005). doi:10.1007/11426639_26
14. Boneh, D., Franklin, M.: Identity-based encryption from the weil pairing. In: Kilian, J. (ed.) CRYPTO 2001. LNCS, vol. 2139, pp. 213–229. Springer, Heidelberg (2001). doi:10.1007/3-540-44647-8_13
15. Boneh, D., Wu, D.J., Zimmerman, J.: Immunizing multilinear maps against zeroizing attacks. IACR Cryptology ePrint Archive 2014, 930 (2014)

28   N. Bitansky et al.

16. Boyen, X.: The uber-assumption family. In: Galbraith, S.D., Paterson, K.G. (eds.) Pairing 2008. LNCS, vol. 5209, pp. 39–56. Springer, Heidelberg (2008). doi:10.1007/978-3-540-85538-5_3

17. Brakerski, Z., Rothblum, G.N.: Virtual black-box obfuscation for all circuits via generic graded encoding. In: Lindell, Y. (ed.) TCC 2014. LNCS, vol. 8349, pp. 1–25. Springer, Heidelberg (2014). doi:10.1007/978-3-642-54242-8_1

18. Bun, M., Zhandry, M.: Order-revealing encryption and the hardness of private learning. In: Kushilevitz, E., Malkin, T. (eds.) TCC 2016. LNCS, vol. 9562, pp. 176–206. Springer, Heidelberg (2016). doi:10.1007/978-3-662-49096-9_8

19. Canetti, R., Kalai, Y.T., Paneth, O.: On obfuscation with random oracles. In: Dodis, Y., Nielsen, J.B. (eds.) TCC 2015. LNCS, vol. 9015, pp. 456–467. Springer, Heidelberg (2015). doi:10.1007/978-3-662-46497-7_18

20. Cheon, J.H., Han, K., Lee, C., Ryu, H., Stehlé, D.: Cryptanalysis of the multi-linear map over the integers. In: Oswald, E., Fischlin, M. (eds.) EUROCRYPT 2015. LNCS, vol. 9056, pp. 3–12. Springer, Heidelberg (2015). doi:10.1007/978-3-662-46800-5_1

21. Coron, J.-S., Gentry, C., Halevi, S., Lepoint, T., Maji, H.K., Miles, E., Raykova, M., Sahai, A., Tibouchi, M.: Zeroizing without low-level zeroes: new MMAP attacks and their limitations. In: Gennaro, R., Robshaw, M. (eds.) CRYPTO 2015. LNCS, vol. 9215, pp. 247–266. Springer, Heidelberg (2015). doi:10.1007/978-3-662-47989-6_12

22. Garg, S., Gentry, C., Halevi, S.: Candidate multilinear maps from ideal lattices. In: Johansson, T., Nguyen, P.Q. (eds.) EUROCRYPT 2013. LNCS, vol. 7881, pp. 1–17. Springer, Heidelberg (2013). doi:10.1007/978-3-642-38348-9_1

23. Garg, S., Gentry, C., Halevi, S., Raykova, M.: Two-round secure MPC from indistinguishability obfuscation. In: Lindell, Y. (ed.) TCC 2014. LNCS, vol. 8349, pp. 74–94. Springer, Heidelberg (2014). doi:10.1007/978-3-642-54242-8_4

24. Garg, S., Gentry, C., Halevi, S., Raykova, M., Sahai, A., Waters, B.: Candidate indistinguishability obfuscation and functional encryption for all circuits. In: 54th Annual Symposium on Foundations of Computer Science, Berkeley, CA, USA, pp. 40–49. IEEE Computer Society Press, 26–29 October 2013

25. Garg, S., Gentry, C., Halevi, S., Zhandry, M.: Functional encryption without obfuscation. In: Kushilevitz, E., Malkin, T. (eds.) TCC 2016. LNCS, vol. 9563, pp. 480–511. Springer, Heidelberg (2016). doi:10.1007/978-3-662-49099-0_18

26. Garg, S., Srinivasan, A.: Unifying security notions of functional encryption. IACR Cryptology ePrint Archive 2016, 524 (2016)

27. Goldwasser, S., Kalai, Y.T., Popa, R.A., Vaikuntanathan, V., Zeldovich, N.: Reusable garbled circuits and succinct functional encryption. In: Boneh, D., Roughgarden, T., Feigenbaum, J. (eds.) 45th Annual ACM Symposium on Theory of Computing, Palo Alto, CA, USA, pp. 555–564. ACM Press, 1–4 June 2013

28. Gorbunov, S., Vaikuntanathan, V., Wee, H.: Functional encryption with bounded collusions via multi-party computation. In: Safavi-Naini, R., Canetti, R. (eds.) CRYPTO 2012. LNCS, vol. 7417, pp. 162–179. Springer, Heidelberg (2012). doi:10.1007/978-3-642-32009-5_11

29. Joux, A.: The weil and tate pairings as building blocks for public key cryptosystems. In: Fieker, C., Kohel, D.R. (eds.) ANTS 2002. LNCS, vol. 2369, pp. 20–32. Springer, Heidelberg (2002). doi:10.1007/3-540-45455-1_3

30. Li, B., Micciancio, D.: Compactness vs collusion resistance in functional encryption. IACR Cryptology ePrint Archive 2016, 561 (2016)

31. Lin, H.: Indistinguishability obfuscation from constant-degree graded encoding schemes. In: Fischlin, M., Coron, J.-S. (eds.) EUROCRYPT 2016. LNCS, vol. 9665, pp. 28–57. Springer, Heidelberg (2016). doi:10.1007/978-3-662-49890-3_2

32. Lin, H.: Indistinguishability obfuscation from DDH on 5-linear maps and locality-5 prgs. IACR Cryptology ePrint Archive 2016, 1096 (2016)

33. Lin, H., Pass, R., Seth, K., Telang, S.: Indistinguishability obfuscation with non-trivial efficiency. In: Cheng, C.-M., Chung, K.-M., Persiano, G., Yang, B.-Y. (eds.) PKC 2016. LNCS, vol. 9615, pp. 447–462. Springer, Heidelberg (2016). doi:10.1007/978-3-662-49387-8_17

34. Lin, H., Pass, R., Seth, K., Telang, S.: Output-compressing randomized encodings and applications. In: Kushilevitz, E., Malkin, T. (eds.) TCC 2016. LNCS, vol. 9562, pp. 96–124. Springer, Heidelberg (2016). doi:10.1007/978-3-662-49096-9_5

35. Lin, H., Vaikuntanathan, V.: Indistinguishability obfuscation from ddh-like assumptions on constant-degree graded encodings. In: IEEE 57th Annual Symposium on Foundations of Computer Science, FOCS 2016 (2016)

36. Mahmoody, M., Mohammed, A., Nematihaji, S.: On the impossibility of virtual black-box obfuscation in idealized models. In: Kushilevitz, E., Malkin, T. (eds.) TCC 2016. LNCS, vol. 9562, pp. 18–48. Springer, Heidelberg (2016). doi:10.1007/978-3-662-49096-9_2

37. Maurer, U.: Abstract models of computation in cryptography. In: Smart, N.P. (ed.) Cryptography and Coding 2005. LNCS, vol. 3796, pp. 1–12. Springer, Heidelberg (2005). doi:10.1007/11586821_1

38. Miles, E., Sahai, A., Zhandry, M.: Annihilation attacks for multilinear maps: cryptanalysis of indistinguishability obfuscation over GGH13. In: Robshaw, M., Katz, J. (eds.) CRYPTO 2016. LNCS, vol. 9815, pp. 629–658. Springer, Heidelberg (2016). doi:10.1007/978-3-662-53008-5_22

39. Pass, R., Shelat, A.: Impossibility of VBB obfuscation with ideal constant-degree graded encodings. In: Kushilevitz, E., Malkin, T. (eds.) TCC 2016. LNCS, vol. 9562, pp. 3–17. Springer, Heidelberg (2016). doi:10.1007/978-3-662-49096-9_1

40. Reingold, O., Trevisan, L., Vadhan, S.: Notions of reducibility between cryptographic primitives. In: Naor, M. (ed.) TCC 2004. LNCS, vol. 2951, pp. 1–20. Springer, Heidelberg (2004). doi:10.1007/978-3-540-24638-1_1

41. Sahai, A., Seyalioglu, H.: Worry-free encryption: functional encryption with public keys. In Al-Shaer, E., Keromytis, A.D., Shmatikov, V. (eds.) ACM CCS 2010: 17th Conference on Computer and Communications Security, Chicago, Illinois, USA, pp. 463–472. ACM Press, 4–8 October 2010

42. Sahai, A., Waters, B.: How to use indistinguishability obfuscation: deniable encryption, and more. In: Shmoys, D.B. (ed.) 46th Annual ACM Symposium on Theory of Computing, pp. 475–484. ACM Press, New York, 31 May–3 June 2014

43. Shoup, V.: Lower bounds for discrete logarithms and related problems. In: Fumy, W. (ed.) EUROCRYPT 1997. LNCS, vol. 1233, pp. 256–266. Springer, Heidelberg (1997). doi:10.1007/3-540-69053-0_18

44. Zimmerman, J.: How to obfuscate programs directly. In: Oswald, E., Fischlin, M. (eds.) EUROCRYPT 2015. LNCS, vol. 9057, pp. 439–467. Springer, Heidelberg (2015). doi:10.1007/978-3-662-46803-6_15

Functional Encryption: Deterministic to Randomized Functions from Simple Assumptions

Shashank Agrawal[1] and David J. Wu[2(✉)]

[1] Visa Research, Palo Alto, USA
shaagraw@visa.com
[2] Stanford University, Stanford, USA
dwu4@cs.stanford.edu

Abstract. Functional encryption (FE) enables fine-grained control of sensitive data by allowing users to only compute certain functions for which they have a key. The vast majority of work in FE has focused on deterministic functions, but for several applications such as privacy-aware auditing, differentially-private data release, proxy re-encryption, and more, the functionality of interest is more naturally captured by a *randomized function.* Recently, Goyal et al. (TCC 2015) initiated a formal study of *FE for randomized functionalities* with security against *malicious encrypters*, and gave a selectively secure construction from indistinguishability obfuscation. To date, this is the only construction of FE for randomized functionalities in the public-key setting. This stands in stark contrast to FE for deterministic functions which has been realized from a variety of assumptions.

Our key contribution in this work is a *generic transformation* that converts any general-purpose, public-key FE scheme for deterministic functionalities into one that supports randomized functionalities. Our transformation uses the underlying FE scheme in a black-box way and can be instantiated using very standard number-theoretic assumptions (for instance, the DDH and RSA assumptions suffice). When applied to existing FE constructions, we obtain several *adaptively-secure*, public-key functional encryption schemes for randomized functionalities with security against malicious encrypters from many different assumptions such as concrete assumptions on multilinear maps, indistinguishability obfuscation, and in the bounded-collusion setting, the existence of public-key encryption, together with standard number-theoretic assumptions.

Additionally, we introduce a new, stronger definition for malicious security as the existing one falls short of capturing an important class

S. Agrawal—Part of this work was done when the author was a graduate student at the University of Illinois, Urbana-Champaign, supported by NSF CNS 12-28856 and the Andrew & Shana Laursen fellowship.

D.J. Wu—This work was supported in part by NSF, DARPA, the Simons foundation, a grant from ONR, and an NSF Graduate Research Fellowship. Opinions, findings and conclusions or recommendations expressed in this material are those of the author(s) and do not necessarily reflect the views of DARPA.

J.-S. Coron and J.B. Nielsen (Eds.): EUROCRYPT 2017, Part II, LNCS 10211, pp. 30–61, 2017.
DOI: 10.1007/978-3-319-56614-6_2

of correlation attacks. In realizing this definition, our compiler combines ideas from disparate domains like related-key security for pseudorandom functions and deterministic encryption in a novel way. We believe that our techniques could be useful in expanding the scope of new variants of functional encryption (e.g., multi-input, hierarchical, and others) to support randomized functionalities.

1   Introduction

Traditionally, encryption schemes have provided an all-or-nothing approach to data access: a user who holds the secret key can completely recover the message from a ciphertext while a user who does not hold the secret key learns nothing at all from the ciphertext. In the last fifteen years, numerous paradigms, such as identity-based encryption [31,45,85], attribute-based encryption [24,66,84], predicate encryption [37,71,75,78], and more have been introduced to enable more fine-grained access control on encrypted data. More recently, the cryptographic community has worked to unify these different paradigms under the general umbrella of functional encryption (FE) [35,79,83].

At a high level, an FE scheme enables delegation of decryption keys that allow users to learn specific functions of the data, and nothing else. More precisely, given a ciphertext for a message x and a secret key for a function f, one can only learn the value $f(x)$. In the last few years, numerous works have explored different security notions [3,4,7,16,23,35,79] as well as constructions from a wide range of assumptions [8,10,50,55,62,64,86]. Until very recently, the vast majority of work in functional encryption has focused on *deterministic functionalities*, i.e., on schemes that issue keys for deterministic functions only. However, there are many scenarios where the functionality of interest is more naturally captured by a *randomized function*. The first two examples below are adapted from those of Goyal et al. [65].

Privacy-aware auditing. Suppose a government agency is tasked with monitoring various financial institutions to ensure that their day-to-day activity is compliant with federal regulations. The financial institutions do not want to give complete access of their confidential data to any external auditor. Partial access is insufficient if the financial institution is able to (adversarially) choose which part of its database to expose. An ideal solution should allow the institutions to encrypt their database before providing access. Next, the government agency can give the external auditors a key that allows them to sample a small number of *randomly chosen* records from each database.

Constructing an encryption scheme that supports this kind of sampling functionality is non-trivial for several reasons. If an auditor obtains two independent keys from the government agency, applying them to the *same* encrypted database should nonetheless generate two *independent* samples from it. On the flip side, if the same key is applied to two distinct databases, the auditor should obtain an *independent* sample from each.

Another source of difficulty that arises in this setting is that the encryption is performed locally by the financial institution. Thus, if malicious institutions are able to construct "bad" ciphertexts such that the auditor obtains correlated or non-uniform samples from the encrypted databases, then they can completely compromise the integrity of the audit. Hence, any encryption scheme we design for privacy-aware auditing must also protect against malicious encrypters.

Differential privacy. Suppose a consortium of hospitals, in an effort to promote medical research, would like to provide restricted access to their patient records to approved scientists. In particular, they want to release information in a differentially-private manner to protect the privacy of their patients. The functionality of interest in this case is the evaluation of some differentially-private mechanism, which is always a randomized function. Thus, the scheme used to encrypt patient data should also support issuing keys for randomized functions. These keys would be managed by the consortium.

Proxy re-encryption. In a proxy re-encryption system, a proxy is able to transform a ciphertext encrypted under Alice's public key into one encrypted under Bob's public key [13]. Such a capability is very useful if, for example, Alice wants to forward her encrypted emails to her secretary Bob while she is away on vacation [27]. We refer to [13] for other applications of this primitive.

Proxy re-encryption can be constructed very naturally from a functional encryption scheme that supports randomized functionalities. For instance, in the above example, Alice would generate a master public/secret key-pair for an FE scheme that supports randomized functionalities. When users send mail to Alice, they would encrypt under her master public key. Then, when Alice goes on vacation, she can delegate her email to Bob by simply giving her mail server a *re-encryption key* that re-encrypts emails for Alice under Bob's public key. Since standard semantically-secure encryption is necessarily randomized, this re-encryption functionality is a randomized functionality. In fact, in this scenario, Alice can delegate an arbitrary decryption capability to other parties. For instance, she can issue a key that only re-encrypts emails tagged with "work" to Bob. Using our solution, the re-encryption function does not require interaction with Bob or knowledge of any of Bob's secrets.

Randomized functional encryption. Motivated by these applications, Alwen et al. [8] and Goyal et al. [65] were the first to formally study the problem of FE for randomized functionalities. In such an FE scheme, a secret key for a randomized function f and an encryption of a message x should reveal *a single sample* from the output distribution of $f(x)$. Moreover, given a collection of secret keys $\mathsf{sk}_{f_1}, \ldots, \mathsf{sk}_{f_n}$ for functions $f_1, \ldots, f_n$, and ciphertexts $\mathsf{ct}_{x_1}, \ldots, \mathsf{ct}_{x_n}$ corresponding to messages $x_1, \ldots, x_n$, where neither the functions nor the messages need to be distinct, each secret key sk_{f_i} and ciphertext ct_{x_j} should reveal an *independent* draw from the output distribution of $f_i(x_j)$, and nothing more.

In supporting randomized functionalities, handling *malicious encrypters* is a central issue: a malicious encrypter may construct a ciphertext for a message x such that when decrypted with a key for f, the resulting distribution differs significantly from that of $f(x)$. For instance, in the auditing application discussed earlier, a malicious bank could manipulate the randomness used to sample records in its database, thereby compromising the integrity of the audit. We refer to [65] for a more thorough discussion on the importance of handling malicious encrypters.

1.1   Our Contributions

To date, the only known construction of public-key FE for randomized functionalities secure against malicious encrypters is due to Goyal et al. [65] and relies on indistinguishability obfuscation ($i\mathcal{O}$) [15,55] together with one-way functions. However, $i\mathcal{O}$ is not a particularly appealing assumption since the security of existing $i\mathcal{O}$ constructions either rely on an exponential number of assumptions [11,14,40,80,87], or on a polynomial set of assumptions but with an exponential loss in the security reduction [58,59]. This shortcoming may even be inherent, as suggested by [57]. Moreover, numerous recent attacks on multilinear maps (the underlying primitive on which all candidate constructions $i\mathcal{O}$ are based) [38,42–44,46,47,69,77] have reduced the community's confidence in the security of existing constructions of $i\mathcal{O}$.

On the other hand, functional encryption for deterministic functions (with different levels of security and efficiency) can be realized from a variety of assumptions such as the existence of public-key encryption [63,83], learning with errors [62], indistinguishability obfuscation [55,86], multilinear maps [56], and more. Thus, there is a very large gap between the assumptions needed to build FE schemes for deterministic functionalities and those needed for randomized functionalities. Hence, it is important to ask:

Does extending public-key FE to support the richer class of randomized functions require strong additional assumptions such as $i\mathcal{O}$?

If there was a general transformation that we could apply to any FE scheme for deterministic functions, and obtain one that supported randomized functions, then we could leverage the extensive work on FE for the former to build FE for the latter with various capabilities and security guarantees. In this paper, we achieve exactly this. We bridge the gap between FE schemes for deterministic and randomized functionalities by showing that any general-purpose, simulation-secure FE scheme for deterministic functionalities can be extended to support randomized functionalities with security against malicious encrypters. Our generic transformation applies to any general-purpose, simulation-secure FE scheme with perfect correctness and only requires fairly mild additional assumptions (e.g., the decisional Diffie-Hellman (DDH) [29] and the RSA [30,82] assumptions suffice). Moreover, our transformation is tight in the sense that it preserves the security of the underlying FE scheme. Because our transformation relies only on simple additional assumptions, future work in constructing general-purpose FE can primarily focus on handling deterministic functions rather than devising specialized constructions to support randomized functions. We now give an informal statement of our main theorem:

Theorem 1.1 (Main theorem, informal). *Under standard number-theoretic assumptions, given any general-purpose, public-key functional encryption scheme for deterministic functions, there exists a general-purpose, public-key functional encryption scheme for randomized functions secure against malicious encrypters.*

In this work, we focus on simulation-based notions of security for FE. As shown by several works [35, 79], game-based formulations of security are inadequate if the function family under consideration has some computational hiding properties. Moreover, as noted by Goyal et al. [65, Remark 2.8], the natural notion of indistinguishability-based security in the randomized setting can potentially introduce circularities in the definition and render it vacuous. Additionally, there are generic ways to boost the security of FE for deterministic functionalities from a game-based notion to a simulation-based notion [50].

We do note though that these generic indistinguishability-to-simulation boosting techniques sometimes incur a loss in expressiveness (due to the lower bounds associated with simulation-based security for FE [5, 7, 35, 79]). For instance, while it is possible to construct a general-purpose FE scheme secure against adversaries that makes an arbitrary (polynomial) number of secret key queries under an indistinguishability-based notion of security, an analogous construction is impossible under a simulation-based notion of security. We leave as an important open problem the development of a generic transformation like the one in Theorem 1.1 that applies to (public-key) FE schemes which satisfy indistinguishability-based notions of security and which does not incur the loss in expressiveness associated with first boosting to a simulation-based notion of security. Such a transformation is known in the secret-key setting [73], though it does not provide security against malicious encrypters.

Concrete instantiations. Instantiating Theorem 1.1 with existing FE schemes such as [55, 56, 64] and applying transformations like [10, 26, 50, 51] to boost correctness and/or security, we obtain several new public-key FE schemes for randomized functionalities with *adaptive* simulation-based security against malicious encrypters. For example, if we start with

- the GVW scheme [63], we obtain a scheme secure under bounded collusions assuming the existence of semantically-secure public-key encryption and low-depth pseudorandom generators.
- the GGHZ scheme [56], we obtain a scheme with best-possible simulation security relying on the polynomial hardness of concrete assumptions on composite-order multilinear maps [36, 48, 49].
- the GGHRSW scheme [55], we obtain a scheme with best-possible simulation security from indistinguishability obfuscation and one-way functions.

The second and third schemes above should be contrasted with the one given by Goyal et al. [65], which achieves *selective* security assuming the existence of $i\mathcal{O}$. We describe these instantiations in greater detail in Sect. 5.

Security definition. We also propose a strong simulation-based definition for security against malicious encrypters, strengthening the one given by Goyal et al. [65]. We first give a brief overview of their definition in Sect. 1.2 and then show why it does not capture an important class of correlation attacks. We also discuss the subtleties involved in extending their definition.

Our techniques. At a very high level, we must balance between two conflicting goals in order to achieve our strengthened security definition. On the one hand, the encryption and key-generation algorithms must be randomized to ensure that the decryption operation induces the correct output distribution, or even more fundamentally, that the scheme is semantically-secure. On the other hand, a malicious encrypter could exploit its freedom to choose the randomness when constructing ciphertexts in order to induce correlations when multiple cipher-texts or keys are operated upon. We overcome this barrier by employing ideas from disparate domains like related-key security for pseudorandom functions and deterministic encryption in a novel way. We discuss our transformation and the tools involved in more detail in Sect. 1.3.

We believe that our techniques could be used to extend the capability of new variants of functional encryption like multi-input FE [32,61], hierarchical or delegatable FE [9,39], and others so that they can support randomized functionalities with security against malicious encrypters as well.

Other related work. Recently, Komargodski et al. [73] studied the same question of extending standard FE to FE for randomized functionalities, but restricted to the private-key setting. They show that starting from any "function-private" secret-key FE scheme for deterministic functionalities, a secret-key FE scheme for randomized functionalities can be constructed (though without robustness against malicious encrypters). However, as we discuss below, it seems challenging to extend their techniques to work in the public-key setting:

- The types of function-privacy that are achievable in the public-key setting are much more limited (primarily because the adversary can encrypt messages of its own and decrypt them in order to learn something about the underlying function keys). For instance, in the case of identity-based and subspace-membership encryption schemes, function privacy is only possible if we assume the function keys are drawn from certain high-entropy distributions [33,34].
- An adversary has limited control over ciphertexts in the private-key setting. For instance, since it cannot construct new ciphertexts by itself, it can only maul honestly-generated ciphertexts. In such a setting, attacks can often be prevented using zero-knowledge techniques.

Concurrent with [65], Alwen et al. [8] also explored the connections between FE for deterministic functionalities and FE for randomized functionalities. Their construction focused only on the simpler case of handling honest encrypters and moreover, they worked under an indistinguishability-based notion of security that has certain circularity problems (see the discussion in [65, Remark 2.8]) which might render it vacuous.

1.2 Security Against Malicious Encrypters

Simulation security. Informally, simulation security for FE schemes supporting randomized functionalities states that the output of any efficient adversary
with a secret key for a randomized function f and an encryption of a message
x can be simulated given only $f(x; r)$, where the randomness r used to evaluate
f is independently and uniformly sampled. Goyal et al. [65] extend this notion
to include security against malicious encrypters by further requiring that the
output of any efficient adversary holding a secret key for a function g and a
(possibly dishonestly-generated) ciphertext $\hat{\mathsf{ct}}$ should be simulatable given only
$g(\hat{x}; r)$, where $\hat{x}$ is a message that is information-theoretically fixed by $\hat{\mathsf{ct}}$, and
the randomness r is uniform and unknown to the adversary. This captures the
notion that a malicious encrypter is unable to influence the randomness used to
evaluate the function during decryption.

More formally, in the simulation-based definitions of security [35,79], an
adversary tries to distinguish its interactions in a real world where ciphertexts
and secret keys are generated according to the specifications of the FE scheme
from its interactions in an ideal world where they are constructed by a simulator
given only a minimal amount of information. To model security against malicious
encrypters, Goyal et al. give the adversary access to a decryption oracle in the
security game (similar to the formulation of IND-CCA2 security [81]) that takes
as input a *single* ciphertext ct along with a function f. In the real world, the
challenger first extracts a secret key sk_f for f and then outputs the decryption
of ct with sk_f. In the ideal world, the challenger invokes the simulator on ct. The
simulator then outputs a value x (or a special symbol $\perp$), at which point the
challenger replies to the adversary with an independently uniform value drawn
from the distribution $f(x)$ (or $\perp$).

Limitations of the existing definition. While the definition in [65] captures
security against dishonest encrypters when dealing with deterministic functionalities, it does not fully capture the desired security goals in the randomized
setting. Notably, the security definition only considers *one* ciphertext. However,
when extending functional encryption to randomized functionalities, we are also
interested in the joint distribution of *multiple* ciphertexts and secret keys. Thus,
while it is the case that in any scheme satisfying the security definition in [65],
the adversary cannot produce any single ciphertext that decrypts improperly, a
malicious encrypter could still produce a collection of ciphertexts such that when
the same key is used for decryption, the outputs are correlated. In the auditing
application discussed before, it is imperative to prevent this type of attack, for
otherwise, the integrity of the audit can be compromised.

Strengthening the definition. A natural way to strengthen Goyal et al.'s definition is to allow the decryption oracle to take in a set of (polynomially-many)
ciphertexts along with a function f. In the real world, the challenger extracts

a single key sk_f for f and applies the decryption algorithm with sk_f to each ciphertext. In the ideal world, the simulator is given the set of ciphertexts and is allowed to query the evaluation oracle $\mathcal{O}_f$ once for each ciphertext submitted. On each query x, the oracle responds with a fresh evaluation of $f(x)$. This direct extension, however, is too strong, and not achievable by any existing scheme. Suppose that an adversary could efficiently find two ciphertexts $ct_1 \neq ct_2$ such that for all secret keys sk, $Decrypt(sk, ct_1) = Decrypt(sk, ct_2)$, then it can easily distinguish the real and ideal distributions. When queried with $(f, (ct_1, ct_2))$, the decryption oracle always replies with two identical values in the real world irrespective of what f is. In the ideal world, however, it replies with two independent values since fresh randomness is used to evaluate f every time.

While we might want to preclude this type of behavior with our security definition, it is also one that arises naturally. For example, in both Goyal et al.'s and our construction, ciphertexts have the form (ct', π) where ct' is the ciphertext component that is actually combined with the decryption key and π is a proof of the well-formedness of ct'. Decryption proceeds only if the proof verifies. Since the proofs are randomized, an adversary can construct a valid ciphertext component ct' and two distinct proofs π_1, π_2 and submit the pair of ciphertexts (ct', π_1) and (ct', π_2) to the decryption oracle. Since π_1 and π_2 do not participate in the decryption process after verification, these two ciphertexts are effectively identical from the perspective of the decryption function. However, as noted above, an adversary that can construct such ciphertexts can trivially distinguish between the real and ideal worlds.

Intuitively, if the adversary submitted the *same* ciphertext multiple times in a decryption query, it does not make sense for the decryption oracle to respond with independently distributed outputs in the ideal experiment. The expected behavior is that the decryption oracle responds with the same value on all identical ciphertexts. In our setting, we allow for this behavior by considering a generalization of "ciphertext equivalence." In particular, when the adversary submits a decryption query, the decryption oracle in the ideal experiment responds consistently on all equivalent ciphertexts that appear in the query. Formally, we capture this by introducing an efficiently-checkable equivalence relation on the ciphertext space of the FE scheme. For example, if the ciphertexts have the form (ct', π), one valid equivalence relation on ciphertexts is equality of the ct' components. To respond to a decryption query, the challenger first groups the ciphertexts according to their equivalence class, and responds consistently for all ciphertexts belonging to the same class. Thus, without loss of generality, it suffices to just consider adversaries whose decryption queries contain at most one representative from each equivalence class. We provide a more thorough discussion of our strengthened definition in Sect. 3.

As far as we understand, the Goyal et al. construction remains secure under our strengthened notion of security against malicious encrypters, but it was only shown to be selectively secure assuming the existence of $i\mathcal{O}$ (and one-way functions).[1] Our transformation, on the other hand, provides a *generic* way of building *adaptively-secure* schemes from both $i\mathcal{O}$ as well as plausibly weaker assumptions such as those on composite-order multilinear maps (Sect. 5). Finally, we note that not all schemes satisfying the Goyal et al. security notion satisfy our strengthened definition. In fact, a simplified version of our transformation yields a scheme secure under their original definition, but not our new definition (Remark 4.2).

Further strengthening the security definition. An important assumption that underlies all existing definitions of FE security against malicious encrypters is that the adversary cannot craft its "malicious" ciphertexts with (partial) knowledge of the secret key that will be used for decryption. More formally, in the security model, when the adversary submits a query to the decryption oracle, the secret key used for decryption is honestly generated and hidden from the adversary. An interesting problem is to formulate stronger notions of randomized FE where the adversary cannot induce correlations within ciphertexts even if it has some (limited) information about the function keys that will be used during decryption. At the same time, we stress that our existing notions already suffice for all of the applications we describe at the beginning of Sect. 1.

1.3   Overview of Our Generic Transformation

Our primary contribution in this work is giving a generic transformation from any simulation-secure general-purpose (public-key) FE scheme[2] for deterministic functionalities to a corresponding simulation-secure (public-key) FE scheme for randomized functionalities. In this section, we provide a brief overview of our generic transformation. The complete construction is given in Sect. 4.

Derandomization. Our starting point is the generic transformation of Alwen et al. [8] who use a pseudorandom function (PRF) to "derandomize" functionalities. In their construction, an encryption of a message x consists of an FE encryption of the pair (x, k) where k is a uniformly chosen PRF key. A secret key for a randomized functionality f is constructed by first choosing a random point t in the domain of the PRF and then extracting an FE secret key for the

---

[1] While there is a generic transformation from selectively-secure FE to adaptively-secure FE [10], it is described in the context of FE for deterministic functions. Though it is quite plausible that the transformation can be applied to FE schemes for randomized functions, a careful analysis is necessary to verify that it preserves security against malicious encrypters. In contrast, our generic transformation allows one to take advantage of the transformation in [10] "out-of-the-box" (i.e., apply it to existing selectively-secure FE schemes for deterministic functions) and directly transform adaptive-secure FE for deterministic functions to adaptively-secure FE for randomized functions.

[2] Our transformation requires that the underlying FE scheme be *perfectly correct*. Using the transformations in [26,51], approximately correct FE schemes can be converted to FE schemes that satisfy our requirement.

derandomized functionality $g_t(x, k) = f(x; \mathsf{PRF}(k, t))$, that is, the evaluation of f using randomness derived from the PRF. Evidently, this construction is not robust against malicious encrypters, since by reusing the same PRF key when constructing the ciphertexts, a malicious encrypter can induce correlations in the function evaluations. In fact, since the PRF key is fully under the control of the encrypter (who needs not sample it from the honest distribution), it is no longer possible to invoke PRF security to argue that $\mathsf{PRF}(k, t)$ looks like a random string.

Secret sharing the PRF key. In our transformation, we start with the same derandomization approach. Since allowing the encrypter full control over the PRF key is problematic, we instead secret share the PRF key across the ciphertext and the decryption key. Suppose the key-space $\mathcal{K}$ of the PRF forms a group under an operation $\diamond$. As before, an encryption of a message x corresponds to an FE encryption of the pair (x, k), but now k is just a single share of the PRF key. To issue a key for f, another random key-share k' is chosen from $\mathcal{K}$. The key sk_f is then an FE key for the derandomized functionality $f(x; \mathsf{PRF}(k \diamond k', x))$. In this scheme, a malicious encrypter is able to influence the PRF key, but does not have full control. However, because the malicious encrypter can induce correlated PRF keys in the decryption queries, the usual notion of PRF security no longer suffices. Instead, we require the stronger property that the outputs of the PRF appear indistinguishable from random even if the adversary observes PRF outputs under *related keys*. Security against related-key attacks (RKA-security) for PRFs has been well-studied [1, 2, 18, 19, 22, 25, 72, 74] in the last few years, and for our particular application, a variant of the Naor-Reingold PRF is related-key secure for the class of group-induced transformations [18].

Applying deterministic encryption. By secret-sharing the PRF key and using a PRF secure against related-key attacks, we obtain robustness against malicious encrypters that only requests the decryption of unique (x, k) pairs (in this case, either k or x is unique, so by related-key security, the output of the PRF appears uniformly random). However, a malicious encrypter can encrypt the same pair (x, k) multiple times, using freshly generated randomness for the base FE scheme each time. Since each of these ciphertexts encrypt the *same* underlying value, in the real world, the adversary receives the same value from the decryption oracle. In the ideal world, the adversary receives independent draws from the distribution $f(x)$. This problem arises because the adversary is able to choose additional randomness when constructing the ciphertexts that does not affect the output of the decryption algorithm. As such, it can construct ciphertexts that induce correlations in the outputs of the decryption process.

To protect against the adversary that encrypts the same (x, k) pair, we note that in the honest-encrypter setting, the messages that are encrypted have high entropy (since the key-share is sampled uniformly at random). Thus, instead of having the adversary choose its randomness for each encryption arbitrarily, we instead force the adversary to derive the randomness from the message. This is similar to what has been done when constructing deterministic public-key encryption [17,20,41,54] and other primitives where it is important to restrict the adversary's freedom when constructing ciphertexts [21]. Specifically, we sample a one-way permutation h on the key-space of the PRF, set the key-share in the ciphertext to $h(k)$ where k is uniform over $\mathcal{K}$, and then derive the randomness used in the encryption using a hard-core function hc of h.[3] In addition, we require the adversary to include a non-interactive zero-knowledge (NIZK) argument that each ciphertext is properly constructed. In this way, we guarantee that for each pair (x, k), there is exactly a *single* ciphertext that is valid. By our admissibility requirement, the adversary is required to submit distinct ciphertexts (since matching ciphertexts belong to the same equivalence class). Thus, the underlying messages encrypted by each ciphertext in a decryption query necessarily differ in either the key-share or the message component. Security then follows by RKA-security.

2   Preliminaries

For $n \geq 1$, we write $[n]$ to denote the set of integers $\{1, \ldots, n\}$. For bit-strings $a, b \in \{0, 1\}^*$, we write $a\|b$ to denote the concatenation of a and b. For a finite set S, we write $x \xleftarrow{\text{R}} S$ to denote that x is sampled uniformly from S. We denote the evaluation of a randomized function f on input x with randomness r by $f(x; r)$. We write $\mathsf{Funs}[\mathcal{X}, \mathcal{Y}]$ to denote the set of all functions mapping from a domain $\mathcal{X}$ to a range $\mathcal{Y}$. We use λ to denote the security parameter. We say a function $f(\lambda)$ is negligible in λ, denoted by $\mathsf{negl}(\lambda)$, if $f(\lambda) = o(1/\lambda^c)$ for all $c \in \mathbb{N}$. We say an algorithm is efficient if it runs in probabilistic polynomial time in the length of its input. We use $\mathsf{poly}(\lambda)$ (or just poly) to denote a quantity whose value is bounded by *some* polynomial in λ.

We now formally define the tools we need to build FE schemes for randomized functionalities with security against malicious encrypters. In the full version of this paper [6], we also review the standard definitions of non-interactive zero-knowledge (NIZK) arguments of knowledge [28,53,67,68] and one-way permutations [60].

2.1   RKA-Secure PRFs

We begin by reviewing the notion of related-key security [1,2,18,19,22,25,72,74] for PRFs.

---

[3] In the deterministic encryption setting of Fuller et al. [54], the hard-core function must additionally be *robust*. This is necessary because $\mathsf{hc}(x)$ is not guaranteed to hide the bits of x, which in the case of deterministic encryption, is the message itself (and precisely what needs to be hidden in a normal encryption scheme!). Our randomized FE scheme does *not* require that the bits of k remain hidden from the adversary. Rather, we only need that $\mathsf{hc}(k)$ does not reveal any information about $h(k)$ (the share of the PRF key used for derandomization). This property follows immediately from the definition of an ordinary hard-core function.

Definition 2.1 (RKA-Secure PRF [18,22]**).** *Let* $\mathcal{K} = \{\mathcal{K}_\lambda\}_{\lambda \in \mathbb{N}}$, $\mathcal{X} = \{\mathcal{X}_\lambda\}_{\lambda \in \mathbb{N}}$, *and* $\mathcal{Y}_\lambda = \{\mathcal{Y}_\lambda\}_{\lambda \in \mathbb{N}}$ *be ensembles where* $\mathcal{K}_\lambda$, $\mathcal{X}_\lambda$, *and* $\mathcal{Y}_\lambda$ *are finite sets and represent the key-space, domain, and range, respectively. Let* $F : \mathcal{K}_\lambda \times \mathcal{X}_\lambda \to \mathcal{Y}_\lambda$ *be an efficiently computable family of pseudorandom functions. Let* $\Phi \subseteq \mathsf{Funs}[\mathcal{K}_\lambda, \mathcal{K}_\lambda]$ *be a family of key derivation functions. We say that* F *is* Φ-*RKA secure if for all efficient, non-uniform adversaries* $\mathcal{A}$,

$$\left| \Pr\left[k \xleftarrow{\text{R}} \mathcal{K}_\lambda : \mathcal{A}^{\mathcal{O}(k,\cdot,\cdot)}(1^\lambda) = 1 \right] - \Pr\left[f \xleftarrow{\text{R}} \mathsf{Funs}[\Phi \times \mathcal{X}_\lambda, \mathcal{Y}_\lambda] : \mathcal{A}^{f(\cdot,\cdot)}(1^\lambda) = 1 \right] \right|$$
$$= \mathsf{negl}(\lambda),$$

where the oracle $\mathcal{O}(k,\cdot,\cdot)$ *outputs* $F(\phi(k),x)$ *on input* $(\phi, x) \in \Phi \times \mathcal{X}_\lambda$.

Definition 2.2 (Group Induced Classes [18,76]**).** *If the key space* $\mathcal{K}$ *forms a group under an operation* $\diamond$, *then the group-induced class* $\Phi_\diamond$ *is the class of functions* $\Phi_\diamond = \{\phi_b : a \in \mathcal{K} \mapsto a \diamond b \mid b \in \mathcal{K}\}$.

2.2   Functional Encryption

The notion of functional encryption was first formalized by Boneh et al. [35] and O'Neill [79]. The work of Boneh et al. begins with a natural indistinguishability-based notion of security. They then describe some example scenarios where these game-based definitions of security are inadequate (in the sense that a trivially insecure FE scheme can be proven secure under the standard game-based definition). To address these limitations, Boneh et al. defined a stronger simulation-based notion of security, which has subsequently been the subject of intense study [7,50,62,63,65]. In this work, we focus on this stronger security notion.

Let $\mathcal{X} = \{\mathcal{X}_\lambda\}_{\lambda \in \mathbb{N}}$ and $\mathcal{Y} = \{\mathcal{Y}_\lambda\}_{\lambda \in \mathbb{N}}$ be ensembles where $\mathcal{X}_\lambda$ and $\mathcal{Y}_\lambda$ are finite sets and represent the input and output domains, respectively. Let $\mathcal{F} = \{\mathcal{F}_\lambda\}_{\lambda \in \mathbb{N}}$ be an ensemble where each $\mathcal{F}_\lambda$ is a finite collection of (deterministic) functions from $\mathcal{X}_\lambda$ to $\mathcal{Y}_\lambda$. A functional encryption scheme FE = (Setup, Encrypt, KeyGen, Decrypt) for a (deterministic) family of functions $\mathcal{F} = \{\mathcal{F}_\lambda\}_{\lambda \in \mathbb{N}}$ with domain $\mathcal{X} = \{\mathcal{X}_\lambda\}_{\lambda \in \mathbb{N}}$ and range $\mathcal{Y} = \{\mathcal{Y}_\lambda\}_{\lambda \in \mathbb{N}}$ is specified by the following four efficient algorithms:

- **Setup:** Setup(1^λ) takes as input the security parameter λ and outputs a public key MPK and a master secret key MSK.
- **Encryption:** Encrypt(MPK, x) takes as input the public key MPK and a message $x \in \mathcal{X}_\lambda$, and outputs a ciphertext ct.
- **Key Generation:** KeyGen(MSK, f) takes as input the master secret key MSK, a function $f \in \mathcal{F}_\lambda$, and outputs a secret key sk.
- **Decryption:** Decrypt(MPK, sk, ct) takes as input the public key MPK, a ciphertext ct, and a secret key SK, and either outputs a string $y \in \mathcal{Y}_\lambda$, or a special symbol $\bot$. We can assume without loss of generality that this algorithm is deterministic.

First, we state the correctness and security definitions for an FE scheme for deterministic functions.

Definition 2.3 (Perfect Correctness). *A functional encryption scheme* FE = (Setup, Encrypt, KeyGen, Decrypt) *for a deterministic function family* $\mathcal{F} = \{\mathcal{F}_\lambda\}_{\lambda \in \mathbb{N}}$ *with message space* $\mathcal{X} = \{\mathcal{X}_\lambda\}_{\lambda \in \mathbb{N}}$ *is perfectly correct if for all* $f \in \mathcal{F}_\lambda$, $x \in \mathcal{X}_\lambda$,

$$\Pr[(\text{MPK}, \text{MSK}) \leftarrow \text{Setup}(1^\lambda);$$
$$\text{Decrypt}(\text{MPK}, \text{KeyGen}(\text{MSK}, f), \text{Encrypt}(\text{MPK}, x)) = f(x)] = 1.$$

Our simulation-based security definition is similar to the one in [7], except that we allow an adversary to submit a vector of messages in its challenge query (as opposed to a single message). Our definition is stronger than the one originally proposed by Boneh et al. [35] because we do not allow the simulator to rewind the adversary. On the other hand, it is weaker than [50,63] since the simulator is allowed to program the public parameters and the responses to the pre-challenge secret key queries.

Definition 2.4 (SIM-Security). *An FE scheme* FE = (Setup, Encrypt, KeyGen, Decrypt) *for a deterministic function family* $\mathcal{F} = \{\mathcal{F}_\lambda\}_{\lambda \in \mathbb{N}}$ *with message space* $\mathcal{X} = \{\mathcal{X}_\lambda\}_{\lambda \in \mathbb{N}}$ *is* (q_1, q_c, q_2)-*SIM-secure if there exists an efficient simulator* $\mathcal{S} = (\mathcal{S}_1, \mathcal{S}_2, \mathcal{S}_3, \mathcal{S}_4)$ *such that for all* PPT *adversaries* $\mathcal{A} = (\mathcal{A}_1, \mathcal{A}_2)$, *where* $\mathcal{A}_1$ *makes at most* q_1 *oracle queries and* $\mathcal{A}_2$ *makes at most* q_2 *oracle queries, the outputs of the following two experiments are computationally indistinguishable:*

Experiment $\text{Real}_{\mathcal{A}}^{\text{FE}}(1^\lambda)$:	**Experiment** $\text{Ideal}_{\mathcal{A}}^{\text{FE}}(1^\lambda)$:
$(\text{MPK}, \text{MSK}) \leftarrow \text{Setup}(1^\lambda)$	$(\text{MPK}, \text{st}') \leftarrow \mathcal{S}_1(1^\lambda)$
$(\mathbf{x}, \text{st}) \leftarrow \mathcal{A}_1^{\mathcal{O}_1(\text{MSK}, \cdot)}(\text{MPK})$ for $\mathbf{x} \in \mathcal{X}_\lambda^{q_c}$	$(\mathbf{x}, \text{st}) \leftarrow \mathcal{A}_1^{\mathcal{O}_1'(\text{st}', \cdot)}(\text{MPK})$ where $\mathbf{x} \in \mathcal{X}_\lambda^{q_c}$
$\text{ct}_i^* \leftarrow \text{Encrypt}(\text{MPK}, x_i)$ for $i \in [q_c]$	• Let $f_1, \ldots, f_{q_1}$ be $\mathcal{A}_1$'s oracle queries
$\alpha \leftarrow \mathcal{A}_2^{\mathcal{O}_2(\text{MSK}, \cdot)}(\text{MPK}, \{\text{ct}_i^*\}_{i \in [q_c]}, \text{st})$	• Let $y_{ij} = f_j(x_i)$ for $i \in [q_c]$, $j \in [q_1]$
Output $(\mathbf{x}, \{f\}, \alpha)$	$(\{\text{ct}_i^*\}_{i \in [q_c]}, \text{st}') \leftarrow \mathcal{S}_3(\text{st}', \{y_{ij}\}_{i \in [q_c], j \in [q_1]})$
	$\alpha \leftarrow \mathcal{A}_2^{\mathcal{O}_2'(\text{st}', \cdot)}(\text{MPK}, \{\text{ct}_i^*\}_{i \in [q_c]}, \text{st})$
	Output $(\mathbf{x}, \{f'\}, \alpha)$

where $\mathcal{O}_1(\text{MSK}, \cdot)$ *and* $\mathcal{O}_1'(\text{st}', \cdot)$ *are pre-challenge key-generation oracles, and* $\mathcal{O}_2(\text{MSK}, \cdot)$ *and* $\mathcal{O}_2'(\text{st}', \cdot)$ *are post-challenge ones. The oracles take a function* $f \in \mathcal{F}_\lambda$ *as input and behave as follows:*

- **Real experiment:** *Oracles* $\mathcal{O}_1(\text{MSK}, \cdot)$ *and* $\mathcal{O}_2(\text{MSK}, \cdot)$ *both implement the key-generation function* KeyGen$(\text{MSK}, \cdot)$. *The set* $\{f\}$ *is the (ordered) set of key queries made to* $\mathcal{O}_1(\text{MSK}, \cdot)$ *in the pre-challenge phase and to* $\mathcal{O}_2(\text{MSK}, \cdot)$ *in the post-challenge phase.*
- **Ideal experiment:** *Oracles* $\mathcal{O}_1'(\text{st}', \cdot)$ *and* $\mathcal{O}_2'(\text{st}', \cdot)$ *are the simulator algorithms* $\mathcal{S}_2(\text{st}', \cdot)$ *and* $\mathcal{S}_4(\text{st}', \cdot)$, *respectively. On each invocation, the post-challenge simulator* $\mathcal{S}_4$ *is also given oracle access to the ideal functionality* KeyIdeal$(\mathbf{x}, \cdot)$. *The functionality* KeyIdeal *accepts key queries* $f' \in \mathcal{F}_\lambda$ *and returns* $f'(x_i)$ *for every* $x_i \in \mathbf{x}$. *Both algorithms* $\mathcal{S}_2$ *and* $\mathcal{S}_4$ *are stateful. In particular, after each invocation, they update their state* st', *which is carried*

over to the next invocation. The (ordered) set $\{f'\}$ denotes the key queries made to $\mathcal{O}'_1(\text{st}', \cdot)$ in the pre-challenge phase, and the queries $\mathcal{S}_4$ makes to KeyIdeal *in the post-challenge phase.*

3   Functional Encryption for Randomized Functionalities

In a functional encryption scheme that supports randomized functionalities, the function class $\mathcal{F}_\lambda$ is expanded to include randomized functions from the domain $\mathcal{X}_\lambda$ to the range $\mathcal{Y}_\lambda$. Thus, we now view the functions $f \in \mathcal{F}_\lambda$ as taking as input a domain element $x \in \mathcal{X}_\lambda$ and randomness $r \in \mathcal{R}_\lambda$, where $\mathcal{R} = \{\mathcal{R}_\lambda\}_{\lambda \in \mathbb{N}}$ is the randomness space. As in the deterministic setting, the functional encryption scheme still consists of the same four algorithms, but the correctness and security requirements differ substantially.

For instance, in the randomized setting, whenever the decryption algorithm is invoked on a fresh encryption of a message x or a fresh key for a function f, we would expect that the resulting output is indistinguishable from evaluating $f(x)$ with fresh randomness. Moreover, this property should hold regardless of the number of ciphertexts and keys one has. To capture this property, the correctness requirement for an FE scheme supporting randomized functions must consider multiple keys and ciphertexts. In contrast, in the deterministic setting, correctness for a single key-ciphertext pair implies correctness for multiple ciphertexts.

Definition 3.1 (Correctness). *A functional encryption scheme* rFE = (Setup, Encrypt, KeyGen, Decrypt) *for a randomized function family* $\mathcal{F} = \{\mathcal{F}_\lambda\}_{\lambda \in \mathbb{N}}$ *over a message space* $\mathcal{X} = \{\mathcal{X}_\lambda\}_{\lambda \in \mathbb{N}}$ *and a randomness space* $\mathcal{R} = \{\mathcal{R}_\lambda\}_{\lambda \in \mathbb{N}}$ *is* correct *if for every polynomial* $n = n(\lambda)$, *every* $\mathbf{f} \in \mathcal{F}_\lambda^n$ *and every* $\mathbf{x} \in \mathcal{X}_\lambda^n$, *the following two distributions are computationally indistinguishable:*

1. **Real***:* $\{\text{Decrypt}\,(\text{MPK}, \text{sk}_i, \text{ct}_j)\}_{i,j \in [n]}$, *where:*
   - $(\text{MPK}, \text{MSK}) \leftarrow \text{Setup}(1^\lambda)$;
   - $\text{sk}_i \leftarrow \text{KeyGen}(\text{MSK}, f_i)$ *for* $i \in [n]$;
   - $\text{ct}_j \leftarrow \text{Encrypt}(\text{MPK}, x_j)$ *for* $j \in [n]$.
2. **Ideal***:* $\{f_i\,(x_j; r_{i,j})\}_{i,j \in [n]}$ *where* $r_{i,j} \xleftarrow{\text{R}} \mathcal{R}_\lambda$.

As discussed in Sect. 1.2, formalizing and achieving security against malicious encrypters in the randomized setting is considerably harder than in the deterministic case. A decryption oracle that takes a *single* ciphertext along with a function f does not suffice in the randomized setting, since an adversary could still produce a *collection* of ciphertexts such that when the same key is used for decryption, the outputs are correlated. We could strengthen the security definition by allowing the adversary to query with multiple ciphertexts instead of just one, but as noted in Sect. 1.2, this direct extension is too strong. In order to obtain a realizable definition, we instead restrict the adversary to submit ciphertexts that do not *behave* in the same way. This is formally captured by defining an *admissible* equivalence relation on the space of ciphertexts.

Definition 3.2 (Admissible Relation on Ciphertext Space). *Let* rFE = (Setup, Encrypt, KeyGen, Decrypt) *be an FE scheme for randomized functions with ciphertext space* $\mathcal{T} = \{\mathcal{T}_\lambda\}_{\lambda \in \mathbb{N}}$. *Let* $\sim$ *be an equivalence relation on* $\mathcal{T}$. *We say that* $\sim$ *is admissible if* $\sim$ *is efficiently checkable and for all* $\lambda \in \mathbb{N}$, *all* (MPK, MSK) *output by* Setup(1^λ), *all secret keys* sk *output by* KeyGen(MSK, $\cdot$), *and all ciphertexts* $\mathsf{ct}_1, \mathsf{ct}_2 \in \mathcal{T}_\lambda$, *if* $\mathsf{ct}_1 \sim \mathsf{ct}_2$, *then one of the following holds:*

- Decrypt(MPK, sk, ct_1) = $\bot$ OR   Decrypt(MPK, sk, ct_2) = $\bot$.
- Decrypt(MPK, sk, ct_1) = Decrypt(MPK, sk, ct_2).

We remark here that there always exists an admissible equivalence relation on the ciphertext space, namely, the equality relation. Next, we define our strengthened requirement for security against malicious encrypters in the randomized setting. Like [65], we build on the usual simulation-based definition of security for functional encryption (Definition 2.4) by providing the adversary access to a decryption oracle. The definition we present here differs from that by Goyal et al. in two key respects. First, the adversary can submit multiple ciphertexts to the decryption oracle, and second, the adversary is allowed to choose its challenge messages adaptively (that is, after seeing the public parameters and making secret key queries).

Definition 3.3 (SIM-security for rFE). *Let* $\mathcal{F} = \{\mathcal{F}_\lambda\}_{\lambda \in \mathbb{N}}$ *be a randomized function family over a domain* $\mathcal{X} = \{\mathcal{X}_\lambda\}_{\lambda \in \mathbb{N}}$ *and randomness space* $\mathcal{R} = \{\mathcal{R}_\lambda\}_{\lambda \in \mathbb{N}}$. *Let* rFE = (Setup, Encrypt, KeyGen, Decrypt) *be a randomized functional encryption scheme for* $\mathcal{F}$ *with ciphertext space* $\mathcal{T}$. *Then, we say that* rFE *is* (q_1, q_c, q_2)-SIM-secure against malicious encrypters *if there exists an admissible equivalence relation* $\sim$ *associated with* $\mathcal{T}$ *and there exists an efficient simulator* $\mathcal{S} = (\mathcal{S}_1, \mathcal{S}_2, \mathcal{S}_3, \mathcal{S}_4, \mathcal{S}_5)$ *such that for all efficient adversaries* $\mathcal{A} = (\mathcal{A}_1, \mathcal{A}_2)$ *where* $\mathcal{A}_1$ *makes at most* q_1 *key-generation queries and* $\mathcal{A}_2$ *makes at most* q_2 *key-generation queries, the outputs of the following experiments are computationally indistinguishable:*[4]

Experiment $\mathsf{Real}_{\mathcal{A}}^{\mathsf{rFE}}(1^\lambda)$:
(MPK, MSK) $\leftarrow$ Setup(1^λ)
$(\mathbf{x}, \mathsf{st}) \leftarrow \mathcal{A}_1^{\mathcal{O}_1(\mathrm{MSK},\cdot), \mathcal{O}_3(\mathrm{MSK},\cdot,\cdot)}(\mathrm{MPK})$
  where $\mathbf{x} \in \mathcal{X}_\lambda^{q_c}$
$\mathsf{ct}_i^* \leftarrow$ Encrypt(MPK, x_i) *for* $i \in [q_c]$
$\alpha \leftarrow \mathcal{A}_2^{\mathcal{O}_2(\mathrm{MSK},\cdot), \mathcal{O}_3(\mathrm{MSK},\cdot,\cdot)}(\mathrm{MPK}, \{\mathsf{ct}_i^*\}, \mathsf{st})$
Output $(\mathbf{x}, \{f\}, \{g\}, \{y\}, \alpha)$

Experiment $\mathsf{Ideal}_{\mathcal{A}}^{\mathsf{rFE}}(1^\lambda)$:
(MPK, st') $\leftarrow \mathcal{S}_1(1^\lambda)$
$(\mathbf{x}, \mathsf{st}) \leftarrow \mathcal{A}_1^{\mathcal{O}_1'(\mathsf{st}',\cdot), \mathcal{O}_3'(\mathsf{st}',\cdot,\cdot)}(\mathrm{MPK})$
  where $\mathbf{x} \in \mathcal{X}_\lambda^{q_c}$
- *Let* $f_1, \ldots, f_{q_1}$ *be* $\mathcal{A}_1$'s *oracle queries to* $\mathcal{O}_1'(\mathsf{st}', \cdot)$
- *Pick* $r_{ij} \xleftarrow{\text{R}} \mathcal{R}_\lambda$, *let* $y_{ij} = f_j(x_i; r_{ij})$ *for all* $i \in [q_c]$, $j \in [q_1]$
$(\{\mathsf{ct}_i^*\}, \mathsf{st}') \leftarrow \mathcal{S}_3(\mathsf{st}', \{y_{ij}\})$
$\alpha \leftarrow \mathcal{A}_2^{\mathcal{O}_2'(\mathsf{st}',\cdot), \mathcal{O}_3'(\mathsf{st}',\cdot,\cdot)}(\mathrm{MPK}, \{\mathsf{ct}_i^*\}, \mathsf{st})$
Output $(\mathbf{x}, \{f'\}, \{g'\}, \{y'\}, \alpha)$

---

[4] In the specification of the experiments, the indices i always range over $[q_c]$ and the indices j always range over $[q_1]$.

where the oracles $\mathcal{O}_1(\text{MSK}, \cdot)$, $\mathcal{O}_1'(\text{st}', \cdot)$, $\mathcal{O}_2(\text{MSK}, \cdot)$, and $\mathcal{O}_2'(\text{st}', \cdot)$ are the analogs of the key-generation oracles from Definition 2.4:

- **Real experiment:** *Oracles $\mathcal{O}_1(\text{MSK}, \cdot)$ and $\mathcal{O}_2(\text{MSK}, \cdot)$ implement* KeyGen $(\text{MSK}, \cdot)$, *and $\{f\}$ is the (ordered) set of key queries made to oracles $\mathcal{O}_1(\text{MSK}, \cdot)$ and $\mathcal{O}_2(\text{MSK}, \cdot)$.*
- **Ideal experiment:** *Oracles $\mathcal{O}_1'(\text{st}', \cdot)$ and $\mathcal{O}_2'(\text{st}', \cdot)$ are the simulator algorithms $\mathcal{S}_2(\text{st}', \cdot)$ and $\mathcal{S}_4(\text{st}', \cdot)$, respectively. The simulator $\mathcal{S}_4$ is given oracle access to* KeyIdeal$(\mathbf{x}, \cdot)$, *which on input a function $f' \in \mathcal{F}_\lambda$, outputs $f'(x_i; r_i)$ for every $x_i \in \mathbf{x}$ and $r_i \xleftarrow{\text{R}} \mathcal{R}_\lambda$. The (ordered) set $\{f'\}$ consists of the key queries made to $\mathcal{O}_1'(\text{st}', \cdot)$, and the queries $\mathcal{S}_4$ makes to* KeyIdeal*.*

Oracles $\mathcal{O}_3(\text{MSK}, \cdot, \cdot)$ and $\mathcal{O}_3'(\text{st}', \cdot, \cdot)$, are the decryption oracles that take inputs of the form (g, C) where $g \in \mathcal{F}_\lambda$ and $C = \{\text{ct}_i\}_{i \in [m]}$ is a collection of $m = \text{poly}(\lambda)$ ciphertexts. For queries made in the post-challenge phase, we additionally require that $\text{ct}_i^* \notin C$ for all $i \in [q_c]$. Without loss of generality, we assume that for all $i, j \in [m]$, if $i \neq j$, then $\text{ct}_i \nsim \text{ct}_j$. In other words, the set C contains at most one representative from each equivalence class of ciphertexts.

- ***Real experiment:*** *On input (g, C), $\mathcal{O}_3$ computes* $\text{sk}_g \leftarrow$ KeyGen(MSK, g). *For $i \in [m]$, it sets $y_i = $* Decrypt$(\text{sk}_g, \text{ct}_i)$ *and replies with the ordered set $\{y_i\}_{i \in [m]}$. The (ordered) set $\{g\}$ denotes the functions that appear in the decryption queries of $\mathcal{A}_2$ and $\{y\}$ denotes the set of responses of $\mathcal{O}_3$.*
- ***Ideal experiment:*** *On input (g', C'), $\mathcal{O}_3'$ does the following:*
  1. *For each $\text{ct}_i' \in C'$, invoke the simulator algorithm $\mathcal{S}_5(\text{st}', \text{ct}_i')$ to obtain a value $x_i \in \mathcal{X}_\lambda \cup \{\perp\}$. Note that $\mathcal{S}_5$ is also stateful.*
  2. *For each $i \in [m]$, if $x_i = \perp$, then the oracle sets $y_i' = \perp$. Otherwise, the oracle choose $r_i \xleftarrow{\text{R}} \mathcal{R}_\lambda$ and sets $y_i' = g'(x_i; r_i)$.*
  3. *Output the ordered set of responses $\{y_i'\}_{i \in [m]}$.*
  The (ordered) set $\{g'\}$ denotes the functions that appear in the decryption queries of $\mathcal{A}_2$ and $\{y'\}$ denotes the outputs of $\mathcal{O}_3'$.

Remark 3.4. Note that the above definition does not put any constraint on the equivalence relation used to prove security. Indeed, *any* equivalence relation—as long as it is admissible—suffices because if two ciphertexts ct_1, ct_2 fall into the same equivalence class, they essentially behave *identically* (for all parameters output by Setup and all keys sk output by KeyGen, decrypting ct_1, ct_2 with sk must either give the same result, or one of the ciphertexts is invalid). Thus, by restricting an adversary to providing at most one ciphertext from each equivalence class in each decryption query, we are only preventing it from submitting ciphertexts which are effectively equivalent to the decryption oracle.

Remark 3.5. One could also consider an ideal model where the adversary is allowed to submit equivalent ciphertexts to the decryption oracle (at the cost of making the security game more cumbersome). In the extreme case where the adversary submits *identical* ciphertexts, it does not make sense for the decryption oracle to respond independently on each of them—rather, it should respond in a consistent way. In constructions of randomized FE that provide malicious security, there naturally arise ciphertexts that are not identical as bit-strings, but are identical from the perspective of the decryption function. In these cases, the expected behavior of the ideal functionality should again be to provide consistent, rather than independent, responses.

Consider now an adversary that submits a function f and a set C of ciphertexts to the decryption oracle, where some ciphertexts in C belong to the same equivalence class. To respond, the challenger can first group these ciphertexts by equivalence class. For each equivalence class C' of ciphertexts in C, the challenger invokes the simulator on C'. On input the collection C', the simulator outputs a *single* value x and indicates which ciphertexts in C', if any, are valid. If C' contains at least one valid ciphertext, the challenger samples a value z from the output distribution of $f(x)$. It then replies with the *same* value z on all ciphertexts marked valid by the simulator, and $\perp$ on all ciphertexts marked invalid. (This is a natural generalization of how we would expect the decryption oracle to behave had the adversary submitted identical ciphertexts to it.)

4   Our Generic Transformation

Let $\mathcal{F} = \{\mathcal{F}_\lambda\}_{\lambda \in \mathbb{N}}$ be a randomized function class over a domain $\mathcal{X} = \{\mathcal{X}_\lambda\}_{\lambda \in \mathbb{N}}$, randomness space $\mathcal{R} = \{\mathcal{R}_\lambda\}_{\lambda \in \mathbb{N}}$ and range $\mathcal{Y} = \{\mathcal{Y}_\lambda\}_{\lambda \in \mathbb{N}}$. We give the formal description of our functional encryption scheme for $\mathcal{F}$ (based on any general-purpose FE scheme for deterministic functionalities) in Fig. 1. All the necessary cryptographic primitives are also shown in Fig. 1.

Theorem 4.1. *If (1)* NIZK *is a simulation-sound extractable non-interactive zero-knowledge argument, (2)* PRF *is a Φ-RKA secure pseudorandom function where Φ is group-induced, (3)* OWP *is a family of one-way permutations with hard-core function* hc, *and (4)* FE *is a perfectly-correct (q_1, q_c, q_2)-SIM secure functional encryption scheme for the derandomized class $\mathcal{G}_\mathcal{F}$, then* rFE *is (q_1, q_c, q_2)-SIM secure against malicious encrypters for the class $\mathcal{F}$ of randomized functions.*

Before proceeding with the proof of Theorem 4.1, we remark that our strengthened definition of security against malicious encrypters (Definition 3.3) is indeed stronger than the original definition by Goyal et al. [65].

Remark 4.2. A simpler version of our generic transformation where we only secret share the RKA-secure PRF key used for derandomization and include a NIZK argument can be shown to satisfy the Goyal et al. [65] definition of security against malicious encrypters, but not our strengthened definition (Definition 3.3). In particular, if the randomness used in the base FE encryption is under the control of the adversary, a malicious encrypter can construct two fresh encryptions (under the base FE scheme) of the same (x, k) pair and submit them to the decryption oracle. In the real world, the outputs are identical (since the ciphertexts encrypt identical messages), but in the ideal world, the oracle

Ingredients:

- A non-interactive zero-knowledge argument system NIZK = (NIZK.Setup, NIZK.Prove, NIZK.Verify) that is simulation-sound extractable.
- A Φ-RKA secure pseudorandom function PRF (Definition 2.1) with key-space $\mathcal{K} = \{\mathcal{K}_\lambda\}_{\lambda \in \mathbb{N}}$, domain $\mathcal{X}$, and range $\mathcal{Y}$, where Φ is group-induced (Definition 2.2). Let $\diamond$ denote the group operation on $\mathcal{K}$.
- A family of one-way permutations OWP = (OWP.Setup, OWP.Eval) over $\mathcal{K}$ with associated hard-core function $\mathsf{hc} : \mathcal{K}_\lambda \to \{0,1\}^\rho$. The number of output bits $\rho = \rho(\lambda)$ is specified below.
- For all $f \in \mathcal{F}_\lambda$ and $k \in \mathcal{K}_\lambda$, let $g_k^f : \mathcal{X}_\lambda \times \mathcal{K}_\lambda \to \mathcal{Y}_\lambda$ be the derandomized function

$$g_k^f(x, k') = f(x; \mathsf{PRF}(k \diamond k', x)). \tag{1}$$

Let $\mathcal{G}_{\mathcal{F},\lambda}$ be the derandomized function class $\left\{ g_k^f \mid f \in \mathcal{F}_\lambda, k \in \mathcal{K}_\lambda \right\}$, and let FE = (FE.Setup, FE.Encrypt, FE.KeyGen, FE.Decrypt) be a functional encryption scheme for the derandomized class $\mathcal{G}_{\mathcal{F}} = \{\mathcal{G}_{\mathcal{F},\lambda}\}_{\lambda \in \mathbb{N}}$. By construction, the message space for FE is $\mathcal{X}_\lambda \times \mathcal{K}_\lambda$. Let $\rho = \rho(\lambda)$ be a bound on the number of bits of randomness FE.Encrypt takes.

A functional encryption scheme rFE = (Setup, Encrypt, KeyGen, Decrypt) for randomized functionalities:

- **Setup:** On input 1^λ, Setup samples $(\textsc{mpk}', \textsc{msk}') \leftarrow$ FE.Setup(1^λ), $t \leftarrow$ OWP.Setup(1^λ), and $\sigma \leftarrow$ NIZK.Setup(1^λ). It sets $h_t(\cdot) =$ OWP.Eval$(t, \cdot)$, and outputs a master public key $\textsc{mpk} = (\textsc{mpk}', t, \sigma)$ and a master secret key $\textsc{msk} = \textsc{msk}'$.
- **Encryption:** On input $\textsc{mpk} = (\textsc{mpk}', t, \sigma)$ and $x \in \mathcal{X}_\lambda$, Encrypt samples $k \xleftarrow{\text{R}} \mathcal{K}_\lambda$ and sets $\mathsf{ct}' =$ FE.Encrypt$(\textsc{mpk}', (x, h_t(k)); \mathsf{hc}(k))$. Then, it runs NIZK.Prove$(\sigma, s, (x, k))$ to obtain an argument π on the following statement s:

$$\exists\, x, k : \mathsf{ct}' = \mathsf{FE.Encrypt}(\textsc{mpk}', (x, h_t(k)); \mathsf{hc}(k)). \tag{2}$$

  Finally, it outputs a ciphertext $\mathsf{ct} = (\mathsf{ct}', \pi)$.
- **Key-generation:** On input $\textsc{msk} = \textsc{msk}'$ and f, KeyGen samples $k \xleftarrow{\text{R}} \mathcal{K}_\lambda$ and outputs a secret key $\mathsf{sk}_f \leftarrow$ FE.KeyGen$(\textsc{msk}', g_k^f)$, where g_k^f is the derandomized function corresponding to f (Eq. (1)).
- **Decryption:** On input $\textsc{mpk} = (\textsc{mpk}', t, \sigma)$, a secret key sk, and a ciphertext $\mathsf{ct} = (\mathsf{ct}', \pi)$, Decrypt first runs NIZK.Verify(σ, s, π) where s is the statement from Eq. (2). If the argument verifies, then it outputs FE.Decrypt$(\mathsf{sk}, \mathsf{ct}')$; otherwise, it outputs $\perp$.

Fig. 1. Generic construction of a functional encryption scheme for any family of randomized functions $\mathcal{F} = \{\mathcal{F}_\lambda\}_{\lambda \in \mathbb{N}}$ over a domain $\mathcal{X} = \{\mathcal{X}_\lambda\}_{\lambda \in \mathbb{N}}$, randomness space $\mathcal{R} = \{\mathcal{R}_\lambda\}_{\lambda \in \mathbb{N}}$ and range $\mathcal{Y} = \{\mathcal{Y}_\lambda\}_{\lambda \in \mathbb{N}}$.

replies with two independent outputs. This is an admissible query because if the underlying FE scheme is secure, one cannot *efficiently* decide whether two FE ciphertexts encrypt the same value without knowing any scheme parameters. But because each *individual* output is still properly distributed (by RKA-security of the PRF), security still holds in the Goyal et al. model.

We now proceed to give a proof of Theorem 4.1 in Sects. 4.1 and 4.2. In the full version [6], we also show that our transformed scheme is correct.

4.1 Proof of Theorem 4.1: Description of Simulator

To prove Theorem 4.1, and show that rFE is secure in the sense of Definition 3.3, we first define an equivalence relation $\sim$ over the ciphertext space $\mathcal{T} = \{\mathcal{T}_\lambda\}_{\lambda \in \mathbb{N}}$. Take two ciphertexts $ct_1, ct_2 \in \mathcal{T}_\lambda$, and write $ct_1 = (ct_1', \pi_1)$ and $ct_2 = (ct_2', \pi_2)$. We say that $ct_1 \sim ct_2$ if $ct_1' = ct_2'$.

Certainly, $\sim$ is an efficiently-checkable equivalence relation over $\mathcal{T}_\lambda$. For the second admissibility condition, take any (MPK, MSK) output by Setup and any sk output by KeyGen(MSK, $\cdot$). Suppose moreover that Decrypt(MPK, sk, ct_1) $\neq \bot \neq$ Decrypt(MPK, sk, ct_2). Then, by definition of Decrypt(MPK, sk, $\cdot$),

$$
\begin{aligned}
\text{Decrypt}(\text{MPK}, \text{sk}, ct_1) &= \text{FE.Decrypt}(\text{MPK}', \text{sk}, ct_1') \\
&= \text{FE.Decrypt}(\text{MPK}', \text{sk}, ct_2') = \text{Decrypt}(\text{MPK}, \text{sk}, ct_2),
\end{aligned}
$$

where MPK' is the master public key for the underlying FE scheme (included in MPK). The second equivalence follows since $ct_1' = ct_2'$.

We now describe our ideal-world simulator $\mathcal{S} = (\mathcal{S}_1, \mathcal{S}_2, \mathcal{S}_3, \mathcal{S}_4, \mathcal{S}_5)$. Let $\mathcal{S}^{(\text{FE})} = (\mathcal{S}_1^{(\text{FE})}, \mathcal{S}_2^{(\text{FE})}, \mathcal{S}_3^{(\text{FE})}, \mathcal{S}_4^{(\text{FE})})$ be the simulator for the underlying FE scheme for deterministic functionalities. Let $\mathcal{S}^{(\text{NIZK})} = (\mathcal{S}_1^{(\text{NIZK})}, \mathcal{S}_2^{(\text{NIZK})})$ and $\mathcal{E}^{(\text{NIZK})} = (\mathcal{E}_1^{(\text{NIZK})}, \mathcal{E}_2^{(\text{NIZK})})$ be the simulation and extraction algorithms, respectively, for the NIZK argument system.

Algorithm $\mathcal{S}_1(1^\lambda)$. $\mathcal{S}_1$ simulates the setup procedure. On input a security parameter 1^λ, it operates as follows:

1. Invoke $\mathcal{S}_1^{(\text{FE})}(1^\lambda)$ to obtain a master public key MPK' and some state $st^{(\text{FE})}$.
2. Invoke $\mathcal{E}_1^{(\text{NIZK})}(1^\lambda)$ to obtain a CRS σ, a simulation trapdoor τ, and an extraction trapdoor ξ.
3. Sample a one-way permutation $t \leftarrow \text{OWP.Setup}(1^\lambda)$ and define $h_t(\cdot) = \text{OWP.Eval}(t, \cdot)$.
4. Set MPK $\leftarrow$ (MPK', t, σ) and st $\leftarrow$ ($st^{(\text{FE})}$, MPK, τ, ξ). Output (MPK, st).

Algorithm $\mathcal{S}_2(st_0, f)$. $\mathcal{S}_2$ simulates the pre-challenge key-generation queries. On input a state $st_0 = (st_0^{(\text{FE})}, \text{MPK}, \tau, \xi)$ and a function $f \in \mathcal{F}_\lambda$, it operates as follows:

1. Choose a random key $k \xleftarrow{\text{R}} \mathcal{K}_\lambda$ and construct the derandomized function g_k^f as defined in Eq. (1).

2. Invoke $\mathcal{S}_2^{(\mathrm{FE})}(\mathrm{st}_0^{(\mathrm{FE})}, g_k^f)$ to obtain a key sk and an updated state $\mathrm{st}_1^{(\mathrm{FE})}$.
3. Output the key sk and an updated state $\mathrm{st}_1 = (\mathrm{st}_1^{(\mathrm{FE})}, \mathrm{MPK}, \tau, \xi)$.

Algorithm $\mathcal{S}_3(\mathrm{st}_0, \{y_{ij}\}_{i\in[q_c], j\in[q_1]})$. $\mathcal{S}_3$ constructs the challenge ciphertexts. Let $\mathbf{x} = (x_1, x_2, \ldots, x_{q_c})$ be the challenge messages the adversary outputs. On input a state $\mathrm{st}_0 = (\mathrm{st}_0^{(\mathrm{FE})}, \mathrm{MPK}, \tau, \xi)$, where $\mathrm{MPK} = (\mathrm{MPK}', t, \sigma)$, and a collection of function evaluations $\{y_{ij}\}_{i\in[q_c], j\in[q_1]}$, $\mathcal{S}_3$ operates as follows:

1. Invoke $\mathcal{S}_3^{(\mathrm{FE})}(\mathrm{st}_0^{(\mathrm{FE})}, \{y_{ij}\}_{i\in[q_c], j\in[q_1]})$ to obtain a set of ciphertexts $\{\mathrm{ct}_i'\}_{i\in[q_c]}$ and an updated state $\mathrm{st}_1^{(\mathrm{FE})}$.
2. For $i \in [q_c]$, let s_i be the statement

$$\exists x, k : \mathrm{ct}_i' = \mathsf{FE.Encrypt}(\mathrm{MPK}', (x, h_t(k)); \mathsf{hc}(k)). \tag{3}$$

   Using the trapdoor τ in st_0, simulate an argument $\pi_i \leftarrow \mathcal{S}_2^{(\mathrm{NIZK})}(\sigma, \tau, s_i)$, and set $\mathrm{ct}_i^* = (\mathrm{ct}_i', \pi_i)$.
3. Output the challenge ciphertexts $\{\mathrm{ct}_i^*\}_{i\in[q_c]}$ and the updated state $\mathrm{st}_1 = (\mathrm{st}_1^{(\mathrm{FE})}, \mathrm{MPK}, \tau, \xi)$.

Algorithm $\mathcal{S}_4(\mathrm{st}_0, f)$. $\mathcal{S}_4$ simulates the post-challenge key-generation queries with help from the ideal functionality $\mathsf{KeyIdeal}(\mathbf{x}, \cdot)$. On input a state $\mathrm{st}_0 = (\mathrm{st}_0^{(\mathrm{FE})}, \mathrm{MPK}, \tau, \xi)$ and a function $f \in \mathcal{F}_\lambda$, it operates as follows:

1. Choose a random key $k \xleftarrow{\mathrm{R}} \mathcal{K}$, and construct the derandomized function g_k^f as defined in Eq. (1).
2. Invoke $\mathcal{S}_4^{(\mathrm{FE})}(\mathrm{st}_0^{(\mathrm{FE})}, g_k^f)$. Here, $\mathcal{S}_4$ also simulates the $\mathsf{FE.KeyIdeal}(\mathbf{x}, \cdot)$ oracle for $\mathcal{S}_4^{(\mathrm{FE})}$. Specifically, when $\mathcal{S}_4^{(\mathrm{FE})}$ makes a query of the form $g_{k'}^{f'}$ to $\mathsf{FE.KeyIdeal}(\mathbf{x}, \cdot)$, $\mathcal{S}_4$ queries its own oracle $\mathsf{KeyIdeal}(\mathbf{x}, \cdot)$ on f' to obtain values z_i for each $i \in [q_c]$.[5] It replies to $\mathcal{S}_4^{(\mathrm{FE})}$ with the value z_i for all $i \in [q_c]$. Let sk and $\mathrm{st}_1^{(\mathrm{FE})}$ be the output of $\mathcal{S}_4^{(\mathrm{FE})}$.
3. Output the key sk and an updated state $\mathrm{st}_1 = (\mathrm{st}_1^{(\mathrm{FE})}, \mathrm{MPK}, \tau, \xi)$.

Algorithm $\mathcal{S}_5(\mathrm{st}, \mathrm{ct})$. $\mathcal{S}_5$ handles the decryption queries. On input a state $\mathrm{st} = (\mathrm{st}^{(\mathrm{FE})}, \mathrm{MPK}, \tau, \xi)$ and a ciphertext ct, it proceeds as follows:[6]

1. Parse MPK as $(\mathrm{MPK}', t, \sigma)$ and ct as (ct', π). Let s be the statement

$$\exists x, k : \mathrm{ct} = \mathsf{FE.Encrypt}(\mathrm{MPK}', (x, h_t(k)); \mathsf{hc}(k)).$$

   If $\mathsf{NIZK.Verify}(\sigma, s, \pi) = 0$, then stop and output $\bot$.

---

[5] The underlying FE scheme is for the derandomized class $\mathcal{G}_\mathcal{F}$, so the only permissible functions $\mathcal{S}_4^{(\mathrm{FE})}$ can issue to $\mathsf{FE.KeyIdeal}$ are of the form $g_{k'}^{f'}$ for some k' and f'.

[6] Recall that in the security definition (Definition 3.3), the decryption oracle accepts *multiple* ciphertexts, and invokes the simulator on each one individually. Thus, the simulator algorithm operates on a single ciphertext at a time.

2. Otherwise, invoke the extractor $\mathcal{E}_2^{(\text{NIZK})}(\sigma, \xi, s, \pi)$ using the extraction trapdoor ξ to obtain a witness $(x, k) \in \mathcal{X}_\lambda \times \mathcal{K}_\lambda$. Output x and state st.

4.2 Proof of Theorem 4.1: Hybrid Argument

To prove security, we proceed via a series of hybrid experiments between an adversary $\mathcal{A}$ and a challenger. Each experiment consists of the following phases:

1. **Setup phase.** The challenger begins by generating the public parameters of the rFE scheme, and sends those to the adversary $\mathcal{A}$.
2. **Pre-challenge queries.** In this phase of the experiment, $\mathcal{A}$ can issue key-generation queries of the form $f \in \mathcal{F}_\lambda$ and decryption queries of the form $(f, C) \in \mathcal{F}_\lambda \times \mathcal{T}_\lambda^m$ to the challenger. For all decryption queries (f, C), we require that for any $\mathsf{ct}_i, \mathsf{ct}_j \in C$, $\mathsf{ct}_i \not\sim \mathsf{ct}_j$ if $i \neq j$. In other words, each set of ciphertexts C can contain at most one representative from each equivalence class.
3. **Challenge phase.** The adversary $\mathcal{A}$ submits a vector of messages $\mathbf{x} \in \mathcal{X}_\lambda^{q_c}$ to the challenger, who replies with ciphertexts $\{\mathsf{ct}_i^*\}_{i \in [q_c]}$.
4. **Post-challenge queries.** In this phase, $\mathcal{A}$ is again allowed to issue key-generation and decryption queries, with a further restriction that no decryption query can contain any of the challenge ciphertexts (i.e., for any query (f, C), $\mathsf{ct}_i^* \notin C$ for all $i \in [q_c]$).
5. **Output.** At the end of the experiment, $\mathcal{A}$ outputs a bit $b \in \{0, 1\}$.

We now describe our sequence of hybrid experiments. Note that in defining a new hybrid, we only describe the phases that differ from the previous one. If one or more of the above phases are omitted, the reader should assume that they are exactly the same as in the previous hybrid.

Hybrid Hyb_0. In this experiment, the challenger responds to $\mathcal{A}$ according to the specification of the real experiment $\mathsf{Real}_{\mathcal{A}}^{\mathsf{rFE}}$.

- **Setup phase.** The challenger samples $(\text{MPK}, \text{MSK}) \leftarrow \mathsf{Setup}(1^\lambda)$ and sends MPK to $\mathcal{A}$.
- **Pre-challenge queries.** The challenger responds to each query as follows:
  - **Key-generation queries.** On a key-generation query $f \in \mathcal{F}_\lambda$, the challenger responds with $\mathsf{KeyGen}(\text{MSK}, f)$.
  - **Decryption queries.** On a decryption query $(f, C) \in \mathcal{F}_\lambda \times \mathcal{T}_\lambda^m$, the challenger samples $\mathsf{sk} \leftarrow \mathsf{KeyGen}(\text{MSK}, f)$. For each $\mathsf{ct}_i \in C$, the challenger sets $y_i = \mathsf{Decrypt}(\mathsf{sk}, \mathsf{ct}_i)$, and sends $\{y_i\}_{i \in [m]}$ to the adversary.
- **Challenge phase.** When the challenger receives a vector $\mathbf{x} \in \mathcal{X}_\lambda^{q_c}$, it sets $\mathsf{ct}_i^* = \mathsf{Encrypt}(\text{MPK}, x_i)$ for each $i \in [q_c]$ and replies to $\mathcal{A}$ with $\{\mathsf{ct}_i^*\}_{i \in [q_c]}$.
- **Post-challenge queries.** This is identical to the pre-challenge phase.

Hybrid Hyb_1. This is the same as Hyb_0, except the challenger simulates the CRS in the setup phase and the arguments in the challenge ciphertexts in the challenge phase. Let $\mathcal{S}^{(\text{NIZK})} = (\mathcal{S}_1^{(\text{NIZK})}, \mathcal{S}_2^{(\text{NIZK})})$ be the simulator for NIZK. Note that we omit the description of the pre- and post-challenge phases in the description below because they are identical to those phases in Hyb_0.

- **Setup phase.** The challenger generates the public parameters as in Hyb_0, except it uses $\mathcal{S}_1^{(\text{NIZK})}$ to generate the CRS. Specifically, it does the following:
    1. Sample $(\text{MPK}', \text{MSK}') \leftarrow \mathsf{FE.Setup}(1^\lambda)$.
    2. Run $\mathcal{S}_1^{(\text{NIZK})}(1^\lambda)$ to obtain a CRS σ and a simulation trapdoor τ.
    3. Sample a one-way permutation $t \leftarrow \mathsf{OWP.Setup}(1^\lambda)$, and define $h_t(\cdot) = \mathsf{OWP.Eval}(t, \cdot)$.
    4. Set $\text{MPK} = (\text{MPK}', t, \sigma)$ and send MPK to $\mathcal{A}$.
- **Challenge phase.** The challenger constructs the challenge ciphertexts as in Hyb_0, except it uses $\mathcal{S}_2^{(\text{NIZK})}$ to simulate the NIZK arguments. Let $\mathbf{x} \in \mathcal{X}_\lambda^{q_c}$ be the adversary's challenge. For $i \in [q_c]$, the challenger samples $k_i^* \xleftarrow{\text{R}} \mathcal{K}_\lambda$ and sets $\text{ct}_i' \leftarrow \mathsf{FE.Encrypt}(\text{MPK}', (x_i, h_t(k_i^*)); \mathsf{hc}(k_i^*))$. It invokes $\mathcal{S}_2^{(\text{NIZK})}(\sigma, \tau, s_i)$ to obtain a simulated argument π_i, where s_i is the statement in Eq. (3). Finally, it sets $\text{ct}_i^* = (\text{ct}_i', \pi_i)$ and sends $\{\text{ct}_i^*\}_{i \in [q_c]}$ to $\mathcal{A}$.

Hybrid Hyb_2. This is the same as Hyb_1, except the challenger uses uniformly sampled randomness when constructing the challenge ciphertexts.

- **Challenge phase.** Same as in Hyb_1, except that for every $i \in [q_c]$, the challenger sets $\text{ct}_i' = \mathsf{FE.Encrypt}(\text{MPK}', (x_i, h_t(k_i^*)); r_i)$ for a randomly chosen $r_i \xleftarrow{\text{R}} \{0,1\}^\rho$.

Hybrid Hyb_3. This is the same as Hyb_2, except the challenger answers the decryption queries by first extracting the message-key pair (m, k) from the NIZK argument and then evaluating the derandomized function on it. Let $\mathcal{E}^{(\text{NIZK})} = (\mathcal{E}_1^{(\text{NIZK})}, \mathcal{E}_2^{(\text{NIZK})})$ be the extraction algorithm for NIZK.

- **Setup phase.** Same as in Hyb_2 (or Hyb_1), except the challenger runs $(\sigma, \tau, \xi) \leftarrow \mathcal{E}_1^{(\text{NIZK})}(1^\lambda)$ to obtain the CRS σ, the simulation trapdoor τ, and the extraction trapdoor ξ.
- **Pre-challenge queries.** The key-generation queries are handled as in Hyb_2, but the decryption queries are handled as follows.
    - **Decryption queries.** On input (f, C), where $C = \{\text{ct}_i\}_{i \in [m]}$,
        1. Choose a random key $k \xleftarrow{\text{R}} \mathcal{K}_\lambda$.
        2. For $i \in [m]$, parse ct_i as (ct_i', π_i), and let s_i be the statement in Eq. (3). If $\mathsf{NIZK.Verify}(\sigma, s_i, \pi_i) = 0$, set $y_i = \bot$. Otherwise, invoke the extractor $\mathcal{E}_2^{(\text{NIZK})}(\sigma, \xi, s_i, \pi_i)$ to obtain a witness (x_i, k_i), and set $y_i = f(x_i; \mathsf{PRF}(k \diamond h_t(k_i), x_i))$.
        3. Send the set $\{y_i\}_{i \in [m]}$ to $\mathcal{A}$.
- **Post-challenge queries.** This is identical to the pre-challenge phase.

Hybrid Hyb_4. This is the same as Hyb_3, except the challenger uses the simulator $\mathcal{S}^{(\mathrm{FE})} = (\mathcal{S}_1^{(\mathrm{FE})}, \mathcal{S}_2^{(\mathrm{FE})}, \mathcal{S}_3^{(\mathrm{FE})}, \mathcal{S}_4^{(\mathrm{FE})})$ for the underlying FE scheme to respond to queries. Let $\mathcal{S} = (\mathcal{S}_1, \mathcal{S}_2, \mathcal{S}_3, \mathcal{S}_4, \mathcal{S}_5)$ be the simulator described in Sect. 4.1.

- **Setup phase.** Same as in Hyb_3, except the challenger invokes the base FE simulator $\mathcal{S}_1^{(\mathrm{FE})}$ to construct MPK. The resulting setup algorithm corresponds to the simulation algorithm $\mathcal{S}_1$. Hence, we can alternately say that the challenger runs $\mathcal{S}_1(1^\lambda)$ to obtain MPK $= (\mathrm{MPK}', t, \sigma)$ and $\mathsf{st} = (\mathsf{st}^{(\mathrm{FE})}, \mathrm{MPK}, \tau, \xi)$, and sends MPK to $\mathcal{A}$.
- **Pre-challenge queries.** The decryption queries are handled as described in Hyb_3, but key-generation queries are handled as follows.
  - **Key-generation queries.** On a key-generation query $f \in \mathcal{F}_\lambda$,
    1. Sample a key $k \xleftarrow{\mathrm{R}} \mathcal{K}_\lambda$. Let g_k^f be the derandomized function corresponding to f.
    2. Run $\mathcal{S}_2^{(\mathrm{FE})}(\mathsf{st}^{(\mathrm{FE})}, g_k^f)$ to obtain a secret key sk and an updated state.
    3. Update st accordingly and send sk to $\mathcal{A}$.

  Note that this is exactly how $\mathcal{S}_2$ behaves when given f and st as inputs.
- **Challenge phase.** The challenger constructs the challenge ciphertexts using the simulation algorithm $\mathcal{S}_3$. Specifically, it does the following on receiving $\mathbf{x} \in \mathcal{X}_\lambda^{q_c}$:
    1. For each $i \in [q_c]$, choose a key $k_i^* \xleftarrow{\mathrm{R}} \mathcal{K}_\lambda$.
    2. Let $f_1, \ldots, f_{q_1} \in \mathcal{F}_\lambda$ be the pre-challenge key-generation queries made by $\mathcal{A}$ and $k_1, \ldots, k_{q_1} \in \mathcal{K}_\lambda$ be the keys chosen when responding to each query. For all $i \in [q_c]$ and $j \in [q_1]$, compute $r_{ij} = \mathsf{PRF}(k_j \diamond h_t(k_i^*), x_i)$ and set $y_{ij} = f_j(x_i; r_{ij})$.
    3. Invoke the simulator algorithm $\mathcal{S}_3(\mathsf{st}, \{y_{ij}\}_{i \in [q_c], j \in [q_1]})$ to obtain a collection of ciphertexts $\{\mathsf{ct}_i^*\}_{i \in [q_c]}$ and an updated state st.
    4. Send $\{\mathsf{ct}_i^*\}_{i \in [q_c]}$ to $\mathcal{A}$.
- **Post-challenge queries.** The decryption queries are handled as in the pre-challenge phase, but key-generation queries are handled differently as follows.
  - **Key-generation queries.** The first step stays the same: a key k is picked at random and g_k^f is defined. The challenger then invokes $\mathcal{S}_4^{(\mathrm{FE})}$ with inputs $\mathsf{st}^{(\mathrm{FE})}$ and g_k^f, instead of $\mathcal{S}_2^{(\mathrm{FE})}$. In invoking $\mathcal{S}_4^{(\mathrm{FE})}$, it simulates the FE.Keyldeal$(\mathbf{x}, \cdot)$ oracle as follows: on input a function of the form $g_{k'}^{f'}$, it computes $y_i = g_{k'}^{f'}(x_i, h_t(k_i^*)) = f'(x_i; \mathsf{PRF}(k' \diamond h_t(k_i^*), x_i))$ and replies with the set $\{y_i\}_{i \in [q_c]}$. The function key returned by $\mathcal{S}_4^{(\mathrm{FE})}$ is given to $\mathcal{A}$, and st is updated appropriately. This is the behavior of $\mathcal{S}_4$.

Hybrid Hyb_5. This is the same as Hyb_4, except the outputs of PRF are replaced by truly random strings. This matches the specification of the ideal experiment $\mathsf{Ideal}_\mathcal{A}^{\mathrm{rFE}}$. We highlight below the differences from the previous hybrid.

- **Pre-challenge queries.** While the key queries are handled as before, the decryption queries are handled as follows.

- **Decryption queries.** Same as in Hyb_4, except the function f is evaluated using uniformly sampled randomness. In other words, on input f and $C = \{\mathsf{ct}_i\}_{i \in [m]}$, the challenger does the following:
    1. For every $\mathsf{ct}_i \in C$, invoke the simulator algorithm $\mathcal{S}_5(\mathsf{st}, \mathsf{ct}_i)$ to obtain a value $x_i \in \mathcal{X}_\lambda \cup \{\bot\}$ and an updated state st.
    2. If $x_i = \bot$, set y_i to $\bot$, else set it to $f(x_i; r_i)$, where $r_i \xleftarrow{\text{R}} \mathcal{R}_\lambda$.
    3. Send the set of values $\{y_i\}_{i \in [m]}$ to $\mathcal{A}$.
- **Challenge phase.** The challenge ciphertexts are constructed as in the ideal experiment. Specifically, instead of using PRF to generate the randomness for evaluating y_{ij} in the first and second steps of the challenge phase, the challenger simply computes $f_j(x_i; r_{ij})$ for $r_{ij} \xleftarrow{\text{R}} \mathcal{R}_\lambda$. The remaining two steps (third and fourth) stay the same.
- **Post-challenge queries.** The decryption queries are handled as in the pre-challenge phase, but key queries are handled as follows:
    - **Key-generation queries.** Same as Hyb_4, except the oracle FE.KeyIdeal $(\mathbf{x}, \cdot)$ is implemented using uniformly sampled randomness as in the ideal experiment. Specifically, if $\mathcal{S}_4^{(\mathrm{FE})}$ makes a query to FE.KeyIdeal$(\mathbf{x}, \cdot)$ with a derandomized function $g_{k'}^{f'}$, the challenger chooses an $r_i \xleftarrow{\text{R}} \mathcal{R}_\lambda$ for every $i \in [q_c]$, and replies with $\{f'(x_i; r_i)\}_{i \in [q_c]}$.

In the full version [6], we complete the hybrid argument by showing that each consecutive pair of experiments are computationally indistinguishable. We also show in the full version that our transformed scheme is correct.

5   Instantiating and Applying the Transformation

In this section, we describe one way to instantiate the primitives (the NIZK argument system, the RKA-secure PRF, and the one-way permutation) needed to apply the generic transformation from Sect. 4, Theorem 4.1. Then, in Sect. 5.2, we show how to obtain new general-purpose functional encryption schemes for randomized functionalities with security against malicious encrypters from a wide range of assumptions by applying our transformation to existing functional encryption schemes.

5.1   Instantiating Primitives

All of the primitives required by our generic transformation can be built from standard number-theoretic assumptions, namely the decisional Diffie-Hellman (DDH) assumption [29], the hardness of discrete log in the multiplicative group $\mathbb{Z}_p^*$ (for prime p), and the RSA assumption [30, 82]. The first two assumptions can be combined by assuming the DDH assumption holds in a prime-order subgroup of $\mathbb{Z}_p^*$, such as the subgroup of quadratic residues of $\mathbb{Z}_p^*$, where p is a safe prime ($p = 2q + 1$, where q is also prime). We describe one such instantiation of our primitives from the DDH and RSA assumptions in the full version [6]. This yields the following corollary to Theorem 4.1:

Corollary 5.1. *Assuming standard number-theoretic assumptions (that is, the DDH assumption in a prime-order subgroup of $\mathbb{Z}_p^*$ and the RSA assumption), and that FE is a perfectly-correct (q_1, q_c, q_2)-SIM secure functional encryption scheme for the derandomized function class $\mathcal{G}_\mathcal{F}$, then rFE is (q_1, q_c, q_2)-SIM secure against malicious encrypters for the class $\mathcal{F}$ of randomized functions.*

5.2   Applying the Transformation

In this section, we give three examples of how our generic transformation from Sect. 4 could be applied to existing functional encryption schemes to obtain schemes that support randomized functionalities. Our results show that functional encryption for randomized functionalities secure against malicious encrypters can be constructed from a wide range of assumptions such as public-key encryption, concrete assumptions over composite-order multilinear maps, or indistinguishability obfuscation, in conjunction with standard number-theoretic assumptions (Corollary 5.1). The examples we present here do not constitute an exhaustive list of the functional encryption schemes to which we could apply the transformation. For instance, the construction of single-key-secure, succinct FE from LWE by Goldwasser et al. [62] and the recent adaptively-secure construction from $i\mathcal{O}$ by Waters [86] are also suitable candidates.

We note that the FE schemes for deterministic functions we consider below are secure (or can be made secure) under a slightly stronger notion of simulation security compared to Definition 2.4. Under the stronger notion (considered in [50,63]), the simulator is not allowed to program the public-parameters (they are generated by the Setup algorithm) or the pre-challenge key queries (they are generated using the KeyGen algorithm). Hence, when our transformation is applied to these schemes, there is a small loss in security. We believe that this loss is inherent because the new schemes are secure under malleability attacks while the original schemes are not. In particular, the construction of Goyal et al. [65] also suffers from this limitation.

The GVW scheme. In [63], Gorbunov et al. give a construction of a general-purpose public-key FE scheme for a bounded number of secret key queries. More formally, they give both a $(q_1, 1, \mathsf{poly})$- and a $(q_1, \mathsf{poly}, 0)$-SIM[7] secure FE scheme for any class of deterministic functions computable by polynomial-size circuits based on the existence of semantically-secure public-key encryption and pseudorandom generators (PRG) computable by low-degree circuits. These assumptions are implied by many concrete intractability assumptions such as factoring.

The GVW scheme can be made perfectly correct if we have the same guarantee from the two primitives it is based on: a semantically-secure public-key encryption scheme and a decomposable randomized encoding scheme [70]. There are many ways to get perfect correctness for the former, like ElGamal [52] or

---

[7] We write poly to denotes that the quantity does not have to be a-priori bounded, and can be any polynomial in λ.

RSA [82]. For the latter, we can use Applebaum et al.'s construction [12, Theorem 4.14]. We can now apply our generic transformation (Corollary 5.1) to the GVW scheme to obtain the following corollary:

Corollary 5.2. *Under standard number-theoretic assumptions, for any polynomial $q_1 = q_1(\lambda)$, there exists a $(q_1, 1, \mathsf{poly})$-SIM and a $(q_1, \mathsf{poly}, 0)$-SIM secure FE scheme for any class of randomized functions computable by polynomial-size circuits with security against malicious encrypters.*

In the full version [6], we describe how to apply our generic transformation from Sect. 4 to the GGHZ [56] and GGHRSW [55] functional encryption schemes to obtains FE schemes supporting randomized functionalities from concrete assumptions over multilinear maps and indistinguishability obfuscation, respectively. We thus obtain the following corollaries:

Corollary 5.3. *Under standard number-theoretic assumptions, and the GGHZ complexity assumptions on composite-order multilinear maps [56, Section 2.3], for any polynomials $q_1 = q_1(\lambda)$ and $q_c = q_c(\lambda)$, there exists a $(q_1, q_c, \mathsf{poly})$-SIM secure functional encryption for all polynomial-sized randomized functionalities with security against malicious encrypters.*

Corollary 5.4. *Under standard number-theoretic assumptions, and the existence of an indistinguishability obfuscator, for any polynomials $q_1 = q_1(\lambda)$ and $q_c = q_c(\lambda)$, there exists a $(q_1, q_c, \mathsf{poly})$-SIM secure functional encryption for all polynomial-sized randomized functionalities with security against malicious encrypters.*

Comparison with the GJKS scheme. We note that $(q_1, q_c, \mathsf{poly})$-SIM security matches the known lower bounds for simulation-based security in the standard model [7,35]. We remark also that the FE schemes from Corollaries 5.3 and 5.4 provide stronger security than the original FE scheme for randomized functionalities by Goyal et al. [65]. Their construction was shown to be selectively rather than adaptively secure. Specifically, in their security model, the adversary must commit to its challenge messages before seeing the master public key. On the contrary, when we apply our generic transformation to both the GGHZ scheme from composite-order multilinear maps as well as the GGHSRW scheme from indistinguishability obfuscation, we obtain an adaptive-secure FE scheme where the adversary can not only see the master public key, but also make secret key queries prior to issuing the challenge query.

6   Open Questions

We conclude with a few interesting open questions for further study:

– Can we construct an FE scheme for a more restrictive class of randomized functionalities (e.g., sampling from a database) without needing to go through our generic transformation? In other words, for simpler classes of randomized functionalities, can we construct a scheme that does not require a general-purpose FE scheme for deterministic functionalities?

- Is it possible to generically convert a public-key FE scheme for deterministic functionalities into one that supports randomized functionalities *without* making any additional assumptions? Komargodski, Segev, and Yogev [73] show that this is possible in the secret-key setting.

Acknowledgments. We thank Venkata Koppula for many helpful conversations and discussions related to this work. We also thank the anonymous reviewers for useful feedback on the presentation.

References

1. Abdalla, M., Benhamouda, F., Passelègue, A.: An algebraic framework for pseudo-random functions and applications to related-key security. In: Gennaro, R., Robshaw, M. (eds.) CRYPTO 2015. LNCS, vol. 9215, pp. 388–409. Springer, Heidelberg (2015). doi:10.1007/978-3-662-47989-6_19

2. Abdalla, M., Benhamouda, F., Passelègue, A., Paterson, K.G.: Related-key security for pseudorandom functions beyond the linear barrier. In: Garay, J.A., Gennaro, R. (eds.) CRYPTO 2014. LNCS, vol. 8616, pp. 77–94. Springer, Heidelberg (2014). doi:10.1007/978-3-662-44371-2_5

3. Agrawal, S., Agrawal, S., Badrinarayanan, S., Kumarasubramanian, A., Prabhakaran, M., Sahai, A.: On the practical security of inner product functional encryption. In: Katz, J. (ed.) PKC 2015. LNCS, vol. 9020, pp. 777–798. Springer, Heidelberg (2015). doi:10.1007/978-3-662-46447-2_35

4. Agrawal, S., Agrawal, S., Prabhakaran, M.: Cryptographic agents: towards a unified theory of computing on encrypted data. In: Oswald, E., Fischlin, M. (eds.) EUROCRYPT 2015. LNCS, vol. 9057, pp. 501–531. Springer, Heidelberg (2015). doi:10.1007/978-3-662-46803-6_17

5. Agrawal, S., Koppula, V., Waters, B.: Impossibility of simulation secure functional encryption even with random oracles. Cryptology ePrint Archive, Report 2016/959 (2016). http://eprint.iacr.org/2016/959

6. Agrawal, S., Wu, D.J.: Functional encryption: Deterministic to randomized functions from simple assumptions. Cryptology ePrint Archive, Report 2016/482 (2016). http://eprint.iacr.org/2016/482

7. Agrawal, S., Gorbunov, S., Vaikuntanathan, V., Wee, H.: Functional encryption: new perspectives and lower bounds. In: Canetti, R., Garay, J.A. (eds.) CRYPTO 2013. LNCS, vol. 8043, pp. 500–518. Springer, Heidelberg (2013). doi:10.1007/978-3-642-40084-1_28

8. Alwen, J., Barbosa, M., Farshim, P., Gennaro, R., Gordon, S.D., Tessaro, S., Wilson, D.A.: On the relationship between functional encryption, obfuscation, and fully homomorphic encryption. In: IMA International Conference on Cryptography and Coding (2013)

9. Ananth, P., Boneh, D., Garg, S., Sahai, A., Zhandry, M.: Differing-inputs obfuscation and applications. Cryptology ePrint Archive, Report 2013/689 (2013). http://eprint.iacr.org/2013/689

10. Ananth, P., Brakerski, Z., Segev, G., Vaikuntanathan, V.: From selective to adaptive security in functional encryption. In: Gennaro, R., Robshaw, M. (eds.) CRYPTO 2015. LNCS, vol. 9216, pp. 657–677. Springer, Heidelberg (2015). doi:10. 1007/978-3-662-48000-7_32

11. Applebaum, B., Brakerski, Z.: Obfuscating circuits via composite-order graded encoding. In: Dodis, Y., Nielsen, J.B. (eds.) TCC 2015. LNCS, vol. 9015, pp. 528–556. Springer, Heidelberg (2015). doi:10.1007/978-3-662-46497-7_21

12. Applebaum, B., Ishai, Y., Kushilevitz, E.: Computationally private randomizing polynomials and their applications. Comput. Complex. **15**(2), 115–162 (2006)

13. Ateniese, G., Fu, K., Green, M., Hohenberger, S.: Improved proxy re-encryption schemes with applications to secure distributed storage. ACM Trans. Inf. Syst. Secur. **9**(1), 1–30 (2006). doi:10.1145/1127345.1127346

14. Barak, B., Garg, S., Kalai, Y.T., Paneth, O., Sahai, A.: Protecting obfuscation against algebraic attacks. In: Nguyen, P.Q., Oswald, E. (eds.) EUROCRYPT 2014. LNCS, vol. 8441, pp. 221–238. Springer, Heidelberg (2014). doi:10.1007/ 978-3-642-55220-5_13

15. Barak, B., Goldreich, O., Impagliazzo, R., Rudich, S., Sahai, A., Vadhan, S., Yang, K.: On the (im)possibility of obfuscating programs. In: Kilian, J. (ed.) CRYPTO 2001. LNCS, vol. 2139, pp. 1–18. Springer, Heidelberg (2001). doi:10. 1007/3-540-44647-8_1

16. Barbosa, M., Farshim, P.: On the semantic security of functional encryption schemes. In: Kurosawa, K., Hanaoka, G. (eds.) PKC 2013. LNCS, vol. 7778, pp. 143–161. Springer, Heidelberg (2013). doi:10.1007/978-3-642-36362-7_10

17. Bellare, M., Boldyreva, A., O'Neill, A.: Deterministic and efficiently searchable encryption. In: Menezes, A. (ed.) CRYPTO 2007. LNCS, vol. 4622, pp. 535–552. Springer, Heidelberg (2007). doi:10.1007/978-3-540-74143-5_30

18. Bellare, M., Cash, D.: Pseudorandom functions and permutations provably secure against related-key attacks. In: Rabin, T. (ed.) CRYPTO 2010. LNCS, vol. 6223, pp. 666–684. Springer, Heidelberg (2010). doi:10.1007/978-3-642-14623-7_36

19. Bellare, M., Cash, D., Miller, R.: Cryptography secure against related-key attacks and tampering. In: Lee, D.H., Wang, X. (eds.) ASIACRYPT 2011. LNCS, vol. 7073, pp. 486–503. Springer, Heidelberg (2011). doi:10.1007/978-3-642-25385-0_26

20. Bellare, M., Fischlin, M., O'Neill, A., Ristenpart, T.: Deterministic encryption: definitional equivalences and constructions without random oracles. In: Wagner, D. (ed.) CRYPTO 2008. LNCS, vol. 5157, pp. 360–378. Springer, Heidelberg (2008). doi:10.1007/978-3-540-85174-5_20

21. Bellare, M., Hoang, V.T.: Resisting randomness subversion: fast deterministic and hedged public-key encryption in the standard model. In: Oswald, E., Fischlin, M. (eds.) EUROCRYPT 2015. LNCS, vol. 9057, pp. 627–656. Springer, Heidelberg (2015). doi:10.1007/978-3-662-46803-6_21

22. Bellare, M., Kohno, T.: A theoretical treatment of related-key attacks: RKA-PRPs, RKA-PRFs, and applications. In: Biham, E. (ed.) EUROCRYPT 2003. LNCS, vol. 2656, pp. 491–506. Springer, Heidelberg (2003). doi:10.1007/3-540-39200-9_31

23. Bellare, M., O'Neill, A.: Semantically-secure functional encryption: possibility results, impossibility results and the quest for a general definition. In: Abdalla, M., Nita-Rotaru, C., Dahab, R. (eds.) CANS 2013. LNCS, vol. 8257, pp. 218–234. Springer, Cham (2013). doi:10.1007/978-3-319-02937-5_12

24. Bethencourt, J., Sahai, A., Waters, B.: Ciphertext-policy attribute-based encryption. In: IEEE Symposium on Security and Privacy (2007)

25. Biham, E.: New types of cryptanalytic attacks using related keys. In: Helleseth, T. (ed.) EUROCRYPT 1993. LNCS, vol. 765, pp. 398–409. Springer, Heidelberg (1994). doi:10.1007/3-540-48285-7_34

26. Bitansky, N., Vaikuntanathan, V.: Indistinguishability obfuscation: from approximate to exact. In: Kushilevitz, E., Malkin, T. (eds.) TCC 2016. LNCS, vol. 9562, pp. 67–95. Springer, Heidelberg (2016). doi:10.1007/978-3-662-49096-9_4

27. Blaze, M., Bleumer, G., Strauss, M.: Divertible protocols and atomic proxy cryptography. In: Nyberg, K. (ed.) EUROCRYPT 1998. LNCS, vol. 1403, pp. 127–144. Springer, Heidelberg (1998). doi:10.1007/BFb0054122

28. Blum, M., Feldman, P., Micali, S.: Non-interactive zero-knowledge and its applications (extended abstract). In: ACM STOC (1988)

29. Boneh, D.: The decision Die-Hellman problem. In: Third Algorithmic Number Theory Symposium (ANTS), vol. 1423 (1998). Invited paper

30. Boneh, D.: Twenty years of attacks on the RSA cryptosystem. Not. Am. Math. Soc. 46(2), 203–213 (1999)

31. Boneh, D., Franklin, M.: Identity-based encryption from the weil pairing. In: Kilian, J. (ed.) CRYPTO 2001. LNCS, vol. 2139, pp. 213–229. Springer, Heidelberg (2001). doi:10.1007/3-540-44647-8_13

32. Boneh, D., Lewi, K., Raykova, M., Sahai, A., Zhandry, M., Zimmerman, J.: Semantically secure order-revealing encryption: multi-input functional encryption without obfuscation. In: Oswald, E., Fischlin, M. (eds.) EUROCRYPT 2015. LNCS, vol. 9057, pp. 563–594. Springer, Heidelberg (2015). doi:10.1007/978-3-662-46803-6_19

33. Boneh, D., Raghunathan, A., Segev, G.: Function-private identity-based encryption: hiding the function in functional encryption. In: Canetti, R., Garay, J.A. (eds.) CRYPTO 2013. LNCS, vol. 8043, pp. 461–478. Springer, Heidelberg (2013). doi:10.1007/978-3-642-40084-1_26

34. Boneh, D., Raghunathan, A., Segev, G.: Function-private subspace-membership encryption and its applications. In: Sako, K., Sarkar, P. (eds.) ASIACRYPT 2013. LNCS, vol. 8269, pp. 255–275. Springer, Heidelberg (2013). doi:10.1007/978-3-642-42033-7_14

35. Boneh, D., Sahai, A., Waters, B.: Functional encryption: definitions and challenges. In: Ishai, Y. (ed.) TCC 2011. LNCS, vol. 6597, pp. 253–273. Springer, Heidelberg (2011). doi:10.1007/978-3-642-19571-6_16

36. Boneh, D., Silverberg, A.: Applications of multilinear forms to cryptography. Cryptology ePrint Archive, Report 2002/080 (2002). http://eprint.iacr.org/2002/080

37. Boneh, D., Waters, B.: Conjunctive, subset, and range queries on encrypted data. In: Vadhan, S.P. (ed.) TCC 2007. LNCS, vol. 4392, pp. 535–554. Springer, Heidelberg (2007). doi:10.1007/978-3-540-70936-7_29

38. Boneh, D., Wu, D.J., Zimmerman, J.: Immunizing multilinear maps against zeroizing attacks. Cryptology ePrint Archive, Report 2014/930 (2014). http://eprint.iacr.org/2014/930

39. Brakerski, Z., Chandran, N., Goyal, V., Jain, A., Sahai, A., Segev, G.: Hierarchical functional encryption. In: ITCS (2017)

40. Brakerski, Z., Rothblum, G.N.: Virtual black-box obfuscation for all circuits via generic graded encoding. In: Lindell, Y. (ed.) TCC 2014. LNCS, vol. 8349, pp. 1–25. Springer, Heidelberg (2014). doi:10.1007/978-3-642-54242-8_1

41. Brakerski, Z., Segev, G.: Better security for deterministic public-key encryption: the auxiliary-input setting. In: Rogaway, P. (ed.) CRYPTO 2011. LNCS, vol. 6841, pp. 543–560. Springer, Heidelberg (2011). doi:10.1007/978-3-642-22792-9_31

42. Cheon, J.H., Fouque, P.-A., Lee, C., Minaud, B., Ryu, H.: Cryptanalysis of the new CLT multilinear map over the integers. In: Fischlin, M., Coron, J.-S. (eds.) EUROCRYPT 2016. LNCS, vol. 9665, pp. 509–536. Springer, Heidelberg (2016). doi:10.1007/978-3-662-49890-3_20

43. Cheon, J.H., Han, K., Lee, C., Ryu, H., Stehlé, D.: Cryptanalysis of the multi-linear map over the integers. In: Oswald, E., Fischlin, M. (eds.) EUROCRYPT 2015. LNCS, vol. 9056, pp. 3–12. Springer, Heidelberg (2015). doi:10.1007/978-3-662-46800-5_1

44. Cheon, J.H., Jeong, J., Lee, C.: An algorithm for NTRU problems and cryptanalysis of the GGH multilinear map without a low level encoding of zero. Cryptology ePrint Archive, Report 2016/139 (2016). http://eprint.iacr.org/2016/139

45. Cocks, C.: An identity based encryption scheme based on quadratic residues. In: Honary, B. (ed.) Cryptography and Coding 2001. LNCS, vol. 2260, pp. 360–363. Springer, Heidelberg (2001). doi:10.1007/3-540-45325-3_32

46. Coron, J.-S., Gentry, C., Halevi, S., Lepoint, T., Maji, H.K., Miles, E., Raykova, M., Sahai, A., Tibouchi, M.: Zeroizing without low-level zeroes: new MMAP attacks and their limitations. In: Gennaro, R., Robshaw, M. (eds.) CRYPTO 2015. LNCS, vol. 9215, pp. 247–266. Springer, Heidelberg (2015). doi:10.1007/978-3-662-47989-6_12

47. Coron, J.-S., Lee, M.S., Lepoint, T., Tibouchi, M.: Cryptanalysis of GGH15 mul-tilinear maps. In: Robshaw, M., Katz, J. (eds.) CRYPTO 2016. LNCS, vol. 9815, pp. 607–628. Springer, Heidelberg (2016). doi:10.1007/978-3-662-53008-5_21

48. Coron, J.-S., Lepoint, T., Tibouchi, M.: Practical multilinear maps over the inte-gers. In: Canetti, R., Garay, J.A. (eds.) CRYPTO 2013. LNCS, vol. 8042, pp. 476–493. Springer, Heidelberg (2013). doi:10.1007/978-3-642-40041-4_26

49. Coron, J.-S., Lepoint, T., Tibouchi, M.: New multilinear maps over the integers. In: Gennaro, R., Robshaw, M. (eds.) CRYPTO 2015. LNCS, vol. 9215, pp. 267–286. Springer, Heidelberg (2015). doi:10.1007/978-3-662-47989-6_13

50. De Caro, A., Iovino, V., Jain, A., O'Neill, A., Paneth, O., Persiano, G.: On the achievability of simulation-based security for functional encryption. In: CRYPTO (2013)

51. Dwork, C., Naor, M., Reingold, O.: Immunizing encryption schemes from decryp-tion errors. In: Cachin, C., Camenisch, J.L. (eds.) EUROCRYPT 2004. LNCS, vol. 3027, pp. 342–360. Springer, Heidelberg (2004). doi:10.1007/978-3-540-24676-3_21

52. ElGamal, T.: A public key cryptosystem and a signature scheme based on discrete logarithms. IEEE Trans. Inf. Theory 31, 469–472 (1985)

53. Feige, U., Lapidot, D., Shamir, A.: Multiple non-interactive zero knowledge proofs based on a single random string (extended abstract). In: FOCS (1990)

54. Fuller, B., O'Neill, A., Reyzin, L.: A unified approach to deterministic encryption: new constructions and a connection to computational entropy. In: Cramer, R. (ed.) TCC 2012. LNCS, vol. 7194, pp. 582–599. Springer, Heidelberg (2012). doi:10.1007/978-3-642-28914-9_33

55. Garg, S., Gentry, C., Halevi, S., Raykova, M., Sahai, A., Waters, B.: Candidate indistinguishability obfuscation and functional encryption for all circuits. In: FOCS (2013)

56. Garg, S., Gentry, C., Halevi, S., Zhandry, M.: Functional encryption without obfus-cation. In: Kushilevitz, E., Malkin, T. (eds.) TCC 2016. LNCS, vol. 9563, pp. 480–511. Springer, Heidelberg (2016). doi:10.1007/978-3-662-49099-0_18

57. Garg, S., Gentry, C., Sahai, A., Waters, B.: Witness encryption and its applications. In: ACM STOC (2013)

60   S. Agrawal and D.J. Wu

58. Gentry, C., Lewko, A., Waters, B.: Witness encryption from instance independent assumptions. In: Garay, J.A., Gennaro, R. (eds.) CRYPTO 2014. LNCS, vol. 8616, pp. 426–443. Springer, Heidelberg (2014). doi:10.1007/978-3-662-44371-2_24
59. Gentry, C., Lewko, A.B., Sahai, A., Waters, B.: Indistinguishability obfuscation from the multilinear subgroup elimination assumption. In: FOCS (2015)
60. Goldreich, O.: The Foundations of Cryptography. Basic Techniques, vol. 1. Cambridge University Press, Cambridge (2001)
61. Goldwasser, S., Gordon, S.D., Goyal, V., Jain, A., Katz, J., Liu, F.-H., Sahai, A., Shi, E., Zhou, H.-S.: Multi-input functional encryption. In: Nguyen, P.Q., Oswald, E. (eds.) EUROCRYPT 2014. LNCS, vol. 8441, pp. 578–602. Springer, Heidelberg (2014). doi:10.1007/978-3-642-55220-5_32
62. Goldwasser, S., Kalai, Y.T., Popa, R.A., Vaikuntanathan, V., Zeldovich, N.: Reusable garbled circuits and succinct functional encryption. In: ACM STOC (2013)
63. Gorbunov, S., Vaikuntanathan, V., Wee, H.: Functional encryption with bounded collusions via multi-party computation. In: Safavi-Naini, R., Canetti, R. (eds.) CRYPTO 2012. LNCS, vol. 7417, pp. 162–179. Springer, Heidelberg (2012). doi:10.1007/978-3-642-32009-5_11
64. Gorbunov, S., Vaikuntanathan, V., Wee, H.: Attribute-based encryption for circuits. In: ACM STOC (2013)
65. Goyal, V., Jain, A., Koppula, V., Sahai, A.: Functional encryption for randomized functionalities. In: Dodis, Y., Nielsen, J.B. (eds.) TCC 2015. LNCS, vol. 9015, pp. 325–351. Springer, Heidelberg (2015). doi:10.1007/978-3-662-46497-7_13
66. Goyal, V., Pandey, O., Sahai, A., Waters, B.: Attribute-based encryption for fine-grained access control of encrypted data. In: ACM CCS (2006). Available as Cryptology ePrint Archive Report 2006/309
67. Groth, J.: Simulation-sound NIZK proofs for a practical language and constant size group signatures. In: Lai, X., Chen, K. (eds.) ASIACRYPT 2006. LNCS, vol. 4284, pp. 444–459. Springer, Heidelberg (2006). doi:10.1007/11935230_29
68. Groth, J., Ostrovsky, R., Sahai, A.: Perfect non-interactive zero knowledge for NP. In: Vaudenay, S. (ed.) EUROCRYPT 2006. LNCS, vol. 4004, pp. 339–358. Springer, Heidelberg (2006). doi:10.1007/11761679_21
69. Hu, Y., Jia, H.: Cryptanalysis of GGH map. In: Fischlin, M., Coron, J.-S. (eds.) EUROCRYPT 2016. LNCS, vol. 9665, pp. 537–565. Springer, Heidelberg (2016). doi:10.1007/978-3-662-49890-3_21
70. Ishai, Y., Kushilevitz, E.: Randomizing polynomials: A new representation with applications to round-efficient secure computation. In: FOCS (2000)
71. Katz, J., Sahai, A., Waters, B.: Predicate encryption supporting disjunctions, polynomial equations, and inner products. In: Smart, N. (ed.) EUROCRYPT 2008. LNCS, vol. 4965, pp. 146–162. Springer, Heidelberg (2008). doi:10.1007/978-3-540-78967-3_9
72. Knudsen, L.R.: Cryptanalysis of LOKI 91. In: Seberry, J., Zheng, Y. (eds.) AUSCRYPT 1992. LNCS, vol. 718, pp. 196–208. Springer, Heidelberg (1993). doi:10.1007/3-540-57220-1_62
73. Goyal, V., Jain, A., Koppula, V., Sahai, A.: Functional encryption for randomized functionalities. In: Dodis, Y., Nielsen, J.B. (eds.) TCC 2015. LNCS, vol. 9015, pp. 325–351. Springer, Heidelberg (2015). doi:10.1007/978-3-662-46497-7_13
74. Lewi, K., Montgomery, H., Raghunathan, A.: Improved constructions of PRFs secure against related-key attacks. In: Boureanu, I., Owesarski, P., Vaudenay, S. (eds.) ACNS 2014. LNCS, vol. 8479, pp. 44–61. Springer, Cham (2014). doi:10.1007/978-3-319-07536-5_4

75. Lewko, A., Okamoto, T., Sahai, A., Takashima, K., Waters, B.: Fully secure functional encryption: attribute-based encryption and (Hierarchical) inner product encryption. In: Gilbert, H. (ed.) EUROCRYPT 2010. LNCS, vol. 6110, pp. 62–91. Springer, Heidelberg (2010). doi:10.1007/978-3-642-13190-5_4

76. Lucks, S.: Ciphers secure against related-key attacks. In: Roy, B., Meier, W. (eds.) FSE 2004. LNCS, vol. 3017, pp. 359–370. Springer, Heidelberg (2004). doi:10.1007/978-3-540-25937-4_23

77. Miles, E., Sahai, A., Zhandry, M.: Annihilation attacks for multilinear maps: cryptanalysis of indistinguishability obfuscation over GGH13. In: Robshaw, M., Katz, J. (eds.) CRYPTO 2016. LNCS, vol. 9815, pp. 629–658. Springer, Heidelberg (2016). doi:10.1007/978-3-662-53008-5_22

78. Okamoto, T., Takashima, K.: Fully secure functional encryption with general relations from the decisional linear assumption. In: Rabin, T. (ed.) CRYPTO 2010. LNCS, vol. 6223, pp. 191–208. Springer, Heidelberg (2010). doi:10.1007/978-3-642-14623-7_11

79. O'Neill, A.: Definitional issues in functional encryption. Cryptology ePrint Archive, Report 2010/556 (2010). http://eprint.iacr.org/2010/556

80. Pass, R., Seth, K., Telang, S.: Indistinguishability obfuscation from semantically-secure multilinear encodings. In: Garay, J.A., Gennaro, R. (eds.) CRYPTO 2014. LNCS, vol. 8616, pp. 500–517. Springer, Heidelberg (2014). doi:10.1007/978-3-662-44371-2_28

81. Rackoff, C., Simon, D.R.: Non-interactive zero-knowledge proof of knowledge and chosen ciphertext attack. In: Feigenbaum, J. (ed.) CRYPTO 1991. LNCS, vol. 576, pp. 433–444. Springer, Heidelberg (1992). doi:10.1007/3-540-46766-1_35

82. Rivest, R.L., Shamir, A., Adleman, L.M.: A method for obtaining digital signature and public-key cryptosystems. Commun. Assoc. Comput. Mach. 21(2), 120–126 (1978)

83. Sahai, A., Seyalioglu, H.: Worry-free encryption: functional encryption with public keys. In: ACM CCS (2010)

84. Sahai, A., Waters, B.: Fuzzy identity-based encryption. In: Cramer, R. (ed.) EUROCRYPT 2005. LNCS, vol. 3494, pp. 457–473. Springer, Heidelberg (2005). doi:10.1007/11426639_27

85. Shamir, A.: Identity-based cryptosystems and signature schemes. In: Blakley, G.R., Chaum, D. (eds.) CRYPTO 1984. LNCS, vol. 196, pp. 47–53. Springer, Heidelberg (1985). doi:10.1007/3-540-39568-7_5

86. Waters, B.: A punctured programming approach to adaptively secure functional encryption. In: Gennaro, R., Robshaw, M. (eds.) CRYPTO 2015. LNCS, vol. 9216, pp. 678–697. Springer, Heidelberg (2015). doi:10.1007/978-3-662-48000-7_33

87. Zimmerman, J.: How to obfuscate programs directly. In: Oswald, E., Fischlin, M. (eds.) EUROCRYPT 2015. LNCS, vol. 9057, pp. 439–467. Springer, Heidelberg (2015). doi:10.1007/978-3-662-46803-6_15

Lattice Attacks and Constructions IV

Random Sampling Revisited: Lattice Enumeration with Discrete Pruning

Yoshinori Aono[1] and Phong Q. Nguyen[2,3($\boxtimes$)]

[1] Security Fundamentals Laboratory, Cybersecurity Research Institute,
National Institute of Information and Communications Technology, Tokyo, Japan
[2] Inria Paris, Paris, France
Phong.Nguyen@inria.fr
[3] CNRS/JFLI and the University of Tokyo, Tokyo, Japan

Abstract. In 2003, Schnorr introduced *Random sampling* to find very short lattice vectors, as an alternative to enumeration. An improved variant has been used in the past few years by Kashiwabara *et al.* to solve the largest Darmstadt SVP challenges. However, the behaviour of random sampling and its variants is not well-understood: all analyses so far rely on a questionable heuristic assumption, namely that the lattice vectors produced by some algorithm are uniformly distributed over certain parallelepipeds. In this paper, we introduce lattice enumeration with discrete pruning, which generalizes random sampling and its variants, and provides a novel geometric description based on partitions of the n-dimensional space. We obtain what is arguably the first sound analysis of random sampling, by showing how discrete pruning can be rigorously analyzed under the well-known Gaussian heuristic, in the same model as the Gama-Nguyen-Regev analysis of pruned enumeration from EUROCRYPT '10, albeit using different tools: we show how to efficiently compute the volume of the intersection of a ball with a box, and to efficiently approximate a large sum of many such volumes, based on statistical inference. Furthermore, we show how to select good parameters for discrete pruning by enumerating integer points in an ellipsoid. Our analysis is backed up by experiments and allows for the first time to reasonably estimate the success probability of random sampling and its variants, and to make comparisons with previous forms of pruned enumeration. Our work unifies random sampling and pruned enumeration and show that they are complementary of each other: both have different characteristics and offer different trade-offs to speed up enumeration.

1 Introduction

With the upcoming NIST standardization of post-quantum cryptography and the development of fully-homomorphic encryption, it is becoming increasingly important to provice convincing security estimates for lattice-based cryptosystems. To do so, we need to understand the best lattice algorithms, such as the ones used to solve the largest numerical challenges. For NTRU challenges [33] and

© International Association for Cryptologic Research 2017
J.-S. Coron and J.B. Nielsen (Eds.): EUROCRYPT 2017, Part II, LNCS 10211, pp. 65–102, 2017.
DOI: 10.1007/978-3-319-56614-6_3

Darmstadt's lattice challenges [18], the largest records were solved (by respectively Ducas-Nguyen and Chen-Nguyen) using algorithms (pruned enumeration [11] with state-of-the-art BKZ [5]) which are reasonably well-understood. However, the seven largest records of Darmstadt's SVP challenges [28] have been solved by Kashiwabara and Teruya (using significant computational power, comparable for the largest challenge to that of RSA-768) using an algorithm which is partially secret: an incomplete description can be found in [8]. The core of the algorithm seems to be an improved variant of Schnorr's random sampling method [30], which was introduced as an alternative to enumeration [31] for finding extremely short lattice vectors.

Unfortunately, our understanding of random sampling and its variants is not satisfactory: all the analyses [4,8,9,20,30] so far rely on several heuristic assumptions, the most questionable being the *Randomness Assumption*, which says that the lattice vectors produced by some algorithm are uniformly distributed over certain parallelepipeds. As noted by Ludwig [20], this assumption cannot hold, and a gap between theoretical analyses and experimental results has been reported [8,20]. In some sense, the situation is reminiscent of enumeration with pruning, which was introduced and partially analyzed in the mid-nineties by Schnorr *et al.* [31,32], but arguably only well-understood in 2010, when Gama *et al.* [11] provided a novel geometric description and presented its first sound analysis, which provided much better parameters.

At this point, we do not know if random sampling is better or worse than pruned enumeration, neither in theory nor in practice, which is rather puzzling, considering their importance for lattice algorithms, which can be used to solve a wide range of problems, such as integer programming [15], factoring polynomials with rational coefficients [16], integer relation finding [13], as well as problems in communication theory (see [1,24] and references therein), and public-key cryptanalysis (see [22] and references therein). Pruned enumeration is used in state-of-the-art implementations of BKZ [2,5].

Our results. We introduce lattice enumeration with discrete pruning, which generalizes naturally Schnorr's random sampling and all its variants, and provides a novel geometric description based on partitions of the n-dimensional space. This new description allows us to rigorously analyze discrete pruning under the well-known Gaussian heuristic, in the same model as the Gama-Nguyen-Regev [11] analysis of pruned enumeration, albeit using different tools. This is the first sound analysis of random sampling and its variants, and our presentation unifies both pruned enumeration and random sampling, by viewing them as two different ways of speeding up the classical enumeration algorithm. In other words, we improve the understanding of random sampling to that of pruned enumeration.

To complement our theoretical analysis, we introduce three technical tools which allow, in practice, to estimate success probabilities and optimize parameters for discrete pruning: this is the most difficult aspect of discrete pruning, because given parameters, estimating the running time of discrete pruning is on the other hand very easy. The first two tools are combined to estimate accurately and efficiently the success probability: the first one computes efficiently the volume of the intersection of an n-dimensional ball with a box, and the second one uses statistical inference to approximate efficiently large sums of such volumes without computing individually each volume. Finally, the third tool is an efficient algorithm to generate nearly-optimal parameters for discrete pruning in practice.

Our analysis is backed up by experiments, and allows us to make concrete comparisons with other forms of pruned enumeration. As an example, our analysis shows that the Fukase-Kashiwabara variant [8] outperforms Schnorr's original algorithm and its variants by Buchmann-Ludwig [4,20]. Experimentally, we find that discrete pruning is complementary with continuous pruning: whether one is more efficient than the other depends on the exact setting, such as what is the lattice dimension, the radius of the enumeration ball, the required time, *etc.*

Technical overview. A *lattice* is the set of all integer combinations of n linearly independent vectors $\mathbf{b}_1, \ldots, \mathbf{b}_n$ in $\mathbb{R}^n$. These vectors are known as a *basis* of the lattice. The most famous computational problem involving lattices is the *shortest vector problem* (SVP), which asks to find a nonzero lattice vector of smallest norm, given a lattice basis as input. A basic approach is *enumeration* which dates back to the early 1980s with work by Pohst [25], Kannan [15], and Fincke-Pohst [7] and is still actively investigated (*e.g.*, [1,11,12,31,35]): given a radius $R > 0$ and a basis $B = (\mathbf{b}_1, \ldots, \mathbf{b}_n)$ of a lattice L, enumeration computes all the points in $L \cap S$ where S is the zero-centered ball of radius R, which allows to find a shortest lattice vector by comparing their norms. Enumeration goes through all lattice vectors $\sum_{i=1}^{n} x_i \mathbf{b}_i$, exhaustively searching in order $x_n, x_{n-1} \ldots x_1 \in \mathbb{Z}$ by projecting L onto suitable subspaces of increasing dimension: for instance, bounds on x_n are found by projecting L onto a line, and intersecting it with S. The running time of enumeration depends on R and the quality of the basis, but is typically super-exponential in n.

To speed up the running time of enumeration, Schnorr, Euchner, and Hörner [31,32] suggested in the 1990s a modification of enumeration called *pruned enumeration*. We follow the geometric presentation of Gama, Nguyen and Regev [11], who revisited pruned enumeration. Essentially, pruned enumeration is a trade-off, defined by a pruning set $P \subseteq \mathbb{R}^n$: P is chosen in such a way that enumerating all the points in $L \cap S \cap P$ is much faster than over $L \cap S$, but this is only useful if $L \cap S \cap P$ is non-trivial $\nsubseteq \{0\}$, in which case we have found a short non-zero lattice vector in $L \cap S$. Under the Gaussian heuristic, $L \cap S \cap P$ is "expected" to be non-trivial when $\mathrm{vol}(S \cap P)$ is sufficiently large with respect to the lattice. Here, the set P is defined as an intersection of n cylinders such that $P \subseteq S$, thus $\mathrm{vol}(S \cap P) = \mathrm{vol}(P)$, and [11] shows how to approximate efficiently $\mathrm{vol}(P)$, which allows to choose good parameters: furthermore, if $\mathrm{vol}(S \cap P)$ turns out to be too small to hope for a solution, one can simply repeat the process with many choices of P, which may still be cheaper overall than the original enumeration.

We introduce *discrete pruning*, which follows the same framework, except that the pruning set P is completely different. Instead of a cylinder-intersection, we consider a set P formed by regrouping finitely many cells of suitable partitions of $\mathbb{R}^n$, related to the lattice L: we show that random sampling and its variants all fall in this category. In practice, P can be rewritten as the union of finitely many non-overlapping boxes, where a box means a cube whose sides are not necessarily of equal length but are still perpendicular. To analyze the algorithm, it therefore suffices to be able to compute $\mathrm{vol}(S \cap H)$ where S is a ball and H is a box: we give exact formulas as infinite series, asymptotical estimations and efficient numerical methods to compute such volumes, which might be of independent interest. However, it actually suffices to approximate a large sum $\sum_i \mathrm{vol}(S \cap H_i)$: we introduce the use of statistical inference to approximate such a sum, without computing each term of the sum.

To select good parameters for the algorithm, we introduce a fast method based on enumerating integer points in an ellipsoid, which allows to select a nearly-optimal pruning set P without resorting to computations of $\mathrm{vol}(S \cap H)$.

Experimental results. Our experiments support our analysis of discrete pruning and explain very clearly why the Fukase-Kashiwabara variant [8] of random sampling outperforms Schnorr's random sampling [30] and its Buchmann-Ludwig variant [4]: for typical parameters, the cells selected by [8] turn out to have a much larger intersection with the ball S than the corresponding cells in [4,30]. Our algorithm for selecting discrete pruning parameters provides parameters which are in practice at least slightly better than [8]: we stress that our method is completely automatic, whereas [8] strongly relies on experiments and do not explain how to select parameters in arbitrary dimension, which makes comparisons difficult. Our experiments suggest that discrete pruning can be slower or faster than continuous pruning [11,31,32], depending on the exact setting. First, the performances are impacted by the lattice dimension, the enumeration radius, and the target running time. For instance, we find that the benefits of discrete pruning increase when the lattice dimension grows, the target running time is small and the basis is not too reduced. Potentially, this might improve BKZ-type algorithms, by speeding up the preprocessing of the enumeration subroutine, or by incorporating high-dimensional discrete pruning in the reduction process itself. Second, our analysis shows that the optimal basis reduction required by discrete pruning is different from the optimal basis reduction required by continuous pruning: roughly speaking, the smaller the sum of squared Gram-Schmidt norms, the better for discrete pruning. This means that to fully take advantage of discrete pruning, one may have to modify the reduction algorithm, and not simply use LLL or BKZ: in fact, some of the secret modifications of [8] may exactly target that, as it is clear that the reduction of [8] is a bit different from BKZ strategies. Third, there are implementation differences between discrete pruning and continuous pruning. It appears that discrete pruning is easier to parallelize, which may make it better suited to special hardware. Finding good parameters for discrete pruning is also a bit easier than for continuous pruning: in particular, there are less parameters to consider, which might be useful for blockwise reduction.

Future work. There are many interesting questions related to discrete pruning. First, one may wonder what is the most efficient form of pruned enumeration, either asymptotically or in practice: we now know continuous pruning [11,31,32] and discrete pruning, and we showed how to compare both in practice, but a theoretical asymptotical comparison is not easy. Can a combination of both, or another form of pruning be more efficient? Second, is it possible to efficiently reduce a basis in such a way that the power of discrete pruning is maximized? For instance, are there better ways to decrease the sum of squared Gram-Schmidt norms?

We presented discrete pruning in the SVP case, *i.e.* to find short lattice vectors. The methodology can be adapted to the CVP case, *i.e.* to find lattice vectors close to a given target: as a special case, [19] already noticed that the Lindner-Peikert algorithm [17] can be viewed as the BDD adaptation of Schnorr's random sampling [30]. However, in the general case of discrete pruning, there appears to be a few differences: the details will be investigated in future work.

Finally, our algorithm to select good discrete pruning parameters is based on enumerating integer points in an ellipsoid: though the algorithm is extremely efficient in practice, we do not know at the moment how to prove it, and it would be interesting to do so.

Related work. Liu and Nguyen [19] also tried to view Schnorr's random sampling as some form of pruning, in the context of bounded distance decoding: however, their formalization does not rely on partitions, and does not capture all the variants of random sampling, including the one of [8].

Roadmap. We start in Sect. 2 with some background and notation on lattices, and continue with a general description of enumeration (with or without pruning) in Sect. 3. In Sect. 4, we present lattice enumeration with discrete pruning based on partitions, show that it generalizes Schnorr's random sampling and its variants, and give our rigorous analysis. In Sect. 5, we address the technical problem of computing the volume of the intersection of a ball with a box, which is required by our discrete pruning analysis. In Sect. 6, we show how to select optimal parameters for discrete pruning. We present experimental results in Sect. 7. In the full version available on the eprint archive, we provide missing proofs and additional information.

2   Preliminaries

General. For any finite set U, we denote by $\#U$ its number of elements. For any measurable subset $S \subseteq \mathbb{R}^n$, we denote by $\text{vol}(S)$ its volume. Throughout the paper, we use row representations of matrices. The Euclidean norm of a vector $\mathbf{v} \in \mathbb{R}^n$ is denoted $\|\mathbf{v}\|$. We denote by $\text{Ball}_n(R)$ the n-dimensional zero-centered Euclidean ball of radius R, whose volume is $\text{vol}(\text{Ball}_n(R)) = R^n \frac{\pi^{n/2}}{\Gamma(n/2+1)}$.

Lattices. A *lattice* L is a discrete subgroup of $\mathbb{R}^m$. Alternatively, we can define a lattice as the set $L(\mathbf{b}_1, \ldots, \mathbf{b}_n) = \{\sum_{i=1}^{n} x_i \mathbf{b}_i : x_i \in \mathbb{Z}\}$ of all integer combinations of n linearly independent vectors $\mathbf{b}_1, \ldots, \mathbf{b}_n \in \mathbb{R}^m$. This sequence of vectors is known as a *basis* of the lattice L. All the bases of L have the same number n of elements, called the dimension or rank of L, and the n-dimensional volume of the parallelepiped $\{\sum_{i=1}^{n} a_i \mathbf{b}_i : a_i \in [0, 1)\}$ they generate. We call this volume the co-volume, or determinant, of L, and denote it by $\mathrm{covol}(L)$. The lattice L is said to be *full-rank* if $n = m$. We denote by $\lambda_1(L)$ the first minimum of L, defined as the length of a shortest nonzero vector of L. The most famous lattice problem is the *shortest vector problem* (SVP), which asks to find a lattice vector of norm $\lambda_1(L)$.

Orthogonalization. For a basis $B = (\mathbf{b}_1, \ldots, \mathbf{b}_n)$ of a lattice L and $i \in \{1, \ldots, n\}$, we denote by π_i the orthogonal projection on $\mathrm{span}(\mathbf{b}_1, \ldots, \mathbf{b}_{i-1})^{\perp}$. The *Gram-Schmidt orthogonalization* of the basis B is defined as the orthogonal sequence of vectors $B^{\star} = (\mathbf{b}_1^{\star}, \ldots, \mathbf{b}_n^{\star})$, where $\mathbf{b}_i^{\star} := \pi_i(\mathbf{b}_i)$. For each i we can write $\mathbf{b}_i$ as $\mathbf{b}_i^{\star} + \sum_{j=1}^{i-1} \mu_{i,j} \mathbf{b}_j^{\star}$ for some unique $\mu_{i,1}, \ldots, \mu_{i,i-1} \in \mathbb{R}$. Thus, we may represent the $\mu_{i,j}$'s by a lower-triangular matrix μ with unit diagonal. The projection of a lattice may not be a lattice, but for all $i \in \{1, \ldots, n\}$, $\pi_i(L)$ is an $n + 1 - i$ dimensional lattice generated by the basis $\pi_i(\mathbf{b}_i), \ldots, \pi_i(\mathbf{b}_n)$, with $\mathrm{covol}(\pi_i(L)) = \prod_{j=i}^{n} \|\mathbf{b}_j^{\star}\|$.

Reduced bases. Lattice reduction algorithms aim to transform an input basis into a "high quality" basis. There are many ways to quantify the quality of bases produced by lattice reduction algorithms. One popular way is to consider the Gram-Schmidt norms $\|\mathbf{b}_1^{\star}\|, \ldots, \|\mathbf{b}_n^{\star}\|$. Intuitively speaking, a good basis is one in which this sequence does not decay too fast. In practice, it turns out that the Gram-Schmidt coefficients of bases produced by the main reduction algorithms (such as LLL or BKZ) have a certain "typical shape", assuming the input basis is sufficiently random. This property was thoroughly investigated in [10,23]. This typical shape is often used to estimate the running time of various algorithms. In particular, many theoretical asymptotic analyses (as introduced by Schnorr [30]) assume for simplicity that this shape is given by $\|\mathbf{b}_i^{\star}\|/\|\mathbf{b}_{i+1}^{\star}\| = q$ where q depends on the reduction algorithm; although less precise, this approximation called the *geometric series assumption (GSA)* is very close to the shape observed in practice.

Gaussian Heuristic. The classical Gaussian Heuristic provides an estimate on the number of lattice points inside a "nice enough" set:

Heuristic 1. *Given a full-rank lattice $L \subseteq \mathbb{R}^n$ and a measurable set $S \subseteq \mathbb{R}^n$, the number of points in $S \cap L$ is approximately $\mathrm{vol}(S)/\mathrm{covol}(L)$.*

If the heuristic holds, we would expect $\lambda_1(L)$ to be close to $\mathrm{GH}(L) = \mathrm{vol}(\mathrm{Ball}_n(1))^{-1/n} \mathrm{covol}(L)^{1/n}$, and that there about α^n points in L which have norm $\leq \alpha \mathrm{GH}(L)$. Some rigorous results along these lines are known. For instance, for a fixed set S, if we consider random lattices of unit-covolume, and scale them to have covol(L), Siegel [34] shows that the average number of points in $S \cap L$ is exactly vol(S)/covol(L). Furthermore, it is known that $\lambda_1(L)$ is in some sense close to GH(L) for a random lattice (see [26]). Note, however, that the heuristic can also be far off; for $L = \mathbb{Z}^n$, for instance, the heuristic estimates the number of lattice points inside a ball of radius $\sqrt{n}/10$ around the origin to be less than 1, yet there are exponentially many lattice points there (see [21] for more such examples). One should therefore experimentally verify the use of the heuristic, as we shall do later. This is particulary necessary, as pruned enumeration relies on strong versions of the heuristic, where the set S is not fixed, but actually depends on a basis of L.

Statistics. We denote by $\mathbb{E}()$ and $\mathbb{V}()$ respectively the expectation and the variance of a random variable.

Lemma 1. *Let X be a random variable uniformly distributed over $[\alpha, \beta]$. Then:*

$$\mathbb{E}(X^2) = \frac{\alpha^2 + \beta^2 + \alpha\beta}{3} \quad \text{and} \quad \mathbb{V}(X^2) = \frac{4}{45}\alpha^4 - \frac{1}{45}\alpha^3\beta - \frac{2}{15}\alpha^2\beta^2 - \frac{1}{45}\alpha\beta^3 + \frac{4}{45}\beta^4$$

Proof. We have:

$$\mathbb{E}(X^2) = \frac{1}{\beta - \alpha}\int_\alpha^\beta x^2 dx = \frac{1}{\beta - \alpha}\left[x^3/3\right]_\alpha^\beta = \frac{\alpha^2 + \beta^2 + \alpha\beta}{3}$$

$$\mathbb{E}(X^4) = \frac{1}{\beta - \alpha}\int_\alpha^\beta x^4 dx = \frac{1}{\beta - \alpha}\left[x^5/5\right]_\alpha^\beta = \frac{\alpha^4 + \alpha^3\beta + \alpha^2\beta^2 + \alpha\beta^3 + \beta^4}{5}$$

Finally, $\mathbb{V}(X^2) = \mathbb{E}(X^4) - \mathbb{E}(X^2)^2$. $\qquad\square$

Corollary 1. *Let $y \in \mathbb{R}$. Let X (resp. X') be a random variable uniformly distributed over $[y - 1/2, y + 1/2]$ (resp. $[y/2, (y + 1)/2]$). Then:*

$$\mathbb{E}(X^2) = y^2 + \frac{1}{12}, \mathbb{E}(X'^2) = y^2/4 + y/4 + \frac{1}{12}, \quad \mathbb{V}(X^2) = \frac{y^2}{3} + \frac{1}{180}, \mathbb{V}(X'^2) = \frac{y^2}{48} + \frac{y}{48} + \frac{1}{180}.$$

In this paper, it is convenient to extend the expectation and variance this to any measurable set C of $\mathbb{R}^n$ by using the squared norm, to measure how short is a random vector of C:

$$\mathbb{E}\{C\} := \mathbb{E}_{\mathbf{x} \in C}(\|\mathbf{x}\|^2) \qquad\qquad \mathbb{V}\{C\} := \mathbb{V}_{\mathbf{x} \in C}(\|\mathbf{x}\|^2).$$

Normal distribution. The CDF of the normal distribution of expectation 0 and variance 1 is the error function

$$\text{erf}(z) := \frac{2}{\sqrt{\pi}}\int_0^z e^{-t^2} dt.$$

3   Enumeration with Pruning

In this section, we give an overview of lattice enumeration and pruning, and revisit the analysis model of [11].

3.1   Enumeration

Let L be a full-rank lattice in $\mathbb{R}^n$. Enumeration [7,15,25] is an elementary algorithm which, given a basis $B = (\mathbf{b}_1, \ldots, \mathbf{b}_n)$ of L and a radius $R > 0$, outputs all the points in $L \cap S$ where $S = \mathrm{Ball}_n(R)$: by comparing all their norms, it is then possible to extract the shortest lattice vectors.

The main idea is to perform a recursive search using projections, which allows to reduce the dimension of the lattice: if $\|\mathbf{v}\| \leq R$, then $\|\pi_k(\mathbf{v})\| \leq R$ for all $1 \leq k \leq n$. We start with the one-dimensional lattice $\pi_n(L)$: it is trivial to enumerate all the points in $\pi_n(L) \cap S$. Assume that we enumerated all the points in $\pi_{k+1}(L) \cap S$ for some $k \geq 1$, then we can derive all the points in $\pi_k(L) \cap S$ by enumerating the intersection of a one-dimensional lattice with a suitable ball, for each point in $\pi_{k+1}(L) \cap S$. Concretely, it can be viewed as a depth first search of a gigantic tree called the enumeration tree. The running-time of enumeration depends on R and the quality of B, but it is typically super-exponential in n, even if $L \cap S$ is small.

We do not need to know more about enumeration: the interested reader is referred to [11] for more details.

3.2   Enumeration with Pruning

We note that in high dimension, enumeration is likely to be unfeasible in general for any radius $R \gg \mathrm{GH}(L)$: indeed, by the Gaussian heuristic, we expect $\#(L \cap \mathrm{Ball}_n(R))$ to have about $(R/\mathrm{GH}(L))^n$ points. For such large radius R, it is therefore more meaningful to just ask for one solution (or say, a bounded number of solutions) in $L \cap \mathrm{Ball}_n(R)$, rather than all the points.

Enumeration with pruning is a natural method to speed up enumeration, which goes back to the work of Schnorr et al. [31,32] in the 90s. We introduce its more general form: pruned enumeration uses an additional parameter, namely a pruning set $P \subseteq \mathbb{R}^n$, and outputs all points in $L \cap S \cap P$. The advantage is that for suitable choices of P, enumerating $L \cap S \cap P$ is much cheaper than enumerating $L \cap S$.

If $L \cap S \cap P \nsubseteq \{0\}$, then pruned enumeration provides non-trivial points in $L \cap S$, which is the first goal of pruned enumeration. Otherwise, it will return nothing or the zero vector, but we can simply repeat the process with many different P's until we find a non-trivial point, provided that there are many choices for P. In fact, by repeating sufficiently many times this process, one might even be able to recover all of $L \cap S$.

In order to analyze the algorithm, we need to predict when $L \cap S \cap P \subseteq \{0\}$: this is especially tricky, since for all choices of P considered in the past, the enumeration of $L \cap S \cap P$ was completely deterministic, which makes the probability space unclear. Gama et al.[11] provided the first sound analysis of enumeration with pruning, by viewing the pruning set P as a random variable: in practice, it depends on the choice of basis B. Then we define the success probability of pruned enumeration as:

$$\Pr_{\mathrm{succ}} = \Pr_P(L \cap S \cap P \nsubseteq \{0\}),$$

that is, the probability that it outputs a non-trivial point in $L \cap S$.

In general, this probability is very hard to compute. To estimate this probability, in the spirit of [11], we make the following heuristic assumption inspired by the Gaussian heuristic applied to the lattice L and the set $S \cap P$:

Heuristic 2. *For reasonable pruning sets P and lattices L, the success probability $\Pr(L \cap S \cap P \not\subseteq \{0\})$ is close to $\min(1, \mathrm{vol}(S \cap P)/\mathrm{covol}(L))$, and when this is close to 1, the cardinal of $L \cap S \cap P$ is close to $\mathrm{vol}(S \cap P)/\mathrm{covol}(L)$.*

We stress that this well-defined heuristic is only a heuristic: it is easy to select pruning sets P for which the heuristic cannot hold. But the experiments of [11] show that the heuristic typically holds for the pruning sets considered by [11], and our experiments show that it also holds for typical pruning sets corresponding to discrete pruning.

If the heuristic holds, then it suffices to be able to compute $\mathrm{vol}(S \cap P)$ to estimate the success probability of pruned enumeration. To estimate the running time of the full algorithm, we need more information:

– An estimate of the cost of enumerating $L \cap S \cap P$.
– An estimate of the cost of computing the (random) reduced basis B.

Gama *et al.* [11] introduced extreme pruning in which $\mathrm{vol}(S \cap P)/\mathrm{covol}(L)$ converges to zero, yet the global running time to find a non-zero short vector is much faster than enumeration, namely exponentially faster asymptotically for the choice of [11].

3.3 Continuous Pruning

Until now, the most general form of pruning set P that has been used is the following generalization [11] of pruned enumeration of [31,32], which was also concurrently used in [35]. There, P is defined by a function $f : \{1, \dots, n\} \to [0, 1]$ and a lattice basis $B = (\mathbf{b}_1, \dots, \mathbf{b}_n)$ as follows:

$$P = \{\mathbf{x} \in \mathbb{R}^n \text{ s.t. } \|\pi_{n+1-i}(\mathbf{x})\| \le f(i)R \text{ for all } 1 \le i \le n\},$$

where the π_i's are the Gram-Schmidt projections defined by the basis B. We call *continuous pruning* this form of pruned enumeration, by opposition with *discrete pruning*, which is the topic of this paper.

By definition, $P \subseteq S$ so $\mathrm{vol}(S \cap P) = \mathrm{vol}(P)$. By suitable rotation, isometry and scaling, $\mathrm{vol}(P)$ can be derived from R, n and the volume of the cylinder intersection defined by f:

$$C_f = \left\{ (x_1, \dots, x_n) \in \mathbb{R}^n \text{ s.t. } \sum_{j=1}^{i} x_j^2 \le f(i)^2 \text{ for all } 1 \le i \le n \right\}.$$

Gama *et al.* [11] showed how to efficiently compute tight lower and upper bounds for $\mathrm{vol}(C_f)$, thanks to the Dirichlet distribution and special integrals. Using Heuristic 2, this allows to reasonably estimate the probability of success.

Using the shape of P, [11] also estimated of the cost of enumerating $L \cap S \cap P$, by using the Gaussian heuristic on projected lattices $\pi_i(L)$, as suggested in [12]: these estimates are usually accurate in practice.

To optimize the whole selection of parameters, one finally needs to take into account the cost of computing the (random) reduced basis of B. For instance, this is done in [2,5], which illustrates the power of continuous pruning in the context of lattice reduction.

Though continuous pruning has proved very successful, it is unknown if it is the most efficient form of pruned enumeration: there might be better choices of pruning sets P, and this is the starting point of discrete pruning, which we introduce next.

4   Enumeration with Discrete Pruning

We now introduce enumeration with discrete pruning, which generalizes Schnorr's random sampling [30] and its variants [4,8], and a provides a novel geometric description based on partitions, which is crucial for our analysis.

4.1   Lattice Partitions

Discrete pruning is based on what we call a *lattice partition*:

Definition 1. *Let L be a full-rank lattice in $\mathbb{Q}^n$. An L-partition is a partition $\mathcal{C}$ of $\mathbb{R}^n$ such that:*

- *The partition is countable: $\mathbb{R}^n = \cup_{t \in T} \mathcal{C}(t)$ where T is a countable set, and $\mathcal{C}(t) \cap \mathcal{C}(t') = \varnothing$ whenever $t \neq t'$.*
- *Each cell $\mathcal{C}(t)$ contains a single lattice point, which can be found efficiently: given any $t \in T$, one can compute in polynomial time the single point of $\mathcal{C}(t) \cap L$. We call this process the cell enumeration.*

We call $t \in T$ the tag of the cell $\mathcal{C}(t)$. Since $\mathbb{R}^n = \cup_{t \in T} \mathcal{C}(t)$, this means that any point in $\mathbb{R}^n$ also has a tag, because it belongs to a unique cell $\mathcal{C}(t)$. Any lattice partition induces a (bijective) encoding of lattice points onto T: any lattice point belongs to $\mathbb{R}^n$, and therefore has a tag; reciprocally, given a tag $t \in T$, one can compute the unique lattice point in $\mathcal{C}(t)$ by the cell enumeration.

The simplest examples of lattice partitions come from fundamental domains of lattices. In particular, one can easily check that any basis $B = (\mathbf{b}_1, \ldots, \mathbf{b}_n)$ of L gives rise to two trivial lattice partitions with $T = \mathbb{Z}^n$ as follows:

- $\mathcal{C}(\mathbf{t}) = \mathbf{t}B + \mathcal{D}$ where $\mathcal{D} = \{\sum_{i=1}^n x_i \mathbf{b}_i \text{ s.t. } -1/2 \leq x_i < 1/2\}$ is a parallelepiped.
- $\mathcal{C}(\mathbf{t}) = \mathbf{t}B + \mathcal{D}$ where $\mathcal{D} = \{\sum_{i=1}^n x_i \mathbf{b}_i^\star \text{ s.t. } -1/2 \leq x_i < 1/2\}$ is a box, *i.e.* a parallelepiped whose axes are pairwise orthogonal.

However, these lattice partitions are not very useful, because $\mathcal{C}(\mathbf{t}) \cap L = \{\mathbf{t}B\}$, which means that we know the lattice point directly from its tag: the cell enumeration is just a matrix/vector product.

The first non-trivial example of lattice partition is the following:

Lemma 2. *Let B be a basis of a full-rank lattice L in $\mathbb{Z}^n$. Let $T = \mathbb{Z}^n$ and for any $\mathbf{t} \in T$, $\mathcal{C}_{\mathbb{Z}}(\mathbf{t}) = \mathbf{t}B^\star + \mathcal{D}$ where $\mathcal{D} = \{\sum_{i=1}^n x_i \mathbf{b}_i^\star \ s.t. \ -1/2 \le x_i < 1/2\}$. Then $(\mathcal{C}_{\mathbb{Z}}(), T)$ with Algorithm 1 is an L-partition, which we call Babai's partition.*

Proof. We already know that $(\mathcal{C}_{\mathbb{Z}}(), T)$ is a $L(B^\star)$-partition because $B^\star$ is a basis of $L(B^\star)$. To show that it is also a L-partition, it suffices to show that $\mathcal{C}_{\mathbb{Z}}(t) \cap L$ is always a singleton, which can be found in polynomial time. To see this, note that Babai's nearest plane algorithm [3] implies that for any $\mathbf{t} \in T$, there is a unique $\mathbf{v} \in L$ such that $\mathbf{v} - \mathbf{t}B^\star \in \mathcal{D}$, and that $\mathbf{v}$ can be found in polynomial time. It follows that $\mathcal{C}_{\mathbb{Z}}(t) \cap L = \{\mathbf{v}\}$. □

The encoding of lattice points induced by Babai's partition is exactly the encoding onto $\mathbb{Z}^n$ introduced in [6]. The paper [8] defines a different encoding of lattice points onto $\mathbb{N}^n$, which implicitly uses a different lattice partition based on $T = \mathbb{N}^n$ rather than $\mathbb{Z}^n$, which we now define:

Algorithm 1. Cell enumeration for Babai's partition from Babai's Nearest Plane algorithm [3]

Input: A tag $\mathbf{t} \in \mathbb{Z}^n$ and a basis $B = (\mathbf{b}_1, \dots, \mathbf{b}_n) \in \mathbb{Q}^n$ of a lattice L, with Gram-Schmidt orthogonalization $B^\star$.
Output: $\mathbf{v} \in L$ such that $\{\mathbf{v}\} = L \cap \mathcal{C}_{\mathbb{Z}}(\mathbf{t})$
1: $\mathbf{v} \leftarrow \mathbf{0}$ and $\mathbf{u} \leftarrow \mathbf{t}B^\star$
2: **for** $i := n$ downto 1 **do**
3:     Compute the integer c closest to $\langle \mathbf{b}_i^\star, \mathbf{u} \rangle / \langle \mathbf{b}_i^\star, \mathbf{b}_i^\star \rangle$
4:     $\mathbf{u} \leftarrow \mathbf{u} - c\mathbf{b}_i$ and $\mathbf{v} \leftarrow \mathbf{v} + c\mathbf{b_i}$
5: **end for**
6: Return $\mathbf{v}$

Algorithm 2. Tagging for the natural partition

Input: A vector $\mathbf{x} = \sum_{i=1}^n x_i \mathbf{b}_i^\star \in \mathbb{R}^n$ where the $\mathbf{b}_i^\star$'s are the Gram-Schmidt vectors of a basis $B = (\mathbf{b}_1, \dots, \mathbf{b}_n)$ of a full-rank lattice L in $\mathbb{R}^n$.
Output: A tag $\mathbf{t} \in \mathbb{N}^n$ such that $\mathbf{x} \in \mathcal{C}_{\mathbb{N}}(\mathbf{t})$.
1: **for** $i := n$ downto 1 **do**
2:     **if** $x_i > 0$ **then**
3:         $t_i \leftarrow \lceil 2x_i \rceil - 1$
4:     **else**
5:         $t_i \leftarrow -\lfloor 2x_i \rfloor$
6:     **end if**
7: **end for**
8: Return $(t_1, \dots, t_n)$

Algorithm 3. Cell enumeration for the natural partition

Input: A tag $\mathbf{t} \in \mathbb{N}^n$ and a basis $B = (\mathbf{b}_1, \ldots, \mathbf{b}_n) \in \mathbb{Q}^n$ of a lattice L, with Gram-Schmidt orthogonalization matrix μ.
Output: $\mathbf{v} \in L$ such that $\{\mathbf{v}\} = L \cap \mathcal{C}_{\mathbb{N}}(\mathbf{t})$
 1: **for** $i := n$ downto 1 **do**
 2:    $y \leftarrow -\sum_{j=i+1}^{n} u_j \mu_{j,i}$
 3:    $u_i \leftarrow \lfloor y + 0.5 \rfloor$
 4:    **if** $u_i < y$ **then**
 5:        $u_i \leftarrow u_i - (-1)^{t_i} \lceil t_i/2 \rceil$
 6:    **else**
 7:        $u_i \leftarrow u_i + (-1)^{t_i} \lceil t_i/2 \rceil$
 8:    **end if**
 9: **end for**
10: Return $\sum_{i=1}^{n} u_i \mathbf{b}_i$

Lemma 3. *Let B be a basis of a full-rank lattice L in $\mathbb{Z}^n$. Let $T = \mathbb{N}^n$ and for any $\mathbf{t} = (t_1, \ldots, t_n) \in T$, $\mathcal{C}_{\mathbb{N}}(\mathbf{t}) = \{\sum_{i=1}^{n} x_i \mathbf{b}_i^{\star}$ s.t. $-(t_i + 1)/2 < x_i \leq -t_i/2$ or $t_i/2 < x_i \leq (t_i + 1)/2\}$. Then $(\mathcal{C}_{\mathbb{N}}(), T)$ with Algorithm 3 is an L-partition, which we call the natural partition.*

Proof. The fact that the cells $\mathcal{C}_{\mathbb{N}}(\mathbf{t})$ form a partition of $\mathbb{R}^n$ is obvious: the cells are clearly disjoint and any point of $\mathbb{R}^n$ belongs to a cell (see Algorithm 2). The only difficulty is to show that any cell contains one and only one lattice point, and that it can be found efficiently: this is achieved by Algorithm 3, which is a variant of Babai's nearest plane algorithm. $\square$

Figure 1 displays Babai's partition and the natural partition (with tags) in dimension two. The encoding of lattice points derived from the natural partition is exactly the encoding introduced by Fukase and Kashiwabara in [8]. What is remarkable is that every cell of the natural partition is not connected, except the zero cell: each cell $\mathcal{C}_{\mathbb{N}}(\mathbf{t})$ is the union of 2^k boxes, where k is the number of non-zero coefficients in $\mathbf{t}$, as illustrated by Fig. 1.

To compare the Babai partition and the natural partition, we study the moments of their cells, which follow from Corollary 1:

Fig. 1. Babai's partition and the natural partition in dimension two: different cells are coloured differently. (Color figure online)

Corollary 2 (Moments of Babai's partition). *Let B be a basis of a full-rank lattice L in $\mathbb{R}^n$. Let $\mathbf{t} = (t_1, \ldots, t_n) \in \mathbb{Z}^n$ and the cell $\mathcal{C}_\mathbb{Z}(\mathbf{t}) = \mathbf{t}B^\star + \mathcal{D}$ where $\mathcal{D} = \{\sum_{i=1}^n x_i \mathbf{b}_i^\star \text{ s.t. } -1/2 \leq x_i < 1/2\}$. Then:*

$$\mathbb{E}\{\mathcal{C}_\mathbb{Z}(\mathbf{t})\} = \sum_{i=1}^n \left(t_i^2 + \frac{1}{12}\right) \|\mathbf{b}_i^\star\|^2 \quad and \quad \mathbb{V}\{\mathcal{C}_\mathbb{Z}(\mathbf{t})\} = \sum_{i=1}^n \left(\frac{t_i^2}{3} + \frac{1}{180}\right) \|\mathbf{b}_i^\star\|^4$$

Corollary 3 (Moments of the natural partition). *Let B be a basis of a full-rank lattice L in $\mathbb{R}^n$. Let $\mathbf{t} = (t_1, \ldots, t_n) \in \mathbb{N}^n$ and the cell $\mathcal{C}_\mathrm{N}(\mathbf{t}) = \{\sum_{i=1}^n x_i \mathbf{b}_i^\star \text{ s.t. } -(t_i+1)/2 < x_i \leq -t_i/2 \text{ or } t_i/2 < x_i \leq (t_i+1)/2\}$. Then:*

$$\mathbb{E}\{\mathcal{C}_\mathrm{N}(\mathbf{t})\} = \sum_{i=1}^n \left(\frac{t_i^2}{4} + \frac{t_i}{4} + \frac{1}{12}\right) \|\mathbf{b}_i^\star\|^2 \quad and \quad \mathbb{V}\{\mathcal{C}_\mathrm{N}(\mathbf{t})\} = \sum_{i=1}^n \left(\frac{t_i^2}{48} + \frac{t_i}{48} + \frac{1}{180}\right) \|\mathbf{b}_i^\star\|^4$$

This suggests that the natural partition is better than Babai's partition: we will return to this topic in Sect. 6.

4.2 Discrete Pruning from Lattice Partitions

Any L-partition $(\mathcal{C}, T)$ defines a partition $\mathbb{R}^n = \cup_{t \in T} \mathcal{C}(t)$. Discrete pruning is simply obtained by choosing a finite number of cells $\mathcal{C}(t)$ to enumerate, as done by Algorithm 4: discrete pruning is parametrized by a finite set $U \subseteq T$, which specifies which cells to enumerate. Discrete pruning is therefore a pruned enumeration with pruning set:

$$P = \cup_{\mathbf{t} \in U} \mathcal{C}(\mathbf{t})$$

Algorithm 4. Discrete Pruning from Lattice Partitions

Input: A lattice partition $(\mathcal{C}(), T)$, a finite subset $U \subseteq T$ and a radius R.
Output: $L \cap S \cap P$ where $S = \mathrm{Ball}_n(R)$ and $P = \cup_{t \in U} \mathcal{C}(t)$.
1: $\mathcal{R} = \varnothing$
2: **for** $t \in U$ **do**
3:    Enumerate $L \cap \mathcal{C}(t)$: if the output vector has norm $\leq R$, add the vector to the set $\mathcal{R}$.
4: **end for**

The algorithm performs exactly k partition-enumerations, where $k = \#U$ is the number of cells of discrete pruning, and each partition-enumeration runs in polynomial time by definition of the lattice partition. So the running time is $\#U$ polynomial-time operations: one can decide how much time should be spent.

Since the running time is easy to evaluate, the only difficulty is to estimate the probability of success. Based on Heuristic 2, the probability can be derived from:

$$\mathrm{vol}(S \cap P) = \sum_{t \in U} \mathrm{vol}(S \cap \mathcal{C}(\mathbf{t})). \tag{1}$$

Thus, the problem is reduced to computing the volume $\mathrm{vol}(S \cap \mathcal{C}(\mathbf{t}))$ of the intersection of a ball with a cell. If we want to maximize the probability of success for a given effort (*i.e.* for a fixed number k of cells), it suffices to select the k cells which maximize $\mathrm{vol}(S \cap \mathcal{C}(\mathbf{t}))$ among all the cells: we will study this topic in Sect. 6 for the natural partition.

The hardness of computing $\mathrm{vol}(S \cap \mathcal{C}(\mathbf{t}))$ depends on the lattice partition, but for Babai's partition and the natural partition, this can be reduced to computing the volume $\mathrm{vol}(S \cap H)$ where S is a ball and H is a box, which is exactly the topic of Sect. 5:

– In the case of Babai's partition, each cell $\mathcal{C}_\mathbb{Z}(\mathbf{t})$ is already a box.
– In the case of the natural partition, each cell $\mathcal{C}_\mathbb{Z}(\mathbf{t})$ is the union of 2^j symmetric (non-overlapping) boxes, where j is the number of non-zero coefficients of $\mathbf{t}$. It follows that $\mathrm{vol}(\mathcal{C}_\mathbb{Z}(\mathbf{t}) \cap S) = 2^j \mathrm{vol}(H \cap S)$, where H is any of these 2^j boxes.

Interestingly, in Sect. 6.3, we show how to approximate well the sum of (1) without computing all the terms, using only a constant number of terms.

4.3   Revisiting Schnorr's Random Sampling and the Fukase-Kashiwabara Variant

Here, we show that Schnorr's random sampling and its variants, including the Fukase-Kashiwabara variant, can all be viewed as special cases of discrete pruning.

Schnorr's Random Sampling. Let $B = (\mathbf{b}_1, \ldots, \mathbf{b}_n)$ be a basis of a full-rank lattice L in $\mathbb{Z}^n$: denote by $B^\star = (\mathbf{b}_1^\star, \ldots, \mathbf{b}_n^\star)$ its Gram-Schmidt orthogonalization. Let $1 \le u \le n-1$ be an integer parameter. Schnorr's random sampling [30] outputs all points $\mathbf{v} \in L$ of the form $\sum_{i=1}^n \mu_i \mathbf{b}_i^\star$ such that $\mu_n = 1$ and the remaining $\mu_i \in \begin{cases} [-1/2, 1/2(& \text{if } i \le n - (u+1) \\ [-1, 1(& \text{if } n - u \le i \le n - 1 \end{cases}$

This is equivalent to pruned enumeration with a pruning set defined as the following (non-centered) box of parameter u:

$$P_u = \left\{ \sum_{i=1}^n x_i \mathbf{b}_i^\star \text{ s.t.} \begin{array}{l} -1/2 \le x_i < 1/2, \text{if } i \le n - (u+1) \\ -1 \le x_i < 1, \text{if } n - u \le i \le n - 1 \\ 1/2 \le x_i < 3/2, \text{if } i = n \end{array} \right\} \tag{2}$$

Clearly $\mathrm{vol}(P_u) = 2^u \mathrm{covol}(L)$. Curiously, the box P_u has slightly bigger moments than $\cup_{\mathbf{t} \in U_u} \mathcal{C}_\mathbb{N}(\mathbf{t})$ where $U_u = \{(0, \ldots, 0, t_{n-u}, \ldots, t_{n-1}, 1) \in \{0,1\}^n\}$ is a finite set defining discrete pruning with the natural partition: the corresponding pruning set also has volume $2^u \mathrm{covol}(L)$. Schnorr's box actually corresponds to discrete pruning with the same finite set U_u of tags, but using a different hybrid lattice partition, which matches the natural partition for the first $n-1$ coordinates, and Babai's partition for the n-th coordinate: namely, $T = \mathbb{N}^{n-1} \times \mathbb{Z}$ and for any $\mathbf{t} \in T$

$$\mathcal{C}_{\text{Schnorr}}(\mathbf{t}) = \left\{ \sum_{i=1}^{n} x_i \mathbf{b}_i^\star \text{ s.t. } \begin{matrix} -(t_i+1)/2 < x_i \leq -t_i/2 \text{ or } t_i/2 < x_i \leq (t_i+1)/2 \text{ if } i \leq n-1 \\ -1/2 \leq x_n - t_n < 1/2 \end{matrix} \right\}$$

Based on Heuristic 2, the success probability of random sampling can be deduced from $\text{vol}(P_u \cap S)$, where P_u is a non-centered box and S is a ball: the computation of such volumes is exactly the topic of Sect. 5. There, the following computations will be useful:

Lemma 4 (Moments of Schnorr's box). *Let $B = (\mathbf{b}_1, \ldots, \mathbf{b}_n)$ be a basis of a full-rank lattice L in $\mathbb{R}^n$, and P_u be Schnorr's box of parameter u defined by (2). Then:*

$$\mathbb{E}\{P_u\} = \frac{\sum_{i=1}^{n-(u+1)} \|\mathbf{b}_i^\star\|^2}{12} + \frac{\sum_{i=n-u}^{n-1} \|\mathbf{b}_i^\star\|^2}{3} + \frac{13\|\mathbf{b}_n^\star\|^2}{12}$$

$$\mathbb{V}\{P_u\} = \frac{\sum_{i=1}^{n-(u+1)} \|\mathbf{b}_i^\star\|^4}{180} + \frac{\sum_{i=n-u}^{n-1} \|\mathbf{b}_i^\star\|^4}{45} + \frac{61\|\mathbf{b}_n^\star\|^4}{180}$$

In some variants of random sampling (see [4, 30]), one actually considers a random subset of U_u of size k for some $k \ll 2^u$: based on Heuristic 2, the success probability of random sampling can be deduced from $\text{vol}(\mathcal{C}(\mathbf{t}) \cap S)$, where $\mathcal{C}(\mathbf{t})$ is the union of symmetric non-overlapping boxes. Again, the problem can be reduced to the volume computation of Sect. 5.

It can easily be checked that the other variants by Buchmann-Ludwig [4] can also be viewed as discrete pruning.

The Fukase-Kashiwabara Variant. As mentioned previously, Fukase and Kashiwabara [8] recently introduced an encoding of lattice points onto $\mathbb{N}^n$, which turns out to be the encoding derived from the natural partition. Their variant proceeds by enumerating all lattice points having certain encodings. In our terminology, this can immediately be rewritten as discrete pruning with the natural partition, where the finite set of tags has size approximately 5×10^7.

They do not provide a general algorithm to select tags: however, they explain which tags they selected to solve the SVP challenges, so this only applies to certain settings and fixed dimensions. Here is an example in dimension n in the range $120 - 140$. The selection proceeds in two stages:

- First, they select a large set of candidates $V \subseteq \mathbb{N}^n$, formed by all tags $(t_1, \ldots, t_n) \in \mathbb{N}^n$ such that:
  - all $t_i \in \{0, 1, 2\}$
  - the total number of indexes i such that $t_i = 1$ is ≤ 13, and the indexes i such that $t_i = 1$ all belong to $\{n - 55 + 1, \ldots, n\}$.
  - the total number of indexes i such that $t_i = 2$ is ≤ 1, and the indexes i such that $t_i = 2$ all belong to $\{n - 15 + 1, \ldots, n\}$.
- For each $\mathbf{t} \in V$, they compute $\mathbb{E}\{\mathcal{C}_{\mathbb{N}}(\mathbf{t})\}$ using the formula of Corollary 3.
- The final set U is formed by the 5×10^7 tags $\mathbf{t} \in V$ which have the smallest $\mathbb{E}\{\mathcal{C}_{\mathbb{N}}(\mathbf{t})\}$.

4.4 Optimizations

If the discrete pruning set U has exactly k elements, then the running time is k polynomial-time operations. However, from a practical point of view, it is important to decrease as much as possible the cost of the polynomial-time operation.

First, one can abort the enumeration of a cell if we realize that the lattice vector inside the cell will be outside the ball S: this is similar to what is done during enumeration.

Second, we can speed up the computation by regrouping cells. A good example is Schnorr's random sampling. We can view Schnorr's pruning set P_u as the union of 2^u cells, but when we want to enumerate all lattice points inside $S \cap P_u$, it is better to view it as a single box: discrete pruning can be rewritten here as some variant of enumeration. More generally, depending on the set U of tags, we can recycle some computations: similar tricks were used for pruned enumeration [11].

Third, we note that all the cell enumerations can be performed in parallel: discrete pruning is easier to parallelize than continuous pruning. It seems that discrete pruning should be better suited to special hardware than continuous pruning.

5 Ball-Box Intersections

In this section, we are interested in computing the volume of the intersection between a ball and a box, either exactly by a formula or approximately by an algorithm. More precisely, we are interested in $\mathrm{vol}(B \cap H)$, where B is the ball of center $\mathbf{c} \in \mathbb{R}^n$ and radius R, and H is the following box:

$$H = \{(x_1, \ldots, x_n) \in \mathbb{R}^n \text{ s.t. } \alpha_i \leq x_i \leq \beta_i\},$$

where the α_i's and β_i's are given. Without loss of generality, we may assume that $\mathbf{c} = 0$ after suitable translation, and $R = 1$ after suitable scaling. Hence, we are interested in:

$$\mathrm{Vol}\left(\mathrm{Ball}_n(1) \cap H\right) = \mathrm{Vol}\left(\mathrm{Ball}_n(1) \cap \prod_{i=1}^{n}[\alpha_i, \beta_i]\right)$$

$$= \prod_{i=1}^{n}(\beta_i - \alpha_i) \Pr_{(x_1,\ldots,x_n) \leftarrow \prod_{i=1}^{n}[\alpha_i,\beta_i]}\left[\sum_{i=1}^{n} x_i^2 \leq 1\right] \quad (3)$$

The section is organized as follows. In Sect. 5.1, we give a rigorous asymptotical estimate of (3) when the box is sufficiently balanced, but we show that it is ill-suited to the case of discrete pruning. In Sect. 5.2, we provide two exact formulas for (3) as infinite series, due to respectively Constales and Tibken [27]: these infinite series give rise to approximation algorithms by truncating the series. In Sect. 5.3, we give a heuristic method to approximate (3), based on fast inverse

Laplace transforms: this method is referred as FILT in the remaining of the paper. Section 5.4 provides an experimental comparison of the running time of the three methods of Sects. 5.2 and 5.3, in the context of discrete pruning: it turns out that in high dimension, the FILT method outperforms the other two.

We note that Buchmann and Ludwig [4, Theorem 1] (more details in [20, Theorem 18]) implicitly adressed the computation of (3): the main part of [4, Algorithm 3] can be viewed as a heuristic approximation algorithm based on the Discrete Fourier transform, but no experimental result seems to be reported in [4, 20].

5.1 Asymptotical Analysis

The following result shows that the volume of the intersection of a ball with a "balanced" box can be asymptotically computed, because the right-hand probability of (3) can be derived from the central limit theorem:

Theorem 3. *Let $C_1, C_2 > 0$ be constants. Let $H = \{\mathbf{x} = (x_1, \ldots, x_n) \in \mathbb{R}^n$ s.t. $\alpha_i \leq x_i \leq \beta_i\}$, where $C_1 \leq \beta_i - \alpha_i$ and $\max(|\alpha_i|, |\beta_i|) \leq C_2$. Then $Y = (\|\mathbf{x}\|^2 - \mathbb{E}\{H\})/\sqrt{\mathbb{V}\{H\}}$ has zero mean and variance one, and converges in distribution to the normal distribution, i.e. for all $y > 0$:*

$$\lim_{n \to \infty} \Pr_{\mathbf{x} \in H} \left(\|\mathbf{x}\|^2 \leq \mathbb{E}\{H\} + y\sqrt{\mathbb{V}\{H\}} \right) = \frac{1}{2} \left(1 + \operatorname{erf}(y/\sqrt{2}) \right),$$

where

$$\mathbb{E}\{H\} = \sum_{i=1}^{n} \left(\frac{\alpha_i^2 + \beta_i^2 + \alpha_i \beta_i}{3} \right) \quad and$$

$$\mathbb{V}\{H\} = \sum_{i=1}^{n} \left(\frac{4}{45}\alpha_i^4 - \frac{1}{45}\alpha_i^3 \beta_i - \frac{2}{15}\alpha_i^2 \beta_i^2 - \frac{1}{45}\alpha_i \beta_i^3 + \frac{4}{45}\beta_i^4 \right).$$

Unfortunately, we cannot apply Theorem 3 when the box H is Schnorr's box which, after suitable rotation, corresponds to:

- If $i \leq n - (u+1)$, then $\beta_i = \|\mathbf{b}_i^\star\|/2$ and $\alpha_i = -\beta_i$;
- If $n - u \leq i \leq n - 1$, then $\beta_i = \|\mathbf{b}_i^\star\|$ and $\alpha_i = -\beta_i$;
- $\alpha_n = \|\mathbf{b}_n^\star\|/2$ and $\beta_n = 3/2\|\mathbf{b}_n^\star\|$

If the basis B is LLL-reduced, then the $\|\mathbf{b}_i^\star\|$'s typically decrease geometrically, which means that the (α_i, β_i)'s do not satisfy the assumptions of Theorem 3. Furthermore, experiments show that in practice, the distribution of the Y defined in Theorem 3 is not normal (see Fig. 2) as its left-tail is below that of the normal distribution and its right-tail is over. Worse, the distance with the normal distribution actually increases with the dimension: see Fig. 3. However, if we fix the dimension and apply stronger and stronger lattice reduction, we expect the box to become more and more "balanced": this is confirmed by Fig. 4, which shows that the more reduced the basis is, the closer Y is to the normal distribution.

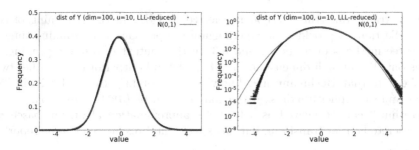

Fig. 2. Comparison of the normal distribution with the experimental distribution of Y defined in Theorem 3 with Schnorr's box (with $u = 10$) over an LLL-reduced basis in dimension 100. Both tails significantly deviate from normal tails. On the right, there is a log-scale.

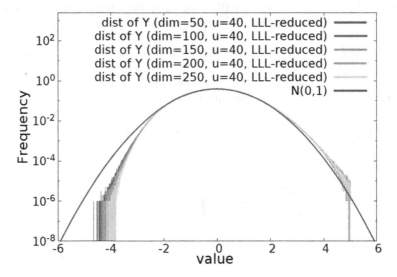

Fig. 3. Evolution of the distribution of Y defined in Theorem 3 with Schnorr's box (with $u = 10$) over an LLL-reduced basis, as the dimension increases: it becomes less and less normal.

If the box is "unbalanced", it is always possible to upper bound and lower bound the right-hand probability of (3). For instance, an upper bound follows from Hoeffding's bound:

Lemma 5. *Let* $H = \{\mathbf{x} = (x_1, \ldots, x_n) \in \mathbb{R}^n \text{ s.t. } \alpha_i \leq x_i \leq \beta_i\}$. *Then for any* $y > 0$:

$$\Pr_{\mathbf{x} \in H} \left(\|\mathbf{x}\|^2 \leq \mathbb{E}\{H\} - y \right) \leq e^{-2y^2 / \sum_{i=1}^n (\beta_i - \alpha_i)^2}.$$

Proof. It suffices to apply Hoeffding's bound with $y > 0$ and the n independent variables $-X_i = -x_i^2$. $\qquad\square$

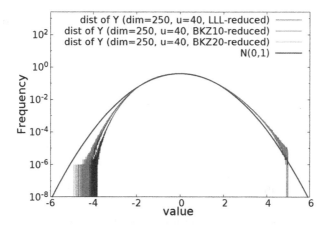

Fig. 4. Evolution of the distribution of Y defined in Theorem 3 with Schnorr's box (with $u = 40$) in dimension 250, with varying reductions: it becomes closer to normal as the basis becomes more reduced

However, these upper bounds are typically very pessimistic for the type of boxes we are interested in. Reciprocally, in the spirit of Schnorr [30] (see also [4,9]), it is also possible to give a lower bound on the right-hand probability of (3), by considering the largest sub-box J of H which is contained in $\mathrm{Ball}_n(1)$, in which case $\mathrm{vol}(\mathrm{Ball}_n(1) \cap H) \geq \mathrm{vol}(J)$. This can lead to an asymptotic lower bound if we are able to conveniently bound the side-lengths of H, *i.e.* the $\|\mathbf{b}_i^\star\|$'s. This is why Schnorr [30] introduced the Geometric Series Assumption (GSA), but this only holds in an approximate sense, which creates some problems (see [4]): for instance, the GSA implies that all $\|\mathbf{b}_i^\star\|$ are always $\leq \|\mathbf{b}_1\|$ for $i \geq 2$, but in practice, it can happen that some $\mathbf{b}_i^\star$ is larger than $\mathbf{b}_1$. To prevent this problem, one can use instead absolute bounds: for instance, Fig. 5 shows that in practice, for a random LLL-reduced basis, $\max_B \|\mathbf{b}_i^\star\|/\|\mathbf{b}_1\|$ can be upper bounded by a geometric sequence indexed by i, with parameters independent of n. However, the lower bound obtained is again typically very pessimistic for the type of boxes we are interested in.

5.2   Exact Formulas

Here, we provide two exact formulas for the intersection volume as infinite series, by sligthly generalizing the works of respectively Constales and Tibken [27], who studied the special case of a zero-centered cube.

Fourier Series. We first generalize Constales' method based on Fourier series. Let $S(x) = \int_0^x \sin(t^2)dt$ and $C(x) = \int_0^x \cos(t^2)dt$ be the Fresnel integrals.

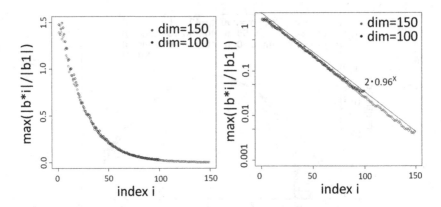

Fig. 5. Evolution of $\max_B \|\mathbf{b}_i^\star\|/\|\mathbf{b}_1\|$ over hundreds thousands LLL-reduced bases, as $i \geq 2$ increases, in dimensions 100 (blue) and 150 (red): the right-hand graph is in log-scale. (Color figure online)

Theorem 4. *Let* $\alpha_j \leq \beta_j$ *for* $1 \leq j \leq n$*. Let* $\ell = \sum_{j=1}^n \max(\alpha_j^2, \beta_j^2)$*. Then:*

$$\text{vol}(\text{Ball}_n(1) \cap \prod_{j=1}^n [\alpha_j, \beta_j])$$

$$= \begin{cases} \prod_{j=1}^n |\beta_j - \alpha_j| & \text{if } \ell \leq 1 \\ \left(\dfrac{1}{2} - \dfrac{\sum_{j=1}^n \alpha_j^2 + \beta_j^2 + \alpha_j \beta_j}{3\ell} + \dfrac{1}{\ell} \right. \\ \quad \left. + \dfrac{1}{\pi} \text{Im} \sum_{k=1}^{\infty} \dfrac{\Phi(-2\pi k/\ell)}{k} e^{2i\pi k/\ell} \right) \prod_{j=1}^n (\beta_j - \alpha_j) \\ \qquad \text{if } \ell > 1 \end{cases} \quad (4)$$

where Φ *is defined as*

$$\Phi(\omega) = \prod_{j=1}^n \frac{(C(\beta_j \sqrt{|\omega|}) - C(\alpha_j \sqrt{|\omega|})) + i\text{sgn}(\omega)(S(\beta_j \sqrt{|\omega|}) - S(\alpha_j \sqrt{|\omega|}))}{(\beta_j - \alpha_j) \sqrt{|\omega|}}.$$

A proof can be found in the full version: it is just a slight generalization of the proof of Constales [27]. An approximation algorithm can be derived by truncating the infinite series of (4) to N terms: computing each term costs approximately n operations, where an operation is dominated by the computation of a constant number of Fresnel integrals. The computation of Fresnel integrals is classical: for instance, it can be done by reduction to erf computations.

Multidimensional Fourier Transforms. We now generalize Tibken's formula based on the n-dimensional Fourier transform.

Theorem 5. *Let $\alpha_j \leq \beta_j$ for $1 \leq j \leq n$. Then:*

$$
\text{vol}(\text{Ball}_n(1) \cap \prod_{j=1}^{n}[\alpha_j, \beta_j]) = \frac{1}{(4\pi)^{n/2}} \left(\frac{I(\frac{1}{2n+4}, 0)}{\Gamma(n/2+1)} + \sum_{k=2}^{\infty} \frac{L_k^{n/2}(n/2+1)I(\frac{1}{2n+4}, k)}{\Gamma(k+n/2+1)(2n+4)^k} \right)
$$
(5)

where $L_k^{\alpha}(x)$ denotes the generalized Laguerre polynomial, Γ is the classical Gamma function and:

$$
I(\lambda, 0) = \pi^n \prod_{j=1}^{n} \left(\text{erf}\left(\frac{\beta_j}{2\sqrt{\lambda}}\right) - \text{erf}\left(\frac{\alpha_j}{2\sqrt{\lambda}}\right) \right) \quad \text{and} \quad I(\lambda, k) = (-1)^k \left(\frac{\partial}{\partial\lambda}\right)^k I(\lambda, 0)
$$

Again, an approximation algorithm can be derived by truncating the infinite series. The first term of (5) is easy to compute from the erf, and turns out to give a much better approximation than the central limit theorem for Schnorr's box. But the right-hand sum terms are trickier to compute (see the full version for details), due to the derivative in the definition of $I(\lambda, k)$ and the presence of Laguerre polynomials.

5.3 Numerical Approximation from Fast Inverse Laplace Transforms (FILT)

We now present the method we used in practice to approximate the volume, which is based on the Laplace transform. The starting point is similar to the Fourier series approach, but it deviates afterwards. For a function $f(x)$ defined over $x \geq 0$, its *Laplace transform* $F = \mathcal{L}\{f\}$ is defined over $\mathbb{C}$ as $F(s) = \int_0^\infty f(t)e^{-st}dt$. The inverse transform is given by the Bromwich integral

$$
f(t) = \frac{1}{2\pi i} \int_{c-\infty i}^{c+\infty i} F(s)e^{st}ds,
$$
(6)

where the integration is done along the vertical line $\text{Re}(s) = c$ in the complex plane such that c is greater than the real part of all singularities of $F(s)$. If $g(t) = \int_0^t f(\tau)d\tau$, then $\mathcal{L}\{g\}(s) = \frac{1}{s}\{f\}$. Thus, if X is a non-negative random variable with probability density function $f(x)$, then its cumulative distribution function $F_X(x)$ satisfies: $F_X(x) = \mathcal{L}^{-1}\{\frac{1}{s}\mathcal{L}\{f\}(s)\}(x)$. Thus, if we denote by $\rho_{\sum_{j=1}^n x_j^2}$ the probability density function of $\sum_{i=1}^n x_i^2$, then the right-hand probability of (3) is given by:

$$
\Pr_{(x_1,\ldots,x_n)\leftarrow\prod_{i=1}^n[\alpha_i,\beta_i]} \left[\sum_{i=1}^n x_i^2 \leq 1 \right] = \mathcal{L}^{-1}\left\{ \frac{1}{s}\mathcal{L}\{\rho_{\sum_{i=1}^n x_i^2}\}(s) \right\}(1).
$$
(7)

When x_i is uniform over $[\alpha_i, \beta_i]$, the p.d.f. of x_i^2 is

$$
\rho_{x_i^2}(z) = \begin{cases} \dfrac{1}{2(\beta_i - \alpha_i)\sqrt{z}} & (\alpha_i^2 \leq z \leq \beta_i^2) \\ 0 & \text{otherwise} \end{cases}.
$$

Thus, the p.d.f. of $x_1^2 + \cdots + x_n^2$ is given by their convolution: $\rho_{x_1^2 + \cdots + x_n^2}(z) = \rho_{x_1^2}(z) * \cdots * \rho_{x_n^2}(z)$, where

$$(f * g)(t) = \int_{\tau=0}^{t} f(\tau)g(t - \tau)d\tau.$$

However, the Laplace transform of a convolution is simply the product of the transforms: if $f_1(t)$ and $f_2(t)$ be two functions with Laplace transform $F_1(s)$ and $F_2(s)$ respectively, then

$$\mathcal{L}\{f_1 * f_2\}(s) = F_1(s) \cdot F_2(s).$$

In our case, the Laplace transform of each individual pdf is given by

$$\mathcal{L}\left\{\rho_{x_i^2}\right\}(s) = \frac{\sqrt{\pi}\left(\operatorname{erf}(\beta_i\sqrt{s}) - \operatorname{erf}(\alpha_i\sqrt{s})\right)}{2(\beta_i - \alpha_i)\sqrt{s}}. \tag{8}$$

Thus,

$$\mathcal{L}\left\{\rho_{\sum_{i=1}^n x_i^2}\right\}(s) = \left(\frac{\pi}{s}\right)^{n/2} \prod_{i=1}^{n} \frac{\left(\operatorname{erf}(\beta_i\sqrt{s}) - \operatorname{erf}(\alpha_i\sqrt{s})\right)}{2(\beta_i - \alpha_i)}, \tag{9}$$

and computing (7) is reduced to computing the inverse Laplace transform of (9) at 1, namely:

$$\frac{\pi^{n/2-1}}{2i} \int_{c-\infty i}^{c+\infty i} \frac{e^s}{s^{n/2+1}} \prod_{j=1}^{n} \frac{\left(\operatorname{erf}(\beta_i\sqrt{s}) - \operatorname{erf}(\alpha_i\sqrt{s})\right)}{2(\beta_i - \alpha_i)} ds \tag{10}$$

for a real number c within a certain range.

We used Hosono's method [14] to invert the Laplace transform: given $F(s)$, we want to compute its inverse $f(t)$. Let $\gamma > \gamma_0$ in the region of convergence. Hosono used the following approximation of the exponential function, for $\gamma_1 > \gamma_0$:

$$e^s \approx E(s, \gamma_1) := \frac{e^{\gamma_1}}{2\cosh(\gamma_1 - s)}$$

which has singularity points at

$$s_m = \gamma_1 + \left(m - \frac{1}{2}\right)\pi i \text{ for } m \in \mathbb{N}.$$

Considering the integral along the sides of the box $\{\gamma_1 + x + yi : |x| < a, |y| < R\}$ for some small number a and large number R. Letting $R \to \infty$, and by the residue theorem, we have

$$f(t) = \frac{1}{2\pi i} \int_{c-\infty i}^{c+\infty i} F(s)e^{st}ds \approx \frac{e^{\gamma_1}}{t} \sum_{m=1}^{\infty} (-1)^m \operatorname{Im}F\left(\frac{\gamma_1 + (m - 1/2)\pi i}{t}\right). \tag{11}$$

Here, we used the fact that $\operatorname{Re}F$ and $\operatorname{Im}F$ are respectively odd and even functions when $f(t)$ is a real function. Thus, by choosing a suitable γ_1 and truncating the sum of (11) to N terms, we can obtain an approximation of the inverse Laplace transform $f(t)$, and therefore approximate (10).

Choice of Preimages. Since our function includes $\sqrt{s}$, we need to specify which square root over $\mathbb{C}$, in such a way that the function is continuous along the path $(c - \infty i, c + \infty i)$. Since the path is in $\text{Re} > 0$, we choose the primal value of $\sqrt{s}$ in the area $|\arg z| < \pi/4$.

Speeding up the convergence by Euler's series transform. To approximate the sum

$$\sum_{m=1}^{\infty} (-1)^m F_m := \sum_{m=1}^{\infty} (-1)^m \text{Im} F\left(\frac{\gamma_1 + (m - 1/2)\pi i}{t}\right)$$

by a small number of terms, we apply the Euler's series transform to its last terms

$$\sum_{m=1}^{\infty} (-1)^m F_m \approx \sum_{m=1}^{k} (-1)^m F_m + (-1)^k \sum_{j=1}^{J} \frac{(-1)^j \Delta^{j-1} F_{k+1}}{2^j}$$

where $\Delta^{j-1} F_{k+1}$ is the forward difference $\sum_{i=0}^{j-1} (-1)^i \binom{j-1}{i} F_{j+k-i}$. The van Wijngaarden transformation gives us a simple algorithm. Let $s_{0,j} = \sum_{m=k+1}^{k+j} (-1)^m F_m$ for $j = 0, 1, \ldots J$ and compute $s_{\ell+1,j} = (s_{\ell,j} + s_{\ell,j+1})/2$ for all $\ell \geq j$. Finally, $s_{J,0}$ is an approximation of the partial sum.

5.4 Experimental Comparison

We give experimental results to compare the running time and accuracy required by the previous three methods (namely, Constales' formula (4), Tibken's formula (5), and our FILT method (11) to approximate the intersection volume (3): in each case, it depends on a number of terms, and only experiments can tell how many terms are required in practice.

Accuracy of the FILT Method. We first report on the accuracy of the FILT method, which turns out to require the least number of terms in practice. To check accuracy, we compared with a very good approximation of the volume, which we denote by "convolution" in all the graphs: it was obtained with the Fourier series method with a huge number of terms. To visualize results more easily, most of the graphs display an approximation of a pdf, instead of a cdf: Fig. 6 shows the accuracy of FILT on (8) in dimension one, when the function is the pdf of x_j^2:

$$\frac{e^{\gamma_1}}{t} \sum_{m=1}^{N} (-1)^m \text{Im} F\left(\frac{\gamma_1 + (m - 1/2)\pi i}{t}\right) \approx \begin{cases} \dfrac{1}{2(\beta_j - \alpha_j)\sqrt{t}} & (\alpha_j^2 \leq z \leq \beta_j^2) \\ 0 & \text{otherwise} \end{cases}. \tag{12}$$

for γ_1 and N, where

$$F(s) = \frac{\sqrt{\pi}\,(\text{erf}(\beta_j\sqrt{s}) - \text{erf}(\alpha_j\sqrt{s}))}{2(\beta_j - \alpha_j)\sqrt{s}}.$$

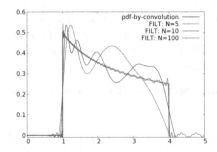

Fig. 6. Accuracy of (12) for (8) as the number N of terms increases.

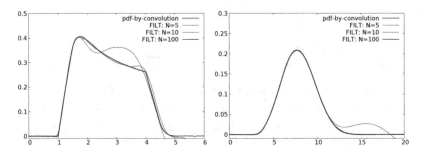

Fig. 7. Accuracy of (11) for evaluating (9) as the number N of terms increases. On the left, $n = 3$. On the right, $n = 10$.

For higher dimensions $n = 3$ and $n = 10$, Fig. 7 shows the accuracy of the method for evaluating (9).

When we apply (10) to (7) to compute the volume of the intersection, the method is very efficient: Fig. 8 shows the accuracy of the method in dimension 140 for the pdf and the CDF of the target random variable, as the number N increases. Here, the box comes from a random natural cell of tag $(0, \ldots, 0, t_{131}, \ldots, t_{140})$ where $t_j \in \{0, 1, 2\}$.

Comparison with the Methods of Constales and Tibken. For each dimension, we generated a random LLL basis and considered the box H corresponding to the tag which as the 1000-th smallest expectation. Then we compute the intersection volume $V = \mathrm{Vol}(Ball_n(0, 1.2 \cdot \mathrm{GH}(L)) \cap H)$ using the FILT method with sufficiently many terms as the reference value r. Table 1 shows what is the number of required terms to achieve a small relative error in practice, that is, the minimum k such that

$$\left| \frac{r - s_k}{r} \right| < 10^{-5}$$

where s_k is the approximation computed with k terms. All computations were done using cpp_dec_float < 50 > of the boost library which can use 50-decimal floating-point numbers. For the computation of the Constales and FILT method,

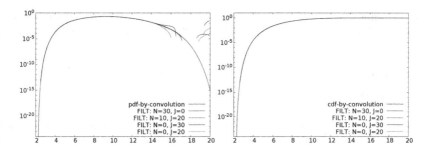

Fig. 8. Accuracy of the van Wijngaarden transformation in dimension 140. On the left, pdf; and on the right, CDF.

Table 1. Required number of terms and running times for the three volume methods

Method/Dimension		40	50	60	80	100	150
Constales	♯ terms	34	34	42	67	109	890
	Time[sec.]	0.55	0.65	0.92	1.9	3.0	45.1
Tibken	♯ terms	34	39	54	-	-	-
	Time[sec.]	5.9	19.5	362.5	-	-	-
FILT	♯ terms	39	46	46	46	43	34
	Time[sec.]	1.06	1.54	1.72	1.96	2.06	1.88

the van Wijngaarden transformation (Sect. 5.3) is used to speed up the convergence. In our FILT method, we used a heuristic value $\gamma_1 = \max(50, 30 + 3\sqrt{dim})$.

From Table 1, we conclude that the FILT method is the best method in practice for discrete pruning: its running time is around 2 s across dimension 40–150, and its required number of terms stays around 40. On the other hand, Tibken's method gets quickly impractical: our implementation requires a huge precomputation table (see the full version for details), which was not taken into account, and we see that the number of terms increases, which creates a sharp increase in the running time. Constales' method is very competitive with FILT until dimension 80, and even faster in dimension ≤ 60, but its number of terms starts to increase from dimension 60, to the point of making the method significantly slower than FILT in dimension 150.

We note that the running time of FILT (resp. Constales' method) is dominated by the computation of $\mathrm{erf}(\cdot)$ (resp. $S(\cdot)$ and $C(\cdot)$): interestingly, in the context of discrete pruning, when we want to compute the intersection volumes for many tags over a fixed basis, we can recycle some computations because $\mathrm{erf}(\beta_i\sqrt{s})$ only depends on t_i. Since tags of interest typically have few non-zero coefficients, and that these non-zero coefficients are small, there is a lot of overlap: given two tags $\mathbf{t}$ and $\mathbf{u}$, there are many indices i such that $t_i = u_i$. In practice, when computing the intersection volumes for sufficiently many tags over a common basis, the total running time is decreased by a factor 10–20.

So the running times of Table 1 can be further decreased when it is applied to discrete pruning: the amortized running time for FILT is a fraction of a second.

6 Optimizing Discrete Pruning with the Natural Partition

We saw in Sect. 4.2 that to optimize discrete pruning for a given effort (*i.e.* for a fixed number M of cells), it suffices to select the M cells which maximize $\mathrm{vol}(S \cap \mathcal{C}(\mathbf{t}))$ among all the cells. In the case of the natural partition, the computation of each $\mathrm{vol}(S \cap \mathcal{C}(\mathbf{t}))$ can be reduced to the computation of $\mathrm{vol}(S \cap H)$ where H is a sub-box of $\mathcal{C}(\mathbf{t})$. And we would like to identify which tags maximize $\mathrm{vol}(S \cap H)$.

Section 5 gave methods to compute $\mathrm{vol}(S \cap H)$ very efficiently, say less than one second. Unfortunately, this is too inefficient to make an online selection, since we may want to process a huge number of cells. Even if we preprocess the computation of tags by using a profile of the Gram-Schmidt norms $\|\mathbf{b}_i^\star\|$, this will be too slow.

In this section, we study practical heuristic methods to select optimal tags for the natural partition: we use our volume computations of Sect. 5 to check the quality of our selection, so the results of Sect. 5 are very useful. The section is organized as follows. In Sect. 6.1, we observe a strong correlation between $\mathrm{vol}(S\cap\mathcal{C}(\mathbf{t}))$ and $\mathbb{E}\{\mathcal{C}_{\mathbb{N}}(\mathbf{t})\}$ for most tags $\mathbf{t}$. This leads us to select the cells which minimize $\mathbb{E}\{\mathcal{C}_{\mathbb{N}}(\mathbf{t})\}$, which avoids any computation of $\mathrm{vol}(S \cap \mathcal{C}(\mathbf{t}))$: Sect. 6.1 explains how to do so efficiently. In Sect. 6.3, we show how to check much more quickly the quality of a discrete pruning set: we show how to speed-up the approximation of a large sum of intersection volumes $\mathrm{vol}(S\cap\mathcal{C}(\mathbf{t}))$, based on statistical inference, which is crucial to estimate the success probability of discrete pruning. Finally, in Sects. 6.4 and 6.5, we compare the quality of cells, depending on the discrete pruning set, and explain how to avoid bad cells.

6.1 Correlation Between Intersection Volumes and Cell Expectations

Inspired by the tag selection of [8] (see Sect. 4.3), we experimentally studied the relationship between the intersection volume $\mathrm{vol}(S \cap \mathcal{C}(\mathbf{t}))$ and the cell expectation $\mathbb{E}\{\mathcal{C}_{\mathbb{N}}(\mathbf{t})\}$. We found that for a fixed-radius centered ball S, for random cells $\mathcal{C}_{\mathbb{N}}(\mathbf{t})$ of the natural partition, there exists a strong negative correlation between $\mathrm{vol}(S\cap\mathcal{C}_{\mathbb{N}}(\mathbf{t}))$ and $\mathbb{E}\{\mathcal{C}_{\mathbb{N}}(\mathbf{t})\}$: roughly speaking, the bigger the expectation $\mathbb{E}\{\mathcal{C}_{\mathbb{N}}(\mathbf{t})\}$, the smaller the volume $\mathrm{vol}(S\cap\mathcal{C}_{\mathbb{N}}(\mathbf{t}))$, as shown in Fig. 9.

More precisely, for a BKZ-20 reduced basis of a 70-dimensional lattice, we computed the best $2^{14} = 16,384$ cells with respect to $\mathbb{E}\{\mathcal{C}_{\mathbb{N}}(\mathbf{t})\}$ (using Corollary 3), and $\mathrm{vol}(S \cap \mathcal{C}_{\mathbb{N}}(\mathbf{t}))$ where S is the ball of radius $\|\mathbf{b}_1\|$. The correlation coefficient of Pearson, Spearman and Kendall are 0.9974191, 0.83618071 and 0.72659780 respectively. The left-hand graph of Fig. 9 shows the intersection volumes of the best cells, sorted by expectation: apart from a few exceptions,

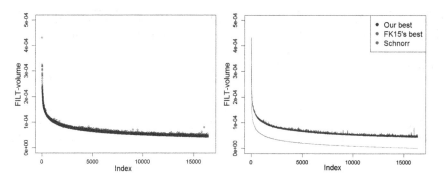

Fig. 9. Graphs of approximated cell-ball intersection volumes sorted by expectation

we see that the curve always decreases, which means that most of the time, the bigger the expectation of the cell, the smaller is the intersection volume.

In other words, if we are only interested in selecting the cells with the largest intersection volumes, we do not need to compute the volume, a near-optimal solution can be found by selecting the cells with the smallest expectation (where the expectation is trivial to compute using Corollary 3), which we will do very efficiently in the next subsection: the volume computation is only necessary when we want to estimate the success probability, not for the selection of tags. We say near-optimal because though the cells with the smallest expectation may not exactly be the cells with the largest intersection volume, most of them are. If we use many cells but miss only a few good cells, it will not affect much the success probability. Note that the selection of cells with the smallest expectation is independent of the radius of the ball S.

As an application of this heuristic supported by experiments, we give evidence that the natural partition is better than Babai's partition: Algorithm 5 shows how to transform efficiently any discrete pruning U for Babai's partition into a discrete pruning V (of the same size) for the natural partition, in such a way that $\sum_{\mathbf{t} \in U} \mathbb{E}(\mathcal{C}_{\mathbb{Z}}(\mathbf{t})) \geq \sum_{\mathbf{t} \in V} \mathbb{E}(\mathcal{C}_{\mathbb{N}}(\mathbf{t}))$. Thus, if we only consider the expectation of cells, we can ignore Babai's partition, and restrict ourselves to the natural partition.

Another interesting application is that for a fixed tag, one can decrease the expectation of its cell by decreasing $\sum_{i=1}^{n} \|\mathbf{b}_i^\star\|^2$: this suggests that the smaller this quantity, the better for discrete pruning.

6.2   Finding the Cells of Lowest Expectations

Recall that the expectation of a cell of the natural partition is given by:

$$\mathbb{E}\{\mathcal{C}_{\mathbb{N}}(\mathbf{t})\} = \sum_{i=1}^{n} \left(\frac{t_i^2}{4} + \frac{t_i}{4} + \frac{1}{12} \right) \|\mathbf{b}_i^\star\|^2$$

We now explain how to compute the tags $\mathbf{t} \in \mathbb{N}^n$ which minimize this expectation.

Algorithm 5. Discrete Pruning: From Babai's partition to the natural partition

---

Input: A finite set $U \subseteq \mathbb{Z}^n$ defining a discrete pruning with Babai's partition $(\mathcal{C}_\mathbb{Z}(), \mathbb{Z}^n)$
over a basis B of a lattice L.

Output: A finite set $V \subseteq \mathbb{N}^n$ of the same cardinal as U, defining a discrete pruning
with the natural partition $(\mathcal{C}_\mathbb{N}(), \mathbb{N}^n)$ over B such that the finite union of cells has
lower $\mathbb{E}$ than for U.

1: Partition the set U as $U_1, \ldots, U_m$ so that m is minimal and within any subset U_i,
   all the tags $\mathbf{t} \in U_i$ are the same in absolute value, $i.e.$ $(|t_1|, \ldots, |t_n|)$ is constant.
2: Define the bijection $\nu : \mathbb{Z} \to \mathbb{N}$ by $\nu(z) = 2|z| - 1$ if z is odd, and $\nu(z) = 2|z|$
   otherwise.
3: **for** $i = 1$ to m **do**
4:    **for** $j = 1$ to n **do**
5:       Compute the number e_j of tags $\mathbf{t} \in U_i$ such that $t_j < 0$
6:       Compute the number f_j of tags $\mathbf{t} \in U_i$ such that $t_j > 0$
7:       **if** $e_j \geq f_j$ **then**
8:          $\varepsilon_j \leftarrow 1$
9:       **else**
10:        $\varepsilon_j \leftarrow -1$
11:       **end if**
12:    **end for**
13:    $V_i \leftarrow \varnothing$
14:    **for** $\mathbf{t} \in U_i$ **do**
15:       **for** $j = 1$ to n **do**
16:          $t'_j \leftarrow \nu(\varepsilon_j t_j)$
17:       **end for**
18:       $V_i \leftarrow V_i \cup (t'_1, \ldots, t'_n)$
19:    **end for**
20: **end for**
21: Return $\cup_{i=1}^n V_i$.

For a fixed sequence $(\|\mathbf{b}_i^\star\|)_{1 \leq i \leq n}$, minimizing $\mathbb{E}\{\mathcal{C}_\mathbb{N}(\mathbf{t})\}$ is equivalent to minimizing the function $q(t_1, \ldots, t_n) = \sum_{i=1}^n (t_i + 1/2)^2 \|\mathbf{b}_i^\star\|^2$ over $\mathbb{N}^n$, which is the same as minimizing $\|\sum_{i=1}^n t_i \mathbf{b}_i^\star + \sum_{i=1}^n \mathbf{b}_i^\star/2\|^2$: the minimal value is $\sum_{i=1}^n \|\mathbf{b}_i^\star\|^2/4$ reached at 0.

Let $L^\star$ be the lattice spanned by the $\mathbf{b}_i^\star$'s, which are pairwise orthogonal. Then finding the M cells with the smallest expectation $\mathbb{E}\{\mathcal{C}_\mathbb{N}(\mathbf{t})\}$ is equivalent to finding the M lattice points $\sum_{i=1}^n t_i \mathbf{b}_i^\star \in L^\star$ with positive coefficients $t_i \in \mathbb{N}$ which are the closest to $\mathbf{u} = -\sum_{i=1}^n \mathbf{b}_i^\star/2$. To solve this problem, we solve a related problem: find all the lattice points $\sum_{i=1}^n t_i \mathbf{b}_i^\star \in L^\star$ whose distance to $\mathbf{u}$ is less than a given bound, with the restriction that all $t_i \in \mathbb{N}$. This is a special case of lattice enumeration in which the coefficients are positive and the input basis vectors are orthogonal: it can be done by slightly modifying enumeration, as shown by Algorithm 6. Given a bound $r > 0$ and $r_1 \leq r_2 \leq \cdots \leq r_n$, Algorithm 6 generates all non-zero $(x_1, \ldots, x_n) \in \mathbb{N}^n$ such that $\sum_{i=1}^n (x_i + 1/2)^2 r_i \leq r$. Algorithm 6 finds all the integer points $\in \mathbb{Z}^n$ which are inside some ellipsoid and the positive orthant (because each $x_i \geq 0$). We will explain later why require that $r_1 \leq r_2 \leq \cdots \leq r_n$, but we note that this is actually not a constraint: if the r_i's are not increasing, sort them by finding a permutation π of $\{1, \ldots, n\}$ such that $r_{\pi(1)} \leq r_{\pi(2)} \leq \cdots \leq r_{\pi(n)}$, then call Algorithm 6 on $(r_{\pi(1)} \leq r_{\pi(2)} \leq \cdots \leq r_{\pi(n)}$ and r, and post-process any output $(x_1, \ldots, x_n)$ of Algorithm 6 by returning instead $(x_{\pi^{-1}(1)}, \ldots, x_{\pi^{-1}(n)})$. Thus, by choosing the r_i's as the $\|\mathbf{b}_i^\star\|^2$'s after suitable reordering, we can indeed use Algorithm 6 to find all the lattice points $\sum_{i=1}^n t_i \mathbf{b}_i^\star \in L^\star$ whose distance to $\mathbf{u}$ is less than a given bound, with the restriction that all $t_i \in \mathbb{N}$.

Now we claim that if we call Algorithm 6 with a suitable value of r we can indeed find the M cells with the smallest expectation $\mathbb{E}\{\mathcal{C}_\mathbb{N}(\mathbf{t})\}$ as follows:

- There are small intervals I such that for any $r \in I$, Algorithm 6 will output only slightly more than M solutions. Then we can sort all the solutions by expectation, and only output the best M tags.
- To find such an interval I slightly modify Algorithm 6 to obtain an algorithm that decides if the number of non-zero $(x_1, \ldots, x_n) \in \mathbb{N}^n$ such that $\sum_{i=1}^n (x_i + 1/2)^2 r_i \leq r$ is larger or less than a given number, with an early-abort as soon as it has found enough solutions. This immediately gives us a suitable I by binary search, using a logarithmic number of calls to the modified version of Algorithm 6. In practice, a non-optimized implementation typically takes a few seconds to find say the best 50 millions tags.

To make this approach practical, it is crucial that Algorithm 6 is very efficient. At first, this looks rather surprising: Algorithm 6 is doing enumeration with a lattice basis $(\mathbf{b}_1^\star, \ldots, \mathbf{b}_n^\star)$ whose Gram-Schmidt norms are identical to that of $(\mathbf{b}_1, \ldots, \mathbf{b}_n)$, with a non-trivial radius beyond the $\mathrm{GH}(L^\star)$. Even if we restrict to positive coefficients, it looks impossible, because the running time of enumeration is typically predicted well by the Gaussian heuristic (see [11]), using values that depend only the Gram-Schmidt norms: this means that, naively, we might expect enumeration in $L^\star$ to be as expensive as enumeration in L, in which case the whole approach would be meaningless, because the goal of discrete pruning is to speed up enumeration. Fortunately, it turns out that the usual predictions for the running time of enumeration do not apply to $L^\star$, because the basis $(\mathbf{b}_1^\star, \ldots, \mathbf{b}_n^\star)$ we use has a very special property: all its vectors are orthogonal, and it is known that in lattices generated by orthogonal vectors, the number of lattice points in a ball can behave significantly differently than usual (see [21]). Yet, that alone would not be sufficient in practice to guarantee efficiency: this is where the constraint $r_1 \leq r_2 \leq \cdots \leq r_n$ matters. In our experiments with that constraint, if ℓ is the number of solutions (*i.e.* the number of $(x_1, \ldots, x_n) \in \mathbb{N}^n$ such that $\sum_{i=1}^n (x_i + 1/2)^2 r_i \leq R$), the running time appears to be polynomial in ℓ. More precisely, like all enumeration algorithms, Algorithm 6 can be viewed as a depth-first search of a tree, and the running time is less than $O(L)$ polynomial-time operations, where L is the total number of nodes of the tree. In practice, at least in the context of discrete pruning, the number L seems to be bounded by $O(\ell \times n)$, and even $\ell \times n$. This is contrast in the usual situation for which the number of nodes of the enumeration tree is exponentially larger than the

---

Algorithm 6. Enumeration of cells of low expectations

Input: $(r_1, \ldots, r_n) \in \mathbb{R}^n$ such that $0 \leq r_1 \leq r_2 \leq \cdots \leq r_n$ and a bound $r > 0$.
Output: All $(x_1, \ldots, x_n) \in \mathbb{N}^n \setminus \{0\}$ such that $\sum_{i=1}^{n} (x_i + 1/2)^2 r_i \leq r$ and all $x_i \geq 0$.
1: $v_2 = \cdots = v_n = \rho_{n+1} = 0$ // *current coefficients*
2: **for** $k = n$ downto 2 **do**
3:    $\rho_k = \rho_{k+1} + (v_k + 1/2)^2 \cdot r_k$ // *partial squared norms*
4: **end for**
5: $k = v_1 = 1$;
6: **while** true **do**
7:    $\rho_k = \rho_{k+1} + (v_k + 1/2)^2 \cdot r_k$ // *compute squared norm of current node*
8:    **if** $\rho_k \leq r$ **then**
9:       **if** $k = 1$ **then**
10:          **return** $(v_1, \ldots, v_n)$; (*solution found*)
11:          $v_k \leftarrow v_k + 1$
12:       **else**
13:          $k \leftarrow k - 1$ // *going down the tree*
14:          $v_k \leftarrow 0$
15:       **end if**
16:    **else**
17:       $k \leftarrow k + 1$ // *going up the tree*
18:       **if** $k = n + 1$ **then**
19:          **exit** (*no more solutions*)
20:       **else**
21:          $v_k \leftarrow v_k + 1$
22:       **end if**
23:    **end if**
24: **end while**

---

number of solutions. We leave it as an open problem to show the efficiency of Algorithm 6.

6.3 Faster Approximation of the Success Probability by Statistical Inference

As seen in Sect. 4.2, estimating the success probability of discrete pruning requires from (1) the computation of $\sum_{\mathbf{t} \in U} \mathrm{vol}(S \cap \mathcal{C}(\mathbf{t}))$, where U is the set of tags. We saw in the previous section how to compute efficiently $\mathrm{vol}(S \cap \mathcal{C}(\mathbf{t}))$ for the natural partition by reducing to the case of $\mathrm{vol}(H \cap S)$ where H is box. Although this computation is efficient, it is not sufficiently efficient to be applied billions of times within a few seconds.

Fortunately, the classical theory of statistical inference allows us to approximate $\sum_{\mathbf{t} \in U} \mathrm{vol}(S \cap \mathcal{C}(\mathbf{t}))$ reasonably well without computing each term of the sum separately: it turns out that even a constant number of terms is sufficient to obtain a good approximation in practice, and a good approximation is sufficient for our heuristic estimate of the success probability of discrete pruning, since Heuristic 2 is only a heuristic estimate. To illustrate the presentation, we report experiments for a 70-dimensional LLL-reduced basis of a random lattice L, radius

$R = 1.2GH(L)$, and a typical set of tags, namely 5,000,000 tags generated by Algorithm 6. Using the FILT method, we find that the sum $\sum_{t \in U} \text{vol}(S \cap \mathcal{C}(t))$ of 5,000,000 volumes is approximately 35.03688covol(L) by computing each volume, and we will show how to approximate this value using only 1,000 volumes.

We want to approximate the sum $\mu = \sum_{t \in U} \text{vol}(S \cap \mathcal{C}(t))$, which is exactly the expectation of the discrete random variable $X = \#U \times \text{vol}(S \cap \mathcal{C}(t))$, when t ranges over the finite set U.

Let $X_1, \ldots, X_m$ be random variables chosen independently from the same distribution as X. Then the sample mean $\overline{X} = (X_1 + \cdots + X_m)/m$. Its expectation $\mathbb{E}(\overline{X})$ is exactly the target $\mu = \sum_{t \in U} \text{vol}(S \cap \mathcal{C}(t))$, and its variance $\mathbb{V}(\overline{X})$ is $\mathbb{V}(X)/m$. We want to know how close is $\overline{X}$ to its expectation μ.

Since the X_i's are discrete, the central limit theorem implies that $\sqrt{m}(\overline{X} - \mu)$ converges in distribution (as m grows) to the centered normal distribution of standard deviation σ: Fig. 10 shows the distribution of X and $\overline{X}$ (with $m = 1000$) for our example. This means that in practice, we can already expect to find a reasonable approximation of μ by simpling selecting m tags $t_1, \ldots, t_m$ independently and uniformly at random from the set U: the estimate would be $\frac{\#U}{m} \sum_{i=1}^m \text{vol}(S \cap \mathcal{C}(t_i))$, with an absolute error of magnitude $\pm \sigma/\sqrt{m}$.

The strategy we described is known as *simple sampling*, because it is the simplest method to estimate the expectation of X, but there are many other methods. In practice, we used a better yet still simple strategy, known as *stratified sampling*, because we can take advantage of properties of the cells we select in practice.

In the simplest form of stratified sampling, the set of tags U can be partitioned into subsets, and one selects a tag uniformly at random inside each subset and "extrapolate", as before. More precisely, if the subsets are $U_1, \ldots U_m$, then the (randomized) estimate is

$$\overline{X'} = \frac{1}{m} \sum_{i=1}^m \#U_i \times \text{vol}(S \cap \mathcal{C}(t_i)), \qquad (13)$$

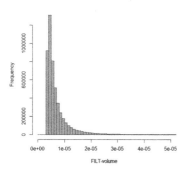

Fig. 10. Distribution of $X/(\#U\text{covol}(L))$ (on the left) and the sample mean $\overline{X}$ (on the right) for $m = 1000$.

where each $\mathbf{t}_i$ is selected uniformly at random from U_i. In our case, assume that we select the set U of tags formed by the best M tags $\mathbf{t}$ with respect to $\mathbb{E}\{\mathcal{C}_{\mathbb{N}}(\mathbf{t})\}$ for some M. Then we sort all these tags by increasing expectation, and split the ordered sequence into m subsets $U_1,\ldots,U_m$ of nearly equal size. Thus, for all $(\mathbf{t}_i,\mathbf{t}_j) \in U_i \times U_j$ and $i < j$, we have: $\mathbb{E}\{\mathcal{C}_{\mathbb{N}}(\mathbf{t}_i)\} \leq \mathbb{E}\{\mathcal{C}_{\mathbb{N}}(\mathbf{t}_j)\}$.

Figure 11 shows that for the same value of m, the distribution of $\overline{X'}$ is narrower than that of $\overline{X}$, and therefore stratified sampling gives a better estimate than simple sampling. Figure 12 shows how accurate is (13), for increasing values of m, where there are 10,000 trials for each value of m. Using these sampling strategies from statistical inference, it is now possible to estimate rather precisely the sum of volumes efficiently in practice. A value of the order $m = 1000$ appears to be sufficient, even for much larger sets of tags. This is consistent with common practice in statistical surveys.

By combining this fast approximation algorithm with Sect. 6.2, we can efficiently find out what is the best trade-off for discrete pruning: by increasing order of magnitude of M, and starting with a small one, we use Sect. 6.2 to identify the

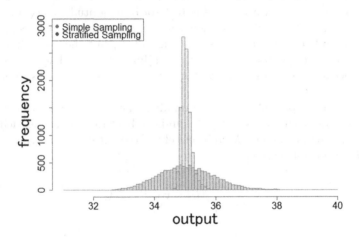

Fig. 11. Distributions of the estimate of simple sampling and stratified sampling.

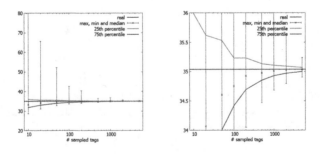

Fig. 12. Accuracy of stratified sampling with $m = 10,\ldots,5000$. The right-hand side zooms in the region $y \in [34, 36]$.

nearly-best M cells, and apply the fast approximation algorithm to estimate the success probability for these M cells. We then deduce which M offers the best trade-off, after taking into account the time of basis reduction, like in continuous pruning [11]. However, this optimization is easier than for continuous pruning: the only parameter is M (or even its order of magnitude).

6.4   Comparison of Cells

As we mentioned in Sect. 4.3, Schnor's random sampling is essentially discrete pruning with the set of tags $U_u = \{(0, ..., 0, t_1, ..., t_u, 1) \in \{0,1\}^n\}$. We have computed the intersection volume and the cell expectations of these tags. The blue curve in Fig. 9 shows the intersection volume sorted by cell expectation. We compared Schnorr's cells with the cells of Fukase-Kashiwabara [8], as selected by the process described in Sect. 4.3. The experiments show that the intersection volumes are much bigger for the FK cells than for Schnorr's cells: the Fukase-Kashiwabara variant [8] outperforms Schnorr's random sampling. We also computed the best cells in terms of expectation, using our algorithm: in this limited experiments, we can see in Fig. 9 that the FK tags are very close to our tags, but that tey are still different. For instance, if we consider a typical BKZ-20 basis of a 250-dimensional random lattice, then among the best 5×10^7 tags selected by Algorithm 6:

– about 70% of the tags have at least one coefficient $\notin \{0, 1\}$ and cannot therefore be selected by Schnorr's random sampling.
– about 25% of the tags have at least two coefficients $\notin \{0, 1\}$ and cannot therefore be selected by the method outlined in [8].

Accordingly, we obtained several experiments in which the shortest vector found by discrete pruning had a tag which did not mach the FK selection nor of course Schnorr's selection.

6.5   Bad Cells

To conclude this section, we note that discrete pruning can be slightly improved by removing what we call bad cells. For instance, the lattice point inside the cell $\mathcal{C}_{\mathbb{N}}(\mathbf{0})$ is zero, which is useless. When selecting a set of tags by lowest expectation, the set usually contains "a trivial" tag of the form of $\mathbf{e}_i = (0, \ldots, 0, 1, 0, \ldots, 0)$. When the basis B is size-reduced, it turns out that the lattice point inside the cell $\mathcal{C}_{\mathbb{N}}(\mathbf{t})$ is simply the trivial vector $\mathbf{b}_i$, which is usually not very short for usual reduction algorithms, and can anyways be tested separately. Although the cells of these tags have small expectation, they are not useful for discrete pruning: they can be removed by precomputation.

7   Experiments

Most of our experiments were performed by a standard server with two Intel Xeon E5-2660 CPUs and 256-GB RAMs. In this section, by random reduced basis, we mean a basis obtained by reducing the Hermite normal form of lattices generated by the SVP challenge generator [28] with different seeds: each seed selects a different lattice, and we only consider one reduced basis per lattice for a given reduction strength. The LLL and BKZ reduced bases are computed by the NTL and progressive BKZ library [2], respectively.

Typical graphs of $\|\mathbf{b}_i^\star\|^2$ of LLL, PBKZ-20 and 60 are shown in Fig. 13.

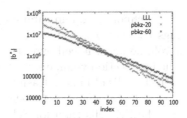

Fig. 13. Typical graphs of $\|\mathbf{b}_i^\star\|$ of LLL, PBKZ-20 and 60 reduced bases of a random 100-dimensional lattice.

For a lattice basis B, a set of tags $T = \{\mathbf{t}_1, \ldots, \mathbf{t}_M\}$, and bounding radius R, define the symbol

$$V(B, T, R) = \sum_{j=1}^{M} \frac{\mathrm{vol}(\mathcal{C}_{\mathbb{N}}(\mathbf{t_i}) \cap \mathrm{Ball}_n(R))}{\mathrm{covol}(L)}$$

which, by Heuristic 1, is a heuristic estimate of the number of lattice points contained in the union of cells.

7.1   Verifying Heuristics on the Success Probability and Number of Solutions

For a lattice basis and a set $\{\mathbf{t}_1, \ldots, \mathbf{t}_M\}$ of tags, consider the union of corresponding cells, that is, the pruning set: $P = \cup_{i=1}^{N} \mathcal{C}_{\mathbb{N}}(\mathbf{t_i})$. Heuristic 1 suggests that the number of lattice points inside P and shorter than R is roughly $\mathrm{vol}(P \cap \mathrm{Ball}_n(R))/\mathrm{covol}(L)$.

We verified this estimate by using random LLL-reduced bases in dimension 100. For each basis B_i, we generated 1,000 tags by Algorithm 6 and selected the best tags so that the total ratio $\mathrm{vol}(P \cap \mathrm{Ball}_n(R))/\mathrm{covol}(L)$ (with radius $R = 1.2GH(L)$) was larger than 0.001. About 300 or 400 tags are selected. For 17,386 bases and 6,281,800 tags generated as above, we display the result in Fig. 14. By the red curve the relation between accumulated volume, that is, for the set of first i tags $T_i = \{\mathbf{t}_1, \ldots, \mathbf{t}_i\}$, $\sum_{\mathbf{t}_i \Leftarrow L_j} V(L, T_i, R = 1.2GH(L_j))$. Here, the notation $\mathbf{t}_i \Leftarrow L_j$ means that $\mathbf{t}_i$ is generated by Algorithm 6 for the basis L_j. The blue curve is the number of non-zero vectors which are contained in the boxes of the tag set T_i.

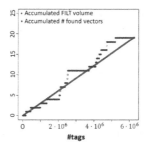

Fig. 14. Comparison between accumulated volume and actual number of found vectors

7.2 Comparison with Classical Pruned Enumeration

We give experiments comparing discrete pruning with continuous pruning [11]. The parameters are: the lattice dimension n, the number of tags M, the enumeration radius $R = \alpha \cdot \mathrm{GH}(L)$ and the blocksize β of BKZ lattice reduction.

The outline of the comparison is as follows: Generate an n-dimensional random lattice and reduce it by BKZ-β. Generate the M best tags $T = \{\mathbf{t}_1, \dots, \mathbf{t}_M\}$ by Algorithm 6 and compute the intersection volume $V(L, T, R = \alpha \cdot \mathrm{GH}(L))$. For $M > 1000$, we use stratified sampling with $m = 1000$.

To compare with continuous pruning, we adapt Gama-Nguyen-Regev's optimization method to minimize the expected number of processed nodes

$$N = \frac{1}{2} \sum_{k=1}^{n} \frac{\mathrm{vol}\{(x_1, \dots, x_k) \in \mathbb{R}^k : \sum_{i=1}^{\ell} x_i^2 < (Rf(\ell) \cdot \alpha GH(L))^2 \text{ for all } \ell \in [k]\}}{\prod_{i=n-k+1}^{n} \|\mathbf{b}_i^\star\|}.$$

subject to the volume of cylinder intersections

$$\frac{\mathrm{vol}\{(x_1, \dots, x_n) \in \mathbb{R}^n : \sum_{i=1}^{\ell} x_i^2 < (R \times f(\ell) \cdot \alpha GH(L))^2 \text{ for } \forall\, \ell \in [n]\}}{\mathrm{covol}(L)}$$

is larger than $V(L, T, R)$ by optimizing the bounding coefficients $f(1), \dots, f(n)$. Note that in the original paper [11], their condition is the probability defined by the surface area of a cylinder intersection.

The cost estimation for discrete pruning is easy: the number of operations is essentially $M \cdot n^2/2$ floating-point operations since one conversion defined in Algorithm 1 requires $n^2/2$ floating-point operations, like Babai's nearest plane algorithm. But the implementation can be massively parallelized (see [29] for the special case of Schnorr's random sampling). On the other hand, the cost estimation for continuous pruning is a bit more tricky, since the actual cost to process one node, i.e., cost to decide whether the node is alive or pruned, is proportional to the depth in the searching tree. To make a rigid comparison, we counted up the actual number N_i of nodes at depth i processed during the enumeration by experiments and we define the cost as $\sum_{i=1}^{n} i \cdot N_i$.

With these settings, we tried 20 random bases for the same parameter set $(n, M, R = \alpha \cdot \mathrm{GH}(L), \beta)$: Fig. 15 shows the average ratio of the costs

$$\frac{1}{20} \sum_{seed=0}^{19} \frac{M \cdot n^2/2}{\sum_{i=1}^{n} i \cdot N_i}$$

which indicates when discrete pruning is faster or slower than continuous pruning, depending on whether the ratio is ≤ 1 or ≥ 1.

The trends are clear from the figures. We find that the discrete pruning is faster than continuous pruning when:

1. The number M of tags is small. This might be useful in extreme enumeration for approximating the shortest vector problem.
2. The lattice dimension is high. Besides the speed factor, it might be useful because it is not easy to run continuous pruning with a very low success probability in high dimension, as it is harder to find suitable optimized bounding functions.
3. The lattice basis is not strongly BKZ-reduced.

On the other hand, the α parameter that sets the enumeration radius does not affect the trends.

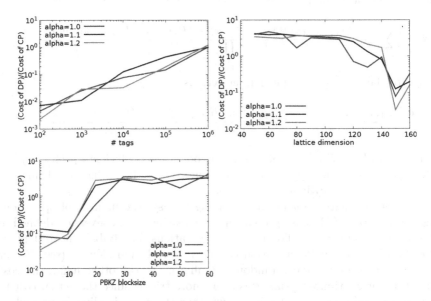

Fig. 15. Discrete pruning vs Continuous pruning: above the virtual horizontal line 10^0, discrete pruning is more expensive. (Upper left) 150-dim LLL-reduced bases with increasing α and M; (Upper right) LLL-reduced bases with increasing dimension and α. $M = 10^4$ is fixed; (Lower left) 150-dim with increasing blocksize β of progressive BKZ and α. $M = 10^4$ is fixed. Note that $\beta = 0$ corresponds to LLL.

Acknowledgements. This work was supported by JSPS KAKENHI Grant Numbers 16H02780 and 16H02830. We thank the reviewers for their comments.

References

1. Agrell, E., Eriksson, T., Vardy, A., Zeger, K.: Closest point search in lattices. IEEE Trans. Info. Theory **48**(8), 2201–2214 (2002)
2. Aono, Y., Wang, Y., Hayashi, T., Takagi, T.: Improved progressive BKZ algorithms and their precise cost estimation by sharp simulator. In: Fischlin, M., Coron, J.-S. (eds.) EUROCRYPT 2016. LNCS, vol. 9665, pp. 789–819. Springer, Heidelberg (2016). doi:10.1007/978-3-662-49890-3_30
3. Babai, L.: On Lovász' lattice reduction and the nearest lattice point problem. In: Mehlhorn, K. (ed.) STACS 1985. LNCS, vol. 182, pp. 13–20. Springer, Heidelberg (1985). doi:10.1007/BFb0023990
4. Buchmann, J., Ludwig, C.: Practical lattice basis sampling reduction. In: Hess, F., Pauli, S., Pohst, M. (eds.) ANTS 2006. LNCS, vol. 4076, pp. 222–237. Springer, Heidelberg (2006). doi:10.1007/11792086_17
5. Chen, Y., Nguyen, P.Q.: BKZ 2.0: better lattice security estimates. In: Lee, D.H., Wang, X. (eds.) ASIACRYPT 2011. LNCS, vol. 7073, pp. 1–20. Springer, Heidelberg (2011). doi:10.1007/978-3-642-25385-0_1
6. Ding, D., Zhu, G., Wang, X.: A genetic algorithm for searching the shortest lattice vector of SVP challenge. In: Proceedings of the Genetic and Evolutionary Computation Conference, GECCO 2015, pp. 823–830 (2015)
7. Fincke, U., Pohst, M.: Improved methods for calculating vectors of short length in a lattice, including a complexity analysis. Math. Comput. **44**(170), 463–471 (1985)
8. Fukase, M., Kashiwabara, K.: An accelerated algorithm for solving SVP based on statistical analysis. JIP **23**(1), 67–80 (2015)
9. Fukase, M., Yamaguchi, K.: Analysis of the extended search space for the shortest vector in lattice. J. Syst. Cybern. Inf. **9**(6), 42–46 (2011)
10. Gama, N., Nguyen, P.Q.: Predicting lattice reduction. In: Smart, N. (ed.) EUROCRYPT 2008. LNCS, vol. 4965, pp. 31–51. Springer, Heidelberg (2008). doi:10.1007/978-3-540-78967-3_3
11. Gama, N., Nguyen, P.Q., Regev, O.: Lattice enumeration using extreme pruning. In: Gilbert, H. (ed.) EUROCRYPT 2010. LNCS, vol. 6110, pp. 257–278. Springer, Heidelberg (2010). doi:10.1007/978-3-642-13190-5_13
12. Hanrot, G., Stehlé, D.: Improved analysis of kannan's shortest lattice vector algorithm. In: Menezes, A. (ed.) CRYPTO 2007. LNCS, vol. 4622, pp. 170–186. Springer, Heidelberg (2007). doi:10.1007/978-3-540-74143-5_10
13. Håstad, J., Just, B., Lagarias, J.C., Schnorr, C.-P.: Polynomial time algorithms for finding integer relations among real numbers. SIAM J. Comput. **18**(5), 859–881 (1989)
14. Hosono, T.: Numerical inversion of laplace transform and some applications to wave optics. Radio Sci. **16**(6), 1015–1019 (1981)
15. Kannan, R.: Improved algorithms for integer programming and related lattice problems. In: Proceedings of the 15th ACM Symposium on Theory of Computing (STOC), pp. 193–206 (1983)
16. Lenstra, A.K., Lenstra Jr., H.W., Lovász, L.: Factoring polynomials with rational coefficients. Math. Ann. **261**, 513–534 (1982)

17. Lindner, R., Peikert, C.: Better key sizes (and Attacks) for LWE-based encryption. In: Kiayias, A. (ed.) CT-RSA 2011. LNCS, vol. 6558, pp. 319–339. Springer, Heidelberg (2011). doi:10.1007/978-3-642-19074-2_21

18. Lindner, R., Rückert, M.: TU Darmstadt lattice challenge. http://www.latticechallenge.org/

19. Liu, M., Nguyen, P.Q.: Solving BDD by enumeration: an update. In: Dawson, E. (ed.) CT-RSA 2013. LNCS, vol. 7779, pp. 293–309. Springer, Heidelberg (2013). doi:10.1007/978-3-642-36095-4_19

20. Ludwig, C.: Practical Lattice Basis Sampling Reduction. Ph.D. thesis, Technische Universität Darmstadt (2005)

21. Mazo, J.E., Odlyzko, A.M.: Lattice points in high dimensional spheres. Monatsh. Math. **17**, 47–61 (1990)

22. Nguyen, P.Q.: Public-key cryptanalysis. In: Luengo, I. (ed.) Recent Trends in Cryptography. Contemporary Mathematics, vol. 477. AMS-RSME (2009)

23. Nguyen, P.Q., Stehlé, D.: LLL on the average. In: Hess, F., Pauli, S., Pohst, M. (eds.) ANTS 2006. LNCS, vol. 4076, pp. 238–256. Springer, Heidelberg (2006). doi:10.1007/11792086_18

24. Nguyen, P.Q., Vallée, B. (eds.): The LLL Algorithm: Survey and Applications. Information Security and Cryptography. Springer, Heidelberg (2009)

25. Pohst, M.: On the computation of lattice vectors of minimal length, successive minima and reduced bases with applications. SIGSAM Bull. **15**(1), 37–44 (1981)

26. Rogers, C.A.: The number of lattice points in a set. Proc. London Math. Soc. **6**(3), 305–320 (1956)

27. Rousseau, C.C., Ruehr, O.G.: Problems and solutions. SIAM Review **39**(4), 779–786 (1997). Subsection: The Volume of the Intersection of a Cube and a Ball in N-space. Two solutions by Bernd Tibken and Denis Constales

28. Schneider, M., Gama, N.: SVP challenge. http://www.latticechallenge.org/svp-challenge/

29. Schneider, M., Göttert, N.: Random sampling for short lattice vectors on graphics cards. In: Preneel, B., Takagi, T. (eds.) CHES 2011. LNCS, vol. 6917, pp. 160–175. Springer, Heidelberg (2011). doi:10.1007/978-3-642-23951-9_11

30. Schnorr, C.P.: Lattice reduction by random sampling and birthday methods. In: Alt, H., Habib, M. (eds.) STACS 2003. LNCS, vol. 2607, pp. 145–156. Springer, Heidelberg (2003). doi:10.1007/3-540-36494-3_14

31. Schnorr, C.-P., Euchner, M.: Lattice basis reduction: improved practical algorithms and solving subset sum problems. Math. Program. **66**, 181–199 (1994)

32. Schnorr, C.P., Hörner, H.H.: Attacking the Chor-Rivest cryptosystem by improved lattice reduction. In: Guillou, L.C., Quisquater, J.-J. (eds.) EUROCRYPT 1995. LNCS, vol. 921, pp. 1–12. Springer, Heidelberg (1995). doi:10.1007/3-540-49264-X_1

33. Innovation, S.: NTRU challenge. https://www.securityinnovation.com/products/ntru-crypto/ntru-challenge

34. Siegel, C.L.: A mean value theorem in geometry of numbers. Ann. Math. **46**(2), 340–347 (1945)

35. Stehlé, D., Watkins, M.: On the extremality of an 80-dimensional lattice. In: Hanrot, G., Morain, F., Thomé, E. (eds.) ANTS 2010. LNCS, vol. 6197, pp. 340–356. Springer, Heidelberg (2010). doi:10.1007/978-3-642-14518-6_27

On Dual Lattice Attacks Against Small-Secret LWE and Parameter Choices in HElib and SEAL

Martin R. Albrecht[✉]

Information Security Group, Royal Holloway,
University of London, Egham, Surrey TW20 0EX, UK
martin.albrecht@royalholloway.ac.uk

Abstract. We present novel variants of the dual-lattice attack against LWE in the presence of an unusually short secret. These variants are informed by recent progress in BKW-style algorithms for solving LWE. Applying them to parameter sets suggested by the homomorphic encryption libraries HElib and SEAL yields revised security estimates. Our techniques scale the exponent of the dual-lattice attack by a factor of $(2\,L)/(2\,L+1)$ when $\log q = \Theta(L \log n)$, when the secret has constant hamming weight h and where L is the maximum depth of supported circuits. They also allow to half the dimension of the lattice under consideration at a multiplicative cost of 2^h operations. Moreover, our techniques yield revised concrete security estimates. For example, both libraries promise 80 bits of security for LWE instances with $n = 1024$ and $\log_2 q \approx 47$, while the techniques described in this work lead to estimated costs of 68 bits (SEAL) and 62 bits (HElib).

1 Introduction

Learning with Errors (LWE), defined in Definition 1, has proven to be a rich source of cryptographic constructions, from public-key encryption and Diffie-Hellman-style key exchange (cf. [Reg09, Pei09, LPR10, DXL12, BCNS15, ADPS16, BCD+16]) to fully homomorphic encryption (cf. [BV11, BGV12, Bra12, FV12, GSW13, CS15]).

Definition 1 (LWE [Reg09]). *Let n, q be positive integers, χ be a probability distribution on $\mathbb{Z}$ and $\mathbf{s}$ be a secret vector in $\mathbb{Z}_q^n$. We denote by $L_{\mathbf{s},\chi,q}$ the probability distribution on $\mathbb{Z}_q^n \times \mathbb{Z}_q$ obtained by choosing $\mathbf{a} \in \mathbb{Z}_q^n$ uniformly at random, choosing $e \in \mathbb{Z}$ according to χ and considering it in $\mathbb{Z}_q$, and returning $(\mathbf{a}, c) = (\mathbf{a}, \langle \mathbf{a}, \mathbf{s} \rangle + e) \in \mathbb{Z}_q^n \times \mathbb{Z}_q$.*

Decision-LWE is the problem of deciding whether pairs $(\mathbf{a}, c) \in \mathbb{Z}_q^n \times \mathbb{Z}_q$ are sampled according to $L_{\mathbf{s},\chi,q}$ or the uniform distribution on $\mathbb{Z}_q^n \times \mathbb{Z}_q$.

Search-LWE is the problem of recovering $\mathbf{s}$ from $(\mathbf{a}, c) = (\mathbf{a}, \langle \mathbf{a}, \mathbf{s} \rangle + e) \in \mathbb{Z}_q^n \times \mathbb{Z}_q$ sampled according to $L_{\mathbf{s},\chi,q}$.

This research was supported by EPSRC grants EP/L018543/1 "Multilinear Maps in Cryptography" and EP/P009417/1 "Bit Security of Learning with Errors for Post-Quantum Cryptography and Fully Homomorphic Encryption".

© International Association for Cryptologic Research 2017
J.-S. Coron and J.B. Nielsen (Eds.): EUROCRYPT 2017, Part II, LNCS 10211, pp. 103–129, 2017.
DOI: 10.1007/978-3-319-56614-6_4

We may write LWE instances in matrix form $(\mathbf{A}, \mathbf{c})$, where rows correspond to samples $(\mathbf{a}_i, c_i)$. In many instantiations, χ is a discrete Gaussian distribution with standard deviation $\alpha q/\sqrt{2\pi}$. Though, in this work, like in many works on cryptanalysis of LWE, the details of the error distribution do not matter as long as we can bound the size of the error under additions.

The bit-security of concrete LWE instances is a prominent area of current cryptographic research, in particular in light of standardisation initiatives for LWE-based schemes and LWE-based (somewhat) homomorphic encryption being proposed for applications such as computation with medical data [KL15]. See [APS15] for a relatively recent survey of known (classical) attacks.

Applications such as [KL15] are enabled by progress in homomorphic encryption in recent years. The two most well-known homomorphic encryption libraries are HElib and SEAL. HElib [GHS12a, HS14] implements BGV [BGV12]. SEAL v2.0 [LP16] implements FV [Bra12, FV12]. Both schemes fundamentally rely on the security of LWE.

However, results on the expected cost of solving generic LWE instances do not directly translate to LWE instances as used in fully homomorphic encryption (FHE). Firstly, because these instances are typically related to the Ring-LWE assumption [LPR10, LPR13] instead of plain LWE. Secondly, because these instances are typically *small-secret* instances. In particular, they typically sample the secret $\mathbf{s}$ from some distribution $\mathcal{B}$ as defined below. We call such instances $\mathcal{B}$-secret LWE instances.

Definition 2. *Let n, q be positive integers. We call*

$\mathcal{B}$ *any distribution on $\mathbb{Z}_q^n$ where each component ≤ 1 in absolute value, i.e. $\|\mathbf{s}_{(i)}\| \leq 1$ for $\mathbf{s} \leftarrow_{\$} \mathcal{B}$.*

$\mathcal{B}^+$ *the distribution on $\mathbb{Z}_q^n$ where each component is independently sampled uniformly at random from $\{0, 1\}$.*

$\mathcal{B}^-$ *the distribution on $\mathbb{Z}_q^n$ where each component is independently sampled uniformly at random from $\{-1, 0, 1\}$.*

$\mathcal{B}_h^+$ *the distribution on $\mathbb{Z}_q^n$ where components are sampled independently uniformly at random from $\{0, 1\}$ with the additional guarantee that at most h components are non-zero.*

$\mathcal{B}_h^-$ *the distribution on $\mathbb{Z}_q^n$ where components are sampled independently uniformly at random from $\{-1, 0, 1\}$ with the additional guarantee that at most h components are non-zero.*

Remark 1. In [BLP+13], instances with $\mathbf{s} \leftarrow_{\$} \mathcal{B}^+$ are referred to as binary-secret; $\mathcal{B}^+$ is used in [FV12]; $\mathcal{B}^-$ is used in Microsoft's SEAL v2.0 library[1] and [LN14]; $\mathcal{B}_{64}^-$ is the default choice in HElib, cf. [GHS12b, Appendix C.1.1] and [HS14].

It is an open question how much easier, if any, $\mathcal{B}$-secret LWE instances are compared to regular LWE instances. On the one hand, designers of FHE schemes

---

[1] cf. KeyGenerator::set_poly_coeffs_zero_one_negone() at https://sealcrypto.codeplex. com/SourceControl/latest#SEAL/keygenerator.h.

typically ignore this issue [GHS12a, LN14, CS16]. This could be considered as somewhat justified by a reduction from [ACPS09] showing that an LWE instance with an arbitrary secret can be transformed into an instance with a secret following the noise distribution in polynomial time and at the loss of n samples. Hence, such instances are not easier than instances with a uniformly random secret, assuming sufficiently many samples are available. As a consequence, LWE with a secret following the noise distribution is considered to be in *normal form*. Given that the noise in homomorphic encryption libraries is also typically rather small—SEAL and HElib use standard deviation $\sigma \approx 3.2$—the distribution $\mathcal{B}^-$ gives rise to LWE instances which could be considered relatively close to normal-form LWE instances. However, considering the actual distributions, not just the standard deviations, it is known that LWE with *error distribution* $\mathcal{B}$ is insecure once sufficiently many samples are available [AG11, ACFP14, KF15].

On the other hand, the best, known reduction from regular LWE to $\mathcal{B}^+$-secret LWE has an expansion factor of $\log q$ in the dimension. That is, [BLP+13] gives a reduction from regular LWE in dimension n to LWE with $\mathbf{s} \leftarrow_{\$} \mathcal{B}^+$ in dimension $n \log q$.

In contrast, even for noise with width $\approx \sqrt{n}$ and $\mathbf{s} \leftarrow_{\$} \mathcal{B}^-$ the best known lattice attacks suggest an expansion factor of at most $\log \log n$ [BG14], if at all. Overall, known algorithms do not perform significantly better for $\mathcal{B}$-secret LWE instances, perhaps reinforcing our confidence in the common approach of simply ignoring the special form of the secret.

One family of algorithms has recently seen considerable progress with regards to $\mathcal{B}$-secret instances: combinatorial algorithms. Already in [Reg09] it was observed that the BKW algorithm, originally proposed for LPN by Blum, Kalai and Wasserman [BKW00], leads to an algorithm in $2^{\Theta(n)}$ time and space for solving LWE. The algorithm proceeds by splitting the components of the vectors $\mathbf{a}_i$ into blocks of k components. Then, it searches for collisions in the first block in an "elimination table" holding entries for (possibly) all q^k different values for that block. This table is constructed by sampling fresh $(\mathbf{a}_i, c_i)$ pairs from the LWE oracle. By subtracting vectors with colliding components in the first block, a vector of dimension $n - k$ is recovered, applying the same subtraction to the corresponding c_i values, produces an error of size $\sqrt{2}\alpha q$. Repeating the process for consecutive blocks reduces the dimension further at the cost of an increase in the noise by a factor $\sqrt{2}$ at each level. This process either continues until all components of $\mathbf{a}_i$ are eliminated or when there are so few components left that exhaustive search can solve the remaining low-dimensional LWE instance.

A first detailed study of this algorithm when applied to LWE was provided in [ACF+15]. Subsequently, improved variants were proposed, for small secret LWE instances via "lazy modulus switching" [AFFP14], via the application of an FFT in the last step of the algorithm [DTV15], via varying the block size k [KF15] and via rephrasing the problem as the coding-theoretic problem of quantisation [GJS15]. In particular, the works [KF15, GJS15] improve the exploitation of a small secret to the point where these techniques improve the cost of solving instances where the secret is as big as the error, i.e. arbitrary

LWE instances. Yet, combinatorial algorithms do not perform well on FHE-style LWE instances because of their large dimension n to accommodate the large modulus q.

1.1 Our Contribution/Outline

We first review parameter choices in HElib and SEAL as well as known algorithms for solving LWE and related problems in Sect. 2.

Then, we reconsider the dual-lattice attack (or "dual attack" in short) which finds short vectors $\mathbf{y}$ such that $\mathbf{y} \cdot \mathbf{A} \equiv 0 \bmod q$ using lattice reduction. In particular, we recast this attack as the lattice-reduction analogue of the BKW algorithm and adapt techniques and lessons learned from BKW-style algorithms. Applying these techniques to parameter sets suggested for HElib and SEAL, we arrive at revised concrete and asymptotic security estimates.

First, in Sect. 3, we recall (the first stage of) BKW as a recursive dimension reduction algorithm for LWE instances. Each step transforms an LWE instance in dimension n to an instance in dimension $n - k$ at the cost of an increase in the noise by a factor of $\sqrt{2}$. This smaller instance is then reduced further by applying BKW again or solved using another algorithm for solving LWE; typically some form of exhaustive search once the dimension is small enough. To achieve this dimension reduction, BKW first produces elimination tables and then makes use of these tables to sample possibly many LWE samples in dimension $n-k$ relatively cheaply. We translate this approach to lattice reduction in the low advantage regime: we perform one expensive lattice reduction step followed by many relatively cheap lattice reductions on rerandomised bases. This essentially reduces the overall solving cost by a factor of m, where m is the number of samples required to distinguish a discrete Gaussian distribution with large standard deviation from uniform modulo q. We note that this approach applies to any LWE instance, i.e. does not rely on an unusually short secret and thus gives cause for a moderate revision of many LWE estimates based on the dual-attack in the low advantage regime. It does, however, rely on the heuristic that these cheap lattice reduction steps produce sufficiently short and random vectors. We give evidence that this heuristic holds.

Second, in Sect. 4, we observe that the normal form of the dual attack— finding short vectors $\mathbf{y}$ such that $\mathbf{y} \cdot \mathbf{A} \equiv \mathbf{x} \bmod q$ is short—is a natural analogue of "lazy modulus switching" [AFFP14]. Then, to exploit the unusually small secret, we apply lattice scaling as in [BG14]. The scaling factor is somewhat analogous to picking the target modulus in modulus switching resp. picking the (dimension of the) code for quantisation. This technique applies to any $\mathcal{B}$-secret LWE instance. For $\mathcal{B}_h^-$-secret instances, it reduces the cost of the dual attack by a factor of $2L/(2L+1)$ in the exponent when $\log q = \Theta(L \log n)$ for L the supported depth of FHE circuits and when h is a constant.

Third, in Sect. 5, we focus on $\mathbf{s} \leftarrow_\$ \mathcal{B}_h^\pm$ and adapt the dual attack to find short vectors which produce zero when multiplied with a *subset* of the columns of $\mathbf{A}$. This, as in BKW, produces a smaller, easier LWE instance which is then solved using another algorithm. In BKW, these smaller instances typically have very small dimension (say, 10). Here, we consider instances with dimension of several hundreds. This is enabled by exploiting the sparsity of the secret and by relaxing the conditions on the second step: we recover a solution only with a small probability of success. The basic form of this attack does not rely on the size of the non-zero components (only on the sparsity) and reduces the cost of solving an instance in dimension n to the cost of solving an instance in dimension $n/2$ multiplied by 2^h where h is the hamming weight of the secret (other trade-offs between multiplicative cost increase and dimension reduction are possible and typically optimal). We also give an improved variant when the non-zero components are also small.

In Sect. 6, we put everything together to arrive at our final algorithm SILKE, which combines the techniques outlined above; inheriting their properties. We also give revised security estimates for parameter sets suggested for HElib and SEAL in Table 1. Table 1 highlights that the techniques described in this work can, despite being relatively simple, produce significantly revised concrete security estimates for both SEAL and HElib.

Table 1. Costs of dual attacks on HElib and SEAL. Rows "$\log_2 q$" give bit sizes for the maximal modulus for a given n, for SEAL it is taken from [LN14], for HElib it is chosen such that the expected cost is 2^{80} resp. 2^{128} s according to [GHS12a]. The rows "dual" give the log cost (in operations) of the dual attack according to our lattice-reduction estimates without taking any of our improvements into account; The row "SILKE$_{\text{small}}$" gives the log cost of Algorithm 3 with "sparse" set to false; The rows "SILKE$_{\text{sparse}}$" give the log cost of Algorithm 3 with "sparse" set to true. The "sparse" flag toggles whether the approach described in Sect. 5 is enabled or not in Algorithm 3.

n	1024	2048	4096	8192	16384
SEAL 80-bit					
$\log_2 q$	47.5	95.4	192.0	392.1	799.6
dual	83.1	78.2	73.7	71.1	70.6
SILKE$_{\text{small}}$	68.1	69.0	68.2	68.4	68.8
HElib 80-bit					
$\log_2 q$	47.0	87.0	167.0	326.0	638.0
dual	85.2	85.2	85.3	84.6	85.5
SILKE$_{\text{sparse}}$	61.3	65.0	67.9	70.2	73.1
HElib 128-bit					
$\log_2 q$	38.0	70.0	134.0	261.0	511.0
dual	110.7	110.1	109.3	108.8	108.9
SILKE$_{\text{sparse}}$	73.2	77.4	81.2	84.0	86.4

Table 2. Logarithms of algorithm costs in operations mod q when applied to example parameters $n = 2048$, $q \approx 2^{63.4}$, $\alpha \approx 2^{-60.4}$ and $\mathbf{s} \leftarrow_\$ \mathcal{B}_{64}^-$. The row "base line" gives the log cost of attacks according to our lattice-reduction estimates without taking any of our improvements into account.

Strategy	Dual	Decode	Embed
HElib	188.9	—	—
Base line	124.2	116.6	114.5
Sect. 4	101.0	—	—
Sect. 5	97.1	111.0	110.9
Sect. 6	83.9	—	—

2 Preliminaries

Logarithms are base 2 if not stated otherwise. We write vectors in bold, e.g. $\mathbf{a}$, and matrices in upper-case bold, e.g. $\mathbf{A}$. By $\mathbf{a}_{(i)}$ we denote the i-th component of $\mathbf{a}$, i.e. a scalar. In contrast, $\mathbf{a}_i$ is the i-th element of a list of vectors. We write $\mathbf{I}_m$ for the $m \times m$ identity matrix over whichever base ring is implied from context. We write $\mathbf{0}_{m \times n}$ for the $m \times n$ zero matrix. A lattice is a discrete subgroup of $\mathbb{R}^n$. It can be represented by a basis $\mathbf{B}$. We write $\Lambda(\mathbf{B})$ for the lattice generated by the rows of the matrix $\mathbf{B}$, i.e. all integer-linear combinations of the rows of $\mathbf{B}$. We write $\Lambda_q(\mathbf{B})$ for the q-ary lattice generated by the rows of the matrix $\mathbf{B}$ over $\mathbb{Z}_q$, i.e. the lattice spanned by the rows $\mathbf{B}$ and multiples of q. We write $\mathbf{A}_{n:m}$ for the rows $n, \ldots, m - 1$ of $\mathbf{A}$. If the starting or end point is omitted it is assumed to be 0 or the number of rows respectively, i.e. we follow Python's slice notation.

2.1 Rolling Example

Throughout, we are going to use Example 1 below to illustrate the behaviour of the techniques described here. See Table 2 for an overview of complexity estimates for solving this set of parameters using the techniques described in this work.

Example 1. The LWE dimension is $n = 2048$, the modulus is $q \approx 2^{63.4}$, the noise parameter is $\alpha \approx 2^{-60.4}$, i.e. we have a standard deviation of $\sigma \approx 3.2$. We have $\mathbf{s} \leftarrow_\$ \mathcal{B}_{64}^-$, i.e. only $h = 64$ components of the secret are ± 1, all other components are zero. This set of parameters is inspired by parameter choices in HElib and produced by calling the function fhe_params(n=2048,L=2) of the LWE estimator from [APS15].

2.2 Parameter Choices in HElib

HElib [GHS12a, HS14] uses the cost of the dual attack for solving LWE to establish parameters. The dual strategy reduces the problem of distinguishing LWE from uniform to the SIS problem [Ajt96]:

Definition 3 (SIS). *Given $q \in \mathbb{Z}$, a matrix $\mathbf{A}$, and $t < q$; find $\mathbf{y}$ with $0 < \|\mathbf{y}\| \leq t$ and*

$$\mathbf{y} \cdot \mathbf{A} \equiv \mathbf{0} \pmod{q}.$$

Now, given samples $\mathbf{A}, \mathbf{c}$ where either $\mathbf{c} = \mathbf{A} \cdot \mathbf{s} + \mathbf{e}$ or $\mathbf{c}$ uniform, we can distinguish the two cases by finding a short $\mathbf{y}$ which solves SIS on $\mathbf{A}$ and by computing $\langle \mathbf{y}, \mathbf{c} \rangle$. On the one hand, if $\mathbf{c} = \mathbf{A} \cdot \mathbf{s} + \mathbf{e}$, then $\langle \mathbf{y}, \mathbf{c} \rangle = \langle \mathbf{y} \cdot \mathbf{A}, \mathbf{s} \rangle + \langle \mathbf{y}, \mathbf{e} \rangle \equiv \langle \mathbf{y}, \mathbf{e} \rangle \pmod{q}$. If $\mathbf{y}$ is short then $\langle \mathbf{y}, \mathbf{e} \rangle$ is also short. On the other hand, if $\mathbf{c}$ is uniformly random, so is $\langle \mathbf{y}, \mathbf{c} \rangle$.

To pick a target norm for $\mathbf{y}$, HElib picks $\|\mathbf{y}\| = q$ which allows distinguishing with good probability because q is not too far from q/σ since $\sigma \approx 3.2$ and q is typically rather large. More precisely, we may rely on the following lemma:

Lemma 1 ([LP11]). *Given an LWE instance characterised by n, α, q and a vector $\mathbf{y}$ of length $\|\mathbf{y}\|$ such that $\mathbf{y} \cdot \mathbf{A} \equiv 0 \pmod{q}$, the advantage of distinguishing $\langle \mathbf{y}, \mathbf{e} \rangle$ from random is close to*

$$\exp(-\pi(\|\mathbf{y}\| \cdot \alpha)^2).$$

To produce a short enough $\mathbf{y}$, we may call a lattice-reduction algorithm. In particular, we may call the BKZ algorithm with block size β. After performing BKZ-β reduction the first vector in the transformed lattice basis will have norm $\delta_0^m \cdot \det(\Lambda)^{1/m}$ where $\det(\Lambda)$ is the determinant of the lattice under consideration, m its dimension and the root-Hermite factor δ_0 is a constant based on the block size parameter β. Increasing the parameter β leads to a smaller δ_0 but also leads to an increase in run-time; the run-time grows at least exponential in β (see below).

In our case, the expression above simplifies to $\|\mathbf{y}\| \approx \delta_0^m \cdot q^{n/m}$ whp, where n is the LWE dimension and m is the number of samples we consider. The minimum of this expression is attained at $m = \sqrt{\frac{n \log q}{\log \delta_0}}$ [MR09].

Explicitly, we are given a matrix $\mathbf{A} \in \mathbb{Z}_q^{m \times n}$, construct a basis $\mathbf{Y}$ for its left kernel modulo q and then consider the q-ary lattice $\Lambda_q(\mathbf{Y})$ spanned by the rows of $\mathbf{Y}$. With high probability $\mathbf{Y}$ is an $(m - n) \times m$ matrix and $\Lambda_q(\mathbf{Y})$ has volume q^n. Let $\mathbf{L}$ be a basis for $\Lambda_q(\mathbf{Y})$, $m' = m - n$ and write $\mathbf{Y} = [\mathbf{I}_{m'}|\mathbf{Y}']$ then we have

$$\mathbf{L} = \begin{pmatrix} \mathbf{I}_{m'} & \mathbf{Y}' \\ 0 & q\,\mathbf{I}_n \end{pmatrix}.$$

In other words, we are attempting to find a short vector $\mathbf{y}$ in the integer row span of $\mathbf{L}$.

Given a target for the norm of $\mathbf{y}$ and hence for δ_0, HElib[2] estimates the cost of lattice reduction by relying on the following formula from [LP11]:

$$\log t_{BKZ}(\delta_0) = \frac{1.8}{\log \delta_0} - 110, \tag{1}$$

---

[2] https://github.com/shaih/HElib/blob/a5921a08e8b418f154be54f4e39a849e74489319/src/FHEContext.cpp#L22.

where $t_{BKZ}(\delta_0)$ is the time in seconds it takes to BKZ reduce a basis to achieve root-Hermite factor δ_0. This estimate is based on experiments with BKZ in the NTL library [Sho01] and extrapolation.

2.3 LP Model

The [LP11] model for estimating the cost of lattice-reduction is not correct.

Firstly, it expresses runtime in seconds instead of units of computation. As Moore's law progresses and more parallelism is introduced, the number of instructions that can be performed in a second increases. Hence, we first must translate Eq. (1) to units of computation. The experiments of Lindner and Peikert were performed on a 2.33 Ghz AMD Opteron machine, so we may assume that about $2.33 \cdot 10^9$ operations can be performed on such a machine in one second and we scale Eq. (1) accordingly.[3]

Secondly, the LP model does not fit the implementation of BKZ in NTL. The BKZ algorithm internally calls an oracle for solving the shortest vector problem in smaller dimension. The most practically relevant algorithms for realising this oracle are enumeration without preprocessing (Fincke-Pohst) which costs $2^{\Theta(\beta^2)}$ operations, enumeration with recursive preprocessing (Kannan) which costs $\beta^{\Theta(\beta)}$ and sieving which costs $2^{\Theta(\beta)}$. NTL implements enumeration without preprocessing. That is, while it was shown in [Wal15] that BKZ with recursive BKZ pre-processing achieves a run-time of $\text{poly}(n) \cdot \beta^{\Theta(\beta)}$, NTL does not implement the necessary recursive preprocessing with BKZ in smaller dimensions. Hence, it runs in time $\text{poly}(n) \cdot 2^{\Theta(\beta^2)}$ for block size β.

Thirdly, the LP model assumes a linear relation between $1/\log(\delta_0)$ and the log of the running time of BKZ, but from the "lattice rule-of-thumb" ($\delta_0 \approx \beta^{1/(2\beta)}$) and $2^{\Theta(\beta)}$ being the complexity of the best known algorithm for solving the shortest vector problem, we get:

Lemma 2 ([APS15]). *The log of the time complexity achieve a root-Hermite factor δ_0 with BKZ is*

$$\Theta\left(\frac{\log(1/\log \delta_0)}{\log \delta_0}\right)$$

if calling the SVP oracle costs $2^{\Theta(\beta)}$.

To illustrate the difference between Lemma 2 and Eq. (1), consider Regev's original parameters [Reg05] for LWE: $q \approx n^2$, $\alpha q \approx \sqrt{n}$. Then, solving LWE with the dual attack and advantage ϵ requires a log root-Hermite factor $\log \delta_0 = \log^2\left(\alpha\sqrt{\ln(1/\varepsilon)/\pi}^{-1}\right)/(4n \log q)$ [APS15]. Picking ε such that $\log \sqrt{\ln(1/\varepsilon)/\pi} \approx 1$, the log root-Hermite factor becomes $\log \delta_0 = \frac{9 \log n}{32 n}$. Plugging this result into Eq. 1, we would estimate that solving LWE for these parameters takes $\log t_{BKZ}(\delta_0) = \frac{32 n}{5 \log n} - 110$ s, which is subexponential in n.

---

[3] The number of operations on integers of size $\log q$ depends on q and is not constant. However, constant scaling provides a reasonable approximation for the number of operations for the parameter ranges we are interested in here.

2.4   Parameter Choices in SEAL 2.0

SEAL v2.0 [LP16] largely leaves parameter choices to the user. However, it provides the ChooserEvaluator::default_parameter_options() function which returns values from [LN14, Table 2].[4] This table gives a maximum $\log q$ for 80 bits of security for $n = 1024, 2048, 4096, 8192, 16384$. We reproduce these values for $\log q$ in Table 1. The default standard deviation is $\sigma = 3.19$.

The values of [LN14, Table 2] are based on enumeration costs and the simulator from [CN11, CN12]. Furthermore, to extrapolate from available enumeration costs from [CN12, LN14] assumes calling the SVP oracle in BKZ grows only exponentially with β, i.e. as $2^{0.64\beta-28}$. Note that this is overly optimistic, as [CN12] calls enumeration with recursive preprocessing to realise the SVP oracle inside BKZ, which has a complexity of $\beta^{\Theta(\beta)}$.

Finally, we note that the SEAL v2.0 manual [LP16] cautions the user against relying on the security provided by the list of default parameters.

2.5   Lattice Reduction

We will estimate the cost of lattice reduction using the following assumptions: BKZ-β produces vectors with $\delta_0 \approx \left(\frac{\beta}{2\pi e}(\pi\beta)^{\frac{1}{\beta}}\right)^{\frac{1}{2(\beta-1)}}$ [Che13]. The SVP oracle in BKZ is realised using sieving and sieving in blocksize β costs $t_\beta = 2^{0.292\,\beta+12.31}$ clock cycles. Here, $0.292\,\beta$ follows from [BDGL16], the additive constant $+12.31$ is based on experiments in [Laa15]. BKZ-β costs $c\,n \cdot t_\beta$ clock cycles in dimension n for some small constant c based on experiments in [Che13]; cf. [Che13, Figure 4.6]. This corresponds roughly to $2\,c$ tours of BKZ. We pick $c = 8$ based on our experiments with [FPL16].

This estimate is more optimistic than the estimate in [APS15], which does not yet take [BDGL16] into account and bases the number of SVP oracle calls on theoretical convergence results [HPS11] instead of experimental evidence. On the other hand, this estimate is more pessimistic than [BCD+16] which assumes *one* SVP call to be sufficient in order to protect against future algorithmic developments. While such developments, amortising costs across SVP calls during one BKZ reduction, are plausible, we avoid this assumption here in order not to "oversell" our results. However, we note that our improvements are somewhat oblivious to the underlying lattice-reduction model used. That is, while the concrete estimates for bit-security will vary depending on which estimate is employed, the techniques described here lead to improvements over the plain dual attack regardless of model. For completeness, we give estimated costs in different cost models in Appendix C.

According to the [LP11] estimate, solving Example 1 costs $2^{157.8}\,\mathrm{s}$ or $2^{188.9}$ operations using the standard dual attack. The estimates outlined in this section predict a cost of $2^{124.2}$ operations for the same standard dual attack.

---

[4] Note that the most recent version of SEAL now recommends more conservative parameters [LCP16], partly in reaction to this work.

2.6   Related Work

LWE. Besides the dual attack, via BKW or lattice-reduction, there is also the primal attack, which solves the bounded distance decoding (BDD) problem directly. That is, given $(\mathbf{A}, \mathbf{c})$ with $\mathbf{c} = \mathbf{A} \cdot \mathbf{s} + \mathbf{e}$ or $\mathbf{c} \leftarrow_\$ \mathcal{U}(\mathbb{Z}_q^m)$ find $\mathbf{s}'$ such that $|\mathbf{w} - \mathbf{c}|$ with $\mathbf{w} = \mathbf{A} \cdot \mathbf{s}'$ is minimised. For this, we may employ Kannan's embedding [AFG14] or variants of Babai's nearest planes after lattice reduction [LP11,LN13]. For Example 1 the cost of the latter approach is $2^{116.6}$ operations, i.e. about a factor 190 faster than the dual attack.

Arora & Ge proposed an asymptotically efficient algorithm for solving LWE [AG11], which was later improved in [ACFP14]. However, these algorithms involve large constants in the exponent, ruling them out for parameters typically considered in cryptography. We, hence, do not consider them further in this work.

Small-Secret LWE. As mentioned in [GHS12b], we can transform instances with an unusually short secret into instances where the secret follows the error distribution, but n samples have the old, short secret as noise [ACPS09].

Given a random $m \times n$ matrix $\mathbf{A} \bmod q$ and an m-vector $\mathbf{c} = \mathbf{A} \cdot \mathbf{s} + \mathbf{e} \bmod q$, let $\mathbf{A}_0$ denotes the first n rows of $\mathbf{A}$, $\mathbf{A}_1$ the next n rows, etc., $\mathbf{e}_0, \mathbf{e}_1, \ldots$ are the corresponding parts of the error vector and $\mathbf{c}_0, \mathbf{c}_1, \ldots$ the corresponding parts of $\mathbf{c}$. We have $\mathbf{c}_0 = \mathbf{A}_0 \cdot \mathbf{s} + \mathbf{e}_0$ or $\mathbf{A}_0^{-1} \cdot \mathbf{c}_0 = \mathbf{s} + \mathbf{A}_0^{-1} \mathbf{e}_0$. For $i > 0$ we have $\mathbf{c}_i = \mathbf{A}_i \cdot \mathbf{s} + \mathbf{e}_i$, which together with the above gives $\mathbf{A}_i \mathbf{A}_0^{-1} \mathbf{c}_0 - \mathbf{c}_i = \mathbf{A}_i \mathbf{A}_0^{-1} \mathbf{e}_0 - \mathbf{e}_i$. The output of the transformation is $\mathbf{z} = \mathbf{B} \cdot \mathbf{e}_0 + \mathbf{f}$ with $\mathbf{B} = (\mathbf{A}_0^{-1} \mid \mathbf{A}_1 \cdot \mathbf{A}_0^{-1} \mid \ldots)$ and $\mathbf{z} = (\mathbf{A}_0^{-1} \mathbf{c}_0 \mid \mathbf{A}_1 \mathbf{A}_0^{-1} \mathbf{c}_1 \mid \ldots)$ and $\mathbf{f} = (\mathbf{s} \mid \mathbf{e}_1 \mid \ldots)$. For Example 1, this reduces α from $2^{-60.4}$ to $\approx 2^{-60.8}$ and marginally improves the cost of solving.

An explicit variant of this approach is given in [BG14]. Consider the lattice

$$\Lambda = \{\mathbf{v} \in \mathbb{Z}^{n+m} \mid [\mathbf{A} \mid \mathbf{I}_m] \cdot \mathbf{v} \equiv 0 \bmod q\}.$$

It has an unusually short vector $(\mathbf{s}\|\mathbf{e})$. When $\|\mathbf{s}\| \ll \|\mathbf{e}\|$, the vector $(\mathbf{s}\|\mathbf{e})$ is uneven in length. To balance the two sides, rescale the first part to have the same norm as the second. When $\mathbf{s} \leftarrow_\$ \mathcal{B}^-$, this scales the volume of the lattice by σ^n. When $\mathbf{s} \leftarrow_\$ \mathcal{B}^+$, this scales the volume of the lattice by $(2\sigma)^n$ because we can scale by 2σ and then re-balance. When $\mathbf{s} \leftarrow_\$ \mathcal{B}_h^\pm$, the volume is scaled depending on h. For our rolling example, this approach costs $2^{114.5}$ operations, i.e. is about a factor 830 faster than the dual attack.

Independently and concurrently to this work, a new key-exchange protocol based on sparse secret LWE was proposed in [CKH+16]. A subset of the techniques discussed here are also discussed in [CKH+16], in particular, ignoring components of the secret and using lattice scaling as in [BG14].

Combinatorial. This work combines combinatorial and lattice-reduction techniques. As such, it has some similarities with the hybrid attack on NTRU [HG07]. This attack was recently adapted to LWE in the $\mathcal{B}$-secret case in [BGPW16] and its complexity revisited in [Wun16].

Rings. Recently, [ABD16] proposed a subfield lattice-attack on the two fully homomorphic encryption schemes YASHE [BLLN13] and LTV [LTV12], showing that NTRU with "overstretched" moduli q is less secure than initially expected. Quickly after, [KF16] pointed out that the presence of subfields is not necessary for attacks to succeed. NTRU can be considered as the homogeneous version of Ring-LWE, but there is currently no indication that these attacks can be translated to the Ring-LWE setting. There is currently no known algorithm which solves Ring-LWE faster than LWE for the parameter choices (ring, error distribution, etc.) typically considered in FHE schemes.

3   Amortising Costs

If the cost of distinguishing LWE from random with probability ε is c, the cost of solving is customary estimated as at least c/ε [LP11]. More precisely, applying Chernoff bounds, we require about $1/\varepsilon^2$ samples to amplify a decision experiment succeeding with advantage ε to a constant advantage. Hence, e.g. in [APS15], the dual attack is costed as the cost of running BKZ-β to achieve the target δ_0 multiplied by the number of samples required to distinguish with the target advantage, i.e. $\approx c/\varepsilon^2$.

In the case of the dual attack, this cost can be reduced by performing rerandomisation on the already reduced basis. If $\mathbf{L}$ is a basis for the lattice $\Lambda_q(\mathbf{Y})$, we first compute $\mathbf{L}'$ as the output of BKZ-β reduction where β is chosen to achieve the target δ_0 required for some given target advantage. Then, in order to produce sufficiently many relatively short vectors $\mathbf{y}_i \in \Lambda_q(\mathbf{Y})$ we repeatedly multiply $\mathbf{L}'$ by a fresh random sparse unimodular matrix with small entries to produce $\mathbf{L}'_i$. As a consequence, $\mathbf{L}'_i$ remains somewhat short. Finally, we run BKZ-β' with $\beta' \le \beta$ on $\mathbf{L}'_i$ and return the smallest non-zero vector as $\mathbf{y}_i$. See Algorithm 1, where ε_d is chosen following Lemma 1 (see below for the expectation of $\|\mathbf{y}\|$) and m is chosen following [SL12].

That is, similar to BKW, which in a first step produces elimination tables which allow sampling smaller dimensional LWE samples in $\mathcal{O}(n^2)$ operations, we first produce a relatively good basis $\mathbf{L}'$ to allow sampling $\mathbf{y}_i$ relatively efficiently.

To produce the estimates in Table 1, we assume the same rerandomisation strategy as is employed in fplll's implementation [FPL16] of extreme pruning for BKZ 2.0.[5] This rerandomisation strategy first permutes rows and then adds three existing rows together using ± 1 coefficients, which would increase norms by a factor of $\sqrt{3} < 2$ when all vectors initially have roughly the same norm. For completeness, we reproduce the algorithm in Appendix A. We then run LLL, i.e. we set $\beta' = 2$, and assume that our $\mathbf{y}_i$ have their norms increased by a factor of two, i.e. $E[\|\mathbf{y}_i\|] = 2 \cdot \delta_0^m q^{n/m}$.

Heuristic. We note that, in implementing this strategy, we are losing statistical independence. To maintain statistical independence, we would consider fresh

---

[5] https://github.com/fplll/fplll/blob/b75fe83/fplll/bkz.cpp#L43.

Data: candidate LWE samples $\mathbf{A}, \mathbf{c} \in \mathbb{Z}_q^{m \times n} \times \mathbb{Z}_q^m$
Data: BKZ block sizes $\beta, \beta' \geq 2$
Data: target success probability ε
$\varepsilon_d \leftarrow \exp(-\pi(\mathrm{E}[\|\mathbf{y}_i\|] \cdot \alpha)^2)$;
$m \leftarrow \lceil 2 \log(2 - 2\varepsilon)/\log(1 - 4\varepsilon_d^2) \rceil$;
$\mathbf{L} \leftarrow$ basis for $\{\mathbf{y} \in \mathbb{Z}^m : \mathbf{y} \cdot \mathbf{A} \equiv 0 \bmod q\}$;
$\mathbf{L}' \leftarrow$ BKZ-β reduced basis for $\mathbf{L}$;
for $i \leftarrow 0$ **to** $m - 1$ **do**
    $\mathbf{U} \leftarrow_{\$}$ a sparse unimodular matrix with small entries;
    $\mathbf{L}_i \leftarrow \mathbf{U} \cdot \mathbf{L}'$;
    $\mathbf{L}_i' \leftarrow$ BKZ-β' reduced basis for $\mathbf{L}_i$;
    $\mathbf{y}_i \leftarrow$ shortest row vector in $\mathbf{L}_i'$;
    $e_i' \leftarrow \langle \mathbf{y}_i, \mathbf{c} \rangle$;
end
if e_i' *follow discrete Gaussian distribution* **then**
    return $\top$;
else
    return $\bot$;
end

Algorithm 1. SILKE₁: Amortising costs in BKW-style SIS strategy for solving LWE

LWE samples and distinguish $\langle \mathbf{y}_i, \mathbf{e}_i \rangle$ from uniform. However, neither HElib nor SEAL provides the attacker with sufficiently many samples to run the algorithm under these conditions. Instead, we are attempting to distinguish $\langle \mathbf{y}_i, \mathbf{e} \rangle$ from uniform. Furthermore, since we are performing only light rerandomisation our distribution could be skewed if our $\mathbf{y}_i$ in $\langle \mathbf{y}_i, \mathbf{e} \rangle$ are not sufficiently random. Just as in BKW-style algorithms [ACF+15] we assume the values $\langle \mathbf{y}_i, \mathbf{e} \rangle$ are distributed closely enough to the target distribution to allow us to ignore this issue.

Experimental Verification. We tested the heuristic assumption of Algorithm 1 by rerandomising a BKZ-60 reduced basis using Algorithm 4 with $d = 3$ followed by LLL reduction several hundred times. In this experiment, we recovered fresh somewhat short vectors in each call, where somewhat short means with a norm at most twice that of the shortest vector of $\mathbf{L}'$. We give further experimental evidence in Sect. 6.

Finally, we note that this process shares some similarities with random sampling reduction (RSR) [Sch03], where random linear combinations are LLL reduced to produce short vectors. While, here, we are only performing sparse sums and accept *larger* norms, the techniques used to analyse RSR might permit reducing our heuristic to a more standard heuristic assumption.

4   Scaled Normal-Form

The line of research improving the BKW algorithm for small secrets starting with [AFFP14] proceeds from the observation that we do not need to find

$\mathbf{y} \cdot \mathbf{A} \equiv 0 \bmod q$, but if the secret is sufficiently small then any $\mathbf{y}$ such that $\mathbf{y} \cdot \mathbf{A}$ is short suffices, i.e. we seek short vectors $(\mathbf{w}, \mathbf{v})$ in the lattice

$$\Lambda = \{(\mathbf{y}, \mathbf{x}) \in \mathbb{Z}^m \times \mathbb{Z}^n : \mathbf{y} \cdot \mathbf{A} \equiv \mathbf{x} \bmod q\}.$$

Note that this lattice is the lattice considered in dual attacks on normal form LWE instances (cf. [ADPS15]).[6] Given a short vector in $(\mathbf{w}, \mathbf{v}) \in \Lambda$, we have

$$\mathbf{w} \cdot \mathbf{c} = \mathbf{w} \cdot (\mathbf{A} \cdot \mathbf{s} + \mathbf{e}) = \langle \mathbf{v}, \mathbf{s} \rangle + \langle \mathbf{w}, \mathbf{e} \rangle.$$

Here, $\mathbf{v}$ corresponds to the noise from "modulus switching" or quantisation in BKW-style algorithms and $\mathbf{w}$ to the multiplicative factor by which the LWE noise increases due to repeated subtractions.

Now, in small secret LWE instances we have $\|\mathbf{s}\| < \|\mathbf{e}\|$. As a consequence, we may permit $\|\mathbf{v}\| > \|\mathbf{w}\|$ such that

$$\| \langle \mathbf{w}, \mathbf{s} \rangle \| \approx \| \langle \mathbf{v}, \mathbf{e} \rangle \|.$$

Hence, we consider the lattice

$$\Lambda_c = \{(\mathbf{y}, \mathbf{x}/c) \in \mathbb{Z}^m \times (1/c \cdot \mathbb{Z})^n : \mathbf{y} \cdot \mathbf{A} \equiv \mathbf{x} \bmod q\}$$

for some constant c, similar to [BG14]. The lattice Λ_c has dimension $m' = m+n$ and whp volume $(q/c)^n$. To construct a basis for Λ_c, assume $\mathbf{A}_{m-n:m}$ has full rank (this holds with high probability for large q). Then $\Lambda_c = \Lambda(\mathbf{L}')$ with

$$\mathbf{L}' = \begin{pmatrix} \frac{1}{c}\mathbf{I}_n & \mathbf{0}_{n \times (m-n)} & \mathbf{A}_{m-n:m}^{-1} \\ & \mathbf{I}_{m-n} & \mathbf{B}' \\ & & q\mathbf{I}_n \end{pmatrix}$$

where $[\mathbf{I}_{m-n}|\mathbf{B}']$ is a basis for the left kernel of $\mathbf{A} \bmod q$.

Remark 2. In our estimates for HElib and SEAL, we typically have $m = n$ and $[\mathbf{I}_{m-n}|\mathbf{B}'] \in \mathbb{Z}^{0 \times n}$.

It remains to establish c. Lattice reduction produces a vector $(\mathbf{w}, \mathbf{v})$ with

$$\|(\mathbf{w}, \mathbf{v})\| \approx \delta_0^{m'} \cdot (q/c)^{n/m'}, \tag{2}$$

which translates to a noise value

$$e = \mathbf{w} \cdot \mathbf{A} \cdot \mathbf{s} + \langle \mathbf{w}, \mathbf{e} \rangle = \langle c \cdot \mathbf{v}, \mathbf{s} \rangle + \langle \mathbf{w}, \mathbf{e} \rangle$$

and we set

$$c = \frac{\alpha \, q}{\sqrt{2 \, \pi \, h}} \equiv \sqrt{m' - n}$$

to equalise the noise contributions of both parts of the above sum.

As a consequence, we arrive at the following lemma, which is attained by combining Eq. (2) with Lemma 1.

---

[6] The strategy seems folklore, we were unable to find a canonical reference for it.

Lemma 3. *Let $m' = 2n$ and $c = \frac{\alpha q}{\sqrt{2\pi h}} \cdot \sqrt{m' - n}$. A lattice reduction algorithm achieving δ_0 such that*

$$\log \delta_0 = \frac{\log \left(\frac{2n \log_2^2 \varepsilon}{\pi \alpha^2 h} \right)}{8n}$$

leads to an algorithm solving decisional LWE with $\mathbf{s} \leftarrow_\$ \mathcal{B}_h^-$ instance with advantage ε and the same cost.

Remark 3. We focus on $m' = 2n$ in Lemma 3 for ease of exposure. For the instances considered in this work, $m' = 2n$ is a good approximation for m' (see Sect. 6).

For Example 1 we predict at a cost of $2^{107.4}$ operations mod q for solving Decision-LWE when applying this strategy. Amortising costs as suggested in Sect. 3 reduces it further to $2^{101.0}$ operations mod q.

Asymptotic Behaviour. The general dual strategy, without exploiting small secrets, requires

$$\log \delta_0 = \frac{\log \left(-\frac{2 \log \varepsilon}{\alpha^2 q} \right)}{4n}$$

according to [APS15]. For HElib's choice of $8 = \alpha q$ and $h = 64$ and setting ε constant, this expression simplifies to

$$\log \delta_0 = \frac{\log q + C_d}{4n},$$

for some constant C_d. On the other hand, Lemma 3 simplifies to

$$\log \delta_0 = \frac{\log q + \frac{1}{2} \log n + C_m}{4n}, \tag{3}$$

for some constant $C_m < C_d$.

For a circuit of depth L, BGV requires $\log q = L \log n + \mathcal{O}(L)$ [GHS12b, Appendix C.2]. Applying Lemma 2, we get that

$$\lim_{\kappa \to \infty} \frac{\mathrm{cost}_m}{\mathrm{cost}_d} = \lim_{n \to \infty} \frac{\mathrm{cost}_m}{\mathrm{cost}_d} = \frac{2L}{2L+1},$$

where cost_d is the log cost of the standard dual attack, cost_m is the log cost under Lemma 3 and κ the security parameter. The same analysis applies to any constant h. Finally, when $h = 2/3\,n$, i.e. $\mathbf{s} \leftarrow_\$ \mathcal{B}^-$, then the term $1/2 \cdot \log n$ vanishes from (3), but $C_m > C_d$.

5   Sparse Secrets

Recall that BKW-style algorithms consist of two stages or, indeed, sub-algorithms. First, in the reduction stage, combinatorial methods are employed to transform an LWE instance in dimension n into an instance of dimension $0 \leq n' \leq n$, typically with increased noise level α. This smaller LWE instance is then, in the solving stage, is solved using some form of exhaustive search over the secret.

Taking the same perspective on the dual attack, write $\mathbf{A} = [\mathbf{A}_0 \mid \mathbf{A}_1]$ with $\mathbf{A}_0 \in \mathbb{Z}_q^{m \times (n-k)}$ and $\mathbf{A}_1 \in \mathbb{Z}_q^{m \times k}$ and find a short vector in the lattice

$$\Lambda = \{\mathbf{y} \in \mathbb{Z}^m : \mathbf{y} \cdot \mathbf{A}_0 \equiv 0 \bmod q\}.$$

Each short vector $\mathbf{y} \in \Lambda$ produces a sample for an LWE instance in dimension k and noise rate $\alpha' = \mathbf{E}[\|\mathbf{y}\|] \cdot \alpha$. Setting $k = 0$ recovers the original dual attack. For $k > 0$, we may now apply our favourite algorithm for solving small dimensional, easy LWE instances. Applying exhaustive search implies $\log_2 k < \kappa$ for $\mathbf{s} \leftarrow_{\$} \mathcal{B}^+$ resp. $\log_3 k < \kappa$ for $\mathbf{s} \leftarrow_{\$} \mathcal{B}^-$ when κ is the target level of security.

The case $\mathbf{s} \leftarrow_{\$} \mathcal{B}_h^{\pm}$ permits much larger k by relaxing the conditions we place on solving the k-dimensional instance. Instead of solving with probability one, we solve with some probability p_k and rerun the algorithm in case of failure.

For this, write $\mathbf{A} \cdot \mathbf{P} = [\mathbf{A}_0 \mid \mathbf{A}_1]$ and $\mathbf{s} \cdot \mathbf{P} = [\mathbf{s}_0 \mid \mathbf{s}_1]$ where $\mathbf{P}$ is a random permutation matrix. Now, over the choice of $\mathbf{P}$ there is a good chance that $\mathbf{s}_1 = 0$ and hence that $\mathbf{A}_1 \cdot \mathbf{s}_1 \equiv 0 \bmod q$. That is, the right choice of $\mathbf{P}$ places all non-zero components of $\mathbf{s}$ in the $\mathbf{s}_0$ part.

In particular, with probability $1 - h/n$ a coordinate $\mathbf{s}_{(i)}$ is zero. More generally, picking k components of $\mathbf{s}$ at random will pick only components such that $\mathbf{s}_{(i)} = 0$ with probability

$$p_k = \prod_{i=0}^{k-1} \left(1 - \frac{h}{n-i}\right) = \frac{\binom{n-h}{k}}{\binom{n}{k}} \approx \left(1 - \frac{h}{n}\right)^k.$$

Hence, simply treating $k > 0$ in the solving stage the same as $k = 0$ succeeds with probability p_k. The success probability can be amplified to close to one by repeating the elimination and solving stages $\approx 1/p_k$ times assuming we distinguish with probability close to 1.

It is clear that the same strategy translates to the primal attack by simply dropping random columns before running the algorithm. However, for the dual attack, the following improvement can be applied. Instead of considering only $\mathbf{s}_1 = 0$, perform exhaustive search over those solutions that occur with sufficiently high probability. In particular, over the choice of $\mathbf{P}$, the probability that $\mathbf{s}_1$ contains $k - j$ components with $\mathbf{s}_{1,(i)} = 0$ and exactly j components with $\mathbf{s}_{1,(i)} \neq 0$ is

$$p_{k,j} = \frac{\binom{n-h}{k-j}\binom{h}{j}}{\binom{n}{k}},$$

i.e. follows the hypergeometric distribution.

Now, assuming $\mathbf{s} \leftarrow_{\$} \mathcal{B}_h^-$, to check if any of those candidates for $\mathbf{s}_1$ is correct, we need to compare $\binom{k}{j} \cdot 2^j$ distributions against the uniform distribution mod q.

Thus, after picking a parameter ℓ we arrive at Algorithm 2 with cost:

1. m calls to BKZ-β in dimension $n - k$.
2. $m \cdot \sum_{i=0}^{\ell} \binom{k}{i} \cdot 2^i \cdot i$ additions mod q to evaluate m samples on all possible solutions up to weight ℓ.

Assuming m is chosen such that distinguishing LWE from uniform succeeds with probability close to one, then Algorithm 2 succeeds with probability $\sum_{j=0}^{\ell} p_{k,j}$.

Data: $m \times n$ matrix $\mathbf{A}$ over $\mathbb{Z}_q$
Data: m vector $\mathbf{c}$ over $\mathbb{Z}_q$
Data: density parameter $0 \le \ell \le 64$
Data: dimension parameter $0 \le k \le n$
$\mathbf{P} \leftarrow_{\$} n \times n$ permutation matrices;
$[\mathbf{A}_0 \mid \mathbf{A}_1] \leftarrow \mathbf{A} \cdot \mathbf{P}$ with $\mathbf{A}_0 \in \mathbb{Z}_q^{m \times (n-k)}$;
$\mathbf{L} \leftarrow$ basis for scaled-dual lattice of $\mathbf{A}_0$;
for $i \leftarrow 0$ **to** $m - 1$ **do**
    $\mathbf{y}_i \leftarrow$ a short vector in the row span of $\mathbf{L}$;
    $e'_i \leftarrow \langle \mathbf{y}_i, \mathbf{c} \rangle$;
end
if e'_i *follow discrete Gaussian distribution* **then**
    return $\top$;
end
foreach $\mathbf{s}'$ *in the set of* $\sum_{i=0}^{\ell} \binom{k}{i} \cdot 2^i$ *candidate solutions* **do**
    for $i \leftarrow 0$ **to** $m - 1$ **do**
        $e''_i = e'_i + \langle \mathbf{y}_i \cdot \mathbf{A}_1, \mathbf{s}' \rangle$;
    end
    if e''_i *follow discrete Gaussian distribution* **then**
        return $\top$;
    end
end
return $\bot$;

Algorithm 2. SILKE2: Sparse secrets in BKW-style SIS strategy for solving LWE.

Asymptotic Behaviour. We arrive at the following simple lemma:

Lemma 4. *Let* $0 \le h < n$ *and* $d > 1$ *be constants,* $p_{h,d}$ *be some constant depending on* h *and* d, $c_{n,\alpha,q}$ *be the cost of solving LWE with parameters* n, α, q *with probability* $\ge 1 - 2^{-p_{h,d}^2}$ *Then, solving LWE in dimension* n *with* $\mathbf{s} \leftarrow_{\$} \mathcal{B}_h^{\pm}$ *costs* $\mathcal{O}(c_{n-n/d,\alpha,q})$ *operations.*

Proof. Observe that $p_{h,d} = \lim_{n\to\infty} \binom{n-h}{n/d} / \binom{n}{n/d}$ is a constant for any constant $0 \le h < n$ and $d > 1$. Hence, solving $\mathcal{O}(1/p_{h,d}) = \mathcal{O}(1)$ instances in dimension $n - n/d$ solves the instance in dimension n. $\qquad\square$

Remark 4. Picking $d = 2$ we get $\lim_{n\to\infty} \binom{n-h}{n/2} / \binom{n}{n/2} = 2^{-h}$ and an overall costs of $\mathcal{O}(2^h \cdot c_{n/2,\alpha,q})$. This improves on exhaustive search, which costs $\mathcal{O}(2^h \cdot \binom{n}{h})$, when $c_{n/2,\alpha,q} \in o\left(\binom{n}{h}\right)$.

6 Combined

Combining the strategies described in this work, we arrive at Algorithm 3 (SILKE). It takes a flag *sparse* which enables the sparse strategy of Algorithm 2. In this case, we enforce that distinguishing LWE from uniform succeeds with probability $1 - 2^{-\kappa}$ when we guessed $\mathbf{s}'$ correctly. Clearly, this parameter can be improved, i.e. this probability reduced, but amplifying the success probability is relatively cheap, so we forego this improvement.

We give an implementation of Algorithm 3 for *sparse* = false in Appendix B. For brevity, we skip the *sparse* = true case. We also tested our implementation on several parameter sets:[7]

1. Considering an LWE instance with $n = 100$ and $q \approx 2^{23}$, $\alpha = 8/q$ and $h = 20$, we first BKZ-50 reduced the basis $\mathbf{L}$ for $c = 16$. This produced a short vector $\mathbf{w}$ such that $|\langle \mathbf{w}, \mathbf{c}\rangle| \approx 2^{15.3}$. Then, running LLL 256 times, we produced short vectors such that $\mathrm{E}[|\langle \mathbf{w}_i, \mathbf{c}\rangle|] = 2^{15.7}$ and standard deviation $2^{16.6}$.
2. Considering an LWE instance with $n = 140$ and $q \approx 2^{40}$, $\alpha = 8/q$ and $h = 32$, we first BKZ-70 reduced the basis $\mathbf{L}$ for $c = 1$. This took 64 hours and produced a short vector $\mathbf{w}$ such that $|\langle \mathbf{w}, \mathbf{c}\rangle| \approx 2^{23.7}$, with $\mathrm{E}[|\langle \mathbf{w}, \mathbf{c}\rangle|] \approx 2^{25.5}$ conditioned on $|\mathbf{w}|$. Then, running LLL 140 times (each run taking about 50 s on average), we produced short vectors such that $\mathrm{E}[|\langle \mathbf{w}_i, \mathbf{c}\rangle|] = 2^{26.0}$ and standard deviation $2^{26.4}$ for $\langle \mathbf{w}_i, \mathbf{c}\rangle$.
3. Considering the same LWE instance with $n = 140$ and $q \approx 2^{40}$, $\alpha = 8/q$ and $h = 32$, we first BKZ-70 reduced the basis $\mathbf{L}$ for $c = 16$. This took 65 hours and produced a short vector $\mathbf{w}$ such that $|\langle \mathbf{w}, \mathbf{c}\rangle| \approx 2^{24.7}$ after scaling by c, cf. $\mathrm{E}[|\langle \mathbf{w}, \mathbf{c}\rangle|] \approx 2^{24.8}$. Then, running LLL 140 times (each run taking about 50 s on average), we produced short vectors such that $\mathrm{E}[|\langle \mathbf{w}_i, \mathbf{c}\rangle|] = 2^{25.5}$ and standard deviation $2^{25.9}$ for $\langle \mathbf{w}_i, \mathbf{c}\rangle$.
4. Considering again the same LWE instance with $n = 140$ and $q \approx 2^{40}$, $\alpha = 8/q$ and $h = 32$, we first BKZ-70 reduced the basis $\mathbf{L}$ for $c = 1$. This took 30 hours and produced a short vector $\mathbf{w}$ such that $|\langle \mathbf{w}, \mathbf{c}\rangle| \approx 2^{25.2}$, cf. $\mathrm{E}[|\langle \mathbf{w}, \mathbf{c}\rangle|] \approx 2^{25.6}$. Then, running LLL 1024 times (each run taking about 50 s on average), we produced 1016 short vectors such that $\mathrm{E}[|\langle \mathbf{w}_i, \mathbf{c}\rangle|] = 2^{25.8}$ and standard deviation $2^{26.1}$ for $\langle \mathbf{w}_i, \mathbf{c}\rangle$.

---

[7] All experiments on "strombenzin" with Intel(R) Xeon(R) CPU E5-2667 v2 @ 3.30 GHz.

Data: candidate LWE samples $\mathbf{A}, \mathbf{c} \in \mathbb{Z}_q^{m \times n} \times \mathbb{Z}_q^m$
Data: BKZ block sizes $\beta, \beta' \geq 2$
Data: target success probability ε
Data: *sparse* flag toggling sparse strategy
Data: scale factor $c \geq 1$
Data: dimension parameter $0 \leq k \leq n$, 0 when *sparse* is set
Data: density parameter $0 \leq \ell \leq k$, 0 when *sparse* is set
`// distinguishing advantage per sample from` β, β'
$\varepsilon_d \leftarrow \exp(-\pi(\mathrm{E}[\|\mathbf{y}_i\|] \cdot \alpha)^2)$;
if *sparse* **then**
> $\varepsilon_t \leftarrow 1 - 1/2^\kappa$; `// for security parameter` κ
> $r \leftarrow \max\left(\lceil \log(1-\varepsilon)/\log(1-\sum_{j=0}^{\ell} p_{k,j})\rceil, 1\right)$;

else
> $\varepsilon_t, r \leftarrow \varepsilon, 1$;

end
`// required number of samples for majority vote`
$m \leftarrow \lceil 2 \log(2 - 2\varepsilon_t)/\log(1 - 4\varepsilon_d^2)\rceil$;
repeat *r* *times*
> $\mathbf{P} \leftarrow_\$ n \times n$ permutation matrices;
> $[\mathbf{A}_0 \mid \mathbf{A}_1] \leftarrow \mathbf{A} \cdot \mathbf{P}$ with $\mathbf{A}_0 \in \mathbb{Z}_q^{m \times (n-k)}$;
> $\mathbf{L} \leftarrow$ basis for $\{(\mathbf{y}, \mathbf{x}/c) \in \mathbb{Z}^m \times (1/c \cdot \mathbb{Z})^n : \mathbf{y} \cdot \mathbf{A}_0 \equiv \mathbf{x} \bmod q\}$;
> $\mathbf{L}' \leftarrow$ BKZ-β reduced basis for $\mathbf{L}$;
> **for** $i \leftarrow 0$ **to** $m - 1$ **do**
>> $\mathbf{U} \leftarrow_\$$ a sparse unimodular matrix with small entries;
>> $\mathbf{L}_i \leftarrow \mathbf{U} \cdot \mathbf{L}'$;
>> $\mathbf{L}_i' \leftarrow$ BKZ-β' reduced basis for $\mathbf{L}_i$;
>> $(\mathbf{w}_i, \mathbf{v}_i) \leftarrow$ shortest row vector in $\mathbf{L}_i'$;
>> $e_i' \leftarrow \langle \mathbf{w}_i, \mathbf{c} \rangle$;
>
> **end**
> **if** e_i' *follow discrete Gaussian distribution* **then**
>> **return** $\top$;
>
> **end**
> **foreach** $\mathbf{s}'$ *in the set of* $\sum_{i=1}^{\ell} \binom{k}{i} \cdot 2^i$ *candidate solutions* **do**
>> **for** $i \leftarrow 0$ **to** $m - 1$ **do**
>>> $e_i'' = e_i' + \langle \mathbf{w}_i \cdot \mathbf{A}_1, \mathbf{s}' \rangle$;
>>
>> **end**
>> **if** e_i'' *follow discrete Gaussian distribution* **then**
>>> **return** $\top$;
>>
>> **end**
>
> **end**

return $\bot$;

Algorithm 3. SILKE: (Sparse) BKW-style SIS Strategy for solving LWE

5. Considering an LWE instance with $n = 180$ and $q \approx 2^{40}$, $\alpha = 8/q$ and $h = 48$, we first BKZ-70 reduced the basis $\mathbf{L}$ for $c = 8$. This took 198 hours[8] and produced a short vector $\mathbf{w}$ such that $| \langle \mathbf{w}, \mathbf{c} \rangle | \approx 2^{26.7}$, cf. $\mathrm{E}[| \langle \mathbf{w}, \mathbf{c} \rangle |] \approx 2^{25.9}$. Then, running LLL 180 times (each run taking about 500 s on average), we produced short vectors such that $\mathrm{E}[| \langle \mathbf{w}_i, \mathbf{c} \rangle |] = 2^{26.6}$ and standard deviation $2^{26.9}$ for $\langle \mathbf{w}_i, \mathbf{c} \rangle$.

All our experiments match our prediction bounding the growth of the norms of our vectors by a factor of two. Note, however, that in the fourth experiment 1 in 128 vectors found with LLL was a duplicate of previously discovered vector, indicating that re-randomisation is not perfect. While the effect of this loss on the running time of the overall algorithm is small, it highlights that further research is required on the interplay of re-randomisation and lattice reduction.

Applying Algorithm 3 to parameter choices from HElib and SEAL, we arrive at the estimates in Table 1. These estimates were produced using the Sage [S+15] code available at http://bitbucket.org/malb/lwe-estimator which optimises the parameters c, ℓ, k, β to minimise the overall cost.

For the HElib parameters in Table 1 we chose the sparse strategy. Here, amortising costs as in Sect. 3 did not lead to a significant improvement, which is why we did not use it in these cases. All considered lattices have dimension $< 2\, n$. Hence, one Ring-LWE sample is sufficient to mount these attacks. Note that this is less than the dual attack as described in [GHS12a] would require (two samples).

For the SEAL parameter choices in Table 1, dimension $n = 1024$ requires two Ring-LWE samples, larger dimensions only require one sample. Here, amortising costs as in Algorithm 1 does lead to a modest improvement and is hence enabled.

Finally, we note that reducing q to $\approx 2^{34}$ resp. $\approx 2^{560}$ leads to an estimated cost of 80 bits for $n = 1024$ resp. $n = 16384$ for $\mathbf{s} \leftarrow_\$ \mathcal{B}_{64}^-$. For $\mathbf{s} \leftarrow_\$ \mathcal{B}^-$, $q \approx 2^{40}$ resp. $q \approx 2^{660}$ leads to an estimated cost of 80 bits under the techniques described here. In both cases, we assume $\sigma \approx 3.2$.

Acknowledgements. We thank Kenny Paterson and Adeline Roux-Langlois for helpful comments on an earlier draft of this work. We thank Hao Chen for reporting an error in an earlier version of this work.

---

[8] We ran 49 BKZ tours until fplll's auto abort triggered. After 16 tours the norm of the then shortest vector was by a factor 1.266 larger than the norm of the shortest vector found after 49 tours.

A   Rerandomisation

Data: $n \times m$ matrix $\mathbf{L}$
Data: density parameter d, default $d = 3$
Result: $\mathbf{U} \cdot \mathbf{L}$ where $\mathbf{U}$ is a sparse, unimodular matrix.
for $i \leftarrow 0$ **to** $4 \cdot n - 1$ **do**
$\quad \mid \quad a \leftarrow_\$ \{0, n-1\};$
$\quad \mid \quad b \leftarrow_\$ \{0, n-1\} \setminus \{a\};$
$\quad \mid \quad \mathbf{L}_{(b)}, \mathbf{L}_{(a)} \leftarrow \mathbf{L}_{(a)}, \mathbf{L}_{(b)} \; ;$
end
for $a \leftarrow 0$ **to** $n - 2$ **do**
$\quad \mid$ **for** $i \leftarrow 0$ **to** $d - 1$ **do**
$\quad \mid \quad \mid \quad b \leftarrow_\$ \{a+1, n-1\};$
$\quad \mid \quad \mid \quad s \leftarrow_\$ \{0, 1\};$
$\quad \mid \quad \mid \quad \mathbf{L}_{(a)} \leftarrow \mathbf{L}_{(a)} + (-1)^s \cdot \mathbf{L}_{(b)};$
$\quad \mid$ **end**
end
return $\mathbf{L}$;
Algorithm 4. Rerandomisation strategy in the fplll library [FPL16].

B   Implementation

```
# -*- coding: utf-8 -*-
from sage.all import shuffle, randint, ceil, next_prime, log, cputime, mean, variance,
set_random_seed, sqrt
from copy import copy
from sage.all import GF, ZZ
from sage.all import random_matrix, random_vector, vector, matrix, identity_matrix
from sage.stats.distributions.discrete_gaussian_integer import DiscreteGaussianDistributionIntegerSampler \
    as DiscreteGaussian
from estimator.estimator import preprocess_params, stddevf


def gen_fhe_instance(n, q, alpha=None, h=None, m=None, seed=None):
    """
    Generate FHE-style LWE instance

    :param n:     dimension
    :param q:     modulus
    :param alpha: noise rate (default: 8/q)
    :param h:     hamming weight of the secret (default: 2/3n)
    :param m:     number of samples (default: n)

    """
    if seed is not None:
        set_random_seed(seed)

    q = next_prime(ceil(q)-1, proof=False)
    if alpha is None:
        alpha = ZZ(8)/q

    n, alpha, q = preprocess_params(n, alpha, q)

    stddev = stddevf(alpha*q)

    if m is None:
        m = n
    K = GF(q, proof=False)
    A = random_matrix(K, m, n)

    if h is None:
        s = random_vector(ZZ, n, x=-1, y=1)
    else:
        S = [-1, 1]
        s = [S[randint(0, 1)] for i in range(h)]
        s += [0 for _ in range(n-h)]
        shuffle(s)
```

```
        s = vector(ZZ, s)
    c = A*s

    D = DiscreteGaussian(stddev)

    for i in range(m):
        c[i] += D()

    return A, c


def dual_instance0(A):
    """
    Generate dual attack basis.

    :param A: LWE matrix A

    """
    q = A.base_ring().order()
    B0 = A.left_kernel().basis_matrix().change_ring(ZZ)
    m = B0.ncols()
    n = B0.nrows()
    r = m-n
    B1 = matrix(ZZ, r, n).augment(q*identity_matrix(ZZ, r))
    B = B0.stack(B1)
    return B


def dual_instance1(A, scale=1):
    """
    Generate dual attack basis for LWE normal form.

    :param A: LWE matrix A

    """
    q = A.base_ring().order()
    n = A.ncols()
    B = A.matrix_from_rows(range(0, n)).inverse().change_ring(ZZ)
    L = identity_matrix(ZZ, n).augment(B)
    L = L.stack(matrix(ZZ, n, n).augment(q*identity_matrix(ZZ, n)))

    for i in range(0, 2*n):
        for j in range(n, 2*n):
            L[i, j] = scale*L[i, j]

    return L


def balanced_lift(e):
    """
    Lift e mod q to integer such that result is between -q/2 and q/2

    :param e: a value or vector mod q

    """
    from sage.rings.finite_rings.integer_mod import is_IntegerMod

    q = e.base_ring().order()
    if is_IntegerMod(e):
        e = ZZ(e)
        if e > q//2:
            e -= q
        return e
    else:
        return vector(balanced_lift(ee) for ee in e)


def apply_short1(y, A, c, scale=1):
    """
    Compute 'y*A', 'y*c' where y is a vector in the integer row span of
    ``dual_instance(A)``

    :param y: (short) vector in scaled dual lattice
    :param A: LWE matrix A
    :param c: LWE vector
    """
    m = A.nrows()
    y = vector(ZZ, 1/ZZ(scale) * y[-m:])
    a = balanced_lift(y*A)
    e = balanced_lift(y*c)
    return a, e


def log_mean(X):
    return log(mean([abs(x) for x in X]), 2)


def log_var(X):
    return log(variance(X).sqrt(), 2)
```

```
def silke(A, c, beta, h, m=None, scale=1, float_type="double"):
    """

    :param A:    LWE matrix
    :param c:    LWE vector
    :param beta: BKW block size
    :param m:    number of samples to consider
    :param scale: scale rhs of lattice by this factor

    """
    from fpylll import BKZ, IntegerMatrix, LLL, GSO
    from fpylll.algorithms.bkz2 import BKZReduction as BKZ2

    if m is None:
        m = A.nrows()

    L = dual_instance1(A, scale=scale)
    L = IntegerMatrix.from_matrix(L)
    L = LLL.reduction(L, flags=LLL.VERBOSE)
    M = GSO.Mat(L, float_type=float_type)
    bkz = BKZ2(M)
    t = 0.0
    param = BKZ.Param(block_size=beta,
                      strategies=BKZ.DEFAULT_STRATEGY,
                      auto_abort=True,
                      max_loops=16,
                      flags=BKZ.VERBOSE|BKZ.AUTO_ABORT|BKZ.MAX_LOOPS)
    bkz(param)
    t += bkz.stats.total_time

    H = copy(L)

    import pickle
    pickle.dump(L, open("L-%d-%d.sobj"%(L.nrows, beta), "wb"))

    E = []
    Y = set()
    V = set()
    y_i = vector(ZZ, tuple(L[0]))
    Y.add(tuple(y_i))
    E.append(apply_short1(y_i, A, c, scale=scale)[1])

    v = L[0].norm()
    v_ = v/sqrt(L.ncols)
    v_r = 3.2*sqrt(L.ncols - A.ncols())*v_/scale
    v_l = sqrt(h)*v_

    fmt = u"{\"t\": %5.1fs, \"log(sigma)\": %5.1f, \"log(|y|)\": %5.1f, \"log(E[sigma]):\"%5.1f}"

    print
    print fmt%(t,
               log(abs(E[-1]), 2),
               log(L[0].norm(), 2),
               log(sqrt(v_r**2 + v_l**2), 2))
    print
    for i in range(m):
        t = cputime()
        M = GSO.Mat(L, float_type=float_type)
        bkz = BKZ2(M)
        t = cputime()
        bkz.randomize_block(0, L.nrows, stats=None, density=3)
        LLL.reduction(L)
        y_i = vector(ZZ, tuple(L[0]))
        l_n = L[0].norm()
        if L[0].norm() > H[0].norm():
            L = copy(H)
        t = cputime(t)

        Y.add(tuple(y_i))
        V.add(y_i.norm())
        E.append(apply_short1(y_i, A, c, scale=scale)[1])
        if len(V) >= 2:
            fmt =  u"{\"i\": %4d, \"t\": %5.1fs, \"log(|e_i|)\": %5.1f, \"log(|y_i|)\": %5.1f,"
            fmt += u"\"log(sigma)\": (%5.1f,%5.1f), \"log(|y|)\": (%5.1f,%5.1f), |Y|: %5d}"
            print fmt%(i+2, t, log(abs(E[-1]), 2), log(l_n, 2), log_mean(E), log_var(E),
                       log_mean(V), log_var(V), len(Y))

    return E
```

C Alternative Cost Models

See Table 3.

Table 3. Costs of dual attacks on HElib and SEAL in the [LP11] cost model resp. assuming SVP in dimension β costs $2^{0.64\beta-28}$ operations as in [LN14] plugged into the estimator from [APS15]; cf. Table 1.

[LP11]					
n	1024	2048	4096	8192	16384
SEAL 80-bit					
q	47.5	95.4	192.0	392.1	799.6
dual	107.9	97.4	88.0	82.0	78.8
small	80.5	81.1	78.2	76.4	75.9
HElib 80-bit					
q	47.0	87.0	167.0	326.0	638.0
dual	111.5	112.4	111.5	111.2	111.2
sparse	58.1	62.6	65.4	69.2	71.5
HElib 128-bit					
q	38.0	70.0	134.0	261.0	511.0
dual	162.0	162.1	160.1	159.1	159.5
sparse	76.3	81.9	85.8	86.2	90.3
[LN14, APS15], 8 − 16 BKZ tours					
n	1024	2048	4096	8192	16384
SEAL 80-bit					
q	47.5	95.4	192.0	392.1	799.6
dual	101.2	91.7	83.1	78.3	76.1
small	74.5	76.0	74.1	73.5	73.2
HElib 80-bit					
q	47.0	87.0	167.0	326.0	638.0
dual	105.1	107.1	106.8	107.7	108.8
sparse	54.1	59.1	62.8	65.8	68.9
HElib 128-bit					
q	38.0	70.0	134.0	261.0	511.0
dual	158.4	159.8	158.6	158.3	160.0
sparse	72.0	77.4	81.4	84.3	87.1

References

[ABD16] Albrecht, M., Bai, S., Ducas, L.: A subfield lattice attack on overstretched NTRU assumptions. In: Robshaw, M., Katz, J. (eds.) CRYPTO 2016. LNCS, vol. 9814, pp. 153–178. Springer, Heidelberg (2016). doi:10.1007/978-3-662-53018-4_6

[ACF+15]   Albrecht, M.R., Cid, C., Faugère, J.-C., Fitzpatrick, R., Perret, L.: On the complexity of the BKW algorithm on LWE. Des. Codes Crypt. **74**, 325–354 (2015)

[ACFP14]   Albrecht, M.R., Cid, C., Faugère, J.-C., Perret, L.: Algebraic algorithms for LWE. Cryptology ePrint Archive, Report 2014/1018 (2014). http:// eprint.iacr.org/2014/1018

[ACPS09]   Applebaum, B., Cash, D., Peikert, C., Sahai, A.: Fast cryptographic primitives and circular-secure encryption based on hard learning problems. In: Halevi, S. (ed.) CRYPTO 2009. LNCS, vol. 5677, pp. 595–618. Springer, Heidelberg (2009). doi:10.1007/978-3-642-03356-8_35

[ADPS15]   Alkim, E., Ducas, L., Pöppelmann, T., Schwabe, P.: Post-quantum key exchange - a new hope. Cryptology ePrint Archive, Report 2015/1092 (2015). http://eprint.iacr.org/2015/1092

[ADPS16]   Alkim, E., Ducas, L., Pöppelmann, T., Schwabe, P.: Post-quantum key exchange - a new hope. In: Holz, T., Savage, S. (eds.) 25th USENIX Security Symposium, USENIX Security, vol. 16, Austin, TX, USA, 10–12 August 2016, pp. 327–343. USENIX Association (2016)

[AFFP14]   Albrecht, M.R., Faugère, J.-C., Fitzpatrick, R., Perret, L.: Lazy modulus switching for the bkw algorithm on LWE. In: Krawczyk, H. (ed.) PKC 2014. LNCS, vol. 8383, pp. 429–445. Springer, Heidelberg (2014). doi:10. 1007/978-3-642-54631-0_25

[AFG14]   Albrecht, M.R., Fitzpatrick, R., Göpfert, F.: On the efficacy of solving LWE by reduction to unique-SVP. In: Lee, H.-S., Han, D.-G. (eds.) ICISC 2013. LNCS, vol. 8565, pp. 293–310. Springer, Cham (2014). doi:10.1007/ 978-3-319-12160-4_18

[AG11]   Arora, S., Ge, R.: New algorithms for learning in presence of errors. In: Aceto, L., Henzinger, M., Sgall, J. (eds.) ICALP 2011. LNCS, vol. 6755, pp. 403–415. Springer, Heidelberg (2011). doi:10.1007/978-3-642-22006-7_34

[Ajt96]   Ajtai, M.: Generating hard instances of lattice problems (extended abstract). In: 28th ACM STOC, pp. 99–108. ACM Press, May 1996

[APS15]   Albrecht, M.R., Player, R., Scott, S.: On the concrete hardness of Learning with Errors. J. Math. Cryptology **9**(3), 169–203 (2015)

[BCD+16]   Bos, J.W., Costello, C., Ducas, L., Mironov, I., Naehrig, M., Nikolaenko, V., Raghunathan, A., Stebila, D.: Frodo: take off the ring! practical, quantum-secure key exchange from LWE. In: Weippl, E.R., Katzenbeisser, S., Kruegel, C., Myers, A.C., Halevi, S. (eds.) ACM CCS 2016, pp. 1006–1018. ACM Press, October 2016

[BCNS15]   Bos, J.W., Costello, C., Naehrig, M., Stebila, D.: Post-quantum key exchange for the TLS protocol from the ring learning with errors problem. In: 2015 IEEE Symposium on Security and Privacy, pp. 553–570. IEEE Computer Society Press, May 2015

[BDGL16]   Becker, A., Ducas, L., Gama, N., Laarhoven, T.: New directions in nearest neighbor searching with applications to lattice sieving. In: Krauthgamer, R. (ed.) 27th SODA, pp. 10–24. ACM-SIAM, January 2016

[BG14]   Bai, S., Galbraith, S.D.: Lattice decoding attacks on binary LWE. In: Susilo, W., Mu, Y. (eds.) ACISP 2014. LNCS, vol. 8544, pp. 322–337. Springer, Cham (2014). doi:10.1007/978-3-319-08344-5_21

[BGPW16]   Buchmann, J., Göpfert, F., Player, R., Wunderer, T.: On the hardness of LWE with binary error: revisiting the hybrid lattice-reduction and meet-in-the-middle attack. In: Pointcheval, D., Nitaj, A., Rachidi, T.

(eds.) AFRICACRYPT 2016. LNCS, vol. 9646, pp. 24–43. Springer, Cham (2016). doi:10.1007/978-3-319-31517-1_2

[BGV12] Brakerski, Z., Gentry, C., Vaikuntanathan, V.: (Leveled) fully homomorphic encryption without bootstrapping. In: Goldwasser, S. (ed.), ITCS 2012, pp. 309–325. ACM, January 2012

[BKW00] Blum, A., Kalai, A., Wasserman, H.: Noise-tolerant learning, the parity problem, and the statistical query model. In: 32nd ACM STOC, pp. 435–440. ACM Press, May 2000

[BLLN13] Bos, J.W., Lauter, K., Loftus, J., Naehrig, M.: Improved security for a ring-based fully homomorphic encryption scheme. In: Stam, M. (ed.) IMACC 2013. LNCS, vol. 8308, pp. 45–64. Springer, Heidelberg (2013). doi:10.1007/978-3-642-45239-0_4

[BLP+13] Brakerski, Z., Langlois, A., Peikert, C., Regev, O., Stehlé, D.: Classical hardness of learning with errors. In: Boneh, D., Roughgarden, T., Feigenbaum, J. (eds.) 45th ACM STOC, pp. 575–584. ACM Press, June 2013

[Bra12] Brakerski, Z.: Fully homomorphic encryption without modulus switching from classical GapSVP. In: In Safavi-Naini and Canetti [SNC12], pp. 868–886

[BV11] Brakerski, Z., Vaikuntanathan, V.: Efficient fully homomorphic encryption from (standard) LWE. In: Ostrovsky, R. (ed.) 52nd FOCS, pp. 97–106. IEEE Computer Society Press, October 2011

[Che13] Chen, Y.: Réduction de réseau et sécurité concrète du chiffrement complètement homomorphe. PhD thesis, Paris 7 (2013)

[CKH+16] Cheon, J.H., Han, K., Kim, J., Lee, C., Son, Y.: A practical post-quantum public-key cryptosystem based on spLWE. In: Hong, S., Park, J.H. (eds.) ICISC 2016. LNCS, vol. 10157, pp. 51–74. Springer, Cham (2017). doi:10.1007/978-3-319-53177-9_3

[CS15] Cheon, J.H., Stehlé, D.: Fully homomphic encryption over the integers revisited. In: Oswald and Fischlin [OF15], pp. 513–536

[CN11] Chen, Y., Nguyen, P.Q.: BKZ 2.0: better lattice security estimates. In: Lee, D.H., Wang, X. (eds.) ASIACRYPT 2011. LNCS, vol. 7073, pp. 1–20. Springer, Heidelberg (2011). doi:10.1007/978-3-642-25385-0_1

[CN12] Chen, Y., Nguyen, P.Q.: BKZ 2.0: Better lattice security estimates (full version) (2012). http://www.di.ens.fr/~ychen/research/Full_BKZ.pdf

[CS16] Costache, A., Smart, N.P.: Which ring based somewhat homomorphic encryption scheme is best? In: Sako, K. (ed.) CT-RSA 2016. LNCS, vol. 9610, pp. 325–340. Springer, Cham (2016). doi:10.1007/978-3-319-29485-8_19

[DXL12] Ding, J., Xie, X., Lin, X.: A simple provably secure key exchange scheme based on the learning with errors problem. Cryptology ePrint Archive, Report 2012/688 (2012). http://eprint.iacr.org/2012/688

[DTV15] Duc, A., Tramèr, F., Vaudenay, S.: Better algorithms for LWE and LWR. In: Oswald and Fischlin [OF15], pp. 173–202

[FPL16] The FPLLL development team. FPLLL 5.0,. a lattice reduction library (2016). https://github.com/fplll/fplll

[FV12] Fan, J., Vercauteren, F.: Somewhat practical fully homomorphic encryption. Cryptology ePrint Archive, Report 2012/144 (2012). http://eprint.iacr.org/2012/144

[GHS12a] Gentry, C., Halevi, S., Smart, N.P.: Homomorphic evaluation of the AES Circuit. In: Safavi-Naini and Canetti [SNC12], pages 850–867

[GHS12b] Gentry, C., Halevi, S., Smart, N.P.: Homomorphic evaluation of the AES circuit. Cryptology ePrint Archive, Report 2012/099 (2012). http://eprint.iacr.org/2012/099

[GJS15] Guo, Q., Johansson, T., Stankovski, P.: Coded-BKW: solving LWE using lattice codes. In: Gennaro and Robshaw [GR15], pp. 23–42

[GR15] Gennaro, R., Robshaw, M. (eds.): CRYPTO 2015. LNCS, vol. 9215. Springer, Heidelberg (2015)

[GSW13] Gentry, C., Sahai, A., Waters, B.: Homomorphic encryption from learning with errors: conceptually-simpler, asymptotically-faster, attribute-based. In: Canetti, R., Garay, J.A. (eds.) CRYPTO 2013. LNCS, vol. 8042, pp. 75–92. Springer, Heidelberg (2013). doi:10.1007/978-3-642-40041-4_5

[HG07] Howgrave-Graham, N.: A hybrid lattice-reduction and meet-in-the-middle attack against NTRU. In: Menezes, A. (ed.) CRYPTO 2007. LNCS, vol. 4622, pp. 150–169. Springer, Heidelberg (2007). doi:10.1007/978-3-540-74143-5_9

[HPS11] Hanrot, G., Pujol, X., Stehlé, D.: Analyzing blockwise lattice algorithms using dynamical systems. In: Rogaway, P. (ed.) CRYPTO 2011. LNCS, vol. 6841, pp. 447–464. Springer, Heidelberg (2011). doi:10.1007/978-3-642-22792-9_25

[HS14] Halevi, S., Shoup, V.: Algorithms in HElib. In: Garay, J.A., Gennaro, R. (eds.) CRYPTO 2014. LNCS, vol. 8616, pp. 554–571. Springer, Heidelberg (2014). doi:10.1007/978-3-662-44371-2_31

[KF15] Kirchner, P., Fouque, P.-A.: An improved BKW algorithm for LWE with applications to cryptography and lattices. In: Gennaro, R., Robshaw, M. (eds.) CRYPTO 2015. LNCS, vol. 9215, pp. 43–62. Springer, Heidelberg (2015). doi:10.1007/978-3-662-47989-6_3

[KF16] Kirchner, P., Fouque, P.-A.: Comparison between subfield and straightforward attacks on NTRU. IACR Cryptology ePrint Archive, 2016: 717 (2016)

[KL15] Kim, M., Lauter, K.: Private genome analysis through homomorphic encryption. BMC Med. Inform. Decis. Mak. $15(5)$, 1–12 (2015)

[Laa15] Laarhoven, T.: Sieving for shortest vectors in lattices using angular locality-sensitive hashing. In: Gennaro, R., Robshaw, M. (eds.) CRYPTO 2015. LNCS, vol. 9215, pp. 3–22. Springer, Heidelberg (2015). doi:10.1007/978-3-662-47989-6_1

[LCP16] Laine, K., Chen, H., Player, R.: Simple Encrypted Arithmetic Library - SEAL (v2.1). Technical report, Microsoft Research, MSR-TR-2016-68, September 2016

[LN13] Liu, M., Nguyen, P.Q.: Solving BDD by enumeration: an update. In: Dawson, E. (ed.) CT-RSA 2013. LNCS, vol. 7779, pp. 293–309. Springer, Heidelberg (2013). doi:10.1007/978-3-642-36095-4_19

[LN14] Lepoint, T., Naehrig, M.: A comparison of the homomorphic encryption schemes FV and YASHE. In: Pointcheval, D., Vergnaud, D. (eds.) AFRICACRYPT 2014. LNCS, vol. 8469, pp. 318–335. Springer, Cham (2014). doi:10.1007/978-3-319-06734-6_20

[LP11] Lindner, R., Peikert, C.: Better key sizes (and attacks) for lwe-based encryption. In: Kiayias, A. (ed.) CT-RSA 2011. LNCS, vol. 6558, pp. 319–339. Springer, Heidelberg (2011). doi:10.1007/978-3-642-19074-2_21

[LP16] Laine, K., Player, R.: Simple Encrypted Arithmetic Library - SEAL (v2.0). Technical report, Microsoft Research, MSR-TR-2016-52, September 2016

[LPR10] Lyubashevsky, V., Peikert, C., Regev, O.: On ideal lattices and learning with errors over rings. In: Gilbert, H. (ed.) EUROCRYPT 2010. LNCS, vol. 6110, pp. 1–23. Springer, Heidelberg (2010). doi:10.1007/978-3-642-13190-5_1

[LPR13] Lyubashevsky, V., Peikert, C., Regev, O.: A toolkit for ring-LWE cryptography. Cryptology ePrint Archive, Report 2013/293 (2013). http://eprint.iacr.org/2013/293

[LTV12] López-Alt, A., Tromer, E., Vaikuntanathan, V.: On-the-fly multiparty computation on the cloud via multikey fully homomorphic encryption. In: Karloff, H.J., Pitassi, T. (eds.) 44th ACM STOC, pp. 1219–1234. ACM Press, May 2012

[MR09] Micciancio, D., Regev, O.: Lattice-based cryptography. In: Bernstein, D.J., Buchmann, J., Dahmen, E. (eds.) Post-Quantum Cryptography, Heidelberg, New York, pp. 147–191 (2009)

[OF15] Oswald, E., Fischlin, M. (eds.): EUROCRYPT 2015. LNCS, vol. 9056. Springer, Heidelberg (2015)

[Pei09] Peikert, C.: Some recent progress in lattice-based cryptography. In: Reingold, O. (ed.) TCC 2009. LNCS, vol. 5444, pp. 72–72. Springer, Heidelberg (2009). doi:10.1007/978-3-642-00457-5_5

[Reg05] Regev, O.: On lattices, learning with errors, random linear codes, and cryptography. In: Gabow, H.N., Fagin, R. (eds.) 37th ACM STOC, pp. 84–93. ACM Press, May 2005

[Reg09] Regev, O.: On lattices, learning with errors, random linear codes, and cryptography. J. ACM **56**(6), 1–40 (2009)

[S+15] Stein, W., et al.: Sage Mathematics Software Version 7.1. The Sage Development Team (2015). http://www.sagemath.org

[Sch03] Schnorr, C.P.: Lattice reduction by random sampling and birthday methods. In: Alt, H., Habib, M. (eds.) STACS 2003. LNCS, vol. 2607, pp. 145–156. Springer, Heidelberg (2003). doi:10.1007/3-540-36494-3_14

[Sho01] Shoup, V.: NTL: A library for doing number theory (2001). http://www.shoup.net/ntl/

[SL12] Sarma, J., Lunawat, P.: IITM-CS6840: Advanced Complexity Theory – Lecture 11: Amplification Lemma (2012). http://www.cse.iitm.ac.in/~jayalal/teaching/CS6840/2012/lecture11.pdf

[SNC12] Safavi-Naini, R., Canetti, R. (eds.): CRYPTO 2012. LNCS, vol. 7417. Springer, Heidelberg (2012)

[Wal15] Walter, M.: Lattice point enumeration on block reduced bases. In: Lehmann, A., Wolf, S. (eds.) ICITS 2015. LNCS, vol. 9063, pp. 269–282. Springer, Cham (2015). doi:10.1007/978-3-319-17470-9_16

[Wun16] Wunderer, T.: Revisiting the hybrid attack: Improved analysis and refined security estimates. Cryptology ePrint Archive, Report 2016/733 (2016). http://eprint.iacr.org/2016/733

Small CRT-Exponent RSA Revisited

Atsushi Takayasu[1,2]([✉]), Yao Lu[1], and Liqiang Peng[3]

[1] The University of Tokyo, Tokyo, Japan
a-takayasu@it.k.u-tokyo.ac.jp
[2] National Institute of Advanced Industrial Science and Technology (AIST),
Tokyo, Japan
[3] State Key Laboratory of Information Security,
Institute of Information Engineering, Chinese Academy of Sciences, Beijing, China

Abstract. Since May (Crypto'02) revealed the vulnerability of the small CRT-exponent RSA using Coppersmith's lattice-based method, several papers have studied the problem and two major improvements have been made. Bleichenbacher and May (PKC'06) proposed an attack for small d_q when the prime factor p is significantly smaller than the other prime factor q; the attack works for $p < N^{0.468}$. Jochemsz and May (Crypto'07) proposed an attack for small d_p and d_q where the prime factors p and q are balanced; the attack works for $d_p, d_q < N^{0.073}$. Even after a decade has passed since their proposals, the above two attacks are still considered to be the state-of-the-art, and no improvements have been made thus far. A novel technique seems to be required for further improvements since the attacks have been studied with all the applicable techniques for Coppersmith's methods proposed by Durfee-Nguyen (Asiacrypt'00), Jochemsz-May (Asiacrypt'06), and Herrmann-May (Asiacrypt'09, PKC'10). In this paper, we propose two improved attacks on the small CRT-exponent RSA: a small d_q attack for $p < N^{0.5}$ (an improvement of Bleichenbacher-May's) and a small d_p and d_q attack for $d_p, d_q < N^{0.091}$ (an improvement of Jochemsz-May's). We use Coppersmith's lattice-based method to solve modular equations and obtain the improvements from a novel lattice construction by exploiting useful algebraic structures of the CRT-RSA key generation. We explicitly show proofs of our attacks and verify the validities by computer experiments. In addition to the two main attacks, we propose small d_q attacks on several variants of RSA.

Keywords: CRT-RSA · Cryptanalysis · Coppersmith's method · Lattices · LLL algorithm

1 Introduction

1.1 Background

Let $N = pq$ be a public RSA modulus whose prime factors p and q are usually the same bit-size. A public exponent e and a secret exponent d satisfy

© International Association for Cryptologic Research 2017
J.-S. Coron and J.B. Nielsen (Eds.): EUROCRYPT 2017, Part II, LNCS 10211, pp. 130–159, 2017.
DOI: 10.1007/978-3-319-56614-6_5

$ed = 1 \mod (p-1)(q-1)$. For encryption/verifying (resp. decryption/signing), the heavy modular exponentiation of e (resp. d) has to be computed. To achieve faster computation, a simple solution is to use a small public or secret exponent. However, Wiener [49] showed that a public RSA modulus is factorized in polynomial time when the secret exponent is too small such that $d < N^{0.25}$. Boneh and Durfee [4] revisited the problem with Coppersmith's lattice-based method [7,17] and improved the bound to $d < N^{0.284}$. Furthermore, in the same work, the bound was improved to $d < N^{0.292}$ by exploiting sublattice structures from the previous one although the proof is involved.

To simultaneously thwart the small secret exponent attack and achieve faster decryption/signing, the Chinese Remainder Theorem (CRT) is often used as described by Quisquater and Couvreur [34]. Instead of the original secret exponent d, there are CRT-exponents d_p and d_q that satisfy

$$ed_p = 1 \mod (p-1) \quad \text{and} \quad ed_q = 1 \mod (q-1).$$

Then a natural question to ask is whether there exist analogous attacks of the Boneh-Durfee [4] to the small CRT-exponents. The first answer was given by May (Crypto'02) [28]. May analyzed the unbalanced RSA whose prime factor p is significantly smaller than the other prime factor q, and proposed an attack for a small d_q with an arbitrary large d_p. The paper contains two attacks where the former attack works for $p < N^{0.382}$. The latter attack works only for smaller p, however, is better than the former attack for $p < N^{0.23}$ in the sense that a larger d_q can be recovered. Since May's attack works only in the unbalanced setting, it is an interesting open question if the attacks can be improved to cover the balanced RSA.

Subsequently, several improved attacks on the small CRT-exponent RSA have been proposed. Bleichenbacher and May (PKC'06) [2] revisited May's work [28] in the same attack scenario and proposed an improved attack. The attack works for a larger p such that $p < N^{0.468}$, and recovers a larger d_q than May's attack for any size of p. However, the balanced prime factors still could not be captured. To capture the balanced RSA, Bleichenbacher and May analyzed other attack scenarios where both d_p and d_q are small in the same work. They proposed an attack which works for $e < N$. Although the same situation was already studied by Galbraith et al. [13], Sun and Wu [39], their attacks only work for a smaller e. Jochemsz and May (Crypto'07) [21] proposed the first attack that works for a full size e when $d_p, d_q < N^{0.073}$.

In the past decade, no improved attacks of Bleichenbacher-May [2] and Jochemsz-May [21] have been proposed. Hence, following these attacks seems to be the best way to study the security of the CRT-RSA. Indeed, until recently, several papers followed the attacks and reported the vulnerabilities of the CRT-RSA, e.g., an attack on Takagi's RSA [38], an attack on the RSA with multiple exponent pairs [33], and partial key exposure attacks [3,26,37,44,46].

1.2   Technical Hardness

Coppersmith introduced two lattice-based methods; to solve a modular equation [7] and an integer equation [6]. May's attack and Bleichenbacher-May's attack used the former method whereas Jochemsz-May's attack used the latter method. Both methods first construct a lattice and then solve equations with a small root in polynomial time. In this research area, constructing better attacks is equivalent to designing better lattices that reflect the more useful algebraic structure of the equation. For the purpose, several useful strategies and techniques for lattice constructions have been introduced thus far. Currently best known small CRT-exponent attacks [2,21,28] are based on the state-of-the-art lattice constructions; the Durfee-Nguyen technique (Asiacrypt'00) [11] and the Jochemsz-May strategy (Asiacrypt'06) [20]. Since the Durfee-Nguyen technique is useful to handle the relation $N = pq$ and the Jochemsz-May construction yields good lattices for arbitrary polynomials, these approaches [2,28] seem appropriate to study the attack. Moreover, to the best of our knowledge, there remained no useful strategies to analyze the attack scenarios at that time. After the proposals of [2,21,28], a new technique called unravelled linearization was introduced by Herrmann and May (Asiacrypt'09) [15]. The technique has been used to study various attack scenarios on RSA, e.g., [1,14,16,18,22,23,41–43,45,47,48], and drastically developed the research area. For example, Herrmann and May [16] showed an elementary proof of Boneh-Durfee's attack [4] to exploit the sublattice structures. However, unfortunately, unravelled linearization could not improve small CRT-exponent attacks. Although Herrmann and May (PKC'10) [16] tried to exploit sublattice structures, they could not obtain better asymptotic bounds. Therefore, to obtain better bounds, a novel technique seems to be developed.

1.3   Our Results

In this paper, we develop a novel lattice construction technique for Coppersmith's modular method where the technique enables us to exploit more useful algebraic structures of the CRT-RSA key generation. A basic application of the technique is an improved small d_q attack for unbalanced prime factors (Sect. 3). As opposed to the previous results by May [28] and Bleichenbacher-May [2], our attack is the first result to reach a meaningful bound, i.e., $p < N^{0.5}$. Hence, we solve one of the major open problems for the security of the small CRT-exponent RSA. Moreover, our attack can recover a larger d_q than [2,28] for any size of p. In addition, our attack requires less lattice dimensions than Bleichenbacher-May's attack [2] since our technique exploits sublattice structures from [2]'s lattice where the approach is similar to Boneh-Durfee [4]. Indeed, our experiments show that Bleichenbacher-May's attack works better than their theoretical analyses.

We claim that our technique is not limited to the small d_q attack. The technique is also applicable to a small d_p and d_q attack (Sect. 4) that improves Jochemsz-May's attack [21]. As we mentioned, small d_q attacks [2,28] and small d_p and d_q attacks [21] were studied with different approaches in previous works; the former attack used Coppersmith's modular method whereas the latter attack used Coppersmith's integer method. However, our powerful technique enables us to improve these attacks in the same manner. Our attack[1] works for $d_p, d_q < N^{0.091}$ with a full size e where the exponent of N is about 25% larger than Jochemsz-May's attack.

Recently, numerous papers [12,19,25,27,32,33,35,36,38,41,45,47] have been studying the security of RSA variants. We further show that we can extend our small d_q attack to the RSA variants (Sect. 5), i.e., the Multi-Prime RSA, Takagi's RSA, and the RSA with multiple exponent pairs. Our attacks significantly improve previous attacks on these variants [33,38].

1.4   Key Technique

We show an overview of our technique. The CRT-RSA key generation for d_q is written as

$$ed_q = 1 + k(q - 1) \tag{1}$$

with some integer k. By multiplying the equation by p, we obtain

$$ed_q p = p + k(N - p) = N + (k - 1)(N - p). \tag{2}$$

Recall in May's and Bleichenbacher-May's attack scenario [2,28], the prime p is significantly smaller than the other prime q. They solved the latter Eq. (2) modulo e to recover unknown $(k-1, p)$. Since the prime p is significantly smaller than the other prime q, to construct better attacks, solving the Eq. (2) is more promising approach than solving the Eq. (1) to recover (k, q). Hence, only the Eq. (2) was used in previous attacks. However, it means that the constructions of previous attacks significantly rely on the fact that p is much smaller than q. As a result, these attacks do not work when p is close to $N^{0.5}$.

What we focus on is a fact that the Eqs. (1) and (2) are essentially the same; there are two representations for the same CRT-RSA key generation. As opposed to previous works, our improved lattice constructions utilize the algebraic structure of both Eqs. (1) and (2) simultaneously not only the Eq. (2). The two representations are compatible in the sense that the combination enables us to exploit more useful algebraic structures. More specifically, we use the Eqs. (1) and (2) where the proportion can be adaptively determined by the sizes of p and q. Then, to solve the modulo e equation as previous works, our framework always yields the better lattices than previous approaches. Our attacks are better than Bleichenbacher-May's attack for any size of p.

At a glance, our lattice construction technique is specialized to the improvement of Bleichenbacher-May's attack. As we pointed out, May's attack and Bleichenbacher-May's attack used Coppersmith's method to solve a modular equation [7,17] whereas Jochemsz-May's attack used the method to solve an integer equation [6,10]. The modular equation for the former attack and the integer equation for the latter attack have completely different algebraic structures.

---

[1] In the full version, we further improve the bound to $d_p, d_q < N^{0.122}$.

However, surprisingly, our powerful technique enables us to construct better lattices and improves Jochemsz-May's attack, too. It suggests that our proposed technique is quite useful to study the security of CRT-RSA over a wide range.

2   Preliminaries

Consider a modular equation $h(x_1, \ldots, x_r) = 0 \pmod{W}$, where all the absolute values of the target solutions $(\tilde{x}_1, \ldots, \tilde{x}_r)$ are bounded above by $X_1, \ldots, X_r$. When $\prod_{j=1}^{r} X_j$ is reasonably smaller than W, Coppersmith's method can find all the solutions in polynomial time. In this section, we recall a simplified reformulation of the method due to Howgrave-Graham [17] and its basis tools, i.e., Howgrave-Graham's lemma and the LLL algorithm.

Let $\|h(x_1, \ldots, x_r)\|$ denote a norm of a polynomial which represents the Euclidean norm of the coefficient vector. The following Howgrave-Graham's lemma reduces the modular equations into integer equations.

Lemma 1 (Howgrave-Graham's Lemma [17]). *Let* $\tilde{h}(x_1, \ldots, x_r) \in \mathbb{Z}[x_1, \ldots, x_r]$ *be a polynomial with at most* n *monomials. Let* $m, W, X_1, \ldots, X_r$ *be positive integers. Suppose that:*

1. $\tilde{h}(\tilde{x}_1, \ldots, \tilde{x}_r) = 0 \pmod{W^m}$, *where* $|\tilde{x}_1| < X_1, \ldots, |\tilde{x}_r| < X_r$,
2. $\|\tilde{h}(x_1 X_1, \ldots, x_r X_r)\| < W^m / \sqrt{n}$.

Then $\tilde{h}(\tilde{x}_1, \ldots, \tilde{x}_r) = 0$ *holds over the integers.*

To solve r-variate modular equations $h(x_1, \ldots, x_r) = 0 \pmod{W}$, it suffices to find r new polynomials $\tilde{h}_1(x_1, \ldots, x_r), \ldots, \tilde{h}_r(x_1, \ldots, x_r)$ whose root is the same as the original one, i.e., $(x_1, \ldots, x_r) = (\tilde{x}_1, \ldots, \tilde{x}_r)$, and whose norms are small enough to satisfy Howgrave-Graham's lemma.

To find such small norm polynomials from the original modular polynomial $h(x_1, \ldots, x_r)$, lattices and the LLL algorithm are used. An n-dimensional lattice is an additive discrete subgroup of $\mathbb{Z}^n$. In other words, a lattice represents all integer linear combinations of its basis vectors. All vectors are row representation throughout the paper. Let $b_1, \ldots, b_m$ be n-dimensional linearly independent vectors in $\mathbb{Z}^n$. A lattice spanned by these vectors as a basis is defined as $L(b_1, \ldots, b_m) := \{\sum_{j=1}^{m} c_j b_j : c_j \in \mathbb{Z} \text{ for all } j = 1, 2, \ldots, n\}$. We also use a matrix representation for the basis. We define a basis matrix B as $m \times n$ matrix which has the basis vectors $b_1, \ldots, b_m$ in each row. A lattice spanned by a basis matrix B is denoted as $L(B)$. We call a lattice full-rank if and only if $n = m$. A determinant of a lattice $\det(L(B))$ is defined as the m-dimensional volume of the fundamental parallelepiped; $\mathcal{P}(B) := \{cB : c \in \mathbb{R}^m, 0 \le c_j < 1, \text{for all } j = 1, 2, \ldots, m\}$. The determinant can be computed as $\det(L(B)) = \sqrt{\det(BB^T)}$ in general and that of a full-rank lattice can be computed as $\det(L(B)) = |\det(B)|$. In this paper, we only use a full-rank lattice. More specifically, we only use a lattice with a triangular basis matrix. Hence, the determinant of the lattice can be computed easily as the absolute value of a product of all diagonals.

Lattice has been used in various ways in cryptographic research. See [8, 9, 29–31] for more information. In cryptanalysis, finding non-zero short lattice vectors is usually an essential operation. In this paper, we recall the LLL algorithm [24] that outputs short lattice vectors in polynomial time.

Proposition 1 (LLL algorithm [24, 29]). *Given linearly independent vectors* $b_1, \ldots, b_n$ *in* $\mathbb{Z}^n$, *the LLL algorithm finds new basis vectors* $\tilde{b}_1, \ldots, \tilde{b}_n$ *for a lattice* $L(b_1, \ldots, b_n)$ *that satisfy*

$$\|\tilde{b}_j\| \leq 2^{n(n-1)/4(n-j+1)} \det(L(\boldsymbol{B}))^{1/(n-j+1)} \quad \text{for } 1 \leq j \leq n,$$

in time polynomial in n *and the maximum input length of* $b_1, \ldots, b_n$.

Again, we explain how to solve the modular equation $h(x_1, \ldots, x_r) = 0$ (mod W). At first, we construct n polynomials $h_1(x_1, \ldots, x_r), \ldots, h_n(x_1, \ldots, x_r)$ that have the root $(\tilde{x}_1, \ldots, \tilde{x}_r)$ modulo W^m with some positive integer m. Then we construct n basis vectors $b_1, \ldots, b_n$ and equivalently its matrix representation $\boldsymbol{B}$. Each elements of a vector b_j for $j = 1, 2, \ldots, n$ consist of coefficients of $h_j(x_1 X_1, \ldots, x_r X_r)$. Since all vectors in a lattice $L(\boldsymbol{B})$ are integer linear combinations of the basis vectors, all polynomials whose coefficients are derived from lattice vectors have the root $(\tilde{x}_1, \ldots, \tilde{x}_r)$ modulo W^m. We apply the LLL algorithm to a lattice basis $\boldsymbol{B}$ and obtain r LLL-reduced vectors $\tilde{b}_1, \ldots, \tilde{b}_r$. Then new polynomials $\tilde{h}_1(x_1, \ldots, x_r), \ldots, \tilde{h}_r(x_1, \ldots, x_r)$ which are derived from the above r LLL-reduced vectors satisfy Howgrave-Graham's lemma provided that $\det(L(\boldsymbol{B}))^{1/n} < W^m$. Here, we omit small terms. When we obtain r polynomials $\tilde{h}_1(x_1, \ldots, x_r), \ldots, \tilde{h}_r(x_1, \ldots, x_r)$, the root $(\tilde{x}_1, \ldots, \tilde{x}_r)$ can easily be recovered by computing resultant or Gröbner bases for the polynomials.

We should note that the method needs heuristic argument for multivariate problems. The polynomials $\tilde{h}_1(x_1, \ldots, x_r), \ldots, \tilde{h}_n(x_1, \ldots, x_r)$ derived from LLL output vectors have no assurance of algebraic independency. In this paper, we assume that the polynomials are algebraic independent as previous works [2, 21, 28] since there exist few negative reports. Moreover, we justify the validity of our attacks by computer experiments.

3   Small d_q Attack

In this section, we propose an attack for small d_q when p is significantly smaller than q. The attack improves Bleichnbacher-May's attack [2].

3.1   An Overview of the Lattice Construction

At first, we explain our strategy for lattice constructions. Since our lattice construction is highly technical, we show toy examples that compare previous lattices [2, 28] and ours. We hope that these examples help readers to understand our technique easily.

Recall the CRT-RSA key generation;

$$ed_q = 1 + k(q-1)$$

with some integer k. If we can solve the following modular equation:

$$f_q(x_q, y_q) = 1 + x_q(y_q - 1) = 0 \pmod e$$

whose root is $(x_q, y_q) = (k, q)$, a public modulus N can be factorized. However, since the prime factor q is significantly larger than the other prime factor p, i.e., $p = N^\beta$ and $q = N^{1-\beta}$ for $\beta \le 1/2$, May [28] multiplied the above equation by p and obtain the following equation:

$$ed_q p = p + k(N-p) = N + (k-1)(N-p).$$

Hence, if the following modular equation can be solved, the public modulus N can be factorized:

$$f_p(x_p, y_p) = N + x_p(N - y_p) = 0 \pmod e$$

whose root is $(x_p, y_p) = (k-1, p)$. Let $e = N^\alpha$ and $d_q = N^\delta$. Then the absolute values of the root (x_p, x_q, y_p, y_q) is bounded above by $X_p := N^{\alpha+\beta+\delta-1}, X_q := N^{\alpha+\beta+\delta-1}, Y_p := N^\beta, Y_q := N^{1-\beta}$ respectively within constant factors. Later we also use a notation $X := X_p = X_q$. In this setting, the other CRT-exponent d_p can be arbitrary large such that $d_p \approx N^\beta$.

May's Matrix. May [28] solved the modular equation $f_p(x_p, y_p) = 0$ under the standard lattice construction which can be captured by Jochemsz-May's strategy [21]. For example, although we omit the detail, he constructed the basis matrix as the following:

$$\begin{pmatrix} e \\ 0 & eX_p \\ N & NX_p & -X_pY_p \\ 0 & 0 & 0 & eY_p \\ 0 & 0 & NX_pY_p & NY_p & -X_pY_p^2 \\ 0 & 0 & 0 & 0 & 0 & eY_p^2 \\ 0 & 0 & 0 & 0 & NX_pY_p^2 & NY_p^2 & -X_pY_p^3 \end{pmatrix}$$

where the rows consist of coefficients of seven polynomials: $e, ex_p, f_p(x_p, y_p), ey_p, y_pf_p(x_p, y_p), ey_p^2, y_p^2f_p(x_p, y_p)$. All the polynomials share the common root as $f_p(x_p, y_p)$ modulo e. In addition to the base polynomials, i.e., $e, ex_p, f_p(x_p, y_p)$, he added extra y_p-shifts, i.e., $ey_p, y_pf_p(x_p, y_p), ey_p^2, y_p^2f_p(x_p, y_p)$. Applying the LLL reduction to the above matrix, polynomials derived from the LLL output vectors satisfy Howgrave-Graham's lemma when

$$X_p^4 Y_p^9 e^4 < e^7 \Leftrightarrow 4(\alpha + \beta + \delta - 1) + 9\beta < 3\alpha$$

$$\Leftrightarrow \delta < 1 - \frac{\alpha + 13\beta}{4}.$$

The core idea of the approach is solving the Eq. (2) not (1) since p is significantly smaller than q. Hence, if p becomes close to q such that $\beta \geq 0.382$, May's attack does not work.

Bleichenbacher-May Matrix. To improve May's attack [28] based on the above matrix, Bleichenbacher and May [2] made use of the relation $y_p y_q = N$ as Durfee and Nguyen [11]. Although the exact solution of y_p is unknown, the relation enables us to reduce powers of Y_p in the diagonals by multiplying powers of y_q to all the polynomials. By optimizing the powers of y_q, Bleichenbacher-May's matrix always offers better results than May's matrix.

To explain our improvement later, we modify Bleichenbacher-May's matrix where the modified matrix offer the same bound as the original Bleichenbacher-May matrix. The modification helps readers to understand the spirit of our improvement. Previous May's matrix used only extra y_p-shifts, however, modified Bleichenbacher-May's matrix used both y_p-shifts and y_q-shifts. Hence, we omit $ey_p^2, y_p^2 f_p(x_p, y_p)$ from the above matrix and add $ey_q, N^{-1} \cdot y_q f_p(x_p, y_p)$ in turn where the new polynomials share the common root as $f_p(x_p, y_p)$ modulo e:

$$\begin{pmatrix} e & & & & & & \\ 0 & eX_p & & & & & \\ N & NX_p & -X_pY_p & & & & \\ 0 & 0 & 0 & eY_p & & & \\ 0 & 0 & NX_pY_p & NY_p & -X_pY_p^2 & & \\ 0 & 0 & 0 & 0 & 0 & eY_q & \\ 0 & -X_p & 0 & 0 & 0 & Y_q & X_pY_q \end{pmatrix}.$$

Although the precise definition of the polynomial selection is slightly different from the one in the original paper, they are essentially the same in the sense that the above matrix yields the same bound as the original Bleichenbacher-May attack. Applying the LLL reduction to the above matrix, polynomials derived from the LLL output vectors satisfy Howgrave-Graham's lemma when

$$X_p^4 Y_p^4 Y_q^2 e^4 < e^7 \Leftrightarrow 4(\alpha + \beta + \delta - 1) + 4\beta + 2(1 - \beta) < 3\alpha$$

$$\Leftrightarrow \delta < \frac{1}{2} - \frac{\alpha + 6\beta}{4}.$$

Compared with May's matrix, the matrix reduces the powers of Y_p by multiplying the powers of Y_q. It means that Bleichenbacher-May's approach tries to control the appearance of Y_p and Y_q. Then the attack works for larger p than May's attack up to $p < N^{0.468}$. By optimizing the selection of y_p-shifts and y_q-shifts, Bleichenbacher-May's attack is always better than May's attack.

Our Matrix. To improve the Bleichenbacher-May attack, what we focus on is the representation of the polynomial. More concretely, previous works used the only one representation, i.e., $f_p(x_p, y_p)$, however, there is the other representation, i.e., $f_q(x_q, y_q)$, for the same polynomial. Indeed, a useful algebraic property can be exploited from the polynomial $f_q(x_q, y_q)$ by making use of the fact that $x_q = x_p + 1$. For the above Bleichenbacher-May matrix to be triangular, the polynomial ey_q is necessary. Since eY_q is larger than the modulus e, the polynomial does not contribute to maximize the solvable root bound as explained in [30, 40]. However, we make use of $f_q(x_q, y_q)$ and show that the matrix becomes triangular without ey_q as follows:

$$
\begin{pmatrix}
e & & & & & \\
0 & eX_p & & & & \\
N & NX_p & -X_pY_p & & & \\
0 & 0 & 0 & eY_p & & \\
0 & 0 & NX_pY_p & NY_p & -X_pY_p^2 & \\
0 & -X_p & 0 & 0 & 0 & X_qY_q
\end{pmatrix}.
$$

Although the above Bleichenbacher-May matrix used $N^{-1} \cdot y_q f_p(x_p, y_p)$ in the bottom row, we use $f_q(x_q, y_q)$ in turn. Notice that $f_q(x_q, y_q) = N^{-1} \cdot y_q f_p(x_p, y_p)$ and we use the same polynomial as the Bleichenbacher-May, however, the algebraic structure of $f_q(x_q, y_q)$, i.e., the relation $x_q = x_p+1$, enables the matrix to be triangular without ey_q. The operation means that Bleichenbacher-May's matrix contains better sublattices. The representation $f_q(x_q, y_q)$, which was not used by Bleichenbacher and May, enables us to exploit the sublattices. Indeed, by construction, our matrix always outperforms the above Bleichenbacher-May matrix with less lattice dimensions. Applying the LLL reduction to our above matrix, polynomials derived from the LLL output vectors satisfy Howgrave-Graham's lemma when

$$
X_p^3 X_q Y_p^4 Y_q e^3 < e^6 \Leftrightarrow 4(\alpha + \beta + \delta - 1) + 4\beta + (1 - \beta) < 3\alpha
$$

$$
\Leftrightarrow \delta < \frac{3}{4} - \frac{\alpha + 7\beta}{4}.
$$

Since $\beta \leq 1/2$, the bound is always better than the above Bleichenbacher-May example.

May's Modulo q Attack. We should notice that our lattice construction technique does not always offer the best attack. More concretely, as we discussed above, our lattice offers better results than all the existing lattices to solve $f_p(x_p, y_p) = 0$ and $f_q(x_q, y_q) = 0$. However, there is the other formulations to attack CRT-RSA, i.e., May's modulo q approach [28]. From the CRT-RSA key generation $ed_q = 1 + k(q - 1)$, May solved a modular equation;

$$
x + ey = 0 \pmod{q}
$$

whose root is $(k - 1, d_q)$. Since the modulo e and the modulo q approach is different, we should check whether which method is the better. Although our modulo e attacks are the better in most cases, we will show in Sect. 5.2 that the modulo p approach outperforms modulo e approach for small d_p attack with a modulus $N = p^r q$.

3.2 Attack for Large e

Although the above discussion handled only toy examples, our approach improves an asymptotic condition of the small CRT-exponent attack. In this section, we propose an improved attack that works when $\alpha > \beta/(1 - \beta)$. The attack is the first result to cover the desired bound, i.e., $\beta < 1/2$ with a full size e.

Theorem 1. *Let $N = pq$ be an RSA modulus where $p = N^\beta$ and $q = N^{1-\beta}$ for $\beta \leq 1/2$. Let $e = N^\alpha$ and $d_q < N^\delta$ be a public/CRT exponent respectively such that $ed_q = 1 \pmod{(q-1)}$. Given public elements N and e, if*

$$\delta < \frac{(1 - \beta)(3 + 2\beta) - 2\sqrt{\beta(1 - \beta)(\alpha\beta + 3\alpha + \beta)}}{3 + \beta} \quad and \quad \alpha > \frac{\beta}{1 - \beta},$$

then N can be factorized in polynomial time by assuming that polynomials which are derived from LLL reduced bases are algebraically independent.

As opposed to previous results, when $\alpha = 1$, the attack works to $\beta < 1/2$. Figure 1 compares our result and the Bleichenbacher-May for $\alpha = 1$. Our attack covers larger δ than the Bleichenbacher-May attack for all β.

Proof of Theorem 1. To solve the modular equation $f_q(x_q, y_q) = 0$ and equivalently $f_p(x_p, y_p) = 0$, we use the following shift-polynomials:

$$g_{[i,j]}(x_p, y_p) := x_p^j f_p^i(x_p, y_p)e^{m-i},$$
$$g'_{[i,j]}(x_p, y_p) := y_p^j f_p^i(x_p, y_p)e^{m-i},$$
$$g''_{[i,j]}(x_p, x_q, y_p, y_q) := f_p^{i-j}(x_p, y_p)f_q^j(x_q, y_q)e^{m-i},$$

with some positive integer m. For non-negative integers i and j, all the shift-polynomials share the same root as $f_p(x_p, y_p)$ and $f_q(x_q, y_q)$ modulo e^m. May [28]

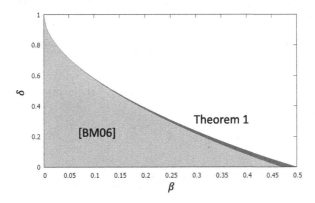

Fig. 1. Comparison between our attack (Theorem 1) and the Bleichenbacher-May for $\alpha = 1$.

used the same shift-polynomials as $g_{[i,j]}(x_p, y_p)$ and $g'_{[i,j]}(x_p, y_p)$. The (modified) Bleichenbacher-May attack used an additional shift-polynomial which used only $f_p(x_p, y_p)$. However, as we showed an example in the previous section, we use the both representations $f_p(x_p, y_p)$ and $f_q(x_q, y_q)$ simultaneously. Then we can construct triangular basis matrices that generalize the toy example as follows.

Lemma 2. *Let all the polynomials be defined as above. Let τ_p and τ_q be constants such that $\tau_p \geq 0$ and $0 \leq \tau_q \leq 1$. Define sets of indices*

$$\mathcal{I}_x := \{i = 0, 1, \ldots, m; j = 0, 1, \ldots, m - i\},$$
$$\mathcal{I}_{y,p} := \{i = 0, 1, \ldots, m; j = 1, 2, \ldots, \lceil \tau_p m \rceil\},$$
$$\mathcal{I}_{y,q} := \{i = 1, 2, \ldots, m; j = 1, 2, \ldots, \lceil \tau_q i \rceil\}.$$

Let $\boldsymbol{B}$ be a matrix whose rows consist of coefficients of $g_{[i,j]}(x_p X_p, y_p Y_p)$, $g'_{[i,j]}(x_p X_p, y_p Y_p)$, and $g''_{[i,j]}(x_p X_p, x_q X_q, y_p Y_p, y_q Y_q)$ with indices in $\mathcal{I}_x$, $\mathcal{I}_{y,p}$, and $\mathcal{I}_{y,q}$, respectively. If the shift-polynomials are ordered as

$$g_{[i,j]} \prec g'_{[i,j]}, g''_{[i,j]},$$
$$g_{[i,j]} \prec g_{[i',j']}, g'_{[i,j]} \prec g'_{[i',j']}, g''_{[i,j]} \prec g''_{[i',j']} \quad \text{for} \ \ i < i',$$
$$g_{[i,j]} \prec g_{[i,j']}, g'_{[i,j]} \prec g'_{[i,j']}, g''_{[i,j]} \prec g''_{[i,j']} \quad \text{for} \ \ j < j',$$

and $N^{-1} \pmod{e^m}$ is multiplied appropriately, then the matrix becomes triangular with diagonals

– $X_p^{i+j} Y_p^i e^{m-i}$ for $g_{[i,j]}(x_p X_p, y_p Y_p)$,
– $X_p^i Y_p^{i+j} e^{m-i}$ for $g'_{[i,j]}(x_p X_p, y_p Y_p)$,
– $X_q^i Y_q^j e^{m-i}$ for $g''_{[i,j]}(x_p Y_p, x_q X_q, y_p Y_p, y_q Y_q)$.

Here, we do not prove the lemma. Later, we prove a more general form of the statement, i.e., Lemma 3.

We compute the resulting condition of Theorem 1. The dimension n and the determinant of the lattice $\det(\boldsymbol{B}) = X^{s_X} Y_p^{s_{Y_p}} Y_q^{s_{Y_q}} e^{s_e}$ can be computed as:

$$n = \sum_{(i,j) \in \mathcal{I}_x} 1 + \sum_{(i,j) \in \mathcal{I}_{y,p}} 1 + \sum_{(i,j) \in \mathcal{I}_{y,q}} 1 = \frac{1 + 2\tau_p + \tau_q}{2} m^2 + o(m^2),$$

$$s_X = \sum_{(i,j) \in \mathcal{I}_x} (i+j) + \sum_{(i,j) \in \mathcal{I}_{y,p}} i + \sum_{(i,j) \in \mathcal{I}_{y,q}} i = \frac{2 + 3\tau_p + 2\tau_q}{6} m^3 + o(m^3),$$

$$s_{Y_p} = \sum_{(i,j) \in \mathcal{I}_x} i + \sum_{(i,j) \in \mathcal{I}_{y,p}} (i+j) = \frac{1 + 3\tau_p + 3\tau_p^2}{6} m^3 + o(m^3),$$

$$s_{Y_q} = \sum_{(i,j) \in \mathcal{I}_{y,q}} j = \frac{\tau_q^2}{6} m^3 + o(m^3),$$

$$s_e = \sum_{(i,j) \in \mathcal{I}_x} (m-i) + \sum_{(i,j) \in \mathcal{I}_{y,p}} (m-i) + \sum_{(i,j) \in \mathcal{I}_{y,q}} (m-i)$$

$$= \frac{2 + 3\tau_p + \tau_q}{6} m^3 + o(m^3).$$

Applying the LLL reduction, the polynomials obtained from the output vectors satisfy Howgrave-Graham's lemma if $X^{s_X} Y_p^{s_{Y_p}} Y_q^{s_{Y_q}} e^{s_e} < e^{nm}$, i.e.,

$$(\alpha + \beta + \delta - 1)\frac{2 + 3\tau_p + 2\tau_q}{6} + \beta\frac{1 + 3\tau_p + 3\tau_p^2}{6}$$
$$+ (1 - \beta)\frac{\tau_q^2}{6} - \alpha\frac{1 + 3\tau_p + 2\tau_q}{6} < 0$$

by omitting low order terms of m. To minimize the left hand side of the inequality, we substitute the parameters $\tau_p = (1 - 2\beta - \delta)/(2\beta)$ and $\tau_q = (1 - \beta - \delta)/(1 - \beta)$, then the condition becomes

$$\delta < \frac{(1 - \beta)(3 + 2\beta) - 2\sqrt{\beta(1 - \beta)(\alpha\beta + 3\alpha + \beta)}}{3 + \beta}$$

as required. To satisfy the restriction $\tau_p \geq 0$, $\alpha > \beta/(1 - \beta)$ should hold. The other parameter τ_q always satisfies $0 \leq \tau_q \leq 1$. $\qquad\square$

3.3 Attack for Small e

The attack of Theorem 1 works only for $\alpha > \beta/(1 - \beta)$. The constraint comes from the fact that the parameter τ_p used in the proof should be non-negative. To capture the other case, i.e., $\alpha \leq \beta/(1 - \beta)$, under the same algorithm construction, we set the parameters $\tau_p = 0$ and $\tau_q = (1 - \beta - \delta)/(1 - \beta)$, then the attack works for $\delta < 2(1 - \beta) - \sqrt{(1 + \alpha)(1 - \beta)}$.

However, by modifying the lattice construction, a better result can be obtained as follows.

Theorem 2. *Let $N = pq$ be an RSA modulus where $p < N^\beta$ and $q \geq N^{1-\beta}$ for $\beta \leq 1/2$. Let $e = N^\alpha$ and $d_q < N^\delta$ be a public/CRT exponent respectively such that $ed_q = 1 \pmod{(q - 1)}$. Given public elements N and e, if*

$$\delta < 1 - \beta - \sqrt{\alpha\beta(1 - \beta)} \text{ for } \beta(1 - \beta) \leq \alpha \leq \frac{\beta}{1 - \beta},$$

then N can be factorized in polynomial time by assuming that polynomials which are derived from LLL reduced bases are algebraically independent.

As we claimed, the bound of Theorem 2 is better than $\delta < 2(1 - \beta) - \sqrt{(1 + \alpha)(1 - \beta)}$ which can be obtained from the same algorithm construction as Theorem 1. We show the proof of Theorem 2. The proof is more technical than that of Theorem 1, however, the spirit is almost the same. In the subsequent sections, lattices which are similar to that of Theorem 2 will be used.

Proof of Theorem 2. To solve the modular equation $f_q(x_q, y_q) = 0$ and equivalently $f_p(x_p, y_p) = 0$, we use the following shift-polynomials:

$$g_{[i,j],\lambda}(x_p, x_q, y_p, y_q) := x_p^j f_p^{\lceil \lambda i \rceil}(x_p, y_p) f_q^{\lfloor (1-\lambda)i \rfloor}(x_q, y_q) e^{m-i},$$
$$g'_{[i,j],\lambda}(x_p, x_q, y_p, y_q) := y_q^j f_p^{\lceil \lambda i \rceil}(x_p, y_p) f_q^{\lfloor (1-\lambda)i \rfloor}(x_q, y_q) e^{m-i},$$

with some positive integer m and a parameter $0 < \lambda \le 1$. For non-negative integers i and j, all the shift-polynomials share the common root as $f_p(x_p, y_p)$ and $f_q(x_q, y_q)$ modulo e^m. Here, notice that $\lceil \lambda i \rceil + \lfloor (1 - \lambda)i \rfloor = i$ for all i. The shift-polynomials $g'_{[i,j]}(x_p, y_p)$ and $g''_{[i,j]}(x_p, y_p)$ used in the proof of Theorem 1 is the special case of $g_{[i,j],\lambda}(x_p, x_q, y_p, y_q)$ and $g'_{[i,j],\lambda}(x_p, x_q, y_p, y_q)$ for $\lambda = 1$. As the attack of Theorem 1, we use both representations $f_p(x_p, y_p)$ and $f_q(x_q, y_q)$ simultaneously for all shift-polynomials. Using these shift-polynomials, we can construct triangular basis matrices as follows.

Lemma 3. *Let all the polynomials be defined as above. Let τ be a constant such that $1 - \lambda < \tau \le 1$. Let m be a positive integer. Define sets of indices as*

$$\mathcal{I}_x := \{i = 0, 1, \dots, m; j = 0, 1, \dots, m - i\},$$
$$\mathcal{I}_{y_q} := \{i = 1, 2, \dots, m; j = 1, 2, \dots, \lceil \tau i \rceil - \lfloor (1 - \lambda)i \rfloor\}.$$

Let $\boldsymbol{B}$ be a matrix whose rows consist of coefficients of $g_{[i,j],\lambda}(x_p X_p, x_q X_q, y_p Y_p, y_q Y_q)$ and $g'_{[i,j],\lambda}(x_p X_p, x_q X_q, y_p Y_p, y_q Y_q)$ with indices in $\mathcal{I}_x$ and $\mathcal{I}_{y,q}$ respectively. If the shift-polynomials are ordered as

$$g_{[i,j],\lambda} \prec g'_{[i,j],\lambda},$$
$$g_{[i,j],\lambda} \prec g_{[i',j'],\lambda}, g'_{[i,j],\lambda} \prec g'_{[i',j'],\lambda} \text{ for } i < i',$$
$$g_{[i,j],\lambda} \prec g_{[i,j'],\lambda}, g'_{[i,j],\lambda} \prec g'_{[i,j'],\lambda} \text{ for } j < j',$$

and $N^{-1} \pmod{e^m}$ is multiplied appropriately, then the matrix becomes triangular with diagonals

- $X_p^{i+j} Y_p^{\lceil \lambda i \rceil} e^{m-i}$ *for $g_{[i,j],\lambda}(x_p X_p, x_q X_q, y_p Y_p, y_q Y_q)$ with i such that $i = 0$ and $\lceil \lambda i \rceil - \lceil \lambda(i - 1) \rceil = 1$,*
- $X_q^{i+j} Y_q^{\lfloor (1-\lambda)i \rfloor} e^{m-i}$ *for $g_{[i,j],\lambda}(x_p X_p, x_q X_q, y_p Y_p, y_q Y_q)$ with i such that $i \ne 0$ and $\lceil \lambda i \rceil - \lceil \lambda(i - 1) \rceil = 0$,*
- $X_q^{i} Y_q^{\lfloor (1-\lambda)i \rfloor + j} e^{m-i}$ *for $g'_{[i,j],\lambda}(x_p X_p, x_q X_q, y_p Y_p, y_q Y_q)$.*

A proof of the lemma is the most technical part of this paper. We prove it in Sect. 3.4.

We compute the resulting condition of Theorem 2. The dimension n and the determinant of the lattice $\det(\boldsymbol{B}) = X^{s_X} Y_p^{s_{Y_p}} Y_q^{s_{Y_q}} e^{s_e}$ can be computed as:

$$n = \sum_{(i,j) \in \mathcal{I}_x} 1 + \sum_{(i,j) \in \mathcal{I}_{y_q}} 1 = \frac{\lambda + \tau}{2} m^2 + o(m^2),$$

$$s_X = \sum_{(i,j) \in \mathcal{I}_x} (i + j) + \sum_{(i,j) \in \mathcal{I}_{y_q}} i = \frac{\lambda + \tau}{3} m^3 + o(m^3),$$

$$s_{Y_p} = \sum_{(i,j) \in \mathcal{I}_x} \lceil \lambda i \rceil = \frac{\lambda^2}{6} m^3 + o(m^3),$$

$$s_{Y_q} = \sum_{(i,j)\in\mathcal{I}_x} \lfloor(1-\lambda)i\rfloor + \sum_{(i,j)\in\mathcal{I}_{yq}} (\lfloor(1-\lambda)i\rfloor + j) = \frac{\tau^2}{6}m^3 + o(m^3),$$

$$s_e = \sum_{(i,j)\in\mathcal{I}_x} (m-i) + \sum_{(i,j)\in\mathcal{I}_{yq}} (m-i) = \frac{1+\lambda+\tau}{6}m^3 + o(m^3).$$

Applying the LLL reduction, the polynomials obtained from the output vectors satisfy Howgrave-Graham's lemma if $X^{s_X}Y_p^{s_{Y_p}}Y_q^{s_{Y_q}}e^{s_e} < e^{nm}$, i.e.,

$$(\alpha+\beta+\delta-1)\frac{\lambda+\tau}{3} + \beta\frac{\lambda^2}{6} + (1-\beta)\frac{\tau^2}{6} - \alpha\frac{-1+2\lambda+2\tau}{6} < 0$$

by omitting low order terms of m. To minimize the left hand side of the inequality, we set the parameters $\lambda = (1-\beta-\delta)/\beta$ and $\tau = (1-\beta-\delta)/(1-\beta)$, then the condition becomes

$$\delta < 1 - \beta - \sqrt{\alpha\beta(1-\beta)}$$

as required. To satisfy the restrictions $0 < \lambda \leq 1$ and $1-\lambda < \tau \leq 1$, $\beta(1-\beta) \leq \alpha \leq \beta/(1-\beta)$ should hold. □

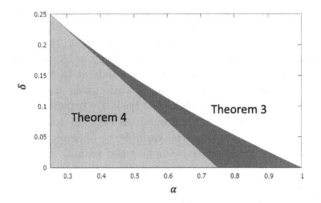

Fig. 2. Comparison between our attack (Theorem 3) and the attack of Lu et al. (Theorem 4) [27].

As opposed to the attack of Theorem 1, that of Theorem 2 is applicable to a balanced RSA, i.e., $\beta = 1/2$, for $\alpha \leq 1$. For a balanced RSA, we substitute $\beta = 1/2$ and the attack becomes as follows.

Theorem 3. *Let $N = pq$ be an RSA modulus where the prime factors p and q are the same bit-size. Let $e = N^\alpha$ and $d_q < N^\delta$ be a public/CRT exponent respectively such that $ed_q = 1 \pmod{(q-1)}$. Given public elements N and e, if*

$$\delta < \frac{1-\sqrt{\alpha}}{2} \text{ for } \alpha \geq \frac{1}{4},$$

then N can be factorized in polynomial time by assuming that polynomials which are derived from LLL reduced bases are algebraically independent.

By construction, the attack always outperforms that under Bleichenbacher-May's lattice construction. We also compare our attack with that of Lu et al. [28] (Theorem 9 of [27]) which follows May's modulo q approach.

Theorem 4 ([27]). *Let $N = pq$ be an RSA modulus where the prime factors p and q are the same bit-size. Let $e = N^\alpha$ and $d_q < N^\delta$ be a public/CRT exponent respectively such that $ed_q = 1 \pmod{(q-1)}$. Given public elements N and e, if*

$$\delta < \frac{3 - 4\alpha}{8},$$

then N can be factorized in polynomial time by assuming that polynomials which are derived from LLL reduced bases are algebraically independent.

Figure 2 compares our attack (Theorem 3) and that of Lu et al. (Theorem 4). Our attack is better for all $1/4 < \alpha < 1$.

3.4   Proof of Lemma 3

In this section, we show a proof of Lemma 3 that is the most technical part of this paper. Before the detailed proof, we explain the spirit of our triangular matrix. The polynomials which we use contains four variables x_p, x_q, y_p, y_q. Furthermore, there are two algebraic relations: $x_q = x_p + 1$ and $y_p y_q = N$. By using the latter relation, i.e., $y_p y_q = N$, we transform all monomials as they do not have both y_p and y_q simultaneously where the same operation was also done in previous works [2,11]. Moreover, we use an additional trick. By using the former relation, i.e., $x_q = x_p + 1$, we transform all monomials as they do not have both x_p and x_q simultaneously. More concretely, the variable x_p appears only in monomials where powers of y_p are non-negative whereas the variable x_q appears only in monomials where powers of y_q are positive. The simple operation is the key technique of this paper.

Then we show the proof of Lemma 3.

Proof of Theorem 3. Since all $g_{[i,j],\lambda}(x_p, x_q, y_p, y_q)$ for $i = 0$ have only one monomial $x_p^j e^m$, these polynomials generate triangular basis matrix with diagonals $X_p^j e^m$. Then remaining proof is inductive; we show that the basis matrix is still triangular with other polynomials.

At first, we assume that polynomials $g_{[i',j'],\lambda}(x_p, x_q, y_p, y_q)$ such that $g_{[i',j'],\lambda}(x_p, x_q, y_p, y_q) \prec g_{[i,j],\lambda}(x_p, x_q, y_p, y_q)$ generate a triangular matrix as stated in Lemma 3. Then, we show that a matrix is still triangular with a new polynomial $g_{[i,j],\lambda}(x_p, x_q, y_p, y_q)$ whose diagonal is $x_p^{i+j} y_p^{\lceil \lambda i \rceil} e^{m-i}$. By definition,

$$g_{[i,j],\lambda}(x_p, x_q, y_p, y_q) = x_p^j f_p^{\lceil \lambda i \rceil}(x_p, y_p) f_q^{\lfloor (1-\lambda)i \rfloor}(x_q, y_q) e^{m-i}$$
$$= x_p^j (N + N x_p - x_p y_p)^{\lceil \lambda i \rceil} (1 - x_q + x_q y_q)^{\lfloor (1-\lambda)i \rfloor} e^{m-i}.$$

From the relation $x_q = x_p + 1$ and equivalently $x_p = x_q - 1$, the polynomial becomes

$$= x_p^j (Nx_q - x_p y_p)^{\lceil \lambda i \rceil} (x_p + x_q y_q)^{\lfloor (1-\lambda)i \rfloor} e^{m-i}.$$

By expanding $(Nx_q - x_p y_p)^{\lceil \lambda i \rceil}$ and $(x_p + x_q y_q)^{\lfloor (1-\lambda)i \rfloor}$,

$$= x_p^j \left(\sum_{i_p=0}^{\lceil \lambda i \rceil} \binom{\lceil \lambda i \rceil}{i_p} (-x_p y_p)^{i_p} \cdot (Nx_q)^{\lceil \lambda i \rceil - i_p} \right) \cdot$$

$$\left(\sum_{i_q=0}^{\lfloor (1-\lambda)i \rfloor} \binom{\lfloor (1-\lambda)i \rfloor}{i_q} (x_q y_q)^{i_q} \cdot x_p^{\lfloor (1-\lambda)i \rfloor - i_q} \right) e^{m-i}$$

$$= \sum_{i_p=0}^{\lceil \lambda i \rceil} \sum_{i_q=0}^{\lfloor (1-\lambda)i \rfloor} (-1)^{i_p} \binom{\lceil \lambda i \rceil}{i_p} \binom{\lfloor (1-\lambda)i \rfloor}{i_q} N^{\lceil \lambda i \rceil - i_p}.$$

$$x_p^{\lfloor (1-\lambda)i \rfloor + i_p - i_q + j} x_q^{\lceil \lambda i \rceil - i_p + i_q} y_q^{i_q} y_p^{i_p} e^{m-i}.$$

From the relation $y_p y_q = N$, the polynomial becomes

$$= \sum_{i_q=0}^{\lfloor (1-\lambda)i \rfloor} \sum_{i_p=i_q}^{\lceil \lambda i \rceil} (-1)^{i_p} \binom{\lceil \lambda i \rceil}{i_p} \binom{\lfloor (1-\lambda)i \rfloor}{i_q} N^{\lceil \lambda i \rceil - i_p + i_q}.$$

$$x_p^{\lfloor (1-\lambda)i \rfloor + i_p - i_q + j} x_q^{\lceil \lambda i \rceil - i_p + i_q} y_p^{i_p - i_q} e^{m-i}$$

$$+ \sum_{i_p=0}^{\lfloor (1-\lambda)i \rfloor - 1} \sum_{i_q=i_p+1}^{\lfloor (1-\lambda)i \rfloor} (-1)^{i_p} \binom{\lceil \lambda i \rceil}{i_p} \binom{\lfloor (1-\lambda)i \rfloor}{i_q} N^{\lceil \lambda i \rceil}.$$

$$x_p^{\lfloor (1-\lambda)i \rfloor + i_p - i_q + j} x_q^{\lceil \lambda i \rceil - i_p + i_q} y_q^{i_q - i_p} e^{m-i}.$$

Notice that there are no monomials that have y_p and y_q simultaneously. The exponents of y_p in the first summation are non-negative whereas the exponents of y_q in the second summation are positive. Hence, as we discussed above, we replace all x_q in the first summation by $x_p + 1$ and replace all x_p in the second summation by $x_q - 1$. Then, the polynomial becomes

$$= \sum_{i_q=0}^{\lfloor (1-\lambda)i \rfloor} \sum_{i_p=i_q}^{\lceil \lambda i \rceil} (-1)^{i_p} \binom{\lceil \lambda i \rceil}{i_p} \binom{\lfloor (1-\lambda)i \rfloor}{i_q} N^{\lceil \lambda i \rceil - i_p + i_q}.$$

$$x_p^{\lfloor (1-\lambda)i \rfloor + i_p - i_q + j} (x_p + 1)^{\lceil \lambda i \rceil - i_p + i_q} y_p^{i_p - i_q} e^{m-i}$$

$$+ \sum_{i_p=0}^{\lfloor (1-\lambda)i \rfloor - 1} \sum_{i_q=i_p+1}^{\lfloor (1-\lambda)i \rfloor} (-1)^{i_p} \binom{\lceil \lambda i \rceil}{i_p} \binom{\lfloor (1-\lambda)i \rfloor}{i_q} N^{\lceil \lambda i \rceil}.$$

$$(x_q - 1)^{\lfloor (1-\lambda)i \rfloor + i_p - i_q + j} x_q^{\lceil \lambda i \rceil - i_p + i_q} y_q^{i_q - i_p} e^{m-i}$$

$$= \sum_{i_q=0}^{\lfloor (1-\lambda)i \rfloor} \sum_{i_p=i_q}^{\lceil \lambda i \rceil} \sum_{i'=0}^{\lceil \lambda i \rceil - i_p + i_q} (-1)^{i_p} \binom{\lceil \lambda i \rceil}{i_p} \binom{\lfloor (1-\lambda)i \rfloor}{i_q} \binom{\lceil \lambda i \rceil - i_p + i_q}{i'}.$$

$$N^{\lceil \lambda i \rceil - i_p + i_q} x_p^{i - i' + j} y_p^{i_p - i_q} e^{m-i}$$

$$+ \sum_{i_p = 0}^{\lfloor (1-\lambda)i \rfloor - 1} \sum_{i_q = i_p + 1}^{\lfloor (1-\lambda)i \rfloor} \sum_{i' = 0}^{\lfloor (1-\lambda)i \rfloor + i_p - i_q + j} (-1)^{i_p + i'} \binom{\lceil \lambda i \rceil}{i_p} \binom{\lfloor (1-\lambda)i \rfloor}{i_q}.$$

$$\binom{\lfloor (1-\lambda)i \rfloor + i_p - i_q + j}{i'} N^{\lceil \lambda i \rceil} x_q^{i - i' + j} y_q^{i_q - i_p} e^{m-i}.$$

The polynomial has monomials for variables

- $x_p^{i_{px}} y_p^{i_{py}}$ for $i_{py} = 0, 1, \ldots, \lceil \lambda i \rceil; i_{px} = i_{py} + \lfloor (1-\lambda)i \rfloor + j, \ldots, i + j,$
- $x_q^{i_{qx}} y_q^{i_{qy}}$ for $i_{qy} = 1, 2, \ldots, \lfloor (1-\lambda)i \rfloor; i_{qx} = i_{qy} + \lceil \lambda i \rceil, \ldots, i + j.$

Then, we show that these variables except $x_p^{i+j} y_p^{\lceil \lambda i \rceil}$ already appeared in the diagonals of a basis matrix. The above variables appeared for diagonals of $g_{[i',j'],\lambda}(x_p X_p, x_q X_q, y_p Y_p, y_q Y_q)$ for

$$i' = 0, 1, \ldots, i - 1 \text{ such that } \lceil \lambda i' \rceil - \lceil \lambda(i' - 1) \rceil = 1;$$
$$j' = \lfloor (1-\lambda)i \rfloor - \lfloor (1-\lambda)i' \rfloor + j, \ldots, i + j - i', \text{ and}$$
$$i' = 1, 2, \ldots, i - 1 \text{ such that } \lceil \lambda i' \rceil - \lceil \lambda(i' - 1) \rceil = 0;$$
$$j' = \lceil \lambda i \rceil - \lceil \lambda i' \rceil, \ldots, i + j - i'.$$

Since $i' < i$, by our definition of the polynomial order,

$$g_{[i',j'],\lambda}(x_p X_p, x_q X_q, y_p Y_p, y_q Y_q) \prec g_{[i,j],\lambda}(x_p X_p, x_q X_q, y_p Y_p, y_q Y_q)$$

holds for all the above i' and j'. All we have to show is that these polynomials are selected in the lattice basis. For the purpose, we show that the indices

$$i' = 0, 1, \ldots, i - 1;$$
$$j' = \min\{\lfloor (1-\lambda)i \rfloor - \lfloor (1-\lambda)i' \rfloor + j, \lceil \lambda i \rceil - \lceil \lambda i' \rceil\}, \ldots, i + j - i',$$

are contained in

$$i' = 0, 1, \ldots, m; j' = 0, 1, \ldots, m - i'.$$

Since $0 < \lambda \le 1, 0 \le i' \le i$, and $j \ge 0$,

$$\lfloor (1-\lambda)i \rfloor - \lfloor (1-\lambda)i' \rfloor + j \ge 0 \quad \text{and} \quad \lceil \lambda i \rceil - \lceil \lambda i' \rceil \ge 0$$

hold. Since $i + j \le m$ holds,

$$i + j - i' \le m - i'$$

holds. Then the statement holds. In the same manner, analogous proof is obtained for the other polynomials $g'_{[i,j],\lambda}(x_p, x_q, y_p, y_q)$. We will show the remaining proof in the full version.   □

To end this section, we briefly show how to deduce Lemma 2 from Lemma 3. The collection of shift-polynomials $g_{[i,j]}(x_p, y_p)$ and $g''_{[i,j]}(x_p, x_q, y_p, y_q)$ in Lemma 2 are essentially the same as $g_{[i,j],\lambda}(x_p, x_q, y_p, y_q)$ and $g'_{[i,j],\lambda}$ (x_p, x_q, y_p, y_q) in Lemma 3 for $\lambda = 1$. Hence, by setting the parameters (λ, τ) in Lemma 3 as $(1, \tau_q)$, Lemma 3 show that $g_{[i,j]}(x_p, y_p)$ and $g''_{[i,j]}(x_p, x_q, y_p, y_q)$ in Lemma 2 generate a triangular matrix. To complete the proof of Lemma 2, we also use May's result [28] that showed that polynomials $g_{[i,j]}(x_p, y_p)$ and $g'_{[i,j]}(x_p, y_p)$ generate a triangular matrix. As a result, $g_{[i,j]}(x_p, y_p)$, $g'_{[i,j]}(x_p, y_p)$, and $g''_{[i,j]}(x_p, x_q, y_p, y_q)$ in Lemma 2 generates a triangular matrix.

Table 1. For 1000-bit RSA moduli, asymptotic and experimental comparisons of small d_q attacks

Bitsize of q	Bleichenbacher-May [2]				Our work			
	Asymptotic	Expt.	dim.	L^3 time	Asymptotic	Expt.	dim.	L^3 time
305	0.210	0.160	63	53 min	0.230	0.170	56	15 min
355	0.140	0.100	63	44 min	0.164	0.100	58	16 min
405	0.075	0.050	63	35 min	0.103	0.055	66	57 min
440	0.033	0.010	63	35 min	0.064	0.012	66	60 min

3.5 Experimental Results

We have implemented the experiment program in Magma 2.10 computer algebra system [5] on a PC with Intel(R) Core(TM) Duo CPU(3.30 GHz, 4.0 GB RAM Windows 7). Table 1 lists some theoretical and experimental results on factoring two 1000-bit RSA moduli with varying bit-size of q. In all experiments, we successfully find the factorization of these RSA moduli.

In [2], the experimental results are much better than their theoretical analysis. For example, for 440-bit factor q, with a lattice dimension of 63, in theory

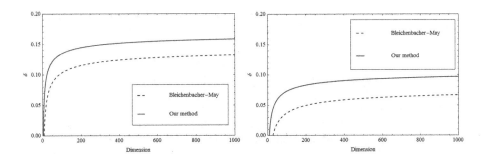

Fig. 3. Comparisons of recoverable bounds depending on lattice dimensions. The left and the right figure is for $\beta = 0.35$ and $\beta = 0.40$, respectively.

Table 2. Asymptotic bounds and lattice dimension for small δ with fixed lattice dimensions.

$\beta = 0.45$					
δ	0.010	0.020	0.030	0.040	0.052
dim.	109	154	340	1055	Asymptotic
$\beta = 0.48$					
δ	0.002	0.005	0.010	0.015	0.020
dim.	486	686	1491	5443	Asymptotic

the attack should not work (we can recover the small private key d_p up to a size of $N^{-0.083}$), however, in practice, we succeed for d_p with bit-size a 0.010-fraction of N. Since our lattice construction captures the underlying sublattice structure of [2]'s desired lattice, we can do better than [2]: with a lattice dimension of 66, experimentally we can reconstruct d_p with a size of $N^{0.012}$.

Note that our result of Theorem 1 is an asymptotic improvement. In Table 2, we present numerical values of δ for different values of β and lattice dimension. Moreover, compared with [2], our method requires smaller lattice dimensions. For $\beta = 0.35$ and $\beta = 0.40$, Fig. 3 shows a comparison of these two approaches in the terms of the bit-size of small secret exponent d_p that can be attacked.

4   Small d_p and d_q Attack

In this section, we propose an attack when both d_p and d_q are small. The attack improves Jochemsz-May's attack [21].

4.1   Our Attack

Recall the CRT-RSA key generation;

$$ed_q = 1 + k_q(q - 1) \quad \text{and} \quad ed_p = 1 + k_p(p - 1)$$

with some integers k_q and k_p. Hence, if we can solve the following simultaneous modular equations, RSA modulus N can be factorized:

$$f_{q,1}(x_{q,1}, y_q) = 1 + x_{q,1}(y_q - 1) = 0 \quad \mod e,$$
$$f_{p,2}(x_{p,2}, y_p) = 1 + x_{p,2}(y_p - 1) = 0 \quad \mod e,$$

where the root is $(x_{q,1}, x_{p,2}, y_q, y_p) = (k_q, k_p, q, p)$.

In addition, by multiplying p and q to the key generation equations respectively, the following representations can be obtained:

$$ed_q p = p + k_q(N - p) = N + (k_q - 1)(N - p),$$
$$ed_p q = q + k_p(N - q) = N + (k_p - 1)(N - q).$$

Then, we can also use the following modular equations:

$$f_{p,1}(x_{p,1}, y_p) = N + x_{p,1}(N - y_p) = 0 \quad \mod e,$$
$$f_{q,2}(x_{q,2}, y_q) = N + x_{q,2}(N - y_q) = 0 \quad \mod e,$$

where the root is $(x_{p,1}, x_{q,2}, y_p, y_q) = (k_q - 1, k_p - 1, p, q)$.

To summarize the above discussion, we want to solve the following simultaneous modular equations:

$$f_{p,1}(x_{p,1}, y_p) = N + x_{p,1}(N - y_p) = 0 \quad \mod e,$$
$$f_{q,1}(x_{q,1}, y_q) = 1 + x_{q,1}(y_q - 1) = 0 \quad \mod e,$$
$$f_{p,2}(x_{p,2}, y_p) = 1 + x_{p,2}(y_p - 1) = 0 \quad \mod e,$$
$$f_{q,2}(x_{q,2}, y_q) = N + x_{q,2}(N - y_q) = 0 \quad \mod e,$$

where the root is $(x_{p,1}, x_{q,1}, x_{p,2}, x_{q,2}, y_p, y_q) = (k_q - 1, k_q, k_p, k_p - 1, p, q)$. Let $e = N^\alpha$, $d_p < N^\delta$, and $d_q < N^\delta$ for a balanced RSA, i.e., $q < p < 2q$. The absolute values of $x_{p,1}, x_{q,1}, x_{p,2}, x_{q,2}$ are bounded above by $X = N^{\alpha+\delta-1/2}$ within constant factors whereas the absolute values of y_p and y_q are bounded above by $Y = N^{1/2}$ within constant factors.

Unfortunately, an approach to solve the above four equations simultaneously does not offer an improvement. The approach gives us only the same bound as Theorem 3. Hence, we use an additional algebraic relation. From the CRT-RSA key generation,

$$ed_q = 1 + k_q(q - 1) \quad \text{and} \quad ed_p = 1 + k_p(p - 1),$$
$$\Leftrightarrow \quad k_q - 1 = k_q q \pmod e \quad \text{and} \quad k_p - 1 = k_p p \pmod e.$$

By multiplying these two equations, we obtain

$$(k_q - 1)(k_p - 1) = k_q k_p N \pmod e.$$

Then the following new equation can be obtained:

$$h(x_{p,1}, x_{q,1}, x_{p,2}, x_{q,2}) = (N - 1)x_{p,1}x_{p,2} + x_{p,1} + Nx_{p,2} = 0 \pmod e$$
$$= (N - 1)x_{q,1}x_{q,2} + Nx_{q,1} + x_{q,2} = 0 \pmod e.$$

The polynomial also has two representations as the previous polynomials. Notice that the same equation as $h(x_{p,1}, x_{q,1}, x_{p,2}, x_{q,2})$ was already used by Galbraith et al. [13]. We make use of these equations and obtain the following result.

Theorem 5. *Let $N = pq$ be an RSA modulus where p and q are the same bit-size. Let $e = N^\alpha$ and $d_p, d_q < N^\delta$ be a public/CRT exponent respectively such that $ed_q = 1 \pmod{(q - 1)}$ and $ed_p = 1 \pmod{(p - 1)}$. Given public elements N and e, if*

$$\delta < \frac{1}{2} - \sqrt{\frac{\alpha}{6}} \quad \text{for} \quad \alpha \geq \frac{3}{8},$$

then N can be factorized in polynomial time by assuming that polynomials which are derived from LLL reduced bases are algebraically independent.

For the full size e, the attack works for $\delta < 1/2 - 1/\sqrt{6} = 0.091\cdots$ which is better than Jochemsz-May's bound [21], i.e., $\delta < 0.073$. Our attack is better than all existing attacks.

Proof of Theorem 5. To solve the above modular equations, we use the following shift-polynomials:

$$g_{[i_1,i_2,j_1,j_2,u]}(x_{p,1}, x_{q,1}, x_{p,2}, x_{q,2}, y_p, y_q)$$
$$:= x_{p,1}^{j_1} x_{p,2}^{j_2} y_q^{\lfloor (i_1+i_2)/2 \rfloor} f_{p,1}^{i_1}(x_{p,1}, y_p) f_{p,2}^{i_2}(x_{p,2}, y_p) h^u(x_{p,1}, x_{p,2}, x_{q,1}, x_{q,2}) \cdot$$
$$e^{m-(i_1+i_2+u)},$$

$$g'_{[i_1,i_2,j_1],p}(x_{p,1}, x_{q,1}, x_{p,2}, x_{q,2}, y_p, y_q)$$
$$:= y_q^{\lfloor (i_1+i_2)/2 \rfloor - j_1} f_{p,1}^{i_1}(x_{p,1}, y_p) f_{p,2}^{i_2}(x_{p,2}, y_p) e^{m-(i_1+i_2+u)},$$

$$g'_{[i_1,i_2,j_2],q}(x_{p,1}, x_{q,1}, x_{p,2}, x_{q,2}, y_p, y_q)$$
$$:= y_q^{\lfloor (i_1+i_2)/2 \rfloor + j_2} f_{p,1}^{i_1}(x_{p,1}, y_p) f_{p,2}^{i_2}(x_{p,2}, y_p) e^{m-(i_1+i_2+u)},$$

with some positive integer m. For non-negative integers i_1, i_2, j_1, i_2, and u, all the shift-polynomials share the common root as $f_{p,1}(x_{p,1}, y_p)$, $f_{p,2}(x_{p,2}, y_p)$, $f_{q,1}(x_{q,1}, y_q)$, $f_{q,2}(x_{q,2}, y_q)$, and $h(x_{p,1}, x_{q,1}, x_{p,2}, x_{q,2})$ modulo e^m. Then we can construct triangular basis matrices as follows.

Lemma 4. *Let all the polynomials be defined as above. Let τ be a constant such that $1/2 \leq \tau \leq 1$. Define sets of indices as*

$$\mathcal{I}_x := \left\{ \begin{array}{l} i_1 = 0, 1, \ldots, m; i_2 = 0, 1, \ldots, m - i_1; j_1 = j_2 = 0; \\ \qquad u = 0, 1, \ldots, \lfloor \frac{m-(i_1+i_2)}{2} \rfloor, \ and \\ i_1 = 0, 1, \ldots, m - 2; i_2 = 1, 2, \ldots, m - 1 - i_1; j_1 = 1; \\ \qquad j_2 = 0; u = 0, 1, \ldots, \lfloor \frac{m-1-(i_1+i_2)}{2} \rfloor, \ and \\ i_1 = 0, 1, \ldots, m; i_2 = 0; j_1 = 1, 2, \ldots, m - i_1; j_2 = 0; \\ \qquad u = 0, 1, \ldots, \lfloor \frac{m-(i_1+j_1)}{2} \rfloor, \ and \\ i_1 = 0; i_2 = 0, 1, \ldots, m; j_1 = 0; j_2 = 1, 2, \ldots, m - i_2; \\ \qquad u = 0, 1, \ldots, \lfloor \frac{m-(i_2+j_2)}{2} \rfloor, \end{array} \right\},$$

$$\mathcal{I}_{y,p} := \left\{ \begin{array}{l} i_1 = 0, 1, \ldots, m; i_2 = 0, 1, \ldots, m - i_1; \\ j_1 = 1, 2, \ldots, \lceil \tau(i_1+i_2) \rceil - \lceil (i_1+i_2)/2 \rceil \end{array} \right\},$$

$$\mathcal{I}_{y,q} := \left\{ \begin{array}{l} i_1 = 0, 1, \ldots, m; i_2 = 0, 1, \ldots, m - i_1; \\ j_2 = 1, 2, \ldots, \lceil \tau(i_1+i_2) \rceil - \lfloor (i_1+i_2)/2 \rfloor \end{array} \right\}.$$

Let B be a matrix whose rows consist of coefficients of $g_{[i_1,i_2,j_1,j_2,u]}(x_{p,1}X_{p,1}, x_{q,1}X_{q,1}, x_{p,2}X_{p,2}, x_{q,2}X_{q,2}, y_pY_p, y_qY_q)$, $g'_{[i_1,i_2,j_1],p}(x_{p,1}X_{p,1}, x_{q,1}X_{q,1}, x_{p,2}X_{p,2}, x_{q,2}X_{q,2}, y_pY_p, y_qY_q)$, and $g'_{[i_1,i_2,j_2],q}(x_{p,1}X_{p,1}, x_{q,1}X_{q,1}, x_{p,2}X_{p,2}, x_{q,2}X_{q,2}, y_pY_p, y_qY_q)$ with indices in $\mathcal{I}_x$, $\mathcal{I}_{y,p}$, and $\mathcal{I}_{y,q}$, respectively. If the shift-polynomials are ordered as

$g_{[i_1,i_2,j_1,j_2,u]} \prec g'_{[i_1,i_2,j_1],p}, g'_{[i_1,i_2,j_2],q},$

$g_{[i'_1,i'_2,j'_1,j'_2,u']} \prec g_{[i_1,i_2,j_1,j_2,u]} \ for \ i'_1 + i'_2 < i_1 + i_2,$

$g_{[i'_1,i'_2,j'_1,j'_2,u']} \prec g_{[i_1,i_2,j_1,j_2,u]} \ for \ i'_1 + i'_2 = i_1 + i_2, u' < u,$

$g_{[i'_1,i'_2,j'_1,0,u]} \prec g_{[i_1,i_2,j_1,0,u]} \ for \ i'_1 + i'_2 = i_1 + i_2, j'_1 < j_1,$

$g_{[i'_1,i'_2,0,j'_2,u]} \prec g_{[i_1,i_2,0,j_2,u]} \ for \ i'_1 + i'_2 = i_1 + i_2, j'_2 < j_2,$

$g'_{[i'_1,i'_2,j'_1]}, g'_{[i'_1,i'_2,j'_2],q} \prec g'_{[i_1,i_2,j_1],p}, g'_{[i_1,i_2,j_2],q} \ for \ i'_1 + i'_2 < i_1 + i_2,$

$g'_{[i'_1,i'_2,j'_1]} \prec g'_{[i_1,i_2,j_1],p} \ for \ i'_1 + i'_2 = i_1 + i_2, j'_1 < j_1,$

$g'_{[i'_1,i'_2,j'_2],q} \prec g'_{[i_1,i_2,j_2],q} \ for \ i'_1 + i'_2 = i_1 + i_2, j'_2 < j_2,$

and N^{-1} (mod e^m) *is multiplied appropriately, then the matrix becomes triangular with diagonals*

$- X_{p,1}^{i_1+j_1+u} X_{p,2}^{i_2+j_2+u} Y_p^{\lceil (i_1+i_2)/2 \rceil} e^{m-(i_1+i_2+u)} \ for \ g_{[i_1,i_2,j_1,j_2,u]} \ if \ i_1 + i_2 \ is \ odd,$

$- X_{q,1}^{i_1+j_1+u} X_{q,2}^{i_2+j_2+u} Y_q^{\lfloor (i_1+i_2)/2 \rfloor} e^{m-(i_1+i_2+u)} \ for \ g_{[i_1,i_2,j_1,j_2,u]} \ if \ i_1 + i_2 \ is \ even,$

$- X_{p,1}^{i_1} X_{p,2}^{i_2} Y_p^{\lceil (i_1+i_2)/2 \rceil + j_1} e^{m-(i_1+i_2)} \ for \ g'_{[i_1,i_2,j_1],p},$

$- X_{q,1}^{i_1} X_{q,2}^{i_2} Y_q^{\lfloor (i_1+i_2)/2 \rfloor + j_2} e^{m-(i_1+i_2)} \ for \ g'_{[i_1,i_2,j_2],q}.$

We do not prove the lemma, however, the proof can be obtained in the same manner as in Sect. 3.4. The polynomials which we use contain six variables $x_{p,1}, x_{p,2}, x_{q,1}, x_{q,2}, y_p, y_q$. Furthermore, there are three algebraic relations, i.e., $x_{q,1} = x_{p,1} + 1$, $x_{p,2} = x_{q,2} + 1$, and $y_p y_q = N$. By using the last relation, i.e., $y_p y_q = N$, we transform all monomials as they do not have both y_p and y_q simultaneously as the proof of Lemma 3. In addition, by using the other relations, i.e., $x_{q,1} = x_{p,1} + 1$ and $x_{p,2} = x_{q,2} + 1$, we transform all monomials as they do not have both $x_{p,1}$ and $x_{q,1}$ simultaneously or both $x_{p,2}$ and $x_{q,2}$ simultaneously. More concretely, the variables $x_{p,1}$ and $x_{p,2}$ appear only in monomials whose exponents of y_p are positive whereas the variables $x_{q,1}$ and $x_{q,2}$ appear only in monomials whose exponents of y_q are non-negative.

We compute the resulting condition of Theorem 5. The dimension n and the determinant of the lattice $\det(\boldsymbol{B}) = X^{s_X} Y^{s_Y} e^{s_e}$ can be computed as:

$$n = \sum_{(i_1,i_2,j_1,j_2,u) \in \mathcal{I}_x} 1 + \sum_{(i_1,i_2,j_1) \in \mathcal{I}_{y,p}} 1 + \sum_{(i_1,i_2,j_2) \in \mathcal{I}_{y,q}} 1$$

$$= \frac{2\tau}{3} m^3 + o(m^3),$$

$$s_X = \sum_{(i_1,i_2,j_1,j_2,u) \in \mathcal{I}_x} (i_1 + i_2 + j_1 + j_2 + 2u) + \sum_{(i_1,i_2,j_1) \in \mathcal{I}_{y,p}} (i_1 + i_2)$$

$$+ \sum_{(i_1,i_2,j_2) \in \mathcal{I}_{y,q}} (i_1 + i_2)$$

$$= \frac{\tau}{2} m^4 + o(m^4),$$

$$s_Y = \sum_{\substack{(i_1, i_2, j_1, j_2, u) \in \\ \mathcal{I}_x \text{ s.t. } i_1 + i_2 \text{ is odd}}} \left\lceil \frac{i_1 + i_2}{2} \right\rceil + \sum_{\substack{(i_1, i_2, j_1, j_2, u) \in \\ \mathcal{I}_x \text{ s.t. } i_1 + i_2 \text{ is even}}} \left\lfloor \frac{i_1 + i_2}{2} \right\rfloor$$

$$+ \sum_{(i_1, i_2, j_1) \in \mathcal{I}_{y,p}} \left(\left\lceil \frac{i_1 + i_2}{2} \right\rceil + j_1 \right) + \sum_{(i_1, i_2, j_2) \in \mathcal{I}_{y,q}} \left(\left\lfloor \frac{i_1 + i_2}{2} \right\rfloor + j_2 \right)$$

$$= \frac{\tau^2}{4} m^4 + o(m^4),$$

$$s_e = \sum_{(i_1, i_2, j_1, j_2, u) \in \mathcal{I}_x} (m - (i_1 + i_2 + u)) + \sum_{(i_1, i_2, j_1) \in \mathcal{I}_{y,p}} (m - (i_1 + i_2))$$

$$+ \sum_{(i_1, i_2, j_2) \in \mathcal{I}_{y,q}} (m - (i_1 + i_2))$$

$$= \frac{2\tau + 1}{12} m^4 + o(m^4).$$

Applying the LLL reduction, the polynomials obtained from the output vectors satisfy Howgrave-Graham's lemma if $X^{s_X} Y^{s_Y} e^{s_e} < e^{nm}$, i.e.,

$$\left(\alpha + \delta - \frac{1}{2} \right) \frac{\tau}{2} + \frac{1}{2} \cdot \frac{\tau^2}{4} + \alpha \cdot \frac{2\tau + 1}{12} < \alpha \cdot \frac{2\tau}{3}$$

by omitting low order terms of m. To minimize the left hand side of the inequality, we set the parameters $\tau = 1 - 2\delta$, then the condition becomes

$$\delta < \frac{1}{2} - \sqrt{\frac{\alpha}{6}}$$

as required. To satisfy the restriction $\tau \geq 1/2$, $\delta \leq 1/4$ and equivalently $\alpha \geq 3/8$ should hold.   □

4.2 Experimental Results

We have implemented the experiment program of Sect. 4.1 in Magma 2.10 computer algebra system [5] on a PC with Intel(R) Core(TM) Duo CPU(3.30 GHz, 4.0 GB RAM Windows 7). Table 3 lists the asymptotic and experimental results on factoring 1000-bit RSA moduli with varying dimension of lattice under small decryption exponents. In all experiments, we successfully find the factorization of these RSA moduli.

5 Attacks on the Variants

In this section, we study small CRT-exponent attacks on the RSA variants, i.e., the Multi-Prime RSA, Takagi's RSA, and the RSA with multiple exponent pairs. We extend our attack of Theorem 2 to the variants.

Table 3. For 1000-bit RSA moduli, asymptotic and experimental comparisons of small d_p and d_q attacks on balanced CRT-RSA

Bitsize of N	Asymptotic	Expt.	$(m,$ dim.$)$	L^3 time (in sec.)
1000	0.091	0.034	(4, 95)	358.787
		0.053	(6, 252)	31390.147

5.1 Multi-Prime RSA

In this section, we extends the small CRT-exponent attack for the Multi-Prime RSA as follows.

Theorem 6. *Let* $N = \prod_{i=1}^{r} p_i$ *be an RSA modulus where* $r \geq 2$ *and all the prime factors* $p_1, \ldots, p_r$ *are the same bit-size. Let* $e = N^\alpha$ *and* $d_{p_i} < N^{\delta_i}$ *be a public/CRT exponent respectively such that* $ed_{p_i} = 1 \pmod{(p_i - 1)}$ *for all* $i = 1, \ldots, r$. *Given public elements* N *and* e, *if*

$$\min_{i \in \{1, \ldots, r\}} \delta_i < \frac{1 - \sqrt{(r-1)\alpha}}{r} \quad for \ \alpha > \frac{r-1}{r^2},$$

then N *can be factorized in polynomial time by assuming that polynomials which are derived from LLL reduced bases are algebraically independent.*

We can successfully extend an attack for the Multi-Prime RSA in the sense that Theorem 6 becomes the same as Theorem 3 for $r = 2$.

We also extend May's modulo p_i attack [28] for the Multi-Prime RSA as follows.

Theorem 7 (Adapted from [27]). *Let* $N = \prod_{i=1}^{r} p_i$ *be an RSA modulus where* $r \geq 2$ *and all the prime factors* $p_1, \ldots, p_r$ *are the same bit-size. Let* $e = N^\alpha$ *and* $d_{p_i} < N^{\delta_i}$ *be a public/CRT exponent respectively such that* $ed_{p_i} = 1 \pmod{(p_i - 1)}$ *for all* $i = 1, \ldots, r$. *Given public elements* N *and* e, *if*

$$\min_{i \in \{1, \ldots, r\}} \delta_i < \frac{r + 1 - r^2\alpha}{2r^2},$$

then N *can be factorized in polynomial time by assuming that polynomials which are derived from LLL reduced bases are algebraically independent.*

We can successfully extend an attack for the Multi-Prime RSA in the sense that Theorem 7 becomes the same as Theorem 4 for $r = 2$. We omit the proof since it is almost the same as Theorem 9 of [27]. The bound of Theorem 6 is always better than or equal to that of Theorem 7. Figure 4 compares the attack condition between Theorems 6 and 7 for $r = 3$ and 4.

5.2 Takagi's RSA

In this section, we extends the small CRT-exponent attack for Takagi's RSA as follows.

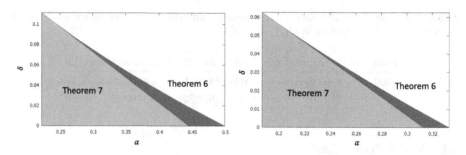

Fig. 4. Comparisons between our attacks of Theorems 6 and 7. The left and the right figure is for $r = 3$ and 4, respectively.

Theorem 8. *Let $N = p^r q$ be an RSA modulus where $r \geq 1$ and the prime factors p and q are the same bit-size. Let $e = N^\alpha$ and $d_p < N^{\delta_p}, d_q < N^{\delta_q}$ be a public/CRT exponent respectively such that $ed_p = 1 \pmod{(p-1)}$ and $ed_q = 1 \pmod{(q-1)}$. Given public elements N and e, if*

$$\min\{\delta_p, \delta_q\} < \frac{1 - \sqrt{r\alpha}}{r+1} \quad for \quad \alpha > \frac{r}{(r+1)^2},$$

then N can be factorized in polynomial time by assuming that polynomials which are derived from LLL reduced bases are algebraically independent.

We can successfully extend an attack for Takagi's RSA in the sense that Theorem 8 becomes the same as Theorem 3 for $r = 1$. Although Shinohara et al. [38] extended Bleichenbacher-May's attack, our attack is always better.

We also extend May's modulo a prime factor attack [28] for Takagi's RSA as follows.

Theorem 9 (Adapted from [28]). *Let $N = p^r q$ be an RSA modulus where $r \geq 1$ and the prime factors p and q are the same bit-size. Let $e = N^\alpha$ and $d_p < N^{\delta_p}, d_q < N^{\delta_q}$ be a public/CRT exponent respectively such that $ed_p = 1 \pmod{(p-1)}$ and $ed_q = 1 \pmod{(q-1)}$. Given public elements N and e, if*

$$\delta_p < \frac{2r + 1 - (r+1)^2 \alpha}{2(r+1)^2} \quad or \quad \delta_q < \frac{r + 2 - (r+1)^2 \alpha}{2(r+1)^2},$$

then N can be factorized in polynomial time by assuming that polynomials which are derived from LLL reduced bases are algebraically independent.

We can successfully extend an attack for the Takagi's RSA in the sense that Theorem 9 becomes the same as Theorem 4 for $r = 1$. We omit the proof since it is almost the same as Theorem 9 of [27]. The bound for δ_q of Theorem 8 is always better than or equal to that of Theorem 9, however, the bound for δ_p of Theorem 9 is better than or equal to that of Theorem 8. Figure 5 compares the attack condition for small d_p between Theorems 8 and 9 for $r = 2$ and 3.

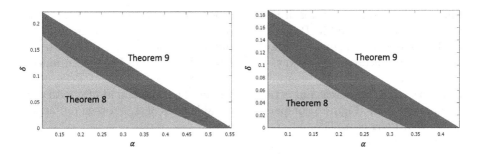

Fig. 5. Comparisons between our attacks of Theorems 8 and 9. The left and the right figure is for $r = 2$ and 3, respectively.

5.3   RSA with Multiple Exponent Pairs

In this section, we extends the small CRT-exponent attack for the RSA with multiple exponent pairs as follows.

Theorem 10. *Let $N = pq$ be an RSA modulus where the prime factors p and q are the same bit-size. Let $e_\ell = N^\alpha$ and $d_{q,\ell} < N^\delta$ for $\ell = 1, \ldots, r$ be a public/CRT exponent respectively such that $e_\ell d_{q,\ell} = 1 \pmod{(q-1)}$. Given public elements N and $e_1, \ldots, e_r$, if*

$$\delta < \frac{1}{2} - \sqrt{\frac{\alpha}{3r+1}},$$

then N can be factorized in time polynomial in input length and exponential in r by assuming that polynomials which are derived from LLL reduced bases are algebraically independent.

We can successfully extend the attack for RSA with multiple exponent pairs in the sense that Theorem 10 becomes the same as Theorem 3 for $r = 1$. We do not

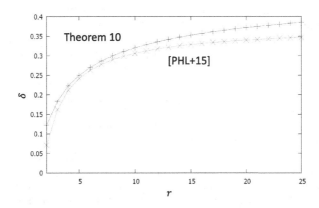

Fig. 6. Comparison between our attack (Theorem 10) and the attack of Peng et al. [33]

think May's modulo q approach is an appropriate way for the attack scenario, hence, we do not extend it. Peng et al. proposed the attack (Theorem 2 of [33]) which extended Bleichenbacher-May's [2] and works when $\delta < (9r-14)/(24r+8)$ for an $\alpha = 1$. Theorem 10 is always better than the attack of Peng et al. Indeed, even if there are infinitely many exponent pairs r, the attack of Peng et al. works for $\delta < 3/8$ whereas our attack works for the same bound of δ with only 21 exponent pairs. Figure 6 compares recoverable sizes of d_q between our attack and that of Peng et al. [33].

6 Concluding Remarks

In this paper, we studied a lattice-based cryptanalysis of the small CRT-exponent RSA. We developed a novel lattice construction technique that is specialized to the CRT-RSA key generation and proposed several improved attacks. When a prime factor p is significantly smaller than the other prime factor q with a small d_q, we solved an open problem which was claimed in [2,28]; we proposed an attack that works for $p < N^{0.5}$. When both d_p and d_q are small, we proposed an attack that works for $d_p, d_q < N^{0.091}$ with a full size e. We also proposed attacks on the RSA variants, i.e., the Multi-Prime RSA, Takagi's RSA, and RSA with multiple exponent pairs.

Acknowledgement. We would like to thank Shuichi Katsumata for his helpful comments. Atsushi Takayasu is supported by a JSPS Fellowship for Young Scientists. This research was supported by CREST, JST, JSPS KAKENHI Grant Number 14J08237, National Key Basic Research Program of China (2013CB834203) and the National Natural Science Foundation of China (Grants 61472417, 61632020, 61472416).

References

1. Bauer, A., Vergnaud, D., Zapalowicz, J.-C.: Inferring sequences produced by nonlinear pseudorandom number generators using Coppersmith's methods. In: Fischlin, M., Buchmann, J., Manulis, M. (eds.) PKC 2012. LNCS, vol. 7293, pp. 609–626. Springer, Heidelberg (2012). doi:10.1007/978-3-642-30057-8_36
2. Bleichenbacher, D., May, A.: New attacks on RSA with small secret CRT-exponents. In: Yung, M., Dodis, Y., Kiayias, A., Malkin, T. (eds.) PKC 2006. LNCS, vol. 3958, pp. 1–13. Springer, Heidelberg (2006). doi:10.1007/11745853_1
3. Blömer, J., May, A.: New partial key exposure attacks on RSA. In: Boneh, D. (ed.) CRYPTO 2003. LNCS, vol. 2729, pp. 27–43. Springer, Heidelberg (2003). doi:10.1007/978-3-540-45146-4_2
4. Boneh, D., Durfee, G.: Cryptanalysis of RSA with private key d less than $N^{0.292}$. IEEE Trans. Inf. Theor. 46(4), 1339–1349 (2000)
5. Bosma, W., Cannon, J.J., Playout, C.: The magma algebra system I: the user language. J. Symb. Comput. 24(3/4), 235–265 (1997)
6. Coppersmith, D.: Finding a small root of a bivariate integer equation; factoring with high bits known. In: Maurer, U. (ed.) EUROCRYPT 1996. LNCS, vol. 1070, pp. 178–189. Springer, Heidelberg (1996). doi:10.1007/3-540-68339-9_16

7. Coppersmith, D.: Finding a small root of a univariate modular equation. In: Maurer, U. (ed.) EUROCRYPT 1996. LNCS, vol. 1070, pp. 155–165. Springer, Heidelberg (1996). doi:10.1007/3-540-68339-9_14

8. Coppersmith, D.: Small solutions to polynomial equations, and low exponent RSA vulnerabilities. J. Cryptology **10**(4), 233–260 (1997)

9. Coppersmith, D.: Finding small solutions to small degree polynomials. In: Silverman, J.H. (ed.) CaLC 2001. LNCS, vol. 2146, pp. 20–31. Springer, Heidelberg (2001). doi:10.1007/3-540-44670-2_3

10. Coron, J.-S.: Finding small roots of bivariate integer polynomial equations revisited. In: Cachin, C., Camenisch, J.L. (eds.) EUROCRYPT 2004. LNCS, vol. 3027, pp. 492–505. Springer, Heidelberg (2004). doi:10.1007/978-3-540-24676-3_29

11. Durfee, G., Nguyen, P.Q.: Cryptanalysis of the RSA schemes with short secret exponent from Asiacrypt '99. In: Okamoto, T. (ed.) ASIACRYPT 2000. LNCS, vol. 1976, pp. 14–29. Springer, Heidelberg (2000). doi:10.1007/3-540-44448-3_2

12. Esgin, M.F., Kiraz, M.S., Uzunkol, O.: A new partial key exposure attack on multi-power RSA. In: Maletti, A. (ed.) CAI 2015. LNCS, vol. 9270, pp. 103–114. Springer, Cham (2015). doi:10.1007/978-3-319-23021-4_10

13. Galbraith, S.D., Heneghan, C., McKee, J.F.: Tunable balancing of RSA. In: Boyd, C., González Nieto, J.M. (eds.) ACISP 2005. LNCS, vol. 3574, pp. 280–292. Springer, Heidelberg (2005). doi:10.1007/11506157_24

14. Herrmann, M.: Lattice-based cryptanalysis using unravelled linearization. Ph.D. thesis, der Ruhr-Universitat Bochum (2011)

15. Herrmann, M., May, A.: Attacking power generators using unravelled linearization: when do we output too much? In: Matsui, M. (ed.) ASIACRYPT 2009. LNCS, vol. 5912, pp. 487–504. Springer, Heidelberg (2009). doi:10.1007/978-3-642-10366-7_29

16. Herrmann, M., May, A.: Maximizing small root bounds by linearization and applications to small secret exponent RSA. In: Nguyen, P.Q., Pointcheval, D. (eds.) PKC 2010. LNCS, vol. 6056, pp. 53–69. Springer, Heidelberg (2010). doi:10.1007/978-3-642-13013-7_4

17. Howgrave-Graham, N.: Finding small roots of univariate modular equations revisited. In: Darnell, M. (ed.) Cryptography and Coding 1997. LNCS, vol. 1355, pp. 131–142. Springer, Heidelberg (1997). doi:10.1007/BFb0024458

18. Huang, Z., Hu, L., Xu, J.: Attacking RSA with a composed decryption exponent using unravelled linearization. In: Lin, D., Yung, M., Zhou, J. (eds.) Inscrypt 2014. LNCS, vol. 8957, pp. 207–219. Springer, Cham (2015). doi:10.1007/978-3-319-16745-9_12

19. Huang, Z., Hu, L., Xu, J., Peng, L., Xie, Y.: Partial key exposure attacks on Takagi's variant of RSA. In: Boureanu, I., Owesarski, P., Vaudenay, S. (eds.) ACNS 2014. LNCS, vol. 8479, pp. 134–150. Springer, Cham (2014). doi:10.1007/978-3-319-07536-5_9

20. Jochemsz, E., May, A.: A strategy for finding roots of multivariate polynomials with new applications in attacking RSA variants. In: Lai, X., Chen, K. (eds.) ASIACRYPT 2006. LNCS, vol. 4284, pp. 267–282. Springer, Heidelberg (2006). doi:10.1007/11935230_18

21. Jochemsz, E., May, A.: A polynomial time attack on rsa with private CRT-exponents smaller than $N^{0.073}$. In: Menezes, A. (ed.) CRYPTO 2007. LNCS, vol. 4622, pp. 395–411. Springer, Heidelberg (2007). doi:10.1007/978-3-540-74143-5_22

22. Kunihiro, N.: On optimal bounds of small inverse problems and approximate GCD problems with higher degree. In: Gollmann, D., Freiling, F.C. (eds.) ISC 2012. LNCS, vol. 7483, pp. 55–69. Springer, Heidelberg (2012). doi:10.1007/978-3-642-33383-5_4

23. Kunihiro, N., Shinohara, N., Izu, T.: A unified framework for small secret exponent attack on RSA. IEICE Trans. **97-A**(6), 1285–1295 (2014)
24. Lenstra, A., Lenstra, H., Lovász, L.: Factoring polynomials with rational coefficients. Math. Ann. **261**, 515–534 (1982)
25. Lu, Y., Zhang, R., Lin, D.: Factoring multi-power RSA modulus $N = p^r q$ with partial known bits. In: Boyd, C., Simpson, L. (eds.) Information Security and Privacy - 18th Australasian Conference, ACISP 2013. LNCS, vol. 7959, pp. 57–71. Springer, Heidelberg (2013)
26. Lu, Y., Zhang, R., Lin, D.: New partial key exposure attacks on CRT-RSA with large public exponents. In: Boureanu, I., Owesarski, P., Vaudenay, S. (eds.) ACNS 2014. LNCS, vol. 8479, pp. 151–162. Springer, Cham (2014). doi:10.1007/978-3-319-07536-5_10
27. Lu, Y., Zhang, R., Peng, L., Lin, D.: Solving linear equations modulo unknown divisors: revisited. In: Iwata, T., Cheon, J.H. (eds.) ASIACRYPT 2015. LNCS, vol. 9452, pp. 189–213. Springer, Heidelberg (2015). doi:10.1007/978-3-662-48797-6_9
28. May, A.: Cryptanalysis of unbalanced RSA with small CRT-exponent. In: Yung, M. (ed.) CRYPTO 2002. LNCS, vol. 2442, pp. 242–256. Springer, Heidelberg (2002). doi:10.1007/3-540-45708-9_16
29. May, A.: New RSA vulnerabilities using lattice reduction methods. Ph.D. thesis, University of Paderborn (2003)
30. May, A.: Using LLL-reduction for solving RSA and factorization problems. In: Nguyen, P.Q., Vallée, B. (eds.) The LLL Algorithm - Survey and Applications. Information Security and Cryptography, pp. 315–348. Springer, Heidelberg (2010)
31. Nguyen, P.Q., Stern, J.: The two faces of lattices in cryptology. In: Silverman, J.H. (ed.) CaLC 2001. LNCS, vol. 2146, pp. 146–180. Springer, Heidelberg (2001). doi:10.1007/3-540-44670-2_12
32. Peng, L., Hu, L., Huang, Z., Xu, J.: Partial prime factor exposure attacks on RSA and its Takagi's variant. In: Lopez, J., Wu, Y. (eds.) ISPEC 2015. LNCS, vol. 9065, pp. 96–108. Springer, Cham (2015). doi:10.1007/978-3-319-17533-1_7
33. Peng, L., Hu, L., Lu, Y., Sarkar, S., Xu, J., Huang, Z.: Cryptanalysis of variants of RSA with multiple small secret exponents. In: Biryukov, A., Goyal, V. (eds.) INDOCRYPT 2015. LNCS, vol. 9462, pp. 105–123. Springer, Cham (2015). doi:10.1007/978-3-319-26617-6_6
34. Quisquater, J.J., Couvreur, C.: Fast decipherment algorithm for RSA public-key cryptosystem. Electron. Lett. **18**, 905–907 (1982)
35. Sarkar, S.: Small secret exponent attack on RSA variant with modulus $N = p^r q$. Des. Codes Crypt. **73**(2), 383–392 (2014)
36. Sarkar, S.: Revisiting prime power RSA. Discrete Appl. Math. **203**, 127–133 (2016)
37. Sarkar, S., Maitra, S.: Partial key exposure attack on CRT-RSA. In: Abdalla, M., Pointcheval, D., Fouque, P.-A., Vergnaud, D. (eds.) ACNS 2009. LNCS, vol. 5536, pp. 473–484. Springer, Heidelberg (2009). doi:10.1007/978-3-642-01957-9_29
38. Shinohara, N., Izu, T., Kunihiro, N.: Small secret CRT-exponent attacks on takagi's RSA. IEICE Trans. **94-A**(1), 19–27 (2011)
39. Sun, H., Wu, M.: An approach towards rebalanced RSA-CRT with short public exponent. IACR Cryptology ePrint Archive 2005, 53 (2005)
40. Takayasu, A., Kunihiro, N.: Better lattice constructions for solving multivariate linear equations modulo unknown divisors. IEICE Trans. **97-A**(6), 1259–1272 (2014)
41. Takayasu, A., Kunihiro, N.: Cryptanalysis of RSA with multiple small secret exponents. In: Susilo, W., Mu, Y. (eds.) ACISP 2014. LNCS, vol. 8544, pp. 176–191. Springer, Cham (2014). doi:10.1007/978-3-319-08344-5_12

42. Takayasu, A., Kunihiro, N.: General bounds for small inverse problems and its applications to multi-prime RSA. In: Lee, J., Kim, J. (eds.) ICISC 2014. LNCS, vol. 8949, pp. 3–17. Springer, Cham (2015). doi:10.1007/978-3-319-15943-0_1

43. Takayasu, A., Kunihiro, N.: Partial key exposure attacks on RSA: achieving the Boneh-Durfee bound. In: Joux, A., Youssef, A. (eds.) SAC 2014. LNCS, vol. 8781, pp. 345–362. Springer, Cham (2014). doi:10.1007/978-3-319-13051-4_21

44. Takayasu, A., Kunihiro, N.: Partial key exposure attacks on CRT-RSA: better cryptanalysis to full size encryption exponents. In: Malkin, T., Kolesnikov, V., Lewko, A.B., Polychronakis, M. (eds.) ACNS 2015. LNCS, vol. 9092, pp. 518–537. Springer, Cham (2015). doi:10.1007/978-3-319-28166-7_25

45. Takayasu, A., Kunihiro, N.: How to generalize RSA cryptanalyses. In: Cheng, C.-M., Chung, K.-M., Persiano, G., Yang, B.-Y. (eds.) PKC 2016. LNCS, vol. 9615, pp. 67–97. Springer, Heidelberg (2016). doi:10.1007/978-3-662-49387-8_4

46. Takayasu, A., Kunihiro, N.: Partial key exposure attacks on CRT-RSA: general improvement for the exposed least significant bits. In: Bishop, M., Nascimento, A.C.A. (eds.) ISC 2016. LNCS, vol. 9866, pp. 35–47. Springer, Cham (2016). doi:10.1007/978-3-319-45871-7_3

47. Takayasu, A., Kunihiro, N.: Partial key exposure attacks on RSA with multiple exponent pairs. In: Liu, J.K., Steinfeld, R. (eds.) ACISP 2016. LNCS, vol. 9723, pp. 243–257. Springer, Cham (2016). doi:10.1007/978-3-319-40367-0_15

48. Takayasu, A., Kunihiro, N.: A tool kit for partial key exposure attacks on RSA. In: Handschuh, H. (ed.) CT-RSA 2017. LNCS, vol. 10159, pp. 58–73. Springer, Cham (2017). doi:10.1007/978-3-319-52153-4_4

49. Wiener, M.J.: Cryptanalysis of short RSA secret exponents. IEEE Trans. Inf. Theor. **36**(3), 553–558 (1990)

Multiparty Computation II

Group-Based Secure Computation: Optimizing Rounds, Communication, and Computation

Elette Boyle[1](✉), Niv Gilboa[2], and Yuval Ishai[3]

[1] IDC Herzliya, Herzliya, Israel
elette.boyle@idc.ac.il
[2] Ben Gurion University, Beersheba, Israel
gilboan@bgu.ac.il
[3] Technion and UCLA, Haifa, Israel
yuvali@cs.technion.ac.il

Abstract. A recent work of Boyle et al. (Crypto 2016) suggests that "group-based" cryptographic protocols, namely ones that only rely on a cryptographically hard (Abelian) group, can be surprisingly powerful. In particular, they present succinct two-party protocols for securely computing branching programs and NC^1 circuits under the DDH assumption, providing the first alternative to fully homomorphic encryption.

In this work we further explore the power of group-based secure computation protocols, improving both their asymptotic and concrete efficiency. We obtain the following results.

– **Black-box use of group.** We modify the succinct protocols of Boyle et al. so that they only make a black-box use of the underlying group, eliminating an expensive non-black-box setup phase.
– **Round complexity.** For any constant number of parties, we obtain 2-round MPC protocols based on a PKI setup under the DDH assumption. Prior to our work, such protocols were only known using fully homomorphic encryption or indistinguishability obfuscation.
– **Communication complexity.** Under DDH, we present a secure 2-party protocol for any NC^1 or log-space computation with n input bits and m output bits using $n + (1 + o(1))m + \mathsf{poly}(\lambda)$ bits of communication, where λ is a security parameter. In particular, our protocol can generate n instances of bit-oblivious-transfer using $(4 + o(1)) \cdot n$ bits of communication. This gives the first constant-rate OT protocol under DDH.
– **Computation complexity.** We present several techniques for improving the computational cost of the share conversion procedure of Boyle et al., improving the concrete efficiency of group-based protocols by several orders of magnitude.

1 Introduction

Gentry's 2009 breakthrough on fully homomorphic encryption (FHE) [18,36] changed the landscape of the theory of secure computation. FHE enables arbitrary computations on encrypted inputs, thereby providing a general-purpose

© International Association for Cryptologic Research 2017
J.-S. Coron and J.B. Nielsen (Eds.): EUROCRYPT 2017, Part II, LNCS 10211, pp. 163–193, 2017.
DOI: 10.1007/978-3-319-56614-6_6

tool for *succinct* secure computation protocols whose communication complexity is smaller than the circuit size of the function being computed. FHE-based protocols were also used to minimize the *round complexity* of secure multiparty computation [2,14,32,33].[1]

On the downside, despite impressive recent progress [13,15,22], the concrete efficiency of current FHE implementations still leaves much to be desired. Moreover, the set of cryptographic assumptions on which FHE can be based is still quite narrow. These two limitations may in fact be related, in that attempts at efficient implementation are curbed by the limited variety of FHE candidates. Indeed, all such candidates rely on similar *lattice-related* algebraic structures and are subject to lattice reduction attacks that have a negative impact on concrete efficiency. In particular, no FHE construction is known under a discrete-log-type assumption or even in the generic group model. This should be contrasted with standard public-key encryption schemes and non-succinct secure computation protocols that can be easily (and unconditionally) realized in the generic group model.

A recent work of Boyle et al. [8] introduced a new technique for succinct secure computation that can be based on any DDH-hard group. (For better concrete efficiency, it is useful to rely on stronger assumptions than DDH, such as the circular security of ElGamal encryption.) While the results obtained using this group-based approach are weaker than corresponding FHE-based results in several important aspects, they do give hope for better concrete efficiency in useful application scenarios. The present work is motivated in part by this hope.

More concretely, the approach of [8] replaces the use of FHE by a 2-party *homomorphic secret sharing* (HSS) primitive, which turns out to be sufficient for the purpose of succinct secure two-party computation. An HSS scheme is a secret sharing scheme that supports homomorphic computations on the shares, such that the output of the computation is compactly shared between the parties. We in fact make the stronger requirement that the output be *additively* shared between the parties over a finite Abelian group. In particular, if the output is a single bit, each output share can be just a single bit. HSS can be viewed as a dual version of *function section sharing* [7], where the roles of the function and the input are reversed, or a weaker version of *additive-spooky encryption* [14].

The main result of [8] is a DDH-based HSS scheme for *branching programs*, which in particular captures logspace and NC^1 computations. We provide a high level overview of this HSS scheme in Sect. 2.2. The HSS scheme of [8] can be used to obtain succinct secure two-party computation protocols for the same classes. One difficulty in applying this HSS scheme towards secure computation is that it has an inverse polynomial error probability, and moreover the event of an error is correlated with the secret input. This difficulty was addressed in [8] by combining error-correcting codes with general-purpose secure two-party

---

[1] As in previous related works, our default notion of secure computation refers to security against *passive* (semi-honest) adversaries. In most cases, similar protocols with security against active (malicious) adversaries can be obtained under the same assumptions by using a suitable version of the GMW compiler [21,24,34].

computation protocols for recovering the correct output from the encoding. This approach has a significant overhead in communication and computation, and requires additional rounds of interaction.

The source of the error in the HSS scheme from [8] is a non-interactive *share conversion* procedure, which converts multiplicative shares into additive shares. To perform this conversion with an error probability bound of δ, the procedure requires $O((1/\delta) \cdot \log(1/\delta))$ (or *expected* $O(1/\delta)$) group multiplications.

1.1   Our Contribution

In this work we further explore the power of group-based secure computation protocols, improving both their asymptotic and concrete efficiency. Following is a detailed overview of our results and the underlying techniques.

Black-box use of group. The group-based succinct protocols from [8] use general-purpose secure computation to distribute the key generation of a "public-key" HSS scheme, namely one that allows joint computation on two or more shared inputs. This procedure leads to poor concrete efficiency, and makes a non-black-box use of the underlying cryptographic group. We present a generic approach for obtaining similar results while only making a black-box use of the underlying group. This approach relies on the plaintext- and key-homomorphism properties of ElGamal encryption (or its circular-secure variant [4]) and can be used for improving the concrete cost of group-based protocols.

Minimizing round complexity. For any constant number of parties, we obtain 2-round MPC protocols based on a Public Key Infrastructure (PKI) setup under the DDH assumption.[2] Prior to our work, such protocols were only known using different flavors of FHE [2,14,33] or indistinguishability obfuscation [14,17]. (Granted, the latter protocols can further support polynomial number of parties, and with milder setup requirements: PKI setup can be relaxed to a CRS setup by using *multi-key FHE,* which can be based on LWE [14,33], or even eliminated by relying on indistinguishability obfuscation [14].)

Our 2-round protocol is obtained in three steps. In the first step, we construct a 1-round (PKI-based) distributed HSS scheme, which can be used to jointly share inputs that originate from multiple clients. This can be used to construct a 2-round protocol in the PKI model that allows m clients to compute a function of their inputs with the help of two servers (of which at most one is corrupted), where in this protocol each client sends a single message to each server and each server sends a single message to each client. The protocol only satisfies a weak notion of $1/\text{poly}$ security (i.e., security with inverse-polynomial simulation error), due to the input-dependent error of the HSS scheme (inherited from the share conversion procedure of [8]). The protocol can be used to succinctly evaluate branching programs. Alternatively, it can be used to evaluate general circuits

---

[2] This implies 3-round protocols in the plain model. Note, however, that unlike the first round in a general 3-round protocol, a PKI setup is *independent of the inputs and the number of parties.*

(at the cost of compromising succinctness) by applying the HSS evaluation to a low-complexity randomized encoding of the circuit [1, 3, 38].

The second step achieves security amplification. That is, we improve the security of the above protocol to hold with negligible simulation error, without increasing the round complexity. This is done by evaluating a *compiled* version of the desired computation, which is resilient to leakage on intermediate computation values. This compilation is obtained by using a virtual "client-server" MPC protocol to make computations locally random, where the initial messages from clients to virtual servers are HSS-shared between the two real servers, and the role of each virtual server is emulated by the two (real) servers via HSS evaluation. This virtual MPC protocol only needs to provide security against a small fraction of corrupted (semi-honest) virtual servers, but additionally needs to be *robust* in the sense that the output can still be computed even when a bounded number of virtual servers fail. The latter feature is important for coping with the error of the underlying HSS.

A technical issue we need to deal with is that the event of failure in the share conversion procedure is correlated not only with the input but also with bits of the secret key. To cope with this type of leakage, we modify the underlying HSS scheme to use a redundant representation of the secret key that makes leakage of a small number of bits harmless.

To make this security amplification step efficient, we need the virtual MPC protocol to have a constant number of rounds, and the next message function computed by each server in each round to be efficiently implementable by branching programs. In particular, we can use 2-round virtual MPC protocols that apply to *constant-degree polynomials* and do not require any server-to-server communication. (Again, general circuits can be handled via randomized encoding.) These protocols are sufficient for our main feasibility result of 2-round MPC from DDH. We can additionally get *succinct* 2-round protocols for NC^1 by applying a different type of virtual MPC protocol that computes NC^1 functions in a constant number of rounds with low client-to-server communication, but additionally requires (a large amount of) server-to-server communication.[3] As a corollary, we get a 2-message 2-party protocol for computing any NC^1 function $f(x, y)$ (with output delivered to one party), where the length of each message is comparable to the length of the corresponding input (and is independent of the complexity of f).

In the third and final step, we use a player virtualization technique [10, 23] to transform the 2-round (m-client) 2-server protocol into a 2-round protocol with m clients and an *arbitrary constant number of servers* k. At a high level, this is done by iteratively emulating the computations of a single server (beginning with a single server in the 2-server protocol) by two separate servers, via another level of 2-round MPC. Because of the complexity blowup in each iteration, this virtualization step can only be applied a constant number of times.

---

[3] Interestingly, this approach does not seem to extend to branching programs using known techniques, since in known constant-round protocols for branching programs the next message function cannot be efficiently computed by branching programs.

Such a client-server protocol readily implies a 2-round (standard) k-party protocol by letting $m = k$ and having each party emulate the corresponding client and server.

Improving communication complexity. Under DDH, we present a secure 2-party protocol for any NC^1 or log-space computation with n input bits and m output bits using $n + (1 + o(1))m + \mathsf{poly}(\lambda)$ bits of communication, where λ is a security parameter. In particular, we generate n instances of $\binom{2}{1}$-oblivious-transfer (OT) of bits using $4n + o(n) + \mathsf{poly}(\lambda)$ bits of communication. This gives the first constant-rate OT protocol under DDH. Constant-rate OT protocols (with a poor concrete rate) could previously be constructed using a polynomial-stretch local pseudorandom generator [27] or the Phi-hiding assumption [28]. A similar result to ours can also be obtained under LWE, via the HSS scheme implied by [14].

The above result is obtained via a new security amplification technique, which provides a simpler and more efficient alternative to the use of virtual MPC in the second step described above. The downside is that this approach is restricted to the 2-party setting and requires an additional round of interaction. The high level idea is as follows. Denote the two parties by P_0, P_1 and assume that the functionality f delivers an output only to P_1. We rely on a Las-Vegas variant of HSS where the shared output is guaranteed to be correct (i.e., the two output shares add up to the correct output) unless P_1 outputs $\bot$, where the latter occurs with small probability. The idea is to have P_1 use $\binom{m}{m-k}$-OT for $m \gg k$ in order to block itself from the k output shares of P_0 that correspond to the positions in which it outputs $\bot$. Note that the $m - k$ selected output shares can be simulated given the correct output and the output shares of P_1, and thus they do not leak any additional information about the input. To make up for the k lost output bits, we use an erasure code to encode the output. Since we can make the number of erasures small, we only need to introduce a small amount of redundancy to the output. A crucial observation which makes this approach useful is that the above form of "punctured OT" can be implemented with only $m + o(m)$ bits of communication by combining general-purpose 2PC with a puncturable pseudo-random function [37].

Improving computation complexity. We present several techniques for reducing the computational cost of the share conversion procedure from [8], improving the concrete efficiency of group-based protocols (both in [8] and the present work) by several orders of magnitude.

First, we present an optimization that improves the asymptotic *worst-case* running time of conversion by an $O(\log(1/\delta))$ factor, where δ is the error probability. In the procedure from [8], a group element h is mapped to the smallest non-negative integer i such that $h \cdot g^i$ (where g is a group generator) belongs to a pseudo-random set of distinguished group elements of density δ. Allowing δ error probability, $O((1/\delta) \cdot \log(1/\delta))$ values of i should be checked, requiring a similar number of group multiplications in the worst case. While the expected number of group multiplications is $O(\log(1/\delta))$, in applications that involve "shallow" computations (where many short sequences of RMS multiplications are performed

168   E. Boyle et al.

in parallel) it is the worst-case time that dominates the overall performance. The alternative approach we propose is to apply an integer-valued hash function ϕ to every group element, and return the (first) value of i in an interval of size $O(1/\delta)$ that minimizes the value of $\phi(h \cdot g^i)$. This requires only $O(1/\delta)$ group multiplications. We can also get an *unconditional* implementation of this alternative share conversion by using explicit constructions of "min-wise independent" hash functions [11,25].

Next, we present several optimization ideas that apply "conversion-friendly groups" towards improving the concrete running time of share conversion by several orders of magnitude. These optimizations rely on discrete-log-type assumptions in multiplicative subgroups of $\mathbb{Z}_p^*$ of a prime order q, where $p = 2q + 1$ is a prime which is close to a power of 2, and where $g = 2$ is a generator of the subgroup. We propose several concrete choices of such p. The advantage of such a group is that multiplying a group element h by the generator g can be done by shifting h by one bit to the left, and adding the difference between p and the closest power of 2 in case that the (removed) leftmost bit is 1. In fact, one can multiply h by g^w, where w is comparable to the machine word size (say, $w = 32$) by using a small constant expected number of machine word operations (64-bit additions or multiplications).

A second observation is that by making a seemingly mild heuristic assumption on the MSB sequence of the powers $h \cdot g^i$ (where h is random), it suffices to search for the first position in the sequence that contains a stretch of 0's of length $\approx \log(1/\delta)$. Concretely we need a *combinatorial* pseudo-randomness assumption asserting that such a stretch occurs roughly as often as expected in a totally random sequence.

By using an optimized "lazy" strategy for finding the first such stretch of 0's, the entire share conversion procedure can be implemented with an amortized cost of less than a single machine word operation per step. Concretely, the amortized cost is roughly 0.03 machine word additions and multiplications and 0.2 masking operations per step. This should be compared to a full group multiplication per step in the procedure of [8]. Combining all the optimizations, one can perform thousands of RMS multiplications per second with error probability that is small enough for performing shallow computations.

We note that the latter optimizations do not apply to Elliptic Curve groups, and hence do not provide the optimal level of succinctness. However, the gain in the computational cost of share conversion is arguably much more significant. We leave open the question of implementing similar optimizations for the case of Elliptic Curve groups.

2   Preliminaries

We give some necessary definitions and provide a high-level overview of the BGI construction of [8]. We refer the reader to the full version for further details.

2.1   Homomorphic Secret Sharing and DEHE

As in [8], we consider the case of 2-out-of-2 secret sharing, where an algorithm Share is used to split a secret $w \in \{0,1\}^n$ into two shares, such that each share computationally hides w. The homomorphic evaluation algorithm Eval is used to locally evaluate a program $P \in \mathcal{P}$ on the two shares, such that the two outputs of Eval add up to $P(w)$ modulo a positive integer β (where $\beta = 2$ by default), except with δ error probability. The running time of Eval is polynomial in the size of P and $1/\delta$. Here we formalize a stronger "Las Vegas" notion of HSS where Eval may output $\perp$ with at most δ probability, and the output is guaranteed to be correct as long as no party outputs $\perp$.

Definition 1 (Homomorphic Secret Sharing: Las Vegas Variant). *A (2-party)* Las Vegas Homomorphic Secret Sharing (HSS) *scheme for a class of programs $\mathcal{P}$ consists of algorithms* (Share, Eval) *with the following syntax:*

- *Share$(1^\lambda, w)$: On security parameter 1^λ and $w \in \{0,1\}^n$, the sharing algorithm outputs a pair of shares* (share$_0$, share$_1$). *We assume that the input length n is included in each share.*
- *Eval$(b,$ share$, P, \delta, \beta)$: On input party index $b \in \{0,1\}$, share* share *(which also specifies an input length n), a program $P \in \mathcal{P}$ with n input bits and m output bits, an error bound $\delta > 0$ and integer $\beta \geq 2$, the homomorphic evaluation algorithm either outputs $y_b \in \mathbb{Z}_\beta^m$, constituting party b's share of an output $y \in \{0,1\}^m$, or alternatively outputs $\perp$ to indicate failure. When β is omitted it is understood to be $\beta = 2$.*

The algorithm Share *is a PPT algorithm, whereas* Eval *can run in time polynomial in its input length and in $1/\delta$. The algorithms* (Share, Eval) *should satisfy the following correctness and security requirements:*

- **Correctness:** *For every polynomial p there is a negligible ν such that for every positive integer λ, input $w \in \{0,1\}^n$, program $P \in \mathcal{P}$ with input length n, error bound $\delta > 0$ and integer $\beta \geq 2$, where $|P|, 1/\delta \leq p(\lambda)$, we have*

$$\Pr[(\text{share}_0, \text{share}_1) \leftarrow \text{Share}(1^\lambda, w); y_b \leftarrow \text{Eval}(b, \text{share}_b, P, \delta, \beta), \ b = 0,1 :$$
$$(y_0 = \perp) \vee (y_1 = \perp)] \leq \delta + \nu(\lambda),$$

  and

$$\Pr[(\text{share}_0, \text{share}_1) \leftarrow \text{Share}(1^\lambda, w); y_b \leftarrow \text{Eval}(b, \text{share}_b, P, \delta, \beta), \ b = 0,1 :$$
$$(y_0 \neq \perp) \wedge (y_1 \neq \perp) \wedge y_0 + y_1 \neq P(w)] \leq \nu(\lambda),$$

  where addition of y_0 and y_1 is carried out modulo β.
- **Security:** *Each share keeps the input semantically secure.*

We will also use a stronger asymmetric *version of Las Vegas HSS where only one party (say, P_1) may output $\perp$. This is defined similarly to the above, except that conditions $y_0 = \perp$ and $y_0 \neq \perp$ in the correctness requirement are removed.*

HSS versus DEHE. We also consider a public-key variant of HSS, known as *distributed-evaluation homomorphic evaluation* (DEHE) [8]. This variant is described and explored in the full version of this work.

Multi-evaluation variant. For our applications of Las Vegas HSS it will sometimes not be enough to consider a single execution of Eval but rather a sequence of such executions following a single execution of Share. In such a case, we will need to assume that the events of outputting $\perp$ in different executions are indistinguishable from being independent. (This will allow us to apply a Chenoff-style bound when analyzing the total number of errors.) To simplify the terminology and notation, we implicitly assume by default that all instances of HSS we use are of the multi-evaluation variant.

2.2 BGI Construction [8]

The work of [8] constructs 2-party HSS (and DEHE) that directly supports homomorphic evaluation of "Restricted-Multiplication Straight-line" (RMS) programs over small integers. Such programs support four operations: Load Input to Memory, Add Values in Memory, Multiply Input by Memory Value, and Output Value. (See full version for formal RMS syntax). We provide here a high-level description of the [8] construction, which serves as a starting point for many of our results. In what follows, let $\mathbb{G}$ be a DDH-hard group of prime order q with generator $g \in \mathbb{G}$, and let $\ell = \lceil \log q \rceil$. We begin with the BGI construction of HSS based on circular-secure ElGamal:

Secret shares: To secret share a (small integer) input w, the BGI construction samples an ElGamal key pair $(c, e = g^c) \in \mathbb{Z}_q \times \mathbb{G}$, and outputs shares as follows: (1) Each party gets an additive secret share over $\mathbb{Z}_q$ of the input w and of the product cw (viewed as an element of $\mathbb{Z}_q$). (2) Each party also gets (copies of the same) $(\ell + 1)$ ElGamal ciphertexts, one encrypting w and one encrypting each product $c^{(t)}w$ of w with the tth bit of the secret key for $t \in [\ell]$.

Homomorphic evaluation: Evaluation maintains the invariant that (after each instruction) for each memory value x in the RMS program execution, the value of x and of cx are each held as an additive secret sharing across the two parties. This directly holds for any "Load Input to Memory" instruction, and can straightforwardly be achieved for each "Add Values in Memory" instruction by linear homomorphism of additive secret shares. "Output Value From Memory" to a target group $\mathbb{Z}_\beta$ (for some integer $\beta \leq q$ specified in the RMS program) is achieved by having each party shift his current share of the relevant memory value by a common rerandomization value and then output this share mod β.

The primary challenge is in supporting "Multiply Input by Value in Memory." Recall in such situation the parties hold additive secret shares of x and cx for the memory value x, and ElGamal ciphertexts of w and $\{c^{(t)}w\}_{t \in [\ell]}$ for the input w. Evaluation takes place in two steps, repeated for each ciphertext; for example, for the ciphertext encrypting w, we convert the common ElGamal ciphertext of w and additive secret shares of x and cx to additive secret shares of wx:

1. Use additive secret shares of x and cx to perform distributed ElGamal decryption via "linear algebra in the exponent," yielding multiplicative secret shares of g^{wx}. For ciphertext (g^r, g^{cr+w}), the multiplicative share of g^{wx} is $(g^r)^{-[\text{share of } cx]}(g^{cr+w})^{[\text{share of } x]}$.

2. To return the computed shares of g^{wx} back to additive shares of wx, the parties execute a share conversion procedure referred to as "Distributed Discrete Log," wherein the parties output the distance (measured by powers of g) of their share value g^{z_b} from the nearest point in an agreed-upon "distinguished set" in $\mathbb{G}$. Error occurs in this step if parties output with respect to *different* distinguished points, which occurs if a distinguished point lies "between" the parties' two shares $g^{z_0}, g^{z_1} = g^{z_0+wx}$.

   A tradeoff between computation and error can be made, by decreasing the density of distinguished points δ, and scaling computation as $1/\delta$; the resulting error probability is roughly δM, where M is the maximal value of the "payload" wx (corresponding to the "distance" between the parties' shares).

By repeating the above 2 steps for w and for each $c^{(t)}w$, the parties receive additive secret shares of wx and of each $c^{(t)}wx$. As a final step, the shares of $\{c^{(t)}wx\}_{t\in[\ell]}$ are combined by the appropriate powers-of-2 linear combination to yield a single set of additive shares of cwx, yielding the desired invariant for the new memory value wx.

Remark 1 (Removing the ElGamal circular security assumption). This can be done by one of two methods: (1) a standard "leveled" approach, using a sequence of secret keys (growing the HSS share size by the depth of computation); alternatively, (2) by replacing ElGamal with the "BHHO" encryption scheme of Boneh, Halevi, Hamburg, and Ostrovsky [4], which is *provably* circular secure based on DDH. Roughly, BHHO ciphertexts are an $O(\lambda)$-element extension of ElGamal, where the first elements are of the form $g_1^r, \dots, g_\ell^r$ (for fixed generators $g_1, \dots, g_\ell$ and encryption randomness r), and the final element contains the message as g^{msg} masked by a subset-product of the previous elements as dictated by the secret key $s \in \{0,1\}^\ell$. In particular, BHHO decryption follows a direct analog of "linear algebra in the exponent" as in ElGamal, and thus can be leveraged in the same manner within homomorphic share evaluation, where the new invariant for each memory value x is holding additive secret shares of x as well as each product $s_t x$, for the secret key bits s_t, $t \in [\ell]$. In addition, BHHO supports the same form of plaintext homomorphism required for DEHE, as discussed above. We refer the reader to [8] for a detailed formal treatment.

2.3   Secure Multiparty Computation

We consider two types of protocols for secure multiparty computation (MPC): standard k-party MPC protocols and client-server protocols. We refer the reader to [12,19] for standard definitions of MPC protocols and only highlight here the aspects that are particularly relevant to this work.

In a standard MPC protocol there are k parties who interact with each other in order to compute a function of their inputs. We say that such a protocol is *secure* if it is computationally secure against a static, passive adversary who may corrupt any strict subset of the parties. We use 2PC to refer to the case $k = 2$.

Client-server protocols. In a client-server protocol there are m clients and k servers. Only the clients have inputs and get an output. Clients and servers can communicate over secure point-to-point channels. We assume protocols in the client-server model to take the following canonical form: in the first round each client sends a message to each server. Then there may $r \geq 0$ rounds of interaction in which each server can send a message to each other server. We assume the servers to be deterministic, so that every message sent by a server in a given round is determined by the messages it received in previous rounds. Finally, there is an output reconstruction round in which each server sends a message to each client, and where each client computes an output by applying a local decoding function to the k messages it received.

We specify such a client-server protocol by $\Pi = (\mathsf{Encode}, \mathsf{NextMsg}, \mathsf{Decode})$, where $\mathsf{Encode}(i, x_i)$ is a randomized function mapping the input of Client i to the k messages it sends in the first round, $\mathsf{NextMsg}(i, \boldsymbol{m})$ is a next message function which determines the messages sent by Server i in the current round given the messages $\boldsymbol{m}$ it received in previous rounds, and $\mathsf{Decode}(i, \boldsymbol{m})$ denote the output of Client i given the messages $\boldsymbol{m}$ it received in the final round. Finally, we will consider by default protocols for functionalities that deliver the same output to all clients. In such a case, we can assume that each server sends the same message to all clients, and $\mathsf{Decode}(i, \cdot)$ is the same for all i.

Security and robustness. We say that Π is a *t-secure* protocol for f if it is secure against a static, passive (semi-honest) adversary who may corrupt any set of parties that includes at most t servers and an arbitrary number of clients. Security is defined by the existence of a simulator $\mathsf{Sim}(1^\lambda, T, 1^n, y)$ that given a security parameter λ (in the computational case), a set T of corrupted parties, an input length n, and an output y of f (in the case at least one client is corrupted) outputs a simulated view of the parties in T. Simulation should be either perfect or computational, depending on the type of security. We assume *computational* $(k-1)$-security by default, but will also consider protocols that offer perfect t-security for smaller values of t. Note that any secure k-client k-server protocol for f implies a standard k-party MPC for f by letting Party i simulate both Client i and Server i.

A *t-robust* protocol for f is a t-secure protocol with the following additional feature: the clients obtain the correct output of f even if t servers fail to send messages. Equivalently, the function Decode outputs the correct output of f at the end of the protocol execution even if up to t of its inputs are replaced by $\perp$.

Succinct MPC. We will consider MPC protocols for a class of programs $\mathcal{P}$, where all parties are given a "program" $P \in \mathcal{P}$ (say, a boolean circuit, boolean formula or branching program) as an input, and their running time should be polynomial in the size of P. See Sect. 4 of [8] for a full definition. We refer to an MPC protocol for $\mathcal{P}$ as being *succinct* if the communication complexity is bounded by a fixed polynomial in the total length of inputs and outputs and the security parameter, independently of the program size.

MPC with PKI setup. For both flavors of MPC protocols, we consider round complexity with a public key infrastructure (PKI) setup. A PKI setup allows a one-time global choice of parameters params $\leftarrow$ ParamGen(1^λ), followed by independent choices of a key pair (sk$_i$, pk$_i$) $\leftarrow$ KeyGen(1^λ, params) by each party P_i.[4] We assume that each party knows the public keys of all parties with whom it wants to interact as well as its own secret key. Note that the public keys are generated independently of any inputs or even the number of other parties in the system. For this reason we do not count the PKI setup towards the round complexity of our protocols.

3   Black-Box Client-Server HSS and MPC

In order to use HSS or its public-key DEHE variant to obtain secure computation, the secret sharing procedure (or DEHE key setup) must be performed in a secure distributed fashion. Applying general-purpose secure computation to do so, as suggested in [8], has poor concrete efficiency and requires non-black-box access to the underlying group.

To avoid this, we introduce the notion of *client-server HSS* (Π, Eval), defined as standard HSS, except that the input is distributed between multiple clients and the centralized sharing algorithm Share is replaced by a distributed protocol Π. That is, Π allows m clients, each holding a secret input w_i, to share the joint input $(w_1, \ldots, w_m)$ between the servers in a way that supports homomorphic computations via Eval. We will be interested in constructing client-server HSS (and DEHE) that only make a black-box access to the underlying group.

The security requirement is that the view of an adversary who corrupts a subset of clients/servers, leaving at least one client and one server uncorrupted, can be simulated given the inputs of corrupted clients, *without* knowledge of the inputs of uncorrupted clients. A formal definition of client-server HSS is deferred to the full version. A "multi-evaluation" version enables independent executions of Eval without re-executing Π.

Intuitively, in our construction of the joint secret sharing protocol Π, each client C_i will generate an independent ElGamal key pair (c_i, e_i), and the joint keys of the system will correspond to $c = \sum c_i \in \mathbb{Z}_q$ and $e = \prod e_i \in \mathbb{G}$, leveraging the key homomorphism of ElGamal. The primary challenge (mirroring the BGI HSS) is how to generate encryptions of the products $c^{(t)} w_i$, where $c^{(t)}$ are the bits of the *joint* secret key $c = \sum c_i$ (where addition is in $\mathbb{Z}_q$). To solve this, we leverage the fact that the BGI construction does not strictly require $\{0, 1\}$ values for this $c^{(t)}$, but rather can support computations on any sufficiently small values

---

[4] We will only use params to specify a group for ElGamal encryption; hence, we can let params be a common random string, or even pick params deterministically under a suitable variant of DDH.

at the expense of greater computation during the share conversion procedure. We will thus use the (possibly non-Boolean) values $\sum_i c_i^{(t)}$ in the place of $c^{(t)}$.

We present the full construction and proof of client-server HSS in the full version. In fact, we achieve the stronger primitive of multi-evaluation client-server DEHE, which directly implies the former.

Remark 2 (ElGamal Circular Security vs. DDH). For simplicity, throughout the present work we describe our constructions based on circular security of ElGamal. However, in each case we may directly remove this circular security assumption, as in [8], by either considering a leveled variant or replacing ElGamal with a circular-secure variant due to BHHO [4], as described in Remark 1. Our theorem statements implicitly apply this transformation directly.

Proposition 1 (Black-box client-server HSS/DEHE). *There exists a multi-evaluation client-server DEHE protocol (and thus also multi-evaluation client-server HSS) for branching programs that makes a black-box access to any DDH-hard group.*

3.1   Black-Box Succinct Secure Computation

Given a black-box m-client 2-server multi-evaluation HSS ($\Pi_{\mathsf{HSS}}, \mathsf{Eval}_{\mathsf{HSS}}$) as above, and an arbitrary general 2PC protocol Π_{MPC}, we obtain succinct secure m-client 2-server computation for branching programs based on DDH which makes only black-box use of the DDH group. Namely, to securely evaluate a program P: (1) the clients and servers interact via Π_{HSS} to share the clients' inputs, (2) the servers homomorphically evaluate λ copies of the desired program P on the resulting shares, and then (3) run the generic protocol Π_{MPC} to securely evaluate the most common combined output.

Note that the procedure for combining evaluated shares and taking the majority (in Step 3) does not require any $\mathbb{G}$ group operations (only operations over the output space $\mathbb{Z}_\beta$), so that general secure computation of this function is still black-box in the DDH group $\mathbb{G}$.

Theorem 1 (Black-box succinct secure computation for branching programs). *There exists a constant-round succinct m-client 2-server protocol Π_{BB} for branching programs that makes only black-box access to any DDH-hard group.*

Remark 3 (1/poly security tradeoff). The round complexity of Π_{BB} is given by the round complexity HSS sharing protocol Π_{HSS} plus that of the generic MPC to evaluate the reconstruction-majority. If one is willing to accept 1/poly security, the MPC reconstruction phase can be replaced by a direct exchange of the output shares computed in the homomorphic evaluation. The corresponding simulator will follow the same simulation strategy, but will fail with inverse-polynomial probability, in the event that a homomorphic evaluation error occurs. The resulting protocol will have $\mathsf{rounds}(\Pi_{\mathsf{HSS}}) + 1$ rounds.

From here on, all of our protocols make a black-box access to the group except for protocols that involve $k \geq 3$ servers (in client server model) or parties (in the MPC model).

4   DDH-Based 2-Round Protocols over PKI

In this section we present a 2-round secure computation protocol in the PKI setup model for a constant number of parties and arbitrary polynomial-size circuits, based on DDH. Our starting point will be the general secure client-server protocol structure given in Theorem 1.

As discussed in the Introduction, our final 2-round solution removes the extra rounds of interaction by means of three main technical steps, which we present in the following three sections: (1) Constructing a Client-Server HSS whose secret sharing protocol Π can be executed in a *single* round of interaction in the PKI model; (2) Amplifying the resulting 2-round client-server protocol (Remark 3) from 1/poly to full security using techniques in leakage resilience; and (3) Compiling from 2 to any constant number k of servers by iteratively emulating a server's computation securely by 2 separate servers.

4.1   Succinct 2-Server Protocol with 1/poly Security

We begin by constructing m-client 2-server HSS whose secret sharing protocol Π takes place via a *single message* from each client within the PKI model.

Our construction takes a similar approach to the black-box client-server HSS of the previous section, where each client owns an independent ElGamal key pair (c_i, e_i). However, the approach does not quite work as is. The primary challenge is in agreeing on common encryptions of the *cross*-products $c_j^{(t)} w_i$ for different clients C_i, C_j. Recall that HSS evaluation requires not only that each party holds an encryption of the same value, but in fact the exact *same* ciphertext.

This remains a problem even if we consider the setting with a public-key infrastructure (PKI). Namely, even given all clients' public keys, it is not clear how in a single message of communication all clients can agree on the same ciphertext of $c_i^{(t)} w_j$ under the joint key $\prod_i e_i$ when $c_i^{(t)}$ and w_j are known by two different clients, and $c_i^{(t)}$ and w_j themselves must remain hidden.

This goal *can* be achieved, however, for the i, j "pairwise" combination of public keys $e_i e_j$, by including an encryption of $c_i^{(t)}$ under key e_i as part of an expanded public key of client C_i. (Note that the value of $c_i^{(t)}$ depends only on C_i's keys themselves and not on inputs or number of parties, hence this is a valid contribution to the PKI setup.) Namely, given an encryption $[\![c_i^{(t)}]\!]_{c_i}$ of $c_i^{(t)}$ (using notation from [8], as per Fig. 1), client C_j can use the homomorphic properties of ElGamal to first shift this to an encryption under e_i of the product $c_i^{(t)} w_j$, and then shift this ciphertext to an encryption of the same value under key $e_i e_j$ by coordinate-wise multiplying in an encryption of 0 under key e_j. (Note that the second step is necessary in order to hide w_j from client C_i.)

We demonstrate that generating these pairwise $c_i^{(t)} w_j$ ciphertexts under the respective pairwise keys is enough to support full homomorphic evaluation capability. The new invariant maintained throughout homomorphic evaluation is that for each memory variable $\hat{y}$, the correct value y of this variable is held as an additive secret sharing $\langle y \rangle$, and as a *collection of m additive secret sharings* $\langle c_i y \rangle$,

---

Secret Sharing Notation. For small $x \in \mathbb{Z}$ (or $x \in \mathbb{Z}_q$ for the case of $\langle x \rangle$).

Items in which *both* parties receive same value.
- $[\![x]\!]_c = (h_1, h_2) \in \mathbb{G}^2$ for which $h_2/(h_1)^c = g^x$. I.e., ElGamal ciphertext of x w.r.t. key c.

Items in which each party receives a separate share.
- $\langle x \rangle$ = Additive secret shares $(x_1, x_2) \in \mathbb{Z}_q^2$ for which $x_1 - x_2 = x \in \mathbb{Z}_q$.
- $\langle\!\langle x \rangle\!\rangle$ = "Multiplicative" secret shares $(h_1, h_2) \in \mathbb{G}^2$ for which $h_1/h_2 = g^x \in \mathbb{G}$.

Pairing Operations.
Let $\phi : \{0,1\}^\lambda \times \mathbb{G} \to \{0,1\}^\ell$ be a given PRF.

- $\mathsf{MultShares}\Big([\![x]\!]_c, \langle y \rangle, \langle cy \rangle\Big) \to \langle\!\langle xy \rangle\!\rangle$.
    1. Denote $[\![x]\!]_c = (h_1, h_2) \in \mathbb{G}^2$.
    2. Compute $\langle\!\langle xy \rangle\!\rangle = h_2^{\langle y \rangle} h_1^{-\langle cy \rangle}$.
- $\mathsf{ConvertShares}(b, \langle\!\langle x \rangle\!\rangle, \mathsf{id}, \delta, M) \to \langle x \rangle$, with party identifier $b \in \{0,1\}$, execution identifier id, error parameter δ and max size bound M.
    1. Denote by $\phi' : \mathbb{G} \to \{0,1\}^{\lceil \log(2M/\delta) \rceil}$ the appropriate prefix output of $\phi(\mathsf{id}, \cdot)$.
    2. Let x_b denote the present party b's share of $\langle\!\langle x \rangle\!\rangle$.
    3. Output $i_b \leftarrow \mathsf{DistributedDLog}_{\mathbb{G},g}(x_b, \delta, M, \phi')$.

Share Conversion Sub-Routine. $\mathsf{DistributedDLog}_{\mathbb{G},g}(h, \delta, M, \phi)$
1: Set $h' \leftarrow h$, $i \leftarrow 0$. Let $T := \lceil 2M \ln(2/\delta) \rceil / \delta$.
2: **while** $(\phi(h') \neq 0^{\lceil \log(2M/\delta) \rceil}$ and $i < T)$ **do**
3:     $h' \leftarrow h' \cdot g$, $i \leftarrow i + 1$.
4: **end while**
5: Output i.

Fig. 1. Notation, pairing operations, and share conversion algorithm, as used in [8]. For simplicity we describe the scheme with *subtractive* (and *division*) secret sharing instead of converting back and forth between additive and subtractive (resp., multiplicative and division) shares; see discussion in full version.

one for the key c_i of each client $i \in [m]$. Whenever we wish to perform an RMS multiplication using a ciphertext $[\![c_i^{(t)} w_j]\!]_{c_i + c_j}$, we can combine the corresponding pair of secret shares $\langle (c_i + c_j)y \rangle = \langle c_i y \rangle + \langle c_j y \rangle$, and then proceed as usual as if the secret key were the sum $c_i + c_j$.

As one additional change (which will be useful in future sections), we replace the bit decomposition $(c^{(t)})_{t \in [\ell]}$ of a key c with a more general, possibly randomized, representation $(\hat{c}^{(t)})_{t \in [\ell']} \leftarrow \mathsf{Decomp}(c)$. The only requirements for correctness are: (1) each value $\hat{c}^{(t)}$ has small magnitude; and (2) there exists a $\mathbb{Z}_q$-*linear* reconstruction procedure Recomp for which $c = \mathsf{Recomp}((\hat{c}^{(t)})_{t \in [\ell']}).$[5]

The formal descriptions of $(\Pi_{1r}, \mathsf{Eval}_{1r})$ are given in Figs. 2 and 3.

---

[5] Note that bit decomposition can be expressed in this form, where $\mathsf{Decomp}(c) := (c^{(t)})_{t \in [\ell]}$ and $\mathsf{Recomp}((c^{(t)})_{t \in [\ell]}) := \sum_{t=1}^{\ell} 2^{t-1} c^{(t)}$.

Lemma 1 (One-Round Client-Server HSS). *Assume hardness of DDH. Then for any polynomial $m = m(\lambda)$, there exists an m-client 2-server HSS $(\Pi_{1r}, \mathsf{Eval}_{1r})$ for which Π_{1r} is a single round in the PKI model.*

Proof. We defer the proof to the full version. We remark that a crucial property for security is that any secret value owned by a client C_i is encrypted under a combination of keys that includes his *own* key, c_i (and distributed as a fresh encryption due to re-randomization). Because of this, semantic security holds for all honest-client values, by the key homomorphism properties of ElGamal.

Plugging in the client-server HSS $(\Pi_{1r}, \mathsf{Eval}_{1r})$ to the framework of Theorem 1, together with the round-savings-for-1/poly tradeoff described in Remark 3, we directly obtain the following proposition.

Proposition 2 (Succinct 2-server protocol with 1/poly security for branching programs). *Assuming PKI setup and DDH, for any polynomial $p(\cdot)$ and $m = m(\lambda)$ there is a (succinct) 2-round m-client 2-server client-server protocol for branching programs with $1/p(\lambda)$ security.*

4.2 Amplifying Security via Leakage Resilience

The 1/poly security loss in the protocol of Sect. 4.1 is due to the noticeable probability of (input-dependent) error in the homomorphic evaluation of the client-server HSS, revealed when evaluated output shares are directly exchanged. We now develop techniques for addressing this information leakage *without* additional communication rounds.

Simulatable Las Vegas HSS. Toward this goal, we first consider and realize two beneficial properties of a client-server HSS:

- *Las Vegas correctness.* In such an HSS scheme, servers can output a special symbol $\bot$ if they identify a possible error situation in the homomorphic evaluation. Las Vegas correctness guarantees that if both servers output a non-$\bot$ value then correct reconstruction will hold.
- *Simulatability of errors.* Unfortunately, it will be the case in constructions that servers do not always agree on whether an error is possible to occur (otherwise error could be removed completely by having each server recompute in such situation), and learning whether the other server reaches $\bot$ may reveal secret information. To address this, we consider a further "simulatability" property which formally characterizes what information is leaked through this process.

We construct simulatable Las Vegas HSS where the information leakage depends *locally* on values of a small number of memory values within the computation of the RMS program and/or symbols $\hat{c}^{(t)}$ of the secret key representation.

In the following two subsections, we present our construction of a simulatable Las Vegas HSS whose secret-sharing protocol is a single round given PKI, and then use this construction as a tool together with leakage-resilient techniques to obtain a (fully) secure 2-round 2-party computation protocol in the PKI model.

One-Round Client-Server HSS (using PKI): m-client secret sharing protocol Π_{1r}.
Global parameters: $\mathbb{G}, g, q$. Let ℓ' be the output size of Decomp.
Inputs: Each client C_i for $i \in [m]$ holds input $w_i \in \{0, 1\}$.
Outputs: Each server S_b for $b \in \{0, 1\}$ learns share_b of all inputs.

Public-Key Infrastructure: Each client C_i's public-key information consists of:
 - An ElGamal private key $c_i \leftarrow \mathbb{Z}_q$, known exclusively by C_i.
 - A public key $\mathsf{pk}_i = \left(e_i, ([\![\hat{c}_i^{(t)}]\!]_{c_i})_{t \in [\ell']} \right)$ consisting of:
    - ElGamal public key $e_i = g^{c_i}$.
    - For $t \in [\ell']$, an ElGamal ciphertext under key e_i of the t'th symbol of $(\hat{c}_i^{(t)})_{t \in [\ell']} \leftarrow \mathsf{Decomp}(c_i)$; i.e., $[\![\hat{c}_i^{(t)}]\!]_{c_i} \leftarrow \mathsf{Enc}_{\mathsf{ElGamal}}(e_i, \hat{c}_i^{(t)})$.

Client Round 1: Each client $i \in [m]$ performs the following:
 1. Generate ciphertexts of "owned" data w_i and $\hat{c}_i^{(t)} w_i$ under self key e_i:
    (a) Encrypt input w_i: i.e., $[\![w_i]\!]_{c_i} \leftarrow \mathsf{Enc}_{\mathsf{ElGamal}}(e_i, w_i)$.
    (b) For each $t \in [\ell']$, encrypt $\hat{c}_i^{(t)} w_i$: i.e., $[\![\hat{c}_i^{(t)} w_i]\!]_{c_i} \leftarrow \mathsf{Enc}_{\mathsf{ElGamal}}(e_i, \hat{c}_i^{(t)} w_i)$.
 2. Generate ciphertexts of "joint" data $\hat{c}_j^{(t)} w_i$ under *pairwise* keys $e_i e_j$ for $j \neq i$:
    For each client $j \in [m], j \neq i$ and key-bit $t \in [\ell']$, generate ciphertext of $\hat{c}_j^{(t)} w_i$ as follows:
    (a) Let e_j and $[\![\hat{c}_j^{(t)}]\!]_{c_j}$ denote the ElGamal public key and tth-key-bit ciphertext within the public key pk_j of client j.
    (b) Sample a fresh encryption of 0 under key $e_i e_j$; i.e., $(h_1^0, h_2^0) \leftarrow \mathsf{Enc}_{\mathsf{ElGamal}}(e_i e_j, 0)$.
    (c) Let $(h_1, h_2) = [\![\hat{c}_j^{(t)}]\!]_{c_j} \in \mathbb{G}^2$ within the public key of Client C_j.
    (d) Compute $[\![\hat{c}_j^{(t)} w_i]\!]_{c_i + c_j}$ as follows:
        i. Let $(h_1', h_2') = (h_1^{w_i}, h_2^{w_i}(h_1^{w_i})^{c_i})$. //Decrypts to $\hat{c}_j^{(t)} w_i$ with key $c_i + c_j$
        ii. Rerandomize using the ciphertext of 0. Namely, take $[\![\hat{c}_j^{(t)} w_i]\!]_{c_i + c_j} = (h_1' h_1^0, h_2' h_2^0)$.
 3. Send all ciphertexts $[\![w_i]\!]_{c_i}, \{[\![\hat{c}_i^{(t)} w_i]\!]_{c_i}\}_{t \in [\ell']}, \{[\![\hat{c}_j^{(t)} w_i]\!]_{c_i + c_j}\}_{i \neq j \in [m], t \in [\ell']}$ to both servers.
 4. Other items:
    (a) Produce an additive secret sharing $\langle c_i \rangle \leftarrow \mathsf{AdditiveShare}(c_i)$ of the key c_i. and send each resulting share $\langle c_i \rangle_b$ to the corresponding server b.
    (b) Sample a random string $r_i \leftarrow \{0, 1\}^\lambda$ (for PRF seed) and send r_i to both servers.

Server Output: Each server $b \in \{0, 1\}$ performs the following:
 1. Take $r = \sum_{i \in [m]} r_i$. Let $\phi = \mathsf{PRFGen}(1^\lambda; r)$ be a PRF from $\{0, 1\}^\lambda \times \mathbb{G} \rightarrow \{0, 1\}^{\ell'}$.
 2. Let $\mathsf{share}_b = $
$$\left(m, \phi, \{\langle c_i \rangle_b\}_{i \in [m]}, \left([\![w_i]\!]_{c_i}, \{[\![\hat{c}_i^{(t)} w_i]\!]_{c_i}\}_{t \in [\ell']}, \{[\![\hat{c}_j^{(t)} w_i]\!]_{c_i + c_j}\}_{i \neq j \in [m], t \in [\ell']} \right)_{i \in [m]} \right).$$

Fig. 2. One-round m-client 2-server HSS secret sharing protocol Π_{1r}. (Decomp, Recomp) refer to a decomposition procedure with low-magnitude shares and linear reconstruction (generalizing bit decomposition).

One-Round Client-Server Homomorphic Evaluation $\mathsf{Eval}_{\mathsf{G},g}(b, \mathsf{share}, P, \delta)$
Inputs: Party identifier $b \in \{0, 1\}$, homomorphic secret share value share, RMS program
description P of size $\leq S$, error bound δ.

Parse share as in Figure 2. Parse P as a magnitude bound 1^M and sequence of RMS
instructions. Take $\delta' = \delta/((\ell'+1)MS)$. We describe here homomorphic evaluation of
multiplication, and defer the other RMS operations to the full version.

Instruction $(\mathsf{id}, \hat{y}_k \leftarrow \hat{w}_\alpha \cdot \hat{y}_j)$:
  1. Produce shares $\langle w_\alpha y_j \rangle$ (using $[\![w_\alpha]\!]_{c_\alpha}$ and $\langle y_j \rangle, \langle c_\alpha y_j \rangle$):
       1: Compute $\langle\!\langle w_\alpha y_j \rangle\!\rangle = \mathsf{MultShares}([\![w_\alpha]\!]_{c_\alpha}, \langle y_j \rangle, \langle c_\alpha y_j \rangle)$, as in Figure 1.
       2: Execute $\langle w_\alpha y_j \rangle = \mathsf{ConvertShares}(b, \langle\!\langle w_\alpha y_j \rangle\!\rangle, (\mathsf{id}, 0), \delta', M, \phi)$, as in Figure 1.
       3: Set $\langle y_k \rangle \leftarrow \langle w_\alpha y_j \rangle$.
  2. Produce shares $\langle c_\gamma w_\alpha y_j \rangle$ for each $\gamma \in [m]$: (using $[\![\hat{c}_\gamma^{(t)} w_\alpha]\!]_{c_\alpha + c_\gamma}$ and
     $\langle y_j \rangle, \langle c_\alpha y_j \rangle, \langle c_\gamma y_j \rangle$)
       1: **for** $\gamma = 1$ to m **do**
       2:     **if** $\alpha = \gamma$ **then** define $c_{\alpha,\alpha} := c_\alpha$. Let $\langle c_{\alpha,\alpha} y_j \rangle = \langle c_\alpha y_j \rangle$.
       3:     **else** define $c_{\alpha,\gamma} := c_\alpha + c_\gamma$. Compute $\langle c_{\alpha,\gamma} y_j \rangle = \langle c_\alpha y_j \rangle + \langle c_\gamma y_j \rangle$.
       4:     **end if**
       5:     **for** $t = 1$ to ℓ' **do**
       6:         Compute $\langle\!\langle \hat{c}_\gamma^{(t)} w_\alpha y_j \rangle\!\rangle = \mathsf{MultShares}([\![\hat{c}_\gamma^{(t)} w_\alpha]\!]_{c_{\alpha,\gamma}}, \langle y_j \rangle, \langle c_{\alpha,\gamma} y_j \rangle)$.
       7:         Execute $\langle \hat{c}_\gamma^{(t)} w_\alpha y_j \rangle = \mathsf{ConvertShares}(b, \langle\!\langle \hat{c}_\gamma^{(t)} w_\alpha y_j \rangle\!\rangle, (\mathsf{id}, t), \delta', M, \phi)$.
       8:     **end for**
       9:     Compute $\langle c_\gamma w_\alpha y_j \rangle = \mathsf{Recomp}((\langle \hat{c}_\gamma^{(t)} w_\alpha y_j \rangle)_{t \in [\ell']})$.
      10:     Set $\langle c_\gamma y_k \rangle \leftarrow \langle c_\gamma w_\alpha y_j \rangle$.
      11: **end for**

Fig. 3. One-round m-client 2-server homomorphic evaluation algorithm Eval. Evaluation maintains the invariant that for each memory value $\hat{y}_i$ the servers hold: (1) additive
shares $\langle y_i \rangle$, and (2) m sets of additive shares $\langle c_\alpha y_i \rangle$, for the secret key c_α of *each* of
the m clients. Here, i, j, k denote memory indices, $t \in [\ell']$ denotes an index of a key
representation, and $\alpha, \gamma \in [m]$ denote client ids.

Defining and Obtaining Simulatable Las Vegas HSS. We define a "simulatable" variant of client-server Las Vegas HSS (LV-HSS), where each server has
a secondary output in Eval that represents its knowledge about the other server's
primary output. The secondary output can either be $\top$, indicating that it is certain that the other server does not output $\bot$, or a predicate Pred (represented
by a circuit) that specifies a function of the clients' inputs $\boldsymbol{w}$ and randomness $\boldsymbol{r}$
such that the other party outputs $\bot$ if and only if $\mathsf{Pred}(\boldsymbol{w}, \boldsymbol{r}) = 1$. We require
that the secondary output is $\top$ except with at most δ probability. Note that
Pred may depend on the program P being homomorphically evaluated.

Definition 2 (Simulatable Client-Server Las Vegas HSS). *A (m-client,
2-server) Simulatable Client-Server Las Vegas HSS scheme for class of programs
$\mathcal{P}$ consists of a distributed protocol Π and PPT algorithm Eval, with syntax:*

- Π specifies an interactive protocol between m clients $C_1, \dots, C_m$ and two servers S_0, S_1, where each client C_i begins with input w_i, and in the end of executing Π the servers S_0, S_1 output homomorphic secret shares $\mathsf{share}_0, \mathsf{share}_1$, respectively, of the joint input $(w_1, \dots, w_m)$.
- Eval has a second output z such that z is either the symbol $\top$ or a predicate $\mathsf{Pred} : \{0,1\}^n \to \{0,1\}$ represented by a boolean circuit.
  We denote by $(\mathsf{share}_0, \mathsf{share}_1) \leftarrow \Pi(\boldsymbol{w}; \boldsymbol{r}, R_0, R_1)$ where $\boldsymbol{w} = (w_1, \dots, w_m)$ and $\boldsymbol{r} = (r_1, \dots, r_m)$ the execution of Π in which each client $i \in [m]$ uses input w_i and randomness r_i, each server $b \in \{0,1\}$ uses randomness R_b, and the output to each server S_b is share_b.

The pair (Π, Eval) should satisfy the correctness of Definition 1 (with respect to the first output of Eval), and the following additional requirements:

- **Security:** There exists a PPT simulator Sim such that for any corrupted set $\mathsf{Corrupt} \subset \{C_1, \dots, C_m\} \cup \{S_0, S_1\}$ of clients and servers for which at least one server and one client are uncorrupted, for every polynomial p, and sequence of input vectors $\boldsymbol{w}^\lambda = (w_1^\lambda, \dots, w_m^\lambda) \in (\{0,1\}^{p(\lambda)})^m$, it holds that
  $$\mathsf{view}(1^\lambda, \mathsf{Corrupt}, \boldsymbol{w}^\lambda) \overset{c}{\cong} \mathsf{Sim}(1^\lambda, \mathsf{Corrupt}, \{w_i\}_{C_i \in \mathsf{Corrupt}}, \{|w_i|\}_{C_i \notin \mathsf{Corrupt}}).$$
- **Error simulation:** For every polynomial p there is a negligible ν such that for every $\lambda \in \mathbb{N}$, input $w \in \{0,1\}^n$, program $P \in \mathcal{P}$ with input length n, error bound $\delta > 0$ and integer $\beta \geq 2$, where $|P|, 1/\delta \leq p(\lambda)$, then for every $b \in \{0,1\}$,
  $$\Pr[(\mathsf{share}_0, \mathsf{share}_1) \leftarrow \Pi(\boldsymbol{w}; \boldsymbol{r}, R_0, R_1);$$
  $$(y_b, z_b) \leftarrow \mathsf{Eval}(b, \mathsf{share}_b, P, \delta, \beta) : z_b \neq \top] \leq \delta + \nu(\lambda),$$
  and for every circuit Pred and $c \in \{0,1\}$:
  $$\Pr[(\mathsf{share}_0, \mathsf{share}_1) \leftarrow \Pi(\boldsymbol{w}; \boldsymbol{r}, R_0, R_1); (y_b, z_b) \leftarrow \mathsf{Eval}(b, \mathsf{share}_b, P, \delta, \beta), \ b = 0, 1 :$$
  $$(z_c = \mathsf{Pred}) \wedge (\chi(y_{1-c} = \bot) \neq \mathsf{Pred}(\boldsymbol{w}, \boldsymbol{r}))] \leq \nu(\lambda),$$
  where $\chi(y_{1-c} = \bot)$ evaluates to 1 if $y_{1-c} = \bot$ and evaluates to 0 otherwise.

Constructing simulatable Las Vegas HSS. Our construction of simulatable (client-server) LV-HSS will be a variant of the 1-round Client-Server HSS construction, with a modified core share-conversion sub-routine DistributedDLog (called within ConvertShares), which enables each party to convert a multiplicative share of $g^z \in \mathbb{G}$ to an additive share of $z \in \mathbb{Z}_q$ (for small z).

Following [8], the procedure DistributedDLog takes as input a share $h \in \mathbb{G}$ and outputs the distance on the cycle generated by $g \in \mathbb{G}$ between h and the first "distinguished" point $h' \in \mathbb{G}$ such that a pseudo-random function (PRF) outputs 0 on h'. Two invocations on inputs h and $h \cdot g^z$ for a small z result, with good probability (over the initial choice of PRF seed), in outputs i and $i - z$ for some $i \in \mathbb{Z}_q$. In such case, the DistributedDLog procedure converts a difference of small z in the cycle generated by g in $\mathbb{G}$ to the same difference over $\mathbb{Z}$.

Algorithm 1. Simulatable SLVDistribDLog$_{\mathbb{G},g}(b,h,\delta,M,\phi)$

1: Let DangerZone $:= \{h, hg^{(-1)^b}, \ldots, hg^{(-1)^b M}\}$.
2: Let SimDangerZone $:= \{hg^{-M+1}, \ldots, h, \ldots, hg^M\}$ and initialize BadValues $\leftarrow \emptyset$.
3: **if** $\exists h' \in$ SimDangerZone with $\phi(h') = 0^{\lceil \log(2M/\delta) \rceil}$ **then** Let BadValues be the set
   of $z \in [M]$ for which $\{hg^{(-1)^b z}, hg^{(-1)^b z+(-1)^{b-1}}, \ldots, hg^{(-1)^b z+(-1)^{b-1}M}\}$ contains
   some h' with $\phi(h') = 0^{\lceil \log(2M/\delta) \rceil}$). If BadValues $= \emptyset$, set BadValues $\leftarrow \top$.
4: **end if**
5: **if** $\exists h' \in$ DangerZone with $\phi(h') = 0^{\lceil \log(2M/\delta) \rceil}$ **then** Let $i = \bot$.
6: **else**
7:     Set $h' \leftarrow h$, $i \leftarrow 0$. Let $T = 2M\lambda/\delta$.
8:     **while** $(\phi(h') \neq 0^{\lceil \log(2M/\delta) \rceil}$ and $i < T)$ **do**
9:         $h' \leftarrow h' \cdot g$, $i \leftarrow i + 1$.
10:    **end while**
11: **end if**
12: Return $(i, \text{BadValues})$.

For any $h, h \cdot g^z \in \mathbb{G}$, DistributedDLog yields an error in two cases:

1. When there exists a distinguished point h' *between* the two inputs h, hg^z: i.e., $h' = hg^i$ for some $i \in \{0, \ldots, z-1\}$.
2. When there does not exist a distinguished point within a fixed polynomial-size range after which the party will abort.

We construct a simulatable Las Vegas version of this sub-routine, SLVDistribDLog, described in Algorithm 1. This algorithm has three primary differences from the original procedure DistributedDLog.

1. For simplicity, the end-case abort threshold T is set large enough $(2M\lambda/\delta)$ so that the probability of abort over the choice of distinguished points (via the PRF ϕ) is negligible. Recall the choice of T gives a tradeoff between error probability and required computation (in [8], and in our complexity-optimized versions in later sections, the threshold is set to a lower value).
2. Given an input share $h \in \mathbb{G}$, maximum magnitude bound M, and "party id" $b \in \{0,1\}$, the algorithm will now output $\bot$ if there is a distinguished point h' within M steps of h in the direction dictated by b. Recall that this sub-routine will be called simultaneously by party P_0 (the "behind" party) holding share h and party P_1 (the "ahead" party) holding share $h \cdot g^z$. In the new procedure, P_0 will output $\bot$ if any of $h \cdot g, \ldots, h \cdot g^{M-1}$ is distinguished, and P_1 will output $\bot$ if any of $h \cdot g^{z-M+1}, \ldots, h \cdot g^{z-1}$ is distinguished. This will guarantee (no matter the value of $z \in [M]$) that if there is a distinguished point between the two parties' shares then both parties will output $\bot$. This zone of values is denoted DangerZone in SLVDistribDLog.
3. SLVDistribDLog now outputs two values: (1) a $\mathbb{Z}_q$-element (or $\bot$) as usual, corresponding to the output additive share, and (2) a subset BadValues $\subset [M]$ of values z such that the other party $1-b$ will have a distinguished point h'

within his DangerZone (and output $\perp$) if and only if he runs SLVDistribDLog with input $hg^{(-1)^b z}$ (i.e., our respective inputs $h, g^{(-1)^b z}$ are multiplicative shares of g^z for some $z \in$ BadValues).

Basically, for each possible share of the other party, we can directly determine if it would result in $\perp$, and record the corresponding secret shared value $z \in [M]$ if it would. In the notation of SLVDistribDLog, the window SimDangerZone is of size $2M$ and captures all possible shifted windows of size M which could be the DangerZone of the other party, depending on which of the M possible values of z is the current offset between shares.

In the full version we present a construction of simulatable Las Vegas HSS, using SLVDistribDLog as a sub-routine. Roughly: At every share conversion step of homomorphic evaluation in $\mathsf{Eval}^{\mathsf{SLV}}$, with some probability there will exist a bad set of plaintext values $z \in [M]$ such that if the newly computed shared value is equal to z then the other party would output $\perp$. These sets of bad values are identified within SLVDistribDLog and are stored as BadValues's within Eval. A pair $(k, \mathsf{BadValues}_k) \in \mathbb{Z} \times 2^{[M]}$ is added to LeakageInfo if partial computation value $y_k = z \in$ BadValues would lead to the other party outputting $\perp$. This corresponds to a share conversion for some $\langle y_k \rangle$. Similarly, a pair $((k, \gamma, t), \mathsf{BadValues}_{k,\gamma,t}) \in (\mathbb{Z} \times [m] \times [\ell]) \times 2^{[M]}$ is added to LeakageInfo if partial computation value $\hat{c}_\gamma^{(t)} y_k = z \in \mathsf{BadValues}_{k,\gamma,t}$ would lead to the other party outputting $\perp$. This corresponds to a share conversion for some $\langle \hat{c}_\gamma^{(t)} y_k \rangle$. Note that the values y_k are defined as a function of the program P and a given input w. The choice of Pred incorporates the P dependency, and operates on input w as well as a subset of (at most λ values of) $\hat{c}^{(t)}$.

Proposition 3. *Assume hardness of DDH. Then for any polynomial $m = m(\lambda)$, the scheme $(\Pi_{\mathsf{SLV}}, \mathsf{Eval}_{\mathsf{SLV}})$ described above is an m-client 2-server simulatable Las Vegas HSS, where Π_{SLV} is a single round in the PKI model. Moreover, with overwhelming probability in λ over the randomness of Π, the predicate Pred depends on at most λ intermediate variables of the evaluation of the RMS program P and values $\hat{c}^{(t)}$.*

Remark 4 (Asymmetric Las Vegas HSS). In some of our later applications (see Sect. 5), it will be advantageous to have an *asymmetric* notion of Las Vegas HSS, where only one of the two parties might output $\perp$. In these applications, simulatability will not be required. We can achieve such notion via a simple tweak of our construction by simply removing the option of outputting $\perp$ for party P_1 within the sub-routine SLVDistribDLog.

Secure 2-server Computation from Simulatable LV-HSS and Leakage Resilience. We now combine the simulatable LV-HSS of Proposition 3, which yields 2-server protocols with partial leakage, together with techniques for protecting computation against this leakage, to obtain a 2-round (m-client 2-server) secure computation protocol (in the PKI model) with *standard* security.

More concretely, the simulatable LV-HSS (Π_{SLV}, $\mathsf{Eval}_{\mathsf{SLV}}$) guaranteed leakage (with high probability) of up to λ intermediate RMS computation memory values y_i and secret-key representation values $\hat{c}^{(t)}$.

To protect against leakage of intermediate computation values, we can replace homomorphic evaluation of the program P with evaluation of a new ("leakage-resilient") program that takes as input *secret shares* $w_i^{(1)}, \ldots, w_i^{(k)}$ of clients' inputs w_i, and emulates a k-server secure computation of the program (whose NextMsg computation is in NC^1) that recombines secret shares and evaluates P, *while guaranteeing correctness and security against λ out of k server corruptions* (referred to as "λ-robustness"). Indeed, the λ leaked/erred intermediate computation values from HSS evaluation now correspond directly to revealing/losing the view of up to λ (virtual) servers in the emulated protocol. For simplicity, we use client-server protocols with no server-server communication, and so we can even emulate servers by *independent* HSS executions. Such protocols are known to exist for secure computation of low-degree polynomials [26]; in turn, this yields a solution for secure computation of general circuits P by instead generating a randomized encoding of the circuit P, computable in low degree [1,38].

To deal with the leakage on the values $\hat{c}^{(t)}$, we further refine the above approach. It will no longer be sufficient to take the $\hat{c}^{(t)}$ directly as the bits of the ElGamal secret key c (as in [8]), since this leakage will compromise the security of the encryptions and thus the HSS. Instead, we take $(\hat{c}^{(t)})_{t \in [\ell']} \leftarrow \mathsf{Decomp}(c)$ defined by first additively secret sharing c over $\mathbb{Z}_q$ into $\lambda+1$ shares, and then taking the $\ell' := (\lambda + 1)\ell$ bits of these separate values. Note that the $\hat{c}^{(t)}$ themselves are bits (in particular, have small magnitude) and reconstruction is linear over $\mathbb{Z}_q$ (first perform powers-of-2 bit reconstruction, then add the resulting values). But, further, any subset of λ values $\hat{c}^{(t)}$ are *statistically independent* of c.

Theorem 2 (Security amplification via virtual client-server protocols). *Let (Π_{SLV}, $\mathsf{Eval}_{\mathsf{SLV}}$) be the one-round simulatable Las Vegas client-server HSS from Proposition 3, and let* (Encode, NextMsg, Decode) *be a λ-robust client-server secure computation protocol with no server-server communication with* NextMsg $\in \mathsf{NC}^1$ *(see Sect. 2.3). Then for any polynomial $m = m(\lambda)$, the protocol Π given in Construction 3 is a secure m-client 2-server protocol for general circuits that executes in 2 rounds in the PKI model.*

Construction 3 (Secure 2-round m-client 2-server protocol (with PKI)).
Input: Each client begins with input w_i.
Tools:

- *("Virtual") 2λ-robust m-client k-server single-round secure computation protocol* (Encode, NextMsg, Decode), *with no server-server interaction (i.e., server computation is a single execution of* NextMsg $\in \mathsf{NC}^1$).
- *One-round simulatable LV-HSS (Π_{SLV}, $\mathsf{Eval}_{\mathsf{SLV}}$) from Proposition 3.*

Protocol:

0. *PKI: The new PKI consists of k independent copies of the PKI distribution from the simulatable LV-HSS; denote each copy by* $\mathsf{PKI}^{(j)}$.
1. *Each client C_i encodes his input as* $(\mathsf{msg}_i^{(1)}, \ldots, \mathsf{msg}_i^{(k)}) \leftarrow \mathsf{Encode}(i, w_i)$.
2. **Communication Round 1:** *In k parallel executions (one for each virtual server in the underlying secure computation protocol), using fresh randomness, the clients each send the corresponding single message as dictated by the one-round sharing protocol Π_{SLV}, where in the j'th execution ($j \in [k]$), client C_i uses $\mathsf{PKI}^{(j)}$ and input $\mathsf{msg}_i^{(j)}$.*
3. *As a result of the previous step, each (real) HSS server S_b learns k shares* $\mathsf{share}_b^{(1)}, \ldots, \mathsf{share}_b^{(k)}$, *one for each virtual server in the secure computation protocol, where $\mathsf{share}_b^{(j)}$ is one share of all clients' messages to virtual server j.*
4. *Each server S_b performs k independent homomorphic evaluations: For each virtual server $j \in [k]$, let* $(\mathsf{output}_b^{(j)}, z_b^{(j)}) = \mathsf{Eval}_{\mathbb{G},g}^{\mathsf{SLV}}(b, \mathsf{share}_b^{(j)}, \mathsf{NextMsg}, 1/2k\lambda)$, *with allowable error probability $1/2k\lambda$. Let $\mathsf{output}_b = (\mathsf{output}_b^{(1)}, \ldots, \mathsf{output}_b^{(k)})$, i.e. S_b's secret share (with possible $\perp$ symbols) of the encoded output of the client-server protocol.*
5. **Communication Round 2:** *Each server $b \in \{0,1\}$ sends his evaluated share, output_b, to all clients.*
6. *Each client outputs $\mathsf{Decode}(\mathsf{output}_0 + \mathsf{output}_1)$: i.e., recombining the HSS output shares (where $\perp + h$ is defined as $\perp$) and running the decoding procedure of the client-server protocol on the resulting output.*

Proof (Sketch). We defer the formal security proof to the full version and briefly outline the simulator $\mathsf{Sim}_{2r}(1^\lambda, \{w_i\}_{C_i \in \mathsf{Corrupt}}, y)$, where $\mathsf{Corrupt} \subset \{C_1, \ldots, C_m\} \cup \{S_0, S_1\}$ is the set of corrupted clients/servers, and y is the output $P(w_1, \ldots, w_m)$ received by the ideal functionality.

Assume wlog that $S_b \in \mathsf{Corrupt}$. Sim_{2r} simulates the HSS shares sent to S_b in the first round on behalf of each honest client C_i, by generating an HSS sharing with respect to $\mathsf{PKI}^{(j)}$ of 0 for each virtual server $j \in [k]$. For $j \in [k]$, Sim_{2r} computes $(\mathsf{output}_b^{(j)}, z_b^{(j)}) = \mathsf{Eval}_{\mathbb{G},g}^{\mathsf{SLV}}(b, \mathsf{share}_b^{(j)}, \mathsf{NextMsg}, 1/2k\lambda)$ on S_b's shares. Let $\mathsf{Corrupt}_S^{\mathsf{Virt}} = \{j \in [k] : z_b^{(j)} = \mathsf{Pred}_b^{(j)} \neq \top\}$ be the virtual servers j for which $\mathsf{output}_{1-b}^{(j)}$ might be $\perp$ (thus leaking information). By Proposition 3, with overwhelming probability $|\mathsf{Corrupt}^{\mathsf{Virt}}| \leq \lambda$ (by correctness and independence of executions) and each $\mathsf{Pred}_b^{(j)}$ depends on the input and at most λ values of $\hat{c}^{(t)}$ for the key c within the corresponding j'th HSS execution.

Sim_{2r} then runs the simulator for the underlying (virtual) m-client k-server protocol, for corrupted clients $\mathsf{Corrupt}_C^{\mathsf{Virt}} = \mathsf{Corrupt} \cap \{C_1, \ldots, C_m\}$ and corrupted (virtual) servers $\mathsf{Corrupt}_S^{\mathsf{Virt}}$, for corrupted inputs $\{w_i\}_{C_i \in \mathsf{Corrupt}}$. The resulting simulated view^{Virt} contains, in particular, the messages $\{\mathsf{msg}_i^{(j)}\}_{C_i \notin \mathsf{Corrupt}}$ received by each corrupt virtual server $j \in \mathsf{Corrupt}_S^{\mathsf{Virt}}$ from honest clients C_i, and all (pre-Decode) values $\mathsf{output}^{(1)}, \ldots, \mathsf{output}^{(k)}$.

For $j \in [k]$, Sim_{2r} simulates the output share $\mathsf{output}_{1-b}^{(j)}$ as follows. Sample λ random bits to serve as the bits $(\hat{c}^{(t)})_{t \in [\lambda]}$ of the jth key that $\mathsf{Pred}_b^{(j)}$ depends on on (if $z_b^{(j)} = \mathsf{Pred}_b^{(j)} \neq \top$). If $j \notin \mathsf{Corrupt}_S^{\mathsf{Virt}}$, or if $\mathsf{Pred}_b^{(j)}(\mathsf{msg}^{(j)}, (\hat{c}^{(t)})_{t \in [\lambda]}) = 0$, then $\mathsf{output}_{1-b}^{(j)} = \mathsf{output}^{(j)} - \mathsf{output}_b^{(j)}$. Otherwise, $\mathsf{output}_{1-b}^{(j)} = \bot$.

Theorem 5 is an application of the above, obtained by using the virtual client-server protocol of [26] for evaluating low-degree polynomials. Our final result follows from generic transformations using low-degree randomized encodings [1].

Theorem 4 (MPC for low-degree polynomials [26]). *For any $t, m, d \in \mathbb{N}$ there is a 2-round, m-client, k-server, perfectly t-robust protocol with no server-server interaction, for the class of degree-d polynomials over $\mathbb{F}_2$, where $k = O(dt)$. When evaluating a vector of ℓ polynomials on n inputs, the computation of each server can be implemented by a circuit of depth $O(\log(n + \ell + k))$.*

Theorem 5 (Succinct 2-server protocol for low-degree polynomials). *Assuming PKI setup and DDH, there is a succinct 2-round 2-server client-server protocol for evaluating degree-d polynomials, for any constant d.*

Corollary 1 (2-server protocol for circuits). *Assuming PKI setup and DDH, there is a (non-succinct) 2-round 2-server client-server protocol for circuits.*

Note that while this solution yields 2 rounds of communication, the amount of information communicated is greater than the program size. In the full version, we describe a more complex solution achieving *succinct* 2-round secure computation for the class of NC^1 programs.

4.3   From 2 to k Servers

As the final step, we compile the 2-round m-client 2-server protocol into a 2-round m-client k-*server* protocol, for any constant $k \in O(1)$. This is achieved by iteratively emulating the role of one server by two servers via the original 2-server protocol. A similar notion of party emulation has appeared within many contexts in the literature (e.g., [10,23]). In each step of this process, the next-message-function computed by the emulated server is realized by using a 2-round client-server protocol involving the m clients and the 2 emulating servers. This increases the number of servers by 1, while still maintaining security as long as only a strict subset of the servers are corrupted. The communication and computation complexity of the protocol increase by a factor of $\mathsf{poly}(\lambda)$ in each such step. Repeating $k - 1$ times, we get the following.

Theorem 6 (2-round k-server client-server protocol). *Assume PKI setup and DDH. Then for any constant $k \geq 2$ there is a 2-round k-server client-server protocol (alternatively, a 2-round k-party MPC protocol) for circuits.*

5   Optimizing Communication

In the previous section, we eliminated the inverse polynomial error and leakage of HSS by using secret-sharing of the inputs and applying virtual client-server MPC protocols to compute on these shares. In this section we describe a simpler alternative approach that has better asymptotic and concrete communication complexity (and better computational complexity as well) at the cost of requiring an additional round of interaction. In contrast to the previous approach, the current approach applies only to the case of 2PC and does not apply to the more general case of client-server MPC.

The high level idea is as follows. Denote the two parties by P_0, P_1 and assume that the functionality f delivers an output only to P_1. We rely on an asymmetric Las-Vegas HSS (see Definition 1) where the output of Eval is guaranteed to be correct (i.e., the two output shares add up to the correct output) unless P_1 outputs $\perp$, where the latter occurs with at most δ probability. The idea is to have P_1 use $\binom{m}{m-k}$-bit-oblivious-transfer (denoted by $\binom{m}{-k}$-OT) in order to block itself from the k output shares of P_0 that correspond to the positions in which it outputs $\perp$. Note that the $m - k$ selected output shares can be simulated given the correct output and the view of P_1, and thus they do not leak any additional information about the input. To make up for the k lost output bits, we use an erasure code to encode the output. Since we can make the number of erasures small, we only need to introduce a small amount of redundancy to the output.

Punctured OT. A key observation is that by setting the error parameter δ to be sufficiently small, we can ensure that the $\binom{m}{-k}$-OT parameters are such that k is much smaller than m. We refer to OT in this parameter regime as *punctured OT* and show how to implement it very efficiently by using a *puncturable PRF*.

A puncturable PRF [37] is a standard PRF family F_K equipped with a puncturing algorithm Puncture that given a set of points $X = \{x, \ldots, x_k\} \subseteq \{0,1\}^d$ produces an evaluation key K_X that allows an evaluation of the PRF on all inputs *except* those in X. Moreover, the PRF values on the inputs in X should be indistinguishable from random given K_X. See full version for a formal definition. As was shown in [5,9,29], the GGM construction [20] of PRFs from a length-doubling PRG can be used to obtain a puncturable PRF for $X = \{x_1, \ldots, x_k\} \subseteq \{0,1\}^d$ with key size $|K_X| = O(\lambda k d)$. The evaluation of F at all points given K or at all non-punctured point given K_X requires $O(2^d)$ invocations of a PRG $G : \{0,1\}^\lambda \rightarrow \{0,1\}^{2\lambda}$. The circuit size required for generating K_X given a λ-bit K and X is $kd \cdot poly(\lambda)$.

A protocol for $\binom{m}{-k}$-OT can be implemented using a puncturable PRF and any general-purpose 2PC protocol (e.g., Yao's protocol [31,38]) in the following natural way.

- Sender's input: $s \in \{0,1\}^m$, where every $i \in [m]$ is represented by a d-bit string.
- Receiver's input: $X \subset [m]$ where $|X| = k$.
- Given primitives: a puncturable PRF $(F_K, \text{Puncture})$, an ideal 2PC oracle Π.

1. Invoke Π on the randomized functionality that, on Receiver input X, delivers a random PRF key K to Sender and constrained PRF key K_X to Receiver.
2. Sender computes and sends $s' \in \{0,1\}^m$ where $s'_i = s_i \oplus F_K(i)$.
3. Receiver outputs $(i, s' \oplus F_{K_X}(i))$ for $i \in [m] \setminus X$.

ANALYSIS. Correctness is straightforward. Security follows from the fact that the values of F_K on all inputs $i \in [m] \setminus X$ are pseudorandom given K_X. Thus, a simulator can simulate the receiver's view given the receiver's output by just running the protocol with an arbitrary s that is consistent with the output. Plugging in Yao's protocol[6] for implementing Π, we get the following theorem.

Theorem 7 (Punctured OT via puncturable PRF). *Suppose a $\binom{2}{1}$-OT protocol exists. Then there is a protocol for $\binom{m}{-k}$-OT with $m + k \cdot \log m \cdot \mathsf{poly}(\lambda)$ bits of communication, where the computational complexity consists of $O(m)$ invocations of a length-doubling PRG $G : \{0,1\}^\lambda \to \{0,1\}^{2\lambda}$ and $\mathsf{poly}(\lambda)$ additional computation.*

We turn to describe our communication-efficient technique for eliminating the inverse polynomial error of HSS. In addition to punctured OT, our second ingredient is a simple randomized erasure correcting code.

Lemma 2 (Erasure correcting code). *There is a randomized linear encoding function $C_r : \{0,1\}^m \to \{0,1\}^{m+m/\lambda}$ that can correct a $1/\lambda^2$ rate of random erasures with all but $m \cdot \mathsf{negl}(\lambda)$ probability.*

Proof. A message $x \in \{0,1\}^m$ is encoded by $(x, y_1, \ldots, y_{m/\lambda})$ where y_i is the parity of a random subset of $\lambda^2/2 - 1$ bits of x. By a Chernoff bound, except with $m \cdot \mathsf{negl}(\lambda)$ probability, every bit of x is involved in at least $\lambda/3$ sets, where every set (including the corresponding parity check) contains an erasure with at most $\frac{\lambda^2/2}{\lambda^2} = 1/2$ error probability. Hence, for any fixed x_i, the probability that all sets involving x_i contain an erasure is at most $2^{-\lambda/3}$. Hence, the probability that some x_i cannot be recovered is bounded by $m \cdot \mathsf{negl}(\lambda)$ as required. □

Finally, we combine punctured OT and erasure codes to give a succinct 2PC protocol for branching programs. This protocol avoids the use of virtual client-server MPC and can thus achieve better communication rate and computational complexity than its counterpart from Sect. 4.2.

The protocol is similar to the protocol for branching programs from [8] (cf. Theorem 4.5 in full version), which evaluates m branching programs on inputs of total length n using $n + m \cdot \mathsf{poly}(\lambda)$ bits of communication, except for the following differences. First, instead of repeating each output bit λ times, the functionality is modified so that the outputs are encoded using the randomized erasure code of Lemma 2 (where a PRG is used to pick the randomness r with

---

[6] We do not attempt here to optimize the concrete efficiency of this secure computation. Given the current speed of secure 2PC protocols for AES, even a naive implementation is expected to be quite efficient.

sublinear communication). Second, instead of applying a standard DEHE to compute shares of the output encoding, we use a (multi-evaluation) asymmetric Las Vegas variant in which P_1 outputs $\perp$ whenever there is a risk of error. We set the error parameter δ to be a sufficiently small $1/\mathsf{poly}(\lambda)$ so that: (1) except with $\mathsf{negl}(\lambda)$ probability, the number of $\perp$ outputs is bounded by $k = m/\lambda^2$, and (2) the communication complexity of $\binom{m'}{-k}$-OT, where $m' = m + m/\lambda$, is $m + o(m)$. Finally, P_1 uses punctured OT to retrieve the output shares of P_0 in the positions where it did not output $\perp$. Note that, by the definition of asymmetric Las Vegas HSS, the shares obtained from P_0 are determined by the shares of P_1 and the output (except with negligible probability), and hence they can be simulated given the output.

The above protocol gives rise to the following theorem.

Theorem 8 (Optimized 2PC for branching programs). *Assuming DDH, there is a constant-round secure 2-party protocol for evaluating any sequence of m branching programs of size S on inputs (x_0, x_1) of total length n, using $n + (1 + o(1))m + \mathsf{poly}(\lambda)$ bits of communication and $\mathsf{poly}(\lambda) \cdot m \cdot S^2$ computation.*

As a corollary, we get the following near-optimal protocol for OT.

Corollary 2 (Constant-rate bit-OT). *Assuming DDH, there is a constant-round secure 2-party protocol for evaluating n instances of bit-OT with $(4 + o(1))n + \mathsf{poly}(\lambda)$ bits of communication and $\mathsf{poly}(\lambda) \cdot n$ computation.*

Combining Corollary 2 with the GMW protocol for secure circuit evaluation using bit-OT [21], we get the following corollary.

Corollary 3 (MPC for general circuits). *Assuming DDH, there is a secure 2-party protocol for evaluating any circuit C of size S with $O(S) + \mathsf{poly}(\lambda)$ bits of communication.*

This should be compared with a similar protocol from the full version of [8] (cf. Theorem 4.10) in which the communication complexity has an additional $(depth + output) \cdot \mathsf{poly}(\lambda)$ term.

6   Optimizing Computation

A bottleneck of the performance of the HSS scheme in [8] and the schemes in this paper is the computation time of homomorphically evaluating RMS multiplications. The time required for the multiplication is almost entirely the result of $\ell + 1$ executions of ConvertShares and $2(\ell + 1)$ executions of MultShares.

We present three optimizations of these procedures. The first optimizes the *worst case* asymptotic running time of the share conversion algorithm by a $\log(1/\delta)$ factor, but does not improve the *expected* running time. The second optimization, which is incompatible with the first, optimizes the concrete running time of the conversion. The third balances the computational complexity of ConvertShares and MultShares to reduce the overall running time of evaluating an RMS multiplication. The first and third of these optimizations (discussed in greater detail in the Introduction) are deferred to the full version of the paper.

6.1 Optimizing the Conversion

A straightforward implementation of the share conversion step in Fig. 1 for a group element $h \in \mathbb{G}$ requires computing the sequence $h, hg, \ldots, hg^x$ for a generator g, computing a pseudo-random function on each element and choosing the first distinguished point (or alternatively the minimal value). A natural strategy for this implementation is to choose the group $\mathbb{G}$ to be a group over elliptic curves, since computing the sequence $h, hg, \ldots, hg^x$ in such groups is more efficient than in other DDH groups.

We explore an alternative implementation to the conversion step which tests whether a sub-sequence of elements $hg^i, \ldots, hg^{i+j}$ includes a distinguished point without explicitly computing each element in the sub-sequence. To achieve this idea we work over groups $\mathbb{Z}_p^*$ with specific structure rather than over an EC group. In addition, this approach requires the distinguished point version of share conversion rather than the min-hash method (described in the full version).

The first idea is to decide if an element $hg^i \in \mathbb{G}$ is distinguished without using a PRF ϕ. We say that an element h' is distinguished if the representation of h' has $d = \lceil \log(1/\delta) \rceil$ leading zeroes, i.e. $h' < 2^{\lceil \log p \rceil - d}$. We conjecture that if $h \in \mathbb{G}$ is chosen randomly then the sequence $h, \ldots, hg^x$ has a distinguished point with essentially the same probability as that of the sequence $\phi(h), \ldots, \phi(hg^x)$. Observe that h can be chosen randomly since the two servers can shift their respective elements h_0, h_1 by a shared random element r maintaining the difference between the elements.

The second idea is to consider pseudo-Mersenne primes, i.e. primes of the form $p = 2^k - \gamma$ for small γ, in which the element 2 generates a large sub-group. We refer to such primes as *conversion friendly*. In this setting, $2h \bmod p$ can be computed by shifting the bit representation of h one bit to the left, removing the most significant bit and adding γ to the result if the removed bit is 1. Therefore, computing the next element of the sequence $h, \ldots, hg^x$ involves little more than a comparison of the bit, an addition, and testing whether the d most significant bits of the result are zero.

Further savings are possible by taking advantage of hardware architectures that enable fast multiplication of w-bit words. If $h = a_1 2^{n-w} + a_0$ for $0 \le a_0 < 2^{n-w}, 0 \le a_1 < 2^w$ then $2^w h \equiv a_0 2^w + a_1 \gamma \bmod p$. Note that if $\gamma << 2^w$ then computing $2^w h$ requires one multiplication of words and with high probability one addition of words.

It is possible to test if any of the w elements $h, 2h \bmod p, \ldots, 2^{w-1} h \bmod p$ are distinguished by checking whether the most significant $2w$ bits of h include the substring 0^d. That can be done efficiently in standard computer architectures by dividing the $2w$ bits into strips of length $d/2$ and checking whether any of the strips is $0^{d/2}$. If none of them are then the sequence $h, 2h, \ldots, 2^{w-1}h$ does not contain a distinguished point and the next element to be examined is $2^w h$. An interesting property of the algorithm is that it is almost independent of the size of the underlying group.

A class of conversion-friendly primes which are relatively common are pseudo-Mersenne primes p which are safe, i.e. $p = 2q + 1$ for a prime q and which satisfy

Table 1. Performance figures for the conversion step over a prime $p = 2^n - \gamma$ with d zero bits determining a distinguished point.

Word size	Multiplications per step	Additions per step	Masking operations per step	No. of Conversion steps per second
32 bits	0.031	0.031	0.22	1.6 billion
w bits	$\frac{1}{w}$	$\frac{1}{w} + \frac{\gamma}{2^w}$	$\frac{2}{w}(\lceil \frac{w}{d} \rceil + \frac{d}{2^{d/2}})$	–

$p \equiv \pm 1 \bmod 8$. For such primes the sub-group $\mathbb{G}$ that includes all the quadratic residues modulo p is of size q. Since q is prime, every element in $\mathbb{G}$ generates the sub-group and one of these elements is 2 since $p \equiv \pm 1 \bmod 8$. Examples for such conversion-friendly primes one can use include $2^{1280} - 7243217$, $2^{1536} - 11510609$ and $2^{2048} - 1942289$.

Assessing the security of DDH over these primes is difficult due to the scarcity of published attacks. Theoretically, the best attack against DDH over pseudo-Mersenne primes is using the Special Number Field Sieve (SNFS) [35] to compute discrete logarithms modulo the prime. The SNFS has been used to factor Mersenne numbers, with the current record being $2^{1199} - 1$ [30]. To account for the speedup offered by SNFS, the bit-length of such special primes needs to be roughly 50% bigger than that of a general prime to provide a similar level of security. For instance, a 2048-bit special p is roughly comparable to a 1340-bit general p [16].

Table 1 presents the average number of basic operations required for one conversion step, i.e. computing $2h \bmod p$ from h and checking whether h is distinguished, and the number of conversion steps per second. The figures in the first row of the table are based on an implementation on a commodity laptop (Dell Latitude 3550, with Intel i7-5500 CPU, running single-threaded at 2.4 GHz and with 8 GByte of RAM) and can be significantly improved given a dedicated hardware and software platform. The implementation used 32-bit words together with multiplications of two 32 bit operands into a 64 bit product. The second row is a general analysis for an architecture with w bit words. The basic operations which are measured in the table are word-sized multiplication, addition and bit level operations (bit-by-bit AND operations and shifts).

The table makes it clear that the conversion step requires on average well below a single instruction, e.g. 0.25 instructions per step in the example in the first row. In the alternative approach for computing a conversion step, each such step includes a group operation over an elliptic curve. Based on [6] Table 3, the fastest elliptic curve multiplication by a scalar for a relatively small, 254-bit, curve requires 196,000 machine instructions (on a somewhat stronger machine than what we used). A multiplication requires on average $254 \cdot (3/2)$ group operations, which means that each group operation, and each conversion, requires at least 2000 times the number of instructions of a conversion step implemented via conversion-friendly primes.

Acknowledgements. We thank Antoine Joux for discussions, suggestions, and pointers that helped improve the results of Sect. 6. We also thank the anonymous reviewers for helpful comments.

First author supported by ISF grant 1861/16, AFOSR Award FA9550-17-1-0069, and ERC starting grant 307952. Second author supported by ISF grant 1638/15, a grant by the BGU Cyber Center, the Israeli Ministry Of Science and Technology Cyber Program and by the European Union's Horizon 2020 ICT program (Mikelangelo project). Third author supported by a DARPA/ARL SAFEWARE award, DARPA Brandeis program under Contract N66001-15-C-4065, NSF Frontier Award 1413955, NSF grants 1619348, 1228984, 1136174, and 1065276, NSF-BSF grant 2015782, ISF grant 1709/14, BSF grant 2012378, a Xerox Faculty Research Award, a Google Faculty Research Award, an equipment grant from Intel, and an Okawa Foundation Research Grant. This material is based upon work supported by the Defense Advanced Research Projects Agency through the ARL under Contract W911NF-15-C-0205. The views expressed are those of the authors and do not reflect the official policy or position of the Department of Defense, the National Science Foundation, or the U.S. Government.

References

1. Applebaum, B., Ishai, Y., Kushilevitz, E.: Computationally private randomizing polynomials and their applications. In: CCC, pp. 260–274 (2005)
2. Asharov, G., Jain, A., López-Alt, A., Tromer, E., Vaikuntanathan, V., Wichs, D.: Multiparty computation with low communication, computation and interaction via threshold FHE. In: Pointcheval, D., Johansson, T. (eds.) EUROCRYPT 2012. LNCS, vol. 7237, pp. 483–501. Springer, Heidelberg (2012). doi:10.1007/978-3-642-29011-4_29
3. Beaver, D., Micali, S., Rogaway, P.: The round complexity of secure protocols (extended abstract). In: STOC, pp. 503–513 (1990)
4. Boneh, D., Halevi, S., Hamburg, M., Ostrovsky, R.: Circular-secure encryption from decision diffie-hellman. In: Wagner, D. (ed.) CRYPTO 2008. LNCS, vol. 5157, pp. 108–125. Springer, Heidelberg (2008). doi:10.1007/978-3-540-85174-5_7
5. Boneh, D., Waters, B.: Constrained pseudorandom functions and their applications. In: Sako, K., Sarkar, P. (eds.) ASIACRYPT 2013. LNCS, vol. 8270, pp. 280–300. Springer, Heidelberg (2013). doi:10.1007/978-3-642-42045-0_15
6. Bos, J.W., Costello, C., Longa, P., Naehrig, M.: Selecting elliptic curves for cryptography: an efficiency and security analysis. J. Cryptographic Eng. **6**(4), 259–286 (2016)
7. Boyle, E., Gilboa, N., Ishai, Y.: Function secret sharing. In: Oswald, E., Fischlin, M. (eds.) EUROCRYPT 2015. LNCS, vol. 9057, pp. 337–367. Springer, Heidelberg (2015). doi:10.1007/978-3-662-46803-6_12
8. Boyle, E., Gilboa, N., Ishai, Y.: Breaking the circuit size barrier for secure computation under DDH. In: Robshaw, M., Katz, J. (eds.) CRYPTO 2016. LNCS, vol. 9814, pp. 509–539. Springer, Heidelberg (2016). doi:10.1007/978-3-662-53018-4_19. Full version: IACR Cryptology ePrint Archive 2016: 585 (2016)
9. Boyle, E., Goldwasser, S., Ivan, I.: Functional signatures and pseudorandom functions. In: Krawczyk, H. (ed.) PKC 2014. LNCS, vol. 8383, pp. 501–519. Springer, Heidelberg (2014). doi:10.1007/978-3-642-54631-0_29
10. Bracha, G.: An asynchronous $[(n-1)/3]$-resilient consensus protocol. In: PODC, pp. 154–162 (1984)

11. Broder, A.Z., Charikar, M., Mitzenmacher, M.: A derandomization using min-wise independent permutations. In: Luby, M., Rolim, J.D.P., Serna, M. (eds.) RAN-DOM 1998. LNCS, vol. 1518, pp. 15–24. Springer, Heidelberg (1998). doi:10.1007/3-540-49543-6_2

12. Canetti, R.: Security and composition of multiparty cryptographic protocols. J. Cryptology **13**, 143–202 (2000)

13. Chillotti, I., Gama, N., Georgieva, M., Izabachène, M.: Faster fully homomorphic encryption: bootstrapping in less than 0.1 seconds. In: Cheon, J.H., Takagi, T. (eds.) ASIACRYPT 2016. LNCS, vol. 10031, pp. 3–33. Springer, Heidelberg (2016). doi:10.1007/978-3-662-53887-6_1

14. Dodis, Y., Halevi, S., Rothblum, R.D., Wichs, D.: Spooky encryption and its applications. In: Robshaw, M., Katz, J. (eds.) CRYPTO 2016. LNCS, vol. 9816, pp. 93–122. Springer, Heidelberg (2016). doi:10.1007/978-3-662-53015-3_4

15. Ducas, L., Micciancio, D.: FHEW: bootstrapping homomorphic encryption in less than a second. In: Oswald, E., Fischlin, M. (eds.) EUROCRYPT 2015. LNCS, vol. 9056, pp. 617–640. Springer, Heidelberg (2015). doi:10.1007/978-3-662-46800-5_24

16. Fried, J., Gaudry, P., Heninger, N., Thomé, E.: A kilobit hidden SNFS discrete logarithm computation. IACR Cryptology ePrint Archive, 2016:961 (2016)

17. Garg, S., Gentry, C., Halevi, S., Raykova, M.: Two-round secure MPC from indistinguishability obfuscation. In: Lindell, Y. (ed.) TCC 2014. LNCS, vol. 8349, pp. 74–94. Springer, Heidelberg (2014). doi:10.1007/978-3-642-54242-8_4

18. Gentry, C.: Fully homomorphic encryption using ideal lattices. In: STOC, pp. 169–178 (2009)

19. Goldreich, O.: Foundations of Cryptography – Basic Applications. Cambridge University Press, Cambridge (2004)

20. Goldreich, O., Goldwasser, S., Micali, S.: How to construct random functions. J. ACM **33**(4), 792–807 (1986)

21. Goldreich, O., Micali, S., Wigderson, A.: How to play any mental game or a completeness theorem for protocols with honest majority. In: STOC, pp. 218–229 (1987)

22. Halevi, S., Shoup, V.: Bootstrapping for HElib. In: Oswald, E., Fischlin, M. (eds.) EUROCRYPT 2015. LNCS, vol. 9056, pp. 641–670. Springer, Heidelberg (2015). doi:10.1007/978-3-662-46800-5_25

23. Hirt, M., Maurer, U.M.: Player simulation and general adversary structures in perfect multiparty computation. J. Cryptology **13**(1), 31–60 (2000)

24. Horvitz, O., Katz, J.: Universally-composable two-party computation in two rounds. In: Menezes, A. (ed.) CRYPTO 2007. LNCS, vol. 4622, pp. 111–129. Springer, Heidelberg (2007). doi:10.1007/978-3-540-74143-5_7

25. Indyk, P.: A small approximately min-wise independent family of hash functions. J. Algorithms **38**(1), 84–90 (2001)

26. Ishai, Y., Kushilevitz, E.: Randomizing polynomials: a new representation with applications to round-efficient secure computation. In: FOCS, pp. 294–304 (2000)

27. Ishai, Y., Kushilevitz, E., Ostrovsky, R., Sahai, A.: Cryptography with constant computational overhead. In: STOC, pp. 433–442 (2008)

28. Ishai, Y., Kushilevitz, E., Ostrovsky, R., Sahai, A.: Zero-knowledge proofs from secure multiparty computation. SIAM J. Comput. **39**(3), 1121–1152 (2009)

29. Kiayias, A., Papadopoulos, S., Triandopoulos, N., Zacharias, T.: Delegatable pseudorandom functions and applications. In: CCS, pp. 669–684 (2013)

30. Kleinjung, T., Bos, J.W., Lenstra, A.K.: Mersenne factorization factory. In: Sarkar, P., Iwata, T. (eds.) ASIACRYPT 2014. LNCS, vol. 8873, pp. 358–377. Springer, Heidelberg (2014). doi:10.1007/978-3-662-45611-8_19

31. Lindell, Y., Pinkas, B.: A proof of security of Yao's protocol for two-party computation. J. Cryptology **22**(2), 161–188 (2009)
32. López-Alt, A., Tromer, E., Vaikuntanathan, V.: On-the-fly multiparty computation on the cloud via multikey fully homomorphic encryption. In: STOC, pp. 1219–1234 (2012)
33. Mukherjee, P., Wichs, D.: Two round multiparty computation via multi-key FHE. In: Fischlin, M., Coron, J.-S. (eds.) EUROCRYPT 2016. LNCS, vol. 9666, pp. 735–763. Springer, Heidelberg (2016). doi:10.1007/978-3-662-49896-5_26
34. Naor, M., Nissim, K.: Communication preserving protocols for secure function evaluation. In: STOC, pp. 590–599 (2001)
35. Pollard J.: Factoring with cubic integers (1988). Unpublished manuscript
36. Rivest, R.L., Adleman, L., Dertouzos, M.L.: On data banks and privacy homomorphisms. In: Foundations of Secure Computation, pp. 169–179 (1978)
37. Sahai, A., Waters, B.: How to use indistinguishability obfuscation: deniable encryption, and more. In: STOC, pp. 475–484 (2014)
38. Yao, A.C.-C.: How to generate and exchange secrets (extended abstract). In: FOCS, pp. 162–167 (1986)

On the Exact Round Complexity
of Self-composable Two-Party Computation

Sanjam Garg[1], Susumu Kiyoshima[2]([✉]), and Omkant Pandey[3]

[1] University of California, Berkeley, USA
sanjamg@berkeley.edu
[2] NTT Secure Platform Laboratories, Tokyo, Japan
kiyoshima.susumu@lab.ntt.co.jp
[3] Stony Brook University, Stony Brook, USA
omkant@cs.stonybrook.edu

Abstract. The round complexity of secure computation has been a fundamental problem in cryptography. Katz and Ostrovsky proved that 5 rounds are both necessary and sufficient for secure computation in the stand alone setting, thus resolving the *exact* round complexity of *standalone* secure computation.

In contrast, round complexity of secure computation in the *concurrent* setting, where several protocols may run simultaneously, is poorly understood. Since standard polynomial time simulation is impossible in the concurrent setting, alternative security notions have been proposed, e.g., *super-polynomial simulation* (SPS). While SPS security can be achieved in constant rounds, the actual constant (> 20) is far from optimal.

In this work, we take the first steps towards studying the exact round complexity of concurrent secure computation. We focus on the two party case and present a new secure computation protocol that achieves SPS security under concurrent self-composition. Our protocol has 5 rounds assuming quasi-polynomially-hard injective one-way functions (or 7 rounds assuming standard polynomially-hard collision-resistant hash functions). We also require other standard assumptions, specifically trapdoor OWPs and lossy TDFs. This matches the rounds for standalone secure computation.

More specifically, our security proof presents a *polynomial time* reduction from SPS security to 3-round public-coin non-malleable commitments with appropriate extractability properties. Such commitments are known based on quasi-polynomially-hard injective OWFs. (The reduction also works with a special 6-round non-malleable commitment to yield the 7-round result under CRHFs.)

S. Garg—Research supported in part from DARPA/ARL SAFEWARE Award W911NF15C0210, AFOSR Award FA9550-15-1-0274, NSF CRII Award 1464397, AFOSR YIP Award and research grants by the Okawa Foundation and Visa Inc. The views expressed are those of the author and do not reflect the official policy or position of the funding agencies.

J.-S. Coron and J.B. Nielsen (Eds.): EUROCRYPT 2017, Part II, LNCS 10211, pp. 194–224, 2017.
DOI: 10.1007/978-3-319-56614-6_7

1   Introduction

Secure computation protocols are protocols that enable mutually distrustful parties to compute a functionality without compromising the correctness of the outputs and the privacy of their inputs. Secure computation protocols have been studied in both two-party case and multi-party case, and it was shown that secure computation protocols for any functionality can be constructed in both cases in a model with malicious adversaries and a dishonest majority [12,36].

The security of secure computation protocols is defined by using simulation paradigm. Specifically, to define the security of a protocol π for computing a function f, we consider the *real world*, where the parties compute f by executing π, and the *ideal world*, where the parties compute f by interacting with a trusted third party. Then, we define the security by requiring that for any adversary in the real world there exists a *simulator* in the ideal world such that whatever an adversary can do in the real world can be "simulated" in the ideal world by the simulator.

Round complexity of secure computation. A fundamental question in this area is to understand how many rounds are necessary and sufficient for securely computing general functionalities. Katz and Ostrovsky [18] proved that five rounds are both necessary and sufficient for secure two-party computation in the *standalone* setting where there is only one protocol execution. These results were further extended in [11,30] w.r.t. black-box constructions and simultaneous message channels. These results completely settle the round complexity of two-party computation in the *standalone* setting.

The concurrent setting. While standalone security is sufficient for many applications, other situations (such as protocol execution over the Internet) require stronger notions of security. This setting where there may be many protocols executions at the same time, is called the *concurrent* setting. Unfortunately, it is known that stand-alone security does not necessarily imply security in the concurrent setting [8].

Secure computation in the concurrent setting is more challenging to define than the standalone setting. Canetti [4] proposed the notion of *universally composable* (UC) security where protocols maintain their strong simulation based security guarantees even in the presence of other arbitrary protocols. However, achieving UC-security in the plain model turned out to be impossible [4,5]. Moreover, Lindell [25,26] proved that even in the special case where only instantiations of the *same* protocol are allowed, standard notion of polynomial-time simulation is impossible to achieve. (This is called "self composition" and corresponds to the setting we are interested in.)

These strong negative results motivated the study of alternative notions for concurrent secure computation, such as *super-polynomial-time simulation* (SPS) security (and the closely related *angel-based* security), input-indistinguishable computation, bounded concurrent composition, and multiple ideal-query model [2,6,10,13,14,19,20,23,28,29,31–33,35].

In this work we focus on SPS security in the two-party setting. In SPS security, the simulator is allowed to run in super-polynomial time; thus, SPS security guarantees that whatever an adversary can do in the real world can also be done in the ideal world *in super-polynomial time*. Although allowing the simulator to run in super-polynomial time weakens the security guarantee, SPS security still guarantees meaningful security for many functionalities. Furthermore, it was shown that under SPS security, concurrent self-composition can be achieved in the plain model. (In what follows, by SPS security we mean SPS-security under concurrent self-composition.)

SPS security has been extensively studied and improved upon in the literature. Prabhakaran and Sahai [35] provided the initial positive result for SPS security. Although, these early results [28, 35] relied on non-standard/subexponential-time assumptions, Canetti, Lin and Pass achieved this (actually, the angel-based) notion under standard polynomial-time assumptions [6] in a polynomial number of rounds. Soon after, Garg et al. [10] presented a *constant round* SPS-secure protocol, thus resolving the *asymptotic* round-complexity of SPS-secure computation (under polynomially-hard assumptions).

Exact round complexity of SPS-secure computation. Although the SPS-secure protocol of [10] has asymptotically constant rounds, its exact round complexity is actually quite large (more than 20). In contrast, the standalone setting only requires five rounds [18]. Is this gap necessary? What is the exact round complexity of SPS-secure protocols for computing general functionalities? To the best of our knowledge, these questions have not been explored before.

1.1   Our Results

In this work, we take the first steps towards studying the exact round complexity of concurrent secure computation. We present a new secure computation protocol whose round complexity matches that of the stand alone setting. More specifically, we present a *five-round* SPS-secure two-party computation protocol. Our protocol guarantees security under concurrent self-composition.

We are interested in basing the security of our protocol on standard, polynomially-hard, assumptions. We do this by providing a *polynomial-time* reduction that reduces the SPS-security of our protocol to that of 3-round public-coin non-malleable commitments with some natural extractability properties. In particular, we want 3-round non-malleable commitments that are extractable *without over-extraction* [19].

One caveat is that such non-malleable commitments, at present, are only known to exist under quasi-polynomially-hard injective OWFs [16].[1] Consequently, we only achieve a result under quasi-polynomially hard injective OWFs. We remark that even under super-polynomially hard assumptions,

---

[1] The 3-round non-malleable commitment of [16] was claimed to be secure under polynomially-hard injective OWFs; however, the public-coin variant of their scheme is proven secure only under quasi-polynomially-hard injective OWFs (see the latest ePrint version [15]).

previous SPS-secure protocols have quite large round complexity (more than 20) [2,14,20,24,33].

While existence of quasi-polynomially-hard injective OWFs is considered a standard assumption, it would be interesting to know if we can rely only on *polynomially*-hard assumptions. Towards this goal, we realize that our construction actually works with a special 6-round non-malleable commitment scheme based on (polynomially-hard) CRHFs. This gives us a 7-round SPS-secure protocol for general functionalities where all underlying assumptions are only polynomially-hard.

1.2   Overview of Techniques

Our overall strategy is to apply the techniques of the constant-round SPS-secure protocol of Garg et al. [10] to the five-round secure two-party computation protocol of Katz and Ostrovsky [18]. In this subsection, we first recall the techniques of Garg et al. and explain the difficulty in applying the techniques of Garg et al. to the protocol of Katz and Ostrovsky. After that, we give an overview of our techniques.

SPS protocol of Garg et al. Like other SPS protocols, the concurrently SPS-secure multi-party computation protocol of Garg et al. [10] has "trapdoor secrets" that enable simulation,[2] and the simulator obtains the trapdoor secrets by breaking cryptographic primitives by brute-force in super-polynomial time. The main technical challenge in the proof of security is to design a *polynomial-time* reduction that reduces the security of the protocol to the security of underlying cryptographic primitives. In fact, since the simulator runs in super-polynomial time, a naive approach that having the reduction emulate the simulator internally can only result in super-polynomial-time reductions.

To obtain a polynomial-time reduction in the proof of security, Garg et al. consider a hybrid experiment in which the brute-force extraction of the trapdoor secrets is replaced with polynomial-time rewinding extraction. With such a hybrid experiment, Garg et al. designs a security proof roughly as follows.

1. First, the indistinguishability between the real and the hybrid experiment is reduced to the security of various protocol components. The reductions run in polynomial time since both the real and the hybrid experiment run in polynomial time.
2. Next, the indistinguishability between the hybrid and the ideal experiment is shown without relying on any cryptographic assumptions. No cryptographic assumption is needed to show this indistinguishability since the two experiments differ only in how the trapdoor secrets are extracted and anyway the same trapdoor secrets are extracted in both experiments except with negligible probability.

---

[2] Concretely, the trapdoor secrets enable the simulator to give "proofs of correct behavior" while executing the protocol incorrectly.

However, since the protocol is executed in the concurrent setting, the use of rewinding extraction causes problems.

The first problem is that rewinding can become recursive in the concurrent setting, which often leads to the necessity of large round complexity of the protocol. Recall that rewinding extraction typically requires the creation of "look-ahead threads," i.e., rewinding the adversary and interacting with it again from an earlier point of the protocol. If the simulator is required to do simulation even on the look-ahead threads, the rewinding can become recursive—if the adversary starts new sessions on look-ahead threads, the simulator need to extract the trapdoor secrets from these newly started sessions, and thus, need to rewind the adversary recursively. A key observation by Garg et al. is that, since the look-ahead threads are created only in the hybrid experiment, the simulator does not need to do "full simulation" on the look-ahead threads. More precisely, Garg et al. observe that in the hybrid experiment, the simulator can use the secret inputs of the honest party to execute newly started sessions honestly on the look-ahead threads, by which the simulator can avoid rewinding the adversary recursively. (The secret inputs of the honest parties are used only on the look-ahead threads, and they are never used on the "main thread.")

The second problem is that the components of the protocol can be rewound in the hybrid experiment due to the rewinding extraction of the trapdoor secrets, which makes it hard to show the indistinguishability between the real and the hybrid experiment. Specifically, since any component in a session can be rewound due to the rewinding extraction of other sessions, and the security of a cryptographic primitive is in general not preserved when it is rewound, it is not clear if the indistinguishability of the real and the hybrid experiment can really be reduced to the security of the components. Garg et al. solved this problem by carefully designing their protocol and a sequence of intermediate hybrid experiments. Specifically:

1. They define the sequence of intermediate hybrids between the real and the hybrid experiment in such a way that the concurrent sessions are switched from honestly executed ones to simulated ones session by session in the order of their special messages—namely, the messages such that the look-ahead threads are created from the rounds of these messages.[3] Switching in this order guarantees that in each intermediate hybrid, rewinding occurs only until special message of the session that has just been switched to simulation.
2. Then, they design the protocol in such a way that all the "rewinding-insecure" components (namely, the components whose security is not preserved when they are rewound) start only after special message of the protocol. A key point is that when the protocol is designed in this way, it is guaranteed that in an intermediate hybrid where a session is switched to simulation (and therefore rewinding occurs only until special message of this session), all the rewinding-insecure components in this session are not rewound and therefore their security can be used in the proof of indistinguishability.

---

[3] In the actual security proof in [10], the sessions are switched to honestly executed ones in a more complex manner since each session has *two* special messages.

Applying techniques of Garg et al. to Katz-Ostrovsky two-party protocol. Unfortunately, the techniques of Garg et al. cannot be applied on the round-optimal two-party secure computation protocol of Katz and Ostrovsky (KO) [18] in a straightforward manner.

The main difficulty is that in the KO protocol, the techniques of Garg et al. is not helpful to solve the second problem described above, i.e., the problem that the components of the protocol can be rewound in the hybrid experiment. Recall that Garg et al. solve this problem by designing their protocol in such a way that the rewinding-insecure components start only after special message. In the KO protocol, however, some components are executed in parallel to compress the round complexity and therefore a rewinding-insecure component starts before special message.

To see the difficulty in more details, let us first recall the KO protocol. (In this overview, we concentrate on the setting where only one party obtains the output. In this setting, the KO protocol has only four rounds.) Roughly speaking, the KO protocol is a semi-honest secure two-party computation protocol that is augmented with proofs of correct behavior. Since the protocol has only four rounds, these proofs are executed somewhat in parallel: One party, P_1, gives a proof in Rounds 1–3 and the other party, P_2, gives in Rounds 1–4. Also, these proofs have the proof-of-knowledge property (and thus are rewinding insecure) and the simulator can extract the implicit input of the adversary from them. When extracting the implicit input, the simulator rewinds the adversary in the last two rounds of the proof; hence, when P_1 is corrupted, special message is the message in Round 2 (since look-ahead threads are created from Round 2), and when P_2 is corrupted, special message is the message in Round 3 (since look-ahead threads are created from Round 3). Notice that when P_2 is corrupted, the proof by P_1 in Rounds 1–3 is executed before special message in Round 4.

Then, the difficulty is the following. Let us consider that we design a sequence of intermediate hybrids following the approach of Garg et al. In the intermediate hybrids, all we can guarantee is that when a session is switched to simulation, no rewinding occurs after special message of this session—hence, when P_2 is corrupted, we can only guarantee that no rewinding occurs after Round 4, and thus, cannot guarantee that the proof by P_1 in Rounds 1–3 is not rewound in this session. Then, since the simulation of the KO protocol is indistinguishable only when the proof by P_1 is secure, it seems hard to prove the indistinguishability among the intermediate hybrids unless the proof by P_1 is rewinding secure. Furthermore, since we require that the proof by P_1 has the proof-of-knowledge property, it seems unlikely that the proof by P_1 can be rewinding secure.

Our techniques. To solve the problem that the components of the protocol are rewound in the hybrid experiment, we use the fact that, as observed by Garg et al., in the SPS setting the look-ahead threads can depend on the inputs of the honest parties. Specifically, we use the fact that we do not need to remove the input of P_1 from the proof of correct behavior on the look-ahead threads.

First, we recall the KO protocol in more details. For simplicity, we assume that each party has only 1-bit input. In this case, P_1 gives the proof of correct behavior using a witness indistinguishable proof of knowledge Π_{WIPOK} for a statement of the form $\mathsf{st}_0 \vee \mathsf{st}_1$, where in the honest execution, only one of st_0 and st_1 is true depending on the input of P_1. In simulation, in a session where P_2 is corrupted, the simulator makes both st_0 and st_1 true and simulates the proof of correct behavior using a witness for st_0. (Notice that the proof no longer depends on P_1's input.) In a session where P_1 is corrupted, the simulator extracts the implicit input of the adversary by extracting a witness from Π_{WIPOK} and checking whether the extracted witness is a witness for st_0 or not.

Then, our idea is the following. We replace Π_{WIPOK} with other cryptographic components—witness-indistinguishable one and extractable one—so that we can have both rewinding-secure witness-indistinguishability and extractability. Specifically, the components we use are:

- a **ZAP system** (namely, a two-round public-coin witness indistinguishable proof system). Since ZAP has only two rounds, it is witness indistinguishable even when it is rewound.
- a **three-round honest-committer extractable commitment** (namely, a commitment scheme such that, as long as the committer behaves honestly, the committed value can be extracted by rewinding the committer). The honest-committer extractable scheme that we use, denoted by ExtCom', is a variant of a three-round challenge-response based extractable scheme. To commit to a message m using ExtCom', the committer commits to many 2-out-of-2 secret shares $\{(\alpha_i^0, \alpha_i^1)\}$ of m using a standard non-interactive commitment scheme in the first round, and after receiving challenge $\{e_i\}$ from the receiver, the committer reveals $\{\alpha_i^{e_i}\}$ in the third round *but does not open the corresponding commitments*. An important property of ExtCom' is that the committer's messages in the first and the third round can be simulated independently of each other. In particular, we can simulate a commitment by committing to all-zero strings in the first round and sending random strings in the last round. (Later, we use this property to say that even though ExtCom' is extractable, it also has some rewinding security.)

We then modify the KO protocol in such a way that P_1 gives two ExtCom' commitments in Rounds 1–3, where one is correctly constructed and the other is simulated, and then proves by ZAP in Rounds 2–3 that either a witness for st_0 is committed in the first ExtCom' commitment or a witness for st_1 is committed in the second one. (Recall that only one of st_0 and st_1 is true in the KO protocol depending on the input of P_1) With this modification, we can solve the problem as follows. When P_2 is corrupted, the simulator makes both st_0 and st_1 true (as the KO simulator does), commits to witnesses for st_0 and st_1 in the two ExtCom' commitments, and completes the ZAP proof using a witness for that st_0 is committed in the first ExtCom' commitment. Then, even though ZAP and ExtCom' can be rewound in the hybrid experiments, we can show the indistinguishability using their security for the following reasons.

1. First, the simulator can switch a simulated ExtCom' commitment to a honest one in an indistinguishable way even under rewinding *as long as the*

commitment does not need to be a honest one on the look-ahead threads. This is because the third message of the simulated commitment consists of just random strings; we can design a reduction that obtains a ExtCom' commitment (either a simulated one or a honest one) on the main thread from an external committer while internally simulating the look-ahead threads by simulating the third message of this ExtCom' commitment with random strings.

2. Second, the witness indistinguishability of ZAP holds even when it is rewound. This is because it has only two rounds.

On the other hand, when P_1 is corrupted, the simulator can extract the implicit input from the adversary by extracting the committed values from the two ExtCom' commitments and checking whether a witness for st_0 or st_1 is extracted. Even though ExtCom' is only honest-committer extractable, the simulator can extract the implicit input in this way since the soundness of ZAP guarantees that at least one of the ExtCom' commitments is constructed correctly.

Other technicalities. To prove security formally, we need to modify the KO protocol further.

First, we need to add non-malleability to the KO protocol because in the concurrent setting with interchangeable roles, the adversary can participate as the first party in a session while participating as the second party in another session. To add non-malleability, we use a non-malleable commitment in a similar manner as Barak et al. [1], who constructed a concurrent non-malleable zero-knowledge argument using a non-malleable commitment. In particular, we modify the KO protocol in such a way that, instead of giving a proof of correct behavior, a party commits to a witness for the correct behavior using a non-malleable commitment and then proves that it committed to a valid witness. As in the protocol of Barak et al. [1], we assume that the non-malleable commitment is extractable and that some components of our protocol are statistically secure. (Roughly, this is for guaranteeing that the non-malleable commitment is non-malleable w.r.t. not only itself but also the other components of our protocol.)

Second, for technical reasons, we augment the KO protocol with a lossy encryption scheme, i.e., an encryption scheme that has a lossy key generation algorithm such that lossy keys statistically hide the plaintexts. Roughly speaking, this is because unlike the SPS-secure protocol of Garg et al. [10], the KO protocol does not have the property that the same information is extracted by rewinding extraction and by brute-force extraction. (Recall that this property is required when the indistinguishability between the hybrid and the ideal experiment is shown.) Specifically, an adversary can make the rewinding simulator obtain a valid implicit input whereas the brute-force simulator obtain an invalid one. We therefore modify the KO protocol so that for such an adversary, all the messages that depend on the extracted implicit input are encrypted under a lossy key (whereas they are encrypted under a normal key in the honest execution).

2 Preliminaries

In this paper, we denote the security parameter by κ. We assume familiarity with the definitions of basic cryptographic schemes and protocols, such as secret-key/public-key encryption schemes, message authentication codes, commitment schemes, and witness-indistinguishable proof/argument of knowledge. We remind the reader that there exists a non-interactive perfectly binding commitment scheme under the existence of injective one-way functions, and there exists a two-round statistically hiding commitment scheme under the existence of collision-resistance hash functions.

2.1 Components of Katz-Ostrovsky 2-Party Computation

We recall the secure two-party computation protocol of Katz and Ostrovsky [18] and its components. Part of the text is taken from [11,18].

Semi-honest two-party computation based on Yao's garbled circuits
We first recall that a semi-honest secure two-party computation protocol can be constructed using Yao's garbled circuit scheme [27,36].

We view Yao's garbled circuit scheme as a tuple of PPT algorithms (GenGC, EvalGC), where GenGC is the "generation procedure" that generates a garbled circuit for a circuit C along with "labels," and EvalGC is the "evaluation procedure" that evaluates the circuit on the "correct" labels. Each individual wire i of the circuit is assigned two labels, $Z_{i,0}, Z_{i,1}$. More specifically, the two algorithms have the following format (here $i \in [\kappa], b \in \{0,1\}$).

- $(\mathsf{GC}_y, \{Z_{i,b}\}) \leftarrow \mathsf{GenGC}(1^\kappa, F, y)$: GenGC takes as input a security parameter κ, a circuit F, and a string $y \in \{0,1\}^\kappa$. It outputs a *garbled circuit* GC_y along with the set of all *input-wire labels* $\{Z_{i,b}\}$.
- $v = \mathsf{EvalGC}(\mathsf{GC}_y, \{Z_{i,x_i}\})$: Given a garbled circuit GC_y and a set of input-wire labels $\{Z_{i,x_i}\}$, where $x = x_1 x_2 \cdots x_\kappa \in \{0,1\}^\kappa$, EvalGC outputs either an invalid symbol $\bot$ or a value $v = F(x,y)$.

The two algorithms have the following properties.

- **Correctness:** $\Pr[\mathsf{EvalGC}(\mathsf{GC}_y, \{Z_{i,x_i}\}) = F(x,y)] = 1$ for all F, x, y, taken over the correct generation of $\mathsf{GC}_y, \{Z_{i,b}\}$ by GenGC.
- **Security:** There exists a PPT simulator SimGC such that for any F, we have $\{(\mathsf{GC}_y, \{Z_{i,x_i}\})\}_{x,y} \approx_c \{\mathsf{SimGC}(1^\kappa, F, v)\}_{x,y}$, where $(\mathsf{GC}_y, \{Z_{i,b}\}) \leftarrow \mathsf{GenGC}(1^\kappa, F, y)$ and $v = F(x,y)$.

Yao's garbled circuit scheme is based on the existence of one-way functions.

Using Yao's garbled circuit scheme and a semi-honest OT protocol, two parties, P_1 and P_2, can compute a function F of their inputs in the semi-honest setting as follows. Let x, y be the inputs of P_1, P_2 respectively. Consider the setting that only one party, say P_1, learns the output of F. Then, P_2 first computes $(\mathsf{GC}_y, \{Z_{i,b}\}) \leftarrow \mathsf{GenGC}(1^\kappa, F, y)$ and sends GC_y to P_1. The two parties then engage in κ parallel instances of OT, where in the i-th instance, P_1 inputs

x_i and P_2 inputs $(Z_{i,0}, Z_{i,1})$ to the OT protocol, and P_1 learns Z_{i,x_i}. Then, P_1 computes $v = \mathsf{EvalGC}(\mathsf{GC}_y, \{Z_{i,x_i}\})$ and outputs $v = F(x, y)$.

We next recall that a three-round semi-honest OT protocol can be constructed from enhanced trapdoor permutations (TDPs).

Definition 1 (Trapdoor permutations). *Let* TDP *be a triple of PPT algorithms* $(\mathsf{TDP.Gen}, \mathsf{TDP.Eval}, \mathsf{TDP.Invert})$ *such that if* $\mathsf{TDP.Gen}(1^\kappa)$ *outputs a pair* (f, td), *then* $\mathsf{TDP.Eval}(f, \cdot)$ *is a permutation over* $\{0,1\}^\kappa$ *and* $\mathsf{TDP.Invert}(f, \mathsf{td}, \cdot)$ *is its inverse.* TDP *is a trapdoor permutation if for any PPT adversary* A, *there exists a negligible function* μ *such that* $\Pr[(f, \mathsf{td}) \leftarrow \mathsf{TDP.Gen}(1^\kappa); y \leftarrow \{0,1\}^\kappa; x \leftarrow A(f, y) : \mathsf{TDP.Eval}(f, x) = y] \leq \mu(\kappa)$.

For convenience, we drop (f, td) from the notation and write $f(\cdot), f^{-1}(\cdot)$ to denote algorithms $\mathsf{TDP.Eval}(f, \cdot), \mathsf{TDP.Invert}(f, \mathsf{td}, \cdot)$ respectively. We assume that TDP satisfies a weak variant of certifiability, namely, given f it is possible to decide in polynomial time whether $\mathsf{TDP.Eval}(f, \cdot)$ is a permutation over $\{0,1\}^\kappa$. Let H be the function that is obtained from a single-bit hardcore function h of $f \in \mathsf{TDP}$ as follows: $\mathsf{H}(z) = h(z) \| h(f(z)) \| \ldots \| h(f^{\kappa-1}(z))$. Informally, $\mathsf{H}(z)$ looks pseudorandom given $f^\kappa(z)$.

The semi-honest OT protocol based on TDP is constructed as follows. Let P_2 hold two strings $Z_0, Z_1 \in \{0,1\}^\kappa$ and P_1 hold a bit b. In the first round, P_2 chooses trapdoor permutation $(f, f^{-1}) \leftarrow \mathsf{TDP.Gen}(1^\kappa)$ and sends f to P_1. Then P_1 chooses two random strings $z_0', z_1' \leftarrow \{0,1\}^\kappa$, computes $z_b = f^\kappa(z_b')$ and $z_{1-b} = z_{1-b}'$, and sends (z_0, z_1) to P_2. In the last round P_2 computes $W_a = Z_a \oplus \mathsf{H}(f^{-\kappa}(z_a))$ for each $a \in \{0,1\}$ and sends (W_0, W_1) to P_1. Finally, P_2 recovers Z_b by computing $Z_b = W_b \oplus \mathsf{H}(z_b)$.

Putting it altogether, we obtain the following three-round semi-honest secure two-party computation protocol for the single-output functionality F:

Protocol Π_{SH}: P_1 holds input $x \in \{0,1\}^\kappa$ and P_2 holds inputs $y \in \{0,1\}^\kappa$. Let TDP be a family of trapdoor permutations and H be its hardcore bit function for κ bits. In the following, i always ranges from 1 to κ and b from 0 to 1.

Round-1: P_2 computes $(\mathsf{GC}_y, \{Z_{i,b}\}) \leftarrow \mathsf{GenGC}(1^\kappa, F, y)$, chooses $\{(f_{i,b}, f_{i,b}^{-1})\}$ using $\mathsf{TDP.Gen}(1^\kappa)$, and sends $(\mathsf{GC}_y, \{f_{i,b}\})$ to P_2.
Round-2: P_1 chooses random strings $\{z_{i,b}'\}$, computes $z_{i,x_i} = f^\kappa(z_{i,x_i}')$ and $z_{i,1-x_i} = z_{i,1-x_i}'$, and sends $\{z_{i,b}\}$ to P_2.
Round-3: P_2 computes $W_{i,b} = Z_{i,b} \oplus \mathsf{H}(f_{i,b}^{-\kappa}(z_{i,b}))$ and sends $\{W_{i,b}\}$ to P_2.
Output: P_1 recovers the labels $Z_{i,x_i} = W_{i,x_i} \oplus \mathsf{H}(z_{i,x_i}')$ and computes $v = \mathsf{EvalGC}(\mathsf{GC}_y, \{Z_{i,x_i}\})$.

Equivocal commitment scheme Eqcom. We next recall the equivocal commitment scheme of [18] that is based on any (standard) non-interactive perfectly binding commitment scheme Com. To commit to a bit $x \in \{0,1\}$, the sender chooses coins ζ_1, ζ_2 and computes $\mathsf{Eqcom}(x; \zeta_1, \zeta_2) \stackrel{\text{def}}{=} \mathsf{Com}(x; \zeta_1) \| \mathsf{Com}(x; \zeta_2)$. It sends $\mathsf{C}_x = \mathsf{Eqcom}(x; \zeta_1, \zeta_2)$ to the receiver along with a zero-knowledge proof that C_x was constructed correctly (i.e., that there exist x, ζ_1, ζ_2 such that $C_x = \mathsf{Eqcom}(x; \zeta_1, \zeta_2)$). To decommit, the sender chooses a bit b at random and reveals x, ζ_b, denoted by open_{C_x}. Note that a simulator can "equivocate" the commitment by setting $C = \mathsf{Com}(x; \zeta_1) \| \mathsf{Com}(1 - x; \zeta_2)$ for a random bit $x \in \{0,1\}$, simulating the zero-knowledge proof, and then revealing ζ_1 or ζ_2 depending on x and the bit to be revealed. This extends to strings by committing bitwise.

Sketch of the Katz-Ostrovsky Two-Party Protocol. The main components of the secure two-party computation protocol of Katz and Ostrovsky [18] are the three-round semi-honest secure two-party computation protocol Π_{SH} and proofs about the correctness of each round. Specifically, the protocol of [18] proceeds as follows. First, both parties commit to their inputs. Then, they run (modified) coin-tossing protocols to guarantee that each party obtains random coins that are committed to the other party. Finally, they run the Π_{SH} protocol together with proofs about the correctness of each round.

Since even a zero-knowledge argument alone requires four rounds, in the protocol of [18] the proof-of-correctness part is executed in parallel with Π_{SH}. To enable such a parallel execution, Katz and Ostrovsky use a zero-knowledge argument system with a "delayed input" property, i.e., a property that the statement to be proven need not be known until the last round. (Specifically, they use a variant of the four-round zero-knowledge proof system by Feige and Samir [9].) Furthermore, for technical reasons, in the protocol of [18] the above equivocal commitment scheme is used to commit to the garbled circuit.

2.2   Component of Our Protocol

Statistical Feige-Shamir zero-knowledge argument Π_{FS}. We use a four-round "delayed-input" statistical zero-knowledge argument Π_{FS} that is based on the four-round zero-knowledge argument system by Feige and Shamir [9]. Recall that the Feige-Shamir zero-knowledge argument for a statement thm consists of the following two (somewhat parallelized) executions of a witness-indistinguishable proof-of-knowledge system: in the first execution (in Rounds 1–3), the verifier proves the knowledge of "simulation trapdoor" σ—namely, selects a one-way function f, sets $x_1 = f(w_1)$ and $x_2 = f(w_2)$, and proves the knowledge of a witness for $\exists w$ s.t. $x_1 = f(w) \lor x_2 = f(w)$; in the second execution (in Rounds 2–4), the prover proves the knowledge of a witness for thm or the simulation trapdoor—i.e., proves the knowledge of a witness for thm $\lor (\exists w$ s.t. $x_1 = f(w) \lor x_2 = f(w))$. We then obtain Π_{FS} by using Blum's three-round witness-indistinguishable proof of knowledge (denoted by Π_{WIPOK}) in the first execution and a four-round statistical witness-indistinguishable version of the "delayed input" witness-indistinguishable argument of Lapidot and Shamir [21] (denoted by Π_{sWIAOK}) in the second execution. It is not hard to see that Π_{FS} has a property that the statement to be proven is not needed until its last round, and it is complete, sound, and zero-knowledge even when the statement is determined in the last round.

Extractable commitment scheme ExtCom'. We use the following commitment scheme ExtCom', which is used in [10]. Let Com be any non-interactive perfectly binding commitment.

Commit Phase: The common input is security parameter 1^κ. The input to the committer is a string $m \in \{0,1\}^{\text{poly}(\kappa)}$.

1. The committer chooses κ independent random pairs $\{\alpha_i^0, \alpha_i^1\}_{i \in [\kappa]}$ such that $\alpha_i^0 \oplus \alpha_i^1 = m$ for every $i \in [\kappa]$. The committer then commits to α_i^b for every $i \in [m]$, $b \in \{0,1\}$ using Com. Let c_i^b be the commitment to α_i^b.
2. The receiver sends uniformly random bits $\{e_i\}_{i \in [\kappa]}$.
3. The committer sends $\alpha_i^{e_i}$ for every $i \in [\kappa]$.
   COMMENT: *The committer just sends $\alpha_i^{e_i}$ and does not decommit $c_i^{e_i}$.*

Open Phase: The committer decommits c_i^b to α_i^b for every $i \in [\kappa], b \in \{0,1\}$.

ExtCom' has extractability in the sense that we can extract the committed value if we can obtain two *correctly constructed* transcripts by rewinding the committer in the last two rounds. We remark that if the commitment is *invalid*, i.e., there is no value to which the commitment can be correctly decommitted, this extracting procedure can output any value. We also remark that in ExtCom', a committer can easily give an invalid commitment by committing to all-zero strings in the first round and sending random strings in the last round. We use such an "fake" execution of ExtCom' in our protocol.

Non-malleable commitment scheme NMCom. Let $\langle C, R \rangle$ be a tag-based commitment scheme (i.e., a commitment scheme that takes a κ-bit string—a *tag*—as an additional input). Informally, $\langle C, R \rangle$ is *non-malleable* if for any man-in-the-middle adversary $\mathcal{M}$, who gives a commitment of $\langle C, R \rangle$ in the "right" interaction while receiving a commitment of $\langle C, R \rangle$ in the "left" interaction, the value committed in the right interaction is "independent" of the value committed in the left interaction as long as the tags in the two interactions are different. See, e.g., [22] for a formal definition.

In our main result, we use a non-malleable commitment scheme such that:

1. The scheme is public coin (i.e., the receiver is public coin) and the round complexity is 3.
2. The scheme has the following extractability: an extractor extracts the committed value from a valid commitment and extracts $\perp$ from an invalid one.

Such a non-malleable commitment exist under quasi-polynomially-hard injective OWFs [15,16]; see Footnote 1. For simplicity, we also assume that the extractor E rewinds the committer in the last·two rounds until it obtains two accepting transcripts. That is, we assume that E interacts with the committer in the same way as the honest receiver on the *main thread* while repeatedly interacting with it from the second round with fresh randomness on the *look-ahead threads*, and when the commitment is accepting on the main thread, E extracts the committed values using the accepting commitment on a look-ahead thread.

Lossy encryption scheme. Informally, a *lossy encryption scheme* [3,17] is a public-key encryption scheme such that, in addition to the standard key generation algorithm, it has a lossy key generating algorithm with the following property: A lossy public key is indistinguishable from a standard public key, and a ciphertext generated under a lossy public key statistically hides the information of the plaintext. More precisely, a lossy public-key encryption scheme is a tuple (LE.Gen, LE.Enc, LE.Dec) of PPT algorithms such that:

– LE.Gen(1^κ, inj) outputs *injective keys* (pk, sk).
– LE.Gen(1^κ, lossy) outputs *lossy keys* ($\mathsf{pk}_{\mathsf{lossy}}, \mathsf{sk}_{\mathsf{lossy}}$).

For a formal security definition, see [3,17].

It is shown in [3] that a lossy encryption scheme can be constructed from *lossy trapdoor functions* [34], which in turn can be realized based on a variety of assumptions including the DDH assumption and the LWE assumption.

ZAP Π_{ZAP}. ZAPs are two-message public-coin witness-indistinguishable proof systems, and can be based on doubly enhanced trapdoor permutations [7].

3   UC Security and Its SPS Variant

We recall the definition of UC security [4] and its SPS variant [2,10,35]. A part of the text below is taken from [10].

3.1   UC Security

We assume familiarity with the UC framework. For full details, see [4].

Recall that in the UC framework, the model for protocol execution consists of the environment $\mathcal{Z}$, the adversary $\mathcal{A}$, and the parties running protocol π. In this paper, we consider static adversaries and assume the existence of authenticated communication channels. Let $\mathrm{EXEC}_{\pi,\mathcal{A},\mathcal{Z}}(\kappa, z)$ denote a random variable for the output of $\mathcal{Z}$ on security parameter $\kappa \in \mathbb{N}$ and input $z \in \{0,1\}^*$ with a uniformly-chosen random tape. Let $\mathrm{EXEC}_{\pi,\mathcal{A},\mathcal{Z}}$ denote the ensemble $\{\mathrm{EXEC}_{\pi,\mathcal{A},\mathcal{Z}}(\kappa, z)\}_{\kappa\in\mathbb{N}, z\in\{0,1\}^*}$.

The security of a protocol π is defined using the *ideal protocol*. In the execution of the ideal protocol, all the parties simply hand their inputs to the *ideal functionality* $\mathcal{F}$. The ideal functionality $\mathcal{F}$ carries out the desired task securely and gives outputs to the parties, and the parties forward these outputs to $\mathcal{Z}$. The adversary $\mathcal{S}$ in the execution of the ideal protocol is often called the *simulator*. Let $\pi(\mathcal{F})$ denote the ideal protocol for functionality $\mathcal{F}$.

We say that a protocol π *emulates* protocol ϕ if for any adversary $\mathcal{A}$ there exists an adversary $\mathcal{S}$ such that no environment $\mathcal{Z}$, on any input, can tell with non-negligible probability whether it is interacting with $\mathcal{A}$ and parties running π or it is interacting with $\mathcal{S}$ and parties running ϕ. We say that π *securely realizes* an ideal functionality $\mathcal{F}$ if it emulates the ideal protocol $\Pi(\mathcal{F})$.

3.2   UC Security with Super-Polynomial Simulation

We next provide a relaxed notion of UC security where the simulator is given access to super-polynomial computational resources.

Definition 2. *Let π and ϕ be protocols. We say that π UC-SPS-emulates ϕ if for any adversary $\mathcal{A}$ there exists a super-polynomial-time adversary $\mathcal{S}$ such that for any environment $\mathcal{Z}$ that obeys the rules of interaction for UC security, we have $\mathrm{EXEC}_{\phi,\mathcal{S},\mathcal{Z}} \approx \mathrm{EXEC}_{\pi,\mathcal{A},\mathcal{Z}}$.*

Definition 3. *Let $\mathcal{F}$ be an ideal functionality and let π be a protocol. We say that π UC-SPS-realizes $\mathcal{F}$ if π UC-SPS-emulates the ideal process $\Pi(\mathcal{F})$.*

The multi-session extension of an ideal functionality. When showing concurrent security of a protocol π under SPS security, we need to construct a simulator in a setting where parties execute π concurrently. (In other words, unlike in UC security, we cannot rely on the *composition theorem* in SPS security.)

To consider the simulator in such a setting, we use a *multi-session extension* of an ideal functionality. Roughly speaking, the multi-session extension $\hat{\mathcal{F}}$ of an ideal functionality $\mathcal{F}$ is a functionally that internally runs multiple copies of $\mathcal{F}$.

4   Our Five-Round Secure Two-Party Protocol

In this section, we prove our main result.

Theorem 1. *Assume the existence of collision-resilient hash function families, trapdoor permutation families[4], lossy encryption schemes, and quasi-polynomially-hard injective one-way functions. Let $\mathcal{F}$ be any well-formed two-party functionality and $\hat{\mathcal{F}}$ be its multi-session extension. Then, there exists a five-round protocol that UC-SPS realizes $\hat{\mathcal{F}}$.*

The other result, a seven-round protocol under polynomially-hard assumptions, is given in the full version of this paper.

Recall that in the UC framework, there are any number of parties $P_1, P_2, \ldots$, and any two of them (say, P_i and P_j) can compute $\mathcal{F}$ using $\hat{\mathcal{F}}$ in each subsession. To simplify the description of the protocol and the proofs, in what follows we denote the first party of $\mathcal{F}$ by P_1 and the second party of $\mathcal{F}$ by P_2 in every subsession. (Equivalently, we consider a setting where two parties P_1, P_2 compute $\mathcal{F}$ any number of times using $\hat{\mathcal{F}}$, and $\mathcal{A}$ corrupts either P_1 or P_2 in each subsession.)

---

[4] Recall that we assume that the trapdoor permutation families satisfy (a weak form of) "certifiability" and their domain/range is $\{0,1\}^\kappa$.

4.1   Our Protocol $\Pi_{2\text{PC}}$

Our protocol is based on the two-party computation protocol of Katz and Ostrovsky [18]; their protocol is described in Sect. 2.1. In our protocol, we use the primitives that are described in Sects. 2.1 and 2.2, and additionally, we use a symmetric-key encryption scheme $\mathsf{SKE} = (\mathsf{SKE.Enc}, \mathsf{SKE.Dec})$ and a message authentication code MAC.

Our protocol $\Pi_{2\text{PC}}$. We denote the two parties by P_1 and P_2. P_1 holds input $x \in \{0,1\}^\kappa$ and P_2 holds input $y \in \{0,1\}^\kappa$. The identities of P_1 and P_2 (i.e., their PIDs) are id_1 and id_2 respectively, where $\mathsf{id}_1 \neq \mathsf{id}_2$. Let $\mathcal{F} = (F_1, F_2) : \{0,1\}^\kappa \times \{0,1\}^\kappa \to \{0,1\}^\kappa \times \{0,1\}^\kappa$ be the functionality to be computed. Let $\mathcal{F}' = (F_1', F_2')$ be a functionality such that:

- $F_1'(x, y') = (F_1(x, y), \mathsf{enc}, \mathsf{mac})$, where $y' = (y, \mathsf{sk}_{\mathsf{ske}}, \mathsf{sk}_{\mathsf{mac}}, \omega_{\mathsf{enc}}) \in \{0,1\}^{4\kappa}$, $\mathsf{enc} = \mathsf{SKE.Enc}_{\mathsf{sk}_{\mathsf{ske}}}(F_2(x, y); \omega_{\mathsf{enc}})$, and $\mathsf{mac} = \mathsf{MAC}_{\mathsf{sk}_{\mathsf{mac}}}(\mathsf{enc})$.
- $F_2'(x, y') = \bot$ for any x and y'.

In the following i always ranges from 1 to κ and b from 0 to 1. We will skip mentioning the SID and SSID to keep the protocol specification simple.

Round 1. P_1 sends a message m_1 that is defined as follows.

1. P_1 commits to 2κ random strings $\{r_{i,b}\}$ using 2κ parallel and independent executions of Com. I.e., it chooses uniformly random strings $r_{i,b}$ and randomness $\omega_{\mathsf{com}}^{i,b}$ and then generates $\mathsf{com}_{i,b} = \mathsf{Com}(r_{i,b}; \omega_{\mathsf{com}}^{i,b})$.
2. P_1 starts committing to κ strings $\{r_{i,1-x_i} \| \omega_{\mathsf{com}}^{i,1-x_i}\}$ using κ parallel and independent executions of ExtCom' and also starts κ "fake" executions of ExtCom'. Concretely, P_1 prepares $\{\mathsf{ext}_1^{i,b}\}$ as follows.
   - For every $i \in [\kappa]$, P_1 prepares $\mathsf{ext}_1^{i,1-x_i}$ by committing to $r_{i,1-x_i} \| \omega_{\mathsf{com}}^{i,1-x_i}$ using ExtCom'. I.e., it generates $\mathsf{ext}_1^{i,1-x_i} \leftarrow \mathsf{ExtCom}'_1(r_{i,1-x_i} \| \omega_{\mathsf{com}}^{i,1-x_i})$, which is the first message of $\mathsf{ExtCom}'(r_{i,1-x_i} \| \omega_{\mathsf{com}}^{i,1-x_i})$.
   - For every $i \in [\kappa]$, P_1 prepares ext_1^{i,x_i} by committing to all-zero strings using Com. (Recall that the first round of ExtCom' consists of 2κ executions of Com.)
3. P_1 prepares the first message fs_1 of Π_{FS}.
4. Message m_1 is the tuple $(\{\mathsf{com}_{i,b}, \mathsf{ext}_1^{i,b}\}, \mathsf{fs}_1)$.

Round 2. P_2 sends a message m_2 that is defined as follows.

1. P_2 samples secret-keys $\mathsf{sk}_{\mathsf{ske}}$ and $\mathsf{sk}_{\mathsf{mac}}$ for SKE and MAC respectively and chooses randomness ω_{enc} for $\mathsf{SKE.Enc}$.
2. P_2 prepares a garbled circuit and labels for F_1' with input $y' = (y, \mathsf{sk}_{\mathsf{ske}}, \mathsf{sk}_{\mathsf{mac}}, \omega_{\mathsf{enc}})$. I.e., it uniformly chooses randomness Ω and generates $(\mathsf{GC}, \{Z_{i,b}\}) = \mathsf{GenGC}(1^\kappa, F_1', y'; \Omega)$.
3. P_2 generates standard commitments to the labels and an equivocal commitment to the garbled circuit. I.e., it uniformly chooses randomness $\{\omega_{\mathsf{lab}}^{i,b}\}$ and ω_{gc} and generates $\mathsf{C}_{\mathsf{lab}}^{i,b} = \mathsf{Com}(Z_{i,b}; \omega_{\mathsf{lab}}^{i,b})$ and $\mathsf{C}_{\mathsf{gc}} = \mathsf{Eqcom}(\mathsf{GC}; \omega_{\mathsf{gc}})$. Let $\mathsf{open}_{\mathsf{C}_{\mathsf{gc}}}$ be the decommitment information that decommits C_{gc} to GC.

4. P_2 samples random strings $\{r'_{i,b}\}$ and $(f_{i,b}, f^{-1}_{i,b}) \leftarrow$ TDP.Gen(1^κ) for the coin tossing and the oblivious transfer executions.
5. P_2 generates the second messages $\{\text{ext}^{i,b}_2\}$ for all the executions of ExtCom$'$ initiated by P_1.
6. P_2 prepares the first message zap_1 of Π_{ZAP}.
7. P_2 prepares the second message fs_2 of Π_{FS} initiated by P_1.
8. P_2 chooses randomness ω_{leEnc} for LE.Enc.
9. Let $\text{wit}_2 := (y', \Omega, \text{GC}, \omega_{\text{gc}}, \text{open}_{C_{\text{gc}}}, \{Z_{i,b}, \omega^{i,b}_{\text{lab}}\})$ and $\text{wit}_4 := (\text{wit}_2, \{f^{-1}_{i,b}\}, \omega_{\text{leEnc}})$. Then, P_2 starts committing to wit_4 using NMCom with identity id_2. I.e., it generates $\text{nm}_1 \leftarrow \text{NMCom}_1(\text{id}_2, \text{wit}_4)$, which is the first message of NMCom$(\text{id}_2, \text{wit}_4)$.
   We remark that wit_2 is a witness for the following statement $\text{st}_2 = (F'_1, C_{\text{gc}}, \{C^{i,b}_{\text{lab}}\})$.
   $\exists\, \text{wit}_2 = (y', \Omega, \text{GC}, \omega_{\text{gc}}, \text{open}_{C_{\text{gc}}}, \{Z_{i,b}, \omega^{i,b}_{\text{lab}}\})$ s.t.
   (a) $(\text{GC}, \{Z_{i,b}\}) = \text{GenGC}(1^\kappa, F'_1, y'; \Omega)$, and
   (b) $C_{\text{gc}} = \text{Eqcom}(\text{GC}; \omega_{\text{gc}})$ and $\forall(i,b) : C^{i,b}_{\text{lab}} = \text{Com}(Z_{i,b}; \omega^{i,b}_{\text{lab}})$, and
   (c) $\text{open}_{C_{\text{gc}}}$ is a valid decommitment that opens C_{gc} to GC.
   COMMENT: *Informally, st_2 is the statement that P_2 performed this step correctly, i.e., generated a garbled circuit and labels correctly and then committed to them in $C_{\text{gc}}, \{C^{i,b}_{\text{lab}}\}$.*
10. Message m_2 is the tuple $(\{C^{i,b}_{\text{lab}}, r'_{i,b}, f_{i,b}, \text{ext}^{i,b}_2\}, C_{\text{gc}}, \text{zap}_1, \text{fs}_2, \text{nm}_1)$.

Round 3. If any of $\{f_{i,b}\}$ is invalid, P_1 aborts. Otherwise, P_1 sends a message m_3 that is defined as follows.

1. P_1 invokes κ parallel executions of oblivious transfer to obtain the input-wire labels corresponding to its input x. More specifically, P_1 does the following for every $i \in [\kappa]$.
   - If $x_i = 0$, sample $z'_{i,0} \leftarrow \{0,1\}^{\kappa/2}$ and then set $z_{i,0} := f^\kappa_{i,0}(\text{PRG}(z'_{i,0}))$ and $z_{i,1} := r_{i,1} \oplus r'_{i,1}$.
   - If $x_i = 1$, sample $z'_{i,1} \leftarrow \{0,1\}^{\kappa/2}$ and then set $z_{i,1} := f^\kappa_{i,1}(\text{PRG}(z'_{i,1}))$ and $z_{i,0} := r_{i,0} \oplus r'_{i,0}$.
2. P_1 prepares $\{\text{ext}^{i,b}_3\}$, where $\{\text{ext}^{i,1-x_i}_3\}$ are the third messages of ExtCom$'$ and $\{\text{ext}^{i,x_i}_3\}$ are random strings.
3. P_1 prepares injective keys $(\text{pk}_{\text{le}}, \text{sk}_{\text{le}})$ of the lossy encryption scheme, i.e., it generates $(\text{pk}_{\text{le}}, \text{sk}_{\text{le}}) \leftarrow$ LE.Gen$(1^\kappa, \text{inj})$.
4. P_1 prepares the second message zap_2 of Π_{ZAP} proving the following statement $\text{st}_3 = (\{\text{com}_{i,b}, \text{ext}^{i,b}_1, \text{ext}^{i,b}_2, \text{ext}^{i,b}_3, r'_{i,b}, z_{i,b}, f_{i,b}\}, \text{pk}_{\text{le}})$:
   $\exists\, \text{wit}_3 = (\{b_i, r_i, \omega^i_{\text{com}}, \omega^i_{\text{ext}}, z'_i\}_{i \in [\kappa]}, \text{sk}_{\text{le}}, \omega_{\text{leGen}})$ s.t. $\forall i$:
   (a) $\text{com}_{i,b_i} = \text{Com}(r_i; \omega^i_{\text{com}})$, and
   (b) ext^{i,b_i}_1 and ext^{i,b_i}_3 are the first and the third messages of ExtCom$'(r_i \| \omega^i_{\text{com}}; \omega^i_{\text{ext}})$ with the second message being ext^{i,b_i}_2, and
   (c) $z_{i,b_i} = r_{i,b_i} \oplus r'_{i,b_i}$, and
   (d) $z_{i,1-b_i} = f^\kappa_{i,1-b_i}(\text{PRG}(z'_i)) \bigvee (\text{pk}_{\text{le}}, \text{sk}_{\text{le}}) = \text{LE.Gen}(1^\kappa, \text{lossy}; \omega_{\text{leGen}})$.
   P_1 uses $(\{1 - x_i, r_{i,1-x_i}, \omega^{i,1-x_i}_{\text{com}}, \omega^{i,1-x_i}_{\text{ext}}, z'_{i,x_i}\}_{i \in [\kappa]}, \bot, \bot)$ as the witness.

COMMENT: *Informally, st_3 is the statement that either P_1 performed this step correctly (i.e., one of $z_{i,0}, z_{i,1}$ is an image of $f_{i,1-b_i}^{\kappa}(\mathsf{PRG}(\cdot))$ and the other is the outcome of the coin-tossing) or $\mathsf{pk}_{\mathsf{le}}$ is a lossy key. Here, PRG is used to make sure that $z_{i,1-b_i} \neq r_{i,1-b_i} \oplus r'_{i,1-b_i}$ holds when $\mathsf{pk}_{\mathsf{le}}$ is an injective key.*

5. P_1 prepares the third message fs_3 of Π_{FS}.
6. P_1 prepares the second message nm_2 of NMCom.
7. Message m_3 is the tuple $(\{z_{i,b}, \mathsf{ext}_3^{i,b}\}, \mathsf{pk}_{\mathsf{le}}, \mathsf{zap}_2, \mathsf{fs}_3, \mathsf{nm}_2)$.

Round 4. If zap_2 or fs_3 is not accepting, P_2 aborts. Otherwise, P_2 sends a message m_4 that is defined as follows.

1. P_2 completes the execution of the oblivious transfers by computing $W_{i,b} = Z_{i,b} \oplus \mathsf{H}(f_{i,b}^{-\kappa}(z_{i,b}))$.
2. P_2 encrypts $\{W_{i,b}\}\|\mathsf{GC}\|\mathsf{open}_{\mathsf{C}_{\mathsf{gc}}}$ using the lossy encryption scheme with public key $\mathsf{pk}_{\mathsf{le}}$ and randomness ω_{leEnc} (which was chosen in Round 2), i.e., it computes $\mathsf{CT}_{\mathsf{gc}} = \mathsf{LE}.\mathsf{Enc}_{\mathsf{pk}_{\mathsf{le}}}(\{W_{i,b}\}\|\mathsf{GC}\|\mathsf{open}_{\mathsf{C}_{\mathsf{gc}}}; \omega_{\mathsf{leEnc}})$.
3. P_2 prepares the final message nm_3 of NMCom.
4. Let $\mathsf{st}_4 = (\{f_{i,b}, z_{i,b}\}, \mathsf{st}_2, \mathsf{CT}_{\mathsf{gc}})$ be the following statement:
   $\exists \mathsf{wit}_4 = (\mathsf{wit}_2, \{g_{i,b}\}, \omega_{\mathsf{leEnc}})$ s.t.
   (a) $\mathsf{wit}_2 = (y', \Omega, \mathsf{GC}, \omega_{\mathsf{gc}}, \mathsf{open}_{\mathsf{C}_{\mathsf{gc}}}, \{Z_{i,b}, \omega_{\mathsf{lab}}^{i,b}\})$ is a valid witness for st_2, and
   (b) $\forall (i,b): f_{i,b}^{\kappa}(g_{i,b}^{\kappa}(z_{i,b})) = z_{i,b}$, and
   (c) $\mathsf{CT}_{\mathsf{gc}} = \mathsf{LE}.\mathsf{Enc}_{\mathsf{pk}_{\mathsf{le}}}(\{W_{i,b}\}\|\mathsf{GC}\|\mathsf{open}_{\mathsf{C}_{\mathsf{gc}}}; \omega_{\mathsf{leEnc}})$, where $W_{i,b} = Z_{i,b} \oplus \mathsf{H}(g_{i,b}^{\kappa}(z_{i,b}))$.
   Then, P_2 prepares the final message fs_4 of Π_{FS} proving the following statement $(\mathsf{nm}_1, \mathsf{nm}_2, \mathsf{nm}_3, \mathsf{st}_4)$.
   $\exists \omega_{\mathsf{nm}}$ and wit_4 s.t.
   (a) nm_1 and nm_3 are the first and the third message of $\mathsf{NMCom}(\mathsf{id}_2, \mathsf{wit}_4; \omega_{\mathsf{nm}})$ with the second message being nm_2, and
   (b) wit_4 is a valid witness for st_4.
   I.e., P_2 proves that it committed to a witness for st_4 using NMCom.
   COMMENT: *Informally, st_4 is the statement that P_1 performed this step and the previous step correctly (in particular, the final messages of the oblivious transfers and the opening of C_{gc} were encrypted in $\mathsf{CT}_{\mathsf{gc}}$).*
5. Message m_4 is the tuple $(\mathsf{CT}_{\mathsf{gc}}, \mathsf{fs}_4, \mathsf{nm}_3)$.

Round 5. If fs_4 or nm_3 is not accepting, P_1 aborts. Otherwise, P_1 sends a message m_5 that is defined as follows.

1. P_1 recovers $\{W_{i,b}\}\|\mathsf{GC}\|\mathsf{open}_{\mathsf{C}_{\mathsf{gc}}}$ by decrypting $\mathsf{CT}_{\mathsf{gc}}$, i.e., it computes $\{W_{i,b}\}\|\mathsf{GC}\|\mathsf{open}_{\mathsf{C}_{\mathsf{gc}}} = \mathsf{LE}.\mathsf{Dec}_{\mathsf{sk}_{\mathsf{le}}}(\mathsf{CT}_{\mathsf{gc}})$. If $(\mathsf{GC}, \mathsf{open}_{\mathsf{C}_{\mathsf{gc}}})$ is not a valid opening of C_{gc}, P_1 aborts. Otherwise, P_1 recovers the garbled labels $\{Z_i := Z_{i,x_i}\}$ from the completion of the oblivious transfer, and then it computes $(F_1(x,y), \mathsf{enc}, \mathsf{mac}) = \mathsf{EvalGC}(\mathsf{GC}, \{Z_i\})$.
2. Message m_5 is the tuple $(\mathsf{enc}, \mathsf{mac})$.

Output computation

P_1's output: P_1 outputs $F_1(x, y)$, which it obtained in Round 5.

P_2's output: If $\mathsf{MAC}_{\mathsf{sk}_{\mathrm{mac}}}(\mathsf{enc}) \neq \mathsf{mac}$, P_2 outputs $\perp$. Otherwise, it outputs $F_2(x, y) = \mathsf{SKE.Dec}_{\mathsf{sk}_{\mathrm{ske}}}(\mathsf{enc})$.

4.2   Description of Simulator $\mathcal{S}$

The simulator $\mathcal{S}$ internally invokes $\mathcal{A}$ and simulates the real-world execution for $\mathcal{A}$ as follows. To simulate the interaction between $\mathcal{A}$ and $\mathcal{Z}$, $\mathcal{S}$ simply forwards messages between $\mathcal{A}$ and $\mathcal{Z}$. To simulate the interaction between P_1 and P_2, $\mathcal{S}$ does the following in each subsession.

Case 1: P_1 is corrupted. $\mathcal{S}$ simulates P_2's messages as follows.

- In Round 1, $\mathcal{S}$ receives $m_1 = (\{\mathsf{com}_{i,b}, \mathsf{ext}_1^{i,b}\}, \mathsf{fs}_1)$ from $\mathcal{A}$.
- In Round 2, $\mathcal{S}$ prepares $m_2 = (\{\mathsf{C}_{\mathrm{lab}}^{i,b}, r'_{i,b}, f_{i,b}, \mathsf{ext}_2^{i,b}\}, \mathsf{C}_{\mathrm{gc}}, \mathsf{zap}_1, \mathsf{fs}_2, \mathsf{nm}_1)$ in the same way as P_2 does except for the following.
  - $\mathcal{S}$ generates $\{\mathsf{C}_{\mathrm{lab}}^{i,b}\}$ by committing to all-zero strings.
  - $\mathcal{S}$ generates C_{gc} in a way that it can be decommitted to any value by using equivocality.
  - $\mathcal{S}$ generates nm_1 by committing to a all-zero string using NMCom.
  Then, $\mathcal{S}$ sends m_2 to $\mathcal{A}$.
- In Round 3, $\mathcal{S}$ receives $m_3 = (\{z_{i,b}, \mathsf{ext}_3^{i,b}\}, \mathsf{pk}_{\mathrm{le}}, \mathsf{zap}_2, \mathsf{fs}_3, \mathsf{nm}_2)$ from $\mathcal{A}$. If m_3 is accepting, $\mathcal{S}$ does the following.
  1. Extracts the committed values of the ExtCom' commitments $\{(\mathsf{ext}_1^{i,b}, \mathsf{ext}_2^{i,b}, \mathsf{ext}_3^{i,b})\}$ by brute force. The extracted values are denoted by $\{\tilde{r}_{i,b} \| \tilde{\omega}_{\mathrm{com}}^{i,b}\}$. (If a commitment is invalid, its committed value is defined to be $\perp$.) If there is $i \in [\kappa]$ such that for any $b_i^* \in \{0, 1\}$, either $(\tilde{r}_{i,b_i^*}, \tilde{\omega}_{\mathrm{com}}^{i,b_i^*})$ is not a valid decommitment of com_{i,b_i^*} or it holds that $z_{i,b_i^*} \neq \tilde{r}_{i,b_i^*} \oplus r'_{i,b_i^*}$, $\mathcal{S}$ aborts the simulation with output Abort_1.
  2. Define $x^* = (x_1^*, \ldots, x_\kappa^*)$ as follows: for each $i \in [\kappa]$, if $(\tilde{r}_{i,0}, \tilde{\omega}_{\mathrm{com}}^{i,0})$ is a valid decommitment of $\mathsf{com}_{i,0}$ and furthermore it holds that $z_{i,0} = \tilde{r}_{i,0} \oplus r'_{i,0}$, define $x_i^* := 1$, and otherwise, define $x_i^* := 0$.
  3. Send x^* to the ideal functionality $\mathcal{F}$ (through $\hat{\mathcal{F}}$) and obtain $v_1 = F_1(x^*, y)$.
  4. Extract the "simulation trapdoor" σ of Π_{FS} by brute force from its first three rounds $(\mathsf{fs}_1, \mathsf{fs}_2, \mathsf{fs}_3)$.
- In Round 4, $\mathcal{S}$ prepares $m_4 = (\mathsf{CT}_{\mathrm{gc}}, \mathsf{fs}_4, \mathsf{nm}_3)$ in the same way as P_2 does except for the following.
  - $\mathcal{S}$ generates $\mathsf{CT}_{\mathrm{gc}}$ as follows. First, $\mathcal{S}$ simulates a garbled circuit and labels by $(\mathsf{GC}^*, \{Z_i^*\}) \leftarrow \mathsf{SimGC}(1^\kappa, F_1', v_1')$, where $v_1' = (v_1, \tilde{\mathsf{enc}}, \tilde{\mathsf{mac}})$, $\tilde{\mathsf{enc}} \leftarrow \mathsf{SKE.Enc}_{\mathsf{sk}_{\mathrm{ske}}}(0^\kappa)$, and $\tilde{\mathsf{mac}} = \mathsf{MAC}_{\mathsf{sk}_{\mathrm{mac}}}(\tilde{\mathsf{enc}})$ for randomly sampled $\mathsf{sk}_{\mathrm{ske}}$ and $\mathsf{sk}_{\mathrm{mac}}$. Second, using the equivocality of Eqcom, $\mathcal{S}$ obtains a decommitment $\mathsf{open}_{\mathsf{C}_{\mathrm{gc}}}^*$ that opens C_{gc} to GC^*. Third, $\mathcal{S}$ generates $\{W_{i,b}\}$ by $W_{i,x_i^*} := Z_i^* \oplus \mathsf{H}(f_{i,x_i^*}^{-\kappa}(z_{i,b}))$ and $W_{i,1-x_i^*} \leftarrow \{0,1\}^\kappa$. Finally, $\mathcal{S}$ generates $\mathsf{CT}_{\mathrm{gc}}$ by $\mathsf{CT}_{\mathrm{gc}} \leftarrow \mathsf{LE.Enc}_{\mathsf{pk}_{\mathrm{le}}}(\{W_{i,b}\} \| \mathsf{GC}^* \| \mathsf{open}_{\mathsf{C}_{\mathrm{gc}}}^*)$.

- $\mathcal{S}$ generates fs_4 by completing Π_{FS} using the simulation trapdoor σ. Then, $\mathcal{S}$ sends m_4 to $\mathcal{A}$.
- In Round 5, $\mathcal{S}$ receives $m_5 = (\mathsf{enc}, \mathsf{mac})$. If m_5 is accepting, $\mathcal{S}$ tells the ideal functionality $\mathcal{F}$ to send the output to P_2.

Case 2: P_2 is corrupted. $\mathcal{S}$ simulates P_1's messages as follows.

- In Round 1, $\mathcal{S}$ generates $m_1 = (\{\mathsf{com}_{i,b}, \mathsf{ext}_1^{i,b}\}, \mathsf{fs}_1)$ in the same way as P_1 except that $\mathcal{S}$ generates $\mathsf{ext}_1^{i,b}$ by committing to $r_{i,b}\|\omega_{\mathrm{com}}^{i,b}$ using ExtCom' correctly for every $i \in [\kappa]$ and $b \in \{0,1\}$. Then, $\mathcal{S}$ sends m_1 to $\mathcal{A}$.
- In Round 2, $\mathcal{S}$ receives $m_2 = (\{\mathsf{C}_{\mathrm{lab}}^{i,b}, r'_{i,b}, f_{i,b}, \mathsf{ext}_2^{i,b}\}, \mathsf{C}_{\mathrm{gc}}, \mathsf{zap}_1, \mathsf{fs}_2, \mathsf{nm}_1)$ from $\mathcal{A}$.
- In Round 3, $\mathcal{S}$ generates $m_3 = (\{z_{i,b}, \mathsf{ext}_3^{i,b}\}, \mathsf{pk}_{\mathrm{le}}, \mathsf{zap}_2, \mathsf{fs}_3, \mathsf{nm}_2)$ in the same way as P_1 except for the following.
  - $\mathcal{S}$ generates $\{z_{i,b}\}$ by $z_{i,b} := r_{i,b} \oplus r'_{i,b}$ for every $i \in [\kappa]$ and $b \in \{0,1\}$.
  - $\mathcal{S}$ generates $\mathsf{ext}_3^{i,b}$ by executing ExtCom' correctly for every $i \in [\kappa]$ and $b \in \{0,1\}$.
  - $\mathcal{S}$ generates $\mathsf{pk}_{\mathrm{le}}$ by $(\mathsf{pk}_{\mathrm{le}}, \mathsf{sk}_{\mathrm{le}}) \leftarrow \mathsf{LE.Gen}(1^\kappa, \mathsf{lossy})$ with randomness ω_{leGen}. (I.e., $\mathcal{S}$ generates a lossy public key rather than an injective one.)
  - When generating zap_2, $\mathcal{S}$ uses $(\{1, r_{i,1}, \omega_{\mathrm{com}}^{i,1}, \omega_{\mathrm{ext}}^{i,1}, \bot\}_{i\in[\kappa]}, \mathsf{sk}_{\mathrm{le}}, \omega_{\mathrm{leGen}})$ as the witness. (I.e., $\mathcal{S}$ proves that $\{(\mathsf{com}_{i,1}, \mathsf{ext}_1^{i,1}, \mathsf{ext}_3^{i,1}, z_{i,1})\}$ are computed correctly and $\mathsf{pk}_{\mathrm{le}}$ is a lossy public key.)

  Then, $\mathcal{S}$ sends m_3 to $\mathcal{A}$.
- In Round 4, $\mathcal{S}$ receives $m_4 = (\mathsf{CT}_{\mathrm{gc}}, \mathsf{fs}_4, \mathsf{nm}_3)$. If m_4 is accepting, $\mathcal{S}$ does the following.
  1. Extract the committed value of the NMCom commitment $(\mathsf{nm}_1, \mathsf{nm}_2, \mathsf{nm}_3)$ by brute force. If the extracted value is not a valid witness for st_4, $\mathcal{S}$ aborts the simulation with output Abort_2. Otherwise, the extracted value is denoted by $\mathsf{wit}_4 = \mathsf{wit}_2\|\{g_{i,b}\}\|\omega_{\mathrm{leEnc}}$, where $\mathsf{wit}_2 = (y', \Omega, \mathsf{GC}, \omega_{\mathrm{gc}}, \mathsf{open}_{\mathsf{C}_{\mathrm{gc}}}, \{Z_{i,b}, \omega_{\mathrm{lab}}^{i,b}\})$.
  2. Parse y' as $(y, \mathsf{sk}_{\mathrm{ske}}, \mathsf{sk}_{\mathrm{mac}}, \omega_{\mathrm{enc}})$, send y to the ideal functionality $\mathcal{F}$, and receive $v_2 = F_2(x, y)$.
- In Round 5, $\mathcal{S}$ generates $m_5 = (\mathsf{enc}, \mathsf{mac})$ by $\mathsf{enc} := \mathsf{SKE.Enc}_{\mathsf{sk}_{\mathrm{ske}}}(v_2; \omega_{\mathrm{enc}})$ and $\mathsf{mac} = \mathsf{MAC}_{\mathsf{sk}_{\mathrm{mac}}}(\mathsf{enc})$.

4.3   Proof of Indistinguishability

Fix any PPT adversary $\mathcal{A}$, and assume for contradiction that there exists a PPT environment $\mathcal{Z}$ and a PPT distinguisher D such that for infinitely many κ:

$$\varepsilon(\kappa) \stackrel{\mathrm{def}}{=} \left| \Pr\left[D(\mathrm{EXEC}_{\Pi_{2\mathrm{PC}}, \mathcal{A}, \mathcal{Z}}(\kappa)) = 1\right] - \Pr\left[D(\mathrm{EXEC}_{\Pi(\mathcal{F}), \mathcal{S}, \mathcal{Z}}(\kappa)) = 1\right] \right| \geq \frac{1}{\mathrm{poly}(\kappa)}. \tag{1}$$

We derive a contradiction by a hybrid argument. Let m be an upper bound on the number of subsessions (e.g., an upper bound on the running time of $\mathcal{Z}$).

Let $N := (10m\kappa/\varepsilon)^2$, which is a parameter that we use in the hybrid experiments. (Roughly speaking, we use N to determine the number of rewinding during extraction procedures in the hybrid experiments. We define N so that the extraction fails with probability that is much smaller than ε.)

Before defining the hybrid experiments, we define the order of the sessions. The order of the sessions is determined by the order of **special messages**, where the message in Round 2 is **special message** when P_1 is corrupted, and the message in Round 3 is **special message** when P_2 is corrupted.

Then, we define the hybrid experiments, $H_{0:17}$, $H_{k:j}$ ($k \in [m], j \in [17]$), and $H_{m+1:1}$, as follows. Hybrid $H_{0:17}$ is identical with the real experiment. Hence, in $H_{0:17}$, several parties (environment $\mathcal{Z}$, adversary $\mathcal{A}$, and two parties P_1, P_2) are invoked and then protocol $\Pi_{2\text{PC}}$ is executed concurrently multiple times among them; we call these executions of $\Pi_{2\text{PC}}$ the *main thread*. Next, for every $k \in [m]$, hybrids $H_{k:1}, \ldots, H_{k:17}$ are defined as follows.

Hybrid $H_{k:1}$ is the same as $H_{k-1:17}$ except that in session k on the main thread, if P_1 is corrupted, then the simulation trapdoor σ and the implicit input x^* are extracted as follows.

1. Just before **special message** of session k appears on the main thread, $2N$ *look-ahead threads* are created. Namely, from **special message** of session k (inclusive), the main thread of $H_{k-1:17}$ is executed $2N$ times with fresh randomness by rewinding all the parties including $\mathcal{Z}$ and $\mathcal{A}$.

   If there are at least two look-ahead threads on which Round 3 of session k is accepting, $\{\tilde{r}_{i,b}, \tilde{\omega}_{\text{com}}^{i,b}\}$ are defined as follows.
   (a) For every u, v such that $1 \le u < v \le 2N$, if Round 3 of session k is accepting both on the u-th look-ahead thread and on the v-th one, and a valid decommitment of $\text{com}_{i,b}$ is extractable from ExtCom' on these threads, then $\tilde{r}_{i,b}$ and $\tilde{\omega}_{\text{com}}^{i,b}$ are defined to be the extracted decommitment.
   (b) If $\tilde{r}_{i,b}$ and $\tilde{\omega}_{\text{com}}^{i,b}$ are not defined by the above process, then $\tilde{r}_{i,b} = \tilde{\omega}_{\text{com}}^{i,b} = \bot$.

2. Then, the main thread is resumed from **special message** of session k, and if Round 3 of session k is accepting on the main thread, the following are done. If there are less than two look-ahead threads on which Round 3 of session k is accepting, the experiment is aborted with output Abort_1. Otherwise, the simulation trapdoor σ is extracted based on the information on the look-ahead threads and the main thread; if a valid simulation trapdoor is not extractable, the experiment is aborted with output Abort_1. Next, $x^* = (x_1^*, \ldots, x_\kappa^*)$ is defined as follows: For every $i \in [\kappa]$, let b_i^* be the bit such that $(\tilde{r}_{i,b_i^*}, \tilde{\omega}_{\text{com}}^{i,b_i^*})$ is a valid decommitment of com_{i,b_i^*} and furthermore it holds that $z_{i,b_i^*} = \tilde{r}_{i,b_i^*} \oplus r'_{i,b_i^*}$; if there is no such b_i^*, the experiment is aborted with output Abort_1, and if b_i^* is not uniquely determined, $b_i^* := 0$; then, define $x_i^* := 1 - b_i^*$.

Hybrid $H_{k:2}$ is the same as $H_{k:1}$ except that in session k on the main thread, if P_1 is corrupted, then Π_{FS} in session k is switched to simulation, i.e., fs_4 is generated by using σ as the witness.

Hybrid $H_{k:3}$ is the same as $H_{k:2}$ except that in session k on the main thread, if P_1 is corrupted, then the value committed by NMCom is switched to a all-zero string.

Hybrid $H_{k:4}$ is the same as $H_{k:3}$ except that in session k on the main thread, if P_1 is corrupted, then $\{C_{\text{lab}}^{i,b}\}$ are generated by committing to all-zero strings by using Com.

Hybrid $H_{k:5}$ is the same as $H_{k:4}$ except that in session k on the main thread, if P_1 is corrupted, then C_{gc} is generated in a way that it can be opened to any value using the equivocality and $\text{open}_{C_{\text{gc}}}$ is computed by the equivocality.

Hybrid $H_{k:6}$ is the same as $H_{k:5}$ except that in session k on the main thread, if P_1 is corrupted, then $\{W_{i,b}\}$ are generated by $W_{i,x_i^*} := Z_{i,x_i^*} \oplus \mathsf{H}(f_{i,x_i^*}^{-\kappa}(z_{i,x_i^*}))$ and $W_{i,1-x_i^*} \leftarrow \{0,1\}^{\kappa}$.

Hybrid $H_{k:7}$ is the same as $H_{k:6}$ except that in session k on the main thread, if P_1 is corrupted, then GC and labels are generated by simulation, i.e., as follows.

1. Compute $v_1 = F_1(x^*, y)$ and $v_2 = F_2(x^*, y)$, where y is the input of P_2 in session k.
2. Compute $(\mathsf{GC}^*, \{Z_i^*\}) \leftarrow \mathsf{SimGC}(1^\kappa, F_1', v_1')$, where $v_1' = (v_1, \tilde{\mathsf{enc}}, \tilde{\mathsf{mac}})$, $\tilde{\mathsf{enc}} \leftarrow \mathsf{SKE.Enc}_{\mathsf{sk}_{\mathsf{ske}}}(v_2)$, and $\tilde{\mathsf{mac}} = \mathsf{MAC}_{\mathsf{sk}_{\mathsf{mac}}}(\tilde{\mathsf{enc}})$ for random $\mathsf{sk}_{\mathsf{ske}}$ and $\mathsf{sk}_{\mathsf{mac}}$.
3. Set $\mathsf{GC} := \mathsf{GC}^*$ and $Z_{i,x_i^*} := Z_i^*$. (Labels $\{Z_{i,1-x_i^*}\}$ are not used in $H_{k:6}$.)

Hybrid $H_{k:8}$ is the same as $H_{k:7}$ except that in session k on the main thread, if P_1 is corrupted, then honest P_2's output v_2 is computed as follows.

1. If $\mathsf{MAC}_{\mathsf{sk}_{\mathsf{mac}}}(\mathsf{enc}) \neq \mathsf{mac}$, P_2 outputs $\bot$. Otherwise, it outputs $F_2(x^*, y)$.

Hybrid $H_{k:9}$ is the same as $H_{k:8}$ except that in session k on the main thread, if P_1 is corrupted, then $\tilde{\mathsf{enc}}$ is generated by $\tilde{\mathsf{enc}} \leftarrow \mathsf{SKE.Enc}_{\mathsf{sk}_{\mathsf{ske}}}(0^\kappa)$ during the generation of GC and labels.

Hybrid $H_{k:10}$ is the same as $H_{k:9}$ except that in session k on the main thread, if P_2 is corrupted, then wit_4 is extracted in session k as follows.

1. Just before special message of session k appears on the main thread, N look-ahead threads are created. Namely, from special message of session k (inclusive), the main thread of $H_{k:9}$ is executed N times with fresh randomness by rewinding all the parties including $\mathcal{Z}$ and $\mathcal{A}$.
2. The main thread is resumed from special message of session k. If Round 4 of session is accepting on the main thread, extract wit_4 from NMCom using the information on the look-ahead threads and the main thread; if the extraction fails or wit_4 is not a valid witness for st_4, the experiment is aborted with output Abort_2.

Hybrid $H_{k:11}$ is the same as $H_{k:10}$ except that in session k on the main thread, if P_2 is corrupted, then honest P_1's output v_1 is computed as follows.

1. Parse the extracted wit_4 as $\mathsf{wit}_2 \| \{g_{i,b}\} \| \omega_{\mathsf{leEnc}}$, where $\mathsf{wit}_2 = (y', \Omega, \mathsf{GC}, \omega_{\mathsf{gc}}, \mathsf{open}_{C_{\mathsf{gc}}}, \{Z_{i,b}, \omega_{\mathsf{lab}}^{i,b}\})$ and $y' = (y, \mathsf{sk}_{\mathsf{ske}}, \mathsf{sk}_{\mathsf{mac}}, \omega_{\mathsf{enc}})$.
2. Set $v_1 := F_1(x, y)$ if the message m_4 in Round 4 is accepting and set $v_1 := \bot$ otherwise.

Hybrid $H_{k:12}$ is the same as $H_{k:11}$ except that in session k on the main thread, if P_2 is corrupted, then $m_5 = (\mathsf{enc}, \mathsf{mac})$ is generated using the keys $\mathsf{sk_{ske}}, \mathsf{sk_{mac}}$ and the randomness ω_{enc} in the extracted wit_4.

Hybrid $H_{k:13}$ is the same as $H_{k:12}$ except that in session k on the main thread, if P_2 is corrupted, then $\mathsf{pk_{le}}$ is switched to a lossy public key, and $\mathsf{CT_{gc}}$ is no longer decrypted in Round 5.

Hybrid $H_{k:14}$ is the same as $H_{k:13}$ except that in session k on the main thread, if P_2 is corrupted, then zap_2 is generated by using $(\{1 - x_i, r_{i,1-x_i}, \omega_{\mathsf{com}}^{i,1-x_i}, \omega_{\mathsf{ext}}^{i,1-x_i}, \bot\}_{i \in [\kappa]}, \mathsf{sk_{le}}, \omega_{\mathsf{leGen}})$ as the witness (i.e., by using a witness for the fact that $\mathsf{pk_{le}}$ is a lossy public key).

Hybrid $H_{k:15}$ is the same as $H_{k:14}$ except that in session k on the main thread, if P_2 is corrupted, then $z_{i,b}$ is generated by $z_{i,b} := r_{i,b} \oplus r'_{i,b}$ for every $i \in [\kappa]$ and $b \in \{0,1\}$.

Hybrid $H_{k:16}$ is the same as $H_{k:15}$ except that in session k on the main thread, if P_2 is corrupted, then $\mathsf{ext}_1^{i,b}$ and $\mathsf{ext}_3^{i,b}$ are generated by committing to $r_{i,b} \| \omega_{\mathsf{com}}^{i,b}$ correctly using ExtCom' for every $i \in [\kappa]$ and $b \in \{0,1\}$.

Hybrid $H_{k:17}$ is the same as $H_{k:16}$ except that in session k on the main thread, if P_2 is corrupted, then zap_2 is generated by using $(\{1, r_{i,1}, \omega_{\mathsf{com}}^{i,1}, \omega_{\mathsf{ext}}^{i,1}, \bot\}_{i \in [\kappa]}, \mathsf{sk_{le}}, \omega_{\mathsf{leGen}})$ as the witness (i.e., by using a witness for the fact that $\{(\mathsf{com}_{i,1}, \mathsf{ext}_1^{i,1}, \mathsf{ext}_3^{i,1}, z_{i,1})\}$ are correctly constructed and $\mathsf{pk_{le}}$ is a lossy public key).

Finally, hybrid $H_{m+1:1}$ is identical with the ideal experiment.

Remark 1. The hybrid experiments $H_{k:1}, \ldots, H_{k:17}$ are designed so that no look-ahead thread is created after special message of session k.

Our goal is to show that the output of the first hybrid $H_{0:17}$ and that of the last hybrid $H_{m+1:1}$ are indistinguishable (more precisely, are distinguishable with advantage at most $\varepsilon/2$.) Toward this goal, we show the indistinguishability among the outputs of the intermediate hybrids. Also, for a technical reason, we show that the following condition holds with high probability in each hybrid: In a session in which P_2 is corrupted, if the session is accepting then the NMCom commitment from P_2 is valid and the committed value wit_4 is a valid witness for st_4. (Notice that if this condition holds, then we can extract the input of P_2 from NMCom.) Formally, for every $k' \in [m]$ let $\mathrm{BAD}_{k'}$ be the event that in the k'-th session on the main thread, P_2 is corrupted, Round 4 is accepting, but the committed value wit_4 of NMCom is not a valid witness for st_4, and let $\rho_{k:j:k'}$ be the probability that $\mathrm{BAD}_{k'}$ occurs in $H_{k:j}$. We first observe that $\rho_{0:17:k'}$ is negligible for every k' (i.e., $\mathrm{BAD}_{k'}$ occurs in the real experiment with negligible probability for every k').

Lemma 1. *For every* $k' \in [m]$, $\rho_{0:17:k'} = \mathsf{negl}(\kappa)$.

Proof. This lemma follows from the soundness of Π_{FS} because P_2 proves in Π_{FS} that a valid witness for st_4 is committed in NMCom. $\square$

Now we are ready to show the indistinguishability among the outputs of the hybrids. Let $\mathsf{H}_{k:j}$ be the random variable representing the output of $H_{k:j}$. We first prove the following lemma.

Lemma 2. *For every $k \in [m]$, the following two inequalities hold.*

1. $\left| \Pr\left[D(\mathsf{H}_{k-1:17}) = 1 \right] - \Pr\left[D(\mathsf{H}_{k:17}) = 1 \right] \right| \leq \dfrac{2\kappa + 1}{\sqrt{N}} + \dfrac{1}{N} + \mathsf{negl}(\kappa).$ (2)

2. $\forall k' \in [m] : \rho_{k:17:k'} \leq \rho_{k-1:17:k'} + \mathsf{negl}(\kappa).$ (3)

Proof. Fix any $k \in [m]$. From Lemma 1, it suffices to show that the above two inequalities hold whenever we have

$$\forall k' \in [m] : \rho_{k-1:17:k'} = \mathsf{negl}(\kappa).$$ (4)

In what follows we show claims about the outputs of each neighboring hybrids.

Claim 1. $\left| \Pr\left[D(\mathsf{H}_{k-1:17}) = 1 \right] - \Pr\left[D(\mathsf{H}_{k:1}) = 1 \right] \right| \leq \frac{2\kappa+1}{\sqrt{N}} + \mathsf{negl}(\kappa).$ *Furthermore, for every $k' \in [m]$, $\rho_{k:1:k'} \leq \rho_{k-1:17:k'}$.*

Proof. We first show the indistinguishability of the outputs of the hybrids. The output of $H_{k:1}$ differs from that of $H_{k-1:17}$ only when it outputs Abort_1 in session k, and $H_{k:1}$ outputs Abort_1 in session k only when one of the following events occur.

Event E_1: Round 3 of session k is accepting on less than two look-ahead threads but it is accepting on the main thread.

Event E_2: The extraction of the simulation trapdoor σ fails.

Event $E_{3,i}$ $(i \in [\kappa])$: There is no b_i^* such that $(\tilde{r}_{i,b_i^*}, \tilde{\omega}_{\mathsf{com}}^{i,b_i^*})$ is a valid decommitment of com_{i,b_i^*} and $z_{i,b_i^*} = \tilde{r}_{i,b_i^*} \oplus r'_{i,b_i^*}$.

From Markov's inequality, E_1 occurs with probability at most $1/\sqrt{N}$, and from the extractability of Π_{WIPOK} (inside Π_{FS}), E_2 occurs with negligible probability. In what follows, we show that for every $i \in [\kappa]$, $E_{3,i}$ occurs with probability at most $2/\sqrt{N} + \mathsf{negl}(\kappa)$. Let prefix be any prefix of the execution of $H_{k:1}$ up until the creation of the look-ahead threads in the k-th session (exclusive). We show that for every i, under that condition that a prefix of the execution of $H_{k:1}$ is prefix, $E_{3,i}$ occurs with probability at most $2/\sqrt{N} + \mathsf{negl}(\kappa)$. For $b \in \{0,1\}$, let us say that session k is (i,b)-*good* if its Round 3 is accepting, a valid decommitment $(r_{i,b}, \omega_{\mathsf{com}}^{i,b})$ of $\mathsf{com}_{i,b}$ is correctly committed in $(\mathsf{ext}_1^{i,b}, \mathsf{ext}_2^{i,b}, \mathsf{ext}_3^{i,b})$, and it holds that $z_{i,b} = r_{i,b} \oplus r'_{i,b}$. From the extractability of ExtCom', one of the following events occurs whenever $E_{3,i}$ occurs.

- Session k is $(i,0)$-good on the main thread, but it is $(i,0)$-good on less than two look-ahead threads. If session k is $(i,0)$-good on the main thread with probability at most $1/\sqrt{N}$, this event occurs with probability at most $1/\sqrt{N}$. Furthermore, even if session k is $(i,0)$-good on the main thread with probability at least $1/\sqrt{N}$, this event occurs with probability at most $1/\sqrt{N}$, since from Markov's inequality, session k is $(i,0)$-good on less than two look-ahead threads with probability at most $1/\sqrt{N}$.

– Session k is $(i,1)$-good on the main thread, but it is $(i,1)$-good on less than two look-ahead threads. From the same argument as above, this event occurs with probability at most $1/\sqrt{N}$.

– On the main thread, Round 3 of session k is accepting but it is neither $(i,0)$-good nor $(i,1)$-good. From the soundness of Π_{ZAP}, this event occurs with negligible probability.

Hence, for every $i \in [\kappa]$, $E_{3,i}$ occurs with probability at most $2/\sqrt{N} + \mathsf{negl}(\kappa)$. From the union bound, the probability that there exists $i \in [\kappa]$ such that $E_{3,i}$ occurs is at most $2\kappa/\sqrt{N} + \mathsf{negl}(\kappa)$. Since prefix is any prefix, we conclude that even without conditioning that a prefix of the execution of $H_{k:1}$ is prefix, the probability that there exists $i \in [\kappa]$ such that $E_{3,i}$ occurs is at most $2\kappa/\sqrt{N} + \mathsf{negl}(\kappa)$. Hence, the indistinguishability follows.

We next observe that we have $\rho_{k:1:k'} \leq \rho_{k-1:17:k'}$. This is because the main thread of $H_{k:1}$ is identical with that of $H_{k-1:17}$ until the experiment outputs Abort_1 in session k, and when it outputs Abort_1, the experiment is aborted immediately and no further NMCom commitment is created. □

Claim 2. $|\Pr[D(H_{k:1}) = 1] - \Pr[D(H_{k:2}) = 1]| \leq \mathsf{negl}(\kappa)$. *Furthermore, for every* $k' \in [m]$, $\rho_{k:2:k'} \leq \rho_{k:1:k'} + \mathsf{negl}(\kappa)$.

Proof. We first show the indistinguishability of the outputs. $H_{k:2}$ differs from $H_{k:1}$ only in that in session k on the main thread, the simulation trapdoor is used in Π_{SWIAOK} (inside Π_{FS}) as the witness. We then observe that, since no look-ahead thread is created after Round 2 of session k on the main thread, Π_{SWIAOK} in session k is not rewound after its second round, and so the indistinguishability follows from the witness indistinguishability of Π_{SWIAOK}.

We next observe that $\rho_{k:2:k'} \leq \rho_{k:1:k'} + \mathsf{negl}(\kappa)$ follows from the statistical witness indistinguishability of Π_{SWIAOK}. Specifically, if $\rho_{k:2:k'}$ differs from $\rho_{k:1:k'}$ with non-negligible amount, we can break the statistical witness indistinguishability of Π_{SWIAOK} by checking whether $\mathrm{BAD}_{k'}$ occurs or not by extracting the committed value of the NMCom commitment by brute force. □

Claim 3. $|\Pr[D(H_{k:2}) = 1] - \Pr[D(H_{k:3}) = 1]| \leq \mathsf{negl}(\kappa)$. *Furthermore, for every* $k' \in [m]$, $\rho_{k:3:k'} \leq \rho_{k:2:k'} + \mathsf{negl}(\kappa)$.

Proof. We first show the indistinguishability of the outputs. $H_{k:3}$ differs from $H_{k:2}$ only in the committed value of NMCom in session k on the main thread. We then observe that, since no look-ahead thread is created after Round 2 of session k on the main thread, NMCom in session k on the main thread is not rewound. Hence, the indistinguishability follows from the hiding property of NMCom.

We next observe that $\rho_{k:3:k'} \leq \rho_{k:2:k'} + \mathsf{negl}(\kappa)$ follows from the non-malleability of NMCom. Specifically, if $\rho_{k:3:k'}$ differs from $\rho_{k:2:k'}$ with non-negligible amount, we can break the non-malleability of NMCom by considering an adversary that internally emulates $H_{k:2}$ while obtaining the NMCom commitment of session k from the external committer and forwarding the NMCom commitment of session k' to the external receiver. We remark that since NMCom is public coin, we can emulate $H_{k:2}$ while forwarding the NMCom commitment of session k' to the external receiver (without worrying that it can be rewound). □

Claim 4. $|\Pr[D(H_{k:3}) = 1] - \Pr[D(H_{k:4}) = 1]| \leq \mathsf{negl}(\kappa)$. *Furthermore, for every* $k' \in [m]$, $\rho_{k:4:k'} \leq \rho_{k:3:k'} + \mathsf{negl}(\kappa)$.

Proof. We first show the indistinguishability of the outputs. $H_{k:4}$ differs from $H_{k:3}$ only in the committed values of Com in session k. Hence, the indistinguishability follows from the hiding properly of Com.

We next observe that $\rho_{k:4:k'} \leq \rho_{k:3:k'} + \mathsf{negl}(\kappa)$ follows from the hiding property of Com and the extractability of NMCom. Specifically, if $\rho_{k:4:k'}$ differs from $\rho_{k:3:k'}$ with non-negligible amount, we can break the hiding property of Com by considering an adversary that internally emulates $H_{k:3}$ while obtaining $\{C^{i,b}_{\mathsf{lab}}\}$ of session k from the external committer and extracting the committed value of the NMCom commitment in session k'. We remark that there are two subtleties:

1. The extraction from NMCom requires rewinding, and hence the Com commitment in session k might be rewound during the extraction from NMCom. Nevertheless, we can use the hiding property of Com since Com is non-interactive (which trivially implies that Com is hiding even when it is rewound).
2. The NMCom commitment in session k' might be rewound in $H_{k:3}$ during the creation of the look-ahead threads. Nevertheless, we can use its extractability since NMCom is public coin (which implies that an adversary can internally emulate $H_{k:3}$ while forwarding NMCom to an external receiver). □

We remark that the statement of Claim 4 also holds w.r.t. $H_{k:j}$ and $H_{k:j+1}$ for $j = 4, \ldots, 8$. The proofs are similar to the proof of Claim 4: the indistinguishability between the outputs of the hybrids is shown by relying on the security of the components (e.g., the equivocality of Eqcom), and the inequality between $\rho_{k:j:k'}$ and $\rho_{k:j+1:k'}$ is shown by additionally using the extractability of NMCom. We therefore have the following claim.

Claim 5. $|\Pr[D(H_{k:4}) = 1] - \Pr[D(H_{k:9}) = 1]| \leq \mathsf{negl}(\kappa)$. *Furthermore, for every* $k' \in [m]$, $\rho_{k:9:k'} \leq \rho_{k:4:k'} + \mathsf{negl}(\kappa)$.

A formal argument for this claim is given in the full version of this paper.

Claim 6. $|\Pr[D(H_{k:9}) = 1] - \Pr[D(H_{k:10}) = 1]| \leq \frac{1}{N} + \mathsf{negl}(\kappa)$. *Furthermore, for every* $k' \in [m]$, $\rho_{k:10:k'} \leq \rho_{k:9:k'}$.

Proof. We first show the indistinguishability of the outputs. The output of $H_{k:10}$ differs from that of $H_{k:9}$ only when it outputs Abort_2 in session k, and $H_{k:10}$ outputs Abort_2 in session k only when one of the following happens.

1. Round 4 of session k does not complete on the look-ahead threads but it completes on the main thread.
2. Even though Round 3 of session k completes on a look-ahead thread and the main thread, a valid witness wit_4 for st_4 is not extractable.

The former occurs with probability at most $1/N$ from the swapping argument. The latter occurs with negligible probability since we have $\rho_{k:9:k'} = \mathsf{negl}(\kappa)$ from Eq. (4). (Notice that when P_2 is corrupted in session k, the main thread of $H_{k:9}$ proceeds identically with that of $H_{k-1:17}$.) Hence, the indistinguishability follows.

We next observe that we have $\rho_{k:10:k'} \leq \rho_{k:9:k'}$. This is because $H_{k:10}$ is identical with $H_{k:9}$ until it outputs Abort_2 in session k, and when it outputs Abort_2, the experiment is immediately aborted.   □

We remark that the statement of Claim 4 also holds w.r.t. $H_{k:j}$ and $H_{k:j+1}$ for $j = 10, \ldots, 14$; the proofs are similar to the proof of Claim 4. We therefore have the following claim.

Claim 7. $|\Pr[D(H_{k:10}) = 1] - \Pr[D(H_{k:15}) = 1]| \leq \mathsf{negl}(\kappa)$. *Furthermore, for every* $k' \in [m]$, $\rho_{k:15:k'} \leq \rho_{k:10:k'} + \mathsf{negl}(\kappa)$.

A formal argument for this claim is given in the full version of this paper.

Claim 8. $|\Pr[D(H_{k:15}) = 1] - \Pr[D(H_{k:16}) = 1]| \leq \mathsf{negl}(\kappa)$. *Furthermore, for every* $k' \in [m]$, $\rho_{k:16:k'} \leq \rho_{k:15:k'} + \mathsf{negl}(\kappa)$.

Proof. We first show the indistinguishability of the outputs. $H_{k:16}$ differs from $H_{k:15}$ only in that ext_1^{i,x_i} and ext_3^{i,x_i} are generated by committing to $r_{i,x_i} \| \omega_{\mathsf{com}}^{i,x_i}$ correctly using ExtCom' (rather than by executing "fake" ExtCom'). Since ext_1^{i,x_i} consists of Com commitments to κ pairs of 2-out-of-2 secret shares and ext_3^{i,x_i} consists of the revealing of the shares that are selected by ext_2^{i,x_i}, we use the hiding property of Com to show the indistinguishability. Assume for contradiction that the output of $H_{k:15}$ and that of $H_{k:16}$ are distinguishable. Then, we consider an adversary $\mathcal{A}_{\mathsf{Com}}$ that internally emulates $H_{k:16}$ honestly except for the following.

– In Round 1 of session k on the main thread, $\mathcal{A}_{\mathsf{Com}}$ obtains $\{\mathsf{ext}_1^{i,x_i}\}$ from the external committer, where each ext_1^{i,x_i} consists of Com commitments whose committed values are either all-zero strings or pairs of 2-out-of-2 secret shares of $r_{i,x_i} \| \omega_{\mathsf{com}}^{i,x_i}$.
– In Round 3 of session k on the main thread, $\mathcal{A}_{\mathsf{Com}}$ computes ext_3^{i,x_i} as in the correct execution of ExtCom' assuming that the values committed in ext_1^{i,x_i} are the pairs of 2-out-of-2 secret shares.

When $\mathcal{A}_{\mathsf{Com}}$ receives Com commitment to the pairs of 2-out-of-2 secret shares, the internally emulated experiment is identical with $H_{k:16}$. When $\mathcal{A}_{\mathsf{Com}}$ receives Com commitments to all-zero strings, the internally emulated experiment is identical with $H_{k:15}$ (since in this case, ext_3^{i,x_i} consists of random strings that are independent of other parts of the experiment). Hence, we derive a contradiction.

Remark 2. ExtCom' in session k might be rewound in $H_{k:15}$ and $H_{k:16}$ since look-ahead threads might be created after Round 1 of session k on the main thread (for simulating other sessions). Nevertheless, $\mathcal{A}_{\mathsf{Com}}$ can emulate $H_{k:16}$

while obtaining ext_1^{i,x_i} from the external committer because (1) the randomness for generating ext_1^{i,x_i} and ext_3^{i,x_i} is not used after Round 1 and (2) ext_3^{i,x_i} on look-ahead thread is a random string (and thus can be simulated trivially).

We next observe that $\rho_{k:16:k'} \leq \rho_{k:15:k'} + \mathsf{negl}(\kappa)$ follows from the indistinguishability of Com. The argument for this statement is similar to the one in the proof of Claim 4. □

Claim 9. $|\Pr[D(\mathsf{H}_{k:16}) = 1] - \Pr[D(\mathsf{H}_{k:17}) = 1]| \leq \mathsf{negl}(\kappa)$. *Furthermore, for every* $k' \in [m]$, $\rho_{k:17:k'} \leq \rho_{k:16:k'} + \mathsf{negl}(\kappa)$.

Proof. We first show the indistinguishability of the outputs. $\mathsf{H}_{k:17}$ differs from $\mathsf{H}_{k:16}$ only in the witness used in Π_{ZAP}. Hence, the indistinguishability follows form the witness indistinguishability of Π_{ZAP}.

We next observe that $\rho_{k:17:k'} \leq \rho_{k:16:k'} + \mathsf{negl}(\kappa)$ follows from the witness indistinguishability of Π_{ZAP} and the extractability of NMCom. The argument for this statement is similar to the one in the proof of Claim 4. □

By combining Claims 1, 2, 3, 4, 5, 6, 7, 8 and 9, we conclude that the two inequalities in the statement of Lemma 2 hold for k. This concludes the proof of Lemma 2. □

We next show that the output of $\mathsf{H}_{m:17}$ and that of the last hybrid $\mathsf{H}_{m+1:1}$ (i.e., the ideal experiment) is indistinguishable.

Lemma 3

$$|\Pr[D(\mathsf{H}_{m:17}) = 1] - \Pr[D(\mathsf{H}_{m+1:1}) = 1]| \leq m\left(\frac{2\kappa+1}{\sqrt{N}} + \frac{1}{N} + \mathsf{negl}(\kappa)\right) + \mathsf{negl}(\kappa).$$

Proof. We consider an intermediate hybrid $\hat{\mathsf{H}}_{m:17}$ that is the same as $\mathsf{H}_{m:17}$ except that the extractions from ExtCom′, Π_{WIPOK}, and NMCom are performed by brute force rather than by rewinding (hence, no look-ahead thread is created in $\hat{\mathsf{H}}_{m:17}$). That is, $\hat{\mathsf{H}}_{m:17}$ is the same as $\mathsf{H}_{m:17}$ except that in a session in which P_1 is corrupted, the simulation trapdoor σ and the committed values $\{\tilde{r}_{i,b} \| \tilde{\omega}_{\mathrm{com}}^{i,b}\}$ of ExtCom′ are extracted by brute force, and in a session in which P_2 is corrupted, the committed value wit_4 of NMCom is extracted by brute force.

First, we observe that the output of $\hat{\mathsf{H}}_{m:17}$ and that of $\mathsf{H}_{m+1:1}$ are identical, that is,

$$\Pr\left[D(\hat{\mathsf{H}}_{m:17}) = 1\right] = \Pr[D(\mathsf{H}_{m+1:1}) = 1]. \tag{5}$$

This can be seen by inspection: in $\hat{\mathsf{H}}_{m:17}$, all the messages of the honest parties are generated in the same way as in $\mathsf{H}_{m+1:1}$ and the outputs of the honest parties are computed in the same way as in $\mathsf{H}_{m+1:1}$.

Next, we show the indistinguishability between the output of $\hat{\mathsf{H}}_{m:17}$ and that of $\mathsf{H}_{m:17}$. We first observe that when $\mathsf{H}_{m:17}$ outputs neither Abort_1 nor Abort_2, the messages and outputs of the honest parties are statistically close to those that would be computed with brute-force extractions (i.e., as in $\hat{\mathsf{H}}_{m:17}$).

- When $H_{m:17}$ does not output Abort_2, a valid witness wit_4 for st_4 is extracted in every session in which P_2 is corrupted, and the same wit_4 would be also extracted by brute-force extraction. (This is because from Lemmas 1 and 2, the probability that $\mathrm{BAD}_{k'}$ occurs in $H_{m:17}$ is negligible for every $k' \in [m]$.)
- When $H_{m:17}$ does not output Abort_1, a valid simulation trapdoor σ is extracted in every session in which P_1 is corrupted, and although a different simulation trapdoor might be extracted as σ by brute-force extraction, the information about σ is statistically hidden because a statistical witness-indistinguishable argument Π_{SWIAOK} is used in Π_{FS}.
- When $H_{m:17}$ does not output Abort_1, an implicit input x^* is defined according to the values extracted from ExtCom' in every session in which P_1 is corrupted. If $\mathsf{pk}_{\mathsf{le}}$ is an injective public key in such a session, the same x^* would be defined by brute-force extraction except with negligible probability. (This is because if $\mathsf{pk}_{\mathsf{le}}$ is an injective public key, the soundness of Π_{ZAP} guarantees that for every $i \in [\kappa]$, there is a unique $b_i^* \in \{0,1\}$ such that $(\mathsf{ext}_1^{i,b_i^*}, \mathsf{ext}_2^{i,b_i^*}, \mathsf{ext}_3^{i,b_i^*})$ is a correct ExtCom' commitment to a valid decommitment $(r_{i,b_i^*}, \omega_{\mathsf{com}}^{i,b_i^*})$ of com_{i,b_i^*} and $z_{i,b_i^*} = r_{i,b_i^*} \oplus r'_{i,b_i^*}$.) If $\mathsf{pk}_{\mathsf{le}}$ is a lossy public key in such a session, a different x^* might be defined by brute-force extraction.[5] However, x^* is used only to compute $\mathsf{CT}_{\mathsf{gc}}$ and the output of honest P_2, where $\mathsf{CT}_{\mathsf{gc}}$ is generated by $\mathsf{LE.Enc}_{\mathsf{pk}_{\mathsf{le}}}(\cdot)$ (which statistically hides the plaintext when $\mathsf{pk}_{\mathsf{le}}$ is lossy) and the output of P_2 is $\bot$ when mac in Round 5 is rejecting (which is almost always the case when $\mathsf{pk}_{\mathsf{le}}$ is lossy because $\mathsf{sk}_{\mathsf{mac}}$ is statistically hidden in this case). Thus, the information about x^* is statistically hidden in this case.

We next analyze the probability that $H_{m:17}$ outputs Abort_1 or Abort_2. From Lemma 2, we have

$$|\Pr[D(\mathsf{H}_{0:17}) = 1] - \Pr[D(\mathsf{H}_{m:17}) = 1]| \leq m\left(\frac{2\kappa + 1}{\sqrt{N}} + \frac{1}{N} + \mathsf{negl}(\kappa)\right).$$

Then, since $H_{0:17}$ (i.e., the real experiment) never output Abort_1 or Abort_2, we have that $H_{m:17}$ outputs Abort_1 or Abort_2 with probability at most

$$m\left(\frac{2\kappa + 1}{\sqrt{N}} + \frac{1}{N} + \mathsf{negl}(\kappa)\right).$$

By combining the above, we obtain

$$\left|\Pr\left[D(\hat{\mathsf{H}}_{m:17}) = 1\right] - \Pr[D(\mathsf{H}_{m:17}) = 1]\right| \leq m\left(\frac{2\kappa + 1}{\sqrt{N}} + \frac{1}{N} + \mathsf{negl}(\kappa)\right) + \mathsf{negl}(\kappa). \tag{6}$$

From Eqs. (5) and (6), we obtain

$$|\Pr[D(\mathsf{H}_{m:17}) = 1] - \Pr[D(\mathsf{H}_{m+1:1}) = 1]| \leq m\left(\frac{2\kappa + 1}{\sqrt{N}} + \frac{1}{N} + \mathsf{negl}(\kappa)\right) + \mathsf{negl}(\kappa).$$

□

---

[5] This is because from an *invalid* ExtCom' commitment, the brute-force extractor always outputs $\bot$ but the rewinding extractor can output any value (in particular, it can output even a valid decommitment of Com).

From Lemmas 2 and 3 and $N = (10m\kappa/\varepsilon)^2$, we have

$$\left| \Pr\left[D(\text{EXEC}_{\Pi_{2\text{PC}},\mathcal{A},\mathcal{Z}}(\kappa)) = 1\right] - \Pr\left[D(\text{EXEC}_{\Pi(\mathcal{F}),\mathcal{S},\mathcal{Z}}(\kappa)) = 1\right]\right|$$
$$= \left|\Pr\left[D(\mathsf{H}_{0:17}) = 1\right] - \Pr\left[D(\mathsf{H}_{m+1:1}) = 1\right]\right|$$
$$\leq 2m\left(\frac{2\kappa+1}{\sqrt{N}} + \frac{1}{N} + \mathsf{negl}(\kappa)\right) + \mathsf{negl}(\kappa) \leq \frac{5m\kappa}{\sqrt{N}} = \frac{\varepsilon}{2}.$$

This contradicts to Eq. (1). This concludes the proof of Theorem 1.

References

1. Barak, B., Prabhakaran, M., Sahai, A.: Concurrent non-malleable zero knowledge. In: 47th FOCS, pp. 345–354. IEEE Computer Society Press, October 2006
2. Barak, B., Sahai, A.: How to play almost any mental game over the net - concurrent composition via super-polynomial simulation. In: 46th FOCS, pp. 543–552. IEEE Computer Society Press, October 2005
3. Bellare, M., Hofheinz, D., Yilek, S.: Possibility and impossibility results for encryption and commitment secure under selective opening. In: Joux, A. (ed.) EUROCRYPT 2009. LNCS, vol. 5479, pp. 1–35. Springer, Heidelberg (2009). doi:10.1007/978-3-642-01001-9_1
4. Canetti, R.: Universally composable security: a new paradigm for cryptographic protocols. In: 42nd FOCS, pp. 136–145. IEEE Computer Society Press, October 2001
5. Canetti, R., Kushilevitz, E., Lindell, Y.: On the limitations of universally composable two-party computation without set-up assumptions. In: Biham, E. (ed.) EUROCRYPT 2003. LNCS, vol. 2656, pp. 68–86. Springer, Heidelberg (2003). doi:10.1007/3-540-39200-9_5
6. Canetti, R., Lin, H., Pass, R.: Adaptive hardness and composable security in the plain model from standard assumptions. In: 51st FOCS, pp. 541–550. IEEE Computer Society Press, October 2010
7. Dwork, C., Naor, M.: Zaps and their applications. In: 41st FOCS, pp. 283–293. IEEE Computer Society Press, November 2000
8. Feige, U., Shamir, A.: Witness indistinguishable and witness hiding protocols. In: 22nd ACM STOC, pp. 416–426. ACM Press, May 1990
9. Feige, U., Shamir, A.: Zero knowledge proofs of knowledge in two rounds. In: Brassard, G. (ed.) CRYPTO 1989. LNCS, vol. 435, pp. 526–544. Springer, Heidelberg (1990). doi:10.1007/0-387-34805-0_46
10. Garg, S., Goyal, V., Jain, A., Sahai, A.: Concurrently secure computation in constant rounds. In: Pointcheval, D., Johansson, T. (eds.) EUROCRYPT 2012, Part II. LNCS, vol. 7237, pp. 99–116. Springer, Heidelberg (2012). doi:10.1007/978-3-642-29011-4_8
11. Garg, S., Mukherjee, P., Pandey, O., Polychroniadou, A.: The exact round complexity of secure computation. In: Fischlin, M., Coron, J.-S. (eds.) EUROCRYPT 2016, Part II. LNCS, vol. 9666, pp. 448–476. Springer, Heidelberg (2016). doi:10.1007/978-3-662-49896-5_16
12. Goldreich, O., Micali, S., Wigderson, A.: How to play any mental game or a completeness theorem for protocols with honest majority. In: Aho, A. (ed.) 19th ACM STOC, pp. 218–229. ACM Press, May 1987

13. Goyal, V., Jain, A.: On concurrently secure computation in the multiple ideal query model. In: Johansson, T., Nguyen, P.Q. (eds.) EUROCRYPT 2013. LNCS, vol. 7881, pp. 684–701. Springer, Heidelberg (2013). doi:10.1007/978-3-642-38348-9_40

14. Goyal, V., Lin, H., Pandey, O., Pass, R., Sahai, A.: Round-efficient concurrently composable secure computation via a robust extraction lemma. In: Dodis, Y., Nielsen, J.B. (eds.) TCC 2015, Part I. LNCS, vol. 9014, pp. 260–289. Springer, Heidelberg (2015). doi:10.1007/978-3-662-46494-6_12

15. Goyal, V., Pandey, O., Richelson, S.: Textbook non-malleable commitments. Cryptology ePrint Archive, Report 2015/1178 (2015). http://eprint.iacr.org/2015/1178

16. Goyal, V., Pandey, O., Richelson, S.: Textbook non-malleable commitments. In: Wichs, D., Mansour, Y. (eds.) 48th ACM STOC, pp. 1128–1141. ACM Press, June 2016

17. Hemenway, B., Libert, B., Ostrovsky, R., Vergnaud, D.: Lossy encryption: constructions from general assumptions and efficient selective opening chosen ciphertext security. In: Lee, D.H., Wang, X. (eds.) ASIACRYPT 2011. LNCS, vol. 7073, pp. 70–88. Springer, Heidelberg (2011). doi:10.1007/978-3-642-25385-0_4

18. Katz, J., Ostrovsky, R.: Round-optimal secure two-party computation. In: Franklin, M. (ed.) CRYPTO 2004. LNCS, vol. 3152, pp. 335–354. Springer, Heidelberg (2004). doi:10.1007/978-3-540-28628-8_21

19. Kiyoshima, S.: Round-efficient black-box construction of composable multiparty computation. In: Garay, J.A., Gennaro, R. (eds.) CRYPTO 2014, Part II. LNCS, vol. 8617, pp. 351–368. Springer, Heidelberg (2014). doi:10.1007/978-3-662-44381-1_20

20. Kiyoshima, S., Manabe, Y., Okamoto, T.: Constant-round black-box construction of composable multi-party computation protocol. In: Lindell, Y. (ed.) TCC 2014. LNCS, vol. 8349, pp. 343–367. Springer, Heidelberg (2014). doi:10.1007/978-3-642-54242-8_15

21. Lapidot, D., Shamir, A.: Publicly verifiable non-interactive zero-knowledge proofs. In: Menezes, A.J., Vanstone, S.A. (eds.) CRYPTO 1990. LNCS, vol. 537, pp. 353–365. Springer, Heidelberg (1991). doi:10.1007/3-540-38424-3_26

22. Lin, H., Pass, R.: Non-malleability amplification. In: Mitzenmacher, M. (ed.) 41st ACM STOC, pp. 189–198. ACM Press, May/June 2009

23. Lin, H., Pass, R.: Black-box constructions of composable protocols without setup. In: Safavi-Naini, R., Canetti, R. (eds.) CRYPTO 2012. LNCS, vol. 7417, pp. 461–478. Springer, Heidelberg (2012). doi:10.1007/978-3-642-32009-5_27

24. Lin, H., Pass, R., Venkitasubramaniam, M.: A unified framework for concurrent security: universal composability from stand-alone non-malleability. In: Mitzenmacher, M. (ed.) 41st ACM STOC, pp. 179–188. ACM Press, May/June 2009

25. Lindell, Y.: Bounded-concurrent secure two-party computation without setup assumptions. In: 35th ACM STOC, pp. 683–692. ACM Press, June 2003

26. Lindell, Y.: Lower bounds for concurrent self composition. In: Naor, M. (ed.) TCC 2004. LNCS, vol. 2951, pp. 203–222. Springer, Heidelberg (2004). doi:10.1007/978-3-540-24638-1_12

27. Lindell, Y., Pinkas, B.: A proof of security of Yao's protocol for two-party computation. J. Cryptology **22**(2), 161–188 (2009)

28. Malkin, T., Moriarty, R., Yakovenko, N.: Generalized environmental security from number theoretic assumptions. In: Halevi, S., Rabin, T. (eds.) TCC 2006. LNCS, vol. 3876, pp. 343–359. Springer, Heidelberg (2006). doi:10.1007/11681878_18

29. Micali, S., Pass, R., Rosen, A.: Input-indistinguishable computation. In: 47th FOCS, pp. 367–378. IEEE Computer Society Press, October 2006

30. Ostrovsky, R., Richelson, S., Scafuro, A.: Round-optimal black-box two-party computation. In: Gennaro, R., Robshaw, M. (eds.) CRYPTO 2015, Part II. LNCS, vol. 9216, pp. 339–358. Springer, Heidelberg (2015). doi:10.1007/978-3-662-48000-7_17
31. Pass, R.: Simulation in quasi-polynomial time, and its application to protocol composition. In: Biham, E. (ed.) EUROCRYPT 2003. LNCS, vol. 2656, pp. 160–176. Springer, Heidelberg (2003). doi:10.1007/3-540-39200-9_10
32. Pass, R.: Bounded-concurrent secure multi-party computation with a dishonest majority. In: Babai, L. (ed.) 36th ACM STOC, pp. 232–241. ACM Press, June 2004
33. Pass, R., Lin, H., Venkitasubramaniam, M.: A unified framework for UC from only OT. In: Wang, X., Sako, K. (eds.) ASIACRYPT 2012. LNCS, vol. 7658, pp. 699–717. Springer, Heidelberg (2012). doi:10.1007/978-3-642-34961-4_42
34. Peikert, C., Waters, B.: Lossy trapdoor functions and their applications. In: Ladner, R.E., Dwork, C. (eds.) 40th ACM STOC, pp. 187–196. ACM Press, May 2008
35. Prabhakaran, M., Sahai, A.: New notions of security: achieving universal composability without trusted setup. In: Babai, L. (ed.) 36th ACM STOC, pp. 242–251. ACM Press, June 2004
36. Yao, A.C.C.: How to generate and exchange secrets (extended abstract). In: 27th FOCS, pp. 162–167. IEEE Computer Society Press, October 1986

High-Throughput Secure Three-Party Computation for Malicious Adversaries and an Honest Majority

Jun Furukawa[1(✉)], Yehuda Lindell[2], Ariel Nof[2], and Or Weinstein[2]

[1] NEC Israel Research Center, Herzliya, Israel
jun.furukawa@necam.com
[2] Bar-Ilan University, Ramat Gan, Israel
lindell@biu.ac.il, nofdinar@gmail.com, oror.wn@gmail.com

Abstract. In this paper, we describe a new protocol for secure three-party computation of any functionality, with an honest majority and a *malicious* adversary. Our protocol has both an information-theoretic and computational variant, and is distinguished by extremely low communication complexity and very simple computation. We start from the recent semi-honest protocol of Araki et al. (ACM CCS 2016) in which the parties communicate only a single bit per AND gate, and modify it to be secure in the presence of malicious adversaries. Our protocol follows the paradigm of first constructing Beaver multiplication triples and then using them to verify that circuit gates are correctly computed. As in previous work (e.g., the so-called TinyOT and SPDZ protocols), we rely on the cut-and-choose paradigm to verify that triples are correctly constructed. We are able to utilize the fact that at most one of three parties is corrupted in order to construct an extremely simple and efficient method of constructing such triples. We also present an improved combinatorial analysis for this cut-and-choose which can be used to achieve improvements in other protocols using this approach.

1 Introduction

1.1 Background

In the setting of secure computation, a set of parties with private inputs wish to compute a joint function of their inputs, without revealing anything but the output. Protocols for secure computation guarantee *privacy* (meaning that the protocol reveals nothing but the output), *correctness* (meaning that the correct function is computed), and more. These security guarantees are provided in the presence of adversarial behavior. There are two classic adversary models that are typically considered: *semi-honest* (where the adversary follows the

---

Supported by the European Research Council under the ERC consolidators grant agreement n. 615172 (HIPS) and by the BIU Center for Research in Applied Cryptography and Cyber Security in conjunction with the Israel National Cyber Bureau in the Prime Minister's Office.

J.-S. Coron and J.B. Nielsen (Eds.): EUROCRYPT 2017, Part II, LNCS 10211, pp. 225–255, 2017.
DOI: 10.1007/978-3-319-56614-6_8

protocol specification but may try to learn more than allowed from the protocol transcript) and *malicious* (where the adversary can run any arbitrary polynomial-time attack strategy). In the *information-theoretic* model, security is obtained unconditionally and even in the presence of computationally unbounded adversaries. In contrast, in the *computational* model, security is obtained in the presence of polynomial-time adversaries and relies on cryptographic hardness assumptions.

Despite its stringent requirements, it has been shown that any polynomial-time functionality can be securely computed with computational security [3,12, 23] and with information-theoretic security [4,8,20]. These results hold both for semi-honest and malicious adversaries, but a two-thirds honest majority must be assumed in order to obtain information-theoretic security (or an honest majority when assuming broadcast).

There are two main approaches for constructing secure computation protocols: the secret-sharing approach (followed by [4,8,12]) works by having the parties interact for every gate of the circuit, whereas the garbled-circuit approach (followed by [3,23]) works by having the parties construct an encrypted version of the circuit which can be computed at once. Both approaches have importance and have settings where they perform better than the other. On the one hand, the garbled-circuit approach yields protocols with a constant number of rounds. Thus, in high-latency networks, they far outperform secret-sharing based protocols which have a number of rounds that is linear in the depth of the circuit being computed. On the other hand, protocols based on secret-sharing typically have low bandwidth, in contrast to garbled circuits that are large and costly in bandwidth. Given that the bandwidth is often the bottleneck, it follows that protocols with low communication have the potential to achieve much higher throughput.

1.2   Our Results

In this paper, we focus on the question of achieving secure computation in the presence of malicious adversaries with *very high throughput* on a *fast network* (without utilizing special-purpose hardware). We start with the recent *three-party* protocol of [1] that achieves security in the presence of semi-honest adversaries. The protocol requires transmitting only a single bit per AND gate, and the computation per gate is very simple. On a cluster of three 20-core servers with a 10 Gbs connection, the protocol of [1] achieves a rate of computation of *7 billion AND gates per second*. This can be used, for example, to securely compute 1.3 million AES block operations per second.

Our approach to achieving malicious security follows the Beaver multiplication triple approach [2] used in [5,9,19] (and many follow-up works). According to this approach, the parties securely generate shares of triples (a, b, c) where a, b are random and $c = ab$ (for the case of Boolean circuits, this is equivalent to $c = a \wedge b$). Such triples can then be used to *verify* that AND gates are computed correctly. In the (difficult) case of no honest majority considered in [5,9,19], there are two major challenges: **(a)** how to generate such triples without malicious parties causing either the shares to be invalid or causing $c \neq ab$, and **(b)** how to force the parties to send their "correct" values in the multiplication triple in the verification stage. The first problem is solved in [5,9,19] by using cut-and-choose: many triples are generated, some are opened, and the others are put in buckets and used to verify each other. The second problem is solved in [5,9,19] by using homomorphic MACs on all of the values. The generation of the triples to start with and the use of MACs adds additional overhead that is very expensive.

In this paper, we heavily utilize the fact that at most one party (out of 3) is corrupted in order to generate triples at very little expense, and to force the parties to send the correct values. In fact, the secret-sharing method used in [1] is such that it is possible to generate shares of random values *without any interaction* (using correlated randomness which is generated by the parties at almost no cost). Furthermore, we show how it is possible to detect if a malicious party sends incorrect values (in the prepared multiplication triples) when there is an honest majority, without requiring MACs of any kind. As a result, generating multiplication triples is very cheap. In turn, this enables us to generate a large number of triples at once, which further improves the parameters of the cut-and-choose step as well.

Overall, our protocol requires very simple computation, and achieves malicious security at very low communication cost. Specifically, with a statistical error of 2^{-40} each party needs to send only 10 bits per AND gate to one other party; for 2^{-80} this rises to only 16 bits per AND gate.

Based on the implementation results in [1], our estimates are that our new protocol should achieve a rate of over 500 million AND gates per second on the same setup as [1]. This is orders of magnitude faster than any other protocol achieving malicious security (see related work below).

1.3 Outline of Our Solution and Organization

In this section, we describe the different subprotocols and constructions that make up our protocol, and provide the high-level ideas behind our constructions.

In Sect. 2.1, we present the 2-out-of-3 secret-sharing scheme used in [1] and some important properties of it. Then, in Sect. 2.2 we describe the semi-honest protocol of [1] for multiplication (AND) gates. In addition, we prove a crucial property of this protocol that we heavily rely on in our construction: for any malicious adversary, the honest parties always hold a *valid sharing* after the multiplication protocol; the shared value may either equal the AND of the input (if the adversary follows the protocol) or its complement (if the adversary cheats).

In Sect. 2.3 we show how to generate correlated randomness (functionality $\mathcal{F}_{cr}$); after an initial exchange of keys for a pseudorandom function, the protocol is non-interactive. This makes it highly efficient, and also secure for malicious adversaries (since there is no interaction at all and so no way to cheat). In Sect. 2.4, we use $\mathcal{F}_{cr}$ to securely compute functionality $\mathcal{F}_{rand}$ that provides random shares to all parties. A very important feature of our protocol is based on the fact that $\mathcal{F}_{rand}$ can be securely computed *non-interactively* using correlated randomness. This means that the first step in generating multiplication triples – generating shares of random a, b via calls to $\mathcal{F}_{rand}$ – can be carried out non-interactively and thus at a very fast rate.

In Sect. 2.5, we use $\mathcal{F}_{rand}$ to carry out secure coin tossing (by generating shares of random values and then just opening them); each coin is generated by sending just a single bit. We explain how this is achieved, since it introduces a key technique that we use throughout. As we have mentioned, shares of a random value can be generated non-interactively and thus this is secure for malicious adversaries by default. However, when opening the shares to obtain the coin, a malicious adversary can cheat by sending an incorrect share value. Here we critically utilize the fact that we have an *honest majority*. In particular, we can simply have all pairs of parties send their shares to each other. Since the sharing is valid and any two parties can reconstruct the secret, each party can reconstruct separately based on the shares received from each other party, and compare. If the adversary cheats, then the result will be *different* reconstructed secrets, which will result in an abort. Our secret-sharing scheme has two bits and so this would cost each party sending 4 bits. However, we observe that in order to open it suffices to send 1 bit of each share only. Furthermore, we observe that if each party sends its bit to only one other party (P_1 to P_2, P_2 to P_3, and P_3 to P_1) then the bit sent by one honest party to another will result in the correct coin (there is always one such pair since only one party is corrupted). Thus, it actually suffices for each party to send its bit to only one other party and to record the result of the coin on a "public view" string. Then, at the end of the entire execution, before any output is revealed, the parties can compare their views by sending a collision-resistant hash of their local public view. If the two honest parties received a different coin at any point then they will have different local public views and so will abort before anything is revealed. As a result, coin-tossing can be achieved by each party sending just a single bit to one other party.

In Sects. 2.6, 2.7 and 2.8, we introduce additional functionalities needed for our protocol. First, in order to carry out the cut-and-choose, a random permutation must be applied to the tuples generated. This is carried out using $\mathcal{F}_{perm}$ (Sect. 2.6) which computes a random permutation of array indices. This functionality is easily realized by the parties just coin tossing the amount of randomness needed to define the permutation. In addition, in order for the parties to share inputs and obtain output, we need a way to deal shares and open shares that is secure for malicious adversaries. These are constructed in Sect. 2.7 ($\mathcal{F}_{reconst}$ for robustly reconstructing a secret to one party) and Sect. 2.8 ($\mathcal{F}_{share}$ for robustly sharing a value).

We now explain how the above subprotocols can be used to generate correct multiplication triples. The parties first call $\mathcal{F}_{rand}$ to generate shares of random values a, b and then run the semi-honest multiplication protocol of [1] to generate shares of c. As we have mentioned above, the semi-honest multiplication protocol has the property that even if the adversary is malicious, the shares of c are valid. However, if the adversary cheats then it may be the case that $c = ab \oplus 1$ instead of equalling ab. In order to prevent this from happening, the parties generate many triples and use some to check the others. Namely, the parties first randomly choose a subset of the triples which are opened to verify that indeed $c = ab$. This uses the subprotocol in Sect. 2.9 which carries out this exact check. Next, the remaining triples are partitioned randomly into buckets of size B (the random division is carried out using $\mathcal{F}_{\text{perm}}$). Then, in each bucket, $B - 1$ of the triples are used to verify that one of the triples is correct, except with negligible probability (without revealing the triple being checked). This uses the subprotocol of Sect. 2.10 which shows how to use one triple to verify that another is correct. This protocol is described in Sect. 3, and it securely computes functionality $\mathcal{F}_{\text{triples}}$ that generates an array of random multiplication triples for the parties.

Finally, we show how to securely compute any functionality f using random multiplication triples. Intuitively, this works by the parties running the semi-honest multiplication protocol for each AND gate and verifying each multiplication using a triple. The verification method, as used in [9,19], has the property that if a multiplication triple is good and the adversary cheats in the gate multiplication, then this is detected by the honest parties. As with all of our protocols, we take care to minimize the communication, and verify each gate by sending only 2 bits (beyond the single bit needed for the multiplication itself).

The efficiency of our construction relies heavily on the cut-and-choose parameters, both with respect to how many triples need to be opened and checked and the bucket size. In Sect. 5 we provide a tight analysis of this cut-and-choose game which yields a significant improvement over previous analyses for similar games in [5,19]. For concrete parameters that are suitable for our protocol, our analysis is approximately 25% better than [5,19].

Caveats. We stress that our protocol is specifically defined for the case of 3 parties only. This case is of interest for outsourced computations, as in the Sharemind business model [22], for two-party setting where a third auxiliary server can be used, and in other settings of interest as described in [1]. The generalization of our protocol to more parties is not straightforward since we rely on replicated secret sharing, and the size of such shares increases *exponentially* in the number of parties. In addition, our protocol is only secure *with abort*; this is unlike other protocols for the honest majority case that achieve fairness. Nevertheless, this is sufficient for many applications. For this setting, we are able to achieve security for *malicious adversaries* with efficiency way beyond any other known protocol.

1.4 Related Work

Most of the work on concretely-efficient secure computation has focused on the dishonest majority case. These protocols are orders of magnitude less efficient than ours, but deal with a much more difficult setting. For example, the best protocols based on garbled circuits for batch executions [17,21] require only sending 4 garbled circuits per execution. Even ignoring all of the additional work and communication (which is very significant), 4 garbled circuits per execution means sending 1000 bits per gate, which is 100 times the cost of our protocol. Likewise, the SPDZ/MASCOT protocol [14] communicates approximately 360 bits per gate for three parties, which is 36 times the cost of our protocol. The same is true for all other dishonest-majority protocols; c.f. [5,9,19].

In the setting of an honest majority, the only highly-efficient protocol with security for malicious adversaries that has been implemented, to the best of our knowledge, is that of [18]. We compare our protocol to [18] in detail in Sect. 6. Our protocol is more than an order of magnitude cheaper both in communication and computation; however, their protocol is constant-round and therefore better suited to slow networks.

1.5   Definition of Security

Our protocols are proven secure under the standard ideal/real simulation paradigm, for the case of malicious adversaries and with abort.

2   Building Blocks and Subprotocols

2.1   The Secret Sharing Scheme

We denote the three parties by P_1, P_2 and P_3. Throughout the paper, in order to simplify notation, when we use an index (say i) to denote the ith party, we will write $i-1$ and $i+1$ to mean the "previous" and "subsequent" party, respectively. That is, when $i = 1$ then P_{i-1} is P_3 and when $i = 3$ then P_{i+1} is P_1.

We use the 2-out-of-3 secret sharing scheme of [1], defined as follows. In order to share a bit v, the dealer chooses three random bits $s_1, s_2, s_3 \in \{0,1\}$ under the constraint that $s_1 \oplus s_2 \oplus s_3 = v$. Then:

- P_1's share is the pair (t_1, s_1) where $t_1 = s_3 \oplus s_1$.
- P_2's share is the pair (t_2, s_2) where $t_2 = s_1 \oplus s_2$.
- P_3's share is the pair (t_3, s_3) and $t_3 = s_2 \oplus s_3$.

It is clear that no single party's share reveals anything about v. In addition, any two parties can obtain v; e.g., given $(t_1, s_1), (t_2, s_2)$ one can compute $v = t_1 \oplus s_2$. We denote by $[v]$ a 2-out-of-3 sharing of the value v according to the above scheme.

Claim 2.1. *The secret v together with the share of one party fully determine the shares of the other parties.*

Proof: By the definition of the secret sharing scheme, it holds that $t_i = s_{i-1} \oplus s_i$. Since (t_i, s_i) for some $i \in \{1, 2, 3\}$ and v are determined, this determines both s_{i-1} and s_{i+1} as well. This follows since $s_{i-1} = t_i \oplus s_i$ and $s_{i+1} = v \oplus s_i \oplus s_{i-1}$. Thus, the shares of the other two parties are determined. ∎

Opening shares. We define a subprocedure, denoted $\mathsf{open}([v])$, for our secret sharing scheme, as follows. Denote the shares of v by $\left\{(t_i, s_i)\right\}_{i=1}^{i=3}$. Then, each party P_i sends t_i to P_{i+1}, and each P_i outputs $v = s_i \oplus t_{i-1}$.

Local operators for shares. We define the following local *operators* on shares:

- *Addition* $[v_1] \oplus [v_2]$: Given a share (t_i^1, s_i^1) of v_1 and a share (t_i^2, s_i^2) of v_2, each party P_i computes: $(t_i^1 \oplus t_i^2, s_i^1 \oplus s_i^2)$.
- *Multiplication by a scalar* $\sigma \cdot [v]$: Given a share (t_i, s_i) of v and a value $\sigma \in \{0,1\}$, each party P_i computes $(\sigma \cdot t_i, \sigma \cdot s_i)$.
- *Addition of a scalar* $[v] \oplus \sigma$: Given a share (t_i, s_i) of v and a value $\sigma \in \{0,1\}$, each party P_i computes $(t_i, s_i \oplus \sigma)$.
- *Complement* $\overline{[v]}$: Given a share (t_i, s_i) of v, each party P_i computes $(t_i, \overline{s_i})$ (where $\overline{b}$ is b's complement)

We stress that when writing $[v_1] \oplus [v_2]$ the symbol "$\oplus$" is an operator on shares and not bitwise XOR, whereas when we write $v_1 \oplus v_2$ the symbol "$\oplus$" *is* bitwise XOR; likewise for the product and complement notation. We now prove that these local operators achieve the expected results.

Claim 2.2. *Let* $[v_1], [v_2]$ *be shares and let* $\sigma \in \{0,1\}$ *be a scalar. Then, the following properties hold:*

1. $[v_1] \oplus [v_2] = [v_1 \oplus v_2]$
2. $\sigma \cdot [v_1] = [\sigma \cdot v_1]$
3. $[v_1] \oplus \sigma = [v_1 \oplus \sigma]$
4. $\overline{[v_1]} = [\overline{v_1}]$

Proof: Denote the shares of v_1 and v_2 by $\left\{(t_i^1, s_i^1)\right\}_{i=1}^{i=3}$ and $\left\{(t_i^2, s_i^2)\right\}_{i=1}^{i=3}$, respectively.

1. We prove that $[v_1] \oplus [v_2] = [v_1 \oplus v_2]$ by showing that $\{(t_i^1 \oplus t_i^2, s_i^1 \oplus s_i^2)\}_{i=1}^{i=3}$ is a valid sharing of $v_1 \oplus v_2$. First, observe that the s-parts are valid since $(s_1^1 \oplus s_1^2) \oplus (s_2^1 \oplus s_2^2) \oplus (s_3^1 \oplus s_3^2) = (s_1^1 \oplus s_2^1 \oplus s_3^1) \oplus (s_1^2 \oplus s_2^2 \oplus s_3^2) = v_1 \oplus v_2$. Furthermore, for every i, $(t_i^1 \oplus t_i^2) = (s_{i-1}^1 \oplus s_i^1) \oplus (s_{i-1}^2 \oplus s_i^2) = (s_{i-1}^1 \oplus s_{i-1}^2) \oplus (s_i^1 \oplus s_i^2)$ as required.
2. We prove that $\sigma \cdot [v_1] = [\sigma \cdot v_1]$ by showing that $\{(\sigma \cdot t_i^1, \sigma \cdot s_i^1)\}_{i=1}^{i=3}$ is a valid sharing of $\sigma \cdot v_1$. This is true since $\sigma \cdot s_1^1 \oplus \sigma \cdot s_2^1 \oplus \sigma \cdot s_3^1 = \sigma \cdot (s_1^1 \oplus s_2^1 \oplus s_3^1) = \sigma \cdot v_1$ and $\sigma \cdot t_i^1 = \sigma \cdot (s_{i-1}^1 \oplus s_i^1) = \sigma \cdot s_{i-1}^1 \oplus \sigma \cdot s_i^1$ as required.
3. We prove that $[v_1] \oplus \sigma = [v_1 \oplus \sigma]$ by showing that $\{(t_i^1, \sigma \oplus s_i^1)\}_{i=1}^{i=3}$ is a valid sharing of $\sigma \oplus v_1$. This is true since $(\sigma \oplus s_1^1) \oplus (\sigma \oplus s_2^1) \oplus (\sigma \oplus s_3^1) = \sigma \oplus (s_1^1 \oplus s_2^1 \oplus s_3^1) = \sigma \oplus v_1$ and $t_i^1 = s_{i-1}^1 \oplus s_i^1 = (\sigma \oplus s_{i-1}^1) \oplus (\sigma \oplus s_i^1)$.
4. We prove that $\overline{[v_1]} = [\overline{v_1}]$ by showing that $\{(t_i^1, \overline{s_i^1})\}_{i=1}^{i=3}$ is a valid sharing of $\overline{v}_1$. This holds since $\overline{s}_1^1 \oplus \overline{s}_2^1 \oplus \overline{s}_3^1 = s_1^1 \oplus s_2^1 \oplus s_3^1 = \overline{v}_1$ and $t_i^1 = s_{i-1}^1 \oplus s_i^1 = \overline{s}_{i-1}^1 \oplus \overline{s}_i^1$. ∎

Consistency. In the setting that we consider here, one of the parties may be maliciously corrupted and thus can behave in an arbitrary manner. Thus, if parties define their shares based on values received, it may be possible that the honest parties hold values that are not a valid sharing of any value. We therefore define the notion of *consistency of shares*. We stress that this definition relates only to the shares held by the honest parties, since the corrupted party can always change its local values. As we will show after the definition, shares are consistent if they define a unique secret v.

Definition 2.3. *Let* $(t_1, s_1), (t_2, s_2)$ *and* (t_3, s_3) *be the shares held by parties* P_1, P_2 *and* P_3 *respectively, and let* P_i *be the corrupted party. We say that the shares are* consistent *if it holds that* $s_{i+1} = s_{i+2} \oplus t_{i+2}$.

In order to understand the definition, recall that in a valid sharing of v it holds that $t_{i+2} = s_{i+1} \oplus s_{i+2}$. Thus, we obtain that $s_{i+1} = s_{i+1} \oplus s_{i+2} \oplus s_{i+2} = s_{i+2} \oplus t_{i+2}$ as the definition requires. The intuition behind this is that, in order to reconstruct the secret, the honest parties P_{i+1} and P_{i+2} need to learn t_i and t_{i+1} respectively. However, since t_i is held by the corrupted party, we use the fact that $t_i = t_{i+1} \oplus t_{i+2}$ to obtain that P_{i+1} can reconstruct the secret using t_{i+1} which it knows and t_{i+2} which is held by the other honest party. The definition says that computing the secret using P_{i+1}'s share and t_{i+2}; i.e., computing $s_{i+1} \oplus t_{i+1} \oplus t_{i+2}$, yields the same value as computing the secret using P_{i+2}'s share and t_{i+1}; i.e., computing $s_{i+2} \oplus t_{i+1}$. We stress that shares may be inconsistent. For example, if P_1 is the corrupted party and the shares of the honest parties P_2, P_3 are $(1,1)$ and $(1,1)$ respectively, then the shares are inconsistent since $s_2 = 1$ whereas $s_3 \oplus t_3 = 1 \oplus 1 = 0$. Thus, these shares cannot be the result of *any* sharing of *any* value.

2.2   Computing AND Gates – One Semi-honest Corrupted Party

We review the protocol for securely computing AND (equivalently, multiplication) gates for semi-honest adversaries from [1] as it will be used in a subprotocol in our protocol for malicious adversaries. This subprotocol requires each party to send a single bit only. The protocol works in two phases: in the first phase the parties compute a simple $\binom{3}{3}$ XOR-sharing of the AND of the input bits, and in the second phase they convert the $\binom{3}{3}$-sharing into the above-defined $\binom{3}{2}$-sharing.

Let $(t_1, s_1), (t_2, s_2), (t_3, s_3)$ be a secret sharing of v_1, and let $(u_1, w_1), (u_2, w_2)$, (u_3, w_3) be a secret sharing of v_2. We assume that the parties P_1, P_2, P_3 hold *correlated randomness* $\alpha_1, \alpha_2, \alpha_3$, respectively, where $\alpha_1 \oplus \alpha_2 \oplus \alpha_3 = 0$. The parties compute $\binom{3}{2}$-shares of $v_1 v_2 = v_1 \wedge v_2$ as follows:

1. **Step 1 – compute $\binom{3}{3}$-sharing:**
   (a) P_1 computes $r_1 = t_1 u_1 \oplus s_1 w_1 \oplus \alpha_1$, and sends r_1 to P_2.
   (b) P_2 computes $r_2 = t_2 u_2 \oplus s_2 w_2 \oplus \alpha_2$, and sends r_2 to P_3.
   (c) P_3 computes $r_3 = t_3 u_3 \oplus s_3 w_3 \oplus \alpha_3$, and sends r_3 to P_1.
   These messages are computed and sent in parallel.

2. **Step 2 – compute $\binom{3}{2}$-sharing:** In this step, the parties construct a $\binom{3}{2}$-sharing from their given $\binom{3}{3}$-sharing and the messages sent in the previous step. This requires local computation only.
   (a) P_1 stores (e_1, f_1) where $e_1 = r_1 \oplus r_3$ and $f_1 = r_1$.
   (b) P_2 stores (e_2, f_2) where $e_2 = r_2 \oplus r_1$ and $f_2 = r_2$.
   (c) P_3 stores (e_3, f_3) where $e_3 = r_3 \oplus r_2$ and $f_3 = r_3$.

It was shown in [1], that $f_1 \oplus f_2 \oplus f_3 = r_1 \oplus r_2 \oplus r_3 = v_1 v_2$. Thus, the obtained sharing is a consistent sharing of $v_1 v_2$. We now show something far stronger; specifically, we show that the above multiplication protocol (for *semi-honest adversaries*) always yields consistent shares, even when run in the presence of a *malicious adversary*. Depending on the adversary, the result is either a consistent sharing of the product or its complement, but it is *always* consistent.

Lemma 2.4. *If $[v_1]$ and $[v_2]$ are consistent and $[v_3]$ was generated by executing the (semi-honest) multiplication protocol on $[v_1]$ and $[v_2]$ in the presence of one malicious party, then $[v_3]$ is a consistent sharing of either $v_1 v_2$ or $v_1 v_2 \oplus 1$.*

Proof: If the corrupted party follows the protocol specification then $[v_3]$ is a consistent sharing of $v_1 v_2$. Else, since the multiplication protocol is symmetric, assume without loss of generality that P_1 is the corrupted party. Then, the only way that P_1 can deviate from the protocol specification is by sending $r_1 \oplus 1$ to the honest P_2 instead of r_1, and in this case P_2 will define its share to be $(e_2, f_2) = (r_2 \oplus r_1 \oplus 1, r_2)$. Meanwhile, P_3 defines its share to be $(e_3, f_3) = (r_3 \oplus r_2, r_3)$, as it receives r_2 from the honest P_2. Thus, $f_3 \oplus e_3 = r_3 \oplus (r_3 \oplus r_2) = r_2 = f_2$ meaning that $[v_3]$ is consistent by Definition 2.3. Furthermore, it is a sharing of $v_1 v_2 \oplus 1$ since $f_3 \oplus e_2 = r_3 \oplus (r_1 \oplus 1 \oplus r_2) = v_1 v_2 \oplus 1$ (utilizing the fact that $r_1 \oplus r_2 \oplus r_3 = v_1 v_2$). ∎

2.3   Generating Correlated Randomness – $\mathcal{F}_{cr}^1 / \mathcal{F}_{cr}^2$

Our protocol relies strongly on the use of random bits which are *correlated*. We define two types of correlated randomness:

- **Type 1:** Consider an ideal functionality $\mathcal{F}_{cr}^1$ that chooses $\alpha_1, \alpha_2, \alpha_3 \in \{0, 1\}$ at random under the constraint that $\alpha_1 \oplus \alpha_2 \oplus \alpha_3 = 0$, and sends α_i to P_i for every i.
- **Type 2:** Consider an ideal functionality $\mathcal{F}_{cr}^2$ that chooses $\alpha_1, \alpha_2, \alpha_3 \in \{0, 1\}$ at random, and sends (α_1, α_2) to P_1, (α_2, α_3) to P_2, and (α_3, α_2) to P_3.

Generating correlated randomness efficiently. It is possible to securely generate type-1 correlated randomness with perfect security by having each party P_j simply choose a random $\rho_j \in \{0, 1\}$ and send it to P_{j+1}. Then, each P_j defines $\alpha_j = \rho_j \oplus \rho_{j-1}$ (observe that $\alpha_1 \oplus \alpha_2 \oplus \alpha_3 = 0$ since each ρ-value appears twice). In order to compute type-2 correlated randomness, each party P_j sends a random ρ_j as before, but now each P_j outputs the pair (ρ_{j-1}, ρ_j). (Formally, the ideal functionalities $\mathcal{F}_{cr}^1 / \mathcal{F}_{cr}^2$ must be defined so that the corrupted party P_i

has some influence, but this suffices.) Despite the elegance and simplicity of this solution, we use a different approach that does not require any communication. Although the above involves sending just a single bit, this would actually *double* the communication per AND gate which is the bottleneck of efficiency.

Protocol 2.5 describes a method for securely compute correlated randomness *computationally* without *any interaction* beyond a short initial setup. Observe that in the output of $\mathcal{F}_{cr}^1$, it holds that $\alpha_1 \oplus \alpha_2 \oplus \alpha_3 = 0$. Furthermore, for every j, P_j does not know k_{j+1} which is used to generate α_{j+1} and α_{j+2}. Thus, α_{j+1} and α_{j+2} are pseudorandom to P_j, under the constraint that $\alpha_2 \oplus \alpha_3 = \alpha_1$. This was proven formally in [1] and the same proof holds for the malicious setting.

PROTOCOL 2.5 (Computing $\mathcal{F}_{cr}^1/\mathcal{F}_{cr}^2$)

- **Auxiliary input**: Each party holds a security parameter κ and a description of a pseudorandom function $F : \{0,1\}^\kappa \times \{0,1\}^\kappa \rightarrow \{0,1\}$.
- **Setup (executed once)**:
    1. Each party P_j chooses a random $k_j \in \{0,1\}^\kappa$.
    2. Each party P_j sends k_j to party P_{j+1}.
- **Generating randomness**: Upon input id,
    - *Computing $\mathcal{F}_{cr}^1$*: each party P_j computes $\alpha_j = F_{k_j}(id) \oplus F_{k_{j-1}}(id)$ and outputs it.
    - *Computing $\mathcal{F}_{cr}^2$*: each party P_j outputs $\left(F_{k_j}(id), F_{k_{j-1}}(id) \right)$.

Formally defining the $\mathcal{F}_{cr}^1/\mathcal{F}_{cr}^2$ ideal functionalities. A naive definition would be to have the ideal functionality choose $\alpha_1, \alpha_2, \alpha_3$ and send α_j to P_j for $= j \in \{1,2,3\}$ (or send α_j, α_{j-1} to P_i in the $\mathcal{F}_{cr}^2$ functionality). However, securely realizing such a functionality would require a full-blown coin tossing protocol. In order to model our non-interactive method, which suffices for our protocol, we need to take into account that the corrupted party P_i can choose its k_i and this influences the output, as P_i's value is generated in a very specific way using a pseudorandom function. In order for the view of the corrupted party to be like in the real protocol, we define the functionality $\mathcal{F}_{cr}^1/\mathcal{F}_{cr}^2$ so that they generate the corrupted party's value in this exact same way.

The functionalities are described formally in Functionality 2.6. The corrupted party chooses two keys k, k' for the pseudorandom function F and sends them to the functionality. These keys are used to generate the values that are influenced by the corrupted party, whereas the other values are chosen uniformly. We denote by κ the computational security parameter, and thus the length of the keys k, k'.

The following is proved in the full version of our paper.

Proposition 2.7. *If F is a pseudorandom function, then Protocol 2.5 securely computes functionalities $\mathcal{F}_{cr}^1$ and $\mathcal{F}_{cr}^2$, respectively, with abort in the presence of one malicious party.*

FUNCTIONALITY 2.6 ($\mathcal{F}_{cr}^1/\mathcal{F}_{cr}^2$ – correlated randomness)

Let $F : \{0,1\}^\kappa \times \{0,1\}^\kappa \to \{0,1\}$ be a keyed function. Upon invocation, the adversary controlling party P_i chooses a pair of keys $k, k' \in \{0,1\}^\kappa$ and sends them to $\mathcal{F}_{cr}^1/\mathcal{F}_{cr}^2$. Then:

- $\mathcal{F}_{cr}^1$: Upon receiving input id from all parties, functionality $\mathcal{F}_{cr}^1$ computes $\alpha_i = F_k(id) \oplus F_{k'}(id)$ and chooses random values $\alpha_{i-1}, \alpha_{i+1} \in \{0,1\}$ under the constraint that $\alpha_1 \oplus \alpha_2 \oplus \alpha_3 = 0$. $\mathcal{F}_{cr}^1$ sends α_j to P_j for every j.
- $\mathcal{F}_{cr}^2$: Upon receiving input id from all parties, functionality $\mathcal{F}_{cr}^2$ computes $\alpha_i = F_k(id)$ and $\alpha_{i-1} = F_{k'}(id)$ and chooses a random value $\alpha_{i+1} \in \{0,1\}$. $\mathcal{F}_{cr}^2$ sends (α_{j-1}, α_j) to P_j for every j.

2.4   Generating Shares of a Random Value – $\mathcal{F}_{rand}$

In this section, we show how the parties can generate a sharing of a random secret value v known to none of them. Formally, we define the functionality $\mathcal{F}_{rand}$ that chooses a random $v \in \{0,1\}$, computes a sharing $[v]$, and sends each party its share of $[v]$. However, $\mathcal{F}_{rand}$ allows the corrupted party to determine its own share, and thus computes the honest parties' shares from the corrupted party's share and the randomly chosen v. $\mathcal{F}_{rand}$ is formally specified in Functionality 2.8.

FUNCTIONALITY 2.8 ($\mathcal{F}_{rand}$ – generating shares of a random value)

- $\mathcal{F}_{rand}$ receives (t_i, s_i) from the corrupted party P_i.
- $\mathcal{F}_{rand}$ chooses a random $v \in \{0,1\}$ and defines the respective shares $(t_{i-1}, s_{i-1}), (t_{i+1}, s_{i+1})$ of P_{i-1}, P_{i+1} based on (t_i, s_i) and v (as described in Claim 2.1).
- $\mathcal{F}_{rand}$ sends (t_{i-1}, s_{i-1}) to P_{i-1}, and sends (t_{i+1}, s_{i+1}) to P_{i+1}.

Protocol 2.9 describes how to securely compute $\mathcal{F}_{rand}$ in the $\mathcal{F}_{cr}^2$-hybrid model, without any interaction.

PROTOCOL 2.9 (Securely computing $\mathcal{F}_{rand}$)

1. The parties call $\mathcal{F}_{cr}^2$ and receive (r_3, r_1), (r_1, r_2), and (r_2, r_3), respectively.
2. P_1 defines $t_1 = r_3 \oplus r_1$ and $s_1 = r_1$.
3. P_2 defines $t_2 = r_1 \oplus r_2$ and $s_2 = r_2$.
4. P_3 defines $t_3 = r_2 \oplus r_3$ and $s_3 = r_3$.

Observe that $t_1 \oplus t_2 \oplus t_3 = 0$. Furthermore, define $v = s_1 \oplus t_3 = r_1 \oplus r_2 \oplus r_3$. Observe that $s_2 \oplus t_1$ and $s_3 \oplus t_2$ also both equal the same v. Thus, this non-interactive protocol defines a valid sharing $[v]$ for a random $v \in \{0,1\}$. The fact that v is random follows from the fact that it equals $r_1 \oplus r_2 \oplus r_3$. Now, by the definition of $\mathcal{F}_{\mathrm{cr}}^2$, a corrupted P_i knows nothing of $r_{i+1} = \alpha_{i+1}$ which is chosen uniformly at random, and thus the defined sharing is of a random value.

Proposition 2.10. *Protocol 2.9 securely computes functionality* $\mathcal{F}_{\mathrm{rand}}$ *with abort in the* $\mathcal{F}_{\mathrm{cr}}^2$*-hybrid model, in the presence of one malicious party.*

Proof: Let $\mathcal{A}$ be a real adversary; we define $\mathcal{S}$ as follows:

- $\mathcal{S}$ receives $\mathcal{A}$'s input k, k' to $\mathcal{F}_{\mathrm{cr}}^2$.
- Upon receiving id from $\mathcal{A}$ as intended for $\mathcal{F}_{\mathrm{cr}}^2$, simulator $\mathcal{S}$ simulates $\mathcal{A}$ receiving back $(r_{i-1}, r_i) = (F_{k'}(id), F_k(id))$ from $\mathcal{F}_{\mathrm{cr}}^2$.
- $\mathcal{S}$ defines $t_i = r_{i-1} \oplus r_i$ and $s_i = r_i$, and externally sends (t_i, s_i) to $\mathcal{F}_{\mathrm{rand}}$.
- $\mathcal{S}$ outputs whatever $\mathcal{A}$ outputs.

We show that the joint distribution of the outputs of $\mathcal{S}$ and the honest parties in an ideal execution is identical to the outputs of $\mathcal{A}$ and the honest parties in a real execution. In order to see this, observe that in a real execution, given a fixed r_{i-1}, r_i (as viewed by the adversary), the value v is fully determined by r_{i+1}. In particular, by the definition of the secret-sharing scheme, $v = s_i \oplus t_{i-1} = r_i \oplus r_{i-2} \oplus r_{i-1} = r_1 \oplus r_2 \oplus r_3$. Since r_{i+1} is randomly generated by $\mathcal{F}_{\mathrm{cr}}^2$, this has the same distribution as $\mathcal{F}_{\mathrm{rand}}$ choosing v randomly (because choosing v randomly, or choosing some r randomly and setting $v = t_i \oplus r$ is identical). Thus, the joint distributions are identical. ∎

2.5   Coin Tossing – $\mathcal{F}_{\mathrm{coin}}$

We now present a highly-efficient three-party coin tossing protocol that is secure in the presence of one malicious adversary. We define the functionality $\mathcal{F}_{\mathrm{coin}}$ that chooses s random bits $v_1, \dots, v_s \in \{0, 1\}$ and sends them to each of the parties. The idea behind our protocol is simply for the parties to invoke s calls to $\mathcal{F}_{\mathrm{rand}}$ and to then open the result (by each P_i sending t_i to P_{i+1}; see Sect. 2.1). Observe that this in itself is not sufficient since a malicious party may send an *incorrect opening*, resulting in the honest parties receiving different output. This can be solved by using a subprocedure called compareview() in which each party P_j sends its output to party P_{j+1}. If any party receives a different output, then it aborts. The reason why this is secure is that the protocol guarantees that if P_i is corrupted then P_{i+2} receives the correct outputs $v_1, \dots, v_s$; this holds because when opening the shares, the only values received by P_{i+2} are sent by the honest P_{i+1} and are not influenced by P_i. Thus, P_{i+2}'s output is guaranteed to be correct, and if P_{i+1} and P_{i+2} have the same output then P_{i+1}'s output is also correct. This is formally described in Protocol 2.11.

PROTOCOL 2.11 (Securely computing $\mathcal{F}_{\text{coin}}$)

1. The parties invoke s calls to $\mathcal{F}_{\text{rand}}$; denote their outputs by $[v_1], ..., [v_s]$.
2. For every $j \in \{1, .., s\}$, the parties run the open($[v_j]$) procedure defined in Sect. 2.1 to obtain v_j.
3. The parties run compareview($v_1, \ldots, v_s$) by each P_j sending the outputs $v_1, \ldots, v_s$ to P_{j+1}. If a party receives different output, then it outputs $\perp$. Otherwise, it outputs $v_1, \ldots, v_s$.

Proposition 2.12. *Protocol 2.11 securely computes functionality $\mathcal{F}_{\text{coin}}$ with abort in the $\mathcal{F}_{\text{rand}}$-hybrid model, in the presence of one malicious party.*

Proof: Let $\mathcal{A}$ be the real adversary controlling P_i; we construct the simulator $\mathcal{S}$:

1. $\mathcal{S}$ receives $v_1, \ldots, v_s$ from the trusted party computing $\mathcal{F}_{\text{coin}}$.
2. $\mathcal{S}$ invokes $\mathcal{A}$ and simulates s calls to $\mathcal{F}_{\text{rand}}$, as follows:
   (a) $\mathcal{S}$ receives P_i's share in every call to $\mathcal{F}_{\text{rand}}$.
   (b) Given $v_1, \ldots, v_s$ and P_i's shares, $\mathcal{S}$ computes the value t_{i-1} that $\mathcal{A}$ should receive from P_{i-1}. (Specifically, for the ℓth value, let (t_i^ℓ, s_i^ℓ) be P_i's share and let v_ℓ be the bit received from $\mathcal{F}_{\text{coin}}$. Then, $\mathcal{S}$ sets $t_{i-1} = v_\ell \oplus s_i^\ell$. This implies that the "opening" is to v_ℓ.)
   (c) If $\mathcal{A}$ sends an incorrect t_i value in any open procedure, then $\mathcal{S}$ sends abort$_{i+1}$ to $\mathcal{F}_{\text{coin}}$ causing P_{i+1} to abort in the ideal model. Otherwise, it sends continue$_{i+1}$ to $\mathcal{F}_{\text{coin}}$. (In all cases it sends continue$_{i-1}$ to $\mathcal{F}_{\text{coin}}$ since P_{i-1} never aborts.)

By the way $\mathcal{F}_{\text{rand}}$ is defined, the output distribution of $\mathcal{A}$ and the honest parties in a real execution is identical to the output distribution of $\mathcal{S}$ and the honest parties in an ideal execution. This is because each v_j is uniformly distributed, and $\mathcal{S}$ can fully determine the messages that $\mathcal{A}$ receives for any fixed $v_1, \ldots, v_s$. ∎

Deferring compareview. Our protocol has the property that nothing is revealed to any party until the end, even if a party behaves maliciously. As such, it will suffice for us to run compareview only once at the very end of the protocol before outputs are revealed. This enables us to have the parties compare their views by only sending a *collision-resistant hash* of their outputs, thereby reducing communication. As we will see below, this method will be used in a number of places, and all compareview operations will be done together.

2.6   Random Shuffle – $\mathcal{F}_{\text{perm}}$

In our protocol, we will need to compute a random permutation of an array of elements (where each element is a "multiplication triple"). Let $\mathcal{F}_{\text{perm}}$ be an ideal functionality that receives a vector $\boldsymbol{d}$ of length M from all parties, chooses a random permutation π over $\{1, ..., M\}$ and returns the vector $\boldsymbol{d}'$ defined by $\boldsymbol{d}'[i] = \boldsymbol{d}[\pi[i]]$ for every $i \in \{1, ..., M\}$. Functionality $\mathcal{F}_{\text{perm}}$ can be securely

---

PROTOCOL 2.13 (Securely computing $\mathcal{F}_{\text{perm}}$)

All parties hold the same input d, and work as follows:

1. For $j = 1$ to M:
   (a) The parties call $\mathcal{F}_{\text{coin}}$ enough times to generate a random index $i \in \{j, \ldots, M\}$.
   (b) Each party swaps $d[j]$ and $d[i]$.
2. Each party output the resulting vector d.

---

computed by the parties running the Fisher-Yates shuffle algorithm [10], and obtaining randomness via $\mathcal{F}_{\text{coin}}$. This is formally described in Protocol 2.13.

The following proposition follows trivially from the security of $\mathcal{F}_{\text{coin}}$:

Proposition 2.14. *Protocol 2.13 securely computes functionality $\mathcal{F}_{\text{perm}}$ with abort in the $\mathcal{F}_{\text{coin}}$-hybrid model, in the presence of one malicious party.*

2.7 Reconstruct a Secret to One of the Parties – $\mathcal{F}_{\text{reconst}}$

In this section we show how the parties can open a consistent sharing $[v]$ of a secret v to *one* of the parties in a secure way. We will use this subprotocol for reconstructing the outputs in our protocol. We remark that we consider security with abort only, and thus the party who should receive the output may abort. We stress that this procedure is fundamentally different to the open procedure of Sect. 2.1 in two ways: first, only one party receives output; second, the open procedure does not guarantee correctness. In contrast, here we ensure that the party either receives the correct value or aborts. We stress, however, that the protocol is only secure if the sharing $[v]$ is consistent; otherwise, nothing is guaranteed. We formally define $\mathcal{F}_{\text{reconst}}$ in Functionality 2.15.

FUNCTIONALITY 2.15 ($\mathcal{F}_{\text{reconst}}$ – secure reconstruction)

Let $\mathcal{S}$ be the adversary and P_i the corrupted party. $\mathcal{F}_{\text{reconst}}$ receives (t_{i+1}, s_{i+1}, j) from P_{i+1} and (t_{i+2}, s_{i+2}, j) from P_{i+2}, and works as follows:

- $\mathcal{F}_{\text{reconst}}$ computes $v = s_{i+2} \oplus t_{i+1}$ and sends v to P_j. In addition, $\mathcal{F}_{\text{reconst}}$ sends (t_i, s_i) to $\mathcal{S}$ (where (t_i, s_i) is P_i's share as defined by the shares received from the honest parties).

Note that $\mathcal{F}_{\text{reconst}}$ also sends P_i's share to $\mathcal{S}$. This is needed technically in the proof to enable simulation; it reveals nothing since this is the corrupted party's share anyway. Also, observe that the output is determined solely by the honest parties' shares; this guarantees that the corrupted party cannot influence the output beyond causing an abort.

We show how to securely compute $\mathcal{F}_{\text{reconst}}$ in Protocol 2.16. Intuitively, the protocol works by the parties sending their shares to P_j who checks that they are consistent, and reconstructs if yes. In order to reduce communication, we actually show that it suffices for the parties to send only the "t" parts of their shares.

PROTOCOL 2.16 (Reconstruct a Secret to One Party)

- **Inputs:** The parties hold a sharing $[v]$ and an index $j \in \{1, 2, 3\}$.
- **The protocol:**
    1. Parties P_{j+1} and P_{j-1} send t_{j+1} and t_{j-1}, respectively, to P_j.
    2. Party P_j checks that $t_j = t_{j+1} \oplus t_{j+2}$. If yes, it outputs $v = s_j \oplus t_{j-1}$; otherwise, it outputs $\perp$.

Proposition 2.17. *If the honest parties' inputs shares are consistent as in Definition 2.3, then Protocol 2.16 securely computes $\mathcal{F}_{\text{reconst}}$ with abort, in the presence of one malicious party.*

Proof: Let $\mathcal{A}$ be the real adversary controlling P_i, and assume that the honest parties' shares are consistent. We first consider the case that P_j is corrupt (i.e., $i = j$). In this case, the simulator $\mathcal{S}$ receives v and $(t_i, s_i) = (t_j, s_j)$ from $\mathcal{F}_{\text{reconst}}$. These values fully define all other shares, and in particular the values t_{j+1} and t_{j-1}. Thus, $\mathcal{S}$ can simulates P_{j+1} and P_{j-1} sending t_{j+1} and t_{j-1} to P_j.

We next consider the case that P_j is honest (i.e., $j \neq i$). In this case, $\mathcal{S}$ receives P_i's share (t_i, s_i) from $\mathcal{F}_{\text{reconst}}$. Then, $\mathcal{S}$ invokes $\mathcal{A}$ and receives the bit t_i' that P_i would send to P_j. If $t_i' = t_i$ (where t_i is the correct share value as received from $\mathcal{F}_{\text{reconst}}$), then $\mathcal{S}$ sends continue$_j$ to $\mathcal{F}_{\text{reconst}}$ so that the honest P_j receives v. Otherwise, $\mathcal{S}$ sends abort$_j$ to $\mathcal{F}_{\text{reconst}}$ so that the honest P_j outputs $\perp$. Observe that in a real protocol P_j aborts unless $t_1 \oplus t_2 \oplus t_3 = 0$ (which is equivalent to $t_j = t_{j+1} \oplus t_{j+2}$). Thus, if the corrupted party sends an incorrect t_i value, then P_j will certainly abort. In contrast, if the adversary controlling P_i sends the correct t_i, then the output will clearly be the correct v, again as in the ideal execution with $\mathcal{S}$. ∎

2.8 Robust Sharing of a Secret – $\mathcal{F}_{\text{share}}$

In this section, we show how to share a secret that is held by one of the parties who may be corrupt. This sub-protocol will be used to share the parties' inputs in the protocol. We define $\mathcal{F}_{\text{share}}$ in Functionality 2.18 We note that the corrupted party always provides its share as input, as in $\mathcal{F}_{\text{rand}}$. In addition, the dealer provides v and the (honest) parties receive their correct shares as defined by these values.

We show how to securely compute $\mathcal{F}_{\text{share}}$ in Protocol 2.19. The idea behind the protocol is to first generate a random sharing via $\mathcal{F}_{\text{rand}}$ which guarantees a *consistent* sharing of a random value (recall that this requires no communication). Next, the parties reconstruct the shared secret to the dealer, who can then send a single bit to "correct" the random share to its actual input. This ensures that the honest parties hold consistent shares, as long as a corrupt dealer sent

---

FUNCTIONALITY 2.18 (Functionality $\mathcal{F}_{\text{share}}$ – sharing a secret)

Let P_j be the party playing the dealer, and let P_i be the corrupted party:

- The corrupted P_i sends (t_i, s_i) to $\mathcal{F}_{\text{share}}$.
- The dealer P_j sends v to $\mathcal{F}_{\text{share}}$.
- $\mathcal{F}_{\text{share}}$ computes (t_{j+1}, s_{j+1}) and (t_{j+2}, s_{j+2}) from (t_i, s_i) and v (as described in Claim 2.1) and sends the honest P_{i-1} and P_{i+1} their respective shares.

PROTOCOL 2.19 (Robust Sharing of a Secret)

- **Inputs:** Party P_j holds a bit $v \in \{0, 1\}$.
- **The protocol:**
  1. The parties call $\mathcal{F}_{\text{rand}}$ to obtain $[a]$ for a random $a \in \{0, 1\}$.
  2. The parties call $\mathcal{F}_{\text{reconst}}$ with $[a]$ and j as its inputs, and so P_j receives a. If P_j receives $\bot$, it sends $\bot$ to all other parties and halts. Else, it proceeds to the next step.
  3. Party P_j sends $b = a \oplus v$ to the other parties.
  4. The parties run compareview(b) by each P_j sending the bit b to P_{j+1}. If any party sees different b values, then it sends $\bot$ to all other parties and halts.
  5. The parties each set their share $[v] = [a] \oplus b$ (using the operator defined in Sect. 2.1).
- **Output**: Each party outputs its share in $[v]$.

---

the same bit to both; this is enforced by the parties comparing to ensure that they received the same bit from the dealer.

The following is proven in the full version of this paper.

Proposition 2.20. *Protocol 2.19 securely computes $\mathcal{F}_{\text{share}}$ with abort in the $(\mathcal{F}_{\text{rand}}, \mathcal{F}_{\text{reconst}})$-hybrid mode, in the presence of one malicious party.*

Deferring compareview. As in Sect. 2.5, the compareview step can be deferred to the end of the execution (before any output is revealed). When using this mechanism, the bits to be compared are simply added to the parties local view and stored, and they are compared at the end.

2.9 Triple Verification with Opening

A multiplication triple is a triple of shares $([a], [b], [c])$ with the property that $c = a \cdot b$. Our protocol works by constructing triples and verifying that indeed $c = a \cdot b$. We begin by defining what it means for such a triple to be correct.

Definition 2.21. $([a], [b], [c])$ *is a* correct *multiplication triple if* $[a], [b]$ *and* $[c]$ *are consistent sharings, and* $c = a \cdot b$.

In our main protocol for secure computation, the parties will generate multiplication triples in two steps:

1. The parties generate random sharings $[a]$ and $[b]$ by calling $\mathcal{F}_{\mathrm{rand}}$ twice.
2. The parties run the semi-honest multiplication protocol described in Sect. 2.2 to obtain $[c]$.

Recall that by Lemma 2.4, the sharing $[c]$ is *always* consistent. However, if one of the parties is malicious, then it may be that $c = ab \oplus 1$. Protocol 2.22 describes a method of verifying that a triple is correct. The protocol is very simple and is based on the fact that if the shares are consistent and $c \neq ab$, then one of the honest parties will detect this in the standard open procedure defined in Sect. 2.1. This protocol is called verification "with opening" since the values a, b, c are revealed.

PROTOCOL 2.22 (Triple Verification With Opening)

- **Inputs:** The parties holds the triple $([a], [b], [c])$.
- **The protocol:**
    1. The parties run the procedures $\mathsf{open}([a]), \mathsf{open}([b])$ and $\mathsf{open}([c])$. Denote the output of party P_j from the three procedures by a_j, b_j and c_j respectively.
    2. Each party P_j checks that $c_j = a_j \cdot b_j$. If no, it sends $\perp$ to both parties and aborts.
    3. If no $\perp$ message is received, each party outputs accept.

Lemma 2.23. *If $[a], [b], [c]$ are consistent shares, but $([a], [b], [c])$ is not a correct multiplication triple, then both honest parties output $\perp$ in Protocol 2.22.*

Proof: Let P_i be the corrupted party, and assume that $[a], [b], [c]$ are consistent shares, but $([a], [b], [c])$ is not a correct multiplication triple. This implies that $c = a \cdot b \oplus 1$. Therefore, in the open procedures, party P_{i+2} will receive values $a_{i+2}, b_{i+2}, c_{i+2}$ such that $c_{i+2} \neq a_{i+2} \cdot b_{i+2}$, and will send $\perp$ to both parties. (This holds since P_{i+2} receives messages only from P_{i+1} that are independent of what P_i sends.) Thus, both honest parties will output $\perp$. ∎

2.10 Triple Verification Using Another (Without Opening)

We have seen how to check a triple by opening it and revealing its values a, b, c. Such a triple can no longer be used in the protocol. In this section, we show how to verify that a multiplication triple is consistent *without opening it*, by using (and wasting) an additional random multiplication triple that is assumed to be consistent. The method is described in Protocol 2.24. The idea behind the protocol is as follows. Given shares of x, y, z and of a, b, c, the parties compute and open shares of $\rho = x \oplus a$ and $\sigma = y \oplus b$; these values reveal nothing about x and y since a, b are both random. As we will show in the proof below, if one of (x, y, z) or (a, b, c) is correct and the other is incorrect (e.g., $x \neq y \cdot z$ but

242 J. Furukawa et al.

$c = a \cdot b$) then $z + c + \sigma \cdot a + \rho \cdot b + \rho \cdot \sigma = 1$. Thus, this value can be computed and opened by the parties. If x, y, z is incorrect and a, b, c is correct, then the honest parties will detect this and abort. In order to save on communication, we observe that if the value to be opened must equal 0 then it must hold that $s_j = t_{j-1}$. Thus, it suffices for the parties to compare a single bit.

PROTOCOL 2.24 (Triple Verif. Using Another Without Opening)

- **Inputs:** The parties hold a triple $([x], [y], [z])$ to verify and an additional triple $([a], [b], [c])$.
- **The protocol:**
    1. Each party locally computes $[\rho] = [x] \oplus [a]$ and $[\sigma] = [y] \oplus [b]$.
    2. The parties run open($[\rho]$) and open($[\sigma]$), as defined in Sect. 2.1. Denote by ρ_j and σ_j the respective output received by P_j in the openings.
    3. The parties run compareview(ρ_j, σ_j) by each P_j sending (ρ_j, σ_j) to P_{j+1}. If a party sees different values, then it sends $\perp$ to all parties and outputs $\perp$.
    4. Each party P_j computes $[z] \oplus [c] \oplus \sigma_j \cdot [a] \oplus \rho_j \cdot [b] \oplus \rho_j \cdot \sigma_j$. Denote by (t_j, s_j) the result of the computation held by party P_j.
    5. The parties run compareview(t_j) by each P_j sending t_j to P_{j+1}. Upon receiving t_{j-1} from P_{j-1}, party P_j checks that $s_j = t_{j-1}$. If yes, it outputs accept; else, it sends $\perp$ to all other parties and outputs $\perp$.
    6. If no abort messages are received, then output accept.

Lemma 2.25. *If $([a], [b], [c])$ is a correct multiplication triple and $[x], [y], [z]$ are consistent shares, but $([x], [y], [z])$ is not a correct multiplication triple, then all honest parties output $\perp$ in Protocol 2.24.*

Proof: Let P_i be the corrupted party. Assume that $([a], [b], [c])$ is a correct multiplication triple, that $[x], [y], [z]$ are consistent sharings, but $([x], [y], [z])$ is not a correct multiplication triple. This implies that all values a, b, c, x, y, z are well defined (from the honest parties' shares) and that $c = ab$ and $z \neq xy$.

Let $\rho = x \oplus a$ and $\sigma = y \oplus b$. If P_{j+1} receives an incorrect bit from P_j in the openings of ρ and σ (i.e., if $\rho_j \neq \rho$ or $\sigma_j \neq \sigma$) then it detects this in compareview of Step 3 with P_{j+2} and thus both honest parties output $\perp$. (Observe that P_{j+2} receives the openings from P_{j+1} who is honest and thus it is guaranteed that $\rho_{j+1} = \rho$ and $\sigma_{j+1} = \sigma$.)

We now show that if P_{i+1} and P_{i+2} did not output $\perp$ in Step 3 (and thus $\sigma_{i+1} = \sigma_{i+2} = \sigma$ and $\rho_{i+1} = \rho_{i+2} = \rho$), then P_{i+1} and P_{i+2} output $\perp$ with probability 1 in Step 5. In order to show this, we first show that in this case, $[z] \oplus [c] \oplus \sigma_j \cdot [a] \oplus \rho_j \cdot [b] \oplus \rho_j \cdot \sigma_j = [1]$. Observe that $z \neq xy$ and thus $z = xy \oplus 1$, and that $\sigma = y \oplus b$ and $\rho = x \oplus a$. Thus, we have:

$[z] \oplus [c] \oplus \sigma[a] \oplus \rho[b] \oplus \rho\sigma$

$\quad = [xy \oplus 1] \oplus [c] \oplus (y \oplus b)[a] \oplus (x \oplus a)[b] \oplus (x \oplus a)(y \oplus b)$

$\quad = [xy \oplus 1] \oplus [c] \oplus [(y \oplus b)a] \oplus [(x \oplus a)b] \oplus (xy \oplus ay \oplus xb \oplus ab)$

$\quad = [xy \oplus 1 \oplus c \oplus (y \oplus b)a \oplus (x \oplus a)b \oplus xy \oplus ay \oplus xb \oplus ab]$

$\quad = [xy \oplus 1 \oplus c \oplus ya \oplus ba \oplus xb \oplus ab \oplus xy \oplus ay \oplus xb \oplus ab]$

$\quad = [1]$

where the last equality follows from simple cancellations and the fact that by the assumption $c = ab$. We therefore have that the honest parties hold a consistent sharing of 1. Denoting the respective shares of P_{i+1} and P_{i+2} by (t_{i+1}, s_{i+1}) and (t_{i+2}, s_{i+2}), by the definition of the secret-sharing scheme we have that $s_{i+2} = t_{i+1} \oplus 1$ and so $s_{i+2} \neq t_{i+1}$. This implies that P_{i+2} sends $\bot$ to all other parties in Step 5 of the protocol, and all output $\bot$. ∎

Deferring compareview: In the first compareview, all parties include ρ_j, σ_j in their view. In the second compareview, P_j includes t_j in its joint view with P_{j+1} and includes s_j in its joint view with P_{j-1}. Observe that the requirement that $s_j = t_{j-1}$ is automatically fulfilled by requiring that the pairwise views be the same. This holds since P_j include s_j in its view with P_{j-1}, whereas P_{j-1} includes t_{j-1} in its view with P_j. As a consequence, in the protocol when compareview is deferred, each party stores two strings – one for its joint view with each of the other parties – and hashes these two strings separately at the end of the protocol. We remark that it is possible to use only a universal hash function by choosing the function after the views have been fixed, if this is desired. Recall that in compareview, it suffices for each party to send its view to one other party. Thus, all communication in our protocol follows the pattern that P_i sends messages to P_{i+1} only, for every $i \in \{1, 2, 3\}$.

3   Secure Generation of Multiplication Triples – $\mathcal{F}_{\text{triples}}$

In this section, we present a three-party protocol for generating an array of correct multiplication triples, as defined in Definition 2.21. Formally, we securely compute the functionality $\mathcal{F}_{\text{triples}}$ defined in Functionality 3.1.

We show how to securely compute $\mathcal{F}_{\text{triples}}$ in Protocol 3.2. The idea behind the protocol is as follows. The parties first use $\mathcal{F}_{\text{rand}}$ to generate many shares of *pairs* of random values $[a_i], [b_i]$. Next, they run the semi-honest multiplication protocol of Sect. 2.2 to compute shares of $c_i = a_i \cdot b_i$. However, since the multiplication protocol is only secure for semi-honest parties, a malicious adversary can cheat in this protocol. We therefore utilize the fact that even if the malicious adversary cheats, the resulting shares $[c_i]$ is *consistent*, but it may be the case that $c_i = a_i b_i \oplus 1$ instead of $c_i = a_i b_i$ (see Lemma 2.4). We therefore use *cut-and-choose* to check that the triples are indeed correct. We do this by opening C triples using Protocol 2.22; this protocol provides a full guarantee that the parties detect any incorrect triple that is opened. Next, the parties randomly divide the remaining

---

FUNCTIONALITY 3.1 ($\mathcal{F}_{\text{triples}}$ – **generating multiplication triples**)
Let P_i be the corrupted party. Upon receiving N from P_1, P_2, P_3, and receiving N triples of pairs $\left\{ (t_{a_i}^j, s_{a_i}^j), (t_{b_i}^j, s_{b_i}^j), (t_{c_i}^j, s_{c_i}^j) \right\}_{j=1}^N$ from P_i, functionality $\mathcal{F}_{\text{triples}}$ works as follows:

- For $j = 1, \ldots, N$, $\mathcal{F}_{\text{triples}}$ chooses random $a_j, b_j \in \{0,1\}$ and computes $c_j = a_j b_j$.
- For $j = 1, \ldots, N$, $\mathcal{F}_{\text{triples}}$ defines a vector of sharings $\boldsymbol{d} = ([a_j], [b_j], [c_j])$. The sharings are computed from $\left[(t_{a_i}^j, s_{a_i}^j), (t_{b_i}^j, s_{b_i}^j), (t_{c_i}^j, s_{c_i}^j) \right]$ provided by P_i and the chosen a_j, b_j, c_j (as in Claim 2.1).
- $\mathcal{F}_{\text{triples}}$ sends each party its shares in all of the generated shares.

---

triples into "buckets" of size B and use Protocol 2.24 to verify that the first triple in the bucket is correct. Recall that in Protocol 2.24, one triple is used to check another without revealing its values. Furthermore, by Lemma 2.25, if the first triple is not correct and the second is, then this is detected by the honest parties. Thus, the only way an adversary can successfully cheat is if **(a)** no incorrect triples are opened, and **(b)** there is no bucket with both correct and incorrect triples. Stated differently, the adversary can only successfully cheat if there exists a bucket where all triples are incorrect. By appropriately choosing the bucket-size B and number of triples C to be opened, the cheating probability can be made negligibly small.

Proposition 3.3. *Let N, B, C be such that $N = CB^2$ and $(B-1)\log_2 C \geq \sigma$. Then, Protocol 3.2 securely computes $\mathcal{F}_{\text{triples}}$ with abort in the $(\mathcal{F}_{\text{rand}}, \mathcal{F}_{\text{perm}})$-hybrid model, with statistical error $2^{-\sigma}$ and in the presence of one malicious party.*

Proof Sketch: Intuitively, the triples generated are to random values since $\mathcal{F}_{\text{rand}}$ is used to generate $[a_i], [b_i]$ for all i (note that in $\mathcal{F}_{\text{triples}}$ the adversary chooses its shares in $[a_i], [b_i]$; this is inherited from its capability in $\mathcal{F}_{\text{rand}}$). Then, Protocol 2.22 is used to ensure that the first C triples are all correct (recall that by Lemma 2.4, $[a_i], [b_i], [c_i]$ are all consistent sharings, and thus by Lemma 2.23 the honest parties output $\perp$ if $c_i \neq a_i b_i$). Finally, Protocol 2.24 is used to verify that all of the triples in $\boldsymbol{d}$ are correct multiplication triples. By Lemma 2.25, if $([a_1], [b_1], [c_1])$ in any of the buckets is not a correct multiplication triple, and there exists a $j \in \{2, \ldots, B\}$ for which $([a_j], [b_j], [c_j])$ *is* a correct multiplication triple, then the honest parties output $\perp$ (note that once again by Lemma 2.4, all of the shares are guaranteed to be consistent). Thus, the only way that $\boldsymbol{d}_i$ for some $i \in [N]$ contains an incorrect multiplication triple is if *all* of the C opened triples were correct and the entire bucket $\boldsymbol{D}_i$ contains incorrect multiplication triples. Denote the event that this happens for some i by bad. By choosing B and C so that $\Pr[\text{bad}]$ is negligible, the protocol is secure. Observe that the triples are all generated and fixed before $\mathcal{F}_{\text{perm}}$ is called, and thus the probability that bad occurs is equal to the balls-and-buckets game of [5,19], where the adversary wins only if there exists no "mixed bucket" (containing both good and bad balls). In [5, Proof of Lemma 12], it is shown that $\Pr[\text{bad}] < 2^{-\sigma}$ when $(B-1)\log_2 C \geq \sigma$ (these parameters are derived from their notation by setting $C = \ell$, $B = b$ and $N = \ell b^2$). For any σ, we therefore choose B and C such that $N = CB^2$ and $(B-1)\log_2 C \geq \sigma$, and the appropriate error probability is obtained.

Observe that simulation is easy; $\mathcal{S}$ receives N triples from the trusted party and simulates the $2(NB + C)$ calls to $\mathcal{F}_{\text{rand}}$. $\mathcal{S}$ places the appropriate values from the N triples in random places, and ensures that they will all be the first triple in each bucket (by setting the output of $\mathcal{F}_{\text{perm}}$ appropriately). Then, $\mathcal{S}$ sends continue to the trusted party if and only if $\mathcal{A}$ did not cheat in any multiplication. The only difference between the protocol execution with $\mathcal{A}$ and the simulation with $\mathcal{S}$ is in the case that bad occurs, which happens with negligible probability.   ∎

Concrete parameters. In our protocol, generation of triples is highly efficient. Thus, we can generate a very large number of triples at once (unlike [19]) which yields better parameters. In [19, Proof of Theorem 8] it was shown that when the probability of a triple being incorrect is $1/2$, the adversary can cheat with probability at most $2^{-\sigma}$ when $(1 + \log_2 N)(B - 1) \geq \sigma$. This implies that for $N = 2^{20}$ and $\sigma = 40$, we can take $B = 3$ because $(1+\log_2 N)(B-1) > 21 \cdot 2 > 40$. In order to make the probability of a triple being incorrect be (close to) $1/2$, we can set $C = 3 \cdot 2^{20}$. This implies that the overall number of triples required is $6 \cdot 2^{20}$.

An improved combinatorial analysis is provided in [5]. They show that when setting $N = CB^2$, the adversary can cheat with probability at most $2^{-\sigma}$ when $(B-1)\log_2 C \geq \sigma$ (in [5], they write ℓ instead of C and b instead of B). In order to minimize the number of triples overall, B must be kept to a minimum. For $N = 2^{20}$ and $\sigma = 40$, one can choose $B = 4$ and $C = 2^{16}$. It then follows that $(B-1)\log_2 C = 3 \cdot 16 > 40$. Thus, the overall number of triples required is $NB + C = CB^3 + C = 2^{22} + 2^{16}$. Observe that these parameters derived from the analysis of [5] yield approximately $2/3$ the cost of $6 \cdot 2^{20}$ as required by the analysis of [19]. (This follows because $6 \cdot 2^{20} = \frac{3}{2} \cdot 2^{22}$.)

In Sect. 5 we provide a *new analysis* showing that it suffices to generate $3 \cdot 2^{20} + 3$ triples. This is approximately 25% less than the analysis of [5]. Concretely, to generate 1 million validated triples, the analysis of [5] requires generating 4,065,536 triples initially, whereas we require only 3,000,003.

Deferring compareview. In the calls to functionalities $\mathcal{F}_{\text{rand}}$ and $\mathcal{F}_{\text{perm}}$ and in the execution of Protocol 2.24, the parties use the compareview subprocedure. As explained before, the parties actually compare only at the end of the entire triple-generation protocol, and compare a hash of the view instead of the entire view, which reduces communication significantly.

Using pseudorandomness in $\mathcal{F}_{\text{perm}}$. In practice, in order to reduce the communication, the calls to $\mathcal{F}_{\text{coin}}$ inside $\mathcal{F}_{\text{perm}}$ are only used in order to generate a short seed. Each party then applies a pseudorandom generator to the seed in order to obtain all of the randomness needed for computing the permutation. This protocol actually no longer securely computes $\mathcal{F}_{\text{perm}}$ since the permutation is not random, and the corrupted party actually knows that it is not random since it has the seed. Nevertheless, by the proof of Proposition 3.3, we have that the only requirement from $\mathcal{F}_{\text{perm}}$ is that the probability of bad happening is negligible. Now, since the triples are fixed before $\mathcal{F}_{\text{perm}}$ is called, the probability of bad happening is simply the probability that a specific subset of permutations occur (that map all of the incorrect triples into the same bucket/s). If this occurs with probability that is non-negligibly higher than when a truly random permutation is used, then this can be used to distinguish the result of the pseudorandom generator from random.

4   Secure Computation of Any Functionality

In this section, we show how to securely compute any three party functionality f. The idea behind the protocol is simple. The parties first use $\mathcal{F}_{\text{triples}}$ to generate a vector of valid multiplication triples. Next, the parties compute the circuit using the semi-honest multiplication protocol of Sect. 2.2 for each AND gate. Recall that by Lemma 2.4, the result of this protocol is a triple of consistent shares $([a], [b], [c])$ where $c = ab$ or $c = ab \oplus 1$, even when one of the parties is malicious. Thus, it remains to verify that for each gate it holds that $c = ab$. Now, utilizing the valid triples generated within $\mathcal{F}_{\text{triples}}$, the parties can use Protocol 2.24 to verify that $([a], [b], [c])$ is a correct multiplication triple (i.e., that $c = ab$) without revealing anything about a, b or c. See Protocol 4.2 for a full specification.

We prove the following theorem:

Theorem 4.1. *Let f be a three-party functionality. Then, Protocol 4.2 securely computes f with abort in the $(\mathcal{F}_{\text{triples}}, \mathcal{F}_{\text{share}}, \mathcal{F}_{\text{reconst}})$-hybrid model, in the presence of one malicious party.*

Proof Sketch: Intuitively, the security of this protocol follows from the following. If the adversary cheats in any semi-honest multiplication, then by Lemma 2.25 the honest parties output $\perp$. This holds because by Lemma 2.4 all shares $[x], [y], [z]$ are consistent (but $z \neq xy$), while $([a_k], [b_k], [c_k])$ is guaranteed to be a correct multiplication triple since it was generated by $\mathcal{F}_{\text{triples}}$. Thus, the adversary must behave honestly throughout, and the security is reduced to the proof of security for the semi-honest case, as proven in [1].

The simulator for $\mathcal{A}$ works by playing the role of the trusted party for $\mathcal{F}_{\text{triples}}$ and $\mathcal{F}_{\text{share}}$, and then by simulating the semi-honest multiplication protocol in the circuit emulation phase for every AND gate. The verification stage involving executions of Protocol 2.24 is then simulated by $\mathcal{S}$ internally handing random ρ, σ values to $\mathcal{A}$ as if sent by P_{i-1}. Since $\mathcal{S}$ plays the trusted party in $\mathcal{F}_{\text{triples}}$ and $\mathcal{F}_{\text{share}}$, it knows all of the values held and therefore can detect if $\mathcal{A}$ tries to cheat. If yes, then it simulates the honest parties aborting, and sends $\perp$ to the trusted party as P_i's input. Otherwise, it sends the input of P_i sent by $\mathcal{A}$ in $\mathcal{F}_{\text{share}}$. Finally, after receiving P_i's output from the trusted party computing f,

$\mathcal{S}$ plays the ideal functionality computing $\mathcal{F}_{\text{reconst}}$. If $\mathcal{A}$ sends abort to $\mathcal{F}_{\text{reconst}}$ then $\mathcal{S}$ sends abort to the trusted party computing f; otherwise, it sends continue. We stress that the above simulation works since the semi-honest multiplication protocol is *private in the presence of a malicious adversary*, meaning that its view can be simulated before any output is revealed. This was shown in [1, Sect. 4]. Thus, the view of a malicious $\mathcal{A}$ in the circuit-emulation phase is simulated as in [1], and then the verification phase is simulated as described above. This completes the proof sketch.   ∎

Generating many triples in the offline. In many cases, the circuit being computed is rather small. However, the highest efficiency in $\mathcal{F}_{\text{triples}}$ is achieved when taking a very large N (e.g., $N = 2^{20}$). We argue that $\mathcal{F}_{\text{triples}}$ can be run once, and the triples used for multiple different computations. This is due to the fact that the honest parties abort if *any* multiplication is incorrect, and this makes no difference whether a single execution utilizing N gates is run, or multiple executions.

5   Improved Combinatorial Analysis

In this section we provide a tighter analysis of the probability that the adversary succeeds in circumventing the computation without being caught. This analysis allows us to reduce both the overall number of triples needed and the number of triples that are opened in the cut-and-bucket process, compared to [5, 19]. In our specific protocol, reducing the number of triples to be opened is of great importance since generation of a triple requires 3 bits of communication, while each opening requires 9 bits.

Loosely speaking, the adversary can succeed if the verification of an incorrect AND gate computation is carried using an incorrect multiplication triple. This event can only happen if no incorrect triples were opened and if the *entire bucket* from which the incorrect triple came contained only incorrect triples. Otherwise, the honest parties would abort in the triples-generation protocol, when running the check phase. Since the triples are randomly assigned to buckets, the probability that this event occurs is small (which is what we need to prove). Clearly, increasing the number of triples checked and the bucket size reduces the success probability of the adversary. However, increasing these parameters raises the computation and communication cost of our protocol. Thus, our goal is to minimize these costs by minimizing the number of triples generated (and opened). We denote by σ the statistical parameter, and our aim to guarantee that the adversary succeeds with probability at most $2^{-\sigma}$. Recall that C is the number of triples opened in the cut-and-bucket process and B is the size of the bucket. We start by defining the following balls-and-buckets game, which is equivalent to our protocol (in the game, a "bad ball" is an incorrect multiplication triple). We say that the adversary "wins", if the output of the game is 1.

PROTOCOL 3.2 (Generating Multiplication triples)

- **Input:** The number N of triples to be generated.
- **Auxiliary input:** Parameters B and C.
- **The Protocol:**
    1. *Generate random sharings:* The parties invoke $2(NB + C)$ calls to $\mathcal{F}_{\text{rand}}$; denote the shares that they receive by $[([a_i], [b_i])]_{i=1}^{NB+C}$.
    2. *Generate multiplication triples:* For $i = 1, \ldots, NB + C$, the parties run the semi-honest multiplication protocol of Sect. 2.2 to compute $[c_i] = [a_i] \cdot [b_i]$. Denote $\boldsymbol{D} = [([a_i], [b_i], [c_i])]_{i=1}^{NB+C}$; observe that $[c_i]$ is the result of the protocol and is not necessarily "correct".
    3. *Cut and bucket:* Let $M = NB+C$. In this stage, the parties perform a first verification that the triples were generated correctly by opening C triples, and then randomly divide the remainder into buckets.
        (a) The parties call $\mathcal{F}_{\text{perm}}$ with vector $\boldsymbol{D}$.
        (b) The parties run Protocol 2.22 (triple verification with opening) for each of the first C triples in $\boldsymbol{D}$, and remove them from $\boldsymbol{D}$. If a party did not output accept in every execution, it sends $\bot$ to the other parties and outputs $\bot$.
        (c) The remaining NB triples in $\boldsymbol{D}$ are divided into N sets of triples $\boldsymbol{D}_1, \ldots, \boldsymbol{D}_N$, each of size B. For $i = 1, \ldots, N$, the bucket $\boldsymbol{D}_i$ contains the triples $([a_{(i-1)\cdot B+1}], [b_{(i-1)\cdot B+1}], [c_{(i-1)\cdot B+1}]), \ldots, ([a_{i \cdot B}], [b_{i \cdot B}], [c_{i \cdot B}])$.
    4. *Check buckets:* The parties initialize a vector $\boldsymbol{d}$ of length N. Then, for $i = 1, \ldots, N$:
        (a) Denote the triples in $\boldsymbol{D}_k$ by $([a_1], [b_1], [c_1]), \ldots, ([a_B], [b_B], [c_B])$.
        (b) For $j = 2, \ldots, B$, the parties run Protocol 2.24 (triple verification using another without opening) on input $([a_1], [b_1], [c_1])$ and $([a_j], [b_j], [c_j])$, to verify $([a_1], [b_1], [c_1])$.
        (c) If a party did not output accept in every execution, it sends $\bot$ to the other parties and outputs $\bot$.
        (d) The parties set $\boldsymbol{d}_i = ([a_1], [b_1], [c_1])$; i.e., they store these shares in the ith entry of $\boldsymbol{d}$.
- **Output:** The parties output $\boldsymbol{d}$.

Game$_1(\mathcal{A}, N, B, C)$:

1. The adversary $\mathcal{A}$ prepares $M = NB + C$ balls. Each ball can be either **bad** or **good**.
2. C random balls are chosen and opened. If one of the C balls is **bad** then output 0. Otherwise, the game proceeds to the next step.
3. The remaining NB balls are randomly divided into N buckets of equal size B. Denote the buckets by $B_1, \ldots, B_N$. We say that a bucket is **fully bad** if all balls inside it are **bad**. Similarly, a bucket is **fully good** if all balls inside it are **good**.
4. The output of the game is 1 if and only if there exists i such that bucket B_i is fully bad, and all other buckets are either **fully bad** or **fully good**.

PROTOCOL 4.2 (Securely Computing a Functionality f)

- **Inputs:** Each party P_j where $j \in \{1, 2, 3\}$ holds an input $x_j \in \{0, 1\}^\ell$.
- **Auxiliary Input:** The parties hold a description of a boolean circuit C that computes f on inputs of length ℓ. Let N be the number of AND gates in C.
- **The protocol – offline phase:** The parties call $\mathcal{F}_{\text{triples}}$ with input N and obtain a vector $\boldsymbol{d}$ of sharings.
- **The protocol – online phase:**
    1. *Sharing the inputs:* For each input wire, the parties call $\mathcal{F}_{\text{share}}$ with the dealer being the party whose input is associated with that wire.
    2. *Circuit emulation:* Let $G_1, ..., G_N$ be a predetermined topological ordering of the gates of the circuit. For $k = 1, ..., N$ the parties work as follows:
        - If G_k is a XOR gate: Given shares $[x]$ and $[y]$ on the input wires, the parties compute $[x] \oplus [y]$ and define the result as their share on the output wire.
        - If G_k is a NOT gate: Given shares $[x]$ on the input wire, the parties compute $\overline{[x]}$ and define the result as their share on the output wire.
        - If G_k is an AND gate: Given shares $[x]$ and $[y]$ on the input wires, the parties run the semi-honest multiplication protocol of Sect. 2.2.
    3. *Verification stage:* *Before* the secrets on the output wires are reconstructed, the parties verify that all the multiplications were carried out correctly, as follows. For $k = 1, \ldots, N$:
        (a) Denote by $([x], [y])$ the shares of the input wires to the kth AND gate, and denote by $[z]$ the shares of the output wire of the kth AND gate.
        (b) The parties run Protocol 2.24 (triple verification using another without opening) on input $([x], [y], [z])$ and $([a_k], [b_k], [c_k])$ to verify $([x], [y], [z])$.
        (c) If a party did not output accept in every execution, it sends $\perp$ to the other parties and outputs $\perp$.
        Observe that all executions of Protocol 2.24 can be run in parallel. In addition, compareview can be run once at the end of all checks, and using a hash of the view as described in Sect. 3.
    4. If any party received $\perp$ in any call to any functionality above, then it outputs $\perp$ and halts.
    5. *Output reconstruction:* For each output wire of the circuit, the parties call $\mathcal{F}_{\text{reconst}}$ with input $([v], j)$ where $[v]$ is the sharing of the value on the output wire, and P_j is the party whose output is on the wire.
    6. If a party received $\perp$ in any call to $\mathcal{F}_{\text{reconst}}$ then it sends $\perp$ to the other parties, outputs $\perp$ and halts.
- **Output:** If a party has not output $\perp$, then it outputs the values it received on its output wires.

Note that the condition in the last step forces the adversary to choose at least one bad ball if it wishes to win. We first show that for $\mathcal{A}$ to win the game, the number of bad balls $\mathcal{A}$ chooses must be a multiple of B, the size of a bucket.

Lemma 5.1. *Let T be the number of bad balls chosen by the adversary $\mathcal{A}$. Then, a necessary condition for $Game_1(\mathcal{A}, N, B, C) = 1$ is that $T = B \cdot t$ for some $t \in \mathbb{N}$.*

Proof: This follows immediately from the fact that the output of the game is 1 only if no bad balls are opened and all buckets are fully bad or good. Thus, if $T \neq B \cdot t$, then either a bad ball is opened or there must be some bucket that is mixed, meaning that it has both bad and good balls inside it. Therefore, in this case, the output of the game will be 0.   ∎

Following Lemma 5.1, we derive a formula for the success probability of the adversary in the game. We say that the adversary $\mathcal{A}$ has chosen to "corrupt" t buckets if $t = \frac{T}{B}$, where T is the number of bad balls generated by the adversary.

Theorem 5.2. *Let t be the number of buckets $\mathcal{A}$ has chosen to corrupt. Then, for every $0 < t \leq N$ it holds that*

$$\Pr[Game_1(\mathcal{A}, N, B, C) = 1] = \binom{N}{t} \binom{NB + C}{tB}^{-1}.$$

Proof: Assume $\mathcal{A}$ has chosen to corrupt t buckets. i.e., $\mathcal{A}$ has generated tB bad balls. Let E_c be the event that no bad balls were detected when opening C random balls. We have:

$$\Pr[E_c] = \frac{\binom{NB+C-tB}{C}}{\binom{NB+C}{C}} = \frac{(NB + C - tB)!(NB)!}{(NB - tB)!(NB + C)!}.$$

Next, let E_B the event that the bad balls are in exactly t buckets after permuting the balls (and so there are t fully bad buckets and all other buckets are fully good). There are $(NB)!$ ways to permute the balls, but if we require the tB bad balls to fall in exactly t buckets then we first choose t buckets out of N, permute the tB balls inside them, and finally permute the other $NB - tB$ balls in the other buckets. Overall, we obtain that

$$\Pr[E_B] = \frac{\binom{N}{t}(tB)!(NB - tB)!}{(NB)!}.$$

Combining the above, we obtain that

$$\begin{aligned}
\Pr[Game_1(\mathcal{A}, N, B, C) = 1] &= \Pr[E_c \wedge E_B] = \Pr[E_c] \cdot \Pr[E_B] \\
&= \binom{N}{t} \frac{(NB + C - tB)!(tB)!}{(NB + C)!} \\
&= \binom{N}{t} \binom{NB + C}{tB}^{-1}.
\end{aligned}$$

∎

Next, we show that if $C \geq B$, then the best strategy of the adversary is to corrupt exactly one bucket. This allows us to derive an upper bound of the success probability of the adversary.

Theorem 5.3. *If $C \geq B$, then for every adversary $\mathcal{A}$, it holds that*

$$\Pr[Game_1(\mathcal{A}, N, B, C) = 1] \leq N\binom{NB+C}{B}^{-1}.$$

Proof: Following Theorem 5.2, we need to show that for *every* $t \geq 1$

$$\binom{N}{t}\binom{NB+C}{tB}^{-1} \leq N\binom{NB+C}{B}^{-1}.$$

First, observe that when $t = 1$, the left side of the inequality is exactly the same as its right side, and thus the theorem holds.

Next, assume that $t \geq 2$; It is suffices to show that:

$$\binom{N}{t}\binom{NB+C}{tB}^{-1} \leq \binom{NB+C}{B}^{-1}$$

which is equivalent to proving that

$$\binom{N}{t}\frac{(tB)!(NB+C-tB)!}{(NB+C)!} \leq \frac{B!(NB+C-B)!}{(NB+C)!}$$

which is in turn equivalent to proving that

$$\binom{N}{t}\frac{(tB)!}{B!} \leq \frac{(NB+C-B)!}{(NB+C-tB)!}.$$

By multiplying both sides of the inequality with $\frac{1}{(tB-B)!}$ we obtain that in order to complete the proof, it suffices to show that

$$\binom{N}{t}\binom{tB}{tB-B} \leq \binom{NB+C-B}{tB-B}. \tag{1}$$

Using the assumption that $C \geq B$, we obtain that instead of proving Eq. (1), it is sufficient to prove that

$$\binom{N}{t}\binom{tB}{tB-B} \leq \binom{NB}{tB-B}. \tag{2}$$

To see that Eq. (2) holds, consider the following two combinatorial processes:

1. Choose t buckets out of N. Then, choose $tB - B$ out of the tB balls in these buckets.
2. Choose $tB - B$ out of NB balls.

Note that since $t \geq 2$, it holds that $tB - B > 0$, and the processes are well defined. Next, observe that both processes end with holding $tB - B$ balls that were chosen from an initial set of NB balls. However, while in the second process we do not place any restriction on the selection process, in the first process we require that t buckets will be chosen first and then the $tB - B$ balls are allowed to be chosen only from the t buckets. Thus, the number of choice options in the second process is strictly larger than in the first process. Finally, since the first process describes exactly the left side of Eq. (2), whereas the second process describes exactly the right side of Eq. (2), we conclude that the inequality indeed holds. ∎

Corollary 5.4. *If $C = B$ and B, N are chosen such that $\sigma \leq \log \left(\frac{(N \cdot B + B)!}{N \cdot B! \cdot (N \cdot B)!} \right)$, then for every adversary $\mathcal{A}$ it holds that $\Pr[Game_1(\mathcal{A}, N, B, C) = 1] \leq 2^{-\sigma}$.*

Proof: This holds directly from Theorem 5.3, which holds if $C \geq B$. Since this holds for any $C \geq B$, we set $C = B$ in the bound of Theorem 5.3, and have that the bound is fulfilled if

$$N \binom{NB + B}{B}^{-1} \leq 2^{-\sigma} \quad \text{and so} \quad \frac{1}{N} \cdot \binom{NB + B}{B} \geq 2^{\sigma}.$$

Taking log of both sides yields the result. ∎

Corollary 5.4 provides us a way of computing the bucket size B for every possible N and σ. For example, setting $N = 2^{20}$ and $\sigma = 40$ (meaning that we want to output 2^{20} triples from the pre-processing protocol with error probability less than 2^{-40}), we obtain that $B = 3$ suffices and we need to generate $NB + C = 3 \cdot 2^{20} + 3$ triples, of which only 3 triples are opened. We performed this computation for $N = 2^{20}, 2^{30}$ and $\sigma = 40, 80, 128$ and compared the results with [5] (recall that according to their analysis, when setting $N = CB^2$, the adversary can cheat with probability at most $2^{-\sigma}$ when $(B - 1) \log_2 C \geq \sigma$). The comparison is presented in Tables 1, 2 and 3. In the tables, we use "M" to denote the number of triples that are initially generated, i.e., $M = NB + C$ in our work whereas $M = CB^3 + C$ in [5].

Table 1. Parameter comparison for $\sigma = 40$

	$N = 2^{20}$			$N = 2^{30}$		
	B	M	C	B	M	C
[5]	4	4,259,840	65,536	3	3,340,530,119	119,304,647
Our work	3	3,145,731	3	3	3,221,225,475	3

As can be seen, in all cases, our combinatorial analysis yields a significant improvement, both in the number of generated triples and in the number of triples needed to be opened. Specifically, although only few triples are opened, we reduce the overall number of triples generated by up to 25%, compared to [5]. Recall that both improvements are important, as each triple less to generate means 1 bit less to send for each party, and each triple less to open means 3 bit less to send for each party.

Table 2. Parameter comparison for $\sigma = 80$

	$N = 2^{20}$			$N = 2^{30}$		
	B	M	C	B	M	C
[5]	7	7,361,432	21,400	5	5,411,658,793	42,949,673
Our work	5	5,242,885	5	4	4,294,967,300	4

Table 3. Parameters comparison for $\sigma = 120$

	$N = 2^{20}$			$N = 2^{30}$		
	B	M	C	B	M	C
[5]	10	10,496,246	10,486	6	6,427,277,106	29,826,162
Our work	7	7,340,039	7	5	5,368,709,125	5

6 Efficiency and Comparison

In this section, we describe the communication and computation complexity of (the computationally secure variant of) our protocol. We compare our protocol to that of [18], since this is the most efficient protocol known for our setting of three parties, malicious adversaries, and an honest majority. The complexity of the protocol in [18] is close to Yao's two-party semi-honest protocol, and thus its communication complexity is dominated by the size of the garbled circuit and its computation complexity is dominated by the amount of work needed to prepare a garbled circuit and evaluate it. The comparison summary is shown in Table 4; for our protocol, the complexity is based on a bucket size of $B = 3$. A detailed explanation appears below.

Communication Complexity. We count the number of bits sent by each party for each AND gate. The semi-honest multiplication protocol requires sending a single bit, and verifying a triple using another without opening requires sending 2 bits (only very few triples are checked with opening and so we ignore this). Now, a single multiplication and a single verification is used for each AND gate (3 bits). Furthermore, each triple is generated from B triples (generated using B multiplications) and $B - 1$ verifications, thereby costing $B + 2(B - 1)$ bits. The overall cost per gate is therefore $B + 2B - 2 + 3 = 3B + 1$. For $B = 3$ (which suffices with $N = 2^{20}$ and error 2^{-40}), we conclude that only 10 bits are sent by each party per gate.

Table 4. Average cost per party (N = number of AND gates in the circuit)

	Communication (bits)	Number of AES computations
Our protocol	$10N$	$\frac{20N + 3N \log(3N)}{128} \approx \frac{N}{5}$
The protocol of [18]	$85N$	$3N$

In contrast, in [18], the communication is dominated by a single garbled circuit. When using the half-gates optimization of [24], such a circuits consists of two ciphertexts per AND gate with a size of 256 bits. Thus, on average, each party sends $256/3 \approx 85$ bits per AND gate.

Number of AES computations. The computations in our protocol are very simple, and the computation complexity is therefore dominated by the AES computations needed to generated randomness (in the multiplication and to compute correlated randomness). Two bits of pseudorandomness are needed for each call to $\mathcal{F}_{cr}$ to generate correlated randomness. In order to generate a triple, 2 calls to $\mathcal{F}_{cr}$ are required and an additional call for the multiplication, for a total of 6 bits. For every AND gate, B triples are generated and one additional multiplication is carried out for the actual gate, resulting in a total of $6B+2$ bits. In addition, $\mathcal{F}_{perm}$ requires an additional $NB\log(NB)$ bits of pseudorandomness. Thus, the total number of pseudorandom bits for N gates equals $(6B+2)N+NB\log(NB)$; taking $B = 3$ as above, we have $20N + 3N(\log 3N)$. Noting that 128-bits of pseudorandomness are generated with a single AES computation, this requires $\frac{20N+3N(\log 3N)}{128} \approx \frac{N}{5}$ calls to AES.

In contrast, in the protocol of [18], two parties need to garble the circuit and one needs to evaluate it. Garbling a circuit requires 4 AES operations per AND gate and evaluating requires 2 operations per AND gate. Thus, the average number of AES operations is $\frac{10}{3} \approx 3$ per party per AND gate.

References

1. Araki, T., Furukawa, J., Lindell, Y., Nof, A., Ohara, K.: High-throughput semi-honest secure three-party computation with an honest majority. In: 23rd ACM CCS (2016, to appear)
2. Beaver, D.: Efficient multiparty protocols using circuit randomization. In: Feigenbaum, J. (ed.) CRYPTO 1991. LNCS, vol. 576, pp. 420–432. Springer, Heidelberg (1992). doi:10.1007/3-540-46766-1_34
3. Beaver, D., Micali, S., Rogaway, P.: The round complexity of secure protocols. In: The 22nd STOC, pp. 503–513 (1990)
4. Ben-Or, M., Goldwasser, S., Wigderson, A.: Completeness theorems for non-cryptographic fault-tolerant distributed computation. In: STOC 1988, pp. 1–10 (1988)
5. Burra, S.S., Larraia, E., Nielsen, J.B., Nordholt, P.S., Orlandi, C., Orsini, E., Scholl, P., Smart, N.P.: High performance multi-party computation for binary circuits based on oblivious transfer. ePrint Cryptology Archive, 2015/472 (2015)
6. Canetti, R.: Security and composition of multiparty cryptographic protocols. J. Cryptol. **13**(1), 143–202 (2000)
7. Canetti, R., Security, U.C.: A new paradigm for cryptographic protocols. In: 42nd FOCS, pp. 136–145 (2001)
8. Chaum, D., Crépeau, C., Damgård, I.: Multi-party unconditionally secure protocols. In: 20th STOC, pp. 11–19 (1988)
9. Damgård, I., Pastro, V., Smart, N., Zakarias, S.: Multiparty computation from somewhat homomorphic encryption. In: Safavi-Naini, R., Canetti, R. (eds.) CRYPTO 2012. LNCS, vol. 7417, pp. 643–662. Springer, Heidelberg (2012). doi:10.1007/978-3-642-32009-5_38

10. Fisher, R.A., Yates, F.: Statistical Tables for Biological, Agricultural and Medical Research, 3rd edn, pp. 26–27. Oliver & Boyd, Edinburgh (1938)
11. Goldreich, O.: Foundations of Cryptography: Volume 2 - Basic Applications. Cambridge University Press, Cambridge (2004)
12. Goldreich, O., Micali, S., Wigderson, A.: How to play any mental game. In: 19th STOC, pp. 218–229 (1987)
13. Goldwasser, S., Lindell, Y.: Secure computation without agreement. J. Cryptol. **18**(3), 247–287 (2005)
14. Keller, M., Orsini, E., Scholl, P.: MASCOT: faster malicious arithmetic secure computation with oblivious transfer. In: The 23rd ACM CCS, pp. 830–842 (2016)
15. Kushilevitz, E., Lindell, Y., Rabin, T.: Information-theoretically secure protocols and security under composition. SIAM J. Comput. **39**(5), 2090–2112 (2010)
16. Lindell, Y., Pinkas, B.: Secure two-party computation via cut-and-choose oblivious transfer. In: Ishai, Y. (ed.) TCC 2011. LNCS, vol. 6597, pp. 329–346. Springer, Heidelberg (2011). doi:10.1007/978-3-642-19571-6_20
17. Lindell, Y., Riva, B.: Blazing fast 2PC in the offline/online setting with security for malicious adversaries. In: ACM Conference on Computer and Communications Security, pp. 579–590 (2015)
18. Mohassel, P., Rosulek, M., Zhang, Y.: Fast and secure three-party computation: the garbled circuit approach. In: ACM Conference on Computer and Communications Security, pp. 591–602 (2015)
19. Nielsen, J.B., Nordholt, P.S., Orlandi, C., Burra, S.S.: A new approach to practical active-secure two-party computation. In: Safavi-Naini, R., Canetti, R. (eds.) CRYPTO 2012. LNCS, vol. 7417, pp. 681–700. Springer, Heidelberg (2012). doi:10.1007/978-3-642-32009-5_40
20. Rabin, T., Ben-Or, M.: Verifiable secret sharing and multi-party protocols with honest majority. In: 21st STOC, pp. 73–85 (1989)
21. Rindal, P., Rosulek, M.: Faster malicious 2-party secure computation with online/offline dual execution. In: USENIX Security, pp. 297–314 (2016)
22. Sharemind, Cybernetica. https://sharemind.cyber.ee
23. Yao, A.: How to generate and exchange secrets. In: The 27th FOCS, pp. 162–167 (1986)
24. Zahur, S., Rosulek, M., Evans, D.: Two halves make a whole. In: Oswald, E., Fischlin, M. (eds.) EUROCRYPT 2015. LNCS, vol. 9057, pp. 220–250. Springer, Heidelberg (2015). doi:10.1007/978-3-662-46803-6_8

Symmetric Cryptanalysis I

Conditional Cube Attack on Reduced-Round Keccak Sponge Function

Senyang Huang[1], Xiaoyun Wang[1,2,3]($\boxtimes$), Guangwu Xu[4], Meiqin Wang[2,3], and Jingyuan Zhao[5]

[1] Institute for Advanced Study, Tsinghua University, Beijing 100084, China
xiaoyunwang@mail.tsinghua.edu.cn
[2] Key Laboratory of Cryptologic Technology and Information Security,
Ministry of Education, Shandong University, Jinan 250100, China
[3] School of Mathematics, Shandong University, Jinan 250100, China
[4] Department of EE and CS, University of Wisconsin-Milwaukee,
Milwaukee, WI 53201, USA
[5] State Key Laboratory of Information Security, Institute of Information
Engineering, Chinese Academy of Sciences, Beijing 100093, China

Abstract. The security analysis of Keccak, the winner of SHA-3, has attracted considerable interest. Recently, some attention has been paid to the analysis of keyed modes of Keccak sponge function. As a notable example, the most efficient key recovery attacks on Keccak-MAC and Keyak were reported at EUROCRYPT'15 where cube attacks and cube-attack-like cryptanalysis have been applied. In this paper, we develop a new type of cube distinguisher, the *conditional cube tester*, for Keccak sponge function. By imposing some bit conditions for certain cube variables, we are able to construct cube testers with smaller dimensions. Our conditional cube testers are used to analyse Keccak in keyed modes. For reduced-round Keccak-MAC and Keyak, our attacks greatly improve the best known attacks in key recovery in terms of the number of rounds or the complexity. Moreover, our new model can also be applied to keyless setting to distinguish Keccak sponge function from random permutation. We provide a searching algorithm to produce the most efficient conditional cube tester by modeling it as an MILP (mixed integer linear programming) problem. As a result, we improve the previous distinguishing attacks on Keccak sponge function significantly. Most of our attacks have been implemented and verified by desktop computers. Finally we remark that our attacks on the reduced-round Keccak will not threat the security margin of Keccak sponge function.

Keywords: Keccak-MAC · Keyak · Cube tester · Conditional cube variable · Ordinary cube variable

1 Introduction

The Keccak sponge function family, designed by Bertoni, Daemen, Peeters, and Giles in 2007 [1], was selected by the U.S. National Institute of Standards and Technology (NIST) in 2012 as the proposed SHA-3 cryptographic hash function.

© International Association for Cryptologic Research 2017
J.-S. Coron and J.B. Nielsen (Eds.): EUROCRYPT 2017, Part II, LNCS 10211, pp. 259–288, 2017.
DOI: 10.1007/978-3-319-56614-6_9

Due to its theoretical and practical importance, cryptanalysis of Keccak has attracted increasing attention. There has been extensive research recently, primarily on the keyless setting. For example, in keyless modes of reduced-round Keccak, many results have been obtained on collision attack [2], preimage attack [3] and second preimage attack [4]. Additionally, there are also some research focused on the distinguishers of Keccak internal permutation, in which the size of input is the full state. In [5], a distinguisher of full 24-round Keccak internal permutation was proposed which takes 2^{1579} Keccak calls. Using the rebound attack and efficient differential trails, Duc et al. [6] derived a distinguisher for 8-round Keccak internal permutation with the complexity 2^{491}. Jérémy et al. [7] provided an 8-round internal differential boomerang distinguisher on Keccak with practical complexity. It should be remarked that these results on Keccak internal permutation seem to be a little far from the security margin of Keccak sponge function, which do not lead any attacks to Keccak hash function. For distinguishing attacks on Keccak sponge function with the bitrate part as its input, some results have been given in [8–10]. These distinguishers are one step closer to the security margin but some of these distinguisher are far from being practical.

By embedding a secret key in a message as an input, Keccak can be used in several settings. For example, Keccak sponge function can produce a pseudorandom binary string of arbitrary length, and hence can serve as a stream cipher. It is also a natural keyed hash function, namely, a message authentication code (MAC). Moreover, an authenticated encryption (AE) scheme based on Keccak was described in [11]. However, there is much less research reported for the keyed modes of the family of Keccak sponge functions. Besides the side channel attack for Keccak-MAC [12], the celebrated paper on key recovery attacks [10] seems to be the only one found in the literature for analysing keyed modes of Keccak. In [10], the authors set cube variables in the column parity (CP) kernel to control the propagation of the mapping θ in the first round. More specifically, the cube dimension can be reduced by carefully selecting cube variables so that they are not multiplied with each other after the first round. The cube sums of output polynomials depend only on a portion of key bits. The dedicated cube-attack-like cryptanalysis uses this property to construct the first key recovery attack on reduced-round Keccak-MAC and Keyak. It is also noted that the cube attack and cube-attack-like are very efficient techniques in analysing Keccak-like cryptosystems in [13,14].

We observe that most of the attacks described in the previously published work deal with propagations of cube variables only after the first round. Thus, it is a natural and interesting question to ask whether and how we can control certain relations of cube variables after the second round of the Keccak sponge function to push this kind of attacks further. The purpose of this paper is to answer this question by proposing the technique of *conditional cube tester* and making the corresponding attacks more efficient. To the best of our knowledge, the results obtained in this paper are currently the best in terms of the number of rounds or the complexity.

1.1   Our Contributions

Conditional Cube Tester for Keccak Sponge Function. Our conditional cube tester model is inspired by the dynamic cube attack on Grain stream cipher [15]. The approach of dynamic cube attack in [15] is to set some bit conditions on the initial value (IV) so that the intermediate polynomials can be simplified and the degree of output polynomial can be reduced. However, this approach cannot be utilized directly in the setting of Keccak sponge function because its structure is very different from that of Grain stream cipher. Additionally, the number of intermediate polynomials related to the ones in the previous round is too large for Keccak, which makes the approach of dynamic cube attack infeasible. The bit-tracing method (see [16]), proposed by one of the authors, is a powerful technique to analyse hash functions. This method has also been used in the cryptanalysis of block ciphers such as Simon family in [17]. Some ideas of our current work are stimulated by the bit-tracing method. In this paper, we propose a new approach by imposing bit conditions on the input to control the propagation of cube variables caused by the nonlinear operation χ. This will be helpful in identifying the cube variables that are not multiplied with each other after the second round of Keccak sponge function. We provide several algorithms for searching the cube variables and imposing the corresponding bit conditions. These algorithms give a base to construct a conditional cube tester. In some cases the dimension of this cube tester is smaller than the cube testers in [10]. Our model is also influenced by the conditional differential cryptanalysis method developed in [18]. Noted that our analysis is algebraic in nature since the attacks are designed by exploring algebraic properties while the previous conditional differential is based on differential bias.

Improved Key Recovery Attack on Reduced-Round Keccak-MAC. We have obtained improved results for Keccak-MAC by applying the conditional cube tester. For 5-round Keccak-MAC-512, our key recovery attack makes 2^{24} Keccak calls. We are also able to recover full key bits for 6-round Keccak-MAC-384 with the complexity of 2^{40}. Furthermore, we prove that a 7-round Keccak-MAC-256 can be broken using 2^{72} Keccak calls. These results greatly improve the current best complexity bounds for key recovery attacks reported in [10]. Notice that in [10] the attacks were performed on 5-round Keccak-MAC-288 and 6-round and 7-round Keccak-MAC-128, with the time complexity of 2^{35}, 2^{66} and 2^{97} respectively. As it is easy to see that an attack on Keccak-MAC-n_1 can be used to break Keccak-MAC-n_2 without increasing its complexity as long as $n_1 \geq n_2$, we conclude that our attacks cover those in [10] with better efficiencies. It is remarked that our attacks on 5-round Keccak-MAC-512 and 6-round Keccak-MAC-384 are practical and have been verified by experiments. In Table 1, we list a comparison of the performance of our key recovery attacks and the existing ones. This table also shows that our attacks save the space complexity significantly.

Table 1. Summary of key recovery attacks on Keccak-MAC

Rounds	Capacity	Time	Data	Memory	Reference
5	576	2^{35}	2^{35}	Negligible	[10]
6	256	2^{66}	2^{64}	2^{32}	[10]
7	256	2^{97}	2^{64}	2^{32}	[10]
5	576/1024	2^{24}	2^{24}	Negligible	Sect. 4
6	256/768	2^{40}	2^{40}	Negligible	Sect. 4
7	256/512	2^{72}	2^{72}	Negligible	Sect. 4

Improved Key Recovery Attack on Reduced-Round Keyak. Keyak is an AE scheme based on Keccak sponge function [11]. In this paper we also use the technique of conditional cube tester to recover the key for reduced-round Keyak. In this situation, we assume that a message is of two blocks and the nonce could be reused. This means that our attacks on Keyak break the properties of authenticity and integrity because the specification of Keyak [11] states that a nonce may not be variable when only authenticity and integrity are required. We perform our attacks on 7-round and 8-round Keyak with the time complexity of 2^{42} and 2^{74} respectively. Under the same assumption on the nonce, [10] proposed a key recovery attack on Keyak, which can work up to 7 rounds with the time complexity of 2^{76}. Table 2 compares our results with the existing attacks on Keyak, and shows a significant reduction of complexity by using our method. It is also interesting to note that the memory complexity in our attacks is negligible.

Table 2. Summary of key recovery attacks on Keyak

Rounds	Capacity	Time	Data	Memory	Reference
7	256	2^{76}	2^{75}	2^{43}	[10]
7	256	2^{42}	2^{42}	Negligible	Sect. 5
8	256	2^{74}	2^{74}	Negligible	Sect. 5

Improved Distinguishing Attack on Keccak Sponge Function. In addition to the cases of keyed modes of Keccak, we use the technique of conditional cube tester in keyless setting as well. To be more specific, we use this technique to carry out distinguishing attacks on Keccak sponge function. With the help of mixed integer linear programming (MILP), we can get a suitable combination of conditional cube variables automatically with good efficiency. As a result, practical distinguishing attacks have been achieved for Keccak sponge function up to seven rounds. There have been several distinguishing attacks on Keccak sponge function reported in the published papers. In [8], Naya-Plasencia et al. put forward a 4-round differential distinguisher over Keccak-256/224. A 6-round distinguisher over Keccak-224 was constructed in [9] by Das et al. Recently, a straightforward distinguisher on n-round Keccak sponge function was given in [10] which invokes $2^{2^{n-1}+1}$ Keccak calls for $n \leq 7$. Table 3 lists these existing distinguishing attacks on Keccak sponge function together with our attacks. It can be seen that our improvements over the previous attacks are quite significant.

Table 3. Summary of distinguishing attacks on Keccak sponge function

Rounds	Capacity	Time	Data	Memory	Reference
4	448/512	2^{25}	2^{24}	Negligible	[8]
6	448	2^{52}	2^{52}	Negligible	[9]
6	448/512/576	2^{33}	2^{33}	Negligible	[10]
7	448/512/576	2^{65}	2^{65}	Negligible	[10]
8	576	2^{129}	2^{129}	Negligible	[10]
5	448/512	2^{9}	2^{9}	Negligible	Sect. 6
6	768/1024	2^{9}	2^{9}	Negligible	Sect. 6
6	448/512/576	2^{17}	2^{17}	Negligible	Sect. 6
7	768	2^{17}	2^{17}	Negligible	Sect. 6
7	448	2^{33}	2^{33}	Negligible	Sect. 6

The remainder of the paper is organized as follows. We introduce some preliminaries needed for the paper in Sect. 2, including Keccak sponge function, two keyed modes of Keccak, and the idea of cube tester. In Sect. 3, we will describe our new model, the conditional cube tester. Key recovery attacks for Keccak-MAC and Keyak based on our new model will be discussed in detail in Sects. 4 and 5. Section 6 is devoted to distinguishing Keccak sponge function from a random permutation using the conditional cube tester. Finally, we conclude the paper in Sect. 7.

2 Preliminaries

In the section, we will briefly introduce some necessary background for this paper. We will describe Keccak sponge function including two keyed modes, namely Keccak-MAC and the AE scheme Keyak. In the later part of the section, the idea of cube tester will be described.

2.1 Keccak Sponge Function

Description of Keccak Sponge Function. We shall just describe the Keccak sponge function in its default version. We refer the readers to [1] for the complete Keccak specification.

The (default) sponge function works on a 1600-bit state A, which is simply a three-dimensional array of bits, namely $A[5][5][64]$. The one-dimensional arrays $A[\ ][y][z], A[x][\ ][z]$ and $A[x][y][\ ]$ are called a column, a row and a lane respectively; the two-dimensional array $A[\ ][\ ][z]$ is called a slice (see Fig. 1). The coordinates are always considered modulo 5 for x and y and modulo 64 for z. Each 1600-bit string a is interpreted as a state A in the following manner: the $(64(5y + x) + z)$th bit of a becomes $A[x][y][z]$.

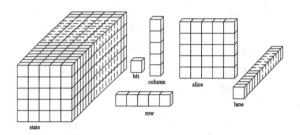

Fig. 1. Terminologies used in Keccak

For each $n \in \{224, 256, 384, 512\}$, the sponge function Keccak-n corresponds to parameters r (bitrate) and $c = 2n$ (capacity) with $r + c = 1600$. Initially, all the 1600 bits are filled with 0s and the message will be split into r-bit blocks. There are two phases in the Keccak sponge function. In the absorbing phase, the next r-bit message block is XORed with its first r-bit segment of the state and then the state is processed by internal permutation which consists of 24 rounds. After all the blocks are absorbed, the squeezing phase starts. In this phase, Keccak-n will return the first r bits as the output of the function with internal permutation iteratively until the n-bit digest is produced.

In the permutation, each round is computed by composing five operations θ, ρ, π, χ and ι as $R = \iota \circ \chi \circ \pi \circ \rho \circ \theta$. Given a round constant RC, the round function can be described by the following pseudo-code, where $r[x, y]$ is the offset of the internal permutation shown in Table 8 and for a lane L, $\text{rot}[L, n]$ means $L >>> n$.

```
R(A, RC)
{
    θ step
    for x in (0...4)
        C[x] = A[x,0] xor A[x,1] xor A[x,2] xor A[x,3] xor A[x,4]
        D[x] = C[x-1] xor rot(C[x+1],1)
    for x in (0...4)
        for y in (0...4)
            A[x,y] = A[x,y] xor D[x]
    ρ step
    for x in (0...4)
```

```
        for y in (0...4)
            A[x,y] = rot[A[x,y],r[x,y]]
    π step
    for x in (0...4)
        for y in (0...4)
            B[y,2*x+3*y] = A[x,y]
    χ step
    for x in (0...4)
        for y in (0...4)
            A[x,y] = B[x,y] xor ((not B[x+1,y]) and B[x+2,y])
    ι step
    A[0,0] = A[0,0] xor RC

    return A
};
```

The purpose of θ is to diffuse the state. If a variable in every column of state has even parity, it will not diffuse to other columns: this is the *column parity kernel (CP kernel)* property. Thus diffusion of some input variables caused by θ can be controlled in the first round. This property has been widely used in cryptanalysis of Keccak. For example, the attacks in [10] use it to decrease the dimension of the cube. The operations ρ and π just change the position of bits. The first three linear operations θ, ρ and π will be called half a round. In the permutation, the only nonlinear operation is χ whose algebraic degree is 2. Therefore, after an n-round Keccak internal permutation, the algebraic degree of the output polynomial is at most 2^n. We will not consider ι since it has no impact on our attacks.

2.2 Keyed Modes of Keccak

MAC Based on Keccak. As an example demonstrated in Fig. 2, one gets a MAC (or a tag) by concatenating a secret key with a message as the input to a hash function. This primitive to ensure data integrity and authentication of a message should satisfy the two following security requirements: no key recovery and resistance of MAC forgery.

Figure 2 shows the construction of Keccak-MAC-n working on a single block. As described in Sect. 2.1, n is half of capacity length. In this paper we will use a single block message and assume that the key and tag are 128 bits long. So there are two significant lanes that consist of key bits. Block sizes may be different based on the variants we analyse.

Authenticated Encryption Scheme Based on Keccak. An AE scheme is used to provide confidentiality, integrity and authenticity of data where decryption is combined with integrity verification. An authenticated encryption scheme based on Keccak is the scheme Keyak [11] which is a third-round candidate algorithm submitted to CAESAR [19]. Figure 3 depicts the construction of Keyak

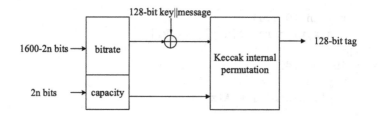

Fig. 2. Construction of Keccak-MAC-n

on two-block message. Both key and nonce are 128 bits. The capacity is 256 bits long and the bitrate is 1344 bits long.

According to the specification of Keyak [11], when confidentiality of data is not required, a nonce can be reused. In this paper, we shall restrict our discussion to the two-block Keyak.

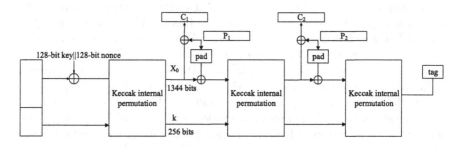

Fig. 3. Construction of Keyak on two blocks

2.3 Cube Tester

Cube tester introduced in [20] is a distinguisher to detect some algebraic property of cryptographic primitives. The idea is to reveal non-random behaviour of a Boolean function with algebraic degree d by summing its values when cube variables of size k $(k \leq d)$ run over all of their 2^k inputs. This cube sum can be taken as higher order derivative [21] of the output polynomial with respect to cube variables. More precisely, we have

Theorem 1. ([10]) *Given a polynomial* $f : \{0,1\}^n \rightarrow \{0,1\}$ *of degree* d. *Suppose that* $0 < k \leq d$ *and* t *is the monomial* $\prod_{i=0}^{k-1} x_i$. *Write* f *as:*

$$f(X) = t \cdot P_t(x_k, \ldots, x_{n-1}) + Q_t(X),$$

where none of the monomials in $Q_t(X)$ *is divisible by* t. *Then the sum of* f *over all values of the cube (cube sum) is*

$$\sum_{x' \in C_t} f(x', x_k, \ldots, x_{n-1}) = P_t(x_k, \ldots, x_{n-1}),$$

where the cube C_t *contains all binary vectors of the length* k.

Some properties for the polynomial P_t, such as its low algebraic degree and highly unbalanced truth table, have been extensively considered in [20,22].

$(n+1)$-**Round Cube Tester on Keccak Sponge Functions.** A cube tester can be constructed based on algebraic properties of Keccak sponge function to distinguish a round-reduced Keccak from a random permutation. An adversary can easily select a combination of $2^n + 1$ cube variables such that they are not multiplied with each other after the first round of Keccak. Note that after n-round Keccak the degree of these cube variables is at most 2^n. So the adversary can sum the output values over a cube of dimension $2^n + 1$ to get zero for a $(n+1)$-round Keccak. This property is also used to perform MAC forgery attack in [10] when $n \leq 6$.

3   Conditional Cube Tester for Keccak Sponge Function

As stated in Sect. 2, the cube attacks against the keyed modes of Keccak in [10] are to select the cube variables that are not multiplied with each other after the first round. Actually, it can be done simply in the context of the differential propagation. Let us consider the following example. In Fig. 4, $A[2][0][0] = A[2][1][0] = v_0$ is set to be a cube variable and it only impacts two bits before the operation χ in the first round. To find $2^n + 1(n \leq 6)$ cube variables to construct an attack, one just needs to trace the positions of these bits.

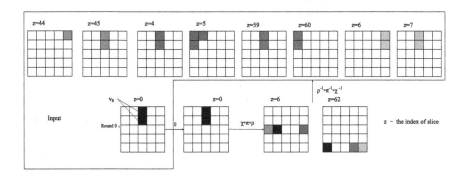

Fig. 4. Overview of bit conditions

In our new model, we develop a strategy to carefully choose the cube variables such that they are either not multiplied with each other after the second round or multiplied within a restrict set of variables.

The idea of our new model—the *conditional cube tester*, is to attach some bit conditions to a cube tester. Figure 4 illustrates how to formulate such conditions. A detailed discussion will be given later in this section. To minimize the possibilities that the cube variable v_0 gets multiplied with other cube variables, we need to slow down the propagation of v_0. This can be done by imposing some additional conditions on the input message so that the coloured input bits of the second round are not related to v_0. Thus these coloured input bits of the second round will not diffuse to other bits in the next round Keccak internal permutation. This is how the propagation of v_0 is controlled.

In the rest of this section, we shall define some types of cube variables in the CP kernel that are involved in the conditional cube tester. An important type is a set of variables that are well behaved through two rounds of Keccak, and we will see that some extra conditions on bits must be satisfied in order to get such variables. Then we will prove a useful result for these cube variables in a conditional cube tester. In the last part of the section, we shall discuss some properties on Keccak sponge function and describe algorithms based on these properties to examine multiplication relation after the second round between every pair of cube variables.

3.1   Conditional and Ordinary Cube Variables in a Conditional Cube Tester

In our discussion, cube variables are the variables in the CP kernel that are not multiplied with each other after the first round of Keccak. Now let us define two types of cube variables for the conditional cube tester.

Definition 1. *Cube variables that have propagation controlled in the first round and are not multiplied with each other after the second round of Keccak are called **conditional cube variables**. Cube variables that are not multiplied with each other after the first round and are not multiplied with any conditional cube variable after the second round are called **ordinary cube variables**.*

An ordinary cube variable has the advantage that it does not need any extra conditions. However, there are no mechanisms to prevent ordinary cube variables from being multiplied with each other after the second round. Thus, in order to get an optimal cube tester for Keccak sponge function, a proper combinations of ordinary cube variables and conditional cube variables should be carefully selected.

To construct an $(n + 2)$-round cube tester, we need to choose p conditional cube variables and q ordinary cube variables. With an appropriate choice of p and q, we have

Theorem 2. *For $(n + 2)$-round Keccak sponge function $(n > 0)$, if there are p $(0 \leq p < 2^n + 1)$ conditional cube variables $v_1, \ldots, v_p$, and $q = 2^{n+1} - 2p + 1$ ordinary cube variables, $u_1, \ldots, u_q$ (If $q = 0$, we set $p = 2^n + 1$), then the term $v_1 v_2 \ldots v_p u_1 \ldots u_q$ will not appear in the output polynomials of $(n + 2)$-round Keccak sponge function.*

Proof. Let $X_1, \cdots, X_s$ be the terms that contain v_i $(i = 1, \ldots, p)$ after the second round. Then by the definition of conditional cube variables, the degree of each X_j is one with respect to some $v_i (i = 1, \ldots, p)$. Similarly, let $Y_1, \cdots, Y_t$ be the terms that contain u_i $(i = 1, \ldots, q)$ after the second round. Then by the definition of ordinary variables, the degree of each Y_j is at most two with respect to some u_is $(i = 1, \ldots, p)$, and no v_i $(i = 1, \ldots, p)$ appears in Y_j.

For output polynomials after another n-round operation, a term with the highest degree with respect to $v_1, \ldots, v_p$ and $u_1, \ldots, u_q$ must be of the following form

$$T_{n+2} = X_{i_1} X_{i_2} \ldots X_{i_k} Y_{j_1} Y_{j_2} \ldots Y_{j_h} \quad \text{with } k + h = 2^n.$$

This implies that there are at most k distinct v_i and $2h$ distinct u_j can appear in T_{n+2}.

If T_{n+2} is divisible by $v_1 v_2 \ldots v_p u_1 \ldots u_q$, then we would have $k \geq p, 2h \geq q+1$ (since q is odd). This yields

$$k + h \geq p + \frac{q+1}{2} = p + 2^n - p + 1 > 2^n,$$

and we have reached a contradiction. $\square$

Let us make some remarks on this theorem. The case that there is no conditional cube variable (i.e., $p = 0$) has been discussed extensively in [10], such as forgery attacks on Keccak-MAC and Keyak. For the case where $1 \leq p \leq 2^n + 1$, we can apply the conditional cube tester to recover the key for the $(n + 2)$-round keyed modes of Keccak based on Theorem 2. The specific methods will be described in Sects. 4 and 5. Furthermore, in Sect. 6, we are able to use the case $p = 2^n + 1$ to implement the distinguishing attacks on Keccak sponge function.

In this paper, we only consider the cases when $n = 3, 4, 5$. If a proper combination of cube variables could be found for $n > 5$, the conditional cube tester still works.

3.2 Properties of Keccak Sponge Function

Before stating three useful properties of Keccak sponge function, we will describe the bitwise derivative of Boolean functions–a tool that helps us to explain our ideas accurately. The bitwise derivative of Boolean functions was proposed by Bo Zhu et al. and used to analyse Boolean algebra based block ciphers [23]. We observe that there is an equivalent relation between the differential characteristic and the bitwise derivatives of Boolean functions. However, it is much more efficient to trace the propagation of a variable by observing the differential characteristic rather than by computing the exact bitwise derivatives of Boolean functions. The bitwise derivative of a Boolean function is defined as follows.

Definition 2. *Given a Boolean function* $f(x_0, x_1, \ldots, x_{n-1})$, *the bitwise derivative of* f *with respect to the variable* x_m *is defined as*

$$\delta_{x_m} f = f_{x_m=1} + f_{x_m=0}$$

The 0-th bitwise derivative is defined to be f *itself. The i-th, where* $i \geq 2$, *bitwise derivative with respect to the variable sequence* $(x_{m_1}, \ldots, x_{m_i})$ *is defined as*

$$\delta^{(i)}_{x_{m_1}, \ldots, x_{m_i}} f = \delta_{x_{m_i}} (\delta^{(i-1)}_{x_{m_1}, \ldots, x_{m_{i-1}}} f)$$

Now let us describe differential properties of χ in the view of bitwise derivative. In this section, we first fix some notations. We will write the input of χ to be
the (vector-valued) Boolean function $F = (f_0, f_1, f_2, f_3, f_4)$. The corresponding
output is written as the (vector-valued) Boolean function $G = (g_0, g_1, g_2, g_3, g_4)$.
The bitwise derivative of a (vector-valued) Boolean function is defined to be the
(vector-valued) Boolean function by taking bitwise derivative in a component-
wise manner.

Property 1 (**Bit Conditions**). If $\delta_{v_0} F = (1, 0, 0, 0, 0)$, then $\delta_{v_0} G = (1, 0, 0, 0, 0)$
if and only if $f_1 = 0$ and $f_4 + 1 = 0$.

Proof. By the structure of χ, the algebraic representation of the output Boolean
function G is given by the following equations:

$$
\begin{aligned}
g_0 &= f_0 + (f_1 + 1)f_2, \\
g_1 &= f_1 + (f_2 + 1)f_3, \\
g_2 &= f_2 + (f_3 + 1)f_4, \\
g_3 &= f_3 + (f_4 + 1)f_0, \\
g_4 &= f_4 + (f_0 + 1)f_1.
\end{aligned}
$$

From the definition of the bitwise derivative, it can be deduced that $\delta_{v_0} G =$
$(1, 0, 0, f_4 + 1, f_1)$. It is clear that $\delta_{v_0} G = (1, 0, 0, 0, 0)$ if and only if $f_1 = 0$ and
$f_4 + 1 = 0$.   □

Fig. 5. Diffusion caused by operation χ

Now we explain the equivalence between the truncated differential characteristic and the bitwise derivatives of Boolean functions when tracing the propagation of a variable by using Fig. 5. Let the input difference for χ be $(1, 0, 0, 0, 0)$
and the truncated output difference is $(1, 0, 0, ?, ?)$ with '?' meaning an unknown
bit. From the view of Boolean functions, the output vector $(1, 0, 0, ?, ?)$ indicates
that $\delta_{v_0} g_0 = 1$, $\delta_{v_0} g_1 = 0$, $\delta_{v_0} g_2 = 0$ and both $\delta_{v_0} g_3, \delta_{v_0} g_4$ are some Boolean functions. From the view of the differential characteristic, if $f_1 = 0$ and $f_4 + 1 = 0$,
then the differential characteristic $(1, 0, 0, 0, 0) \rightarrow (1, 0, 0, 0, 0)$ holds with probability 1. This also implies that g_0 is related to v_0 but g_i (for $1 \leq i \leq 4$) are
independent of v_0. Therefore, the truncated differential characteristics and the
bitwise derivatives of Boolean functions are equivalent representations.

Table 4. Summary of conditions for bitwise derivative of χ

Input/Output bitwise derivative (Difference)	Conditions
$(1,0,0,0,0) \rightarrow (1,0,0,0,0)$	$f_1 = 0, f_4 = 1$
$(0,1,0,0,0) \rightarrow (0,1,0,0,0)$	$f_2 = 0, f_0 = 1$
$(0,0,1,0,0) \rightarrow (0,0,1,0,0)$	$f_3 = 0, f_1 = 1$
$(0,0,0,1,0) \rightarrow (0,0,0,1,0)$	$f_4 = 0, f_2 = 1$
$(0,0,0,0,1) \rightarrow (0,0,0,0,1)$	$f_0 = 0, f_3 = 1$

We summarize all of the five input bitwise derivative cases in Table 4 where each input bitwise derivative has only one non-zero bit. Each case can be proved in a similar manner as Property 1. As discussed before, in each case, the input and output have the same vector of bitwise derivatives so that the propagation of v_0 by χ is under control. This will be used in constructing our conditional cube tester.

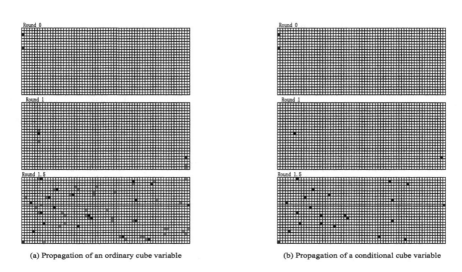

(a) Propagation of an ordinary cube variable   (b) Propagation of a conditional cube variable

Fig. 6. 1.5-round differential of an ordinary and a conditional cube variable

In order to show the advantage of a conditional cube variable over an ordinary cube variable, we consider the propagation of variable $A[2][0][0] = A[2][1][0] = v_0$ in two views: as an ordinary cube variable in the view of truncated differential characteristic (Fig. 6(a)) and as a conditional cube variable in the view of differential characteristic (Fig. 6(b)).

It is obvious to see that the two active bits at the beginning of the second round will affect 22 bits caused by the step θ. Thus, the conditional cube variable in Fig. 6(b) only relates to 22 active bits after 1.5-round Keccak internal permutation. However, not only bits with black colour but also those with gray colour after 1.5-round Keccak involve the ordinary cube variable in Fig. 6(a). In total, there are 62 bits related to v_0 after 1.5-round Keccak. So it is more likely for a ordinary cube variable to get multiplied with other cube variables after the second round Keccak.

The pattern of the conditional cube variable v_0 in Fig. 6(b) will be called a 2-2-22 pattern to reflect the number of active bits in three states (the input state, the output state of the first round and the output state of the first 1.5 rounds).

During the process of searching more cube variables, we need to determine whether candidate variables get multiplied after the second round of Keccak and eliminate conditional cube variable candidates that require conflicting conditions. We observe that the following two properties with respect to the operation χ will be useful in dealing with these situations.

Property 2 (**Multiplication**). Assume that $\delta_{v_0} F = (\delta_{v_0} f_0, 0, 0, 0, 0)$ and $\delta_{v_1} F = (0, \delta_{v_1} f_1, 0, 0, 0)$ with $\delta_{v_0} f_0 \cdot \delta_{v_1} f_1 \neq 0$, then the term $v_0 v_1$ will be in the output of χ.

Proof. As mentioned in the proof of the Property 1, the component g_4 of the output $G = (g_0, g_1, g_2, g_3, g_4)$ is $f_4 + (f_0 + 1)f_1$. From

$$\delta^{(2)}_{v_0, v_1} g_4 = \delta_{v_1}(\delta_{v_0} g_4) = \delta_{v_1}(\delta_{v_0} f_0) \cdot f_1 + \delta_{v_0} f_0 \cdot \delta_{v_1} f_1 = \delta_{v_0} f_0 \cdot \delta_{v_1} f_1$$

we see that $\delta^{(2)}_{v_0, v_1} g_4 \neq 0$ and hence g_4 contains the term $v_0 v_1$. In particular, if $\delta_{v_0} f_0 = \delta_{v_1} f_1 = 1$, then $g_4 = v_0 v_1 + h$, where h is a Boolean function not divisible by $v_0 v_1$. □

Property 3 (**Exclusion**). If $\delta_{v_0} F = (1, 0, 0, 0, 0)$ and $\delta_{v_1} F = (0, 0, 1, 0, 0)$, then at least one of $\delta_{v_0} G = (1, 0, 0, 0, 0)$ and $\delta_{v_1} G = (0, 0, 1, 0, 0)$ is false.

Proof. From the Property 1 as well as the Table 4, the conditions $\delta_{v_0} F = (1, 0, 0, 0, 0)$ and $\delta_{v_0} G = (1, 0, 0, 0, 0)$ would imply $f_1 = 0, f_4 = 1$. Under the assumption $\delta_{v_1} F = (0, 0, 1, 0, 0)$, if $\delta_{v_1} G = (0, 0, 1, 0, 0)$ also holds true, then we would have $f_1 = 1, f_3 = 0$. This is a contradiction. □

For a version of Keccak sponge function, many positions in the plaintext space can be set as cube variables. For example, as shown in Fig. 7, we can set the bits in the same colour as a cube variable for the version Keccak-512. There are 256 such cases in 64 slices. Each of these cases in a version of Keccak is called a cube variable candidate.

Before searching for a proper combination of cube variables from these candidates to construct a conditional cube tester, we need to know the relation between every pair of cube variable candidates, namely, whether they are multiplied after the second round of Keccak. This problem could be solved directly by examining exact intermediate polynomials after the second round. However, it is very time-consuming to derive such an exact representation for the polynomials after the second round. Our approach with the application of truncated differentials can determine the (multiplication) relation between two cube variables efficiently. The precise procedures will be given in Algorithms 1, 2 and 3. These three algorithms are based on Properties 2 and 3.

Fig. 7. Cube variable candidates in a slice for Keccak-512

In the three algorithms, v_0 and v_1 are assumed to be two cube variable candidates in a Keccak version. We use $\delta_{v_0}A$ ($\delta_{v_1}A$) to denote the positions of v_0 (v_1) in the input state, which means to apply bitwise derivative on each entry of A. For example $\delta_{v_0}A[i][j][k] = 1$ means $A[i][j][k] = v_0 + h$, where h is a Boolean function independent of v_0. For a cube variable candidate v, we shall use '0', '1' and '2' to denote the inactive bit, the active bit and the unknown bit respectively. To be more specific, v is of type '0' if $\delta_v A[i][j][k] = 0$, v is of type '1' if $\delta_v A[i][j][k] = 1$ and type '2' if $\delta_v A[i][j][k]$ is a Boolean function. In this way, the truncated differences or differences in the algorithms can be used to interpret the bitwise derivatives on the state with respect to v.

Now we include three algorithms in this subsection for determining whether two possible cube variables (conditional or ordinary) have a multiplication relation after first round and the second round. The first algorithm is restricted to the case of two ordinary cube variable candidates. They should not be multiplied together after the first round.

The second algorithm is to test the relation between a conditional cube variable candidate and an ordinary cube variable candidate, whose multiplication is not allowed after the second round. The third algorithm is to examine the relation between two conditional cube variable candidates, whose multiplication is not allowed after the second round either.

4   Key Recovery Attack on Reduced-Round Keccak-MAC

In this section, we will use conditional cube testers to perform key recovery attacks against Keccak-MAC. First, we will discuss the general procedure for key recovery attack, including the attack process, complexity analysis and searching algorithm for suitable combinations of conditional and ordinary cube variables. Then we will describe conditional cube attacks to different variants of Keccak-MAC, including Keccak-MAC-512, Keccak-MAC-384 and Keccak-MAC-224.

274   S. Huang et al.

Algorithm 1. Determine Relation of Two Ordinary Cube Variable Candidates

Input: $\delta_{v_0} A$ and $\delta_{v_1} A$ for two ordinary cube variable candidates v_0 and v_1
Output: multiplication relation of v_0 and v_1
1: compute the 0.5-round output difference B_0 (B_1) based on $\delta_{v_0} A(\delta_{v_1} A)$;
2: flag=0
3: **for** each integer $i \in [0, 63]$, each integer $j \in [0, 4]$, each integer $k \in [0, 4]$ **do**
4:    **if** $B_0[k][j][i] \cdot B_1[k+1][j][i] = 1$ **then**
5:       flag=1;   ▷ Property 2.
6:    **end if**
7: **end for**
8: **if** (flag) **then**
9:    **return** multiplied after the first round;
10: **else**
11:    **return** not multiplied after the first round;
12: **end if**

Algorithm 2. Determine Relation of a Conditional Cube Variable Candidate and an Ordinary Cube Variable Candidate

Input: $\delta_{v_0} A$ and $\delta_{v_1} A$ for the conditional cube variable candidate v_0 and the ordinary cube variable candidate v_1
Output: multiplication relation of v_0 and v_1
1: flag=[0,0]
2: compute the 0.5-round output difference B_0 (B_1) based on $\delta_{v_0} A(\delta_{v_1} A)$;
3: compute the 1.5-round truncated output difference C_0 (C_1) based on $\delta_{v_0} A(\delta_{v_1} A)$;
4: **for** each integer $i \in [0, 63]$, each integer $j \in [0, 4]$, each integer $k \in [0, 4]$ **do**
5:    **if** $B_0[k][j][i] \cdot B_1[k+1][j][i] = 1$ **then**
6:       flag[0]=1;   ▷ Property 2.
7:    **end if**
8:    **if** $C_0[k][j][i] \cdot C_1[k+1][j][i] \neq 0$ **then**
9:       flag[1]=1;   ▷ Property 2.
10:    **end if**
11: **end for**
12: **if** (flag[0]) **then**
13:    **return** multiplied after the first found;
14: **else if** (flag[1]) **then**
15:    **return** multiplied after the second round;
16: **end if**
17: **return** not multiplied after the second round;

Algorithm 3. Determine Relation of Two Conditional Cube Variable Candidates

---

Input: $\delta_{v_0} A$ and $\delta_{v_1} A$ for two conditional cube variable candidates v_0 and v_1
Output: multiplication relation of v_0 and v_1
1: flag=[0,0,0]
2: compute the 0.5-round output difference B_0 (B_1) based on $\delta_{v_0} A(\delta_{v_1} A)$;
3: compute the 1.5-round output difference C_0 (C_1) based on $\delta_{v_0} A(\delta_{v_1} A)$;
4: **for** each integer $i \in [0, 63]$, each integer $j \in [0, 4]$, each integer $k \in [0, 4]$ **do**
5:     **if** $B_0[k][j][i] \cdot B_1[k + 2][j][i] = 1$ **then**
6:        flag[0]=1;   ▷ Property 3.
7:     **end if**
8:     **if** $B_0[k][j][i] \cdot B_1[k + 1][j][i] = 1$ **then**
9:        flag[1]=1;   ▷ Property 2.
10:   **end if**
11:   **if** $C_0[k][j][i] \cdot C_1[k + 1][j][i] = 1$ **then**
12:      flag[2]=1;   ▷ Property 2.
13:   **end if**
14: **end for**
15: **if** (flag[0]) **then**
16:   **return** contradiction;
17: **else if** (flag[1]) **then**
18:   **return** multiplied after the first round;
19: **else if** (flag[2]) **then**
20:   **return** multiplied after the second round;
21: **end if**
22: **return** not multiplied by the second round;

4.1 General Process for Key Recovery Attack on Keccak-MAC

Given a cube tester with p conditional cube variables and $q = 2^{n+1} - 2p + 1$ ordinary cube variables ($1 \leq p \leq 2^n + 1$), we can construct a key recovery attack on $(n + 2)$-round Keccak-MAC. In order to explain the general attack process clearly, we need to define some types of variables other than cube variables. As we know, a bit condition is an equality with a single variable on the left hand side and a Boolean function on the right hand side. The variable on the left hand side is called a **conditional variable**. Other public variables (that can be assigned to arbitrary values) are called **free variables**. Thus, a bit condition is a relation between conditional variable, equivalent key bit and free variables. It is assumed that s equivalent key bits are related to the bit conditions derived from conditional cube variables. The general attack process is described as follows.

Step 1. Assign free variables with random values.
Step 2. Guess values of the s equivalent key bits.
Step 3. Calculate the values of conditional variables under the guess of key bits.
Step 4. For each possible set of values of cube variables, compute the corresponding tag and then sum all of the 128-bit tags over the $(2^{n+1} - p + 1)$-dimension cube.

Step 5. If the sum is zero, the guess of these s key bits is probable correct and the process terminates; otherwise the guess is invalid, go back to Step 2.

After one execution of the above process, which takes $2^{2^{n+1}-p+1} \cdot 2^s$ Keccak calls at most, the values of s key bits can be recovered. To recover the remaining $128 - s$ key bits, we just shift the positions of all the cube variables equally to the right along the z-direction and repeat the process for $128/s$ times. In this case, the bitwise derivatives with respect to the cube variables are rotated equally along the z axis as well. This rotation, known as translation invariance in the direction of the z axis, will change the equivalent key bits in the bit conditions but not the relations between the cube variables. Therefore, the time and data complexity of the key recovery attack are both $\frac{1}{s} \cdot 2^{2^{n+1}-p+s+8} = \frac{2^{s-p}}{s} \cdot 2^{2^{n+1}+8}$. Thus, for an $(n+2)$-round conditional cube attack, the complexity is determined by $\frac{2^{s-p}}{s}$. We would like this term to be small to achieve a better performance. Notice that when the number of conditional cube variables gets larger, more key bits will be involved in the bit conditions and hence more guesses will be required. So p can not be too large to make the attack better. In our case, we use one conditional cube variable and $2^{n+1} - 1$ ordinary cube variables to construct our key recovery attack on Keccak-MAC.

We choose $A[2][0][0] = A[2][1][0] = v_0$ as the conditional cube variable in our attacks. As shown in Fig. 4, bit conditions are derived from $\delta_{v_0} A[2][0][6] = \delta_{v_0} A[2][4][6] = \delta_{v_0} A[4][3][62] = \delta_{v_0} A[4][4][62] = 0$, where A is the intermediate state after 1.5-round Keccak. This procedure could be done efficiently with the help of SAGE [24], a software on symbol computation. We fix $A[2][0][0] = A[2][1][0] = v_0$ as the conditional cube variable because there are only two equivalent key bits involved in the bit conditions. But if we choose other positions to set the conditional cube variable, the number of key bits involved in the bit conditions may be greater than two. Thus, $A[2][0][0] = A[2][1][0] = v_0$ is the cube variable and we find the corresponding ordinary cube variables using Algorithm 4.

In the discussion later, we will see that $2^{n+1} - 1$ ordinary cube variables can be always found for $n = 3, 4$ and 5. So in these cases, the cube tester with v_0 and $2^{n+1} - 1$ ordinary cube variables can be constructed to perform key recovery attacks on different variants of Keccak-MAC.

4.2 Key Recovery on 5/6/7-Round Keccak-MAC

We first discuss 5-round Keccak-MAC-512. In this case, $n = 3$ and full key bits can be recovered with one conditional cube variable and 15 ordinary cube variables. The block size of this version is $1600 - 2 \cdot 512 = 576$ bits. As discussed in Sect. 4.1, we set $A[2][0][0] = A[2][1][0] = v_0$ to be the conditional cube variable. $A[4][0][44]$, $A[2][0][4]$, $A[2][0][59]$ and $A[2][0][27]$ are the conditional variables assigned with Boolean functions and a set of the corresponding ordinary cube variables is produced by Algorithm 4 (see Table 5). To recover the remaining key bits, the positions of the conditional cube variable shall be shifted to

---

Algorithm 4. Searching Ordinary Cube Variables along with the conditional cube variable $A[2][0][0] = A[2][1][0] = v_0$ for Keccak-MAC

Output: a set of ordinary cube variables;
1: $m=\#\{$ordinary cube variable candidates in bitrate part$\}$
2: $S = \varnothing$
3: **for** each integer $i \in [0, m - 1]$ **do**
4:     **execute** Algorithm 2 with v_0 and the i-th ordinary cube variable candidate u_i as the input;
5:     **if** Algorithm 2 returns 'not multiplied by the second round' **then**
6:         $S \leftarrow S \cup \{u_i\}$
7:     **end if**
8: **end for**
9: Choose the maximum number of variables in S which will not be multiplied with each other after the first round and put these variables into T
10: **return** T

$A[2][0][i] = A[2][1][i] = v_0(1 \le i \le 63)$ and the positions of ordinary cube variables shall be rotated at the same time. The key is recovered in 2^{24} time and data, which is very practical. On a desktop computer, the process of recovering a key only costs a few minutes.

The next example is a simple illustration of the attack where the key was generated randomly. For the convenience of statement, all the free variables are fixed to be zero, but they can be random bits. It is obvious that the correct key can be easily distinguished.

128-bit key:
11100001000101000010110100100010111111100000011001011100111 00101
110001110001011110100011111111101000010101100000001 1000100100010
correct value: $k_5 + k_{69} = 1, k_{60} = 0$
guessed value:00, cube sum: 0xe93169ae5c86d086, 0xf6ec898c859bea1a
guessed value:01, cube sum: 0xc7d0bc36dc141c5e, 0x523a33c8753eb171
guessed value:10, cube sum: 0x0,0x0
guessed value:11, cube sum: 0x2ee1d5988092ccd8, 0xa4d6ba44f0a55b6b

To perform a conditional cube attack on 6-round Keccak-MAC-384, we use one conditional cube variable and 31 ordinary cube variables to recover full 128-bit key with 2^{40} Keccak calls. Fixing the conditional cube variable, we collect the corresponding ordinary cube variables by applying Algorithm 4. The parameters for this attack can be found in Table 9. It takes just a few days to run this attack on a desktop with four $i5$ processors. An instance for attacking 6-round Keccak-MAC-384 is summarized below, with randomly generated key and free variables are fixed to be zero:

128-bit key:
11110111110010010001110100101001111000111100011101111001000 00010
011100001001010001010111011011111101000010101010001110111001100011
correct value: $k_5 + k_{69} = 1, k_{60} = 0$
guessed value:00, cube sum: 0x3f9d5fa4e143f779, 0x26607b3ce1c56f2b

Table 5. Parameters set for attack on 5-round Keccak-MAC-512

Ordinary cube variables	$A[2][0][8]=A[2][1][8]=v_1$, $A[2][0][12]=A[2][1][12]=v_2$, $A[2][0][20]=A[2][1][20]=v_3$, $A[2][0][28]=A[2][1][28]=v_4$, $A[2][0][41]=A[2][1][41]=v_5$, $A[2][0][43]=A[2][1][43]=v_6$, $A[2][0][45]=A[2][1][45]=v_7$, $A[2][0][53]=A[2][1][53]=v_8$, $A[2][0][62]=A[2][1][62]=v_9$, $A[3][0][3]=A[3][1][3]=v_{10}$, $A[3][0][4]=A[3][1][4]=v_{11}$, $A[3][0][9]=A[3][1][9]=v_{12}$, $A[3][0][13]=A[3][1][13]=v_{13}$, $A[3][0][23]=A[3][1][23]=v_{14}$, $A[3][0][30]=A[3][1][30]=v_{15}$
Conditional cube variable	$A[2][0][0]=A[2][1][0]=v_0$
Bit conditions	$A[4][0][44]=0$, $A[2][0][4]= k_5 + k_{69} + A[0][1][5] + A[2][1][4] + 1$, $A[2][0][59]= k_{60} + A[0][1][60] + A[2][1][59] + 1$, $A[2][0][7]= A[4][0][6] + A[2][1][7] + A[3][1][7]$
Guessed key bits	$k_{60}, k_5 + k_{69}$

guessed value:01, cube sum: 0x99bbf2ae6b93a7fb, 0xdbbb864fcc563747
guessed value:10, cube sum: 0x0,0x0
guessed value:11, cube sum: 0x398b37a846e81e42, 0x691cf4345e2164ee

For 7-round Keccak-MAC-256, our conditional cube attack takes 2^{72} Keccak calls to recover full 128-bit key, with a cube of dimension 64. We include the parameters of this attack in Table 10.

5   Key Recovery Attacks on Reduced-Round Keyak

Similar to the key recovery attack on Keyak in [10], we also deal with two-block messages (as depicted in Fig. 3) and allow the reuse of a nonce. In this way, we can use the first block to control the input of the second permutation and the second block to get the output of the second permutation. The attack described here is in fact a state recovery attack. We are able to get the bitrate part X_0 (see Fig. 3) but not the 256 bits in the capacity part. Denoting the capacity part as $k = (k_0, k_1, \cdots, k_{255})$, we will first recover k, then get the master key by performing the inverse of the first Keccak internal permutation.

In the attack, cube variables are set in the input state of the second Keccak internal permutation by choosing the values of P_1 while the second message block P_2 is set to zero bits. This implies that the second ciphertext block C_2 is the output of Keccak internal permutation. The attack procedure is almost identical to the general process described in Sect. 4.1 except for the bit conditions and the inverse process on the output. For 1344 output bits of Keyak, the operation χ of the last round on the most significant 1280 bits can be reversed. Note that the linear operations of the final round do not increase the degree of output polynomials, so the previous $(n + 2)$-round cube tester can be used for $(n + 3)$-round. In other words, conditional cube attack can be extended by one more round forward without increasing the dimension of cube.

For 7-round Keyak, conditional cube attack is built with the same cube as in Table 9 except for a different set of bit conditions as shown in Table 6. Note that in Table 6 A denotes the input state to the second internal permutation. By shifting the positions of cube variables and repeating the attack for $192/4 = 48$ times, three lanes of secret values, i.e. $k_0, \ldots, k_{191}$, can be recovered with $2^{36} \cdot 48 = 2^{41.58}$ Keyak calls. The other lane of key bits can be recovered by changing the conditional cube variable to $A[3][0][i] = A[3][1][i] = v_0$ and a set of the corresponding ordinary cube variables could be produced similarly by Algorithm 4. Since only one key bit is involved in the bit conditions after recovering three lanes of secret values, the remaining lane of secret values can be identified with $2^{33} \cdot 2^6 = 2^{39}$ Keyak calls. In total, the time complexity to recover the full 128-bit master key is about 2^{42} Keyak calls.

For 8-round Keyak, cube variables in Table 10 and bit conditions in the Table 6 are used in the conditional cube attack. Using a similar analysis as that to 7-round Keyak, the data and time complexities for 8-round attack are 2^{74}.

Finally, we remark that the memory complexity for both attacks can be neglected.

Table 6. Parameters for attacking 7-round and 8-round Keyak

Bit conditions for 8(7)-round Keyak	$A[4][0][44]=k_{169}$ $(+A[4][1][44])$ + $A[2][2][45]$ + $A[3][2][45]$ + $A[4][2][44]$ + $A[2][3][45]$ + $A[4][3][44]$, $A[0][0][5]= k_{128}$ + $A[1][0][5]$ + $A[2][0][4]$ + $A[0][1][5]$ + $A[2][1][4]$ + $A[0][2][5]$ + $A[2][2][4]$ + $A[0][3][5]$ + $A[2][3][4]$ + $A[0][4][5]$ + 1, $A[0][0][60]= k_{56}$ + k_{183} + $A[2][0][59]$ + $A[0][1][60]$ + $A[2][1][59]$ + $A[0][2][60]$ + $A[2][2][59]$ + $A[0][3][60]$ + $A[2][3][59]$+ $A[0][4][60]$ + 1, $A[2][0][7]= k_{131}$ + $A[4][0][6]$ + $A[2][1][7]$ + $A[3][1][7]$ + $A[4][1][6]$ + $A[2][2][7]$+ $A[4][2][6]$ + $A[2][3][7]$ + $A[4][3][6]$
Guessed key bits	$k_{169}, k_{128}, k_{56} + k_{183}, k_{131}$

6   Distinguishing Attacks on Keccak Sponge Function

In this section, conditional cube tester will be applied to establish distinguishing attacks on Keccak sponge function with practical complexity. By Theorem 2, if we use $2^n + 1$ conditional cube variables, the monomial containing these $2^n + 1$ conditional cube variables will not appear in the output polynomials of $(n + 2)$-round Keccak sponge function. This means that the dimension of the cube to distinguish $(n + 2)$-round Keccak is reduced to $2^n + 1$ from higher numbers reported in [10]. In some cases like Keccak-512 and Keccak-384, the distinguishing attacks could be extended one more round forward.

Our construction of the cube tester includes two parts:

- Find a combination of sufficiently many conditional cube variables;
- Derive the corresponding bit conditions for the chosen conditional cube variables.

6.1   Constructing Conditional Cube Tester with MILP

A mixed-integer linear programming (MILP) problem is a linear programming problem with some variables taking integer values. MILP has been used to find the best differential characteristic in [25]. In this section, we model the problem of finding a combination of sufficiently many conditional cube variables as an MILP problem.

In this new model, each conditional cube variable candidate is assigned with a variable x_i ($1 \leq i \leq m$) where x_i takes value from $\{0, 1\}$. The i-th conditional cube variable candidate is selected as a conditional cube variable if and only if $x_i = 1$. To find sufficiently many conditional cube variables, we need to find an assignment $X = \{(x_1, x_2, \ldots, x_m) | x_i \in \{0, 1\}, 1 \leq i \leq m\}$ of hamming weight larger than $2^n + 1$. From earlier analysis, we know that in some cases two conditional cube variable candidates can not be selected simultaneously. We will first generate such constraints in terms of X. The precise generation procedure is the following.

Algorithm 5. Generating Constraints on X

Input: m conditional cube variable candidates;
Output: A set F of constrains on X
 1: $F = \varnothing$
 2: **for** each integer $i \in [1, m - 1]$ **do**
 3:     **for** each integer $j \in [i + 1, m]$ **do**
 4:         excute Algorithm 3 on the i-th and j-th conditional cube variable candidates;

 5:         **if** Algorithm 3 does not return 'Not Multiplied after the Second Round' **then**
 6:             $F \leftarrow F \cup \{x_i + x_j \leq 1\}$
 7:         **end if**
 8:     **end for**
 9: **end for**
10: **return** F

With the constraint set F, the selection problem for conditional cube variables is modeled into a binary linear programming problem as follow:

$$\sum_{i=1}^{m} x_i \geq 2^n + 1$$

$$\text{s.t.} A_0 X \leq b, X = \{(x_1, x_2, \ldots, x_m) | x_i \in \{0, 1\}, \quad 1 \leq i \leq m\}$$

where A_0 is a binary matrix and b a binary vector such that $A_0 X \leq b$ describes the constraint set F. Although MILP is proved to be NP-hard, our problem is a special (and small) instance and can be solved by the programming solver Gurobi Optimizer [26] based on branch and cut algorithm.

We can get a desired combination of conditional cube variables by solving the MILP problem. In the rest of the section, we will construct distinguishing

Table 7. Four differential characteristics in double kernel pattern

NO	δ_i	Differential
0	δ_0	. . . . 1 8- . . . .
		. . . . 1 . -. . . . . 4- . . . .
		. . . . 8- . . . . 4- . . . .
		. . . .
0	δ_1	. . . . 1 . . . . 2- . . . .
		. . . . 1 . - . . . . 1- . . . .
		. . . . 1- . . . . 2- . . . .
0	$\delta_{1.5}$	. . . . 1 . . . . 8 . . . . 2 . . . .
		. . . . 4- . . . . 1 . . . .
		. . . . 8 . . . .
1	δ_0	. . . . 8 . . . . 1 .
		. . . . 8 . . . . 8 - . . . . 1 .
		. . . . 8 . . . . 8 - . . . .
1	δ_1	. . . . 1 . . . .
		. . . . 2- . . . .
		. . . . 8 - . . . . 2- . . . .
		. . . . 8 - . . . . 1 . . . .
1	$\delta_{1.5}$	. . . . 1 . . . .
		. . . . 1 . . . .
		. . . . 1 . . . . 1 . . . .
		. . . . 8 . . . . 2 . . . .
2	δ_0	. . . . 1 . . . . 4 . . . . 4 . . .
		. . . . 1 . . . . 4 . . . .
		. . . . 1 . . . . 1 . . . . 4 . . .
2	δ_1	. . . . 2- . . . .
		. . . . 2 . . . . 1 . . . . 2 . . . .
		. . . . 2 . . . . 1 . . . . 2- . . . . 2 . . . .
2	$\delta_{1.5}$	. . . . 1 . . . . 2 . . . . 4 -
		. . . .
		. . . . 8 - . . . 1 . . . . 4 . . . . 4 . . . . 2 . . . .
3	δ_0	. . . . 1 . . . . 4 - .
		. . . . 8 . . . . 1 . . . .
		. . . . 8 . . . . 1 . . . 1 . . . . 4 . . . .
3	δ_1	. . . . 4 - . . . .
		. . . . 2 . . . . 2 - . . . .
		. . . . 2 . . . . 8 . . . . 4 - . . . .
		. . . . 8 . . . . 2 - . . . .
3	$\delta_{1.5}$	. . . . 1 - . . . . 1 . . . . . . 1 . . . . 4 -
		. . . . 1 . . . . 2 . . . .
		. . . . 4 . . . . 2 . . . .

attacks on Keccak sponge function by solving the MILP problems and deriving the corresponding conditions for these conditional cube variables.

6.2   Distinguishing Attack on Keccak-512 and Keccak-384

As depicted in Fig. 7, there are 4 conditional cube variables candidates in one slice for Keccak-512. There are total 256 such candidates in 64 slices. Applying Algorithm 5 to generate all of the constraints with respect to these 256 candidates and solving the problem with Gurobi Optimizer, we get a set of 9 conditional cube variables. The bit conditions can be derived directly from $\delta_{v_m} A[i][j][k] = 0(0 \leq m \leq 9)$, where $\delta_{v_m} A[i][j][k] \neq 1$ and A is the 1.5-round intermediate state. We then construct a 5-round conditional cube tester. Note that the algebraic degree of output polynomial of 5-round Keccak-512 is at most 8, the cube sum of 5-round Keccak-512 output is zero.

The most significant 320 bits of Keccak-512 output can be reversed so that the distinguishing attack can be extended one more round further without increasing

Table 8. Offsets r[x,y] in operation ρ

0	1	62	28	27
36	44	6	55	20
3	10	43	25	39
41	45	15	21	8
18	2	61	56	14

Table 9. Parameters set for attack on 6-round Keccak-MAC-384

Ordinary cube variables	A[2][0][12]=A[2][1][12]=v_1, A[2][0][20]=A[2][1][20]=v_2, A[2][0][28]=A[2][1][28]=v_3, A[2][0][41]=A[2][1][41]=v_4, A[2][0][43]=A[2][1][43]=v_5, A[2][0][45]=A[2][1][45]=v_6, A[2][0][53]=A[2][1][53]=v_7, A[2][0][61]=A[2][1][61]=v_8, A[2][0][62]=A[2][1][62]=v_9, A[3][0][3]=A[3][1][3]=v_{10}, A[3][0][9]=A[3][1][9]=v_{11}, A[3][0][13]=A[3][1][13]=v_{12}, A[3][0][15]=A[3][1][15]=v_{13}, A[3][0][23]=A[3][1][23]=v_{14}, A[3][0][30]=A[3][1][30]=v_{15}, A[3][0][40]=A[3][1][40]=v_{16}, A[3][0][46]=A[3][1][46]=v_{17}, A[3][0][56]=A[3][1][56]=v_{18}, A[3][0][57]=A[3][1][57]=v_{19}, A[4][0][5]=A[4][1][5]=v_{20}, A[4][0][10]=A[4][1][10]=v_{21}, A[4][0][12]=A[4][1][12]=v_{22}, A[4][0][14]=A[4][1][14]=v_{23}, A[4][0][47]=A[4][1][47]=v_{24}, A[4][0][58]=A[4][1][58]=v_{25}, A[4][0][62]=A[4][1][62]=v_{26}, A[4][0][63]=A[4][1][63]=v_{27}, A[0][1][28]=A[0][2][28]=v_{28}, A[0][1][34]=A[0][2][34]=v_{29}, A[0][1][37]=A[0][2][37]=v_{30}, A[0][1][46]=A[0][2][46]=v_{31}
Conditional cube variable	A[2][0][0]=A[2][1][0]=v_0
Bit conditions	A[4][0][44]= A[4][1][44] + A[2][2][45], A[2][0][4]= $k_5 + k_{69}$ + A[0][1][5] + A[2][1][4] + A[0][2][5] + A[2][2][4] + 1, A[2][0][59]= k_{60} + A[0][1][60] + A[2][1][59] + A[0][2][60] + A[2][2][59] + 1, A[2][0][7]= A[4][0][6] + A[2][1][7] + A[4][1][6] + A[2][2][7] + A[3][1][7].
Guessed key bits	$k_{60}, k_5 + k_{69}$

the complexity. The time complexity for the distinguishing attack on 6-round Keccak-512 with the conditional cube tester is thus 2^9 Keccak calls and the data complexity is also 2^9. From the fact that a distinguishing attack on Keccak with the capacity c_1 also works on Keccak with the capacity c_2 with the same complexity as long as $c_1 > c_2$, this attack can also distinguish Keccak-224, Keccak-256 up to 5 rounds and Keccak-384 up to 6 rounds.

We can find a combination of 17 conditional cube variables for Keccak-384 and construct a 7-round conditional cube tester in a similar manner with a complexity of 2^{17}.

The conditions for these two conditional cube tester are shown in Tables 11 and 12. We have verified both of these two conditional cube testers by experiments.

Table 10. Parameters set for attack on 7-round Keccak-MAC-256

Ordinary cube variables	$A[2][0][8]=A[2][1][8]=v_1$, $A[2][0][12]=A[2][1][12]=v_2$, $A[2][0][20]=A[2][1][20]=v_3$, $A[2][0][28]=A[2][1][28]=v_4$, $A[2][0][41]=A[2][1][41]=v_5$, $A[2][0][43]=A[2][1][43]=v_6$, $A[2][0][45]=A[2][1][45]=v_7$, $A[2][0][53]=A[2][1][53]=v_8$, $A[2][0][62]=A[2][1][62]=v_9$, $A[3][0][3]=A[3][1][3]=v_{10}$, $A[3][0][9]=A[3][1][9]=v_{11}$, $A[3][0][13]=A[3][1][13]=v_{12}$, $A[3][0][30]=A[3][1][30]=v_{13}$, $A[3][0][40]=A[3][1][40]=v_{14}$, $A[3][0][46]=A[3][1][46]=v_{15}$, $A[3][0][56]=A[3][1][56]=v_{16}$, $A[4][0][5]=A[4][1][5]=v_{17}$, $A[4][0][10]=A[4][1][10]=v_{18}$, $A[4][0][12]=A[4][1][12]=v_{19}$, $A[4][0][14]=A[4][1][14]=v_{20}$, $A[4][0][31]=A[4][1][31]=v_{21}$, $A[4][0][47]=A[4][1][47]=v_{22}$, $A[4][0][58]=A[4][1][58]=v_{23}$, $A[4][0][62]=A[4][1][62]=v_{24}$, $A[4][0][63]=A[4][1][63]=v_{25}$, $A[0][1][37]=A[0][2][37]=v_{26}$, $A[0][1][47]=v_{27}$, $A[0][2][47]=v_{27}+v_{28}$, $A[0][3][47]=v_{28}$, $A[0][1][46]=A[0][2][46]=v_{29}$, $A[0][1][59]=A[0][2][59]=v_{30}$, $A[1][1][7]=A[1][2][7]=v_{31}$, $A[1][1][15]=A[1][2][15]=v_{32}$, $A[1][1][20]=A[1][2][20]=v_{33}$, $A[1][1][26]=A[1][2][26]=v_{34}$, $A[1][1][30]=A[1][2][30]=v_{35}$, $A[1][1][38]=A[1][2][38]=v_{36}$, $A[1][1][39]=A[1][2][39]=v_{37}$, $A[1][1][40]=A[1][2][40]=v_{38}$, $A[1][1][52]=A[1][2][52]=v_{39}$, $A[1][1][54]=A[1][2][54]=v_{40}$, $A[2][1][11]=A[2][2][11]=v_{41}$, $A[2][1][15]=A[2][2][15]=v_{42}$, $A[2][1][19]=A[2][2][19]=v_{43}$, $A[2][1][24]=A[2][2][24]=v_{44}$, $A[2][1][52]=A[2][2][52]=v_{45}$, $A[2][1][58]=A[2][2][58]=v_{46}$, $A[2][1][61]=A[2][2][61]=v_{47}$, $A[3][1][23]=A[3][2][23]=v_{48}$, $A[3][1][29]=A[3][2][29]=v_{49}$, $A[3][1][58]=A[3][2][58]=v_{50}$, $A[4][1][1]=A[4][2][1]=v_{51}$, $A[4][1][28]=A[4][2][28]=v_{52}$, $A[4][1][44]=A[4][2][44]=v_{53}$, $A[4][1][50]=A[4][2][50]=v_{54}$, $A[4][1][61]=A[4][2][61]=v_{55}$, $A[0][2][17]=A[0][3][17]=v_{56}$, $A[0][2][28]=A[0][3][28]=v_{57}$, $A[0][2][34]=A[0][3][34]=v_{58}$, $A[0][2][56]=A[0][3][56]=v_{59}$, $A[1][2][44]=A[1][3][44]=v_{60}$, $A[1][2][49]=A[1][3][49]=v_{61}$, $A[1][2][57]=A[1][3][57]=v_{62}$, $A[2][0][5]=A[2][2][5]=v_{63}$
Conditional cube variable	$A[2][0][0]=A[2][1][0]=v_0$
Bit conditions	$A[4][0][44]= A[2][2][45] + A[3][2][45]$, $A[2][0][4]= k_5 + k_{69} + A[0][1][5] + A[2][1][4] + A[0][2][5] + A[2][2][4] + A[0][3][5] + 1$, $A[2][0][59]= k_{60} + A[0][1][60] + A[2][1][59] + A[0][2][60] + A[2][2][59] + A[0][3][60] + 1$, $A[2][0][7]= A[3][1][7] + A[4][1][6] + A[2][1][7] + A[4][1][6] + A[2][2][7] + A[4][2][6]$
Guessed key bits	$k_{60}, k_5 + k_{69}$

6.3 Distinguishing Attack on Keccak-224

For Keccak-224, the same process can be applied with the conditional cube variables candidates in a 2-2-22 pattern. But with 1536 conditional cube variable candidates, the searching problem becomes difficult to solve. So we turn to consider the conditional cube variable candidates in double kernel patterns. The bitwise derivatives of such a chosen variable are still invariant with respect to the operation θ in the second round.

Table 11. Conditions to distinguish Keccak-512

Conditional cube variables	$A[2][0][0]=A[2][1][0]=v_0$, $A[2][0][1]=A[2][1][1]=v_1, A[2][0][2]=A[2][1][2]=v_2$, $A[2][0][3]=A[2][1][3]=v_3, A[2][0][22]=A[2][1][22]=v_4$, $A[2][0][23]=A[2][1][23]=v_5, A[2][0][44]=A[2][1][44]=v_6$, $A[2][0][45]=A[2][1][45]=v_7, A[3][0][15]=A[3][1][15]=v_8$
Bit conditions	$A[2][0][4]= A[0][0][5]+ A[1][0][5]+ A[0][1][5]+ A[2][1][4]+ 1$, $A[2][0][5]= A[0][0][6]+ A[1][0][6]+ A[0][1][6]+ A[2][1][5]+ 1$, $A[2][0][6]= A[0][0][7]+ A[1][0][7]+ A[0][1][7]+ A[2][1][6]+ 1$, $A[2][0][7]= A[0][0][8]+ A[1][0][8]+ A[0][1][8]+ A[2][1][7]+ 1$, $A[2][0][8]= A[4][0][7]+ A[2][1][8]+ A[3][1][8]$, $A[2][0][9]=$ $A[4][0][8]+ A[2][1][9]+ A[3][1][9]$, $A[2][0][10]= A[4][0][9]+$ $A[2][1][10]+ A[3][1][10]$, $A[2][0][17]= A[0][0][18]+ A[0][1][18]+$ $A[2][1][17]+ 1$, $A[2][0][25]= A[4][0][24]+ A[2][1][25]$, $A[2][0][26]= A[0][0][27]+ A[1][0][27]+ A[0][1][27]+$ $A[2][1][26]+ 1$, $A[2][0][27]= A[0][0][28]+ A[1][0][28]+$ $A[0][1][28]+ A[2][1][27]+ 1$, $A[2][0][29]= A[4][0][28]+$ $A[2][1][29]+ A[3][1][29]$, $A[2][0][30]= A[4][0][29]+ A[2][1][30]+$ $A[3][1][30]$, $A[2][0][40]= A[0][0][41]+ A[0][1][41]+ A[2][1][40]+$ 1, $A[2][0][46]= A[4][0][45]+ A[2][1][46]$, $A[2][0][47]=$ $A[4][0][46]+ A[2][1][47]$, $A[2][0][48]= A[4][0][47]+ A[2][1][48]$, $A[2][0][49]= A[0][0][50]+ A[1][0][50]+ A[0][1][50]+$ $A[2][1][49]+ 1$, $A[2][0][51]= A[4][0][50]+ A[2][1][51]+$ $A[3][1][51]$, $A[2][0][52]= A[4][0][51]+ A[2][1][52]+ A[3][1][52]$, $A[2][0][59]= A[0][0][60]+ A[0][1][60]+ A[2][1][59]+ 1$, $A[2][0][60]= A[0][0][61]+ A[0][1][61]+ A[2][1][60]+ 1$, $A[2][0][61]= A[0][0][62]+ A[0][1][62]+ A[2][1][61]+ 1$, $A[2][0][62]= A[0][0][63]+ A[0][1][63]+ A[2][1][62]+ 1$, $A[3][0][23]= A[0][0][22]+ A[0][1][22]+ A[3][1][23]$, $A[3][0][31]=$ $A[0][0][30]+ A[0][1][30]+ A[3][1][31]$, $A[3][0][45]= A[1][0][46]+$ $A[1][1][46]+ A[3][1][45]+ 1$, $A[4][0][3]= A[0][0][5]+$ $A[1][0][5]+ A[0][1][5]+ 1$, $A[4][0][6]= A[0][0][8]+ A[1][0][8]+$ $A[0][1][8]+ A[3][1][7]+ 1$, $A[4][0][25]= A[0][0][27]+$ $A[1][0][27]+ A[0][1][27]+ 1$, $A[0][1][49]= A[0][0][49]+$ $A[1][0][49]+ A[4][0][47]+ 1$, $A[4][0][44]=0$, $A[4][0][2] = 1$.

Four differential characteristics in double kernel pattern are shown in Table 7 in hexadecimal format with '-' denoting zero. The rows labeled with δ_0 and δ_1 are the input difference of the first round and the second round respectively; $\delta_{1.5}$ is the output difference after 1.5-round Keccak. The first two differential characteristics can be found in [9] in 6-6-6 pattern and the other two are found using the method in [8] in 8-8-8 pattern. As an example, a conditional cube variable can be set as

$$A[0][0][0]=A[0][1][0]=A[2][1][30]=A[2][2][30]= A[1][0][63]=A[1][2][63]=v_0.$$

This variable only impacts 6 bits after 1.5 round, which reduces the possibilities for the conditional cube variables to multiply with each other. Because of translation invariance in the direction of the z axis, we have 256 conditional cube variable candidates to build the MILP problem by applying Algorithm 5.

Table 12. Conditions to distinguish Keccak-384

Conditional cube variables	$A[0][0][14]=A[0][1][14]=v_0$, $A[2][0][23]=A[2][1][23]=v_1$, $A[2][0][24]=A[2][1][24]=v_2$, $A[2][0][43]=A[2][1][43]=v_3$, $A[2][0][44]=v_4$, $A[2][1][44]=v_4+v_5$, $A[2][2][44]=v_5$, $A[3][0][56]=A[3][1][56]=v_6$, $A[3][0][58]=A[3][1][58]=v_7$, $A[0][1][57]=A[0][2][57]=v_8$, $A[0][1][58]=A[0][2][58]=v_9$, $A[1][1][49]=A[1][2][49]=v_{10}$, $A[1][1][50]=A[1][2][50]=v_{11}$, $A[2][1][41]=A[2][2][41]=v_{12}$, $A[0][0][20]=A[0][2][20]=v_{13}$, $A[1][0][13]=A[1][2][13]=v_{14}$, $A[2][0][0]=A[2][2][0]=v_{15}$, $A[2][0][16]=A[2][2][16]=v_{16}$
Bit conditions	$A[0][0][1]= A[3][0][2]+ A[0][1][1]+ A[3][1][2]+ A[4][1][2]+ A[0][2][1]$
	$A[0][0][2]= A[3][0][3]+ A[0][1][2]+ A[3][1][3]+ A[4][1][3]+ A[0][2][2]+ 1$
	$A[0][0][5]= A[3][0][6]+ A[0][1][5]+ A[3][1][6]+ A[0][2][5]+ 1$
	$A[0][0][7]= A[3][0][8]+ A[0][1][7]+ A[3][1][8]+ A[0][2][7]$
	$A[0][0][9]= A[3][0][10]+ A[0][1][9]+ A[3][1][10]+ A[0][2][9]$
	$A[0][0][12]= A[2][0][11]+ A[0][1][12]+ A[2][1][11]+ A[0][2][12]+ A[2][2][11]+ 1$
	$A[0][0][15]= A[2][0][14]+ A[0][1][15]+ A[2][1][14]+ A[0][2][15]+ A[2][2][14]$
	$A[0][0][16]= A[2][0][15]+ A[0][1][16]+ A[2][1][15]+ A[0][2][16]+ A[2][2][15]$
	$A[0][0][19]= A[2][0][18]+ A[0][1][19]+ A[2][1][18]+ A[0][2][19]+ A[2][2][18]+ 1$
	$A[0][0][22]= A[3][0][23]+ A[0][1][22]+ A[3][1][23]+ A[4][1][23]+ A[0][2][22]+ A[2][2][24]+ 1$
	$A[0][0][28]= A[1][0][29]+ A[2][0][27]+ A[2][0][28]+ A[4][0][29]+ A[0][1][28]+ A[0][1][29]+ A[1][1][28]+ A[2][1][27]+ A[2][1][28]+ A[4][1][29]+ A[0][2][28]+ A[0][2][29]+ A[1][2][28]+ A[2][2][27]+ A[2][2][28]+ 1$
	$A[0][0][29]= A[1][0][29]+ A[2][0][28]+ A[0][1][29]+ A[2][1][28]+ A[0][2][29]+ A[2][2][28]+ 1$
	$A[0][0][30]= A[1][0][29]+ A[4][0][30]+ A[1][1][29]+ A[4][1][30]+ A[1][2][29]+ 1$
	$A[0][0][34]= A[2][0][33]+ A[0][1][34]+ A[1][1][34]+ A[2][1][33]+ A[0][2][34]+ A[2][2][33]$
	$A[0][0][39]= A[4][0][37]+ A[0][1][39]+ A[4][1][37]+ A[0][2][39]+ 1$
	$A[0][0][40]= A[3][0][41]+ A[0][1][40]+ A[3][1][41]+ A[4][1][41]+ A[0][2][40]+ 1$
	$A[0][0][42]= A[2][0][41]+ A[0][1][42]+ A[0][2][42]$
	$A[0][0][43]= A[2][0][42]+ A[0][1][43]+ A[1][1][43]+ A[2][1][42]+ A[0][2][43]+ A[2][2][42]+ 1$
	$A[0][0][46]= A[1][2][46]+ A[2][0][45]+ A[0][1][46]+ A[2][1][45]+ A[0][2][46]+ A[2][2][45]+ 1$
	$A[0][0][48]= A[1][0][48]+ A[2][0][47]+ A[0][1][48]+ A[2][1][47]+ A[0][2][48]+ A[2][2][47]+ 1$
	$A[0][0][49]= A[2][0][48]+ A[2][0][49]+ A[3][0][48]+ A[0][1][49]+ A[2][1][48] + A[3][1][48]+ A[0][2][49]+ A[2][2][48]$
	$A[0][0][60]= A[2][0][59]+ A[0][1][60]+ A[2][1][59]+ A[0][1][60]+ A[2][2][59]+ 1$
	$A[0][0][63]= A[3][0][0]+ A[0][1][63]+ A[3][1][0]+ A[0][2][63]+ 1$
	$A[1][0][8]= A[3][0][7]+ A[1][1][8]+ A[3][1][7]+ A[1][2][8]$
	$A[1][0][22]= A[0][1][23]+ A[1][1][22]+ A[1][2][22]+ A[2][2][24]+ 1$
	$A[1][0][23]= A[4][0][24]+ A[0][1][24]+ A[1][1][23]+ A[4][1][24]+ A[1][2][23]+ 1$
	$A[1][0][25]= A[3][0][24]+ A[1][1][25]+ A[3][1][24]+ A[1][2][25]+ 1$
	$A[1][0][28]= A[1][0][29]+ A[2][0][28]+ A[4][0][29]+ A[0][1][29]+ A[1][1][28]+ A[2][1][28]+ A[4][1][29]+ A[0][2][29]+ A[1][2][28]+ A[2][2][28]$
	$A[1][0][44]= A[3][0][43]+ A[1][1][44]+ A[3][1][43]+ A[1][2][44]$
	$A[1][0][45]= A[3][0][44]+ A[1][1][45]+ A[3][1][44]+ A[1][2][45]$
	$A[1][0][49]= A[2][0][49]+ A[3][0][48]+ A[3][1][48]+ 1$
	$A[1][0][50]= A[4][0][51]+ A[0][1][51]+ A[4][4][51]+ 1$
	$A[1][0][51]= A[3][0][50]+ A[1][1][51]+ A[3][1][50]+ A[1][2][51]+ A[2][2][51]$
	$A[1][0][59]= A[4][0][60]+ A[1][1][59]+ A[4][1][60]+ A[1][2][59]$
	$A[2][0][2]= A[4][0][1]+ A[2][1][2]+ A[4][1][1]+ A[2][2][2]$
	$A[2][0][4]= A[2][1][4]+ A[2][2][4]$
	$A[2][0][5]= A[4][0][4]+ A[2][1][5]+ A[4][1][4]+ A[2][2][5]$
	$A[2][0][7]= A[4][0][6]+ A[2][1][7]+ A[3][1][7]+ A[4][1][6]+ A[2][2][7]$
	$A[2][0][22]= A[4][0][21]+ A[2][1][22]+ A[4][1][21]+ A[2][2][22]$
	$A[2][0][25]= A[4][0][24]+ A[2][1][25]+ A[4][0][24]+ A[2][2][25]$
	$A[2][0][30]= A[4][0][29]+ A[2][1][30]+ A[3][1][30]+ A[4][1][29]+ A[2][2][30]$
	$A[2][0][31]= A[4][0][30]+ A[2][1][31]+ A[3][1][31]+ A[4][1][30]+ A[2][2][31]$
	$A[2][0][38]= A[4][0][37]+ A[2][1][38]+ A[4][1][37]+ A[2][2][38]$
	$A[2][0][39]= A[3][0][41]+ A[2][1][39]+ A[3][1][41]+ A[4][1][41]+ A[2][2][39]$
	$A[2][0][50]= A[4][0][49]+ A[2][1][50]+ A[3][1][50]+ A[4][1][49]+ A[2][2][50]$
	$A[2][0][51]= A[2][2][51]+ 1$
	$A[2][0][62]= A[3][0][0]+ A[1][1][63]+ A[2][1][62]+ A[3][1][0]+ A[2][2][62]$
	$A[2][0][63]= A[4][0][62]+ A[2][1][63]+ A[4][1][62]+ A[2][2][63]$
	$A[3][0][22]= A[4][0][24]+ A[0][1][24]+ A[3][1][22]+ A[4][1][24]$
	$A[3][0][40]= A[4][0][37]+ A[3][1][40]+ A[4][1][37]+ A[4][1][40]$
	$A[3][0][49]= A[4][0][51]+ A[0][1][51]+ A[3][1][49]+ A[4][1][51]+ A[2][2][50]+ 1$
	$A[4][0][2]= A[4][1][2]+ 1$
	$A[4][0][22]= A[3][1][23]+ A[4][1][22]+ A[2][2][23]$
	$A[4][0][23]= A[4][1][23]+ A[2][2][24]$
	$A[4][0][50]= A[2][1][51]+ A[3][1][51]+ A[4][1][50]+ 1$
	$A[0][1][20]= A[2][0][19]+ A[2][1][19]+ A[2][2][19]+ 1$
	$A[1][2][40]= 1$ $A[4][1][0]= 1$ $A[1][2][19]= 1$
	$A[1][2][20]= 1$ $A[1][1][40]= 1$ $A[2][1][8]= 0$ $A[1][1][15]= 1$

Table 13. Conditional cube variables to distinguish Keccak-224

Conditional cube variables
A[0][0][3]=A[1][2][2]=v_0, A[0][1][3]=A[2][1][33]=A[2][2][33]=v_0+v_{25}, A[1][0][2]=v_0+v_{17},
A[0][0][6]=A[1][0][5]=A[1][2][5]=v_1, A[0][1][6]=A[2][1][36]=A[2][2][36]=v_1+v_{17},
A[0][0][9]=A[0][1][9]=A[2][1][39]=A[2][2][39]=A[1][0][8]=A[1][2][8]=v_2, A[0][0][11]=A[1][0][10]=v_3,
A[0][1][11]=A[2][1][41]=A[2][2][41]=v_3+v_{18}, A[1][2][10]=v_3+v_{16},
A[0][0][14]=A[2][1][44]=A[2][2][44]=A[1][0][13]=A[1][2][13]=v_4, A[0][1][14]=$v_4+v_{16}+v_{26}$, A[0][0][16]=v_5,
A[0][1][16]=A[2][1][46]=A[2][2][46]=v_5+v_{19}, A[1][0][15]=$v_5+v_{20}+v_{27}$, A[1][2][15]=v_5+v_{27},
A[0][0][19]=A[1][0][18]=A[1][2][18]=v_6, A[0][1][19]=A[2][1][49]=A[2][2][49]=v_6+v_{20},
A[0][0][21]=A[0][1][21]=A[2][1][51]=A[2][2][51]=v_7, A[1][0][20]=$v_7+v_{21}+v_{28}$, A[1][2][20]=v_7+v_{28},
A[0][0][22]=A[2][1][52]=A[2][2][52]=A[1][0][21]=v_8, A[0][1][22]=A[1][2][21]=v_8+v_{14},
A[0][0][24]=A[1][0][23]=A[1][2][23]=v_9, A[0][1][24]=A[2][1][54]=A[2][2][54]=v_9+v_{21},
A[0][0][27]=A[2][1][57]=A[2][2][57]=A[1][0][26]=A[1][2][26]=v_{10}, A[0][1][27]=$v_{10}+v_{28}$,
A[0][0][29]=A[1][0][28]=v_{11}, A[0][1][29]=A[2][1][59]=A[2][2][59]=$v_{11}+v_{22}$, A[1][2][28]=$v_{11}+v_{15}$,
A[0][1][32]=$v_{12}+v_{15}+v_{29}$, A[0][0][32]=A[2][1][62]=A[2][2][62]=A[1][0][31]=A[1][2][31]=v_{12},
A[0][0][62]=A[1][2][61]=v_{13}, A[0][1][62]=A[2][1][28]=A[2][2][28]=$v_{13}+v_{23}$, A[1][0][61]=$v_{13}+v_{24}$,
A[3][1][6]=A[3][2][6]=A[1][3][21]=A[0][1][25]=A[0][3][25]=v_{14}, A[3][1][13]=A[3][2][13]=$v_{15}+v_{29}$,
A[1][3][28]=A[0][3][32]=v_{15}, A[3][1][59]=A[3][2][59]=$v_{16}+v_{26}$, A[1][3][10]=A[0][3][14]=v_{16}
A[1][3][2]=A[4][0][40]=A[4][2][40]=A[0][3][6]=v_{17}, A[1][0][7]=A[4][0][45]=A[4][2][45]=$v_{18}+v_{26}$,
A[1][3][7]=A[0][3][11]=v_{18} A[1][0][12]=A[1][3][12]=A[4][0][50]=A[4][2][50]=A[0][3][16]=v_{19},
A[1][3][15]=A[0][3][19]=v_{20}, A[4][0][53]=A[4][2][53]=$v_{20}+v_{27}$, A[1][3][20]=A[0][3][24]=v_{21},
A[4][0][58]=A[4][2][58]=$v_{21}+v_{28}$, A[1][0][25]=A[4][0][63]=A[4][2][63]=$v_{22}+v_{29}$,
A[1][3][25]=A[0][3][29]=v_{22}, A[1][0][58]=A[1][3][58]=A[4][0][32]=A[4][2][32]=A[0][3][62]=v_{23},
A[1][3][61]=A[4][0][35]=A[4][2][35]=A[0][1][1]=A[0][3][1]=A[2][1][31]=A[2][2][31]=v_{24},
A[1][0][63]=A[1][3][63]=A[4][0][37]=A[4][2][37]=A[0][3][3]=v_{25}, A[1][2][7]=A[0][2][14]=v_{26},
A[0][2][22]=A[3][1][3]=A[3][2][3]=v_{27} A[0][2][27]=A[3][1][8]=A[3][2][8]=v_{28}, A[1][2][25]=A[0][2][32]=v_{29},
A[2][0][55]=A[2][1][55]=v_{30} A[0][2][60]=A[0][3][60]=v_{31}, A[0][1][37]=A[0][3][37]=v_{32}

With Gurobi Optimizer, we can find a combination of 30 conditional cube variables. Three conditional cube variables in 2-2-22 pattern have been added to the combination to get 33 independent conditional cube variables. Refer to Table 13 for the list of the conditional cube variables. The bit conditions can be derived exactly from the conditional cube variables, but they will be listed in the auxiliary supporting material due to the space limitation. Thus, a 7-round cube tester on Keccak-224 is constructed.

The time complexity of this distinguishing attack is 2^{33}. Memory complexity is negligible. This distinguishing attack can be performed on a desktop computer in several hours.

7   Conclusion

In this paper, we propose the conditional cube tester for Keccak sponge function with the advantage of having smaller dimensions compared to the previous cube tester in some cases. Our approach is based on a novel idea to add some conditions for certain cube variables, so that the multiplication between cube variables are under control after the second round of Keccak sponge function. More specifically, using a conditional cube tester to round-reduced Keccak-MAC and Keyak, our key recovery attacks are more efficient than the currently best known attacks according to the number of rounds or the complexity. Another application of our conditional cube tester is to construct distinguishing attacks on Keccak sponge function. Our distinguishing attacks are much faster and improve the existing attacks. Most of our attacks are very practical and implementations and experiments have been conducted on desktop computers. We should also remark that our proposed conditional cube testers may be used to analyse Keccak-like cryptosystems. Implementations of our methods are available at http://people.uwm.edu/gxu4uwm/eurocrypt17_code/.

Acknowledgement. This work is supported by 973 Program (No. 2013CB834205), and the Strategic Priority Research Program of the Chinese Academy of Sciences (No. XDB01010600) and the National Natural Science Foundation of China (No. 61133013).

References

1. Guido, B., Joan, D., Michaël, P., Van Assche, G.: Keccak Sponge Function Family Main Document. http://Keccak.noekeon.org/Keccak-main-2.1.pdf
2. Dinur, I., Dunkelman, O., Shamir, A.: Improved practical attacks on round-reduced keccak. J. Cryptology **27**(2), 183–209 (2014)
3. Pawel, M., Josef, P., Marian, S., Michal, S.: Preimage Attacks on the Round-Reduced Keccak with the Aid of Differential Cryptanalysis. Cryptology ePrint Archive, Report 2013/561 (2013). http://eprint.iacr.org/
4. Bernstein, D.J.: Second Preimages for 6 (7 (8??)) Rounds of Keccak. NIST mailing list (2010)
5. Duan, M., Lai, X.J.: Improved zero-sum distinguisher for full round $keccak$-f permutation. Chin. Sci. Bull. **57**(6), 694–697 (2012)
6. Duc, A., Guo, J., Peyrin, T., Wei, L.: Unaligned rebound attack: application to keccak. In: Canteaut, A. (ed.) FSE 2012. LNCS, vol. 7549, pp. 402–421. Springer, Heidelberg (2012). doi:10.1007/978-3-642-34047-5_23
7. Jean, J., Nikolić, I.: Internal differential boomerangs: practical analysis of the round-reduced keccak-f permutation. In: Leander, G. (ed.) FSE 2015. LNCS, vol. 9054, pp. 537–556. Springer, Heidelberg (2015). doi:10.1007/978-3-662-48116-5_26
8. Naya-Plasencia, M., Röck, A., Meier, W.: Practical analysis of reduced-round KECCAK. In: Bernstein, D.J., Chatterjee, S. (eds.) INDOCRYPT 2011. LNCS, vol. 7107, pp. 236–254. Springer, Heidelberg (2011). doi:10.1007/978-3-642-25578-6_18
9. Das, S., Meier, W.: Differential biases in reduced-round keccak. In: Pointcheval, D., Vergnaud, D. (eds.) AFRICACRYPT 2014. LNCS, vol. 8469, pp. 69–87. Springer, Cham (2014). doi:10.1007/978-3-319-06734-6_5
10. Dinur, I., Morawiecki, P., Pieprzyk, J., Srebrny, M., Straus, M.: Cube attacks and cube-attack-like cryptanalysis on the round-reduced keccak sponge function. In: Oswald, E., Fischlin, M. (eds.) EUROCRYPT 2015. LNCS, vol. 9056, pp. 733–761. Springer, Heidelberg (2015). doi:10.1007/978-3-662-46800-5_28
11. Guido, B., Joan, D., Michaël, P., Van Assche, G.: Keyak. http://keyak.noekeon.org
12. Taha, M., Schaumont, P.: Differential power analysis of MAC-keccak at any key-length. In: Sakiyama, K., Terada, M. (eds.) IWSEC 2013. LNCS, vol. 8231, pp. 68–82. Springer, Heidelberg (2013). doi:10.1007/978-3-642-41383-4_5
13. Dobraunig, C., Eichlseder, M., Mendel, F., Schläffer, M.: Cryptanalysis of Ascon. In: Nyberg, K. (ed.) CT-RSA 2015. LNCS, vol. 9048, pp. 371–387. Springer, Cham (2015). doi:10.1007/978-3-319-16715-2_20
14. Pawel, M., Josef, P., Michal, S., Marian, S.: Applications of Key Recovery Cube-attack-like. Cryptology ePrint Archive, Report 2015/1009 (2015). http://eprint.iacr.org/

15. Dinur, I., Shamir, A.: Breaking Grain-128 with dynamic cube attacks. In: Joux, A. (ed.) FSE 2011. LNCS, vol. 6733, pp. 167–187. Springer, Heidelberg (2011). doi:10. 1007/978-3-642-21702-9_10

16. Wang, X., Yu, H.: How to break MD5 and other hash functions. In: Cramer, R. (ed.) EUROCRYPT 2005. LNCS, vol. 3494, pp. 19–35. Springer, Heidelberg (2005). doi:10.1007/11426639_2

17. Chen, H., Wang, X.: Improved linear hull attack on round-reduced SIMON with dynamic key-guessing techniques. In: Peyrin, T. (ed.) FSE 2016. LNCS, vol. 9783, pp. 428–449. Springer, Heidelberg (2016). doi:10.1007/978-3-662-52993-5_22

18. Knellwolf, S., Meier, W., Naya-Plasencia, M.: Conditional differential cryptanalysis of NLFSR-based cryptosystems. In: Abe, M. (ed.) ASIACRYPT 2010. LNCS, vol. 6477, pp. 130–145. Springer, Heidelberg (2010). doi:10.1007/978-3-642-17373-8_8

19. CAESAR: Competition for Authenticated Encryption: Security, Applicability, and Robustness. http://competitions.cr.yp.to/caesar.html

20. Aumasson, J.-P., Dinur, I., Meier, W., Shamir, A.: Cube testers and key recovery attacks on reduced-round MD6 and trivium. In: Dunkelman, O. (ed.) FSE 2009. LNCS, vol. 5665, pp. 1–22. Springer, Heidelberg (2009). doi:10.1007/ 978-3-642-03317-9_1

21. Lai, X.: Higher order derivatives and differential cryptanalysis. In: Blahut, R.E., Costello, D.J., Maurer, U., Mittelholzer, T. (eds.) Communications and Cryptography: Two Sides of One Tapestry, pp. 227–233. Springer (1994)

22. Dinur, I., Shamir, A.: Cube attacks on tweakable black box polynomials. In: Joux, A. (ed.) EUROCRYPT 2009. LNCS, vol. 5479, pp. 278–299. Springer, Heidelberg (2009). doi:10.1007/978-3-642-01001-9_16

23. Zhu, B., Chen, K., Lai, X.: Bitwise higher order differential cryptanalysis. In: Chen, L., Yung, M. (eds.) INTRUST 2009. LNCS, vol. 6163, pp. 250–262. Springer, Heidelberg (2010). doi:10.1007/978-3-642-14597-1_16

24. Stein, W., Joyner, D.: SAGE: System for algebra and geometry experimentation. ACM SIGSAM Bull. **39**(2), 61–64 (2005)

25. Sun, S., Hu, L., Wang, M., Yang, Q., Qiao, K., Ma, X., Song, L., Shan, J.: Extending the applicability of the mixed-integer programming technique in automatic differential cryptanalysis. In: Lopez, J., Mitchell, C.J. (eds.) ISC 2015. LNCS, vol. 9290, pp. 141–157. Springer, Cham (2015). doi:10.1007/978-3-319-23318-5_8

26. Gurobi optimization. Gurobi: Gurobi optimizer reference manual (2015). http:// www.gurobi.com

A New Structural-Differential Property
of 5-Round AES

Lorenzo Grassi[1]([✉]), Christian Rechberger[1,3], and Sondre Rønjom[2,4]

[1] IAIK, Graz University of Technology, Graz, Austria
{lorenzo.grassi,christian.rechberger}@iaik.tugraz.at
[2] Nasjonal sikkerhetsmyndighet, Oslo, Norway
[3] DTU Compute, DTU, Kongens Lyngby, Denmark
[4] Department of Informatics, University of Bergen, Bergen, Norway
Sondre.Ronjom@ii.uib.no

Abstract. AES is probably the most widely studied and used block cipher. Also versions with a reduced number of rounds are used as a building block in many cryptographic schemes, e.g. several candidates of the SHA-3 and CAESAR competition are based on it.

So far, non-random properties which are independent of the secret key are known for up to 4 rounds of AES. These include differential, impossible differential, and integral properties.

In this paper we describe a *new structural property for up to 5 rounds of AES*, differential in nature and which is independent of the secret key, of the details of the MixColumns matrix (with the exception that the branch number must be maximal) and of the SubBytes operation. It is very simple: By appropriate choices of difference for a number of input pairs it is possible to make sure that the number of times that the difference of the resulting output pairs lie in a particular subspace is *always* a multiple of 8.

We not only observe this property experimentally (using a small-scale version of AES), we also give a detailed proof as to why it has to exist. As a first application of this property, we describe a way to distinguish the 5-round AES permutation (or its inverse) from a random permutation with only 2^{32} chosen texts that has a computational cost of $2^{35.6}$ look-ups into memory of size 2^{36} bytes which has a success probability greater than 99%.

Keywords: Block cipher · Permutation · AES · Secret-key distinguisher

1 Introduction

Block ciphers play an important role in symmetric cryptography providing the basic tool for encryption. They are the oldest and most scrutinized

---

The extended version of this paper can be found in [13]. It includes a more formal description of the main result of this paper which exploits the subspace trail notation [14] recently introduced at FSE 2017.

© International Association for Cryptologic Research 2017
J.-S. Coron and J.B. Nielsen (Eds.): EUROCRYPT 2017, Part II, LNCS 10211, pp. 289–317, 2017.
DOI: 10.1007/978-3-319-56614-6_10

cryptographic tools. Consequently, they are the most trusted cryptographic algorithms that are often used as the underlying tool to construct other cryptographic algorithms, whose proofs of security are performed under the assumption that the underlying block cipher is ideal.

While the security of public-key encryption schemes are related to the hardness of well-defined mathematical problems, informally a block cipher is considered secure if an (efficient) adversary, with access to the encryptions of messages of its choice, cannot tell apart those encryptions from the values of a truly random permutation. In other words, this means that an (efficient) adversary, with access to the encryptions of messages of its choice, cannot tell the difference between the block cipher (equipped with a random key) and a truly random permutation. This notion of block cipher security was introduced and formally modeled by Luby and Rackoff [19] in 1988, and it was motivated by the design of DES. To be a bit more precise (but without going into the details), a secret key distinguisher is one of the weakest cryptographic attacks that can be launched against a secret-key cipher. In this attack, there are two oracles: one that simulates the cipher for which the cryptographic key has been chosen at random and the other simulates a truly random permutation. The adversary can query both oracles and his task is to decide which oracle is the cipher and which is the random permutation. The attack is considered to be successful if the number of queries required to make a correct decision is below a well defined level.

The Rijndael block cipher [8] has been designed by Daemen and Rijmen in 1997 and was chosen as the AES (Advanced Encryption Standard) by NIST in 2000. Nowadays, it is probably the most used and studied block cipher. The possibility to set up a *secret key distinguisher for 5-round of AES* that exploits a property which is *independent of the secret key* was already considered in [21] and improved in [14]. However, only partial solutions have been proposed and the problem is still open. As we will argue below, the solutions so far are partial because the distinguishers are derived from a key-recovery attack and they actually exploit as property the existence of a sub-key for which a property on 4 rounds holds.

In this paper, we present (and practical verify) the *first secret-key distinguisher for 5-round AES* which exploits a new structural/differential property which is independent of the secret key, that is a property that can be practically verified without needing to know or to get to know any information of the secret key. As we are going to show, it requires 2^{33} chosen plaintexts/ciphertexts and has a computational cost of $2^{36.6}$ table look-ups.

1.1 Secret-Key Distinguishers for AES-128

In the usual security model, the adversary is given a *black box* (oracle) access to an instance of the encryption function associated with a random secret key and its inverse. The goal is to find the key or more generally to efficiently distinguish the encryption function from a random permutation.

More formally, a block cipher is a family of functions $E : \mathcal{K} \times \mathcal{S} \rightarrow \mathcal{S}$, with $\mathcal{K}$ a finite set called the key space and $\mathcal{S}$ a finite set called the domain or message space. For every $k \in \mathcal{K}$, the function $E_k(\cdot) = E(k, \cdot)$ is a permutation. The inverse of the block cipher E is defined as a function $E^{-1} : \mathcal{K} \times \mathcal{S} \to \mathcal{S}$ that satisfies $E_k^{-1}(E_k(s)) = s$ for each $k \in \mathcal{K}$ and for each $s \in \mathcal{S}$. A block cipher $E_k(\cdot)$ with key space $\mathcal{K}$ is a (q, t, ε)-pseudorandom permutation (PRP) if any adversary making at most q oracle queries and running in time at most t can distinguish E_k (for a random key k) from a uniformly random permutation with advantage at most ε.

Definition 1. *Let E be block cipher defined as before, and $Perm(\mathcal{S})$ be the set of all permutations of $\mathcal{S}$. Let D be a distinguisher with oracle access to a permutation and its inverse, and returning a single bit. The (Strong PseudoRandom Permutation) SPRP-advantage of D against E is defined as*

$$\mathbf{Adv}_E^{sprp}(D) = |Prob(\pi \leftarrow Perm(\mathcal{S}) : D^{\pi(\cdot), \pi^{-1}(\cdot)} = 1)$$
$$- Prob(k \leftarrow \mathcal{K} : D^{E_k(\cdot), E_k^{-1}(\cdot)} = 1)|.$$

For integers q and t, the SPRP-advantage of E is defined as

$$\mathbf{Adv}_E^{sprp}(q, t) = \max_D \mathbf{Adv}_E^{sprp}(D),$$

where the maximum is taken over all distinguishers making at most q oracle queries and running in time at most t. E is a (q, t, ε)-SPRP if $\mathbf{Adv}_E^{sprp}(q, t) \leq \varepsilon$.

Note that if $Adv_E(D) \simeq 0$, then the $E_k(\cdot)$ behaves (exactly) like a random permutation from the distinguisher point of view.

Before we focus on the 5-round distinguisher, we briefly summarize the properties exploited by the secret key distinguisher on AES-like permutations up to 4 rounds. We stress that, even if a key-recovery attack can also be used as a secret key distinguisher in this paper we focus only on secret-key distinguisher that are independent of the secret key.

The most competitive secret-key distinguishers up to 3-round are based on the differential [5] and on the truncated differential cryptanalysis [17]. These distinguishers exploit the fact that some r-round differential characteristics exist with higher probability for an AES permutation than for a random one. In [7], Daemen *et al.* proposed an attack vector that uses a 3-round distinguisher to attack up to 6 rounds of the cipher and later became known as integral attacks. In an integral distinguisher, given inputs with particular properties, one exploits the fact that the sum of the corresponding ciphertexts is zero with probability 1 for an AES permutation, while this happens with a (much) lower probability for a random permutation. Finally, another possible distinguisher exploits the impossible-differential cryptanalysis, which was independently proposed by Knudsen [18] and by Biham *et al.* [3]. In impossible-differential cryptanalysis, the idea is to exploit the fact that some differential trails hold with probability 0 for an AES permutation (i.e. impossible differential trails), while they have probability greater than 0 for a random permutation.

5-Round "Distinguisher" for AES-128: State of Art. A distinguisher for five rounds of AES-128 has been recently proposed by Sun, Liu, Guo, Qu, and Rijmen at Crypto 2016 [21]. This distinguisher - which requires the *whole* input-output space to work - has been improved in [14], where authors set up a secret key distinguisher in the same setting of the one proposed in [21], but which requires only $2^{98.2}$ chosen plaintexts.

Both these two distinguishers are derived by a key-recovery attack on AES-128 with a secret S-Box. In particular, they are able to distinguish a random permutation from an AES one exploiting the existence of a (secret) key for which a property on 4-round is verified. In more details, the property on 4-round used in [21] is the balance property, while the one used in [14] is the impossible differential one. With respect to a classical key-recovery attack, these distinguishers require the knowledge only of a single byte of the secret subkey to distinguish an AES permutation with a secret S-Box from a random one.

For a complete comparison with the distinguisher presented in this paper, we briefly recall how they are set up, and we refer to [14,21] for a complete discussion. In both cases, authors first assume to know the difference of two bytes (i.e. 1 byte) of one secret subkey. Using this knowledge, they are able to extend a four rounds distinguisher to five rounds. In order to turn these distinguishers into secret-key ones, the idea is simply to iterate these distinguishers on all the 2^8 possible values of the difference of these two bytes of the secret subkey. The idea is that for an AES permutation there exists one difference of these two bytes for which a property (which is independent of the secret key) on 4-round is satisfied, while for a random permutation this property on 4-round is never satisfied (with high probability) for any of the 2^8 possible values.

We stress that both these distinguishers require to find part of the secret key in order to verify a property on 4-round, i.e. they work as key-recovery attacks. Note that the research of a secret-key distinguisher which is independent of the secret key is of particular interest and importance since it (theoretically) allows to set up key recovery attacks, as it already happened for the secret-key distinguishers up to 4 rounds just described. Moreover, we highlight that both these distinguishers are independent of the details of the S-Box, but they depend on the details of the MixColumns matrix (in particular, they exploit the fact that for at least one column of the MixColumns matrix or its inverse two elements are identical).

1.2 Our Result: The First 5-Round Secret-Key Distinguisher for AES-128 Independent of the Secret Key

The results presented in the previous two papers don't solve the problem to set up *a 5-round secret key distinguisher of AES which exploits a property which is independent of the secret key*. In Sect. 3 of this paper, we provide a solution to this problem, that is we propose the *first* secret-key distinguisher on 5-round of AES which exploits a new property which is independent of the secret key and of the details of the S-Box.

The high-level idea is very easily described. By appropriate choices of difference for a number of input pairs it is possible to make sure that the number of times that the difference of the resulting output pairs assumes certain values is *always* a multiple of 8. More concretely, given a set of plaintexts which are equal in certain bytes, consider the corresponding ciphertexts after 5 rounds. The idea is to count the total number of different ciphertext pairs with zero-difference in certain bytes. As we show in detail in the paper, for an AES permutation this number can only be a multiple of 8, while it does not have any particular property for the case of a random permutation. As we will see in the comparison, the resulting distinguisher proposed in this paper is much more efficient than those proposed earlier, it *works both in the encryption and in the decryption mode of AES and it does not depend on the details of the MixColumns matrix (with the exception that the branch number must be five) or/and of the SubBytes operation.* A formal statement of this property used by our distinguisher is given in Theorem 1 in Sect. 3.1, and its detailed proof is given in Sect. 4.

Comparison with 4-Round Secret-Key Distinguishers. These last properties also highlight a difference between our new distinguisher and the others currently used in literature. In most cases, especially in the cryptanalysis of AES, one does not have the necessity to investigate the details of the S-Boxes. Consider for example the 4-round secret-key distinguishers, based on the integral [12] and on the impossible-differential [4] properties. In the first one, given a set of chosen plaintexts of which part is held constant and another part varies through all possibilities, it is possible to prove that their XOR-sum after 4-round is always equal to 0. In the second one, given the same set of chosen plaintexts, it is possible to prove that the difference of each possible pair of ciphertexts after 4-round can not take some values (some differences have prob. 0, i.e. they are impossible). In both cases, the corresponding results are independent of the key and of the non-linear components. That is, if some other S-Boxes with similar differential/linear properties are chosen in a cipher, the corresponding cryptanalytic results remain the same.

Although there are already 4-round impossible differentials and zero-correlation linear hulls for AES, the effort to find new impossible differentials and zero-correlation linear hulls that could cover more rounds has never been stopped. In Eurocrypt 2016, Sun *et al.* [22] proved that, unless the details of the S-Boxes are exploited, one cannot find any impossible differential or zero-correlation linear hull of the AES that covers 5 or more rounds. Moreover, due to the link among impossible differential, integral and zero correlation linear cryptanalysis [23], an analogous result holds also for the integral case. On the other hand, our new property presented in this paper holds up to 5-round of AES independently of the key and of the details of the S-Box (and of the MixColumns operation), and allows to answer an almost 20-year old problem: given a set of chosen plaintexts similar to the one used by the integral and impossible differential distinguishers just recalled, is there any property which is independent of the secret key after 5-round AES?

Table 1. *5-round Secret-Key Distinguishers for AES with a Single Secret S-Box.* In this table, we limit ourselves to consider the distinguishers that exploit a property which is independent of the key, or which are derived by a key-recovery attack but are independent of the S-Box and require the knowledge of only part of the key. The complexity is measured in minimum number of chosen plaintexts CP or/and chosen ciphertexts CC which are needed to distinguish the AES permutation from a random one with probability higher than 99%. Time complexity is measured in memory accesses (M) or XOR operations (XOR). The case in which the final MixColumns operation is omitted is denoted by "$r.5$ round", i.e. r full rounds and the final one. "Key-Independence" denotes a distinguisher which is able to distinguish 5-round AES from a random permutation without discovering any information of the secret key or of part of it.

Property	Rounds	Data	CP	CC	Cost	Key-Independence	Ref.
Structural Diff.	$4.5-5$	2^{33}	✓	✓	$2^{36.6}$ M	✓	**Sect. 3**
Impossible Diff.	$4.5-5$	$2^{98.2}$	✓		2^{107} M		[14]
Integral	5	2^{128}		✓	2^{128} XOR		[21]

Comparison of 5-Round Secret-Key Distinguishers. For a better comparison between this new secret-key distinguisher proposed in this paper and earlier ones, we propose to classify the secret-key distinguishers in the following way (from strongest to weakest):

1. a distinguisher which is completely independent of the secret key (e.g., it exploits properties that are not related to the existence of a key) and independent of the details of the S-Box;
2. a distinguisher which depends on the existence of a key and is derived by a key-recovery attack.

A comparison between our new distinguisher and the ones proposed in [14,21] is given in Table 1, where "Key-Independence" denotes a secret-key distinguisher which is derived by a key-recovery attack, i.e. that does not exploit a property which is independent of the secret key. Moreover, with respect to the previous classification, a complete comparison of all the secret key distinguishers and key-recovery attacks (used as distinguishers) for 5-round AES is provided in Table 2 - Appendix C of the full version of the paper [13].

2 Preliminary - Description of AES

The Advanced Encryption Standard [8] is a *Substitution-Permutation network* that supports key size of 128, 192 and 256 bits. The 128-bit plaintext initializes the internal state as a 4×4 matrix of bytes as values in the finite fields $\mathbb{F}_{256}$, defined using the irreducible polynomial $x^8 + x^4 + x^3 + x + 1$. Depending on the version of AES, N_r round are applied to the state: $N_r = 10$ for AES-128, $N_r = 12$ for AES-192 and $N_r = 14$ for AES-256. An AES round applies four operations to the state matrix:

- *SubBytes* (S-Box) - applying the same 8-bit to 8-bit invertible S-Box 16 times in parallel on each byte of the state (it provides non-linearity in the cipher);
- *ShiftRows* (*SR*) - cyclic shift of each row to the left;
- *MixColumns* (*MC*) - multiplication of each column by a constant 4×4 invertible matrix M_{MC} (*MC* and *SR* provide diffusion in the cipher[1]);
- *AddRoundKey* (*ARK*) - XORing the state with a 128-bit subkey.

One round of AES can be described as $R(x) = K \oplus MC \circ SR \circ$ S-Box(x). In the first round an additional AddRoundKey operation (using a whitening key) is applied, and in the last round the MixColumns operation can be omitted. For the following, we *assume that the last MixColumns operation is always omitted.* In the case in which the last MixColumns is not omitted, it is sufficient to exchange the order of the last MixColumns operation and of the AddRoundKey operation - they are linear.

Finally, as we don't use the details of the AES key schedule in this paper, we refer to [8] for a complete description.

The Notation Used in the Paper. Let x denote a plaintext, a ciphertext, an intermediate state or a key. Then $x_{i,j}$ with $i, j \in \{0, ..., 3\}$ denotes the byte in the row i and in the column j. We denote by k^r the key of the r-th round, where k^0 is the secret key. If only the key of the final round is used, then we denote it by k to simplify the notation. Finally, we denote by R one round of AES, while we denote r rounds of AES by R^r. We sometimes use the notation R_K instead of R to highlight the round key K. If the MixColumns operation is omitted in the last round, then we denote it by R_f.

2.1 Differential Trail over 2-round AES

For the following, we recall a 2-round truncated differential trail of AES (see [9] or [10] for details), largely used in the paper and illustrated in Fig. 1.

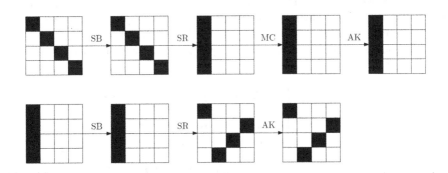

Fig. 1. Differential Trail over 2-round AES.

---

[1] *SR* makes sure column values are spread, *MC* makes sure each column is mixed.

Let $R^2(\cdot)$ denote two AES rounds with fixed random round keys. Consider two plaintexts which are equal in all bytes except for the ones in the i-th diagonal for a certain $i = 0, 1, 2, 3$, i.e. for the bytes in row j and column $i + j$ for each $j = 0, 1, 2, 3$ (the index $i + j$ is taken modulo 4). After one round, the two texts are equal in all bytes except for the ones in the i-th column, i.e. for the bytes in row j and column i for each j. After the second and last round - assuming the final MixColumns is omitted, the two texts are equal in all bytes except for the ones in the i-th anti-diagonal, that is for the bytes in row j and column $i - j$ for each j (the index $i - j$ is taken modulo 4) by definition of anti-diagonal.

For the following, we work with *diagonal sets* of 2^{32} plaintexts, defined as sets of texts which are equal in 3 diagonals, i.e. texts with active bytes in the i-th diagonal for a certain $i = 0, 1, 2, 3$ and with constant bytes in the other three:

$$
\begin{bmatrix} A & C & C & C \\ C & A & C & C \\ C & C & A & C \\ C & C & C & A \end{bmatrix} \xrightarrow{R(\cdot)} \begin{bmatrix} A & C & C & C \\ A & C & C & C \\ A & C & C & C \\ A & C & C & C \end{bmatrix} \xrightarrow{R_f(\cdot)} \begin{bmatrix} A & C & C & C \\ C & C & C & A \\ C & C & A & C \\ C & A & C & C \end{bmatrix},
$$

where A denotes an active byte (i.e. a byte in which every value in $\mathbb{F}_{2^8}$ appears the same number of times) and C denotes a constant byte (i.e. a byte in which the value is fixed to a constant for all texts). For completeness, we label the last set by *inverse-diagonal set*, i.e. a set of texts where the bytes in one (or more) anti-diagonal(s) are active while the others are constant.

If the final MixColumns is not omitted, certain linear relations - which are given by the definition of the MixColumns matrix - hold between the bytes of the texts that lie in the same column:

$$
\begin{bmatrix} A & C & C & C \\ C & A & C & C \\ C & C & A & C \\ C & C & C & A \end{bmatrix} \xrightarrow{R(\cdot)} \begin{bmatrix} A & C & C & C \\ A & C & C & C \\ A & C & C & C \\ A & C & C & C \end{bmatrix} \xrightarrow{R(\cdot)} MC \times \begin{bmatrix} A & C & C & C \\ C & C & C & A \\ C & C & A & C \\ C & A & C & C \end{bmatrix},
$$

In this case, we label the last set by *mixed set*. As an example, consider two plaintexts p^1 and p^2 which are equal in all bytes except for the ones in the 0-th diagonal, i.e. except for the bytes in positions (j, j) for each $j = 0, 1, 2, 3$. After 2 (complete) rounds, there exist $x, y, z, w \in F_{2^8}$ such that their difference $R^2(p^1) \oplus R^2(p^2)$ can be re-written as:

$$
R^2(p^1) \oplus R^2(p^2) = \begin{bmatrix} 0 \times 02 \cdot x & y & z & 0 \times 03 \cdot w \\ x & y & 0 \times 03 \cdot z & 0 \times 02 \cdot w \\ x & 0 \times 03 \cdot y & 0 \times 02 \cdot z & w \\ 0 \times 03 \cdot x & 0 \times 02 \cdot y & z & w \end{bmatrix}. \tag{1}
$$

Finally, the same truncated differential analysis of 2-round can be generalized to the cases of an initial diagonal set with more than a single active diagonal, i.e. a set of plaintexts which are equal in all bytes except for the ones that lie in two or three diagonals (instead of only one).

3   New 5-round Secret Key Distinguisher for AES

3.1   Statement of the Property

Consider a diagonal set of plaintexts - i.e. a set of 2^{32} plaintexts which are equal in all bytes except for the ones in i-diagonal for a certain $i = 0, 1, 2, 3$, and the corresponding ciphertexts after 5 rounds. Assume the final MixColumns operation is omitted. In order to set up the distinguisher on 5 rounds of AES, the idea is to count the number of different pairs of ciphertexts which are equal in d anti-diagonals for a certain $1 \leq d \leq 3$ - that is the number of pairs of ciphertexts with zero-difference in the bytes in positions $(i, j - i)$ for all $i = 0, 1, 2, 3$ and $j \in J$ for a certain $J \subseteq \{0, 1, 2, 3\}$ with $|J| = d$ - and to exploit the property that for an AES-like permutation this number is a multiple of 8 with prob. 1.

In more detail, given a set of plaintexts/ciphertexts (p^i, c^i) for $i = 0, ..., 2^{32} - 1$ - where all the plaintexts are in the same diagonal set, the idea is to construct all the possible pairs of ciphertexts (c^i, c^j) for $i \neq j$ and to count the number of different pairs[2] of ciphertexts (c^i, c^j) for which the bytes of the difference $c^i \oplus c^j$ that lie in d anti-diagonals are equal to zero (where $1 \leq d \leq 3$ and the anti-diagonals are fixed in advance). It is possible to prove that for 5-round AES this number has the special property to be a multiple of 8 independently of d - that is on the number of considered anti-diagonals. Instead, for a random permutation the same number does not have any special property (e.g. it has the same probability to be even or odd). This allows to distinguish 5-round AES from a random permutation.

Theorem 1. *Given 2^{32} plaintexts in the same diagonal set defined as before, consider the corresponding ciphertexts after 5 rounds, that is (p^i, c^i) for $i = 0, ..., 2^{32} - 1$ where $c^i = R^5(p^i)$ The number n of different pairs of ciphertexts (c^i, c^j) for $i \neq j$ for which the bytes of the difference $c^i \oplus c^j$ that lie in d anti-diagonals are equal to zero (where $1 \leq d \leq 3$ and the anti-diagonals are fixed in advance) is a multiple of 8, that is $\exists n' \in \mathbb{N}$ such that $n = 8 \cdot n'$.*

Idea of the Proof - Lemma 1. As we have seen in the previous section, a diagonal set is always mapped after two rounds into a mixed set. In other words, if two plaintexts have equal bytes expect for the ones in one diagonal, then after two rounds some particular linear relationships (given in (1)) hold among the bytes of the difference of these two texts that lie in the same column with probability 1. In the same way, if two ciphertexts have equal bytes in d anti-diagonals, then these two texts have equal bytes in d diagonals two rounds before (due to the 2-round differential trail described in Sect. 2.1). In other words, an inverse-diagonal set is mapped into a diagonal set two rounds before (assuming the final MixColumns operation is omitted).

Assume for simplicity that the 2^{32} plaintexts are chosen in a diagonal set with the active bytes in the first diagonal (analogous for the other cases).

---

[2] The two pairs (c^i, c^j) and (c^j, c^i) are considered equivalent. To formalize this concept, one can consider the number of ciphertexts (c^i, c^j) *with* $i < j$ for which the bytes of the difference $c^i \oplus c^j$ that lie in d anti-diagonals are equal to zero.

Due to these two previous considerations, Theorem 1 on 5 rounds of AES (and its proof) is strongly related to the following lemma on 1-round AES.

Lemma 1. *Given 2^{32} plaintexts in a mixed set of the form*

$$MC \cdot \begin{bmatrix} A\,C\,C\,C \\ C\,C\,C\,A \\ C\,C\,A\,C \\ C\,A\,C\,C \end{bmatrix}, \tag{2}$$

consider the corresponding ciphertexts after 1 round, that is $(\hat{p}^i, \hat{c}^i)$ for $i = 0, ..., 2^{32} - 1$ where $\hat{c}^i = R(\hat{p}^i)$. The number n of different pairs of ciphertexts $(\hat{c}^i, \hat{c}^j)$ for $i \neq j$ for which the bytes of the difference $c^i \oplus c^j$ that lie in d diagonals are equal to zero (where $1 \leq d \leq 3$ and the diagonals are fixed in advance) is a multiple of 8, that is $\exists\, n' \in \mathbb{N}$ s.t. $n = 8 \cdot n'$.

The complete proof is provided in the next section - Sect. 4. We emphasize that the proof of Theorem 1 follows immediately by the proof of Lemma 1, due to the 2-round truncated differential trail described in Sect. 2.1. In particular, note that considering 2^{32} plaintexts in the same diagonal set (that is 2^{32} plaintexts which are equal in three diagonals and with active bytes in the other one) is equivalent to consider 2^{32} texts in the same mixed set as defined in (2) after two rounds. In other words, all 2^{32} plaintexts of Lemma 1 are definitely reachable in 2 rounds from the initial plaintext (diagonal) structure defined in Theorem 1.

To prove the lemma, the idea is show that given one pair of ciphertexts for which the bytes that lie in d diagonals are equal, then also other pairs of ciphertexts have the same property with probability 1. The complete proof is given in Sect. 4. We highlight that the statement given in Theorem 1 (or Lemma 1) does not depend on the details of the MixColumns matrix (with the exception that the branch number must be five) or/and of the SubBytes operation. In other words, the only property that the proof - given in the next section - exploits is the branch number of the MixColumns matrix.

3.2   Setting up the Distinguisher

Our 5-round distinguisher exploits the property just described that the above defined number n is a multiple of 8 for 5-round AES, while it can take any possible value in the case of a random permutation. In the following we show how to set up the previous distinguisher in an efficient way for the case $d = 1$ (analougos for the other cases).

To implement the distinguisher, one has to count the number of pairs of ciphertexts for which the difference in $d = 1$ anti-diagonal is equal to zero (where this anti-diagonal is fixed in advance). First of all, since the probability that two ciphertexts satisfy this property is 2^{-32} (in general, $2^{-32 \cdot d}$ for d anti-diagonals), we expect that *on average*

$$\binom{2^{32}}{2} \cdot 2^{-32} = 2^{31} \cdot (2^{32} - 1) \cdot 2^{-32} \simeq 2^{31}$$

different pairs of ciphertexts have difference zero in one fixed anti-diagonal both for an AES permutation and for a random one. However, while for an AES permutation this number is a multiple of 8 with probability 1, for a random permutation this happens only with probability $0.125 \equiv 2^{-3}$. In particular, consider s initial arbitrary diagonal sets of plaintexts and for each of them count the number of different pairs of ciphertexts that have difference zero in d anti-diagonals. For an AES permutation, each of these numbers is a multiple of 8, while the probability that this happens for a random permutation is only $2^{-3 \cdot s}$. In order to distinguish the AES permutation from the random one with probability at least pr, it is sufficient that for a random permutation at least one of these numbers is not a multiple of 8, which happens with probability pr:

$$pr = 1 - 2^{-3 \cdot s}.$$

Thus, the probability of success of this distinguisher is greater than 99% (i.e. $pr \geq 0.99$) for $s \geq 3$. Note that for each initial diagonal set, one can count the above defined number n for each one of the four possible anti-diagonals. In other words, there are four different anti-diagonals for which one can count the number n of pairs of ciphertexts with zero difference in that anti-diagonal. It follows that using a single initial diagonal set, it is possible to distinguish 5-round AES from a random permutation with a probability of success of approximately $1 - (2^{-12}) = 99.975\%$.

In conclusion, 2^{32} chosen plaintexts in a single initial arbitrary diagonal set - i.e. a set of 2^{32} plaintexts which are equal in all bytes except for the ones in the i-th diagonal for a certain $i = 0, 1, 2, 3$ - are sufficient to distinguish a random permutation from an AES one. An approximation of the computational cost is given in the following. For completeness, it is also possible to set up a distinguisher for the cases $d = 2$ and $d = 3$ - i.e. the cases in which one count the number n of pairs of ciphertexts for which the bytes in $d = 2, 3$ anti-diagonals are equal. However, it should be noticed that the average number of collisions in these cases are respectively $2^{31} \cdot (2^{32} - 1) \cdot 2^{-64} \simeq 2^{-1}$ and $2^{31} \cdot (2^{32} - 1) \cdot 2^{-96} \simeq 2^{-33}$. As a consequence, the data and computational cost of these cases is not lower than for the case $d = 1$.

3.3   The Computational Cost

We have just seen that 2^{32} chosen plaintexts in a single diagonal set are sufficient to distinguish a random permutation from 5 rounds of AES, simply counting the number of pairs of ciphertexts with equal bytes in d anti-diagonal and checking if it is a multiple of 8 or not. Here we give an estimation of the computational cost of the distinguisher, which is approximately given by the sum of the cost to construct all the pairs and of the cost to count the number of pairs of ciphertexts with the previous property. As a result, the total computational cost can be well approximated by $2^{35.6}$ table look-ups.

Assume the final MixColumns operation is omitted. As we have just said, for each initial diagonal set the two steps of the distinguisher are (1) construct all the possible pairs of ciphertexts and (2) count the number of collisions. First of all, given pair of ciphertexts, note that the cost to check that the bytes in d anti-diagonals are equal corresponds to the cost of a XOR operation[3]. As we are going to show, the major cost of this distinguisher regards the construction of all the possible different pairs, which corresponds to step (1). Since it is possible to construct approximately 2^{63} pairs for each initial diagonal set, the simplest way to do it requires 2^{63} table look-ups. In the following, we present a way to reduce the total cost to approximately $2^{35.6}$ table look-ups, where the used tables are of size 2^{32} texts (or equivalently $2^{32} \cdot 16 = 2^{36}$ byte).

The basic idea is to implement the distinguisher using a *data structure*. The goal is to count the number of pairs of ciphertexts (c^1, c^2) for which the bytes in one of the anti-diagonal are equal, that is such that for a fixed $j \in \{0,1,2,3\}$ the following condition is satisfied:

$$c^1_{i,j-i} = c^2_{i,j-i} \qquad \forall i = 0,1,2,3 \qquad (3)$$

where the index is computed modulo 4. To do this, consider an array A of 2^{32} elements completely initialized to zero. The element of A in position x for $0 \leq x \leq 2^{32} - 1$ - denote by $A[x]$ - represents the number of ciphertexts c that satisfy the following equivalence (in the integer field $\mathbb{N}$):

$$x = c_{0,0-j} + 256 \cdot c_{1,1-j} + c_{2,2-j} \cdot 256^2 + c_{3,3-j} \cdot 256^3.$$

It's simple to observe that if two ciphertexts c^1 and c^2 satisfy (3), then they increment the same element x of the array A. It follows that given $r \geq 0$ texts that increment the same element x of the array A, then it is possible to construct

$$\binom{r}{2} = \frac{r \cdot (r-1)}{2}$$

different pairs of texts that satisfy (3). The complete pseudo-code of such an algorithm is given in Algorithm 1.

What is the total computational cost of this procedure? Given a set of 2^{32} (plaintexts, ciphertexts) pairs, one has first to fill the array A using the strategy just described, and then to compute the number of total of pairs of ciphertexts that satisfy the property, for a cost of $3 \cdot 2^{32} = 2^{33.6}$ table look-ups - each one of these three operations require 2^{32} table look-ups. Since one has to repeat this algorithm 4 times - one time for each one of the four anti-diagonal, the total cost is of $4 \cdot 2^{33.6} = 2^{35.6}$ table look-ups, or equivalently 2^{29} five-round encryptions of AES (using the approximation[4] 1 table look-up $\approx$ 1 round of AES).

---

[3] As example, let $J \subseteq \{0,1,2,3\}$ with $d = |J|$. Given a pair (c^1, c^2), this operation can be reduced to check that $\tilde{c}_{k,j-k} = 0$ for each $k = 0, ..., 3$ and $j \in J$, where $\tilde{c} \equiv c^1 \oplus c^2$.

[4] We highlight that even if this approximation is not formally correct - the size of the table of an S-Box look-up is lower than the size of the table used for our proposed distinguisher, it allows to give a comparison between our proposed distinguisher and the others currently present in literature. At the same time, we note that the same approximation is largely used in literature.

Data: 2^{32} (plaintext, ciphertext) pairs (p^i, c^i) for $i = 0, ..., 2^{32} - 1$ in a single diagonal set.

Result: 1 for an AES permutation, 0 otherwise (prob. $\geq 99\%$)

Let (p^i, c^i) for $i = 0, ..., 2^{32} - 1$ the (plaintext, ciphertext) pairs;

for *all* $j \in \{0, 1, 2, 3\}$ **do**

> Let $A[0, ..., 2^{32} - 1]$ an array initialized to zero;
>
> **for** i from *0 to* $2^{32} - 1$ **do**
>
>> $x \leftarrow 0$;
>>
>> **for** k from *0 to 3* **do**
>>
>>> $x \leftarrow x + c^i_{k, j-k} \cdot 256^k$; // $c^i_{k, j-k}$ `denotes the byte of` c^i `in row` k
>>> `and column` $j - k$ `mod 4`
>>
>> **end**
>>
>> $A[x] \leftarrow A[x] + 1$;   // $A[x]$ `denotes the value stored in the` x`-th`
>> `address of the array` A
>
> **end**
>
> $n \leftarrow 0$;
>
> **for** i from *0 to* $2^{32} - 1$ **do**
>
>> $n \leftarrow n + A[i] \cdot (A[i] - 1)/2$;
>
> **end**
>
> **if** $(n \bmod 8) \neq 0$ **then**
>
>> **return** 0;
>
> **end**

end

return 1.

Algorithm 1. *Secret-Key Distinguisher for 5 rounds of AES* which exploits a property which is independent of the secret key - probability of success: $\geq 99\%$.

Another possible way to implement our distinguisher exploits a re-ordering algorithm. The goal is again to count the number of pairs of ciphertexts for which the bytes that lie in d fixed anti-diagonals are equal. In this case, the idea is to re-order the texts using a particular numerical order which depends - in a "certain way" - on these d anti-diagonals. Then, given a set of ordered texts, the idea is to work only on two consecutive elements in order to count the total number of pairs of ciphertexts with the required property. In other words, given ordered ciphertexts, one can work only on approximately 2^{32} different pairs (composed of consecutive elements with respect to the used order) instead of 2^{63} for each initial diagonal set. All the details of this method are given in Appendix D of [13]. This second implementation could be in some cases more efficient than the one proposed in details in this section when e.g. it is required to do further operations on the pairs of ciphertexts which are equal in the d fixed anti-diagonals.

3.4 Practical Verification

Using a C/C++ implementation[5], we have practically verified the distinguisher on a small scale variant of AES, as presented in [6]. While in "real" AES, each

---

[5] The source code is available at https://github.com/Krypto-iaik/AES_5round_SKdistinguisher.

word is composed of 8 bits, in this variant each word is composed of 4 bits. We refer to [6] for a complete description of this small-scale AES, and we limit ourselves to describe the results of our 5-round distinguisher in this case.

First of all, note that Theorem 1 holds exactly in the same way also for this small-scale variant of AES (the proof is independent by the fact that each word of AES is of 4 or 8 bits). Thus, our verification on the small-scale variant of AES is strong evidence for it to hold for the real AES.

We have verified the theorem for each possible value of d (i.e. for $1, 2, 3$). For the verification of the secret-key distinguisher, we have chosen plaintexts in the diagonal sets with a single active diagonal and $d = 1$. As result, we have verified that for 5-round AES the number of collisions is a multiple of 2, while this number does not have any particular property for a random permutation. Moreover, we have found that 2 initial diagonal sets are largely sufficient to distinguish a random permutation from an AES permutation also from a practical point of view, as predicted.

The differences between this small-scale AES and the real AES regard the total number of pairs of ciphertexts that satisfy the required property (equal bytes in 1 fixed diagonal), which in this case is well approximated by $2^{15} \cdot (2^{16} - 1) \cdot 2^{-16} \approx 2^{15}$ for each diagonal set, and the lower computational cost, which can be approximated by $2^{17.6} \cdot 4 \approx 2^{19.6}$ memory look-ups for each initial diagonal set, besides the memory costs. The *average* practical results of our experiments are in accordance with these numbers.

3.5   Generalizations of the Central Theorem

Until now we have considered only a particular case in order to set up our distinguisher. However, here we show that it is possible to generalize Theorem 1 as follows.

Firstly, note that the same distinguisher works also in the reverse direction (i.e. in the decryption mode) with the same complexity. Assume that the final MixColumns operation is omitted. In this case the strategy is to choose 2^{32} ciphertexts in the same inverse-diagonal set, i.e. a set of 2^{32} ciphertexts which are equal in all the bytes expect for the ones in the i-th anti-diagonal for a certain $i = 0, 1, 2, 3$ (similar definition of the diagonal set). As before, the idea is to count the number of different pairs of plaintexts for which the bytes that lie in d diagonals are equal, for d fixed diagonals with $1 \le d \le 3$. This number has the same properties given in Theorem 1, while for a random permutation it can take any possible value.

Theorem 2. *Assume the final MixColumns operation is omitted. Given 2^{32} ciphertexts in the same inverse-diagonal set - that is, a set of texts with equal bytes expect the ones in the i-th anti-diagonal for a certain $i \in \{0, 1, 2, 3\}$, consider the corresponding plaintexts 5 rounds before, that is (p^i, c^i) for $i = 0, ..., 2^{32} - 1$ where $p^i = R^{-5}(c^i)$ The number n of different pairs of plaintexts (p^i, p^j) for $i \ne j$ for which the bytes of the difference $p^i \oplus p^j$ that lie in d diagonals are equal to zero (where $1 \le d \le 3$ and the diagonals are fixed in advance) is a multiple of 8, that is $\exists n' \in \mathbb{N}$ such that $n = 8 \cdot n'$.*

A complete proof of this Theorem can be found in Appendix A of the full version of the paper [13].

Secondly, Theorem 1 can be generalized for the cases of diagonal sets in which more than a single diagonal is active. As an example, diagonal sets with 2 or 3 active diagonals can be

$$\begin{bmatrix} A\ A\ C\ C \\ C\ A\ A\ C \\ C\ C\ A\ A \\ A\ C\ C\ A \end{bmatrix} \quad \text{or} \quad \begin{bmatrix} A\ A\ A\ C \\ C\ A\ A\ A \\ A\ C\ A\ A \\ A\ A\ C\ A \end{bmatrix}.$$

It is possible to prove that the result given in Theorem 1 is completely independent of the number of active diagonals. In other words, independently of the number of active diagonals of the initial diagonal set of the plaintexts, then the number of pairs of ciphertexts for which the bytes that lie in d anti-diagonals are equal (for d fixed anti-diagonals with $1 \le d \le 3$) is a multiple of 8. A formal statement is the following:

Theorem 3. *Given $2^{32 \cdot D}$ plaintexts in the same diagonal set with $1 \le D \le 3$ active diagonals defined as before, consider the corresponding ciphertexts after 5 rounds, that is (p^i, c^i) for $i = 0, ..., 2^{32} - 1$ where $c^i = R^5(p^i)$ The number n of different pairs of ciphertexts (c^i, c^j) for $i \ne j$ for which the bytes of the difference $c^i \oplus c^j$ that lie in d anti-diagonals are equal to zero (where $1 \le d \le 3$ and the anti-diagonals are fixed in advance) is a multiple of 8, that is $\exists\, n' \in \mathbb{N}$ such that $n = 8 \cdot n'$.*

The proof of this theorem is given in Appendix A - it is simply a generalization of the proof of Theorem 1 given in the next section.

4   A Detailed Proof of Theorem 1 - Lemma 1

In this section we give a detailed and formal proof of Theorem 1. As we have already said, since it is sufficient to prove Lemma 1 in order to prove the Theorem, we focus on this Lemma, which is recalled in the following. Moreover, we assume that for simplicity that the 2^{32} plaintexts are chosen in a diagonal set with the active bytes in the first diagonal (analogous for the other cases).

Lemma 1. *Given 2^{32} plaintexts in the same mixed set of the form (2)*

$$MC \cdot \begin{bmatrix} A\ C\ C\ C \\ C\ C\ C\ A \\ C\ C\ A\ C \\ C\ A\ C\ C \end{bmatrix},$$

consider the corresponding ciphertexts after 1 round, that is $(\hat{p}^i, \hat{c}^i)$ for $i = 0, ..., 2^{32} - 1$ where $\hat{c}^i = R(\hat{p}^i)$. The number n of different pairs of ciphertexts $(\hat{c}^i, \hat{c}^j)$ for which the bytes of the difference $c^i \oplus c^j$ that lie in d diagonals are equal to zero (where $1 \le d \le 3$ and the diagonals are fixed in advance) is a multiple of 8, that is $\exists\, n' \in \mathbb{N}$ s.t. $n = 8 \cdot n'$.

Proof. Consider two elements p^1 and p^2 in the set just defined. By definition, there exist $x, y, z, w \in \mathbb{F}_{2^8}$, $x', y', z', w' \in \mathbb{F}_{2^8}$ and a constant $a \in \mathbb{F}_{2^8}^{4 \times 4}$ such that:

$$p^1 = a \oplus \begin{bmatrix} 2 \cdot x & y & z & 3 \cdot w \\ x & y & 3 \cdot z & 2 \cdot w \\ x & 3 \cdot y & 2 \cdot z & w \\ 3 \cdot x & 2 \cdot y & z & w \end{bmatrix}, \qquad p^2 = a \oplus \begin{bmatrix} 2 \cdot x' & y' & z' & 3 \cdot w' \\ x' & y' & 3 \cdot z' & 2 \cdot w' \\ x' & 3 \cdot y' & 2 \cdot z' & w' \\ 3 \cdot x' & 2 \cdot y' & z' & w' \end{bmatrix}$$

where $2 \equiv 0 \times 02$ and $3 \equiv 0 \times 03$. For the following, we say that p^1 is "generated" by the variables $\langle x, y, z, w \rangle$ and that p^2 is "generated" by the variables $\langle x', y', z', w' \rangle$.

First Case. First, we consider the case in which three variables are equal. W.l.o.g. we assume for example that $y = y'$, $z = z'$, $w = w'$ and $x \neq x'$ (the other cases are analogous). As we are going to show, in this case it is not possible that after one round the bytes of one diagonal (e.g. the j-th diagonal for $j \in \{0, 1, 2, 3\}$) of the two texts are equal). In other words, it is not possible that $(R(p^1) \oplus R(p^2))_{i,j+i} = 0$ for each $i = 0, ..., 3$ (i.e. the four bytes of the j-th diagonal of $R(p^1) \oplus R(p^2)$ are equal to zero), where the indexes are taken modulo 4. As we are going to show, this is due to the given hypothesis of this case and to the fact that the branch number of the MixColumns operation is equal to five.

By simple computation, the first column (analogues for the other ones) of $SR \circ$ S-Box$(p^1) \oplus SR \circ$ S-Box(p^2) - denoted by $(SR \circ$ S-Box$(p^1) \oplus SR \circ$ S-Box$(p^2))_{.,0}$ - is equal to:

$$(SR \circ \text{ S-Box}(p^1) \oplus SR \circ \text{ S-Box}(p^2))_{.,0} = \begin{bmatrix} \text{S-Box}(2 \cdot x \oplus a_{0,0}) \oplus \text{S-Box}(2 \cdot x' \oplus a_{0,0}) \\ 0 \\ 0 \\ 0 \end{bmatrix}.$$

After the MixColumns operation (note $R(p^1) \oplus R(p^2) = MC(SR \circ$ S-Box$(p^1) \oplus SR \circ$ S-Box$(p^2)) = MC \circ SR \circ$ S-Box$(p^1) \oplus MC \circ SR \circ$ S-Box$(p^2))$, since only one input byte[6] is different from zero, it follows that at least four output bytes must be different from zero, that is all the output bytes are different from zero. This simply implies that it is not possible that the bytes of one or more diagonals of $R(p^1) \oplus R(p^2)$ are equal to zero. As a consequence, this case does not contribute to the number n.

Second Case. Secondly, we consider the case in which two variables are equal, that is w.l.o.g. we assume for example that $z = z'$ and $w = w'$, while $x \neq x'$ and $y \neq y'$ (the other cases are analogous).

Assume there exist two elements p^1 (generated by $\langle x, y \rangle$) and p^2 (generated by $\langle x', y' \rangle$) defined as before such that they have zero-difference in the j-th

---

[6] Note that S-Box$(2 \cdot x \oplus a_{0,0}) \oplus$ S-Box$(2 \cdot x' \oplus a_{0,0}) = 0$ if and only if $x = x'$, which can never happen for hypothesis.

diagonal after one round. In other words, let $j \in \{0, 1, 2, 3\}$ and assume that there exist x, y and x', y' such that the generated elements p^1 and p^2 satisfy $(R(p^1) \oplus R(p^2))_{i,i+j} = 0$ for each $i = 0, 1, 2, 3$, where the indexes are taken modulo 4.

This implies that *other* two elements $\hat{p}^1$ (generated by $\langle x, y' \rangle$) and $\hat{p}^2$ (generated by $\langle x, y' \rangle$), that is

$$
\hat{p}^1 = a \oplus \begin{bmatrix} 2 \cdot x' & y & 0 & 0 \\ x' & y & 0 & 0 \\ x' & 3 \cdot y & 0 & 0 \\ 3 \cdot x' & 2 \cdot y & 0 & 0 \end{bmatrix} \quad \text{and} \quad \hat{p}^2 = a \oplus \begin{bmatrix} 2 \cdot x & y' & 0 & 0 \\ x & y' & 0 & 0 \\ x & 3 \cdot y' & 0 & 0 \\ 3 \cdot x & 2 \cdot y' & 0 & 0 \end{bmatrix},
$$

satisfy the condition $(R(\hat{p}^1) \oplus R(\hat{p}^2))_{i,i+j} = 0$ for each $i = 0, 1, 2, 3$ and for a certain j after one round. To prove this fact, it is sufficient to compute $R(p^1) \oplus R(p^2)$ and $R(\hat{p}^1) \oplus R(\hat{p}^2)$, and to prove that they are equal, i.e.

$$
R(p^1) \oplus R(p^2) = R(\hat{p}^1) \oplus R(\hat{p}^2).
$$

Since $(R(p^1) \oplus R(p^2))_{i,i+j} = 0$ for each $i = 0, 1, 2, 3$, it also follows that $(R(\hat{p}^1) \oplus R(\hat{p}^2))_{i,i+j} = 0$ for each i. In particular, by simple computation the first column of $R(p^1) \oplus R(p^2)$ is given by:

$$
(R(p^1) \oplus R(p^2))_{0,0} = 2 \cdot (\text{S-Box}(2 \cdot x \oplus a_{0,0}) \oplus \text{S-Box}(2 \cdot x' \oplus a_{0,0})) \oplus \\
\oplus 3 \cdot (\text{S-Box}(y \oplus a_{1,1}) \oplus \text{S-Box}(y' \oplus a_{1,1})),
$$
$$
(R(p^1) \oplus R(p^2))_{1,0} = \text{S-Box}(2 \cdot x \oplus a_{0,0}) \oplus \text{S-Box}(2 \cdot x' \oplus a_{0,0}) \oplus \\
\oplus 2 \cdot (\text{S-Box}(y \oplus a_{1,1}) \oplus \text{S-Box}(y' \oplus a_{1,1})),
$$
$$
(R(p^1) \oplus R(p^2))_{2,0} = \text{S-Box}(2 \cdot x \oplus a_{0,0}) \oplus \text{S-Box}(2 \cdot x' \oplus a_{0,0}) \oplus \\
\oplus \text{S-Box}(y \oplus a_{1,1}) \oplus \text{S-Box}(y' \oplus a_{1,1}),
$$
$$
(R(p^1) \oplus R(p^2))_{3,0} = 3 \cdot (\text{S-Box}(2 \cdot x \oplus a_{0,0}) \oplus \text{S-Box}(2 \cdot x' \oplus a_{0,0})) \oplus \\
\oplus \text{S-Box}(y \oplus a_{1,1}) \oplus \text{S-Box}(y' \oplus a_{1,1}).
$$

Due to the definition of $\hat{p}^1$ and $\hat{p}^2$, it follows immediately that $(R(p^1) \oplus R(p^2))_{\cdot,0} = (R(\hat{p}^1) \oplus R(\hat{p}^2))_{\cdot,0}$. The same happens for the other columns. Note that the two elements $\hat{p}^1$ and $\hat{p}^2$ exist for sure since we are working with all the 2^{32} plaintexts in the same mixed set (2). This implies that the number of collisions must be even, that is a multiple of 2.

Question: given p^1 and p^2 as before, is it possible that x, y, x', y' exist such that $(R(p^1) \oplus R(p^2))_{i,i+j} = 0$ for each $i = 0, 1, 2, 3$? Yes, again because the branch number of the MixColumns operation is five. Indeed, compute $SR \circ \text{S-Box}(p^1) \oplus SR \circ \text{S-Box}(p^2)$ and analyze the first column (the others are analogous):

$$
(SR \circ \text{S-Box}(p^1) \oplus SR \circ \text{S-Box}(p^2))_{\cdot,0} = \begin{bmatrix} \text{S-Box}(2 \cdot x \oplus a_{0,0}) \oplus \text{S-Box}(2 \cdot x' \oplus a_{0,0}) \\ \text{S-Box}(y \oplus a_{1,1}) \oplus \text{S-Box}(y' \oplus a_{1,1}) \\ 0 \\ 0 \end{bmatrix}.
$$

After the MixColumns operation (note $R(p^1) \oplus R(p^2) = MC(SR \circ \text{S-Box}(p^1) \oplus SR \circ \text{S-Box}(p^2)))$, since two input bytes[7] are different from zero, it follows that at least three output bytes must be different from zero, or at most one output byte could be equal to zero (similar for the other columns). Moreover, this also implies that it is not possible that two or more output bytes in the same column are equal to zero.

Moreover, observe that $(R(p^1) \oplus R(p^2))_{i,i+j} = 0$ for each i if and only if four bytes (one per column) of $R(p^1) \oplus R(p^2)$ are equal to zero. Since there are four "free" variables (i.e. x, y, x', y') and a system of four equations, such a system can have a non-negligible solution.

Finally, since the previous result is independent of the values of $z = z'$ and $w = w'$, it follows that the number of collisions for this case must be a multiple of 2^{17}. Indeed, assume that for certain $\hat{z}$ and $\hat{w}$ there exist x, y, x', y' such that the two elements p^1 and p^2 generated respectively by $\langle x, y \rangle$ by $\langle x', y' \rangle$ satisfy the condition that $R(p^1) \oplus R(p^2)$ has zero-difference in the j-th diagonal. By simple computation, the difference $R(p^1) \oplus R(p^2)$ doesn't depend on $z = z'$ and on $w = w'$, that is for each byte of $(R(p^1) \oplus R(p^2))_{k,l}$ for $k, l = 0, 1, 2, 3$ there exist constant A_i, B_i, C_i for $i = 0, 1, 2, 3$ - that depend only on the coefficients of the MixColumns matrix or/and of the secret-key - such that

$$
\begin{aligned}
(R(p^1) \oplus R(p^2))_{k,l} =& A_0 \cdot (\text{S-Box}(B_0 \cdot x \oplus C_0) \oplus \text{S-Box}(B_0 \cdot x' \oplus C_0)) \oplus \\
& \oplus A_1 \cdot (\text{S-Box}(B_1 \cdot y \oplus C_1) \oplus \text{S-Box}(B_1 \cdot y' \oplus C_1)) \oplus \\
& \oplus A_2 \cdot (\text{S-Box}(B_2 \cdot z \oplus C_2) \oplus \text{S-Box}(B_2 \cdot z' \oplus C_2)) \oplus \\
& \oplus A_3 \cdot (\text{S-Box}(B_3 \cdot w \oplus C_3) \oplus \text{S-Box}(B_3 \cdot w' \oplus C_3)) = \\
=& A_0 \cdot (\text{S-Box}(B_0 \cdot x \oplus C_0) \oplus \text{S-Box}(B_0 \cdot x' \oplus C_0)) \oplus \\
& \oplus A_1 \cdot (\text{S-Box}(B_1 \cdot y \oplus C_1) \oplus \text{S-Box}(B_1 \cdot y' \oplus C_1)).
\end{aligned}
$$

It follows that - under the previous hypothesis - each pair of elements p^1 and p^2 respectively generated by (1) $\langle x, y, z, w \rangle$ and by $\langle x', y', z, w \rangle$ or (2) $\langle x, y', z, w \rangle$ and by $\langle x', y, z, w \rangle$ *for each possible value of z and w* satisfy the condition that $R(p^1) \oplus R(p^2)$ has zero-difference in the j-th diagonal. Thus, the number of collisions for this case must be a multiple of $2 \cdot (2^8)^2 = 2^{17}$. As before, the existence of all these elements is guaranteed by the fact that we are working with all the 2^{32} plaintexts in the same mixed set (2).

Third Case. Thirdly, we consider the case in which only one variable is equal, that is w.l.o.g. we assume for example $w = w'$, while $x \neq x'$, $y \neq y'$ and $z \neq z'$ (the other cases are analogous).

Assume there exist two elements p^1 (generated by $\langle x, y, z \rangle$) and p^2 (generated by $\langle x', y', z' \rangle$) defined as before and $J \subseteq \{0, 1, 2, 3\}$ with $1 \leq d = |J| \leq 2$ such that the bytes of the two texts are equal after one round in the j-th diagonals

---

[7] Note that $\text{S-Box}(2 \cdot x \oplus a_{0,0}) \oplus \text{S-Box}(2 \cdot x' \oplus a_{0,0}) = 0$ if and only if $x = x'$, which can never happen for hypothesis. In the same way, $\text{S-Box}(y \oplus a_{1,1}) \oplus \text{S-Box}(y' \oplus a_{1,1}) = 0$ if and only if $y = y'$, which can never happen for hypothesis.

for $j \in J$. In other words, assume there exist x, y, z and x', y', z' such that the generated elements p^1 and p^2 satisfy $(R(p^1) \oplus R(p^2))_{i,j+i}$ for $j \in J$ with $1 \leq |J| \leq 2$. Similar to before, it follows that also the following three pairs of plaintexts generated by:

- $\langle x', y, z \rangle$ and $\langle x, y', z' \rangle$
- $\langle x, y', z \rangle$ and $\langle x', y, z' \rangle$
- $\langle x, y, z' \rangle$ and $\langle x', y', z \rangle$

have the same property (that is, the bytes in the j-th diagonals for $j \in J$ are equal after one round), for a total of four different pairs. As before, in order to prove this fact it is sufficient to show that $R(p^1) \oplus R(p^2) = R(\hat{p}^1) \oplus R(\hat{p}^2)$, where $\hat{p}^1$ and $\hat{p}^2$ are generated by the previous combinations of variables. Note that the two elements $\hat{p}^1$ and $\hat{p}^2$ exist for sure since we are working with all the 2^{32} plaintexts in the same mixed set (2). This implies that the number of collisions must be a multiple of 4.

Finally, we have only to prove that such x, y, z and x', y', z' can exist. As before, we compute $SR \circ$ S-Box$(p^1) \oplus SR \circ$ S-Box(p^2) and analyze the first column (the others are analogous):

$$(SR \circ \text{S-Box}(p^1) \oplus SR \circ \text{S-Box}(p^2))_{.,0} = \begin{bmatrix} \text{S-Box}(2 \cdot x \oplus a_{0,0}) \oplus \text{S-Box}(2 \cdot x' \oplus a_{0,0}) \\ \text{S-Box}(y \oplus a_{1,1}) \oplus \text{S-Box}(y' \oplus a_{1,1}) \\ \text{S-Box}(2 \cdot z \oplus a_{2,2}) \oplus \text{S-Box}(2 \cdot z' \oplus a_{2,2}) \\ 0 \end{bmatrix}.$$

After the MixColumns operation, since three input bytes[8] are different from zero, it follows that at least two output bytes must be different from zero, or at most two output bytes could be equal to zero. This implies that the event $(R(p^1) \oplus R(p^2))_{i,j+i} = 0$ for all $i = 0, 1, 2, 3$ and $j \in J$ with $1 \leq |J| \leq 2$ is possible. Moreover, this also implies that it is not possible that three output bytes (of the same column) are equal to zero, or in other words that $(R(p^1) \oplus R(p^2))_{i,j+i} = 0$ for all i and all $j \in J$ with $d = |J| = 3$ is not possible Also in this case, variables x, y, z and x', y', z' can exist since the number of equations is less or equal than the number of variables.

Finally, since the previous result is independent of the values of $w = w'$, it follows that the number of collisions for this case must be a multiple of $4 \cdot 2^8 = 2^{10}$. As before, assume that for a certain $\hat{w}$ there exist x, y, z, x', y', z' such that the two elements p^1 and p^2 generated respectively by $\langle x, y, z \rangle$ and by $\langle x', y', z' \rangle$ satisfy the condition that $R(p^1) \oplus R(p^2)$ has zero-difference in the j-th diagonals for $j \in J$. Also in this case, the idea is to show that the difference $R(p^1) \oplus R(p^2)$ doesn't depend on $w = w'$, that is for each byte of $(R(p^1) \oplus R(p^2))_{i,j}$ there exist constant A_i, B_i, C_i for $i = 0, 1, 2$ - that depend only on the coefficients of the MixColumns matrix or/and of the secret-key - such that

---

[8] Note that S-Box$(2 \cdot x \oplus a_{0,0}) \oplus$ S-Box$(2 \cdot x' \oplus a_{0,0}) =$ S-Box$(y \oplus a_{1,1}) \oplus$ S-Box$(y' \oplus a_{1,1}) =$ S-Box$(2 \cdot z \oplus a_{2,2}) \oplus$ S-Box$(2 \cdot z' \oplus a_{2,2}) = 0$ if and only if $x = x'$, $y = y'$ and $z = z'$, which can never happen for hypothesis.

$$(R(p^1) \oplus R(p^2))_{i,j} = A_0 \cdot (\text{S-Box}(B_0 \cdot x \oplus C_0) \oplus \text{S-Box}(B_0 \cdot x' \oplus C_0)) \oplus$$
$$\oplus A_1 \cdot (\text{S-Box}(B_1 \cdot y \oplus C_1) \oplus \text{S-Box}(B_1 \cdot y' \oplus C_1) \oplus$$
$$\oplus A_2 \cdot (\text{S-Box}(B_2 \cdot z \oplus C_2) \oplus \text{S-Box}(B_2 \cdot z' \oplus C_2).$$

It follows that - under the previous hypothesis - each pair of elements p^1 and p^2 respectively generated by one of the four different combinations of the variables $\langle x, y, z, w \rangle$ and $\langle x', y', z', w \rangle$ *for each possible value of w* satisfy the condition that $R(p^1) \oplus R(p^2)$ has zero-difference in the j-th diagonals for $j \in J$. As before, the existence of all these elements is guaranteed by the fact that we are working with all the 2^{32} plaintexts in the same mixed set (2).

Fourth Case. Fourthly, we consider the case in which all the variables are different, that is w.l.o.g. we assume that $x \neq x'$, $y \neq y'$, $z \neq z'$ and $w \neq w'$.

Assume there exist two elements p^1 (generated by $\langle x, y, z, w \rangle$) and p^2 (generated by $\langle x', y', z', w' \rangle$) defined as before and $J \subseteq \{0, 1, 2, 3\}$ with $1 \leq d = |J| \leq 3$ such that the bytes of the two texts are equal after one round in the j-th diagonals for $j \in J$. In other words, assume there exist x, y, z, w and x', y', z', w' such that the generated elements p^1 and p^2 satisfy $(R(p^1) \oplus R(p^2))_{i,j+i} = 0$ for all $i = 0, 1, 2, 3$ and for all $j \in J$ with $1 \leq d = |J| \leq 3$. Similar to before, it follows that also the following seven pairs of plaintexts generated by:

- $\langle x', y, z, w \rangle$ and $\langle x, y', z', w' \rangle$
- $\langle x, y', z, w \rangle$ and $\langle x', y, z', w' \rangle$
- $\langle x, y, z', w \rangle$ and $\langle x', y', z, w' \rangle$
- $\langle x, y, z, w' \rangle$ and $\langle x', y', z', w \rangle$
- $\langle x', y', z, w \rangle$ and $\langle x, y, z', w' \rangle$
- $\langle x', y, z', w \rangle$ and $\langle x, y', z, w' \rangle$
- $\langle x', y, z, w' \rangle$ and $\langle x, y', z', w \rangle$

have the same property (thta is, the bytes in the j-th diagonals for $j \in J$ are equal after one round), for a total of eight different pairs. As before, in order to prove this fact it is sufficient to show that $R(p^1) \oplus R(p^2) = R(\hat{p}^1) \oplus R(\hat{p}^2)$. Moreover, as before note that the two elements $\hat{p}^1$ and $\hat{p}^2$ exist for sure since we are working with all the 2^{32} plaintexts in the same mixed set (2). This implies that the number of collisions must be a multiple of 8.

Finally, we have only to prove that such x, y, z, w and x', y', z', w' can exist. As before, we compute $SR \circ \text{S-Box}(p^1) \oplus SR \circ \text{S-Box}(p^2)$ and analyze the first column (the others are analogous):

$$(SR \circ \text{S-Box}(p^1) \oplus SR \circ \text{S-Box}(p^2))_{\cdot,0} = \begin{bmatrix} \text{S-Box}(2 \cdot x \oplus a_{0,0}) \oplus \text{S-Box}(2 \cdot x' \oplus a_{0,0}) \\ \text{S-Box}(y \oplus a_{1,1}) \oplus \text{S-Box}(y' \oplus a_{1,1}) \\ \text{S-Box}(2 \cdot z \oplus a_{2,2}) \oplus \text{S-Box}(2 \cdot z' \oplus a_{2,2}) \\ \text{S-Box}(w \oplus a_{3,3}) \oplus \text{S-Box}(w' \oplus a_{3,3}) \end{bmatrix}.$$

After the MixColumns operation, since four input bytes[9] are different from zero, it follows that at least one output byte must be different from zero, or at most three output bytes could be equal to zero. This implies that the event $(R(p^1) \oplus R(p^2))_{i,j+i} = 0$ for all $i = 0, 1, 2, 3$ and for all $j \in J$ with $1 \leq |J| \leq 3$ is possible. Also in this case, variables x, y, z, w and x', y', z', w' can exist since the number of equations is less or equal than the number of variables.

Conclusion. We summarize the previous results and we prove the lemma. Given a set (2), we analyze the number of pairs of texts for which the bytes of d diagonals are equal after one round.

If $d = 3$, it is possible to have a collision only in the case in which all the variables that generate the two texts are different, that is $x \neq x'$, $y \neq y'$, and so on. In this case, the number of collisions n must be a multiple of 8, that is there exists $n' \in \mathbb{N}$ such that $n = 8 \cdot n'$.

If $d = 2$, it is possible to have a collision only if at least three variables that generate the two texts are different (i.e. at most one variable can be equal). If all the variables are different, the number of collisions is a multiple of 8, while if one is equal then the number of collisions is a multiple of $1024 \equiv 2^{10}$. In other words, there exist $n', n'_2 \in \mathbb{N}$ such that the total number of collisions n is equal to $n = 8 \cdot n' + 1024 \cdot n'_2 = 8 \cdot (n' + 128 \cdot n'_2)$, i.e. it is a multiple of 8.

If $d = 3$, it is possible to have a collision only if at least two variables that generate the two texts are different (i.e. at most two variables can be equal). If all the variables are different, the number of collisions is a multiple of 8, if one is equal then the number of collisions is a multiple of $1024 \equiv 2^{10}$, while if two are equal then the number of collisions is a multiple of $131072 \equiv 2^{17}$. In other words, there exist $n', n'_2, n'_3 \in \mathbb{N}$ such that the total number of collisions n is equal to $n = 8 \cdot n' + 2^{10} \cdot n'_2 + 2^{17} \cdot n'_3 = 8 \cdot (n' + 2^7 \cdot n'_2 + 2^{14} \cdot n'_3)$, i.e. it is a multiple of 8.

This proves the lemma. □

For completeness, we briefly recall why the proof of Lemma 1 implies Theorem 1. As we have already seen, if two plaintexts are in the same diagonal set, then after two rounds some particular linear relationships (given in (1)) hold among the bytes of the two texts that lie in the same column with probability 1. In the same way, if two ciphertexts have equal bytes in d anti-diagonals, then two rounds before - assuming the final MixColumns operation is omitted - the the two texts have equal bytes in d diagonals (due to the 2-round differential trail described in Sect. 2.1). Thus, it is sufficient to prove that given a mixed set of the form (2), the number of pairs of texts for which the bytes of d diagonals are equal after one round is a multiple of 8, which is the statement of Lemma 1. This finally proves the theorem.

---

[9] Note that S-Box$(2 \cdot x \oplus a_{0,0}) \oplus$ S-Box$(2 \cdot x' \oplus a_{0,0}) =$ S-Box$(y \oplus a_{1,1}) \oplus$ S-Box$(y' \oplus a_{1,1}) =$ S-Box$(2 \cdot z \oplus a_{2,2}) \oplus$ S-Box$(2 \cdot z' \oplus a_{2,2}) =$ S-Box$(w \oplus a_{3,3}) \oplus$ S-Box$(w' \oplus a_{3,3}) = 0$ if and only if $x = x'$, $y = y'$, $z = z'$ and $w = w'$, which can never happen for hypothesis.

5   Conclusion, Applications and Open Problems

In this paper, we have presented a new non-random property for 5 rounds of AES. Additionally, we showed how to set up an efficient 5-round secret-key distinguisher for AES which exploits this property, which is independent of the secret key, improving the very recent results [21] and providing answers to the questions posed in [21]. This distinguisher is structural in the sense that it is independent of the details of the MixColumns matrix (with the exception that the branch number must be five) and also independent of the SubBytes operation. As such it will be straightforward to apply to many other AES-like constructions. Starting from our results, a range of new questions arise for future investigations:

Application to Schemes that directly use round-reduced AES. Round-reduced AES is a popular construction to build different schemes. For example, in the on-going "Competition for Authenticated Encryption: Security, Applicability, and Robustness" (CAESAR) [1], which is currently at its third round, several candidates are designed based on an AES-like SPN structure. Focusing only on the third-round candidates[10], among many others, AEGIS [15] uses four AES round-functions in the state update functions while ELmD [20] recommends to use round-reduced AES including 5-round AES to partially encrypt the data. Although the security of these candidates does not completely depend on the underlying primitives, we believe that a better understanding of the security of round-reduced AES can help get insights to both the design and cryptanalysis of authenticated encryption algorithms.

Further Extensions. Is it possible to set up a secret-key distinguisher for 6-round of AES which exploits a property which is independent of the secret key? Is it possible to set up efficient key recovery attacks for 6- or more rounds of AES that exploits this new 5-round secret-key distinguisher proposed in this paper or a modified version of it?

Permutation and Known-Key Distinguishers. The new 5-round property (or its approach to derive it) might find applications to permutation distinguishers or known-key distinguishers. Permutation distinguisher are usually set up by combining two secret-key distinguishers in an inside-out fashion. It is not immediately clear how the 5-round secret-key distinguisher presented in this paper used in an inside-out approach would be able to maintain the property in both directions simultaneously, but it seems interesting to investigate this direction also.

Acknowledgments. The work in this paper has been partially supported by the Austrian Science Fund (project P26494-N15).

---

[10] Among previous-round candidates, it is also possible to include PRIMATEs [11] which design is based on an AES-like SPN structure, while 4-round AES is adopted by Marble [16] and used to build the AESQ permutation in PAEQ [2].

A   Generalization of Theorem 1

In Theorem 1 given in Sect. 3, we only considered the case of chosen plaintexts in the same diagonal set with a single active diagonal - i.e. $D = 1$. A natural question arises: is it possible to generalize the theorem also for $D = 2$ or/and $D = 3$, that is for chosen plaintexts in the same diagonal set with two or three active diagonals? The answer is yes, and it is given in Theorem 3 recalled in the following. In particular, we prove in this section that the result obtained in Theorem 1 is independent of the number of initial active diagonals D, or, in other words, the property of n to be a multiple of 8 is independent of D.

Theorem 1. *Given $2^{32 \cdot D}$ plaintexts in the same diagonal set with $1 \leq D \leq 3$ active diagonals defined as before, consider the corresponding ciphertexts after 5 rounds, that is (p^i, c^i) for $i = 0, ..., 2^{32} - 1$ where $c^i = R^5(p^i)$ The number n of different pairs of ciphertexts (c^i, c^j) for $i \neq j$ for which the bytes of the difference $c^i \oplus c^j$ that lie in d anti-diagonals are equal to zero (where $1 \leq d \leq 3$ and the anti-diagonals are fixed in advance) is a multiple of 8, that is $\exists n' \in \mathbb{N}$ such that $n = 8 \cdot n'$.*

Since the proof for the case $D = 1$ is given in Sect. 4, we focus on the cases $D = 2$ and $D = 3$. Also for these cases, the idea is to analyze the middle round and to study each possible case, as done in Sect. 4. Thus, given pair of texts of the form

$$MC \cdot \begin{bmatrix} A\,A\,C\,C \\ A\,C\,C\,A \\ C\,C\,A\,A \\ C\,A\,A\,C \end{bmatrix} \quad \text{or} \quad MC \cdot \begin{bmatrix} A\,A\,A\,C \\ A\,A\,C\,A \\ A\,C\,A\,A \\ C\,A\,A\,A \end{bmatrix}, \tag{4}$$

we analyze the property of the number of pairs of texts which are equal in d diagonals after one round.

Since the idea of the proof for $D = 2$ and $D = 3$ is analogous to that given for $D = 1$, we limit ourselves to do some considerations which justify the theorem. A complete proof can be easily obtained exploiting the following considerations and using the same strategy proposed in Sect. 4.

First Consideration. As first consideration, note that we are considering pairs of plaintexts/ciphertexts (p^1, c^1) and (p^2, c^2) such that the plaintexts are in the same diagonal set with at least 2 active diagonals. On the other hand, such a set can be seen as a collection of diagonal set with only 1 active diagonal. Since Theorem 1 holds for each one of these sets, it follows that if n is a multiple of 2^m then m must satisfy $m \leq 3$. This follows immediately by Theorem 1 and the corresponding proof of Sect. 4.

Thus, we have to prove that n is a multiple of 2^m and that $m = 3$ also for the cases $D = 2$ and $D = 3$.

A.1 Case $D = 2$

We start studying the case $D = 2$. As we show in details in the following, the same analysis can be simply modified and adapted for the case $D = 3$.

Consider two texts p^1 and p^2 in the same set (4) (the other cases are analogous). By definition, there exist $x_0, x_1, y_0, y_1, z_0, z_1, w_0, w_1 \in \mathbb{F}_{2^8}$, x_0', x_1', y_0', $y_1', z_0', z_1', w_0', w_1' \in \mathbb{F}_{2^8}$ and $a \in \mathbb{F}_{2^8}^{4 \times 4}$ such that:

$$p^1 = a \oplus MC \cdot \begin{bmatrix} x_0 & y_0 & 0 & 0 \\ x_1 & 0 & 0 & w_0 \\ 0 & 0 & z_0 & w_1 \\ 0 & y_1 & z_1 & 0 \end{bmatrix}, \qquad p^2 = a \oplus MC \cdot \begin{bmatrix} x_0' & y_0' & 0 & 0 \\ x_1' & 0 & 0 & w_0' \\ 0 & 0 & z_0' & w_1' \\ 0 & y_1' & z_1' & 0 \end{bmatrix}.$$

For the following, let $2 \equiv 0 \times 02$ and $3 \equiv 0 \times 03$.

Following the same strategy of Sect. 4, the idea is to consider all the possible cases in which some or no-one variables of p^1 are equal to the ones of p^2. Note that the case $x_1 = x_1'$, $y_1 = y_1'$, $z_1 = z_1'$ and $w_1 = w_1'$ (i.e. two texts that belong into the same set (2)) has already been considered. In particular, by Theorem 1 it follows that in this case the number n is a multiple of 8.

First Case. W.l.o.g. we consider the case $y_1 = y_1'$, $w_i = w_i'$ and $z_i = z_i'$ for $i = 0, 1$, while $y_0 \neq y_0'$ and $x_i \neq x_i'$ for $i = 0, 1$ (the other cases are analogous).

Assume there exist x_0, x_1, y_0 and x_0', x_1', y_0' such that the generated elements p^1 and p^2 satisfy the condition $(R(p^1) \oplus R(p^2))_{i, j+i} = 0$ for all $i = 0, 1, 2, 3$ and for a certain $j \in 0, 1, 2, 3$ - i.e. the bytes of one diagonal of the two texts are equal after one round. First of all, we show that such variables can exist. The condition $(R(p^1) \oplus R(p^2))_{i, j+i} = 0$ for all i and a certain $j \in \{0, 1, 2, 3\}$ implies that four bytes (one per column) of $R(p^1) \oplus R(p^2)$ must be equal to 0. Since there are six independent variables, a solution can exist (note that the number of variables is higher than the number of equations, so two variables are still "free"). Moreover, this is also due to the branch number of the MixColumns operation, which is five. Indeed, by simple computation the first column of $SR(\text{S-Box}(p^1) \oplus \text{S-Box}(p^2))$ (analogous for the others) is given by:

$$SR(\text{S-Box}(p^1) \oplus \text{S-Box}(p^2))_{0,0} = \text{S-Box}(2 \cdot x_0 \oplus 3 \cdot x_1 \oplus a_{0,0} \oplus a_{1,0})$$
$$\oplus \text{S-Box}(2 \cdot x_0' \oplus 3 \cdot x_1' \oplus a_{0,0} \oplus a_{1,0}),$$
$$SR(\text{S-Box}(p^1) \oplus \text{S-Box}(p^2))_{1,0} = \text{S-Box}(y_0 \oplus a_{1,1}) \oplus \text{S-Box}(y_0' \oplus a_{1,1}),$$
$$SR(\text{S-Box}(p^1) \oplus \text{S-Box}(p^2))_{2,0} = SR(\text{S-Box}(p^1) \oplus \text{S-Box}(p^2))_{3,0} = 0.$$

Thus, if we compute $MC \circ SR(\text{S-Box}(p^1) \oplus \text{S-Box}(p^2))$ (that is, $R(p^1) \oplus R(p^2)$), since at most two input bytes are different from zero, then it follows that at least three output bytes must be different from zero, or equivalently at most one output byte can be equal to zero. As a consequence, it is possible that $(R(p^1) \oplus R(p^2))_{i, j+i} = 0$ for all i and a certain $j \in \{0, 1, 2, 3\}$. Note that the same can not happen for $d \geq 2$ diagonals. We emphasize that with respect to the case $D = 1$, it is possible that one input byte of the MixColumns operation can be equal to zero. Indeed, it is possible that exist x_0 and x_0' such that $SR(\text{S-Box}(p^1) \oplus \text{S-Box}(p^2))_{0,0}$ (analogous for the others columns).

As before, the idea is to consider the pairs of texts generated by all the possible combinations of these six variables, as for example $\langle x_0, x_1, y_0' \rangle$ and $\langle x_0', x_1', y_0 \rangle$, $\langle x_0, x_1', y_0 \rangle$ and $\langle x_0', x_1, y_0' \rangle$, $\langle x_0', x_1, y_0 \rangle$ and $\langle x_0, x_1', y_0' \rangle$, $\langle x_1, x_0, y_0' \rangle$ and $\langle x_0', x_1', y_0 \rangle$ (note that the elements generated by $\langle x_0, x_1, y_0' \rangle$ and by $\langle x_1, x_0, y_0' \rangle$ are different) and so on.

We analyze these cases. It is simple to observe that if p^1 generated by $\langle x_0, x_1, y_0 \rangle$ and p^2 generated by $\langle x_0', x_1', y_0 \rangle$ satisfy the condition that $(R(p^1) \oplus R(p^2))_{i,j+i} = 0$ for all i and a certain $j \in \{0, 1, 2, 3\}$ - i.e. one diagonal of the two texts are equal after one round, then also the elements generated by $\langle x_0, x_1, y_0' \rangle$ and $\langle x_0', x_1', y_0' \rangle$ have the same property. To prove this fact, it is sufficient to show that $R(p^1) \oplus R(p^2) = R(\hat{p}^1) \oplus R(\hat{p}^2)$. As an example, by simple computation, it is simple to observe that for the first column:

$$SR(\text{S-Box}(\hat{p}^1) \oplus \text{S-Box}(\hat{p}^2))_{i,0} = SR(\text{S-Box}(p^1) \oplus \text{S-Box}(p^2))_{i,0} \qquad \forall i,$$

which implies the statement.

Consider now the elements $\hat{p}^1$ generated by $\langle x_0, x_1', y_0 \rangle$ and $\hat{p}^2$ generated by $\langle x_0', x_1, y_0 \rangle$ (similar for the elements generated by $\langle x_0', x_1, y_0 \rangle$ and $\langle x_0, x_1', y_0 \rangle$). By simple computation, the first column of $SR(\text{S-Box}(\hat{p}^1) \oplus \text{S-Box}(\hat{p}^2))$ (analogous for the others) is given by:

$$SR(\text{S-Box}(\hat{p}^1) \oplus \text{S-Box}(\hat{p}^2))_{0,0} = \text{S-Box}(2 \cdot x_0 \oplus 3 \cdot x_1' \oplus a_{0,0} \oplus a_{1,0})$$
$$\oplus \text{S-Box}(2 \cdot x_0' \oplus 3 \cdot x_1 \oplus a_{0,0} \oplus a_{1,0})$$

and for $i = 1, 2, 3$

$$SR(\text{S-Box}(\hat{p}^1) \oplus \text{S-Box}(\hat{p}^2))_{i,0} = SR(\text{S-Box}(p^1) \oplus \text{S-Box}(p^2))_{i,0}.$$

Since the S-Box is a non-linear operation, three different cases can happen:

1. $SR(\text{S-Box}(\hat{p}^1) \oplus \text{S-Box}(\hat{p}^2))_{0,0} = 0$;
2. $SR(\text{S-Box}(\hat{p}^1) \oplus \text{S-Box}(\hat{p}^2))_{0,0} \neq 0$ and the elements $\hat{p}^1$ and $\hat{p}^2$ satisfy the condition $(R(\hat{p}^1) \oplus R(\hat{p}^2))_{i,j+i} = 0$ for all i and a certain $j \in \{0, 1, 2, 3\}$;
3. $SR(\text{S-Box}(\hat{p}^1) \oplus \text{S-Box}(\hat{p}^2))_{0,0} \neq 0$ and the elements $\hat{p}^1$ and $\hat{p}^2$ don't satisfy the condition $(R(\hat{p}^1) \oplus R(\hat{p}^2))_{i,j+i} = 0$ for all i and a certain $j \in \{0, 1, 2, 3\}$.

We analyze in details these three cases, starting from the first one. As first thing, note that this case can happen since the condition $(R(p^1) \oplus R(p^2))_{i,j+i} = 0$ for all i and a certain $j \in \{0, 1, 2, 3\}$ imposes a condition only on four out of six variables, that is two variables are still "free". If $SR(\text{S-Box}(\hat{p}^1) \oplus \text{S-Box}(\hat{p}^2))_{0,0} = 0$, it follows that only one byte (i.e. the second one) of the first column of $SR(\text{S-Box}(\hat{p}^1) \oplus \text{S-Box}(\hat{p}^2))$ is different from 0 (since $y_0 \neq y_0'$). Thus, since MixColumns operation has branch number 5, all the bytes of the first column of $R(\hat{p}^1) \oplus R(\hat{p}^2)$ must be different from zero, that is no diagonals of $R(\hat{p}^1)$ and $R(\hat{p}^2)$ can be equal. However, note that also in this case it is possible to deduce something. Indeed, by the previous consideration, it follows that the elements generated by $\langle x_0, x_1', y_0' \rangle$ and by $\langle x_0', x_1, y_0 \rangle$ don't satisfy the condition $(R(p^1) \oplus R(p^2))_{i,j+i} = 0$ for all i and a certain $j \in \{0, 1, 2, 3\}$

Consider now the other two cases. Since the S-Box is a non-linear operation, it is not possible to guarantee that

$$SR(\text{S-Box}(\hat{p}^1) \oplus \text{S-Box}(\hat{p}^2))_{0,0} = SR(\text{S-Box}(p^1) \oplus \text{S-Box}(p^2))_{0,0}.$$

In other words, they can be equal (which implies that the condition $(R(\hat{p}^1) \oplus R(\hat{p}^2))_{i,j+i} = 0$ for all i and a certain $j \in \{0, 1, 2, 3\}$ - the same j of p^1 and p^2 - holds) or different. In this second case, one can not say anything about the fact that the elements $\hat{p}^1$ and $\hat{p}^2$ satisfy or not the condition $(R(\hat{p}^1) \oplus R(\hat{p}^2))_{i,j+i} = 0$ for all i and a certain $j \in \{0, 1, 2, 3\}$ (the same j of p^1 and p^2). However, suppose that $\hat{p}^1$ and $\hat{p}^2$ satisfy it after one round for the same j of p^1 and p^2 (which is independent by the previous condition). In the same way of before, note that also the elements generated by $\langle x_0, x_1', y_0' \rangle$ and $\hat{p}^2$ generated by $\langle x_0', x_1, y_0 \rangle$ have the same property.

Thus, assume that p^1 generated by $\langle x_0, x_1, y_0 \rangle$ and p^2 generated by $\langle x_0', x_1', y_0' \rangle$ satisfy or not the condition $(R(p^1) \oplus R(p^2))_{i,j+i} = 0$ for all i and a certain $j \in \{0, 1, 2, 3\}$ after one round. By previous considerations, it follows that also the $\hat{p}^1$ generated by $\langle x_0, x_1', y_0 \rangle$ and $\hat{p}^2$ generated by $\langle x_0', x_1, y_0' \rangle$ have the same property. Thus, even if we can not do any claim for the other texts generated by a different combination of these six variables, it is possible to conclude that - for fixed $y_1 = y_1'$, $w_i = w_i'$ and $z_i = z_i'$ for $i = 0, 1$ - the number of collisions must be a multiple of 2 for this case.

Finally, since we are working with the entire set of the form (4) - that is, $y_1 = y_1'$, $w_i = w_i'$ and $z_i = z_i'$ for $i = 0, 1$ can take any possible value - and due to the same considerations of Sect. 4, it follows that the number of collisions must be a multiple of $2 \cdot (2^8)^5 = 2^{41}$ for this case.

Second Case. Similar considerations can be done for the case $w_i = w_i'$ and $z_i = z_i'$ for $i = 0, 1$, while $x_i \neq x_i'$ and $y_i \neq y_i'$ for $i = 0, 1$ (the other cases are analogous).

Assume there exist x_0, x_1, y_0, y_1 and x_0', x_1', y_0', y_1' such that the generated elements p^1 and p^2 satisfy the condition $(R(p^1) \oplus R(p^2))_{i,j+i} = 0$ for all i and a certain $j \in \{0, 1, 2, 3\}$. As before, note that this is possible since this implies that four bytes of $R(p_1) \oplus R(p^2)$ (one per column) must be equal to 0. Since there are eight independent variables, a solution can exist (note that the number of variables is higher than the number of equations, so four variables are still "free"). Due to the branch number of the MixColumns operation, even if four variables are still "free" it is not possible that the condition $(R(p^1) \oplus R(p^2))_{i,j+i} = 0$ for all i holds for two different j. Indeed, the first column of $SR(\text{S-Box}(p^1) \oplus \text{S-Box}(p^2))$ (analogous for the others) is given by:

$$SR(\text{S-Box}(p^1) \oplus \text{S-Box}(p^2))_{0,0} = \text{S-Box}(2 \cdot x_0 \oplus 3 \cdot x_1 \oplus a_{0,0} \oplus a_{1,0})$$
$$\oplus \text{S-Box}(2 \cdot x_0' \oplus 3 \cdot x_1' \oplus a_{0,0} \oplus a_{1,0}),$$
$$SR(\text{S-Box}(p^1) \oplus \text{S-Box}(p^2))_{1,0} = \text{S-Box}(y_0 \oplus y_1 \oplus a_{0,1} \oplus a_{3,0})$$

$$\oplus \text{S-Box}(y_0' \oplus y_1' \oplus a_{0,1} \oplus a_{3,0}),$$

$$SR(\text{S-Box}(p^1) \oplus \text{S-Box}(p^2))_{2,0} = SR(\text{S-Box}(p^1) \oplus \text{S-Box}(p^2))_{3,0} = 0.$$

After the MixColumns operation $MC \circ SR(\text{S-Box}(p^1) \oplus \text{S-Box}(p^2))$, since at most two input bytes are different from zero, then it follows that at least three output bytes must be different from zero.

Thus, given x_0, x_1, y_0, y_1 and x_0', x_1', y_0', y_1', the idea is to consider all the possible combinations as before. Also in this case, we can do a claim only on one of them. In particular, if two elements p^1 generated by $\langle x_0, x_1, y_0, y_1 \rangle$ and p^2 generated by $\langle x_0', x_1', y_0', y_1' \rangle$ satisfies the condition $(R(p^1) \oplus R(p^2))_{i,j+i} = 0$ for all i and a certain $j \in \{0, 1, 2, 3\}$, we can only claim that also the elements $\hat{p}^1$ generated by $\langle x_0', x_1', y_0, y_1 \rangle$ and $\hat{p}^2$ generated by $\langle x_0, x_1, y_0', y_1' \rangle$ have the same property. Considerations for the other combinations are similar to the previous case. Thus, we can claim that - *for fixed* $w_i = w_i'$ and $z_i = z_i'$ for $i = 0, 1$ - also for this case the number of collisions is a multiple of 2.

Finally, since we are working with the entire set of the form (4) - that is, $w_i = w_i'$ and $z_i = z_i'$ for $i = 0, 1$ can take any possible value - and due to the same considerations of Sect. 4, it follows that the number of collisions must be a multiple of $2 \cdot (2^8)^4 = 2^{33}$ for this case.

Second Consideration. What can we deduce by the previous two cases? Suppose to have two texts p^1 generated by $\langle x \equiv (x_0, x_1), y \equiv (y_0, y_1) \rangle$ and p^2 generated by $\langle x' \equiv (x_0', x_1'), y' \equiv (y_0', y_1') \rangle$ that satisfy the condition $(R(\hat{p}^1) \oplus R(\hat{p}^2))_{i,j+i} = 0$ for all i and a certain $j \in \{0, 1, 2, 3\}$ and where $x, y \in \mathbb{F}_{2^8} \times \mathbb{F}_{2^8} \equiv \mathbb{F}_{2^8}^2$. We have seen that given these two elements, one can only claim that also the texts $\hat{p}^1$ generated by $\langle x' \equiv (x_0', x_1'), y \equiv (y_0, y_1) \rangle$ and $\hat{p}^2$ generated by $\langle x \equiv (x_0, x_1), y' \equiv (y_0', y_1') \rangle$ have the same property, that is the condition $(R(\hat{p}^1) \oplus R(\hat{p}^2))_{i,j+i} = 0$ for all i and a certain $j \in \{0, 1, 2, 3\}$ for the same j of p^1 and p^2.

As a consequence, the idea for the case $D = 2$ is not to consider the variables that generate the texts and that are in the same column as independent. In other words, the idea is to work with variables in $\mathbb{F}_{2^8}^2$ and not in $\mathbb{F}_{2^8}$, i.e. to consider only all the possible combinations of $x \equiv (x_0, x_1), y \equiv (y_0, y_1)$ and $x' \equiv (x_0', x_1'), y' \equiv (y_0', y_1')$, and not of x_0, x_1, y_0, y_1 and x_0', x_1', y_0', y_1'. Using this strategy and working in the same way of Sect. 4, it is possible to analyze all the possible cases.

For example, consider the case in which $w_i = w_i'$ for $i = 0, 1$ and $x \equiv (x_0, x_1) \neq x' \equiv (x_0', x_1')$, $y \equiv (y_0, y_1) \neq y' \equiv (y_0', y_1')$ and $z \equiv (z_0, z_1) \neq z' \equiv (z_0', z_1')$. In the same way of before, it is only possible to prove that if there exist p^1 generated by $\langle x, y, z \rangle$ and p^2 generated by $\langle x', y', z' \rangle$ such that $(R(p^1) \oplus R(p^2))_{i,j+i} = 0$ for all i and certain $j \in J$ where $J \subseteq \{0, 1, 2, 3\}$ and $|J| = 2$ - i.e. two diagonals are equal, then a total of four elements generated by

- $\langle x, y, z \rangle$ and $\langle x', y', z' \rangle$
- $\langle x', y, z \rangle$ and $\langle x, y', z' \rangle$

- $\langle x, y', z \rangle$ and $\langle x', y, z' \rangle$
- $\langle x, y, z' \rangle$ and $\langle x', y', z \rangle$

have the same property. No claim can be made about other combinations of variables (as before, this is due to the fact that the S-Box is non-linear). It follows that - for fixed $w_i = w'_i$ for $i = 0, 1$- the number of collisions must be a multiple of 4 for this case. As before, since we are working with the entire set of the form (4) it follows that the number of collisions must be a multiple of $4 \cdot (2^8)^2 = 2^{18}$. Moreover, since the branch number of the MixColumns operation is five, note that it is not possible that $(R(p^1) \oplus R(p^2))_{i,j+i} = 0$ for all i and certain $j \in \{0, 1, 2, 3\}$ if $w_l = w'_l$ for $l = 0, 1$ (even if $(R(p^1) \oplus R(p^2))_{i,j+i} = 0$ for all i and certain $j \in J$ where $J \subseteq \{0, 1, 2, 3\}$ imposes only 8 conditions while the number of variables is 12, so 4 variables are still "free").

Similar considerations can be done for the case in which all the variables are different. As a consequence, the theorem is proved for the case $|I| = 2$.

A.2   Case $D = 3$

The case $D = 3$ is analogous to the case $D = 2$ and to the proof given in Sect. 4. For this reason, we limit ourselves to show how to adapt the proof of the case $D = 2$ for this case.

W.l.o.g consider two texts p^1 and p^2 in the same set (4) (the other cases are analogous). By definition, there exist $x_0, x_1, x_2, y_0, y_1, y_2, z_0, z_1, z_2, w_0, w_1, w_2 \in \mathbb{F}_{2^8}$, $x'_0, x'_1, x'_2, y'_0, y'_1, y'_2, z'_0, z'_1, z'_2, w'_0, w'_1, w'_2 \in \mathbb{F}_{2^8}$ and $a \in \mathbb{F}_{2^8}^{4\times4}$ such that:

$$p^1 = a \oplus MC \cdot \begin{bmatrix} x_0 & y_0 & z_0 & 0 \\ x_1 & y_1 & 0 & w_0 \\ x_2 & 0 & z_1 & w_1 \\ 0 & y_2 & z_2 & w_2 \end{bmatrix}, \quad p^2 = a \oplus MC \cdot \begin{bmatrix} x'_0 & y'_0 & z'_0 & 0 \\ x'_1 & y'_1 & 0 & w'_0 \\ x'_2 & 0 & z'_1 & w'_1 \\ 0 & y'_2 & z'_2 & w'_2 \end{bmatrix}.$$

Similarly to the case $D = 2$, the idea is to work with variables in $\mathbb{F}_{2^8}^3 \equiv \mathbb{F}_{2^8} \times \mathbb{F}_{2^8} \times \mathbb{F}_{2^8}$, e.g. $x \equiv (x_0, x_1, x_2), y \equiv (y_0, y_1, y_2)$ and so on. In other words, the idea is to consider the variables in the same column as not independent, that is to consider the possible combinations only of variables in $\mathbb{F}_{2^8}^3$ and not in $\mathbb{F}_{2^8}$.

References

1. CAESAR: Competition for Authenticated Encryption: Security, Applicability, and Robustness. http://competitions.cr.yp.to/caesar.html
2. Biryukov, A., Khovratovich, D.: PAEQ v1. http://competitions.cr.yp.to/round1/paeqv1.pdf
3. Biham, E., Biryukov, A., Shamir, A.: Cryptanalysis of skipjack reduced to 31 rounds using impossible differentials. In: Stern, J. (ed.) EUROCRYPT 1999. LNCS, vol. 1592, pp. 12–23. Springer, Heidelberg (1999). doi:10.1007/3-540-48910-X_2
4. Biham, E., Keller, N.: Cryptanalysis of Reduced Variants of Rijndael, unpublished (2001). http://csrc.nist.gov/archive/aes/round2/conf3/papers/35-ebiham.pdf

5. Biham, E., Shamir, A.: Differential Cryptanalysis of the Data Encryption Standard. Springer, New York (1993)
6. Cid, C., Murphy, S., Robshaw, M.J.B.: Small scale variants of the AES. In: Gilbert, H., Handschuh, H. (eds.) FSE 2005. LNCS, vol. 3557, pp. 145–162. Springer, Heidelberg (2005). doi:10.1007/11502760_10
7. Daemen, J., Knudsen, L., Rijmen, V.: The block cipher Square. In: Biham, E. (ed.) FSE 1997. LNCS, vol. 1267, pp. 149–165. Springer, Heidelberg (1997). doi:10.1007/BFb0052343
8. Daemen, J., Rijmen, V.: The Design of Rijndael: AES - The Advanced Encryption Standard. Information Security and Cryptography. Springer, Heidelberg (2002)
9. Daemen, J., Rijmen, V.: Two-round aes differentials. Cryptology ePrint Archive, Report 2006/039 (2006). http://eprint.iacr.org/2006/039
10. Daemen, J., Rijmen, V.: Understanding two-round differentials in AES. In: Prisco, R., Yung, M. (eds.) SCN 2006. LNCS, vol. 4116, pp. 78–94. Springer, Heidelberg (2006). doi:10.1007/11832072_6
11. Andreeva, E., Bilgin, B., Bogdanov, A., Luykx, A., Mendel, F., Mennink, B., Mouha, N., Wang, Q., Yasuda, K.: PRIMATEs v1.02 Submission to the CAESAR Competition. http://competitions.cr.yp.to/round2/primatesv102.pdf
12. Ferguson, N., Kelsey, J., Lucks, S., Schneier, B., Stay, M., Wagner, D., Whiting, D.: Improved cryptanalysis of Rijndael. In: Goos, G., Hartmanis, J., Leeuwen, J., Schneier, B. (eds.) FSE 2000. LNCS, vol. 1978, pp. 213–230. Springer, Heidelberg (2001). doi:10.1007/3-540-44706-7_15
13. Grassi, L., Rechberger, C., Rønjom, S.: A New Structural-Differential Property of 5-Round AES. IACR Cryptology ePrint Archive, vol. 2017 (2017). http://eprint.iacr.org/2017
14. Grassi, L., Rechberger, C., Rønjom, S.: Subspace trail cryptanalysis and its applications to AES. IACR Trans. Symmetric Cryptology 2016(2), 192–225 (2017). http://ojs.ub.rub.de/index.php/ToSC/article/view/571
15. Wu, H., Preneel, B.: A Fast Authenticated Encryption Algorithm. http://competitions.cr.yp.to/round1/aegisv1.pdf
16. Guo, J.: Marble Version 1.1. https://competitions.cr.yp.to/round1/marblev11.pdf
17. Knudsen, L.R.: Truncated and higher order differentials. In: Preneel, B. (ed.) FSE 1994. LNCS, vol. 1008, pp. 196–211. Springer, Heidelberg (1995). doi:10.1007/3-540-60590-8_16
18. Knudsen, L.R.: DEAL - a 128-bit block cipher. Technical report 151, Department of Informatics, University of Bergen, Norway, February 1998
19. Luby, M., Rackoff, C.: How to construct pseudorandom permutations from pseudorandom functions. SIAM J. Comput. 17(2), 373–386 (1988)
20. Datta, N., Nandi, M.: ELmD v2.0. http://competitions.cr.yp.to/round2/elmdv20.pdf
21. Sun, B., Liu, M., Guo, J., Qu, L., Rijmen, V.: New insights on AES-Like SPN ciphers. In: Robshaw, M., Katz, J. (eds.) CRYPTO 2016. LNCS, vol. 9814, pp. 605–624. Springer, Heidelberg (2016). doi:10.1007/978-3-662-53018-4_22
22. Sun, B., Liu, M., Guo, J., Rijmen, V., Li, R.: Provable security evaluation of structures against impossible differential and zero correlation linear cryptanalysis. In: Fischlin, M., Coron, J.-S. (eds.) EUROCRYPT 2016. LNCS, vol. 9665, pp. 196–213. Springer, Heidelberg (2016). doi:10.1007/978-3-662-49890-3_8
23. Sun, B., Liu, Z., Rijmen, V., Li, R., Cheng, L., Wang, Q., Alkhzaimi, H., Li, C.: Links among impossible differential, integral and zero correlation linear cryptanalysis. In: Gennaro, R., Robshaw, M. (eds.) CRYPTO 2015. LNCS, vol. 9215, pp. 95–115. Springer, Heidelberg (2015). doi:10.1007/978-3-662-47989-6_5

Zero Knowledge II

Removing the Strong RSA Assumption from Arguments over the Integers

Geoffroy Couteau[1(✉)], Thomas Peters[2], and David Pointcheval[1]

[1] CNRS, INRIA, ENS/PSL Research University, Paris, France
geoffroy.couteau@ens.fr
[2] FNRS and UCLouvain, Louvain-la-Neuve, Belgium

Abstract. Committing integers and proving relations between them is an essential ingredient in many cryptographic protocols. Among them, range proofs have been shown to be fundamental. They consist in proving that a committed integer lies in a public interval, which can be seen as a particular case of the more general Diophantine relations: for the committed vector of integers x, there exists a vector of integers w such that $P(x, w) = 0$, where P is a polynomial.

In this paper, we revisit the security strength of the statistically hiding commitment scheme over the integers due to Damgård-Fujisaki, and the zero-knowledge proofs of knowledge of openings. Our first main contribution shows how to remove the Strong RSA assumption and replace it by the standard RSA assumption in the security proofs. This improvement naturally extends to generalized commitments and more complex proofs without modifying the original protocols.

As a second contribution, we design an interactive technique turning commitment scheme over the integers into commitment scheme modulo a prime p. Still under the RSA assumption, this results in more efficient proofs of relations between committed values. Our methods thus improve upon existing proof systems for Diophantine relations both in terms of performance and security. We illustrate that with more efficient range proofs under the sole RSA assumption.

Keywords: Public-key cryptography · Commitment schemes · Interactive arguments of knowledge · Zero-knowledge proofs · RSA assumption

1 Introduction

Commitment Schemes. Commitments are one of the most fundamental and widely used tools in cryptography. A commitment scheme allows a committer $\mathscr{C}$ holding a secret value s to send a *commitment* c of s to a verifier $\mathscr{V}$, and later on to *open* this commitment to reveal the value s. Such a commitment should *hide* the committed value s to the verifier, but *binds* the committer in opening only s. A famous example of commitment scheme, that perfectly hides its input, is the Pedersen commitment scheme [38], whose binding property relies on the discrete

© International Association for Cryptologic Research 2017
J.-S. Coron and J.B. Nielsen (Eds.): EUROCRYPT 2017, Part II, LNCS 10211, pp. 321–350, 2017.
DOI: 10.1007/978-3-319-56614-6_11

logarithm assumption: let $\mathbb{G}$ be a group of prime order p with two generators (g, h). To commit to $m \in \mathbb{Z}_p$, $\mathscr{C}$ picks at random $r \in \mathbb{Z}_p$ and sends $c = g^m h^r$.

Fujisaki and Okamoto introduced the first *integer commitment scheme* [23], which was later generalized in [20]. Unlike classical commitment schemes, an integer commitment scheme allows $\mathscr{C}$ to commit to any $m \in \mathbb{Z}$. Intuitively, this is done by committing to m in a group $\mathbb{Z}_\tau$ of unknown order τ, where division by units cannot be performed in general.

Interactive Proofs of Knowledge. An interactive proof of knowledge is a two-party protocol in which a prover $\mathscr{P}$ wants to convince a verifier $\mathscr{V}$ of his knowledge of some values satisfying a public statement. It should be *knowledge-extractable*, which means that an extractor can get values satisfying the statement when interacting with a successful prover, and *zero-knowledge*, which means that no information about these values leaks to the verifier (except that they satisfy the statement). Such proofs of knowledge are useful in many cryptographic constructions. Commitment schemes are a core component of zero-knowledge proofs of knowledge. In particular, integer commitment schemes have been extensively used in various interactive protocols involving zero-knowledge proofs of knowledge.

Assumptions for Proofs on Integer Commitments. The binding property of the Damgård-Fujisaki commitment scheme relies on the hardness of factoring composite integers. Even though the intractability of factoring is widely considered as a mild computational assumption, the knowledge-extractability of the proofs using these commitments relies on the Strong-RSA assumption [3,23], which is a much stronger assumption than the classical RSA assumption. This assumption states that, given a composite integer n and a random element $u \in \mathbb{Z}_n^*$, it is hard to find a pair (v, e) such that $u = v^e \bmod n$. Unlike the RSA assumption [43], where the exponent $e > 1$ is imposed, there are exponentially many solutions to a given instance of the Strong-RSA problem, the problem is thus easier to solve. However, these commitments still provide the best solution to prove relations over integers.

Range Proof. The most widespread reason to work over the integers is to prove that a committed value x lies in a public integer range $[\![a\,;b]\!]$. Indeed, working over the integers allows to show that $x - a$ and $b - x$ are positive by decomposing them as sum of four squares, following the well-known Lagrange's result. Boudot in his Eurocrypy'00 talk, and Lipmaa [36], were the first to propose such a method by relying on a commitment over the integers. As a consequence, the knowledge extractability of this range proof requires the Strong-RSA assumption.

1.1   Our Contribution

First, we revisit the Damgård-Fujisaki integer commitment scheme and show that the security of arguments of knowledge of openings can be based on the standard RSA assumption, instead of the Strong-RSA assumption. In the reduction, we use the rewinding technique in another way than in [20] as well as the splitting lemma [39,40]. Our result extends to any protocols involving arguments or relations between committed integers which first prove the knowledge of the inputs before proving that the relations are satisfied. This implies that the security of numerous protocols, such as two-party computation [18,32], e-cash [12], e-voting [25], secure generation of RSA keys [21,33], zero-knowledge primality tests [14], password-protected secret sharing [31], and range proofs [36], among many others, can be proven under the RSA assumption instead of the Strong-RSA assumption at no computational cost. In addition, we believe that the ideas on which our proof relies could be used in several other constructions whose security was proven under the Strong-RSA assumption, and might allow to replace the Strong-RSA assumption by the standard RSA assumption as well.

Second, we revisit a commitment scheme which was formally introduced in [24]: $c = g^m R^\pi \bmod n$, for a message $m \in \mathbb{Z}_\pi$ and $R \in \mathbb{Z}_n^*$. It is perfectly hiding, and the binding property relies on the RSA assumption (with modulus n, exponent π, and challenge g). We prove, as for the Damgård-Fujisaki commitment scheme, that the security of an argument of knowledge of an opening can also be based on the classical RSA assumption. Therefore, we identify an interesting property that is satisfied by this commitment, which corresponds informally to the possibility to see this commitment scheme either as an integer commitment scheme (i.e., $c = g^m h^r \bmod n$), or, after some secret has been revealed, as a commitment scheme over $\mathbb{Z}_\pi$ for some prime π (i.e., $c = g^m R^\pi \bmod n$). Without additional assumption, we show how the unpredictability of π allows improving the efficiency of zero-knowledge arguments over the integers as the knowledge of the order π is *delayed* in the protocol. This method allows to save communication and greatly reduces the work of the verifier, compared with a classical zero-knowledge argument for the same statement. We illustrate our method on range proofs [36], a zero-knowledge argument of knowledge of an input to a commitment such that the input belongs to some public interval.

Taken together, our contributions allow us to enhance both the security, by removing the Strong-RSA assumption, and the efficiency of numerous cryptographic protocols relying on integer commitment schemes.

1.2 Related Works

The Damgård-Fujisaki commitment scheme [20,23] is the only known homomorphic statistically-hiding commitment scheme over the integers. Arguments of knowledge over the integers were studied in [16,34,36].

Range proofs were introduced in [10]. They are a core component in numerous cryptographic protocols, including e-cash [12], e-voting [25], private auctions [37], group signatures [15], and anonymous credentials [13], among many others. There are two classical methods for performing a range proof:

- Writing the number in binary notation [10,27] or u-ary notation [11], committing to its decomposition and performing a specific proof for each of these commitments For example, membership to $[\![0\,; 2^\ell]\!]$ is proven in communication $O(\ell/(\log \ell - \log\log \ell))$ in the protocol of [11], and in communication $O(\ell^{1/3})$ in the protocol of [27] (only counting the number of group elements).
- Using an integer commitment scheme [8,25,36].

Note that protocols such as [17] do also allow to prove that a committed integer x lies in a given interval $[\![0\,;a]\!]$ up to an accuracy parameter δ: actually only membership to $[\![0\,;(1+\delta)a]\!]$ is proved.

Eventually, several papers have proposed signatures based on the standard RSA assumption [7,29,30] as alternatives to classical signature schemes based on the Strong-RSA assumption. Our work is in the same vein as these papers, replacing the Strong-RSA assumption by the RSA assumption in arguments over the integers. However, note that we do not actually propose a new argument system to get rid of the Strong-RSA assumption, but rather show that the security of the classical argument system is implied by the RSA assumption. As a consequence, the schemes using arguments over the integers do not need to be modified to benefit from our security analysis.

1.3   Organization

Section 2 introduces the necessary background for what follows, and namely some useful facts on the RSA groups. Section 3 recalls the Damgård-Fujisaki commitment scheme, its properties, and the argument of knowledge of [20]. A new security proof of the latter, under the standard RSA assumption, is given in details in Sect. 4. Section 5 illustrates some extensions of our result. First, we show how one can commit to vectors at once with generalized commitments. And then, we show how one can make range proofs under the standard RSA assumption. Section 6 revisits the commitment scheme of [24] and shows how, by switching from the previous commitment to this one, we can get a new interactive proof system for performing zero-knowledge arguments over the integers, that is more efficient. Eventually, Sect. 7 illustrates our method on range proofs, with concrete efficiency comparisons.

For the sake of completeness, in the full version [19] we exhibit a flaw in the optimized version of Lipmaa's range proof [36, Annex B]. We then propose a fix as well as security proof.

2   Backgrounds

Throughout this paper, κ denotes the security parameter. An algorithm is *efficient* when it runs in polynomial time in the (implicit) security parameter κ. A positive function f is *negligible* if for any polynomial p there exists a bound $B > 0$ such that, for any integer $k \geq B$, $f(k) \leq 1/|p(k)|$. An event depending on κ occurs with *overwhelming probability* when its probability is at least $1 - \varepsilon(\kappa)$ for a negligible function ε.

2.1   Notations

Given a finite set S, the notation $x \leftarrow_R S$ means a uniformly random assignment of an element of S to the variable x, then for any $s \in S$ we have $\Pr_S[x = s] = 1/\#S$ where $\#S$ denotes the cardinality of S. When an element s is represented by an integer, $|s|_b$ is the bit-length of the integer, and $|s|$ denotes its absolute value (or norm). Bold variables denote vectors. For a vector $\boldsymbol{x} = (x_1, \cdots, x_\ell)$, $g^{\boldsymbol{x}}$ denotes $(g^{x_1}, \cdots, g^{x_\ell})$ and $\|\boldsymbol{x}\|_\infty = \max_{1 \le i \le \ell} |x_i|$.

The integer range $[\![a\,;b]\!]$ stands for $\{x \in \mathbb{Z} \mid a \le x \le b\}$. For any integers $a \le b$, the statistical distance between two uniform distributions, over $U_a = [\![1\,;a]\!]$ and $U_b = [\![1\,;b]\!]$ respectively, is given by $\sum_{i=1}^{b} |\Pr_{U_a}[x = i] - \Pr_{U_b}[x = i]| = \sum_{i=1}^{a}(1/a - 1/b) + \sum_{i=a+1}^{b} 1/b = 2(b-a)/b$.

2.2 Commitment Scheme

We first recall the basic definition of a commitment scheme on the message space $\mathcal{M}$. This is an essential primitive in cryptography, that is used to lock a value in a box, so that the sender cannot change at the opening time (the *binding* property) but still the receiver has no information about the value before the opening (the *hiding* property). A *non-interactive* commitment scheme is defined by three algorithms (Setup, Commit, Verify):

- Setup(1^κ), generates the public parameters pp, which also specifies the message space $\mathcal{M}$, the commitment space $\mathcal{C}$, the opening space $\mathcal{D}$, and the random source $\mathcal{R}$;
- Commit(pp, $m; r$), given the message $m \in \mathcal{M}$ and some random coins $r \in \mathcal{R}$, outputs a commitment-opening pair (c, d). When there is no ambiguity, we will abuse the notation $(c, d) \leftarrow_R$ Commit(m), for pp and $r \leftarrow_R \mathcal{R}$;
- Verify(pp, c, d, m), outputs a bit whose value depends on the validity of the opening (m, d) with respect to the commitment c.

A commitment scheme *must* be

Correct. For any public parameters pp $\leftarrow_R$ Setup(1^κ), any message $m \in \mathcal{M}$, and any random coin $r \in \mathcal{R}$, if $(c, d) \leftarrow$ Commit(pp, $m; r$), then we necessarily have Verify(pp, c, d, m) = 1.

Hiding. No probabilistic polynomial-time adversary $\mathcal{A}$, that is first given pp $\leftarrow_R$ Setup(1^κ), can distinguish commitments on two messages (m_0, m_1) of its choice. The commitment scheme is said *statistically hiding* if the indistinguishability holds even for unbounded adversaries.

Binding. No probabilistic polynomial-time adversary $\mathcal{A}$ can open a commitment c on two different messages $m_0 \ne m_1$. The commitment scheme is said *statistically binding* if this is infeasible even for unbounded adversaries.

A commitment scheme can also be *homomorphic*, if for a group law $\oplus$ on the message space $\mathcal{M}$, from pp, $(c_0, d_0) \leftarrow$ Commit(pp, $m_0; r_0$) and $(c_1, d_1) \leftarrow$ Commit(pp, $m_1; r_1$), one can generate c and d so that Verify(pp, $c, d, m_0 \oplus m_1$) = 1.

2.3 Interactive Proof Systems

We now recall the second tool we will use in this paper, the zero-knowledge proofs of knowledge, and their variants.

Zero-Knowledge Proofs and Arguments. Let R be an NP-relation over a set $\mathfrak{X}$ defining an NP-language $\mathscr{L} = \{x \in \mathfrak{X} \mid \exists w, \mathsf{R}(x, w) = 1\}$, where a w such that $\mathsf{R}(x, w) = 1$ is called a *witness* for the statement $x \in \mathscr{L}$.

A *zero-knowledge proof of knowledge* (ZKPoK) for a relation R and a word $x \in \mathfrak{X}$ is an interactive protocol $\langle \mathscr{P}(w), \mathscr{V} \rangle (x \in \mathscr{L})$ between a *prover* $\mathscr{P}$ holding a *witness* w for the statement $x \in \mathscr{L}$, and a verifier $\mathscr{V}$, both modeled as interactive probabilistic polynomial-time Turing machines. The purpose of a ZKPoK is for $\mathscr{P}$ to convince $\mathscr{V}$ of its knowledge of some witness w of the statement $x \in \mathscr{L}$, without revealing any information about this witness. More formally, let $\mathrm{VIEW}_{\mathscr{V}}[\langle \mathscr{P}(w), \mathscr{V} \rangle (x \in \mathscr{L})]$ be the view of $\mathscr{V}$ during the execution of the interactive protocol (i.e., all the messages it received when interacting with $\mathscr{P}$). A ZKPoK must be:

Correct. For every $x \in \mathscr{L}$, if $\mathscr{P}$ knows a witness w, and both $\mathscr{P}$ and $\mathscr{V}$ behave honestly, $\langle \mathscr{P}(w), \mathscr{V} \rangle (x \in \mathscr{L})$ is accepted by $\mathscr{V}$ with overwhelming probability.

Knowledge Extractable. For any prover $\mathscr{P}'$ which succeeds in convincing $\mathscr{V}$ of $x \in \mathscr{L}$ with non-negligible probability, there exists a simulator $\mathscr{S}im_{\mathsf{KE}}$, running in expected polynomial time, which extracts a witness w for $x \in \mathscr{L}$ from $\mathscr{P}'$.

Zero-Knowledge. For any verifier $\mathscr{V}'$, there exists a simulator $\mathscr{S}im_{\mathsf{ZK}}$ such that for every $x \in \mathscr{L}$, $\mathscr{S}im_{\mathsf{ZK}}(x)$ and $\mathrm{VIEW}_{\mathscr{V}'}[\langle \mathscr{P}(w), \mathscr{V}' \rangle (x \in \mathscr{L})]$, where w is a witness for $x \in \mathscr{L}$, are indistinguishable.

If the knowledge-extractability holds only for a computationally-bounded $\mathscr{P}'$, the protocol is a zero-knowledge *argument* of knowledge (ZKAoK). If the verifier is restricted to being honest in the zero-knowledge property, the proof is an *honest-verifier* zero-knowledge proof.

Zero-Knowledge Arguments from Diophantine Relations. A *Diophantine set* $S \subseteq \mathbb{Z}^k$ is a set of vectors over $\mathbb{Z}^k$ defined by a *representing polynomial* $P_S(X, W)$ with $X = (X_1, \cdots, X_k)$ and $W = (Y_1, \cdots, Y_\ell)$, i.e. a set of the form $S = \{x \in \mathbb{Z}^k \mid \exists w \in \mathbb{Z}^\ell, P_S(x, w) = 0\}$ for some polynomial P_S. It was shown in [22] that any recursively enumerable set is Diophantine. An interesting class for cryptographic applications is the class **D** of Diophantine sets S such that each $x \in S$ has at least one witness w satisfying $\|w\|_\infty \leq \|x\|_\infty^{O(1)}$. It is widely conjectured that $\mathbf{D} = \mathsf{NP}$, as **D** contains several NP-complete problems, and it was shown in [41] that if $\mathsf{co\text{-}NLOGTIME} \subseteq \mathbf{D}$, then $\mathbf{D} = \mathsf{NP}$. The class **D** was introduced in [1] and its cryptographic relevance was pointed out in [36]. For example, the set $\mathbb{Z}_+$ of positive integers is in **D**, as by a well-known result of Lagrange, it can be defined as $\mathbb{Z}_+ = \{x \in \mathbb{Z} \mid \exists (w_1, w_2, w_3, w_4) \in \mathbb{Z}^4, x - (w_1^2 + w_2^2 + w_3^2 + w_4^2) = 0\}$. In addition, each w_i is of bounded size $|w_i| \leq |x|$.

Lipmaa [36] has shown that zero-knowledge arguments of membership to a set $S \in \mathbf{D}$, with representing polynomial P over k-vector inputs and ℓ-vector witnesses, can be constructed using an integer commitment scheme, such as [20]. The size of the argument (the communication between $\mathscr{P}$ and $\mathscr{V}$) depends on k, ℓ, and $\deg(P)$, the degree of P. As noted in [36], intervals, unions of intervals, exponential relations (i.e., set of tuples (x, y, z) such that $z = x^y$) and gcd relation (i.e., set of tuples (x, y, z) such that $z = \gcd(x, y)$) are all in $\mathbf{D}$, with parameters (k, ℓ and $\deg(P)$) small enough for cryptographic applications.

2.4   RSA Group Structure

In this paper we focus on $\mathbb{Z}_n^*$ for a strong RSA modulus $n = pq$ where p, q are distinct safe primes. That means that $p = 2p' + 1$ and $q = 2q' + 1$ for two other primes so that p, p', q, q' are all distinct, and $\varphi(n) = 4p'q'$. We write $a = b \bmod n$ to specify that $a = b$ in $\mathbb{Z}_n$ and we write $a \leftarrow [b \bmod n]$ to affect the smallest positive integer to a so that $a = b \bmod n$.

By GenMod(1^κ), we denote a probabilistic efficient algorithm that, given the security parameter κ, generates a strong RSA modulus n and secret parameters (p, q) of at least κ bits each with the specification that $n = pq$. In the following, we write $(n, (p, q)) \leftarrow_R$ GenMod(1^κ). We will sometimes abuse the notation $n \leftarrow_R$ GenMod(1^κ) to say that the modulus n has been generated according to this distribution.

The RSA Assumption. The RSA assumption states, informally, that given an exponent e prime to $\varphi(n)$, it is hard for any probabilistic polynomial-time algorithm to find the e-th root modulo n of a random $y \leftarrow_R \mathbb{Z}_n^*$. More formally, let P_n be the subset of $\mathbb{Z}_n$ of elements prime to $\varphi(n)$. The RSA assumption does in fact refer to a class of assumptions, depending of the distribution $\mathscr{D}_n$ over P_n from which the exponent e is drawn.

$\mathscr{D}_n$-**RSA Assumption** [43]. For $n \leftarrow_R$ GenMod(1^κ) and $e \leftarrow_R \mathscr{D}_n$, it is hard for any probabilistic polynomial-time algorithm to find the e-th root modulo n of a random $y \leftarrow_R \mathbb{Z}_n^*$. The triple (n, e, y) is the RSA instance.

Various flavours of the RSA assumption are standard in the literature. In particular, the RSA assumption with a fixed small exponent (the most common being 65537) is widely used in practical implementations. In theoretical papers, it is common to consider the RSA assumption for exponents picked from the uniform distribution over P_n (see [30] for example). In this paper, we use a flavour of the RSA assumption which is somewhat intermediate between these two standard variants: we will consider the RSA assumption for exponents picked from the uniform distribution over $[\![3\,;a]\!] \cap \mathsf{P}_n$ for a value a polynomial in κ (hence, we consider random small exponents). To simplify the notations, we will denote by a-RSA this variant of the RSA assumption[1].

Other Computational Assumptions. Other famous computational assumptions over RSA groups are the intractability of the factorization and the Strong-RSA assumption:

---

[1] It should be noted that in our proof, the bound a will depend on the success probability of the adversary.

Integer Factorization Assumption. It states that finding a non-trivial factor of $n \leftarrow_R \mathsf{GenMod}(1^\kappa)$ is hard for any probabilistic polynomial-time algorithm.

Strong-RSA Assumption [3,23]. It lets the choice of e to the algorithm: It states that, for $n \leftarrow_R \mathsf{GenMod}(1^\kappa)$, this is hard to find the e-th root modulo n, for a random $y \leftarrow_R \mathbb{Z}_n^*$, for any probabilistic polynomial-time algorithm, for an exponent $e > 1$ *of its choice.*

It is well-known that breaking the integer factorization assumptions allows to break both the RSA and the Strong-RSA assumption. From the definition, it is clear that the Strong-RSA assumption gives more degree of freedom to the adversary, so it is seemingly much stronger. Indeed, for the RSA assumption, the exponent is not chosen by the adversary, but can be fixed in any way in advance by the challenger.

Properties of *Strong* RSA Groups. One can note that in groups modulo n, where $n = pq$ is a strong RSA modulus, p and q are Blum primes: $p = q = 3 \bmod 4$. If we denote QR_n the subgroup of the squares, $\mathsf{QR}_n = \{a \in \mathbb{Z}_n^* \mid \exists b \in \mathbb{Z}_n^*, a = b^2 \bmod n\}$, this is a cyclic subgroup of $\mathbb{Z}_n^*$ of order $\varphi(n)/4 = p'q'$.

Proposition 1. *The following facts hold:*

1. $-1 \notin \mathsf{QR}_n$;
2. *any square* $h \in \mathsf{QR}_n$ *has four square roots, with exactly one in* QR_n;
3. *for a random element* $h \in \mathsf{QR}_n$, *finding a square root of* h *is equivalent to factoring the modulus* n;
4. *for random elements* $g, h \in \mathsf{QR}_n$, *finding non-zero integers* a, b *such that* $g^a = h^b \bmod n$ *is equivalent to factoring the modulus* n;
5. *for an RSA instance* (n, e, y), *finding* $x \in \mathbb{Z}_n^*$ *and* e' *prime to* e *such that* $x^e = y^{e'} \bmod n$ *is equivalent to finding an* e-th root of y modulus n.

Proof. Let us briefly explain why these facts hold, using the Jacobi symbol function $J_n(x) = J_p(x) \times J_q(x)$ in $\mathbb{Z}_n^*$, as the extension of the Legendre symbol on $\mathbb{Z}_p^*$ for prime p, $J_p(x) = (x)^{(p-1)/2}$, which determines whether x is a square or not in $\mathbb{Z}_p^*$. Since p and q are Blum primes, $J_p(-1) = J_q(-1) = -1$, and so $J_n(-1) = 1$, but -1 is not in QR_n, hence the Fact 1. The four square roots of 1, in $\mathbb{Z}_n^*$ are 1 and -1, both with Jacobi symbol $+1$, but respectively $(+1, +1)$ and $(-1, -1)$ for the Legendre symbols in $\mathbb{Z}_p^*$ and $\mathbb{Z}_q^*$, and α, and $-\alpha$, both with Jacobi symbol -1, but respectively $(+1, -1)$ and $(-1, +1)$ for the Legendre symbols in $\mathbb{Z}_p^*$ and $\mathbb{Z}_q^*$. As a consequence, given a square $h \in \mathsf{QR}_n$, and a square root u, the four square roots are $u, -u$, and $\alpha u, -\alpha u$, one of which being in QR_n, since the four kinds of Legendre symbols are represented. This leads to the Fact 2.

For Fact 3, if one chooses a random $u \in \mathbb{Z}_n^*$ and sets $h = u^2 \bmod n$, $J_n(u)$ is completely hidden. Another square root v has probability one-half to have $J_n(v) = -J_n(u)$. This means that $u^2 = v^2 \bmod n$, but $u \neq \pm v \bmod n$. Then, $\gcd(u - v, n)$ gives a non-trivial factor of n.

For Fact 4, if one chooses a random $u \in \mathbb{Z}_n^*$ and a large random scalar α and sets $h = u^2 \bmod n$ and $g = h^\alpha \bmod n$, h is likely a generator of QR_n, and then

$g^a = h^b \bmod n$ means that $m = b - a\alpha$ is a multiple of $p'q'$, the order of the subgroup of the squares. Let us note $m = 2^v \cdot t$, for an odd t, then $p'q'$ divides t: let us choose a random element $u \in \mathbb{Z}_n^*$, with probability close to one-half, $J_n(u) = -1$, and so $J_n(u^t) = -1$ while u^t is a square root of 1. As in the proof of the previous Fact 3, we can obtain a non-trivial factor of n.

Eventually, for Fact 5, using Bézout relation $ue + ve' = 1$, then $x^{ve} = y^{ve'} = y^{1-ue} \bmod n$. So $y = (x^v y^u)^e \bmod n$. □

3   Commitment of Integers Revisited

In [23], Okamoto and Fujisaki proposed a statistically-hiding commitment scheme allowing commitment to arbitrary-size integers. Their commitment was later generalized in [20]. It relies on the fact that when the factorization is unknown, it is infeasible to know the order of the sub-group QR_n of the squares in $\mathbb{Z}_n^*$, where n is a strong RSA modulus. Hence, the only way for a computationally-bounded committer to open a commitment is to do it over the integers.

In addition, [23] gave an argument of knowledge of an opening of a commitment and proved that the knowledge extractability of the argument is implied by the Strong-RSA assumption. A flaw in the original proof was later identified and corrected in [20]. We will revisit the argument of knowledge of an opening due to Damgård-Fujisaki [20] and provide a new proof for its knowledge extractability, in order to remove the requirement of the Strong-RSA assumption. Our proof requires the standard RSA assumption only, with an exponent randomly chosen in a polynomially-bounded set.

3.1   Commitments over the Integers

Description. Let us recall the commitment of one integer m:

- Setup(1^κ) runs $(n, (p, q)) \leftarrow_R \mathsf{GenMod}(1^\kappa)$, and picks two random generators g, h of QR_n. It returns $\mathsf{pp} = (n, g, h)$;
- Commit($\mathsf{pp}, m; r$), for $\mathsf{pp} = (n, g, h)$, a message $m \in \mathbb{Z}$, and some random coins $r \leftarrow_R [\![0; n]\!]$, computes $c = g^m h^r \bmod n$, and returns (c, d) with $d = r$;
- Verify(pp, c, d, m) parses pp as $\mathsf{pp} = (n, g, h)$ and outputs 1 if $c = \pm g^m h^d \bmod n$ and 0 otherwise.

One should note that an honest user will always open such that $c = g^m h^d \bmod n$. But the knowledge-extractability of the next ZKAoK of opening cannot exclude the change of sign. In this description, we provide a trusted setup algorithm. But as we see below, the guarantees for the committer (the hiding property of the commitment) just rely on the existence of α such that $g = h^\alpha \bmod n$. For the verifier to be convinced, one can just let him generate the parameters (n, g, h), and prove the existence of such an α to the committer.

Security Analysis. The above commitment scheme is obviously *correct*. The *hiding* property relies on the existence of α such that $g = h^\alpha \bmod n$ (they are both generators of the same subgroup QR_n), and so, for any $m' \in \mathbb{Z}$,

$$c = \mathsf{Commit}(\mathsf{pp}, m; r) = g^m h^r = h^{r+\alpha m} = h^{(r+\alpha(m-m'))+\alpha m'}$$

$$= g^{m'} h^{r+\alpha(m-m')} = \mathsf{Commit}(\mathsf{pp}, m'; r'),$$

with $r' \leftarrow [r + \alpha(m - m') \bmod p'q']$, that is smaller than n and follows a distribution statistically close to the distribution of r. The binding property relies on the Integer Factorization assumption: indeed, from two different openings m_0, d_0, m_1, d_1 for a commitment c, with $d_1 > d_0$, the validity checks show that $g^{m_0} h^{d_0} = \pm g^{m_1} h^{d_1} \bmod n$, and so $g^{m_0-m_1} = \pm h^{d_1-d_0} \bmod n$. Since g and h are squares, and -1 is not a square, necessarily $g^{m_0-m_1} = h^{d_1-d_0} \bmod n$. The Fact 4 from Proposition 1 leads to a non-trivial factor of n.

3.2 Zero-Knowledge Argument of Opening

Let us now study the argument of knowledge of a valid opening for such a commitment. The common inputs are the public parameters pp and the commitment $c = g^x h^r \bmod n$, together with the bit-length k_x of the message x, that is then assumed to be in $[\![-2^{k_x} ; 2^{k_x}]\!]$, while $r \in [\![0 ; n]\!]$ and x are the private inputs, i.e. the witness of the prover. We stress that k_x is chosen by the prover, since this reveals some information about the integer x, while r is always in the same set, whatever the committed element x is.

Description of the Protocol. The protocol works as follows:

Initialize: $\mathscr{P}$ and $\mathscr{V}$ decide to run the protocol on input $(\mathsf{pp}, \kappa, c, k_x)$;
Commit: $\mathscr{P}$ computes $d = g^y h^s \bmod n$, for randomly chosen $y \leftarrow_R [\![0 ; 2^{k_x+2\kappa}]\!]$
    and $s \leftarrow_R [\![0 ; 2^{|n|_b+2\kappa}]\!]$, and sends d to the $\mathscr{V}$;
Challenge: $\mathscr{V}$ outputs $e \leftarrow_R [\![0 ; 2^\kappa]\!]$;
Response: $\mathscr{P}$ computes and outputs the integers $z = ex + y$ and $t = er + s$;
Verify: $\mathscr{V}$ accepts the proof and outputs 1 if $c^e d = g^z h^t \bmod n$. Otherwise, $\mathscr{V}$ rejects the proof and outputs 0.

In the rest of this section, we prove this protocol is indeed a zero-knowledge argument of knowledge of an opening. Which means it is correct, zero-knowledge, and knowledge-extractable.

Correctness. First, the correctness is quite obvious: if $c = g^x h^r \bmod n$, with $z = ex + y$ and $t = er + s$, we have $g^z h^t = (g^x h^r)^e \cdot g^y h^s = c^e d \bmod n$.

Zero-Knowledge. For the zero-knowledge property, in the honest-verifier setting, the simulator $\mathscr{S}im$ (that is $\mathscr{S}im_{\mathsf{ZK}}$ in this case) can simply do as follows:

1. $\mathscr{S}im$ chooses a random challenge $e \leftarrow_R [\![0 ; 2^\kappa]\!]$;

$\mathscr{D}_0$	$y \leftarrow_R [\![0\,;2^{k_x+2\kappa}]\!], s \leftarrow_R [\![0\,;2^{	n	_b+2\kappa}]\!],$
	$e \leftarrow_R [\![0\,;2^{\kappa}]\!], z = xe+y, t = re+s, d = g^y h^s \bmod n$
$\mathscr{D}_1$	$z \leftarrow_R [\![xe\,;2^{k_x+2\kappa}+xe]\!], t \leftarrow_R [\![re\,;2^{	n	_b+2\kappa}+re]\!],$
	$e \leftarrow_R [\![0\,;2^{\kappa}]\!], d = g^{z-xe} h^{t-re} \bmod n$
$\mathscr{D}_2$	$z \leftarrow_R [\![xe\,;2^{k_x+2\kappa}+xe]\!], t \leftarrow_R [\![re\,;2^{	n	_b+2\kappa}+re]\!],$
	$e \leftarrow_R [\![0\,;2^{\kappa}]\!], d = g^z h^t c^{-e} \bmod n$
$\mathscr{D}_3$	$z \leftarrow_R [\![0\,;2^{k_x+2\kappa}]\!], t \leftarrow_R [\![0\,;2^{	n	_b+2\kappa}]\!],$
	$e \leftarrow_R [\![0\,;2^{\kappa}]\!], d = g^z h^t c^{-e} \bmod n$

Fig. 1. Distributions for the zero-knowledge property

2. $\mathscr{S}im$ chooses random responses $z \leftarrow_R [\![0\,;2^{k_x+2\kappa}]\!]$ and $t \leftarrow_R [\![0\,;2^{|n|_b+2\kappa}]\!]$;
3. $\mathscr{S}im$ sets $d = g^z h^t c^{-e} \bmod n$.

The simulated transcript is the tuple $(d, e, (z, t))$, where the elements follow the distribution $\mathscr{D}_3$ from Fig. 1, while the real transcript follows the distribution $\mathscr{D}_0$.

However, it is clear that $\mathscr{D}_0 = \mathscr{D}_1 = \mathscr{D}_2$, while the distance between $\mathscr{D}_2$ and $\mathscr{D}_3$ is the sum of the distances between the distributions of z and t, respectively in $\mathscr{Z}_2 = [\![xe\,;2^{k_x+2\kappa}+xe]\!]$ and $\mathscr{Z}_3 = [\![0\,;2^{k_x+2\kappa}]\!]$, and $\mathscr{T}_2 = [\![re\,;2^{|n|_b+2\kappa}+re]\!]$ and $\mathscr{T}_3 = [\![0\,;2^{|n|_b+2\kappa}]\!]$:

$$\Delta_z = \sum_{Z=0}^{2^{k_x+2\kappa}+xe} |\Pr[z \leftarrow_R \mathscr{Z}_2 : z = Z] - \Pr[z \leftarrow_R \mathscr{Z}_3 : z = Z]|$$

$$= \sum_{Z=0}^{xe-1} 2^{-k_x-2\kappa} + \sum_{Z=2^{k_x+2\kappa}+1}^{2^{k_x+2\kappa}+xe} 2^{-k_x-2\kappa} = 2 \cdot xe \cdot 2^{-k_x-2\kappa} \leq 2 \cdot 2^{k_x+\kappa} \cdot 2^{-k_x-2\kappa}$$

that is bounded by $2 \cdot 2^{-\kappa}$. Similarly, $\Delta_t \leq 2 \cdot 2^{-\kappa}$. Hence the statistical zero-knowledge property, since the real distribution $\mathscr{D}_0$ and the simulated distribution $\mathscr{D}_3$ have a negligible distance bounded by $2^{-\kappa+2}$.

Knowledge-Extractability. The last property is the most intricate, and this is the one that required the Strong-RSA assumption in the original proof of Damgård and Fujisaki [20]. In the next section, we present a detailed proof of the following theorem:

Theorem 2. *Given a prover $\mathscr{P}'$ able to convince a verifier $\mathscr{V}$ of its knowledge of an opening of c for random system parameters $\mathsf{pp} = (n, g, h)$ with probability greater than ε within time t, one either breaks the $4/\varepsilon$-RSA assumption with expected time upper-bounded by $256t/\varepsilon^3$, or extracts a valid opening with expected time upper-bounded by $16t/\varepsilon^2$.*

4 Proof of Theorem 2

Since this proof is the main technical contribution of the paper, with many practical applications, we provide it in details. We start with some preliminaries, and then discuss various cases.

4.1   Preliminaries

The proof will make use of the splitting lemma [39,40], that we recall below:

Lemma 3. *Let $A \subset X \times Y$ such that $\Pr[(x,y) \in A] \geq \varepsilon$. For any $\varepsilon' < \varepsilon$, if one defines $B = \{(x,y) \in X \times Y \mid \Pr_{y' \in Y}[(x,y') \in A] \geq \varepsilon - \varepsilon'\}$, then it holds that:*

(i) $\Pr[B] \geq \varepsilon'$ *(ii)* $\forall (x,y) \in B, \underset{y' \in Y}{\Pr}[(x,y') \in A] \geq \varepsilon - \varepsilon'$ *(iii)* $\Pr[B \mid A] \geq \varepsilon'/\varepsilon$.

In the proof, we will consider an adversary with a random tape R who succeeds with some probability ε in any run of the full argument. Our proof will make use of rewinding: we will rewind the adversary several times to get several transcripts of the protocol for the *same* random tape R, and various challenges. The purpose of the splitting lemma is therefore to get a bound on the probability of getting valid transcripts when we fix R and run the adversary on various challenges.

4.2   Detailed Proof

Let us suppose the extractor $\mathscr{S}im$ (that is $\mathscr{S}im_{KE}$ in this case) is given a $4/\varepsilon$-RSA challenge (n,e,u), which means that the exponents e is randomly chosen prime to $\varphi(n)$ but also in the set $[1, 4/\varepsilon]$. It sets $h \leftarrow u^2 \bmod n$ and $g \leftarrow h^\alpha \bmod n$ for a random exponent $\alpha \leftarrow_R \mathbb{Z}_{n^2}$. It sets $\mathsf{pp} = (n,g,h)$. Note that as u is random in $\mathbb{Z}_n^*$, (g,h) are actually distributed as in the real protocol. We consider an adversary $\mathscr{A}$ that provides a convincing proof of knowledge of an opening of c (an accepted transcript) with probability ε, with the parameters $(\mathsf{pp} = (n,g,h), \kappa, c, k_x)$.

Note that the probability distribution of a protocol execution is $D = (R, e)$, where R is the adversary's random tape that determines d, and e is the random challenge from the honest verifier. Then, we can assume that on a random pair (R, e_0), its probability to output an accepted transcript (d, e_0, z_0, t_0) is greater than ε. We apply the splitting lemma with $\varepsilon' = \varepsilon/2$ for the distribution $D = \{R\} \times \{e\}$: after one execution, with probability greater than ε, we obtain an accepted transcript (d, e_0, z_0, t_0). In such a case, with probability greater than $1/2$, R is a good random tape, which means that another execution with the same R but a random challenge e_1 will lead to another accepted transcript (d, e_1, z_1, t_1) with probability $\varepsilon' = \varepsilon/2$. Note that since R is kept unchanged, d is the same. Globally, with probability greater than $\varepsilon^2/4$, after 2 executions of the protocol, one gets two related accepted transcripts: (d, e_0, z_0, t_0) and (d, e_1, z_1, t_1).

Without loss of generality, we may assume $e_0 \geq e_1$. Writing $e_1' \leftarrow e_0 - e_1$, $z_1' \leftarrow z_0 - z_1$, and $t_1' \leftarrow t_0 - t_1$, the two valid tuples lead to the relation $c^{e_1'} = g^{z_1'} h^{t_1'} \bmod n$.

Then, with our adversary $\mathscr{A}$ and a rewind, with random (R, e_0, e_1), we have at least one of the two statements below that is true after a first execution of $\mathscr{A}$ with (R, e_0) and a rewind with (R, e_1):

- **Statement 1.** one gets two related accepted transcripts (d, e_0, z_0, t_0) and (d, e_1, z_1, t_1), and e_1' divides both z_1' and t_1' (with above notations) with probability greater than $\varepsilon^2/8$;

– **Statement 2.** one gets two related accepted transcripts (d, e_0, z_0, t_0) and (d, e_1, z_1, t_1), and e'_1 *does not* divide both z'_1 and t'_1 (with above notations) with probability greater than $\varepsilon^2/8$.

Statement 1: One gets two related accepted transcripts and e'_1 divides both z'_1 and t'_1 with probability greater than $\varepsilon^2/8$. $\mathscr{S}im$ simply outputs the pair of integers $(x_1, r_1) \leftarrow (z'_1/e'_1, t'_1/e'_1)$. If e'_1 is odd, and thus prime to $\varphi(n)$, we have $c = g^{x_1} h^{r_1} \bmod n$. However, if $e'_1 = 2^v \rho$ for an odd ρ and $v \geq 1$, $(c^{-1} g^{x_1} h^{r_1})^{2^v} = 1 \bmod n$: from the Fact 2 from Proposition 1, $(c^{-1} g^{x_1} h^{r_1})^2 = 1 \bmod n$:

– either $c^{-1} g^{x_1} h^{r_1} = \pm 1 \bmod n$, and so $c = \pm g^{x_1} h^{r_1} \bmod n$ (valid opening);
– or we have a non-trivial square root of 1 modulo n, which leads to the factorization of n (see Proposition 1). As the RSA assumption is stronger than the factorization, when we solve the factorization, we can compute the solution to the RSA challenge.

Statement 2: One gets two related accepted transcripts and e'_1 does not divide both z'_1 and t'_1 with probability greater than $\varepsilon^2/8$. We first show that, with reasonable probability, e'_1 does not divide $\alpha z'_1 + t'_1$ either (this is exactly the case 2 from [20]). The intuition behind this argument is that the only information that $\mathscr{A}$ can get about α is from $g = h^\alpha \bmod n$. However, this leaks only $\alpha \bmod p'q'$, while α was taken at random in $\mathbb{Z}_{n^2}$: all the information on its most significant bits is *perfectly* hidden. We recall below the proof given by Damgård and Fujisaki, for completeness.

Let Q be a prime factor of e'_1 and j be the integer such that Q^j divides e'_1 but Q^{j+1} does not divide e'_1, and at least one of z'_1 or t'_1 is non-zero modulo Q^j. Since e'_1 does not divide both z'_1 and t'_1, such a pair (Q, j) does necessarily exist. Actually, if Q^j divides z'_1, as it divides e'_1, it must also divide $\alpha z'_1 + t'_1$ and therefore t'_1, which was excluded (at least one of z'_1 or t'_1 is non-zero modulo Q^j). Therefore, $z'_1 \neq 0 \bmod Q^j$.

We can write $\alpha = [\alpha \bmod p'q'] + \lambda p'q'$ for some λ. Let us denote $\mu = [\alpha \bmod p'q']$. The tuple (n, g, h) uniquely determines μ, whereas λ is perfectly unknown to the prover. As Q^j divides e'_1, it also divides $\alpha z'_1 + t'_1$:

$$\alpha z'_1 + t'_1 = \lambda z'_1 p'q' + \mu z'_1 + t'_1 = 0 \bmod Q^j.$$

Note that $p'q' \neq 0 \bmod Q$, since p' and q' are κ-bit primes and the challenges are less than 2^κ. And from the view of the adversary, λ is uniformly distributed in $\mathbb{Z}_n$, while it should satisfy the above equation. But since this equation has at most $\gcd(z'_1 p'q', Q^j)$ solutions, which is a power of Q (and at most Q^{j-1}), and since n is larger than Q^j by a factor (far) bigger than 2^κ, the distribution of $\lambda \bmod Q^j$ is statistically close to uniform in $\mathbb{Z}_{Q^j}$, and the probability that λ satisfies the above equation is bounded by $1/Q - 2^{-\kappa} \leq 1/2$, independently of the actions of $\mathscr{A}$. Hence, when Statement 2 holds (the global probability is greater than $\varepsilon^2/8$), e'_1 cannot divide $\alpha z'_1 + t'_1$ more than half the time. As a consequence, we necessarily have a stronger statement

**One gets two related accepted transcripts and e_1' does not divide
$\alpha z_1' + t_1'$ with probability greater than $\varepsilon^2/16$.**

This allows $\mathscr{S}im$ to solve an RSA instance, which is the difference with the
original proof. Let $\beta_1 = \gcd(e_1', \alpha z_1' + t_1')$. Since e_1' does not divide $\alpha z_1' + t_1'$, we
necessarily have $1 \le \beta_1 < e_1'$. Let $\Gamma_1 \leftarrow e_1'/\beta_1$ and $F_1 \leftarrow (\alpha z_1' + t_1')/\beta_1$: F_1/Γ_1 is
the irreducible fraction form of $(\alpha z_1' + t_1')/e_1'$ and $e_1' \ge \Gamma_1 > 1$. We now consider
the following statements, among which at least one holds:

- **Statement 2.a.** One gets two related accepted transcripts, e_1' does not divide
  $\alpha z_1' + t_1'$, and $\Gamma_1 \le 8/\varepsilon$ with probability at least $\varepsilon^2/32$;
- **Statement 2.b.** One gets two related accepted transcripts, e_1' does not divide
  $\alpha z_1' + t_1'$, and $\Gamma_1 > 8/\varepsilon$ with probability at least $\varepsilon^2/32$.

**Statement 2.a: One gets two related accepted transcripts, e_1' does not
divide $\alpha z_1' + t_1'$, and $\Gamma_1 \le 8/\varepsilon$ with probability at least $\varepsilon^2/32$.** If $\Gamma_1 \le 8/\varepsilon$,
since $\beta_1 < e_1'$, we must have $\Gamma_1 \in [\![2\,;8/\varepsilon]\!]$. Let us recall we have (e_1', z_1', t_1') so
that $c^{e_1'} = g^{z_1'} h^{t_1'} \bmod n$ and $\beta_1 = \gcd(e_1', \alpha z_1' + t_1')$ with $1 < \Gamma_1 = e_1'/\beta_1 \le 8/\varepsilon$.

So we have $e_1' = \beta_1 \Gamma_1$ and $\alpha z_1' + t_1' = \beta_1 F_1$ for relatively prime integers Γ_1
and F_1. Since $h = u^2 \bmod n$, we have $c^{e_1'} = u^{2(\alpha z_1' + t_1')} \bmod n$, which reduces
to $c^{\Gamma_1} = c^{e_1'/\beta_1} = \pm u^{2(\alpha z_1' + t_1')/\beta_1} = \pm u^{2F_1} \bmod n$, unless one finds a non-trivial
square root of 1 modulo n (which allows to solve any RSA instance modulo n,
see above). We now consider two additional statements, among which at least
one holds:

- **Statement 2.a.1.** One gets two related accepted transcripts, e_1' does not
  divide $\alpha z_1' + t_1'$, $\Gamma_1 \le 8/\varepsilon$, and $\Gamma_1 = 2^a$ with $a \ge 1$, with probability at least
  $\varepsilon^2/64$.
  We thus have, with probability $\varepsilon^2/64$, an odd k_1 such that $c^{2^a} = u^{2F_1} \bmod n$:
  $c^{2^{a-1}}$ and u^{F_1} are two square roots of the same value. Since no information
  leaks about the actual square roots $\{u, -u\}$ known for h, nor for $h^{F_1} \bmod n$, so
  $c^{2^{a-1}} \ne \pm u^{F_1} \bmod n$ with probability $1/2$, which leads to the factorization of
  n with probability $1/2$ (see Proposition 1). Hence, we solve the RSA challenge
  with probability at least $\varepsilon^2/128$.
- **Statement 2.a.2.** One gets two related accepted transcripts, e_1' does not
  divide $\alpha z_1' + t_1'$, $\Gamma_1 \le 8/\varepsilon$, and $\Gamma_1 = 2^a v$ with $a \ge 0$ and an odd $v > 1$, with
  probability at least $\varepsilon^2/64$.
  It thus holds, with probability $\varepsilon^2/64$ (unless one finds a non-trivial square
  root of 1 modulo n, which allows to solve any RSA instance modulo n, see
  above), that $C^v = u^{2F_1} \bmod n$, for $C = \pm c^{2^a}$ and $\gcd(v, 2F_1) = 1$, since
  $v \mid \Gamma_1$ and v is odd. Using the Fact 5 from Proposition 1, one gets the v-th
  root of u modulo n, for $v \in [\![3\,;8/\varepsilon]\!] \cap \mathsf{P}_n$. Since our simulation that uses the
  RSA challenge (n, u, e) does not leak *any* information about e, then $v = e$
  with probability greater than $\varepsilon/4$, if the exponent e is randomly chosen in
  $[\![2\,;8/\varepsilon]\!] \cap \mathsf{P}_n$ (this set being exactly the set of odd integers smaller than $8/\varepsilon$, it
  contains approximately $4/\varepsilon$ elements). Hence, we solve an RSA challenge with
  probability at least $\varepsilon^2/64 \times \varepsilon/4 = \varepsilon^3/256$.

Statement 2.b: One gets two related accepted transcripts, e_1' does not divide $\alpha z_1' + t_1'$, and $\Gamma_1 > 8/\varepsilon$ with probability at least $\varepsilon^2/32$. When $\Gamma_1 > 8/\varepsilon$, the simulator rewinds the protocol once more, with a third challenge e_2. Let us consider all the possible challenges e_2 for this rewinding (independently of any success). Among all the possible challenges e_2, and so the differences $e_2' = |e_0 - e_2|$, the number of differences that Γ_1 divides is at most $(2^\kappa + 1)/\Gamma_1 < 8(2^\kappa + 1)/\varepsilon$. A given e_2' appears with probability at most $2/2^\kappa$ (since $0 \le e_2' \le \max\{e_0, 2^\kappa - e_0\}$). Therefore, the probability that Γ_1 divides e_2' for a random e_2 is less than $\varepsilon/4$. Recall that, from the splitting lemma (with a good R), one gets a third related accepted transcript with probability greater than $\varepsilon/2$. Hence globally, we get three related accepted transcripts, such that e_1' does not divide $\alpha z_1' + t_1'$, $\Gamma_1 > 8/\varepsilon$, and Γ_1 does not divide e_2', with probability at least $\varepsilon^3/128$.

As above, for the third transcript (d, e_2, z_2, t_2), we assume $e_0 \ge e_2$, and define $e_2' \leftarrow e_0 - e_2$, $z_2' \leftarrow z_0 - z_2$ (otherwise we change the signs). We also define $\beta_2 = \gcd(e_2', \alpha z_2' + t_2')$. Note that we do not require that e_2' does not divide $\alpha z_2' + t_2'$. We also set $\Gamma_2 \leftarrow e_2'/\beta_2$ and $F_2 \leftarrow (\alpha z_2' + t_2')/\beta_2$: F_2/Γ_2 is the irreducible fraction form of $(\alpha z_2' + t_2')/e_2'$. Since Γ_2 divides e_2', it cannot be equal to Γ_1.

Since these are all accepted transcripts, so $c^{e_1'} = g^{z_1'} h^{t_1'} \bmod n$ and $c^{e_2'} = g^{z_2'} h^{t_2'} \bmod n$, and then $c^{e_1'e_2'} = g^{e_2'z_1'}h^{e_2't_1'} = g^{e_1'z_2'}h^{e_1't_2'} \bmod n$. This leads, for $\Delta_z = e_2'z_1' - e_1'z_2'$ and $\Delta_t = e_2't_1' - e_1't_2'$, to

$$g^{\Delta_z} = g^{e_2'z_1' - e_1'z_2'} = h^{e_1't_2' - e_2't_1'} = h^{-\Delta_t} \bmod n.$$

If $\Delta_z = \Delta_t = 0$, then it holds that $z_2'/e_2' = z_1'/e_1'$ and $t_2'/e_2' = t_1'/e_1'$:

$$\frac{F_2}{\Gamma_2} = \frac{\alpha z_2' + t_2'}{e_2'} = \alpha \cdot \frac{z_2'}{e_2'} + \frac{t_2'}{e_2'} = \alpha \cdot \frac{z_1'}{e_1'} + \frac{t_1'}{e_1'} = \frac{\alpha z_1' + t_1'}{e_1'} = \frac{F_1}{\Gamma_1}.$$

Since they are both the irreducible notations of the same fraction, we necessarily have $\Gamma_1 = \Gamma_2$ and $F_1 = F_2$, which contradicts the above remark that $\Gamma_2 \ne \Gamma_1$. Hence, the pair (Δ_z, Δ_t) is non-trivial, which leads to the factorization of n with probability $1/2$, from the Fact 4 from Proposition 1. Overall, we get a solution to the RSA challenge with probability at least $\varepsilon^3/128 \times 1/2 = \varepsilon^3/256$ (after getting the factorization).

Overall Success Probability. All in all, if Statement 2 is true, we get a solution to the RSA challenge with probability at least $\varepsilon^3/256$. On the other hand, if Statement 1 holds, there are two complementary situations: either we get a valid opening with probability at least $\varepsilon^2/16$, or we get a non-trivial square root of 1 modulo n with probability at least $\varepsilon^2/16$. Overall, we either get a valid opening with probability at least $\varepsilon^2/16$, or we solve an RSA challenge modulo n with probability at least $\varepsilon^3/256$. $\qquad\square$

5   Classical Extensions and Applications

We revisit the natural implications of the commitment scheme of Sect. 3 and its argument of knowledge. More precisely, we generalize the results of previous sections while we commit to vectors of integers. Then, we also show the security of Lipmaa's range proofs [36] under the RSA assumption to illustrate how the result of Sect. 4 extends to more general arguments over the integers.

5.1 Generalized Commitment of Integers

The following commitment scheme allows committing to a vector of integers $(m_1, \ldots, m_\ell)$ with a single element of the form $c = g_1^{m_1} \cdots g_\ell^{m_\ell} h^r \bmod n$:

- Setup($1^\kappa, \ell$) runs $(n, (p, q)) \leftarrow_R \mathsf{GenMod}(1^\kappa)$, and picks $\ell+1$ random generators $(g_1, \ldots, g_\ell, h)$ of QR_n. It returns $\mathsf{pp} = (n, g_1, \ldots, g_\ell, h)$;
- Commit($\mathsf{pp}, \boldsymbol{m}; r$), for $\mathsf{pp} = (n, g_1, \ldots, g_\ell, h)$, a vector $\boldsymbol{m} = (m_1, \ldots, m_\ell) \in \mathbb{Z}^\ell$, and some random coins $r \leftarrow_R [\![0 \,;\, n]\!]$, computes $c = g_1^{m_1} \cdots g_\ell^{m_\ell} h^r \bmod n$, and returns (c, d) with $d = r$;
- Verify($\mathsf{pp}, c, d, \boldsymbol{m}$) parses pp as $\mathsf{pp} = (n, g_1, \ldots, g_\ell, h)$ and outputs 1 if $c = \pm g_1^{m_1} \cdots g_\ell^{m_\ell} h^d \bmod n$ and 0 otherwise.

Again, the above commitment scheme is obviously *correct*. The *hiding* property relies on the existence of α_i such that $g_i = h^{\alpha_i} \bmod n$ for $i = 1, \ldots, \ell$, and so

$$c = \mathsf{Commit}(\mathsf{pp}, \boldsymbol{m}; r) = g_1^{m_1} \cdots g_\ell^{m_\ell} h^r = h^{r + \sum \alpha_i m_i}$$
$$= h^{(r + \sum \alpha_i(m_i - m_i')) + \sum \alpha_i m_i'} = g_1^{m_1'} \cdots g_\ell^{m_\ell'} h^{r + \sum \alpha_i(m_i - m_i')}$$
$$= \mathsf{Commit}(\mathsf{pp}, \boldsymbol{m}'; r'),$$

for any $\boldsymbol{m}' = (m_1', \ldots, m_\ell') \in \mathbb{Z}$, with $r' \leftarrow [r + \sum \alpha_i(m_i - m_i') \bmod p'q']$, that is smaller than n.

The binding property relies on the Integer Factorization assumption: indeed, from two different openings $(\boldsymbol{m}, d)$ and $(\boldsymbol{m}', d')$ for a commitment c, with $d' > d$, the validity checks show that $g_1^{m_1} \cdots g_\ell^{m_\ell} h^d = g_1^{m_1'} \cdots g_\ell^{m_\ell'} h^{d'} \bmod n$, and so, if one has chosen β_i such that $g_i = g^{\beta_i} \bmod n$, for a random square g, then one knows $g^{\sum \beta_i(m_i - m_i')} = h^{d'-d} \bmod n$. The Fact 4 from Proposition 1 leads to the conclusion.

To avoid a trusted setup, one can note that the guarantees for the prover (the hiding property) just rely on the existence of α_i such that $g_i = h^{\alpha_i} \bmod n$ for $i = 1, \ldots, \ell$. The well-formedness of the RSA modulus is for the security guarantees against the verifier. It is important for him that the prover cannot break the RSA assumption. So the setup can be run by the verifier, with an additional proof of existence of α_i such that $g_i = h^{\alpha_i} \bmod n$ for $i = 1, \ldots, \ell$ to the prover.

5.2 Zero-Knowledge Argument of Opening

An argument of knowledge of an opening of a commitment $c = g_1^{x_1} \cdots g_\ell^{x_\ell} h^r \bmod n$ in the general case can be easily adapted from the normal case leading to a transcript of the form $(d, e, (z_1, \ldots, z_\ell, t))$ with $d = g_1^{y_1} \cdots g_\ell^{y_\ell} h^s$, and $c^e d = g_1^{z_1} \cdots g_\ell^{z_\ell} h^t \bmod n$. As above, the knowledge-extractor rewinds the execution for the same d, but two different challenges $e_0 \neq e_1$. Doing the quotient of the two relations, d cancels out: $c^{e'} = g_1^{z_1'} \cdots g_\ell^{z_\ell'} h^{t'} \bmod n$. Let us assume that one would have set $g_i = g^{a_i} h^{b_i} \bmod n$, we would have

$$c^{e'} = g^{\sum a_i z_i'} h^{\sum b_i z_i' + t'} \bmod n.$$

Under the RSA assumption, the above Statement 1 (from the proof, in Sect. 4) holds: e' divides both $\sum a_i z_1'$ and $\sum b_i z_i' + t'$ with non-negligible probability. Since the coefficients a_i's and b_i's are random, this means that e' divides all the z_i''s and t'. Hence, one can set $\mu_i = z_i'/e'$, for $i = 1, \ldots, \ell$ and $\tau = t'/e'$, and $c = \pm g_1^{\mu_1} \cdots g_\ell^{\mu_\ell} h^\tau \bmod n$ is a valid opening of c, unless one finds a non-trivial square-root of 1 modulo n.

5.3 Efficient Range Proofs from RSA

We show that Lipmaa's range proof [36] also benefits from our technique as the Strong-RSA assumption can also be avoided in the security analysis.

Range Proof from Integer Commitment Scheme. Let $c = g^x h^r \bmod n$ be a commitment of a value x and $[a \, ; b]$ be a public interval. As the commitment is homomorphic, one can efficiently compute a commitment c_a of $x - a$ and a commitment c_b of $b - x$ from c. To prove that $x \in [a \, ; b]$, this is enough to show that c_a and c_b commit to positive values. Let us focus on the proof that $c_a = g^{x-a} h^r \bmod n$ commits to a positive value, since the same method applies for c_b. To do so, the prover computes (x_1, x_2, x_3, x_4) such that $x - a = \sum_{i=1}^4 x_i^2$. By a famous result from Lagrange, such a decomposition exists if and only if $x - a \geq 0$. Moreover, this decomposition can be efficiently computed by the Rabin-Shallit algorithm [42], for which Lipmaa [36] also suggested some optimizations. The prover commits to (x_1, x_2, x_3, x_4) in (c_1, c_2, c_3, c_4), where $c_i = g^{x_i} h^{r_i} \bmod n$ for each $i = 1$ to 4. Now, the prover proves his knowledge of openings $x - a$, x_1, x_2, x_3, x_4 (along with random coins r, r_1, r_2, r_3, r_4) of c_a, c_1, c_2, c_3, c_4 satisfying $\sum_{i=1}^4 x_i^2 = x - a$ over the integers.

The reason allowing to solely rely on the RSA assumption in the range proof comes from the fact that the first part of the argument reduces to an argument of knowledge of openings x_1, x_2, x_3, x_4 of c_1, c_2, c_3, c_4 while the remaining part simply ensures the relation $\sum_{i=1}^4 x_i^2 = x - a$ to hold. Indeed, once the witnesses are extracted, this is implied by the representation $c_a = \prod_{i=1}^4 c_i^{x_i} h^{r - \sum x_i r_i} \bmod n$ which can be seen as generalized commitment scheme with basis (c_1, c_2, c_3, c_4, h) from which the opening cannot change. Therefore, the argument can be seen as five parallel arguments of knowledge, the fifth one being an argument of knowledge for a generalized commitment, where the opening for the last argument is the vector of the openings for the other arguments. A formal proof of an optimized version of this protocol under the intractability of the RSA assumption is presented in the full version [19].

Extension. Since most of the arguments of knowledge of a solution to a system of equations over the integers [16] can be split into parallel arguments of knowledge of values assigned to the variables and a proof of membership (in the language composed of all the solutions of the system), which is expressed as representations corresponding to generalized commitments, our analysis extends to all "discrete-logarithm relation set" (see [34]): the description of the protocol is unchanged but the security only relies on the standard RSA assumption.

6   Commitment with Knowledge-Delayed Order

Arguments of knowledge of openings for the Damgård-Fujisaki commitment scheme can rely on the RSA assumption rather than the Strong-RSA assumption. In this section, we show that this scheme can be efficiently combined with another RSA-based commitment scheme which, as far as we know, was proposed by Gennaro [24]: we show how Damgård-Fujisaki commitments (which are homomorphic over the integers) can be converted into Gennaro commitments (which are homomorphic over $\mathbb{Z}_\pi$ for some prime π). We rely on this feature to design a method to improve the efficiency of zero-knowledge arguments over the integers on several aspects, by allowing the players to perform some of the computations over $\mathbb{Z}_\pi$ rather than over the integers. We then illustrate our technique on the famous example of range proofs.

6.1   RSA-Based Commitments with Known Order

We recall the homomorphic commitment scheme over $\mathbb{Z}_\pi$ of [24]. The order of the commitment is a known prime $\pi > 2^\kappa$.

Description of the Generalized Commitment Scheme. Let us describe the commitment of vectors of integers $(m_1, \ldots, m_\ell)$:

- Setup(1^κ) runs $(n, (p, q)) \leftarrow_R \mathsf{GenMod}(1^\kappa)$, and picks ℓ random generators $g_1, \ldots, g_\ell$ of QR_n. Then, it picks a random prime $\pi \in [\![2^{\kappa+1} ; 2^{\kappa+2}]\!]$, and returns $\mathsf{pp} = (n, g_1, \ldots, g_\ell, \pi)$;
- Commit($\mathsf{pp}, \boldsymbol{m}; r$), for $\mathsf{pp} = (n, g_1, \ldots, g_\ell, \pi)$, a vector $\boldsymbol{m} = (m_1, \ldots, m_\ell) \in \mathbb{Z}_\pi^\ell$, and some random coins $r \leftarrow_R \mathbb{Z}_n$, computes $c = g_1^{m_1} \cdots g_\ell^{m_\ell} r^\pi \bmod n$, and returns (c, d) with $d = r$;
- Verify($\mathsf{pp}, c, d, \boldsymbol{m}$) parses pp as $\mathsf{pp} = (n, g_1, \ldots, g_\ell, \pi)$ and outputs 1 if $c = g_1^{m_1} \cdots g_\ell^{m_\ell} r^\pi \bmod n$, and 0 otherwise.

The above commitment scheme is obviously *correct*. The *hiding* property relies on the bijectivity of the π-th power modulo n (as π is prime): for any message $\boldsymbol{m}' = (m_1', \ldots, m_\ell') \in \mathbb{Z}_\pi^\ell$, we have $c = g_1^{m_1'} \cdots g_\ell^{m_\ell'} \times g_1^{m_1-m_1'} \cdots g_\ell^{m_\ell-m_\ell'} \times r^\pi \bmod n$. By noting s the π-th root of $g_1^{m_1-m_1'} \cdots g_\ell^{m_\ell-m_\ell'}$, $c = \mathsf{Commit}(\mathsf{pp}, \boldsymbol{m}'; rs)$. The *binding* property uses an extension of the Fact 5 from Proposition 1: if one has chosen β_i such that $g_i = u^{2\beta_i}$, for a challenge RSA $u \in \mathbb{Z}_n^*$, two distinct openings $(\boldsymbol{m}, r) \neq (\boldsymbol{m}', s)$ satisfy $g_1^{m_1} \cdots g_\ell^{m_\ell} r^\pi = g_1^{m_1'} \cdots g_\ell^{m_\ell'} s^\pi \mod n$, and so $(s/r)^\pi = u^{2a} \mod n$, where $a = \sum \beta_i(m_i - m_i') = a_1\pi + a_0$, with $0 \leq a_0 < \pi$. Let us note α and β the integers such that $\alpha\pi + \beta 2a_0 = \gcd(\pi, 2a_0) = 1$, and output $u_0 := u^{\alpha - 2a_1\beta} \cdot (s/r)^\beta \mod n$, then

$$u_0^\pi = u^{\alpha\pi - 2a_1\beta\pi} \cdot (s/r)^{\beta\pi} = u^{1-2(a_0+a_1\pi)\beta} \cdot u^{2a\beta} = u \mod n.$$

This breaks the RSA assumption with exponent π.

Homomorphic-Opening. In addition, this commitment scheme is homomorphic in $\mathbb{Z}_\pi$: given $c = g_1^{m_1} \cdots g_\ell^{m_\ell} r^\pi \mod n$ and $d = g_1^{m_1'} \cdots g_\ell^{m_\ell'} s^\pi \mod n$ with known openings, we can efficiently open the commitment $c \cdot d \mod n$ to $\bar{\boldsymbol{m}} = (\bar{m}_1, \ldots, \bar{m}_\ell)$, with $\bar{m}_i = m_i + m_i' \mod \pi$ for $1 \leq i \leq \ell$, and a random coin $rs \prod g_i^{(m_i+m_i')\div\pi} \mod n$, where $a \div b$ is the quotient of the Euclidean division. We emphasize this property to be essential to avoid working with long integers in the arguments of knowledge of an opening: the prover can "reduce" its openings since π is known.

Argument of Opening. Given $\mathsf{pp} = (n, g_1, \ldots, g_\ell, \pi)$ and $c = g_1^{x_1} \cdots g_\ell^{x_\ell} r^\pi \mod n$, with witness $(x_1, \ldots, x_\ell, r)$, we can describe a standard argument of knowledge of an opening:

Initialize: $\mathscr{P}$ and $\mathscr{V}$ decide to run the protocol on input (pp, κ, c);
Commit: $\mathscr{P}$ computes $d = g_1^{y_1} \cdots g_\ell^{y_\ell} s^\pi$, for $y_i \leftarrow_R \mathbb{Z}_\pi$, and $s \leftarrow_R \mathbb{Z}_n^*$, and sends d to $\mathscr{V}$;
Challenge: $\mathscr{V}$ outputs $e \leftarrow_R [\![0\, ; 2^\kappa]\!]$;
Response: $\mathscr{P}$ computes k_i, z_i, t such that $ex_i + y_i = k_i\pi + z_i$, with $0 \leq z_i < \pi$, and $t = g_1^{k_1} \cdots g_\ell^{k_\ell} \cdot r^e s \mod n$. $\mathscr{P}$ outputs $(z = (z_i)_i, t)$;
Verify: $\mathscr{V}$ accepts the proof and outputs 1 if, for each i, $0 \leq z_i < \pi$, and $c^e d = g_1^{z_1} \cdots g_\ell^{z_\ell} t^\pi \mod n$. Otherwise, $\mathscr{V}$ rejects the proof and outputs 0.

Completeness and *zero-knowledge* are straightforward. Then, let us focus on the *knowledge-extractability*: From two related valid transcripts, for the same d, we get as usual $c^{e-e'} = g_1^{z_1-z_1'} \cdots g_\ell^{z_\ell-z_\ell'} \cdot (t/t')^\pi \mod n$. Since the prime $\pi > 2^\kappa \geq |e - e'|$, the simulator can compute $\alpha(e - e') + \beta\pi = 1$ and we have

$$c^{1-\beta\pi} = c^{\alpha(e-e')} = g_1^{\alpha(z_1-z_1')} \cdots g_\ell^{\alpha(z_\ell-z_\ell')} \cdot (t/t')^{\alpha\pi} \mod n.$$

Then, for $\alpha(z_i - z_i') = l_i\pi + x_i'$ with $0 \leq x_i' < \pi$, and $T = c^\beta \cdot g_1^{l_1} \cdots g_\ell^{l_\ell} \cdot (t/t')^\alpha \mod n$, we have a valid opening $(x_1', \ldots, x_\ell', T)$ of c.

6.2 Commitment with Knowledge-Delayed Order

The above commitment scheme with known prime order π can temporarily pass itself off as a commitment scheme of Sect. 3 with hidden order.

Description of the Commitment Scheme. The verifier sets up the parameter pp of the commitment scheme with hidden order but hides a prime order π in pp during this execution. To guarantee the hiding property, in the setup the verifier also adds a proof that $g = h^\alpha \bmod n$ for some α.

- Setup(1^κ) runs $(n, (p,q)) \leftarrow_R$ GenMod(1^κ), and picks $h_0 \leftarrow_R$ QR$_n$ and a random prime $\pi \in [\![2^{\kappa+1}; 2^{\kappa+2}]\!]$. Then, it picks $\rho \leftarrow_R [\![0; n^2]\!]$, relatively prime to π, and sets $g \leftarrow h_0^\rho \bmod n$ and $h \leftarrow h_0^\pi \bmod n$. Finally, it returns pp $= (n, g, h)$ and keeps sk $= (\pi, h_0)$. Actually, we have $h^\rho = g^\pi \bmod n$. So, if one sets $\alpha = \rho \cdot \pi^{-1} \bmod \varphi(n)$, one has $g = h^\alpha \bmod n$, and proves it;
- Commit(pp, $m; r$) parses pp as above and commits to $m \in \mathbb{Z}$ by picking $r \leftarrow_R \mathbb{Z}_n$ and computing $c = g^m h^r \bmod n$. It returns (c, r);
- Verify(pp, c, m, r) parses pp $= (n, g, h)$ and outputs 1 if $c = \pm g^m h^r \bmod n$ and 0 otherwise;
- Reveal(pp, sk) returns sk $= (\pi, h_0)$;
- Adapt(pp, sk, c, m, r) first parses sk $= (\pi, h_0)$ and checks whether $h = h_0^\pi \bmod n$. Then, it adapts the opening by computing $m = k\pi + \bar{m}$ for $0 \le \bar{m} < \pi$ and $t = g^k h_0^r \bmod n$. It outputs $(\bar{m}, t)$;
- Verify'(pp, $\pi, c, \bar{m}, t$) outputs 1 if $c = g^{\bar{m}} t^\pi \bmod n$, and 0 otherwise.

This construction easily extends to commitments of vectors. Note that from $g^{\bar{m}} t^\pi = c = g^{\bar{m}'} t'^\pi \bmod n$, with $\bar{m} \ne \bar{m}' \bmod \pi$, setting $h_0 = y^2$ from an RSA challenge (n, y) of exponent $\pi > 2^\kappa$, we obtain $y^{2\rho(\bar{m}-\bar{m}')} = (t'/t)^\pi \bmod n$, with $2\rho(\bar{m} - \bar{m}') \ne 0 \bmod \pi$, which leads to the π-th root of y modulo n (using Fact 5 from Proposition 1).

Switching Between Commitments. Let com denote the Damgård-Fujisaki integer commitment scheme, such that $\text{com}(m; r) = g^m h^r \bmod n$, and com_π denote the Gennaro commitment scheme, such that $\text{com}_\pi(m; R) = g^m R^\pi \bmod n$. On the one-hand, only com leads to proof of relations over the integers. On the other hand, com_π leads to much more efficient proofs of relation modulo π. The above commitment with knowledge-delayed order allows generating pp $= (n, g, h)$ so that $c = \text{com}(m; r) = g^m h^r \bmod n$ can be switched into

$$c = \text{com}_\pi(\mathfrak{r}_\pi(m); g^{\mathfrak{q}_\pi(m)} h_0^r), \tag{1}$$

where $\mathfrak{q}_\pi(m)$ and $\mathfrak{r}_\pi(m)$ denote the quotient and remainder of the euclidean division of m by π. This switching allows to keep some good properties over the integers and working modulo π since pp gives no information about π until the verifier reveals (π, h_0).

6.3   Improving Zero-Knowledge Arguments over the Integers

The commitment with knowledge-delayed order provides a new technique to zero-knowledge arguments for statements over the integers, while working modulo π. This technique leads to more efficient membership arguments, with a lower communication and a smaller verifier computation (some part of the cost is *delegated* to the prover). We restrict our attention to statements that can be expressed as membership to a set $S \in \mathbf{D}$. The protocol we describe is *honest-verifier zero-knowledge*. At the end of the section we recall standard methods to get full-fledged zero-knowledge.

Membership Argument for D. Given $S \in \mathbf{D}$, let P_S be a representing polynomial with k-vector input and ℓ-vector witness (e.g., if S is the set of positive integers, $P_S : (x, w_1, w_2, w_3, w_4) = x - (\sum_i w_i^2)$). We assume $\mathscr{P}$ and $\mathscr{V}$ agreed on a bound t such that each $\boldsymbol{x} \in S$ has a witness $\boldsymbol{w}$ such that $\|\boldsymbol{w}\|_\infty \leq \|\boldsymbol{x}\|_\infty^t$ ($S \in \mathbf{D}$, so there is always such a t. As shown in [36], $t < 2$ is sufficient for most cryptographic applications).

Let $\boldsymbol{x}$ be a secret vector held by $\mathscr{P}$, and $\boldsymbol{w}$ be a *witness* for $\boldsymbol{x} \in S$, meaning that $P_S(\boldsymbol{x}, \boldsymbol{w}) = 0$. Zero-knowledge argument for polynomial relations over committed inputs usually demands committing to intermediate values, and proving additive and multiplicative relationships with the inputs, see e.g. [9]. To prove a multiplicative relationship $z = xy$ between values (x, y, z) committed in (c_x, c_y, c_z), $\mathscr{P}$ proves knowledge of inputs (x, y, z) and random coins (r_x, r_y, r_z) such that $c_x = g^x r_x^\pi \bmod n$, $c_y = g^y r_y^\pi \bmod n$, and $c_z = c_x^y r_z^\pi$.

We almost follow this principle except that we use the commitment scheme of Sect. 6.2 to switch from com to com_π once $\mathscr{P}$ proved knowledge of both $\boldsymbol{x}$ and $\boldsymbol{w}$ over the integers. Proving $P_S(\boldsymbol{x}, \boldsymbol{w}) = 0$ over the integers is then replaced by proving $P_S(\boldsymbol{x}, \boldsymbol{w}) = 0$ modulo π.

Argument of knowledge of the inputs and witnesses.

1. $\mathscr{V}$ runs the setup from the Sect. 6.2, which generates $\mathsf{pp} = (n, g, h)$ and $\mathsf{sk} = (\pi, h_0)$: this defines com : $(x; r) \mapsto g^x h^r \bmod n$. It additionally proves the existence of α such that $g = h^\alpha \bmod n$;
2. $\mathscr{P}$ picks random coins (r_x, r_w) and commits to $(\boldsymbol{x}, \boldsymbol{w})$ with (r_x, r_w) as $(c_x, c_w) \leftarrow (\mathsf{com}(\boldsymbol{x}; r_x), \mathsf{com}(\boldsymbol{w}; r_w))$;
3. $\mathscr{P}$ performs a $\mathsf{ZKAoK}\{(\boldsymbol{x}, \boldsymbol{w}, r_x, r_w) \mid c_x = g^x h^{r_x} \wedge c_w = g^w h^{r_w}\}$, we thereafter refer to ZK_1, with $\mathscr{V}$. If the argument fails, $\mathscr{V}$ aborts the protocol.

Argument of knowledge of $(\boldsymbol{x}', \boldsymbol{w}')$ such that $P_S(\boldsymbol{x}', \boldsymbol{w}') = 0 \bmod \pi$.

1. $\mathscr{V}$ reveals (π, h_0) to $\mathscr{P}$ who checks whether $h = h_0^\pi \bmod n$ or not, to switch to com_π. Let $(\boldsymbol{x}', \boldsymbol{w}') = (\mathfrak{r}_\pi(\boldsymbol{x}), \mathfrak{r}_\pi(\boldsymbol{w})) = (\boldsymbol{x}, \boldsymbol{w}) \bmod \pi$.
2. $\mathscr{P}$ performs a $\mathsf{ZKAoK}\{(\boldsymbol{x}', \boldsymbol{w}', R_x, R_w)\}$, we thereafter refer to ZK_2, such that $(c_x, c_w) = (\mathsf{com}_\pi(\boldsymbol{x}; R_x), \mathsf{com}_\pi(\boldsymbol{w}; R_w))$ and $P_S(\boldsymbol{x}, \boldsymbol{w}) = 0 \bmod \pi$. Note that (c_x, c_w) are now seen as commitments over $\mathbb{Z}_\pi$, using the fact that $\mathsf{com}(\boldsymbol{x}; r_x) = \mathsf{com}_\pi(\mathfrak{r}_\pi(\boldsymbol{x}); R_x)$ and $\mathsf{com}(\boldsymbol{w}; r_w) = \mathsf{com}_\pi(\mathfrak{r}_\pi(\boldsymbol{w}); R_w)$, with appropriate (R_x, R_w). If the argument succeeds, $\mathscr{V}$ returns accept.

Theorem 4. *Under the* **RSA** *assumption, the above protocol is a statistical zero-knowledge argument of knowledge of openings of (c_x, c_w) to vectors of integers $(\boldsymbol{x}, \boldsymbol{w})$ such that $P_S(\boldsymbol{x}, \boldsymbol{w}) = 0$: which proves that $\boldsymbol{x} \in S$.*

Proof. The intuition behind Theorem 4 is that ZK_1 proves that $\mathscr{P}$ knows $(\boldsymbol{x}, \boldsymbol{w})$ in $(\boldsymbol{c_x}, \boldsymbol{c_w})$, and ZK_2 proves that $P_S(\boldsymbol{x}, \boldsymbol{w}) = 0 \bmod \pi$ for a κ-bit prime π which was revealed *after* $(\boldsymbol{x}, \boldsymbol{w})$ *were committed*. Hence, $\mathscr{P}$ knew vectors of integer $(\boldsymbol{x}, \boldsymbol{w})$ such that $P_S(\boldsymbol{x}, \boldsymbol{w}) = 0 \bmod \pi$ for a random κ-bit prime π. This has a negligible probability to happen unless $P_S(\boldsymbol{x}, \boldsymbol{w}) = 0$ holds over the integers, since P_S is a polynomial. The full proof consists of the three properties: correctness, zero-knowledge, and knowledge-extractability.

Correctness. It easily follows from the correctness of ZK_1 and ZK_2: if $\mathscr{P}$ knows $(\boldsymbol{x}, \boldsymbol{w}, \boldsymbol{r_x}, \boldsymbol{r_w})$ such that $(\boldsymbol{c_x}, \boldsymbol{c_w}) = (\mathsf{com}(\boldsymbol{x}; \boldsymbol{r_x}), \mathsf{com}(\boldsymbol{w}; \boldsymbol{r_w}))$ and $P_S(\boldsymbol{x}, \boldsymbol{w}) = 0$, then the argument of knowledge of $(\boldsymbol{x}, \boldsymbol{r_x})$ such that $\boldsymbol{c_x} = \mathsf{com}(\boldsymbol{x}; \boldsymbol{r_x})$ will succeed, and it holds that $(\boldsymbol{c_x}, \boldsymbol{c_w}) = (\mathsf{com}_\pi(\boldsymbol{x} \bmod \pi; v^{q_\pi(\boldsymbol{x})} \tilde{h}^{\boldsymbol{r_x}}), \mathsf{com}_\pi(\boldsymbol{w} \bmod \pi; v^{q_\pi(\boldsymbol{x})} \tilde{h}^{\boldsymbol{r_x}}))$. Moreover, as P_S is a polynomial, the modular reduction applies, and leads to $P_S(\boldsymbol{x} \bmod \pi, \boldsymbol{w} \bmod \pi) = P_S(\boldsymbol{x}, \boldsymbol{w}) = 0 \bmod \pi$.

Zero-Knowledge. It also follows from the zero-knowledge of ZK_1 and ZK_2, and the hiding property of the commitments. Let $\mathscr{S}im_{\mathsf{ZK}}$ be the following simulator: one first generates dummy commitments $(\boldsymbol{c_x}, \boldsymbol{c_w})$, which does not make any difference under the hiding property, and runs the simulator of ZK_1. Once (π, h_0) is revealed, $\mathscr{S}im_{\mathsf{ZK}}$ runs the simulator of ZK_2.

Since the commitment is statistically hiding, ZK_1 is our statistically zero-knowledge argument of knowledge of opening from Sect. 3 and ZK_2 is an argument of relations on commitments with known order π (since $h = h_0^\pi \bmod n$) that is possible in statistical zero-knowledge, the full protocol is statistically zero-knowledge.

Knowledge Extractability. Let $\mathscr{P}'$ outputing a convincing argument with probability ε, i.e. $\mathscr{P}'$ succeeds in ZK_1 and ZK_2 with probability greater than ε.

Under the RSA assumption, there is an extractor of ZK_1 which computes $(\boldsymbol{x}, \boldsymbol{w}, \boldsymbol{r_x}, \boldsymbol{r_w})$ such that $\boldsymbol{c_x} = g^{\boldsymbol{x}} h^{\boldsymbol{r_x}}$ and $\boldsymbol{c_w} = g^{\boldsymbol{w}} h^{\boldsymbol{r_w}}$. Then, (π, h_0) is revealed in the protocol and still under the RSA assumption, there is another extractor of ZK_2 which computes $(\boldsymbol{x}', \boldsymbol{w}', \boldsymbol{R_x}, \boldsymbol{R_w})$ such that both relations $(\boldsymbol{c_x}, \boldsymbol{c_w}) = (\mathsf{com}_\pi(\boldsymbol{x}'; \boldsymbol{R_x}), \mathsf{com}_\pi(\boldsymbol{w}'; \boldsymbol{R_w}))$ and $P_S(\boldsymbol{x}', \boldsymbol{w}') = 0 \bmod \pi$ are satisfied. Now, let us consider two situations:

– If $\boldsymbol{x}' = \boldsymbol{x} \bmod \pi$ and $\boldsymbol{w}' = \boldsymbol{w} \bmod \pi$, then the value committed over the integers, *before π was revealed*, satisfy $P_S(\boldsymbol{x}, \boldsymbol{w}) = 0 \bmod \pi$, for a random $\pi \in [\![2^{\kappa+1}; 2^{\kappa+2}]\!]$. We stress that the view of (n, g, h) does not reveal any information on the prime π.

   Since there are approximately $2^{\kappa+1}/\kappa$ primes in this set, and this extraction works with probability greater than ε^2, $P_S(\boldsymbol{x}, \boldsymbol{w}) = 0 \bmod Q$, for $Q \geq 2^{2^\kappa/\varepsilon^2}$, which is much larger than the values that can be taken in the integers, since the inputs and the witnesses have a size polynomial in κ, and the polynomial P_S has a bounded degree.

– If $x' \neq x \bmod \pi$ or $w' \neq w \bmod \pi$, wlog, we can assume that $x' \neq x \bmod \pi$:
  - we get (x, r_x) such that $c_x = \pm g^x h^{r_x} = g^{\tau_\pi(x)}(\pm g^{q_\pi(x)} h_0^{r_x})^\pi \bmod n$;
  - and (x', R_x) such that $c_x = g^{x'} R_x^\pi \bmod n$.

Hence, $g^{\tau_\pi(x)}(\pm g^{q_\pi(x)} h_0^{r_x})^\pi = g^{x'} R_x^\pi \bmod n$, and so $g^{\tau_\pi(x)-x'} = S^\pi \bmod n$, for $S = R_x/(\pm g^{q_\pi(x)} h_0^{r_x}) \bmod n$. If one would have set $h_0 = y^2$ from an RSA challenge (n, y, π) of exponent $\pi > 2^\kappa$, and thus $g = y^{2\rho}$, using Fact 5 from Proposition 1, one gets the π-th root of y modulo n.

This concludes the proof of the knowledge-extractability of the protocol, under the RSA assumption over $\mathbb{Z}_n$.   □

On the Efficiency of the Method. The advantages of this method compared to the classical method are twofold. First, most of the work in the protocol comes from the computation of exponentiations; and our technique transfers most of this work from $\mathcal{V}$ to $\mathcal{P}$. This comes from the fact that verifying an equation such as $c = \mathrm{com}(x; r)$ involves exponentiations by integers of size $O(\log n + \kappa)$ while verifying the equation $c = \mathrm{com}_\pi(x \bmod \pi; R)$ involves only two exponentiations by κ-bit values, which greatly reduces $\mathcal{V}$'s work. However, to switch from com to com_π $\mathcal{P}$ has to adapt the opening as in (1) of Sect. 6.2, which costs exponentiations by integers of size $O(\log n + \kappa)$ to compute the random coin R. $\mathcal{V}$ will still need to perform exponentiations by integers during ZK_1, but his work during this step can be made essentially independent of the number N of inputs and witnesses (up to a small $\log N$ additive term) and completely independent of the degree of the representing polynomial.

Second, our method separates the argument of *knowledge* of inputs to a Diophantine equation from the argument that they do indeed satisfy the equation. The arguments of knowledge of an opening of a commitment can be very efficiently batched: if $\mathcal{P}$ commits to $(x_1, \cdots, x_N)$ with random coins $(r_1, \cdots, r_N)$ as $(c_1, \cdots, c_N)$, the verifier can simply send a random seed $\lambda \leftarrow_R \{0,1\}^\kappa$ from which both players compute $(\lambda_1, \cdots, \lambda_N)$ using a pseudo-random generator[2]. Then, $\mathcal{P}$ performs a *single* argument of knowledge of an opening $(\sum_i \lambda_i x_i; \sum_i \lambda_i r_i)$ of the commitment $\prod_i c_i^{\lambda_i}$ (see [5,6] for more details). Therefore, when performing multiple membership arguments, $\mathcal{P}$ and $\mathcal{V}$ will have to perform a single argument for ZK_1 (of size essentially independent of the number of committed values).

In general, the higher the degree of the representing polynomial is, the lower will be the communication with our method. Still, we show in the next section that even for the case of range proofs, which is a membership proof to a Diophantine set whose representing polynomial is of degree 2, our method provides efficiency improvements.

Further Improvements. $\mathcal{V}$ can set h to $h_0^{\prod_i \pi_i}$ for several primes π_i instead of h_0^π. For some integer i, let $p_i \leftarrow \prod_{j \neq i} \pi_j$. Doing so allows $\mathcal{V}$ to reveal $(h_0^{p_i}, \pi_i)$

---

[2] The classical trick that consists of using $\lambda_i = \lambda^i$ is not efficient here since we are in the integers, and so no reduction can be applied.

instead of (h_0, π) in our method. Hence, in addition to allowing arbitrary parallel arguments with a single prime π, a single setting is sufficient to perform a polynomial number of sequential arguments (fixed in advance) with different primes π_i. In addition, we explained that commitments with knowledge-delayed order allow splitting the arguments of knowledge of the witnesses, denoted ZK_1, and the argument that they indeed belong to a Diophantine set, denoted ZK_2. The arguments ZK_1 can be batched as described above but, for efficiency reason, we should not generate $(\lambda_1, \lambda_2 \ldots, \lambda_N)$ as $(\lambda, \lambda^2, \ldots, \lambda^N)$. Indeed, $|\lambda^j|_b$ grows linearly with j over the integers. However, for the argument ZK_2, the order of the commitment has been revealed. Hence, we can now use batching technique with such $\lambda_j = \lambda^j$ since the prover is able to reduce the exponents modulo π at this stage. That means that our technique consisting of efficiently revealing the order of the commitment between ZK_1 and ZK_2 allows to use any method that crucially relies on batching coefficients expressed as powers of some λ, that were only available for discrete-log based proofs of statement over (pairing-free) known-order groups. For instance, we can get a *sub-linear* size argument to show that a committed matrix is the Hadamard products *over the integers* of two other committed matrices. Indeed, we can commit the rows of the matrices using a generalized commitment and make a batch proof for ZK_1, which remain sub-linear in the number of entries, and then we can import the results of [4,26] to ZK_2, preserving its sub-linearity.

Full-Fledged Zero-Knowledge. With an honest verifier, there is no need to prove the existence of α such that $g = h^\alpha$. In the malicious setting, this proof guarantees the hiding property of the commitments to the prover, who additionally checks $h = h_0^\pi \bmod n$ when they are revealed. Then we can use classical techniques to convert the HVZK protocol into a ZK protocol, such as an equivocable commitment of the challenge by the verifier, before the commitments from the prover.

7   Application to Range Proofs

7.1   Lipmaa's Compact Argument for Positivity

As explained before, Lipmaa [36] proposed an efficient argument for positivity, using generalized Damgård-Fujisaki commitments, and the proof that an integer is positive if and only if it can be written as the sum of four squares. However, it appears that the explicit construction given in [36, annex B] is flawed — although the high-level description is correct: any prover can provide a convincing argument for positivity, regardless of the sign of the committed integer, and so without holding valid witnesses. This might raise some concerns as the protocol of Lipmaa is the "textbook" range proof based on hidden order groups. Hence the protocol is suggested in several papers, and was implemented in e.g. [2]. In the full version [19], we recall the argument of [36], identify its flaw, and provide a correct optimized version with a full proof of security.

In the following, we describe a range proof in the same vein as the positivity argument of Lipmaa: an integer x belongs to an interval $[\![a\,;b]\!]$ if and only if $(x - a)(b - x) \geq 0$. In addition, we take into account the following improvement suggested by Groth [25]: x is positive if and only if $4x + 1$ can be written as the sum of three squares, and such a decomposition can be computed in polynomial time by the prover. We view this range proof (we call the three-square range proof, and denote it 3SRP) as an optimized version of the textbook range proof with integer commitments, to which we will compare our new method with knowledge-delayed order commitments (denoted 3SRP-KDO).

7.2   Three-Square Range Proof

To prove that $x \in [\![a\,;b]\!]$, for x committed with an integer commitment scheme, we prove that $4(x - a)(b - x) + 1$ can be written as the sum of three squares. Let (n, g, h) be the public parameters of the Damgård-Fujisaki commitment scheme, generated by the verifier. The three-square range proof (3SRP) is described in full details on Fig. 2. Basically, both $\mathscr{P}$ and $\mathscr{V}$ know that c_a contains $4(x - a)$ and c_0 contains $(b - x)$. The latter, with c_1, c_2, c_3 containing respectively x_1, x_2, x_3, is proven in a classical way, and the last part of the proof shows that $c_a^{x_0} g$, which implicitly contains $4(x - a)(b - x) + 1$ also contains $x_1^2 + x_2^2 + x_3^2$.

We then illustrate the technique introduced in Sect. 6.3 on this 3SRP protocol. The full converted protocol, denoted 3SRP-KDO, is described on Fig. 3.

7.3   Results

Let $B = \log(b - a)$. Note that for all $i \in \{0, 1, 2, 3\}$, $x_i^2 \leq (b - a)^2$ hence $\log x_i \leq B$. An exponentiation by a t-bit value takes on average $1.5t$ multiplications using a

---

For $\mathsf{pp} = (n, g, h)$ generated by $\mathscr{V}$, $\mathscr{P}$ has sent c, for which he knows (x, r) such that $c = g^x h^r \bmod n$ and $x \in [\![a\,;b]\!]$. Let $H : \mathbb{Z}_n^5 \mapsto \{0, 1\}^{2\kappa}$ be a collision-resistant hash function. $\mathscr{V}$ compute $c_a = (cg^{-a})^4 \bmod n$ and $c_0 = c^{-1}g^b \bmod n$; $\mathscr{P}$ computes c_a.

1. $\mathscr{P}$ computes $(x_i)_{1 \leq i \leq 3}$ such that $4(b - x)(x - a) + 1 = \sum_{i=1}^{3} x_i^2$. $\mathscr{P}$ commits to $(x_i)_{1 \leq i \leq 3}$ with random coins $(r_i)_{1 \leq i \leq 3} \leftarrow_R [\![0\,;n]\!]^3$ as $(c_i = g^{x_i} h^{r_i} \bmod n)_{1 \leq i \leq 3}$. Let $x_0 \leftarrow (b - x)$ and $r_0 \leftarrow r$.
2. $\mathscr{P}$ picks $(m_0, \cdots, m_3) \leftarrow_R [\![0\,;2^{B+2\kappa}]\!]^4$, $(s_0, \cdots, s_3) \leftarrow_R [\![0\,;2^{2\kappa}n]\!]^4$, $\sigma \leftarrow_R [\![0\,;2^{B+2\kappa}n]\!]$, and sends $\Delta = H((g^{m_i} h^{s_i} \bmod n)_{0 \leq i \leq 3}, h^\sigma c_a^{m_0} \prod_{i=1}^{3} c_i^{-m_i} \bmod n)$.
3. $\mathscr{V}$ picks a challenge $e \leftarrow_R [\![0\,;2^\kappa]\!]$ and sends it to $\mathscr{P}$.
4. $\mathscr{P}$ computes and sends $z_i = ex_i + m_i$ and $t_i = er_i + s_i$ for $i \in \{0, 1, 2, 3\}$, and $\tau = \sigma + e(x_0 r_0 - \sum_{i=1}^{3} x_i r_i)$.
5. $\mathscr{V}$ accepts the argument if

$$\Delta = H\left((g^{z_i} h^{t_i} c_i^{-e} \bmod n)_{0 \leq i \leq 3}, h^\tau g^e c_a^{z_0} (\prod_{i=1}^{3} c_i^{-z_i}) \bmod n\right).$$

Fig. 2. Three-square range proof (3SRP)

For $pp = (n, g, h)$ and $sk = (\pi, h_0)$ generated by $\mathcal{V}$, $\mathcal{P}$ has sent c, for which he knows (x, r) such that $c = g^x h^r \bmod n$ and $x \in [a\,;b]$. Let $H : \mathbb{Z}_n^6 \mapsto \{0,1\}^{2\kappa}$ be a collision-resistant hash function. $\mathcal{V}$ compute $c_a = (cg^{-a})^4 \bmod n$ and $c_0 = c^{-1}g^b \bmod n$; $\mathcal{P}$ computes c_a.

1. $\mathcal{P}$ computes $(x_i)_{1 \leq i \leq 3}$ such that $4(b-x)(x-a)+1 = \sum_{i=1}^3 x_i^2$. $\mathcal{P}$ commits to $(x_i)_{1 \leq i \leq 3}$ with random coins $(r_i)_{1 \leq i \leq 3} \leftarrow_R [0\,;n]^3$ as $(c_i = g^{x_i} h^{r_i} \bmod n)_{1 \leq i \leq 3}$. Let $x_0 \leftarrow (b-x)$ and $r_0 \leftarrow r$.
2. $\mathcal{P}$ picks $m \leftarrow_R [0\,;2^{B+3\kappa}]$, $(m_0, \cdots, m_3) \leftarrow_R [0\,;2^\kappa]^4$, $s \leftarrow_R [0\,;2^{3\kappa}n]$, $(s_0, \cdots, s_3) \leftarrow_R [0\,;n]^4$, $\sigma \leftarrow_R [0\,;2^{B+2\kappa}n]$, and sends $\Delta = H(g^m h^s \bmod n, (g^{m_i} h^{s_i} \bmod n)_{0 \leq i \leq 3}, h^\sigma c_a^{m_0} \prod_{i=1}^3 c_i^{-m_i} \bmod n)$.
3. $\mathcal{V}$ picks a challenge $e' \leftarrow_R [0\,;2^\kappa]$ and sends (e', π, h_0) to $\mathcal{P}$.
4. $\mathcal{P}$ extends the challenge e' into $(e, (\lambda_i)_{0 \leq i \leq 3}) \in [0\,;2^\kappa]^5$, computes and sends $z = e \sum \lambda_i x_i + m$ and $t = e \sum \lambda_i r_i + s$, as well as $z_i = \mathfrak{r}_\pi(ex_i + m_i)$ and $T_i = h_0^{er_i+s_i} g^{q_\pi(ex_i+m_i)} \bmod n$ for $i \in \{0,1,2,3\}$, and $T = h_0^{\sigma + e(x_0 r_0 - \sum_{i=1}^3 x_i r_i)} c_a^{q_\pi(ex_0+m_0)} \prod_{i=1}^3 c_i^{-q_\pi(ex_i+m_i)} \bmod n$.
5. $\mathcal{V}$ accepts the argument if

$$\Delta = H\left(g^z h^t (\prod_{i=0}^3 c_i^{\lambda_i})^{-e} \bmod n, (g^{z_i} T_i^\pi c_i^{-e} \bmod n)_{i=0}^3, T^\pi g^e c_a^{z_0} (\prod_{i=1}^3 c_i^{-z_i}) \bmod n\right)$$

Fig. 3. Three-square range proof with knowledge-delayed order (3SRP-KDO)

square-and-multiply algorithm; we do not take into account possible optimizations from multi-exponentiation algorithms.

Table 1 sums up the communication complexity and the computational complexity of both the 3SRP and the 3SRP-KDO arguments for the execution of N parallel range proofs on the interval $[a\,;b]$, as classical batch techniques [5,6] allow to batch arguments of knowledge. Note that we omit constant terms. The communication is given in bits, while the work is given as a number of multiplications of elements of QR_n. When comparing the work of the prover, we also omit the cost of the decomposition in sum of squares, as it is the same in both protocols. Similarly, we omit the cost of the initial proof of $g = h^\alpha \bmod n$ by the verifier to the prover.

Table 1. Complexities of 3SRP and 3SRP-KDO

	3SRP	3SRP-KDO
Communication	$N(8 \log n + 18\kappa + 5B) + 3\kappa$	$N(8 \log n + 4\kappa) + 10\kappa + 2 \log n + B + \log N$
Prover's work	$1.5N(8 \log n + 12B + 26\kappa + \log a)$	$1.5(N(13 \log n + 13B + 18\kappa + \log a) + \log n + B + 6\kappa + \log N)$
Verifier's work	$1.5(N(5 \log n + 9B + 30\kappa + \log a + \log b) + \kappa)$	$1.5(N(12\kappa + \log a + \log b) + \log n + B + 10\kappa + \log N)$

Efficiency Analysis. We now provide a detailed comparison between the 3SRP and the 3SRP-KDO protocols. We set the order of the modulus n to 2048 bits and the security parameter κ to 128. As the communication of the protocols does also depend on the bound 2^B on the size of the interval, we consider various bounds in our estimation. For the sake of simplicity, we assume $B = \log b$. We evaluate the overhead of the 3SRP-KDO with respect to 3SRP, computed as $100 \times (\text{cost(3SRP-KDO)} - \text{cost(3SRP)})/\text{cost(3SRP)}$, cost being either a number of bits exchanged, or a number of exponentiations.

Small Intervals and Large Intervals. As pointed out in [11], several practical applications of range proofs, such as e-voting [25] and e-cash [12], involve quite small intervals (say, of size at most 2^{30}, and so $B \leq 30$). However, in numerous cryptographic schemes, range proofs on very large intervals are involved. Examples include anonymous credentials [13], mutual private set intersection protocols [35], secure generation of RSA keys [21,33], zero-knowledge primality tests [14], and some protocols for performing non-arithmetic operations on Paillier ciphertexts [18,28]. In such protocols, B typically range from 1024 to 8000. We note that such intervals are exactly the ones for which range proofs based on groups of hidden order are likely to be used, since for small intervals, protocols based on some u-ary decomposition of the input [11,27] will in general have better performances (essentially because they avoid the need of the Rabin-Shallit algorithm, which is computationally involved).

Comparisons. Table 2 gives a summary of our results. As already noted, the overhead of the work of the prover in 3SRP-KDO is measured by comparing the works *without considering the cost of the Rabin-Shallit algorithm*; the latter one, however, is by far the dominant cost when B is large (as it runs in expected $O(B^2 \log B \cdot M(\log B))$ time, where $M(\log B)$ is the time taken to perform a multiplication of $(\log B)$-bit integers). Therefore, for a large B, the overhead of the work of the prover in 3SRP-KDO is very small, whereas there is a huge gain for the verifier. As expected, the 3SRP-KDO protocol provides interesting performances in settings where the verifier is computationally weak (e.g. in secure

Table 2. Comparison between the 3SRP and the 3SRP-KDO

	Communication overhead	Prover's work overhead	Verifier's work overhead
$B = 30, N = 1$	+16%	+60.2%	−66%
$B = 1024, N = 1$	−3.7%	+44%	−71.7%
$B = 2048, N = 1$	−17%	+36.4%	−74.1%
$B = 30, N = 10$	−7.6%	+47.5%	−86.8%
$B = 1024, N = 10$	−26.5%	+33.2%	−87.7%
$B = 2048, N = 10$	−39.1%	+26.5%	−88%

This is for various interval sizes (2^B) and numbers N of parallel executions

Cloud computing), and/or multiples range proofs are likely to be used in parallel, and/or the intervals are large.

Acknowledgments. This work has been partially done while the second author was at ENS. This work was supported in part by the European Research Council under the European Community's Seventh Framework Programme (FP7/2007-2013 Grant Agreement no. 339563 – CryptoCloud).

References

1. Adleman, L., Manders, K.: Diophantine complexity. In: Proceedings of the 17th Annual Symposium on Foundations of Computer Science, SFCS 1976, pp. 81–88 (1976). http://dx.doi.org/10.1109/SFCS.1976.13
2. Adelsbach, A., Rohe, M., Sadeghi, A.-R.: Non-interactive watermark detection for a correlation-based watermarking scheme. In: Dittmann, J., Katzenbeisser, S., Uhl, A. (eds.) CMS 2005. LNCS, vol. 3677, pp. 129–139. Springer, Heidelberg (2005). doi:10.1007/11552055_13
3. Barić, N., Pfitzmann, B.: Collision-free accumulators and fail-stop signature schemes without trees. In: Fumy, W. (ed.) EUROCRYPT 1997. LNCS, vol. 1233, pp. 480–494. Springer, Heidelberg (1997). doi:10.1007/3-540-69053-0_33
4. Bayer, S., Groth, J.: Efficient zero-knowledge argument for correctness of a shuffle. In: Pointcheval, D., Johansson, T. (eds.) EUROCRYPT 2012. LNCS, vol. 7237, pp. 263–280. Springer, Heidelberg (2012). doi:10.1007/978-3-642-29011-4_17
5. Bellare, M., Garay, J.A., Rabin, T.: Batch verification with applications to cryptography and checking. In: Lucchesi, C.L., Moura, A.V. (eds.) LATIN 1998. LNCS, vol. 1380, pp. 170–191. Springer, Heidelberg (1998). doi:10.1007/BFb0054320
6. Bellare, M., Garay, J.A., Rabin, T.: Fast batch verification for modular exponentiation and digital signatures. In: Nyberg, K. (ed.) EUROCRYPT 1998. LNCS, vol. 1403, pp. 236–250. Springer, Heidelberg (1998). doi:10.1007/BFb0054130
7. Böhl, F., Hofheinz, D., Jager, T., Koch, J., Seo, J.H., Striecks, C.: Practical signatures from standard assumptions. In: Johansson, T., Nguyen, P.Q. (eds.) EUROCRYPT 2013. LNCS, vol. 7881, pp. 461–485. Springer, Heidelberg (2013). doi:10.1007/978-3-642-38348-9_28
8. Boudot, F.: Efficient proofs that a committed number lies in an interval. In: Preneel, B. (ed.) EUROCRYPT 2000. LNCS, vol. 1807, pp. 431–444. Springer, Heidelberg (2000). doi:10.1007/3-540-45539-6_31
9. Bresson, E., Stern, J.: Proofs of knowledge for non-monotone discrete-log formulae and applications. In: Chan, A.H., Gligor, V. (eds.) ISC 2002. LNCS, vol. 2433, pp. 272–288. Springer, Heidelberg (2002). doi:10.1007/3-540-45811-5_21
10. Brickell, E.F., Chaum, D., Damgård, I., van de Graaf, J.: Gradual and verifiable release of a secret. In: Pomerance, C. (ed.) CRYPTO 1987. LNCS, vol. 293, pp. 156–166. Springer, Heidelberg (1988)
11. Camenisch, J., Chaabouni, R., shelat, A.: Efficient protocols for set membership and range proofs. In: Pieprzyk, J. (ed.) ASIACRYPT 2008. LNCS, vol. 5350, pp. 234–252. Springer, Heidelberg (2008). doi:10.1007/978-3-540-89255-7_15
12. Camenisch, J., Hohenberger, S., Lysyanskaya, A.: Compact e-cash. In: Cramer, R. (ed.) EUROCRYPT 2005. LNCS, vol. 3494, pp. 302–321. Springer, Heidelberg (2005). doi:10.1007/11426639_18

13. Camenisch, J., Lysyanskaya, A.: An efficient system for non-transferable anonymous credentials with optional anonymity revocation. In: Pfitzmann, B. (ed.) EUROCRYPT 2001. LNCS, vol. 2045, pp. 93–118. Springer, Heidelberg (2001). doi:10.1007/3-540-44987-6_7

14. Camenisch, J., Michels, M.: Proving in zero-knowledge that a number is the product of two safe primes. In: Stern, J. (ed.) EUROCRYPT 1999. LNCS, vol. 1592, pp. 107–122. Springer, Heidelberg (1999). doi:10.1007/3-540-48910-X_8

15. Camenisch, J., Michels, M.: Separability and efficiency for generic group signature schemes. In: Wiener, M. (ed.) CRYPTO 1999. LNCS, vol. 1666, pp. 413–430. Springer, Heidelberg (1999). doi:10.1007/3-540-48405-1_27

16. Canard, S., Coisel, I., Traoré, J.: Complex zero-knowledge proofs of knowledge are easy to use. In: Susilo, W., Liu, J.K., Mu, Y. (eds.) ProvSec 2007. LNCS, vol. 4784, pp. 122–137. Springer, Heidelberg (2007). doi:10.1007/978-3-540-75670-5_8

17. Chan, A.H., Frankel, Y., Tsiounis, Y.: Easy come - easy go divisible cash. In: Nyberg, K. (ed.) EUROCRYPT 1998. LNCS, vol. 1403, pp. 561–575. Springer, Heidelberg (1998)

18. Couteau, G., Peters, T., Pointcheval, D.: Encryption switching protocols. In: Robshaw, M., Katz, J. (eds.) CRYPTO 2016. LNCS, vol. 9814, pp. 308–338. Springer, Heidelberg (2016). doi:10.1007/978-3-662-53018-4_12

19. Couteau, G., Peters, T., Pointcheval, D.: Removing the strong RSA assumption from arguments over the integers. Cryptology ePrint Archive, Report 2016/128 (2016). http://eprint.iacr.org/2016/128

20. Damgård, I., Fujisaki, E.: A statistically-hiding integer commitment scheme based on groups with hidden order. In: Zheng, Y. (ed.) ASIACRYPT 2002. LNCS, vol. 2501, pp. 125–142. Springer, Heidelberg (2002). doi:10.1007/3-540-36178-2_8

21. Damgård, I., Mikkelsen, G.L.: Efficient, robust and constant-round distributed rsa key generation. In: Micciancio, D. (ed.) TCC 2010. LNCS, vol. 5978, pp. 183–200. Springer, Heidelberg (2010). doi:10.1007/978-3-642-11799-2_12

22. Davis, M., Putnam, H., Robinson, J.: The decision problem for exponential diophantine equations. Ann. Math. **72**, 425–436 (1961)

23. Fujisaki, E., Okamoto, T.: Statistical zero knowledge protocols to prove modular polynomial relations. In: Kaliski, B.S. (ed.) CRYPTO 1997. LNCS, vol. 1294, pp. 16–30. Springer, Heidelberg (1997). doi:10.1007/BFb0052225

24. Gennaro, R.: Multi-trapdoor commitments and their applications to proofs of knowledge secure under concurrent man-in-the-middle attacks. In: Franklin, M. (ed.) CRYPTO 2004. LNCS, vol. 3152, pp. 220–236. Springer, Heidelberg (2004). doi:10.1007/978-3-540-28628-8_14

25. Groth, J.: Non-interactive zero-knowledge arguments for voting. In: Ioannidis, J., Keromytis, A., Yung, M. (eds.) ACNS 2005. LNCS, vol. 3531, pp. 467–482. Springer, Heidelberg (2005). doi:10.1007/11496137_32

26. Groth, J.: Linear algebra with sub-linear zero-knowledge arguments. In: Halevi, S. (ed.) CRYPTO 2009. LNCS, vol. 5677, pp. 192–208. Springer, Heidelberg (2009). doi:10.1007/978-3-642-03356-8_12

27. Groth, J.: Efficient zero-knowledge arguments from two-tiered homomorphic commitments. In: Lee, D.H., Wang, X. (eds.) ASIACRYPT 2011. LNCS, vol. 7073, pp. 431–448. Springer, Heidelberg (2011). doi:10.1007/978-3-642-25385-0_23

28. Guajardo, J., Mennink, B., Schoenmakers, B.: Modulo reduction for paillier encryptions and application to secure statistical analysis. In: Sion, R. (ed.) FC 2010. LNCS, vol. 6052, pp. 375–382. Springer, Heidelberg (2010). doi:10.1007/978-3-642-14577-3_32

29. Hofheinz, D., Jager, T., Kiltz, E.: Short signatures from weaker assumptions. In: Lee, D.H., Wang, X. (eds.) ASIACRYPT 2011. LNCS, vol. 7073, pp. 647–666. Springer, Heidelberg (2011). doi:10.1007/978-3-642-25385-0_35

30. Hohenberger, S., Waters, B.: Short and stateless signatures from the RSA assumption. In: Halevi, S. (ed.) CRYPTO 2009. LNCS, vol. 5677, pp. 654–670. Springer, Heidelberg (2009). doi:10.1007/978-3-642-03356-8_38

31. Jarecki, S., Kiayias, A., Krawczyk, H.: Round-optimal password-protected secret sharing and T-PAKE in the password-only model. In: Sarkar, P., Iwata, T. (eds.) ASIACRYPT 2014. LNCS, vol. 8874, pp. 233–253. Springer, Heidelberg (2014). doi:10.1007/978-3-662-45608-8_13

32. Jarecki, S., Shmatikov, V.: Efficient two-party secure computation on committed inputs. In: Naor, M. (ed.) EUROCRYPT 2007. LNCS, vol. 4515, pp. 97–114. Springer, Heidelberg (2007). doi:10.1007/978-3-540-72540-4_6

33. Juels, A., Guajardo, J.: RSA key generation with verifiable randomness. In: Naccache, D., Paillier, P. (eds.) PKC 2002. LNCS, vol. 2274, pp. 357–374. Springer, Heidelberg (2002). doi:10.1007/3-540-45664-3_26

34. Kiayias, A., Tsiounis, Y., Yung, M.: Traceable signatures. In: Cachin, C., Camenisch, J.L. (eds.) EUROCRYPT 2004. LNCS, vol. 3027, pp. 571–589. Springer, Heidelberg (2004). doi:10.1007/978-3-540-24676-3_34

35. Kim, M., Lee, H.T., Cheon, J.H.: Mutual private set intersection with linear complexity. In: Jung, S., Yung, M. (eds.) WISA 2011. LNCS, vol. 7115, pp. 219–231. Springer, Heidelberg (2012). doi:10.1007/978-3-642-27890-7_18

36. Lipmaa, H.: On diophantine complexity and statistical zero-knowledge arguments. In: Laih, C.-S. (ed.) ASIACRYPT 2003. LNCS, vol. 2894, pp. 398–415. Springer, Heidelberg (2003). doi:10.1007/978-3-540-40061-5_26

37. Lipmaa, H., Asokan, N., Niemi, V.: Secure vickrey auctions without threshold trust. Cryptology ePrint Archive, Report 2001/095 (2001). http://eprint.iacr.org/2001/095

38. Pedersen, T.P.: Non-interactive and information-theoretic secure verifiable secret sharing. In: Feigenbaum, J. (ed.) CRYPTO 1991. LNCS, vol. 576, pp. 129–140. Springer, Heidelberg (1992). doi:10.1007/3-540-46766-1_9

39. Pointcheval, D., Stern, J.: Security proofs for signature schemes. In: Maurer, U. (ed.) EUROCRYPT 1996. LNCS, vol. 1070, pp. 387–398. Springer, Heidelberg (1996). doi:10.1007/3-540-68339-9_33

40. Pointcheval, D., Stern, J.: Security arguments for digital signatures and blind signatures. J. Cryptology **13**(3), 361–396 (2000)

41. Pollett, C.: On the bounded version of hilbert's tenth problem. Arch. Math. Log. **42**(5), 469–488 (2003). http://dx.doi.org/10.1007/s00153-002-0162-y

42. Rabin, M.O., Shallit, J.O.: Randomized algorithms in number theory. Commun. Pure Appl. Math. **39**(S1), S239–S256 (1986). http://dx.doi.org/10.1002/cpa.3160390713

43. Rivest, R.L., Shamir, A., Adleman, L.M.: A method for obtaining digital signature and public-key cryptosystems. Commun. Assoc. Comput. Mach. **21**(2), 120–126 (1978)

Magic Adversaries Versus Individual Reduction: Science Wins Either Way

Yi Deng[1,2(✉)]

[1] SKLOIS, Institute of Information Engineering, CAS,
Beijing, People's Republic of China
deng@iie.ac.cn
[2] State Key Laboratory of Cryptology, P.O. Box 5159, Beijing 100878, China

Abstract. We prove that, assuming there exists an injective one-way function f, *at least* one of the following statements is true:

- (Infinitely-often) Non-uniform public-key encryption and key agreement exist;
- The Feige-Shamir protocol instantiated with f is distributional concurrent zero knowledge for a large class of distributions over any OR NP-relations with small distinguishability gap.

The questions of whether we can achieve these goals are known to be subject to black-box limitations. Our win-win result also establishes an unexpected connection between the complexity of public-key encryption and the round-complexity of concurrent zero knowledge.

As the main technical contribution, we introduce a dissection procedure for concurrent adversaries, which enables us to transform a magic concurrent adversary that breaks the distributional concurrent zero knowledge of the Feige-Shamir protocol into non-black-box constructions of (infinitely-often) public-key encryption and key agreement.

This dissection of complex algorithms gives insight into the fundamental gap between the known *universal* security reductions/simulations, in which a single reduction algorithm or simulator works for *all* adversaries, and the natural security definitions (that are sufficient for almost all cryptographic primitives/protocols), which switch the order of qualifiers and only require that for every adversary there *exists* an *individual* reduction or simulator.

1 Introduction

The seminal work of Impagliazzo and Rudich [IR89] provides a methodology for studying the limitations of black-box reductions. Following this methodology, plenty of black-box barriers, towards building cryptographic systems on simpler primitives/assumptions and achieving more efficient constructions, have

---

A full version of this work appeared in [Den16]. "Science wins either way" is credited to Silvio Micali. This work was supported by the National Natural Science Foundation of China (Grant No. 61379141), and the Open Project Program of the State Key Laboratory of Cryptology.

J.-S. Coron and J.B. Nielsen (Eds.): EUROCRYPT 2017, Part II, LNCS 10211, pp. 351–377, 2017.
DOI: 10.1007/978-3-319-56614-6_12

been found in the last three decades. These findings have long challenged us to develop new reduction methods and get around the limitations of black-box reduction, however, the progress towards this goal is quite slow, and for most of the known black-box barriers, it is still unclear whether they even hold for arbitrary reductions.

We revisit two seemingly unrelated fundamental problems, for both of which the black-box impossibility results are well known.

The first problem is to identify the weakest complexity assumptions required for public-key encryption. Ever since the invention of public key cryptography by Diffie and Hellman [DH76], the complexity of public-key cryptography, i.e., lowering the underlying complexity assumptions for cryptographic primitives/protocols, is one of the most basic problems. In the past four decades, for some primitives, including pseudorandom generators, signatures and statistically-hiding commitments, we witnessed huge success on this line of research and can now base them on the existence of one-way functions [Rom90, HILL99, HR07], which is the minimum assumption in the sense that, as showed by [IL89], almost all cryptographic primitives/protocols imply the existence of one-way functions.

But for public-key encryption and key agreement–the concepts that were conceived in the original paper of Diffie and Hellman, we did not make that successful progress yet. Impagliazzo and Rudich proved in their seminal work [IR89] that there is no black-box reduction of one-way permutations to key agreement, and since public-key encryption implies key agreement, their result also separates one-way permutations from public-key encryption with respect to black-box reduction.

In [Imp95] Impagliazzo describes five possible worlds of complexity theory. The top two worlds among them are Cryptomania, where public-key cryptography exists, and Minicrypt where there are one-way functions but no public-key cryptography. Though the above black-box separation provides some strong negative evidences, they do not rule out the possibility of constructing public-key encryption from one-way functions, i.e., do not prove that we live in Minicrypt.

The other fundamental problem we consider is that of the round-complexity of concurrent zero knowledge. The notion of concurrent zero-knowledge, put forward by Dwork, Naor and Sahai [DNS98], extends the standard-alone zero-knowledge security notion [GMR89] to the case where multiple concurrent executions of the same protocol take place and an adversarial verifier may corrupt multiple verifiers and control the scheduling of the messages.

As observed in [DNS98], the traditional black-box simulator does not work for the classic constant-round protocols (including the Feige-Shamir type protocol [FS89] and the Goldreich-Kahan type protocol [GK96]) in the concurrent setting. Indeed, Canetti et al. [CKPR01] proved that concurrent zero-knowledge with black-box simulation requires a logarithmic number of rounds for languages outside BPP. Prabhakaran et al. [PRS02] later refined the analysis of the Kilian and Petrank's [KP01] recursive simulator and gave an (almost) logarithmic round concurrent zero knowledge protocol for NP.

In his breakthrough work, Barak [Bar01] introduced a non-black-box simulation technique based on PCP mechanism and constructed a constant-round public-coin *bounded-concurrent* zero knowledge protocol for NP, which breaks several known lower bounds for black-box zero knowledge. There has been a vast body of work (see Sect. 1.4) since then on developing new non-black-box techniques and reducing the round-complexity of zero knowledge protocol in the concurrent setting. However, The problem of whether we can achieve constant-round concurrent zero knowledge based on standard assumptions is still left open.

Note also that the known constructions that beat the lower bound on the black-box round-complexity are rather complicated and therefore impractical. Given the current state of the art, a more ambitious question is whether we can prove the concurrent zero knowledge property of the classic 4-round protocols (such as the Feige-Shamir protocol), although it is known to be impossible to give such a proof for these simple and elegant constructions via black-box simulations.

1.1 Universal Simulator "$\exists S \forall A$" Versus Individual Simulator "$\forall a \exists S$"

We observe that almost all known reduction and simulation techniques are *universal* in the sense that, in the security proof of a protocol/premitive, the reduction R (or simulator S) works for all possible efficient adversaries and turn the power of a given adversary A into the power of breaking the underlying assumptions (i.e., "$\exists R$ or $S \ \forall A$"). However, for most natural security definitions, it is only required that for every adversary A there *exists* an *individual* reduction R (or a simulator S) that works for A (i.e., "$\forall A \exists R$ or S").

This motivates us to step back and look at the concurrent security of the simplest Feige-Shamir protocol. We will show that there is an *individual* simulator for the specific adversarial verifier (and thus it is not a concrete "attacker") constructed by Canetti et al. [CKPR01], though it was shown that for such a adversary the known black-box simulator fails. Sure, showing the existence of a simulator for a specific verifier does not mean that the Feige-Shamir protocol is concurrent zero knowledge, but this example does reveal a gap between the universal simulation "$\exists S \forall A$" and the individual simulation "$\forall A \exists S$".

The Feige-Shamir protocol for proving $x \in L$ proceeds as follows. In the first phase, the verifier picks two random strings α_1 and α_2, computes two images, $\beta_1 = f(\alpha_1)$, $\beta_2 = f(\alpha_2)$, of a one-way function f, and then proves to the prover via a constant-round witness indistinguishability protocol that he knows either α_1 or α_2; in the second phase, the prover proves that either $x \in L$ or he knows one of α_1, α_2. The adversary V^* constructed in [CKPR01] adopts a delicate scheduling strategy, and when computing a verifier message, it applies a hash function h with high independence to the history hist sofar and generates the randomness $r = h(\text{hist})$ for computing the current message. In our case, the randomness for the first verifier step of a session includes the two pre-images α_1 and α_2.

Canetti et al. showed that it is impossible for an efficient simulator to simulate V^*'s view when treating it as a black-box[1]. However, as mentioned before, the natural concurrent zero knowledge condition does not require a universal (or black-box) simulator that works for all adversarial verifiers, but just requires that for every specific V^* there *exists* an *individual* simulator.

Note that the individual simulator may depends on the specific verifier, and more importantly, since we are only required to show the *mere existence* of such a simulator, we can assume that the individual simulator knows (or equivalently, takes as input) the verifier's *functionality, randomness*, etc.

Indeed, for the adversary V^* of [CKPR01], there *exists, albeit* probably not efficiently constructible from a given (possibly obfuscated) code of V^*, a simple simulator for the above specific V^*: Note that there *exists* an adversary V' that acts exactly in the same way as V^* except that at each step V' outputs $r = h(\text{hist})$ together with the current message, and thus a trivial simulator $\text{Sim}(V')$, incorporating V' and using the fake witness (one of α_1 and $\alpha_2{}^2$) output by V' at the first verifier step of each session, can easily generate a transcript that is indistinguishable from the real interaction between V^* and honest provers.

1.2   Our Work

We prove an unexpected connection between the complexity of public-key encryption and the round-complexity of concurrent zero knowledge. Specifically, we show how to transform an attacker that can break a weak version of distributional concurrent zero knowledge of the Feige-Shamir protocol instantiated with injective one-way functions into (infinitely-often) constructions of public-key encryption and key agreement. This means at least one of the two problems (with respect to infinitely-often version and distributional version respectively) mentioned above has a *positive* answer.

A formal statement of our result. Let L and R_L be an arbitrary NP language and its associated NP relation respectively. The OR language $L \vee L^3$ and the corresponding relation $R_{L_{OR}}$ are defined in a natural way.

Given an arbitrary efficiently samplable distribution ensemble $D = \{D_n\}_{n \in N}$ over R_L (each D_n is over $R_L^n := \{(x,w) : (x,w) \in R_L \wedge |x| = n\}$), and an arbitrary efficiently samplable distribution Z_n over $\{0,1\}^{*4}$, we define the joint distribution $\{(X_n, W_n, Z_n)\}_{n \in N}$ over $R_{L_{OR}} \times \{0,1\}^*$ in the following way: Sample $(x_1, w_1) \leftarrow D_n, (x_2, w_2) \leftarrow D_n, z \leftarrow Z_n, b \leftarrow \{1,2\}$, and output $((x_1, x_2), w_b)$.

---

[1] I.e., the simulator is given only oracle access to V^*, and does not have knowledge about its code, running time, etc.

[2] Note that α_1 and α_2 are part of the randomness r used in the first verifier message of a session.

[3] For simplicity, we consider only the OR composition of the same NP language L, but our result holds with respect to the OR composition of any two NP languages.

[4] The element z from Z_n will be given as auxiliary input to the verifier of Feige-Shamir protocol.

Theorem 1. *Assume that there exists an injective one-way function f. Then, at least one of the following statements is true:*

- *(Infinitely-often) Non-uniform public-key encryption and key agreement exist;*
- *For every inverse polynomial ϵ, the Feige-Shamir protocol instantiated with f is distributional concurrent zero knowledge on $\{(X_n, W_n, Z_n)\}_{n \in N}$ defined as above with distinguishability gap bounded by ϵ.*

In an infinitely-often version of a primitive, the correctness and security of a construction are required to hold only for infinitely many security parameter n. The notion of ϵ-*distributional* concurrent zero knowledge (first defined in [CLP15b]) differs from the traditional zero knowledge in that its zero knowledge property holds on average (i.e., holds for distributions over the statements), and that the indistinguishability gap for any efficient distinguisher is bounded by an arbitrary inverse polynomial (instead of a negligibly function).

Very roughly, Theorem 1 says the Feige-Shamir protocol is concurrent secure in the Minicrypt: In the world where there are injective one-way functions but no public-key encryption, the Feige-Shamir protocol satisfies certain version of concurrent zero knowledge.

Remark 1. We note that the black-box lower bounds [IR89,CKPR01] also hold for the infinitely-often version of public-key encryption and the ϵ-*distributional* concurrent zero knowledge[5]. We stress that our public-key encryption (and the key agreement) is based on the injective one-way function f and the specific attacker against the Feige-Shamir protocol, and is non-uniform and non-black-box in nature: The key generation, encryption and decryption algorithms in our public-key encryption scheme are all non-uniform, and make non-black-box usage of the underlying function f and the attacker.

Dissecting a complex adversary: Revealing the Creation of a Trapdoor. The basic proof strategy of Theorem 1 is to transform a magic verifier against the Feige-Shamir protocol into constructions for (infinitely-often) public-key encryption and key agreement. This proof idea is somewhat similar in spirit to the one appeared in [DNRS03] but still quite unusual in cryptography. In our setting, formalizing such a proof idea is very complicated and requires substantially new techniques.

To deal with the complex concurrent adversary, we introduce a dissection procedure to pinpoint where a supposed successful adversary magically endow a set of images of the injective one-way function f with a trapdoor, which is the key step towards our construction of public-key encryption via the Goldreich-Levin Theorem. On the very high level, if an adversarial verifier V^* that can

---

[5] Our result holds with respect to distributions that are not always over YES instances. By applying the lower-bound proof strategy of [CKPR01], we conclude that the Feige-Shamir protocol cannot be ϵ-distributional concurrent *black-box* zero knowledge for any non-trivial distribution over hard problems, see the full version of this work for more details.

break concurrent zero knowledge of the Feige-Shamir protocol, then in the real interaction there must exist a step i (verifier steps are ordered according to their appearance in the concurrent setting) such that:

- With high probability, V^* will output a pair of images β_1 and β_2, i.e., the first verifier message of some session j at this step i, and at a later time it will reach its second step of session j, i.e., completes its 3-round proof that it knows one pre-image of β_1 and β_2 under f.
- But for any efficient algorithm T, even taking the code of V^* and the history prefix up to its i-th step, the probability that T inverts any one of these two images β_1 and β_2 is bounded away from 1.

The intuition behind this observation is as follows. If the above two items does not hold simultaneously, then at each verifier step, either V^* does not output a pair of images of a session, or it outputs a pair of images of session j but will never reach its second message of session j, or there is an efficient algorithm that can find one of the corresponding pre-images. In each case we will have a simple simulator that can simulate the view of the V^*, which leads to a contradiction.

Thus, for a given successful adversary V^* the above two items must hold simultaneously. This means V^* magically endow the images β_1 and β_2 output at its step i with a trapdoor (i.e., the witness w to the common input x): With the trapdoor w, one can play the role of honest prover until V^* completes his 3-round proof, then using standard rewinding technique to obtain one of the pre-images; while, without the knowledge of w, no efficient algorithm can invert any one of β_1 and β_2 with overwhelming probability. This is the key observation that enables us to construct public key encryption and key agreement from the injective one-way f.

The major challenge in the actual dissection is to show the existence of *infinitely many* security parameter n for each of which the above conditions hold (as required by infinitely-often public key encryption and key agreement). To cope with this difficulty, we develop a set of techniques that convert concrete security into asymptotic security, which may be of independent interest.

An overview of the proof. We divide the proof into four steps, which will be presented in Sects. 3, 4, 5 and 6 respectively. Roughly, the proof proceeds as follows.

Step I: We introduce a dissection procedure and prove that there must be infinitely many n, for each of which there exists a step i of V^*, such that the above two items hold simultaneously. This illustrates the power of V^* that magically endows the images of f output by V^* at its step i with a sort of trapdoor.

Step II: Note that V^* outputs a pair of images of f at its step i. To avoid that the sender and the receiver (both with a witness to x) may recover different pre-images from V^*, we construct a pair of non-interactive algorithms C and E from the code of V^* such that for each (n, i) obtained in the above step:

- C (with knowledge of a witness w to x) outputs a *single* image β of f with high probability;
- E (with knowledge of a witness w to x) will extract the pre-image of β output by C;
- No efficient algorithm can compute the pre-image of β with probability close to 1.

Step III: Using standard techniques, we amplify the gap between the success probability of E and the success probability of any efficient inverting algorithm without knowing a witness to x, and obtain two algorithms M and Find, where M takes a sequence of (x, w) as input and outputs a sequence of images β of f, and Find takes the same sequence of (x, w) and outputs all pre-images corresponding to the sequence of images β, both with probability negligibly close to 1; further, there is no efficient algorithm that can invert all the images output by M simultaneously with non-negligible probability.

Step IV: Note that the Feige-Shamir protocol is concurrent witness indistinguishable, and thus the above holds when M and Find use different witnesses. Starting with a magic adversary V^* that breaks the distributional concurrent zero knowledge of the Feige-Shamir protocol for distribution over OR NP-statements of the form $(x_1 \vee x_2)$, we construct the public-key encryption scheme (and key-exchange scheme) in a natural way: The receiver generates a sequence of (x_1, w_1) as the public/secret key pair; to encrypt a bit, the sender generates a sequence of (x_2, w_2) and runs M on input the sequence of OR statements $(x_1 \vee x_2)$ and their corresponding witnesses w_2 to generate a set of images of f, computes the hard-core of the corresponding pre-images and XOR the plaintext bit with the hardcore; to decrypt, the receiver runs Find on input the ciphertext and the sequence of witnesses w_1 to obtain the corresponding pre-images, and then computes the hardcore and gets the plaintext.

Remark 2. We use the code of V^* in our final construction of public-key encryption. However, what we actually need to construct public-key encryption is the *functionality* of V^*, that is, we can replace the code of V^* with *any* code[6] of the *same* functionality in the intermediate algorithms in each of above steps along the way.

1.3   A Wide Perspective on Reductions

As mentioned, the mostly common used security proof techniques–black-box techniques (see [RTV04, BBF13] for refined treatments) and the known non-black-box techniques [Bar01, DGS09, BP15]–are universal, where a single universal reduction algorithm works for all possible adversaries. Here in this section we abuse the term *reduction* and view *simulation* as a type of reduction. Note that the description of an adversary that the reduction algorithm has access to probably is an obfuscated code. This causes a trouble in cases where the *functionality*

---

[6] As long as it is of polynomial size.

of the adversary is crucial for the reduction to go through (as showed in the above example of simulation for the adversary in [CKPR01], and see also [DGL+16]), since we cannot expect the efficient reduction algorithm to figure out the functionality from a given obfuscated code of an arbitrary adversary.

However, in almost all cases, in a security proof the reduction can be *arbitrary*. This means the reduction is allowed to depend not only on the code of the adversary, but also on any "nice" properties of the adversary (if exist), such as functionality, good random tapes, etc. Furthermore, to show the mere existence of such an arbitrary reduction, we do not need to care about whether such properties can be efficiently extracted from the code of the adversary, but just assume that the reduction takes these properties as input. We refer to an arbitrary reduction as *individual* reduction, which is also called non-constructive reduction or non-uniform reduction in some previous work [BU08, CLMP13]. We stress that it is not always possible to turn an individual reduction into a universal reduction with a non-uniform advice because, in many cases, even if we can prove all possible adversaries share a certain property, this property may not have a short description. (This will be clear in the following example.)

Recall that, to complete a security proof, we have to show for *every* adversary there is an individual reduction. This would be impossible unless we can prove that all possible adversaries have certain properties *in common*. Indeed, we observe that a few exceptional individual reductions in complexity (e.g., [Adl78]) and hardness amplification (e.g., [GNW95, CHS05, HS11]) literature are based on a property–the existence of "good" random tapes–shared by all possible adversaries. Let's take the reduction for BPP $\subseteq$ P/poly [Adl78] as an example. The first step of the proof of [Adl78] is to show a common property that every machine deciding a language $L \in$ BPP must have at least one good random tape on which this machine will make correct decisions on all instances of a given size. Using the mere existence of a good random tape, we can then simply hardwire this good random tape into the circuit family that decide the language L deterministically. This circuit family can be thought of as a reduction, which varies depending on the specific BPP machine since different machines may have different good random taps.

Besides the structure (success/failure) of the random tapes, there seems to be a more important structure of the adversaries, i.e., the structure of the adversary's computation, that would empower the individual reduction greatly. In cryptography, we actually already exploited structures of this type, such as the knowledge of exponent assumption and extractable one-way functions [Dam91, BCPR14], but most of them are viewed as just non-standard assumption. Our work seems to raise some hope that we may be able to prove highly non-trivial structures of the adversary's computation in some settings under standard assumptions in the future.

1.4   Related Work

There have been numerous efficient constructions ([RSA78, Rab79, GM82, CS99, Reg09, HKS03], to name a few) for public-key encryption with various security notions based on specific assumptions with various algebraic structures, and some less efficient constructions [NY90, BHSV98, Sah99, Lin03a] based on more abstract assumptions–enhanced trapdoor permutations or trapdoor functions with polynomial pre-image size. Since public-key encryption implies key agreement (secure against eavesdropping adversaries), the same assumptions are sufficient for the latter. On the negative side, the recent work of [DS16] strengthens the black-box separation of public-key encryption and general one-way functions in [IR89] by allowing the reduction to take the code of the underlying primitive as input.

In the line of research on concurrent zero knowledge, Goyal [Goy13] extended Barak's idea to achieve fully concurrent zero knowledge in polynomial rounds. In the globe hash model, Canetti et al. [CLP13a] showed that public-coin concurrent zero knowledge can be obtained with logarithmic round-complexity. Recently, Chung et al. [CLP15a] (based on [CLP13b]) presented the first constant-round concurrent zero knowledge protocol based on indistinguishability obfuscation with super-polynomial security. Assuming the existence public-coin input-differing obfuscation, Pandey et al. [PPS15] presented a 4-round concurrent zero knowledge protocol. Over the last two decades, concurrent zero knowledge protocols have been used as a key building block in the construction of generally composable cryptographic protocols [CLOS02, PR03, Lin03b, PR05, Pas04, Lin08, GGJ13, GGJS12, GGS15, GLP+15].

2 Preliminaries

In this section we mainly present the definition of ϵ-distributional concurrent zero knowledge and some related new notions and definitions that we will use, and refer readers to [Gol01, KL07] for some other standard notions and definitions.

If D is a distribution (or random variable), we denote by $x \leftarrow D$ the process of sampling x according to D, and by $\{x_i\}_{i=1}^{k} \leftarrow D^{\otimes k}$ the process of sampling k times x from D independently. Similarly, for a function $f : \{0,1\}^n \rightarrow \{0,1\}^{\ell(n)}$, $f^{\otimes k}$ denotes the function that maps $(x_1, x_2, ..., x_k)$ to $(f(x_1), f(x_2), ..., f(x_k))$.

We abbreviate probabilistic polynomial-time with PPT. Throughout this paper, all PPT algorithms/Turing machines are allowed to be non-uniform, and we use non-uniform PPT algorithms/Turing machines interchangeably with circuit families of polynomial size. In our default setting, the circuit families are also probabilistic.

Given a two-party protocol $\Pi = (P_1, P_2)$, for $i \in \{1, 2\}$, we denote by $\mathsf{Trans}_{P_i}(P_1(x), P_2(y))$ the transcript of an execution of Π (including the input to P_i) when P_1's input is x and P_2's input is y. For a joint distribution (X, Y) over the two parties' inputs, $\mathsf{Trans}_{P_i}(P_1(X), P_2(Y))$ naturally defines the distribution over all possible view of P_i.

Throughout the paper, we let n be the security parameter and denote by $negl(n)$ a negligible function. We write $\{X_n\}_{n\in\mathbb{N}} \overset{c}{\approx} \{Y_n\}_{n\in\mathbb{N}}$ to indicate that the two distribution ensembles $\{X_n\}_{n\in\mathbb{N}}$ and $\{Y_n\}_{n\in\mathbb{N}}$ are computationally distinguishable.

A zero knowledge argument system is an interactive argument for which the view of the (even malicious) verifier in an interaction can be efficiently reconstructed. In this paper, we consider *distributional* zero knowledge, defined by Goldreich [Gol93], for which the indistinguishability between the real interaction and the simulation is only required to hold for any distribution over the inputs to each party, rather than to hold for every individual inputs. We follow the definition of [CLP15b], which departs from the one of [Gol93] in that it only requires that for each distribution over the inputs there exists an efficient simulator[7], and consider the case (following [DNRS03, CLP15b]) where the indistinguishability gap between the simulation and the real interaction is less than any inverse polynomial ϵ (instead of a negligible function). As we will show, the size of encryption algorithm of our encryption scheme is polynomial in the value $\frac{1}{\epsilon}$, which needs to be upper-bounded by a fixed (but arbitrary) polynomial.

Steps of the Concurrent Verifier and Steps of a Session. We also allow the adversary V^* to launch a *concurrent* attack [DNS98, PRS02] in which it interacts with a polynomial number of independent provers over an asynchronous network, and fully controls over the scheduling of all messages in these interactions.

We refer to the action of sending a message by V^* as a step (of V^*). In a real concurrent interaction, we order the steps of V^* according to their appearance. Note that in the concurrent setting, sessions of the Feige-Shamir protocol are executed in interleaving way, and thus, "the second verifier step of a session" refers to the second verifier step that appears in this specific session, not to the second step of V^* in the real concurrent interaction.

Definition 1 (ϵ-Distributional Concurrent zero knowledge). *We say that an interactive argument (P, V) for language L is ϵ-distributional concurrent zero knowledge if for every concurrent adversary V^*, and every distribution ensemble $\{(X_n, W_n, Z_n)\}_{n \in \mathbb{N}}$ over $R_L^n \times \{0,1\}^*$, there exists a non-uniform PPT Sim such that for all non-uniform PPT D and sufficient large n it holds that*

$$\Pr[D(\mathit{Trans}_{V^*}(P(X_n, W_n), V^*(Z_n)), Z_n) = 1]$$
$$- \Pr[D(\mathit{Sim}(V^*, X_n, Z_n), Z_n) = 1] < \epsilon(n),$$

where both distributions are over (X_n, W_n, Z_n) and the random tapes of P and V^*.

The Feige-Shamir ZK Argument for NP. We here describe the Feige-Shamir constant-round[8] zero knowledge argument for NP based on an injective one-way function $f : \{0,1\}^n \to \{0,1\}^{\ell(n)}$.

PROTOCOL FEIGE-SHAMIR
Common input: $x \in L$.

---

[7] Instead, the definition of [Gol93] requires an efficient simulator for all distributions over the inputs.

[8] By merging the first and the second prover messages, one can obtain a 4-round Feige-Shamir protocol.

The prover P's input: w such that $(x, w) \in R_L$.

The verifier V's (auxiliary) input:z

First phase:

Execute the n-parallel-repetition of the 3-round Blum's protocol in which V plays the role of the prover:

$V \longrightarrow P$: Choose $\alpha_1, \alpha_2 \leftarrow \{0, 1\}^n$ independently and at random, compute $\beta_1 = f(\alpha_1)$, $\beta_2 = f(\alpha_2)$, and compute the first prover message a of the 3-round n-parallel-repetition of the Blum's protocol in which V proves to P that he knows one of α_1, α_2.
Send β_1, β_2 and a.

$P \longrightarrow V$: Send a random challenge $e \leftarrow \{0, 1\}^n$.

$V \longrightarrow P$: Send t.

Second phase:

P and V execute the n-parallel-repetition of the 3-round Blum's protocol in which P proves to V that either $x \in L$ or he knows one of α_1, α_2.

3 The Dissection of a Concurrent Verifier

In this section we develop a technique to dissect concurrent verifiers that reveals where a supposed concrete attacker against the Feige-Shamir protocol magically endows some images of an injective one-way function with a trapdoor. This is the key step towards constructing public-key encryption (and key agreement) from an injective one-way function.

As mentioned in the introduction, we show that a magic adversary V^* will endow a set of images of f with a trapdoor in the following sense: there are infinitely many n, for each of which there exists a step index i_n, such that the images (β_1, β_2) output by V^* at its step i_n can *only* be inverted by PPT algorithms with the trapdoor knowledge of a witness to the common input x with overwhelming probability.

3.1 The Main Lemma

We need the following notations to give a formal statement of our main lemma:

- Trans^{i_n} and $h \leftarrow \mathsf{Trans}^{i_n}$: The former denotes the distribution of the history prefix in the view of V^* up to its i_n-th step in the real concurrent interaction $\mathsf{Trans}_{V^*}(P(X_n, W_n), V^*(Z_n))$; the latter denotes the event of drawing a history prefix h from Trans^{i_n}, i.e., the event of generating h in the real concurrent interaction between honest prover(s) and V^*, where h consists of the statement x, the auxiliary input z to V^* and the interaction history prefix upto the step i_n of the verifier.
- $V^*|_h \rightsquigarrow (j, 2)$ denotes the event that, conditioned on the given history prefix h, V^* reaches the second verifier step of session j in the real concurrent interaction, i.e., V^* completes its proof of knowledge of one pre-image in session j.

– PartR$_h$ consists of the randomness used by V^* and the *partial* randomness used by honest provers in those *incomplete* sessions in h (i.e., sessions in which the last prover message does not appear in h) in a real concurrent interaction. Observe that in a session of the Feige-Shamir protocol, the honest prover uses the knowledge of corresponding witness w *only* in its last step, and the transcript of a session before the prover last step is independent of w. Thus, the transcript of an *incomplete* session together with the prover's randomness used do not help reveal the witness w, but this is not the case for a *complete* session.

In the real concurrent interaction, given a history prefix h up to the i_n-th step of V^*, we denote by $h = h'||(\beta_1^j, \beta_2^j, a^j)$ the event that V^* outputs the first verifier message $(\beta_1^j, \beta_2^j, a^j)$ of some session j at its i_n-th step, where "$||$" denotes concatenation of messages.

Let ϵ be an arbitrary inverse polynomial, and poly($\cdot$) be an arbitrary polynomial. Define

$$p(\cdot) := \frac{\epsilon(\cdot)}{2\text{poly}^2(\cdot)}.$$

Lemma 1. *(Main Lemma) Let ϵ, p, poly be as above, and f be the one-way function used in the Feige-Shamir protocol. Assume that there is a non-uniform PPT verifier V^*, running in at most* poly(n) *steps, that breaks ϵ-distributional concurrent zero knowledge of the Feige-Shamir protocol on a joint distribution ensemble $\{(X_n, W_n, Z_n)\}_{n \in N}$ over a NP relation R_L[9] and auxiliary inputs. Then, there exists an infinite set $I = \{(n, i_n)\}$ for which the following two conditions simultaneously hold:*

1. *For a random history prefix generated in the real concurrent interaction,*

$$\Pr\left[h \leftarrow \text{Trans}^{i_n} : \begin{array}{c} h = h'||(\beta_1^j, \beta_2^j, a^j) \wedge \\ \Pr[V^* |_h \rightsquigarrow (j, 2)] \geq p(n) \end{array}\right] \geq p(n).$$

2. *For every circuit family T of polynomial size, there is N_0 such that for every $n > N_0$ (s.t. $(n, \cdot) \in I$) it holds that,*

$$\Pr\left[T(h, \text{PartR}_h) \in \{f^{-1}(\beta_1^j), f^{-1}(\beta_2^j)\} \,\middle|\, \begin{array}{c} h'||(\beta_1^j, \beta_2^j, a^j) = h \leftarrow \text{Trans}^{i_n} \\ \wedge \Pr[V^* |_h \rightsquigarrow (j, 2)] \geq p(n) \end{array}\right]$$
$$\leq 1 - p(n).$$

Remark 3. Note that if, conditioned on outputting the first verifier message $(\beta_1^j, \beta_2^j, a^j)$ of session j at its i_n-th step, V^* reaches the second verifier step of session j (i.e., completes the proof of knowledge of one pre-image) in the real concurrent interaction with probability greater than an inverse polynomial,

---

[9] Though in our final construction of public-key encryption we need to assume a magic adversarial verifier against the Feige-Shamir protocol for a distribution $\{(X_n, W_n)\}_{n \in N}$ over some *OR* NP-relation, Lemma 1 and the results in Sects. 4 and 5 hold with respect to distribution $\{(X_n, W_n)\}_{n \in N}$ over *any* NP relation.

we can construct an efficient algorithm, taking the corresponding witness w as input and playing the role of the honest prover, that extracts one of pre-images of (β_1^j, β_2^j) from V^* by rewinding it with probability negligibly close to 1. The first condition of our lemma asserts that it is relatively easy to obtain images of f for which there is an efficient algorithm with knowledge of w can invert one of them with overwhelming probability, while the second condition of the above lemma guarantees that for any efficient algorithm without knowledge of w the success probability of inversion is bounded away from 1. This illustrates the magic power that the supposed adversary V^* endows the images output at its step i_n with a sort of trapdoor.

As we shall see later, in the final construction of public key encryption, the partial randomness PartR_h together with some images of f will be part of ciphertext, and to ensure the semantic security it is naturally required that for any efficient algorithm with PartR_h as input the success probability of inverting the images of f is small. This is guaranteed by the second condition of the above lemma.

Remark 4. (On the role of the value ϵ) The main reason we deal only with ϵ-distributional concurrent zero knowledge, rather than the standard one, is that, as we will see later, our approach will yield encryption algorithm that runs in time $poly(\frac{1}{\epsilon})$, and thus the value $\frac{1}{\epsilon}$ has to be upper-bounded by a fixed (but arbitrarily) polynomial.

3.2   The Dissection Procedure Leading to a Proof of Lemma 1

Formally, if for an arbitrary inverse polynomial ϵ, V^* breaks ϵ-distributional concurrent zero knowledge of Feige-Shamir protocol over distribution $\{(X_n, W_n, Z_n)\}_{n \in \mathbb{N}}$, then $\forall$ Sim $\exists$ D and infinitely many n, such that

$$\Pr[\mathsf{D}(\mathsf{Trans}_{V^*}(P(X_n, W_n), V^*(Z_n)), Z_n) = 1]$$
$$- \Pr[\mathsf{D}(\mathsf{Sim}(V^*, X_n, Z_n), Z_n) = 1] > \epsilon(n). \tag{1}$$

As mentioned, the intuition behind Lemma 1 is quite straightforward: For a successful V^*, there must exist a step i at which V^* outputs a pair of images and will complete the proof of knowledge of one pre-image at a later time in the real concurrent interaction with high probability, but without knowledge of the corresponding witness no efficient algorithm can invert one of the images, since otherwise, if for every step of V^* there is an efficient algorithm that can extract the target pre-images with overwhelming probability, we are able to show that there *exists* a simulator, incorporating all these efficient inverting algorithms as its subroutines, that will simulate the view of V^* successfully.

To formalize this intuition in the asymptotic setting, we view the behaviour of V^* as an infinite table, in which the entry in the i-th row and n-th column represents the i-th step of V^* (followed immediately by the response from the honest prover) in its concurrent interaction on input the security parameter n (c.f. Fig. 1).

$(P(w), V^*)$

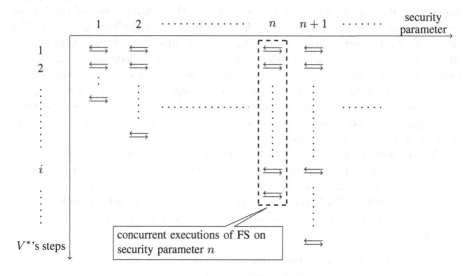

Fig. 1. V^*'s behaviour.

With this table, we dissect V^* and examine its every step *across all security parameters* $n \in \mathbb{N}$, i.e., examine the set of entries $\{(n, i_n = i)\}_{n \in \mathbb{N}}$. A few terminologies follow.

Imaginary steps. Note that for the i-th row of the table (i.e., V^*'s step i), if a security parameter n satisfies $\text{poly}(n) < i$, V^* on the input security parameter n will never reach step i. To simplify the presentation, we think of the step i in every n-th column with $\text{poly}(n) < i$ as an *imaginary step* of V^* with

$$\Pr\left[h \leftarrow \text{Trans}^i : \frac{h = h'||(\beta_1^j, \beta_2^j, a^j) \wedge}{\Pr[V^* |_h \rightsquigarrow (j, 2)] \geq p(n)}\right] = 0.$$

Significant/insignificant entries. Given a (possibly infinite) set K of security parameters, and a set $K' = \{(n, i_n)\}_{n \in K}$, we say the entry $(n, i_n) \in K'$ is *significant* if for which the first condition of Lemma 1 holds, i.e.,

$$\Pr\left[h \leftarrow \text{Trans}^{i_n} : \frac{h = h'||(\beta_1^j, \beta_2^j, a^j) \wedge}{\Pr[V^* |_h \rightsquigarrow (j, 2)] \geq p(n)}\right] > p(n).$$

Otherwise, we call it *insignificant*.

Solving a set of entries. Given a set (possibly infinite) K of security parameters, and a set $K' = \{(n, i_n)\}_{n \in K}$, we say a circuit family T of size $\mathbb{P}$ *solves* the set K', if for every *significant* entry $(n, i_n) \in K'$, T breaks the second condition of Lemma 1 on (n, i_n), i.e., for all $n \in K$,

$$\Pr\left[T(h, \mathrm{PartR}_h) \in \{f^{-1}(\beta_1^j), f^{-1}(\beta_2^j)\} \,\middle|\, \begin{array}{l} h'||(\beta_1^j, \beta_2^j, a^j) = h \leftarrow \mathsf{Trans}^{i_n} \\ \wedge \ \Pr[V^* \,|_h \leadsto (j, 2)] \geq p(n) \end{array}\right]$$
$$> 1 - p(n). \tag{2}$$

Otherwise, we say T fails to solve the set K', i.e., there are *some* entries in K' on which the above inequality does not hold for T. When we say T of size $\mathbb{P}$ fails to solve *any* entry in the set K', we mean that every entry in K' is significant and T cannot solve even a single entry in K'.

Note that we don't make any requirement on T for those *insignificant* entries K' (i.e., those entries for which the first condition of Lemma 1 does not hold). To take an extreme example, if for *all* $(n, i_n) \in K'$ the first condition of Lemma 1 fails to hold, i.e.,

$$\Pr\left[h \leftarrow \mathsf{Trans}^{i_n} : \begin{array}{l} h = h'||(\beta_1^j, \beta_2^j, a^j) \wedge \\ \Pr[V^* \,|_h \leadsto (j, 2)] \geq p(n) \end{array}\right] < p(n),$$

then, by definition, any circuit family can solve the set K'. For simplicity, we let the circuit family that solves such a set K' to be a special dummy circuit family denoted by ϕ, which is of size 0.

With these definitions, we observe the following fact.

Fact 1. Fix a verifier step i. If for any polynomial $\mathbb{P}$, there does not exist a circuit family of size $\mathbb{P}$ that solves the set $\{(n, i_n = i)\}_{n \in \mathbb{N}}$, then there is an infinite set I on which both conditions of Lemma 1 hold.

Proof. Observe first that if for any polynomial $\mathbb{P}$, there is no $\mathbb{P}$-size circuit family that solves the set $\{(n, i)\}_{n \in \mathbb{N}}$, then for every $\mathbb{P}$-size circuit family T, there exists an *infinite* set K of security parameters such that T cannot solve any entry in the set $\{(n, i)\}_{n \in K}$. To see this, suppose for the sake of contradiction that, there is a $\mathbb{P}$-size circuit family T for which there is a *finite* set K such that T solves the set $\{(n, i_n = i)\}_{n \in \mathbb{N} \setminus K}$. Let c_k be the largest security parameter in K, and the circuit family T' be the inverting algorithm that, upon receiving a pair of images, inverts one of them by exhausting all possible pre-images. We now have a new circuit family of size $\mathbb{P}(n) + 2^{c_k}$, denoted by T_i, which applies T on the security parameters $n \in \mathbb{N} \setminus K$ and T' on $n \in K$, can solve the set $\{(n, i)\}_{n \in \mathbb{N}}$, which contradicts the hypothesis of this fact since $\mathbb{P}(n) + 2^{c_k}$ is still a polynomial in n.

We now fix a polynomial (monomial) n^c, and construct a *best possible* n^c-size circuit family $T := \{T^n\}$: Each circuit T_n is of size n^c and achieves the highest success probability of inverting. It follows from the observation above that there is an infinite set K_c of security parameters such that T cannot solve *any* entry in $\{(n, i)\}_{n \in K_c}$.

Since for each security parameter n, the circuit T^n is best possible, we conclude that, for any n^c-size circuit family $T' := \{T'^n\}$, T' cannot solve any entry in $\{(n, i)\}_{n \in K_c}$ (note that the success probability of the inverting circuit T'^n is less than the one of T^n).

Note that $K_c \subseteq K_{c-1}$ for all $c \in \mathbb{N}$. The desired infinite set I can be constructed as follows. Let $n_0 = 0$ and $n_c := \min\{K_c \setminus \{n_{c-1}, n_{c-1}, \cdots, n_0\}\}^{10}$ for each $c \in \mathbb{N}$. We define I to be

$$I := \{(n_c, i)\}_{c \in \mathbb{N}}.$$

It is easy to verify that the first condition of Lemma 1 holds on I.[11] Consider an arbitrary polynomial size circuit family T, say, of size $\mathbb{P}^\dagger$, and suppose that $\mathbb{P}^\dagger(n) \le n^{c'}$[12]. Then T cannot solve *any* entry $(n_c, i) \in I$ for any $c > c'$. Note that $c > c'$ implies $n_c > n_{c'}$, we have that T cannot solve any entry $(n_c, i) \in I$ for any $n_c > n_{c'}$. This establishes the second condition of Lemma 1.   □

The following dissection procedure (c.f. Fig. 2) will yield an infinite set I as desired.

The dissection procedure. Initially set $I_0 := \{(n_0 = 0, i_{n_0} = 0)\}$, $S_0 := \{(T_0 = \phi, \mathbb{P}_0 = 0)\}$.

For $i = 1, 2, ...$, given $I_{i-1} = \{(n_0, i_{n_0}), ..., (n_{k-1}, i_{n_{k-1}})\}^{13}$, $S_{i-1} = \{(T_0, \mathbb{P}_0), ..., (T_{i-1}, \mathbb{P}_{i-1})\}$ and $\mathbb{P} = \max\{\mathbb{P}_0, \mathbb{P}_1, ..., \mathbb{P}_{i-1}\}$, we check the i-th step of V^* for all $n \in \mathbb{N}$ and do the following:

1. If for any polynomial $\mathbb{P}'$ there is no $\mathbb{P}'$-size circuit family that solves the set $\{(n, i_n = i)\}_{n \in \mathbb{N}}$, let I be as defined in the above Fact 1, and stop this process;
2. If there are a polynomial $\mathbb{P}_i$ such that $\mathbb{P}_i \le \mathbb{P}$, and a $\mathbb{P}_i$-size circuit family T_i that solves the set $\{(n, i_n = i)\}_{n \in \mathbb{N}}$, set $S_i \leftarrow S_{i-1} \cup (T_i, \mathbb{P}_i)$, and $I_i \leftarrow I_{i-1}$ (Note that we do not update the set I_{i-1});
3. If there are a polynomial $\mathbb{P}_i$ such that $\mathbb{P}_i > \mathbb{P}$, and a $\mathbb{P}_i$-size circuit family T_i that solves the set $\{(n, i_n = i)\}_{n \in \mathbb{N}}$, but no circuit family of size less than $\mathbb{P}$ that can solve the set $\{(n, i_n = i)\}_{n \in \mathbb{N}}$, then
   (a) set $S_i \leftarrow S_{i-1} \cup \{(T_i, \mathbb{P}_i)\}$, and,
   (b) if $i > \text{poly}(n_{k-1})$[14], find a $n_k > n_{k-1}$ on which the first condition of Lemma 1 holds, but no circuit family of size less than $\mathbb{P}$ can solve the set $I_{i-1} \cup \{(n_k, i_{n_k} = i)\}$[15]. Set $I_i \leftarrow I_{i-1} \cup \{(n_k, i_{n_k} = i)\}$.

Denote by I the set resulted from the above dissection procedure, which is either of the form $\{(n_c, i)\}_{c \in \mathbb{N}}$ (when we encounter the first case during the dissection procedure), or of the form $\{(n_k, i_{n_k})\}$ (otherwise).

Lemma 1 follows from the following two claims. Due to space limitations, we provide detailed proofs of these claims in the full version of this work [Den16].

---

[10] Note that in case K_c is identical to K_{c-1}, then $n_{c-1} \in K_c$.

[11] Note that for every $c \in \mathbb{N}$, for any entry (n, i) in $\{(n, i)\}_{n \in K_c}$, the first condition of Lemma 1 holds for (n, i), since otherwise the entry (n, i) is insignificant, and by definition can be solved by any circuit family.

[12] A little bit oversimplified. In case that, for some N, $\mathbb{P}^\dagger(n) \le n^{c'}$ only when $n > N$, we should set N_0 to be $\max\{N, n_{c'}\}$ and conclude that T cannot solve any entry $(n_c, i) \in I$ for any $n_c > N_0$.

[13] Here $k \le i - 1$. Note that we may not update the set I at each step i.

[14] This means that the current i-step is an imaginary step of V^* for those $n \le n_{k-1}$.

[15] As will be showed in proof of claim 1 in the next section, we can always find such a n_k.

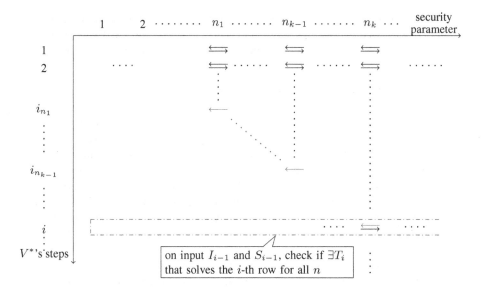

Fig. 2. The dissection procedure. For a magic adversary V^* there must exist either a single row (a step of V^*) from which we find the desired infinite set I, or infinite many rows from each of which we add a new entry to the set I.

Claim 1. If we encounter the first case during the above dissection, or there is no polynomial $\mathbb{P}$ s.t. $\mathbb{P} = \sup\{\mathbb{P}_i : i \in \mathbb{N}\}$, i.e., there is no polynomial upper-bound on the infinite set $\{\mathbb{P}_i : i \in \mathbb{N}\}$, then the set I is infinite and on which both conditions of Lemma 1 hold.

Claim 2. If we will never encounter the first case during the above dissection, and there is a polynomial $\mathbb{P}$ s.t. $\mathbb{P} = \sup\{\mathbb{P}_i : i \in \mathbb{N}\}$, then there is a non-uniform PPT simulator that breaks the inequality (1).

Remark 5. (On the mere existence of T_i and the dependence between T_i's) Note that at each step of the dissection procedure we only ask if there *exists* a good extractor T_i, and that these algorithms may depend on a specific verifier. It may be the case that these T_i exist but we cannot construct them from the code V^* efficiently, as we showed for the concrete adversary from [CKPR01].

However, the mere existence of *good* extractors T_i, satisfying that all of them have size upper-bounded by a fixed polynomial as in Claim 2, helps us show the *existence* of a simulator for V^* under the natural security definition of "$\forall V^* \exists S$".

We stress that the dependence between the possible algorithms T_i's is irrelevant here. Note that at each step i, we set a clear bar $\mathbb{P}$ and check if there exists a circuit family T_i of size less than $\mathbb{P}$ that can solve all those significant entries in the i-th row. If there exists a circuit family T_i that solves this row but the minimal size $\mathbb{P}_i$ required is strictly greater than $\mathbb{P}$, we record this new $\mathbb{P}_i$ and when we enter the next step $(i+1)$, we have a higher bar on the circuit size for checking the existence of T_{i+1}.

Nevertheless, if one can construct a verifier V^* for which there is a deep dependence between these T_i's such that, say, the size of T_{i-1} is twice that of T_i for many i, then we will soon find a desired set I as required by Lemma 1.

4   Tuning in to the Same Channel

As showed in the previous section, the real concurrent interaction between the honest prover and a successful adversary V^* will magically generate a history prefix of the form $h'||(\beta_1, \beta_2, a)$ for which only algorithms with knowledge of the corresponding witness can extract one of the pre-images of (β_1, β_2) with overwhelming probability. However, different algorithms using different witnesses/randomness may recover different pre-images from this history. Thus, to exploit the power of V^* in our setting, we first need to make sure that all parties are in the same channel, i.e., recover the same pre-image from a given history.

In this section we construct non-interactive algorithms C and E from the magic adversary V^* such that, taking as input the witness to x, C generates a β and E can obtain the pre-image of the *same* β. Detailed analyses of these two algorithms can be found in [Den16].

Lemma 2. *Let* p, f, $\{(X_n, W_n, Z_n)\}_{n \in N}$, *the infinite set* I, *and* V^* *be as in Lemma 1. Then there exist two non-uniform PPT algorithms* C *and* E *such that for every* $(n, i_n) \in I$ *the following conditions hold:*

1. C *generates* β, α *and an auxiliary string* aux *satisfying* $\beta = f(\alpha)$ *with probability*

$$\Pr[(x, w, z) \leftarrow (X_n, W_n, Z_n) : C(x, w, z) = (\beta, \alpha, aux)] \geq p^2 - negl(n).$$

2. *It is easy for* E *with knowledge of* w *to invert the image output by* C *with probability*

$$\Pr\left[(x, w, z) \leftarrow (X_n, W_n, Z_n) : E(\beta, aux, w) = f^{-1}(\beta) \,\middle|\, \begin{array}{l} C(x, w, z) \\ = (\beta, \alpha, aux) \end{array}\right]$$
$$\geq 1 - negl(n).$$

3. *For any polynomial-size circuit family* T *without knowing* w, *there is* N_0 *such that for every* $n > N_0$ *(s.t.* $(n, \cdot) \in I$*) it holds that:*

$$\Pr\left[(x, w, z) \leftarrow (X_n, W_n, Z_n) : T(\beta, aux) = f^{-1}(\beta) \,\middle|\, \begin{array}{l} C(x, w, z) \\ = (\beta, \alpha, aux) \end{array}\right]$$
$$\leq 1 - p.$$

Fix $(n, i) \in I$ (from here on we drop the n on i_n for simplicity). Incorporating V^* and the honest prover P, (n, i) and the inverse polynomial p, the algorithm C, on input (x, w, z), plays the role of the honest prover and extracts (by rewinding) one-pre-image of the pair images of f output by V^* at its i-th step, and then

---

The Algorithm C

---

input : $(x, w, z) \leftarrow (X_n, W_n, Z_n)$

1. Run P and V^* on input (x, w, z) until obtain the history prefix h up to the step i of V^*. If the V^*'s step i message v_i is the first verifier message of the form (β_1, β_2, a) in a session, say, session j, then continue; otherwise, return $\bot$.

2. Resume the interaction between P and V^* until V^* terminates. If the second accepting verifier message t in session j appears in this interaction, continue; otherwise, return $\bot$.

3. Repeat the following two steps $\frac{n}{p}$ times (there are at most $\frac{n^2}{p^2}$ iterations of step 2 within this step):

    (a) Run the above step 2 using fresh randomness (based on the same history prefix h) until either the second accepting verifier message in session j appears *twice* or the $\frac{n}{p}$-th iteration is reached. If two accepting transcripts of the first phase in session j of the Feige-Shamir protocol are obtained within these $\frac{n}{p}$ iterations (for the purpose of simplifying the analysis of the algorithm E, here we don't use the transcript obtained in step 2), compute α such that $\beta_b = f(\alpha)$ from them; otherwise, return $\bot$.

    (b) Store (β_b, α) in a list.

4. Set β to be β_b for which the corresponding pair (β_b, α) appears most often in the above list, and aux to be $(h, \text{PartR}_h, x, z)$, where PartR_h includes only the randomness used by V^* and the randomness used by honest provers in those *incomplete* sessions in producing h.
    output: (β, α, aux).

---

outputs the pre-image extracted and the corresponding image (together with some auxiliary information). To make sure that different algorithms can extract the same pre-image, we have C repeat the extraction procedure many times and output the image corresponding to the *most-often* extracted pre-image. See below for the detailed description of C.

The algorithm E, taking (β, aux, w) as input, simply repeats $\frac{n}{p}$ times the step 3(a) of the algorithm C to extract the pre-image of β.

The Algorithm E

---

input : (β, aux, w)

1. Parse aux into $(h, \text{PartR}_h, x, z)$, and parse the last message v_i in h into (β_1, β_2, a).

2. Suppose that $\beta = \beta_b$. Repeat the step 3(a) of C until the pre-image α of β_b is extracted or the $\frac{n}{p}$-th iteration is reached, and if all iterations fail, return $\bot$.
    output: α.

---

5 Hardness Amplification and a Tailored Hard-Core Lemma

For our applications, we need to increase the success probability of the algorithm C significantly while decreasing T's success probability (as in the third condition of Lemma 2) to a negligible level. In addition, if the statement x has multiple witnesses, we also want the algorithm E to work when given an arbitrary one (not necessarily the same as the one given as input to C) as input.

Our basic strategy for achieving these goals is to use classic hardness amplification method with some careful modifications. Let p be as in Lemma 1, and define

$$q_1 := \frac{n}{(p)^2}, q_2 := \frac{n}{p} \text{ and } q := q_1 q_2.$$

Given as input a $q_1 \times q_2$ matrix of simples from (X_n, W_n, Z_n), M runs C on each column and outputs a vector of q_2 number of images of f (together with the corresponding pre-images and some auxiliary strings). The formal descriptions of algorithms M and Find are given below.

The Algorithm M

input : $\{(x_k, w_k, z_k)\}_{k=1}^q$

1. Arrange $\{(x_k, w_k, z_k)\}_{k=1}^q$ into $q_1 \times q_2$ tuples, denoted by $\{(x_i^j, w_i^j, z_i^j)\}_{i,j=1}^{q_2, q_1}$.
2. For $i = 1, 2, ..., q_2$, run C on each (x_i^j, w_i^j, z_i^j), $j \in [1, q_1]$, until C outputs (β, α, aux). If for some i all these q_1 runs of C fail, return $\bot$; otherwise, set $(\beta_i, \alpha_i, aux_i)$ to be (β, α, aux).

output: $\{(\beta_i, \alpha_i, aux_i)\}_{i=1}^{q_2}$.

The Algorithm Find

input : $\{(x_k, w_k, z_k)\}_{k=1}^q$, $\{(\beta_i, aux_i)\}_{i=1}^{q_2}$

1. Arrange $\{(x_k, w_k, z_k)\}_{k=1}^q$ in the same way as M and obtain $\{(x_i^j, w_i^j, z_i^j)\}_{i,j=1}^{q_2, q_1}$.
2. For $i = 1, 2, ..., q_2$, obtain the statement x_i from aux_i, find the j-th entry (x_i^j, w_i^j, z_i^j) from $\{(x_i^j, w_i^j, z_i^j)\}_{j=1}^{q_1}$ such that $x_i^j = x_i$ and fetch the corresponding w_i^j, set $w_i = w_i^j$ and run E on input (β_i, aux_i, w_i). If E fails, output $\bot$, otherwise, set α_i to be the output of E.

output: $\{\alpha_i\}_{i=1}^{q_2}$.

It easily follows from Lemma 2 that the algorithms M and Find enjoys the following security properties.

Lemma 3. *The following properties hold for algorithms M and Find:*

1. *The probability that M outputs $\{(\beta_i, \alpha_i, aux_i)\}_{i=1}^{q_2}$ such that $\beta_i = f(\alpha_i)$ holds for each i is negligibly close to 1.*

2. *Conditioned on* M *outputting* $\{(\beta_i, \alpha_i, aux_i)\}_{i=1}^{q_2}$, *the probability that* Find *inverts all these* β_i's *successfully is negligibly close to* 1.
3. *Conditioned on* M *outputting* $\{(\beta_i, \alpha_i, aux_i)\}_{i=1}^{q_2}$, *for any polynomial-size circuit family* T, *given as input only* $(\{(x_k, z_k)\}_{k=1}^{q}, \{(\beta_i, aux_i)\}_{i=1}^{q_2})$ *(without any witnesses to the* x_k's*), the probability that* T *inverts all these* β_i's *successfully is negligible.*
4. *For any two inputs to* Find *with different witnesses,* $(\{(x_k, w_k, z_k)\}_{k=1}^{q}, \{(\beta_i, aux_i)\}_{i=1}^{q_2})$ *and* $(\{(x_k, w'_k, z_k)\}_{k=1}^{q}, \{(\beta_i, aux_i)\}_{i=1}^{q_2})$ *such that* $\{w_k\}_{k=1}^{q} \neq \{w'_k\}_{k=1}^{q}$, Find *succeeds on each input with almost (negligibly close to each other) the same probability.*

The algorithm M generates q_2 number of images $(\beta_1, \beta_2, ..., \beta_{q_2})$ of one-way function $f : \{0,1\}^n \rightarrow \{0,1\}^{\ell(n)}$ in a way such that they are hard for any polynomial-size circuit family (without knowing the corresponding witnesses) to invert simultaneously. This enables us to apply Goldreich-Levin hard-core predicate for the function of $f^{\otimes q_2}$ with respect to the distribution on $(\beta_1, \beta_2, ..., \beta_{q_2})$ generated by M. Formally, we need the following form of the Goldreich-Levin theorem.

Lemma 4 (Goldreich-Levin). *Let* $f : \{0,1\}^n \rightarrow \{0,1\}^{\ell(n)}$ *be a function computable in polynomial time,* G *be a PPT algorithm. If for every polynomial-size circuit family* T,

$$\Pr[(f(x), aux) \leftarrow G(1^n) : T(1^n, f(x), aux) \in f^{-1}(f(x))] \leq negl(n),$$

then, the inner product of x *and a random* r *modulo* 2, *denoted by* $\langle x, r \rangle$, *is a hardcore predicate for* f, *i.e., for every polynomial-size circuit family* T'

$$\Pr[(f(x), aux) \leftarrow G(1^n), r \leftarrow \{0,1\}^n : T'(1^n, f(x), r, aux) = \langle x, r \rangle]$$
$$\leq \frac{1}{2} + negl(n).$$

The Goldreich-Levin theorem typically states for the distribution $f(U)$, i.e., for x being drawn from uniform distribution, but its proof ignores the distribution on the images of f and the auxiliary input (as long as both T and T' are given the same auxiliary string as input) completely, so the same proof applies to Lemma 4 (c.f. [Gol01]).

In our setting, this means that the inner product (modulo 2) $\langle(\alpha_1, \alpha_2, ..., \alpha_{q_2}), r \leftarrow \{0,1\}^{n \times q_2}\rangle$ is a hard core predicate for $f^{\otimes q_2} : \{0,1\}^{n \times q_2} \rightarrow \{0,1\}^{\ell(n) \times q_2}$ against arbitrary circuit families of polynomial size that takes as auxiliary input $(\{(x_k, z_k)\}_{k=1}^{q}, \{(\beta_i, aux_i)\}_{i=1}^{q_2})$.

6 Constructions for Public-Key Encryption and Key Agreement

In this section, we construct semantic secure (under chosen-plaintext-attack) public-key encryption and key agreement from a supposed adversary V^* against

372   Y. Deng

the Feige-Shamir protocol and an injective one-way function. This completes the proof of Theorem 1.

Let ϵ, q, q_2, M, Find and the infinite set I be as defined in previous sections. The final construction of public-key encryption scheme proceeds as follows. The key generation algorithm generates q number of YES instances together with their corresponding witnesses, $\{(x_{1,k}, w_{1,k})\}_{k=1}^{q}$, where $\{w_{1,k}\}_{k=1}^{q}$ is kept secret and $\{x_{1,k}\}_{k=1}^{q}$ is made public. To encrypt a bit m, the encryption algorithm generates $\{(x_{2,k}, w_{2,k})\}_{k=1}^{q}$, prepares a sequence of OR statements $\{(x_{1,k} \vee x_{2,k})\}_{k=1}^{q}$ (thus each $\{w_{b,k}\}_{k=1}^{q}$, $b \in [1,2]$, are valid witnesses), and then applies M on $\{w_{2,k}\}_{k=1}^{q}$ to generate an image of $f^{\otimes q_2}$ and encrypts m using Goldreich-Levin; to decrypt the cipher-text, the decryption algorithm applies Find on $\{w_{1,k}\}_{k=1}^{q}$ as witnesses to obtain the corresponding pre-image and then computes the plain-text.

Formally, we need to assume the following for our constructions of public-key encryption and key agreement:

- An arbitrary *injective* one-way function $f : \{0,1\}^n \rightarrow \{0,1\}^{\ell(n)}$ (used in the Feige-Shamir protocol). The injectiveness will be used for one party to recover the same hardcore bit that generated by the other party.
- An arbitrary efficiently samplable distribution ensemble $D = \{D_n\}_{n \in N}$ over R_L for an arbitrary NP language L.
- An arbitrary efficiently samplable distribution ensemble $\{Z_n\}_{n \in N}$ over $\{0,1\}^*$.
- A joint distribution ensemble $\{(X_n, W_n, Z_n)\}_{n \in N}$ on which the adversary V^* breaks the ϵ-distributional concurrent zero knowledge of Feige-Shamir protocol, where each distribution (X_n, W_n, Z_n) defined in the following way: Sample $(x_1, w_1) \leftarrow D_n$, $(x_2, w_2) \leftarrow D_n$, $z \leftarrow Z_n$, $b \leftarrow \{1,2\}$, and output $((x_1, x_2), w_b)$.

We now construct public-key encryption for a single bit message on each security parameter n s.t. $(n, \cdot) \in I$.

Key generation $\mathsf{Gen}(1^n)$: $\{(x_{1,k}, w_{1,k})\}_{k=1}^{q} \leftarrow D_n^{\otimes q}$, and set $pk = \{x_{1,k}\}_{k=1}^{q}$, $sk = \{w_{1,k}\}_{k=1}^{q}$.

Encryption $\mathsf{Enc}(pk = \{x_{1,k}\}_{k=1}^{q}, m)$ $(m \in \{0,1\})$:

1. $\{(x_{2,k}, w_{2,k})\}_{k=1}^{q} \leftarrow D_n^{\otimes q}$, $\{z_k\}_{k=1}^{q} \leftarrow Z_n^{\otimes q}$.
2. for $k \in [1, q]$, set x_k to be a random order of the pair $(x_{1,k}, x_{2,k})$.
3. $\{(\beta_i, \alpha_i, aux_i)\}_{i=1}^{q_2} \leftarrow \mathsf{M}(\{(x_k, w_{2,k}, z_k)\}_{k=1}^{q})$.
4. $r \leftarrow \{0,1\}^{n \times q_2}$, $h \leftarrow \langle (\alpha_1, \alpha_2, ..., \alpha_{q_2}), r \rangle \in \{0,1\}$.
5. Output $c = (\{(x_k, z_k)\}_{k=1}^{q}, \{(\beta_i, aux_i)\}_{i=1}^{q_2}, r, h \oplus m)$.

Decryption $\mathsf{Dec}(sk = \{w_{1,k}\}_{k=1}^{q}, c)$:

1. Parse c into $\{(x_k, z_k)\}_{k=1}^{q} || \{(\beta_i, aux_i)\}_{i=1}^{q_2} || r || c'$.
2. $\{\alpha_i\}_{i=1}^{q_2} \leftarrow \mathsf{Find}(\{(x_k, w_{1,k}, z_k)\}_{k=1}^{q}, \{(\beta_i, aux_i)\}_{i=1}^{q_2})$.
3. $h \leftarrow \langle (\alpha_1, \alpha_2, ..., \alpha_{q_2}), r \rangle$.
4. Output $m = h \oplus c'$.

Notice that the input to M in the encryption algorithm can be viewed as being drawn from (X_n, W_n, Z_n) defined above. The correctness of this scheme follows from properties 1, 2, 4 of algorithms M and Find presented in the previous section. It should be noted that our scheme is not perfectly correct since it is possible for M/Find to fail during the encryption/decryption process. However, this happens only with negligible probability.

It is also easy to verify the semantic security under chosen-plaintext-attack, which is essentially due to the property 3 of M, together with the security of the hardcore bit for $f^{\otimes q_2}$.

Following the well-known paradigm, one can transform a semantic secure (under chosen-plaintext-attack) public-key encryption scheme into a key agreement protocol (A, B) with security against eavesdropping adversary in a simple way: the party A generates a public/secrete key pair and send the public-key to B, and then B sends back a ciphertext of the session secret key under A's public key to A. This establishes a common session secret key between A and B.

Extensions to Multiparty Key Agreement. Our key agreement protocol can be easily extended to the multiparty setting. Roughly, if V^* is able to break ϵ-distributional concurrent zero knowledge of the Feige-Shamir protocol on a distribution over instances of the form $(x_1 \vee x_2 \vee ... \vee x_n)$, then the n parties can establish a session secret key as follows. Each party A_i generates a sequence of pairs $\{(x_{i,k}, w_{i,k})\}_{k=1}^q$). In their first round the parties $A_1, A_2, ..., A_{n-1}$ send their sequences of $\{(x_{i,k})\}_{i,k=1}^{n-1,q}$ to the n-th party, then the n-th party uses these sequences as a public key of the above public-key encryption scheme to encrypt the session secret key and send the ciphertext to all $n-1$ parties. Upon receiving the ciphertext, each A_i, $i = [1, n-1]$, decrypts it and obtains the session secret key using their own $\{(w_{i,k})\}_{k=1}^q$.

7 Concluding Remarks

We prove a win-win result regarding the complexity of public-key encryption and the round-complexity of concurrent zero knowledge. We believe that when one can prove one of these two statements listed in Theorem 1, one might obtain a much stronger result (e.g., result with respect to the (nicer) standard definition) than the ones stated therein. The ideas and techniques used here may be applied to investigate some other black-box lower bounds in cryptography.

Our result can be viewed as a step toward breaking the known black-box or universal reduction barriers, and a proof (or disproof) of either one of the two statements in Theorem 1 will be exciting. A construction of public-key encryption (key agreement) from general one-way functions will, borrowing from the Impagliazzo's terminology [Imp95], rule out the world Minicrypt and build for the first time the world Cryptomania from (trapdoor/algebraic) structure-free hardness assumption, which definitely is a major achievement in cryptography.

On the other hand, a concurrent security proof of the Feige-Shamir protocol will also be an exciting breakthrough, both technically and conceptually. On the technical level, such a proof will reveal a fascinating fact that all possible efficient adversaries against the Feige-Shamir protocol have *in common* a highly non-trivial structure of computation–e.g., the existence of those good extractors $\{T_i\}_{i\in\mathbb{N}}$ from the second claim in Sect. 3.2, which might shed light on the longstanding open problem of constructing extractable one-way functions from standard assumptions; on the conceptual level, it will bring a new individual reduction/simulation for cryptography and refute the impression that a new reduction technique always gives more complicated and inefficient constructions.

Acknowledgement. We are grateful to Nir Bitansky, Omer Paneth, Alon Rosen and anonymous reviewers for constructive comments and suggestions. We thank Yu Chen and Jiang Zhang for helpful discussions, and Yanyan Liu, Shunli Ma, Hailong Wang, Bo Wu, Zhenbin Yan and Jingyue Yu for careful proofreading.

References

[Adl78] Adleman, L.M.: Two theorems on random polynomial time. In: Proceedings of the 19th Annual Symposium on Foundations of Computer Science, FOCS 1978, pp. 75–83. IEEE Computer Society (1978)

[Bar01] Barak, B.: How to go beyond the black-box simulation barrier. In: Proceedings of the 42th Annual IEEE Symposium on Foundations of Computer Science, FOCS 2001, pp. 106–115. IEEE Computer Society (2001)

[BBF13] Baecher, P., Brzuska, C., Fischlin, M.: Notions of black-box reductions, revisited. In: Sako, K., Sarkar, P. (eds.) ASIACRYPT 2013. LNCS, vol. 8269, pp. 296–315. Springer, Heidelberg (2013). doi:10.1007/978-3-642-42033-7_16

[BCPR14] Bitansky, N., Canetti, R., Paneth, O., Rosen, A.: On the existence of extractable one-way functions. In: Proceedings of the 45th Annual ACM Symposium on the Theory of Computing, STOC 2014, pp. 505–514. ACM Press (2014)

[BHSV98] Bellare, M., Halevi, S., Sahai, A., Vadhan, S.: Many-to-one trapdoor functions and their relation to public-key cryptosystems. In: Krawczyk, H. (ed.) CRYPTO 1998. LNCS, vol. 1462, pp. 283–298. Springer, Heidelberg (1998). doi:10.1007/BFb0055735

[BP15] Bitansky, N., Paneth, O.: On non-black-box simulation and the impossibility of approximate obfuscation. SIAM J. Comput. **44**(5), 1325–1383 (2015)

[BU08] Backes, M., Unruh, D.: Limits of constructive security proofs. In: Pieprzyk, J. (ed.) ASIACRYPT 2008. LNCS, vol. 5350, pp. 290–307. Springer, Heidelberg (2008). doi:10.1007/978-3-540-89255-7_18

[CHS05] Canetti, R., Halevi, S., Steiner, M.: Hardness amplification of weakly verifiable puzzles. In: Kilian, J. (ed.) TCC 2005. LNCS, vol. 3378, pp. 17–33. Springer, Heidelberg (2005). doi:10.1007/978-3-540-30576-7_2

[CKPR01] Canetti, R., Kilian, J., Petrank, E., Rosen, A.: Black-box concurrent zero-knowledge requires omega(log n) rounds. In: Proceedings of the 33rd Annual ACM Symposium Theory of Computing, STOC 2001, pp. 570–579. ACM Press (2001)

[CLMP13] Chung, K.-M., Lin, H., Mahmoody, M., Pass, R.: On the power of nonuniformity in proofs of security. In: ITCS 2013, pp. 389–400 (2013)

[CLOS02] Canetti, R., Lindell, Y., Ostrovsky, R., Sahai, A.: Universally composable two-party and multi-party computation. In: Proceedings of the 34th Annual ACM Symposium on the Theory of Computing, STOC 2002, pp. 494–503. ACM Press (2002)

[CLP13a] Canetti, R., Lin, H., Paneth, O.: Public-coin concurrent zero-knowledge in the global hash model. In: Sahai, A. (ed.) TCC 2013. LNCS, vol. 7785, pp. 80–99. Springer, Heidelberg (2013). doi:10.1007/978-3-642-36594-2_5

[CLP13b] Chung, K.-M., Lin, H., Pass, R.: Constant-round concurrent zero knowledge from p-certificates. In: Proceedings of the 54th Annual Symposium on Foundations of Computer Science, FOCS 2013, pp. 50–59. IEEE Computer Society (2013)

[CLP15a] Chung, K.-M., Lin, H., Pass, R.: Constant-round concurrent zero-knowledge from indistinguishability obfuscation. In: Gennaro, R., Robshaw, M. (eds.) CRYPTO 2015. LNCS, vol. 9215, pp. 287–307. Springer, Heidelberg (2015). doi:10.1007/978-3-662-47989-6_14

[CLP15b] Chung, K.-M., Lui, E., Pass, R.: From weak to strong zero-knowledge and applications. In: Dodis, Y., Nielsen, J.B. (eds.) TCC 2015. LNCS, vol. 9014, pp. 66–92. Springer, Heidelberg (2015). doi:10.1007/978-3-662-46494-6_4

[CS99] Cramer, R., Shoup, V.: Signature schemes based on the strong RSA assumption. In: ACM Conference on Computer and Communications Security, CCS 1999, pp. 46–52. ACM Press (1999)

[Dam91] Damgård, I.: Towards practical public key systems secure against chosen ciphertext attacks. In: Feigenbaum, J. (ed.) CRYPTO 1991. LNCS, vol. 576, pp. 445–456. Springer, Heidelberg (1992). doi:10.1007/3-540-46766-1_36

[Den16] Deng, Y.: Magic adversaries versus individual reduction: science wins either way. Cryptology ePrint Archive, Report 2016/1107 (2016)

[DGL+16] Deng, Y., Garay, J., Ling, S., Wang, H., Yung, M.: On the implausibility of constant-round public-coin zero-knowledge proofs. In: Zikas, V., Prisco, R. (eds.) SCN 2016. LNCS, vol. 9841, pp. 237–253. Springer, Cham (2016). doi:10.1007/978-3-319-44618-9_13

[DGS09] Deng, Y., Goyal, V., Sahai, A.: Resolving the simultaneous resettability conjecture and a new non-black-box simulation strategy. In: Proceedings of the 50th Annual Symposium on Foundations of Computer Science, FOCS 2009, pp. 251–260. IEEE Computer Society (2009)

[DH76] Diffie, W., Hellman, M.E.: New directions in cryptography. IEEE Trans. Inf. Theor. 22(6), 644–654 (1976)

[DNRS03] Dwork, C., Naor, M., Reingold, O., Stockmeyer, L.J.: Magic functions. J. ACM 50(6), 852–921 (2003)

[DNS98] Dwork, C., Naor, M., Sahai, A.: Concurrent zero-knowledge. In: Proceedings of the 30rd Annual ACM Symposium Theory of Computing, STOC 1998, pp. 409–418. ACM Press (1998)

[DS16] Dachman-Soled, D.: Towards non-black-box separations of public key encryption and one way function. In: Hirt, M., Smith, A. (eds.) TCC 2016. LNCS, vol. 9986, pp. 169–191. Springer, Heidelberg (2016). doi:10.1007/978-3-662-53644-5_7.

[FS89] Feige, U., Shamir, A.: Zero knowledge proofs of knowledge in two rounds. In: Brassard, G. (ed.) CRYPTO 1989. LNCS, vol. 435, pp. 526–544. Springer, New York (1990). doi:10.1007/0-387-34805-0_46

[GGJ13] Goyal, V., Gupta, D., Jain, A.: What information is leaked under concurrent composition? In: Canetti, R., Garay, J.A. (eds.) CRYPTO 2013. LNCS, vol. 8043, pp. 220–238. Springer, Heidelberg (2013). doi:10.1007/978-3-642-40084-1_13

[GGJS12] Garg, S., Goyal, V., Jain, A., Sahai, A.: Concurrently secure computation in constant rounds. In: Pointcheval, D., Johansson, T. (eds.) EUROCRYPT 2012. LNCS, vol. 7237, pp. 99–116. Springer, Heidelberg (2012). doi:10.1007/978-3-642-29011-4_8

[GGS15] Goyal, V., Gupta, D., Sahai, A.: Concurrent secure computation via non-black box simulation. In: Gennaro, R., Robshaw, M. (eds.) CRYPTO 2015. LNCS, vol. 9216, pp. 23–42. Springer, Heidelberg (2015). doi:10.1007/978-3-662-48000-7_2

[GK96] Goldreich, O., Kahan, A.: How to construct constant-round zero-knowledge proof systems for NP. J. Cryptology 9(3), 167–190 (1996)

[GLP+15] Goyal, V., Lin, H., Pandey, O., Pass, R., Sahai, A.: Round-efficient concurrently composable secure computation via a robust extraction lemma. In: Dodis, Y., Nielsen, J.B. (eds.) TCC 2015. LNCS, vol. 9014, pp. 260–289. Springer, Heidelberg (2015). doi:10.1007/978-3-662-46494-6_12

[GM82] Goldwasser, S., Micali, S.: Probabilistic encryption and how to play mental poker keeping secret all partial information. In: Proceedings of the 14rd Annual ACM Symposium Theory of Computing, STOC 1982, pp. 365–377. ACM Press (1982)

[GMR89] Goldwasser, S., Micali, S., Rackoff, C.: The knowledge complexity of interactive proof systems. SIAM J. Comput. 18(1), 186–208 (1989)

[GNW95] Goldreich, O., Nisan, N., Wigderson, A.: On yao's xor-lemma. In: Electronic Colloquium on Computational Complexity, TR95-050 (1995)

[Gol93] Goldreich, O.: A uniform-complexity treatment of encryption and zero-knowledge. J. Cryptology 6(1), 21–53 (1993)

[Gol01] Goldreich, O.: Foundations of Cryptography, Basic Tools. Cambridge University Press, Cambridge (2001)

[Goy13] Goyal, V.: Non-black-box simulation in the fully concurrent setting. In: Proceedings of the 45th Annual ACM Symposium on the Theory of Computing, STOC 2013, pp. 221–230. ACM Press (2013)

[HILL99] Hastad, J., Impagliazzo, R., Levin, L.A., Luby, M.: A pseudorandom generator from any one-way function. SIAM J. Comput. 28(4), 1364–1396 (1999)

[HKS03] Hofheinz, D., Kiltz, E., Shoup, V.: Practical chosen ciphertext secure encryption from factoring. J. Cryptology 26(1), 102–118 (2003)

[HR07] Haitner, I., Reingold, O.: Statistically-hiding commitment from any one-way function. In: Proceedings of the 39rd Annual ACM Symposium Theory of Computing, STOC 2007, pp. 1–10. ACM Press (2007)

[HS11] Holenstein, T., Schoenebeck, G.: General hardness amplification of predicates and puzzles. In: Ishai, Y. (ed.) TCC 2011. LNCS, vol. 6597, pp. 19–36. Springer, Heidelberg (2011). doi:10.1007/978-3-642-19571-6_2

[IL89] Impagliazzo, R., Luby, M.: One-way functions are essential for complexity based cryptography. In: Proceedings of the 30th Annual Symposium on Foundations of Computer Science, FOCS 1989, pp. 230–235. IEEE Computer Society (1989)

[Imp95] Impagliazzo, R.: A personal view of average-case complexity. In: Proceedings of the 10th Annual IEEE Structure in Complexity Theory Conference, pp. 134–147. IEEE Computer Society (1995)

[IR89] Impagliazzo, R., Rudich, S.: Limits on the provable consequences of one-way permutations. In: Proceedings of the 21th Annual ACM Symposium on the Theory of Computing, STOC 1989, pp. 44–61. ACM Press (1989)

[KL07] Katz, J., Lindell, Y.: Introduction to Modern Cryptography. Chapman and Hall/CRC Press (2007)

[KP01] Kilian, J., Petrank, E.: Concurrent and resettable zero-knowledge in polyloalgorithm rounds. In: Proceedings of the 33rd Annual ACM Symposium Theory of Computing, STOC 2001, pp. 560–569. ACM Press (2001)

[Lin03a] Lindell, Y.: Bounded-concurrent secure two-party computation without setup assumptions. In: Proceedings of the 35rd Annual ACM Symposium Theory of Computing, STOC 2003, pp. 683–692. ACM Press (2003)

[Lin03b] Lindell, Y.: General composition and universal composability in secure multi-party computation. In: Proceedings of the 44th Annual Symposium on Foundations of Computer Science, FOCS 2003, pp. 394–403. IEEE Computer Society (2003)

[Lin08] Lindell, Y.: Lower bounds and impossibility results for concurrent self composition. J. Cryptology 21(2), 200–249 (2008)

[NY90] Naor, M., Yung, M.: Public-key cryptosystems provably secure against chosen ciphertext attacks. In: Annual ACM Symposium on the Theory of Computing, STOC 1990, pp. 427–437. ACM Press (1990)

[Pas04] Pass, R.: Bounded-concurrent secure multi-party computation with a dishonest majority. In: Proceedings of the 36th Annual ACM Symposium on the Theory of Computing, STOC 2004, pp. 232–241. ACM Press (2004)

[PPS15] Pandey, O., Prabhakaran, M., Sahai, A.: Obfuscation-based non-black-box simulation and four message concurrent zero knowledge for NP. In: Dodis, Y., Nielsen, J.B. (eds.) TCC 2015. LNCS, vol. 9015, pp. 638–667. Springer, Heidelberg (2015). doi:10.1007/978-3-662-46497-7_25

[PR03] Pass, R., Rosen, A.: Bounded-concurrent secure two-party computation in a constant number of rounds. In: Proceedings of the 44th Annual Symposium on Foundations of Computer Science, FOCS 2003, pp. 404–413. IEEE Computer Society (2003)

[PR05] Pass, R., Rosen, A.: Concurrent non-malleable commitments. In: Proceedings of the 46th Annual IEEE Symposium on Foundations of Computer Science, FOCS 2005, pp. 563–572. IEEE Computer Society (2005)

[PRS02] Prabhakaran, M., Rosen, A., Sahai, A.: Concurrent zero knowledge with logarithmic round-complexity. In: Proceedings of the 43th Annual IEEE Symposium on Foundations of Computer Science, FOCS 2002, pp. 366–375. IEEE Computer Society (2002)

[Rab79] Rabin, M.: Digitalized signatures and public-key encryptions as intractable as factorization. Technical Report MIT/LCS/TR-212, MIT Laboratory for Computer Science (1979)

[Reg09] Regev, O.: On lattices, learning with errors, random linear codes, and cryptography. J. ACM 56(6), 1–40 (2009)

[Rom90] Rompel, J.: One-way functions are necessary and sufficient for secure signatures. In: Proceedings of the 22rd Annual ACM Symposium Theory of Computing, STOC 1990, pp. 387–394. ACM Press (1990)

[RSA78] Rivest, R.L., Shamir, A., Adleman, L.M.: A method for obtaining digital signatures and public-key cryptosystems. Commun. ACM 21(2), 120–126 (1978)

[RTV04] Reingold, O., Trevisan, L., Vadhan, S.: Notions of reducibility between cryptographic primitives. In: Naor, M. (ed.) TCC 2004. LNCS, vol. 2951, pp. 1–20. Springer, Heidelberg (2004). doi:10.1007/978-3-540-24638-1_1

[Sah99] Sahai, A.: Non-malleable non-interactive zero knowledge and adaptive chosen-ciphertext security. In: Proceedings of the 40th Annual Symposium on Foundations of Computer Science, FOCS 1999, pp. 543–553. IEEE Computer Society (1999)

Provable Security for Symmetric Cryptography I

The Multi-user Security of Double Encryption

Viet Tung Hoang[1]([⊠]) and Stefano Tessaro[2]

[1] Department of Computer Science, Florida State University, Tallahassee, USA
hviettung@gmail.com
[2] Department of Computer Science, University of California Santa Barbara,
Santa Barbara, USA

Abstract. It is widely known that double encryption does not substantially increase the security of a block cipher. Indeed, the classical meet-in-the middle attack recovers the $2k$-bit secret key at the cost of roughly 2^k off-line enciphering operations, in addition to very few known plaintext-ciphertext pairs. Thus, essentially as efficiently as for the underlying cipher with a k-bit key.

This paper revisits double encryption under the lens of multi-user security. We prove that its security degrades only very mildly with an increasing number of users, as opposed to single encryption, where security drops linearly. More concretely, we give a tight bound for the multi-user security of double encryption as a pseudorandom permutation in the ideal-cipher model, and describe matching attacks.

Our contribution is also conceptual: To prove our result, we enhance and generalize the generic technique recently proposed by Hoang and Tessaro for lifting single-user to multi-user security. We believe this technique to be broadly applicable.

Keywords: Symmetric security · Provable security · Multi-user security · Double encryption

1 Introduction

A classical problem in cryptography is that of stretching the key length of a block cipher. Namely, from a block cipher E with block length n and key length k, we want to obtain a new one with key length $k' > k$ which is *more* secure than E. The problem was naturally motivated by legacy designs – in particular, DES – with inherently too-short keys (e.g., 56 bits), and the desire to stretch this key length generically without resorting to designing a new cipher.

The common wisdom is that *double encryption* is not useful for key-stretching purposes. Here, by double encryption, we mean the construction that, given an n-bit plaintext M and two k-bit keys K_1, K_2, outputs $E_{K_1}(E_{K_2}(M))$. Indeed, there is a well-known meet-in-the-middle attack recovering the key with only marginally more than 2^k operations given (very few) valid plaintext-ciphertext pairs. This weakness has led to the widespread deployment (which continues to

© International Association for Cryptologic Research 2017
J.-S. Coron and J.B. Nielsen (Eds.): EUROCRYPT 2017, Part II, LNCS 10211, pp. 381–411, 2017.
DOI: 10.1007/978-3-319-56614-6_13

date in some niche areas) of Triple-DES [1], as well as a number of works on analyzing the theory of triple and multiple encryption [7,13–17,20], and alternative constructions with extra whitening steps (and key material) [15,16,18–20].

In this paper, we revisit double encryption in the context of multi-user security, where we give tight bounds, and show that it constitutes a sound and simple method to mitigate multi-user attacks on block ciphers. However, this problem will also serve as an application for a generic framework to provide good multi-user security bounds, and which we hope to be of wider applicability.

Double Encryption in the Single User Setting. As in previous works, we study the security of double encryption in the ideal-cipher model as a (strong) pseudorandom permutation (PRP). The attacker A is given access to an ideal cipher E to which it can issue p forward or backward queries for any chosen key (these are usually referred to as "offline queries"), and up to q queries (in either direction) to $E_{K_1} \circ E_{K_2}$ (for random secret keys K_1, K_2) or a truly random permutation on the n-bit strings (this being usually called "online queries"). The attacker's goal is to decide which of the two it is accessing. In this model, Aiello *et al.* [2] proved that A's distinguishing advantage satisfies

$$\mathsf{Adv}^{\mathrm{prp}}_{\mathrm{DE}[E]}(A) \leq \left(\frac{p}{2^k}\right)^2 . \tag{1}$$

where $\mathrm{DE}[E]$ denotes double encryption. Note that for *single* encryption, the bound is easily shown to be $\mathsf{Adv}^{\mathrm{prp}}_E(A) \leq \frac{p}{2^k}$. Both advantages become non-negligible for the same $p \approx 2^k$, although (1) is smaller when $p \ll 2^k$.

The Multi-user Setting. In the *multi-user (mu) setting*, originally proposed by Bellare, Boldyreva, and Micali [5] for public-key encryption, the attacker can distribute its online queries adaptively across multiple independent key pairs (in the real world) or independent permutations (in the ideal world). A few recent block-cipher analyses [19,24,29] have focused on mu security, and the notion has established itself as a more realistic security target.

One expects security to degrade as the number of users increases, and this loss can be linear in the worst case. For example, for single-encryption, we do have

$$\mathsf{Adv}^{\pm \mathrm{mu\text{-}prp}}_E(A) \leq \frac{u\,(p+u)}{2^k} \leq \frac{q\,(p+q)}{2^k} , \tag{2}$$

where u is a bound on the number of users A queries, and this bound is tight, i.e., there is a matching attack [10]. Also, we can only guarantee $u \leq q$, as the attacker can decide to only issue one query per user. However, for *double* encryption, we can use a simple hybrid argument to show that

$$\mathsf{Adv}^{\pm \mathrm{mu\text{-}prp}}_{\mathrm{DE}[E]}(A) \leq u\left(\frac{p+2q}{2^k}\right)^2 \leq q\left(\frac{p+2q}{2^k}\right)^2 . \tag{3}$$

This bound is already better than the one from (2) – for instance, for roughly $p = q = 2^{k/2}$, this latter bound is still $O(2^{-k/2})$, but (2) gives $\Omega(1)$. However, contrary to the single-encryption case, it is not clear that the bound is tight. We will indeed show a much better bound.

Our Bounds. Our main result shows that the security of double encryption does not degrade substantially in the multi-user setting, and that the bound from (3) is overly pessimistic. In particular, we prove that

$$\mathsf{Adv}^{\pm\mathrm{mu\text{-}prp}}_{\mathrm{DE}[E]}(A) \leq \frac{1}{2^n} + \frac{5q}{2^{k+n/2}} + \frac{6qB^2 + 222BQ^2}{2^{2k}}$$

where $Q = \max\{p, q\}$ and $B = 5\max\{n + k/2, 2q/2^n\}$. This bound is rather cumbersome, but the key observation is that third-degree monomials in p and q all appear with denominator 2^{2k+n}, whereas any term with denominator 2^{2k} is at most *quadratic* in p, q – very similar to the single-user case.

Recall that the meet-in-the-middle attack on the single user security of double encryption succeeds with advantage $p^2/2^{2k}$, and Biham's key-collision attack [10] achieves advantage $q^2/2^{2k}$. Therefore for the setting that $n \geq k$ (such as DES or AES), our bound is tight. For the setting $n \ll k$ (which occurs in Format-Preserving Encryption [6], and several block-cipher designs), finding matching attacks is difficult, and we leave it as an open problem. However, as an intermediate step, we note that most proofs are in models where the keys are revealed to the distinguisher at the end of the execution. In this model, we can give a matching attack (based on the meet-in-the-middle paradigm) that achieves distinguishing advantage

$$\max\{\lfloor n/8 \lg(n) \rfloor, q/2^n\} \cdot \frac{p^2}{3 \cdot 2^{2k}} \,.$$

We discuss attacks below in Sect. 6.

A Disclaimer. We stress that the common wisdom that there is no security increase is obviously still in place. However, the envisioned application is to ciphers whose key length is not an issue in the single-user setting, but becomes too short in a multi-user regime. For instance, a multi-user attack reduces the security of (single) AES128 to 64 bits. Our result shows that iterating AES128 *twice* substantially mitigates the impact of a multi-user attack, and that in fact we obtain almost optimal multi-user security, namely around 115 bits for a total key length of 256 bits. (Also see Fig. 2.)

Techniques. Our result is obtained using new techniques we introduce and that we believe to be of broad applicability in lifting existing analyses from the single-user (su) to the mu setting.

384   V.T. Hoang and S. Tessaro

Hoang and Tessaro (HT) [19] already proposed a generic approach for this purpose. It is illustrative to briefly review it, and see why it fails for double encryption. HT's idea is to show that the construction (e.g., double encryption) satisfies, in the su case, a property called *point-wise proximity*, a stronger property than indistinguishability, already used in previous works (e.g., in [9]). Concretely, this means that there exists a function $\epsilon = \epsilon(p, q)$ of the query parameters p and q, such that for all transcripts τ containing p offline and q online queries, we have

$$\mathsf{ps}_{\mathrm{ideal}}(\tau) - \mathsf{ps}_{\mathrm{real}}(\tau) \le \epsilon(p, q) \cdot \mathsf{ps}_{\mathrm{ideal}}(\tau), \tag{4}$$

where $ps_{\mathrm{ideal}}(\tau)$ and $ps_{\mathrm{real}}(\tau)$ are the so-called ideal and real *interpolation probabilities*. Namely, they describe the probability that the real ($\mathsf{ps}_{\mathrm{real}}$) and the ideal ($\mathsf{ps}_{\mathrm{ideal}}$) worlds behave consistently with the transcript when the queries the transcript contains are asked in that order.

HT show that then point-wise proximity is achieved in the multi-user experiment, where $\epsilon(p, q)$ is replaced by $\epsilon(p+qt, q)$, where t is the number of calls made by the construction to the underlying primitive (in the case of double encryption, $t = 2$). This implies that the distinguishing advantage is also at most $\epsilon(p + qt, q)$. For this argument to hold, however, ϵ needs to be super-additive, i.e., $\epsilon(x, y) + \epsilon(x, z) \le \epsilon(x, x + y)$, and moreover, $\epsilon(\cdot, y)$ and $\epsilon(x, \cdot)$ need to be non-decreasing functions for all $x, y \in \mathbb{N}$. For double encryption, no such ϵ can be established. For instance, the natural candidate $\epsilon(p, q) = \left(\frac{p}{2^k}\right)^2$ is not super-additive, as $\epsilon(x, y) + \epsilon(x, z) = 2\epsilon(x, y + z)$.

We take a different approach, by introducing a relaxed notion of *almost proximity*, which in particular akin to the H-coefficient method (cf. e.g. [12, 26]), introduced a partition the set of *single-user* transcripts into good and bad transcripts, and proximity guarantees are shown only on the former. Our main technical insight is the introduction of a precise framework to mitigate the effects of the growth of the probability of a bad transcript when increasing the number of users. We dispense with a formulation here – the conditions are not concise – and refer the reader to Sect. 3. We note that we also provide simplifications of the framework in Sect. 4, one of which is in particular sufficient for analyzing double encryption. We finally apply it in Sect. 5.

Further Related Work. Multiple encryption has been studied also in the standard computational model, with respect to the question of how it amplifies (weak) PRP security. Luby and Rackoff [21] initially studied double encryption, and bounds for multiple encryption were later provided by Myers [25] and Tessaro [28].

Also, while above we have focused on block cipher analyses, recent works have studied mu security in different contents, in particular for authentication encryption [8] and message-authentication codes [3,4].

2   Preliminaries

Notation. For a finite set S, we let $x \leftarrow_\$ S$ denote the uniform sampling from S and assigning the value to x. Let $|x|$ denote the length of the string x, and for $1 \leq i < j \leq |x|$, let $x[i,j]$ denote the substring from the ith bit to the jth bit (inclusive) of x. If A is an algorithm, we let $y \leftarrow A(x_1, \ldots; r)$ denote running A with randomness r on inputs $x_1, \ldots$ and assigning the output to y. We let $y \leftarrow_\$ A(x_1, \ldots)$ be the resulting of picking r at random and letting $y \leftarrow A(x_1, \ldots; r)$.

Multi-user PRP Security of Blockciphers. Let $\Pi : \mathcal{K} \times \{0,1\}^n \to \{0,1\}^n$ be a blockcipher, which is built on another blockcipher $E : \{0,1\}^k \times \mathcal{M} \to \mathcal{M}$. We associate with Π a key-sampling algorithm Sample. Let A be an adversary. Define

$$\mathsf{Adv}^{\pm\mathrm{mu\text{-}prp}}_{\Pi[E],\mathrm{Sample}}(A) = \Pr[\mathrm{Real}^A_{\Pi[E],\mathrm{Sample}} \Rightarrow 1] - \Pr[\mathrm{Rand}^A_{\Pi[E],\mathrm{Sample}} \Rightarrow 1]$$

where games Real and Rand are defined in Fig. 1. If Sample is the uniform sampling of $\mathcal{K}$ then we only write $\mathsf{Adv}^{\pm\mathrm{mu\text{-}prp}}_{\Pi[E]}(A)$.

proc INITIALIZE() $\mathrm{Real}^A_{\Pi[E],\mathrm{Sample}}$	**proc** INITIALIZE() $\mathrm{Rand}^A_{\Pi[E],\mathrm{Sample}}$
for $i = 1, 2, \ldots$ **do** $K_i \leftarrow_\$ \mathrm{Sample}()$	**for** $i = 1, 2, \ldots$ **do** $f_i \leftarrow_\$ \mathrm{Perm}(\{0,1\}^n)$
proc ENC(i, x) {**return** $\Pi_{K_i}[E](x)$}	**proc** ENC(i, x) {**return** $f_i(x)$}
proc DEC(i, y) {**return** $\Pi^{-1}_{K_i}[E](y)$}	**proc** DEC(i, y) {**return** $f_i^{-1}(y)$}
proc PRIM(J, u) {**return** $E_J(u)$}	**proc** PRIM(J, u) {**return** $E_J(u)$}
proc PRIMINV(J, v) {**return** $E_J^{-1}(v)$ }	**proc** PRIMINV(J, v) {**return** $E_J^{-1}(v)$}

Fig. 1. Games defining the multi-user security of a blockcipher $\Pi : \mathcal{K} \times \{0,1\}^n \to \{0,1\}^n$. This blockcipher is based on another blockcipher $E : \{0,1\}^k \times \{0,1\}^n \to \{0,1\}^n$. The game is associated with a key-sampling algorithm Sample. Here $\mathrm{Perm}(\{0,1\}^n)$ denotes the set of all permutations on $\{0,1\}^n$.

In the games above, we first use Sample to sample keys $K_1, K_2, \ldots \in \mathcal{K}$ for Π, and independent, random permutations $f_1, f_2, \ldots$ on $\mathcal{M}$. The adversary is given four oracles PRIM, PRIMINV, ENC, and DEC. In both games, the oracles PRIM and PRIMINV always give access to the primitive E and its inverse respectively. The ENC and DEC oracles give access to $f_1(\cdot), f_2(\cdot), \ldots$ and their inverses respectively in game Rand, and access to $\Pi[E](K_1, \cdot), \Pi[E](K_2, \cdot), \ldots$ and their inverses in game Real. The adversary finally needs to output a bit to tell which game it is interacting with.

Single and Double Encryption. Let $k, n \in \mathbb{N}$ and let $E : \{0,1\}^k \times \{0,1\}^n \to \{0,1\}^n$ be a blockcipher. The Single Encryption of E is the blockcipher E itself. The Double Encryption $\mathrm{DE}[E]$ of E is a blockcipher with keyspace $(\{0,1\}^k)^2$ and message space $\{0,1\}^n$. On key $K = (J_1, J_2)$ and message $x \in \{0,1\}^n$, $\mathrm{DE}_K[E](x)$ returns $E_{J_2}(E_{J_1}(x))$.

Systems and Transcripts. Following up the notation from [19] (which was in turn inspired by Maurer's framework [22]), it is convenient to consider interactions of a distinguisher A with an abstract system $\mathbf{S}$ which answers A's queries. The resulting interaction then generates a transcript $\tau = ((X_1, Y_1), \ldots, (X_q, Y_q))$ of query-answer pairs. It is well known that $\mathbf{S}$ is entirely described by the probabilities $\mathsf{p}_{\mathbf{S}}(\tau)$ that if we make queries in τ to system $\mathbf{S}$, we will receive the answers as indicated in τ. We say in particular that $\mathbf{S}$ is *stateless* if $\mathsf{p}_{\mathbf{S}}(\tau)$ is invariant under permuting the orders of the input-output pairs it contains.

We will generally describe systems informally, or more formally in terms a set of oracles they provide, and only use the fact that they define a corresponding probabilities $\mathsf{p}_{\mathbf{S}}(\tau)$ without explicitly giving these probabilities.

The Expectation Method. In this paper, we shall use the expectation method of Hoang and Tessaro [19]. For a pair of systems $\mathbf{S}_{\mathrm{real}}$ and $\mathbf{S}_{\mathrm{ideal}}$, this method aims to bound the gap $\mathsf{p}_{\mathbf{S}_{\mathrm{ideal}}}(\tau) - \mathsf{p}_{\mathbf{S}_{\mathrm{real}}}(\tau)$, for a fixed (su) transcript τ such that $\mathsf{p}_{\mathbf{S}_{\mathrm{ideal}}}(\tau) > 0$. Under this method, one extends the transcript with a random variable S. In $\mathbf{S}_{\mathrm{real}}$, this S is often a part of the key and suppose that it has marginal distribution μ. In $\mathbf{S}_{\mathrm{ideal}}$, we pick S of the same marginal distribution μ, but independent of τ. Let $\mathsf{p}_{\mathbf{S}_{\mathrm{real}}}(\tau, s)$ denote the probability that $\mathbf{S}_{\mathrm{real}}$ behaves according to τ, and S agrees with s. Let $\mathsf{p}_{\mathbf{S}_{\mathrm{ideal}}}(\tau, s)$ denote the probability that $\mathbf{S}_{\mathrm{ideal}}$ behaves according to τ, and $S \leftarrow_{\$} \mu$ agrees with s. Under the expectation method, one partitions the range of S into two sets, Γ_{good} and Γ_{bad}. For s such that $\mathsf{p}_{\mathbf{S}_{\mathrm{ideal}}}(\tau, s) > 0$, if $s \in \Gamma_{\mathrm{bad}}$ then we say that s is *bad*; otherwise s is *good*. We write $\Pr[S \in \Gamma_{\mathrm{bad}}]$ to denote the probability that $S \leftarrow_{\$} \mu$ independent of τ is bad. Hoang and Tessaro give the following result.

Lemma 1 (The expectation method). *[19] Fix a su transcript τ such that $\mathsf{p}_{\mathbf{S}_{\mathrm{ideal}}}(\tau) > 0$. Assume that there is a partition Γ_{good} and Γ_{bad} of the range $\mathcal{U}$ of S, as well as a function $g : \mathcal{U} \to [0, \infty)$ such that $\Pr[S \in \Gamma_{\mathrm{bad}}] \leq \delta$ and for all $s \in \Gamma_{\mathrm{good}}$,*

$$1 - \frac{\mathsf{p}_{\mathbf{S}_{\mathrm{real}}}(\tau, s)}{\mathsf{p}_{\mathbf{S}_{\mathrm{ideal}}}(\tau, s)} \leq g(s).$$

Then

$$\mathsf{p}_{\mathbf{S}_{\mathrm{ideal}}}(\tau) - \mathsf{p}_{\mathbf{S}_{\mathrm{real}}}(\tau) \leq (\delta + \mathbf{E}[g(S)]) \cdot \mathsf{p}_{\mathbf{S}_{\mathrm{ideal}}}(\tau). \qquad \square$$

Note that in Lemma 1, the expectation is taken over all possible (good or bad) values of S.

3   A Generic Method to Bound Multi-user Security

In this section we present a generic method to prove information-theoretic mu security bounds, based (mostly) on upper bounding single-user quantities. The framework is very general, and in fact generalizes the approach by Hoang and Tessaro [19] based on pointwise proximity.

The Generic Setting. We consider two (stateless) systems $\mathbf{S}_{\text{real}}$ and $\mathbf{S}_{\text{ideal}}$, called the *real* and *ideal* systems, respectively. Each of these two systems can be invoked via two oracles CONS and PRIM, allowing for *construction* and *primitive* queries, respectively. First off, PRIM gives access to an ideal primitive (for example, an ideal cipher, a random function or permutation), whereas CONS's role depends on the context, but always answers queries of the form (i, X), where i is the index of a user and X is the query for that user. More specifically:

1. In $\mathbf{S}_{\text{real}}$, the oracle CONS upon a query (i, X) invokes a construction Π which makes calls to PRIM, and additionally depends on some local, initially chosen randomness (or key) K_i. That is, the output is $\Pi^{\text{PRIM}}(K_i, X)$.
2. In $\mathbf{S}_{\text{ideal}}$, the oracle CONS samples independent functions $f_1, f_2, \ldots$ from some distribution, and answers a query (i, X) as $f_i(X)$.

For example, the game from Fig. 1 can be described as suitable systems $\mathbf{S}_{\text{real}}$ and $\mathbf{S}_{\text{ideal}}$: We would simply handle inversion queries (to DEC and PRIMINV) by specifying the direction of the query in the input given to CONS and PRIM, i.e., $X = (+, x)$ or $X = (-, y)$. Also, we can model more complex scenarios, like the security of authenticated encryption schemes, as long as we can map the security notion to suitable $\mathbf{S}_{\text{real}}$ and $\mathbf{S}_{\text{ideal}}$.

We generally will assume that there exists a metric of *data complexity* associated with queries made to CONS. For instance, if CONS takes variable-length inputs, σ could be number bits queried to it, whereas if the input length is fixed, this could just be the number of queries. We assume that there exists a parameter t indicating that when answering multiple queries with overall data complexity σ, Π makes at most $t \cdot \sigma$ queries to PRIM.

The Distinguishing Problem. For any adversary A and a system $\mathbf{S}$, we let $\text{Script}(A, \mathbf{S})$ denote the random variable for the transcript of the interaction of A and $\mathbf{S}$. Recall that the advantage of the adversary in distinguishing two systems $\mathbf{S}_{\text{real}}$ and $\mathbf{S}_{\text{ideal}}$ is at most the statistical distance between the distributions of the adversary's transcript in the real and ideal games, which is

$$\text{Adv}^{\text{dist}}_{\mathbf{S}_{\text{real}}, \mathbf{S}_{\text{ideal}}}(A) \leq \sum_{\tau} \max\{0, \mathsf{ps}_{\mathbf{S}_{\text{ideal}}}(\tau) - \mathsf{ps}_{\mathbf{S}_{\text{real}}}(\tau)\}, \tag{5}$$

where the sum is taken over all τ such that $\Pr[\text{Script}(A, \mathbf{S}_{\text{ideal}}) = \tau] > 0$.

Note that there might be some context-dependent constraints on the adversary's queries. For example, if part of the inputs to CONS include nonces to a nonce-based authenticated encryption, then one might require that the nonces will not repeat. This is easy to handle, since it will only restrict the set of valid transcripts to be considered. We will usually capture the complexity of A in terms of the number of PRIM queries, p, the number of CONS queries q, and the overall data complexity σ for the queries made to CONS. A security bound ϵ is then viewed as a function $\epsilon(p, q, \sigma)$. We say that a function $\epsilon(\cdot, \cdot, \cdot) : \mathbb{N}^3 \to [0, 1]$ is *monotonic* if $\epsilon(\cdot, y, z)$, $\epsilon(x, \cdot, z)$, and $\epsilon(x, y, \cdot)$ are increasing functions, for any $x, y, z \in \mathbb{N}$. Often security bounds are monotonic functions, since increasing the adversary's resources can only help it.

Almost Proximity. We now establish a condition on $\mathbf{S}_{\mathrm{real}}$ that we call *almost proximity*, which will allow us to establish mu security from a number of functions, δ_0, δ_1 and δ_2, we define next. In particular, some of these functions (δ_1 and δ_2) are defined with respect to single-user (su) transcript, i.e., transcripts were all queries to CONS are of the form (i, X) *for one single i*.

One begins by defining a context-dependent, undesirable property on su transcripts that we call *bad*, and if a su transcript is not bad then it is *good*. We partition in particular the set of bad transcripts into two sets, $\mathcal{S}$ and $\mathcal{S}'$. In many cases (such as our Double Encryption application below), one of the two sets $\mathcal{S}$ and $\mathcal{S}'$ is simply the empty set, but we envision more general application scenarios.

Further, we will assume that there exists a function Rate such that for any good su transcript τ,

$$\mathsf{ps}_{\mathrm{ideal}}(\tau) - \mathsf{ps}_{\mathrm{real}}(\tau) \le \mathrm{Rate}(\tau) \cdot \mathsf{ps}_{\mathrm{ideal}}(\tau) \,,$$

where Rate is in particular an *increasing* function mapping a transcript to a number in $[0, 1]$, meaning that for any transcripts τ and τ' such that τ' contains all the query-answer pairs of τ (possibly in a different order), we have $\mathrm{Rate}(\tau') \ge \mathrm{Rate}(\tau)$.

Then, we also assume that there is a monotonic function δ_2 such that for any adversary B attacking a single user via p PRIM queries, q ENC queries with overall data complexity σ, we have

$$\Pr[\mathrm{Script}(B, \mathbf{S}_{\mathrm{ideal}}) \in \mathcal{S}] \le \delta_2(p, q, \sigma).$$

Note that the bound above is with respect to the ideal system, $\mathbf{S}_{\mathrm{ideal}}$, and thus often easy to compute.

We also define another, context-dependent, desired property on mu transcripts that we call *nice* — we let $\mathcal{N}$ be the set of all nice mu transcripts. (We stress that niceness is with respect to mu transcripts, whereas being good/bad is only with respect to su ones.) The notion of niceness involves only the CONS query-answer pairs: for any two transcripts τ and τ' that have the same CONS query-answer pairs (possibly in different orders), if $\tau \in \mathcal{N}$ then so is τ'. Also, for a mu transcript τ involving queries to exactly r users, and for each $i \in \{1, \ldots, r\}$, let $\mathrm{Map}(i, \tau)$ denote the su transcript obtained by deleting the CONS$(j, \cdot)$ queries and answers for any $j \ne i$. We require the following conditions:

- For any transcript $\tau \in \mathcal{N}$ and all i, $\mathrm{Map}(i, \tau) \notin \mathcal{S}'$.
- There is a monotonic function δ_0 such that for any mu adversary A making p PRIM queries, q CONS queries, and data complexity σ,

$$\Pr[\mathrm{Script}(A, \mathbf{S}_{\mathrm{ideal}}) \notin \mathcal{N}] \leq \delta_0(p, q, \sigma).$$

- There is a monotonic function δ_1 such that for any $\tau \in \mathcal{N}$ of r users that contains p PRIM, q CONS queries of total data complexity at most σ,

$$\sum_{i=1}^{r} \mathrm{Rate}(\mathrm{Map}(i, \tau)) \leq \delta_1(p, q, \sigma). \tag{6}$$

We refer to this last property as mu-boundedness.

We refer to the existence of suitable functions $\delta_0, \delta_1, \delta_2$ for corresponding Rate, Map, $\mathcal{S}$, $\mathcal{S}'$ and $\mathcal{N}$ as meeting the *almost-proximity* conditions.

Mu Security via Almost Proximity. The following result bounds the mu advantage in distinguishing $\mathbf{S}_{\mathrm{real}}$ and $\mathbf{S}_{\mathrm{ideal}}$, granted the almost-proximity conditions defined above are met.

Lemma 2 (Mu-security via almost proximity). *Assume that the almost-proximity conditions above are met, for some δ_2, δ_0 and δ_1. Then for any adversary A that makes at most q CONS queries of total data complexity σ, and p PRIM queries, we have*

$$\mathsf{Adv}^{\mathrm{dist}}_{\mathbf{S}_{\mathrm{real}}, \mathbf{S}_{\mathrm{ideal}}}(A) \leq \delta_0(p, q, \sigma) + 2\delta_1(p + t\sigma, q, \sigma) + 2q \cdot \delta_2(p + t\sigma, q, \sigma).$$

Discussion. A meaningful question is why we need to separate the set of bad su transcripts into $\mathcal{S}$ and $\mathcal{S}'$. The reason is that, when we move from su to mu setting, under our method, the term δ_2 will blow up to $q\delta_2$, which is similar to the hybrid argument. To avoid an inferior mu bound, we would like to minimize the term δ_2 as much as possible, by carving out $\mathcal{S}'$ from the set of bad su transcripts. Due to the requirement that $\mathrm{Map}(i, \tau) \notin \mathcal{S}'$ for every nice mu transcript τ and every i, the set $\mathcal{S}'$ and the notion of niceness needs to be chosen in tandem to minimize $q\delta_2 + \delta_0(p, q, \sigma)$. Bounding $\Pr[\mathrm{Script}(A, \mathbf{S}_{\mathrm{ideal}}) \notin \mathcal{N}]$ requires working directly in the mu setting, but recall that we are in the ideal game, which is often simple to deal with.

Proof (of Lemma 2). Since we consider a computationally unbounded adversary, without loss of generality, assume that the adversary is deterministic. For simplicity, from this point, we will write δ_2 and δ_1 instead of $\delta_2(p + t\sigma, q, \sigma)$ and $\delta_1(p + t\sigma, q, \sigma)$. Without loss of generality, assume that $\delta_1 < 1/2$; otherwise the the claimed bound in the statement of this lemma is moot. We also assume that the adversary's transcript involves at most r users.

Restricting to Nice Transcripts. Recall that in the ideal system, the probability that the adversary A can produce a mu transcript that is not nice is at most $\delta_0(p, q, \sigma)$. From Eq. (5), what is left is to show that

$$\sum_\tau \mathsf{ps}_{\mathrm{ideal}}(\tau) - \mathsf{ps}_{\mathrm{real}}(\tau) \leq 2\delta_1 + 2q\delta_2, \tag{7}$$

where the sum in the left hand side is taken over all nice transcripts τ in the support $\mathrm{supp}(\mathrm{Script}(A, \mathbf{S}_{\mathrm{ideal}}))$ of $\mathrm{Script}(A, \mathbf{S}_{\mathrm{ideal}})$ such that $\mathsf{ps}_{\mathrm{ideal}}(\tau) > \mathsf{ps}_{\mathrm{real}}(\tau)$. Below, when we talk about a *valid* transcript τ, this means that τ meets the constraint above.

Building Hybrids. For each $i \in \{0, \ldots, r\}$, consider the hybrid system $\mathbf{S}_i$ that provides the interface compatible with the real and ideal systems, but queries for user u_j are answered via the actual construction $\Pi^{\mathrm{PRIM}}(K_j, \cdot)$ for $j > i$, and via an independent, perfect simulation of the $\mathrm{CONS}(j, \cdot)$ oracle of the ideal game if $j \leq i$. Then $\mathbf{S}_0 = \mathbf{S}_{\mathrm{real}}$ and $\mathbf{S}_r = \mathbf{S}_{\mathrm{ideal}}$ and thus for any valid transcript τ,

$$\mathsf{ps}_{\mathrm{ideal}}(\tau) - \mathsf{ps}_{\mathrm{real}}(\tau) = \sum_{i=1}^r \mathsf{ps}_{\mathbf{S}_i}(\tau) - \mathsf{ps}_{\mathbf{S}_{i-1}}(\tau). \tag{8}$$

Let B_i be the following hybrid su adversary. It samples key K_j for Π^{PRIM} for every $i < j \leq r$, and then runs A. Queries for user u_j are answered via $\Pi^{\mathrm{PRIM}}(K_j, \cdot)$ if $j > i$, and via the $\mathrm{CONS}(1, \cdot)$ oracle of B_i if $j = i$, and via an independent, perfect simulation of the $\mathrm{CONS}(j, \cdot)$ oracle of the ideal game if $j < i$. In other words, adversary B_i simulates system $\mathbf{S}_{i-1}$ in its su real game, and simulates system $\mathbf{S}_i$ in its su ideal game. It makes at most q CONS queries of total data complexity σ and at most $p + t\sigma$ PRIM queries.

Reducing to Transcript-Wise Gap. Fix a valid transcript τ. Let $\mathcal{T}(i, \tau)$ denote the set of extended transcripts of B_i in its su ideal game that are enhanced with the simulated CONS queries and answers as well as the simulated keys K_j, such that the corresponding simulated transcript for A is τ. For each $\tau_i \in \mathcal{T}(i, \tau)$, let $\mathrm{Tr}(\tau_i)$ be the transcript of B_i derived from τ_i. For $\mathbf{S} \in \{\mathbf{S}_{\mathrm{real}}, \mathbf{S}_{\mathrm{ideal}}\}$, let $\mathsf{ps}(\tau_i)$ denote the probability that, when B_i interacts with $\mathbf{S}$, its enhanced transcript is τ_i. Note that compared to $\mathrm{Tr}(\tau_i)$, the additional information τ_i contains is the keys K_j, and the queries/answers on the simulated oracle $\mathrm{CONS}(j, \cdot)$ of the ideal game for users $j < i$. Since this information is independent of $\mathbf{S}_{\mathrm{real}}$ and $\mathbf{S}_{\mathrm{ideal}}$,

$$\frac{\mathsf{ps}_{\mathrm{real}}(\tau_i)}{\mathsf{ps}_{\mathrm{ideal}}(\tau_i)} = \frac{\mathsf{ps}_{\mathrm{real}}(\mathrm{Tr}(\tau_i))}{\mathsf{ps}_{\mathrm{ideal}}(\mathrm{Tr}(\tau_i))}. \tag{9}$$

Let $\mathcal{S}_i$ be the set of extended transcripts τ_i of B_i such that $\mathrm{Tr}(\tau_i) \in \mathcal{S}$. We claim that

$$\mathsf{ps}_{\mathrm{ideal}}(\tau) - \mathsf{ps}_{\mathrm{real}}(\tau) \le 2\Big(\sum_{i=1}^{r} \sum_{\tau_i \in \mathcal{T}(i,\tau) \cap \mathcal{S}_i} \mathsf{ps}_{\mathrm{ideal}}(\tau_i)\Big) + 2\delta_1 \sum_{\tau_1 \in \mathcal{T}(1,\tau)} \mathsf{ps}_{\mathrm{real}}(\tau_1)$$

$$= 2\Big(\sum_{i=1}^{r} \sum_{\tau_i \in \mathcal{T}(i,\tau) \cap \mathcal{S}_i} \mathsf{ps}_{\mathrm{ideal}}(\tau_i)\Big) + 2\delta_1 \cdot \mathsf{ps}_{\mathrm{real}}(\tau)$$

$$\le 2\Big(\sum_{i=1}^{r} \sum_{\tau_i \in \mathcal{T}(i,\tau) \cap \mathcal{S}_i} \mathsf{ps}_{\mathrm{ideal}}(\tau_i)\Big) + 2\delta_1 \cdot \mathsf{ps}_{\mathrm{ideal}}(\tau), \qquad (10)$$

where the last inequality is due to the assumption that τ is valid. This claim will be justified later. By summing both sides of Eq. (10) over all valid τ, we can bound the left-hand side of Eq. (7) by

$$2\Big(\sum_{i=1}^{r} \Pr[\mathrm{Script}(B_i, \mathbf{S}_{\mathrm{ideal}}) \in \mathcal{S}]\Big) + 2\delta_1 \le 2q \cdot \delta_2 + 2\delta_1$$

which is the right-hand side of Eq. (7). To justify Eq. (10), note that

$$\mathsf{ps}_{\mathrm{ideal}}(\tau) - \mathsf{ps}_{\mathrm{real}}(\tau) = \sum_{i=1}^{r} \mathsf{ps}_i(\tau) - \mathsf{ps}_{i-1}(\tau).$$

Moreover, for each $i \le r$,

$$\mathsf{ps}_i(\tau) = \sum_{\tau_i \in \mathcal{T}(i,\tau)} \mathsf{ps}_{\mathrm{ideal}}(\tau_i),$$

whereas

$$\mathsf{ps}_{i-1}(\tau) \ge \sum_{\tau_i \in \mathcal{T}(i,\tau)} \mathsf{ps}_{\mathrm{real}}(\tau_i),$$

because (a) the left-hand side is the chance that adversary B_i in its real world (recall that the real world of B_i is the ideal world of B_{i-1}) can generate τ, which is $\sum_{\tau'} \mathsf{ps}_{\mathrm{real}}(\tau')$ over all enhanced transcripts τ' that B_i can witness such that the corresponding transcript for A is τ, and (b) the right-hand side is $\sum_{\tau'} \mathsf{ps}_{\mathrm{real}}(\tau')$ over *some* (but probably not all) such τ'. Hence

$$\mathsf{ps}_{\mathrm{ideal}}(\tau) - \mathsf{ps}_{\mathrm{real}}(\tau) \le \sum_{i=1}^{r} \sum_{\tau_i \in \mathcal{T}(i,\tau)} \mathsf{ps}_{\mathrm{ideal}}(\tau_i) - \mathsf{ps}_{\mathrm{real}}(\tau_i)$$

$$= \Big(\sum_{i=1}^{r} \sum_{\tau_i \in \mathcal{T}(i,\tau) \cap \mathcal{S}_i} \mathsf{ps}_{\mathrm{ideal}}(\tau_i) - \mathsf{ps}_{\mathrm{real}}(\tau_i)\Big) + \sum_{i=1}^{r} \sum_{\tau_i \in \mathcal{T}(i,\tau) \setminus \mathcal{S}_i} \mathsf{ps}_{\mathrm{ideal}}(\tau_i) - \mathsf{ps}_{\mathrm{real}}(\tau_i)$$

$$\le \Big(\sum_{i=1}^{r} \sum_{\tau_i \in \mathcal{T}(i,\tau) \cap \mathcal{S}_i} \mathsf{ps}_{\mathrm{ideal}}(\tau_i)\Big) + \sum_{i=1}^{r} \sum_{\tau_i \in \mathcal{T}(i,\tau) \setminus \mathcal{S}_i} \mathsf{ps}_{\mathrm{ideal}}(\tau_i) - \mathsf{ps}_{\mathrm{real}}(\tau_i).$$

What is left is to prove that

$$\sum_{i=1}^{r} \sum_{\tau_i \in \mathcal{T}(i,\tau) \backslash \mathcal{S}_i} \mathsf{ps}_{\mathrm{ideal}}(\tau_i) - \mathsf{ps}_{\mathrm{real}}(\tau_i)$$

$$\leq \left(\sum_{i=1}^{r} \sum_{\tau_i \in \mathcal{T}(i,\tau) \cap \mathcal{S}_i} \mathsf{ps}_{\mathrm{ideal}}(\tau_i) \right) + 2\delta_1 \sum_{\tau_1 \in \mathcal{T}(1,\tau)} \mathsf{ps}_{\mathrm{real}}(\tau_1). \tag{11}$$

Now, recall that for each $\tau_i \in \mathcal{T}(i,\tau) \backslash \mathcal{S}_i$, the su transcript $\mathrm{Tr}(\tau_i)$ is good. Since the two systems satisfy the almost proximity condition,

$$\mathsf{ps}_{\mathrm{ideal}}(\mathrm{Tr}(\tau_i)) - \mathsf{ps}_{\mathrm{real}}(\mathrm{Tr}(\tau_i)) \leq \mathrm{Rate}(\mathrm{Tr}(\tau_i)) \cdot \mathsf{ps}_{\mathrm{ideal}}(\mathrm{Tr}(\tau_i)).$$

Recall that from Eq. (9), the ratio between $\mathsf{ps}_{\mathrm{ideal}}(\mathrm{Tr}(\tau_i))$ and $\mathsf{ps}_{\mathrm{real}}(\mathrm{Tr}(\tau_i))$ is exactly that between $\mathsf{ps}_{\mathrm{ideal}}(\tau_i)$ and $\mathsf{ps}_{\mathrm{real}}(\tau_i)$. Then

$$\mathsf{ps}_{\mathrm{ideal}}(\tau_i) - \mathsf{ps}_{\mathrm{real}}(\tau_i) \leq \mathrm{Rate}(\mathrm{Tr}(\tau_i)) \cdot \mathsf{ps}_{\mathrm{ideal}}(\tau_i). \tag{12}$$

This in turn implies that

$$\mathsf{ps}_{\mathrm{ideal}}(\tau_i) \leq \frac{\mathsf{ps}_{\mathrm{real}}(\tau_i)}{1 - \mathrm{Rate}(\mathrm{Tr}(\tau_i))}. \tag{13}$$

To justify that the denominator of the right-hand side is nonzero so that Eq. (13) above is well-defined, let τ' be the mu transcript that has the same CONS queries/answers as τ, and the same PRIM queries/answers as τ_i. Since τ is nice, so is τ'. Thus, $1 - \mathrm{Rate}(\mathrm{Tr}(\tau_i)) = 1 - \mathrm{Rate}(\mathrm{Map}(i,\tau')) \geq 1 - \delta_1 > 0$. From Eq. (12), to justify Eq. (11), we need to bound each sum

$$\sum_{\tau_i \in \mathcal{T}(i,\tau) \backslash \mathcal{S}_i} \mathrm{Rate}(\mathrm{Tr}(\tau_i)) \cdot \mathsf{ps}_{\mathrm{ideal}}(\tau_i),$$

for every $i \in \{1,\ldots,r\}$. For $\ell \leq i$, define $\mathrm{Rate}(i,\tau_\ell)$ as follows. Let τ' be the su transcript induced by τ_ℓ in which we only keep CONS queries/answers for user u_i, and all PRIM queries/answers. Let $\mathrm{Rate}(i,\tau_\ell) = \mathrm{Rate}(\tau')$. The special case $\mathrm{Rate}(i,\tau_i)$ coincides with $\mathrm{Rate}(\mathrm{Tr}(\tau_i))$. We claim that for each i, the sum above is at most

$$\sum_{\tau_1 \in \mathcal{T}(1,\tau)} 2\mathrm{Rate}(i,\tau_1) \cdot \mathsf{ps}_{\mathrm{real}}(\tau_1) + \sum_{s=1}^{i} \sum_{\tau_s \in \mathcal{T}(s,\tau) \cap \mathcal{S}_s} 2\mathrm{Rate}(i,\tau_s) \cdot \mathsf{ps}_{\mathrm{ideal}}(\tau_s). \tag{14}$$

Note that for any $s \geq 1$ and any $\tau_s \in \mathcal{T}(s,\tau)$, if we let τ' be the mu transcript that has the same CONS queries/answers as τ, and the same PRIM queries/answers as τ_s, then τ' is also nice, because τ is nice. Then

$$\sum_{i=s}^{r} \mathrm{Rate}(i,\tau_s) = \sum_{i=s}^{r} \mathrm{Rate}(\mathrm{Map}(i,\tau')) \leq \delta_1. \tag{15}$$

From Eq. (15),

$$\sum_{i=1}^{r}\sum_{\tau_1\in\mathcal{T}(1,\tau)} 2\mathrm{Rate}(i,\tau_1)\cdot \mathsf{ps}_{\mathrm{real}}(\tau_1) \le \sum_{\tau_1\in\mathcal{T}(1,\tau)} 2\delta_1\cdot \mathsf{ps}_{\mathrm{real}}(\tau_1), \qquad (16)$$

and

$$\sum_{i=1}^{r}\sum_{s=1}^{i}\sum_{\tau_s\in\mathcal{T}(s,\tau)\cap\mathcal{S}_s} 2\mathrm{Rate}(i,\tau_s)\cdot \mathsf{ps}_{\mathrm{ideal}}(\tau_s)$$

$$=\sum_{s=1}^{r}\sum_{\tau_s\in\mathcal{T}(s,\tau)\cap\mathcal{S}_s}\sum_{i=s}^{r} 2\mathrm{Rate}(i,\tau_s)\cdot \mathsf{ps}_{\mathrm{ideal}}(\tau_s) \qquad (17)$$

$$\le \sum_{s=1}^{r}\sum_{\tau_s\in\mathcal{T}(s,\tau)\cap\mathcal{S}_s} 2\delta_1\cdot \mathsf{ps}_{\mathrm{ideal}}(\tau_s) \le \sum_{s=1}^{r}\sum_{\tau_s\in\mathcal{T}(s,\tau)\cap\mathcal{S}_s} \mathsf{ps}_{\mathrm{ideal}}(\tau_s). \qquad (18)$$

Combining Eqs. (12), (14), (16), and (18) gives us Eq. (11).

To justify Eq. (14), fix $i \in \{1, \ldots, r\}$. We create a binary tree whose weight at the root is exactly the sum above for i. In this tree, for any two children of a node, the left one must be a leaf node. Moreover, we will put weights on the nodes so that the weight of a parent node is bounded by the sum of the weights of its children. Hence the weight at the root is bounded by the total weight of the leaves.

Starting at the root, from Eq. (13), we can bound the weight at the root by a linear combination of $\mathsf{ps}_{\mathrm{real}}(\tau_i)$, where $\tau_i \in \mathcal{T}(i,\tau)\backslash\mathcal{S}_i$. For each such τ_i, if we enhance it with the key of user u_i and the internal PRIM queries/answers due to the CONS queries of user u_i then we will get an extended transcript τ_{i-1} for adversary B_{i-1}. (Recall that the real world of B_i is the ideal world of B_{i-1}.) Hence the linear combination of $\mathsf{ps}_{\mathrm{real}}(\tau_i)$ becomes a linear combination of $\mathsf{ps}_{\mathrm{ideal}}(\tau_{i-1})$, for $\tau_{i-1} \in \mathcal{T}(i-1,\tau)$. We divide this into two parts, one for $\tau_{i-1} \in \mathcal{S}_{i-1}$, and another for $\tau_{i-1} \notin \mathcal{S}_{i-1}$. The first partial sum will be the weight of the left child of the root, and the second the weight of the right child. So far, we have placed the weights up to the second level of the tree. We will repeat the process above, starting at the right child of the root, until we reach the i-th level. At that point, the weight of the right-most leaf is a linear combination of $\mathsf{ps}_{\mathrm{ideal}}(\tau_1)$, for $\tau_1 \in \mathcal{T}(1,\tau)$.

Recall that the weight of each node of the binary tree above is a linear combination. We now specify the coefficients. At the root, each coefficient for $\mathsf{ps}_{\mathrm{ideal}}(\tau_i)$ is $\mathrm{Rate}(i,\tau_i)$. We will have to bound $\mathsf{ps}_{\mathrm{ideal}}(\tau_i)$ via $\mathsf{ps}_{\mathrm{real}}(\tau_i)$ by Eq. (13), so the coefficients for the left and right children of the root are at most

$$\frac{\mathrm{Rate}(i,\tau_i)}{1-\mathrm{Rate}(i,\tau_i)} \le \frac{\mathrm{Rate}(i,\tau_{i-1})}{1-\mathrm{Rate}(\tau_{i-1})},$$

where the inequality is due to the fact that Rate is increasing and τ_{i-1} contains all queries/answers of τ_i, and thus $\mathrm{Rate}(i-1,\tau_{i-1}) \ge \mathrm{Rate}(i,\tau_i)$. By repeating this process, for nodes at the $(i+1-s)$-th level, the coefficients are at most

$$\frac{\text{Rate}(i, \tau_s)}{\prod_{\ell=s+1}^{i} \big(1 - \text{Rate}(\ell, \tau_s)\big)}.$$

Now, for the right most leaf, its weight is currently a linear combination of $\text{ps}_{\text{ideal}}(\tau_1)$, but we want to have its weight as a linear combination of $\text{ps}_{\text{real}}(\tau_1)$ instead. To achieve this, we will again use Eq. (13) (but i replaced by 1), and the new coefficients for this leaf are at most

$$\frac{\text{Rate}(i, \tau_1)}{\prod_{\ell=1}^{i} \big(1 - \text{Rate}(\ell, \tau_1)\big)}.$$

Hence the coefficients for a leaf at the $(i + 1 - s)$-th level of the tree are at most

$$\frac{\text{Rate}(i, \tau_s)}{\prod_{\ell=s}^{i} \big(1 - \text{Rate}(\ell, \tau_s)\big)} \leq \frac{\text{Rate}(i, \tau_s)}{1 - \sum_{\ell=s}^{i} \text{Rate}(\ell, \tau_s)} \leq \frac{\text{Rate}(i, \tau_s)}{1 - \delta_1} \leq 2\text{Rate}(i, \tau_s),$$

where the first inequality is due to the fact that $(1 - x)(1 - y) \geq 1 - x - y$ for any $0 \leq x, y < 1$, and the second inequality is due to Eq. (15). The total weight of the leaves therefore is at most

$$\sum_{\tau_1 \in \mathcal{T}(1, \tau)} 2\text{Rate}(i, \tau_1) \cdot \text{ps}_{\text{real}}(\tau_1) + \sum_{s=1}^{i} \sum_{\tau_s \in \mathcal{T}(s, \tau) \cap \mathcal{S}_s} 2\text{Rate}(i, \tau_s) \cdot \text{ps}_{\text{ideal}}(\tau_s).$$

This concludes the proof.   □

4   Simplification of the Framework for Specific Settings

Since the framework in Sect. 3 aims to provide an umbrella for *all* settings, it appears unnecessarily complex in many important settings. To improve the usability of our framework, in this section, we consider some simplified treatments of our general framework for specific settings. Each such specialized result is somewhat more limited in scope, but simpler to use.

4.1   A Simple Specialization of the Framework

We now describe a specialization of the framework that is very simple, but might be powerful enough for typical real-world cryptographic schemes, such as the authenticated encryption scheme GCM [23]. This simple treatment however is not enough for Double Encryption, and thus in the next subsection, we will consider another specialized result of the general framework to handle Double Encryption.

The Setting. Here we still use the generic setting as stated in Sect. 3, but make an assumption on the metric σ. For a mu transcript τ and each user u_i of τ, let $\text{Map}(i, \tau)$ be the induced su transcript for user u_i that consists of the $\text{Cons}(i, \cdot)$ queries/answers and $\text{Prim}(\cdot)$ queries/answers of τ. We require that for any mu transcript τ, if the CONS queries in τ have data complexity σ, and those in each Map(i, τ) have data complexity σ_i, then

$$\sum_i \sigma_i \leq \sigma.$$

This requirement obviously holds if we let, for example, σ be the total length of the CONS queries.

Super-Additivity. For a function $\delta : (\mathbb{N})^3 \to [0, 1]$, we say that it is *super-additive* if

$$\delta(x, y_0, z_0) + \delta(x, y_1, z_1) \leq \delta(x, y_0 + y_1, z_0 + z_1)$$

for every $x, y_0, y_1, z_0, z_1 \in \mathbb{N}$. In many schemes, the desired bounds (such as $\delta(p, q, \sigma) = \sigma^2/2^n$) are often super-additive.

The Technique. One begins by defining an undesirable property on su transcripts that involves only CONS queries/answers. If a su transcript has this property then we say that it is *bad*, otherwise it is *good*.[1] A mu transcript τ is *nice* if there is no user u_i such that its induced su transcript Map(i, τ) is bad. Let $\mathcal{N}$ be the set of nice mu transcripts. We require that there be a monotonic function δ such that for any adversary A making p PRIM queries and q Cons queries of data complexity σ,

$$\Pr[\text{Script}(A, \mathbf{S}_{\text{ideal}}) \notin \mathcal{N}] \leq \delta(p, q, \sigma), \tag{19}$$

where for any system $\mathbf{S}$, Script$(A, \mathbf{S})$ denotes the random variable for the transcript of the interaction of A and $\mathbf{S}$. Moreover, we require that there be a monotonic function ϵ' and a super-additive, monotonic function ϵ such that for any good su transcript τ of p PRIM queries and q CONS queries of data complexity σ,

$$\mathsf{ps}_{\text{ideal}}(\tau) - \mathsf{ps}_{\text{real}}(\tau) \leq (\epsilon(p, q, \sigma) + \epsilon'(p, q, \sigma)) \cdot \mathsf{ps}_{\text{ideal}}(\tau). \tag{20}$$

Lemma 3. *Assume that the systems* $\mathbf{S}_{\text{real}}$ *and* $\mathbf{S}_{\text{ideal}}$ *meet the conditions in Eqs.* (19) *and* (20). *Then*

$$\mathsf{Adv}^{\text{dist}}_{\mathbf{S}_{\text{real}}, \mathbf{S}_{\text{ideal}}}(A) \leq \delta(p, q, \sigma) + 2\epsilon(p + t\sigma, q, \sigma) + 2q \cdot \epsilon'(p + t\sigma, q, \sigma).$$

Proof. For a su transcript τ of p PRIM queries and q CONS queries of data complexity σ, let

$$\text{Rate}(\tau) = \epsilon(p, q, \sigma) + \epsilon'(p, q, \sigma).$$

This function Rate is increasing, in the sense that if τ' contains all the query-answer pairs of τ, then Rate$(\tau') \geq$ Rate(τ). To use Lemma 2, we need to establish

---

[1] In Sect. 3, we partitioned the set of bad su transcripts into $\mathcal{S}$ and $\mathcal{S}'$, and required that it is unlikely for the adversary to produce a bad transcript in $\mathcal{S}$. Here $\mathcal{S}$ is simply the empty set.

the mu-boundedness of Rate. We claim that for any nice mu transcript τ of r users, using p PRIM queries and q CONS queries of data complexity σ,

$$\sum_{i=1}^{r} \text{Rate}(\text{Map}(i,\tau)) \le \epsilon(p,q) + q\epsilon'(p,q).$$

To justify this, suppose that τ_i contains q_i CONS queries of data-complexity σ_i. Then

$$\sum_{i=1}^{r} \text{Rate}(\text{Map}(i,\tau)) = \sum_{i=1}^{r} \epsilon(p,q_i,\sigma_i) + \epsilon'(p,q_i,\sigma_i)$$

$$\le \sum_{i=1}^{r} \epsilon(p,q_i,\sigma_i) + \epsilon'(p,q,\sigma)$$

$$\le \epsilon(p,q,\sigma) + r \cdot \epsilon'(p,q,\sigma) \le \epsilon(p,q,\sigma) + q \cdot \epsilon'(p,q,\sigma).$$

Finally, applying Lemma 2 for $\delta_0 = \delta$, $\delta_1 = \epsilon + q\epsilon'$, and $\delta_2 = 0$, leads to the claimed advantage. □

4.2   The Specialized Framework for Double Encryption and Beyond

We now specialize the general framework into a more specific result that covers the case of Single Encryption, Double Encryption, and Key-Alternating Cipher (KAC) [11]. This result explains why these constructions, despite being somewhat similar in the structure, have different blowups when we move from su setting to mu one.

The Setting. Let $\Pi[E] : \mathcal{K} \times \{0,1\}^n \times \{0,1\}^n$ be a blockcipher construction built on top of an ideal blockcipher $E : \{0,1\}^k \times \{0,1\}^n \to \{0,1\}^n$ such that a single call to Π/Π^{-1} makes at most t calls to E/E^{-1}. Let $\mathbf{S}_{\text{real}}$ and $\mathbf{S}_{\text{ideal}}$ be two stateless systems implementing games $\text{Real}_{\Pi[E],\text{Sample}}$ and $\text{Rand}_{\Pi[E],\text{Sample}}$ in Fig. 1, respectively. We will measure adversaries' resources in terms of q (the number of ENC/DEC queries) and p (the number of PRIM/PRIMINV queries). A transcript recording the interaction between an adversary and a system $\mathbf{S} \in \{\mathbf{S}_{\text{ideal}}, \mathbf{S}_{\text{real}}\}$ contains the following:

- ENC/DEC queries: A query to ENC(i,x) returning y is associated with an entry $(\text{enc}, +, i, x, y)$. Likewise, a query to DEC(i,y) returning x is associated with an entry $(\text{enc}, -, i, x, y)$.
- PRIM/PRIMINV queries: A query to PRIM(J,u) returning v is recorded in the transcript as $(\text{prim}, +, J, u, v)$. Likewise, a query to PRIMINV(J,v) returning u is associated with an entry $(\text{prim}, -, J, u, v)$.

Super-Additivity and Beyond. For a function $\delta : (\mathbb{N})^2 \to [0,1]$, we say that it is *super-additive* if $\delta(x,y) + \delta(x,z) \le \delta(x,y+z)$, for every $x,y,z \in \mathbb{N}$. For real numbers $M > 0$ and $z \ge 0$, let $\text{Cost}(M,z) = \max\{M,z\}$ if $z > 1$, and $\text{Cost}(M,z) = M/\lg(M)$ if $z \le 1$.

The Technique. One begins by defining an undesirable property on su transcripts, which can involve both ENC/DEC and PRIM/PRIMINV queries/answers. If a su transcript has this property, we say that it is *bad*; otherwise it is *good*. Let $\mathcal{S}$ be the set of all bad su transcripts.[2] If a su transcript is not bad, we say that it is *good*. We demand that there be a monotonic function ϵ^* such that for any su adversary A that makes at most q ENC/DEC queries and p PRIM/PRIMINV queries,

$$\Pr[\text{Script}(A, \mathbf{S}_{\text{ideal}}) \in \mathcal{S}] \leq \epsilon^*(p, q) \tag{21}$$

where for any system $\mathbf{S}$, $\text{Script}(A, \mathbf{S})$ denotes the random variable for the transcript of the interaction of A and $\mathbf{S}$.

For any transcript τ in which the adversary attacks just a single user, let $\text{Ent}(\tau)$ be the number of entries $(\texttt{prim}, \cdot, \cdot, u, v)$ such that τ contains either an entry $(\texttt{enc}, +, 1, \cdot, x)$ or an entry $(\texttt{enc}, -, 1, x, \cdot)$, for some $x \in \{u, v\}$. Suppose that there are monotonic functions ϵ', ϵ'' and a monotonic, super-additive function ϵ such that, for any good su transcript of q queries to ENC/DEC, and p queries to PRIM/PRIMINV,

$$\mathsf{ps}_{\text{ideal}}(\tau) - \mathsf{ps}_{\text{real}}(\tau) \leq (\epsilon(p, q) + \epsilon'(p, q) \cdot \text{Ent}(\tau) + \epsilon''(p, q)) \cdot \mathsf{ps}_{\text{ideal}}(\tau). \tag{22}$$

If Eqs. (21) and (22) are met, then we say that $\Pi[E]$ has the $(\epsilon, \epsilon', \epsilon'', \epsilon^*)$-proximity property.

Note that $\text{Ent}(\tau) \leq \min\{p, 2^{k+2}q\}$, where k is the key length of the primitive E. Thus $(\epsilon, \epsilon', \epsilon'', \epsilon^*)$-proximity immediately implies that for any adversary attacking a single user via q ENC/DEC queries and p PRIM/PRIMINV queries, its su advantage is at most $\epsilon(p, q) + \epsilon'(p, q) \cdot \min\{p, q \cdot 2^{k+2}\} + \epsilon''(p, q) + \epsilon^*(p, q)$. The following result bounds the mu security of $\Pi[E]$.

Lemma 4. *Assume that $\Pi[E]$ has the $(\epsilon, \epsilon', \epsilon'', \epsilon^*)$-proximity property. Then for any adversary A that makes at most q ENC/DEC queries, and p PRIM/PRIMINV queries,*

$$\text{Adv}^{\pm\text{mu-prp}}_{\Pi[E], \text{Sample}}(A) \leq 2^{-n} + 2\epsilon + 2q(\epsilon'' + \epsilon^*) + \text{Cost}(4n, 8q/2^n) \cdot 10(p + qt)\epsilon',$$

where t is the number of calls to E/E^{-1} that a single call to Π/Π^{-1} makes, and functions $\epsilon, \epsilon', \epsilon'', \epsilon^*$ all take arguments $p + qt$ and q.   □

Discussion. Recall that our technique dissects a su bound into three components: $\epsilon, \epsilon' \cdot \min\{p, q \cdot 2^{k+2}\}$, and $(\epsilon'' + \epsilon^*)$. Lemma 4 above then lifts those to ϵ, $\text{Cost}(4n, 8q/2^n) \cdot (p + qt) \cdot \epsilon'$, and $q \cdot (\epsilon'' + \epsilon^*)$, respectively, for the corresponding mu bound. This trisection captures different possibilities of security loss when one moves from su to mu security: (i) Key-Alternating Cipher (where ϵ is the dominant term in both the su and mu bounds) [19], (ii) Single Encryption (where $\epsilon'' + \epsilon^*$ and $q \cdot (\epsilon'' + \epsilon^*)$ are the dominant term in the su and mu

---

[2] In Sect. 3, we factored the set of bad su transcripts into two disjoint sets $\mathcal{S}$ and $\mathcal{S}'$, and required that it is unlikely for the adversary to produce a bad transcript in $\mathcal{S}$. Here $\mathcal{S}'$ is simply the empty set.

bounds respectively), and (iii) Double Encryption (where $\epsilon' \cdot \min\{p, q \cdot 2^{k+2}\}$ and $\mathrm{Cost}(4n, 8q/2^n) \cdot (p + qt)\epsilon'$ are the dominant term in the su and mu bounds respectively).

Given a su analysis, there might be multiple choices for ϵ and ϵ''. However, recall that when we move from su to mu security, the former term remains the same, whereas the latter blows up with a factor q. Therefore, when we need to pinpoint ϵ and ϵ'', we will shift as much weight to ϵ as possible, and the optimal choice of ϵ will often be clear from the context and the best mu attacks. On the other hand, due to the q-blowup of ϵ'', one may need a very fine-grained su analysis to obtain a good mu bound.

The Proof of Lemma 4. We want to show that Lemma 2 implies the claimed result. In order to do that, we need to define (i) function $\mathrm{Rate}(\tau)$ for su transcripts τ, and (ii) a *niceness* property for mu transcripts. The former is obvious: for a su transcript τ of p PRIM/PRIMINV queries and q ENC/DEC queries, let

$$\mathrm{Rate}(\tau) = \epsilon(p, q) + \epsilon'(p, q) \cdot \mathrm{Ent}(\tau) + \epsilon''(p, q).$$

This function Rate is increasing, in the sense that if τ' contains all the query-answer pairs of τ then $\mathrm{Rate}(\tau') \geq \mathrm{Rate}(\tau)$. Next, let $d = \frac{5}{4}\mathrm{Cost}(4n, 8q/2^n)$. We say that a mu transcript τ in the support of $\mathrm{Script}(A, \mathbf{S}_{\mathrm{ideal}})$ is *nice* if it satisfies the following constraints:

– There are no d entries in τ of the form $(\mathsf{enc}, +, \cdot, \cdot, y), \ldots, (\mathsf{enc}, +, \cdot, \cdot, y)$.
– There are no d entries in τ of the form $(\mathsf{enc}, -, \cdot, x, \cdot), \ldots, (\mathsf{enc}, -, \cdot, x, \cdot)$.

Clearly, the definition of niceness involves only ENC/DEC query-answer pairs of τ. Let $\mathcal{N}$ be the set of nice mu transcripts. The following bounds the chance that A's transcript is not nice; the proof is in the full version of this paper.

Lemma 5. *For any adversary A that makes at most q ENC/DEC queries, and p PRIM/PRIMINV queries,*

$$\Pr[\mathrm{Script}(A, \mathbf{S}_{\mathrm{ideal}}) \notin \mathcal{N}] \leq \frac{1}{2^n}. \qquad \square$$

To use Lemma 2, we need to establish the mu-boundedness of Rate. Specifically, we claim that, for any nice mu transcript τ of r users, using p PRIM/PRIMINV queries and q ENC/DEC queries,

$$\sum_{i=1}^{r} \mathrm{Rate}(\mathrm{Map}(i, \tau)) \leq \epsilon(p, q) + q\epsilon''(p, q) + 4dp\epsilon'(p, q). \qquad (23)$$

Then using Lemma 2 for $\delta_0 = 2^{-n}$ and $\delta_1 = \epsilon + q\epsilon'' + 4dp\epsilon'$ and $\delta_2 = \epsilon^*$ leads to our claimed result.

We now justify Eq. (23). Suppose that in τ, the adversary uses q_i ENC/DEC queries for the i-th user. Then

$$\sum_{i=1}^{r} \text{Rate}(\text{Map}(i,\tau)) = \sum_{i=1}^{r}\left(\epsilon(p,q_i) + \epsilon''(p,q_i) + \text{Ent}(\text{Map}(i,\tau)) \cdot \epsilon'(p,q_i)\right)$$

$$\leq \epsilon(p,q) + r\epsilon''(p,q) + \sum_{i=1}^{r}\text{Ent}(\text{Map}(i,\tau)) \cdot \epsilon'(p,q)$$

$$\leq \epsilon(p,q) + q\epsilon''(p,q) + \sum_{i=1}^{r}\text{Ent}(\text{Map}(i,\tau)) \cdot \epsilon'(p,q),$$

where the first inequality is due to the superadditivity of ϵ and the monotone of ϵ' and ϵ''. Thus to justify (23), what's left is to prove that

$$\sum_{i=1}^{r}\text{Ent}(\text{Map}(i,\tau)) \leq 4dp.$$

Since τ is nice, for each entry $(\texttt{prim}, \cdot, \cdot, u, v)$, there are at most $4d$ entries $(\texttt{enc}, \cdot, \cdot, \cdot, x)$ or $(\texttt{dec}, \cdot, \cdot, x, \cdot)$, for $x \in \{u, v\}$. Since each ENC/DEC entry belongs to exactly one user, for each PRIM/PRIMINV entry of τ, there are at most $4d$ indices i such that $\text{Ent}(\text{Map}(i,\tau))$ counts this entry, and thus summing over p PRIM/PRIMINV entries of τ gives us

$$\sum_{i=1}^{r}\text{Ent}(\text{Map}(i,\tau)) \leq 4dp$$

as claimed.

5   Exact Multi-user Security of Double Encryption

5.1   Results and Discussion

Results. In this section, we give an exact mu security bound of Double Encryption via the specialized framework in Sect. 4.2; the key-sampling algorithm is uniform. While it is relatively easy to give an exact su security bound of Double Encryption [2,14], giving a good $(\epsilon, \epsilon', \epsilon'', \epsilon^*)$-proximity bound, as in Lemma 6 below, requires a much more fine-grained analysis. The proof, given in Sect. 5.2, is based on the expectation method of Hoang and Tessaro [19].

Lemma 6. *Let $n \geq 16$ and $k \geq 1$ be integers, and let $E : \{0,1\}^k \times \{0,1\}^n \to \{0,1\}^n$ be a blockcipher. Then $\text{DE}[E]$ satisfies the $(\epsilon, \epsilon', \epsilon'', \epsilon^*)$-proximity property, with $\epsilon(p,q) = \frac{2q}{2^{k+n/2}} + \frac{3qB^2+2Bpq}{2^{2k}}$, $\epsilon'(p,q) = \frac{2p}{2^{2k}}$, $\epsilon''(p,q) = \frac{5Bp}{2^{2k}}$, and $\epsilon^*(p,q) = \frac{1}{2^{k+n}}$, where $B = \frac{5}{4} \cdot \text{Cost}(4n + 2k, 8q/2^n)$.* ☐

From Lemmas 4 and 6, we immediately obtain the following result.

Theorem 1 (Mu security of Double Encryption). *Let $n, k \in \mathbb{N}$ be integers, and let $E : \{0,1\}^k \times \{0,1\}^n \to \{0,1\}^n$ be a blockcipher. Then for any adversary making only q ENC/DEC queries and p PRIM/PRIMINV queries,*

$$\mathsf{Adv}_{\mathrm{DE}[E]}^{\pm\mathrm{mu\text{-}prp}}(A) \leq \frac{1}{2^n} + \frac{5q}{2^{k+n/2}} + \frac{6qB^2 + 222BQ^2}{2^{2k}}$$

where $B = \frac{5}{4} \cdot \mathrm{Cost}(4n + 2k, 8q/2^n)$ and $Q = \max\{p, q\}$. $\square$

Discussion. Admittedly, the bound in Theorem 1 looks complicated. However, for the "usual" setting $n \geq k \geq 16$ and $q \leq \frac{2^k}{8}$, the bound can be simplified to $\mathsf{Adv}_{\mathrm{DE}[E]}^{\pm\mathrm{mu\text{-}prp}}(A) \leq \frac{1}{2^n} + \frac{(n+5)q}{2^{1.5k}} + \frac{1554nQ^2}{\lg(4n)\cdot 2^{2k}}$. On the other hand, recall that the classical su bound of $\mathrm{DE}[E]$ by Aiello et al. [2] is $\mathsf{Adv}_{\mathrm{DE}[E]}^{\pm\mathrm{prp}}(A) \leq \frac{p^2}{2^{2k}}$. If we apply the hybrid argument to this, we will get the following inferior bound $\mathsf{Adv}_{\mathrm{DE}[E]}^{\pm\mathrm{mu\text{-}prp}}(A) \leq \frac{q(p+2q)^2}{2^{2k}}$. While this bound is enough to show that Double Encryption squarely beats Single Encryption in mu security,[3] it is much worse than the bound in Theorem 1, as illustrated in Fig. 2.

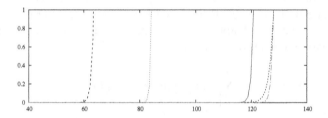

Fig. 2. Mu and su security of Single and Double Encryption on AES. From left to right: the mu bound of Single Encryption, the naive mu bound of Double Encryption via the hybrid argument, the mu bound of Double Encryption via Theorem 1, the su bound of Single Encryption, and the classical su bound of Double Encryption by Aiello et al. [2]. We set $p = q$ and $n = k = 128$. The x-axis gives the log (base 2) of p, and the y-axis gives the security bounds.

5.2 Proof of Lemma 6

It is convenient to assume without loss of generality that the adversary doesn't make redundant queries. Our proof borrows the overall approach used by Hoang and Tessaro [19] for key-alternating ciphers. We begin with some high-level setup.

Assumptions on the Transcript. We consider an arbitrary fixed transcript τ which contains q ENC/DEC queries and p PRIM/PRIMINV queries. Moreover, for a transcript τ, we also denote (following [14])

---

[3] Recall that $\mathsf{Adv}_E^{\pm\mathrm{prp}}(A) \leq \frac{p}{2^k}$ and $\mathsf{Adv}_E^{\pm\mathrm{mu\text{-}prp}}(A) \leq \frac{p(p+q)}{2^k}$ for an adversary A making only q ENC/DEC queries and p PRIM/PRIMINV queries.

$$\mathrm{Fwd}(\tau) = \max_{y \in \{0,1\}^n} \left| \{(J, x) \mid (\mathtt{prim}, +, J, x, y) \in \tau\} \right| ,$$

$$\mathrm{Bwd}(\tau) = \max_{x \in \{0,1\}^n} \left| \{(J, y) \mid (\mathtt{prim}, -, J, x, y) \in \tau\} \right| .$$

Recall that to establish $(\epsilon, \epsilon', \epsilon'', \epsilon^*)$-proximity, we have to define bad transcripts. A transcript is bad if either $\mathrm{Fwd}(\tau) > B$ or $\mathrm{Bwd}(\tau) > B$, where

$$B := \frac{5}{4} \cdot \mathrm{Cost}(4n + 2k, 8p/2^n) .$$

Let $\mathcal{S}$ be the set of all bad transcripts. The following bounds the chance that the adversary produces a bad transcript; the proof is in the full version of this paper.

Lemma 7. *For any adversary A that makes p PRIM/PRIMINV queries and q ENC/DEC queries,*

$$\Pr[\mathrm{Script}(A, \mathbf{S}_{\mathrm{ideal}}) \in \mathcal{S}] \leq \frac{1}{2^{n+k}}.$$
□

From now on, we assume that additionally $\tau \notin \mathcal{S}$. We shall use the expectation method to prove the claimed bound of the gap $\mathsf{ps}_{\mathrm{ideal}}(\tau) - \mathsf{ps}_{\mathrm{real}}(\tau)$. We begin with some combinatorial results on the transcript.

Type-1 Chains. Consider a pair of entries $(\mathtt{prim}, \cdot, \cdot, x_1, y_1), (\mathtt{prim}, \cdot, \cdot, x_2, y_2)$ in τ such that $y_1 = x_2$. We say that it is a *positive type-1 chain* if there's an entry $(\mathtt{enc}, +, x_1, \cdot)$ in τ. We say that it is a *negative type-1 chain* if there's an entry $(\mathtt{enc}, -, \cdot, y_2)$. The following lemma bounds the number of type-1 chains; the proof is in Appendix A.

Lemma 8. *The number of type-1 chains is at most $4Bp + 2B^2q + 2Bpq$.* □

Type-2 Chains. Consider a pair of entries $(\mathtt{prim}, \cdot, \cdot, x_1, y_1), (\mathtt{prim}, \cdot, \cdot, x_2, y_2)$. We say that it is a *positive type-2 chain* if there's an entry $(\mathtt{enc}, +, x_1, y_2)$ in τ. We say that it is a *negative type-2 chain* if there's an entry $(\mathtt{enc}, -, x_1, y_2)$ in τ. The following lemma bounds the number of type-2 chains; the proof is in Appendix B.

Lemma 9. *The number of type-2 chains is at most $2p \cdot \mathrm{Ent}(\tau)$.* □

Good and Bad Keys. We shall use the expectation method. Let S be the random variable for the key. The key-space $\mathcal{K}$ is $(\{0,1\}^k)^2$ and S is uniformly distributed over $\mathcal{K}$. For each key vector $s = (K_1, K_2) \in \mathcal{K}$ and each $i \in \{1, 2\}$, let $p_i[s]$ be the number of queries $(\mathtt{prim}, \cdot, K_i, \cdot, \cdot)$ in τ.

Definition 1 (Good and bad keys). We say that a key $s = (K_1, K_2)$ is *bad* if one the following happens:

(i) $K_1 = K_2$ and $p_1[s] \geq 1$, or
(ii) $K_1 \neq K_2$, $p_1[s] \geq 1$ and $p_2[s] \geq 2^n/4$, or
(iii) $K_1 \neq K_2$, $p_1[s] \geq 2^n/4$ and $p_2[s] \geq 1$, or
(iv) $K_1 \neq K_2$ and there's a (type-1 or 2) chain $(\mathtt{prim}, \cdot, K_1, \cdot, \cdot)$, $(\mathtt{prim}, \cdot, K_2, \cdot, \cdot)$.

If a key is not bad then we say that it is *good*. Let Γ_{bad} be the set of bad keys, and let $\Gamma_{\mathrm{good}} = \mathcal{K} \backslash \Gamma_{\mathrm{bad}}$.

We first bound the probability that S is bad. First, the chance that S satisfies condition (i) above is at most $\frac{p}{2^{2k}}$. Next, we say that a subkey $J \in \{0,1\}^k$ is *heavy* if there are at least $2^n/4$ entries $(\mathtt{prim}, \cdot, J, \cdot, \cdot)$ in τ. Since there are at most $4p/2^n$ heavy subkeys, the chance that S satisfies condition (ii) above is at most $\frac{4p/2^n}{2^k} \cdot \frac{p}{2^k} = \frac{4p^2}{2^{2k+n}}$. Likewise, the chance that S satisfies condition (ii) above is at most $\frac{4p^2}{2^{2k+n}}$. From Lemmas 8 and 9, there are at most $2p \cdot \mathrm{Ent}(\tau) + 4Bp + 2qB^2 + 2Bpq$ chains, and thus the chance that S satisfies condition (iii) above is at most $\frac{2p \cdot \mathrm{Ent}(\tau) + 4Bp + 2qB^2 + 2Bpq}{2^{2k}}$. Summing up

$$\Pr[S \text{ is bad}] \leq \frac{p + 8p^2/2^n}{2^{2k}} + \frac{2p \cdot \mathrm{Ent}(\tau) + 4Bp + 2qB^2 + 2Bpq}{2^{2k}}$$
$$\leq \frac{2p \cdot \mathrm{Ent}(\tau) + 5Bp + 2qB^2 + 2Bpq}{2^{2k}}.$$

Next, recall that in the expectation method, one needs to find a non-negative function $g : \mathcal{K} \rightarrow [0, \infty)$ such that $g(s)$ bounds $1 - \mathsf{ps}_0(\tau, s)/\mathsf{ps}_1(\tau, s)$ for all $s \in \Gamma_{\mathrm{good}}$. Let U be the subset of Γ_{good} such that for any $(K_1, K_2) \in U$, we have $K_1 = K_2$. We will define $g(s)$ such that $g(s) = 2q/2^{n/2}$ for every $s \in U$, and $g(s) = \frac{4q \cdot p_1[s] \cdot p_2[s]}{2^{2n}}$ for every $s \in \mathcal{K} \backslash U$. We will show that $g(s)$ bounds $1 - \mathsf{ps}_0(\tau, s)/\mathsf{ps}_1(\tau, s)$ later. Then

$$\mathbf{E}[g(S)] = \frac{1}{2^{2k}} \left(\sum_{s \in U} g(s) + \sum_{s \in \mathcal{K} \backslash U} g(s) \right)$$
$$= \frac{1}{2^{2k}} \left(\sum_{s \in U} \frac{q}{2^{n/2}} + \sum_{s \in \mathcal{K} \backslash U} \frac{4qp_1[s]p_2[s]}{2^{2n}} \right)$$
$$\leq \frac{1}{2^{2k}} \left(\frac{q2^k}{2^{n/2}} + \frac{4qp^2}{2^{2n}} \right) \leq \frac{q}{2^{k+n/2}} + \frac{qB^2}{2^{2k}}.$$

Then from Lemma 1,

$$\mathsf{ps}_{\mathrm{ideal}}(\tau) - \mathsf{ps}_{\mathrm{real}}(\tau) \leq \left(\Pr[S \text{ is bad}] + \mathbf{E}[g(S)] \right) \mathsf{ps}_{\mathrm{ideal}}(\tau)$$
$$\leq \left(\frac{2q}{2^{k+n/2}} + \frac{2p \cdot \mathrm{Ent}(\tau) + 5Bp + 3qB^2 + 2Bpq}{2^{2k}} \right) \mathsf{ps}_{\mathrm{ideal}}(\tau).$$

We now show that $g(s)$ indeed bounds $1 - \mathsf{ps}_0(\tau, s)/\mathsf{ps}_1(\tau, s)$ for every $s \in \Gamma_{\mathrm{good}}$. We consider the following cases, depending on whether $s \in \Gamma_{\mathrm{good}} \backslash U$ or $s \in U$.

CASE 1: $s \in \Gamma_{\text{good}} \backslash U$. For this case, we have to consider two sub-cases, depending on whether $q \leq N/4$ or not.

CASE 1.1: $q \leq N/4$. Let $s = (K_1, K_2)$. Since $s \in \Gamma_{\text{good}} \backslash U$, we must have $K_1 \neq K_2$. We now use the following result of Chen and Steinberger [12]. (Their proof considered key-alternating ciphers, but we note that we are restricting ourselves to the setting $K_1 \neq K_2$, and and their proof also applies to the special case that all sub-keys of the key-alternating cipher are 0^n, which is equivalent to our setting here.)

Lemma 10. [12] *Assume that* $p_1[s], p_2[s], q < 2^n/2$. *Then*

$$1 - \frac{\mathsf{ps}_0(\tau, s)}{\mathsf{ps}_1(\tau, s)} \leq \frac{q \cdot p_1[s] \cdot p_2[s]}{(2^n - q - p_1[s])(2^n - q - p_2[s])}. \qquad \square$$

From Lemma 10, since $p_1[s], p_2[s], q \leq 2^n/4$,

$$1 - \frac{\mathsf{ps}_0(\tau, s)}{\mathsf{ps}_1(\tau, s)} \leq \frac{4q \cdot p_1[s] \cdot p_2[s]}{2^{2n}} = g(s).$$

CASE 1.2: $N/4 < q \leq N$. Let Z be the random variable for the additional $(N - q)$ ENC queries that τ lacks. For we write $\mathsf{ps}_{\mathbf{S}_{\text{real}}}(\tau, s, z)$ to be the probability that $\mathbf{S}_{\text{real}}$ answers queries according to τ, and that $S = s$ and $Z = z$. In this case $\mathsf{ps}_{\mathbf{S}_{\text{ideal}}}(\tau, s, z)$ is the probability that $\mathbf{S}_{\text{ideal}}$ behaves according to the entries in (τ, z), and $S \leftarrow_{\$} \{0, 1\}^{2k}$ agrees with s. We now show that $\mathsf{ps}_{\mathbf{S}_{\text{ideal}}}(\tau, s, z) \leq \mathsf{ps}_{\mathbf{S}_{\text{real}}}(\tau, s, z)$ for all choices of z such that $\mathsf{ps}_{\mathbf{S}_{\text{ideal}}}(\tau, s, z) > 0$, and thus

$$\mathsf{ps}_{\mathbf{S}_{\text{ideal}}}(\tau, s) - \mathsf{ps}_{\mathbf{S}_{\text{real}}}(\tau, s) \leq \sum_z \mathsf{ps}_{\mathbf{S}_{\text{ideal}}}(\tau, s, z) - \mathsf{ps}_{\mathbf{S}_{\text{real}}}(\tau, s, z) \leq 0 \leq g(s).$$

Let $s = (K_1, K_2)$ and $a = p_1[s]$ and $b = p_1[s] + p_2[s] < 2^n$. As $s \in \Gamma_{\text{good}} \backslash U$, the entries in (τ, z) consist of the following categories:

(1) $(\text{enc}, \cdot, 1, x_1, y_1), \ldots, (\text{enc}, \cdot, 1, x_{2^n}, y_{2^n})$,
(2) $(\text{prim}, \cdot, K_1, x_1, u_1), \ldots, (\text{prim}, \cdot, K_1, x_a, u_a)$ and $(\text{prim}, \cdot, K_2, u_{a+1}, y_{a+1})$, $\ldots, (\text{prim}, \cdot, K_2, u_b, y_b)$, and
(3) $(\text{prim}, \cdot, J, \cdot, \cdot)$, with $J \notin \{K_1, K_2\}$.

Hence $\mathsf{ps}_{\mathbf{S}_{\text{real}}}(\tau, s, z)$ is the probability of the following events:

(i) If we make queries in category (3) above, we get the answers provided by τ.
(ii) $S \leftarrow_{\$} \{0, 1\}^{2k}$ agrees with s.
(iii) For any $i \in \{1, \ldots, a + b\}$, querying $\text{PRIM}(K_1, x_i)$ in $\mathbf{S}_{\text{real}}$ yields u_i, and querying $\text{PRIMINV}(K_2, y_i)$ in $\mathbf{S}_{\text{real}}$ yields u_i. Moreover, for any $j \in \{b + 1, \ldots, 2^n\}$, in $\mathbf{S}_{\text{real}}$, the output of $\text{PRIM}(K_1, x_j)$ is the same as the output of $\text{PRIMINV}(K_2, y_j)$.

Note that the three events above are independent, and the first two are independent of the system. On the other hand, $\mathsf{ps}_{\mathbf{S}_{\text{ideal}}}(\tau, s, z)$ is likewise the probability of events (i), (ii), and the following

(iv) For any $i \in \{1, \ldots, a\}$, querying $\mathrm{PRIM}(K_1, x_i)$ in $\mathbf{S}_{\mathrm{ideal}}$ yields u_i. For any $i \in \{a+1, \ldots, b\}$, querying $\mathrm{PRIMINV}(K_2, y_i)$ in $\mathbf{S}_{\mathrm{ideal}}$ yields u_i. Moreover, for any $j \in \{1, \ldots, 2^n\}$, querying $\mathrm{ENC}(1, x_j)$ yields y_j.

Again, note that events (i), (ii), and (iv) are independent. Hence we need only show that the probability that event (iii) happens is at least the probability that event (iv) happens. The chance that event (iii) is

$$\frac{1}{(2^n)! \cdot 2^n(2^n - 1)(2^n - a - b)}$$

whereas the chance that event (iv) happens is

$$\frac{1}{(2^n)! \cdot 2^n(2^n - 1)\cdots(2^n - a) \cdot 2^n(2^n - 1)\cdots(2^n - b)}.$$

Hence the probability that event (iii) happens is indeed at least the probability that event (iv) happens.

CASE 2: $s \in U$. Then $p_1[s] = 0$. Clearly if $q \geq 2^{n/2-1}$ then the claim vacuously holds. Assume that $q < 2^{n/2-1}$. Let $s = (K_1, K_1)$. Let the ENC/DEC entries in τ be $(\mathrm{enc}, \cdot, 1, x_1, y_1), \ldots, (\mathrm{enc}, \cdot, 1, x_q, y_q)$. Note that τ doesn't contain any entry $(\mathrm{prim}, \cdot, K_1, \cdot, \cdot)$. Then $\mathsf{ps}_{\mathbf{S}_{\mathrm{ideal}}}(\tau, s)$ is the probability of the following events:

(a) $S \twoheadleftarrow \{0,1\}^{2k}$ agrees with s.
(b) If we make PRIM/PRIMINV queries in τ, we get the answers provided by τ.
(c) $\mathbf{S}_{\mathrm{ideal}}$ behaves according to the ENC/DEC queries in τ.

Note that the three events above are independent, and the first two are independent of the system. On the other hand, $\mathsf{ps}_{\mathbf{S}_{\mathrm{real}}}(\tau, s)$ is at least the probability of events (a), (b), and the following:

(d) For every $i \in \{1, \ldots, q\}$, if we query $\mathrm{PRIM}(K, x_i)$, we will get an answer $z_i \notin \{x_1, y_1, \ldots, x_q, y_q\}$, and the strings $z_1, \ldots, z_q$ are distinct. Moreover, if we query $\mathrm{PRIM}(K, z_i)$, we will get y_i.

Again, events (a), (b), and (d) are independent. Hence we only need to show that, $\Pr[\text{Event (d)}] \geq (1 - 2q/2^{n/2})\Pr[\text{Event (c)}]$. Note that event (c) happens with probability

$$\frac{1}{2^n(2^n - 1)\cdots(2^n - q + 1)},$$

whereas event (d) happens with probability

$$\left(\prod_{i=0}^{q-1} \frac{2^n - 2q - i}{2^n - i}\right) \frac{1}{(2^n - q)\cdots(2^n - 2q + 1)}.$$

Hence

$$\frac{\Pr[\text{Event (d)}]}{\Pr[\text{Event (c)}]} = \prod_{i=0}^{q-1} \frac{2^n - 2q - i}{2^n - q - i} = \prod_{i=0}^{q-1} \left(1 - \frac{q}{2^n - q - i}\right)$$

$$\geq 1 - \sum_{i=0}^{q-1} \frac{q}{2^n - q - i} \geq 1 - \frac{q^2}{2^n - 2q} \geq 1 - \frac{2q}{2^{n/2}},$$

where the first inequality is due to the fact that $(1-x)(1-y) \geq 1 - x - y$ for any $x, y \geq 0$, and the last inequality is due to the assumption that $q < 2^{n/2-1}$. This concludes the proof.

6 Matching Attacks

In this section, we give matching attacks for both Single Encryption and Double Encryption, in which the adversary uses $\Theta(q)$ ENC/DEC queries and $\Theta(p)$ PRIM/PRIMINV queries. Our attack on Single Encryption generalizes Biham's work [10] for all choices of the parameters p and q. For Double Encryption, recall that one can launch a su attack with advantage $\frac{p^2}{2^{2k}}$, and Biham's key-collision attack [10] gives a mu attack with advantage $\frac{q^2}{2^{2k}}$. Thus those attacks already give matching bounds in the usual case $n \geq k$ (such as DES or AES). Hence for Double Encryption, the only interesting setting to find matching attacks is where $n \ll k$ (such as Format-Preserving Encryption or MISTY-1). We however only know how to give matching attacks for this setting if the adversary is given all the keys after it finishes querying, which is the model in our security proof and many prior works [14,16]. Our attack yields the bound around $p^2 s/2^{2k}$, where $s = \max\{\lfloor n/8 \lg(n) \rfloor, q/2^n\}$, which is much better than the two known attacks above. We leave as an open problem to find matching attacks for $n \ll k$ without key revelation.

A Useful Inequality. In the attacks, we often need to make use of the following technical result.

Lemma 11. *Let $r \geq 1$ be an integer and $0 < a \leq 1/r$. Then $(1-a)^r \leq 1 - ar/2$.*

Proof. Clearly the claimed inequality holds for $r = 1$, and thus we need only consider $r \geq 2$. Let $f(x) = xr/2 + (1-x)^r - 1$. Our goal is to show that $f(a) \leq 0$. The derivative and second derivative of the function f are $f'(x) = \frac{r}{2} - r(1-x)^{r-1}$ and $f''(x) = \frac{1}{2} + r(r-1)(1-x)^{r-2}$ respectively. Since $f''(x) > 0$ for all $x \in [0, 1/r]$, the function $f'(x)$ is strictly increasing. We claim that $f(a) \leq \max\{f(0), f(1/r)\}$. Since $f(0) = 0$ and

$$f(1/r) = \frac{1}{2} + (1 - 1/r)^r - 1 \leq \frac{1}{e} - \frac{1}{2} < 0,$$

we have $f(a) \leq 0$. To justify the claim above, note that if $f'(1/r) < 0$ then function f is decreasing, and thus $f(a) \leq f(0) = \max\{f(0), f(1/r)\}$. If $f'(1/r) \geq 0$, since function f' is strictly increasing and $f'(0) = -r/2 < 0$, there exists $b \in [0, 1/r]$ such that $f'(x) < 0$ for every $x \in [0, b)$ and $f'(x) \geq 0$ for every $x \in [b, 1/r]$. Hence function f is decreasing in $[0, b)$ and increasing in $[b, 1/r]$, and thus $f(a) \leq \max\{f(0), f(1/r)\}$. □

6.1   Attacking Single Encryption

Let $d = \lceil \frac{k+2}{n-1} \rceil$ and assume that $d \leq 2^{n-1}$, which holds for all practical values of n and k. Then

$$2^n(2^n - 1) \cdots (2^n - d + 1) \geq (2^{n-1})^d \geq 2^{k+2}.$$

For all practical values of n and k, the value d will be very small. For example, if $n = 64$ and $k = 56$ (meaning DES parameters), we have $d = 1$. Or if $n = k = 128$ (meaning AES parameters), we have $d = 2$. Let $p, q \in \mathbb{N}$ such that $pq \leq 2^k$. Consider the following adversary A. It picks arbitrary distinct $x_1, \ldots, x_d \in \{0, 1\}^n$ and queries $\text{ENC}(i, x_\ell)$ to get answer $y_{i,\ell}$, for every $i \in \{1, \ldots, q\}$ and $\ell \in \{1, \ldots, d\}$. It then picks p arbitrary distinct keys $K_1, \ldots, K_p \in \{0, 1\}^k$ and queries $E(K_j, x_\ell)$ to get answer $z_{j,\ell}$, for every $j \in \{1, \ldots, p\}$ and $\ell \in \{1, \ldots, d\}$. If there are i and j such that $y_{i,\ell} = z_{j,\ell}$ for every $\ell \in \{1, \ldots, d\}$ then the adversary outputs 1, otherwise it outputs 0. In the real game, from Lemma 11, the chance that the adversary outputs 1 is

$$1 - \left(1 - \frac{p}{2^k}\right)^q \geq \frac{pq}{2^{k+1}}.$$

In the ideal game, the chance that it outputs 1 is

$$\frac{pq}{2^n(2^n - 1) \cdots (2^n - d + 1)} \leq \frac{pq}{2^{k+2}}.$$

Hence $\text{Adv}_E^{\pm\text{mu-prp}}(A) \geq \frac{pq}{2^{k+2}}$, and the adversary uses dq ENC queries and dp PRIM queries.

6.2   Attacking Double Encryption

Here we assume that $16 \leq n < k$, and aim to achieve advantage $p^2 s/2^{2k}$, where $s = \max\{\lfloor n/8 \lg(n) \rfloor, q/2^n\}$. We have the following restrictions on the parameters p and q:

- Since there are attacks with advantage $Q^2/2^{2k}$, where $Q = \max\{p, q\}$, we need only consider $2^n/n \leq q \leq 2^k$.
- Since using $p \approx 2^k/\sqrt{s}$ is already enough to get a constant advantage, without loss of generality, we can assume that $p \leq 2^k/\sqrt{s}$.

Moreover, recall that we are in the model where the keys are given to the adversary after it finishes querying.

The Attack. For every $i \in \{1, \ldots, q\}$, query $(i, 0^n)$ to ENC to receive answer y_i. View each string in $\{0, 1\}^n$ as a bin, and querying $\text{ENC}(i, 0^n)$ means throwing a ball to those 2^n bins at random. Let y be the bin of the most balls, and let S be the set of indices i such that the corresponding ball falls into bin y. The following lemma gives a strong concentration bound on $|S|$ in both the real and ideal games; see Appendix C for the proof.

Lemma 12. *Let $n \geq 16$ and $q \geq 2^n/n$ be integers. Consider throwing q balls to 2^n bins at random. Let X denote the random variable for the number of balls in the bin of most balls. Then*

$$\Pr\left[X \geq \max\{\lfloor n/8 \lg(n) \rfloor, q/2^n\}\right] \geq 1 - 2^{-n/3}.$$
□

Next, if $|S| < s$ then the output a random guess to get advantage 0. If $|S| \geq s$, which happens with probability at least $1 - 2^{-n/3}$ according to Lemma 12, then adapt the meet-in-the-middle attack as follows. Pick distinct keys $J_1, \ldots, J_{2p} \in \{0, 1\}^k$, and query $\text{PRIM}(J_i, x)$ and $\text{PRIMINV}(J_{i+p}, y)$ to get answer u_i and v_i respectively. When the keys are given, check if there are some $i, j \in \{1, \ldots, p\}$ and $\ell \in S$ such that (J_i, J_{j+p}) is the key of user ℓ. If such a triple (i, j, ℓ) exists then output 1 if and only if $u_i = v_j$.

Analyses. Suppose that $|S| \geq s$. Then the chance that there are $i, j \in \{1, \ldots, p\}$ and $\ell \in S$ such that (J_i, J_{j+p}) is the key of user ℓ is

$$1 - (1 - p^2/2^{2k})^{|S|} \geq 1 - (1 - p^2/2^{2k})^s \geq \frac{p^2 s}{2^{2k+1}},$$

where the last inequality is due to Lemma 11. If this pair exists then in the ideal game, the conditional probability that $v_i = u_i$ is $1/2^n$, whereas in the real game, $v_i = u_i$ with conditional probability 1. Putting this all together, the adversary wins with advantage at least

$$(1 - 2^{-n/3})(1 - 2^{-n}) \cdot \frac{p^2 s}{2^{2k+1}} \geq \max\{\lfloor n/8 \lg(n) \rfloor, q/2^n\} \cdot \frac{p^2}{3 \cdot 2^{2k}}.$$

Discussion. What's the problem if we are not given keys at the end of the querying process? Now we have many pairs (i, j) such that $u_i = v_i$. One such pair will yield the key (J_i, J_{j+p}) for some user, but we don't know which user. Moreover, there are too many pairs (i, j)—one can show that in the ideal world, there are on average $O(p^2/2^n)$ such pairs—and most of them are just false positives.

Acknowledgments. This research was partially supported by NSF grants CNS-1423566, CNS-1528178, CNS-1553758 (CAREER), IIS-152804, by a Hellman Fellowship, and by the Glen and Susanne Culler Chair.

A   Proof of Lemma 8

We claim that the number of positive type-1 chains is at most $2Bp + B^2q + Bpq$. By symmetry, the number of negative type-1 chains is also at most $2Bp + B^2q + Bpq$. Hence the total number of type-1 chains is at most $4Bp + 2B^2q + 2Bpq$.

To justify the claim above, consider a positive type-1 chain $(\texttt{prim}, \cdot, \cdot, x_1, y_1)$, $(\texttt{prim}, \cdot, \cdot, x_2, y_2)$. There are four ways to assign the signs $+/-$ to the entries. Fix a specific way to assign the signs. We consider the following cases.

Case 1: Both entries have sign $-$. Then there are at most Bq choices for the first entry, since $\tau \notin S$. Moreover, once the first entry is fixed, there are only B choices for the second entry. Thus in this case, the total number of positive type-1 chains is at most B^2q.

Case 2: Both entries have sign $+$. There are at most p choices for the last entry. Moreover, once the last entry is fixed, there are at most B choices for the first entry. Thus in this case, the total number of positive type-1 chains is at most Bp.

Case 3: The first entry has sign $-$ and the second sign $+$. There are at most Bq choices for the first entry and p choices for the last one. Thus in this case, the total number of positive type-1 chains is at most Bpq.

Case 4: The first entries has sign $+$ and the second sign $-$. Then there are at most p choices for the first entry. Moreover, once the first entry is fixed, there are at most B choices for the last entry. Thus in this case, the total number of positive type-1 chains is at most Bp.

Summing up, the total number of positive type-1 chains is at most $2Bp + B^2q + Bpq$.

B   Proof of Lemma 9

We claim that the number of negative type-2 chains is at most $p \cdot \mathrm{Ent}(\tau)$. By symmetry, the number of positive type-2 chains is also at most $p \cdot \mathrm{Ent}(\tau)$. Hence the total number of type-2 chains is at most $2p \cdot \mathrm{Ent}(\tau)$.

To justify the claim above, consider a negative type-2 chain $(\texttt{prim}, \cdot, \cdot, x_1, y_1)$, $(\texttt{prim}, \cdot, \cdot, x_2, y_2)$. Then there are at most $\mathrm{Ent}(\tau)$ choices for the first entry, and p choices for the last entry. Thus the total number of negative type-2 chains is at most $p \cdot \mathrm{Ent}(\tau)$.

C   Proof of Lemma 12

Let $s = \lfloor n/8 \lg(n) \rfloor$. Clearly $X \geq q/2^n$, hence we only need to consider the case that $q/2^n \leq s$. Our proof will closely follow the second-moment method in classic balls-into-bins papers [27]. For any $i \in \{1, \ldots, 2^n\}$, the chance that the i-th bin has at least s balls is

$$\binom{q}{s}\frac{1}{(2^n)^s}\left(1-\frac{1}{2^n}\right)^{q-s} \geq \left(\frac{q}{s}\right)^s\frac{1}{(2^n)^s}\left(1-\frac{1}{2^n}\right)^q \geq \left(\frac{q/2^n}{s}\right)^s \cdot e^{-q/2^n} \geq n^{-2s}\cdot e^{-q/2^n}.$$

Moreover,

$$n^{-2s} \cdot e^{-q/2^n} \geq 2^{-2s \lg(n)} \cdot 2^{-1.5q/2^n} \geq 2^{-n/4 - \frac{1.5n}{8 \lg(n)}} \geq 2^{-n/3}.$$

Let Y_i be the Bernoulli random variable such that $Y_i = 1$ if and only if the i-th bin has at least s balls. Then $\mathbf{E}[Y_i] = \Pr[Y_i = 1] \geq 2^{-n/3}$. Let $Y = Y_1 + \cdots + Y_{2^n}$, and thus

$$\mathbf{E}[Y] = \mathbf{E}[Y_1] + \cdots + \mathbf{E}[Y_{2^n}] \geq 2^{2n/3}.$$

Since

$$\Pr[X \geq s] = \Pr[Y \geq 1] = 1 - \Pr[Y = 0] \geq 1 - \Pr\Big[|Y - \mathbf{E}[Y]| \geq \mathbf{E}[Y]\Big],$$

what's left is to show that $\Pr\Big[|Y - \mathbf{E}[Y]| \geq \mathbf{E}[Y]\Big] \leq 2^{-n/3}$. By Chebyshev's inequality,

$$\Pr\Big[|Y - \mathbf{E}[Y]| \geq \mathbf{E}[Y]\Big] \leq \frac{\mathbf{Var}[Y]}{(\mathbf{E}[Y])^2} \leq \frac{\mathbf{Var}[Y]}{2^{4n/3}}.$$

It then suffices to show that $\mathbf{Var}[Y] \leq 2^n$. On the one hand, for any $i \neq j$, each Y_i and Y_j are negatively correlated, as some bin having more balls means that it is less likely for another bin to be so. Therefore, each covariance $\mathbf{Cov}(Y_i, Y_i)$ is at most 0. On the other hand, since each Y_i is a Bernoulli random variable, $(Y_i)^2 = Y_i$, and thus

$$\mathbf{Var}[Y_i] \leq \mathbf{E}[(Y_i)^2] = \mathbf{E}[Y_i] \leq 1.$$

Hence

$$\mathbf{Var}[Y] = \sum_{i=1}^{2^n} \mathbf{Var}[Y_i] + \sum_{i \neq j} \mathbf{Cov}(Y_i, Y_j) \leq 2^n.$$

References

1. ANSI X9.52: Triple data encryption algorithm modes of operation (1998)
2. Aiello, W., Bellare, M., Crescenzo, G., Venkatesan, R.: Security amplification by composition: the case of doubly-iterated, ideal ciphers. In: Krawczyk, H. (ed.) CRYPTO 1998. LNCS, vol. 1462, pp. 390–407. Springer, Heidelberg (1998). doi:10. 1007/BFb0055743
3. Andreeva, E., Daemen, J., Mennink, B., Van Assche, G.: Security of keyed sponge constructions using a modular proof approach. In: Leander, G. (ed.) FSE 2015. LNCS, vol. 9054, pp. 364–384. Springer, Heidelberg (2015). doi:10.1007/ 978-3-662-48116-5_18
4. Bellare, M., Bernstein, D.J., Tessaro, S.: Hash-function based PRFs: AMAC and its multi-user security. In: Fischlin, M., Coron, J.-S. (eds.) EUROCRYPT 2016. LNCS, vol. 9665, pp. 566–595. Springer, Heidelberg (2016). doi:10.1007/ 978-3-662-49890-3_22
5. Bellare, M., Boldyreva, A., Micali, S.: Public-key encryption in a multi-user setting: security proofs and improvements. In: Preneel, B. (ed.) EUROCRYPT 2000. LNCS, vol. 1807, pp. 259–274. Springer, Heidelberg (2000). doi:10.1007/3-540-45539-6_18

6. Bellare, M., Ristenpart, T., Rogaway, P., Stegers, T.: Format-preserving encryption. In: Jacobson, M.J., Rijmen, V., Safavi-Naini, R. (eds.) SAC 2009. LNCS, vol. 5867, pp. 295–312. Springer, Heidelberg (2009). doi:10.1007/978-3-642-05445-7_19

7. Bellare, M., Rogaway, P.: The security of triple encryption and a framework for code-based game-playing proofs. In: Vaudenay, S. (ed.) EUROCRYPT 2006. LNCS, vol. 4004, pp. 409–426. Springer, Heidelberg (2006). doi:10.1007/11761679_25

8. Bellare, M., Tackmann, B.: The multi-user security of authenticated encryption: AES-GCM in TLS 1.3. In: Robshaw, M., Katz, J. (eds.) CRYPTO 2016. LNCS, vol. 9814, pp. 247–276. Springer, Heidelberg (2016). doi:10.1007/978-3-662-53018-4_10

9. Bernstein, D.J.: How to stretch random functions: the security of protected counter sums. J. Cryptology **12**(3), 185–192 (1999)

10. Biham, E.: How to decrypt or even substitute DES-encrypted messages in 2^{28} steps. Inf. Process. Lett. **84**(3), 117–124 (2002)

11. Bogdanov, A., Knudsen, L.R., Leander, G., Standaert, F.-X., Steinberger, J., Tischhauser, E.: Key-alternating ciphers in a provable setting: encryption using a small number of public permutations. In: Pointcheval, D., Johansson, T. (eds.) EUROCRYPT 2012. LNCS, vol. 7237, pp. 45–62. Springer, Heidelberg (2012). doi:10.1007/978-3-642-29011-4_5

12. Chen, S., Steinberger, J.: Tight security bounds for key-alternating ciphers. In: Nguyen, P.Q., Oswald, E. (eds.) EUROCRYPT 2014. LNCS, vol. 8441, pp. 327–350. Springer, Heidelberg (2014). doi:10.1007/978-3-642-55220-5_19

13. Dai, Y., Lee, J., Mennink, B., Steinberger, J.: The security of multiple encryption in the ideal cipher model. In: Garay, J.A., Gennaro, R. (eds.) CRYPTO 2014. LNCS, vol. 8616, pp. 20–38. Springer, Heidelberg (2014). doi:10.1007/978-3-662-44371-2_2

14. Dai, Y., Steinberger, J.: Tight security bounds for multiple encryption. Cryptology ePrint Archive, Report 2014/096 (2014). http://eprint.iacr.org/2014/096

15. Gaži, P.: Plain versus randomized cascading-based key-length extension for block ciphers. In: Canetti, R., Garay, J.A. (eds.) CRYPTO 2013. LNCS, vol. 8042, pp. 551–570. Springer, Heidelberg (2013). doi:10.1007/978-3-642-40041-4_30

16. Gaži, P., Lee, J., Seurin, Y., Steinberger, J., Tessaro, S.: Relaxing full-codebook security: a refined analysis of key-length extension schemes. In: Leander, G. (ed.) FSE 2015. LNCS, vol. 9054, pp. 319–341. Springer, Heidelberg (2015). doi:10.1007/978-3-662-48116-5_16

17. Gaži, P., Maurer, U.: Cascade encryption revisited. In: Matsui, M. (ed.) ASIACRYPT 2009. LNCS, vol. 5912, pp. 37–51. Springer, Heidelberg (2009). doi:10.1007/978-3-642-10366-7_3

18. Gaži, P., Tessaro, S.: Efficient and optimally secure key-length extension for block ciphers via randomized cascading. In: Pointcheval, D., Johansson, T. (eds.) EUROCRYPT 2012. LNCS, vol. 7237, pp. 63–80. Springer, Heidelberg (2012). doi:10.1007/978-3-642-29011-4_6

19. Hoang, V.T., Tessaro, S.: Key-alternating ciphers and key-length extension: exact bounds and multi-user security. In: Robshaw, M., Katz, J. (eds.) CRYPTO 2016. LNCS, vol. 9814, pp. 3–32. Springer, Heidelberg (2016). doi:10.1007/978-3-662-53018-4_1

20. Lee, J.: Towards key-length extension with optimal security: cascade encryption and xor-cascade encryption. In: Johansson, T., Nguyen, P.Q. (eds.) EUROCRYPT 2013. LNCS, vol. 7881, pp. 405–425. Springer, Heidelberg (2013). doi:10.1007/978-3-642-38348-9_25

21. Luby, M., Rackoff, C.: Pseudo-random permutation generators and cryptographic composition. In: Hartmanis, J. (ed.) Proceedings of the 18th Annual ACM Symposium on Theory of Computing, Berkeley, California, USA, 28–30 May 1986, pp. 356–363. ACM (1986)

22. Maurer, U.: Indistinguishability of random systems. In: Knudsen, L.R. (ed.) EUROCRYPT 2002. LNCS, vol. 2332, pp. 110–132. Springer, Heidelberg (2002). doi:10.1007/3-540-46035-7_8

23. McGrew, D.A., Viega, J.: The security and performance of the Galois/Counter Mode (GCM) of Operation. In: Canteaut, A., Viswanathan, K. (eds.) INDOCRYPT 2004. LNCS, vol. 3348, pp. 343–355. Springer, Heidelberg (2004). doi:10.1007/978-3-540-30556-9_27

24. Mouha, N., Luykx, A.: Multi-key security: the Even-Mansour construction revisited. In: Gennaro, R., Robshaw, M. (eds.) CRYPTO 2015. LNCS, vol. 9215, pp. 209–223. Springer, Heidelberg (2015). doi:10.1007/978-3-662-47989-6_10

25. Myers, S.A.: On the development of block-ciphers and pseudo-random function generators using the composition and xor operators (1999)

26. Patarin, J.: The "Coefficients H" technique. In: Avanzi, R.M., Keliher, L., Sica, F. (eds.) SAC 2008. LNCS, vol. 5381, pp. 328–345. Springer, Heidelberg (2009). doi:10.1007/978-3-642-04159-4_21

27. Raab, M., Steger, A.: "Balls into Bins" — a simple and tight analysis. In: Luby, M., Rolim, J.D.P., Serna, M. (eds.) RANDOM 1998. LNCS, vol. 1518, pp. 159–170. Springer, Heidelberg (1998). doi:10.1007/3-540-49543-6_13

28. Tessaro, S.: Security amplification for the cascade of arbitrarily weak PRPs: tight bounds via the interactive hardcore lemma. In: Ishai, Y. (ed.) TCC 2011. LNCS, vol. 6597, pp. 37–54. Springer, Heidelberg (2011). doi:10.1007/978-3-642-19571-6_3

29. Tessaro, S.: Optimally secure block ciphers from ideal primitives. In: Iwata, T., Cheon, J.H. (eds.) ASIACRYPT 2015. LNCS, vol. 9453, pp. 437–462. Springer, Heidelberg (2015). doi:10.1007/978-3-662-48800-3_18

Public-Seed Pseudorandom Permutations

Pratik Soni$^{(\boxtimes)}$ and Stefano Tessaro

University of California, Santa Barbara, USA
{pratik_soni,tessaro}@cs.ucsb.edu

Abstract. This paper initiates the study of standard-model assumptions on permutations – or more precisely, on families of permutations indexed by a *public* seed. We introduce and study the notion of a *public-seed pseudorandom permutation* (psPRP), which is inspired by the UCE notion by Bellare, Hoang, and Keelveedhi (CRYPTO '13). It considers a two-stage security game, where the first-stage adversary is known as the source, and is restricted to prevent trivial attacks – the security notion is consequently parameterized by the class of allowable sources. To this end, we define in particular unpredictable and reset-secure sources analogous to similar notions for UCEs.

We first study the relationship between psPRPs and UCEs. To start with, we provide efficient constructions of UCEs from psPRPs for both reset-secure and unpredictable sources, thus showing that most applications of the UCE framework admit instantiations from psPRPs. We also show a converse of this statement, namely that the five-round Feistel construction yields a psPRP for reset-secure sources when the round function is built from UCEs for reset-secure sources, hence making psPRP and UCE equivalent notions for such sources.

In addition to studying such reductions, we suggest generic instantiations of psPRPs from both block ciphers and (keyless) permutations, and analyze them in ideal models. Also, as an application of our notions, we show that a simple modification of a recent highly-efficient garbling scheme by Bellare et al. (S&P '13) is secure under our psPRP assumption.

Keywords: Symmetric cryptography · UCE · Permutation-based cryptography · Assumptions · Indifferentiability

1 Introduction

Many symmetric cryptographic schemes are built generically from an underlying component, like a hash function or a block cipher. For several recent examples (e.g., hash functions [15,39], authenticated-encryption schemes [4], PRNGs [16], garbling schemes [10]), this component is a (keyless) *permutation*, which is either designed from scratch to meet certain cryptanalytic goals (as in the case of SHA-3 and derived algorithms based on the sponge paradigm) or is instantiated by fixing the key in a block cipher like AES (as in the garbling scheme of [10]).

© International Association for Cryptologic Research 2017
J.-S. Coron and J.B. Nielsen (Eds.): EUROCRYPT 2017, Part II, LNCS 10211, pp. 412–441, 2017.
DOI: 10.1007/978-3-319-56614-6_14

The security of these schemes is usually proved in the *ideal-permutation model*, that is, the permutation is randomly chosen, and all parties are given (black-box) access to it. Essentially no non-tautological assumptions on permutations are known which are sufficient to imply security.[1] This situation is in sharp contrast to that of hash functions, where despite the popularity of the random-oracle model, we have a good understanding of plausible security assumptions that can be satisfied by these functions. This is particularly important – not so much because we want to put ideal models in question, but because we would like to assess what is really expected from a good permutation or hash function that makes these schemes secure.

OUR CONTRIBUTIONS, IN A NUTSHELL. This paper initiates the study of computational assumptions for permutation-based cryptography. Akin to the case of hash functions, we extend permutations with a *public* seed, that is, π_s is used in lieu of π, where s is a public parameter of the scheme. We introduce a new framework – which we call *public-seed pseudorandom permutations*, or psPRPs, for short – which we investigate, both in terms of realizability, as well as in terms of applications. Our approach takes inspiration from Bellare, Hoang, and Keelveedhi's UCE framework [8], which we extend to permutations. As we will see, psPRPs are both useful and interesting objects of study.

Beyond definitions, we contribute in several ways. First off, we build UCEs from psPRPs via efficient permutation-based hash functions, and show conversely how to build psPRPs from UCEs using the Feistel construction. We also discuss generic instantiations of psPRPs from block ciphers and keyless permutations. Finally, we show a variant of the garbling scheme from [10] whose security can be based on a psPRP assumption on the underlying block cipher, without compromising efficiency. Our reductions between psPRPs and UCEs are established by general theorems that connect them with a weak notion of indifferentiability, which is of independent interest. We explain all of this in detail in the remainder of this introduction; an overview of the results is in Fig. 1.

THE UCE FRAMEWORK: A PRIMER. Bellare, Hoang, and Keelveedhi (BHK) [8] introduced the notion of a *universal computational extractor* (UCE). For a seeded hash function $H : \{0,1\}^s \times \{0,1\}^* \to \{0,1\}^h$, the UCE framework considers a two-stage security game. First, a *source* S is given oracle access to either $H(s, \cdot)$ (for a random, and for now secret, seed s), or a random function $\rho : \{0,1\}^* \to \{0,1\}^h$. After a number of queries, the source produces some leakage $L \in \{0,1\}^*$. In the second stage, the distinguisher D learns *both* L and the seed s, and needs to decide whether S was interacting with $H(s, \cdot)$ or ρ – a task we would like to be hard. Clearly, this is unachievable without restrictions on S, as it can simply set $L = y^*$, where y^* is the output of the oracle on a fixed input x^*, and D then checks whether $H(s, x^*) = y^*$, or not.

---

[1] A notable exception is the line of work on establishing good bounds on the PRF-security of MACs derived from sponge-based constructions, as e.g. in [3,26,37], where one essentially assumes that the underlying permutation yields a secure Even-Mansour [25] cipher.

BHK propose to restrict the set of allowable sources – the security notion UCE[$\mathcal{S}$] corresponds to a function H being secure against all sources within a class $\mathcal{S}$. For example, *unpredictable* sources are those for which a predictor P, given the leakage $L \xleftarrow{\$} S^\rho$, cannot guess any of S's queries. They further distinguish between the class of *computationally* unpredictable sources $\mathcal{S}^{\mathsf{cup}}$ and the class of *statistically* unpredictable sources $\mathcal{S}^{\mathsf{sup}}$, depending on the powers of P. A somewhat stronger notion – referred to as *reset-security* – demands that a distinguisher R given $L \xleftarrow{\$} S^\rho$ accessing the random function ρ cannot tell whether it is given access to the same oracle ρ, or to a completely independent random oracle ρ'. One denotes as $\mathcal{S}^{\mathsf{srs}}$ and $\mathcal{S}^{\mathsf{crs}}$ the classes of statistical and computational reset-secure sources, respectively.

While UCE[$\mathcal{S}^{\mathsf{cup}}$]-security (even under meaningful restrictions) was shown impossible to achieve in the standard model [14,17] assuming indistinguishability obfuscation (IO) [5], there is no evidence of impossibility for UCE[$\mathcal{S}^{\mathsf{sup}}$] and UCE[$\mathcal{S}^{\mathsf{srs}}$], and several applications follow from them. Examples include providing standard-model security for a number of schemes and applications previously only secure in the random-oracle model, including deterministic [8] and hedged PKE [7], immunizing backdoored PRGs [24], message-locked encryption [8], hardcore functions [8], point-function obfuscation [8,13] simple KDM-secure symmetric encryption [8], adaptively-secure garbling [8], and CCA-secure encryption [34]. Moreover, as also pointed out by Mittelbach [36], and already proved in the original BHK work, UCE[$\mathcal{S}^{\mathsf{crs}}$] and UCE[$\mathcal{S}^{\mathsf{cup}}$] are achievable in *ideal* models, and act as useful intermediate security notions for two-stage security games, where indifferentiability does not apply [38].

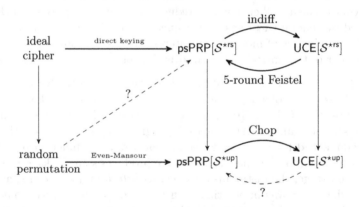

Fig. 1. Relations established in this paper. Here, $\star$ is set consistently everywhere either to c or to s. Lack of arrow indicates a separation, dashed lines indicate implications that are open and which we conjecture to hold true. Also note that in the ideal-cipher model, a random permutation is obtained by fixing the cipher key (e.g., to the all-zero string). We do not know whether the converse is true generically – indifferentiable constructions of ideal ciphers from random permutations (e.g., [1]) do not apply here [38].

PUBLIC-SEED PRPs. We extend the UCE approach to the case of a *seeded* permutation $\pi : \{0,1\}^s \times \{0,1\}^n \rightarrow \{0,1\}^n$, that is, $\pi_s = \pi(s, \cdot)$ is an efficiently invertible permutation on n-bit strings. As in the UCE case, the security game will involve a source making queries to a permutation P *and its inverse* P^{-1}. In the real case, P/P^{-1} give access to π_s and π_s^{-1}, whereas in the ideal case they give access to a random permutation ρ and its inverse ρ^{-1}. Then, S passes on the leakage L to the distinguisher D, which additionally learns s. The psPRP$[\mathcal{S}]$ security notion demands indistinguishability of the real and ideal cases for all PPT D and all $S \in \mathcal{S}$.

This extension is straightforward, but it is not clear that it is useful at all. For instance, UCEs naturally generalize the notion of an extractor, yet no such natural "generalization" exists here, except that of extending the PRP game (played by the source) with a public-seed stage (and hence, the name psPRP). In addition, necessary source restrictions are somewhat less intuitive than in the UCE case. For instance, for psPRPs, for statistically/computationally unpredictable sources (we abuse notation, and denote the corresponding source classes also as $\mathcal{S}^{\mathsf{sup}}$ and $\mathcal{S}^{\mathsf{cup}}$) it must be hard for a predictor to guess an input *or* an output of the queries made by S.

UCEs FROM psPRPs. We first show that psPRPs are not only useful, but essentially allow to recover *all* previous applications of UCEs through simple constructions of UCEs from psPRPs.

Our first result shows that all permutation-based constructions which are *indifferentiable* from a random oracle [20,35] transform a psPRP$[\mathcal{S}^{\star\mathsf{rs}}]$-secure permutation into UCE$[\mathcal{S}^{\star\mathsf{rs}}]$-secure hash function, where $\star \in \{\mathsf{c}, \mathsf{s}\}$.[2] In particular, this implies that the sponge paradigm by Bertoni *et al.* [15], which underlies the SHA-3 hash function, can be used for such transformation, thus giving extra validation for the SHA-3 standard. We note that the permutation underlying SHA-3 is not seeded, but under the assumption that the underlying permutation is psPRP$[\mathcal{S}^{\star\mathsf{rs}}]$-secure when seeded via the Even-Mansour construction [25], it is easy to enhance the sponge construction with a seed.

Note that $\mathcal{S}^{\star\mathsf{rs}}$ is a strictly larger class than $\mathcal{S}^{\star\mathsf{up}}$ (for both psPRP and UCE). Therefore, when an application only needs UCE$[\mathcal{S}^{\star\mathsf{up}}]$-secure hashing, one may ask whether the assumption on the underlying psPRP can also be reduced. We will prove that this is indeed the case, and show that whenever π is psPRP$[\mathcal{S}^{\star\mathsf{up}}]$-secure, then the simple construction that on input X outputs $\pi_s(X)[1 \ldots r]$, that is, the first r bits of $\pi_s(X)$ is a secure fixed-input length UCE$[\mathcal{S}^{\star\mathsf{up}}]$ as long as $r < n - \omega(\log \lambda)$. This result can be combined with the domain extender of [9] to obtain a variable-input-length UCE$[\mathcal{S}^{\star\mathsf{up}}]$-secure hash function.[3]

---

[2] We note that the computational case, by itself, is not that useful, given we know that UCE$[\mathcal{S}^{\mathsf{cup}}]$ and hence also UCE$[\mathcal{S}^{\mathsf{crs}}]$ security is unachievable, unless IO does not exist. However, we may want to occasionally apply these results in ideal models, where the notion is achievable, and thus they are worth stating.

[3] Their construction pre-processes the arbitrary-long input with an almost universal hash function, as e.g. one based on polynomial evaluation.

CP-SEQUENTIAL INDIFFERENTIABILITY. The technique behind the above results is inspired by Bellare, Hoang, and Keelveedhi's work [9] on UCE domain extension. They show that every construction that transforms a fixed-input length random oracle into a variable-input length one in the sense of indifferentiability [20,35] is a good domain extender for UCEs.

We extend their result along three axes. First off, we show that it applies to arbitrary pairs of ideal primitives – e.g., a fixed-input-length or variable-input length random function or a random permutation. For example, a construction using a permutation which is indifferentiable from a random oracle transforms psPRP[$\mathcal{S}^{\star rs}$]-secure permutations into UCE[$\mathcal{S}^{\star rs}$]-secure functions. Through such a general treatment, our above result on sponges is a corollary of the indifferentiability analysis of [15].

Second, we show that a weaker version of indifferentiability, which we call *CP-sequential indifferentiability*, suffices. Recall that indifferentiability of a construction M transforming an ideal primitive **I** into an ideal primitive **J** means that there exists a simulator Sim such that (M^I, I) and (J, Sim^J) are indistinguishable. CP-sequential indifferentiability only demands this for distinguishers that make all of their *construction* queries to M^I/J *before* they proceed to *primitive* queries to I/Sim^J. As we will see, this significantly enlarges the set of constructions this result applies to. For example, truncating the permutation output to $r < n$ bits does not achieve indifferentiability, because a simulator on an inverse query Y needs to fix $\pi^{-1}(Y)$ to some X such that, for the given random function $\rho : \{0,1\}^n \to \{0,1\}^r$, $\rho(X)$ is consistent with Y on the first r bits, which is infeasible. Yet, the same construction *is* CP-sequentially indifferentiable, intuitively because there is no way for a distinguisher to catch an inconsistent random X, as this would require an extra query to ρ. CP-sequential indifferentiability is dual to the sequential indifferentiability notion of Mandal, Patarin, and Seurin [33], which considers the opposite order of construction and primitive queries. In fact, the two notions are incomparable, as we explain below.

Finally, we will also show that under suitable restrictions on the construction M, the result extends from reset-secure sources to unpredictable ones. This will allow to lift our result for truncation to unpredictable sources.

CONSTRUCTING PSPRPS. Obviously, a central question is whether the assumption of being a psPRP is, by itself, attainable. Our general theorem answers this question already – existing indifferentiability result for Feistel constructions [21–23,29] imply already that the 8-round Feistel construction transforms a function which is UCE[$\mathcal{S}^{\star rs}$]-secure into a psPRP[$\mathcal{S}^{\star rs}$]-secure permutation.

It is important however to assess whether simpler constructions achieve this result. *Here, we show that the five-round Feistel construction suffices.* Our proof heavily exploits our connection to CP-indifferentiability. Indeed, the six-round lower bound of [21] does not apply for CP-indifferentiabiliy, as it requires the ability to ask construction queries *after* primitive queries, and we show that CP-indifferentiability is achieved at five rounds already. Our result is not merely a simplification of earlier ones, and our simulation strategy is novel. In particular, while we still follow the chain-completion approach of previous works, due to the low number of rounds, we need to introduce new techniques to bound the complexity of the simulator. To our rescue will come the fact that no construction queries can be made after primitive queries, and hence only a limited number of chain types will need to be completed.

We also note that the we are not aware of any obvious lower bound that shows that more than four rounds are really necessary – four rounds are necessary alone to reach PRP security in the eyes of the source. We leave it as an open problem to show whether four rounds are sufficient. We also note that our result only deals with reset-secure sources, and we leave it as an open problem to find a similar result for unpredictable sources. For reasons we explain in the body, it seems reasonable to conjecture that a heavily unbalanced Feistel network with $\Omega(n)$ rounds achieves this transformation.

CONSTRUCTING psPRPs, IN IDEAL MODELS. While the main purpose of the psPRP framework is that of removing ideal model assumptions, it is still valuable to assess how psPRPs are built in the first place. To this end, we also show how to *heuristically* instantiate psPRPs from existing cryptographic primitives, and here validation takes us necessarily back to ideal models. Plus, for ideal-model applications that require psPRP security as an intermediate notion (for instance, because we are analyzing two-stage games), these provide instantiations.

We validate two strategies: (1) Using a block cipher, and seed it through the key, and (2) Using a *keyless* permutation, and seeding via the Even-Mansour construction [25]. We prove that the first approach is psPRP[$\mathcal{S}^{crs}$]-secure in the ideal-cipher model, and prove the second psPRP[$\mathcal{S}^{cup}$]-secure in the random permutation model.[4]

FIXED-KEY BLOCK-CIPHER BASED GARBLING FROM psPRPs. As a benchmark for psPRPs, we revisit the garbling schemes from [10] based on fixed-key block ciphers, which achieve high degrees of efficiency by eliminating re-keying costs. Their original security analysis was in the ideal-cipher model, and their simplicity is unmatched by schemes with standard-model reductions.

We consider a simple variant of their Ga scheme and prove it secure under the assumption the underlying block cipher, when seeded through its key input, is psPRP[$\mathcal{S}^{sup}$]-secure. Our construction is slightly less efficient than the scheme from [10], since a different seed/key is used for every garbling. However, we still gain from the fact that no re-keying is necessary throughout a garbling operation, or the evaluation of a garbled circuit. We also note that our approach also extends to the GaX scheme of [10] with further optimizations.

EXTRA RELATED WORK. A few works gave UCE constructions. Brzuska and Mittelbach [18] gave constructions from auxiliary-input point obfuscation (AIPO) and iO. In a recent paper, under the exponential DDH assumption, Zhandry [41] built a primitive (called an AI-PRG) which is equivalent to a UCE for a subset of $\mathcal{S}^{cup}$ which is sufficient for instantiating point obfuscators. (The observation is not made explicit in [41], but the definitions are equivalent.)

---

[4] Again, recall that IO-based impossibility for $\mathcal{S}^{cup}$ and $\mathcal{S}^{crs}$ do not apply because we are in ideal models.

None of these results is sufficiently strong to instantiate our Feistel-based construction of psPRPs.

The cryptanalysis community has studied block-cipher security under known keys, albeit with a different focus. For example, Knuden and Rijmen [30] gave attacks against Feistel networks and reduced-round versions of AES that find input-output pairs with specific properties (given the key) in time faster than it should be possible if the block cipher were a random permutation. Several such attacks were later given for a number of block ciphers. We are not aware of these attacks however invalidating psPRP security. Andreeva, Bogdanov, and Mennink [2] gave formal models for known-key security in ideal models based on a weak form of indifferentiability, where construction queries are to the construction *under a known random key*. These are however unrelated.

OUTLINE. Section 2 proposes a general framework for public-seed pseudorandom notions, and Sect. 3 puts this to use to provide general reduction theorems between pairs of such primitives, and defines in particular CP-sequential indifferentiability. UCE constructions from psPRPs are given in Sect. 4, whereas Sect. 5 presents our main result on building psPRPs via the Feistel construction. Heuristic constructions are presented in Sect. 6, and finally we apply psPRPs to the analysis of garbling schemes in Sect. 7.

NOTATIONAL PRELIMINARIES. Throughout this paper, we denote by Funcs (X, Y) the set of functions $X \to Y$, and in particular use the shorthand Funcs(m, n) whenever $X = \{0, 1\}^m$ and $Y = \{0, 1\}^n$. We also denote by Perms(X) the set of permutations on the set X, and analogously, Perms(n) denotes the special case where $X = \{0, 1\}^n$. We say that a function $f : \mathbb{N} \to \mathbb{R}_{\geq 0}$ is *negligible* if for all $c \in \mathbb{N}$, there exists a λ_0 such that $f(\lambda) \leq \lambda^{-c}$ for all $\lambda \geq \lambda_0$.

Our security definitions and proofs will often use games, as formalized by Bellare and Rogaway [12]. Typically, our games will have boolean outputs – that is, either true or false – and we use the shorthand Pr [G] to denote the probability that a certain game outputs the value true, or occasionally 1 (when the output is binary, rather than boolean).

2   Public-Seed Pseudorandomness

We present a generalization of the UCE notion [8], which we term *public-seed pseudorandomness*. We apply this notion to define psPRPs as a special case, but the general treatment will be useful to capture transformations between UCEs and psPRPs in Sect. 3 via one single set of theorems.

2.1   Ideal Primitives and Their Implementations

We begin by formally defining *ideal primitives* using notation inspired by [6,27].

IDEAL PRIMITIVES. An *ideal primitive* is a pair $\mathbf{I} = (\Sigma, \mathcal{T})$, where $\Sigma = \{\Sigma_\lambda\}_{\lambda \in \mathbb{N}}$ is a family of sets of functions (such that all functions in Σ_λ have the same domain and range), and $\mathcal{T} = \{\mathcal{T}_\lambda\}_{\lambda \in \mathbb{N}}$ is a family of probability distributions, where $\mathcal{T}_\lambda$'s range is a subset of Σ_λ for all $\lambda \in \mathbb{N}$. The ideal primitive $\mathbf{I}$, once the security parameter λ is fixed, should be thought of as an oracle that initially samples a function I as its initial state according to $\mathcal{T}_\lambda$ from Σ_λ. We denote this sampling as $I \leftarrow_{\$} \mathbf{I}_\lambda$. Then, $\mathbf{I}$ provides access to I via queries, that is, on input $\mathbf{x}$ it returns $I(\mathbf{x})$.[5]

EXAMPLES. We give a few examples of ideal primitives using the above notation. In particular, let $\kappa, m, n : \mathbb{N} \to \mathbb{N}$ be functions.

Example 1. The *random function* $\mathbf{R}_{m,n} = (\Sigma^{\mathbf{R}}, \mathcal{T}^{\mathbf{R}})$ is such that for all $\lambda \in \mathbb{N}$, $\Sigma^{\mathbf{R}}_\lambda = \mathsf{Funcs}(m(\lambda), n(\lambda))$, and $\mathcal{T}^{\mathbf{R}}_\lambda$ is the uniform distribution on $\Sigma^{\mathbf{R}}_\lambda$. We also define $\mathbf{R}_{*,n}$ to be the same for $\mathsf{Funcs}(*, n(\lambda))$, that is, when the domain is extended to arbitrary length input strings.[6]

Example 2. The *random permutation* $\mathbf{P}_n = (\Sigma^{\mathbf{P}}, \mathcal{T}^{\mathbf{P}})$ is such that for all $\lambda \in \mathbb{N}$,

$$\Sigma^{\mathbf{P}}_\lambda = \Big\{ P : \{+,-\} \times \{0,1\}^{n(\lambda)} \to \{0,1\}^{n(\lambda)} \mid$$
$$\exists \pi \in \mathsf{Perms}(n(\lambda)) : \ P(+,x) = \pi(x), P(-,x) = \pi^{-1}(x) \Big\},$$

and moreover, $\mathcal{T}^{\mathbf{P}}_\lambda$ is the uniform distribution on $\Sigma^{\mathbf{P}}_\lambda$.

Example 3. The *ideal cipher* $\mathbf{IC}_{\kappa,n} = (\Sigma^{\mathbf{IC}}, \mathcal{T}^{\mathbf{IC}})$ is such that

$$\Sigma^{\mathbf{IC}}_\lambda = \Big\{ E : \{0,1\}^{\kappa(\lambda)} \times \{+,-\} \times \{0,1\}^{n(\lambda)} \to \{0,1\}^{n(\lambda)} \mid$$
$$\forall k \in \{0,1\}^{\kappa(\lambda)} \exists \pi_k \in \mathsf{Perms}(n(\lambda)) : \ E(k,+,x) = \pi_k(x), E(k,-,x) = \pi_k^{-1}(x) \Big\},$$

and $\mathcal{T}^{\mathbf{IC}}_\lambda$ is the uniform distribution on $\Sigma^{\mathbf{IC}}_\lambda$.

EFFICIENCY CONSIDERATIONS. Usually, for an ideal primitive $\mathbf{I} = (\Sigma, \mathcal{T})$, the bit-size of the elements of Σ_λ grows exponentially in λ, and thus one would not implement a primitive $\mathbf{I}$ by sampling I from Σ_λ, but rather using techniques such as lazy sampling. An implementation of a primitive $\mathbf{I}$ is a *stateful* randomized PPT algorithm A such that $A(1^\lambda, \cdot)$ behaves as $I \xleftarrow{\$} \mathbf{I}_\lambda$ for all $\lambda \in \mathbb{N}$. We say that $\mathbf{I}$ is *efficiently implementable* if such an A exists. All the above examples – $\mathbf{R}_{m,n}, \mathbf{R}_{*,n}, \mathbf{P}_n$, and $\mathbf{IC}_{\kappa,n}$ – are efficiently implementable as long as m, n, κ are polynomially bounded functions.

Σ-COMPATIBLE FUNCTION FAMILIES. A *function family* $\mathsf{F} = (\mathsf{Kg}, \mathsf{Eval})$ consists of a *key (or seed) generation algorithm* $\mathsf{F.Kg}$ and an *evaluation algorithm* $\mathsf{F.Eval}$. In particular, $\mathsf{F.Kg}$ is a randomized algorithm that on input the unary representation of the security parameter λ returns a *key* k, and we let $[\mathsf{F.Kg}(1^\lambda)]$ denote

---

[5] The reader may wonder whether defining Σ is necessary, but this will allow us to enforce a specific format on valid implementations below.

[6] Note that this requires some care, because Σ_λ is now uncountable, and thus sampling from it requires a precise definition. We will not go into formal details, similar to many other papers, but it is clear that this can easily be done.

MAIN $\mathsf{psPR}_{\mathsf{F},\mathbf{I}}^{S,D}(\lambda)$:	ORACLE $\mathcal{O}(i,\mathbf{x})$:
$(1^n, t) \twoheadleftarrow\!\!\!\!{}_{\$}\, S(1^\lambda, \varepsilon)$	if $b = 1$ then
$b \leftarrow\!\!{}_{\$}\, \{0,1\}$	$\quad$ return $\mathsf{F}.\mathsf{Eval}(1^\lambda, k_i, \mathbf{x})$
$k_1, \ldots, k_n \leftarrow\!\!{}_{\$}\, \mathsf{F}.\mathsf{Kg}(1^\lambda)$	else
$f_1, \ldots, f_n \leftarrow\!\!{}_{\$}\, \mathbf{I}_\lambda$	$\quad$ return $f_i(\mathbf{x})$
$L \leftarrow\!\!{}_{\$}\, S^{\mathcal{O}}(1^\lambda, t)$	
$b' \leftarrow\!\!{}_{\$}\, D(1^\lambda, k_1, \ldots, k_n, L)$	
return $b' = b$	

Fig. 2. Game psPR used to define pspr-security for a primitive F that is Σ-compatible with **I**. Here, S is the source and D is the distinguisher. Recall that the notation $f \leftarrow\!\!{}_{\$}\, \mathbf{I}_\lambda$ indicates picking a function from Σ_λ using $\mathcal{T}_\lambda$.

the set of all possible outputs of $\mathsf{F}.\mathsf{Kg}(1^\lambda)$. Moreover, $\mathsf{F}.\mathsf{Eval}$ is a deterministic algorithm that takes three inputs; the security parameter in unary form 1^λ, a key $k \in [\mathsf{F}.\mathsf{Kg}(1^\lambda)]$ and a query $\mathbf{x}$ such that $\mathsf{F}.\mathsf{Eval}(1^\lambda, k, \cdot)$ implements a function that maps queries $\mathbf{x}$ to $\mathsf{F}.\mathsf{Eval}(1^\lambda, k, \mathbf{x})$. We say that F is *efficient* if both Kg and Eval are polynomial-time algorithms.

Definition 1 (Σ-compatibility). *A function family* F *is* Σ-*compatible with* $\mathbf{I} = (\Sigma, \mathcal{T})$ *if* $\mathsf{F}.\mathsf{Eval}(1^\lambda, k, \cdot) \in \Sigma_\lambda$ *for all* $\lambda \in \mathbb{N}$ *and* $k \in [\mathsf{F}.\mathsf{Kg}(1^\lambda)]$.

2.2 Public-Seed Pseudorandomness, PsPRPs, and Sources

We now define a general notion of public-seed pseudorandom implementations of ideal primitives.

THE GENERAL DEFINITION. Let $\mathsf{F} = (\mathsf{Kg}, \mathsf{Eval})$ be a function family that is Σ-compatible with an ideal primitive $\mathbf{I} = (\Sigma, \mathcal{T})$. Let S be an adversary called the *source* and D an adversary called the *distinguisher*. We associate to them, F and **I**, the game $\mathsf{psPR}_{\mathsf{F},\mathbf{I}}^{S,D}(\lambda)$ depicted in Fig. 2. The source initially chooses the number of keys n. Then, in the second stage, it is given access to an oracle $\mathcal{O}$ and we require any query $(i, \mathbf{x})$ made to this oracle be valid, that is, $\mathbf{x}$ is a valid query for any $f_i \in \Sigma_\lambda$ and $i \in [n]$, for n output by the first stage of the source. When the challenge bit $b = 1$ ("real") the oracle responds via $\mathsf{F}.\mathsf{Eval}$ under the key k_i ($\mathsf{F}.\mathsf{Eval}(1^\lambda, k_i, \cdot)$) that is chosen by the game and *not* given to the source. When $b = 0$ ("ideal") it responds via f_i where $f_i \leftarrow\!\!{}_{\$}\, \mathbf{I}_\lambda$. After its interaction with the oracle $\mathcal{O}$, the source S communicates the *leakage* $L \in \{0,1\}^*$ to D. The distinguisher is given access to the keys $k_1, \ldots, k_n$ and must now guess $b' \in \{0,1\}$ for b. The game returns true iff $b' = b$ and we describe the pspr-advantage of (S, D) for $\lambda \in \mathbb{N}$ as

$$\mathsf{Adv}_{\mathsf{F},S,D}^{\mathsf{pspr}[\mathbf{I}]}(\lambda) = 2 \Pr\left[\mathsf{psPR}_{\mathsf{F},\mathbf{I}}^{S,D}(\lambda)\right] - 1. \tag{1}$$

In the following, we are going to use the shorthands $\mathsf{UCE}[m, n]$ for $\mathsf{pspr}[\mathbf{R}_{m,n}]$, $\mathsf{UCE}[n]$ for $\mathsf{pspr}[\mathbf{R}_{*,n}]$, and $\mathsf{psPRP}[n]$ for $\mathsf{pspr}[\mathbf{P}_n]$.

Note that our security game captures the multi-key version of the security notions, also considered in past works on UCE, as it is not known to be implied by the single-key version, which is recovered by having the source initially output $n = 1$.

RESTRICTING SOURCES. One would want to define F as secure if $\mathsf{Adv}^{\mathsf{pspr[I]}}_{\mathsf{F},S,D}(\lambda)$ is negligible in λ for all polynomial time sources S and distinguishers D. However, as shown already in the special case of UCEs [8], this is impossible, as one can always easily construct (at least for non-trivial $\mathbf{I}$'s) a simple source S which leaks the evaluation of $\mathcal{O}$ on a given point, and D can check consistency given k.

Therefore to obtain meaningful and non-empty security definitions we restrict the considered sources to some class $\mathcal{S}$, without restricting the distinguisher class. Consequently, we denote by $\mathsf{psPR}[\mathbf{I}, \mathcal{S}]$ the security notion that asks $\mathsf{Adv}^{\mathsf{pspr[I]}}_{\mathsf{F},S,D}(\lambda)$ to be negligible for all polynomial time distinguishers D and all sources $S \in \mathcal{S}$. Following [8], we also use $\mathsf{psPR}[\mathbf{I}, \mathcal{S}]$ to denote the set of F's which are $\mathsf{psPR}[\mathbf{I}, \mathcal{S}]$-secure. Note that obviously, if $\mathcal{S}_1 \subseteq \mathcal{S}_2$, then $\mathsf{psPR}[\mathbf{I}, \mathcal{S}_2] \subseteq \mathsf{psPR}[\mathbf{I}, \mathcal{S}_1]$ where $\mathcal{S}_1$ and $\mathcal{S}_2$ are source classes for the ideal primitive $\mathbf{I}$. We will use the shorthands $\mathsf{psPRP}[n, \mathcal{S}]$ for $\mathsf{psPR}[\mathbf{P}_n, \mathcal{S}]$ and $\mathsf{UCE}[m, n, \mathcal{S}]$ for $\mathsf{psPR}[\mathbf{R}_{m,n}, \mathcal{S}]$, where $m = *$ if the domain is unbounded.

Below, we discuss two important classes of restrictions, which are fundamental for the remainder of this paper – unpredictable and reset-secure sources.

MAIN $\mathsf{Pred}^P_{\mathbf{I},S}(\lambda)$:	MAIN $\mathsf{Reset}^R_{\mathbf{I},S}(\lambda)$:
done $\leftarrow$ false; $Q \leftarrow \emptyset$; $(1^n, t) \leftarrow_{\$} S(1^\lambda, \varepsilon)$	done $\leftarrow$ false; $(1^n, t) \leftarrow_{\$} S(1^\lambda, \varepsilon)$
$f_1, \ldots, f_n \leftarrow_{\$} \mathbf{I}_\lambda$	$f_1^0, f_1^1, \ldots, f_n^0, f_n^1 \leftarrow_{\$} \mathbf{I}_\lambda$
$L \leftarrow_{\$} S^{\mathcal{O}}(1^\lambda, t)$; done $\leftarrow$ true	$L \leftarrow_{\$} S^{\mathcal{O}}(1^\lambda, t)$; done $\leftarrow$ true
$Q' \leftarrow_{\$} P^{\mathcal{O}}(1^\lambda, 1^n, L)$	$b \leftarrow_{\$} \{0, 1\}$; $b' \leftarrow_{\$} R^{\mathcal{O}}(1^\lambda, 1^n, L)$
return $(Q \cap Q' \neq \emptyset)$	return $b' = b$
ORACLE $\mathcal{O}(i, \mathbf{x})$:	ORACLE $\mathcal{O}(i, \mathbf{x})$:
if $\neg$done then $Q \leftarrow Q \cup \{\mathbf{x}\}$	if $\neg$done then return $f_i^0(\mathbf{x})$
return $f_i(\mathbf{x})$	else return $f_i^b(\mathbf{x})$

Fig. 3. Games Pred and Reset are used to define the unpredictability and reset-security of the source S respectively against the ideal primitive $\mathbf{I}$. Here, S is the source, P is the predictor and R is the reset adversary.

UNPREDICTABLE SOURCES. Let S be a source. Consider the game $\mathsf{Pred}^P_{\mathbf{I},S}(\lambda)$ of Fig. 3 associated to S and an adversary P called the predictor. Given the leakage, the latter outputs a set Q'. It wins if this set contains any $\mathcal{O}$-query of the source. For $\lambda \in \mathbb{N}$ we let

$$\mathsf{Adv}^{\mathsf{pred[I]}}_{S,P}(\lambda) = \Pr\left[\mathsf{Pred}^P_{\mathbf{I},S}(\lambda)\right]. \tag{2}$$

We say that P is a *computational predictor* if it is polynomial time, and it is a *statistical* predictor if there exists polynomials q, q' such that for all $\lambda \in \mathbb{N}$, predictor P makes at most $q(\lambda)$ oracle queries and outputs a set Q' of size at most $q'(\lambda)$ in game $\mathsf{Pred}_{\mathbf{I},S}^P(\lambda)$. We stress that in this case the predictor need not be polynomial time, even though it makes a polynomial number of queries. We say S is *computationally unpredictable* if $\mathsf{Adv}_{S,P}^{\mathsf{pred}[\mathbf{I}]}(\lambda)$ is negligible for all computational predictors P. We say S is *statistically unpredictable* if $\mathsf{Adv}_{S,P}^{\mathsf{pred}[\mathbf{I}]}(\lambda)$ is negligible for all statistical predictors P. We let $\mathcal{S}^{\mathsf{cup}}$ be the class of all polynomial time, computationally unpredictable sources and $\mathcal{S}^{\mathsf{sup}} \subseteq \mathcal{S}^{\mathsf{cup}}$ be the class of all polynomial time statistically unpredictable sources.[7]

<u>RESET-SECURE SOURCES.</u> Let S be a source. Consider the game $\mathsf{Reset}_{\mathbf{I},S}^R(\lambda)$ of Fig. 3 associated to S and an adversary R called the reset adversary. The latter wins if given the leakage L it can distinguish between f^0 used by the source S and an independent f^1 where $f^0, f^1 \leftarrow_{\$} \mathbf{I}_\lambda$. For $\lambda \in \mathbb{N}$ we let

$$\mathsf{Adv}_{S,R}^{\mathsf{reset}[\mathbf{I}]}(\lambda) = 2\Pr\left[\mathsf{Reset}_{\mathbf{I},S}^R(\lambda)\right] - 1. \tag{3}$$

We say that R is a *computational reset adversary* if it is polynomial time, and it is a statistical reset adversary if there exists a polynomial q such that for all $\lambda \in \mathbb{N}$, reset adversary R makes at most $q(\lambda)$ oracle queries in game $\mathsf{Reset}_{\mathbf{I},S}^R(\lambda)$. We stress that in this case the reset adversary need not be polynomial time. We say S is *computationally reset-secure* if $\mathsf{Adv}_{S,R}^{\mathsf{reset}[\mathbf{I}]}(\lambda)$ is negligible for all computational reset adversaries R. We say S is *statistically reset-secure* if $\mathsf{Adv}_{S,R}^{\mathsf{reset}[\mathbf{I}]}(\lambda)$ is negligible for all statistical reset adversaries R. We let $\mathcal{S}^{\mathsf{crs}}$ be the class of all polynomial time, computationally reset-secure sources and $\mathcal{S}^{\mathsf{srs}} \subseteq \mathcal{S}^{\mathsf{crs}}$ the class of all polynomial time statistically reset-secure sources.

<u>RELATIONSHIPS.</u> For the case of psPRPs, we mention the following fact, which is somewhat less obvious than in the UCE case, and in particular only holds if the permutation's domain grows with the security parameter.

Proposition 1. *For all $n \in \omega(\log \lambda)$, we have* $\mathsf{psPRP}[n, \mathcal{S}^{\star\mathsf{rs}}] \subseteq \mathsf{psPRP}[n, \mathcal{S}^{\star\mathsf{up}}]$ *where $\star \in \{\mathsf{c}, \mathsf{s}\}$.*

Proof (Sketch). In the reset secure game, consider the event that R queries its oracle $\mathcal{O}$ on input (i, σ, x) which was queried by S already as an $\mathcal{O}(i, \sigma, x)$ query, or it was the answer to a query $\mathcal{O}(i, \overline{\sigma}, y)$. Here (like elsewhere in the paper), we use the notational convention $\overline{+} = -$ and $\overline{-} = +$. The key point here is proving that as long as this bad event does not happen, the $b = 0$ and $b = 1$ case are hard to distinguish. A difference with the UCE case is that due to the permutation property, they will not be perfectly indistinguishable, but a fairly standard (yet somewhat tedious) birthday argument suffices to show that indistinguishability still holds as long as the overall number of $\mathcal{O}$ queries (of S and R) is below $2^{n(\lambda)/2}$, which is super-polynomial for $n(\lambda) = \omega(\log \lambda)$.   □

---

[7] We note that computational unpredictability is only meaningful for sufficiently restricted classes of sources or in ideal models, as otherwise security against $\mathcal{S}^{\mathsf{cup}}$ is not achievable assuming IO, using essentially the same attack as [17].

MAIN $\mathsf{CP}[\mathbf{I} \to \mathbf{J}]_{\mathsf{M},\mathsf{Sim}}^{A}(\lambda)$:	ORACLE $\mathsf{Func}(\mathbf{x})$:	ORACLE $\mathsf{Prim}(\mathbf{u})$:
$b \leftarrow_\$ \{0,1\}; f \leftarrow_\$ \mathbf{I}_\lambda; g \leftarrow_\$ \mathbf{J}_\lambda$	if $b = 1$ then	if $b = 1$ then
$\mathsf{st} \leftarrow_\$ A_1^{\mathsf{Func}}(1^\lambda)$	return $\mathsf{M}^f(\mathbf{x})$	return $f(\mathbf{u})$
$b' \leftarrow_\$ A_2^{\mathsf{Prim}}(1^\lambda, \mathsf{st})$	else	else
return $b' = b$	return $g(\mathbf{x})$	return $\mathsf{Sim}^g(\mathbf{u})$

Fig. 4. Game CP used to define cpi-security for a construction M implementing the primitive **J** using primitive **I**. Here, Sim is the simulator and $A = (A_1, A_2)$ is the two-stage distinguisher.

3 Reductions and Indifferentiability

We present general theorems that we will use to obtain reductions between psPRPs and UCEs. Our general notation for public-seed pseudorandom primitives allows us to capture the reductions through two general theorems.

CP-SEQUENTIAL INDIFFERENTIABILITY. Indifferentiability was introduced in [35] by Maurer, Renner, and Holenstein to formalize reductions between ideal primitives. Following their general treatment, it captures the fact that a (keyless) construction M using primitive **I** (which can be queried by the adversary directly) is *as good* as another primitive **J** by requiring the existence of a simulator that can simulate **I** consistently by querying **J**.

Central to this paper is a weakening of indifferentiability that we refer to as *CP-sequential indifferentiability*, where the distinguisher A makes all of its *construction* queries to $\mathsf{M}^\mathbf{I}$ (or **J**) *before* moving to making *primitive queries* to **I** (or $\mathsf{Sim}^\mathbf{J}$, where Sim is the simulator). Note that this remains a non-trivial security goal, since Sim does not learn the construction queries made by A, but needs to simulate correctly nonetheless. However, the hope is that because A has committed to its queries before starting its interaction with Sim, the simulation task will be significantly easier. (We will see that this is indeed the case.)

More concretely, the notion is concerned with constructions which implement **J** from **I**, and need to at least satisfy the following syntactical requirement.

Definition 2 ($(\mathbf{I} \to \mathbf{J})$-compatibility). *Let* $\mathbf{I} = (\mathbf{I}.\Sigma, \mathbf{I}.\mathcal{T})$ *and* $\mathbf{J} = (\mathbf{J}.\Sigma, \mathbf{J}.\mathcal{T})$ *be ideal primitives. A construction* M *is called* $(\mathbf{I} \to \mathbf{J})$-compatible *if for every* $\lambda \in \mathbb{N}$, *and every* $f \in \mathbf{I}.\Sigma_\lambda$, *the construction* M *implements a function* $x \mapsto \mathsf{M}^f(1^\lambda, x)$ *which is in* $\mathbf{J}.\Sigma_\lambda$.

The game CP is described in Fig. 4. For ideal primitives **I**, **J**, a two-stage adversary $A = (A_1, A_2)$, an $(\mathbf{I} \to \mathbf{J})$-compatible construction M, and simulator Sim, as well as security parameter $\lambda \in \mathbb{N}$, we define

$$\mathsf{Adv}_{\mathsf{M},\mathsf{Sim},A}^{\mathsf{cpi}[\mathbf{I} \to \mathbf{J}]}(\lambda) = 2 \cdot \mathsf{Pr}\left[\mathsf{CP}[\mathbf{I} \to \mathbf{J}]_{\mathsf{M},\mathsf{Sim}}^{A}(\lambda)\right] - 1. \tag{4}$$

We remark that the CP-sequential indifferentiability notion is the exact dual of sequential indifferentiability as introduced by Mandal, Patarin, and

Seurin [33], which postpones construction queries *to the end.* As we will show below in Sect. 4.2, there are CP-indifferentiable constructions which are not sequentially indifferentiable in the sense of [33].

REDUCTIONS. We show that CP-sequential indifferentiability yields a reduction between public-seed pseudorandomness notions. A special case was shown in [9] by Bellare, Hoang, and Keelvedhi for domain extension of UCEs. Our result goes beyond in that: (1) It is more general, as it deals with arbitrary ideal primitives, (2) It only relies on CP-sequential indifferentiability, as opposed to full indifferentiability, and (3) The reduction of [9] only considered reset-secure sources, whereas we show that under certain conditions on the construction, the reduction also applies to unpredictable sources. Nonetheless, our proofs follow the same approach of [9], and the main contribution is conceptual.

We let $\mathsf{F} = (\mathsf{F.Kg}, \mathsf{F.Eval})$ be a function family which is Σ-compatible with an ideal primitive $\mathbf{I}$. Further, let M be an $(\mathbf{I} \to \mathbf{J})$-compatible construction. Then, overloading notation, we define the new function family $\mathsf{M}[\mathsf{F}] = (\mathsf{M.Kg}, \mathsf{M.Eval})$, where $\mathsf{M.Kg} = \mathsf{F.Kg}$, and for every $k \in [\mathsf{M.Kg}(1^\lambda)]$, we let

$$\mathsf{M.Eval}(1^\lambda, k, x) = \mathsf{M}^O(1^\lambda, x), \qquad (5)$$

where $O(z) = \mathsf{F.Eval}(1^\lambda, k, z)$.

RESET-SECURE SOURCES. The following is our general reduction theorem for the case of reset-secure sources. Its proof follows similar lines as the one in [9] and we refer the reader to the full version for details.

Theorem 1 (Composition theorem, reset-secure case). *Let* $\mathsf{M}, \mathsf{F}, \mathbf{I},$ *and* $\mathbf{J}$ *be as above. Fix any simulator* Sim. *Then, for every source-distinguisher pair* (S, D), *where* S *requests at most* $N(\lambda)$ *keys, there exists a source-distinguisher pair* $(\overline{S}, \overline{D})$, *and a further distinguisher* A, *such that*

$$\mathsf{Adv}^{\mathsf{pspr}[\mathbf{J}]}_{\mathsf{M}[\mathsf{F}], S, D}(\lambda) \le \mathsf{Adv}^{\mathsf{pspr}[\mathbf{I}]}_{\mathsf{F}, \overline{S}, \overline{D}}(\lambda) + N(\lambda) \cdot \mathsf{Adv}^{\mathsf{cpi}[\mathbf{I} \to \mathbf{J}]}_{\mathsf{M}, \mathsf{Sim}, A}(\lambda). \qquad (6)$$

Here, in particular: The complexities of D *and* $\overline{D}$ *are the same. Moreover, if* S, D, *and* M *are polynomial time, and* $\mathbf{I}, \mathbf{J}$ *are efficiently implementable, then* A, $\overline{S}$ *and* $\overline{D}$ *are also polynomial-time.*

Moreover, for every reset adversary R, *there exists a reset adversary* R' *and a distinguisher* B *such that*

$$\mathsf{Adv}^{\mathsf{reset}[\mathbf{I}]}_{\overline{S}, R}(\lambda) \le \mathsf{Adv}^{\mathsf{reset}[\mathbf{J}]}_{S, R'}(\lambda) + 3N(\lambda) \cdot \mathsf{Adv}^{\mathsf{cpi}[\mathbf{I} \to \mathbf{J}]}_{\mathsf{M}, \mathsf{Sim}, B}(\lambda), \qquad (7)$$

where R' *makes a polynomial number of query/runs in polynomial time if* R *and* Sim *make a polynomial number of queries/run in polynomial time, and* $\mathbf{I}, \mathbf{J}$ *are efficiently implementable.* ∎

QUERY EXTRACTABLE CONSTRUCTIONS. Next, we show that under strong conditions on the construction M, Theorem 1 extends to the case of unpredictability.

GAME $\text{EXT}^{S,P}_{M,I,\text{Ext}}(\lambda)$:	ORACLE $\mathcal{O}(i,x)$:
done $\leftarrow$ false	if $\neg$done then $Q_I \xleftarrow{\cup} \{x\}$
$Q_I, Q_M \leftarrow \emptyset$	return $f_i(x)$
$(1^n, \text{st}) \leftarrow_\$ S(1^\lambda, \varepsilon)$	
$f_1, \ldots, f_n \leftarrow_\$ I_\lambda$	ORACLE $\mathcal{O}_M(i,x)$:
$L \leftarrow_\$ S^{\mathcal{O}_M}(1^\lambda, 1^n, \text{st})$	if $\neg$done then $Q_M \xleftarrow{\cup} \{x\}$
done $\leftarrow$ true	$y \leftarrow M^{\mathcal{O}(i,\cdot)}(x)$
$Q \leftarrow_\$ P^{\mathcal{O}}(1^\lambda, 1^n, L); Q^* \leftarrow \text{Ext}^{\mathcal{O}}(Q)$	return y
return $((Q \cap Q_I \neq \emptyset) \wedge (Q^* \cap Q_M = \emptyset))$	

Fig. 5. Game $\text{EXT}^{S,P}_{M,I,\text{Ext}}(\lambda)$ in the definition of query extractability.

In particular, we consider constructions which we term *query extractable*. Roughly, what such constructions guarantee is that every query made by M to an underlying ideal primitive **I** can be assigned to a (small) set of possible inputs to M that would result in this query during evaluation. Possibly, this set of inputs may be found by making some additional queries to **I**. We define this formally through the game $\text{EXT}^{S,P}_{M,I,\text{Ext}}(\lambda)$ in Fig. 5. It involves a *source* S and a *predictor* P, as well as an extractor Ext. Here, S selects an integer n, which results in n instances $f_1, \ldots, f_n$ of **I** being spawned, and then makes queries to n instances of M^{f_i}, gives some leakage to the predictor P, and the predictor makes further query to the **I**-instances, until it outputs a set Q. Then, we run the extractor Ext on Q, and the extractor can also make additional queries to the **I**-instances, and outputs an additional set Q^*. We are interested in the event that Q contains one of queries made to the f_i's by M in the first stage of the game, yet Q^* does not contain any of S's queries to M^{f_i} for some i. In particular, we are interested in

$$\text{Adv}^{\text{ext}[I]}_{M,S,P,\text{Ext}}(\lambda) = \Pr\left[\text{EXT}^{S,P}_{M,I,\text{Ext}}(\lambda)\right].$$

We say that M is *query extractable with respect to* **I** if there exists a polynomial time Ext such that $\text{Adv}^{\text{ext}[I]}_{M,S,P,\text{Ext}}(\lambda)$ is negligible for all PPT P and S. We say it is *perfectly* query extractable if the advantage is 0, rather than simply negligible.

The next theorem provides an alternative to Theorem 1 for the case of unpredictable sources whenever M guarantees query extractability.

Theorem 2 (Composition theorem, unpredictable case). *Let* M, F, **I**, *and* **J** *be as before. Fix any simulator* Sim. *Then, for every source-distinguisher pair* (S, D), *where* S *requests at most* $N(\lambda)$ *keys, there exists a source-distinguisher pair* $(\overline{S}, \overline{D})$, *and a further distinguisher* A, *such that*

$$\text{Adv}^{\text{pspr}[J]}_{M[F],S,D}(\lambda) \leq \text{Adv}^{\text{pspr}[I]}_{F,\overline{S},\overline{D}}(\lambda) + N(\lambda) \cdot \text{Adv}^{\text{cpi}[I\to J]}_{M,\text{Sim},A}(\lambda). \tag{8}$$

Here, in particular: The complexities of D *and* $\overline{D}$ *are the same. Moreover, if* S, D, *and* M *are polynomial time, and* **I**, **J** *are efficiently implementable, then* A, $\overline{S}$ *and* $\overline{D}$ *are also polynomial-time.*

Moreover, for every predictor P and extractor Ext, *there exists a predictor adversary P′ and a distinguisher B such that*

$$\mathsf{Adv}_{\overline{S},P}^{\mathsf{pred}[\mathbf{I}]}(\lambda) \leq \mathsf{Adv}_{S,P'}^{\mathsf{pred}[\mathbf{J}]}(\lambda) + \mathsf{Adv}_{\mathsf{M},S,P,\mathsf{Ext}}^{\mathsf{ext}[\mathbf{I}]}(\lambda) + N(\lambda) \cdot \mathsf{Adv}_{\mathsf{M},\mathsf{Sim},B}^{\mathsf{cpi}[\mathbf{I}\to\mathbf{J}]}(\lambda), \qquad (9)$$

where P′ makes a polynomial number of query/runs in polynomial time if P, Sim *and* Ext *make a polynomial number of queries/run in polynomial time, and* **I, J** *are efficiently implementable.* ∎

4   From psPRPs to UCEs

We consider the problem of building UCEs from psPRPs. On the one hand, we want to show that all applications of UCEs can be recovered modularly by instantiating the underlying UCE with a psPRP-based construction. Second, we want to show that practical permutation-based designs can be instantiated by assuming the underlying permutation (when equipped with a seed) is a psPRP.

4.1   Reset-Secure Sources and Sponges

The case of reset-secure sources follows by a simple application of Theorem 1: A number of constructions from permutations have been proved indifferentiable from a random oracle, and all of these yield a construction of a UCE for $S^{\star\mathsf{rs}}$ when the underlying permutation is a psPRP for $S^{\star\mathsf{rs}}$, where $\star \in \{\mathsf{c},\mathsf{s}\}$.[8]

SPONGES. A particular instantiation worth mentioning is the sponge construction by Bertoni et al. [15], which underlies KECCAK/SHA-3. In particular, let $\mathsf{Sponge}_{n,r}$ be the $(\mathbf{P}_n \to \mathbf{R}_{*,r})$-compatible construction which operates as follows, on input 1^λ, $M \in \{0,1\}^*$, and given oracle access to a permutation $\rho : \{0,1\}^{n(\lambda)} \to \{0,1\}^{n(\lambda)}$. The message M is split into r-bit blocks $M[1],\ldots,M[\ell]$, and the computation keeps a state $S_i\|T_i$, where $S_i \in \{0,1\}^r$ and $T_i \in \{0,1\}^{n-r}$. Then, $\mathsf{Sponge}_{n,r}^\rho(1^\lambda, M) = S_\ell[1..r]$, where

$$S_0\|T_0 \leftarrow 0^n, \quad S_i\|T_i \leftarrow \rho((S_{i-1} \oplus M[i])\|T_{i-1}) \text{ for} i = 1,\ldots,\ell.$$

Then, the following theorem follows directly from Theorem 1 and the indifferentiability analysis of [15]. (We state here only the asymptotic version, but concrete parameters can be obtained from these theorems.)

Theorem 3 (UCE-security for Sponges). *For $\star \in \{\mathsf{c},\mathsf{s}\}$ and $n(\lambda)$ polynomially bounded in λ, if* $F \in \mathsf{psPRP}[n, S^{\star\mathsf{rs}}]$, *then* $\mathsf{Sponge}_{n,r}[F] \in \mathsf{UCE}[*, r, S^{\star\mathsf{rs}}]$ *whenever $n(\lambda) - r(\lambda) = \omega(\log \lambda)$.* ∎

---

[8] One caveat is that some of these constructions use a few independent random permutations, whereas Theorem 1 assumes only one permutation is used. We point out in passing that Theorem 1 can easily be adapted to this case.

HEURISTIC INSTANTIATION. We wish to say this validates SHA-3 as being a good UCE. One caveat of Theorem 3 is that the actual sponge construction (as used in SHA-3) uses a seedless permutation π. We propose the following assumption on such a permutation π that – if true – implies a simple way to modify an actual Sponge construction to be a secure UCE using Theorem 3. In particular, we suggest using the Even-Mansour [25] paradigm to add a seed to π. Given a family of permutations $\Pi = \{\pi_\lambda\}_{\lambda \in \mathbb{N}}$, where $\pi_\lambda \in \mathsf{Perms}(n(\lambda))$, define then $\mathsf{EM}[\Pi] = (\mathsf{EM.Kg}, \mathsf{EM.Eval})$ where $\mathsf{EM.Kg}$ outputs a random $n(\lambda)$-bit string s on input 1^λ, and

$$\mathsf{EM.Eval}(1^\lambda, s, (+, x)) = s \oplus \pi_\lambda(x \oplus s), \quad \mathsf{EM.Eval}(1^\lambda, s, (-, y)) = s \oplus \pi_\lambda^{-1}(y \oplus s)$$

for all $s, x \in \{0, 1\}^{n(\lambda)}$. Now, if Π is such that $\mathsf{EM}[\Pi]$ is $\mathsf{psPRP}[n, \mathcal{S}^{\mathsf{srs}}]$-secure, then $\mathsf{Sponge}[\mathsf{EM}[\Pi]]$ is $\mathsf{UCE}[*, r, \mathcal{S}^{\mathsf{srs}}]$-secure by Theorem 3. We discuss the conjecture that EM is $\mathsf{psPRP}[n, \mathcal{S}^{\mathsf{srs}}]$-secure further below in Sect. 6.

The attractive feature of $\mathsf{Sponge}[\mathsf{EM}[\Pi]]$ is that it can be implemented in a (near) black-box way from $\mathsf{Sponge}[\Pi]$, that is, the original sponge construction run with fixed oracle Π, by setting (1) The initial state $S_0 \| T_0$ to the seed s (rather than $0^{n(\lambda)}$), and (2) xoring the first r bits $s[1 \dots r]$ of the seed s to the output. The other additions of the seed s to the inner states are unnecessary, as they cancel out. (A similar observation was made by Chang et al. [19] in the context of keying sponges to obtain PRFs.)

4.2 Unpredictable Sources

Many UCE applications only require (statistical) unpredictability. In this section, we see that for this weaker target a significantly simpler construction can be used. In particular, we will first build a $\mathsf{UCE}[n, r, \mathcal{S}^{*\mathsf{up}}]$-secure *compression function* from a $\mathsf{psPRP}[n, \mathcal{S}^{*\mathsf{up}}]$-secure permutation, where $n(\lambda) - r(\lambda) = \omega(\log \lambda)$ and $\star \in \{\mathsf{c}, \mathsf{s}\}$. Combined with existing domain extension techniques [9], this can be enhanced to a variable-input-length UCE for the same class of sources.

THE CHOP CONSTRUCTION. Let $r, n : \mathbb{N} \to \mathbb{N}$ be polynomially bounded functions of the security parameter λ, where $r(\lambda) \leq n(\lambda)$ for all $\lambda \in \mathbb{N}$. We consider the following construction $\mathsf{Chop}[n, r]$ which is $(\mathbf{P}_n \to \mathbf{R}_{n,r})$-compatible. On input 1^λ, it expects a permutation $\pi : \{0, 1\}^n \to \{0, 1\}^n$ for $n = n(\lambda)$, and given additionally $x \in \{0, 1\}^{n(\lambda)}$, it returns

$$\mathsf{Chop}[n, r]^\pi(1^\lambda, x) = \pi(x)[1 \dots r(\lambda)], \tag{10}$$

that is, the first $r = r(\lambda)$ bits of $\pi(x)$. It is not hard to see that the construction is (perfectly) query extractable using the extractor Ext which given oracle access to $\mathcal{O}$ and a set Q of queries of the form $(+, x)$ and $(-, y)$, returns a set consisting of all x such that $(+, x) \in Q$, and moreover adds x' to the set obtained by querying $\mathcal{O}(i, -, y)$ for every $i \in [n]$ and $(-, y) \in Q$.

CP-SEQUENTIAL INDIFFERENTIABILITY. The following theorem establishes CP-sequential indifferentiability of the Chop construction. We refer the reader to the full version for the proof but give some intuition about it after the theorem.

Theorem 4 (CP-indifferentiability of Chop). *Let $r, n : \mathbb{N} \to \mathbb{N}$ be such that $r(\lambda) \leq n(\lambda)$ for all $\lambda \in \mathbb{N}$. Let $\mathbf{P} = \mathbf{P}_n$ and $\mathbf{R} = \mathbf{R}_{n,r}$ be the random permutation and random function, respectively. Then, there exists a simulator Sim such that for all distinguishers A making at most q construction and p primitive queries,*

$$\mathsf{Adv}^{\mathsf{cpi}[\mathbf{P} \to \mathbf{R}]}_{\mathsf{Chop}[n,r],\mathsf{Sim},A}(\lambda) \leq \frac{(q+p)^2}{2^n} + \frac{p \cdot q}{2^{n-r}}. \tag{11}$$

Here, Sim makes at most one oracle query upon each invocation, and otherwise runs in time polynomial in the number of queries answered. ■

The dependence on r is necessary, as otherwise the construction becomes invertible and cannot be CP-sequentially indifferentiable. Also, note that we cannot expect full indifferentiability to hold for the Chop construction, and in fact, not even sequential indifferentiability in the sense of [33]. Indeed, a distinguisher A can simply first query $\mathsf{Prim}(-, y)$, obtaining x, and then query $\mathsf{Func}(x)$, that yields y'. Then, A just checks that the first r bits of y equals y', and if so outputs 1, and otherwise outputs 0. Note that in the real world, A always outputs 1, by the definition of Chop. However, in the ideal world, an arbitrary simulator Sim needs, on input y, to return an x for which the random oracle (to which it access) returns the first r bits of y. This is however infeasible if $n - r = \omega(\log \lambda)$, unless the simulator can make roughly 2^r queries.

The proof in full version shows this problem vanishes for CP-sequential indifferentiability. Indeed, our simulator will respond to queries $\mathsf{Sim}(-, y)$ with a random (and inconsistent) x. The key point is that due to the random choice, it is unlikely that the distinguisher has already issued a prior query $\mathsf{Func}(x)$. Moreover, it is also unlikely (in the real world) that the distinguisher, after a query $\mathsf{Func}(x)$, makes an inverse query on $\pi(x)$. The combination of these two facts will be enough to imply the statement.

UCE SECURITY. We can now combine Theorem 4 with the fact that the Chop construction is (perfectly) query extractable, and use Theorem 2:

Corollary 1. *For all n, r such that $n(\lambda) - r(\lambda) = \omega(\log \lambda)$, if F is $\mathsf{psPRP}[n, \mathcal{S}^{\star\mathsf{up}}]$-secure, then $\mathsf{Chop}[\mathsf{F}]$ is $\mathsf{UCE}[n, r, \mathcal{S}^{\star\mathsf{up}}]$-secure, where $\star \in \{\mathsf{c}, \mathsf{s}\}$.*

The construction of [9] can be used to obtain variable-input-length UCE: It first hashes the arbitrary-long input down to an $n(\lambda)$-bit long input using an almost-universal hash function, and then applies $\mathsf{Chop}[\mathsf{F}]$ to the resulting value.

5   Building psPRPs from UCEs

This section presents our main result on building psPRPs from UCEs, namely that the five-round Feistel construction, when its round functions are instantiated from a $\mathsf{UCE}[\mathcal{S}^{\star\mathsf{rs}}]$-secure function family (for $\star \in \{\mathsf{c}, \mathsf{s}\}$), yields a $\mathsf{psPRP}[\mathcal{S}^{\star\mathsf{rs}}]$-secure permutation family.

CP-INDIFFERENTIABILITY OF FEISTEL. Let $n : \mathbb{N} \to \mathbb{N}$ be a (polynomially bounded) function. We define the following construction Ψ_5, which, for security parameter λ, implements an invertible permutation on $2n(\lambda)$-bit strings, and makes calls to an oracle $f : [5] \times \{0,1\}^{n(\lambda)} \to \{0,1\}^{n(\lambda)}$. In particular, on input 1^λ and $X = X_0 \| X_1$, where $X_0, X_1 \in \{0,1\}^{n(\lambda)}$, running $\Psi_5^f(1^\lambda, (+, X))$ outputs $X_5 \| X_6$, where

$$X_{i+1} \leftarrow X_{i-1} \oplus f(i, X_i) \text{ for all } i = 1, \ldots, 5. \tag{12}$$

Symmetrically, upon an inverse query, $\Psi_5^f(1^\lambda, (-, Y = X_5 \| X_6))$ simply computes the values backwards, and outputs $X_0 \| X_1$. Construction Ψ_5 is clearly $(\mathbf{R}_{n,n}^5 \to \mathbf{P}_{2n})$-compatible, where we use the notation $\mathbf{R}_{n,n}^k$ to denote the k-fold combination of independent random functions which takes queries of the form (i, x) that are answered by evaluating on x the i-th function.

The following theorem establishes CP-indifferentiability for Ψ_5. We discuss below its consequences, and give a detailed description of our simulation strategy. The full analysis of the simulation strategy – which employs the randomness-mapping technique of [29] – is found in the full version.

Theorem 5 (CP-indifferentiability of Feistel). *Let* $\mathbf{R} = \mathbf{R}_{n,n}^5$ *and* $\mathbf{P} = \mathbf{P}_{2n}$. *Then, there exists a simulator* Sim *(described in Fig. 6) such that for all distinguisher A making at most* $q(\lambda)$ *queries,*

$$\mathsf{Adv}_{\Psi_5, \mathsf{Sim}, A}^{\mathsf{cpi}[\mathbf{R} \to \mathbf{P}]}(\lambda) \leq \frac{360 q(\lambda)^6}{2^{n(\lambda)}}. \tag{13}$$

Here, Sim *makes at most* $2q(\lambda)^2$ *queries, and otherwise runs in time polynomial in the number of queries answered, and* n.   ∎

This, together with Theorem 1, gives us immediately the following corollary: Given a keyed function family $\mathsf{F} = (\mathsf{F.Kg}, \mathsf{F.Eval})$, where for all $\lambda \in \mathbb{N}$, $k \in [\mathsf{F.Kg}(1^\lambda)]$, $\mathsf{F.Eval}(1^\lambda, k, \cdot)$ is a function from $n(\lambda)+3$ bits to $n(\lambda)$ bits, interpreted as a function $[5] \times \{0,1\}^{n(\lambda)} \to \{0,1\}^{n(\lambda)}$, then define the keyed function family $\Psi_5[\mathsf{F}] = (\Psi.\mathsf{Kg}, \Psi.\mathsf{Eval})$ obtained by instantiating the round function using F.

Corollary 2. *For any polynomially bounded* $n = \omega(\log \lambda)$, *if* $\mathsf{F} \in \mathsf{UCE}[n + 3, S^{\star\mathsf{rs}}]$, *then* $\Psi_5[\mathsf{F}] \in \mathsf{psPRP}[2n, S^{\star\mathsf{rs}}]$, *where* $\star \in \{\mathsf{c}, \mathsf{s}\}$.   ∎

REMARKS. Theorem 5 is interesting in its own right, as part of the line of works on (full-fledged) indifferentiability of Feistel constructions. Coron et al. [21] show that six rounds are necessary for achieving indifferentiability, and proofs of indifferentiability have been given for 14, 10, and 8 rounds, respectively [21–23, 29]. Thus, our result shows that CP-indifferentiability is a strictly weaker goal in terms of round-complexity of the Feistel construction. (Also for sequential indifferentiability as in [33], six rounds are necessary.) As we will see in the next paragraph, our simulation strategy departs substantially from earlier proofs.

Two obvious problems remain open. First off, we know four rounds are necessary (as they are needed for indistinguishability alone [32]), but we were unable to make any progress on whether CP-sequential indifferentiability (or psPRP security) is achievable. The second is the case of unpredictable sources. We note that a heavily unbalanced Feistel construction (where each round function

```
PROCEDURE Sim(k, X):
1:  if G_k[X] = ⊥ then
2:     if k = 2 then
3:        F^inner(k, X)
4:        foreach (X_1, X_2) ∈ G_1 × {X} do
5:           if (X_1, X_2, 1) ∉ CompletedChains then
6:              X_0 ← F^inner(1, X_1) ⊕ X_2
7:              (X_5, X_6) ← Func(+, X_0||X_1)
8:              C ← (X_1, X_2, 1)
9:              if G_5[X_5] ≠ ⊥ then                          // Immediate Completion
10:                 Complete(C, (X_5, X_6))
11:             else                                          // Completion is delayed
12:                 X_3 ← F^inner(2, X_2) ⊕ X_1
13:                 Chains[3, X_3] ← (5, X_5), Chains[5, X_5] ←∪ {(C, (X_5, X_6))}
14:    elseif k = 4 then
15:       F^inner(k, X)
16:       foreach (X_4, X_5) ∈ {X} × G_5 do
17:          if (X_4, X_5, 4) ∉ CompletedChains then
18:             X_6 ← F^inner(4, X_4) ⊕ X_5
19:             (X_0, X_1) ← Func(-, X_5||X_6)
20:             C ← (X_4, X_5, 4)
21:             if G_1[X_1] ≠ ⊥ then                          // Immediate Completion
22:                Complete(C, (X_0, X_1))
23:             else                                          // Completion is delayed
24:                X_3 ← F^inner(4, X_4) ⊕ X_5
25:                Chains[3, X_3] ← (1, X_1), Chains[1, X_1] ←∪ {(C, (X_0, X_1))}
26:    elseif k ∈ {1, 5} then
27:       F^inner(k, X)
28:       foreach (C, (U, V)) ∈ Chains[k, X] do
29:          if C ∉ CompletedChains then                      // Delayed Completion
30:             Complete(C, (U, V))
31:    elseif Chains[3, X] ≠ ⊥ then
32:       Sim(Chains[k, X])
33:    return F^inner(k, X)
```

Fig. 6. The code for simulator Sim. Sim has access to the Func oracle and maintains data structures G_k, Chains and CompletedChains as global variables.

outputs one bit) would be query extractable, as the input of the round function leaves little uncertainty on the inner state, and the extractor can evaluate the round functions for other rounds to infer the input/output of the construction. Thus, if we could prove CP-indifferentiabilty, we could combine this with Theorem 2. Unfortunately, such a proof appears beyond our current understanding.

SIMULATOR DESCRIPTION. We explain now our simulation strategy, which is described formally in Fig. 6. We note that our approach inherits the chain-completion technique from previous proofs, but it will differ substantially in how and when chains are completed.

PROCEDURE $\mathsf{F}^{\text{inner}}(i, X_i)$:
34:    **if** $G_i[X_i] = \bot$ **then**
35:      $G_i[X_i] \leftarrow_{\$} \{0,1\}^n$
36:    **return** $G_i[X_i]$

PROCEDURE $\mathsf{ForceVal}(X, Y, l)$:
37:    $G_l[X] \leftarrow Y$

PROCEDURE $\mathsf{Complete}(C, (U, V))$:
38:    $(X, Y, i) \leftarrow C$
39:    **if** $i = 1$ **then**
40:      $X_1 \leftarrow X, X_2 \leftarrow Y, X_3 \leftarrow \mathsf{F}^{\text{inner}}(2, X_2) \oplus X_1$
41:      $(X_5, X_6) \leftarrow (U, V)$
42:      $X_4 \leftarrow \mathsf{F}^{\text{inner}}(5, X_5) \oplus X_6$
43:      $\mathsf{ForceVal}(X_3, X_4 \oplus X_2, 3), \mathsf{ForceVal}(X_4, X_5 \oplus X_3, 4)$
44:    **elseif** $i = 4$ **then**
45:      $X_4 \leftarrow X, X_5 \leftarrow Y, X_3 \leftarrow \mathsf{F}^{\text{inner}}(4, X_4) \oplus X_5$
46:      $(X_0, X_1) \leftarrow (U, V)$
47:      $X_2 \leftarrow \mathsf{F}^{\text{inner}}(1, X_1) \oplus X_0$
48:      $\mathsf{ForceVal}(X_3, X_4 \oplus X_2, 3), \mathsf{ForceVal}(X_2, X_1 \oplus X_3, 2)$
49:    $\mathsf{CompletedChains} \overset{\cup}{\leftarrow} \{(X_1, X_2, 1), (X_4, X_5, 4)\}$

Fig. 7. The code for subroutines used by simulator Sim (continuation of Fig. 6).

Recall that in the ideal case, in the first stage of the CP-indifferentiability game, A_1 makes queries to Func implementing a random permutation, and then passes the control of the game to A_2 which interacts with Sim. Our Sim maintains tables G_k for $k \in [5]$ to simulate the round functions. We denote by $G_k[X] = \bot$ that the table entry for X is undefined, and we assume all values are initially undefined. Also, we refer to a tuple (X_k, X_{k+1}, k) as a *partial chain* where $G_k[X_k] \neq \bot$ and $G_{k+1}[X_{k+1}] \neq \bot$ for $k \in \{1, 4\}$, $X_k, X_{k+1} \in \{0,1\}^n$.

For any query (k, X) by A_2, Sim checks if $G_k[X] = \bot$. If not then the image $G_k[X]$ is returned. Otherwise, depending on the value of k, Sim takes specific steps as shown in Figs. 6 and 7. If $k \in \{2, 4\}$ then Sim sets $G_k[X]$ to a uniformly random n-bit string by calling the procedure $\mathsf{F}^{\text{inner}}$. At this point, Sim considers newly formed tuples $(X_1, X_2) \in G_1 \times \{X\}$ (when $k = 2$) and detects partial chains $C = (X_1, X_2, 1)$. The notation $X_1 \in G_1$ is equivalent to $G_1[X_1] \neq \bot$. For every partial chain C that Sim detects, it queries Func on (X_0, X_1) and receives (X_5, X_6) where $X_0 = G_1[X_1] \oplus X_2$. If (X_0, X_1) does not appear in one of the queries/responses by/to A_1 then it is unlikely for A_2 to guess the corresponding (X_5, X_6) pair. Therefore, if $G_5[X_5] \neq \bot$ then Sim assumes that C is a chain that most likely corresponds to a query by A_1. We refer to partial chains that correspond to the queries by A_1 as *relevant* chains. In this case, Sim *immediately* completes C by calling the procedure Complete. C is completed by forcing the values of $G_3[X_3]$ and $G_4[X_4]$ to be consistent with the Func query where $X_3 \leftarrow G_2[X_2] \oplus X_1$ and $X_4 \leftarrow G_5[X_5] \oplus X_6$.

If $G_5[X_5] = \perp$ then either C is not a relevant chain or C is a relevant chain but A_2 has not queried $(5, X_5)$ yet. An aggressive strategy would be to complete C, thereby asking Sim to complete every partial chain ever detected. The resulting simulation strategy will however end up potentially managing an exponential number of partial chains, contradicting our goal of efficient simulation. Hence, Sim *delays* the completion and only completes C on A_2's query to either $(3, X_3)$ or $(5, X_5)$ where $X_3 = G_2[X_2] \oplus X_1$. The completion is delayed by storing information about X_3 and X_5, that fall on the chain C, in the table Chains. In particular, Sim stores a pointer to $(5, X_5)$ at Chains$[3, X_3]$. The inputs $((X_1, X_2, 1), (X_5, X_6))$ to the Complete call on C are stored in Chains$[5, X_5]$. As many chains can share the same X_5, we allow Chains$[5, X_5]$ to be a set. The idea of delaying the chain completions is unique to our simulation strategy and it translates to an efficient Sim which consistently completes chains in the eyes of A. Sim works symmetrically when $k = 4$.

For queries of the form (k, X) where $k \in \{1, 5\}$, Sim always assigns $G_k[X]$ to a uniform random n-bit string by calling $\mathsf{F}^{\mathrm{inner}}$. Moreover as discussed earlier, X could be on previously detected partial chains whose completion was delayed. Therefore after the assignment, Sim picks up all partial chains C' (if any) stored in Chains$[k, X]$ and completes them. This is where Sim captures a relevant partial chain which was delayed for completion. Finally for queries $(3, X)$, Sim checks if this X was on a partial chain that was detected but not completed. If Chains$[3, X] = \perp$ then Sim assigns $G_3[X]$ a uniform random n-bit string otherwise it follows the pointer to Chains$[3, X]$ to complete the chain X was on. Since Chains$[3, X]$ just stores a tuple (instead of a set) there can be at most one chain C that Chains$[3, X]$ can point to at any time. In the execution, Chains$[3, X]$ can get overwritten which may lead to inconsistencies in chain completions. However, we show that there are no overwrites in either tables G_k or the data-structure Chains, except with negligible probability. This allows Sim to Complete chains consistently in the eyes of A. Furthermore, to avoid completing the same chains again, Sim maintains a set of all CompletedChains and completes any chain if it is not in CompletedChains. A pictorial description of Sim is found in Fig. 8.

6   Ideal-Model PsPRP Constructions

We discuss two natural approaches to instantiate psPRPs. One is by taking any block cipher, and using its key as a (now public) seed. The second is by using a key-less permutation (e.g., the one within SHA-3), and adding the seed through the Even-Mansour [25] construction. While the purpose of our psPRP framework is to remove ideal-model assumptions, the only obvious way to validate (heuristically) these methods *is* via ideal-model proofs, and this is what we do here. Also, note that such ideal-model proofs are useful since, as in the case of UCE, psPRP security can become a powerful intermediate notion within ideal-model proofs for multi-stage security games [36].

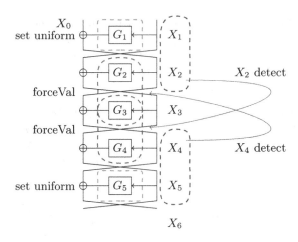

Fig. 8. The 5-round Feistel where Sim sets $G_1[X_1]$ and $G_5[X_5]$ uniformly at random (green). Sim detects chains at either (X_1, X_2) or (X_4, X_5) (blue) and adapts at (X_3, X_4) and (X_2, X_3) respectively (red). (Color figure online)

PsPRPs FROM BLOCK CIPHERS. Given a family of block ciphers E_λ : $\{0,1\}^{s(\lambda)} \times \{0,1\}^\lambda \to \{0,1\}^\lambda$, we consider the construction $F = (F.Kg, F.Eval)$, where $F.Kg(1^\lambda)$ outputs a random $k \leftarrow_\$ \{0,1\}^{s(\lambda)}$, whereas

$$F.Eval(1^\lambda, (k, +, x)) = E(k, x), \quad F.Eval(1^\lambda, (k, -, y)) = E^{-1}(k, y). \quad (14)$$

The following theorem establishes its security in the ideal cipher model, that is, we assume (without overloading notation) that all parties (i.e., the source, the distinguisher, and the reset adversary in the proof) are given access to a randomly chosen block cipher E, which is also used within F. We refer the reader to the full version for the proof which closely follows the proof from [8] that a random oracle is UCE-secure.

Theorem 6 (Ideal Cipher as a psPRP). *Let* P *be the random permutation with input length* $n(\lambda) = \lambda$. *For every source-distinguisher pair* S, D, *where the source* S, *in its first stage, outputs* n *which is at most* $N(\lambda)$ *and makes* q Prim *queries to its oracle, there exists* R *(described in the proof) such that*

$$\mathsf{Adv}^{\mathsf{psPRP}[n]}_{F,S,D}(\lambda) \leq \mathsf{Adv}^{\mathsf{reset}[P]}_{S,R}(\lambda) + \frac{2qN(\lambda)}{2^{s(\lambda)}} + \frac{2N(\lambda)^2}{2^{s(\lambda)}}. \quad (15)$$

In particular, if D *is polynomial time, then so is* R. ∎

EVEN-MANSOUR. We find it practically valuable to assess whether simple constructions can work. To this end, here, we show that the Even-Mansour construction [25] yields a $\mathsf{psPRP}[\mathcal{S}^{\mathsf{cup}}]$-secure permutation family in the random permutation model. In particular, we assume we are given a family of permutations $\Pi = \{\pi_\lambda\}_{\lambda \in \mathbb{N}}$, where $\pi_\lambda : \{0,1\}^\lambda \to \{0,1\}^\lambda$, and consider the construction

EM = (EM.Kg, EM.Eval) as defined in Sect. 4.1. Similar to the above, the following theorem implicitly assumes all parties are given oracle access to a random permutation which is used to sample the permutation inside EM.

Below, we give a few remarks on why the above approach to show psPRP$[\mathcal{S}^{crs}]$ security does not extend to the case of Even-Mansour.

Theorem 7 (Even-Mansour as a psPRP). *Let* **P** *be the random permutation with input length* $n(\lambda) = \lambda$. *For every source-distinguisher pair* S, D, *where the source* S, *in its first stage, outputs* n *which is at most* $N(\lambda)$ *and where* S *and* D *jointly make at most* q *queries to their oracles, there exists* P *(described in the proof) such that*

$$\mathsf{Adv}_{\mathsf{EM},S,D}^{\mathsf{psPRP}[n]}(\lambda) \leq \mathsf{Adv}_{S,P}^{\mathsf{pred}[\mathbf{P}]}(\lambda) + \frac{3q^2}{2^\lambda} + \frac{2N(\lambda)q^2}{2^\lambda}. \qquad (16)$$

In particular, if D *runs in polynomial time, then so does* P. ∎

The proof of Theorem 7 can be found in the full version. It resembles the original indistinguishability proof from [25], which bounds the advantage via the probability of an intersection query, that is, a direct query (by the source or by the distinguisher) to the random permutation that overlaps with one of the queries to the random permutation made internally by oracle $\mathcal{O}$ invoked by the source. Bounding the probability that S makes an intersection query proceeds as in [25] (exploiting lack of knowledge of the seed), whereas bounding the probability that D makes such a query requires a reduction to unpredictability.

WHY NOT RESET-SECURE SOURCES? We would like to extend Theorem 7 to $\mathcal{S}^{crs}$, as this would provide validation for the assumption from Sect. 4.1. While we conjecture this to be true, the statement seems to evade a simple proof. The proof approach behind Theorem 6 fails in particular, as it heavily exploits the property that for each distinct seed, the construction F queries a disjoint portion of the domain of the ideal cipher, which is not true for EM.

7   Efficient Garbling from PsPRPs

As an application of the psPRP framework, we study the security of the efficient garbling schemes of Bellare, Hoang, Keelveedhi, and Rogaway [10], and in particular, their simplest scheme (called Ga). It follows Yao's general garbling paradigm [40], but proposes a particular gadget to garble individual gates that only relies on evaluating the underlying block cipher on a fixed key. In terms of efficiency, this has been shown to be advantageous, as it avoids higher re-keying costs. However, its security has only been proved in the ideal-cipher model, and recent work by Gueron et al. [28] has debated this. Here, we show that a minor variant of Ga (which still largely benefits from the lack of re-keying) is secure assuming the underlying block cipher is psPRP$[\mathcal{S}^{sup}]$-secure. While this assumption is undoubtedly strong, it makes it clear what is expected from the permutation. In particular, the main concern of [28] is the existence of fixed-key distinguishers (as in [30]), but these do not seem to affect psPRP-security, while they may invalidate the permutation being ideal.

SIMPLE CIRCUIT DESCRIPTION. For representing circuits we adopt the SCD notation of [11]. A circuit is a 6-tuple $f = (n, m, q, W_1, W_2, G)$ where $n \geq 2$ is the number of inputs, $m \geq 1$ is the number of outputs, $q \geq 1$ is the number of gates, and $n + q$ is the number of wires. We let $\mathsf{Inputs} = [1, \ldots, n]$, $\mathsf{Wires} = [1, \ldots, n+q]$, $\mathsf{OutputWires} = [n + q - m + 1, \ldots, n + q]$ and $\mathsf{Gates} = [n + 1, \ldots, n + q]$. Then $W_1 : \mathsf{Gates} \rightarrow \mathsf{Wires} \setminus \mathsf{OutputWires}$ is a function to identify each gate's first incoming wire and $W_2 : \mathsf{Gates} \rightarrow \mathsf{Wires} \setminus \mathsf{OutputWires}$ to identify each gate's second incoming wire. Finally $G : \mathsf{Gates} \times \{0,1\}^2 \rightarrow \{0,1\}$ is a function that determines the functionality of each gate i.e. G_g is a table storing the output of gate g with input i and j at $G_g[i, j]$. We require that $W_1(g) < W_2(g) < g$ for all gates g.

Following [10], our definitions will be parameterized by the side information about the circuit obtained from its garbled counterpart. We consider the *topology side information* ϕ_{topo} which maps f to its topology $\phi_{\mathsf{topo}}(f) = (n, m, q, W_1, W_2)$. Another example is ϕ_{xor}, which maps f to a circuit $\phi_{\mathsf{xor}}(f) = (n, m, q, W_1, W_2, G')$ which obscures the functionality of non-xor gates. As shown in [11] and [10], an important property is that ϕ_{topo} and ϕ_{xor} are efficiently invertible, i.e., there exists an efficient algorithm which given $\phi(f)$ and y, outputs (f', x') such that $\phi(f) = \phi(f')$ and $y = \mathsf{ev}(f', x')$.

GARBLING SCHEMES AND THEIR SECURITY. To describe a garbling scheme we use the notation from [11]. A *garbling scheme* is a tuple of algorithms $\mathcal{G} = (\mathsf{Gb}, \mathsf{En}, \mathsf{De}, \mathsf{Ev}, \mathsf{ev})$. The algorithm Gb is probabilistic and others are determinisitic. Gb takes as inputs a circuit $f = (n, m, q, W_1, W_2, G)$ represented in the SCD notation and a security parameter 1^λ and returns a tuple of strings (F, e, d) where F is the garbled circuit, e is the input encoding information and d is the output decoding information. $\mathsf{En}(e, \cdot) : \{0,1\}^n \rightarrow \{0,1\}^*$ transforms the n-bit input x to the garbled input X. $\mathsf{Ev}(F, \cdot) : \{0,1\}^* \rightarrow \{0,1\}^*$ runs the garbled circuit F on garbled input X and returns the garbled output Y. $\mathsf{De}(d, \cdot) : \{0,1\}^* \rightarrow \{0,1\}^m \cup \{\bot\}$ decodes the garbled output Y to return $y \in \{0,1\}^m$. The algorithm ev is the canonical circuit-evaluation function where $\mathsf{ev}(f, x)$ is the m-bit output one gets by feeding x to f. Finally, we require that $\mathcal{G}$ is *correct*, that is, if $f \in \{0,1\}^*$, $\lambda \in \mathbb{N}$, $x \in \{0,1\}^n$ and $(F, e, d) \in [\mathsf{Gb}(1^\lambda, f)]$, then $\mathsf{De}(d, \mathsf{Ev}(F, \mathsf{En}(e, x))) = \mathsf{ev}(f, x)$. We require that all algorithms run in time polynomial in the security parameter λ.

In this work, we are only concerned with indistinguishability-based privacy, as defined in Game $\mathsf{PrvInd}^A_{\mathcal{G},\phi}$ in Fig. 9. Since both ϕ_{topo} and ϕ_{xor} are efficiently invertible, [11] show that it is sufficient to focus on this target, since simulation-based security is implied. We say that $\mathcal{G}$ is prvind-*secure over side information function* ϕ if for all PPT adversaries A,

$$\mathsf{Adv}^{\mathsf{prvind}[\phi]}_{\mathcal{G},A}(\lambda) = 2 \Pr \left[\mathsf{PrvInd}^A_{\mathcal{G},\phi}(\lambda) \right] - 1 \qquad (17)$$

MAIN $\mathsf{PrvInd}^A_{\mathcal{G},\phi}(1^\lambda)$	PROCEDURE $\mathsf{Garble}(f_0, f_1, x_0, x_1)$
$b \leftarrow_\$ \{0,1\}$	if $\phi(f_0) \neq \phi(f_1)$ then return $\perp$
$b' \leftarrow_\$ A^{\mathsf{Garble}}(1^\lambda)$	if $\{x_0, x_1\} \not\subseteq \{0,1\}^{f_0 \cdot n}$ then return $\perp$
return $(b' = b)$	if $\mathsf{ev}(f_0, x_0) \neq \mathsf{ev}(f_1, x_1)$ then return $\perp$
	$(F, e, d) \leftarrow_\$ \mathsf{Gb}(1^\lambda, f_b), X \leftarrow \mathsf{En}(e, x_b)$
	return (F, X, d)

Fig. 9. $\mathsf{PrvInd}^A_{\mathcal{G},\phi}$ game for $\mathcal{G}$ with adversary A.

is negligible (in λ). Also, it is not hard to see (by a simple hybrid argument) that it is sufficient to prove this for adversaries which make one single query to their oracle.

GARBLING SCHEME $\mathsf{Ga}[\mathsf{P}]$. Our garbling scheme resembles heavily that of [10]. The only modification is that we assume it uses a function family P meant to be $\mathsf{psPRP}[\mathcal{S}^{\mathsf{sup}}]$-secure (which could be instantiated from a block cipher, by letting the key take the role of the seed.). During the garbling procedure, a fresh seed for P is chosen and made part of the garbled circuit. Clearly, re-keying costs are still largely avoided (especially for large circuits), even though re-keying is necessary when garbling multiple circuits.

Concretely, the garbling scheme $\mathsf{Ga}[\mathsf{P}]$ (Fig. 10) is a tuple of algorithms $\mathsf{Ga}[\mathsf{P}] = (\mathsf{Gb}, \mathsf{En}, \mathsf{De}, \mathsf{Ev}, \mathsf{ev})$ where $\mathsf{P} \in \mathsf{psPRP}[k(\lambda), \mathcal{S}^{\mathsf{sup}}]$ for some polynomial $k(\lambda)$. Algorithm Gb transforms the input circuit f to a tuple of strings (F, e, d) where the seed s for the permutation P sampled independently for each input f is now part of F.

Though the Ga scheme of [10] comes in several variants, where each variant is defined by the dual-key cipher used, we focus on a specific dual-key cipher (namely A1 in [10]) that leads to the most efficient implementation of Ga.

In the following theorem we prove the prvind-security of $\mathsf{Ga}[\mathsf{P}]$ and later discuss about the more efficient schemes GaX and GaXR from [10].

Theorem 8 (Garbling from psPRPs). *Let* $\mathsf{P} \in \mathsf{psPRP}[k(\lambda), \mathcal{S}^{\mathsf{sup}}]$ *then* $\mathsf{Ga}[\mathsf{P}]$ *is* prvind-*secure over* ϕ_{topo}. ∎

Proof. Let us assume that $\mathsf{Ga}[\mathsf{P}]$ is not prvind-secure then there exists a PPT adversary A that issues circuits with at most $q(\lambda)$ gates and achieves a non-negligible advantage $\epsilon(\lambda)$ in the $\mathsf{PrvInd}^A_{\mathsf{Ga}[\mathsf{P}],\phi}(\lambda)$ [9] game. Using A we construct a pair (S, D) (Fig. 11) breaking the psPRP security of P, where S is a statistically unpredictable source. Without loss of generality, we can assume that A queries its oracle exactly once.

Let c be the challenge bit in the psPR game for P, and let Perm be the oracle called by S. We allow S to sample the challenge bit b for the $\mathsf{PrvInd}^A_{\mathsf{Ga}[\mathsf{P}],\phi}$ game. Further, for syntactic reasons we decompose A into (A_0, A_1) where A_0 on input 1^λ outputs (f_0, x_0, f_1, x_1) (inputs for Garble) and forwards a state st

[9] We drop the subscript topo from ϕ for ease of notation.

PROCEDURE $\mathsf{En}(e, x)$
$(X_1^0, X_1^1, \ldots, X_n^0, X_n^1) \leftarrow e$
$x_1, \ldots, x_n \leftarrow x$
$X \leftarrow (X_1^{x_1}, \ldots, X_n^{x_n})$
return X

PROCEDURE $\mathsf{Ev}(F, X)$
$(n, m, q, W_1, W_2, T, s) \leftarrow F$
$(X_1, \ldots, X_n) \leftarrow X$

foreach $g \in [n + 1, \ldots, n + q]$ **do**
$\quad w_1 \leftarrow W_1(g), w_2 \leftarrow W_2(g)$
$\quad \alpha \leftarrow \mathrm{lsb}(X_{w_1}), \beta \leftarrow \mathrm{lsb}(X_{w_2})$
$\quad K \leftarrow X_{w_1} \oplus X_{w_2} \oplus g$
$\quad X_g \leftarrow T[g, \alpha, \beta] \oplus \mathsf{P.Eval}(s, K) \oplus K$
return $(X_{n+q-m+1}, \ldots, X_{n+q})$

PROCEDURE $\mathsf{De}(d, Y)$
$(d_1, \ldots, d_m) \leftarrow d$
$(Y_1, \ldots, Y_m) \leftarrow Y$
foreach $i \in [1, \ldots, m]$ **do**
$\quad y_i \leftarrow \mathrm{lsb}(Y_i) \oplus d_i$
return $y \leftarrow y_1, \ldots, y_m$

PROCEDURE $\mathsf{Gb}(1^\lambda, f)$
$s \leftarrow_\$ \mathsf{P.Kg}(1^\lambda)$
$(n, m, q, A', B', G) \leftarrow f$

foreach $i \in [1, \ldots, n + q]$ **do**
$\quad t \leftarrow_\$ \{0, 1\}$
$\quad X_i^0 \leftarrow_\$ \{0, 1\}^{k-1} t$
$\quad X_i^1 \leftarrow_\$ \{0, 1\}^{k-1} \bar{t}$

foreach $g \in [n + 1, \ldots, n + q]$ **do**
$\quad w_1 \leftarrow W_1(g), w_2 \leftarrow W_2(g)$
$\quad$ **foreach** $(i, j) \in \{0, 1\}^2$ **do**
$\quad\quad A \leftarrow X_{w_1}^i, \alpha \leftarrow \mathrm{lsb}(A)$
$\quad\quad B \leftarrow X_{w_2}^j, \beta \leftarrow \mathrm{lsb}(B)$
$\quad\quad K \leftarrow X_{w_1}^i \oplus X_{w_2}^j \oplus g$
$\quad\quad T[g, \alpha, \beta] \leftarrow \mathsf{P.Eval}(s, K) \oplus K \oplus X_g^{G_g[i,j]}$

$F \leftarrow (n, m, q, W_1, W_2, T, s)$
$e \leftarrow (X_1^0, X_1^1, \ldots, X_n^0, X_n^1)$
$d \leftarrow (\mathrm{lsb}(X_{n+1-m+1}^0), \ldots, \mathrm{lsb}(X_{n+q}^0))$
return (F, e, d)

Fig. 10. Scheme $\mathsf{Ga[P]}$.

to A_1. The result of Garble i.e. (F, X, d) is forwarded to A_1 to guess the challenge bit b in the $\mathsf{PrvInd}_{\mathsf{Ga[P]}, \phi}^A$ game. The source S nearly acts as Gb on input f_b. To satisfy unpredictability, the leakage L must give no information about the queries made by S. Therefore, S refrains from compiling the rows in the garbled table T which can be opened by A. S outputs this partially garbled circuit F^- as leakage in addition to (b, d, st). Moreover, since (S, D) must perfectly simulate the $\mathsf{PrvInd}_{\mathsf{Ga[P]}, \phi}^A$ game for A, leakage also contains the vector X^+ which is the set of all visible tokens (one for each wire). Given s and L, D completes the garbled circuit and invokes A_1 with appropriate inputs. D then outputs $b' \oplus b$ where b' was the guess of A in the $\mathsf{PrvInd}_{\mathsf{Ga[P]}, \phi}^A$ game.

It is easy to see that when $c = 1$, (S, D) simulate the game $\mathsf{PrvInd}_{\mathsf{Ga[P]}, \phi}^A$ for A. Furthermore, when $c = 0$ the leakage L can be transformed to be independent of the bit b by modifying Perm to act like a random function. This allows rows in the garbled table to be independent of the tokens $X_g^{G_g[i,j]}$ which might depend on bit b. Therefore, in this modified game A can do no better than guessing. For a detailed analysis, we direct the reader to the full version.

To prove that S is statistically unpredictable we need to show that any (possibly unbounded) predictor P making at most $p(\lambda)$ number of queries to the oracle Perm is unlikely to predict a query made by S given $L = (F^-, X^+, b, d, \mathsf{st})$. The idea is to swiftly transition to a game where L is independent of the queries made

SOURCE $S^{\mathsf{Perm}}(1^\lambda)$:
$b \leftarrow_\$ \{0,1\}$
$(f_0, x_0, f_1, x_1, \mathsf{st}) \leftarrow A_0(1^\lambda)$
$(n, m, q, W_1, W_2) \leftarrow \phi(f_b)$
$G \leftarrow f_b.G$
foreach $i \in [1, \ldots, n+q]$ **do**
$\quad v_i \leftarrow \mathsf{ev}(f_b, x_b, i); \; t_i \leftarrow_\$ \{0,1\}$
$\quad X_i^{v_i} \leftarrow_\$ \{0,1\}^{k-1} t_i$
$\quad X_i^{\overline{v_i}} \leftarrow_\$ \{0,1\}^{k-1} \overline{t_i}$
foreach $g \in [n+1, \ldots, n+q]$ **do**
$\quad$ **foreach** $(i,j) \in \{0,1\}^2$ **do**
$\quad\quad w_1 \leftarrow W_1(g), \; w_2 \leftarrow W_2(g)$
$\quad\quad A \leftarrow X_{w_1}^i, \; \alpha \leftarrow \mathsf{lsb}(A)$
$\quad\quad B \leftarrow X_{w_2}^j, \; \beta \leftarrow \mathsf{lsb}(B)$
$\quad\quad K \leftarrow A \oplus B \oplus g$
$\quad\quad$ **if** $i \neq v_{w_1} \vee j \neq v_{w_2}$ **then**
$\quad\quad\quad T[g, \alpha, \beta] \leftarrow \mathsf{Perm}(K) \oplus K \oplus X_g^{G_g[i,j]}$
$F^- \leftarrow (n, m, q, W_1, W_2, T)$
$X^+ \leftarrow (X_1^{v_1}, \ldots, X_{n+q}^{v_{n+q}})$
return $(F^-, X^+, d, b, \mathsf{st})$

DISTINGUISHER $D(1^\lambda, s, L)$:
$(F^-, X^+, d, b, \mathsf{st}) \leftarrow L$
$(n, m, q, W_1, W_2, T) \leftarrow F^-$
$X_1, \ldots, X_{n+q} \leftarrow X^+$
for $g \in [n+1, \ldots, n+q]$ **do**
$\quad w_1 \leftarrow W_1(g), \; w_2 \leftarrow W_2(g)$
$\quad \alpha \leftarrow \mathsf{lsb}(X_{w_1}), \; \beta \leftarrow \mathsf{lsb}(X_{w_2})$
$\quad K \leftarrow X_{w_1} \oplus X_{w_2} \oplus g$
$\quad T[g, \alpha, \beta] \leftarrow \mathsf{P.Eval}(s, K) \oplus K \oplus X_g$

$F \leftarrow (n, m, q, W_1, W_2, T, s)$
$X \leftarrow (X_1, \ldots, X_n)$
$b' \leftarrow A_1(1^\lambda, \mathsf{st}, F, X, d)$
return $(b' \oplus b)$

Fig. 11. (S, D) in the psPR game of P where A_0's inputs are honest.

by S to Perm. This then reduces P to merely guess the queries. To achieve this, we take a similar path as the psPRP game of P ($c = 0$). We transition to a game G_1 where F^- is independent of bit b. However, unlike the psPRP case, P and S share the same oracle Perm (to which P can also make inverse queries), and therefore it is non-trivial to argue about the independence of L and queries of S as we desire. Therefore, we make a final transition to a game G_2 where Perm returns random strings for queries by S and refrains from storing any information about the queries made by S. The resulting leakage can be viewed as being constructed by S without making any queries to Perm. Then we exploit the fact X^+ information theoretically hides X^- and hence queries by S are hidden from any P making only polynomially many queries to Perm. Again we direct the reader to the full version for a rigorous argument. (We also note that this argument is implicitly contained in the original security proof in the ideal-cipher model.)   □

RELATED SCHEMES. Along with Ga, [10] propose another scheme GaX which achieves faster garbling and evaluation times using the free-xor technique [31]. We consider a variant GaX[P] of GaX where (like Ga[P]) the permutation is replaced by a psPRP[S^{sup}]-secure permutation P and the seed for P is sampled freshly for every new instantiation of Gb. The security proof of GaX[P] almost readily follows from the security proof of GaX from [10] with slight modifications as done in the proof of Ga[P].

The third scheme proposed by [10], called GaXR, further improves over GaX in the size of the garbled table due to the use of row reduction technique at the cost of slower garbling and evaluation times. This means that for every gate, GaXR serves only three rows in the garbled table. Here, we note that adapting GaXR to be proved secure under a suitable psPRP assumption does not appear to have a simple and clear solution, and we leave this as an open problem.

Acknowledgments. We wish to thank John Retterer-Moore for his involvement in an earlier stage of this project. This research was partially supported by NSF grants CNS-1423566, CNS-1528178, CNS-1553758 (CAREER), and IIS-152804, and by a Hellman Fellowship.

References

1. Andreeva, E., Bogdanov, A., Dodis, Y., Mennink, B., Steinberger, J.P.: On the indifferentiability of key-alternating ciphers. In: Canetti, R., Garay, J.A. (eds.) CRYPTO 2013. LNCS, vol. 8042, pp. 531–550. Springer, Heidelberg (2013). doi:10.1007/978-3-642-40041-4_29

2. Andreeva, E., Bogdanov, A., Mennink, B.: Towards understanding the known-key security of block ciphers. In: Moriai, S. (ed.) FSE 2013. LNCS, vol. 8424, pp. 348–366. Springer, Heidelberg (2014). doi:10.1007/978-3-662-43933-3_18

3. Andreeva, E., Daemen, J., Mennink, B., Van Assche, G.: Security of keyed sponge constructions using a modular proof approach. In: Leander, G. (ed.) FSE 2015. LNCS, vol. 9054, pp. 364–384. Springer, Heidelberg (2015). doi:10.1007/978-3-662-48116-5_18

4. Aumasson, J.-P., Jovanovic, P., Neves, S.: NORX8 and NORX16: authenticated encryption for low-end systems. Cryptology ePrint Archive, Report 2015/1154 (2015). http://eprint.iacr.org/2015/1154

5. Barak, B., Goldreich, O., Impagliazzo, R., Rudich, S., Sahai, A., Vadhan, S., Yang, K.: On the (Im)possibility of obfuscating programs. In: Kilian, J. (ed.) CRYPTO 2001. LNCS, vol. 2139, pp. 1–18. Springer, Heidelberg (2001). doi:10.1007/3-540-44647-8_1

6. Bellare, M., Bernstein, D.J., Tessaro, S.: Hash-function based PRFs: AMAC and its multi-user security. In: Fischlin, M., Coron, J.-S. (eds.) EUROCRYPT 2016. LNCS, vol. 9665, pp. 566–595. Springer, Heidelberg (2016). doi:10.1007/978-3-662-49890-3_22

7. Bellare, M., Hoang, V.T.: Resisting randomness subversion: fast deterministic and hedged public-key encryption in the standard model. In: Oswald, E., Fischlin, M. (eds.) EUROCRYPT 2015. LNCS, vol. 9057, pp. 627–656. Springer, Heidelberg (2015). doi:10.1007/978-3-662-46803-6_21

8. Bellare, M., Hoang, V.T., Keelveedhi, S.: Instantiating random Oracles via UCEs. In: Canetti, R., Garay, J.A. (eds.) CRYPTO 2013. LNCS, vol. 8043, pp. 398–415. Springer, Heidelberg (2013). doi:10.1007/978-3-642-40084-1_23

9. Bellare, M., Hoang, V.T., Keelveedhi, S.: Cryptography from compression functions: the UCE bridge to the ROM. In: Garay, J.A., Gennaro, R. (eds.) CRYPTO 2014. LNCS, vol. 8616, pp. 169–187. Springer, Heidelberg (2014). doi:10.1007/978-3-662-44371-2_10

10. Bellare, M., Hoang, V.T., Keelveedhi, S., Rogaway, P.: Efficient garbling from a fixed-key blockcipher. In: 2013 IEEE Symposium on Security and Privacy, pp. 478–492. IEEE Computer Society Press, May 2013

11. Bellare, M., Hoang, V.T., Rogaway, P.: Foundations of garbled circuits. In: Yu, T., Danezis, G., Gligor, V.D. (eds.), ACM CCS 2012, pp. 784–796. ACM Press, October 2012
12. Bellare, M., Rogaway, P.: The security of triple encryption and a framework for code-based game-playing proofs. In: Vaudenay, S. (ed.) EUROCRYPT 2006. LNCS, vol. 4004, pp. 409–426. Springer, Heidelberg (2006). doi:10.1007/11761679_25
13. Bellare, M., Stepanovs, I.: Point-function obfuscation: a framework and generic constructions. In: Kushilevitz, E., Malkin, T. (eds.) TCC 2016. LNCS, vol. 9563, pp. 565–594. Springer, Heidelberg (2016). doi:10.1007/978-3-662-49099-0_21
14. Bellare, M., Stepanovs, I., Tessaro, S.: Contention in cryptoland: obfuscation, leakage and UCE. In: Kushilevitz, E., Malkin, T. (eds.) TCC 2016. LNCS, vol. 9563, pp. 542–564. Springer, Heidelberg (2016). doi:10.1007/978-3-662-49099-0_20
15. Bertoni, G., Daemen, J., Peeters, M., Assche, G.: On the indifferentiability of the sponge construction. In: Smart, N. (ed.) EUROCRYPT 2008. LNCS, vol. 4965, pp. 181–197. Springer, Heidelberg (2008). doi:10.1007/978-3-540-78967-3_11
16. Bertoni, G., Daemen, J., Peeters, M., Assche, G.: Sponge-based pseudo-random number generators. In: Mangard, S., Standaert, F.-X. (eds.) CHES 2010. LNCS, vol. 6225, pp. 33–47. Springer, Heidelberg (2010). doi:10.1007/978-3-642-15031-9_3
17. Brzuska, C., Farshim, P., Mittelbach, A.: Indistinguishability obfuscation and UCEs: the case of computationally unpredictable sources. In: Garay, J.A., Gennaro, R. (eds.) CRYPTO 2014. LNCS, vol. 8616, pp. 188–205. Springer, Heidelberg (2014). doi:10.1007/978-3-662-44371-2_11
18. Brzuska, C., Mittelbach, A.: Using indistinguishability obfuscation via UCEs. In: Sarkar, P., Iwata, T. (eds.) ASIACRYPT 2014. LNCS, vol. 8874, pp. 122–141. Springer, Heidelberg (2014). doi:10.1007/978-3-662-45608-8_7
19. Chang, D., Dworkin, M., Hong, S., Kelsey, J., Nandi, M.: A keyed sponge construction with pseudorandomness in the standard model. In: Proceedings of the Third SHA-3 Candidate Conference (2012)
20. Coron, J.-S., Dodis, Y., Malinaud, C., Puniya, P.: Merkle-Damgård revisited: how to construct a hash function. In: Shoup, V. (ed.) CRYPTO 2005. LNCS, vol. 3621, pp. 430–448. Springer, Heidelberg (2005). doi:10.1007/11535218_26
21. Coron, J.-S., Holenstein, T., Künzler, R., Patarin, J., Seurin, Y., Tessaro, S.: How to build an ideal cipher: the indifferentiability of the Feistel construction. J. Cryptol. 29(1), 61–114 (2016)
22. Dachman-Soled, D., Katz, J., Thiruvengadam, A.: 10-round Feistel is indifferentiable from an ideal cipher. In: Fischlin, M., Coron, J.-S. (eds.) EUROCRYPT 2016. LNCS, vol. 9666, pp. 649–678. Springer, Heidelberg (2016). doi:10.1007/978-3-662-49896-5_23
23. Dai, Y., Steinberger, J.: Indifferentiability of 8-round Feistel networks. In: Robshaw, M., Katz, J. (eds.) CRYPTO 2016. LNCS, vol. 9814, pp. 95–120. Springer, Heidelberg (2016). doi:10.1007/978-3-662-53018-4_4
24. Dodis, Y., Ganesh, C., Golovnev, A., Juels, A., Ristenpart, T.: A formal treatment of backdoored pseudorandom generators. In: Oswald, E., Fischlin, M. (eds.) EUROCRYPT 2015. LNCS, vol. 9056, pp. 101–126. Springer, Heidelberg (2015). doi:10.1007/978-3-662-46800-5_5
25. Even, S., Mansour, Y.: A construction of a cipher from a single pseudorandom permutation. J. Cryptol. 10(3), 151–162 (1997)
26. Gaži, P., Pietrzak, K., Tessaro, S.: The exact PRF security of truncation: tight bounds for keyed sponges and truncated CBC. In: Gennaro, R., Robshaw, M. (eds.) CRYPTO 2015. LNCS, vol. 9215, pp. 368–387. Springer, Heidelberg (2015). doi:10.1007/978-3-662-47989-6_18

27. Gazi, P., Tessaro, S.: Secret-key cryptography from ideal primitives: a systematic overview. In: IEEE Information Theory Workshop, ITW, Jerusalem, Israel, 26 April–1 May 2015, pp. 1–5. IEEE (2015)

28. Gueron, S., Lindell, Y., Nof, A., Pinkas, B.: Fast garbling of circuits under standard assumptions. In: Ray, I., Li, N., Kruegel, C. (eds.), ACM CCS 2015, pp. 567–578. ACM Press, October 2015

29. Holenstein, T., Künzler, R., Tessaro, S.: The equivalence of the random Oracle model and the ideal cipher model, revisited. In: Fortnow, L., Vadhan, S.P. (eds.), 43rd ACM STOC, pp. 89–98. ACM Press, June 2011

30. Knudsen, L.R., Rijmen, V.: Known-key distinguishers for some block ciphers. In: Kurosawa, K. (ed.) ASIACRYPT 2007. LNCS, vol. 4833, pp. 315–324. Springer, Heidelberg (2007). doi:10.1007/978-3-540-76900-2_19

31. Kolesnikov, V., Schneider, T.: Improved garbled circuit: free XOR gates and applications. In: Aceto, L., Damgård, I., Goldberg, L.A., Halldórsson, M.M., Ingólfsdóttir, A., Walukiewicz, I. (eds.) ICALP 2008. LNCS, vol. 5126, pp. 486–498. Springer, Heidelberg (2008). doi:10.1007/978-3-540-70583-3_40

32. Luby, M., Racko, C.: How to construct pseudorandom permutations from pseudorandom functions. SIAM J. Comput. $17(2)$, 373–386 (1988)

33. Mandal, A., Patarin, J., Seurin, Y.: On the public indifferentiability and correlation intractability of the 6-round Feistel construction. In: Cramer, R. (ed.) TCC 2012. LNCS, vol. 7194, pp. 285–302. Springer, Heidelberg (2012). doi:10.1007/978-3-642-28914-9_16

34. Matsuda, T., Hanaoka, G.: Chosen ciphertext security via UCE. In: Krawczyk, H. (ed.) PKC 2014. LNCS, vol. 8383, pp. 56–76. Springer, Heidelberg (2014). doi:10.1007/978-3-642-54631-0_4

35. Maurer, U., Renner, R., Holenstein, C.: Indifferentiability, impossibility results on reductions, and applications to the random Oracle methodology. In: Naor, M. (ed.) TCC 2004. LNCS, vol. 2951, pp. 21–39. Springer, Heidelberg (2004). doi:10.1007/978-3-540-24638-1_2

36. Mittelbach, A.: Salvaging indifferentiability in a multi-stage setting. In: Nguyen, P.Q., Oswald, E. (eds.) EUROCRYPT 2014. LNCS, vol. 8441, pp. 603–621. Springer, Heidelberg (2014). doi:10.1007/978-3-642-55220-5_33

37. Mouha, N., Mennink, B., Herrewege, A., Watanabe, D., Preneel, B., Verbauwhede, I.: Chaskey: an efficient MAC algorithm for 32-bit microcontrollers. In: Joux, A., Youssef, A. (eds.) SAC 2014. LNCS, vol. 8781, pp. 306–323. Springer, Heidelberg (2014). doi:10.1007/978-3-319-13051-4_19

38. Ristenpart, T., Shacham, H., Shrimpton, T.: Careful with composition: limitations of the indifferentiability framework. In: Paterson, K.G. (ed.) EUROCRYPT 2011. LNCS, vol. 6632, pp. 487–506. Springer, Heidelberg (2011). doi:10.1007/978-3-642-20465-4_27

39. Rogaway, P., Steinberger, J.: Security/efficiency tradeoffs for permutation-based hashing. In: Smart, N. (ed.) EUROCRYPT 2008. LNCS, vol. 4965, pp. 220–236. Springer, Heidelberg (2008). doi:10.1007/978-3-540-78967-3_13

40. Yao, A.C.-C.: How to generate and exchange secrets (extended abstract). In: 27th FOCS, pp. 162–167. IEEE Computer Society Press, October 1986

41. Zhandry, M.: The magic of ELFs. In: Robshaw, M., Katz, J. (eds.) CRYPTO 2016. LNCS, vol. 9814, pp. 479–508. Springer, Heidelberg (2016). doi:10.1007/978-3-662-53018-4_18

Security Models I

Cryptography with Updates

Prabhanjan Ananth[1]([⊠]), Aloni Cohen[2], and Abhishek Jain[3]

[1] UCLA, Los Angeles, USA
prabhanjan@cs.ucla.edu
[2] MIT, Cambridge, USA
aloni@mit.edu
[3] Johns Hopkins University, Baltimore, USA
abhishek@cs.jhu.edu

Abstract. Starting with the work of Bellare, Goldreich and Goldwasser [CRYPTO'94], a rich line of work has studied the design of updatable cryptographic primitives. For example, in an updatable signature scheme, it is possible to efficiently transform a signature over a message into a signature over a related message without recomputing a fresh signature.

In this work, we continue this line of research, and perform a systematic study of updatable cryptography. We take a unified approach towards adding updatability features to recently studied cryptographic objects such as attribute-based encryption, functional encryption, witness encryption, indistinguishability obfuscation, and many others that support non-interactive computation over inputs. We, in fact, go further and extend our approach to classical protocols such as zero-knowledge proofs and secure multiparty computation.

To accomplish this goal, we introduce a new notion of *updatable randomized encodings* that extends the standard notion of randomized encodings to incorporate updatability features. We show that updatable randomized encodings can be used to generically transform cryptographic primitives to their updatable counterparts.

We provide various definitions and constructions of updatable randomized encodings based on varying assumptions, ranging from one-way functions to compact functional encryption.

P. Ananth—Work done in part while visiting the Simons Institute for Theoretical Computer Science, supported by the Simons Foundation and and by the DIMACS/Simons Collaboration in Cryptography through NSF grant #CNS-1523467. This work was partially supported by grant #360584 from the Simons Foundation.

A. Cohen—Supported in part by NSF MACS CNS-1413920, DARPA IBM W911NF-15-C-0236, and Simons Investigator Award Agreement Dated 6-5-12.

A. Jain—Work done in part while visiting the Simons Institute for Theoretical Computer Science, supported by the Simons Foundation and and by the DIMACS/Simons Collaboration in Cryptography through NSF grant #CNS-1523467. Supported in part by a DARPA/ARL Safeware Grant W911NF-15-C-0213 and NSF CNS-1414023.

J.-S. Coron and J.B. Nielsen (Eds.): EUROCRYPT 2017, Part II, LNCS 10211, pp. 445–472, 2017.
DOI: 10.1007/978-3-319-56614-6_15

1   Introduction

The last decade has seen the advent of a vast array of advanced cryptographic primitives such as attribute-based encryption [45,55], predicate encryption [20, 43,47,57], fully homomorphic encryption [36], fully homomorphic signatures [7, 17,44], functional encryption [19,41,54,55], constrained pseudorandom functions [21,22,48], witness encryption [34,38], witness PRFs [60], indistinguishability obfuscation [9,32], and many more. Most of these primitives can be viewed as "cryptographic circuit compilers" where a circuit C can be compiled into an encoding $\langle C \rangle$ and an input x can be encoded as $\langle x \rangle$ such that they can be evaluated together to compute $C(x)$. For example, in a functional encryption scheme, circuit compilation corresponds to the key generation process whereas input encoding corresponds to encryption. Over the recent years, cryptographic circuit compilers have revolutionized cryptography by providing non-interactive means of computing over inputs/data.

A fundamental limitation of these circuit compilers is that they only support *static* compilation. That is, once a circuit is compiled, it can no longer be modified. In reality, however, compiled circuits may need to undergo several updates over a period of time. For example, consider an organization where each employee is issued a decryption key SK_P of an attribute-based encryption scheme where the predicate P corresponds to her access level determined by her employment status. However, if her employment status later changes, then we would want to update the predicate P associated with her decryption key. Known schemes, unfortunately, do not support this ability.

Motivated by the necessity of supporting updates in applications, in this work, we study and build *dynamic* circuit compilers. In a dynamic circuit compiler, it is possible to update a compiled circuit $\langle C \rangle$ into another compiled circuit $\langle C' \rangle$ by using an *encoded update string* whose size only depends on the "difference" between the plaintext circuits C and C'. For example, if the difference between C and C' is simply a single gate change, then this should be reflected in the size of the encoded update. Note that this rules out the trivial solution of simply releasing a new compiled circuit at the time of update.

Background: Incremental Cryptography. The study of cryptography with updates was initiated by Bellare, Goldreich and Goldwasser [10] under the umbrella of *incremental cryptography*. They studied the problem of incremental digital signatures, where given a signature of a message m, it should be possible to efficiently compute a signature of a related message m', without having to recompute the signature of m' from scratch. Following their work, the study of incremental cryptography was extended to other basic cryptographic primitives such as encryption and hash functions [10–12,24,31,52,53], and more recently, indistinguishability obfuscation [5,35].

Our Goal. In this work, we continue this line of research, and perform a systematic study of updatable cryptographic primitives. We take a unified approach towards adding updatability features to recently studied primitives such as attribute-based encryption, functional encryption and more generally, cryptographic circuit compilers. We, in fact, go further and also study updatability for classical protocols such as zero-knowledge proofs and secure multiparty computation.

To accomplish this goal, we introduce a new notion of *updatable randomized encodings* that extends the standard notion of randomized encoding [46] to incorporate updatability features. We show that updatable randomized encodings can be used to generically transform cryptographic primitives (discussed above) to their updatable counterparts.

Updatable Randomized Encodings. The notion of randomized encoding [46] allows one to encode a "complex" computation $C(x)$ into a "simple" randomized function $\mathsf{Encode}(C, x; r)$ such that given the output $\langle C(x) \rangle$ of the latter, it is possible to recover the value $C(x)$ (by running a public Decode algorithm) but it is impossible to learn anything else about C or x. The typical measure of complexity studied in the literature is parallel-time complexity or circuit depth. Such randomized encodings are known to exist for general circuits based on only the existence of one-way functions [6] (also referred to as Yao's garbled circuits [59], where $\mathsf{Encode}(C, x; r)$ is in $\mathbf{NC}^1$).

In this work, we study *updatable* randomized encodings (URE): given a randomized encoding $\langle C(x) \rangle$ of $C(x)$, we want the ability to update it to an encoding $\langle C'(x') \rangle$ of $C'(x')$, where C' and x' are derived from C and x by applying some "update" $\mathbf{u}$. For now, we may think of this update as some small modification to the circuit or input (e.g., change the output gate of C to AND and the second bit of x to 1). We require that the update $\mathbf{u}$ can be encoded as $\langle \mathbf{u} \rangle$ which can then be used to transform $\langle C(x) \rangle$ into $\langle C'(x') \rangle$, a randomized encoding of $C'(x')$. A bit more precisely, a URE scheme consists of the following algorithms:

– $\mathsf{Encode}(C, x)$ takes as input a circuit C and an input x, and outputs an encoding $\langle C(x) \rangle$ and a secret state st.
– $\mathsf{GenUpd}(\mathsf{st}, \mathbf{u})$ takes as input an update $\mathbf{u}$, and outputs an encoded update $\langle \mathbf{u} \rangle$ and a possibly updated state st'.
– $\mathsf{ApplyUpd}(\langle C(x) \rangle, \langle \mathbf{u} \rangle)$ takes as input a randomized encoding $\langle C(x) \rangle$ and an update encoding $\langle \mathbf{u} \rangle$, and outputs an (updated) encoding $\langle C'(x') \rangle$.
– $\mathsf{Decode}(\langle C(x) \rangle)$ takes as input a (possibly updated) randomized encoding $\langle C(x) \rangle$, and outputs the value $y = C(x)$.

If we make no additional requirements, the above could be easily achieved. For instance, let Encode output the state $\mathsf{st} = (C, x)$, and let GenUpd – which now has access to C and x from st in addition to the update $\mathbf{u}$ – compute the updated C' and x' directly and output a as the encoded update $\langle \mathbf{u} \rangle$ the standard randomized encoding of $\langle C'(x') \rangle$. $\mathsf{ApplyUpd}$ would correspondingly output $\langle \mathbf{u} \rangle = \langle C'(x') \rangle$. The drawback of this approach is that a fresh randomized encoding is computed during every evaluation of GenUpd, irrespective of whether $\mathbf{u}$ constitutes a minute or significant change to the underlying C and x.

Our key efficiency requirement is that the running time of the GenUpd algorithm must be a fixed polynomial size of the update (and a security parameter), and independent of the size of the circuit and input being updated. This, in particular, implies that the size of an update encoding $\langle \mathbf{u} \rangle$ is also a fixed polynomial in the size of $\mathbf{u}$ (and the security parameter).

The above discussion immediately generalizes to the setting of *multiple sequential updates*.[1] Let $\langle C_0(x_0) \rangle$ denote an initial randomized encoding. Let $\mathbf{u}_1, \ldots, \mathbf{u}_n$ denote a sequence of updates and let $\langle \mathbf{u}_i \rangle$ denote an encoding of $\mathbf{u}_i$. In a URE scheme for multiple updates, $\langle C_0(x_0) \rangle$ can be updated to $\langle C_1(x_1) \rangle$ using $\langle u_1 \rangle$; the result can then be updated into $\langle C_2(x_2) \rangle$ using $\langle u_2 \rangle$, and so on, until we obtain $\langle C_n(x_n) \rangle$. We allow the number of updates n to be an arbitrary polynomial in the security parameter.

Within this framework, two distinct notions naturally arise.

URE with multiple evaluations: Every intermediate encoding $\langle C_i(x_i) \rangle$ can be decoded to obtain $C_i(x_i)$. For security, we require that given an initial randomized encoding $\langle C_0(x_0) \rangle$ and a sequence of encoded updates $\{\langle \mathbf{u}_i \rangle\}_{i=1}^n$, an adversary can learn only the outputs $\{C_i(x_i)\}_{i=0}^n$, and nothing else.

URE with single evaluation: Only the final encoding $\langle C_n(x_n) \rangle$ can be decoded. To enable this, we will consider an augmented decoding algorithm that additionally requires an "unlocking key."[2] This unlocking key is provided after all the updates are completed, allowing the user to decode the final encoding, but preventing her from decoding any intermediate values. For security, we require that given an initial randomized encoding $\langle C_0(x_0) \rangle$ and a sequence of encoded updates $\{\langle \mathbf{u}_i \rangle\}_{i=1}^n$), an adversary can only learn the final output $C_n(x_n)$, and nothing else.

Except where otherwise specified, we use URE to mean the multiple-evaluation variant. For both conceptual reasons and to minimize confusion, we in fact consider an alternative but equivalent formulation of single-evaluation URE which we call **updatable garbled circuits** (UGC). A garbled circuit [59] is a "decomposable" randomized encoding, where a circuit C and an input x can be encoded separately. In an updatable garbled circuit scheme, given an encoding $\langle C_0 \rangle$ of a circuit C_0 and a sequence of update encodings $\langle \mathbf{u}_1 \rangle, \ldots, \langle \mathbf{u}_n \rangle$, it is possible to compute updated circuit encodings $\langle C_1 \rangle, \ldots, \langle C_n \rangle$, where C_i is derived from C_{i-1} using $\mathbf{u}_i$. Once all the updates are completed, an encoding $\langle x \rangle$ for an input x is released. This input encoding can then be used to decode the final circuit encoding $\langle C_n \rangle$ and learn $C_n(x_n)$. Intuitively, the input encoding can be viewed as the unlocking key in single-evaluation URE.

It is easy to see that UGC is a weaker notion than multi-evaluation URE. In particular, since UGC only allows for decoding "at the end," it remains *single-use*, while multi-evaluation URE captures *reusability*.

---

[1] One may also consider an alternative notion of *parallel* updates, where every update $\langle \mathbf{u}_i \rangle$ is applied to the *original* encoding $\langle C_0(x_0) \rangle$. It turns out that URE with parallel updates is closely connected to the notion of reusable garbled circuits [42]. We refer the reader to the full version [2] for further discussion on this subject.

[2] In the setting of bounded updates, this modification is unnecessary. We focus primarily on the unbounded setting.

We find the notions of URE and UGC to be of interest from a purely complexity-theoretic perspective. Further, as we discuss later, they have powerful applications to updatable cryptography.

1.1   Our Results

In this work, we initiate the study of updatable randomized encodings. We study both simulation and indistinguishability-based security definitions and obtain general positive results. We showcase URE as a central object for the study of updatable cryptography by demonstrating applications to other updatable cryptographic primitives. The technical ideas we develop for our constructions are quite general, and may be applicable to future works on updatable cryptography.

Multi-evaluation URE for General Updates. Before stating our positive results for multi-evaluation URE, we first informally describe which classes of updates we can support. An update Update $\in \mathcal{U}$ represents some way to modify any circuit C and an input x to some modified circuit C' and input x'. We denote by $\mathbf{u}$ the procedure $(C', x') \leftarrow \mathbf{u}(C, x, \text{Update})$ which applies the update to C and x. We consider all $\mathcal{U}$ and Update subject to two restrictions: (1) Update is computed by a (family) of circuits, one for every circuit size $|C|$, and (2) Update preserves circuit size (i.e., $|C| = |C'|$). We refer to this very broad class of updates as *general circuit updates*.

For general circuit updates, we construct URE from *compact* functional encryption. The summary below focuses on indistinguishability-based security, and concludes with a remark on achieving simulation-based security.

Theorem 1 (Informal). *Assuming the existence of secret-key, compact functional encryption supporting a single key query and B ciphertexts, there exists a multi-evaluation* URE *scheme supporting B sequential general circuit updates.*

A compact functional encryption is one where the running time of the encryption algorithm for a message m is a fixed polynomial in the size of m and the security parameter, and independent of the complexity of the function family supported by the FE scheme.

For the case of **unbounded** updates, a recent work of Bitansky et al. [14] shows that secret-key compact functional encryption with unbounded-many ciphertexts implies exponentially-efficient indistinguishability obfuscation (XIO) [49]. Put together with the results of [49] and [3,15], it shows that *sub-exponentially* secure secret-key compact FE that supports a single function key query together with the learning with errors (LWE) assumption implies indistinguishability obfuscation.

In contrast, in Theorem 1, we require secret-key compact FE with only *polynomial security*. Such an FE scheme can be based on polynomial-hardness assumptions on multilinear maps using the results of [33] and [4,15].

For the case of **polynomially-bounded** updates, we can, in fact, relax our assumption to only *one-way functions*. We obtain this result by using a *stateful* single-key compact secret-key FE scheme for an a priori bounded number B

of ciphertexts. A stateful single-key compact secret-key FE scheme can be constructed from garbled circuits: a functional key consists of B garbled circuits, i^{th} ciphertext consists of garbled wire keys corresponding to the ith garbled circuit. This FE scheme is stateful since the encryption algorithm needs to store how many messages it has encrypted so far.

Plugging in such an FE scheme in Theorem 1 yields the following corollary.

Corollary 1 (Informal). *Assuming one-way functions, for any fixed polynomial B, there exists a multi-evaluation URE scheme supporting B sequential general circuit updates.*

ON THE NECESSITY OF FUNCTIONAL ENCRYPTION. It is natural to ask whether secret-key compact FE is necessary for building multi-evaluation URE with unbounded updates. We show that if a (multi-evaluation) URE scheme is *output compact*, then it implies XIO. Put together with the result of [49], we have that a URE scheme with output compactness together with LWE implies a public-key compact FE scheme that supports a single key query.

Theorem 2 (Informal). *Assuming LWE, a multi-evaluation URE scheme with unbounded output-compact updates implies a public-key compact FE scheme that supports a single key query.*

In an output-compact URE scheme, the running time of the GenUpd algorithm is independent of the output length of the updated circuit. We remark that the URE scheme obtained from Theorem 1 is, in fact, output compact. Our construction in Theorem 1 is in this sense tight.

ON OUTPUT COMPACTNESS. We study both indistinguishability and simulation-based security notions for URE. In the context of FE, it is known from [1,26] that simulation-secure FE with output compactness is impossible for general functions. We observe that the same ideas as in [1,26] can be used to establish impossibility of simulation-secure URE with output compact updates.

However, when we consider indistinguishability-based security, URE with output compact updates is indeed possible. The results in Theorem 1 and Corollary 1 are stated for this case. Furthermore, using the trapdoor circuits technique of [26], one can generically transform output-compact URE with indistinguishability security to non-output-compact URE with simulation-based security.

Updatable garbled circuits with gate-wise updates. We now turn to updatable garbled circuits, an alternate formulation of single-evaluation URE. We consider the family of gate-wise updates, where an update $\mathbf{u}$ can modify a single gate of a circuit or add or delete a gate. Below, we consider the case of unbounded updates and bounded updates separately.

UGC WITH UNBOUNDED UPDATES FROM LATTICE ASSUMPTIONS. Our first result is a construction of UGC for general circuits that supports an unbounded number of sequential updates from the family of gate-wise updates. We build such a scheme from worst-case lattice assumptions.

Theorem 3 (Informal). *Let C be a family of general circuits. Assuming the hardness of approximating either GapSVP or SIVP to within sub-exponential factors, there exists a* UGC *scheme for C that supports an unbounded polynomial number of sequential gate-wise updates.*

We defer the proof of this theorem to the full version. [2] At the heart of this result is a new notion of *puncturable symmetric proxy re-encryption scheme* that extends the well-studied notion of proxy re-encryption [16]. In a symmetric proxy re-encryption scheme, for any pair of secret keys SK_1, SK_2, it is possible to construct a re-encryption key $\mathsf{RK}_{1\to 2}$ that can be used to publicly transform a ciphertext w.r.t. SK_1 into a ciphertext w.r.t. SK_2. In our new notion of puncturable proxy re-encryption, re-encryption keys can be "disabled" on ciphertexts CT^* (w.r.t. SK_1) s.t. the semantic security of CT^* holds even if the adversary is given the punctured key $\mathsf{RK}_{1\to 2}^{\mathsf{CT}^*}$ and SK_2. We give a construction of such a scheme based on the hardness of approximating either GapSVP or SIVP to within sub-exponential factors.

Given the wide applications of proxy re-encryption (see, e.g., [8] for a discussion), we believe that our notion of puncturable proxy re-encryption is of independent interest and likely to find new applications in the future.

UGC with Bounded Updates from One-Way Functions. For the case of a polynomially-bounded number of updates, we can relax our assumption to only *one-way functions*. We obtain this result by using a puncturable PRF scheme that can be based on one-way functions [40,56].

Theorem 4 (Informal). *Let C be a family of general circuits, and λ be a security parameter. Assuming one-way functions, for any fixed polynomial p, there exists a* UGC *scheme for C that supports $p(\lambda)$ sequential gate-wise updates. The size of the initial garbled circuit as well as each update encoding is independent of p. However, the initial circuit garbling time and update generation time grows with p.*

The construction of this scheme is quite simple and does not require a puncturable proxy re-encryption scheme. We provide an informal description of this scheme in the technical overview Sect. 2.1.

Applications. We next discuss applications of our results.

Updatable Primitives with IND security. We start by discussing application of multi-evaluation URE to dynamic circuit compilers. Here, we demonstrate our main idea by a concrete example, namely, by showing how to use URE to transform any (key-policy) attribute-based encryption (ABE) scheme into *updatable ABE*. The same idea can be used in a generic way to build dynamic circuit compilers and obtain updatable functional encryption, updatable indistinguishability obfuscation, and so on. We refer the reader to the full version [2] for the general case.

We briefly describe a generic transformation from any ABE scheme to one where the policies associated with secret keys can be updated. The setup and encryption algorithms for the updatable ABE scheme are the same as in the underlying ABE scheme. The key generation algorithm in the updatable ABE scheme works as follows: to compute an attribute key for a function f, we compute a URE $\langle C_f \rangle$ of a circuit C_f where C runs the key generation algorithm of the underlying ABE scheme using function f and outputs a key SK_f. To decrypt a ciphertext, a user can first decode $\langle C_f \rangle$ to compute SK_f and then use it to decrypt the ciphertext.

In order to update an attribute key for a function f to another key for function f', we can simply issue an update encoding $\langle \mathbf{u} \rangle$ for $\langle C_f \rangle$ where $\mathbf{u}$ captures the modification from f to f'. To compute the updated attribute key, a user can first update $\langle C_f \rangle$ using $\langle \mathbf{u} \rangle$ to obtain $\langle C_{f'} \rangle$, and then decode it to obtain an attribute key $\mathsf{SK}_{f'}$ for f'.

Let us inspect the efficiency of updates in the above updatable ABE scheme. As in URE, we would like the size (as well as the generation time) of an update encoding here to be independent of the size of the updated function. Note, however, that the output of the updated function $C_{f'}$ is very large – an entire attribute key $\mathsf{SK}_{f'}$! Thus, in order to achieve the aforementioned efficiency, we require that the URE scheme has updates with output compactness.

Recall that URE with output compact updates is only possible with indistinguishability based security. As such, the above idea is only applicable to cryptographic primitives with indistinguishability-based security.

UPDATABLE PRIMITIVES WITH SIM SECURITY. Next, we discuss applications of URE to cryptographic primitives with simulation-based security. In the main body of the paper, we describe two concrete applications, namely, *updatable non-interactive zero-knowledge proofs* (UNIZK) and *updatable multiparty computation* (UMPC). A notable feature of these constructions is that they only require a URE scheme with *non-output-compact* updates and simulation-based security. Below, we briefly describe our main idea for constructing UNIZKs.

Let (x, w) denote an instance and witness pair for an **NP** language L. Let $\mathbf{u}$ denote an update that transforms (x, w) to another valid instance and witness pair (x', w'). In a UNIZK proof system for L, it should be possible for a prover to efficiently compute an encoding $\langle \mathbf{u} \rangle$ of $\mathbf{u}$ that allows a verifier to transform a valid proof π for x into a proof π' for x' and verify its correctness.

We now briefly describe our transformation. A proof π for (x, w) in the UNIZK scheme is computed as follows: we first compute a URE $\langle C_{x,w} \rangle$ for a circuit $C_{x,w}$ that checks whether (x, w) satisfies the **NP** relation associated with L and outputs 1 or 0 accordingly. Furthermore, we also compute a regular NIZK proof ϕ to prove that $\langle C_{x,w} \rangle$ is computed "honestly." To verify $\pi = (\langle C_{x,w} \rangle, \phi)$, a verifier first verifies ϕ and if the check succeeds, it decodes $\langle C_{x,w} \rangle$ and outputs its answer.

In order to update a proof π, we can simply issue an update encoding $\langle \mathbf{u} \rangle$ for the randomized encoding $\langle C_{x,w} \rangle$, along with a regular NIZK proof ϕ' that $\langle \mathbf{u} \rangle$ was computed honestly. Upon receiving the update $(\langle \mathbf{u} \rangle, \phi')$, a verifier can first verify ϕ' and then update $\langle C_{x,w} \rangle$ using $\langle \mathbf{u} \rangle$ to obtain $\langle C_{x',w'} \rangle$. Finally, it can decode the updated URE $\langle C_{x',w'} \rangle$ to learn whether x' is in the language L or not.

It should be easy to see that the above idea can, in fact, be also used to make *interactive* zero-knowledge proofs updatable. Finally, we note that the above is a slightly oversimplified description and we refer the reader to the full version [2] for further details on UNIZK and UMPC, respectively.

1.2   Related Work

Incremental Cryptography. The area of incremental cryptography was pioneered by Bellare, Goldreich and Goldwasser [10]. While their work dealt with signature schemes, the concept of incremental updates has been subsequently studied for other basic cryptographic primitives such as hash functions, semantically-secure encryption and deterministic encryption [11,24,31,52,53]. To the best of our knowledge, all of these works only consider bit-wise updates, in which a single bit of the message is modified.

While our work shares much in spirit with these works, we highlight one important difference. In incremental cryptography, update operation is performed "in house," e.g., in the case of signatures, the entity who produces the original signature also performs the update. In contrast, we consider a *client-server* scenario where the client simply produces an update encoding, and the actual updating process is performed by the server. This difference stipulates different efficiency and security requirements. On the one hand, incremental cryptography necessarily requires efficient updating time for the notion to be non-trivial, while we consider the weaker property of efficient update encoding generation time. On the other hand, our security definition is necessarily stronger since we allow the adversary to view the update encodings – a property not necessary when the updating is done "in house."

Incremental/Patchable Obfuscation. Recently, [35] and [5] study the notion of updatability in the context of *indistinguishability obfuscation*. The work of [35] considers incremental (i.e., bit-wise) updates, while [5] allow for arbitrary updates, including those that may increase the size of the program (modeled as a Turing machine).

We note that one of our results, namely, URE with unbounded updates can be derived from [5] at the cost of requiring sub-exponentially secure iO. In contrast, we obtain our result by using *polynomially secure* secret-key compact FE.

Malleable NIZKs. Our notion of updatable NIZKs should be contrasted with the notion of malleable NIZKs proposed by Chase et al. [28]. In a malleable NIZK, it is possible to publicly "maul" a proof string π for a statement x into a a proof string π' for a related statement x'. In contrast, our notion of UNIZK only allows for privately generated updates. To the best of our knowledge, malleable NIZKs are only known either for a limited class of update relations from standard assumptions [28], or for general class of update relations based on non-falsifiable assumptions such as succinct non-interactive arguments [29]. In contrast, we show how to build UNIZK for unbounded number of general updates from compact secret-key FE and regular NIZKs, and for a bounded number of general updates from regular NIZKs.

454   P. Ananth et al.

Updatable Codes. The concept of updating was also studied in the context of error correcting codes by [27]. In this context, it is difficult to model the problem of updating – we should be able to change few bits of the code to correspond to a codeword of a different message and at the same time we want the distance between codewords of different messages to be far apart. We refer the reader to their work for discussion on this seemingly contradictory requirement. In a subsequent work, [30] studied this problem in the context of non-malleable codes.

2   Our Techniques

We start with the construction of UGC and present the main ideas underlying the construction. We then build upon the intuition developed in the construction of UGC, to construct (multi-evaluation) URE.

2.1   Construction of UGC

Below, we restrict our discussion to updates that correspond to a gate change.

A *Lock-and-Release* Mechanism for Single Update. Let us first start with the simpler goal of building a UGC scheme that supports updating a *single* gate. Let C be a circuit comprised of s-many gates $C^1, \ldots, C^s$. Our starting idea towards a UGC scheme with single update is as follows: in order to garble C, we simply compute a garbling of C using a standard gate-by-gate garbling scheme such as [59].[3] We denote by $\langle C \rangle_{\mathsf{gc}}$ the garbled circuit for C, and by $\langle C \rangle_{\mathsf{gc}}^i$ the garbled gate corresponding to gate C^i. Encrypt each garbled gate, and output the resulting ciphertexts $\mathsf{CT}_1, \ldots, \mathsf{CT}_s$.

Now, suppose we wish to update garbling of C to garbling of C' where C' only differs from C in the first gate. Then, a natural idea is to release a decryption key that only decrypts the ciphertexts $\mathsf{CT}_2, \ldots, \mathsf{CT}_s$ (but not CT_1). The encoding of the update is consists of these $s-1$ keys and, along with a garbled-version of the new gate $\langle C' \rangle_{\mathsf{gc}}^1$. Using this information, the receiver can decrypt and recover the garbled gates $\langle C \rangle_{\mathsf{gc}}^2, \ldots, \langle C \rangle_{\mathsf{gc}}^s$. Together with $\langle C \rangle_{\mathsf{gc}}^1$, this forms a complete garbled circuit for C'.

The main remaining question is how to implement the aforementioned *conditional decryption* mechanism. A naive way to achieve this is to encrypt each ciphertext with an independent encryption key, and then release the decryption key for every position $i \neq 1$. However, note that in this naive solution, the size of the update encoding is proportional to $s = |C|$. In terms of size, this is no better than garbling C' from scratch.

To address this, we could instead use a (secret key) *puncturable* encryption scheme. In such a scheme, for any ciphertext CT, it is possible to compute "punctured decryption key" that enables one to decrypt all ciphertexts except CT.

---

[3] In gate-by-gate garbling schemes such as [59], each boolean gate can be garbled knowing only the circuit topology and the gate's functionality, independently of the remainder of the circuit.

In order to be non-trivial, the size of punctured decryption keys must be independent of the number of ciphertexts generated. Such an encryption scheme can be built from puncturable pseudorandom functions [21,22,48,56] (c.f. Waters [58]) which in turn can be based on any one-way function. It is easy to verify that given such an encryption scheme, we can efficiency we desire in above construction for UGC supporting a single update.

We find it instructive to abstract the above idea as a **lock-and-release** mechanism. Roughly speaking, the encryption of the wire keys corresponding to C constitutes the *locking* step, while the dissemination of the punctured decryption key constitutes the (conditional) *release* step. We find this abstraction particularly useful going forward, in order to develop our full solution for an unbounded number of updates.

Multiple Updates: Main Challenges. The above solution does not offer any security for multiple sequential updates – even for two. If the two updates for two different gates would allow an adversary to recover a garbling of the original circuit Additionally, the above scheme does not "connect" the two updates in any manner; an adversary could choose to apply none, one, or both of the updates before evaluating the circuit.

A *Layered Lock-and-Release* Mechanism for Bounded Updates. We next consider the case of an a priori *bounded* number of updates (the setting of Theorem 4). The key idea, in a nutshell, is to use *layered* punctured encryption, or alternatively, a **layered lock-and-release** mechanism.

Suppose we wish to handle p-many of updates. When garbling the circuit C, instead of encrypting the garbled gates a single time, we instead use p "onion" layers of encryption scheme, each using a punctured encryption scheme. Let $\mathbf{u}_1,\ldots,\mathbf{u}_p$ be a sequence of gate updates, each consisting of a gate $g \in [s]$ to change and a new gate type. To generate an updatable garbled circuit for C, first garble $\tilde{C}$ using a traditional gate-by-gate scheme. Sample p keys $SK_1,\ldots,SK_p$ for a puncturable encryption scheme. Encrypt each garbled gate $\langle C \rangle^i_{\mathsf{gc}}$ of the garbled circuit in p layers, yielding a ciphertext $\mathsf{CT}_i = \mathsf{Enc}(SK_1, \mathsf{Enc}(SK_2, \ldots \mathsf{Enc}(SK_p, \langle C \rangle^i_{\mathsf{gc}})))$.

The encoding of the first update $\mathbf{u}_1 = (g_1, \mathsf{gateType}_1)$ simply corresponds to releasing a decryption key for the outermost encryption layer that is punctured at CT_{g_1}, along with a layer $(p-1)$ encryption $\mathsf{CT}'_{g_i} = \mathsf{Enc}(SK_2, \mathsf{Enc}(SK_3, \ldots \mathsf{Enc}(SK_p, \langle C' \rangle^{g_i}_{\mathsf{gc}})))$, where $\langle C' \rangle^{g_i}_{\mathsf{gc}}$ is the new garbled gate corresponding to replacing gate g_i of C with $\mathsf{gateType}$. Likewise, an encoding of the i-th update $\mathbf{u}_i$ corresponds to releasing a punctured decryption key for SK_i, $(i-1)$ encryption of the new garbled gate.

The above idea of layered (punctured) encryption ensures that the receiver cannot "skip" any update, and instead must apply all the updates one-by-one to "peel-off" all the encryption layers from the garbled gates. Furthermore, since the encryption layers can only be removed in a prescribed order, the receiver must applies the updates *in order*. Finally, after all the decryption operations, the receiver only obtains a single garbled gate every location in the (updated) circuit.

We now briefly argue that the above construction satisfies the efficiency properties stated in Theorem 4. We first note that punctured encryption scheme in the above construction can simply correspond to a one-time pad where the randomness for computing the ith ciphertext, for every $i \in |C|$, is generated by evaluating a puncturable PRF over the index i. The PRF key (i.e., the secret key for the punctured encryption scheme) is different for every layer. With this instantiation, note that the size of the initial garbled circuit as well as every update is independent of the total number of updates p; however, the garbling time as well as update generation time depends on p.

A *Relock-and-Eventual-Release* Mechanism for Unbounded Updates. The above solution is that it inherently requires the number of updates to be a priori bounded. To support multiple updates, our main insight is to develop a **relock-and-eventual-release** mechanism as opposed to the layered lock-and-release mechanism discussed above. That is, instead of removing a lock at every step, our idea is to *change the lock* at every step. In encryption terminology, our idea is to replace the layered encryption in the above approach with a symmetric *re-encryption* scheme [16]. In a symmetric re-encryption scheme, given two encryption keys SK_1 and SK_2, it is possible to issue a re-encryption key $RK_{1 \to 2}$ that transforms any ciphertext w.r.t. SK_1 into a ciphertext w.r.t. SK_2. In order to allow for updates, we, require the re-encryption scheme to support key puncturing. That is, we require that it is possible to compute a punctured re-encryption key $RK_{1 \to 2}^{CT^*}$ that allows one to transform any ciphertext w.r.t. SK_1 into a ciphertext w.r.t. SK_2, *except the ciphertext* CT^* (computed under SK_1). From a security viewpoint, we require that the semantic security of CT^* should hold even if the adversary is given $RK_{1 \to 2}^{CT^*}$ *and the terminal secret key* SK_2. We refer to such an encryption scheme as a puncturable symmetric re-encryption scheme. While the above description only refers to a "single-hop" puncturable re-encryption scheme, we in fact consider a "multi-hop" scheme.

Armed with the above insight, we modify the previous solution template as follows: the garbling of a circuit C consists of $\tilde{U}$ as before. The main difference is that the wire keys $w_C = \{w_{C_1}, \ldots, w_{C_n}\}$ corresponding to the circuit C are now encrypted w.r.t. a puncturable re-encryption scheme. Let SK_0 denote the secret key used to encrypt the wire keys. In order to issue an update encoding for an update $\mathbf{u}_i$, we release (a) a re-encryption key $RK_{i-1 \to i}^{CT}$ that is punctured at ciphertext CT, where CT is the encryption of w_{C_ℓ} w.r.t. SK_{i-1} and ℓ is the position associated with update $\mathbf{u}_i$, along with (b) an encryption of $w_{\bar{C}_\ell}$ w.r.t. SK_i. For the final update L, we simply release the Lth secret key SK_L.

We argue the security of the construction by using the security of the puncturable re-encryption scheme and the garbling scheme (see the technical sections for details). We note, however, that this construction does not hide the location of the updates. Indeed, the correctness of the above scheme requires the evaluator to know the locations that are being updated. To address this, we provide a generic transformation from any UGC scheme (or in fact, any URE scheme) that does not achieve update hiding into one that achieves update hiding. Our transformation uses non-interactive oblivious RAM in the same manner as in [35]. Finally, we note that while the above only discusses single-bit updates, our construction handles multi-bit updates as well.

The only missing piece in the above solution is a construction of a puncturable symmetric re-encryption scheme. We discuss it next.

Puncturable Symmetric Re-encryption from Worst-case Lattice Assumptions. The work of [18] constructs re-encryption schemes from *key homomorphic PRFs*, which have the property that for all x, K_1, and K_2, $\mathsf{PRF}(K_1, x) + \mathsf{PRF}(K_2, x) = \mathsf{PRF}(K_1 + K_2, x)$, where the keys and outputs of the PRF lie in appropriate groups. A secret key for the encryption scheme is simply a PRF key, and the encryption of a message m with secret key K_1 and randomness r is $\mathsf{CT} = (r, m + \mathsf{PRF}(K_1, r))$.

A re-encryption key between between secret keys K_1 and K_2 is simply their difference: $\mathsf{RK}_{1 \to 2} = K_2 - K_1$. The key-homomorphism suggests a natural way to re-encrypt ciphertexts, as $(r, m + \mathsf{PRF}(K_1, r) + \mathsf{PRF}(\mathsf{RK}_{1 \to 2}, r)) = (r, m + (K_2, r))$ is a ciphertext w.r.t K_2. Observe that successful re-encryption of a ciphertext with randomness r relies on the ability to compute $\mathsf{PRF}(\mathsf{RK}_{1 \to 2}, r)$.

We construct *puncturable* proxy re-encryption scheme following the above approach, but instantiated with *constrained* key-homomorphic PRFs [23]. A punctured re-encryption key $\mathsf{RK}_{1 \to 2}^{\mathsf{CT}^*}$ for a ciphertext CT^* with randomness r^* is the PRF key $K_2 - K_1$ punctured at the input r^*. This key, which can be used to evaluate $\mathsf{PRF}(K_2 - K_1, r)$ for all $r \neq r^*$, enables the re-encryption of all ciphertexts except for the ciphertext CT^*.

For security, we require that the semantic security of CT^* holds given both $\mathsf{RK}_{1 \to 2}^{\mathsf{CT}^*}$ and K_2. We reduce to the security of the constrained PRF, which guarantees that $y^* := \mathsf{PRF}(K_2 - K_1, r^*)$ is pseudorandom. The key idea is that (partial information about) y^* can be computed given CT^*, K_2, and (partial information about) the message m.

2.2 Construction of URE

We now shift our focus on building multi-evaluation URE.

Relock-and-Release Mechanism. Recall that the main difference between UGC and URE is that UGC only allows for a single evaluation after a sequence of updates, while URE allows for evaluation after *every* update. As such, the relock-and-*eventual*-release mechanism that we discussed above does not suffice for building URE. Our starting insight is to instead develop a **relock-and-release** mechanism that performs both relocking and release at every step. Intuitively, relocking allows us to "carry over" the updates, while the release mechanism allows us to evaluate the updated randomized encoding at every step.

Starting Idea: Garbled RAM with Persistent Memory. A natural starting approach to implement such a relock-and-release mechanism is via the use of garbled RAMs with persistent memory [37,51]. In a garbled RAM scheme, it is possible to encode a database D_0 and later issue encodings for RAM programs $M_1, \ldots, M_n$. Each RAM program encoding $\widetilde{M_i}$ updates the database encoding from $\widetilde{D}_{i-1}$ to $\widetilde{D}_i$, and outputs the result of some computation on D_i.

Given this description, it is not difficult to see why such a notion is useful for our purpose. Starting from a garbled RAM scheme and a standard randomized encodings scheme without updates [59], we can build a candidate construction of multi-evaluation URE as follows:

- We set the initial database D_0 in garbled RAM to the initial circuit and input pair (C_0, x_0) in the URE scheme. The initial updatable randomized encoding of (C_0, x_0) is an encoding of D_0, computed under garbled RAM scheme, along with an encoding of (C_0, x_0) computed under the standard randomized encoding scheme.
- In order to compute an encoding $\langle \mathbf{u}_i \rangle$ for an update $\mathbf{u}_i$, we compute an encoding $\widetilde{M_i}$ of a machine M_i w.r.t. the garbled RAM scheme where the machine M_i has $\mathbf{u}_i$ hardcoded in it. The machine M_i on input $D_{i-1} = (C_{i-1}, x_{i-1})$ first updates the database to $D_i = (C_i, x_i)$, where $(C_i, x_i) \leftarrow$ Update$(C_{i-1}, x_{i-1}; \mathbf{u}_i)$, and outputs a fresh standard randomized encoding of (C_i, x_i).

Let us inspect the above solution closely; specifically, the complexity of the machine M_i corresponding to an update $\mathbf{u}_i$. Since M_i computes a fresh (standard) randomized encoding "on-the-fly," in order to achieve the necessary efficiency guarantee for URE, we will require that the encoding time for M_i is independent of its running time. Such a garbled RAM scheme is called a *succinct* garbled RAM scheme [13,25]. Furthermore, since the output of M_i consists of a fresh randomized encoding, we will also require that the time of encode M_i is independent of its output length. Such a garbled RAM scheme is referred to as *output-compressing* [3,50].

Recent works [3,50] show that output-compressing succinct garbled RAM (with sub-exponential security) imply indistinguishability obfuscation (iO). Furthermore, the only known constructions for such a garbled RAM scheme are based on iO, which, in turn seems to require sub-exponential hardness assumptions. Our goal, however, is to obtain a solution for URE using *polynomial hardness assumptions*. As such, the above is not a viable solution for us.

Garbled RAM meets Delegation of Computation. Towards that end, our next idea is to instantiate the above approach using a non-succinct garbled RAM scheme where the size of the encoding of a machine M_i depends on the running time and the output length of M_i. Such garbled RAM schemes are known to exist based on only one-way functions. At first, it is not clear how to make this approach work since the efficiency requirements of URE are immediately violated.

Towards that end, our next idea is to delegate the computation of the encoding of M_i to the receiver. We implement this idea by using secret-key functional encryption [19,54,55]. Roughly speaking, the initial encoding of $C_0(x_0)$ now corresponds to a database encoding of $D_0 = (C_0, x_0)$ w.r.t. a non-succinct garbled RAM scheme along with FE functional key for a circuit P that takes as input an update string $\mathbf{u}_i$ and outputs an encoding $\widetilde{M}_i$ of the machine M_i (as described before). Encoding of an update $\mathbf{u}_i$ now corresponds to an FE encryption of $\mathbf{u}_i$.

In order to achieve the necessary efficiency guarantee of URE, we require that the secret-key FE scheme used above is *compact*, i.e., where the running time of the encryption algorithm on a message m is a fixed polynomial in the length of m and the security parameter, and in particular, independent of the size complexity of any function f in the function family supported by the FE scheme. Indeed, if this were not the case, then the encoding time for an update $\mathbf{u}_i$ in the above solution would depend on the size of the circuit C, which in turn depends on the running time and output length of M_i. Therefore, if the FE scheme were not compact, then the efficiency requirements of URE would once again be violated.

As discussed earlier, a secret-key compact FE scheme with polynomial hardness can be built from polynomial hardness assumptions on multilinear maps using using the results of [33] and [4,15].

Challenges in Proving Security. While the above construction seems to achieve correctness, it is not immediately clear how to argue security. Note that the circuit P computed by an FE key in the above construction contains the garbling key of the garbled RAM scheme hardwired inside it. Indeed, this is necessary for it to compute the encodings corresponding to machines M_i as discussed above. In order to leverage security of garbled RAM, one approach is to remove the garbling key from the FE function key. However, in order to maintain functionality, this would require hardwiring the output of P, either in the FE key, or in the FE ciphertext. We cannot afford to hardwire the output in the ciphertext since that would violate the efficiency requirements of URE. Thus, our only option is to hardwire the output in the FE key. Note, however, that in the setting of multiple updates, we have to deal with multiple outputs. In particular, the above approach would require hardwiring all the outputs (one corresponding to each update) in the FE key. Doing so "at once" would require putting a bound on the number of updates.

A better option is to hardwire the outputs "one-at-a-time," analogous to many proofs in the iO literature (see, e.g., [3,23,39]). Implementing this idea, however, would require puncturing the garbling key. Such a notion of key puncturing is not supported by standard garbled RAM schemes.

Using Cascaded Garbled Circuits. Towards that end, we take a step back and revisit our requirements from the garbled RAM scheme. Our first observation is that in the above solution template, machine M_i need not be a RAM since we are already requiring it to read the entire database! Instead, the key property of garbled RAM with persistent memory that is used in the above template is its ability to maintain *updated state* in the form of encoded database.

We now discuss how to implement this property in a more direct manner by "downgrading" the garbled RAM to a cascaded garbled circuit. Along the way, we will also address the security issues discussed above. Very briefly, we modify the above construction as follows: consider a circuit Q_i that has an update string $\mathbf{u}_i$ hardwired in its description. It takes as input (C_{i-1}, x_{i-1}) and outputs two values. The first value is a fresh randomized encoding of $C_i(x_i)$ where $(C_i, x_i) \leftarrow \mathsf{Update}(C_{i-1}, x_{i-1}; \mathbf{u}_i)$, and the second value is a set of wire keys for the string (C_i, x_i) corresponding to a garbling of the circuit Q_{i+1} (that is defined analogously to Q_i). The initial encoding of $C_0(x_0)$ now corresponds to the input wire keys for the string (C_0, x_0) corresponding to a garbling of circuit Q_1 as defined above, as well as an FE key for a function f that takes as input $\mathbf{u}_i$ and outputs a garbling a circuit Q_i. The encoding of an update $\mathbf{u}_i$ now corresponds to an FE encryption of $\mathbf{u}_i$ as before.

We prove the security of the above construction with respect to indistinguishability based security definition. Simulation-based security can be argued via a generic transformation following [26]. Let C_0^0, C_0^1, x be the initial circuits and input submitted by the adversary in the security proof. And let, $(\mathbf{u}_1^0, \mathbf{u}_1^1), \ldots, (\mathbf{u}_q^0, \mathbf{u}_q^1)$ be the tuple of updates. There are two "chains" of updating processes with the 0th chain starting from C_0^0 and 1st chain starting from C_1^1. The ith "bead" on 0th (resp., 1st) chain corresponds to update $\mathbf{u}_i^0$ (resp., $\mathbf{u}_i^1$).

In the security proof, we start with the real experiment where challenge bit 0 is used. That is, the 0th chain is active in the experiment. In the next step, we introduce the 1st chain, along with the already present 0th chain, into the experiment. However even in this step, 0th chain is still active – that is, generating the randomized encoding at every step is performed using the 0th chain. In the next intermediate hybrids, we slowly switch from 0th chain being activated to 1st chain being activated. In the ith intermediate step, the first i beads on 1st chain are active and on the 0th chain, all except the first i beads are active – this means that the first i updated randomized encodings are computed using the 1st chain and the rest of them are computed using 0th chain. At the end of these intermediate hybrids, we have the 1st chain to be active and 0th chain to be completely inactive. At this stage, we can remove the 0th chain and this completes the proof.

The two chains described above are implemented in a sequence of garbled circuits, that we call *cascaded* garbled circuits. That is, every i^{th} garbled circuit in this sequence produces wire keys for the next garbled circuit. Every garbled circuit in this sequence is a result of $\mathsf{ApplyUpd}$ procedure and encapsulates, for some i, the ith beads on both the chains. In order to move from the ith intermediate step to $(i+1)^{th}$ intermediate step, we use the security of garbled circuits. But since these garbled circuits are not given directly, but instead produced by a FE key, we need to make use of security of FE to make this switch work.

3   Preliminaries

We denote the security parameter by λ. The background for randomized encodings and private key (function hiding) functional encryption can be found in the full version [2].

3.1 Updatable Circuits

A boolean circuit C is an directed acyclic graph of in-degree at most 2 with the non-leaf nodes representing $\vee$ (OR), $\wedge$ (AND) and $\neg$ (NOT) gates and the leaf nodes representing the input variables and constants 0 and 1. The nodes with no outgoing edges are designated to be output gates. The size of a circuit $|C|$ is the number of nodes in the graph. Each node is labeled with a different index between 1 and $|C|$. The evaluation of C on input x is performed by first substituting the leaf nodes with the value x and then evaluating gate-by-gate till we reach the output gates. The joint value of all the output gates determine the output of the circuit. Circuit C is said to represent a function $f : \{0,1\}^\lambda \to \{0,1\}^{\ell(\lambda)}$ if for every $x \in \{0,1\}^\lambda$ we have $C(x) = f(x)$. We assume that the class of all boolean circuits for every fixed size $|C|$ and n inputs has an efficient binary representation $\mathsf{binary}(C) \in \{0,1\}^{O(|C|)}$. That is, there is an efficient algorithm that computes $C \mapsto (n, |C|, \mathsf{binary}(C))$, and its inverse.

We define the notion of updatable circuits next. A family of updatable circuits $\mathcal{C}$ has associated with it a class of updates $\mathcal{U}$. Given any circuit $C \in \mathcal{C}$ we can transform this circuit into another circuit $C' \in \mathcal{C}$ with the help of an update $\mathbf{u} \in \mathcal{U}$. The updating process could, for instance, change one of the output gates from $\vee$ to $\neg$, change all the gates to $\wedge$ gates and so on. Formally,

Definition 1 (Updatable Circuits). *Consider a circuit family $\mathcal{C} = \{\mathcal{C}_\lambda\}_{\lambda \in \mathbb{N}}$, where $\mathcal{C}_\lambda$ contains $\mathrm{poly}(\lambda)$-sized boolean circuits $C : \{0,1\}^\lambda \to \{0,1\}^{\ell(\lambda)}$. Consider a set system of strings $\mathcal{U} = \{\mathcal{U}_\lambda\}_{\lambda \in \mathbb{N}}$, where $\mathcal{U}_\lambda$ is a set of strings of length $\mathrm{poly}(\lambda)$. We define $\mathcal{C}$ to be $(\mathsf{Upd}, \mathcal{U})$-updatable if $C' \leftarrow \mathsf{Upd}(C, \mathbf{u} \in \mathcal{U}_\lambda)$ is also a boolean circuit with input domain $\{0,1\}^\lambda$ and output domain $\{0,1\}^{\ell(\lambda)}$.*

The size of update $\mathbf{u}$ could potentially be much smaller than the size of the circuit C. For instance, the length of the instruction to change all the gates in C to $\wedge$ gate is in fact independent of $|C|$.

4 Updatable Randomized Encodings

We define the notion of updatable randomized encodings (URE) next. Since this notion deals with transforming circuits, this notion will be associated to a class of updatable circuits. But to also capture the joint updatability of both the circuit and the input together, we introduce the notion of hardwired circuits below.

Hardwired Circuits. A hardwired circuit, associated to a circuit C and input x, takes no input but upon evaluation yields a fixed output $C(x)$.

We provide the formal definition of hardwired circuits below.

Definition 2 (Hardwired Circuit). *Consider a circuit $C : \{0,1\}^\lambda \to \{0,1\}^{\ell(\lambda)}$ and $x \in \{0,1\}^\lambda$. We define a **hardwired circuit**, denoted by $C[x]$, to be a circuit such that,*

- *it takes no input.*
- *upon evaluation (always) outputs $C(x)$.*

We interchangeably use $C[x]$ to denote the circuit as well as the output $C(x)$ it computes.

Two hardwired circuits $C_0[x_0]$ and $C_1[x_1]$ are *equivalent* if and only if $C_0(x_0) = C_1(x_1)$ and $|C_0| = |C_1|$. If $C_0[x_0]$ and $C_1[x_1]$ are equivalent then they are denoted by $C_0[x_0] \equiv C_1[x_1]$. We can generalize this notion and define a class of hardwired circuits as stated below.

Definition 3 *Consider a circuit family $\mathcal{C} = \{\mathcal{C}_\lambda\}_{\lambda \in \mathbb{N}}$. We define a **hardwired circuit family** $\{\mathcal{C}[X]_\lambda\}_{\lambda \in \mathbb{N}}$ where $\mathcal{C}[X]_\lambda$ comprises of hardwired circuits of fixed input length and is associated with a bijective function $\phi : \mathcal{C}_\lambda \times \{0,1\}^\lambda \to \mathcal{C}[X]_\lambda$ such that if $\phi(C \in \mathcal{C}_\lambda, x) = \mathbf{C}$ then the output of the hardwired circuit $\mathbf{C}$ is $C(x)$.*

We can now talk about updatability of hardwired circuits. Note that this captures joint updating of both the circuit as well as the input hardwired into it.

Definition 4 (Updatable Hardwired Circuits). *Consider a family of hardwired circuits $\{\mathcal{C}[X]_\lambda\}_{\lambda \in \mathbb{N}}$, where $\mathcal{C}[X]_\lambda$ contains $\text{poly}(\lambda)$-sized boolean circuits $C[X] : \bot \to \{0,1\}^{\ell(\lambda)}$. Consider a set system of strings $\mathcal{U} = \{\mathcal{U}_\lambda\}_{\lambda \in \mathbb{N}}$, where $\mathcal{U}_\lambda$ contains a set of strings of length $\text{poly}(\lambda)$. We define $\mathcal{C}[X]$ to be $(\mathsf{Upd}, \mathcal{U})$-**updatable** if $\mathbf{C} \leftarrow \mathsf{Upd}(C[x], \mathbf{u})$, where $C[x] \in \mathcal{C}[X]_\lambda, \mathbf{u} \in \mathcal{U}_\lambda$, then $\mathbf{C}$ is also a hardwired circuit.*

We now proceed to give a formal definition of URE.

Syntax. A scheme $\mathsf{URE} = (\mathsf{Encode}, \mathsf{GenUpd}, \mathsf{ApplyUpd}, \mathsf{Decode})$ for a $(\mathsf{Upd}, \mathcal{U})$-updatable class of circuits $\mathcal{C} = \{\mathcal{C}_\lambda\}_{\lambda \in \mathbb{N}}$ is defined below. We denote $\mathcal{C}[X]$ to be the corresponding updatable *hardwired* circuit family.

- **Encode,** $(\langle C[x]\rangle_{\mathsf{ure}}, \mathsf{st}) \leftarrow \mathsf{Encode}(1^\lambda, C, x)$: On input security parameter λ, circuit $C \in \mathcal{C}_\lambda$, input $x \in \{0,1\}^\lambda$, it outputs the joint encoding $\langle C[x]\rangle_{\mathsf{ure}}$ and state st.
- **Generating Secure Update,** $(\langle \mathbf{u}\rangle_{\mathsf{ure}}, \mathsf{st}') \leftarrow \mathsf{GenUpd}(\mathsf{st}, \mathbf{u})$: On input state st, update $\mathbf{u} \in \mathcal{U}_\lambda$, output the secure update $\langle \mathbf{u}\rangle_{\mathsf{ure}}$ along with the new state st'.
- **Apply Secure Update,** $\langle C'[x']\rangle_{\mathsf{ure}} \leftarrow \mathsf{ApplyUpd}(\langle C[x]\rangle_{\mathsf{ure}}, \langle \mathbf{u}\rangle_{\mathsf{ure}})$: On input randomized encoding $\langle C[x]\rangle_{\mathsf{ure}}$, secure update $\langle \mathbf{u}\rangle_{\mathsf{ure}}$, output the updated randomized encoding $\langle C'[x']\rangle_{\mathsf{ure}}$.
- **Evaluation,** $\alpha \leftarrow \mathsf{Decode}(\langle C[x]\rangle_{\mathsf{ure}})$: On input randomized encoding $\langle C[x]\rangle_{\mathsf{ure}}$, output the decoded value α.

We associate the above scheme with efficiency, correctness and security properties. We first talk about the efficiency requirement. Modeling of correctness and security properties is tricky and we will deal with them in a separate subsection.

Efficiency. We lay out different efficiency properties associated with the above scheme.

- *Encoding Time*: This property requires that the encoding time of (C, x) is significantly "simpler" than computing $C(x)$. The efficiency aspect can be quantified in many ways – in this work, we define encoding to be efficient if the depth of Encode circuit is smaller than C.
- *Secure Update Generation Time*: This property requires that the runtime of GenUpd $(\mathsf{st}, \mathbf{u})$ is $p(\lambda, |\mathbf{u}|)$, where p is an a priori fixed polynomial. In other words, the update generation time is independent of the size of the encoded circuit.
- *State Size*: This property requires that the size of the state maintained by the authority is a fixed polynomial in the security parameter. That is, the size of st output by Encode and GenUpd is always poly(λ) independent of the size of the machines and the update sizes.
- *Secure Update Size*: This property states that the size of the secure version of the update should solely depend on the size of the update. Formally, we have the size of the secure update to be $|\langle \mathbf{u} \rangle_{\mathsf{ure}}| = p(\lambda, |\mathbf{u}|)$, where $(\langle \mathbf{u} \rangle_{\mathsf{ure}}, \mathsf{st}') \leftarrow$ GenUpd $(\mathsf{st}, \mathbf{u})$. Note that any URE scheme that satisfies the above secure update generation time property also satisfies this property.
- *Runtime of Update*: Informally, this property states that the time to update the secure encoding incurs a polynomial overhead in the time to update the plaintext circuit. Formally, the runtime of ApplyUpd$(\langle C[x] \rangle_{\mathsf{ure}}, \langle \mathbf{u} \rangle_{\mathsf{ure}})$ is $p(\lambda, t, |\mathbf{u}|)$, where t is the time taken to execute Upd$(C[x], \mathbf{u})$.

Our constructions achieve a restricted version of the above properties. On the positive side, our construction in Sect. 5 achieves the 'Encoding Time' property and 'Secure Update Generation Time' properties. We use a term to define a URE scheme that satisfies the secure update generation time property – we call it *output compact URE*.

Definition 5 (Output Compact URE). *An URE scheme that is said to be* **output compact** *if it satisfies 'Secure update generation time' property.*

In the case of indistinguishability security, our construction will be output-compact, i.e., the updates will be independent of the output length of the circuit. In the case of simulation-based security, our construction will not achieve output compactness. This is, in fact, inherent and a formal lower bound to this effect can be established along the same lines as in [1,26]. On the flip side, our construction does not satisfy 'Runtime of Update' property.

In the full version [2], we provide a transformation from any URE scheme that satisfies the 'Secure Update Generation Time' property to one that additionally satisfies the 'State Size' property. This transformation uses non-succinct garbled RAMs, and assumes only one-way functions.

4.1 Sequential Updating

We first consider sequential updating process that will be the main focus of this work. For alternate updating processes, refer to the full version [2]. Sequential

Updating process allows for updating a randomized encoding using multiple patches in a sequential manner. That is, given secure updates $\langle \mathbf{u}_1 \rangle_{\text{ure}}, \ldots, \langle \mathbf{u}_\ell \rangle_{\text{ure}}$, we can update a randomized encoding $\langle C[x] \rangle_{\text{ure}}$ by first applying $\langle \mathbf{u}_1 \rangle_{\text{ure}}$ on $\langle C[x] \rangle_{\text{ure}}$ to obtain the updated encoding $\langle C_1[x_1] \rangle_{\text{ure}}$; next we apply $\langle \mathbf{u}_2 \rangle_{\text{ure}}$ on $\langle C_1[x_1] \rangle_{\text{ure}}$ to obtain the updated encoding $\langle C_2[x_2] \rangle_{\text{ure}}$ and so on. After all the updates, we end up with the updated encoding $\langle C_\ell[x_\ell] \rangle_{\text{ure}}$.

Correctness of Sequential Updating. Intuitively, the correctness property states that computing the randomized encoding $\langle C[x] \rangle_{\text{ure}}$, applying the secure updates $\langle \mathbf{u}_1 \rangle_{\text{ure}}, \ldots, \langle \mathbf{u}_\ell \rangle_{\text{ure}}$ sequentially and finally decoding yields the same result as the output of the circuit obtained by updating the hardwired circuit $C[x]$ by applying the updates $\mathbf{u}_1, \ldots, \mathbf{u}_\ell$ sequentially. We give the formal description below.

Consider a circuit $C \in \mathcal{C}_\lambda$, input $x \in \{0,1\}^\lambda$. Consider a vector of updates $\mathbf{U} \in (\mathcal{U}_\lambda)^q$, where $q(\lambda)$ is a polynomial in λ. Consider the following two processes:

Secure updating process:

1. $(\langle C[x] \rangle_{\text{ure}}, \mathsf{st}_0) \leftarrow \mathsf{Encode}\,(1^\lambda, C, x)$.
2. For every $i \in [q]$; $(\langle \mathbf{u}_i \rangle_{\text{ure}}, \mathsf{st}_i) \leftarrow \mathsf{GenUpd}\,(\mathsf{st}_{i-1}, \mathbf{u}_i)$, where $\mathbf{u}_i$ is the ith entry in $\mathbf{U}$.
3. Let $\langle C_0[x_0] \rangle_{\text{ure}} := \langle C[x] \rangle_{\text{ure}}$. For every $i \in [q]$;

$$\langle C_i[x_i] \rangle_{\text{ure}} \leftarrow \mathsf{ApplyUpd}\,(\langle C_{i-1}[x_{i-1}] \rangle_{\text{ure}}, \langle \mathbf{u}_i \rangle_{\text{ure}}).$$

Insecure updating process:

1. Let $(C_0, x_0) := (C, x)$. For every $i \in [q]$, we have $C_i[x_i] \leftarrow \mathsf{Upd}(C_{i-1}[x_{i-1}], \mathbf{u}_i)$. The output of $C_q[x_q]$ is $C_q(x_q)$.

We have,

$$\mathsf{Decode}\Big(\langle C_q[x_q] \rangle_{\text{ure}}\Big) = C_q(x_q)$$

Security of Sequential Updating. We consider two different security notions of sequential updatable RE. First, we consider simulation-based notion and then we consider the weaker indistinguishability-based notion.

Our security notions attempt to capture the intuition that an updateable randomized encoding $\langle C_0[x_0] \rangle_{\text{ure}}$ and a sequence of updates $\langle \mathbf{u}_1 \rangle_{\text{ure}}, \ldots, \langle \mathbf{u}_q \rangle_{\text{ure}}$ should reveal only the outputs $C_0(x_0), C_1(x_1), \ldots C_q(x_q)$ where C_i and X_i are defined as in the preceding correctness definition. In addition to hiding the circuits and inputs as in traditional randomized encodings, a URE additionally hides the sequence of updates. Our URE construction satisfies this update-hiding property.

We could instead consider a relaxed notion, in which updates are partially or wholly revealed (modifying the definitions appropriately). Indeed, this is what we will do in the context of updatable garbled circuits. In the full version [2], we provide a generic transformation from an update-revealing URE scheme to an update-hiding URE scheme, assuming only the existence of one-way functions.

Simulation-Based Security. We adopt the *real world/ideal world* paradigm in formalizing the simulation-based security definition of sequential updatable RE. In the real world, the adversary receives a randomized encoding and encodings of updates. All the encodings are generated honestly as per the description of the scheme. In the ideal world, the adversary is provided simulated randomized encodings and encodings of updates. These simulated encodings are generated as a function of the outputs and in particular, the simulation process does not receive as input the circuit, input or the plaintext updates. A sequential updatable RE scheme is secure if an efficient adversary cannot tell apart real world from the ideal world.

The ideal world is formalized by considering a simulator Sim that runs in probabilistic polynomial time. Sim gets as input the output of circuit $C(x)$, the length of C and produces a simulated randomized encoding. We emphasize that Sim does not receive as input C or x. After this, Sim simulates the update encodings. On input length of update $\mathbf{u}_i$, value $C_i(x_i)$, it generates a simulated encoding of $\mathbf{u}_i$. Here, $C_i(x_i)$ is obtained by first updating $C_{i-1}[x_{i-1}]$ using $\mathbf{u}_i$ to obtain $C_i[x_i]$, whose output is $C_i(x_i)$ and also, $C_0[x_0]$ is initialized with $C[x]$. For this discussion, we consider the scenario where the circuit, input along with the updates are fixed at the beginning of the experiment. This is termed as the *selective* setting. We describe the formal experiment in Fig. 1.

We present the formal security definition below.

Definition 6 (SIM-secure Sequential URE). *A sequential URE scheme* URE *for* (Upd, $\mathcal{U}$)*-updatable class of circuits* $\mathcal{C} = \{\mathcal{C}_\lambda\}_{\lambda \in \mathbb{N}}$ *is said to be* **SIM-secure** *if for every PPT adversary* $\mathcal{A}$*, for every circuit* $C \in \mathcal{C}_\lambda$*, updates* $\mathbf{u}_1, \ldots, \mathbf{u}_q \in \mathcal{U}_\lambda$*, there exists a PPT simulator* Sim *such that the following holds for sufficiently large* $\lambda \in \mathbb{N}$*,*

$$\left| \Pr\left[0 \leftarrow \mathsf{IdealExpt}^{\mathcal{A}} \left(1^\lambda, C, x, \{\mathbf{u}_i\}_{i \in [q]} \right) \right] \right.$$
$$\left. - \Pr\left[0 \leftarrow \mathsf{RealExpt}^{\mathcal{A}} \left(1^\lambda, C, x, \{\mathbf{u}_i\}_{i \in [q]} \right) \right] \right| \leq \mathsf{negl}(\lambda),$$

where negl *is a negligible function.*

We also define indistinguishability-based security notion. We show a transformation from indistinguishability-based security notion to simulation based security notion. This can be found in the full version.

5 Output-Compact URE from FE

In this section, we present our construction of updatable randomized encodings satisfying output compactness properties.

5.1 Construction

Our goal is to construct an updatable randomized encoding scheme, URE = (Encode, GenUpd, ApplyUpd, Decode) for $\mathcal{C}$. The main tools we use in our construction are the following. We refer the reader to the preliminaries for the definitions of these primitives.

$\boxed{\begin{array}{ll}
\underline{\text{IdealExpt}^{\mathcal{A}}(1^\lambda, C, x, \{\mathbf{u}_i\}_{i\in[q]}):} & \underline{\text{RealExpt}^{\mathcal{A}}(1^\lambda, C, x, \{\mathbf{u}_i\}_{i\in[q]}):}
\end{array}}$

IdealExpt$^{\mathcal{A}}(1^\lambda, C, x, \{\mathbf{u}_i\}_{i\in[q]})$:

$(\langle C[x]\rangle_{\mathsf{ure}}, \mathsf{st}_0) \leftarrow \mathsf{Sim}(1^\lambda, 1^{|C|}, C(x))$.

$C_0[x_0] :=$ hardwired circuit of (C, x).

$\forall i \in [q],\ C_i[x_i] \leftarrow \mathsf{Upd}(C_{i-1}[x_{i-1}], \mathbf{u}_i)$.

$\forall i \in [q],\ (\langle \mathbf{u}_i\rangle_{\mathsf{ure}}, \mathsf{st}_i) \leftarrow \mathsf{Sim}(\mathsf{st}_{i-1}, 1^{|\mathbf{u}_i|}, C_i(x_i))$.

Output $\mathcal{A}\Big(\langle C[x]\rangle_{\mathsf{ure}}, \langle \mathbf{u}_1\rangle_{\mathsf{ure}}, \ldots, \langle \mathbf{u}_q\rangle_{\mathsf{ure}}\Big)$.

RealExpt$^{\mathcal{A}}(1^\lambda, C, x, \{\mathbf{u}_i\}_{i\in[q]})$:

$(\langle C[x]\rangle_{\mathsf{ure}}, \mathsf{st}_0) \leftarrow \mathsf{Encode}\left(1^\lambda, C, x\right)$.

$\forall i \in [q],\ (\langle \mathbf{u}\rangle_{\mathsf{ure}}, \mathsf{st}_i) \leftarrow \mathsf{GenUpd}\left(\mathsf{st}_{i-1}, \mathbf{u}_i\right)$.

Output $\mathcal{A}\Big(\langle C[x]\rangle_{\mathsf{ure}}, \langle \mathbf{u}_1\rangle_{\mathsf{ure}}, \ldots, \langle \mathbf{u}_q\rangle_{\mathsf{ure}}\Big)$.

Fig. 1. Selective simulation-based definition of sequential URE.

- Randomized Encoding scheme, $\mathsf{RE} = (\mathsf{RE.Enc}, \mathsf{RE.Dec})$ for the same class of circuits $\mathcal{C}$.
- Compact, Function-private, Single-Key, Secret-key functional encryption (FE) scheme, $\mathsf{FE} = (\mathsf{FE.Setup}, \mathsf{FE.KeyGen}, \mathsf{FE.Enc}, \mathsf{FE.Dec})$.
- Garbling Scheme for circuits, $\mathsf{GC} = (\mathsf{GrbCkt}, \mathsf{GrbInp}, \mathsf{EvalGC})$.

Remark 1. In the full version [2], we show that compact secret-key functional encryption is necessary for our construction of updatable randomized encodings if we believe that learning with errors assumption holds true.

We assume, without loss of generality, that all randomized algorithms require only λ-many random bits. We use the above tools to design the algorithms of URE as given below.

The updatable randomized encoding of (C, x) will consist of a (standard) randomized encoding (C, x) and some additional information necessary to carry out the updating process. This additional information consists of a garbled input encoding of C and x with respect to GC, and a FE secret key for a function that takes as input an update and outputs a garbled circuit mapping C and x to a new randomized encoding and new garbled circuit input encodings of C' and x', which are the updated values. Henceforth, we denote by s the size of the representation of the harwired circuit $C[x]$.

$\underline{\mathsf{Encode}\left(1^\lambda, C, x\right)}$: On input security parameter λ, perform the following operations.

1. Execute the setup of FE, $\mathsf{FE.MSK} \leftarrow \mathsf{FE.Setup}(1^\lambda)$.
2. Compute a functional key $\mathsf{FE.SK}_{\mathsf{RRGarbler}} \leftarrow \mathsf{FE.KeyGen}(\mathsf{FE.MSK}, \mathsf{RRGarbler})$, where RRGarbler is as defined in Fig. 3.
3. In the next step, generate a randomized encoding of input (C, x). That is, compute $\mathsf{RE.Enc}(1^\lambda, C, x)$ to obtain $\langle C[x]\rangle_{\mathsf{re}}$.

4. As stated earlier, let s be the size of the representation of $C[x]$. Generate a garbled circuit input encoding of $(C[x], \perp)$ by evaluating $\langle C[x], \perp \rangle_{gc} \leftarrow$ GrbInp$(C[x], \perp; r_{gc})$, where r_{gc} is the randomness used to garble the input. Here we view $(C[x], \perp)$ as an *input* (to the circuit RelockRelease).
5. Output as the randomized encoding the tuple,

$$\langle C[x] \rangle_{ure} = \Big(\text{FE.SK}_{\text{RRGarbler}}, \ \langle C[x] \rangle_{re}, \ \langle C[x], \perp \rangle_{gc} \Big)$$

and set the state to be $st = (\text{FE.MSK}, r_{gc})$.

GenUpd $(st_i, \ \mathbf{u}_{i+1})$: It takes as input the state $st_i = (\text{FE.MSK}, r_{gc,i})$ and update $\mathbf{u}_{i+1}$.

1. Sample random coins $r_{re,i+1}$ and $r_{gc,i+1}$. Let mode $= 0$.
2. Generate the FE ciphertext,

$$\text{CT}_{i+1} \leftarrow \text{FE.Enc}(\text{FE.MSK}, (\mathbf{u}_{i+1}, \ \perp, \ r_{gc,i}, \ r_{gc,i+1}, \ r_{re,i+1}, \ \text{mode}))$$

3. Set the new state $st_{i+1} = (\text{FE.MSK}, r_{gc,i+1})$.
4. Output $\langle \mathbf{u}_{i+1} \rangle_{ure} = \text{CT}_{i+1}$.

ApplyUpd $(\langle C_i[x_i] \rangle_{ure}, \langle \mathbf{u}_{i+1} \rangle_{ure})$: On input circuit encoding $\langle C_i[x_i] \rangle_{ure}$ and update encoding $\langle \mathbf{u}_{i+1} \rangle_{ure} = \text{CT}_{i+1}$, execute the following (Fig. 2).

1. Parse the circuit encoding as:

$$\langle C_i[x_i] \rangle_{ure} = \Big(\text{FE.SK}_{\text{RRGarbler}}, \ \langle C_i[x_i] \rangle_{re}, \ \langle C_i[x_i], \perp \rangle_{gc} \Big)$$

RelockRelease$_{i+1}$

Input: $C_i^0[x_i^0], C_i^1[x_i^1]$
Hard-coded values: $\mathbf{u}_{i+1}^0, \mathbf{u}_{i+1}^1, r_{gc,i+1}, r_{re,i+1}$, and mode

- Update both the hardwired circuits $C_i^b[x_i^b]$ using $\mathbf{u}_{i+1}^b$:

$$C_{i+1}^b[x_{i+1}^b] \leftarrow \text{Upd}(C_i^b[x_i^b], \mathbf{u}_{i+1}^b)$$

- Encode the updated hardwired circuit $C_{i+1}^{\text{mode}}[x_{i+1}^{\text{mode}}]$:

$$\langle C_{i+1}^{\text{mode}}[x_{i+1}^{\text{mode}}] \rangle_{re} \leftarrow \text{RE.Enc}\Big(C_{i+1}^{\text{mode}}[x_{i+1}^{\text{mode}}]; \ r_{re,i+1}\Big)$$

- Compute the randomized encoding of the input $(C_{i+1}^0[x_{i+1}^0], C_{i+1}^1[x_{i+1}^1])$:

$$\langle C_{i+1}^0[x_{i+1}^0], C_{i+1}^1[x_{i+1}^1] \rangle_{gc} \leftarrow \text{GrbInp}\Big((C_{i+1}^0[x_{i+1}^0], C_{i+1}^1[x_{i+1}^1]); \ r_{gc,i+1}\Big)$$

- Output $\Big(\langle C_{i+1}^{\text{mode}}[x_{i+1}^{\text{mode}}] \rangle_{re}, \ \langle C_{i+1}^0[x_{i+1}^0], C_{i+1}^1[x_{i+1}^1] \rangle_{gc}\Big)$

Fig. 2. RelockRelease$_{i+1}$.

RRGarbler

Input: $(\mathbf{u}_{i+1}^0, \mathbf{u}_{i+1}^1, r_{\mathsf{gc},i}, r_{\mathsf{gc},i+1}, r_{\mathsf{re},i+1}, \mathsf{mode})$

Compute the garbled circuit encoding of $\mathsf{RelockRelease}_{i+1}$, *which is defined in Figure 2:*

$$\langle \mathsf{RelockRelease}_{i+1} \rangle_{\mathsf{gc}} \leftarrow \mathsf{GrbCkt}\left(\mathsf{RelockRelease}_{i+1}; \ r_{\mathsf{gc},i} \right)$$

Output $\langle \mathsf{RelockRelease}_{i+i} \rangle_{\mathsf{gc}}$.

Fig. 3. RRGarbler.

2. Execute the FE decryption, $\mathsf{FE.Dec}(\mathsf{FE.SK}_{\mathsf{RRGarbler}}, \mathsf{CT}_{i+1})$ to obtain:

$$\langle \mathsf{RelockRelease}_{i+1} \rangle_{\mathsf{gc}}.$$

3. Execute the decode algorithm of the garbling scheme,

$$(\langle C_{i+1}[x_{i+1}] \rangle_{\mathsf{re}}, \langle C_{i+1}[x_{i+1}], \perp \rangle_{\mathsf{gc}}) \leftarrow \mathsf{EvalGC}(\langle \mathsf{RelockRelease}_{i+1} \rangle_{\mathsf{gc}}, \langle C_i[x_i] \rangle_{\mathsf{gc}})$$

That is, the decode algorithm outputs the randomized encoding of updated hardwired circuit $C_{i+1}[x_{i+1}]$ and also wire keys of $(C_{i+1}[x_{i+1}], \perp)$ that will be input to the next level garbled circuit.

4. Output $\left(\mathsf{FE.SK}_{\mathsf{RRGarbler}}, \langle C_{i+1}[x_{i+1}] \rangle_{\mathsf{re}}, \langle C_{i+1}[x_{i+1}], \perp \rangle_{\mathsf{gc}} \right)$.

<u>Decode</u> $(\langle C_i[x_i] \rangle_{\mathsf{ure}})$: On input encoding

$$\langle C_i[x_i] \rangle_{\mathsf{ure}} = (\mathsf{FE.SK}_{\mathsf{RRGarbler}}, \langle C_i[x_i] \rangle_{\mathsf{re}}, \langle C_i[x_i], \perp \rangle_{\mathsf{gc}}),$$

decode the encoding $\langle C_i[x_i] \rangle_{\mathsf{re}}$ by executing $\mathsf{RE.Dec}(\langle C_i[x_i] \rangle_{\mathsf{re}})$ to obtain α. Output the value α.

In the full version [2], we show that the above scheme satisfies all the properties of an updatable randomized encodings scheme.

References

1. Agrawal, S., Gorbunov, S., Vaikuntanathan, V., Wee, H.: Functional encryption: new perspectives and lower bounds. In: Canetti, R., Garay, J.A. (eds.) CRYPTO 2013, Part II. LNCS, vol. 8043, pp. 500–518. Springer, Heidelberg (2013). doi:10.1007/978-3-642-40084-1_28
2. Ananth, P., Cohen, A., Jain, A.: Cryptography with updates. Cryptology ePrint Archive, Report 2016/934 (2016). http://eprint.iacr.org/2016/934
3. Ananth, P., Jain, A.: Indistinguishability obfuscation from compact functional encryption. In: Gennaro, R., Robshaw, M. (eds.) CRYPTO 2015. LNCS, vol. 9215, pp. 308–326. Springer, Heidelberg (2015). doi:10.1007/978-3-662-47989-6_15

4. Ananth, P., Jain, A., Sahai, A.: Indistinguishability obfuscation from functional encryption for simple functions. Technical report, Cryptology ePrint Archive, Report 2015/730 (2015)
5. Ananth, P., Jain, A., Sahai, A.: Patchable indistinguishability obfuscation: io for evolving software. IACR Cryptology ePrint Archive, 2015:1084 (2015)
6. Applebaum, B., Ishai, Y., Kushilevitz, E.: Cryptogaphy in nc0. SIAM J. Comput. **36**(4), 845–888 (2007)
7. Ateniese, G., Burns, R., Curtmola, R., Herring, J., Kissner, L., Peterson, Z., Song, D.: Provable data possession at untrusted stores. In: Proceedings of the 14th ACM Conference on Computer and Communications Security, pp. 598–609. ACM (2007)
8. Ateniese, G., Fu, K., Green, M., Hohenberger, S.: Improved proxy re-encryption schemes with applications to secure distributed storage. In: Proceedings of the Network and Distributed System Security Symposium, NDSS 2005, San Diego, California, USA (2005)
9. Barak, B., Goldreich, O., Impagliazzo, R., Rudich, S., Sahai, A., Vadhan, S., Yang, K.: On the (im)possibility of obfuscating programs. In: Kilian, J. (ed.) CRYPTO 2001. LNCS, vol. 2139, pp. 1–18. Springer, Heidelberg (2001). doi:10.1007/3-540-44647-8_1
10. Bellare, M., Goldreich, O., Goldwasser, S.: Incremental cryptography: the case of hashing and signing. In: Desmedt, Y.G. (ed.) CRYPTO 1994. LNCS, vol. 839, pp. 216–233. Springer, Heidelberg (1994). doi:10.1007/3-540-48658-5_22
11. Bellare, M., Goldreich, O., Goldwasser, S.: Incremental cryptography and application to virus protection. In: Proceedings of the Twenty-Seventh Annual ACM Symposium on Theory of Computing, pp. 45–56. ACM (1995)
12. Bellare, M., Micciancio, D.: A new paradigm for collision-free hashing: incrementality at reduced cost. In: Fumy, W. (ed.) EUROCRYPT 1997. LNCS, vol. 1233, pp. 163–192. Springer, Heidelberg (1997). doi:10.1007/3-540-69053-0_13
13. Bitansky, N., Garg, S., Lin, H., Pass, R., Telang, S.: Succinct randomized encodings and their applications. In: STOC (2015)
14. Bitansky, N., Nishimaki, R., Passelègue, A., Wichs, D.: From cryptomania to obfustopia through secret-key functional encryption. In: Hirt, M., Smith, A. (eds.) TCC 2016. LNCS, vol. 9986, pp. 391–418. Springer, Heidelberg (2016). doi:10.1007/978-3-662-53644-5_15
15. Bitansky, N., Vaikuntanathan, V.: Indistinguishability obfuscation from functional encryption. In: 2015 IEEE 56th Annual Symposium on Foundations of Computer Science (FOCS), pp. 171–190. IEEE (2015)
16. Blaze, M., Bleumer, G., Strauss, M.: Divertible protocols and atomic proxy cryptography. In: Nyberg, K. (ed.) EUROCRYPT 1998. LNCS, vol. 1403, pp. 127–144. Springer, Heidelberg (1998). doi:10.1007/BFb0054122
17. Boneh, D., Freeman, D.M.: Homomorphic signatures for polynomial functions. In: Paterson, K.G. (ed.) EUROCRYPT 2011. LNCS, vol. 6632, pp. 149–168. Springer, Heidelberg (2011). doi:10.1007/978-3-642-20465-4_10
18. Boneh, D., Lewi, K., Montgomery, H., Raghunathan, A.: Key homomorphic PRFs and their applications. In: Canetti, R., Garay, J.A. (eds.) CRYPTO 2013. LNCS, vol. 8042, pp. 410–428. Springer, Heidelberg (2013). doi:10.1007/978-3-642-40041-4_23
19. Boneh, D., Sahai, A., Waters, B.: Functional encryption: definitions and challenges. In: Ishai, Y. (ed.) TCC 2011. LNCS, vol. 6597, pp. 253–273. Springer, Heidelberg (2011). doi:10.1007/978-3-642-19571-6_16

20. Boneh, D., Waters, B.: Conjunctive, subset, and range queries on encrypted data. In: Vadhan, S.P. (ed.) TCC 2007. LNCS, vol. 4392, pp. 535–554. Springer, Heidelberg (2007). doi:10.1007/978-3-540-70936-7_29

21. Boneh, D., Waters, B.: Constrained pseudorandom functions and their applications. In: Sako, K., Sarkar, P. (eds.) ASIACRYPT 2013. LNCS, vol. 8270, pp. 280–300. Springer, Heidelberg (2013). doi:10.1007/978-3-642-42045-0_15

22. Boyle, E., Goldwasser, S., Ivan, I.: Functional Signatures and pseudorandom functions. In: Krawczyk, H. (ed.) PKC 2014. LNCS, vol. 8383, pp. 501–519. Springer, Heidelberg (2014). doi:10.1007/978-3-642-54631-0_29

23. Brakerski, Z., Vaikuntanathan, V.: Constrained key-homomorphic PRFs from standard lattice assumptions. In: Dodis, Y., Nielsen, J.B. (eds.) TCC 2015. LNCS, vol. 9015, pp. 1–30. Springer, Heidelberg (2015). doi:10.1007/978-3-662-46497-7_1

24. Buonanno, E., Katz, J., Yung, M.: Incremental unforgeable encryption. In: Matsui, M. (ed.) FSE 2001. LNCS, vol. 2355, pp. 109–124. Springer, Heidelberg (2002). doi:10.1007/3-540-45473-X_9

25. Canetti, R., Holmgren, J., Jain, A., Vaikuntanathan, V.: Indistinguishability obfuscation of iterated circuits and RAM programs. In: STOC (2015)

26. De Caro, A., Iovino, V., Jain, A., O'Neill, A., Paneth, O., Persiano, G.: On the achievability of simulation-based security for functional encryption. In: Canetti, R., Garay, J.A. (eds.) CRYPTO 2013. LNCS, vol. 8043, pp. 519–535. Springer, Heidelberg (2013). doi:10.1007/978-3-642-40084-1_29

27. Chandran, N., Kanukurthi, B., Ostrovsky, R.: Locally updatable and locally decodable codes. In: Lindell, Y. (ed.) TCC 2014. LNCS, vol. 8349, pp. 489–514. Springer, Heidelberg (2014). doi:10.1007/978-3-642-54242-8_21

28. Chase, M., Kohlweiss, M., Lysyanskaya, A., Meiklejohn, S.: Malleable proof systems and applications. In: Pointcheval, D., Johansson, T. (eds.) EUROCRYPT 2012. LNCS, vol. 7237, pp. 281–300. Springer, Heidelberg (2012). doi:10.1007/978-3-642-29011-4_18

29. Chase, M., Kohlweiss, M., Lysyanskaya, A., Meiklejohn, S.: Succinct malleable NIZKs and an application to compact shuffles. In: Sahai, A. (ed.) TCC 2013. LNCS, vol. 7785, pp. 100–119. Springer, Heidelberg (2013). doi:10.1007/978-3-642-36594-2_6

30. Dachman-Soled, D., Liu, F.-H., Shi, E., Zhou, H.-S.: Locally decodable and updatable non-malleable codes and their applications. In: Dodis, Y., Nielsen, J.B. (eds.) TCC 2015. LNCS, vol. 9014, pp. 427–450. Springer, Heidelberg (2015). doi:10.1007/978-3-662-46494-6_18

31. Fischlin, M.: Incremental cryptography and memory checkers. In: Fumy, W. (ed.) EUROCRYPT 1997. LNCS, vol. 1233, pp. 393–408. Springer, Heidelberg (1997). doi:10.1007/3-540-69053-0_27

32. Garg, S., Gentry, C., Halevi, S., Raykova, M., Sahai, A., Waters, B.: Candidate indistinguishability obfuscation and functional encryption for all circuits. In: 54th Annual IEEE Symposium on Foundations of Computer Science, FOCS 2013, Berkeley, CA, USA, 26–29 October 2013, pp. 40–49. IEEE Computer Society (2013)

33. Garg, S., Gentry, C., Halevi, S., Zhandry, M.: Fully secure functional encryption without obfuscation. IACR Cryptology ePrint Archive, 2014:666 (2014)

34. Garg, S., Gentry, C., Sahai, A., Waters, B.: Witness encryption and its applications. In: Proceedings of the Forty-Fifth Annual ACM Symposium on Theory of Computing, pp. 467–476. ACM (2013)

35. Garg, S., Pandey, O.: Incremental program obfuscation. IACR Cryptology ePrint Archive, 2015:997 (2015)

36. Gentry, C.: Fully homomorphic encryption using ideal lattices. In: Proceedings of the 41st Annual ACM Symposium on Theory of Computing, STOC 2009, Bethesda, MD, USA, May 31-June 2, 2009, pp. 169–178 (2009)
37. Gentry, C., Halevi, S., Lu, S., Ostrovsky, R., Raykova, M., Wichs, D.: Garbled RAM revisited. In: Nguyen, P.Q., Oswald, E. (eds.) EUROCRYPT 2014. LNCS, vol. 8441, pp. 405–422. Springer, Heidelberg (2014). doi:10.1007/978-3-642-55220-5_23
38. Gentry, C., Lewko, A., Waters, B.: Witness encryption from instance independent assumptions. In: Garay, J.A., Gennaro, R. (eds.) CRYPTO 2014. LNCS, vol. 8616, pp. 426–443. Springer, Heidelberg (2014). doi:10.1007/978-3-662-44371-2_24
39. Gentry, C., Lewko, A.B., Sahai, A., Waters, B.: Indistinguishability obfuscation from the multilinear subgroup elimination assumption. IACR Cryptology ePrint Archive, 2014:309 (2014)
40. Goldreich, O., Goldwasser, S., Micali, S.: How to construct random functions. J. ACM (JACM) 33(4), 792–807 (1986)
41. Goldwasser, S., Dov Gordon, S., Goyal, V., Jain, A., Katz, J., Liu, F.-H., Sahai, A., Shi, E., Zhou, H.-S.: Multi-input functional encryption. In: Nguyen, P.Q., Oswald, E. (eds.) EUROCRYPT 2014. LNCS, vol. 8441, pp. 578–602. Springer, Heidelberg (2014). doi:10.1007/978-3-642-55220-5_32
42. Goldwasser, S., Kalai, Y.T., Popa, R.A., Vaikuntanathan, V., Zeldovich, N.: Reusable garbled circuits and succinct functional encryption. In: Boneh, D., Roughgarden, T., Feigenbaum, J. (eds.) Symposium on Theory of Computing Conference, STOC 2013, Palo Alto, CA, USA, 1–4 June 2013, pp. 555–564. ACM (2013)
43. Gorbunov, S., Vaikuntanathan, V., Wee, H.: Predicate encryption for circuits from LWE. In: Gennaro, R., Robshaw, M. (eds.) CRYPTO 2015. LNCS, vol. 9216, pp. 503–523. Springer, Heidelberg (2015). doi:10.1007/978-3-662-48000-7_25
44. Gorbunov, S., Vaikuntanathan, V., Wichs, D.: Leveled fully homomorphic signatures from standard lattices. In: Proceedings of the Forty-Seventh Annual ACM on Symposium on Theory of Computing, pp. 469–477. ACM (2015)
45. Goyal, V., Pandey, O., Sahai, A., Waters, B.: Attribute-based encryption for fine-grained access control of encrypted data. In: Proceedings of the 13th ACM Conference on Computer and Communications Security, CCS 2006, Alexandria, VA, USA, October 30 - November 3, 2006, pp. 89–98 (2006)
46. Ishai, Y., Kushilevitz, E.: Randomizing polynomials: a new representation with applications to round-efficient secure computation. In: Proceedings of 41st Annual Symposium on Foundations of Computer Science, pp. 294–304. IEEE (2000)
47. Katz, J., Sahai, A., Waters, B.: Predicate encryption supporting disjunctions, polynomial equations, and inner products. In: Smart, N. (ed.) EUROCRYPT 2008. LNCS, vol. 4965, pp. 146–162. Springer, Heidelberg (2008). doi:10.1007/978-3-540-78967-3_9
48. Kiayias, A., Papadopoulos, S., Triandopoulos, N., Zacharias, T.: Delegatable pseudorandom functions and applications. In: Proceedings of the 2013 ACM SIGSAC Conference on Computer & Communications Security, pp. 669–684. ACM (2013)
49. Lin, H., Pass, R., Seth, K., Telang, S.: Indistinguishability obfuscation with nontrivial efficiency. In: Cheng, C.-M., Chung, K.-M., Persiano, G., Yang, B.-Y. (eds.) PKC 2016, Part II. LNCS, vol. 9615, pp. 447–462. Springer, Heidelberg (2016). doi:10.1007/978-3-662-49387-8_17
50. Lin, H., Pass, R., Seth, K., Telang, S.: Output-compressing randomized encodings and applications. In: Kushilevitz, E., Malkin, T. (eds.) TCC 2016. LNCS, vol. 9562, pp. 96–124. Springer, Heidelberg (2016). doi:10.1007/978-3-662-49096-9_5

51. Lu, S., Ostrovsky, R.: How to garble RAM programs? In: Johansson, T., Nguyen, P.Q. (eds.) EUROCRYPT 2013. LNCS, vol. 7881, pp. 719–734. Springer, Heidelberg (2013). doi:10.1007/978-3-642-38348-9_42

52. Micciancio, D.: Oblivious data structures: applications to cryptography. In: Proceedings of the Twenty-Ninth Annual ACM Symposium on Theory of Computing, pp. 456–464. ACM (1997)

53. Mironov, I., Pandey, O., Reingold, O., Segev, G.: Incremental deterministic public-key encryption. In: Pointcheval, D., Johansson, T. (eds.) EUROCRYPT 2012. LNCS, vol. 7237, pp. 628–644. Springer, Heidelberg (2012). doi:10.1007/978-3-642-29011-4_37

54. O'Neill, A.: Definitional issues in functional encryption. IACR Cryptology ePrint Archive, 2010:556 (2010)

55. Sahai, A., Waters, B.: Fuzzy identity-based encryption. In: Cramer, R. (ed.) EUROCRYPT 2005. LNCS, vol. 3494, pp. 457–473. Springer, Heidelberg (2005). doi:10.1007/11426639_27

56. Sahai, A., Waters, B.: How to use indistinguishability obfuscation: deniable encryption, and more. In: Shmoys, D.B. (ed.) Symposium on Theory of Computing, STOC 2014, New York, NY, USA, May 31–June 03, pp. 475–484. ACM (2014)

57. Shi, E., Bethencourt, J., Hubert Chan, T.H., Song, D., Perrig, A.: Multidimensional range query over encrypted data. In: 2007 IEEE Symposium on Security and Privacy (SP 2007), pp. 350–364. IEEE (2007)

58. Waters, B.: A punctured programming approach to adaptively secure functional encryption. In: Gennaro, R., Robshaw, M. (eds.) CRYPTO 2015. LNCS, vol. 9216, pp. 678–697. Springer, Heidelberg (2015). doi:10.1007/978-3-662-48000-7_33

59. Yao, A.C.-C.: How to generate and exchange secrets (extended abstract). In: FOCS, pp. 162–167 (1986)

60. Zhandry, M.: How to avoid obfuscation using witness PRFs. In: Kushilevitz, E., Malkin, T. (eds.) TCC 2016. LNCS, vol. 9563, pp. 421–448. Springer, Heidelberg (2016). doi:10.1007/978-3-662-49099-0_16

Fixing Cracks in the Concrete:
Random Oracles with Auxiliary Input, Revisited

Yevgeniy Dodis[1], Siyao Guo[2($\boxtimes$)], and Jonathan Katz[3]

[1] New York University, New York, USA
dodis@cs.nyu.edu
[2] Simons Institute, UC Berkeley, Berkeley, USA
siyao.guo@berkeley.edu
[3] University of Maryland, College Park, USA
jkatz@cs.umd.edu

Abstract. We revisit the security of cryptographic primitives in the random-oracle model against attackers having a bounded amount of *auxiliary information* about the random oracle. This situation arises most naturally when an attacker carries out offline preprocessing to generate state (namely, auxiliary information) that is later used as part of an on-line attack, with perhaps the best-known example being the use of rainbow tables for function inversion. The resulting model is also critical to obtain accurate bounds against *non-uniform* attackers when the random oracle is instantiated by a concrete hash function.

Unruh (Crypto 2007) introduced a generic technique (called presampling) for analyzing security in this model: a random oracle for which S bits of arbitrary auxiliary information can be replaced by a random oracle whose value is fixed in some way on P points; the two are distinguishable with probability at most $O(\sqrt{ST/P})$ by attackers making at most T oracle queries. Unruh conjectured that the distinguishing advantage could be made negligible for a sufficiently large polynomial P. We show that Unruh's conjecture is *false* by proving that the distinguishing probability is at least $\Omega(ST/P)$.

Faced with this negative general result, we establish new security bounds, — which are nearly optimal and beat pre-sampling bounds, — for specific applications of random oracles, including one-way functions, pseudorandom functions/generators, and message authentication codes. We also explore the effectiveness of *salting* as a mechanism to defend against offline preprocessing, and give quantitative bounds demonstrating that salting provably helps in the context of one-wayness, collision-resistance, pseudorandom generators/functions, and message authentication codes. In each case, using (at most) n bits of salt, where n is the length of the secret key, we get the same security $O(T/2^n)$ in the random oracle model with auxiliary input as we get without auxiliary input.

Y. Dodis—Work done while visiting the University of Maryland. Partially supported by gifts from VMware Labs and Google, and NSF grants 1619158, 1319051, 1314568.
S. Guo—Work done while at NYU and visiting the University of Maryland.
J. Katz—Work supported in part by NSF award #1223623.

J.-S. Coron and J.B. Nielsen (Eds.): EUROCRYPT 2017, Part II, LNCS 10211, pp. 473–495, 2017.
DOI: 10.1007/978-3-319-56614-6_16

At the heart of our results is the compression technique of Gennaro and Trevisan, and its extensions by De, Trevisan and Tulsiani.

1   Introduction

The random-oracle model [4] often provides a simple and elegant way of analyzing the concrete security of cryptographic schemes based on hash functions. To take a canonical example, consider (naïve) password hashing where a password pw is stored as $H(pw)$, for H a cryptographic hash function, and we are interested in the difficulty of recovering pw from $H(pw)$ (i.e., we are interested in understanding the *one-wayness* of H). It seems difficult to formalize a concrete assumption about H that would imply the difficulty of recovering pw for *all* high-entropy distributions on pw; it would be harder still to come up with a natural assumption implying that for all distributions on pw with min-entropy k, recovering pw requires $O(2^k)$ work. If we model H as a random oracle, however, then both these statements can be proven easily—and this matches the best known attacks for many cryptographic hash functions.

Importantly, the above discussion assumes that no preprocessing is done. That is, we imagine an attacker who does no work prior to being given $H(pw)$ or, more formally, we imagine that the attacker is fixed before the random oracle H is chosen. In that case, the only way an attacker can learn information about H is by making explicit queries to an oracle for H, and the above-mentioned bounds hold. In practice, however, H is typically a standardized hash function that is known in advance, and *offline preprocessing attacks*—during which the attacker can query and store arbitrary information about H—can be a significant threat.

Concretely, let $H : [N] \to [N]$ and assume that pw is uniform in $[N]$. The obvious attack to recover pw from $H(pw)$ is an exhaustive-search attack which uses time $T = N$ in the online phase (equating time with the number of queries to H) to recover pw. But an attacker could also generate the entire function table for H during an offline preprocessing phase; then, given $H(pw)$ in the on-line phase, the attacker can recover pw in $O(1)$ time using a table lookup. The data structure generated during the offline phase requires $S = O(N)$ space (ignoring $\log N$ factors), but Hellman [11] showed a more clever construction of a data structure which, in particular, gives an attack using $S = T = O(N^{2/3})$ (see [12, Sect. 5.4.3] for a self-contained description). *Rainbow tables* implementing this approach along with later improvements (most notably by Oechslin [14]), are widely used in practice, and must be taken into account in any practical analysis of password security. Further work has explored improving these result and proving rigorous versions of them, as well as showing bounds on how well such attacks can perform [2,6,8,9,14,18].

The above discussion in the context of function inversion gives a practical example of where *auxiliary information* about a random oracle (in this case, in the form of rainbow tables generated using the random oracle) can quantitatively change the security of a given application that uses the random oracle. For a more dramatic (but less practical) example, consider the case of collision finding.

Given a random function $H : [N] \to [N]$, one can show that $O(\sqrt{N})$ queries are needed in order to find a collision in H (i.e., distinct points x, x' with $H(x) = H(x')$). But clearly we can find a collision in H during an offline pre-processing phase and store that collision using $O(1)$ space, after which it is trivial to output that collision in an online phase in $O(1)$ time. The conclusion is that in settings where offline preprocessing is a possibility, security proofs in the random-oracle model must be interpreted carefully. (We refer the reader to [5,16], as well as many of the references below, for further discussion).

From a different viewpoint, another motivation for studying auxiliary information comes from the desire for obtaining accurate security bounds against *non-uniform attackers* when instantiating random oracle by a concrete hash function. Indeed, non-uniform attackers are allowed to have some arbitrary 'advice' before attacking the system. Translated to the random oracle model, this would require the attacker to be able to compute some arbitrary function of the *entire random oracle*, which cannot be done using only bounded number T of oracle queries. This mismatch already led to considerable confusion among both theoreticians and practitioners. We refer to [5,15] for some in-depth discussion, here only mentioning two most well-known examples. (1) In the standard (non-uniform) model, no single function can be collision-resistant, while a single random oracle is trivially collision-resistant (without preprocessing); this is why in the standard model one considers a *family* of CRHFs, whose public key (which we call *salt*) is chosen *after* the attacker gets his non-uniform advice. To the best of our knowledge, prior to our work no meaningful CRHF bound was given for *salted* random oracle if (salt-independent) preprocessing was allowed. (2) In the standard (non-uniform) model, it is well known [1,5,7] that no pseudorandom generator (PRG) $H(x)$ can have security better than $2^{-n/2}$ even against linear-time attackers, where n is the seed-length of x. In contrast, an expanding random oracle can be trivially shown to be $(T/2^n)$-secure PRG in the traditional random oracle model, easily surpassing the $2^{-n/2}$ barrier in the standard model (even for huge T up to $2^{n/2}$, let alone polynomial T).

Random Oracle with Auxiliary Input. While somewhat different, the two motivating applications above effectively reduce to the following identical extension of the traditional random oracle model (ROM). A (computationally unbounded) attacker A can compute arbitrary S bits of information $z = z(\mathcal{O})$ about the random oracle $\mathcal{O}$ *before* attacking the system, and then use additional T *oracle* queries to $\mathcal{O}$ *during* the attack. Following Unruh [16], we call this the *Random Oracle Model with Auxiliary Input* (ROM-AI), and this is the model we thoroughly study in this work. As we mentioned, while the traditional ROM only uses one parameter T, the ROM-AI is parameterized by two parameters, S and T which roughly correspond to space (during off-line pre-processing) and time (during on-line attack). For the application to non-uniform security, one can also use the ROM-AI to get good estimates for non-uniform security against (non-uniform) circuits of size C by setting $S = T = C$.[1]

---

[1] Since circuit of size C can encode up to $S = \Omega(C)$ bits of information about a given hash function H, as well as evaluate it close to $T = \Omega(C)$ times, assuming H is efficient.

1.1   Handling Random Oracles with Auxiliary Input

Broadly speaking, there are three ways one can address the issue of preprocessing/auxiliary input in the random-oracle model: (1) by using a generic approach to analyze existing or proposed schemes, (2) by using an application-specific approach to analyze an existing or proposed scheme, or (3) by modifying existing schemes in an attempt to defeat preprocessing/non-uniform attacks. We discuss limited prior work on these three approaches below, before stating our results.

A generic approach. Unruh [16] was the first to propose a generic approach for dealing with auxiliary input in the random-oracle model. We give an informal overview of his results (a formal statement is given in Sect. 2). Say we wish to bound the success probability ϵ (in some experiment) of an online attacker making T random-oracle queries, and relying on S bits of (arbitrary) auxiliary information about the random oracle. Unruh showed that it suffices to analyze the success probability $\epsilon'(P)$ of the attack in the presence of a "pre-sampled" random oracle that is chosen uniformly subject to its values being fixed in some adversarial way on P adversarial points (where P is a parameter), and no other auxiliary information is given; ϵ is then bounded by $\epsilon'(P) + O(\sqrt{ST/P})$, while P is then chosen optimally as to balance out the resulting two terms (see an example below).

This is an impressive result, but it falls short of what one might hope for. In particular, P must be super-polynomial in order to make the "security loss" $O(\sqrt{ST/P})$ negligible, but in many applications if P is too large then the bound $\epsilon'(P)$ one can prove on an attacker's success probability in the presence of a "pre-sampled" random oracle with P fixed points becomes too high. Unruh conjectured that his bound was not tight, and that it might be possible to bound the "security loss" by a negligible quantity for P a sufficiently large polynomial.

An application-specific approach. Given that the generic approach might lead to very sub-optimal bounds, one might hope to develop a much tighter application-specific approach to get concrete bounds. To the best of our knowledge, no such work was done for the random oracle model with preprocessing. Indirectly, however, De et al. [6] adapted the beautiful compression "compression paradigm" introduced by Gennaro and Trevisan [9,10] to show nearly tight security bounds for inverting inverting one-way *permutations* as well as *specific* PRGs (based on one-way permutations and hardcore bits). This was done not for the sake of analyzing security of these constructions,[2] but rather to show limitations of generic inversion/distinguishing attacks *all* one-way functions or PRGs. Still, this elegant theoretical approach suggests that application-specific techniques, such as the compression paradigm, might be useful in the analysis of schemes based on real-world hash functions, such as SHA.

"Salting." Even with optimal application-specific techniques, we have already discussed how preprocessing attacks can be effective for tasks like function inversion

---

[2] For which we currently have no real-world candidates, since we do not have any candidates for efficient *uninvertible* "random permutations".

and collision finding, as well as non-trivial distinguishing attacks against pseudo-random generators/functions.

A natural defense against preprocessing attacks, which has been explicitly suggested [13] and is widely used to defeat such attacks in the context of password hashing, is to use *salting*. Roughly, this involves choosing a random but public value a and including it in the input to the hash function. Thus, in the context of password hashing we would choose a uniform salt a and store $(a, H(a, pw))$; in the context of collision-resistant hashing we would choose and publish a and then look at the hardness of finding collisions in the function $H(a, \cdot)$; and in the context of pseudorandom generators we would choose a and then look at the pseudorandomness of $H(a, x)$ (for uniform x) given a.

De et al. [6] briefly study the effect of salting for inverting one-way *permutations* as well as *specific* PRGs (based on one-way permutations and hardcore bits), but beyond that we are aware of no analysis of the effectiveness of salting for defeating preprocessing in any other contexts, including the use of hash functions which are not permutations.[3] We highlight that although it may appear "obvious" that salting defeats, say, rainbow tables, it is not at all clear what is the *quantitative* security benefit of salting, and it is not clear whether rainbow tables can be adapted to give a (possibly different) online/offline tradeoff when salting is used.

1.2   Our Results

We address all three approaches outlined in the previous section. First, we investigate the generic approach to proving security in the random-oracle model with auxiliary input, and specifically explore the extent to which Unruh's pre-sampling technique can be improved. Here, our result is largely negative: disproving Unruh's conjecture, we show that there is an attack for which the "security loss" stemming from Unruh's approach is at least $\Omega(ST/P)$. Although there remains a gap between our lower bound and Unruh's upper bound that will be interested to close, as we discuss next the upshot is that Unruh's technique is not sufficient (in general) for proving strong concrete-security bounds in the random-oracle model when preprocessing is a possibility.

Consider, e.g., the case of function inversion. One can show that the probability of inverting a random oracle $H : [N] \rightarrow [N]$ for which P points have been "pre-sampled" is $O(P/N + T/N)$. Combined with the security loss of $O(\sqrt{ST/2P})$ resulting from Unruh's technique and plugging in the optimal value of P, we obtain a security bound of $O((ST/N)^{1/3} + T/N)$ for algorithms making T oracle queries and using S bits of auxiliary input about H. And our negative result shows that the best bound one could hope to achieve by using Unruh's approach is $O((ST/N)^{1/2} + T/N)$. Both bounds fall short of the best known attacks, which succeed with probability $\Omega\left(\min\left\{\frac{T}{N}, (\frac{S^2 T}{N^2})^{1/3}\right\} + \frac{T}{N}\right)$. Similar gaps exist for other cryptographic primitives.

---

[3] Bellare et al. [3] study security of salting for the purposes of multi-instance security, but they do not address the issue of preprocessing.

Faced with this, we turn to studying a more direct approach for proving tighter bounds for specific important applications of hash functions, such as their use as one-way functions, pseudorandom generators/functions (PRGs/PRFs) or message authentication codes (MACs).[4] Here we show much tighter, and in many cases optimal bounds for all of these primitives, which always beat the provable version of Unruh's pre-sampling (see Table 1 with value $K = 1$). Not surprisingly, our bounds are not as good as what is possible to show without pre-processing, since those bounds are no longer true once pre-processing is allowed. In particular, setting $S = T = C$ we now get meaningful non-uniform security bounds against circuits of size C for all of the above primitives, which often match the existing limitations known for non-uniform attacks. (For example, when $C = S = T$ is polynomial in n, we get that the optimal non-uniform PRG/PRF security is lower bounded by $2^{-n/2}$, matching existing attacks).

Given these inherent limitation as compared to the traditional ROM without preprocessing, we formally examine the effects of "salting" as a way of mitigating or even defeating the effects of pre-processing/non-uniformity. As before, we look at the natural, "salted" constructions of one-way functions, PRGs, PRFs and MACs, but now can also examine collision-resistant hash functions (CRHFs), which can be potentially secure against pre-processing, once the salt is long-enough. In all these case we analyze the security of these constructions in the presence of auxiliary information about the random oracle. In fact, the "unsalted" results for one-way functions, PRGs, PRFs and MACs mentioned above are simply special cases of salted result with the cardinality K of the salting space is $K = 1$.

Our results are summarized in Table 1, where they are compared to the best known attacks using preprocessing. Our bounds for inverting one-way functions and distinguishing PRGs matches the bounds De et al. [6] for inverting one-way permutations and distinguishing PRGs based on one-way permutations and hardcore bits, but apply to real-world candidates for these primitives based on existing hash functions. In the case of CRHFs, our bound is tight and matches the best known attack of storing explicit collisions for roughly S distinct salts. In the remaining cases, although our bounds are not tight (but close), it is interesting to note that, assuming $N \geq T \geq S$, our results show that setting the length of the salt equal to the length of the secret (i.e., setting $K = N$) yields the same security bound $O(T/N)$ that is achieved for constructions in the standard random-oracle model *without* preprocessing. Summarizing a bit informally: using an n-bit salt and an n-bit secret gives n-bit security *even in the presence of preprocessing*. Namely, salts provably defeats pre-processing in these settings.

All our new bounds are proven using the "compression paradigm" introduced by Gennaro and Trevisan [9,10]. The main idea is to argue that if some attacker succeeds with "high" probability, then that attacker can be used to reversibly encode (i.e., compress) a random oracle beyond what is possible from

---

[4] As we mentioned, collision-resistance is impossible without salting, which we discuss shortly.

Table 1. Security bounds and best known attacks using space S and time T for "salted" constructions of primitives based on a random oracle. The first three (unkeyed) primitives are constructed from a random oracle $\mathcal{O} : [K] \times [N] \to [M]$, where $[K]$ is the domain of the salt and $[N]$ is the domain of the secret; the final two (keyed) primitives are constructed from a random oracle $\mathcal{O} : [K] \times [N] \times [L] \to [M]$, where $[L]$ is the domain of the input. For simplicity, logarithmic factors and constant terms are omitted.

	Security bounds (here)	Best known attacks
OWFs	$\frac{ST}{KN} + \frac{T}{N}$	$\min\left\{ \frac{ST}{KN}, \left(\frac{S^2T}{K^2N^2}\right)^{1/3} \right\} + \frac{T}{N}$
CRHFs	$\frac{S}{K} + \frac{T^2}{M}$	$\frac{S}{K} + \frac{T^2}{M}$
PRGs	$\left(\frac{ST}{KN}\right)^{1/2} + \frac{T}{N}$	$\left(\frac{S}{KN}\right)^{1/2} + \frac{T}{N}$
PRFs	$\left(\frac{ST}{KN}\right)^{1/2} + \frac{T}{N}$	$\left(\frac{S}{KN}\right)^{1/2} + \frac{T}{N}$
MACs	$\frac{ST}{KN} + \frac{T}{N} + \frac{T}{M}$	$\min\left\{ \frac{ST}{KN}, \left(\frac{S^2T}{K^2N^2}\right)^{1/3} \right\} + \frac{T}{N} + \frac{1}{M}$

an information-theoretic point of view. Since we are considering attackers who perform preprocessing, our encoding must include the S-bit auxiliary information produced by the attacker. Thus, the main technical challenge we face is to ensure that our encoding compresses by (significantly) more than S bits.

Outlook. In this work we thoroughly revisited the ROM with auxiliary input, as we believe it has not gotten enough attention from the cryptographic community, despite being simultaneously important for the variety of reasons detailed above, and also much more interesting than the traditional ROM from a technical point in view. Indeed, even the most trivial one-line proof in the traditional ROM is either completely false once preprocessing is allowed (e.g., CRHFs), or becomes an interesting technical challenge (OWFs, PRGs, MACs) that requires new techniques, and usually teaches us something new about the primitive in question in relation to pre-processing.

Of course, given an abundance of works using random oracle, we hope our work will generate a lot of follow-up research analyzing the effects of pre-processing and non-uniformity for many other important uses of hash functions, as well as other idealized primitives (e.g., ideal ciphers).

2   Limits on the Power of Preprocessing

For two distributions D_1, D_2 over universe Ω, we use $\Delta(D_1, D_2)$ to denote their statistical distance $\frac{1}{2} \cdot \sum_{y \in \Omega} |\Pr[D_1 = y] - \Pr[D_2 = y]|$.

In this section, we revisit the result of Unruh [16] that allows one to replace arbitrary (bounded-length) auxiliary information about a random oracle $\mathcal{O}$ with a (bounded-size) set fixing the value of the random oracle on some fraction of points. For a set of tuples $Z = \{(x_1, y_1), \ldots\}$, we let $\mathcal{O}'[Z]$ denote a random oracle chosen uniformly subject to the constraints $\mathcal{O}'(x_i) = y_i$.

Theorem 1 ([16]). *Let $P, S, T \geq 1$ be integers, and let A_0 be an oracle algorithm that outputs state of length at most S bits. Then there is an oracle algorithm* Pre *outputting a set containing at most P tuples such that for any oracle algorithm A_1 that makes at most T oracle queries,*

$$\Delta(A_1^{\mathcal{O}}(A_0^{\mathcal{O}}),\ A_1^{\mathcal{O}'[\mathsf{Pre}^{\mathcal{O}}]}(A_0^{\mathcal{O}})) \leq \sqrt{\frac{ST}{2P}}.$$

This theorem enables proving various results in the random-oracle model even in the presence of auxiliary input by first replacing the auxiliary input with a fixed set of input/output pairs and then using standard lazy-sampling techniques for the value of the random oracle at other points. However, applying this theorem incurs a cost of $\sqrt{ST/2P}$, and so super-polynomial P is required in order to obtain negligible advantage overall. It is open whether one can improve the bound in Theorem 1; Unruh conjectures [16, Conjecture 14] that for all polynomials S, T there is a polynomial P such that the statistical difference above is negligible. We disprove this conjecture by showing that the bound in the theorem cannot be improved (in general) below $O(ST/P)$. That is,

Theorem 2. *Consider random oracles $\mathcal{O} : [N] \rightarrow \{0,1\}$, and let $S, T, P \geq 1$ be integers with $4P^2/ST + ST \leq N$. Then there is an oracle algorithm A_0 that outputs S-bit state and an oracle algorithm A_1 that makes T oracle queries such that for any oracle algorithm Pre outputting a set containing at most P tuples,*

$$\Delta(A_1^{\mathcal{O}}(A_0^{\mathcal{O}}),\ A_1^{\mathcal{O}'[\mathsf{Pre}^{\mathcal{O}}]}(A_0^{\mathcal{O}})) \geq \frac{ST}{24P}.$$

Proof. Pick S disjoint sets $X_1, \ldots, X_S \subset [N]$, where each set is of size $t = T \cdot (4(P/ST)^2 + 1)$. Partition each set X_i into $t/T = 4(P/ST)^2 + 1$ disjoint blocks $X_{i,1}, \ldots, X_{i,t/T}$, each of size T. Algorithm $A_1^{\mathcal{O}}$ outputs an S-bit state where the ith bit is equal to $\mathrm{maj}(\oplus_{x \in X_{i,1}} \mathcal{O}(x), \ldots, \oplus_{x \in X_{i,t/T}} \mathcal{O}(x))$ where maj is the majority function. Algorithm $A_1^{\mathcal{O}}(b_1, \ldots, b_S)$ chooses a uniform block $X_{i,j}$ and outputs 1 iff $\oplus_{x \in X_{i,j}} \mathcal{O}(x) = b_i$.

We have

$$\Pr[A_1^{\mathcal{O}}(A_0^{\mathcal{O}}) = 1]$$

$$= \Pr_{\mathcal{O},i,j}\left[\mathrm{maj}(\oplus_{x \in X_{i,1}} \mathcal{O}(x), \ldots, \oplus_{x \in X_{i,t/T}} \mathcal{O}(x)) = \oplus_{x \in X_{i,j}} \mathcal{O}(x)\right]$$

$$= \Pr_{z_1,\ldots,z_{t/T} \leftarrow \{0,1\}, j \leftarrow [t/T]}\left[\mathrm{maj}(z_1, \ldots, z_{t/T}) = z_j\right]$$

$$= \mathbb{E}_j\left[\Pr\left[\sum_{i \neq j} z_i = \frac{t/T - 1}{2}\right] + \frac{1}{2} \cdot \Pr\left[\sum_{i \neq j} z_i \neq \frac{t/T - 1}{2}\right]\right]$$

$$= \frac{1}{2} + \frac{1}{2} \cdot \Pr\left[\sum_{i > 1} z_i = \frac{t/T - 1}{2}\right]$$

$$= \frac{1}{2} + \binom{t/T - 1}{\frac{t/T-1}{2}} \cdot 2^{-t/T}$$

$$\geq \frac{1}{2} + \frac{1}{3\sqrt{t/T - 1}} = \frac{1}{2} + \frac{ST}{6P},$$

where the inequality uses $\sqrt{2\pi n}\,(n/e)^n \le n! \le e\sqrt{n}\,(n/e)^n$ so that

$$\binom{n}{n/2} \ge \frac{\sqrt{2\pi n}\,(n/e)^n}{(e\sqrt{n/2}\,(n/2e)^{n/2})^2} = \frac{2\sqrt{2\pi}}{e^2\sqrt{n}} \cdot 2^n \ge \frac{2}{3} \cdot \frac{2^n}{\sqrt{n}}.$$

On the other hand, for any algorithm Pre we have

$$\Pr[A_1^{\mathcal{O}'[\mathsf{Pre}^{\mathcal{O}}]}(A_0^{\mathcal{O}}) = 1]$$

$$= \Pr_{i,j,\mathcal{O},\mathcal{O}'}[\mathrm{maj}(\oplus_{x\in X_{i,1}}\mathcal{O}(x),\ldots,\oplus_{x\in X_{i,t/T}}\mathcal{O}(x)) = \oplus_{x\in X_{i,j}}\mathcal{O}'(x)]$$

$$\le \frac{P/T}{St/T} + \frac{1}{2}\cdot\left(1 - \frac{P/T}{St/T}\right)$$

$$= \frac{1}{2} + \frac{P}{2St} \le \frac{1}{2} + \frac{ST}{8P}.$$

The first inequality above holds since, for any fixed $i, j, \mathcal{O}$,

$$\Pr_{\mathcal{O}'}\left[\mathrm{maj}(\oplus_{x\in X_{i,1}}\mathcal{O}(x),\ldots,\oplus_{x\in X_{i,t/T}}\mathcal{O}(x)) = \oplus_{x\in X_{i,j}}\mathcal{O}'(x)\right] = 1/2$$

unless the value of $\mathcal{O}'$ is fixed by $\mathsf{Pre}^{\mathcal{O}}$ at every point in $X_{i,j}$. But $\mathsf{Pre}^{\mathcal{O}}$ can ensure that the value of $\mathcal{O}'$ is fixed in that way for at most P/T out of the St/T blocks defined by i, j. This concludes the proof. ∎

3   Function Inversion

For natural number n, we define $[n] = \{1,\ldots,n\}$. In this section, we prove bounds on the hardness of inverting "salted" random oracles in the presence of preprocessing. That is, consider choosing a random function $\mathcal{O} : [K]\times[N] \to [M]$ and then allowing an attacker A_0 (with oracle access to $\mathcal{O}$) to perform arbitrary preprocessing to generate an S-bit state st. We then look at the hardness of inverting $\mathcal{O}(a,x)$, given st and a, for algorithms A_1 making up to T oracle queries, where $a \in [K]$ and $x \in [N]$ are uniform. We consider two notions of inversion: computing x itself, or the weaker goal of finding any x' such that $\mathcal{O}(a,x') = \mathcal{O}(a,x)$. Assuming $N = M$ for simplicity in the present discussion, we show that in either case the probability of successful inversion is $O(\frac{ST}{KN} + \frac{T\log N}{N})$. We remark that the best bound one could hope to prove via a generic approach (i.e., using Theorem 1 with best-possible bound $O(ST/P)$) is[5] $O(\sqrt{ST/KN}+T/N)$.

By way of comparison, rainbow tables [2,6,8,11,14] address the case $K = 0$ (i.e., no salt), and give success probability $O(\min\{ST/N, (S^2T/N^2)^{1/3}\} + T/N)$. One natural way to adapt rainbow tables to handle salt is to compute K independent rainbow tables, each using space S/K, for the K reduced functions $\mathcal{O}(a,\cdot)$.

---

[5] Any such bound would take the form $O(ST/P + P/KN + T/N)$, where the first term is from application of the theorem, the second is the probability that the input to A_1 is from the set of fixed points, and the third is the success probability of a trivial brute-force search. Setting $P = \sqrt{ST/KN}$ optimizes this bound.

Using this approach gives success probability $O(\min\{ST/KN, (S^2T/K^2N^2)^{1/3}\} + T/N)$. This shows that our bound is tight when $ST^2 < KN$.

We begin with some preliminary lemmas that we will rely on in this and the following sections.

Lemma 1. *Say there exist encoding and decoding procedures* (Enc, Dec) *such that for all* $m \in M$ *we have* Dec(Enc(m)) = m. *Then* $\mathbb{E}_m[|\mathsf{Enc}(m)|] \geq \log |M|$.

Proof. For $m \in M$, let $s_m = |\mathsf{Enc}(m)|$. Define $C = \sum_m 2^{-s_m}$, and for $m \in M$ let $q_m = 2^{-s_m}/C$. Then $\mathbb{E}_m[|\mathsf{Enc}(m)|] = -\mathbb{E}_m[\log q_m] - \log C$. By Jensen's inequality, $\mathbb{E}_m[\log q_m] \leq \log \mathbb{E}_m[q_m] = -\log |M|$, and by Kraft's inequality $C \leq 1$. The lemma follows. ∎

Following De et al. [6], we also consider randomized encodings (Enc, Dec) for a set M. We say that an encoding has *recovery probability* δ if for all $m \in M$,

$$\Pr_r[\mathsf{Dec}(\mathsf{Enc}(m, r), r) = m] \geq \delta.$$

(Note that Dec is given the randomness used by Enc). The *encoding length* of (Enc, Dec) is defined to be $\max_{m,r}\{|\mathsf{Enc}(m, r)|\}$.

Lemma 2 ([6]). *Suppose there exist randomized encoding and decoding procedures* (Enc, Dec) *for a set* M *with recovery probability* δ. *Then the encoding length of* (Enc, Dec) *is at least* $\log |M| - \log 1/\delta$.

Proof. By a standard averaging argument, there exists an r and a set $M' \subseteq M$ with $|M'| \geq \delta \cdot |M|$ such that Dec(Enc(m, r), r) = m for all $m \in M'$. Let Enc', Dec' be the deterministic algorithms obtained by fixing the randomness to r. By Lemma 1, $\mathbb{E}_{m'}[|\mathsf{Enc}'(m')|] \geq |M'| \geq |M| - \log 1/\delta$, and hence there exists an m' with $|\mathsf{Enc}'(m')| \geq |M| - \log 1/\delta$. ∎

We now state and prove the main results of this section. Let $\mathsf{Func}(A, B)$ denote the set of all functions from A to B.

Theorem 3. *Consider random oracles* $\mathcal{O} \in \mathsf{Func}([K] \times [N], [M])$. *For any oracle algorithms* (A_0, A_1) *such that* A_0 *outputs* S-bit state and A_1 *makes at most* T *oracle queries,*

$$\Pr_{\mathcal{O}, a, x}[A_1^{\mathcal{O}}(A_0^{\mathcal{O}}, a, \mathcal{O}(a, x)) = x] = O\left(\frac{ST}{KN} + \frac{T \log N}{N}\right).$$

Theorem 4. *Consider random oracles* $\mathcal{O} \in \mathsf{Func}([K] \times [N], [M])$. *For any oracle algorithms* (A_0, A_1) *such that* A_0 *outputs* S-bit state and A_1 *makes at most* T *oracle queries,*

$$\Pr_{\mathcal{O}, a, x}[A_1^{\mathcal{O}}(A_0^{\mathcal{O}}, a, \mathcal{O}(a, x)) = x' : \mathcal{O}(a, x) = \mathcal{O}(a, x')] = \varepsilon,$$

if $\varepsilon = \Omega(\log MN/N)$, *then*

$$\varepsilon = O\left(\frac{ST}{K \cdot \alpha} + \frac{T \log N}{\alpha}\right)$$

where $\alpha = \min\{N/\log M, M\}$

To prove Theorem 3, we first prove the following lemma:

Lemma 3. *Consider random oracles $\mathcal{O} \in \mathsf{Func}\,([K] \times [N], [M])$. Assume there exist oracle algorithms (A_0, A_1) such that A_0 outputs S-bit state and A_1 makes at most T oracle queries, and such that*

$$\Pr_{\mathcal{O},a,x}[A_1^{\mathcal{O}}(A_0^{\mathcal{O}}, a, \mathcal{O}(a,x)) = x] = \epsilon.$$

Then there exists a randomized encoding for a set $\mathcal{F} \subseteq \mathsf{Func}\,([K] \times [N], [M])$ of size at least $\frac{\epsilon}{2} \cdot M^{KN}$, with recovery probability at least 0.9 and encoding length (in bits) at most

$$KN \log M + S + K \log N - \frac{\varepsilon KN}{100T} \log\left(\frac{\varepsilon N}{100eT}\right).$$

Proof. By an averaging argument, there is a set $\mathcal{F} \subseteq \mathsf{Func}\,([K] \times [N], [M])$ of size at least $\varepsilon/2 \cdot |\mathsf{Func}\,([K] \times [N], [M])| = \frac{\varepsilon}{2} \cdot M^{KN}$ such that for all $\mathcal{O} \in \mathbb{F}$

$$\Pr_{a,x}[A_1^{\mathcal{O}}(A_0^{\mathcal{O}}, a, \mathcal{O}(a,x)) = x] \geq \epsilon/2.$$

Fix arbitrary $\mathcal{O} \in \mathcal{F}$. We encode $\mathcal{O}$ as follows. Let $\mathsf{st}_{\mathcal{O}}$ be the output of $A_0^{\mathcal{O}}$ and, for $a \in [K]$, let $U_a \subseteq [N]$ be the points x on which $A_1^{\mathcal{O}}(\mathsf{st}_{\mathcal{O}}, a, \mathcal{O}(a,x)) = x$. The high-level idea is that rather than encode the mapping $\{(x, \mathcal{O}(a,x))\}_{x \in U_a}$ explicitly, we will encode the set of points $\{\mathcal{O}(a,x)\}_{x \in U_a}$ and then use A_1 to recover the mapping. If we attempt this in the straightforward way, however, then it may happen that A_1 queries its oracle on a point for which the mapping is not yet known. To get around this issue, we instead use this approach for a random subset of U_a so that this only happens with small probability.

Specifically, the encoder uses randomness r to pick a set $R \subseteq [K] \times [N]$, where each $(a,x) \in [K] \times [N]$ is included in R with probability $1/10T$. For $a \in [K]$, let $G_a \subseteq R$ be the set of $(a,x) \in R$ such that $A_1^{\mathcal{O}}(\mathsf{st}_{\mathcal{O}}, a, \mathcal{O}(a,x)) = x$ and moreover A_1 does not query $\mathcal{O}$ on any $(a',x') \in R$ (except possibly (a,x) itself). Let $G = \bigcup_a G_a$. Define $V_a = \{\mathcal{O}(a,x)\}_{x \in G_a}$, and note that $|V_a| = |G_a|$.

As in De et al. [6], with probability at least 0.9 the size of G is at least $\varepsilon KN/100T$. To see this, note that by a Chernoff bound, R has at least $\varepsilon KN/40T$ points with probability at least 0.95. The expected number of points $(a,x) \in R$ for which $A_1^{\mathcal{O}}(\mathsf{st}_{\mathcal{O}}, a, \mathcal{O}(a,x)) = x$ but A_1 queries $\mathcal{O}$ on some point $(a',x') \in R$ (besides (a,x) itself) is at most $\frac{\varepsilon KN}{2} \cdot \frac{1}{10T} \cdot \left(1 - (1 - 1/10T)^T\right) \leq \frac{\varepsilon KN}{2000T}$. By Markov's inequality, with probability at least 0.95 the number of such points is at most $\frac{\varepsilon KN}{100T}$. So with probability at least 0.9, we have $|G| \geq \frac{3\varepsilon KN}{200T} \geq \frac{\varepsilon KN}{100T}$.

Assuming $|G| \geq \varepsilon KN/100T$, we encode $\mathcal{O}$ as follows:

1. Include $\mathsf{st}_{\mathcal{O}}$ and, for each $a \in [K]$, include $|V_a|$ and a description of V_a. This uses a total of $S + K \log N + \sum_{a \in [K]} \log \binom{M}{|G_a|}$ bits.
2. For each a and $y \in V_a$ (in lexicographic order), run $A_1^{\mathcal{O}}(\mathsf{st}_{\mathcal{O}}, a, y)$ and include in the encoding the answers to all the oracle queries made by A_1 that have not been included in the encoding so far, except for any queries in R. (By definition of G_a, there will be at most one such query and, if so, it will be the query (a,x) such that $\mathcal{O}(a,x) = y$.)

3. For each $(a, x) \in ([K] \times [N]) \setminus G$ (in lexicographic order) for which $\mathcal{O}(a, x)$ has not been included in the encoding so far, add $\mathcal{O}(a, x)$ to the encoding.

Steps 2 and 3 explicitly include in the encoding the value of $\mathcal{O}(a, x)$ for each $(a, x) \in ([K] \times [N]) \setminus G$. Thus, the total number of bits added to the encoding by those steps is $(KN - \sum_a |G_a|) \log M$.

To decode, the decoder first uses r to recover the set R defined above. Then it does the following:

1. Recover $\mathsf{st}_\mathcal{O}$, $\{|V_a|\}_{a \in K}$, and $\{V_a\}_{a \in K}$.
2. For each a and $y \in V_a$ (in lexicographic order), run $A_1(\mathsf{st}_\mathcal{O}, a, y)$ while answering the oracle queries of A_1 using the values stored in the encoding. The only exception is if A_1 ever makes a query $(a, x) \in R$, in which case y itself is returned as the answer. The output x of A_1 will be such that $\mathcal{O}(a, x) = y$.
3. For each $(a, x) \in [K] \times [N]$ (in lexicographic order) for which $\mathcal{O}(a, x)$ is not yet defined, recover the value of $\mathcal{O}(a, x)$ from the remainder of the encoding.

Assuming $|G| \geq \varepsilon KN/100T$, the encoding is not empty and the decoding procedure recovers $\mathcal{O}$. The encoding length is

$$S + K \log N + \sum_{a \in [K]} \log \binom{M}{|G_a|} + \left(KN - \sum_{a \in K} |G_a| \right) \log M.$$

Because $\binom{M}{|G_a|} \leq \left(\frac{eM}{|G_a|} \right)^{|G_a|}$, the encoding length is bounded by

$$S + K \log N + KN \log M - \sum_a |G_a| \log \left(\frac{|G_a|}{e} \right)$$

$$\leq S + K \log N + KN \log M - |G| \log \left(\frac{|G|}{eK} \right)$$

$$\leq S + K \log N + KN \log M - \frac{\varepsilon KN}{100T} \log \left(\frac{\varepsilon N}{100eT} \right),$$

where the second line uses concavity of the function $f(y) = -y \log (y/e)$, and the last line is because $|G| \geq \frac{\varepsilon KN}{100T}$. ∎

Lemma 3 gives an encoding for a set of size $\frac{\varepsilon}{2} \cdot M^{KN}$ with recovery probability 0.9, and encoding length at most $NK \log M + S + K \log N - \frac{\varepsilon KN}{100T} \log \left(\frac{\varepsilon N}{100eT} \right)$ bits. But Lemma 2 shows that any such encoding must have encoding length at least $NK \log M - \log \frac{2}{\varepsilon} - \log \frac{10}{9}$ bits. We thus conclude that

$$S + K \log N + \log \frac{20}{9\varepsilon} \geq \frac{\varepsilon KN}{100T} \log \left(\frac{\varepsilon N}{100eT} \right).$$

This implies Theorem 3 since either $\varepsilon < \frac{200eT}{N}$, or else it must be the case that $\varepsilon \leq \left(\frac{100T}{KN} \right) \cdot (S + K \log N + \log N)$.

We now prove Theorem 4. For fixed $\mathcal{O}$ and $a \in [K]$, let $Y_{\mathcal{O},a} \subseteq [M]$ be the set of points A_1 successfully inverts, i.e.,

$$Y_{\mathcal{O},a} = \{y \ : \ A_1^{\mathcal{O}}(A_0^{\mathcal{O}}, a, y) = x' : \mathcal{O}(a, x') = y\}.$$

Let $X_{\mathcal{O},a} \subseteq [N]$ be the pre-images of the points in $Y_{\mathcal{O},a}$. That is,

$$X_{\mathcal{O},a} = \{x \ : \ \mathcal{O}(a, x) \in Y_{\mathcal{O},a}\}.$$

We show a deterministic encoding for $\mathsf{Func}([K] \times [N], [M])$. Given a function $\mathcal{O}$, we encode it by including for each $a \in [K]$ the following information:

1. The set $X_{\mathcal{O},a}$ (along with its size), using $\log N + \binom{N}{|X_{\mathcal{O},a}|}$ bits.
2. The set $Y_{\mathcal{O},a}$ (along with its size), using $\log M + \binom{M}{|Y_{\mathcal{O},a}|}$ bits.
3. For each $x \in X_{\mathcal{O},a}$, the value $\mathcal{O}(a, x) \in Y_{\mathcal{O},a}$ encoded using $\log |Y_{\mathcal{O},a}|$ bits.
4. For each $x \notin X_{\mathcal{O},a}$, the value $\mathcal{O}(a, x)$ encoded using $\log M$ bits.

Decoding is done in the obvious way. The encoding length of $\mathcal{O}$ (in bits) is

$$K \log N + K \log M$$
$$+ \sum_{a \in [K]} \log \binom{N}{|X_{\mathcal{O},a}|} + \log \binom{M}{|Y_{\mathcal{O},a}|} + |X_{\mathcal{O},a}| \cdot \log |Y_{\mathcal{O},a}| + (N - |X_{\mathcal{O},a}|) \cdot \log M.$$

Using the inequality $\log \binom{A}{B} \leq B \cdot \log \frac{eA}{B}$ and the log-sum[6] inequality, the encoding length of $\mathcal{O}$ (in bits) is at most

$$K \log N + K \log M + \left(\sum_{a \in [K]} |X_{\mathcal{O},a}| \right) \cdot \log \frac{eN \sum_{a \in [K]} |Y_{\mathcal{O},a}|}{M \sum_{a \in [K]} |X_{\mathcal{O},a}|}$$
$$+ \left(\sum_{a \in [K]} |Y_{\mathcal{O},a}| \right) \cdot \log \frac{eKM}{\sum_{a \in [K]} |Y_{\mathcal{O},a}|} + KN \log M. \qquad (1)$$

Let $\epsilon' \overset{\text{def}}{=} \Pr_{\mathcal{O},a,x}[A_1^{\mathcal{O}}(A_0^{\mathcal{O}}, a, \mathcal{O}(a, x)) = x]$, and note that $\mathbb{E}_{\mathcal{O}}[\sum_a |X_{\mathcal{O},a}|] = \epsilon NK$ and $\mathbb{E}_{\mathcal{O}}[\sum_a |Y_{\mathcal{O},a}|] = \epsilon'NK$. By averaging over $\mathcal{O}$ and log-sum inequality, the average encoding length of $\mathcal{O}$ is upper bounded by replacing $\sum_{a \in K} X_{\mathcal{O},a}$ by $\mathbb{E}_{\mathcal{O}}[\sum_{a \in K} |X_{\mathcal{O},a}|]$ and $\sum_{a \in K} Y_{\mathcal{O},a}$ by $\mathbb{E}_{\mathcal{O}}[\sum_{a \in K} |Y_{\mathcal{O},a}|]$ in (1), namely

$$K \log N + K \log M + (\varepsilon NK) \cdot \log \frac{eN\varepsilon'NK}{M\varepsilon NK} + (\varepsilon'NK) \cdot \log \frac{eKM}{\varepsilon'NK} + KN \log M.$$

Using the fact that (by Lemma 1) the encoding length must be at least $KN \log M$ bits and rearranging the inequality, we obtain

$$\frac{\log N + \log M}{N} + \varepsilon' \cdot \log \frac{eM}{\varepsilon'N} \geq \varepsilon \cdot \log \frac{M\varepsilon}{eN\varepsilon'}.$$

---

[6] The log-sum inequality states that for nonnegative $t_1, \ldots, t_n$ and $w_1, \ldots, w_n$, it holds that $\sum_{i=1}^n t_i \log(w_i/t_i) \leq (\sum_{i=1}^n t_i) \cdot \log(\sum_{i=1}^n w_i / \sum_{i=1}^n t_i)$. It also implies the average of $t_1 \log(w_1/t_1), \ldots, t_n \log(w_n/t_n)$ is less that $\bar{t} \log(\overline{w}/\bar{t})$ where $\bar{t}$ is the average of $t_1, \ldots, t_n$ and $\overline{w}$ is the average of $w_1, \ldots, w_n$.

If $\varepsilon = \Omega((\log MN)/N)$, then there exists a sufficiently large constant C such that $\varepsilon N \geq (\log MN)/C$. If $M\varepsilon/(eN\varepsilon') \leq 2^{C+1}$, then $\varepsilon = O(\varepsilon'N/M)$. Otherwise, $(M\varepsilon)/(eN\varepsilon') \geq 2^{C+1}$, then

$$\varepsilon' \log \frac{eM}{\varepsilon'N} \geq \varepsilon(C+1) - (\log MN)/N \geq \varepsilon,$$

which implies $\varepsilon = O(\varepsilon' \log M)$ (here we assume $\varepsilon'N \geq 1$). Overall we get $\varepsilon = O(\varepsilon' \max(\log M, N/M))$. By the bound on ε' from Theorem 3, we obtain the desired bound on ε.

4   Collision-Resistant Hash Functions

In this section, we prove the following theorem.

Theorem 5. *Consider random oracles $\mathcal{O} \in \mathsf{Func}([K] \times [N], [M])$. For any oracle algorithms (A_0, A_1) such that A_0 outputs S-bit state and A_1 makes at most T oracle queries,*

$$\Pr_{\mathcal{O},a}[(x,x') := A_1^{\mathcal{O}}(A_0^{\mathcal{O}}, a) : x \neq x' \wedge \mathcal{O}(a,x) = \mathcal{O}(a,x')] = O\left(\frac{S + \log K}{K} + \frac{T^2}{M}\right).$$

The bound in the above theorem matches (up to the $K^{-1} \log K$ term) the parameters achieved by the following: A_0 outputs collisions in $\mathcal{O}(a_i, \cdot)$ for each of $a_1, \ldots, a_S \in [K]$. Then A_1 outputs the appropriate collision if $a = a_i$, and otherwise performs a birthday attack in an attempt to find a collision.

To prove Theorem 5, we first prove the following lemma:

Lemma 4. *Consider random oracles $\mathcal{O} \in \mathsf{Func}([K] \times [N], [M])$. Assume there exist oracle algorithms (A_0, A_1) such that A_0 outputs S-bit state and A_1 makes at most T oracle queries, and such that*

$$\Pr_{\mathcal{O},a}[(x,x') := A_1^{\mathcal{O}}(A_0^{\mathcal{O}}, a) : x \neq x' \wedge \mathcal{O}(a,x) = \mathcal{O}(a,x')] = \epsilon.$$

Then there exists a deterministic encoding for the set $\mathsf{Func}([K] \times [N], [M])$ with expected encoding length (in bits) at most

$$S + KN \log M + \log K - \frac{\varepsilon K}{2} \log\left(\frac{\varepsilon M}{8eT^2}\right).$$

Proof. Fix $\mathcal{O} \colon [K] \times [N] \to [M]$, and let $\mathsf{st}_{\mathcal{O}} = A_0^{\mathcal{O}}$. Let $G_{\mathcal{O}}$ be the set of $a \in [K]$ such that $A_1^{\mathcal{O}}(\mathsf{st}_{\mathcal{O}}, a)$ outputs a collision in $\mathcal{O}(a, \cdot)$. We assume, without loss of generality, that if $A_1^{\mathcal{O}}(\mathsf{st}_{\mathcal{O}}, a)$ outputs x, x', then it must have queried $\mathcal{O}(a, x)$ and $\mathcal{O}(a, x')$ at some point in its execution. The basic observation is that we can use this to compress $\mathcal{O}(a, \cdot)$ for $a \in G_{\mathcal{O}}$. Specifically, rather than store both $\mathcal{O}(a, x)$ and $\mathcal{O}(a, x')$ (using $2 \log M$ bits), where x, x' is the collision in $\mathcal{O}(a, \cdot)$ output by A_1, we instead store the value $\mathcal{O}(a, x) = \mathcal{O}(a, x')$ once, along with the indices i, j of the oracle queries $\mathcal{O}(a, x)$ and $\mathcal{O}(a, x')$ made by A_1 (using a total of $\log M + 2 \log T$ bits). This is a net savings if $2 \log T < \log M$. Details follow.

A simple case. To illustrate the main idea, we first consider a simple case where $A_1^{\mathcal{O}}(\mathsf{st}_{\mathcal{O}}, a)$ never makes oracle queries $\mathcal{O}(a', x)$ with $a' \neq a$. Under this assumption, we encode $\mathcal{O}$ as follows:

1. Encode $\mathsf{st}_{\mathcal{O}}$, $|G_{\mathcal{O}}|$, and $G_{\mathcal{O}}$. This requires $S + \log K + \log \binom{K}{|G_{\mathcal{O}}|}$ bits.
2. For each $a \in G_{\mathcal{O}}$ (in lexicographic order), run $A_1^{\mathcal{O}}(\mathsf{st}_{\mathcal{O}}, a)$ and let the second components of the oracle queries of A_1 be $x_1, \ldots, x_T$. (We assume without loss of generality these are all distinct.) If x, x' are the output of A_1, let $i < j$ be such that $\{x, x'\} = \{x_i, x_j\}$. Encode i and j, along with the answers to each of A_1's oracle queries (in order) except for the jth. Furthermore, encode $\mathcal{O}(a, x)$ for all $x \in [N] \setminus \{x_1, \ldots, x_T\}$ (in lexicographic order). This requires $(N-1) \cdot \log M + 2 \log T$ bits for each $a \in G_{\mathcal{O}}$.
3. For each $a \notin G_{\mathcal{O}}$ and $x \in [N]$ (in lexicographic order), store $\mathcal{O}(a, x)$. This uses $N \log M$ bits for each $a \notin G_{\mathcal{O}}$.

Decoding is done in the obvious way.

The encoding length of $\mathcal{O}$ (in bits) is

$$S + \log K + \log \binom{K}{|G_{\mathcal{O}}|} + KN \log M - |G_{\mathcal{O}}| \cdot (\log M - 2 \log T).$$

Using the inequality $\binom{K}{|G_f|} \leq (\frac{eK}{|G_f|})^{|G_f|}$, the expected encoding length (in bits) is thus

$$
\begin{aligned}
S + \log K + &\mathbb{E}_{\mathcal{O}}\left[|G_{\mathcal{O}}| \cdot \log \frac{eK}{|G_{\mathcal{O}}|}\right] \\
&+ KN \log M - \mathbb{E}_{\mathcal{O}}[|G_{\mathcal{O}}|] \cdot (\log M - 2 \log T) \\
\leq S + \log K + &\mathbb{E}_{\mathcal{O}}[|G_{\mathcal{O}}|] \cdot \log \frac{eK}{\mathbb{E}_{\mathcal{O}}[|G_{\mathcal{O}}|]} \\
&+ KN \log M - \mathbb{E}_{\mathcal{O}}[|G_{\mathcal{O}}|] \cdot (\log M - 2 \log T) \\
= S + \log K + &KN \log M - \varepsilon K \log \left(\frac{\varepsilon M}{eT^2}\right),
\end{aligned}
$$

where the inequality uses concavity of the function $y \cdot \log 1/y$, and the third line uses $\mathbb{E}_{\mathcal{O}}[|G_{\mathcal{O}}|] = \varepsilon K$.

The general case. In the general case, we need to take into account the fact that A_1 may make arbitrary queries to $\mathcal{O}$. This affects the previous approach because $A_1(\mathsf{st}_{\mathcal{O}}, a)$ may query $\mathcal{O}(a', x)$ for a value x that is output as part of a collision by $A_1(\mathsf{st}_{\mathcal{O}}, a')$.

To deal with this, consider running $A_1^{\mathcal{O}}(\mathsf{st}_{\mathcal{O}}, a)$ for all $a \in G_{\mathcal{O}}$. There are at most $T \cdot |G_{\mathcal{O}}|$ distinct oracle queries made overall. Although several of them may share the same prefix $a \in [K]$, there are at most $|G_{\mathcal{O}}|/2$ values of a that are used as a prefix in more than $2T$ queries. In other words, there is a set $G'_{\mathcal{O}} \subseteq G_{\mathcal{O}}$ of size at least $|G_{\mathcal{O}}|/2$ such that each $a \in G'_{\mathcal{O}}$ is used in at most $2T$ queries when running $A_1^{\mathcal{O}}(\mathsf{st}_{\mathcal{O}}, a)$ for all $a \in G'_{\mathcal{O}}$.

To encode $\mathcal{O}$ we now proceed in a manner similar to before, but using $G'_{\mathcal{O}}$ in place of $G_{\mathcal{O}}$. Moreover, we run $A_1^{\mathcal{O}}(\mathsf{st}_{\mathcal{O}}, a)$ for all $a \in G'_{\mathcal{O}}$ (in lexicographic order) and consider all the distinct oracle queries made. For each $a \in G'_{\mathcal{O}}$, let $i_a < j_a \leq 2T$ be such that the i_ath and j_ath oracle queries that use prefix a are distinct but yield the same output. (There must exist such indices by assumption on A_1.) We encode (i_a, j_a) for all $a \in G'_{\mathcal{O}}$, along with the answers to all the (distinct) oracle queries made with the exception of the j_ath oracle query made using prefix a for all $a \in G'_{\mathcal{O}}$. The remainder of $\mathcal{O}(\cdot, \cdot)$ is then encoded in the trivial way as before. Decoding is done in the natural way.

Arguing as before, but with ϵK replaced by $\epsilon K/2$ and T replaced by $2T$, we see that the expected encoding length (in bits) is now at most

$$ S + \log K + KN \log M - \frac{\varepsilon K}{2} \log \left(\frac{\varepsilon M}{8eT^2} \right), $$

as claimed.   ∎

Lemma 4 gives an encoding for $\mathsf{Func}\,([K] \times [N], [M])$ with expected length at most

$$ S + \log K + KN \log M - \frac{\varepsilon K}{2} \log \left(\frac{\varepsilon M}{8eT^2} \right) $$

bits. But Lemma 1 shows that any such encoding must have expected length at least $NK \log M$ bits. We thus conclude that

$$ S + \log K \geq \frac{\varepsilon K}{2} \log \left(\frac{\varepsilon M}{8eT^2} \right). $$

This implies Theorem 5 since either $\varepsilon \leq \frac{16eT^2}{M}$ or else $\varepsilon \leq \frac{2S + 2 \log K}{K}$.

5   Pseudorandom Generators and Functions

In this section, we prove the following theorems.

Theorem 6. *Consider random oracles $\mathcal{O} \in \mathsf{Func}\,([K] \times [N], [M])$ where it holds that $M > N$. For any oracle algorithms (A_0, A_1) such that A_0 outputs S-bit state and A_1 makes at most T oracle queries,*

$$ \left| \Pr_{\mathcal{O}, a, x}[A_1^{\mathcal{O}}(A_0^{\mathcal{O}}, a, \mathcal{O}(a, x)) = 1] - \Pr_{\mathcal{O}, a, y}[A_1^{\mathcal{O}}(A_0^{\mathcal{O}}, a, y) = 1] \right| $$

$$ = O\left(\log M \cdot \left(\sqrt{\frac{ST}{KN}} + \frac{T \log N}{N} \right) \right). $$

Theorem 7. *Consider random oracles $\mathcal{O} \in \mathsf{Func}\,([K] \times [N] \times [L], \{0,1\})$. For any oracle algorithms (A_0, A_1) such that A_0 outputs S-bit state and A_1 makes at most T oracle queries to $\mathcal{O}$ and at most q queries to its other oracle,*

$$\left| \Pr_{\mathcal{O},a,k}[A_1^{\mathcal{O},\mathcal{O}(a,k,\cdot)}(A_0^{\mathcal{O}}, a) = 1] - \Pr_{\mathcal{O},a,f}[A_1^{\mathcal{O},f}(A_0^{\mathcal{O}}, a) = 1] \right|$$

$$= O\left(q \cdot \left(\sqrt{\frac{ST}{KN}} + \frac{T \log N}{N} \right) \right),$$

where f is uniform in $\mathsf{Func}\,([L], \{0,1\})$.

Note that in both cases, an exhaustive-search attack (with $S = 0$) achieves distinguishing advantage $\Theta(T/N)$. With regard to pseudorandom generators (Theorem 6), De et al. [6] show an attack with $T = 0$ that achieves distinguishing advantage $\Omega(\sqrt{\frac{S}{KN}})$. Their attack can be extended to the case of pseudorandom functions (assuming $q > \log KN$) to obtain distinguishing advantage $\Omega(\sqrt{\frac{S}{KN}})$ in that case as well.

In proving the above, we rely on the following [6, Lemma 8.4]:

Lemma 5. *Fix a parameter ϵ, and oracle algorithms (A_0, A_1) such that A_0 outputs S-bit state and A_1 makes at most T queries to $\mathcal{O}$ but may not query its input. Let $\mathcal{F} \subseteq \mathsf{Func}\,([K] \times [N], \{0,1\})$ be such that if $\mathcal{O} \in \mathcal{F}$ then*

$$\Pr_{a,x}[A_1^{\mathcal{O}}(A_0^{\mathcal{O}}, a, x) = \mathcal{O}(a, x)] \geq \frac{1}{2} + \varepsilon.$$

Then there is a randomized encoding for $\mathcal{F}$ with recovery probability $\Omega(\varepsilon/T)$ and encoding length (in bits) at most $KN + S - \Omega\left(\frac{\varepsilon^2 NK}{T}\right) + O(1)$.

We now prove Theorem 6.

Proof. Let

$$\epsilon = \left| \Pr_{\mathcal{O},a,x}[A_1^{\mathcal{O}}(A_0^{\mathcal{O}}, a, \mathcal{O}(a, x)) = 1] - \Pr_{\mathcal{O},a,y}[A_1^{\mathcal{O}}(A_0^{\mathcal{O}}, a, y) = 1] \right|.$$

We assume for simplicity that M is a power of 2. By Yao's equivalence of distinguishability and predictability [17], there exist $i \in [\log M]$ and oracle algorithms (B_0, B_1) such that B_0 outputs at most $S+1$ bits and B_1 makes at most T oracle queries, and such that

$$\Pr_{\mathcal{O},a,x}[B_1^{\mathcal{O}}(B_0^{\mathcal{O}}, a, \mathcal{O}_1(a, x), \ldots, \mathcal{O}_{i-1}(a, x)) = \mathcal{O}_i(a, x)] \geq 1/2 + \varepsilon/\log M,$$

where $\mathcal{O}_i(a, x)$ denotes the ith bit of $\mathcal{O}(a, x)$. If B_1 queries (a, x) with probability at least $\varepsilon/2 \log M$, we can turn B_1 into an algorithm that inverts $\mathcal{O}(a, x)$ with at least that probability; Theorem 3 then implies

$$\varepsilon = O\left(\log M \cdot \left(\frac{ST}{KN} + \frac{T \log N}{N} \right) \right). \tag{2}$$

Otherwise, we may construct algorithms (C_0, C_1) such that

- C_1 makes at most T oracle queries, and never queries its own input;
- C_0 runs B_0 and also outputs as part of its state the truth table of a function mapping $[K] \times [N]$ to outputs of length at most $(\log M - 1)$ bits;

and such that

$$\Pr_{\mathcal{O}_i, a, x}[C_1^{\mathcal{O}_i}(C_0^{\mathcal{O}_i}, a, x) = \mathcal{O}_i(a, x)] \geq 1/2 + \varepsilon/2 \log M.$$

This means that for at least an $(\varepsilon/4 \log M)$-fraction of $\mathsf{Func}\,([K] \times [N], \{0, 1\})$ it holds that

$$\Pr_{a, x}[C_1^{\mathcal{O}_i}(C_0^{\mathcal{O}_i}, a, x) = \mathcal{O}_i(a, x)] \geq 1/2 + \varepsilon/4 \log M.$$

Lemma 5 thus implies that we can encode that set of functions using at most $KN + KN \cdot (\log M - 1) + S - \Omega \left(\frac{(\varepsilon/\log M)^2 KN}{T} \right) + O(1)$ bits. By Lemma 2, this means we must have

$$\Omega \left(\frac{(\varepsilon/\log M)^2 KN}{T} \right) - \log \left(\frac{\varepsilon}{T} \right) - \log \left(\frac{\varepsilon}{4 \log M} \right) \leq S + O(1),$$

which in turn implies $\varepsilon = O \left(\log M \cdot \sqrt{\frac{ST}{KN}} \right)$. This, combined with (2), implies the theorem. ∎

As intuition for the proof of Theorem 7, note that we may view a pseudorandom function as a pseudorandom generator mapping a key to the truth table for a function, with the main difference being that the distinguisher is not given the entire truth table as input but instead may only access parts of the truth table via queries it makes. We may thus apply the same idea as in the proof of Theorem 6, with the output length (i.e., $\log M$) replaced by the number of queries the distinguisher makes. However in this case, Lemma 5 cannot be directly applied and a slightly more involved compression argument is required.

With this in mind, we turn to the proof of Theorem 7:

Proof. Let

$$\epsilon = \left| \Pr_{\mathcal{O}, a, k}[A_1^{\mathcal{O}, \mathcal{O}(a, k, \cdot)}(A_0^{\mathcal{O}}, a) = 1] - \Pr_{\mathcal{O}, a, f}[A_1^{\mathcal{O}, f}(A_0^{\mathcal{O}}, a) = 1] \right|.$$

By Yao's equivalence of distinguishability and predictability [17], there exist $i \in [q]$ and oracle algorithms (B_0, B_1) such that B_0 outputs at most $S + 1$ bits and B_1 makes at most T oracle queries to $\mathcal{O}$ and $i \leq q$ distinct queries to the second oracle, such that

$$\Pr_{\mathcal{O}, a, k}[B_1^{\mathcal{O}, \mathcal{O}(a, k, \cdot)}(B_0^{\mathcal{O}}, a) \text{ outputs } (x, b), \text{ s.t. } \mathcal{O}(a, k, x) = b] \geq \frac{1}{2} + \varepsilon/q,$$

where it is required that B_1 not query x to its second oracle. If B_1 queries $\mathcal{O}$ on any query with prefix (a, k), with probability at least $\varepsilon/2q$, we can turn B_1 into an algorithm that inverts random oracle $\mathcal{O}'$ from $[K] \times [N]$ to $\{0, 1\}^L$ with that probability where the output of $\mathcal{O}'(a,k)$ is the truth table of $\mathcal{O}(a,k,\cdot)$. Theorem 3 then implies

$$\varepsilon = O\left(q \cdot \left(\frac{ST}{KN} + \frac{T\log N}{N}\right)\right). \tag{3}$$

Otherwise, we may construct algorithms C_1 which behaves as B_1 except when B_1 queries $\mathcal{O}$ on any query with prefix (a,k), C_1 outputs a random guess. C_1 satisfies that

$$\Pr_{\mathcal{O},a,k}[C_1^{\mathcal{O},\mathcal{O}(a,k,\cdot)}(B_0^{\mathcal{O}},a) \text{ outputs } (x,b) \text{ s.t. } \mathcal{O}(a,k,x) = b] \geq 1/2 + \varepsilon/2q.$$

This means that for at least an $(\varepsilon/4q)$-fraction of $\mathsf{Func}\left([K] \times [N], \{0,1\}^{[L]}\right)$, it holds that

$$\Pr_{\mathcal{O},a,k}[C_1^{\mathcal{O},\mathcal{O}(a,k,\cdot)}(B_0^{\mathcal{O}},a) \text{ outputs } (x,b) \text{ s.t. } \mathcal{O}(a,k,x) = b] \geq 1/2 + \varepsilon/4q.$$

We can encode the set of functions using randomized encoding. Specifically, the encoder uses randomness r to pick a set $R \subseteq [K] \times [N]$, where each $(a,k) \in [K] \times [N]$ is included in R with probability $1/10T$. For $a \in [K]$, let $G \subseteq R$ be the set of $(a,k) \in R$ such that $C_1^{\mathcal{O},\mathcal{O}(a,k,\cdot)}(B_0^{\mathcal{O}},a)$ does not query $\mathcal{O}$ on any point with prefix $(a',k') \in G$. Let G_0 be the subset of G such that the output of C_1 is correct and $G_1 = G \setminus G_0$.

As in De et al. [6], with probability at least $\varepsilon/160qT$, $|G_0| - |G_1| \geq \frac{\varepsilon KN}{80qT}$ and $|G| = \Omega(\frac{KN}{T})$ hold. To see this, note by a Chernoff bound, G has $\Omega(\frac{KN}{T})$ points with probability at least $1 - e^{-\frac{2KN}{T}}$. The expected difference between $|G_0|$ and $|G_1|$ is at least $\frac{\varepsilon KN}{40qT}$. By averaging argument, with probability at least $\frac{\varepsilon}{80qT}$ their difference is at least $\frac{\varepsilon KN}{80qT}$. So with probability at least $\frac{\varepsilon}{80qT} - e^{-\frac{2KN}{T}} \geq \frac{\varepsilon}{160qT}$, both events happen. Conditioned on that, we encode $\mathcal{O}$ as follows (otherwise we output empty string):

1. Include $B_0^{\mathcal{O}}$. This uses at most $S + 1$ bits.
2. For each $(a,k) \in ([K] \times [N]) \backslash R$ (in lexicographic order), include the truth table of $\mathcal{O}(a,k,\cdot)$. Then for each $(a,k) \in R \backslash G$ (in lexicographic order), include the truth table of $\mathcal{O}(a,k,\cdot)$. This uses a total of $(KN - |G|) \cdot L$ bits.
3. Include a description of G_0. This uses $\log\binom{|G|}{|G_0|}$ bits.
4. For each $(a,k) \in G$ (in lexicographic order), include in the encoding the answers to all the oracle queries made by C_1 to the second oracle $\mathcal{O}(a,k,\cdot)$, and for every x such that (a,k,x) is not queried by C_1 to $\mathcal{O}(a,k,\cdot)$ and x is not the output of C_1, add $\mathcal{O}(a,k,x)$ to the encoding. This uses a total of $|G|(L-1)$ bits.

To decode, the decoder first uses r to recover the set R defined above. Then it does the following:

1. Recover $B_0^{\mathcal{O}}$.
2. For each $(a, k) \in ([K] \times [N]) \setminus R$, recover the truth table of $\mathcal{O}(a, k, \cdot)$. Identify set G by running C_1 with $B_0^{\mathcal{O}}$ on $(a, k) \in R$ because if C_1 on (a, k) only makes query outside R, then $(a, k) \in G$. Go over $(a, k) \in R \setminus G$, and recover the truth table of $\mathcal{O}(a, k, \cdot)$.
3. Recover G_0.
4. For each $(a, k) \in G$, run $C_1(B_0^{\mathcal{O}}, a)$ while answering the oracle queries to the first oracle using recovered values and to the second oracle using the values stored in the encoding. Suppose C_1 outputs x, b, if $(a, k) \in G_0$, recover $\mathcal{O}(a, k, x) = b$ otherwise $\mathcal{O}(a, k, x) = 1 - b$. After that for which $\mathcal{O}(a, k, x)$ is not yet defined, recover the value of $\mathcal{O}(a, k, x)$ from the remainder of the encoding.

Because we condition on $|G| \leq KN/T$ and $|G_0| - |G_1| \geq \varepsilon KN/80qT$ which implies $\log \binom{|G|}{|G_0|} \leq |G| H(1/2 + \varepsilon KN/80T |G|) \leq |G| - \Omega((\varepsilon/q)^2 KN/T)$, where H is the binary entropy function. The maximal length is at most

$$KNL + S + 1 + \log \binom{|G|}{|G_0|} - |G| \leq KNL + S + O(1) - \Omega((\varepsilon/q)^2 KN/T).$$

By Lemma 2, we have

$$S \geq \Omega((\varepsilon/q)^2 KN/T) - \log \Omega(\frac{\varepsilon}{160qT}) - \log(\frac{\varepsilon}{4q}).$$

which implies $\varepsilon \leq O(q \cdot \sqrt{\frac{ST}{KN}})$. Overall we obtain $\varepsilon \leq O(q \cdot (\sqrt{\frac{ST}{KN}} + \frac{T}{N} \cdot \log N))$.

∎

6   Message Authentication Codes (MACs)

In this section, we prove the following theorem.

Theorem 8. *Consider random oracles $\mathcal{O} \in \mathsf{Func}([K] \times [N] \times [L], [M])$. For any oracle algorithms (A_0, A_1) such that A_0 outputs S-bit state and A_1 makes at most T queries to $\mathcal{O}$,*

$$\Pr_{\mathcal{O}, a, k} \left[(m, t) := A_1^{\mathcal{O}, \mathcal{O}(a, k, \cdot)}(A_0^{\mathcal{O}}, a) : \mathcal{O}(a, k, m) = t \right]$$

$$= O \left(\frac{ST}{KN} + \frac{T}{M} + \frac{T \log N}{N} \right),$$

where it is required that A_1 not query m to its second oracle.

Note that any generic inversion attack can be used to attack the above construction of a MAC by fixing some $m \in [L]$ and then inverting the function $\mathcal{O}(a, \cdot, m)$ given a; in this sense, it is perhaps not surprising that the bound above contains terms $\mathcal{O}\left(\frac{ST}{KN} + \frac{T \log N}{N}\right)$ as in Theorem 3. There is, of course, also a trivial guessing attack that achieves advantage $1/M$.

Proof. If A_1 queries $\mathcal{O}$ on any query with prefix (a, k), with probability at least $\varepsilon/2$, we can turn A_1 into an algorithm that inverts random oracle $\mathcal{O}'$ from $[K] \times [N]$ to $[M^L]$ with that probability where the output of $\mathcal{O}'(a, k)$ is the truth table of $\mathcal{O}(a, k, \cdot)$. Then by Theorem 3, we obtain $\varepsilon \leq O(\frac{ST}{KN} + \frac{T \log N}{N})$. Otherwise, we may construct algorithms B_1 which behaves as A_1 except when B_1 queries $\mathcal{O}$ on any query with prefix (a, k), B_1 outputs a random guess. B_1 satisfies that

$$\Pr_{\mathcal{O}, a, k}[B_1^{\mathcal{O}, \mathcal{O}(a, k, \cdot)}(A_0^{\mathcal{O}}, a) \text{ outputs } (m, t) \text{ s.t. } \mathcal{O}(a, k, m) = t] \geq \varepsilon/2.$$

where it is required that B_1 not query m to its second oracle.

Fix $\mathcal{O}: [K] \times [N] \times [L] \to [M]$. Let $U_{\mathcal{O}}$ be the set of (a, k) such that B_1 succeeds on (a, k). Let $G_{\mathcal{O}}$ be the subset of $U_{\mathcal{O}}$ such that for every $(a, k) \in G_{\mathcal{O}}$, $B_1^{\mathcal{O}, \mathcal{O}(a, k, \cdot)}$ does not query its first oracle with any query with prefix $(a', k') \in G_{\mathcal{O}}$. Because B_1 makes at most T queries, there exists $G_{\mathcal{O}}$ with size at least $|U_{\mathcal{O}}|/(T+1)$.

We can encode $\mathcal{O}$ as follows.

1. Include $A_0^{\mathcal{O}}$, $|G_{\mathcal{O}}|$ and a description of $G_{\mathcal{O}}$. This uses a total of $S + \log KN + \log \binom{NK}{|G_{\mathcal{O}}|}$ bits.
2. For each $(a, k) \in ([K] \times [N]) \setminus G_{\mathcal{O}}$ (in lexicographic order), include the truth table of $\mathcal{O}(a, k, \cdot)$. This uses a total of $(KN - |G_{\mathcal{O}}|) \cdot L \log M$ bits.
3. For each $(a, k) \in G_{\mathcal{O}}$ (in lexicographic order), include in the encoding the answers to all the oracle queries made by B_1 to the second oracle $\mathcal{O}(a, k, \cdot)$, and then for every m such that (a, k, m) is not queried by C_1 to $\mathcal{O}(a, k, \cdot)$ and m is not the output of C_1, add $\mathcal{O}(a, k, m)$ to the encoding. This uses a total of $|G| (L - 1) \log M$ bits.

Decoding is done in the obvious way. The encoding length is at most

$$KNL \log M + S + \log KN + \log \binom{KN}{|G_{\mathcal{O}}|} - |G_{\mathcal{O}}| \log M$$

By $\log \binom{KN}{|G_{\mathcal{O}}|} \leq |G_{\mathcal{O}}| \log \frac{eKN}{|G_{\mathcal{O}}|}$ and log-sum inequality, the average length over all possible $\mathcal{O}$ is at most

$$KNL \log M + S + \log KN + \mathbb{E}[|G_{\mathcal{O}}|] \log \frac{eKN}{M \cdot \mathbb{E}_f[|G_{\mathcal{O}}|]}.$$

But Lemma 1 shows that any such encoding must have expected length at least $KNL \log M$ bits. We thus conclude that

$$S + \log KN \geq \mathbb{E}[|G_{\mathcal{O}}|] \log \frac{M \mathbb{E}[|G_{\mathcal{O}}|]}{eKN} \geq \frac{\varepsilon NK}{2(T+1)} \log \frac{M\varepsilon}{2e(T+1)}.$$

where the second inequality is due to the monotonicity of $y \log y$ for $y \geq 1$ and $\mathbb{E}[\|G_{\mathcal{O}}\|] \geq \mathbb{E}[\frac{|U_{\mathcal{O}}|}{T+1}] \geq \frac{\varepsilon NK}{2(T+1)}$. This implies Theorem 8 since either $\varepsilon \leq \frac{4e(T+1)}{M}$ or else $\varepsilon \leq \frac{2(S+\log KN)(T+1)}{NK} = O(\frac{ST}{NK} + \frac{T \log N}{N})$.   ∎

Acknowledgments. Jonathan Katz thanks Christine Evangelista, Aaron Lowe, Jordan Schneider, Lynesia Taylor, Aishwarya Thiruvengadam, and Ellen Vitercik, who explored problems related to salting and rainbow tables as part of an NSF-REU program in the summer of 2014.

References

1. Alon, N., Goldreich, O., Håstad, J., Peralta, R.: Simple constructions of almost k-wise independent random variables. Random Struct. Algorithms **3**(3), 289–304 (1992)
2. Barkan, E., Biham, E., Shamir, A.: Rigorous Bounds on Cryptanalytic Time/Memory Tradeoffs. In: Dwork, C. (ed.) CRYPTO 2006. LNCS, vol. 4117, pp. 1–21. Springer, Heidelberg (2006). doi:10.1007/11818175_1
3. Bellare, M., Ristenpart, T., Tessaro, S.: Multi-instance security and its application to password-based cryptography. In: Safavi-Naini, R., Canetti, R. (eds.) CRYPTO 2012. LNCS, vol. 7417, pp. 312–329. Springer, Heidelberg (2012). doi:10.1007/978-3-642-32009-5_19
4. Bellare, M., Rogaway, P.: Random oracles are practical: a paradigm for designing efficient protocols. In: 1st ACM Conference on Computer and Communications Security, pp. 62–73. ACM Press (1993)
5. Bernstein, D.J., Lange, T.: Non-uniform cracks in the concrete: the power of free precomputation. In: Sako, K., Sarkar, P. (eds.) ASIACRYPT 2013. LNCS, vol. 8270, pp. 321–340. Springer, Heidelberg (2013). doi:10.1007/978-3-642-42045-0_17
6. De, A., Trevisan, L., Tulsiani, M.: Time space tradeoffs for attacks against one-way functions and PRGs. In: Rabin, T. (ed.) CRYPTO 2010. LNCS, vol. 6223, pp. 649–665. Springer, Heidelberg (2010). doi:10.1007/978-3-642-14623-7_35
7. Dodis, Y., Steinberger, J.: Message authentication codes from unpredictable block ciphers. In: Halevi, S. (ed.) CRYPTO 2009. LNCS, vol. 5677, pp. 267–285. Springer, Heidelberg (2009). doi:10.1007/978-3-642-03356-8_16
8. Fiat, A., Naor, M.: Rigorous time/space trade-offs for inverting functions. SIAM J. Comput. **29**(3), 790–803 (1999)
9. Gennaro, R., Gertner, Y., Katz, J., Trevisan, L.: Bounds on the efficiency of generic cryptographic constructions. SIAM J. Comput. **35**(1), 217–246 (2005)
10. Gennaro, R., Trevisan, L.: Lower bounds on the efficiency of generic cryptographic constructions. In: 41st Annual Symposium on Foundations of Computer Science (FOCS), pp. 305–313. IEEE (2000)
11. Hellman, M.: A cryptanalytic time-memory trade-off. IEEE Trans. Inf. Theory **26**(4), 401–406 (1980)
12. Katz, J., Lindell, Y.: Introduction to Modern Cryptography, 2nd edn. Chapman & Hall/CRC Press (2014)
13. Morris, R., Thompson, K.: Password security: a case history. Commun. ACM **22**(11), 594–597 (1979)
14. Oechslin, P.: Making a faster cryptanalytic time-memory trade-off. In: Boneh, D. (ed.) CRYPTO 2003. LNCS, vol. 2729, pp. 617–630. Springer, Heidelberg (2003). doi:10.1007/978-3-540-45146-4_36

15. Rogaway, P.: Formalizing human ignorance. In: Nguyen, P.Q. (ed.) VIETCRYPT 2006. LNCS, vol. 4341, pp. 211–228. Springer, Heidelberg (2006). doi:10.1007/11958239_14

16. Unruh, D.: Random oracles and auxiliary input. In: Menezes, A. (ed.) CRYPTO 2007. LNCS, vol. 4622, pp. 205–223. Springer, Heidelberg (2007). doi:10.1007/978-3-540-74143-5_12

17. Yao, A.C.: Theory and applications of trapdoor functions. In: 23rd Annual Symposium on Foundations of Computer Science (FOCS), pp. 80–91. IEEE (1982)

18. Yao, A.C.-C.: Coherent functions and program checkers. In: 22nd Annual ACM Symposium on Theory of Computing (STOC), pp. 84–94. ACM Press (1990)

Provable Security for Symmetric Cryptography II

Modifying an Enciphering Scheme After Deployment

Paul Grubbs[1]([✉]), Thomas Ristenpart[1], and Yuval Yarom[2]

[1] Cornell Tech, New York, USA
pag225@cornell.edu
[2] CSIRO, University of Adelaide and Data61, Adelaide, Australia

Abstract. Assume that a symmetric encryption scheme has been deployed and used with a secret key. We later must change the encryption scheme in a way that preserves the ability to decrypt (a subset of) previously encrypted plaintexts. Frequent real-world examples are migrating from a token-based encryption system for credit-card numbers to a format-preserving encryption (FPE) scheme, or extending the message space of an already deployed FPE. The ciphertexts may be stored in systems for which it is not easy or not efficient to retrieve them (to re-encrypt the plaintext under the new scheme).

We introduce methods for functionality-preserving modifications to encryption, focusing particularly on deterministic, length-preserving ciphers such as those used to perform format-preserving encryption. We provide a new technique, that we refer to as the Zig-Zag construction, that allows one to combine two ciphers using different domains in a way that results in a secure cipher on one domain. We explore its use in the two settings above, replacing token-based systems and extending message spaces. We develop appropriate security goals and prove security relative to them assuming the underlying ciphers are themselves secure as strong pseudorandom permutations.

1 Introduction

We explore the ability to modify a deployed symmetric encryption scheme in a way that preserves some of its previous input-output mappings. This may prove useful in a variety of settings, but we are motivated and will focus on addressing two specific ones that arise in the increasing deployment of format-preserving encryption (FPE) schemes.

Modifying deployed FPE schemes. In a variety of settings conventional symmetric encryption is difficult or impossible to utilize, due to unfortunate constraints imposed by legacy software systems. A common problem is that encryption produces ciphertexts whose format are ruled out by restrictive application programming interfaces (APIs) and/or pre-existing database schema. This problem prevented, for example, encryption of credit-card numbers (CCNs) in a variety of settings. Format-preserving encryption (hereafter FPE) is a technique aimed at such problems, allowing one to encrypt a plaintext item of some format

© International Association for Cryptologic Research 2017
J.-S. Coron and J.B. Nielsen (Eds.): EUROCRYPT 2017, Part II, LNCS 10211, pp. 499–527, 2017.
DOI: 10.1007/978-3-319-56614-6_17

to a ciphertext of the same format (16 digit CCN). It has seen wide academic study [4,6,11,16,17,20] as well as widespread use in industry [12,18,23,25].

Before the advent of strong FPE schemes, companies often instead used what are called *tokenization systems* to solve the format-constrained ciphertext problem. One generates a random *token* using generic techniques for creating random strings with a certain format, i.e., sampling a token C uniformly from some set $\mathcal{M}$ that defines the set of strings matching the format. A *token table* containing plaintext-to-token mappings is stored in a database, and applications which need access to data in the clear ask the database to do a lookup in this table for the plaintext corresponding to a particular token. Often applications reside in other organizations that have outsourced CCN management to a payments service. This technique can be viewed as a particularly inefficient FPE implementing a permutation $F_{K_o} : \mathcal{M} \rightarrow \mathcal{M}$ for a "key" K_o that is a lazily constructed map of plaintexts to random ciphertexts (tokens).

Now that we have better approaches to FPE, a common problem faced by companies is upgrading from tokenization to an FPE scheme. This can be challenging when tokens have been distributed to various systems and users; there may be no way to recall the old tokens. The challenge here is therefore to build a new cipher that "completes" the domain of the cipher partially defined by the token table thus far.

A second example arises in the use of FPE for encryption of data before submission to restrictive cloud computing APIs. An instance of this arises with Salesforce, a cloud provider that performs customer relations management — at core they maintain on behalf of other companies databases of information about the companies' customers. As such, companies using Salesforce and desiring encryption of data before uploading have a large number of data fields with various format restrictions: email addresses, physical addresses, CCNs, names, phone numbers, etc. While we now have in-use solutions for defining formats via easy-to-use regular expressions [4,11,16], it is often the case that formats must change later. As a simple example: one may have thought only 16-digit CCNs were required, but later realized that 15-digit cards will need to be handled as well. Here we would like to, as easily as possible, modify an FPE $F_{K_o} : \mathcal{D} \rightarrow \mathcal{D}$ to one that works for an extended message space $\mathcal{M} \supset \mathcal{D}$. As with tokens and for similar reasons, it would be useful to maintain some old mappings under F in the new cipher.

Functionality-preserving modifications to encryption. The core challenge underneath both examples above is to take an existing cipher F_{K_o} operating on some domain and, using knowledge of K_o, build a new cipher E_K such that $E_K(M) = F_{K_o}(M)$ for $M \in \mathcal{T} \subset \mathcal{M}$. We refer to $\mathcal{T}$ as the *preservation set*. In the tokenization example $\mathcal{T}$ could be the full set of messages for which entries in the table exist, and in the format-extension example $\mathcal{T}$ could be a subset of $\mathcal{D}$.

We note that trivial solutions don't seem to work. As already explained, the simplest solution of replacing old ciphertexts with new ones won't work when the old ciphertexts are unavailable (e.g., because other organizations have stored them locally). Furthermore, even when old ciphertexts can be revoked, the cost of decrypting and re-encrypting the whole database may be prohibitive.

Alternatively, consider encrypting a new point M in the following way. First check if $M \in \mathcal{T}$ and if so use the old cipher $F_{K_{\circ}}(M)$. Otherwise use a fresh key K for a new cipher E and apply $E_K(M)$. But this doesn't define a correct cipher, because different messages may encrypt to the same ciphertext: there will exist $M \notin \mathcal{T}$ and $M' \in \mathcal{T}$ for which $E_K(M) = F_{K_{\circ}}(M')$.

Our contributions. We explore for the first time functionality-preserving modifications to deployed ciphers. A summary of the settings and our constructions is given in Fig. 1. Our main technical contribution is a scheme that we call the Zig-Zag construction. It can be used both in the tokenization setting and, with simple modifications, in the expanded format setting. It uses a new kind of cycle walking to define the new cipher on $\mathcal{M}$ using the old cipher $F_{K_{\circ}}$ and a helper cipher $\overline{E}_{\overline{K}} \colon \mathcal{M} \to \mathcal{M}$. The old mappings on points in $\mathcal{T}$ are preserved.

We analyze security of Zig-Zag in two cases, corresponding to the two situations discussed: (1) in domain completion, F has ciphertexts in the range $\mathcal{M}$ and (2) in domain extension, F works on a smaller domain $\mathcal{D} \subset \mathcal{M}$. For the first case, we show that the Zig-Zag construction is provably a strong pseudorandom permutation (SPRP) assuming that F and $\overline{E}$ both are themselves SPRPs. Extending to deal with tweaks [13] is straightforward.

Setting	Description	Achievable security	Construction
Domain completion	Preserve partially defined cipher $\mathcal{T} \to \mathcal{M}$ in new cipher $\mathcal{M} \to \mathcal{M}$	SPRP	Zig-Zag
Domain extension	Extend cipher $\mathcal{D} \to \mathcal{D}$ to $\mathcal{M} \to \mathcal{M}$	SEPRP	Zig-Zag
	Extend cipher $\mathcal{D} \to \mathcal{D}$ to $\mathcal{M} \to \mathcal{M}$	SPRP (unknown $\mathcal{T}$)	Recursive Zig-Zag

Fig. 1. Summary of different settings, security goals, and constructions. The set $\mathcal{M}$ is the new cipher's domain, the set $\mathcal{T}$ is the set of preserved points, and $\mathcal{D} \subset \mathcal{M}$ is the original domain in the extension setting.

The second case is more nuanced. We first observe that *no* scheme can achieve SPRP security when adversaries know $\mathcal{T}$. The attack is straightforward: query a point from $\mathcal{T}$ and see if the returned ciphertext is in $\mathcal{D}$ or not. Because it is functionality preserving, the construction must always have a ciphertext in $\mathcal{D}$, whereas a random permutation will do so only with probability $|\mathcal{D}|/|\mathcal{M}|$. Since we expect this ratio to usually be small, the attack will distinguish with high probability.

This begs the question of what security level is possible in this context. Investigating the attack ruling out SPRP security, we realize that the main issue is that ciphers that preserve points will necessarily leak to adversaries that the underlying plaintext is in $\mathcal{T}$. We formalize a slightly weaker security goal, called strong extended pseudorandom permutation (SEPRP) security in which a cipher must be indistinguishable from an ideal extended random permutation. Roughly this formalizes the idea that attackers should learn nothing but the fact that the distribution of points in T is slightly different from those in $M \setminus T$. We show that the Zig-Zag construction meets this new notion.

SEPRP security does leak more information to adversaries than does traditional SPRP security, and so we investigate the implications of this for applications. We formally relate SEPRP security to two security notions for FPE schemes from Bellare, Ristenpart, Rogaway, and Stegers [4], message recovery (MR) and message privacy (MP). We highlight the main results regarding MR here, and leave MP to the body. MR requires that an adversary, given the encryption of some challenge message as well as a chosen-plaintext encryption oracle, cannot recover the message with probability better than a simulator can, given no ciphertext and instead a test oracle that only returns one if the queried message equals the challenge. We show that there exist settings for which SEPRP security does *not* imply MR security, by way of an adversary whose success probability is 1, but any simulator succeeds with probability at most $1/2$. The reason is that the adversary can exploit knowledge of membership in T, whereas the simulator cannot.

This result may lead us to pessimistically conclude that SEPRP provides very weak security, but intuition states otherwise: an SEPRP-secure cipher should not leak more than one bit of information about a plaintext (whether or not it is in T). The best MR attack one can come up with should therefore only have a factor of two speedup over attacking an SPRP-secure cipher. The gap between intuition and formalism lies in the strict way MR security was defined in [4]: simulators can only make as many queries as adversaries make and simulators receive nothing to aid in their attack. We therefore introduce a more general MR security notion that we uncreatively call generalized MR (gMR) security. The definition is now parameterized by both an auxiliary information function on the challenge plaintext as well as a query budget for simulators. We show that SEPRP security implies gMR security when the auxiliary information indicates whether the challenge is in T or not. We then show a general result that gMR security with this auxiliary information implies gMR security without any auxiliary information, as long as the simulator can make twice as many queries as the adversary. This makes precise the security gap between SEPRP and SPRP, and the interpretation is simply that adversaries get at most a factor of two speedup in message recovery attacks.

Security and side-channel attacks. The Zig-Zag construction is not inherently constant time, which suggests it may be vulnerable to timing or other side-channel attacks. We prove in the body that timing leaks only whether a message is in T or not, and nothing further. We also discuss possible implementation approaches that avoid even this timing attack.

The Recursive Zig-Zag construction. Above we argued that in the domain extension setting SPRP security is unachievable should the adversary know (at least one) point in T. In some scenarios, the attacker may be unable to learn which points are in T, but is able to learn some information on T such as its size. This might arise, for example, should an attacker learn the size of a database but not have direct access to its contents. In this context the attack discussed above showing SPRP insecurity for all schemes no longer applies. We therefore explore feasibility of SPRP security in the domain extension setting when attackers know $|\mathcal{T}|$ but do not know $\mathcal{T}$.

First we show that SPRP security is still ruled out if the gap between the size of the old domain and the target domain is smaller than the number of new points by which the domain was extended, namely $|\mathcal{D}| - |\mathcal{T}| < |\mathcal{M} \setminus \mathcal{D}|$. To gain some intuition, consider the minimum number of points from $\mathcal{D}$ that map back to points in $\mathcal{D}$ for both an extended cipher and for a random permutation. For an extended cipher, at least $|\mathcal{T}|$ points are necessarily preserved, and so map to points in $\mathcal{D}$. For a random permutation, if the number of added points is large enough there is a probability that no point in $\mathcal{D}$ is mapped back to $\mathcal{D}$. Consequently, for a large enough $\mathcal{T}$ or when we add many points, the distribution of the number of points in $\mathcal{D}$ that map back to $\mathcal{D}$ differs sufficiently between extended ciphers and random permutations to give an adversary distinguishing advantage. For other ranges of parameters, however, with overwhelming probability a random permutation will have a subset of inputs that maps back to $\mathcal{D}$.

Unfortunately, the Zig-Zag construction does not meet SPRP security in this unknown $\mathcal{T}$ setting. Intuitively, the reason is that the construction biases the number of sets of size $|\mathcal{T}|$ that map to $\mathcal{D}$, with this bias growing as $|\mathcal{T}|$ grows.

We therefore provide a domain extension construction in the unknown $\mathcal{T}$ setting. It starts with a helper cipher $\overline{E}$ on $\mathcal{M}$, and utilizes the basic structure of the Zig-Zag to patch it in order preserve the points of $\mathcal{T}$. The patching occurs by replacing mappings for a t-size subset of $\mathcal{M}$ that $\overline{E}$ maps to $\mathcal{T}$. By patching those points in that set, as opposed to arbitrary points as in done in Zig-Zag, we preserve the distribution of sets of size $|\mathcal{T}|$ that are mapped to $\mathcal{D}$. To make this efficient we perform the patching recursively, handling the points in $\mathcal{T}$ one at a time, hence the name Recursive Zig-Zag for the construction. We prove that the construction works in expected time and space proportional (with a small constant) to $|\mathcal{T}|$, making it feasible for an application where $\mathcal{T}$ would need to be stored anyway, and analyze its SPRP security.

A ranking-based approach. An anonymous reviewer pointed out a potential alternative to our Zig-Zag construction that takes advantage of ranking functions, which are efficiently computable and invertible bijections from a domain $\mathcal{M}$ to $\mathbb{Z}_{|\mathcal{M}|}$. Ranking underlies many FPE constructions, and in some ways the reviewer's construction is simpler than Zig-Zag. The reviewer gave us permission to present the idea and discuss it in comparison to Zig-Zag. See Sect. 4.

Limitations and open problems. The approach we explore, of modifying a scheme after deployment, has several limitations. First, it requires the ability to perform membership tests against $\mathcal{T}$ and requires the old key K_{o} for the lifetime of the updated cipher. These must both be protected (in the former case, since it may leak some information about how people were using the cipher). In the case that $\mathcal{T}$ is an explicit list, one could cryptographically hash each point to provide some partial protection of plaintext data in case of key compromise, but this would only provide marginal benefit in case of exposure since dictionary attacks would be possible.

Second, as points in $\mathcal{T}$ are submitted it would be convenient to gracefully remove points from it and "refresh" them with new mappings. This would be useful in the tokenization scenario should the client be able to update its token after a query. But there is no way to make "local" modifications to a cipher as any changed mapping necessarily affects at least one other domain point. We leave how to modify schemes gradually over time as an interesting open problem.

Our work only considered updating ciphers, but it could be that other cryptographic primitives might benefit from functionality-preserving updates. Future work could determine whether compelling scenarios exist, and what solutions could be brought to bear.

Full version. Due to space constraints, we had to omit a number of proofs. These will be available in the full version, which will be available from the authors' websites.

2 Preliminaries

Let $\mathcal{M}$ be a set, called the domain, and $\mathcal{K}$ be a set called the key space. Later we abuse notation and use sets to also denote efficient representations of them. A cipher is a family of permutations $E : \mathcal{K} \times \mathcal{M} \rightarrow \mathcal{M}$. This means that $E_K(\cdot) = E(K, \cdot)$ is a permutation for all $K \in \mathcal{K}$. Both E_K and its inverse E_K^{-1} must be efficient to compute. Block ciphers are a special case in which $\mathcal{M} = \{0, 1\}^n$ for some integer n, and format-preserving encryption [4] is a generalization of ciphers that allows multiple lengths as well as tweaks [13]. Our results translate to that more general setting as well, but for the sake of simple exposition we focus on only a single domain, and use the term cipher instead of FPE.

For a function f and set $\mathcal{X}$ that is a subset of its domain, we write $\mathsf{Img}_f(\mathcal{X})$ to denote the image of $\mathcal{X}$ under f, i.e., the set $\{f(x) \mid x \in \mathcal{X}\}$.

SPRP security. Ciphers are considered secure if they behave like strong pseudorandom permutations (SPRPs). Let $\mathrm{Perm}(\mathcal{M})$ be the set of all permutations on any set $\mathcal{M}$. Consider a cipher $E : \mathcal{K} \times \mathcal{M} \rightarrow \mathcal{M}$. We define the *advantage* of an adversary $\mathcal{A}$ in distinguishing E and its inverse from a random permutation and its inverse as $\mathbf{Adv}_E^{\mathrm{sprp}}(\mathcal{A}) = \left|\Pr\left[\mathrm{SPRP1}_E^{\mathcal{A}} \Rightarrow 1\right] - \Pr\left[\mathrm{SPRP0}_E^{\mathcal{A}} \Rightarrow 1\right]\right|$. The two games SPRP1 and SPRP0 are defined in Fig. 2 and the probabilities are taken over the random coins used in the games. We will give a concrete security treatment, meaning we will explicitly relay the running time (in some RAM

main $\mathrm{SPRP1}_E^{\mathcal{A}}$	**main** $\mathrm{SPRP0}_E^{\mathcal{A}}$
$K \leftarrow_\$ \mathcal{K}$	$\pi \leftarrow_\$ \mathrm{Perm}(\mathcal{D})$
$b' \leftarrow \mathcal{A}^{\mathrm{Enc,Dec}}$	$b' \leftarrow \mathcal{A}^{\mathrm{Enc,Dec}}$
return b'	return b'
proc $\mathrm{Enc}(M)$	**proc** $\mathrm{Enc}(M)$
return $E_K(M)$	return $\pi(M)$
proc $\mathrm{Dec}(C)$	**proc** $\mathrm{Dec}(C)$
return	return
$E_K^{-1}(C)$	$\pi^{-1}(C)$

Fig. 2. SPRP security games for a cipher E.

model of computation) and number of queries made by adversaries.

We assume that adversaries do not repeat any oracle queries and do not ask queries to which they already knows the answer, like querying a decryption oracle with the result of a previous encryption oracle query.

The hypergeometric distribution. A hypergeometrically distributed random variable X has probability mass function

$$\Pr[X = k] = \frac{\binom{K}{k}\binom{N-K}{n-k}}{\binom{N}{n}}$$

where N is the total number of samples, K is the number of marked samples, n is the number of samples drawn, and k is the number of marked samples of the n total samples.

The hypergeometric distribution has very strong tail bounds. We formalize this as the following lemma. A full proof of this lemma can be found in [8,22] but is omitted here.

Lemma 1. *Let X be a hypergeometrically distributed random variable and n be the number of samples. Then for any real number t, $\Pr[E[X]+tn \le X] \le e^{-2t^2 n}$, where e is the base of the natural logarithm.*

3 Extending Partially Used Message Spaces

We start by considering how to replace an existing cipher $F : \mathcal{K} \times \mathcal{M} \to \mathcal{M}$ with a new one $E : \mathcal{K} \times \mathcal{M} \to \mathcal{M}$, while maintaining backwards compatibility with the subset of the message space $\mathcal{T} \subset \mathcal{M}$ for which messages have already been encrypted. Our motivation for this originates with the following situation that arises in practice.

Updating tokenization deployments. Tokenization [27] is a set of techniques whose purpose is to provide confidentiality for relatively small data values (e.g., government ID numbers or credit card numbers). Usually tokenization is employed to meet regulatory requirements imposed by governments or industry standards bodies like PCI [10].

A tokenization system usually consists of a few parts: a server front-end which accepts tokenize/detokenize requests from authenticated clients, a cryptographic module that produces tokens for plaintext values, and a database backend that stores the plaintext/token mapping table. Each time a new tokenize request occurs for a plaintext M, a randomly generated value from $\mathcal{M}$ is chosen to

Fig. 3. Tokenization system after choosing random values C_1, C_2, C_3 for plaintexts $\mathcal{T} = \{M_1, M_2, M_3\}$.

be $F_{K_o}(M)$. Here K_o is just an explicitly stored table of message, token pairs. The token $F_{K_o}(M)$ is given back to the requester and stored for later use. Let $T \subseteq M$ be the set of all points for which tokens have been distributed. A diagram is shown in Fig. 3.

Such tokenization systems are a bit clumsy. Primarily they do not scale very well, requiring protected storage linear in the number of plaintexts encrypted compared to FPE schemes that achieve this with just a small 128 bit key. (One cannot just store a key for a pseudorandom number generator and recreate values; this doesn't allow efficient decryption.) Companies therefore often want to move from tokenization to a modern solution using an FPE E.

One could choose a new key for E but the problem is then that the previously returned tokens will be invalidated, and this may cause problems down the road when clients make use of these tokens. Hence the desire to perform what we call *domain completion*: define $E_K(M)$ such that $E_K(M) = F_{K_o}(M)$ for all $M \in T$. This ensures that previously distributed tokens are still valid even under the new functionality. In deployment, any method for domain completion would most likely retain the token table, but the crucial difference in terms of performance is the immutability of the table. After switching to the keyed cipher, the table can be made read-only and distributed with the FPE software as a file with no expensive and complicated database needed to ensure consistency and availability. In most contexts, one will want to keep the file secret since it may leak information about previous use of F.

Domain completion, formally. A domain completion setting is defined to be a tuple (F, M, T) consisting of a cipher with domain M and the preservation set $T \subseteq M$. Relative to some fixed domain completion setting (that later will always be made clear from context), a domain-completed cipher DCC $= (KT, E)$ consists of an algorithm and a cipher. The algorithm is called a domain-completion key transform. It is randomized, takes as input a key K_o for F and the preservation set T, and outputs a key for the cipher E. The cipher is assumed to have a key space compatible with the output of KT. For some preservation set T, the induced key generation algorithm for E consists of choosing a random key K_o for F, running $KT(K_o, T)$ and returning the result.

A domain-completed cipher DCC preserves a point M if $E_K(M) = F_{K_o}(M)$ with probability one over the experiment defined by running the induced key generation for E. We say that KT preserves a set T if it preserves each $M \in T$. The ability of a key transformation to achieve preservation implies that K must somehow include (an encoding of) K_o. In the case where F is a tokenization system, then K_o is a table of at least $t = |T|$ points.

We measure security for a domain-completed cipher via the SPRP advantage of the cipher E using its induced key generation algorithm. We will quantify over all preservation sets or that security should hold even if the adversary knows T.

4 Domain Completion via Rank-Encipher-Unrank

An anonymous reviewer pointed out an approach to domain completion for schemes constructed using the rank-encipher-unrank approach of [4]. With permission we reproduce it here. Recall that a rank-encipher-unrank construction uses a ranking function $rank : \mathcal{M} \to \mathbb{Z}_m$, which is a bijection with inverse $unrank : \mathbb{Z}_m \to \mathcal{M}$. Both must be efficiently computable. One additionally uses cipher $\overline{E}$ that operates on domain $\mathbb{Z}_m$. (This is referred to as an integer FPE in [4]). Then one enciphers a point $X \in \mathcal{M}$ via $unrank(\overline{E}_K(rank(X)))$ and decrypts a point Y via $unrank(\overline{D}_K(rank(Y)))$.

Now consider a domain completion setting $(F, \mathcal{M}, \mathcal{T})$. Let $\mathcal{D} = \mathcal{M} \setminus \mathcal{T}$ be the set of domain points not in the preservation set. Let $\mathcal{R} = \mathcal{M} \setminus \mathsf{Img}_{F_{K_o}}(\mathcal{T})$, where $\mathsf{Img}_{F_{K_o}}(\mathcal{T}) = \{F_{K_o}(X) \mid X \in \mathcal{T}\}$, be the set of range points not in the image of F_{K_o} on $\mathcal{T}$. The construction uses a cipher $\overline{E} : \mathbb{Z}_d \to \mathbb{Z}_d$ and a ranking function $rank : \mathcal{M} \to \mathbb{Z}_m$ with inverse $unrank$. The construction builds from $rank$ rankings $rank_{\mathcal{D}} : \mathcal{D} \to \mathbb{Z}_d$ and $rank_{\mathcal{R}} : \mathcal{R} \to \mathbb{Z}_d$. It then encrypts by checking if a point X is in $\mathcal{T}$ and, if so, outputting $F_{K_o}(X)$ and otherwise outputting $unrank_{\mathcal{R}}(\overline{E}_{\overline{K}}(rank_{\mathcal{D}}(X)))$.

In detail, the domain-completed cipher RTE $= (KT^{rte}, E^{rte})$ is defined as follows. The domain-completion key transform $KT^{rte}(K_o, \mathcal{T})$ first computes the set $\mathsf{Img}_{F_{K_o}}(\mathcal{T})$. Then it computes the set $\{rank(X) \mid X \in \mathcal{T}\}$ and sorts it to obtain a list $\boldsymbol{x} = (\boldsymbol{x}_1, \ldots, \boldsymbol{x}_t) \in \mathbb{Z}_m^t$. Similarly it computes $\{rank(Y) \mid Y \in \mathsf{Img}_{F_{K_o}}(\mathcal{T})\}$ and sorts it to obtain a list $\boldsymbol{y} = (\boldsymbol{y}_1, \ldots, \boldsymbol{y}_t) \in \mathbb{Z}_m^t$. It also chooses a new key $\overline{K}$ for a helper cipher $\overline{E}$ on $\mathbb{Z}_d$ and outputs $K = (K_o, \overline{K}, \boldsymbol{x}, \boldsymbol{y})$.

Enciphering is performed via $E_K^{rte}(X) = unrank_{\mathcal{R}}(\overline{E}_{\overline{K}}(rank_{\mathcal{D}}(X)))$ for rankings defined as follows. The first ranking, $rank_{\mathcal{D}}(X)$, works for $X \in \mathcal{D}$ by computing $x \leftarrow rank(X)$, then determining, via a binary search, the largest index i such that $\boldsymbol{x}_i < x$, and finally outputting $x - i$. The inverse of $rank_{\mathcal{D}}$ is $unrank_{\mathcal{D}}(x')$. It works for $x' \in \mathbb{Z}_d$ by using a binary search to determine the largest index j such that $\boldsymbol{x}_j - j + 1 \le x'$, and then outputting $X \leftarrow unrank(x' + j)$. The construction of $rank_{\mathcal{R}}$ is similar, using $\boldsymbol{y}$ instead of $\boldsymbol{x}$.

This domain-completed cipher can be shown to be SPRP secure and, looking ahead, one can simply adapt it to the domain extension case to achieve SEPRP security (as defined in Sect. 6). This approach relies on having a ranking for $\mathcal{M}$. While not all languages have efficient rankings [4], efficient rankings can be built for most formats of practical interest [4,11,16,17]. The additional overhead of removing the $\mathcal{T}$ (resp. $\mathsf{Img}_{F_{K_o}}(\mathcal{T})$) points requires time proportional to $\log t$ and space equal to $2t$ multiplied by some representation-specific constant.

Our construction, to be presented in the next section, avoids the extra space requirements and the binary search. It only requires the ability to determine membership in $\mathcal{T}$, which affords various flexibilities such as using an API to check membership or representing $\mathcal{T}$ via a compact Bloom filter. It also allows, via precomputation, a constant-time implementation using table look-up (assuming constant time implementations of F, $\overline{E}$).

We note that the straightforward implementation of both approaches leaks, via timing side-channels, whether a domain point is in $\mathcal{T}$. The ranking-based approach may leak more with a naive implementation of binary search. Ranking itself may be in some cases tricky to implement without side-channels, if one uses the table-based constructions for ranking regular languages [4,11,16] or context-free grammars [17].

5   The Zig-Zag Construction for Domain Completion

In this section we will introduce an algorithm that achieves SPRP security in the domain completion setting. Fix a domain completion setting $(F, \mathcal{M}, \mathcal{T})$. Then the Zig-Zag domain-completed cipher $ZZ = (KT^{zz}, E^{zz})$ is defined as follows. The key transform KT^{zz} takes inputs K_0 and $\mathcal{T}$ and outputs the tuple $(\mathcal{T}, K_0, \overline{K})$ where $\overline{K}$ is a randomly chosen key for a cipher $\overline{E}$ on domain $\mathcal{M}$. We refer to $\overline{E}$ as the helper cipher. The triple $(\mathcal{T}, K_0, \overline{K})$ define a key for the enciphering algorithm E^{zz} and deciphering algorithm D^{zz}, detailed in Fig. 4.

$E^{zz}_{\mathcal{T}, K_0, \overline{K}}(M)$:	$D^{zz}_{\mathcal{T}, K_0, \overline{K}}(C)$:
if ($M \in \mathcal{T}$):	if ($F^{-1}_{K_0}(C) \in \mathcal{T}$):
$\quad$ return $F_{K_0}(M)$	$\quad$ return $F^{-1}_{K_0}(C)$
else	else
$\quad i \leftarrow 1$	$\quad i \leftarrow 1$
$\quad M_0 \leftarrow M$	$\quad M_i \leftarrow \overline{D}_{\overline{K}}(C)$
$\quad$ while ($M_{i-1} \in \mathcal{T}$)	$\quad$ while ($M_i \in \mathcal{T}$ or $i = 1$):
$\quad\quad Y_i \leftarrow \overline{E}_{\overline{K}}(M_{i-1})$	$\quad\quad Y_{i+1} \leftarrow F_{K_0}(M_i)$
$\quad\quad M_i \leftarrow F^{-1}_{K_0}(Y_i)$	$\quad\quad M_{i+1} \leftarrow \overline{D}_{\overline{K}}(Y_{i+1})$
$\quad\quad i \leftarrow i + 1$	$\quad\quad i \leftarrow i + 1$
$\quad$ return Y_{i-1}	$\quad$ return M_i

Fig. 4. Zig-Zag encryption and decryption algorithms.

Towards building intuition about the Zig-Zag construction, we start by discussing why traditional cycle walking will not work for our context. Cycle walking is a generic method for achieving format-preserving encryption on a set by re-encrypting an input point until it falls in a desired subset of the domain of the cipher [6]. Cycle-walking could ostensibly be used instead of zig-zagging as follows. Consider an input point $X \in \mathcal{M} \setminus \mathcal{T}$, and suppose that $Y = \overline{E}_{\overline{K}}(X) \in \mathsf{Img}_{F_{K_0}}(\mathcal{T})$. Then since Y is already a point required for the preservation set, we can't map it to X, and instead cycle walk by computing $Y' = \overline{E}_{\overline{K}}(Y)$, $Y'' = \overline{E}_{\overline{K}}(Y')$ and so on, stopping the first time we find a point not in the image of $\mathcal{T}$ and having X map to that final value. But the problem is that, unlike with traditional cycle walking, the intermediate points Y, Y', Y'' are themselves valid *inputs* to the cipher, and using them for the cycle walk obviates using the obvious mapping of, e.g., $\overline{E}_{\overline{K}}(Y)$.

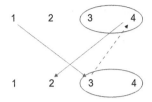

Fig. 5. Graphical depiction of a zig-zag. The domain (first row) has a target set $T = \{3, 4\}$. We encrypt 1 to 3, which collides with the ciphertext in the image of T. We then decrypt to get $4 \in T$ and re-encrypt 4 to get 2.

The Zig-Zag avoids this problem by only trying a different point of T each time $\overline{E}$ returns a point in $\mathsf{Img}_F(T)$. This ensures that as we do our search for a point to which we will map the input X, we are only using $\overline{E}$ on points $Y, Y', Y'' \in T$. A diagram depicting this process appears in the diagram of Fig. 5. There $\mathcal{M} = \{1, 2, 3, 4\}$ and $T = \{3, 4\}$. The solid red lines signify encryption by $\overline{E}_{\overline{K}}$ and the dashed black line represents $F_{K_\circ}^{-1}$. We start by calling $\overline{E}_{\overline{K}}(1)$, which (say) gives us the image of a point in T. We call F^{-1} and find that the preimage of this point is 4. We then call $\overline{E}_{\overline{K}}(4)$, which gives us $E_{\overline{K}}^{zz}(1) = \overline{E}_{\overline{K}}(4) = 2$.

5.1   Running Time of the Zig-Zag Construction

The Zig-Zag construction, in the worst-case, requires $|T|$ iterations of the while loop. First, we note that if the algorithm enters the while loop, it must exit — otherwise permutivity of $\overline{E}_{\overline{K}}$ would be violated. In the worst case, then, the while loop will hit every point in T. In the full version we provide a formal proof of this. We also show that, when encrypting the entire domain $\mathcal{M}$ under E_{K}^{zz}, $\overline{E}$ is executed at most once per point in $\mathcal{M}$. Roughly speaking, the aggregate cost of enciphering the entire domain under Zig-Zag is almost the same as enciphering with $\overline{E}$ (assuming $|T| \ll |\mathcal{M}|$, otherwise it's at most twice the cost).

This doesn't mean that for individual points the running time is not significantly delayed (in the worst case, requiring $2 \cdot |T|$ underlying cipher calls). But in fact the expected running time for an arbitrary point is small, as captured by the next theorem, and assuming the underlying ciphers are random permutations.

Theorem 1. *Let E^{zz} be implemented as in Fig. 4 except with F replaced with Π_1 and $\overline{E}$ replaced with Π_2, where Π_1 and Π_2 are random permutations over $\mathcal{M}$. Let I be a random variable denoting the number of iterations of the inner 'while' loop of E^{zz} with domain $\mathcal{M}$ and preservation set T taken when enciphering an arbitrary point in $\mathcal{M} \setminus T$. Let $|T| = t$. Then, if $t \leq |\mathcal{M}|/2$, it holds that $E[I] \leq 2$.*

Proof. Consider an arbitrary $M \in \mathcal{M}$. First if $M \in T$ then the number of iterations of the while loop is zero. Consider otherwise, and let the transcript of points defined in the while loop be $\mathcal{P} = \{M_0, Y_1, M_1, Y_2, M_2, \ldots, Y_i, M_i\}$. Then the size of this transcript is a random variable, over the coins used to choose Π_1, Π_2, and we denote it by I. We have that $\Pr[I > t] = 0$ by our arguments of correctness discussed above. Then for any $1 \leq j \leq t$, because Π_1, Π_2 are random permutations independent of M, we have that

$$\Pr[I = j] = \left[\prod_{i=1}^{j-1} \frac{(t-i+1)}{(m-i+1)} \right] \cdot \left(1 - \frac{t-j+1}{m-j+1} \right) .$$

Letting $P_j = \prod_{i=1}^{j-1} \frac{t-i+1}{m-i+1}$, we can plug into the definition of expectation to get that

$$\mathrm{E}[I] = \sum_{j=1}^{t} j \left[P_j - P_j \cdot \frac{t-j+1}{m-j+1} \right] = \sum_{j=1}^{t} j P_j - \sum_{j=1}^{t} j P_{j+1}$$

where we've used the fact $P_j \cdot \frac{t-j+1}{m-j+1} = P_{j+1}$. Investigating the right-hand side of the equation, we have that the left summand is one factor of P_j larger than the right summand when the index of summation on the left is one greater than on the right. Thus, for every P_j there will be a jP_j term in the overall summation and a $-(j-1)P_j$ term, so every term of the right summation is cancelled by a term of the left summation except for the final one, tP_{t+1}. Thus we can rewrite the equation to get

$$\mathrm{E}[I] = \sum_{j=1}^{t} P_j - t P_{t+1} \leq 1 + \sum_{j=2}^{t} \frac{1}{2^{j-1}} .$$

To justify the final inequality, observe that the first term of the summation is the empty product, which is by definition equal to 1. For the second summation, noticing that for $2 \leq j \leq t$, plugging $t = \frac{m}{2}$ into P_j gives us a summand which is upper-bounded by $\frac{1}{2^{j-1}}$. The rightmost term is bounded above by 1 so $\mathrm{E}[I] \leq 2$ for $t \leq \frac{m}{2}$. ∎

For simplicity in the above we restricted to $t \leq m/2$. For larger t, i.e. $\frac{m}{2} \leq t \leq m - 1$, a similar analysis can be done using the sum of a geometric series with ratio $\frac{1}{2} \leq \frac{t}{m} < 1$ in the final step instead of $\frac{1}{2}$. Finally, a similar analysis can be done for the run time of D^{zz}.

Security of the Zig-Zag construction. In the domain completion setting, our Zig-Zag construction achieves SPRP security. We prove the following theorem in the full version. The proof proceeds via standard reductions to move to an information-theoretic setting in which F and $\overline{E}$ are replaced by random permutations. Then one performs an analysis to show that the Zig-Zag domain-completed cipher, when using random permutations, exactly defines a random permutation.

Theorem 2. *Fix a domain completion setting $(F, \mathcal{M}, T)$ and let $ZZ = (KT^{zz}, E^{zz})$ be the Zig-Zag domain-completed cipher using helper cipher $\overline{E}$. Let $\mathcal{A}$ be an SPRP adversary making at most q queries to its oracles. Then the proof gives explicit adversaries $\mathcal{B}$ and $\mathcal{C}$ such that*

$$\mathbf{Adv}^{\mathrm{sprp}}_{E^{zz}}(\mathcal{A}) \leq \mathbf{Adv}^{\mathrm{sprp}}_{\overline{E}}(\mathcal{B}) + \mathbf{Adv}^{\mathrm{sprp}}_{F}(\mathcal{C}) .$$

Adversaries $\mathcal{B}$ and $\mathcal{C}$ each run in time that of $\mathcal{A}$ plus negligible overhead and each make at most $q + |\mathcal{T}|$ queries.

6   Domain Extenders for Deployed Ciphers

We now look at a distinct but related setting in which we want to extend the message space of a cipher after it has been deployed. Suppose we have an FPE F_{K_0} for some message space $\mathcal{D}$ that has already been used to encrypt a number of plaintexts. We later learn that we need to be able to encrypt as well plaintexts from a larger format $\mathcal{M} = \mathcal{N} \cup \mathcal{D}$.

Practical motivations for domain extension. While perhaps odd at first, message space extension arises frequently in deployment. An example is the use of encryption schemes in settings with constrained formatting on ciphertexts, such as the traditional credit card number example. Say we have deployed an FPE scheme for 16-digit credit card numbers only to later realize we must handle 15-digit credit card numbers as well. In this case it might be that $|\mathcal{D}| = 10^{15}$, $|\mathcal{N}| = 10^{14}$ and $|\mathcal{M}| = 10^{15} + 10^{14}$. (Recall that the last digit of a credit card number is a checksum, so a 15 digit CCN is only 14 digits of information.)

In deployment, such a format change is often precipitated by one of two things: a change in customer requirements or a change in application behavior. Changes in customer requirements often occur when businesses adopt FPE incrementally, rather than all at once. For example, a business might deploy FPE for users in the United States first, then later deploy it for users in China as well. If the format of the FPE was English-only initially, the inclusion of Chinese users will necessitate a change to this format. Sometimes customer requirements change because of external changes in their industries. When computerized gift cards gained widespread adoption, the credit card industry had to modify its standard for assigning credit card numbers to include a reserved range for numbers corresponding to temporary gift cards.

Changes in application behavior are problematic for businesses that use FPE in conjunction with cloud-hosted software. When FPE is deployed in such a setting (in which, it is important to note, the users have no control over application behavior) the formats are chosen to hew as closely as possible to the format validation used by the application. If the software vendor changes the way formats are validated, the FPE format must change as well or leave businesses with an unusable application.

We can achieve the desired security trivially by using an FPE on $\mathcal{M}$ with a fresh key. But this requires retrieving, decrypting, and re-encrypting all ciphertexts already stored under the old format and key. In many contexts this is rather expensive, and may not even be feasible should the ciphertexts be unavailable for update (e.g., because they've been handed back to some client's systems as a token and no API exists for recalling them).

One way to handle this extension would be to use a separate FPE for 16-digit credit card numbers and for 15-digit credit card numbers. The security of this kind of solution is, a priori, unclear, as such a ciphertext later accessible to adversaries will trivially leak which portion of the message space a plaintext sits. We will analyze the security of this formally below.

We might also be able to do better in the case that we have only used the old FPE on a small portion $T \subset D$ of the old message space. Ideally we'd like to preserve the decryptability of the points in T while otherwise picking mappings that are indistinguishable from a random permutation. We will formalize this goal below.

It may seem odd to assume that a list of already-encrypted points T is obtainable. After all, if we can extract a list of previously encrypted values, why can't we simply download and re-encrypt them? As discussed above, it is often difficult to authoritatively change any value in a complex software system after it has been created. It's also possible a description of T (like a regular expression) may be stored in a concise form in some application metadata that is stored on the encryption server.

Domain extension, formally. A domain extension setting is defined as a tuple (F, D, T, M) consisting of a cipher F on domain D, an extended domain M, and a preservation set $T \subseteq D$. A domain extended cipher DEC = (KT, E) is an algorithm and a cipher. The randomized algorithm KT, called a domain extension key transformation, takes as input a key K_o for F, the preservation set T, and outputs a key K for the cipher E whose domain is M. The cipher E is assumed to have a key space compatible with the output of KT. For some preservation set T, the induced key generation algorithm for E consists of choosing a random key K_o for F, running $KT(K_o, T)$ and returning the result.

A domain-extended cipher DEC preserves a point M if $E_K(M) = F_{K_o}(M)$ with probability one over the experiment defined by running the induced key generation for E. We say that DEC preserves a set T if it preserves each $M \in T$.

Impossibility of SPRP security. We can measure security for a domain-completed cipher DEC = (KT, E) via the SPRP advantage of the cipher E using its induced key generation algorithm. As before, we quantify over all preservation sets, meaning that security must hold even if the adversary knows T.

This definition proves too strong for most domain extension settings of interest. Roughly speaking, unless m is very close to d, the $|T|$ is small, and d is large, one can give simple SPRP adversaries successful against any construction. The following theorem captures the negative result.

Theorem 3. *Fix a domain extension setting (F, D, T, M). Let DEC = (KT, E) be a domain-extended cipher that preserves T. Let $d = |D|$ and $m = |M|$ and $t = |T|$. Then we give a fast, specific SPRP adversary A that makes $q \leq t$ queries for which*

$$\mathbf{Adv}_E^{\mathrm{sprp}}(A) = 1 - \frac{d! \cdot (m-q)!}{m! \cdot (d-q)!} .$$

Proof. Our adversary picks any size q subset of T and queries each point in the subset to its encryption oracle. If any of the resulting ciphertexts are in $N = M \setminus D$ it outputs 0, because this violates the definition of preserving T.

The adversary thus knows its oracle is a random permutation. If all q queries are in $\mathcal{D}$ it outputs 1.

In the real world, it is obvious the adversary outputs 1 with probability 1. In the case where the adversary's oracle is a random permutation, we have to treat the possibility of all the queries to the encryption oracle landing in $\mathcal{D}$ by chance. If this happens, the adversary is fooled into thinking its oracle is an extended FPE even though it's actually a random permutation.

The probability that the first query's ciphertext is in $\mathcal{D}$ is $\frac{d}{m}$. The probability that the next one is also in $\mathcal{D}$ is $\frac{d(d-1)}{m(m-1)}$, because there are $d-1$ remaining points in $\mathcal{D}$ and $m-1$ points left in m. We multiply the probability of the first query also being in $\mathcal{D}$ because this probability is conditioned on that also happening. If we carry out this argument to q queries, we'll get

$$\frac{d(d-1)\cdots(d-q)}{m(m-1)\cdots(m-q)} = \frac{d! \cdot (m-q)!}{m! \cdot (d-q)!} \, .$$

SEPRP security. Given the negative result about SPRP security, we turn to weaker, but still meaningful, security notions. The first is a relaxation of SPRP in which we do not seek to hide from an adversary that an extension has taken place. For an domain extension setting $(F, \mathcal{D}, \mathcal{T}, \mathcal{M})$, an adversary $\mathcal{A}$ and domain-extended cipher DEC $= (KT, E)$, the SEPRP$0^{\mathcal{A}}_{\text{DEC}}$ and SEPRP$1^{\mathcal{A}}_{\text{DEC}}$ games in Fig. 6 capture what we call "indistin-

main SEPRP1$^{\mathcal{A}}_{\text{DEC},\mathcal{T}}$	main SEPRP0$^{\mathcal{A}}_{\text{DEC},\mathcal{T}}$
$K_o \leftarrow\!\!\$\ \mathcal{K}$	$\pi \leftarrow\!\!\$\ \text{Perm}(\mathcal{D})$
$K \leftarrow\!\!\$\ KT(K_o, \mathcal{T})$	$\tilde{\pi} \leftarrow\!\!\$\ \text{ExtPerm}(\mathcal{D},\mathcal{T},\pi)$
$b' \leftarrow \mathcal{A}^{\text{Enc,Dec}}(\mathcal{T})$	$b' \leftarrow \mathcal{A}^{\text{Enc,Dec}}(\mathcal{T})$
return b'	return b'
proc $\text{Enc}(M)$	**proc** $\text{Enc}(M)$
return $E_K(M)$	return $\tilde{\pi}(M)$
proc $\text{Dec}(C)$	**proc** $\text{Dec}(C)$
return $E_K^{-1}(C)$	return $\tilde{\pi}^{-1}(C)$

Fig. 6. Games defining SEPRP security.

guishability from an extended random permutation". (The games are implicitly parameterized by the domain extension setting). There $\text{ExtPerm}(\mathcal{D}, \mathcal{T}, \pi)$ is the set of all possible permutations $\tilde{\pi}$ such that for all $X \in \mathcal{T}$ it is the case that $\tilde{\pi}(X) = \pi(X)$. An adversary $\mathcal{A}$'s SEPRP advantage against DEC is defined as

$$\mathbf{Adv}^{\text{seprp}}_{\text{DEC}}(\mathcal{A}) = \left| \Pr\left[\, \text{SEPRP1}^{\mathcal{A}}_{\text{DEC}} \Rightarrow 1 \,\right] - \Pr\left[\, \text{SEPRP0}^{\mathcal{A}}_{\text{DEC}} \Rightarrow 1 \,\right] \right| .$$

Zig-Zag for domain extension. We now consider the security of the Zig-Zag construction in the domain extension setting. Fixing a setting $(F, \mathcal{D}, \mathcal{T}, \mathcal{M})$, observe that the construction ZZ $= (KT^{zz}, E^{zz})$ from Sect. 5 provides a domain-extended cipher. The next theorem captures its SEPRP security. The proof appears in the full version of the paper.

Theorem 4. *Fix a domain-extension setting $(F, \mathcal{D}, \mathcal{T}, \mathcal{M})$ and let ZZ $=$ (KT^{zz}, E^{zz}) be the Zig-Zag domain-extended cipher using helper cipher $\overline{E}$.*

Let $\mathcal{A}$ be an SEPRP adversary making at most q queries to its oracles. Then the proof gives explicit adversaries $\mathcal{B}$ and $\mathcal{C}$ for which

$$\mathbf{Adv}_{E^{zz}}^{seprp}(\mathcal{A}) \le \mathbf{Adv}_F^{\mathrm{sprp}}(\mathcal{B}) + \mathbf{Adv}_{\overline{E}}^{\mathrm{sprp}}(\mathcal{C}) .$$

The adversaries $\mathcal{B}, \mathcal{C}$ each run in time at most that of $\mathcal{A}$ plus a negligible overhead and each make at most $q + |\mathcal{T}|$ queries.

7 Understanding SEPRP Security

In this section we study SEPRP security in more detail, in particular understanding its relationship with prior security definitions. In particular we'll explore the relationship between SEPRP and the notions of message recovery and message privacy security for ciphers introduced by Bellare et al. [4]. Throughout this section we fix a domain extension setting $(F, \mathcal{D}, \mathcal{T}, \mathcal{M})$, which is known to the adversary.

7.1 Message Recovery Security

The weakest definition from [4] is message-recovery security. At a high level it states that an attacker, given the encryption of some unknown message, should not be able to recover that message with probability better than that achieved by a simulator given no ciphertext. The adversary is additionally given an encryption oracle to which it can submit queries; the simulator is given access to an equality oracle that checks if the submitted message equals the target one. This latter reflects the fact that chosen-message attacks against an FPE scheme can always rule out messages one at a time by obtaining encryptions and comparing it to the challenge ciphertext.

We'll present a generalization of the standard message recovery definition called "generalized message recovery". We use the games gMR and gMRI to specify a simulation-style security target. The "real" game gMR tasks an adversary $\mathcal{A} = (\mathcal{A}_1, \mathcal{A}_2)$ with recovering a message chosen by $\mathcal{A}_1$ given its encryption under the domain-extended cipher DEC. We emphasize that $\mathcal{A}_1$ and $\mathcal{A}_2$ do not share any state (otherwise the definition would be vacuous). The adversary $\mathcal{A}_2$ has the ability to obtain encryptions on messages of its choosing. The "ideal" game gMRI tasks a simulator $\mathcal{S}$ to recover an identically distributed message X^* given some leakage $\mathrm{aux}(X^*)$ about it and the ability to query an equality oracle $\mathbf{Eq}$ that returns whether or not the submitted message equals X^*.

The generalized MR-advantage of an adversary $\mathcal{A} = (\mathcal{A}_1, \mathcal{A}_2)$ against a domain-extended cipher DEC is defined as

$$\mathbf{Adv}_{\mathrm{DEC}}^{\mathrm{gmr}}(\mathcal{A}, q', \mathrm{aux}) = \Pr\left[\, \mathrm{gMR}_{\mathrm{DEC}}^{\mathcal{A}} \Rightarrow \mathsf{true} \,\right] - \max_{\mathcal{S} \in \mathbb{S}_{q'}} \Pr\left[\, \mathrm{gMRI}^{\mathcal{S},\mathrm{aux}} \Rightarrow \mathsf{true} \,\right]$$

where the rightmost term is defined over $\mathbb{S}_{q'}$, the set of all simulators making at most q' queries to their $\mathbf{Eq}$ oracles. In what follows, the simulator can depend

main gMR$_{\text{DEC}}$	main gMRI	main gMP$_{\text{DEC}}$	main gMPI
$K_0 \leftarrow\!\!\$\ \mathcal{K}$	$X^* \leftarrow\!\!\$\ \mathcal{A}_1(\mathcal{T})$	$K_0 \leftarrow\!\!\$\ \mathcal{K}$	$X^* \leftarrow\!\!\$\ \mathcal{A}_1(\mathcal{T})$
$K \leftarrow\!\!\$\ KT(K_0,\mathcal{T})$	$X \leftarrow \mathcal{S}^{\text{Eq}}(\mathcal{T},\text{aux}(X^*))$	$K \leftarrow\!\!\$\ KT(K_0,\mathcal{T})$	$X \leftarrow \mathcal{S}^{\text{Eq}}(\mathcal{T},\text{aux}(X^*))$
$X^* \leftarrow\!\!\$\ \mathcal{A}_1(\mathcal{T})$	return $(X = X^*)$	$X^* \leftarrow\!\!\$\ \mathcal{A}_1(\mathcal{T})$	return $(X = \mathcal{A}_3(X^*))$
$Y^* \leftarrow E_K(X^*)$		$Y^* \leftarrow E_K(X^*)$	
$X \leftarrow \mathcal{A}_2^{\text{Enc}}(\mathcal{T},Y^*)$		$Z \leftarrow \mathcal{A}_2^{\text{Enc}}(\mathcal{T},Y^*)$	
return $(X = X^*)$	**Eq(X)**	return $(Z = \mathcal{A}_3(X^*))$	**Eq(X)**
	return $(X = X^*)$		return $(X = X^*)$
Enc(X)		**Enc(X)**	
return $E_K(X)$		return $E_K(X)$	

Fig. 7. Generalized Message-recovery and message privacy games.

on an adversary $\mathcal{A}$. The string aux is the description of a function which takes a point of $\mathcal{M}$ and outputs either some information about it or $\bot$.

The value q' is a function of q, the number of queries the adversary makes in its experiment. Below, q' will be some small constant like 1 or 2 times q.[1] When $q' > q$, it means that the security provided is weaker because the simulator needs more queries to its ideal functionality to achieve the same probability of success in its game. Intuitively, this means that the real oracle **Enc** leaks more information than the ideal oracle. All reductions below are tight up to some small constant factors (Fig. 7).

Since MR security is shown not to imply SPRP security in [4], we expect that it does not imply SEPRP security. To demonstrate this, imagine we take an MR-secure cipher E over a size-d domain and add one bit to its domain, making it $d+1$ bits. Define a new cipher $E'(X)$ on this domain by calling E on the first d bits of X and concatenating the $d+1$st bit (in the clear) to make the ciphertext of X under E'. The MR-security of E' is reducible to the MR-security of E by a simple argument. However, this new cipher E' does not meet SEPRP security, because (with M and $E'(M)$ interpreted as integers) the quantity $|M - E'(M)|$ is the same whether the top bit of M is 1 or 0.

We can also show that SEPRP does not imply MR security. Take a similar setting in which the new domain $\mathcal{M}$ has $|\mathcal{M}| = m = 2 \cdot d$ where $|\mathcal{D}| = d$ and every point in $\mathcal{D}$ is preserved. We claim that for an SEPRP E, $\mathbf{Adv}_E^{\text{gmr}}(\mathcal{A},q,\text{B}) \geq \frac{1}{2}$, where B is the function that always outputs $\bot$ (meaning no information is leaked). To see this, take $\mathcal{A} = (\mathcal{A}_1,\mathcal{A}_2)$ and have $\mathcal{A}_2$ first check if the point it was given is in $\mathcal{D}$. If so, it queries every point in $\mathcal{D}$ until it finds the right one. Likewise for $\mathcal{N}$. $\mathcal{A}$ wins the gMR game with probability 1, but any simulator wins the gMRI game with probability at most $\frac{1}{2}$ because it doesn't receive any information about the hidden point and only has q queries.

This is troubling, because we seem to have a separation in two directions when $q = q'$: generalized MR does not imply SEPRP, and SEPRP does not imply generalized MR. However, we can prove that SEPRP does imply generalized MR

---

[1] Note that when $q' = q$ and aux is the function that always outputs $\bot$, this definition corresponds exactly to the message recovery definition from [4].

when the simulator is given some auxiliary information about the hidden point, namely whether or not it is in $\mathcal{T}$.

Theorem 5. *Fix a domain-extension setting $(F, \mathcal{D}, \mathcal{T}, \mathcal{M})$. For any domain-extended cipher DEC and adversary $\mathcal{A}$ making q oracle queries, we give in the proof an adversary $\mathcal{B}$ making q oracle queries such that*

$$\mathbf{Adv}^{gmr}_{DEC}(\mathcal{A}, q, aux) \leq \mathbf{Adv}^{seprp}_{DEC}(\mathcal{B})$$

where $aux(X)$ returns 1 if $X \in \mathcal{T}$ and 0 otherwise, for all $X \in \mathcal{M}$.

Proof. Our adversary $\mathcal{B}$ is given the description of $\mathcal{A} = (\mathcal{A}_1, \mathcal{A}_2)$. It runs $\mathcal{A}_1(\mathcal{T})$ and gets a point X^*, then runs $\mathcal{A}_2(\mathcal{T}, \mathrm{Enc}(X^*))$, simulating $\mathcal{A}_2$'s **Enc** oracle using its own encryption oracle for the SEPRP game. When $\mathcal{A}_2$ outputs its guess X, $\mathcal{B}$ returns 1 if $(X = X^*)$ and 0 otherwise. By construction

$$\Pr\left[\,\mathrm{gMR}^{\mathcal{A}}_{DEC} \Rightarrow \mathrm{True}\,\right] = \Pr\left[\,\mathrm{SEPRP1}^{\mathcal{B}}_{DEC} = 1\,\right]$$

To complete the proof we must show that

$$\max_{\mathcal{S} \in \mathbb{S}_q} \Pr\left[\,\mathrm{gMRI}^{\mathcal{S},\mathrm{aux}} \Rightarrow \mathrm{true}\,\right] \geq \Pr\left[\,\mathrm{SEPRP0}^{\mathcal{B}}_{DEC} = 1\,\right]$$

Construct a simulator $\mathcal{S}$ by giving it the target set $\mathcal{T}$ and the leakage bit that indicates whether the hidden point is in $\mathcal{T}$. $\mathcal{S}$ runs $\mathcal{A}_2(\mathcal{T}, X')$, where X' is a random point of $\mathcal{D}$ if its leakage bit indicates that the hidden point is preserved, and a random point of $\mathcal{M}$ otherwise. $\mathcal{S}$ simulates $\mathcal{A}$'s **Enc** oracle by taking each of $\mathcal{A}_2$'s queries and checking it against its own **Eq** oracle. If the **Eq** oracle returns true, $\mathcal{S}$ returns X'. Otherwise, $\mathcal{S}$ returns a random (subject to permutivity) point of $\mathcal{D}$ if $\mathcal{A}_2$'s query is in $\mathcal{T}$, and a random point of $\mathcal{M}$ otherwise. $\mathcal{S}$ returns whatever $\mathcal{A}_2$ does. By inspection, $\mathcal{S}$ is simulating the same environment for $\mathcal{A}_2$ as $\mathcal{B}$ does in the ideal SEPRP game, because in either case the environment is lazy-sampling a random ideal extended permutation. Thus, the probability of $\mathcal{S}$ winning is exactly the probability of $\mathcal{B}$ guessing 1 in the SEPRP0 game. The max value of the left-hand side is at least the success probability of this simulator, so the inequality holds. ∎

We can also prove the following relationship between different parameterizations of the generalized MR games. Intuitively, this theorem says that leaking whether the hidden point is in $\mathcal{T}$ is roughly equivalent to speeding up a guessing attack by a factor of two.

Theorem 6. *Fix a domain-extension setting $(F, \mathcal{D}, \mathcal{T}, \mathcal{M})$. For any domain-extended cipher DEC and adversary $\mathcal{A}$ making q queries,*

$$\mathbf{Adv}^{gmr}_{DEC}(\mathcal{A}, 2(q+1), B) \leq \mathbf{Adv}^{gmr}_{DEC}(\mathcal{A}, q, aux)$$

where aux and B are as above.

Proof. First, observe that

$$\Pr\left[\,\mathrm{gMR}_{\mathrm{DEC}}^{\mathcal{A}} \Rightarrow \mathrm{true}\,\right] = \Pr\left[\,\mathrm{gMR}_{\mathrm{DEC}}^{\mathcal{A}} \Rightarrow \mathrm{true}\,\right]$$

This is tautological because the gMR game is the same in either case; only the gMRI game changes.

To complete the proof we need to show that

$$\max_{\mathcal{S} \in \mathbb{S}_{2q+2}} \Pr\left[\,\mathrm{gMRI}^{\mathcal{S},\mathrm{B}} \Rightarrow \mathrm{true}\,\right] \geq \max_{\mathcal{S}' \in \mathbb{S}_q} \Pr\left[\,\mathrm{gMRI}^{\mathcal{S}',\mathrm{aux}} \Rightarrow \mathrm{true}\,\right]$$

The simulator $\mathcal{S}$ is given a description of the $\mathcal{S}'$ that maximizes the right-hand side and runs it twice — once with the leakage bit set to 0 and once with the bit set to 1. $\mathcal{S}$ answers $\mathcal{S}'$'s Eq queries with its own Eq oracle. If one of $\mathcal{S}'$'s guesses is correct, $\mathcal{S}$ returns that as its guess. Since $\mathcal{S}$ runs $\mathcal{S}'$ with the leakage bit set to both possible values, if $\mathcal{S}'$ wins in either case, $\mathcal{S}$ wins as well. Thus, the success probability of $\mathcal{S}$ is at least the success probability of $\mathcal{S}'$. ∎

7.2   Message Privacy

In [4] a (strictly stronger) definition than message recovery is proposed. They refer to this definition as message privacy. It says, roughly, that no adversary can compute any function of the message given only its ciphertext. Message-recovery security is a special case of message privacy where the function the adversary wants to compute is the equality function. We define the generalized MP-advantage of an adversary $\mathcal{A}$ as

$$\mathbf{Adv}_E^{\mathrm{gmp}}(\mathcal{A}, q', \mathrm{aux}) = \Pr\left[\,\mathrm{gMP}_{\mathrm{DEC}}^{\mathcal{A}} \Rightarrow \mathrm{true}\,\right] - \max_{\mathcal{S} \in \mathbb{S}_{q'}} \Pr\left[\,\mathrm{gMPI}^{\mathcal{S},\mathrm{aux}} \Rightarrow \mathrm{true}\,\right]$$

We will use this generalized definition instead of the one used in [4] because extended permutations leak more information to adversaries than standard SPRPs. To demonstrate the necessity of this generalized definition, we'll prove that SEPRP security does not imply the standard message privacy definition from [4], which corresponds to our generalized definition when $q = q'$ and $\mathrm{aux} = \perp$.

Theorem 7. *Fix a domain-extension setting $(F, \mathcal{D}, \mathcal{T}, \mathcal{M})$ for which $\mathcal{T} = \mathcal{D}$, i.e., every point is preserved. For any domain-extended cipher DEC, the proof gives a specific adversary $\mathcal{A} = (\mathcal{A}_1, \mathcal{A}_2, \mathcal{A}_3)$ in the message privacy game such that*

$$\mathbf{Adv}_{DEC}^{\mathrm{gmp}}(\mathcal{A}, 0, \perp) = 1 - \max(\frac{d}{m}, \frac{n}{m})$$

Proof. The algorithm $\mathcal{A}_1$ samples uniformly from its input space. The function represented by $\mathcal{A}_3$ is

$$\mathcal{A}_3(m) = \begin{cases} 1 & \text{The message is in } \mathcal{D} \\ 0 & \text{The message is in } \mathcal{M} \setminus \mathcal{D} \end{cases}$$

$\mathcal{A}$ wins with probability 1 by checking whether the point Y^* it is given is in $\mathcal{D}$ or $\mathcal{M} \setminus \mathcal{D}$. Because every point is preserved, $\mathcal{A}$ always computes the function correctly. The simulator $\mathcal{S}$ does not get any **Eq** queries because $\mathcal{A}$ used no encryption queries, and its auxiliary function always outputs $\perp$, so the simulator's optimal strategy is to output 1 if $d > n$ and 0 otherwise. A point from the larger of the two sets is more likely, so the simulator wins with probability $\max(\frac{d}{m}, \frac{n}{m})$. $\qquad\blacksquare$

We can, however, prove that SEPRP does imply generalized message privacy when an oracle for membership in $\mathcal{T}$ is given to the simulator.

Theorem 8. *Fix a domain-extension setting* $(F, \mathcal{D}, \mathcal{T}, \mathcal{M})$. *For any domain-extended cipher DEC and adversary $\mathcal{A}$ making q oracle queries, we give in the proof an adversary $\mathcal{B}$ making q queries such that*

$$\mathbf{Adv}_{DEC}^{gmp}(\mathcal{A}, q, aux) \leq \mathbf{Adv}_{DEC}^{seprp}(\mathcal{B})$$

where aux returns 1 if its input is in $\mathcal{T}$ and 0 otherwise.

Proof. Our adversary $\mathcal{B}$ is given the description of $\mathcal{A} = (\mathcal{A}_1, \mathcal{A}_2, \mathcal{A}_3)$. It runs $\mathcal{A}_1(\mathcal{T})$ and gets a point X^*, then runs $\mathcal{A}_2(\text{Enc}(X^*))$, simulating $\mathcal{A}_2$'s **Enc** oracle using its own encryption oracle for the SEPRP game. When $\mathcal{A}_2$ outputs its guess Z, $\mathcal{B}$ returns 1 if $(Z = \mathcal{A}_3(X^*))$ and 0 otherwise. By construction

$$\Pr\left[\, \text{gMP}_{DEC}^{\mathcal{A}} \Rightarrow \text{True} \,\right] = \Pr\left[\, \text{SEPRP1}_{DEC}^{\mathcal{B}} = 1 \,\right]$$

It suffices to show that

$$\max_{\mathcal{S} \in \mathbb{S}_q} \Pr\left[\, \text{gMPI}^{\mathcal{S}, aux} \Rightarrow \text{true} \,\right] \geq \Pr\left[\, \text{SEPRP0}_{DEC}^{\mathcal{B}} = 1 \,\right]$$

Define a simulator $\mathcal{S}$ that takes $\mathcal{T}$ and the value of $aux(X^*)$. The simulator $\mathcal{S}$ runs $\mathcal{A}_2(X')$ where X' is a random point of $\mathcal{D}$ if the simulator's leakage from aux indicates the hidden point is in $\mathcal{T}$ and X' is a random point of $\mathcal{M}$ otherwise. The simulator simulates $\mathcal{A}_2$'s **Enc** oracle by first using its own **Eq** oracle to check if $\mathcal{A}_2$'s guess is equal to the hidden point. If $\mathcal{S}$'s oracle returns true, $\mathcal{S}$ returns X' in response to $\mathcal{A}_2$'s query. If it returns false, $\mathcal{S}$ checks if the queried point is in $\mathcal{T}$. If it is, $\mathcal{S}$ returns a random point of $\mathcal{D}$, else $\mathcal{S}$ returns a random point of $\mathcal{M}$. The simulator makes both choices subject to permutivity. When $\mathcal{A}_2$ returns its guess for $\mathcal{A}_3(X^*)$, $\mathcal{S}$ outputs the same guess. Since $\mathcal{S}$ is simulating the same environment for $\mathcal{A}_2$ as $\mathcal{B}$ does in the case where $\mathcal{B}$'s oracle is an ideal extended permutation, the probability of this simulator $\mathcal{S}$ winning is exactly the probability of $\mathcal{B}$ guessing 1 in the SEPRP0 game. The true max of the left-hand side is at least the probability that this $\mathcal{S}$ we've constructed wins the gMPI game, so the inequality holds. $\qquad\blacksquare$

8 The Zig-Zag Construction and Side-Channel Resistance

One important question when designing any encryption scheme that contains branches on secret data or has variable timing for different points (e.g. the Zig-Zag construction) is whether this gives rise to any kind of side-channel attack. Timing side-channels have proven particularly dangerous in applications [5,7] so we would like to prove the Zig-Zag construction does not give rise to a timing side-channel. In this section, we will prove that the time taken to encrypt or decrypt with the Zig-Zag construction does not leak useful information to an adversary about the encrypted message.

Fix some domain extension setting $(F, \mathcal{D}, \mathcal{T}, \mathcal{M})$ and let $ZZ = (KT^{zz}, E^{zz})$ be the Zig-Zag construction for it. We define two games, detailed in Fig. 8. The first, $\mathrm{Real}_{ZZ}^{\mathcal{A}}$, gives the adversary $\mathcal{A}$ access to a Zig-Zag enciphering oracle that additionally reveals the number of iterations of the inner loop of Zig-Zag. The second, $\mathrm{Ideal}^{\mathcal{A}}$ gives the adversary $\mathcal{A}$ access to an oracle that returns random permutation applied to the message as well as a simulated while-loop count that only uses whether $M \in \mathcal{T}$. Define the SPRP-with-timing advantage of an adversary $\mathcal{A}$ against an ZZ as

$$\mathbf{Adv}_{ZZ}^{\mathrm{sprp+t}}(\mathcal{A}) = \left| \Pr\left[\mathrm{Real}_{ZZ}^{\mathcal{A}} \Rightarrow 1 \right] - \Pr\left[\mathrm{Ideal}^{\mathcal{A}} \Rightarrow 1 \right] \right|$$

The interpretation is that efficient adversaries should not be able to distinguish E^{zz} from a random permutation, even with this additional information. In the Sample procedures, the function $B(x,y)$ generates a random bit that is 1 with probability $\frac{x}{y}$.

main Real_{ZZ}	**Main** Ideal
$K_0 \leftarrow\!\!\$\ \mathcal{K}$	$z \leftarrow 0$
$(\mathcal{T}, K_0, \overline{K}) \leftarrow\!\!\$\ KT(K_0, \mathcal{T})$	$q \leftarrow 0$
$b \leftarrow \mathcal{A}^{\mathrm{Enc}}$	$\pi \leftarrow \mathrm{GetPerm}(\mathcal{T})$
	$b \leftarrow \mathcal{A}^{\mathrm{Enc}}(\mathcal{T})$
proc Enc(M)	
$c \leftarrow 0$	**proc** Sample$(M, \mathcal{T})$
If $M \in \mathcal{T}$ then	If $(M \in \mathcal{T})$ then return 0
$\quad$ return $(F_{K_0}(M), c)$	$c \leftarrow 0$
else	$b \leftarrow\!\!\$\ B(t - z - c, m - q - z - c)$
$\quad X \leftarrow \overline{E}_{\overline{K}}(M)$	while $b \neq 0$:
$\quad$ while $F_{K_0}^{-1}(X) \in \mathcal{T}$:	$\quad b \leftarrow\!\!\$\ B(t - z - c, m - q - z - c)$
$\quad\quad X \leftarrow \overline{E}_{\overline{K}}(F_{K_0}^{-1}(X))$	$\quad c \leftarrow c + 1$
$\quad\quad c \leftarrow c + 1$	$z \leftarrow z + c$
$\quad$ return (X, c)	$q \leftarrow q + 1$
	return c
	proc Enc(M)
	$c \leftarrow\!\!\$\ \mathrm{Sample}(M, \mathcal{T})$
	return $(\pi(M), c)$

Fig. 8. Games defining SPRP-with-timing advantage for Zig-Zag.

The following theorem captures Zig-Zag's security in this new model.

Theorem 9. *Fix a domain extension setting $(F, \mathcal{D}, \mathcal{T}, \mathcal{M})$ and let $ZZ = (KT^{zz}, E^{zz})$ be the Zig-Zag domain-extended cipher for it built using an underlying helper cipher $\overline{E}$. Then for any $\mathcal{A}$ making at most q queries the proof gives specific adversaries $\mathcal{B}$ and $\mathcal{C}$ such that*

$$\mathbf{Adv}^{\mathrm{sprp+t}}_{ZZ}(\mathcal{A}) \leq \mathbf{Adv}^{\mathrm{sprp}}_{\overline{E}}(\mathcal{B}) + \mathbf{Adv}^{\mathrm{sprp}}_{F}(\mathcal{C})$$

Adversaries $\mathcal{B}, \mathcal{C}$ each run in time at most that of $\mathcal{A}$ plus a negligible overhead and each make at most $q + |\mathcal{T}|$ queries.

We defer the proof to the full version. This theorem lets us say with that information on how long encryption of a particular point takes leaks only whether it is in $\mathcal{T}$ or not. Since in a chosen-plaintext attack the adversary already knows whether $M \in \mathcal{T}$, this means the adversary learns nothing. Intuitively this is because, for every point $M \notin \mathcal{T}$ the distribution of M's zig-zag lengths is the same.

8.1   Other Sources of Side Information

Now that we have shown formally that the timing side-channel of the Zig-Zag construction's inner loop does not leak information to an adversary other than whether or not a point is in $\mathcal{T}$, we will discuss more coarse-grained side channels. We first look at remote timing attacks, where the adversary learns how long it takes to perform the encryption or decryption and from that deduces secret information. For convenience, we only discuss leaks in the encryption algorithm. Similar leaks exist in the decryption algorithm.

The main source of secret-dependent timing variations is the zig-zag operation. Each time the algorithm iterates through the while loop it performs two more encryptions. Thus, timing information discloses the number of times the algorithm iterates. Knowing that the algorithm iterates through the while loop is an indication that $M \notin \mathcal{T}$. We prove above that this is the only information about M leaked to an adversary by this timing information.

Other sources of timing variations that may leak secret information include: the different code path taken for $M \in \mathcal{T}$, the test for $M \in \mathcal{T}$ and potential timing variations in the implementations of F and $\overline{E}$.

A common technique for protecting against timing channels is to pad the computation time. The implementation is modified to ensure that the time between the start of the computation and the delivery of the result is fixed [3,9]. To avoid any side-channel information, this fixed time must be long enough to accommodate any possible length of computation. Askarov et al. [3] suggest an adaptive approach that ensures that only a small amount of information leaks while adapting to the execution time of the computation.

As discussed in Sect. 5.1, the Zig-Zag construct iterates, on average, less than one time per encryption. The worst-case scenario, however, is that it iterates t times. Padding to the worst-case scenario incurs a significant performance loss. Yet, failure to pad to the worst-case scenario may result in information leaks.

To avoid both excessive padding and information leaks, we can pre-compute the value of the new cipher E^{zz} on all the points that require zig-zagging. That is, on the points of the set:

$$\left\{ \overline{D}_{\overline{K}}(F_{K_\circ}(M)) | M \in \mathcal{T} \wedge \overline{D}_{\overline{K}}(F_{K_\circ}(M)) \notin \mathcal{T} \right\}$$

Informally, these are the points not in $\mathcal{T}$ whose encryptions under $\overline{E}$ are in $\mathrm{Im}_{F_{K_\circ}}(\mathcal{T})$.

Storing the precomputed values takes a space linear in t. However, as the Zig-Zag construction needs to store $\mathcal{T}$, the pre-computation only increases the space requirements by a constant factor and, at the same time, guarantees that the computation of E^{zz} requires at most a single application of either F or of $\overline{E}$. With the pre-computation, padding can provide an efficient countermeasure for remote timing attacks.

Padding is not an efficient countermeasure for local side-channel attacks. Local adversaries can monitor traces that software execution leaves in the cache or in other microarchitectural components [1,2,15,19]. Constant-time implementations that perform no secret-dependent branches or memory accesses can provide protection for ciphers against local side channel attacks [5,19]. However, such implementations need to access every table element when performing a table access. Thus, for the Zig-Zag construction, the check whether $M \in \mathcal{T}$ would require a time linear in t. Rather than using a constant-time implementation of the cipher, implementer can rely on hardware or operating system based measure to provide protection against local side-channel attacks [14,21,24,26].

9   Domain Extension When Adversaries Do Not Know $\mathcal{T}$

In a previous section we demonstrated that we cannot achieve SPRP security while preserving a subset of the original domain if the adversary knows which subset is preserved. One can naturally ask, then, if there are weaker adversarial settings in which SPRP security can be attained. In particular, we may want to know what the strongest "weaker" adversary is — namely, how much information can we reveal about $\mathcal{T}$ before SPRP becomes provably impossible. In this section we provide a constructive partial answer to this question by building an SPRP-secure scheme in the setting where the adversary only knows $|\mathcal{T}| = t$, the size of the preserved set, but does not know which elements it contains. This weakening of the adversary is motivated not only by theoretical questions, but by practical settings in which the attacker, through application logs or other non-sensitive information, is able to infer the number of ciphertexts in the database before an extension has occurred.

In terms of security goal, we target SPRP security in a setting where the adversary knows $\mathcal{D}$ but the preserved set is chosen uniformly from the subsets of size t of the original domain $\mathcal{D}$, and the random coins used to make this choice are hidden from the adversary. (We leave treating other cases as an open question).

Observe that if $t > d - n$, a random permutation has a nonzero probability of having fewer than t elements of $\mathcal{D}$ mapped into $\mathcal{D}$. When a permutation maps a point of $\mathcal{D}$ back to $\mathcal{D}$, we say that it "domain-preserves" that point. Since one of our goals for domain extension is to preserve mappings for points in $\mathcal{T}$ (which are domain-preserved mappings) there must be at least t domain-preserved points in our permutation. To make this more intuitive, consider how few points can possibly be domain-preserved in any permutation. This occurs when (for $d > n$) as many points as possible are mapped from $\mathcal{D}$ to $\mathcal{N}$. Since this is a permutation, only n points can have ciphertexts in $\mathcal{N}$. The rest *have* to be domain-preserved. If this happens, n points of $\mathcal{D}$ are not domain-preserved, so $d - n$ points are. If t is indeed greater than this strict lower bound, we are excluding some nonzero number of possible permutations (i.e., the ones that domain-preserve between $t-1$ and $d-n$ points). This will give a distinguishing advantage to an adversary.

The Recursive Zig-Zag. For the case that $t \leq d - n$, any permutation on $\mathcal{M}$ domain-preserves at least t elements of $\mathcal{D}$. We will use this fact to construct a Zig-Zag algorithm that achieves SPRP security for domain extension. Since the key transformation acts in a recursive fashion on its state, we will call the algorithm the "Recursive Zig-Zag" (RZZ). The key transformation works by selecting a set of points that are domain-preserved under the helper cipher and, for each point τ in $\mathcal{T}$, "swapping" the image of one of these points with the image of τ if τ is *not* domain-preserved. This is done so the number of domain-preserved points in the resulting permutation is unchanged. Points that are swapped are stored in a lookup table Emap. Below, we will prove SPRP security of the RZZ and demonstrate that the expected amortized cost of the KT^{rzz} is constant for each point of $\mathcal{T}$.

To motivate the RZZ, it may be useful to give a concrete example of why the previous Zig-Zag construction cannot be SPRP-secure for domain extension when only t is known. Take $d = 99$, $t = 98$ and $n = 1$. In this case, Zig Zag will have a 50% probability that the newly added element maps to itself. However, the probability of that happening in a random permutation is 1%. The main cause of the problem is that standard Zig-Zag may change the size of the domain-preserved set. In KT^{rzz} we guarantee that does not happen.

The construction. Figure 9 shows the encryption and decryption algorithms. They consult the lookup table Emap for the existence of the value, returning it if found. Otherwise they return the value of the helper cipher $\overline{E}$. The new key $K = \langle \text{Emap}, \overline{K} \rangle$ output by KT^{rzz} contains the lookup table Emap which

$E^{rzz}_{\text{Emap},\overline{K}}(M)$:	$D^{rzz}_{\text{Emap},\overline{K}}(M)$:
if (Emap$[M] \neq \perp$):	if (Emap$^{-1}[M] \neq \perp$):
return Emap$[M]$	return Emap$^{-1}[M]$
else	else
return $\overline{E}_{\overline{K}}(M)$	return $\overline{E}^{-1}_{\overline{K}}(M)$

Fig. 9. Recursive Zig-Zag encryption and decryption algorithms

```
KT^{rzz}(K_o, T)

K ←$ K
for i from 0 to t:
    τ_old ← E^{rzz}_{Emap,K̄}(τ_i)
    if ( τ_old = F_{K_o}(τ_i)) :
        // Case 1: Do nothing
    if ( τ_old ∈ D) :
        // Case 2: set mapping for preimage of F_{K_o}(τ_i))
        τ_m ← D^{rzz}_{Emap,K̄}(F_{K_o}(τ_i))
        Emap[τ_m] ← τ_old
    else:
        // Case 3: chose a p_i and swap through it
        τ_m ← D^{rzz}_{Emap,K̄}(F_{K_o}(τ_i))
        //Select a random p_i
        //that is domain-preserved under E^{rzz}_{Emap,K̄}
        do:
            p_i ←$ D \ {τ_1, . . . , τ_{i-1}}
            p_i^{old} ← E^{rzz}_{Emap,K̄}(p_i)
        while (p_i^{old} ∈ N)
        Emap[p_i] ← τ_old
        Emap[τ_m] ← p_i^{old}
    endif
    // Always record τ_i
    Emap[τ_i] ← F_{K_o}(τ_i)
endfor
return (Emap, K̄)
```

Fig. 10. The Recursive Zig-Zag key transformation. The original cipher is F. The helper cipher is $\overline{E}$.

is pre-calculated by the key transformation algorithm in Fig. 10. We use the notation Emap[x] to refer to the mapping of the element x under Emap. The notation Emap^{-1}[y] returns the value X such that Emap[X] = y. If Emap does not provide a mapping for x, the value of Emap[x] is ⊥. If there is no point mapped to y in Emap, Emap^{-1}[y] will likewise output ⊥.

The KT^{rzz} algorithm. The key transformation algorithm records all the values modified during the t iterations. At the start of the ith iteration of KT^{rzz}, Emap contains the values changed in all previous iterations. We begin the iteration by computing $τ_{old} ← E^{rzz}_{Emap,K}(τ_i)$ where $E^{rzz}_{Emap,K}$ is computed as in Fig. 9. There are then three cases. We will explain each in turn, referring to the case numbers given in Fig. 10.

Case 1 ($τ_{old} = F_{K_o}(τ_i)$)): This occurs if E^{rzz} already contains the correct mapping for $τ_i$. That would happen if the helper cipher $\overline{E}_{\overline{K}}$ maps $τ_i$ to $F_{K_o}(τ_i)$. We simply update Emap and continue.

Case 2 ($τ_{old} ∈ D$)): This occurs if E^{rzz} domain-preserves $τ_i$. Here we do not need to worry about biasing the number of domain-preserved points by preserving $τ_i$'s mapping to $F_{K_o}(τ_i)$ because both $F_{K_o}(τ_i)$ and $τ_{old}$ are in D. In this case we can do a zig-zag as above, assigning $τ_{old}$ to the decryption of $F_{K_o}(τ_i)$ under D^{rzz} and $τ_i$ to $F_{K_o}(τ_i)$, as desired.

Case 3 ($\tau_{\text{old}} \in \mathcal{N}$)): This occurs if E^{rzz} does *not* domain-preserve τ_i. This is the case that requires special handling, since if we patch E^{rzz} to map τ_i to $F_{K_o}(\tau_i)$ we may increase the number of domain-preserved points of E^{rzz} and give a distinguishing advantage to an adversary. We use rejection sampling on points of $\mathcal{D} \setminus \{\tau_1, \ldots, \tau_{i-1}\}$ to find a point not in $\mathcal{T}$ that is domain-preserved under E^{rzz}. Such a point is guaranteed to exist by our assumption that $t \leq d - n$. When we find such a point p_i, we record it with its new image τ_{old} in Emap. Finally, we assign its old image under E^{rzz}, p_i^{old}, to be the image of $D^{rzz}_{\text{Emap},\overline{K}}(F_{K_o}(\tau_i))$ to preserve permutivity. Once we select p_i it cannot be selected in a subsequent iteration, since it will no longer be domain-preserved under E^{rzz}.

Note that we do not need special handling of the case that on the ith iteration of KT^{rzz} we assign $E^{rzz}_{\text{Emap},\overline{K}}(\tau_i) = \tau_j$ for some $j > i$. We will change the value of $E^{rzz}_{\text{Emap},\overline{K}}(\tau_j)$ on the ith iteration, but we will fix it in the jth iteration.

The number of points changed in each of the t transformations is at most 3. Consequently, the number of points we need to pre-calculate is at most $3t$ and with the result of precalculation we need to encode at most $6t$ values—6 times as much as we need to encode to remember $\mathcal{T}$.

9.1 Security of the Construction

To analyze the construction, we will assume tables for the adjusted points have not been created, for ease of exposition. To begin, let $\overline{E}: \mathcal{M} \to \mathcal{M}$ be a uniformly random permutation. For a preserved domain set $\mathcal{T} = \{\tau_1, \ldots, \tau_t\}$ and a uniformly random permutation $F: \mathcal{D} \to \mathcal{D}$ we construct a sequence of permutations $\mathcal{R}_0, \ldots, \mathcal{R}_t$, such that $\mathcal{R}_0 = \overline{E}$, $\mathcal{R}_i(\tau_j) = F(\tau_j)$ for all $1 \leq j \leq i \leq t$. Each $\mathcal{R}_i$ corresponds to the lookup table Emap for E^{rzz} after the ith iteration of KT^{rzz}. Note that we will abuse the notation slightly below, since if $\overline{E}$ and F are random permutations there will be no keys generated in KT^{rzz}; we refer to the straightforward modification of KT^{rzz} with random permutations.

We will now state the theorems showing $\mathcal{R}_t$, the cipher $E^{rzz}_{\text{Emap},\overline{K}}$ obtained after the t iterations of KT^{rzz}, is an SPRP. We defer their proofs to the full version. First, we state the information-theoretic step. Intuitively, this theorem shows that if $\overline{E}$ and F are uniformly random permutations, the permutation E^{rzz}_{Emap} resulting from constructing Emap from F and $\overline{E}$ as in KT^{rzz} is also uniformly random.

Theorem 10. *Let $\mathcal{T}$ be a randomly-chosen subset of $\mathcal{D}$, $|\mathcal{T}| = t$, and $t \leq d - n$. Let $\mathcal{R}_t$ be a random variable denoting the permutation over $\mathcal{M}$ induced by the RZZ algorithm after all t iterations of KT^{rzz}, instantiated with $\overline{E}$ and F as uniformly random permutations over $\mathcal{M}$ and $\mathcal{D}$, respectively. For any fixed permutation Π on $\mathcal{M}$,*

$$\Pr[\mathcal{R}_t = \Pi] = \frac{1}{m!}$$

To complete the proof that $E^{rzz}_{\text{Emap},\overline{K}}$ is an SPRP, we need to transition to the computational setting. Our current proof establishing this uses a reduction that requires exponential time in the worst case, but is efficient in expectation. This is due to the rejection sampling in case 3 of KT^{rzz}, which can take as many as n queries in the worst case but this happens with small probability.

Theorem 11. *Assume that $\overline{E}$ and F are ciphers on domains $\mathcal{M}$ and $\mathcal{D}$, respectively. Let t be a non-zero number, and let $\mathcal{T}$ be a random size t subset $\mathcal{D}$. Let $\mathcal{A}$ be an SPRP adversary against $E^{rzz}_{\text{Emap},\overline{K}}$ making at most q queries to its oracles. Then the proof gives an adversary $\mathcal{B}$ and an adversary $\mathcal{C}$ such that*

$$\mathbf{Adv}^{\text{sprp}}_{E^{rzz}_{\text{Emap},\overline{K}}}(\mathcal{A}(t)) \leq \mathbf{Adv}^{\text{sprp}}_{F}(B) + \mathbf{Adv}^{\text{sprp}}_{\overline{E}}(C)$$

Adversary $\mathcal{B}$ makes at most q queries and runs in time that of $\mathcal{A}$ plus a negligible overhead. Adversary $\mathcal{C}$ runs in expected time $c(q+8t)$ for a small constant c and and makes $q + 8t$ queries in expectation.

9.2   Efficiency of Recursive Zig-Zag's KT

Now that we know our construction meets the desired security, we turn to analyzing the efficiency of the key transformation KT. Examining the algorithm, we can see that with the exception of the random selection of p_i, the algorithm requires $O(1)$ steps for each iteration, or a time linear in t for the whole calculation. Selecting the p_i's is, however, a bit more involved. As described above, the p_i's are randomly-selected domain-preserved points in $\mathcal{D} \setminus \{\tau_1, \ldots, \tau_{i-1}\}$.

One way of selecting a p_i is to keep picking random points in $\mathcal{D} \setminus \{\tau_1, \ldots, \tau_{i-1}\}$ until a domain-preserved point is found. For a small i, there are many points in $\mathcal{D} \setminus \{\tau_1, \ldots, \tau_{i-1}\}$ and we expect to find an acceptable p_i within very few tries. However, as i approaches $d - n$, for some permutations, the number of domain-preserved elements in $\mathcal{D} \setminus \{\tau_1, \ldots, \tau_{i-1}\}$ may be very small, requiring a large number of tries.

The next theorem shows that we can limit the number of points we need to encrypt in order to generate the p_i's. More specifically, we show that with a high probability, the number of encryptions required for finding the p_i's is linear in t, with a reasonably small factor. Hence, the expected amortized cost of finding each of the p_i is constant.

Theorem 12. *Let X be a random variable (over the probability space defined by the keys of F and $\overline{E}$ and the random choices of the p_i values) whose value is the number of encryptions required to select all the p_i values in the t transformations of KT. Let $d = |\mathcal{D}|$. Then $\Pr[X > 8t] \leq e^{-d/8}$.*

Proof. If $t > d/8$, where d is the size of the original domain $\mathcal{D}$, we can enumerate $\mathcal{S}_0$ by calculating $\mathcal{R}(x)$ for every $x \in \mathcal{D}$. This requires $d < 8t$ encryptions. Once enumerated, we can calculate $\mathcal{S}_i$ during the generation of the extended permutation without requiring any further encryptions.

For smaller t we look at $s' = d - s$ as a random variable with a hypergeometric distribution whose mean is $E[s'] = d - d^2/(d+n) < d/2$. Hypergeometric distributions are concentrated around the mean. Hence, by Lemma 1, we have

$$\Pr[d/4 > s] = \Pr[d/2 + d/4 < s'] < \Pr[E[s'] + d/4 < s'] < e^{-d/8}$$

Thus, with a high probability, at the ith step in the construction we have a choice of at least $s - i \geq d/4 - t \geq d/8$ elements of $\mathcal{D}$ as candidates for p_i. We can, now, repeatedly pick a random $p_i \in \mathcal{D}$ and check whether $p_i \in \mathcal{S}_i$. Each such try requires one encryption. Because $s - i \geq d/8$ the expected number of encryption required is less than 8. Thus, with a high probability, the expected number of encryption required to select the p_i's is less than $8t$.

Acknowledgments. The authors thank Terence Spies for suggesting this problem to them, and for discussions about how it impacts industry practice. The authors also thank the anonymous reviewer whose observations led to Sect. 4, and for his or her permission to include it. This work was supported in part by NSF grants CNS-1514163, CNS-1330308, and CNS-1558500 as well as a generous gift by Microsoft.

References

1. Acıiçmez, O., Koç, Ç.K., Seifert, J.-P.: On the power of simple branch prediction analysis. In: 2nd ACM Symposium on Information, Computer and Communications Security, Singapore (2007)
2. Acıiçmez, O., Seifert, J.-P.: Cheap hardware parallelism implies cheap security. In: Fourth International Workshop on Fault Diagnosis and Tolerance in Cryptography, Vienna, AT, pp. 80–91 (2007)
3. Askarov, A., Zhang, D., Myers, A.C.: Predictive black-box mitigation of timing channels. In: Al-Shaer, E., Keromytis, A.D., Shmatikov, V. (eds.) ACM CCS 2010, Chicago, Illinois, USA, October 4–8, pp. 297–307. ACM Press (2010)
4. Bellare, M., Ristenpart, T., Rogaway, P., Stegers, T.: Format-preserving encryption. In: Jacobson, M.J., Rijmen, V., Safavi-Naini, R. (eds.) SAC 2009. LNCS, vol. 5867, pp. 295–312. Springer, Heidelberg (2009). doi:10.1007/978-3-642-05445-7_19
5. Bernstein, D.J.: Cache-timing attacks on AES (2005). Preprint available at http://cr.yp.to/papers.html#cachetiming
6. Black, J., Rogaway, P.: Ciphers with arbitrary finite domains. In: Preneel, B. (ed.) CT-RSA 2002. LNCS, vol. 2271, pp. 114–130. Springer, Heidelberg (2002). doi:10.1007/3-540-45760-7_9
7. Brumley, D., Boneh, D.: Remote timing attacks are practical. Comput. Netw. **48**(5), 701–716 (2005)
8. Chavátal, V.: The tail of the hypergeometric distribution. Discrete Math. **25**(3), 285–287 (1979)
9. Cock, D., Ge, Q., Murray, T.C., Heiser, G.: The last mile: an empirical study of timing channels on seL4. In: Ahn, G.-J., Yung, M., Li, N. (eds.) ACM CCS 2014, Scottsdale, AZ, USA, November 3–7, pp. 570–581. ACM Press (2014)
10. PCI Security Standards Council. The payment card industry data security standard specification. https://www.pcisecuritystandards.org/security_standards/documents.php?agreements=pcidss&association=pcidss

11. Dyer, K.P., Coull, S.E., Ristenpart, T., Shrimpton, T.: Protocol misidentification made easy with format-transforming encryption. In: Proceedings of the 20th ACM Conference on Computer and Communications Secuirty (CCS 2013), November 2013
12. Protegrity Inc., Protegrity data protection methods. http://www.protegrity.com/data-security-platform/#protection-methods
13. Liskov, M., Rivest, R.L., Wagner, D.: Tweakable block ciphers. In: Yung, M. (ed.) CRYPTO 2002. LNCS, vol. 2442, pp. 31–46. Springer, Heidelberg (2002). doi:10.1007/3-540-45708-9_3
14. Liu, F., Ge, Q., Yarom, Y., Mckeen, F., Rozas, C., Heiser, G., Lee, R.B.: CATalyst: Defeating last-level cache side channel attacks in cloud computing. In: IEEE Symposium on High-Performance Computer Architecture, Barcelona, Spain, March 2016
15. Liu, F., Yarom, Y., Ge, Q., Heiser, G., Lee, R.B.: Last-level cache side-channel attacks are practical. In: 2015 IEEE Symposium on Security and Privacy, San Jose, CA, US, pp. 605–622, May 2015
16. Luchaup, D., Dyer, K.P., Jha, S., Ristenpart, T., Shrimpton, T.: LibFTE: A user-friendly toolkit for constructing practical format-abiding encryption schemes. In: Proceedings of the 14th Conference on USENIX Security Symposium (2014)
17. Luchaup, D., Shrimpton, T., Ristenpart, T., Jha, S.: Formatted encryption beyond regular languages. In: Proceedings of the 2014 ACM SIGSAC Conference on Computer and Communications Security, CCS 2014, New York, USA, pp. 1292–1303. ACM (2014)
18. Mihir Bellare, T.S., Rogaway, P.: The FFX mode of operation for format-preserving encryption. http://csrc.nist.gov/groups/ST/toolkit/BCM/documents/proposedmodes/ffx/ffx-spec.pdf
19. Osvik, D.A., Shamir, A., Tromer, E.: Cache attacks and countermeasures: the case of AES. Cryptology ePrint Archive, Report 2005/271, 2005. http://eprint.iacr.org/2005/271
20. Ristenpart, T., Yilek, S.: The mix-and-cut shuffle: small-domain encryption secure against N queries. In: Canetti, R., Garay, J.A. (eds.) CRYPTO 2013. LNCS, vol. 8042, pp. 392–409. Springer, Heidelberg (2013). doi:10.1007/978-3-642-40041-4_22
21. Shi, J., Song, X., Chen, H., Zang, B.: Limiting cache-based side-channel in multi-tenant cloud using dynamic page coloring. In: International Conference on Dependable Systems and Networks Workshops (DSN-W), HK, pp. 194–199, June 2011
22. Skala, M.: Hypergeometric tail inequalities: ending the insanity. http://arxiv.org/pdf/1311.5939v1.pdf
23. Inc., Skyhigh Networks. Skyhigh for Salesforce. https://www.skyhighnetworks.com/product/salesforce-encryption/
24. Varadarajan, V., Ristenpart, T., Swift, M.: Scheduler-based defenses against cross-VM side-channels. In: Proceedings of the 24th USENIX Security Symposium, San Diego, CA, US (2014)
25. HP Security Voltage. Hp format-preserving encryption. https://www.voltage.com/technology/data-encryption/hp-format-preserving-encryption/
26. Wang, Z., Lee, R.B.: New cache designs for thwarting software cache-based side channel attacks. In: Proceedings of the 34th International Symposium on Computer Architecture, San Diego, CA, US (2007)
27. Wikipedia. Tokenization. https://en.wikipedia.org/wiki/Tokenization_(data_security)

Separating Semantic and Circular Security for Symmetric-Key Bit Encryption from the Learning with Errors Assumption

Rishab Goyal[(✉)], Venkata Koppula, and Brent Waters

University of Texas at Austin, Austin, USA
{rgoyal,kvenkata,bwaters}@cs.utexas.edu

Abstract. In this work we separate private-key semantic security from 1-circular security for bit encryption using the Learning with Error assumption. Prior works used the less standard assumptions of multilinear maps or indistinguishability obfuscation. To achieve our results we develop new techniques for obliviously evaluating branching programs.

1 Introduction

Over the past several years the cryptographic community has given considerable attention to the notion of key-dependent message security. In key dependent security we consider an attacker that gains access to ciphertexts that encrypt certain functions of the secret key(s) of the user(s). Ideally, a system should remain semantically secure even in the presence of this additional information.

One of the most prominent problems in key dependent message security is the case of circular security. A circular secure system considers security in the presence of key cycles. A key cycle of k users consists of k encryptions where the i-th ciphertext ct_i is an encryption of the $i+1$ user's secret key under user i's public key. That is $\mathsf{ct}_1 = \mathsf{Encrypt}(\mathrm{PK}_1, \mathrm{SK}_2), \mathsf{ct}_2 = \mathsf{Encrypt}(\mathrm{PK}_2, \mathrm{SK}_3)\ldots, \mathsf{ct}_k = \mathsf{Encrypt}(\mathrm{PK}_k, \mathrm{SK}_1)$. If a system is k circular secure, then such a cycle should be indistinguishable from an encryption of k arbitrary messages. The notion also applies to secret key encryption systems.

One reason that circular security has received significant attention is that the problem has arisen in multiple applications [2,16,27], the most notable is that Gentry [22] showed how a circular secure leveled homomorphic encryption can be bootstrapped to homomorphic encryption that works for circuits of unbounded depth. Stemming from this motivation there have been several positive results [4–6,8,11,13,14,29] that have achieved circular and more general notions of key dependent messages security from a variety of cryptographic assumptions.

On the flip side several works have sought to discover if there exist separations between IND-CPA security and different forms of circular security. That is they

B. Waters—Supported by NSF CNS-1228599 and CNS-1414082, DARPA SafeWare, Microsoft Faculty Fellowship, and Packard Foundation Fellowship.

J.-S. Coron and J.B. Nielsen (Eds.): EUROCRYPT 2017, Part II, LNCS 10211, pp. 528–557, 2017.
DOI: 10.1007/978-3-319-56614-6_18

sought to develop a system that was *not* circular secure, but remained IND-CPA secure. For the case of 1-circular security achieving such a separation is trivial. The (secret key) encryption system simply tests if the message to be encrypted is equal to the secret key SK, if so it gives the message in the clear; otherwise it encrypts as normal. (This example can be easily extended to public key encryption.) Clearly, such a system is not circular secure and it is easy to show it maintains IND-CPA security. More work is required, however, to achieve separations of length greater than one. Separations were first shown for the case of $k = 2$ length cycles using groups with bilinear maps [1,17] and later [10] under the Learning with Errors assumption [34]. Subsequently, there existed works that achieved separations for arbitrary length cycles [25,28], however, these required the use obfuscation. All current candidates of general obfuscation schemes rely on the relatively new primitive of multilinear maps, where many such multilinear map candidates have suffered from cryptanalysis attacks [18,19]. Most recently and Alamati and Peikert [3] and Koppula and Waters [26] showed separations of arbitrary length cycles from the much more standard Learning with Errors assumption.

Another challenging direction in achieving separations for circular security is to consider encryptions systems where the message consist of a single bit. Separating from IND-CPA is difficult even in the case of cycles of length 1 (i.e. someone encrypts their own secret key). Consider a bit encryption system with keys of length $\ell = \ell(\lambda)$. Suppose an attacker receives an encryption of the secret key in the form of ℓ successive bit by bit encryptions. Can this be detected?

We observe that encrypting bit by bit seems to make detection harder. Our trivial counterexample from above no longer applies since the single bit message cannot be compared to the much longer key. The first work to consider such a separation was due to Rothblum [35] who showed that a separation could be achieved from multilinear maps under certain assumptions. One important caveat, however, to his result was that the level of multilinearlity must be greater than $\log(q)$ where q is the group order. This restriction appears to be at odds with current multilinear map/encoding candidates which are based off of "noisy cryptography" and naturally require a bigger modulus whose log is greater than the number of multiplications allowed. Later, Koppula, Ramchen and Waters [25] showed how to achieve a separation from bit encryption using indistinguishability obfuscation. Again, such a tool is not known from standard assumptions.

In this work we aim to separate semantic security from 1-circular security for bit encryption systems under the Learning with Errors assumption. Our motivation to study this problem is two fold. First, achieving such a separation under a standard assumption will significantly increase our confidence compared to obfuscation or multilinear map-based results. Second, studying such a problem presents the opportunity for developing new techniques in the general area of computing on encrypted data and may lead to other results down the line.

To begin with, we wish to highlight some challenges presented by bit encryption systems that were not addressed in prior work. First, the recent results of [3,26] both use a form of telescoping cancellation where the encryption algorithm takes in a message and uses this as a 'lattice trapdoor' [24,30]; if the message contained the needed secret key then it cancels out the public key of an "adjacent" ciphertext. We observe that such techniques require an encryption algorithm that receives the entire secret key at once, and there is no clear path to leverage this in the case where an encryption algorithm receives just a single bit message. Second, while the level restriction in Rothblum's result [35] appeared in the context of multilinear maps, the fundamental issue will transcend to our Learning with Errors solution. Looking ahead we will need to perform a computation where the number of multiplication steps is restricted to be less than $\log(q)$, where here q is the modulus we work in.

1.1 Separations from Learning with Errors

We will now describe our bit encryption scheme that is semantically secure but not circular secure. Like previous works [3,10,26], we will take decryption out of the picture, and focus on building an IND-CPA secure encryption scheme where one can distinguish between an encryption of the secret key and encryptions of zeroes.

The two primary ingredients of our construction are low-depth pseudorandom functions (PRFs) and lattice trapdoors. In particular, we require a PRF which can be represented using a permutation branching program of polynomial length and polynomial width.[1] Banerjee, Peikert and Rosen [7] showed how to construct LWE based PRFs that can be represented using $\mathbf{NC^1}$ circuits, and using Barrington's theorem [9], we get PRFs that can be represented using branching programs of polynomial length and width 5.

Next, let us recall the notion of lattice trapdoors. A lattice trapdoor generation algorithm outputs a matrix $\mathbf{A}$ together with a trapdoor $T_{\mathbf{A}}$. The matrix looks uniformly random, while the trapdoor can be used to compute, for any matrix $\mathbf{U}$, a low norm matrix $\mathbf{S} = \mathbf{A}^{-1}(\mathbf{U})$ such that $\mathbf{A} \cdot \mathbf{S} = \mathbf{U}$.[2] As a result, the matrix $\mathbf{S}$ can be used to 'transform' the matrix $\mathbf{A}$ to another matrix $\mathbf{U}$. In this work, we will be interested in *oblivious sequence transformation*: we want a sequence of matrices $\mathbf{B}_1, \ldots, \mathbf{B}_w$ such that for any sequence of matrices $\mathbf{U}_1, \ldots, \mathbf{U}_w$, we can compute a low norm matrix $\mathbf{S}$ such that $\mathbf{B}_i \cdot \mathbf{S} = \mathbf{U}_i$. Note that the same matrix $\mathbf{S}$ should be able to transform any $\mathbf{B}_i$ to $\mathbf{U}_i$; that is, $\mathbf{S}$ is oblivious of i. This obliviousness property will be important for our solution, and together with the telescoping products/cascading cancellations idea of [3,23,26], we get our counterexample.

---

[1] Recall, a permutation branching program of length L and width w has w states at each level, an accepting and rejecting state at the top level. Each level $j \le L$ has two permutations $\sigma_{j,0}$ and $\sigma_{j,1}$ associated, and there is an input-selector function which determines the input read at each level. The program execution starts at state 1 of level 0. Suppose, at level j, the state is $\mathsf{st} \in [w]$. Let b be the input read at level j. Then, the state at level $j+1$ is $\sigma_{j,b}(\mathsf{st})$. Proceeding this way, the program terminates at level L in either the accepting state or rejection state.

[2] For simplicity, we use the notation $\mathbf{A}^{-1}(\cdot)$ to represent the pre-image $\mathbf{S}$. In the formal description of our algorithms, we use the pre-image sampling algorithm SamplePre.

Oblivious Sequence Transformation. We first observe that one can easily obtain oblivious sequence transformation, given standard lattice trapdoors. Consider the following matrix $\mathbf{B}$:

$$\mathbf{B} = \begin{bmatrix} \mathbf{B}_1 \\ \vdots \\ \mathbf{B}_w \end{bmatrix}.$$

Let T denote the trapdoor of $\mathbf{B}$ (we will refer to T as the 'joint trapdoor' of $\mathbf{B}_1, \ldots, \mathbf{B}_w$). Now, given any sequence $\mathbf{U}_1, \ldots, \mathbf{U}_w$, we similarly define a new matrix $\mathbf{U}$ which has the $\mathbf{U}_i$ stacked together, and set $\mathbf{S} = \mathbf{B}^{-1}(\mathbf{U})$. Clearly, this satisfies our oblivious sequence transformation requirement.

Our Encryption Scheme. As mentioned before, we will only focus on the setup, encryption and testing algorithms. Let PRF be a pseudorandom function family with keys and inputs of length λ, and output being a single bit. For any input i, we require that the function $\mathrm{PRF}(\cdot, i)$ can be represented using a branching program of length L and width 5 (we choose 5 for simplicity here; our formal description works for any polynomial width w). The setup algorithm chooses a PRF key s. Let nbp be a parameter which represents the number of points at which the PRF is evaluated, and let $t_i = \mathrm{PRF}(s, i)$ for $i \leq$ nbp. Finally, for each $i \leq$ nbp, let $\mathsf{BP}^{(i)}$ denote the branching program that evaluates $\mathrm{PRF}(\cdot, i)$. Each branching program $\mathsf{BP}^{(i)}$ has L levels and 5 possible states at each level. At the last level, there are only two valid states — $\mathsf{acc}^{(i)}$ and $\mathsf{rej}^{(i)}$, i.e. the accepting and rejecting state. For each branching program $\mathsf{BP}^{(i)}$ and level j, there are two state transition functions $\sigma_{j,0}^{(i)}, \sigma_{j,1}^{(i)}$ that decide the transition between states depending upon the input bit read. The setup algorithm also chooses, for each branching program $\mathsf{BP}^{(i)}$, level $j \leq L$ and state $k \leq 5$, a matrix $\mathbf{B}_{j,k}^{(i)}$. At all levels $j \neq L$, the matrices $\mathbf{B}_{j,1}^{(i)}, \ldots, \mathbf{B}_{j,5}^{(i)}$ have a joint trapdoor. At the top level, the matrices satisfy the following relation:

$$\sum_{i \,:\, t_i=0} \mathbf{B}_{L,\mathsf{rej}^{(i)}}^{(i)} + \sum_{i \,:\, t_i=1} \mathbf{B}_{L,\mathsf{acc}^{(i)}}^{(i)} = \mathbf{0}.$$

The secret key consists of the PRF key s and nbp $\cdot\, L$ trapdoors $T_j^{(i)}$.

The encryption algorithm is designed specifically to distinguish key encryptions from encryptions of zeros. Each ciphertext consists of L sub-ciphertexts, one for each level, and each sub-ciphertext consists of nbp sub-sub-ciphertexts. The sub-sub-ciphertext corresponding to $\mathsf{BP}^{(i)}$ at level j can be used to transform $\mathbf{B}_{j,k}^{(i)}$ to $\mathbf{B}_{j+1,\sigma_{j,0}^{(i)}(k)}^{(i)}$ or $\mathbf{B}_{j+1,\sigma_{j,1}^{(i)}(k)}^{(i)}$, depending on the bit encrypted. This is achieved via oblivious sequence transformation. Let b denote the bit encrypted, and let $\mathbf{D}$ be the matrix constructed by stacking $\{\mathbf{B}_{j,1}^{(i)}, \ldots, \mathbf{B}_{j,5}^{(i)}\}$ according to the permutation $\sigma_{j,b}^{(i)}$. The sub-sub-ciphertext $\mathsf{ct}_j^{(i)}$ for program $\mathsf{BP}^{(i)}$ at level j is simply (a noisy approximation of) $\mathbf{B}_j^{(i)^{-1}}(\mathbf{D})$. The ciphertext also includes the base matrices $\{\mathbf{B}_0^{(i)}\}$ for each program.

The testing algorithm is used to distinguish between an encryption of the secret key and encryptions of zeros. It uses the first $|s| = \lambda$ ciphertexts, which are either encryptions of the PRF key s, or encryptions of zeros. Let us consider the case where the λ ciphertexts are encryptions of s. At a high level, the testing algorithm combines the ciphertext components appropriately, such that for each $i \leq$ nbp, the result is $\mathbf{B}^{(i)}_{L,\text{rej}^{(i)}}$ if $\text{PRF}(s,i) = 0$, and $\mathbf{B}^{(i)}_{L,\text{acc}^{(i)}}$ otherwise. Once the testing algorithm gets these matrices, it can sum them to check if it is (close to) the zero matrix. The testing algorithm essentially mimics the program evaluation on s using the encryption of s. Let us fix a program $\text{BP}^{(i)}$, and say it reads bit positions $p_1, \ldots, p_L$. At step 1, the program goes from state 1 at level 0 to state $\text{st}_1 = \sigma^{(i)}_{1,s_{p_1}}$ at level 1. The test algorithm has $\mathbf{B}^{(i)}_{0,1}$. It combines this with the $(i,1)$ component of the p_1^{th} ciphertext to get $\mathbf{B}^{(i)}_{1,\text{st}_1}$. Next, the program reads the bit at position p_2 and goes to state st_2 at level 2. The test algorithm, accordingly, combines $\mathbf{B}^{(i)}_{1,\text{st}_1}$ with the $(i,2)$ sub-sub-component of the p_2^{th} ciphertext to compute $\mathbf{B}^{(i)}_{2,p_2}$. Proceeding this way, the actual program evaluation reaches either $\text{acc}^{(i)}$ or $\text{rej}^{(i)}$, and the test algorithm accordingly reaches either $\mathbf{B}^{(i)}_{L,\text{acc}^{(i)}}$ or $\mathbf{B}^{(i)}_{L,\text{rej}^{(i)}}$.

The solution described above, however, is not IND-CPA secure. To hide the encrypted bit without affecting the above computation, we will have to add some noise to each sub-sub-ciphertext. In particular, instead of outputting $\mathbf{B}_j^{(i)^{-1}}(\mathbf{D})$ for some matrix $\mathbf{D}$, we will now have $\mathbf{B}_j^{(i)^{-1}}(\mathbf{S} \cdot \mathbf{D} + \text{noise})$,[3] where $\mathbf{S}$ is a low norm matrix. To prove IND-CPA security, we first switch the top level matrices to uniformly random matrices. Once we've done that, we can use LWE, together with the properties of lattice trapdoors, to argue that the top level sub-sub-ciphertexts look like random matrices from a low-norm distribution. As a result, we don't need trapdoors for the matrices at level $L - 1$, and hence, they can be switched to uniformly random matrices. Using LWE with trapdoor properties, we can then switch the sub-sub-ciphertexts at level $L - 1$ to random matrices. Proceeding this way, all sub-sub-ciphertexts can be made random Gaussian matrices. This concludes our proof.

Separation from Chosen Ciphertext Security. One interesting question is whether achieving chosen ciphertext security (as opposed to IND-CPA security) makes a bit encryption system more likely to be resistant to circular security attacks. Here we show generically that achieving a bit encryption system that is IND-CCA secure, but not circular secure is no more difficult than our original separation problem. In particular, we show generically how to combine a IND-CPA secure, but not circular secure bit encryption with multi-bit CCA secure encryption to achieve a single bit encryption system that is IND-CPA secure. We note that

---

[3] Strictly speaking, if $\mathbf{D}$ consists of 5 components $\mathbf{D}_1, \ldots, \mathbf{D}_5$ stacked together, then our sub-sub-ciphertext will be $\mathbf{B}_j^{(i)^{-1}}(\mathbf{D}' + \text{noise})$ where $\mathbf{D}'$ consists of 5 components $\mathbf{S} \cdot \mathbf{D_k}$ for $k \leq 5$.

Rothblum addressed CCA security, but used the more specific assumption of trapdoor permutations to achieve NIZKs.

Our transformation is fairly simple and follows in a similar manner to how an analogous theorem in Bishop, Hohenberger and Waters [10].

Relation to GGH15 Graph Based Multilinear Maps. Our counterexample construction bears some similarities to the graph-induced multilinear maps scheme of Gentry, Gorbunov and Halevi [23]. In a graph induced multilinear maps scheme, we have an underlying graph G, and encodings of elements are relative to pairs of connected nodes in in the graphs. Given encodings of s_1 and s_2 relative to connected nodes $u \rightsquigarrow v$, one can compute an encoding of $s_1 + s_2$ relative to $u \rightsquigarrow v$. Similarly, given an encoding of s_1 relative to $u \rightsquigarrow v$ and an encoding of s_2 relative to $v \rightsquigarrow w$, one can compute an encoding of $s_1 \cdot s_2$ relative to $u \rightsquigarrow w$. Finally, one is allowed to zero-test corresponding to certain source-destination pairs. Gentry et al. gave a lattice based construction for graph-induced encoding scheme, where each vertex u has an associated matrix $\mathbf{A}_u$ (together with a trapdoor T_u). The encoding of an element s corresponding to the edge (u, v) is simply $\mathbf{A}_u^{-1}(s\mathbf{A}_v + \mathsf{noise})$.

At a high level, our construction looks similar to the GGH15 multilinear maps construction. In particular, while GGH15 uses the cascading cancellations property to prove correctness, we use it for proving that the testing algorithm succeeds with high probability. Our security requirements, on the other hand, are different from that in multilinear maps. However, we believe that the ideas used in this work can be used to prove security of GGH15 mmaps for special graphs/secret distributions (note that GGH15 gave a candidate multilinear maps construction, and it did not have a proof of security for general graphs).

Summary and Conclusions. To summarise, we show how to perform computation using an *outside primitive* by means of our oblivious sequence transformation approach. This allows us to show a separation between private-key semantic security and circular security for bit encryption schemes. While such counterexamples are contrived and do not give much insight into the circular security of existing schemes, we see this as a primitive of its own. The tools/techniques used for developing such counterexamples might have other applications. In particular, these counterexamples share certain features with more advanced cryptographic primitives such as witness encryption and code obfuscation.

2   Preliminaries

Notations. We will use lowercase bold letters for vectors (e.g. $\mathbf{v}$) and uppercase bold letters for matrices (e.g. $\mathbf{A}$). For any finite set S, $x \leftarrow S$ denotes a uniformly random element x from the set S. Similarly, for any distribution $\mathcal{D}$, $x \leftarrow \mathcal{D}$ denotes an element x drawn from distribution $\mathcal{D}$. The distribution $\mathcal{D}^n$ is used to represent a distribution over vectors of n components, where each component is drawn independently from the distribution $\mathcal{D}$.

Min-Entropy and Randomness Extraction. The min-entropy of a random variable X is defined as $\mathbf{H}_\infty(X) \overset{\text{def}}{=} -\log_2(\max_x \Pr[X = x])$. Let $\mathsf{SD}(X, Y)$ denote the statistical distance between two random variables X and Y. Below we state the Leftover Hash Lemma (LHL) from [20, 21].

Theorem 1. *Let $\mathcal{H} = \{h : X \to Y\}_{h \in \mathcal{H}}$ be a universal hash family, then for any random variable W taking values in X, the following holds*

$$\mathsf{SD}\left((h, h(W)), (h, U_Y)\right) \le \frac{1}{2}\sqrt{2^{-\mathbf{H}_\infty(W)} \cdot |Y|}.$$

We will use the following corollary, which follows from the Leftover Hash Lemma.

Corollary 1. *Let $\ell > m \cdot n \log_2 q + \omega(\log n)$ and q a prime. Let $\mathbf{R}$ be an $k \times m$ matrix chosen as per distribution $\mathcal{R}$, where $k = k(n)$ is polynomial in n and $\mathbf{H}_\infty(\mathcal{R}) = \ell$. Let $\mathbf{A}$ and $\mathbf{B}$ be matrices chosen uniformly in $\mathbb{Z}_q^{n \times k}$ and $\mathbb{Z}_q^{n \times m}$, respectively. Then the statistical distance between the following distributions is negligible in n.*

$$\{(\mathbf{A}, \mathbf{A} \cdot \mathbf{R})\} \approx_s \{(\mathbf{A}, \mathbf{B})\}$$

Proof. The proof of above corollary follows directly from the Leftover Hash Lemma. Note that for a prime q the family of hash functions $h_\mathbf{A} : \mathbb{Z}_q^{k \times m} \to \mathbb{Z}_q^{n \times m}$ for $\mathbf{A} \in \mathbb{Z}_q^{n \times k}$ defined by $h_\mathbf{A}(\mathbf{X}) = \mathbf{A} \cdot \mathbf{X}$ is universal. Therefore, if $\mathcal{R}$ has sufficient min-entropy, i.e. $\ell > m \cdot n \log_2 q + \omega(\log n)$, then the Leftover Hash Lemma states that statistical distance between the distributions $(\mathbf{A}, \mathbf{A} \cdot \mathbf{R})$ and $(\mathbf{A}, \mathbf{B})$ is at most $2^{-\omega(\log n)}$ which is negligible in n as desired.

2.1   Lattice Preliminaries

This section closely follows [26].

Given positive integers n, m, q and a matrix $\mathbf{A} \in \mathbb{Z}_q^{n \times m}$, we let $\Lambda_q^\perp(\mathbf{A})$ denote the lattice $\{\mathbf{x} \in \mathbb{Z}^m : \mathbf{A} \cdot \mathbf{x} = \mathbf{0} \mod q\}$. For $\mathbf{u} \in \mathbb{Z}_q^n$, we let $\Lambda_q^\mathbf{u}(\mathbf{A})$ denote the coset $\{\mathbf{x} \in \mathbb{Z}^m : \mathbf{A} \cdot \mathbf{x} = \mathbf{u} \mod q\}$.

Discrete Gaussians. Let σ be any positive real number. The Gaussian distribution $\mathcal{D}_\sigma$ with parameter σ is defined by the probability distribution function $\rho_\sigma(\mathbf{x}) = \exp(-\pi \cdot ||\mathbf{x}||^2/\sigma^2)$. For any set $\mathcal{L} \subset \mathcal{R}^m$, define $\rho_\sigma(\mathcal{L}) = \sum_{\mathbf{x} \in \mathcal{L}} \rho_\sigma(\mathbf{x})$. The discrete Gaussian distribution $\mathcal{D}_{\mathcal{L}, \sigma}$ over $\mathcal{L}$ with parameter σ is defined by the probability distribution function $\rho_{\mathcal{L}, \sigma}(\mathbf{x}) = \rho_\sigma(\mathbf{x})/\rho_\sigma(\mathcal{L})$ for all $\mathbf{x} \in \mathcal{L}$.

The following lemma (Lemma 4.4 of [24, 31]) shows that if the parameter σ of a discrete Gaussian distribution is small, then any vector drawn from this distribution will be short (with high probability).

Lemma 1. *Let m, n, q be positive integers with $m > n$, $q \ge 2$. Let $\mathbf{A} \in \mathbb{Z}_q^{n \times m}$ be a matrix of dimensions $n \times m$, $\sigma = \tilde{\Omega}(n)$ and $\mathcal{L} = \Lambda_q^\perp(\mathbf{A})$. Then*

$$\Pr[||\mathbf{x}|| > \sqrt{m} \cdot \sigma : \mathbf{x} \leftarrow \mathcal{D}_{\mathcal{L}, \sigma}] \le negl(n).$$

Learning with Errors (LWE). The Learning with Errors (LWE) problem was introduced by Regev [34]. The LWE problem has four parameters: the dimension of the lattice n, the number of samples m, the modulus q and the error distribution $\chi(n)$.

Assumption 1 (Learning with Errors). *Let n, m and q be positive integers and χ a noise distribution on $\mathbb{Z}$. The Learning with Errors assumption (n, m, q, χ)-LWE, parameterized by n, m, q, χ, states that the following distributions are computationally indistinguishable:*

$$\left\{ (\mathbf{A}, \mathbf{s}^\top \cdot \mathbf{A} + \mathbf{e}) \ : \ \begin{array}{l} \mathbf{A} \leftarrow \mathbb{Z}_q^{n \times m}, \\ \mathbf{s} \leftarrow \mathbb{Z}_q^n, \mathbf{e} \leftarrow \chi^m \end{array} \right\} \approx_c \left\{ (\mathbf{A}, \mathbf{u}) \ : \ \begin{array}{l} \mathbf{A} \leftarrow \mathbb{Z}_q^{n \times m}, \\ \mathbf{u} \leftarrow \mathbb{Z}_q^m \end{array} \right\}$$

Under a quantum reduction, Regev [34] showed that for certain noise distributions, LWE is as hard as worst case lattice problems such as the decisional approximate shortest vector problem (GapSVP) and approximate shortest independent vectors problem (SIVP). The following theorem statement is from Peikert's survey [33].

Theorem 2 ([34]). *For any $m \leq \mathsf{poly}(n)$, any $q \leq 2^{\mathsf{poly}(n)}$, and any discretized Gaussian error distribution χ of parameter $\alpha \cdot q \geq 2 \cdot \sqrt{n}$, solving (n, m, q, χ)-LWE is as hard as quantumly solving GapSVP_γ and SIVP_γ on arbitrary n-dimensional lattices, for some $\gamma = \tilde{O}(n/\alpha)$.*

Later works [15,32] showed classical reductions from LWE to GapSVP_γ. Given the current state of art in lattice algorithms, GapSVP_γ and SIVP_γ are believed to be hard for $\gamma = \tilde{O}(2^{n^\epsilon})$, and therefore (n, m, q, χ)-LWE is believed to be hard for Gaussian error distributions χ with parameter $2^{-n^\epsilon} \cdot q \cdot \mathsf{poly}(n)$.

LWE with Short Secrets. In this work, we will be using a variant of the LWE problem called *LWE with Short Secrets*. In this variant, introduced by Applebaum et al. [6], the secret vector is also chosen from the noise distribution χ. They showed that this variant is as hard as LWE for sufficiently large number of samples m.

Assumption 2 (LWE with Short Secrets). *Let n, m and q be positive integers and χ a noise distribution on $\mathbb{Z}$. The LWE with Short Secrets assumption (n, m, q, χ)-LWE-ss, parameterized by n, m, q, χ, states that the following distributions are computationally indistinguishable[4]:*

$$\left\{ (\mathbf{A}, \mathbf{S} \cdot \mathbf{A} + \mathbf{E}) \ : \ \begin{array}{l} \mathbf{A} \leftarrow \mathbb{Z}_q^{n \times m}, \\ \mathbf{S} \leftarrow \chi^{n \times n}, \mathbf{E} \leftarrow \chi^{n \times m} \end{array} \right\} \approx_c \left\{ (\mathbf{A}, \mathbf{U}) \ : \ \begin{array}{l} \mathbf{A} \leftarrow \mathbb{Z}_q^{n \times m}, \\ \mathbf{U} \leftarrow \mathbb{Z}_q^{n \times m} \end{array} \right\}.$$

---

[4] Applebaum et al. showed that $\{(\mathbf{A}, \mathbf{s}^\top \cdot \mathbf{A} + \mathbf{e}) : \mathbf{A} \leftarrow \mathbb{Z}_q^{n \times m}, \mathbf{s} \leftarrow \chi^n, \mathbf{e} \leftarrow \chi^m\} \approx_c$ $\{(\mathbf{A}, \mathbf{u}) : \mathbf{A} \leftarrow \mathbb{Z}_q^{n \times m}, \mathbf{u} \leftarrow \mathbb{Z}_q^m\}$, assuming LWE is hard. However, by a simple hybrid argument, we can replace vectors $\mathbf{s}, \mathbf{e}, \mathbf{u}$ with matrices $\mathbf{S}, \mathbf{E}, \mathbf{U}$ of appropriate dimensions.

Lattices with Trapdoors. Lattices with trapdoors are lattices that are statistically indistinguishable from randomly chosen lattices, but have certain 'trapdoors' that allow efficient solutions to hard lattice problems.

Definition 1. *A trapdoor lattice sampler consists of algorithms* TrapGen *and* SamplePre *with the following syntax and properties:*

- TrapGen$(1^n, 1^m, q) \rightarrow (\mathbf{A}, T_{\mathbf{A}})$: *The lattice generation algorithm is a randomized algorithm that takes as input the matrix dimensions n, m, modulus q, and outputs a matrix $\mathbf{A} \in \mathbb{Z}_q^{n \times m}$ together with a trapdoor $T_{\mathbf{A}}$.*
- SamplePre$(\mathbf{A}, T_{\mathbf{A}}, \mathbf{u}, \sigma) \rightarrow \mathbf{s}$: *The presampling algorithm takes as input a matrix $\mathbf{A}$, trapdoor $T_{\mathbf{A}}$, a vector $\mathbf{u} \in \mathbb{Z}_q^n$ and a parameter $\sigma \in \mathcal{R}$ (which determines the length of the output vectors). It outputs a vector $\mathbf{s} \in \mathbb{Z}_q^m$.*

These algorithms must satisfy the following properties:

1. Correct Presampling: *For all vectors $\mathbf{u}$, parameters σ, $(\mathbf{A}, T_{\mathbf{A}}) \leftarrow$* TrapGen$(1^n, 1^m, q)$, *and $\mathbf{s} \leftarrow$* SamplePre$(\mathbf{A}, T_{\mathbf{A}}, \mathbf{u}, \sigma)$, $\mathbf{A} \cdot \mathbf{s} = \mathbf{u}$ *and* $\|\mathbf{s}\|_\infty \leq \sqrt{m} \cdot \sigma$.
2. Well Distributedness of Matrix: *The following distributions are statistically indistinguishable:*

$$\{\mathbf{A} : (\mathbf{A}, T_{\mathbf{A}}) \leftarrow \text{TrapGen}(1^n, 1^m, q)\} \approx_s \{\mathbf{A} : \mathbf{A} \leftarrow \mathbb{Z}_q^{n \times m}\}.$$

3. Well Distributedness of Preimage: *For all $(\mathbf{A}, T_{\mathbf{A}}) \leftarrow$* TrapGen$(1^n, 1^m, q)$, *if $\sigma = \omega(\sqrt{n \cdot \log q \cdot \log m})$, then the following distributions are statistically indistinguishable:*

$$\{\mathbf{s} : \mathbf{u} \leftarrow \mathbb{Z}_q^n, \mathbf{s} \leftarrow \text{SamplePre}(\mathbf{A}, T_{\mathbf{A}}, \mathbf{u}, \sigma)\} \approx_s \mathcal{D}_{\mathbb{Z}^m, \sigma}.$$

These properties are satisfied by the gadget-based trapdoor lattice sampler of [30].

2.2 Branching Programs

Branching programs are a model of computation used to capture space-bounded computations [9,12]. In this work, we will be using a restricted notion called *permutation branching programs.*

Definition 2 (Permutation Branching Program). *A permutation branching program of length L, width w and input space $\{0,1\}^n$ consists of a sequence of $2L$ permutations $\sigma_{i,b} : [w] \rightarrow [w]$ for $1 \leq i \leq L, b \in \{0,1\}$, an input selection function* inp $: [L] \rightarrow [n]$, *an accepting state* acc $\in [w]$ *and a rejection state* rej $\in [w]$. *The starting state* st_0 *is set to be 1 without loss of generality. The branching program evaluation on input $x \in \{0,1\}^n$ proceeds as follows:*

- *For $i = 1$ to L,*
    - *Let* pos $=$ inp(i) *and $b = x_{\text{pos}}$. Compute* $\text{st}_i = \sigma_{i,b}(\text{st}_{i-1})$.
- *If* $\text{st}_L =$ acc, *output 1. If* $\text{st}_L =$ rej, *output 0, else output $\perp$.*

In a remarkable result, Barrington [9] showed that any circuit of depth d can be simulated by a permutation branching program of width 5 and length 4^d.

Theorem 3 ([9]). *For any boolean circuit C with input space $\{0,1\}^n$ and depth d, there exists a permutation branching program BP of width 5 and length 4^d such that for all inputs $x \in \{0,1\}^n$, $C(x) = \mathsf{BP}(x)$.*

Looking ahead, the permutation property is crucial for our construction in Sect. 4. We will also require that the permutation branching program has a fixed input-selector function inp. In our construction, we will have multiple branching programs, and all of them must read the same input bit at any level $i \le L$.

Definition 3. *A permutation branching program with input space $\{0,1\}^n$ is said to have a fixed input-selector $\mathsf{inp}(\cdot)$ if for all $i \le L$, $\mathsf{inp}(i) = i \bmod n$.*

Any permutation branching program of length L and input space $\{0,1\}^n$ can be easily transformed to a fixed input-selector branching program of length nL. In this work, we only require that all branching programs share the same input selector function $\mathsf{inp}(\cdot)$. The input selector which satisfies $\mathsf{inp}(i) = i \bmod n$ is just one possibility, and we stick with it for simplicity.

2.3   Symmetric Key Encryption and Pseudorandom Functions

Symmetric Key Encryption. A symmetric key encryption scheme SKBE with message space $\mathcal{M}$ consists of algorithms Setup, Enc, Dec with the following syntax.

- Setup$(1^\lambda) \to \mathsf{sk}$. The setup algorithm takes as input the security parameter and outputs secret key sk.
- Enc$(\mathsf{sk}, m \in \mathcal{M}) \to \mathsf{ct}$. The encryption algorithm takes as input a secret key sk and a message $m \in \mathcal{M}$. It outputs a ciphertext ct.
- Dec$(\mathsf{sk}, \mathsf{ct}) \to y \in \mathcal{M}$. The decryption algorithm takes as input a secret key sk, ciphertext ct and outputs a message $y \in \mathcal{M}$.

A symmetric key encryption scheme must satisfy correctness and IND-CPA security.

Correctness: For any security parameter λ, message $m \in \mathcal{M}$, $\mathsf{sk} \leftarrow \mathsf{Setup}(1^\lambda)$,

$$\Pr[\mathsf{Dec}(\mathsf{sk}, \mathsf{Enc}(\mathsf{sk}, m)) \ne m] < \mathsf{negl}(\lambda)$$

where the probability is over the random coins used during encryption and decryption.

Security: In this work, we will be using the IND-CPA security notion.

Definition 4. *Let SKBE = (Setup, Enc, Dec) be a symmetric key encryption scheme. The scheme is said to be IND-CPA secure if for all security parameters λ, all PPT adversaries $\mathcal{A}$, $\mathsf{Adv}^{\mathsf{ind\text{-}cpa}}_{\mathsf{SKBE},\mathcal{A}}(\lambda) = |\Pr[\mathcal{A} \text{ wins the IND-CPA game }] - 1/2|$ is negligible in λ, where the IND-CPA experiment is defined below:*

- *The challenger chooses* sk ← Setup(1^λ), *and bit* $b \leftarrow \{0,1\}$.
- *The adversary queries the challenger for encryptions of polynomially many messages* $m_i \in \mathcal{M}$, *and for each query* m_i, *the challenger sends ciphertext* ct$_i$ ← Enc(sk, m_i) *to* $\mathcal{A}$.
- *The adversary sends two challenge messages* m_0^*, m_1^* *to the challenger. The challenger sends* ct* ← Enc(sk, m_b^*) *to* $\mathcal{A}$.
- *Identical to the pre-challenge phase, the adversary makes polynomially many encryption queries and the challenger responds as before.*
- $\mathcal{A}$ *sends its guess* b' *and wins if* $b = b'$.

Pseudorandom Functions. A family of keyed functions PRF = $\{\text{PRF}_\lambda\}_{\lambda \in \mathbb{N}}$ is a pseudorandom function family with key space $\mathcal{K} = \{\mathcal{K}_\lambda\}_{\lambda \in \mathbb{N}}$, domain $\mathcal{X} = \{\mathcal{X}_\lambda\}_{\lambda \in \mathbb{N}}$ and co-domain $\mathcal{Y} = \{\mathcal{Y}_\lambda\}_{\lambda \in \mathbb{N}}$ if function $\text{PRF}_\lambda : \mathcal{K}_\lambda \times \mathcal{X}_\lambda \to \mathcal{Y}_\lambda$ is efficiently computable, and satisfies the pseudorandomness property defined below.

Definition 5. *A pseudorandom function family* PRF *is secure if for every PPT adversary* $\mathcal{A}$, *there exists a negligible function* negl($\cdot$) *such that*

$$\left| \Pr[\mathcal{A}^{\text{PRF}_\lambda(s,\cdot)}(1^\lambda) = 1] - \Pr[\mathcal{A}^{\mathcal{O}(\cdot)}(1^\lambda) = 1] \right| < negl(\lambda),$$

where $\mathcal{O}$ *is a random function and the probability is taken over the choice of seeds* $s \in \mathcal{K}_\lambda$ *and the random coins of the challenger and adversary.*

Theorem 4 *(PRFs in* **NC**^{1} *[7]). For some* $\sigma > 0$, *suitable universal constant* $C > 0$, *modulus* $p \geq 2$, *any* $m = \text{poly}(n)$, *let* $\chi = \mathcal{D}_{\mathbb{Z},\sigma}$ *and* $q \geq p \cdot k(C\sigma\sqrt{n})^k \cdot n^{\omega(1)}$, *assuming hardness of* (n, m, q, χ)-LWE, *there exists a function family* PRF *consisting of functions from* $\{0,1\}^k$ *to* $\mathbb{Z}_p^{m \times n}$ *that satisfies pseudorandomness property as per Definition 5 and the entire function can be computed in* **TC**$^0 \subseteq$ **NC**^{1}.

From Theorems 3 and 4, the following corollary is immediate.

Corollary 2. *Assuming hardness of* (n, m, q, χ)-LWE *with parameters as in Theorem 4, there exists a family of branching programs* BP = $\{\text{BP}_\lambda\}_{\lambda \in \mathbb{N}}$ *with input space* $\{0,1\}^\lambda \times \{0,1\}^\lambda$ *of width* 5 *and length* poly(λ) *that computes a pseudorandom function family.*

3   Circular Security for Symmetric-Key Bit Encryption and Framework for Generating Separations

In this section, we define the notion of circular security for symmetric-key bit-encryption schemes. We also extend the BHW framework [10] to separate IND-CPA and circular security for bit-encryption in the symmetric-key setting. Informally, the circular security definition requires that it should be infeasible for any adversary to distinguish between encryption of the secret key and encryption of all-zeros string. In the bit-encryption case, each secret key bit is encrypted separately and independently.

Definition 6 *(1-Circular Security for Bit Encryption). Let* SKBE = *(*Setup, Enc, Dec*) be a symmetric-key bit-encryption scheme. Consider the following security game:*

- *The challenger chooses* sk ← Setup(1^λ) *and* b ← $\{0, 1\}$.
- *The adversary is allowed to make following queries polynomially many times:*
    1. **Encryption Query.** *It queries the challenger for encryption of message* $m \in \{0, 1\}$.
    2. **Secret Key Query.** *It queries the challenger for encryption of* i^{th} *bit of the secret key* sk.
- *The challenger responds as follows:*
    1. **Encryption Query.** *For each query* m, *it computes the ciphertext* ct ← Enc(sk, m), *and sends* ct *to the adversary.*
    2. **Secret Key Query.** *For each query* $i \leq |$sk$|$, *if* $b = 0$, *it sends the ciphertext* ct* ← Enc(sk, sk$_i$), *else it sends* ct* ← Enc(sk, 0).
- *The adversary sends its guess* b' *and wins if* $b = b'$.

The scheme SKBE *is said to be circular secure if it satisfies semantic security (Definition 4), and for all security parameters* λ, *all PPT adversaries* $\mathcal{A}$, Adv$^{\text{bit-circ}}_{\text{SKBE},\mathcal{A}}(\lambda) = |\Pr[\mathcal{A} \text{ wins}] - 1/2|$ *is negligible in* λ.

Next, we extend the BHW cycle tester framework for bit-encryption schemes.

3.1   Bit-Encryption Cycle Tester Framework

In a recent work, Bishop et al. [10] introduced a generic framework for separating IND-CPA and circular security. In their cycle tester framework, there are four algorithms - Setup, KeyGen, Encrypt and Test. The setup, key generation and encryption algorithms behave same as in any standard encryption scheme. However, the cycle tester does not contain a decryption algorithm, but provides a special testing algorithm. Informally, the testing algorithm takes as input a sequence of ciphertexts, and outputs 1 if the sequence corresponds to an encryption cycle, else it outputs 0. The security requirement is identical to semantic security for encryption schemes.

The BHW cycle tester framework is a useful framework for separating IND-CPA and n-circular security as it allows us to focus on building the core testing functionality without worrying about providing decryption. The full decryption capability is derived by generically combining a tester with a normal encryption scheme. The BHW framework does not directly work for generating circular security separations for bit-encryption. Below we provide a *bit-encryption cycle tester framework* for symmetric-key encryption along the lines of BHW framework.

Definition 7 *(Bit-Encryption Cycle Tester). A symmetric-key cycle tester* $\Gamma =$ (Setup, Enc, Test) *for message space* $\{0, 1\}$ *and secret key space* $\{0, 1\}^s$ *is a tuple of algorithms (where* $s = s(\lambda)$*) specified as follows:*

- Setup(1^λ) → sk. *The setup algorithm takes as input the security parameter λ, and outputs a secret key* sk ∈ $\{0,1\}^s$.
- Enc(sk, m ∈ $\{0,1\}$) → ct. *The encryption algorithm takes as input a secret key* sk *and a message* m ∈ $\{0,1\}$, *and outputs a ciphertext* ct.
- Test(ct) → $\{0,1\}$. *The testing algorithm takes as input a sequence of s ciphertexts* ct = (ct$_1$, . . . , ct$_s$), *and outputs a bit in* $\{0,1\}$.

The algorithms must satisfy the following properties.

1. *(Testing Correctness) There exists a polynomial $p(\cdot)$ such that for all security parameters λ, the Test algorithm's advantage in distinguishing sequence of encryptions of secret key bits from encryptions of zeros, denoted by* $\mathsf{Adv}^{\text{bit-circ}}_{\mathsf{SKBE,Test}}(\lambda)$ *(Definition 6), is at least $1/p(\lambda)$.*

2. *(IND-CPA security) Let Π = (Setup, Enc, $\cdot$) be an encryption scheme with empty decryption algorithm. The scheme Π must satisfy the IND-CPA security definition (Definition 4).*

Next, we prove that given a cycle tester, we can transform any semantically secure bit-encryption scheme to another semantically secure bit-encryption scheme that is *circular insecure*.

3.2 Circular Security Separation from Cycle Testers

In this section, we prove the following theorem.

Theorem 5 *(Separation from Cycle Testers). If there exists an IND-CPA secure symmetric-key bit-encryption scheme Π for message space $\{0,1\}$ and secret key space $\{0,1\}^{s_1}$ and symmetric-key bit-encryption cycle tester Γ for message space $\{0,1\}$ and secret key space $\{0,1\}^{s_2}$ (where $s_1 = s_1(\lambda)$ and $s_2 = s_2(\lambda)$), then there exists an IND-CPA secure symmetric-key bit-encryption scheme Π' for message space $\{0,1\}$ and secret key space $\{0,1\}^{s_1+s_2}$ that is circular insecure.*

The proof of the above theorem is provided in the full version of the paper.

4   Private Key Bit-Encryption Cycle Tester

In this section, we present our Bit-Encryption Cycle Tester $\mathcal{E}$ = (Setup, Enc, Test) satisfying Definition 7. Before describing the formal construction, we will give an outline of our construction and describe intuitively how the cycle testing algorithm works.

Outline of Our Construction: To begin with, let us first discuss the tools required for our bit-encryption cycle tester. The central primitive in our construction is a low depth pseudorandom function family. More specifically, we require a pseudorandom function PRF : $\{0,1\}^\lambda \times \{0,1\}^\lambda \to \{0,1\}$ (the first input is the PRF key, and the second input is the PRF input) such that for all $i < 2^\lambda$, PRF$(\cdot, i)$[5]

---

[5] Here i is represented as a binary string of length λ.

can be computed using a permutation branching program of polynomial length and polynomial width. Recall, from Corollary 2, there exist PRF constructions [7] that satisfy this requirement. Let $\mathsf{BP}^{(i)}$ denote a branching program of length L and width w computing $\mathrm{PRF}(\cdot, i)$. Each program $\mathsf{BP}^{(i)}$ has an accept state $\mathsf{acc}^{(i)} \in [w]$ and a reject state $\mathsf{rej}^{(i)} \in [w]$. We will also require that at each level $j \leq L$, all branching programs $\mathsf{BP}^{(i)}$ read the same input bit.

The setup algorithm first chooses the LWE parameters: the matrix dimensions n, m, LWE modulus q and noise χ. It also chooses a parameter nbp which is sufficiently larger than n, m and denotes the number of branching programs. Next, it chooses a PRF key s. Finally, for each state of each branching program, it chooses a 'random looking' matrix. In particular, it chooses matrices $\mathbf{B}_{j,k}^{(i)}$ for the state k at level j in $\mathsf{BP}^{(i)}$, and all these matrices have certain 'trapdoors'. The top level matrices corresponding to the accept/reject state satisfy a special constraint: for each branching program $\mathsf{BP}^{(i)}$, choose the matrix $\mathbf{B}_{L,\mathsf{acc}^{(i)}}^{(i)}$ if $\mathrm{PRF}(s, i) = 1$, else choose $\mathbf{B}_{L,\mathsf{rej}^{(i)}}^{(i)}$, and these chosen matrices must sum to 0. The secret key consists of the PRF key s and the matrices, together with their trapdoors.

Next, we describe the encryption algorithm. The ciphertexts are designed such that given an encryption of the secret key, we can combine the components appropriately in order to compute, for each $i \leq \mathsf{nbp}$, a noisy approximation of either $\mathbf{B}_{L,\mathsf{acc}^{(i)}}^{(i)}$ or $\mathbf{B}_{L,\mathsf{rej}^{(i)}}^{(i)}$ depending on $\mathrm{PRF}(s, i)$. If $\mathrm{PRF}(s, i) = 1$, then the output of this combination procedure is $\mathbf{B}_{L,\mathsf{acc}^{(i)}}^{(i)}$, else it is $\mathbf{B}_{L,\mathsf{rej}^{(i)}}^{(i)}$. As a result, adding these matrices results in the zero matrix. On the other hand, the same combination procedure with encryptions of zeroes gives us a matrix with large entries, thereby allowing us to break circular security. Let us now consider a simple case where we have two branching programs $\mathsf{BP}^{(1)}, \mathsf{BP}^{(2)}$, each of length $L = 4$, width $w = 3$ and reading two bit inputs (see Fig. 1).

Let us consider an encryption of a bit b. Each ciphertext consists of 4 sub-ciphertexts, one for each level. At each level, each sub-ciphertext consists of 2 sub-sub-ciphertexts, one for each branching program. The sub-sub-ciphertext $\mathsf{ct}_j^{(i)}$ at level j for program $\mathsf{BP}^{(i)}$ has the following 'propagation' property: for any state matrix $\mathbf{B}_{j-1,k}^{(i)}$ corresponding to state k at level $j-1$ in program $\mathsf{BP}^{(i)}$, $\mathbf{B}_{j-1,k}^{(i)} \cdot \mathsf{ct}_j^{(i)} = \mathbf{B}_{j,\sigma_b(k)}^{(i)}$. In our example (see Fig. 1), if

$$\mathsf{ct} = \left(\left(\mathsf{ct}_1^{(1)}, \mathsf{ct}_1^{(2)} \right), \left(\mathsf{ct}_2^{(1)}, \mathsf{ct}_2^{(2)} \right), \left(\mathsf{ct}_3^{(1)}, \mathsf{ct}_3^{(2)} \right), \left(\mathsf{ct}_4^{(1)}, \mathsf{ct}_4^{(2)} \right) \right)$$

is an encryption of 0, then $\mathbf{B}_{2,3}^{(1)} \cdot \mathsf{ct}_3^{(1)} = \mathbf{B}_{3,1}^{(1)}$. To achieve this, we use the lattice trapdoors. Finally, the ciphertext also contains the base level starting matrices $\{\mathbf{B}_{0,1}^{(i)}\}$.

To see how the test algorithm works, let us consider an encryption of the secret key. Recall, due to the cancellation property of the top level matrices, all we need is a means to compute $\mathbf{B}_{L,\mathsf{acc}^{(i)}}^{(i)}$ if $\mathsf{BP}^{(i)}(x) = 1$, else $\mathbf{B}_{L,\mathsf{rej}^{(i)}}^{(i)}$

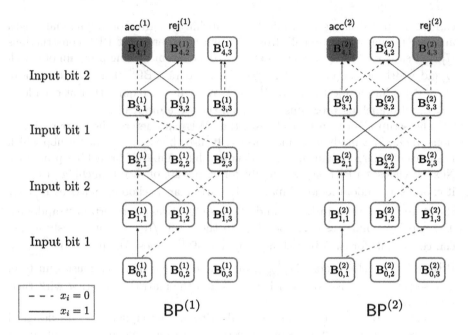

Fig. 1. Branching programs $BP^{(1)}$ and $BP^{(2)}$.

if $BP^{(i)}(x) = 0$. Let us consider $BP^{(2)}$ in our example, and suppose we have encryptions $ct[1]$ and $ct[2]$ of bits 0 and 1 respectively. Now, from the propagation property, it follows that $\mathbf{B}_{0,1}^{(2)} \cdot ct[1]_1^{(2)} = \mathbf{B}_{1,3}^{(2)}$. Similarly, $\mathbf{B}_{1,3}^{(2)} \cdot ct[2]_2^{(2)} = \mathbf{B}_{2,3}^{(2)}$. Continuing this way, we can see that $\mathbf{B}_{0,1}^{(2)} \cdot ct[1]_1^{(2)} \cdot ct[2]_2^{(2)} \cdot ct[1]_3^{(2)} \cdot ct[2]_4^{(2)} = \mathbf{B}_{4,3}^{(2)}$. As a result, we have our desired $\mathbf{B}_{4,\mathsf{rej}^{(2)}}^2$. We can add the matrices computed for each $i \le \mathsf{nbp}$, and see if they sum up to the zero matrix.

For proving security under LWE, we need to make some changes. Instead of having an exact propagation property, we will have an approximate version, where for any state matrix $\mathbf{B}_{j,k}^{(i)}$, $\mathbf{B}_{j,k}^{(i)} \cdot ct_{j+1}^{(i)} \approx \mathbf{S}_{j+1} \cdot \mathbf{B}_{j+1,\sigma_b(k)}^{(i)}$. Here $\mathbf{S}_{j+1}$ is a random low norm matrix chosen during encryption, and is common for all sub-sub-ciphertexts at level $j+1$. As a result, given an encryption of the secret key, at the top level, we either get an approximation of $\mathbf{T} \cdot \mathbf{B}_{L,\mathsf{acc}^{(i)}}^{(i)}$ or $\mathbf{T} \cdot \mathbf{B}_{L,\mathsf{rej}^{(i)}}^{(i)}$. Since $\mathbf{T}$ is a low norm matrix, adding the top-level outputs will be a low norm matrix if we have an encryption of the secret key.

4.1 Our Construction

Let $\mathrm{PRF} = \{\mathrm{PRF}_\lambda\}_{\lambda \in \mathbb{N}}$ be a family of secure pseudorandom functions, where $\mathrm{PRF}_\lambda : \{0,1\}^\lambda \times \{0,1\}^\lambda \to \{0,1\}$ and for all $i \in \{0,1\}^\lambda$, $\mathrm{PRF}_\lambda(\cdot, i)$ can be computed by a fixed-input selector permutation branching program $BP^{(i)}$ of length $L = \ell\text{-bp}(\lambda)$ and width $w = \mathsf{w}\text{-bp}(\lambda)$, where $\ell\text{-bp}(\cdot)$ and $\mathsf{w}\text{-bp}(\cdot)$ are fixed polynomials and

$$\mathsf{BP}^{(i)} = \left(\left\{ \sigma_{j,b}^{(i)} : [w] \to [w] \right\}_{j \in [L], b \in \{0,1\}}, \mathsf{acc}^{(i)} \in [w], \mathsf{rej}^{(i)} \in [w] \right).$$

Note that $\mathsf{BP}^{(i)}$ are fixed-input selector permutation branching programs, therefore they share the same input selector function $\mathsf{inp}(\cdot)$ defined as $\mathsf{inp}(i) = i \bmod n$ (see Definition 3). For simplicity of notation, we will drop the dependence on security parameter λ when it is clear from the context. Fix any $\epsilon < 1/2$. Below we describe our construction.

- $\mathsf{Setup}(1^\lambda) \to \mathsf{sk}$. The setup algorithm first chooses the following parameters: matrix dimensions n, m, LWE modulus q, parameter σ for the Gaussian noise distribution χ and an additional parameter nbp (which denotes the number of branching programs). Let $L = \ell\text{-}\mathsf{bp}(\lambda)$ and $w = \mathsf{w}\text{-}\mathsf{bp}(\lambda)$. Let $\mathsf{params} = (n, m, q, \sigma, \mathsf{nbp})$. The different parameters must satisfy the following constraints:
  - $n \geq \lambda$ (for LWE security)
  - $m = \Omega(n \cdot w \cdot \log q)$ (for TrapGen)
  - $\chi = \mathcal{D}_{\mathbb{Z}, \sigma}$ and $\sigma/q \geq \mathsf{poly}(n)/2^{n^\epsilon}$ (for LWE noise/modulus ratio to be greater than $\mathsf{poly}(n)/2^{n^\epsilon}$)
  - $\mathsf{nbp} \cdot L \cdot (m \cdot \sigma)^L < q/4$ (for the correctness of our Test algorithm)
  - $\mathsf{nbp} = \Omega(m \cdot n \cdot \log q)$ (for applying Leftover Hash Lemma)

  One possible setting of parameters is as follows: set n such that $w \cdot L \leq n^{\epsilon/2}$, $m = n \cdot w \cdot \log q \cdot \log n$, $\sigma = n^c$ for some constant c, $q = 2^{n^\epsilon}/n^c$ and $\mathsf{nbp} = m \cdot n \cdot \log q \cdot \log n$.

  Next, it chooses a random string $s \leftarrow \{0,1\}^\lambda$ and computes, for $i = 1$ to nbp, $t_i = \mathrm{PRF}(s, i)$.[6] It then samples $\mathsf{nbp} \cdot L$ matrices of dimensions $(w \cdot n) \times m$ along with their trapdoors (independently) as $(\mathbf{B}_j^{(i)}, T_j^{(i)}) \leftarrow \mathsf{TrapGen}(1^{w \cdot n}, 1^m, q)$ for $i = 1, \ldots, \mathsf{nbp}$ and $j = 0, \ldots, L-1$.

  It also chooses nbp uniformly random matrices $\mathbf{B}_L^{(i)}$ of dimensions $(w \cdot n) \times m$, such that the following constraint is satisfied

$$\sum_{i \,:\, t_i = 0} \mathbf{B}_{L,\mathsf{rej}^{(i)}}^{(i)} + \sum_{i \,:\, t_i = 1} \mathbf{B}_{L,\mathsf{acc}^{(i)}}^{(i)} = \mathbf{0}.$$

  Each matrix $\mathbf{B}_j^{(i)} \in \mathbb{Z}_q^{w \cdot n \times m}$ can be parsed as follows

$$\mathbf{B}_j^{(i)} = \begin{bmatrix} \mathbf{B}_{j,1}^{(i)} \\ \vdots \\ \mathbf{B}_{j,w}^{(i)} \end{bmatrix}$$

  where matrices $\mathbf{B}_{j,k}^{(i)} \in \mathbb{Z}_q^{n \times m}$ for $k \leq w$. Intuitively, the matrix $\mathbf{B}_{j,k}^{(i)}$ corresponds to state k at level j of branching program $\mathsf{BP}^{(i)}$. The algorithm sets secret key as $\mathsf{sk} = \left(s, \left\{ \mathbf{B}_j^{(i)}, T_j^{(i)} \right\}_{i,j}, \mathsf{params} \right)$.

---

[6] Here, i is represented as a λ bit string.

544   R. Goyal et al.

- Encrypt(sk, $m \in \{0,1\}$) $\rightarrow$ ct. The encryption algorithm takes as input the secret key sk and message m, where sk $= \left(s, \left\{\mathbf{B}_j^{(i)}, T_j^{(i)}\right\}_{i,j}, \text{params}\right)$. It runs the sub-encryption algorithm L times (SubEncrypt is defined in Fig. 2) to compute L sub-ciphertexts.

SubEncrypt

Input: Secret key sk $= (s, \{\mathbf{B}_j^{(i)}, T_j^{(i)}\}_{i\leq \text{nbp}, j\leq L}, \text{params})$, message $m \in \{0,1\}$, level level $\in [L]$
Output: Sub-ciphertext ct_{level}.

1. Choose matrices $\mathbf{S} \leftarrow \chi^{n\times n}$ and $\mathbf{E}^{(i)} \leftarrow \chi^{w\cdot n\times m}$ for $i \leq$ nbp.
2. Set matrix $\mathbf{D}^{(i)}$ as a permutation of the matrix blocks of $\mathbf{B}_{\text{level}}^{(i)}$ according to the permutation $\sigma_{\text{level},m}^{(i)}(\cdot)$. More formally, for $i \leq$ nbp, set

$$\mathbf{D}^{(i)} = \begin{bmatrix} \mathbf{B}_{\text{level},\sigma_{\text{level},m}^{(i)}(1)}^{(i)} \\ \vdots \\ \mathbf{B}_{\text{level},\sigma_{\text{level},m}^{(i)}(w)}^{(i)} \end{bmatrix}.$$

3. Set $\mathbf{C}^{(i)} = (\mathbf{I}_w \otimes \mathbf{S}) \cdot \mathbf{D}^{(i)} + \mathbf{E}^{(i)}$ for $i \leq$ nbp.
4. Compute $\text{ct}^{(i)} \leftarrow \text{SamplePre}(\mathbf{B}_{\text{level}-1}^{(i)}, T_{\text{level}-1}^{(i)}, \sigma, \mathbf{C}^{(i)})$ for $i \leq$ nbp.
5. Output $\text{ct}_{\text{level}} = \left(\text{ct}^{(1)}, \ldots, \text{ct}^{(\text{nbp})}\right)$.

Fig. 2. Routine SubEncrypt

For level $= 1$ to L, it computes the sub-ciphertexts at level level as

$$\text{ct}_{\text{level}} = \left(\text{ct}_{\text{level}}^{(1)}, \ldots, \text{ct}_{\text{level}}^{(\text{nbp})}\right) \leftarrow \text{SubEncrypt}(\text{sk}, m, \text{level}), \quad \forall \text{ level} \in \{1, \ldots, L\}.$$

Finally, it outputs the ciphertext as ct $= \left(\left\{\mathbf{B}_{0,1}^{(i)}\right\}_i, \left\{\text{ct}_j^{(i)}\right\}_{i,j}\right)$.

- Test(ct[1], ..., ct[λ], ..., ct[|sk|]) $\rightarrow \{0,1\}$. The testing algorithm takes as input a sequence of |sk| ciphertexts (ct[1], ..., ct[λ], ...). We will assume the algorithm also knows the LWE modulus q. It parses the first λ ciphertexts as ct[k] $= \left(\left\{\mathbf{B}_{0,1}^{(i)}\right\}_i, \left\{\text{ct}[k]_j^{(i)}\right\}_{i,j}\right)$ for $k \leq \lambda$. Next, it computes the following

$$\text{sum} = \sum_{i=1}^{\text{nbp}} \mathbf{B}_{0,1}^{(i)} \cdot \prod_{j=1}^{L} \text{ct}[\text{inp}(j)]_j^{(i)}.$$

If each component of sum lies in $(-q/4, q/4)$, then the algorithm outputs 1 to indicate a cycle. Otherwise it outputs 0. We would like to remind the reader that the starting state st_0 of each branching program $BP^{(i)}$ is 1 (assumed w.l.o.g. in Sect. 2.2), therefore the testing algorithm only requires the matrices $\mathbf{B}_{0,1}^{(i)}$ to start oblivious evaluation of each branching program.

4.2   Proof of Correctness

In this section, we will prove correctness of our bit-encryption cycle tester. Concretely, we show that the Test algorithm distinguishes between a sequence of $|sk|$ ciphertexts where k^{th} ciphertext encrypts k^{th} bit of the secret key, and a sequence of encryptions of zeros with non-negligible probability. First, we show that if Test algorithm is given encryptions of secret key bits, then it outputs 1 with all-but-negligible probability. Next, we show that if Test algorithm is run on encryptions of zeros, then it outputs 0 with all-but-negligible probability. Using these two facts, correctness of our cycle tester follows.

Testing Encryptions of Key Bits. Let $\mathbf{ct} = (ct[1], \ldots, ct[\lambda], \ldots)$ be the sequence of $|sk|$ ciphertexts where k^{th} ciphertext encrypts bit sk_k, and it can be parsed as $ct[k] = \left(\left\{ \mathbf{B}_{0,1}^{(i)} \right\}_i, \left\{ ct[k]_j^{(i)} \right\}_{i,j} \right)$. Recall that the first λ bits of secret key sk correspond to the PRF key s. Therefore, $ct[k]$ is an encryption of the bit s_k for $k \leq \lambda$. Also, i^{th} branching program $BP^{(i)}$ computes the function $PRF_\lambda(\cdot, i)$. This could be equivalently stated as

$$\forall \; i \leq \mathsf{nbp}, \quad \sigma_{L,b_L}^{(i)} \left(\cdots \left(\sigma_{1,b_1}^{(i)}(1) \right) \cdots \right) = \begin{cases} \mathsf{rej}^{(i)} & \text{if } PRF(s,i) = 0, \\ \mathsf{acc}^{(i)} & \text{if } PRF(s,i) = 1 \end{cases}$$

where $b_j = s_{\mathsf{inp}(j)}$ for $j \leq L$. Let $st_j^{(i)}$ denote the state of the ith branching program after j steps. The initial state $st_0^{(i)}$ is 1 for all programs, and j^{th} state can be computed as $st_j^{(i)} = \sigma_{j,s_{\mathsf{inp}(j)}}^{(i)}(st_{j-1}^{(i)})$.

Note that every ciphertext $ct[k]$ consists of L sub-ciphertexts $ct[k]_j$ for each level $j \leq L$, and each sub-ciphertext consists of nbp short matrices, each for a separate branching program. For constructing each sub-ciphertext, exactly one short secret matrix $\mathbf{S}_j$ is chosen, and it is shared across all nbp branching programs for generating LWE-type samples. It is crucial for testability that $\mathbf{S}_j$'s stay same for all branching programs.

First, we will introduce some notations for this proof.

- $\mathbf{S}[k]_j$: matrix chosen at level j for computing $ct[k]_j^{(i)}$
- $\mathbf{E}[k]_j^{(i)}$: error matrix chosen at level j, program i for computing $ct[k]_j^{(i)}$
- $\mathsf{inp}_j = \mathsf{inp}(j)$: the input bit read at level j of the branching program
- $\mathbf{S}_j = \mathbf{S}[\mathsf{inp}_j]_j$, $\mathbf{E}_j^{(i)} = \mathbf{E}[\mathsf{inp}_j]_j^{(i)}$, $CT_j^{(i)} = ct[\mathsf{inp}_j]_j^{(i)}$

- $\Gamma_{j^*} = \prod_{j=1}^{j^*} \mathbf{S}_j$
- $\boldsymbol{\Delta}_{j^*}^{(i)} = \mathbf{B}_{0,1}^{(i)} \cdot \left(\prod_{j=1}^{j^*} \mathsf{CT}_j^{(i)} \right), \quad \widetilde{\boldsymbol{\Delta}}_{j^*}^{(i)} = \Gamma_{j^*} \cdot \mathbf{B}_{j^*,\mathrm{st}_{j^*}^{(i)}}^{(i)}, \quad \mathbf{Err}_{j^*}^{(i)} = \boldsymbol{\Delta}_{j^*}^{(i)} - \widetilde{\boldsymbol{\Delta}}_{j^*}^{(i)}.$

The Test algorithm checks that $\left\| \sum_{i=1}^{\mathrm{nbp}} \boldsymbol{\Delta}_L^{(i)} \right\|_{\infty} < q/4$. Also, note that

$$\sum_{i=1}^{\mathrm{nbp}} \widetilde{\boldsymbol{\Delta}}_L^{(i)} = \sum_{i=1}^{\mathrm{nbp}} \Gamma_L \cdot \mathbf{B}_{L,\mathrm{st}_L^{(i)}}^{(i)} = \Gamma_L \cdot \sum_{i=1}^{\mathrm{nbp}} \mathbf{B}_{L,\mathrm{st}_L^{(i)}}^{(i)} = \mathbf{0}.$$

Thus, it would be sufficient to show that, with high probability, $\mathbf{Err}_L^{(i)} = \boldsymbol{\Delta}_L^{(i)} - \widetilde{\boldsymbol{\Delta}}_L^{(i)}$ is bounded. We will show that for all $i \leq \mathrm{nbp}$, $j^* \leq L$, $\mathbf{Err}_{j^*}^{(i)}$ is bounded.

Lemma 2. $\forall\, i \in \{1, \ldots, \mathrm{nbp}\}, j^* \in \{1, \ldots, L\}, \quad \left\| \mathbf{Err}_{j^*}^{(i)} \right\|_{\infty} \leq j^* \cdot (m \cdot \sigma)^{j^*}$ with overwhelming probability.

Proof. The above lemma is proven by induction over j^*, and all arguments hold irrespective of the value of i. Therefore, for simplicity of notation, we will drop the dependence on i. We will slightly abuse the notation and use $\mathbf{B}_{j,\sigma_{j,m}^{(i)}}^{(i)}$ to denote the following matrix.

$$\mathbf{B}_{j,\sigma_{j,m}^{(i)}}^{(i)} = \begin{bmatrix} \mathbf{B}_{j,\sigma_{j,m}^{(i)}(1)}^{(i)} \\ \vdots \\ \mathbf{B}_{j,\sigma_{j,m}^{(i)}(w)}^{(i)} \end{bmatrix}.$$

Before proceeding to our inductive proof, we would like to note the following fact.

Fact 1. *For all* $j \leq L$, $\mathsf{CT}_j^{(i)} \leftarrow \mathsf{SamplePre}(\mathbf{B}_{j-1}^{(i)}, T_{j-1}^{(i)}, \sigma, \mathbf{C}_j^{(i)})$, *where* $\mathbf{C}_j^{(i)} = (\mathbf{I}_w \otimes \mathbf{S}_j) \cdot \mathbf{B}_{j,\sigma_{j,m}^{(i)}}^{(i)} + \mathbf{E}_j^{(i)}$ *and* $m = s_{\mathrm{inp}_j}$.

Base case $(j^* = 1)$. We know that $\boldsymbol{\Delta}_1 = \mathbf{B}_{0,1} \cdot (\mathsf{CT}_1)$. Therefore, using Fact 1, we can say that $\boldsymbol{\Delta}_1 = \mathbf{S}_1 \cdot \mathbf{B}_{1,\mathrm{st}_1} + \mathbf{E}_{1,1} = \widetilde{\boldsymbol{\Delta}}_1 + \mathbf{E}_{1,1}$. Note that $\mathbf{E}_{1,1}$ is an $n \times m$ submatrix consisting of first n rows of $\mathbf{E}_1$. Thus, we could write the following

$$\|\mathbf{Err}_1\|_{\infty} = \left\| \boldsymbol{\Delta}_1 - \widetilde{\boldsymbol{\Delta}}_1 \right\|_{\infty} = \|\mathbf{E}_{1,1}\|_{\infty} \leq m \cdot \sigma.$$

This completes the proof of base case. For the induction step, we assume that the above lemma holds for $j^* - 1$, and show that it holds for j^* as well.

Induction Step. We know that $\mathbf{\Delta}_{j*} = \mathbf{\Delta}_{j*-1} \cdot (\mathsf{CT}_{j*})$. Also, $\mathbf{\Delta}_{j*-1} = \widetilde{\mathbf{\Delta}}_{j*-1} + \mathbf{Err}_{j*-1}$. So, we could write the following

$$
\begin{aligned}
\mathbf{\Delta}_{j*} &= \widetilde{\mathbf{\Delta}}_{j*-1} \cdot \mathsf{CT}_{j*} + \mathbf{Err}_{j*-1} \cdot \mathsf{CT}_{j*} \\
&= \mathbf{\Gamma}_{j*-1} \cdot \left(\mathbf{B}_{j*-1,\mathsf{st}_{j*-1}} \cdot \mathsf{CT}_{j*} \right) + \mathbf{Err}_{j*-1} \cdot \mathsf{CT}_{j*} \\
&= \mathbf{\Gamma}_{j*-1} \cdot \left(\mathbf{S}_{j*} \cdot \mathbf{B}_{j*,\mathsf{st}_{j*}} + \mathbf{E}_{j*,\mathsf{st}_{j*-1}} \right) + \mathbf{Err}_{j*-1} \cdot \mathsf{CT}_{j*} \\
&= \widetilde{\mathbf{\Delta}}_{j*}^{(i)} + \mathbf{\Gamma}_{j*-1} \cdot \mathbf{E}_{j*,\mathsf{st}_{j*-1}} + \mathbf{Err}_{j*-1} \cdot \mathsf{CT}_{j*}
\end{aligned}
$$

Here, $\mathbf{E}_{j*,\mathsf{st}_{j*-1}}$ is an $n \times m$ submatrix of $\mathbf{E}_{j*}$. Finally, we can bound $\mathbf{Err}_{j*}$ as follows

$$
\begin{aligned}
\left\| \mathbf{Err}_{j*} \right\|_\infty &= \left\| \mathbf{\Delta}_{j*} - \widetilde{\mathbf{\Delta}}_{j*} \right\|_\infty = \left\| \mathbf{\Gamma}_{j*-1} \cdot \mathbf{E}_{j*,\mathsf{st}_{j*-1}} + \mathbf{Err}_{j*-1} \cdot \mathsf{CT}_{j*} \right\|_\infty \\
&\leq \left\| \mathbf{\Gamma}_{j*-1} \cdot \mathbf{E}_{j*,\mathsf{st}_{j*-1}} \right\|_\infty + \left\| \mathbf{Err}_{j*-1} \cdot \mathsf{CT}_{j*} \right\|_\infty \\
&\leq (n \cdot \sigma)^{j*-1} \cdot m \cdot \sigma + (j* - 1) \cdot (m \cdot \sigma)^{j*-1} \cdot m \cdot \sigma \leq j* \cdot (m \cdot \sigma)^{j*}
\end{aligned}
$$

This completes the proof.

Using Lemma 2, we can claim that for all $i \leq \mathsf{nbp}$, $\left\| \mathbf{\Delta}_L^{(i)} - \widetilde{\mathbf{\Delta}}_L^{(i)} \right\|_\infty \leq L \cdot (m \cdot \sigma)^L$. Therefore,

$$
\left\| \mathsf{sum} \right\|_\infty = \left\| \sum_{i=1}^{\mathsf{nbp}} \mathbf{\Delta}_L^{(i)} \right\|_\infty = \left\| \sum_{i=1}^{\mathsf{nbp}} \mathbf{\Delta}_L^{(i)} - \sum_{i=1}^{\mathsf{nbp}} \widetilde{\mathbf{\Delta}}_L^{(i)} \right\|_\infty \leq \mathsf{nbp} \cdot L \cdot (m \cdot \sigma)^L < q/4
$$

Therefore, for our setting of parameters, if ciphertexts encrypt the secret key bit-by-bit, then Test algorithm outputs 1 with high probability.

Testing Encryptions of Zeros

Lemma 3. *If* PRF *is a family of secure pseudorandom functions and challenge ciphertexts are encryptions of zeros, then* Test *outputs 0 with all-but-negligible probability.*

Proof. Since the ciphertexts are encryptions of zeros, each branching program $\mathsf{BP}^{(i)}$ computes the value $t_i' = \mathrm{PRF}_\lambda(0, i)$. Also, with high probability, t_i' and t_i can not be equal for all $i \leq \lambda$ as otherwise PRF_λ will not be a secure pseudorandom function. Therefore, with high probability,

$$
\widetilde{\mathsf{sum}} = \sum_{i=1}^{\mathsf{nbp}} \widetilde{\mathbf{\Delta}}_L^{(i)} = \left(\prod_{j=1}^L \mathbf{S}_j \right) \cdot \sum_{i=1}^{\mathsf{nbp}} \mathbf{B}_{L,\mathsf{st}_L^{(i)}}^{(i)} \neq \mathbf{0}.
$$

Now, $\sum_{i=1}^{\mathsf{nbp}} \mathbf{B}_{L,\mathsf{st}_L^{(i)}}^{(i)}$ will be a uniformly random matrix in $\mathbb{Z}_q^{n \times m}$ as $t' \neq t$ and $\mathbf{B}_{L,\mathsf{st}_L^{(i)}}^{(i)}$ are randomly chosen for $i \leq \mathsf{nbp}$. Let $\mathbf{S}$ denote the product $\prod_{j=1}^L \mathbf{S}_j$ and

$\mathbf{B}$ denote the sum $\sum_{i=1}^{\mathsf{nbp}} \mathbf{B}_{L,\mathsf{st}_L^{(i)}}^{(i)}$. We can write $\widetilde{\mathsf{sum}}$ as $\widetilde{\mathsf{sum}} = \mathbf{S} \cdot \mathbf{B}$, where $\mathbf{B}$ is a random $n \times m$ matrix. Thus, $\widetilde{\mathsf{sum}}$ is a random $n \times m$ matrix as $\mathbf{S}$, product of L full rank matrices, is also full rank. So, with high probability, at least one entry in matrix sum will have absolute value $> q/4$ which implies that Test outputs 0.

4.3   IND-CPA Proof

We will now show that the construction described above is IND-CPA secure. The adversary queries for ciphertexts, and each ciphertext consists of $L \cdot \mathsf{nbp}$ sub-sub-ciphertexts. In our proof, we will gradually switch the sub-sub-ciphertexts to random low-norm (Gaussian) matrices, starting with the top-level sub-ciphertext and moving down. Once all sub-ciphertexts are switched to Gaussian matrices, the adversary has no information about the challenge message.

Our proof proceeds via a sequence of hybrid games. First, we switch the PRF evaluation to a truly random nbp bit string. Next, we switch the top level matrices to truly random matrices. This is possible since nbp is much larger than n, m, and as a result, we can use Leftover Hash Lemma. Once all top level matrices are truly random, we can make the top-level sub-sub-ciphertexts to be random low norm (Gaussian) matrices. This follows from the LWE security, together with the Property 3 of lattice trapdoors. Once the top level sub-sub-ciphertexts are Gaussian, we do not require the trapdoors at level $L - 1$. As a result, we can choose uniformly random matrices at level $L-1$. This will allow us to switch the sub-sub-ciphertexts at level $L-1$ to Gaussian matrices. Proceeding this way, we can switch all sub-sub-ciphertexts to Gaussian matrices.

We will first define the sequence of hybrid games, and then show that they are computationally indistinguishable. The first hybrid corresponds to the original security game. In the subsequent hybrids, we only show the steps that are modified.

Sequence of Hybrid Games
Game 0: This corresponds to the original security game.

- **Setup Phase**
    1. The challenger first chooses the LWE parameters n, m, q, σ, χ and nbp. Recall $L = \ell\text{-bp}(\lambda)$ and $w = \mathsf{w\text{-}bp}(\lambda)$.
    2. Next, it chooses a uniformly random string $s \leftarrow \{0,1\}^\lambda$ and sets $t_i = \mathsf{PRF}(s, i)$ for $i \le \mathsf{nbp}$.
    3. For $i = 1$ to nbp and $j = 0$ to $L - 1$, it chooses $(\mathbf{B}_j^{(i)}, T_j^{(i)}) \leftarrow \mathsf{TrapGen}(1^{w \cdot n}, 1^m, q)$.
    4. It chooses nbp uniformly random matrices $\mathbf{B}_L^{(i)}$ of dimensions $w \cdot n \times m$, such that the following constraint is satisfied

$$\sum_{i \,:\, t_i = 0} \mathbf{B}_{L,\mathsf{rej}^{(i)}}^{(i)} + \sum_{i \,:\, t_i = 1} \mathbf{B}_{L,\mathsf{acc}^{(i)}}^{(i)} = \mathbf{0}.$$

5. Finally, the challenger sets $\mathsf{sk} = \left(s, \left\{ \mathbf{B}_j^{(i)}, T_j^{(i)} \right\}_{i,j} \right)$.

- **Pre-Challenge Query Phase**
    1. The adversary requests polynomially many encryption queries. The challenger responds to each encryption query as follows.
    For $j = 1$ to L, the challenger computes $\mathsf{ct}_j \leftarrow \mathsf{SubEncrypt}(\mathsf{sk}, m, j)$ and sends $\mathsf{ct} = \left(\left\{ \mathbf{B}_{0,1}^{(i)} \right\}_i, (\mathsf{ct}_1, \ldots, \mathsf{ct}_L) \right)$.
- **Challenge Phase.** The challenger chooses a bit $b \leftarrow \{0,1\}$, and computes the challenge ciphertext identical to any pre-challenge query ciphertext for bit b.
- **Post-Challenge Query Phase.** This is identical to the pre-challenge query phase.
- **Guess.** The adversary finally sends the guess b', and wins if $b = b'$.

Game 1: This hybrid experiment is similar to the previous one, except that the string $t = (t_1, \ldots, t_{\mathsf{nbp}})$ is a uniformly random nbp bit string. Also, in place of the PRF key in the secret key, we have an empty string $\bot$. Note that this does not affect the encryption algorithm since it works oblivious to the PRF key (the PRF key is not used during encryption).

- **Setup Phase**
    1. The challenger first chooses the LWE parameters n, m, q, σ, χ and nbp. Recall $L = \ell\text{-}\mathsf{bp}(\lambda)$ and $w = \mathsf{w}\text{-}\mathsf{bp}(\lambda)$.
    2. Next, it chooses $t \leftarrow \{0,1\}^{\mathsf{nbp}}$.
    3. For $i = 1$ to nbp and $j = 0$ to $L-1$, it chooses $(\mathbf{B}_j^{(i)}, T_j^{(i)}) \leftarrow \mathsf{TrapGen}(1^{w \cdot n}, 1^m, q)$.
    4. It chooses nbp uniformly random matrices $\mathbf{B}_L^{(i)}$ of dimensions $w \cdot n \times m$, such that the following constraint is satisfied
    $$\sum_{i \,:\, t_i = 0} \mathbf{B}_{L,\mathsf{rej}^{(i)}}^{(i)} + \sum_{i \,:\, t_i = 1} \mathbf{B}_{L,\mathsf{acc}^{(i)}}^{(i)} = 0.$$
    5. Finally, the challenger sets $\mathsf{sk} = \left(\bot, \left\{ \mathbf{B}_j^{(i)}, T_j^{(i)} \right\}_{i,j} \right)$.

Game 2: In this hybrid experiment, the challenger chooses the top-level matrices $\mathbf{B}_L^{(i)}$ uniformly at random.

- **Setup Phase**
    1. The challenger first chooses the LWE parameters n, m, q, σ, χ and nbp. Recall $L = \ell\text{-}\mathsf{bp}(\lambda)$ and $w = \mathsf{w}\text{-}\mathsf{bp}(\lambda)$.
    2. Next, it chooses $t \leftarrow \{0,1\}^{\mathsf{nbp}}$.
    3. For $i = 1$ to nbp and $j = 0$ to $L-1$, it chooses $(\mathbf{B}_j^{(i)}, T_j^{(i)}) \leftarrow \mathsf{TrapGen}(1^{w \cdot n}, 1^m, q)$.
    4. For $i = 1$ to nbp, it chooses uniformly random matrices $\mathbf{B}_L^{(i)} \leftarrow \mathbb{Z}_q^{w \cdot n \times m}$ of dimensions $w \cdot n \times m$.

5. Finally, the challenger sets $\mathsf{sk} = \left(\bot, \left\{ \mathbf{B}_j^{(i)}, T_j^{(i)} \right\}_{i,j} \right)$.

Next, we have a sequence of $3L$ hybrid experiments Game 2.level.$\{1,2,3\}$ for level $= L$ to 1.

Game 2.level.1: In hybrids Game 2.level.1, the sub-ciphertexts corresponding to levels greater than level are Gaussian matrices. At level level, the sub-ciphertext computation does not use SubEncrypt routine. Instead, it chooses a uniformly random matrix and computes the SamplePre of the uniformly random matrix. Also, for levels greater than level $- 1$, matrices $\mathbf{B}_j^{(i)}$ are chosen uniformly at random instead of being sampled using TrapGen.

- **Pre-Challenge Query Phase**
    1. The adversary requests polynomially many encryption queries. The challenger responds to each encryption query as follows.
    2. For $j = 1$ to level-1, the challenger computes $\mathsf{ct}_j \leftarrow \mathsf{SubEncrypt}(\mathsf{sk}, m, j)$.
    3. For $j = $ level, the challenger chooses uniformly random matrix $\mathbf{C}_j^{(i)} \leftarrow \mathbb{Z}_q^{w \cdot n \times m}$ and sets $\mathsf{ct}_j^{(i)} \leftarrow \mathsf{SamplePre}(\mathbf{B}_{j-1}^{(i)}, T_{j-1}^{(i)}, \sigma, \mathbf{C}_j^{(i)})$. It sets $\mathsf{ct}_j = (\mathsf{ct}_j^{(1)}, \ldots, \mathsf{ct}_j^{(\mathsf{nbp})})$.
    4. For $i = 1$ to nbp and $j = $ level $+ 1$ to L, the challenger chooses $\mathsf{ct}_j^{(i)} \leftarrow \chi^{m \times m}$. It sets $\mathsf{ct}_j = (\mathsf{ct}_j^{(1)}, \ldots, \mathsf{ct}_j^{(\mathsf{nbp})})$.
    5. Finally, it sets $\mathsf{ct} = \left(\left\{ \mathbf{B}_{0,1}^{(i)} \right\}_i, (\mathsf{ct}_1, \ldots, \mathsf{ct}_L) \right)$ and sends ct to the adversary.

Game 2.level.2: In hybrids Game 2.level.2, the sub-ciphertexts corresponding to levels greater than level $- 1$ are Gaussian matrices.

- **Pre-Challenge Query Phase**
    1. The adversary requests polynomially many encryption queries. The challenger responds to each encryption query as follows.
    2. For $j = 1$ to level $- 1$, the challenger computes $\mathsf{ct}_j \leftarrow \mathsf{SubEncrypt}(\mathsf{sk}, m, j)$.
    3. For $i = 1$ to nbp and $j = $ level to L, the challenger chooses $\mathsf{ct}_j^{(i)} \leftarrow \chi^{m \times m}$. It sets $\mathsf{ct}_j = (\mathsf{ct}_j^{(1)}, \ldots, \mathsf{ct}_j^{(\mathsf{nbp})})$.
    4. Finally, it sets $\mathsf{ct} = \left(\left\{ \mathbf{B}_{0,1}^{(i)} \right\}_i, (\mathsf{ct}_1, \ldots, \mathsf{ct}_L) \right)$ and sends ct to the adversary.

Game 2.level.3: In hybrids Game 2.level.3, matrices $\mathbf{B}_j^{(i)}$ are chosen uniformly at random instead of being sampled using TrapGen for levels greater than level $- 2$.

- **Setup Phase**
    1. The challenger first chooses the LWE parameters n, m, q, σ, χ and nbp. Recall $L = \ell\text{-bp}(\lambda)$ and $w = \mathsf{w\text{-}bp}(\lambda)$.
    2. Next, it chooses $t \leftarrow \{0,1\}^{\mathsf{nbp}}$.

3. For $i = 1$ to nbp and $j = 0$ to level $- 2$, it chooses $(\mathbf{B}_j^{(i)}, T_j^{(i)}) \leftarrow$ TrapGen$(1^{w \cdot n}, 1^m, q)$.
4. For $i = 1$ to nbp and $j =$ level $- 1$ to L, it chooses uniformly random $\mathbf{B}_j^{(i)} \leftarrow \mathbb{Z}_q^{w \cdot n \times m}$ of dimensions $w \cdot n \times m$.
5. Finally, the challenger sets sk $= \left(\perp, \left\{ \mathbf{B}_j^{(i)}, T_j^{(i)} \right\}_{i,j} \right)$.

Indistinguishability of Hybrid Games. We now establish via a sequence of lemmas that no PPT adversary can distinguish between any two adjacent games with non-negligible advantage. To conclude, we show that the advantage of any PPT adversary in the last game is 0.

Let $\mathcal{A}$ be a PPT adversary that breaks the security of our construction in the IND-CPA security game (Definition 4). In Game i, advantage of $\mathcal{A}$ is defined as $\mathsf{Adv}_{\mathcal{A}}^i = |\Pr[\mathcal{A}\text{ wins}] - 1/2|$. We show via a sequence of claims that $\mathcal{A}$'s advantage is distinguishing between any two consecutive games must be negligible, otherwise there will be a poly-time attack on the security of some underlying primitive. Finally, in last game, we show that $\mathcal{A}$'s advantage in the last game is 0.

Lemma 4. *If* PRF *is a family of secure pseudorandom functions, then for any PPT adversary* $\mathcal{A}$, $|\mathsf{Adv}_{\mathcal{A}}^0 - \mathsf{Adv}_{\mathcal{A}}^1| \leq negl(\lambda)$ *for some negligible function* $negl(\cdot)$.

Proof. We describe a reduction algorithm $\mathcal{B}$ which plays the indistinguishability based game with PRF challenger. $\mathcal{B}$ runs the Setup Phase as in Game 0, except it does not choose a string $s \leftarrow \{0,1\}^{\lambda}$. $\mathcal{B}$ makes nbp queries to the PRF challenger, where in the ith query it sends i to the PRF challenger and sets t_i as the challenger's response. $\mathcal{B}$ performs remaining steps as in Game 0, and sends 1 to the PRF challenger if $\mathcal{A}$ guesses the bit correctly, otherwise it sends 0 to the PRF challenger as its guess.

Note that when PRF challenger honestly evaluates the PRF on each query, then $\mathcal{B}$ exactly simulates the view of Game 0 for $\mathcal{A}$. Otherwise if PRF challenger behaves as a random function, then $\mathcal{B}$ exactly simulates the view of Game 1. Therefore, if $|\mathsf{Adv}_{\mathcal{A}}^0 - \mathsf{Adv}_{\mathcal{A}}^1|$ is non-negligible, then PRF is not secure pseudo-random function family.

Lemma 5. *For any adversary* $\mathcal{A}$, $|\mathsf{Adv}_{\mathcal{A}}^1 - \mathsf{Adv}_{\mathcal{A}}^2| \leq negl(\lambda)$ *for some negligible function* $negl(\cdot)$.

Proof. The proof of this lemma follows from Corollary 1 which itself follows from the Leftover Hash Lemma Theorem 1. Note that the difference between Game 1 and 2 is the way top level matrices $\mathbf{B}_L^{(i)}$ are sampled during Setup Phase. In Game 1, matrix $\mathbf{B}_{L, \mathsf{st}_L^{(i)}}^{(\mathsf{nbp})}$ is chosen as

$$\mathbf{B}_{L, \mathsf{st}_L^{(\mathsf{nbp})}}^{(\mathsf{nbp})} = - \left(\sum_{i \leq \mathsf{nbp}-1 \,:\, t_i = 0} \mathbf{B}_{L, \mathsf{rej}^{(i)}}^{(i)} + \sum_{i \leq \mathsf{nbp}-1 \,:\, t_i = 1} \mathbf{B}_{L, \mathsf{acc}^{(i)}}^{(i)} \right),$$

where $\mathsf{st}_L^{(\mathsf{nbp})}$ is $\mathsf{acc}^{(\mathsf{nbp})}$ if $t_{\mathsf{nbp}} = 1$, and $\mathsf{rej}^{(\mathsf{nbp})}$ otherwise. It can be equivalently written as follows

$$\mathbf{B}_{L,\mathsf{st}_L^{(\mathsf{nbp})}}^{(\mathsf{nbp})} = -\mathbf{A}\cdot\mathbf{R}, \qquad \mathbf{A} = \left[\mathbf{B}_{L,\mathsf{rej}^{(1)}}^{(1)} \,||\, \mathbf{B}_{L,\mathsf{acc}^{(1)}}^{(1)} \,||\cdots||\, \mathbf{B}_{L,\mathsf{rej}^{(\mathsf{nbp}-1)}}^{(\mathsf{nbp}-1)} \,||\, \mathbf{B}_{L,\mathsf{acc}^{(\mathsf{nbp}-1)}}^{(\mathsf{nbp}-1)}\right]$$

where $\mathbf{R} = \mathbf{u}\otimes\mathbf{I}_m \in \mathbb{Z}_q^{2m(\mathsf{nbp}-1)\times m}$, $\mathbf{u} = (u_1,\dots,u_{2\mathsf{nbp}-2})^\top \in \{0,1\}^{2\mathsf{nbp}-2}$ and for all $i \leq \mathsf{nbp}-1$, $u_{2i} = t_i$ and $u_{2i-1} = 1 - t_i$. That is, matrix $\mathbf{R}$ consists of $2\mathsf{nbp}-2$ submatrices where if $t_i = 1$, then its $2i^{th}$ submatrix is identity and $(2i-1)^{th}$ submatrix is zero, otherwise it is the opposite. Let $\mathcal{R}$ denote the distribution of matrix $\mathbf{R}$ as described above with t drawn uniformly from $\{0,1\}^{\mathsf{nbp}}$. Note that $\mathbf{H}_\infty(\mathcal{R}) = \mathsf{nbp} - 1$ (min-entropy of $\mathcal{R}$), and $\mathsf{nbp} > m \cdot n\log_2 q + \omega(\log n)$. Therefore, it follows (from Corollary 1) that

$$\left\{\left(\mathbf{A}, \mathbf{B}_{L,\mathsf{st}_L^{(\mathsf{nbp})}}^{(\mathsf{nbp})} = -\mathbf{A}\cdot\mathbf{R}\right) : \mathbf{A} \leftarrow \mathbb{Z}_q^{n\times 2m(\mathsf{nbp}-1)}, \mathbf{R} \leftarrow \mathcal{R}\right\}$$

$$\approx_s$$

$$\left\{\left(\mathbf{A}, \mathbf{B}_{L,\mathsf{st}_L^{(\mathsf{nbp})}}^{(\mathsf{nbp})}\right) : \mathbf{A} \leftarrow \mathbb{Z}_q^{n\times 2m(\mathsf{nbp}-1)}, \mathbf{B}_{L,\mathsf{st}_L^{(\mathsf{nbp})}}^{(\mathsf{nbp})} \leftarrow \mathbb{Z}_q^{n\times m}\right\}$$

Thus, $|\mathsf{Adv}_{\mathcal{A}}^1 - \mathsf{Adv}_{\mathcal{A}}^2|$ is negligible in the security parameter for all PPT adversaries $\mathcal{A}$.

Lemma 6. *If $(n, \mathsf{nbp}\cdot w\cdot m, q, \chi)$-LWE-ss assumption holds (Assumption 2), then for any PPT adversary $\mathcal{A}$, $|\mathsf{Adv}_{\mathcal{A}}^2 - \mathsf{Adv}_{\mathcal{A}}^{2.L.1}| \leq negl(\lambda)$ for some negligible function $negl(\cdot)$.*

Proof. The difference between Game 2 and 2.L.1 is the way top-level sub-ciphertexts (ct_L) are created for all encryption queries (including challenge query). Recall that ct_L contains nbp short matrices $\mathsf{ct}_L^{(i)}$, and each $\mathsf{ct}_L^{(i)}$ is sampled as $\mathsf{ct}_L^{(i)} \leftarrow \mathsf{SamplePre}(\mathbf{B}_{L-1}^{(i)}, T_{L-1}^{(i)}, \sigma, \mathbf{C}_L^{(i)})$. In Game 2, matrix $\mathbf{C}_L^{(i)}$ is computed as $\mathbf{C}_L^{(i)} = (\mathbf{I}_w \otimes \mathbf{S}_L)\cdot\mathbf{D}_L^{(i)} + \mathbf{E}_L^{(i)}$, where $\mathbf{D}_L^{(i)}$ is a permutation of $\mathbf{B}_L^{(i)}$ and $\mathbf{E}_L^{(i)}$ is chosen as $\mathbf{E}_L^{(i)} \leftarrow \chi^{w\cdot n\times m}$. On the other hand, in Game 2.L.1, it is chosen as $\mathbf{C}_L^{(i)} \leftarrow \mathbb{Z}_q^{w\cdot n\times m}$.

For proving indistinguishability of Game 2 and 2.L.1, we need to sketch q intermediate hybrids, where q is the total number of queries made by $\mathcal{A}$.[7] In k^{th} hybrid, the challenger proceeds as Game 2.L.1 while answering first k queries, and proceeds as in Game 2 for answering remaining queries. Indistinguishability between any two consecutive intermediate hybrids follows directly from LWE-ss assumption. Below we describe a reduction algorithm $\mathcal{B}$ which plays the LWE-ss indistinguishability game.

First, $\mathcal{B}$ receives as LWE-ss challenge two $n\times(\mathsf{nbp}\cdot w\cdot m)$ matrices $(\mathbf{F}, \mathbf{G})$. It parses $\mathbf{F}$ into nbp submatrices of dimensions $n \times (w \cdot m)$ as $[\mathbf{F}^{(1)} ||\dots|| \mathbf{F}^{(\mathsf{nbp})}] = \mathbf{F}$.

---

[7] Here q includes the challenge query as well.

Further, each matrix $\mathbf{F}^{(i)}$ is parsed into w matrices of dimensions $n \times m$ as $[\mathbf{F}_1^{(i)} \| \ldots \| \mathbf{F}_w^{(i)}] = \mathbf{F}^{(i)}$. Similarly, it parses $\mathbf{G}$ as well. Next, it runs the Setup phase as in Game 2, except instead of choosing matrices $\mathbf{B}_L^{(i)}$ uniformly at random, it sets them as $\mathbf{B}_{L,v}^{(i)} = \mathbf{F}_v^{(i)}$ for $i \leq \mathsf{nbp}$ and $v \leq w$.

$\mathcal{B}$ answers the first $i-1$ queries as in Game 2.L.1. On receiving k^{th} query m_k, it computes $L-1$ sub-ciphertexts ct_j $(j \leq L-1)$ honestly using sub-encrypt routine.[8] For computing sub-ciphertext ct_L, it first sets matrices $\mathbf{C}_L^{(i)}$ for $i \leq \mathsf{nbp}$ as follows

$$\mathbf{C}_L^{(i)} = \begin{bmatrix} \mathbf{G}_{\sigma_{L,m_k}^{(i)}(1)}^{(i)} \\ \vdots \\ \mathbf{G}_{\sigma_{L,m_k}^{(i)}(w)}^{(i)} \end{bmatrix}.$$

Next, it computes $\mathsf{ct}_L^{(i)} \leftarrow \mathsf{SamplePre}(\mathbf{B}_{L-1}^{(i)}, T_{L-1}^{(i)}, \sigma, \mathbf{C}_L^{(i)})$ for $i \leq \mathsf{nbp}$, and sets $\mathsf{ct}_L = (\mathsf{ct}_L^{(1)}, \ldots, \mathsf{ct}_L^{(\mathsf{nbp})})$. $\mathcal{B}$ answers k^{th} query as $\mathsf{ct} = (\mathsf{ct}_1, \ldots, \mathsf{ct}_L)$. Now, $\mathcal{B}$ answers remaining queries as in Game 2. Finally, $\mathcal{A}$ sends b' as its guess to $\mathcal{B}$. If $b = b'$, then $\mathcal{B}$ sends 1 to LWE-ss challenger to indicate that $\mathbf{G}$ consists of LWE samples, otherwise it sends 0.

Since, LWE-ss chooses $\mathbf{F}$ uniformly at random, therefore $\mathcal{B}$ simulates the distribution of $\mathbf{B}_L^{(i)}$ for $i \leq \mathsf{nbp}$ exactly. Next, if $\mathbf{G} = \mathbf{S} \cdot \mathbf{F} + \mathbf{E}$ for some matrices $\mathbf{S} \leftarrow \chi^{n \times n}$ and $\mathbf{E} \leftarrow \chi^{n \times (\mathsf{nbp} \cdot w \cdot m)}$, then $\mathcal{B}$ simulates the view of Game 2 for $\mathcal{A}$, otherwise it simulates the view of Game 2.L.1. Therefore, if $|\mathsf{Adv}_{\mathcal{A}}^2 - \mathsf{Adv}_{\mathcal{A}}^{2.L.1}|$ is non-negligible, then LWE-ss assumption does not hold.

Lemma 7. *If the* preimage well-distributedness property *of lattice trapdoor sampler* (TrapGen, SamplePre) *holds (Definition 1), then for every adversary $\mathcal{A}$, for any level* level $\in [L]$, $|\mathsf{Adv}_{\mathcal{A}}^{2.\mathsf{level}.1} - \mathsf{Adv}_{\mathcal{A}}^{2.\mathsf{level}.2}| \leq \mathsf{negl}(\lambda)$ *for some negligible function* $\mathsf{negl}(\cdot)$.

Proof. To prove indistinguishability of Game 2.level.1 and 2.level.2, we need to sketch q intermediate hybrids as in Lemma 6. In k^{th} intermediate hybrid, the challenger proceeds as Game 2.level.2 while answering first k queries, and proceeds as in Game 2.level.1 for answering remaining queries. Indistinguishability between any two consecutive intermediate hybrids follows from preimage well-distributedness property of lattice trapdoor sampler.

Observe that in $(k-1)^{th}$ intermediate hybrid, $\mathsf{ct}_{\mathsf{level}}^{(i)}$ is chosen as $\mathsf{ct}_{\mathsf{level}}^{(i)} \leftarrow \mathsf{SamplePre}(\mathbf{B}_{\mathsf{level}-1}^{(i)}, T_{\mathsf{level}-1}^{(i)}, \sigma, \mathbf{C}_{\mathsf{level}}^{(i)})$ for $i \leq \mathsf{nbp}$, where $\mathbf{C}_{\mathsf{level}}^{(i)} \leftarrow \mathbb{Z}_q^{w \cdot n \times m}$. On the other hand, in k^{th} intermediate hybrid, they are chosen as $\mathsf{ct}_{\mathsf{level}}^{(i)} \leftarrow \chi^{m \times m}$ for $i \leq \mathsf{nbp}$. By a simple hybrid argument, we canrestate the preimage

---

[8] If k^{th} query is the challenge query, then $m_k = b$. In other words, m_k will be the random challenge bit.

well-distributedness property for matrices instead of vectors such that for all $i \leq \mathsf{nbp}$, the following holds

$$\left\{ \mathsf{ct}_{\mathsf{level}}^{(i)} : \begin{array}{c} (\mathbf{B}_{\mathsf{level}-1}^{(i)}, T_{\mathsf{level}-1}^{(i)}) \leftarrow \mathsf{TrapGen}(1^{w \cdot n}, 1^m, q), \ \mathbf{C}_{\mathsf{level}}^{(i)} \leftarrow \mathbb{Z}_q^{w \cdot n \times m}, \\ \mathsf{ct}_{\mathsf{level}}^{(i)} \leftarrow \mathsf{SamplePre}(\mathbf{B}_{\mathsf{level}-1}^{(i)}, T_{\mathsf{level}-1}^{(i)}, \sigma, \mathbf{C}_{\mathsf{level}}^{(i)}) \end{array} \right\}$$

$$\approx_s$$

$$\left\{ \mathsf{ct}_{\mathsf{level}}^{(i)} : \mathsf{ct}_{\mathsf{level}}^{(i)} \leftarrow \chi^{w \cdot n \times m} \right\}.$$

Thus, by a hybrid argument over i, we can switch the nbp short matrices in sub-ciphertext $\mathsf{ct}_{\mathsf{level}}$ from being sampled using SamplePre to Gaussian matrices. Therefore, intermediate hybrid $k-1$ and k are statistically indistinguishable. Hence, using nbp intermediate hybrids between Game 2.level.1 and 2.level.2, we can switch level level sub-ciphertexts to low-norm Gaussian matrices for all queries such that if preimage well-distributedness property of lattice trapdoor sampler holds, then Game 2.level.1 and 2.level.2 are statistically indistinguishable for all level $\leq L$.

Lemma 8. *If the matrix well-distributedness property of lattice trapdoor sampler* (TrapGen, SamplePre) *holds (Definition 1), then for every adversary $\mathcal{A}$, for any level* level $\in [L]$, $|\mathsf{Adv}_{\mathcal{A}}^{2.\mathsf{level}.2} - \mathsf{Adv}_{\mathcal{A}}^{2.\mathsf{level}.3}| \leq negl(\lambda)$ *for some negligible function* $negl(\cdot)$.

Proof. The proof of this lemma follows directly from the matrix well-distributedness property of lattice trapdoor sampler. First, note that in both Games 2.level.2 and 2.level.3 sub-ciphertexts at level level (for all queries) consist of nbp random low-norm Gaussian matrices. Thus, the challenger does not need to know the trapdoor of matrices at level (level -1), that is matrices $\mathbf{B}_{\mathsf{level}-1}^{(i)}$ for all $i \leq \mathsf{nbp}$ can be sampled without trapdoor. The matrix well-distributedness property states that for all $i \leq \mathsf{nbp}$

$$\{\mathbf{B}_{\mathsf{level}-1}^{(i)} : (\mathbf{B}_{\mathsf{level}-1}^{(i)}, T_{\mathsf{level}-1}^{(i)}) \leftarrow \mathsf{TrapGen}(1^{w \cdot n}, 1^m, q)\}$$

$$\approx_s$$

$$\{\mathbf{B}_{\mathsf{level}-1}^{(i)} : \mathbf{B}_{\mathsf{level}-1}^{(i)} \leftarrow \mathbb{Z}_q^{w \cdot n \times m}\}.$$

Therefore, by a simple hybrid argument over i, we can move from Game 2.level.2 to 2.level.3 using matrix well-distributedness property of lattice trapdoor sampler with only negligible drop in the advantage.

Lemma 9. *If $(n, \mathsf{nbp} \cdot w \cdot m, q, \chi)$-LWE-ss assumption holds (Assumption 2), then for any PPT adversary $\mathcal{A}$, for any level* level $\in [L-1]$, $|\mathsf{Adv}_{\mathcal{A}}^{2.(\mathsf{level}+1).3} - \mathsf{Adv}_{\mathcal{A}}^{2.\mathsf{level}.1}| \leq negl(\lambda)$ *for some negligible function* $negl(\cdot)$.

Proof. The proof of this lemma is similar to that of Lemma 6.

Lemma 10. *For any PPT adversary $\mathcal{A}$,* $\mathsf{Adv}_{\mathcal{A}}^{2.1.3} = 0$.

Proof. The proof of this lemma follows from the fact that in Game 2.1.3, each ciphertext contains nbp$L$ random low-norm Gaussian matrices irrespective of the message bit being encrypted. Therefore, the distribution of ciphertexts when 0 is encrypted is identical to the distribution of ciphertexts when 1 is encrypted, thus they do not contain any information about the encrypted message bit. Hence, the advantage of any adversary is this game is exactly 0.

References

1. Acar, T., Belenkiy, M., Bellare, M., Cash, D.: Cryptographic agility and its relation to circular encryption. In: Gilbert, H. (ed.) EUROCRYPT 2010. LNCS, vol. 6110, pp. 403–422. Springer, Heidelberg (2010). doi:10.1007/978-3-642-13190-5_21
2. Adão, P., Bana, G., Herzog, J., Scedrov, A.: Soundness and completeness of formal encryption: the cases of key cycles and partial information leakage. J. Comput. Secur. **17**, 737–797 (2009)
3. Alamati, N., Peikert, C.: Three's compromised too: circular insecurity for any cycle length from (ring-)LWE. In: Robshaw, M., Katz, J. (eds.) CRYPTO 2016. LNCS, vol. 9815, pp. 659–680. Springer, Heidelberg (2016). doi:10.1007/978-3-662-53008-5_23
4. Alperin-Sheriff, J., Peikert, C.: Circular and KDM security for identity-based encryption. In: Fischlin, M., Buchmann, J., Manulis, M. (eds.) PKC 2012. LNCS, vol. 7293, pp. 334–352. Springer, Heidelberg (2012). doi:10.1007/978-3-642-30057-8_20
5. Applebaum, B.: Key-dependent message security: generic amplification and completeness. In: Paterson, K.G. (ed.) EUROCRYPT 2011. LNCS, vol. 6632, pp. 527–546. Springer, Heidelberg (2011). doi:10.1007/978-3-642-20465-4_29
6. Applebaum, B., Cash, D., Peikert, C., Sahai, A.: Fast cryptographic primitives and circular-secure encryption based on hard learning problems. In: Halevi, S. (ed.) CRYPTO 2009. LNCS, vol. 5677, pp. 595–618. Springer, Heidelberg (2009). doi:10.1007/978-3-642-03356-8_35
7. Banerjee, A., Peikert, C., Rosen, A.: Pseudorandom functions and lattices. In: Pointcheval, D., Johansson, T. (eds.) EUROCRYPT 2012. LNCS, vol. 7237, pp. 719–737. Springer, Heidelberg (2012). doi:10.1007/978-3-642-29011-4_42
8. Barak, B., Haitner, I., Hofheinz, D., Ishai, Y.: Bounded key-dependent message security. In: Gilbert, H. (ed.) EUROCRYPT 2010. LNCS, vol. 6110, pp. 423–444. Springer, Heidelberg (2010). doi:10.1007/978-3-642-13190-5_22
9. Barrington, D.A.: Bounded-width polynomial-size branching programs recognize exactly those languages in NC^1. In: STOC 1986 (1986)
10. Bishop, A., Hohenberger, S., Waters, B.: New circular security counterexamples from decision linear and learning with errors. In: Iwata, T., Cheon, J.H. (eds.) ASIACRYPT 2015. LNCS, vol. 9453, pp. 776–800. Springer, Heidelberg (2015). doi:10.1007/978-3-662-48800-3_32
11. Boneh, D., Halevi, S., Hamburg, M., Ostrovsky, R.: Circular-secure encryption from decision Diffie-Hellman. In: Wagner, D. (ed.) CRYPTO 2008. LNCS, vol. 5157, pp. 108–125. Springer, Heidelberg (2008). doi:10.1007/978-3-540-85174-5_7
12. Borodin, A., Dolev, D., Fich, F.E., Paul, W.J.: Bounds for width two branching programs. SIAM J. Comput. **15**(2), 549–560 (1986)
13. Brakerski, Z., Goldwasser, S.: Circular and leakage resilient public-key encryption under subgroup indistinguishability. In: Rabin, T. (ed.) CRYPTO 2010. LNCS, vol. 6223, pp. 1–20. Springer, Heidelberg (2010). doi:10.1007/978-3-642-14623-7_1

14. Brakerski, Z., Goldwasser, S., Kalai, Y.T.: Black-box circular-secure encryption beyond affine functions. In: Ishai, Y. (ed.) TCC 2011. LNCS, vol. 6597, pp. 201–218. Springer, Heidelberg (2011). doi:10.1007/978-3-642-19571-6_13

15. Brakerski, Z., Langlois, A., Peikert, C., Regev, O., Stehlé, D.: Classical hardness of learning with errors. In: STOC 2013 (2013)

16. Camenisch, J., Lysyanskaya, A.: An efficient system for non-transferable anonymous credentials with optional anonymity revocation. In: Pfitzmann, B. (ed.) EUROCRYPT 2001. LNCS, vol. 2045, pp. 93–118. Springer, Heidelberg (2001). doi:10.1007/3-540-44987-6_7

17. Cash, D., Green, M., Hohenberger, S.: New definitions and separations for circular security. In: Fischlin, M., Buchmann, J., Manulis, M. (eds.) PKC 2012. LNCS, vol. 7293, pp. 540–557. Springer, Heidelberg (2012). doi:10.1007/978-3-642-30057-8_32

18. Cheon, J.H., Han, K., Lee, C., Ryu, H., Stehlé, D.: Cryptanalysis of the multilinear map over the integers. In: Oswald, E., Fischlin, M. (eds.) EUROCRYPT 2015. LNCS, vol. 9056, pp. 3–12. Springer, Heidelberg (2015). doi:10.1007/978-3-662-46800-5_1

19. Coron, J.-S., Gentry, C., Halevi, S., Lepoint, T., Maji, H.K., Miles, E., Raykova, M., Sahai, A., Tibouchi, M.: Zeroizing without low-level zeroes: new MMAP attacks and their limitations. In: Gennaro, R., Robshaw, M. (eds.) CRYPTO 2015. LNCS, vol. 9215, pp. 247–266. Springer, Heidelberg (2015). doi:10.1007/978-3-662-47989-6_12

20. Dodis, Y., Ostrovsky, R., Reyzin, L., Smith, A.D.: Fuzzy extractors: how to generate strong keys from biometrics and other noisy data. SIAM J. Comput. **38**, 97–139 (2008)

21. Dodis, Y., Reyzin, L., Smith, A.: Fuzzy extractors: how to generate strong keys from biometrics and other noisy data. In: Cachin, C., Camenisch, J.L. (eds.) EUROCRYPT 2004. LNCS, vol. 3027, pp. 523–540. Springer, Heidelberg (2004). doi:10.1007/978-3-540-24676-3_31

22. Gentry, C.: Fully homomorphic encryption using ideal lattices. In: STOC (2009)

23. Gentry, C., Gorbunov, S., Halevi, S.: Graph-induced multilinear maps from lattices. In: Dodis, Y., Nielsen, J.B. (eds.) TCC 2015. LNCS, vol. 9015, pp. 498–527. Springer, Heidelberg (2015). doi:10.1007/978-3-662-46497-7_20

24. Gentry, C., Peikert, C., Vaikuntanathan, V.: Trapdoors for hard lattices and new cryptographic constructions. In: STOC, pp. 197–206 (2008)

25. Koppula, V., Ramchen, K., Waters, B.: Separations in circular security for arbitrary length key cycles. In: Dodis, Y., Nielsen, J.B. (eds.) TCC 2015. LNCS, vol. 9015, pp. 378–400. Springer, Heidelberg (2015). doi:10.1007/978-3-662-46497-7_15

26. Koppula, V., Waters, B.: Circular security separations for arbitrary length cycles from LWE. In: Robshaw, M., Katz, J. (eds.) CRYPTO 2016. LNCS, vol. 9815, pp. 681–700. Springer, Heidelberg (2016). doi:10.1007/978-3-662-53008-5_24

27. Laud, P.: Encryption cycles and two views of cryptography. In: NORDSEC 2002 (2002)

28. Marcedone, A., Orlandi, C.: Obfuscation $\Rightarrow$ (IND-CPA security $\not\Rightarrow$ circular security). In: Abdalla, M., Prisco, R. (eds.) SCN 2014. LNCS, vol. 8642, pp. 77–90. Springer, Cham (2014). doi:10.1007/978-3-319-10879-7_5

29. Marcedone, A., Pass, R., Shelat, A.: Bounded KDM security from iO and OWF. In: Zikas, V., Prisco, R. (eds.) SCN 2016. LNCS, vol. 9841, pp. 571–586. Springer, Cham (2016). doi:10.1007/978-3-319-44618-9_30

30. Micciancio, D., Peikert, C.: Trapdoors for lattices: simpler, tighter, faster, smaller. In: Pointcheval, D., Johansson, T. (eds.) EUROCRYPT 2012. LNCS, vol. 7237, pp. 700–718. Springer, Heidelberg (2012). doi:10.1007/978-3-642-29011-4_41

31. Micciancio, D., Regev, O.: Worst-case to average-case reductions based on Gaussian measures. SIAM J. Comput. **37**(1), 267–302 (2007)

32. Peikert, C.: Public-key cryptosystems from the worst-case shortest vector problem: extended abstract. In: STOC 2009 (2009)

33. Peikert, C.: A decade of lattice cryptography. Found. Trends Theor. Comput. Sci. **10**, 283–424 (2016)

34. Regev, O.: On lattices, learning with errors, random linear codes, and cryptography. In: STOC, 2005 (2005)

35. Rothblum, R.D.: On the circular security of bit-encryption. In: Sahai, A. (ed.) TCC 2013. LNCS, vol. 7785, pp. 579–598. Springer, Heidelberg (2013). doi:10. 1007/978-3-642-36594-2_32

Security Models II

Toward Fine-Grained Blackbox Separations Between Semantic and Circular-Security Notions

Mohammad Hajiabadi[1,2]($\boxtimes$) and Bruce M. Kapron[1,2]

[1] Department of Computer Science, University College London, London, UK
m.hajiabadi@ucl.ac.uk
[2] Department of Computer Science, University of Victoria, Victoria, Canada
bmkapron@uvic.ca

Abstract. We address the problems of whether t-circular-secure encryption can be based on $(t-1)$-circular-secure encryption or on semantic (CPA) security, if $t = 1$. While for $t = 1$ a folklore construction, based on CPA-secure encryption, can be used to build a 1-circular-secure encryption with the same secret-key and message space, no such constructions are known for the bit-encryption case, which is of particular importance in fully-homomorphic encryption. Also, all constructions of t-circular encryption (bitwise or otherwise) are based on specific assumptions.

We make progress toward these problems by ruling out all fully-blackbox constructions of

- 1-*seed-circular-secure* bit encryption from CPA-secure encryption;
- t-seed-circular-secure encryption from $(t-1)$-seed-circular secure encryption, for any $t > 1$.

Informally, seed-circular security is a variant of the circular security notion in which the seed of the key-generation algorithm, instead of the secret key, is encrypted. We also show how to extend our first result to rule out a large and non-trivial class of constructions of 1-circular-secure bit encryption, which we dub *key-isolating constructions*. Our separations follow the model of Gertner, Malkin and Reingold (FOCS'01), which is a weaker separation model than that of Impagliazzo and Rudich.

1 Introduction

A public-key encryption scheme is 1-circular secure if it is CPA secure in the presence of an encryption of the secret key under its corresponding public key. A more general notion is that of t-circular security under which CPA security under t public keys $pk_0, \ldots, pk_{t-1}$ should be maintained even when each pk_i is used to encrypt the secret key of $pk_{(i+1 \bmod t)}$. These notions are a special case of the general notion of key-dependent-message (KDM) security, under which more general functions of the secret key(s) may be encrypted.

Work supported in part by the NSERC Discovery Grant "Foundational Studies in Privacy and Security". Part of this work completed while the first author was at University College London and received funding from the European Research Council under the ERC Grant Agreement no. 307937.

© International Association for Cryptologic Research 2017
J.-S. Coron and J.B. Nielsen (Eds.): EUROCRYPT 2017, Part II, LNCS 10211, pp. 561–591, 2017.
DOI: 10.1007/978-3-319-56614-6_19

A primary foundational application of the notion of circular security (for any t) is in the context of fully homomorphic encryption. Currently, with the exception of [11], all constructions of *pure fully homomorphic* encryption go through a *bootstrapping* procedure, requiring a circular-security assumption on a *bootstrappable* scheme built along the way.

When discussing circular security for an encryption scheme with secret-key space $\{0,1\}^\tau$ and plaintext space $\{0,1\}^\eta$, an important feature is the relation between τ and η: we call a scheme *full-length* if $\tau = \eta$. It is straightforward to build a full-length 1-circular-secure scheme from any CPA-secure scheme.[1] Informally, this folklore construction is based on the idea that the underlying plaintext m and public key pk can "communicate" to see if m is pk's secret key. Attempts in extending this idea to the t-circular security setting (for $t > 1$) have so far met with less success and in fact to date all constructions of t-circular secure schemes (full-length or otherwise) are based on assumptions with certain algebraic properties or obfuscation assumptions [3,7,9,30,31,43]. One of the goals of our work is to explain this state of difficulty.

Unfortunately, the full-length assumption is not the end of the story since in many applications of circular security, the secret key is indeed encrypted bit-by-bit or block-by-block, where the size of each block is considerably smaller than the secret-key size (e.g., [16,42]). In such cases the above folklore construction (for $t = 1$) no longer applies: the main difficulty is that since the secret key is no longer encrypted as a whole, but as short blocks, we cannot perform the simple check described above. Of particular importance in such settings is the notion of circular security for *single-bit* encryption schemes (which we call *bit-circular security*), which, beyond FHE applications, is fundamental for the following reason: as shown by Applebaum [2], *projection security*, a notion slightly extending bit-circular security by also allowing for encryptions of negated secret-key bits, is sufficient to obtain KDM security w.r.t. any (*a priori* fixed) function family. Thus, understanding basic forms of KDM security in the bitwise setting is essential for the general understanding of KDM security.

Toward understanding the notion of circular security, several papers based on various specific assumptions have given schemes that are CPA secure, but not t-circular secure (for various values of t), [1,6,12,27]. We remark that although these works provide evidence that t-circular security of any scheme cannot be reduced to the CPA security of the same scheme, they do not shed light on the impossibility of positive constructions.

Finally, we mention that despite the foundational importance of the notion of bit-circular security, our understanding of what it takes to obtain this notion (without relying on specific assumptions) is still lacking, and there is little previous addressing the problem. Haitner and Holenstein [21] rule out fully-blackbox constructions of KDM-secure encryption w.r.t. quite large function families from

---

[1] Assume, w.l.o.g, the CPA-secure scheme (G, E, D) has plaintext space $\{0,1\}^n$ and that G uses an n-bit seed, which is also the outputted secret key. Briefly, the idea is to modify E so that $E(pk, m)$ will first check whether $G(m)$ produces pk as the public key, in which case it returns an encryption of an innocuous message.

trapdoor permutations. Rothblum [39] shows no fully-blackbox reduction can prove that CPA security of a bit-encryption scheme implies circular security of the same scheme. We stress that the result of [39] only considers reductions to and from the same scheme, as opposed to the results of this paper which are concerned with constructions.

Before moving on, we remind the reader of the simple fact that bit t-circular security implies full-length t-circular security. Briefly, the state of knowledge regarding circular security can be summarized as follows:

- Full-length t-circular security based on CPA security: we have a simple construction for $t = 1$, but no known constructions for $t > 1$.
- For bit t-circular security: all constructions (for $t = 1$ or beyond) are based on specific assumptions [7,9] and there is a preliminary separation for $t = 1$ from CPA security [39].

In this work we ask the following two questions

(1) Can bit 1-circular security be based on CPA security?
(2) Can full-length t-circular security (for $t > 1$) be based on CPA security?

1.1 Our Contributions and Discussion

In this paper we make progress toward answering both questions above in the negative, by considering the stronger notion of *seed-circular* security. In its simplest form, an encryption scheme is 1-seed-circular secure if it is CPA secure in the presence of an encrypted version of the seed (of the key-generation algorithm) under its corresponding public key. Similarly, we may define bit/full-length) t-seed circular security. Note that the assumption of t-seed-circular security is indeed at least as strong as that of t-circular security since any scheme meeting the former can slightly be changed to meet the latter by altering the key-generation algorithm to return the underlying seed as its secret-key output. We first describe our main results and then discuss them in detail.

1. We prove there exists no fully blackbox construction (in the sense of [36]) of 1-*seed-circular-secure* bit encryption from CPA-secure encryption (Theorem 6.) We also show that this separation holds so long as the constructed scheme has plaintext space $\{0,1\}^{c \log n}$ for any constant c (Sect. 5.6).
2. We prove that full-length $(t + 1)$-seed-circular security cannot be based in a fully-blackbox way on *bit* t-seed circular security, for any $t \geq 1$ (Theorem 9).

Our first result already rules out certain types of constructions for 1-circular-secure encryption, namely those in which seeds and secret keys are the same. We show how to adapt this result to the setting of circular security, to rule out a large and non-trivial class of constructions of circular-secure encryption that we call *key-isolating constructions*. Due to technicalities involved we refer the reader to Sect. 7 for this notion. (A similar adaptation may be given for the second result, but we do not pursue it in this paper.)

For our second result, choosing the target notion to be full-length $(t+1)$-seed circular security (as opposed to bit $(t+1)$-seed circular security) and the base notion to be bit t-seed circular security (as opposed to full-length t-seed-circular security) only makes our result stronger.

Discussion of results and notions. We first start by discussing the significance of the second result. We note that the folklore CPA-security-based construction alluded to earlier indeed results in a full-length 1-seed-circular secure scheme, since the constructed scheme has the same seed and secret-key space. This shows that the notion of seed-circular security (at least for the full-length case) is not so far fetched, reinforcing the significance of the separation result and providing partial justification for the lack of success in basing full-length t-circular security, for $t > 1$, on CPA security. In fact, it suggests that a less ambitious goal than that of Question (2), namely of basing t-circular security on $(t-1)$-circular security, may still be too much to hope for.

As for the first result, we mention the following fact regarding the notion of bit 1-seed-circular security. Since one of the main applications of this notion is in the context of FHE, it is worth mentioning that if $\mathcal{E}$ is fully homomorphic (or homomorphic enough to evaluate G), then if $\mathcal{E}$ is 1-seed-circular secure it is also 1-circular secure, since one can use the homomorphic properties of $\mathcal{E}$ to evaluate G homomorphically, thereby producing an encrypted secret key from an encrypted seed. (This simple proof is, however, non-blackbox.)

From a practical point of view, the notion of seed circular security for specific schemes is not very natural since such schemes typically come with *public parameters* (e.g., a group), and it is not very meaningful to talk about, say, encrypting the bits used to generate those parameters. Nevertheless, if public-parameter generation is thought of as a separate process, many specific schemes have the property that their secret keys are just the same as their seeds (e.g., ElGamal). For example, both circular-secure schemes of [7,9] have the property that w.r.t. fixed public parameters (which are a group plus l group elements), their secret keys are just random l-bit-strings, being the same as their seeds. Thus, as a step toward proving full blackbox impossibility for circular-secure encryption, it may be worthwhile to formulate a notion of encryption with public parameters, and investigate whether our results extend to this case. We have not, however, carried this out at this moment.

We conclude the discussion with the following observation. Our first result leaves us with an unexplained gap, namely to what extent the plaintext size of the constructed scheme could be made bigger before obtaining a positive (seed-)circular security result? For example, what happens if the construction is allowed to have plaintexts of $\omega(\log n)$ bits long? We believe that filling this gap will further improve our understanding of the notion of 1-(seed-)circular security.

Our separation model. All our separations follow the model of [19]. We discuss the model for the first result. For any candidate 1-seed-circular-secure construction $\mathcal{E} = (G, E, D)$ we show the existence of two oracles $\mathbf{O} = (\mathbf{g}, \mathbf{e}, \mathbf{d})$ and $\mathbf{T}$ such that (a) there exists a PPT oracle adversary $\mathcal{A}^{\mathbf{O},\mathbf{T}}$ that breaks the (supposed) seed-circular security of $\mathcal{E}^\mathbf{O}$ and (b) no PPT oracle adversary $\mathcal{B}^{\mathbf{O},\mathbf{T}}$ can break the CPA security of $\mathbf{O}$. This immediately implies that there exists no fully-blackbox reduction. As common in separation models we show the existence of $\mathbf{O}$ and $\mathbf{T}$ non-constructively by proving results w.r.t. randomly chosen $\mathbf{O}$ and $\mathbf{T}$. We give an overview of our techniques and separation model in Sect. 4.

Most separation results in the literature indeed rule out the existence of *relativizing reductions*, e.g., [8,17,25,40,41], which constitute a broader class of constructions than fully-blackbox ones. We stress that our results do not rule out relativizing reductions. Nonetheless, we are not aware of any "natural" cryptographic construction that is relativizing but not fully-blackbox. Finally, we mention that there exists separation results in the literature that also only rule out fully-blackbox reductions, e.g., [21,24,28,29].

Blackbox versus non-blackbox techniques. We note that there are non-blackbox reductions in cryptography, for which a blackbox-counterpart may or may not (both provably and ostensibly) exist. (Here by non-blackbox we are referring to the construction, not to the security proof.) We mention [14,26] as two examples of blackbox constructions that replaced their earlier non-blackbox counterparts [20,35]. Classical examples of non-blackbox constructions with no known blackbox counterparts are [15,34], giving non-blackbox constructions of CCA1- and CCA2-secure encryption from *enhanced trapdoor permutations*. The state of our knowledge regarding the blackbox status of CCA-secure encryption versus other "classical" public-key primitives is arguably limited, and the only known works are the work of Gertner *et al.* [18], ruling out *sheilding* blackbox constructions of CCA1-secure encryption from CPA-secure encryption, and that of Myers and shelat [33] proving equivalence of one-bit and many-bit CCA2 secure encryption. Finally, we mention that the work of Mahmoody and Pass [29] shows the existence of a non-blackbox construction (that of non-interactive commitment schemes from so called *hitting one-way functions*) for which provably no blackbox counterpart exists.

Other related work. The question of what "general" assumptions may be used to obtain KDM security is addressed in [23], where it is shown that projection-secure public-key encryption (PKE) can be built from any CPA-secure PKE with some structural properties. The power of circular-secure encryption is addressed in [22], where it is shown that in combination with a so-called reproducibility property, bit circular security implies the existence of powerful primitives including correlation-secure trapdoor functions [38], CCA2-secure encryption and deterministic encryption. The body of work on blackbox separations is extensive, some of which were mentioned earlier. We also mention the progress that has been made in understanding the limitations of some of the common non-blackbox techniques, e.g., [4,10].

Open Problems. The main open problem is to extend our impossibility results to the circular-security setting. We explain in Sect. 8 why we were not able to do this. Another interesting problem is to see to what extent our techniques extend to obtain separations based on other classical public-key primitives.

Note on proofs. Due to space constraints, proofs for some results have been omitted. In all cases, proofs for these results appear in the full version.

2   Preliminaries

If $R(x_1, \ldots, x_i; r)$ is a randomized algorithm using randomness r, by $R(a_1, \ldots, a_i)$ we mean the random variable obtained by sampling r uniformly at random and returning $R(a_1, \ldots, a_i; r)$. If $\mathcal{D}$ is a distribution $x \in \mathcal{D}$ means $x \in support(\mathcal{D})$.

The notion of a public-key encryption scheme (PKE) (G, E, D) is standard. The only convention we make is that the order of keys produced by G is as a secret/public key pair (as opposed to a public/secret key pair). We refer to the randomness space of G as the *seed* space of the scheme. We assume the decryption algorithm is deterministic, and always decrypts correctly, and refer to this as the *correctness* condition. (Our separation results will hold even if the constructed scheme is allowed to make a small decryption error. However, for the sake of simplicity we assume the stated condition.) All schemes in this paper are many-bit or single-bit encryption schemes. If E's plaintext space is $\{0,1\}^\eta$ by $E(PK, M)$ for $M \in \{0,1\}^*$ we mean that M is encrypted in blocks of size η, augmenting M with enough zero bits to make $|M|$ a multiple of η, if necessary. In particular, when $\eta = 1$, this will denote the bit-by-bit encryption of M.

We shall use lowercase letters $(\mathbf{g}, \mathbf{e}, \mathbf{d})$ to denote base (i.e., blackbox) schemes and uppercase letters (G, E, D) to denote constructions.

Oracle convention. Whenever we talk about an oracle adversary/algorithm $\mathcal{A}$ we adopt the following conventions: we say $\mathcal{A}$ is *efficient* (or PPT) if $\mathcal{A}$ can be implemented as a PPT oracle algorithm; we say $\mathcal{A}$ is *query-efficient* if $\mathcal{A}$ always makes at most a poly-number of oracle queries (but unlimited otherwise, and may run exponential local computations). Whenever we put no restriction on an adversary it means that it is not restricted in any way.

We define when an adversary breaks the (seed-)circular security of a bit-encryption scheme. The definition naturally extends to the many-bit case.

Definition 1. *Let $\mathcal{E} = (G, E, D)$ be a bit PKE with seed space $\{0,1\}^n$. Let*

$$\mathbf{InpSeed} = (PK_1, \ldots, PK_t, E_{PK_1}(S_2), \ldots, E_{PK_{t-1}}(S_t), E_{PK_t}(S_1))$$
$$\mathbf{InpSec} = (PK_1, \ldots, PK_t, E_{PK_1}(SK_2), \ldots, E_{PK_{t-1}}(S_t), E_{PK_t}(SK_1))$$
$$b \leftarrow \{0,1\}, C \leftarrow E_{PK_1}(b),$$

where $S_i \leftarrow \{0,1\}^n$ and $(SK_i, PK_i) = G(S_i)$, for $1 \le i \le t$. Then we say

- $\mathcal{A}$ *breaks the t-seed-circular security of $\mathcal{E}$ if $\Pr[\mathcal{A}(\mathbf{InpSeed}, C) = b]$ is non-negligibly greater than $1/2$.*
- $\mathcal{A}$ *breaks the t-circular security of $\mathcal{E}$ if $\Pr[\mathcal{A}(\mathbf{InpSec}, C) = b]$ is non-negligibly greater than $1/2$.*

We now define the assumptions underlying our results in this paper.

Toward a Separation of Circular Security and Semantic Security   567

Terminology 1. The assumption of "bit t-seed-circular security" refers to the existence of a t-seed-circular secure single-bit PKE. Also, "full-length t-seed-circular security" refers to the existence of a t-seed-circular secure PKE with the same seed and plaintext space. We have the following simple implications: (a) CPA security $\Rightarrow$ full-length 1-seed circular security and (b) bit t-seed-circular security $\Rightarrow$ full-length t-seed-circular security.

We define a notion of blackbox reductions between encryption primitives. See [5,36] for more general notions of blackbox reductions.

Definition 2. *A fully-blackbox reduction of P-secure (e.g., circular-secure) encryption to Q-secure (e.g., CPA-secure) encryption consists of two PPT oracle algorithms $(\mathcal{E}, Red)$, satisfying the following: for any PKE $\mathbf{O} = (\mathbf{g}, \mathbf{e}, \mathbf{d})$,*

1. *$\mathcal{E}^{\mathbf{O}} = (G^{\mathbf{O}}, E^{\mathbf{O}}, D^{\mathbf{O}})$ forms a PKE, and*
2. *for any adversary $\mathcal{A}$ breaking the P-security of $(G^{\mathbf{O}}, E^{\mathbf{O}}, D^{\mathbf{O}})$, the oracle algorithm $Red^{\mathcal{A}, \mathbf{O}}$ breaks the Q-security of $\mathbf{O}$.*

3   PKE Oracle Distribution

Convention. Whenever we say a function $f \colon D \to R$ with property P (e.g., injectivity) is a randomly chosen function we mean f is chosen uniformly at random from the space of all functions from D to R having property P.

We describe a distribution under which a PKE oracle (with some auxiliary oracles) is sampled. These oracles will be used to model "ideal" base primitives in our separations. We largely follow the notational style of [18]. As notation, if f is a function whose output is a tuple, say a pair, we write $f(x) = (*, y)$ to indicate that $f(x) = (y', y)$, for some y'.

Definition 3. *We define an oracle distribution Ψ which produces a PKE oracle with certain length parameters, plus two auxiliary oracles. Formally, Ψ produces an ensemble of oracles $\mathcal{O}_n = (\mathbf{O}_n, \mathbf{u}_n, \mathbf{w}_n)_{n \in \mathbb{N}}$, where for every $n \in \mathbb{N}$, $\mathbf{O}_n = (\mathbf{g}_n, \mathbf{e}_n, \mathbf{d}_n)$ and $(\mathbf{u}_n, \mathbf{w}_n)$ are chosen as follows.*

- *$\mathbf{g}_n \colon \{0,1\}^n \to \{0,1\}^{5n}$ is a random one-to-one function, mapping a secret key to a public key.*
- *$\mathbf{e}_n \colon \{0,1\}^{5n} \times \{0,1\} \times \{0,1\}^n \to \{0,1\}^{7n}$ is a function, where for every $pk \in \{0,1\}^{5n}$, $\mathbf{e}_n(pk, \cdot, \cdot)$ is a random one-to-one function.*
- *$\mathbf{d}_n \colon \{0,1\}^n \times \{0,1\}^{7n} \to \{0,1\} \cup \{\bot\}$ is defined by letting $\mathbf{d}_n(sk, c) = b$ if and only if $\mathbf{e}_n(\mathbf{g}_n(sk), b, r) = c$, for some $r \in \{0,1\}^n$; otherwise, $\mathbf{d}_n(sk, c) = \bot$.*
- *$\mathbf{u}_n \colon \{0,1\}^{5n} \times \{0,1\}^{7n} \to (\{0,1\} \times \{0,1\}^n) \cup \{\bot\}$ is defined as $\mathbf{u}_n(pk, c) = (b, r)$ if $\mathbf{e}_n(pk, b, r) = c$, and $\mathbf{u}_n(pk, c) = \bot$ if for no (b, r) does it hold that $\mathbf{e}_n(pk, b, r) = c$. That is, $\mathbf{u}_n(pk, c)$ decrypts c relative to pk, and if successful, also returns the unique randomness used to produce c. (The oracle $\mathbf{u}$ is not typically allowed to be freely used. See Definition 4.)*
- *$\mathbf{w}_n \colon \{0,1\}^{5n} \to \{\bot, \top\}$ is defined as $\mathbf{w}_n(pk) = \top$ if for some sk $\mathbf{g}_n(sk) = pk$, and $\mathbf{w}_n(pk) = \bot$, otherwise. That is, $\mathbf{w}_n(pk)$ checks whether pk is a valid public key.*

Definition 4. *In all settings where access to* **u** *is granted this access is limited and is determined based on the underlying challenge inputs. Specifically, we call* $\mathcal{A}^{\mathbf{g},\mathbf{e},\mathbf{d},\mathbf{u},\mathbf{w}}$ *CCA-valid if* $\mathcal{A}^{\mathbf{g},\mathbf{e},\mathbf{d},\mathbf{u},\mathbf{w}}$ *on input* (pk, c) *never calls* $\langle \mathbf{u}, (pk, c) \rangle$. *This definition naturally generalizes to the case in which* $\mathcal{A}$*'s input consists of several challenge public keys with several challenge ciphertexts for each public key, e.g., the t-seed circular security setting.*

Omitting the security parameter. We define $\mathbf{g}(sk) = \mathbf{g}_n(sk)$, for every n and $sk \in \{0,1\}^n$, and use a similar convention for other functions in Definition 3. Sometimes when we need to emphasize under what security parameter a query is made, we put in the sub-index n; in other places we typically omit the sub-index.

Ψ-valid oracles. We call a triple of functions $(\mathbf{g}, \mathbf{e}, \mathbf{d})$ *Ψ-valid* if $(\mathbf{g}, \mathbf{e}, \mathbf{d})$ is part of a possible output of Ψ, i.e., the domains and ranges of $\mathbf{g}$, $\mathbf{e}$ and $\mathbf{d}$ are as specified in Definition 3, and also all the corresponding injectivity conditions hold. Similarly, we may use the same convention to call, say, $\mathbf{g}$, Ψ-valid.

Notation. For oracles $O = (O_1, \ldots, O_m)$ and an oracle algorithm $\mathcal{A}^O$, we let $qry = \langle O_i, q \rangle$ denote an $\mathcal{A}$'s query q to oracle O_i; if $u = O_i(q)$ we use $((\langle O_i, q \rangle), u)$ to indicate that $\mathcal{A}$ calls O_i on q and receives u; we also define $O(qry) = u$. If Que is a set of such *query/response pairs* we use shorthands like $((\langle O_j, * \rangle), u) \in$ Que to mean that for some q, $((\langle O_j, q \rangle), u) \in$ Que. Thus, $((\langle O_j, * \rangle), u) \notin$ Que indicates that for no q, we have $((\langle O_j, q \rangle), u) \in$ Que.

Symbolic representation of oracle queries. Sometimes we need to talk about sets containing query/response pairs generated under some oracle, and later on check them against another oracle. For this reason, we may sometimes talk about *symbolic* query/response pairs. For example, the symbolic form of a concrete query/response pair $((\langle \mathbf{g}, sk \rangle), pk)$ is denoted $((\langle \mathfrak{g}, sk \rangle), pk)$.

4 General Overview of Techniques

We give an overview of our approaches for the two main results: separating bit 1-seed circular security (see Terminology 1) from CPA security and separating full-length $(t+1)$-seed-circular security from bit t-seed-circular security.

4.1 CPA Security $\not\Rightarrow$ Bit 1-seed-circular Security

Summary of approach. First, note a random $\mathbf{O} = (\mathbf{g}, \mathbf{e}, \mathbf{d})$, chosen as $(\mathbf{O}, \mathbf{u}, \mathbf{w}) \leftarrow \Psi$ will be "ideally" secure w.r.t. all notions security discussed in this paper. One idea for proving separations is to add some weakening components $\mathbf{v}$ to $\mathbf{O}$ and show that relative to $(\mathbf{O}, \mathbf{v})$ the base primitive exists, but not the target primitive. We could not make this approach work. Instead, we follow the model of [19], by defining a weakening oracle $\mathbf{T}$, for every candidate construction (G, E, D), in such a way that $\mathbf{T}$ breaks the claimed security of $(G^{\mathbf{O}}, E^{\mathbf{O}}, D^{\mathbf{O}})$, for a random $\mathbf{O}$, but not the base security of $\mathbf{O}$. We emphasize that $\mathbf{T}$ depends on (G, E, D).

Let $\mathcal{E} = (G, E, D)$ be a candidate bit-encryption construction, $(\mathbf{g}, \mathbf{e}, \mathbf{d}, \mathbf{u}, \mathbf{w}) \leftarrow \Psi$ and $\mathbf{O} = (\mathbf{g}, \mathbf{e}, \mathbf{d})$. Our goal is to give an oracle $\mathbf{T}$ s.t. (I) $\mathbf{T}$ is helpful in breaking the (alleged) seed-circular-security of $\mathcal{E}^{\mathbf{O}}$ and (II) $\mathbf{T}$ is not helpful in breaking the CPA security of $\mathbf{O}$. The most obvious idea is that on inputs of the form $(PK, C_1, \ldots, C_n)$, an alleged public key PK and a bit-by-bit encryption of PK's seed under $\mathcal{E}^{\mathbf{O}}(PK, \cdot)$, $\mathbf{T}$ will check whether PK is a *valid* public key under $G^{\mathbf{O}}$ and if so decrypt $C_1, \ldots, C_n$ under a secret key corresponding to PK to get some string S and return S if $G^{\mathbf{O}}(S)$ produces PK.

There are two problems with the above raw approach. First, even doing a simple check, namely whether PK is a valid public key, can potentially grant a CPA adversary against $\mathbf{O}$ much power, violating Condition (II) above. (It is not hard to think of contrived constructions $\mathcal{E}$ for which this is the case.) Second, even if we assume a CPA-adversary $\mathcal{A}^{\mathbf{O}, \mathbf{T}}(pk, c)$—against $\mathbf{O}$—always calls $\mathbf{T}$ on valid PK's, we still have to make sure that $\mathcal{A}$ cannot come up with a clever query $\mathbf{T}(PK, C_1, \ldots, C_n)$ whose response leaks information about $\mathbf{g}^{-1}(pk)$ or about c's plaintext bit.

Our approach starts by resolving the first problem, using an idea from [19] (also used in some subsequent works [18,41]): the oracle $\mathbf{T}$ performs the decryption of $(C_1, \ldots, C_n)$ not relative to $\mathbf{O}$, but relative to some $\widetilde{\mathbf{O}} = (\widetilde{\mathbf{g}}, \widetilde{\mathbf{e}}, \widetilde{\mathbf{d}})$, under which PK is indeed a valid public key (i.e., $\mathbf{T}$ decrypts using $D^{\widetilde{\mathbf{O}}}(SK', C_1 \ldots C_n)$, where $(SK', PK) \in G^{\widetilde{\mathbf{O}}}$). Without further restrictions on $\widetilde{\mathbf{O}}$ the result of decryption is most likely a random noise, as $\widetilde{\mathbf{e}}$ and $\widetilde{\mathbf{d}}$ can behave arbitrarily. Thus, we need to ensure that w.h.p. over a random R, $E^{\mathbf{O}}(PK, b; R) = E^{\widetilde{\mathbf{O}}}(PK, b; R)$, for any bit b. This would ensure that $D^{\widetilde{\mathbf{O}}}(SK', C_1 \ldots C_n)$ w.h.p. will be the real output, if $(PK, C_1, \ldots, C_n)$ were "honestly" generated, showing that $\mathbf{T}$ is useful in breaking 1-seed-circular security of $(G^{\mathbf{g}}, E^{\mathbf{e}}, D^{\mathbf{d}})$.

Specifically, we construct $\widetilde{\mathbf{O}}$ by *super-imposing* a poly number of query/response pairs $\mathbf{Q_s}$, which serve as a *certificate* of PK's validity, on $\mathbf{O}$. More precisely, we first sample (offline) a set of query/response pairs $\mathbf{Q_s}$ in such a way that $G^{\mathbf{Q_s}} = (*, PK)$; then, we super-impose (Definition 5) $\mathbf{Q_s}$ on $\mathbf{O}$ to obtain $\widetilde{\mathbf{O}}$. (Sometimes $\mathbf{Q_s}$ needs to also agree with some previous information.)

To resolve the second problem the oracle $\mathbf{T}$ will refuse to decrypt queries deemed "dangerous": those that can be issued by a CPA adversary $\mathcal{A}$ against $\mathbf{O}$, and whose responses may leak information about $\mathcal{A}$'s challenge secrets. The main challenge is to formulate these dangerous queries in such a way that $\mathbf{T}$ is provably of no use to any CPA adversary against $\mathbf{O}$, while guaranteeing that $\mathbf{T}$ still decrypts w.h.p. a randomly encrypted random seed chosen relative to $E^{\mathbf{O}}$.

Concrete overview. We now give a concrete overview of the above approach for a simple class of constructions. We first start by defining the task of super-imposing a set of $\mathfrak{g}$-type query/response pairs on an oracle $(\mathbf{g}, \mathbf{e}, \mathbf{d})$.

Definition 5. *We define the following procedure we call KeyImpose.*

- **Input:** $(\mathbf{g}, \mathbf{e}, \mathbf{d})$ *and a set* $\mathbf{Q_s} = \{((\langle \mathfrak{g}, sk_1 \rangle, pk_1), \ldots, ((\langle \mathfrak{g}, sk_w \rangle, pk_w)\}$, *satisfying* $sk_i \neq sk_j$ *for all distinct i and j.*

- **Output:** $(\widetilde{\mathbf{g}}, \widetilde{\mathbf{d}})$, where

$$\widetilde{\mathbf{g}}(sk) = \begin{cases} \mathbf{g}(sk) & \text{if } sk \notin \{sk_1, \ldots, sk_w\} \\ pk_i & \text{if } sk = sk_i \text{ for some } 1 \le i \le w \end{cases} \quad (1)$$

$\widetilde{\mathbf{d}}(sk, c)$ is defined as follows: if there exist b and r such that $\mathbf{e}(\widetilde{\mathbf{g}}(sk), b, r) = c$ then $\widetilde{\mathbf{d}}(sk, c) = b$; otherwise, $\widetilde{\mathbf{d}}(sk, c) = \bot$.

Note that in the above definition if $(\mathbf{g}, \mathbf{e}, \mathbf{d})$ is a valid PKE scheme and $\mathsf{Q_s}$ satisfies the required condition then $(\widetilde{\mathbf{g}}, \mathbf{e}, \widetilde{\mathbf{d}})$ is also a valid PKE scheme. The resulting $\widetilde{\mathbf{g}}$, however, will not be injective if there are "collisions" between $\mathsf{Q_s}$ and $\mathbf{g}$. Nonetheless, the resulting $(\widetilde{\mathbf{g}}, \mathbf{e}, \widetilde{\mathbf{d}})$ is still both well-defined and valid.

We will use the following fact over and over again in the paper. Informally, it shows one particular situation where $\widetilde{\mathbf{d}}$ queries, defined as above, can be handled using full access to $(\mathbf{g}, \mathbf{e}, \mathbf{d})$ and partial access to $\mathbf{u}$.

Fact 1. *Let $(\mathbf{g}, \mathbf{e}, \mathbf{d}, \mathbf{u}, \mathbf{w})$ be a Ψ-valid oracle and let $\mathcal{B}^{\mathbf{g}, \mathbf{e}, \mathbf{d}, \mathbf{u}, \mathbf{w}}(pk, \ldots)$ be a CCA-valid adversary (Definition 4) with a challenge public key pk. (The set of $\mathcal{B}$'s challenge ciphertexts is not important for this discussion.) Let $\mathsf{Q_s}$ be a set of query/response pairs meeting the condition of Definition 5 and $(\widetilde{\mathbf{g}}, \widetilde{\mathbf{d}}) = KeyImpose(\mathbf{g}, \mathbf{e}, \mathbf{d}, \mathsf{Q_s})$. Assuming $((\mathfrak{g}, *), pk) \notin \mathsf{Q_s}$, then $\mathcal{B}^{\widetilde{\mathbf{g}}, \mathbf{e}, \widetilde{\mathbf{d}}, \mathbf{u}, \mathbf{w}}(pk, \ldots)$, by having $\mathsf{Q_s}$ as a separate input, can efficiently compute $\widetilde{\mathbf{d}}(sk', c')$, for all sk' and c' without violating the CCA condition.*

Proof. For any query $qu = \langle \widetilde{\mathbf{d}}, (sk', c') \rangle$, either (i) $(\langle \mathfrak{g}, sk' \rangle, *) \notin \mathsf{Q_s}$ or (ii) for some $pk' \ne pk$, $(\langle \mathfrak{g}, sk' \rangle, pk') \in \mathsf{Q_s}$. If (i) holds then $\widetilde{\mathbf{d}}(sk', c') = \mathbf{d}(sk', c')$ and so $\mathcal{B}$ can reply to qu by calling $\langle \mathbf{d}, (sk', c') \rangle$. If (ii) holds, the answer to qu can be determined by calling $\langle \mathbf{u}, (pk', c') \rangle$, which is a valid query for $\mathcal{B}$ as $pk' \ne pk$. $\square$

We make the following two assumptions for any construction (G, E, D) discussed throughout.

Assumption 1. For any Ψ-valid $\mathbf{O} = (\mathbf{g}, \mathbf{e}, \mathbf{d})$ we assume $G^{\mathbf{O}}$, $E^{\mathbf{O}}$ and $D^{\mathbf{O}}$, on inputs corresponding to security parameter n make exactly n^{ϑ} oracle calls (for $\vartheta \ge 1$) and that $G^{\mathbf{O}}(1^n)$ uses exactly n random bits.[2]

Assumption 2. We assume G, E and D, on inputs relative to security parameter 1^n only call their oracles under the same security parameter 1^n. This assumption is only made to simplify our exposition. Indeed, we are not aware of any construction that does not satisfy this assumption.

---

[2] Note that we do not claim that there exists a universal ϑ that works for all constructions $\mathcal{E} = (G, E, D)$. Rather, for any fixed construction (G, E, D) which we want to rule out (i.e., define a breaking oracle **T** for), we fix a ϑ that satisfies the stated conditions. Also, the assumption that G relative to any Ψ valid oracle uses n coins is not necessary; it can indeed be any fixed $p(n)$ number of coins, but assuming it to be n allows us to dispense with an additional parameter p.

We now describe our techniques for a simple class of constructions, those with oracle access of the form $(G^{\mathbf{g}}, E^{\mathbf{e}}, D^{\mathbf{d}})$. We first give the oracle $\mathbf{T}$, defined w.r.t. a fixed $(\mathbf{g}, \mathbf{e}, \mathbf{d})$ and a fixed construction (G, E, D), which helps us to break the seed-circular security of $(G^{\mathbf{g}}, E^{\mathbf{e}}, D^{\mathbf{d}})$. Fix (G, E, D) throughout this section, so we make the dependence of $\mathbf{T}$ on (G, E, D) implicit below. The oracle $\mathbf{T}$ is selected from a class of oracles, but it is convenient to define the output distribution of a randomly chosen $\mathbf{T}$ on an arbitrary given input, as we do below.

Description of T:
Oracles: $(\mathbf{g}, \mathbf{e}, \mathbf{d}, \mathbf{w})$
Input: $(1^n, PK, C_1, \ldots, C_n)$

1. Choose $(\mathbf{g}', S')$ uniformly at random from the set of all pairs satisfying (a) $\mathbf{g}'$ is Ψ-valid and (b) $G^{\mathbf{g}'}(1^n, S') = (*, PK)$. If no such a pair exists return $\perp$. Otherwise, let SK' be the secret key output by $G^{\mathbf{g}'}(1^n, S')$.
2. Let Q_s contain the symbolic versions of all query/response pairs made in the execution of $G^{\mathbf{g}'}(1^n, S')$. Define $(\widetilde{\mathbf{g}}, \widetilde{\mathbf{d}}) = KeyImpose(\mathbf{g}, \mathbf{e}, \mathbf{d}, \mathsf{Q}_\mathsf{s})$. Let QPub include any pk such that $\mathbf{w}(pk) = \top$ and $(\langle \mathfrak{g}, * \rangle, pk) \in \mathsf{Q}_\mathsf{s}$.
3. Compute $S_{out} = D^{\widetilde{\mathbf{d}}}(SK', C_1 \ldots C_n)$. Execute $G^{\mathbf{g}}(S_{out})$ and if for all $pk \in \mathsf{QPub}$ the query/response $(\langle \mathbf{g}, * \rangle, pk)$ is made during the execution, then return S_{out}; otherwise, return $\perp$.

We now informally discuss why $\mathbf{T}$ provides the "desired" properties.

4.1.1   T Does Not Break CPA Security of a Random $(\mathbf{g}, \mathbf{e}, \mathbf{d})$

We show that any adversary $\mathcal{A}^{\mathbf{O}, \mathbf{T}}(1^n, pk, c)$ against the CPA-security of $\mathbf{O}$ can be fully simulated without $\mathbf{T}$, by a CCA-valid adversary $\mathcal{B}^{\mathbf{O}, \mathbf{u}, \mathbf{w}}(1^n, pk, c)$ (See Definition 4). We then show any such $\mathcal{B}$ has a very small chance of breaking the security of $\mathbf{O}$, by relying on a special case of the following lemma which shows a random $\mathbf{O}$ is t-seed-circular secure in a strong sense. As notation whenever we write $f_1(n) \leq f_2(n)$ we mean that this holds asymptotically.

Lemma 1. Let $t = t(n)$ be a poly. Let $\mathcal{B}$ be a CCA-valid oracle adversary (Definition 4), which has access to some Ψ-valid oracle $(\mathbf{g}, \mathbf{e}, \mathbf{d}, \mathbf{u}, \mathbf{w})$, and which makes at most $2^{n/4}$ queries and outputs a bit. It then holds that

$$\Pr\left[\mathcal{B}^{\mathbf{O}, \mathbf{u}, \mathbf{w}}(1^n, pk_1, \ldots, pk_t, \mathbf{e}(pk_1, sk_2), \ldots, \mathbf{e}(pk_t, sk_1), \mathbf{e}(pk_1, b)) = b\right] \leq \frac{1}{2} + \frac{1}{2^{n/4}},$$

where $\mathcal{O} = (\mathbf{g}, \mathbf{e}, \mathbf{d}, \mathbf{u}, \mathbf{w}) \leftarrow \Psi$, $\mathbf{O} = (\mathbf{g}, \mathbf{e}, \mathbf{d})$, $b \leftarrow \{0, 1\}$, $sk_i \leftarrow \{0, 1\}^n$ and $pk_i = \mathbf{g}(sk_i)$ for $1 \leq i \leq t$.

Fix a Ψ-valid oracle $(\mathbf{O}, \mathbf{u}, \mathbf{w})$. We show that any adversary $\mathcal{A}^{\mathbf{O}, \mathbf{T}}(1^n, pk, c)$, against the CPA-security of $\mathbf{O} = (\mathbf{g}, \mathbf{e}, \mathbf{d})$, can be perfectly simulated by a CCA-valid adversary $\mathcal{B}^{\mathbf{O}, \mathbf{u}, \mathbf{w}}(1^n, pk, c)$ that makes a poly-related number of queries. The crux of our techniques lies in showing how $\mathcal{B}$, using $\mathbf{u}$ and $\mathbf{w}$, can simulate $\mathcal{A}$'s $\mathbf{T}$ access.

Specifically, $\mathcal{B}^{\mathbf{O},\mathbf{u},\mathbf{w}}(1^n, pk, c)$ starts running $\mathcal{A}^{\mathbf{O},\mathbf{T}}(1^n, pk, c)$ and forwards all $\mathcal{A}$'s $\mathbf{O} = (\mathbf{g}, \mathbf{e}, \mathbf{d})$ queries to its own corresponding oracles.

To respond to a $\mathbf{T}$ query of the form $Tqu \stackrel{\text{def}}{=} \langle \mathbf{T}, (1^{n_1}, PK, C_1, \ldots, C_{n_1}) \rangle$ made by $\mathcal{A}$, $\mathcal{B}$ acts as follows (note it may be that $n_1 \neq n$, as $\mathcal{A}$ can make queries under different security parameters): $\mathcal{B}$ forms SK' and Q_s exactly as in Steps 1 and 2 of $\mathbf{T}$'s computation. It is able to do so since during these two computations no queries are made to the real oracles (though, a massive offline search is involved). Next, $\mathcal{B}$ starts simulating $D^{\tilde{\mathbf{d}}}(SK', C_1, \ldots, C_{n_1})$, where $(\tilde{\mathbf{g}}, \tilde{\mathbf{d}}) = KeyImpose(\mathbf{g}, \mathbf{e}, \mathbf{d}, Q_s)$. Since it is not clear how $\mathcal{B}$ can perform this decryption by only making a polynomial number of queries and without ever calling $\langle \mathbf{u}, (pk, c) \rangle$, we consider two possible cases:

(A) $(\langle \mathbf{g}, * \rangle, pk) \notin Q_s$: In this case $\mathcal{B}$ can fully execute $D^{\tilde{\mathbf{d}}}(SK', C_1, \ldots, C_{n_1})$, since by Fact 1 $\mathcal{B}$ can handle all encountered queries, which are all of type $\tilde{\mathbf{d}}$. (Recall that pk is $\mathcal{B}$'s challenge public key.) Now if $S_{out} = D^{\tilde{\mathbf{d}}}(SK', C_1, \ldots, C_{n_1})$, then $\mathcal{B}$ performs the rest of Step 3 of $\mathbf{T}$, which $\mathcal{B}$ can fully do since the rest only involves making $\mathbf{g}$ and $\mathbf{w}$ queries. Thus, $\mathcal{B}$ can find the answer to Tqu.

(B) $(\langle \mathbf{g}, * \rangle, pk) \in Q_s$: In this case, recalling the definition of Qpub, we have $pk \in \mathsf{QPub}$, since pk is $\mathcal{B}$'s challenge public key and so by definition $\mathbf{w}(pk) = \top$. Thus, by the condition given in Step 3 of $\mathbf{T}$'s description, if $S_{out} = D^{\tilde{\mathbf{d}}}(SK', C_1, \ldots, C_{n_1})$ then at least one of the following holds:
 (a) The answer to Tqu is $\bot$; or
 (b) The query/response pair $(\langle \mathbf{g}, * \rangle, pk)$ will show up during $G^{\mathbf{g}}(S_{out})$, i.e., $\mathbf{g}^{-1}(pk)$, $\mathcal{B}$'s challenge secret key, is revealed during $G^{\mathbf{g}}(S_{out})$.
We now claim that $\mathcal{B}$ can find two strings S_0 and S_1 such that $S_{out} \in \{S_0, S_1\}$. If this is the case, $\mathcal{B}$ can execute both $G^{\mathbf{g}}(S_0)$ and $G^{\mathbf{g}}(S_1)$; if during either execution a query/response $(\langle \mathbf{g}, * \rangle, pk)$ is observed, $\mathcal{B}$ has learned $\mathbf{g}^{-1}(pk)$, winning the game; otherwise, $\mathcal{B}$ in response to Tqu returns $\bot$, which is indeed the correct response.
It remains to demonstrate the claim. To find S_0 and S_1, $\mathcal{B}$ attempts to simulate $D^{\tilde{\mathbf{d}}}(SK', C_1, \ldots, C_{n_1})$. For any query $qu = \langle \tilde{\mathbf{d}}, (sk', c') \rangle$ encountered in the simulation, one of the following holds
 (i) $(\langle \mathbf{g}, sk' \rangle, *) \notin Q_s$; or
 (ii) $(\langle \mathbf{g}, sk' \rangle, pk') \in Q_s$ for $pk' \neq pk$; or
 (iii) $(\langle \mathbf{g}, sk' \rangle, pk) \in Q_s$ and $c' \neq c$; or
 (iv) $(\langle \mathbf{g}, sk' \rangle, pk) \in Q_s$ and $c' = c$.
$\mathcal{B}$ can find the answer to qu by querying $\langle \mathbf{d}, (sk', c') \rangle$ for Case (i), querying $\langle \mathbf{u}, (pk', c') \rangle$ for Case (ii), and querying $\langle \mathbf{u}, (pk, c') \rangle$ for Case (iii). The latter two are legitimate $\mathbf{u}$ queries (see Definition 4).

For Case (iv) $\mathcal{B}$ continues the execution of $D^{\tilde{\mathbf{d}}}(SK', C_1, \ldots, C_{n_1})$ in two parallel branches BR_0 and BR_1, where $\mathcal{B}$ replies to qu with b on BR_b. On both branches $\mathcal{B}$ replies to queries for which Cases (i), (ii) and (iii) hold exactly as above. If on some branch $BR_{b'}$, still during the execution of $D^{\tilde{\mathbf{d}}}(SK', C_1, \ldots, C_{n_1})$, for a query qu' Case (iv) holds again (i.e.,

$qu' = \langle \widetilde{\mathbf{d}}, (sk', c) \rangle$ and $(\langle \mathfrak{g}, sk' \rangle, pk) \in \mathsf{Q_s})$ $\mathcal{B}$ replies to qu' with b', making it consistent with the previous reply. Thus, these two branches result in two strings S_0, S_1 satisfying the claim.

By invoking Lemma 1 we deduce that for random $\mathbf{O} = (\mathbf{g}, \mathbf{e}, \mathbf{d})$ and $\mathbf{T}$, any $\mathcal{A}^{\mathbf{O},\mathbf{T}}(pk, c)$ that makes at most, say, $2^{n/5}$ queries (basically any number m of queries where $2^{n/4}/m$ is super-polynomial) has advantage at most $\frac{1}{2} + \frac{1}{2^{n/4}}$ of computing b, where $sk \leftarrow \{0,1\}^n$, $pk = \mathbf{g}(sk)$ and $c \leftarrow \mathbf{e}(pk, b)$.

4.1.2   T Breaks Seed-Circular Security of (G, E, D)

We show

CLAIM A. T is useful if used honestly: For $\mathcal{O} = (\mathbf{g}, \mathbf{e}, \mathbf{d}, \mathbf{u}, \mathbf{w}) \leftarrow \Psi$, $S \leftarrow \{0,1\}^n$, $(SK, PK) = G^{\mathbf{g}}(S)$ and $(C_1, \ldots, C_n) \leftarrow E^{\mathbf{e}}(PK, S)$, the probability that $\mathbf{T}(1^n, PK, C_1, \ldots, C_n)$ does not return S is exponentially small.

To prove CLAIM A, we need the following simple information-theoretic lemma, showing that the probability that an adversary can "forge" a public key is small.

Lemma 2. *Let $\mathcal{B}$ be an oracle adversary, which has access to some Ψ-valid oracle $\mathcal{O} = (\mathbf{g}, \mathbf{e}, \mathbf{d}, \mathbf{u}, \mathbf{w})$, and which on input 1^n makes a list Que of at most 2^n queries and outputs a public key $pk_{out} \in \{0,1\}^{5n}$. It then holds that*

$$\Pr_{\mathcal{O} \leftarrow \Psi}[\mathbf{w}(pk_{out}) = \top \text{ and } (\langle \mathbf{g}, * \rangle, pk_{out}) \notin \mathsf{Que}] \leq \frac{1}{2^{2n}}.$$

Proof of CLAIM A. *Let the variables $\mathbf{g}', S', \mathsf{Q_s}, SK', \widetilde{\mathbf{g}}, \widetilde{\mathbf{d}}$ and S_{out} be sampled as in $\mathbf{T}(1^n, PK, C_1, \ldots, C_n)$. Recall that $(SK', PK) = G^{\mathsf{Q_s}}(1^n, S')$, that $(\widetilde{\mathbf{g}}, \widetilde{\mathbf{d}}) = KeyImpose(\mathbf{g}, \mathbf{e}, \mathbf{d}, \mathsf{Q_s})$ and that $S_{out} = D^{\widetilde{\mathbf{d}}}(SK', C_1 \ldots C_n)$.*

Recall the way $S, PK, C_1, \ldots, C_n$ are chosen in the claim. We first claim $S_{out} = S$. This follows since (a) $(C_1, \ldots, C_n) \leftarrow E^{\mathbf{e}}(PK, S)$, (b) $(\widetilde{\mathbf{g}}, \mathbf{e}, \widetilde{\mathbf{d}})$ is a correct PKE, and (c) $(SK', PK) = G^{\widetilde{\mathbf{g}}}(1^n; S')$, since $\widetilde{\mathbf{g}}$ agrees with $\mathsf{Q_s}$. Thus, by the correctness of (G, E, D), $S_{out} = S$. Thus, $\mathbf{T}(1^n, PK, C_1, \ldots, C_n)$ either returns S or $\bot$.

Let $Fail$ be the event $\mathbf{T}(PK, C_1, \ldots, C_n) = \bot$. We show how to successfully forge a public $pk \in \{0,1\}^{5n}$ whenever $Fail$ holds. By Lemma 2 we will then have $\Pr[Fail] \leq \frac{1}{2^{2n}}$, implying that with probability at least $1 - \frac{1}{2^{2n}}$, $\mathbf{T}(PK, C_1, \ldots, C_n)$ returns S. We first start with some intuition behind the forgery.

Recall that $S_{out} = S$. By definition, $Fail$ occurs if there exists pk s.t. (a) $\mathbf{w}(pk) = \top$, (b) pk is embedded in $\mathsf{Q_s}$ (i.e., $(\langle \mathfrak{g}, * \rangle, pk) \in \mathsf{Q_s}$) and (c) the query/response $(\langle \mathbf{g}, * \rangle, pk)$ does not show up during $G^{\mathbf{g}}(1^n, S)$. Now the forgery is enabled by the fact that $\mathsf{Q_s}$ is produced based on PK in offline mode. (Recall that $(SK, PK) = G^{\mathbf{g}}(1^n, S)$). The only thing left is to prove that the forged pk is indeed in $\{0,1\}^{5n}$. The reason is the following: since $(\langle \mathfrak{g}, * \rangle, pk) \in \mathsf{Q_s}$, the public key pk shows up as the response to a $\mathbf{g}'$ query during $G^{\mathbf{g}'}(1^n, S')$. Recalling that $\mathbf{g}'$ is Ψ-valid (see Step 1 of $\mathbf{T}$'s computation), by Assumption 2 we have $pk \in \{0,1\}^{5n}$. Given this intuition, the forging adversary $\mathcal{A}$ works as follows.

$\mathcal{A}^{\mathcal{O}}(1^n)$ generates $S \leftarrow \{0,1\}^n$ and $(SK, PK) = G^{\mathfrak{g}}(S)$; it then samples a Ψ-valid function $\mathbf{g}'$ and a seed S' in such a way that $G^{\mathfrak{g}'}(1^n, S') = (*, PK)$ and lets Q_s contain the symbolic versions of all query/response pairs made to $\mathbf{g}'$. Denoting by Que the set of all query/response pairs of $\mathcal{A}$ so far (which was populated only during $G^{\mathfrak{g}}(S)$), for all pk s.t. $(\langle \mathfrak{g}, * \rangle, pk) \in \mathsf{Q}_s$ and $(\langle \mathbf{g}, * \rangle, pk) \notin \mathsf{Que}$, $\mathcal{A}$ calls $\langle \mathbf{w}, pk \rangle$: as soon as $\mathcal{A}$ receives $\top$ in response, it returns pk.   □

We can now, using standard techniques, combine the two facts above about $\mathbf{T}$ to rule out fully-blackbox reductions for the construction type considered.

We conclude this subsection with a remark. The separation proved in this subsection will hold even if the candidate construction $\mathcal{E}$ is full length. This can easily be checked, considering nowhere in our analysis do we use the fact that E is a single-bit encryption algorithm. This may briefly be thought of as contradicting the positive construction basing full-length 1-seed circular security on CPA security! However, the catch here is that the positive construction alluded to earlier does not belong in the class of constructions ruled out here, since the constructed E calls the base key-generation algorithm. When discussing the general separation result in Sect. 5 we will point out exactly where our separation fails if the constructed scheme is full-length.

4.2   Bit t-seed-circular Security $\not\Rightarrow$ Full-Length $(t+1)$-seed-circular Security

For simplicity, we show how to separate full-length 2-seed-circular security from bit 1-seed circular security, as this case already captures most of the underlying techniques. Since in this case the seed in the constructed scheme is encrypted as a whole we denote a seed encryption as $C \leftarrow E_{PK}(S)$. Fix the proposed full-length encryption construction (G, E, D), for which we will define a weakening oracle $\mathbf{T}_2$ in such a way that $\mathbf{T}_2$ breaks the 2-seed circular security of $(G^{\mathfrak{g}}, E^{\mathbf{e}}, D^{\mathbf{d}})$, but not the 1-seed circular security of $(\mathbf{g}, \mathbf{e}, \mathbf{d})$.

For this new setting, we cannot use the previous approach, mainly because the analysis there for showing that $\mathbf{T}$ is simulatable by a CCA-valid adversary against $(\mathbf{g}, \mathbf{e}, \mathbf{d})$ (Subsect. 4.1.1) heavily relies on the fact that only one challenge ciphertext is present, whose value can be guessed in two branches. That simulation trick will fail here because an adversary against the bit 1-seed circular security of $(\mathbf{g}, \mathbf{e}, \mathbf{d})$, which will have access to $\mathbf{T}_2$ and which we want to simulate without $\mathbf{T}_2$, is provided with $n + 1$ ciphertexts. Thus, we need some new ideas for the oracle $\mathbf{T}_2$, outlined below. We also use some of the previous ideas.

$\mathbf{T}_2$ accepts inputs of the form (PK_0, PK_1, C_0, C_1), where purportedly C_i, for $i = 0, 1$, is the encryption of PK_{1-i}'s seed under $E(PK_i, \cdot)$. Intuitively, $\mathbf{T}_2$ will decrypts C_0 and C_1 relative to, respectively, oracles $\widetilde{\mathbf{O}_0}$ and $\widetilde{\mathbf{O}_1}$, obtained by superimposing two sampled sets Q_s^0 and Q_s^1, meeting a certain condition, on $\mathbf{O}$. Specifically, $\mathbf{T}_2$ samples two sets of query/response pairs Q_s^0 and Q_s^1 in such a way that for $i = 0, 1$ (a) $G^{\mathsf{Q}_s^i} = (SK_i', PK_i)$ for some SK_i' and (b) the sets of embedded public keys in Q_s^0 and Q_s^1 are disjoint, namely for all pk: if $(\langle \mathfrak{g}, * \rangle, pk) \in \mathsf{Q}_s^0$ then $(\langle \mathfrak{g}, * \rangle, pk) \notin \mathsf{Q}_s^1$. (If such Q_s^0 and Q_s^1 cannot be found,

$\mathbf{T_2}$ returns $\bot$.) Then, $\mathbf{T_2}$ forms $(\widetilde{g}_i, \widetilde{d}_i) = KeyImpose(\mathbf{g}, \mathbf{e}, \mathbf{d}, Q_s^i)$, and $S_{out}^0 = D^{\widetilde{d}_1}(SK_1', C_1)$ and $S_{out}^1 = D^{\widetilde{d}_0}(SK_0', C_0)$. Finally, $\mathbf{T_2}$ returns S_{out}^0 if for both $i = 0, 1$ all embedded public keys in Q_s^i appear during the execution of $G^{\mathbf{g}}(S_{out}^i)$.

The check (b) above is aimed at making $\mathbf{T_2}$ simulatable using $\mathbf{u}$ and $\mathbf{w}$ oracles: namely, to make any 1-seed circular security adversary $\mathcal{A}^{\mathbf{O}, \mathbf{T_2}}(pk, c_1, \ldots, c_n, c)$ against $\mathbf{O} = (\mathbf{g}, \mathbf{e}, \mathbf{d})$ simulatable by CCA-valid adversary $\mathcal{B}^{\mathbf{O}, \mathbf{u}, \mathbf{w}}(pk, c_1, \ldots, c_n, c)$. The main idea behind the simulation is that, for any query $\langle \mathbf{T_2}, (PK_0, PK_1, C_0, C_1) \rangle$ of $\mathcal{A}$, the adversary $\mathcal{B}$ will be able to decrypt at least one of C_i's, specifically the one for which $(\langle \mathfrak{g}, * \rangle, pk) \notin Q_s^i$. This follows by Fact 1. (Recall pk is $\mathcal{B}$'s challenge public key.) If for the other index (i.e., $1 - i$) it holds that $(\langle \mathfrak{g}, * \rangle, pk) \in Q_s^{1-i}$ then as before we can show that either the answer to the underlying $\mathbf{T_2}$ query is $\bot$, or $\mathcal{B}$ will learn its challenge secret key (i.e., $\mathbf{g}^{-1}(pk)$) along the way.

The check (b) however may make the oracle $\mathbf{T_2}$ too weak to break the 2-seed circular security of $(G^{\mathbf{O}}, E^{\mathbf{O}}, D^{\mathbf{O}})$. In particular, if there are pk's that occur w.h.p. as responses to $\mathbf{g}$ queries during a random execution of $G^{\mathbf{g}}(1^n)$, then $\mathbf{T_2}$, even on "honest" inputs, may return $\bot$ too often. To resolve this problem, we first sample a large number of executions of $G^{\mathbf{O}}$, record all the query/response pairs and make Q_s^0 and Q_s^1 be consistent with this information.

We now describe the oracle $\mathbf{T_2}$.

Description of $\mathbf{T_2}$:
Oracles: $(\mathbf{g}, \mathbf{e}, \mathbf{d}, \mathbf{w})$
Input: $(1^n, PK_0, PK_1, C_0, C_1)$

1. **Learning heavy key-generation queries:** Execute $G^{\mathbf{g}}(1^n)$ ϱ times independently at random and record all query/response pairs to Freq. (We instantiate ϱ later.) For any $(\langle \mathbf{g}, * \rangle, pk) \in$ Freq add pk to FreqPub.
2. **Sampling oracles/secret keys consistent with Freq, PK_1 and PK_2.** For $i = 0, 1$:
   - choose $(\mathbf{g}_i', S_i')$ uniformly at random from the set of all pairs satisfying (a) $\mathbf{g}_i'$ is Ψ-valid and is consistent with Freq and (b) $G^{\mathbf{g}_i'}(1^n, S_i') = (*, PK_i)$. (If no such a pair exists return $\bot$.) Let SK_i' be the secret key output by $G^{\mathbf{g}_i'}(1^n, S_i')$.
   - Let Q_s^i contain the symbolic versions of all query/response pairs made in the execution of $G^{\mathbf{g}_i'}(1^n, S_i')$. Define $(\widetilde{g}_i, \widetilde{d}_i) = KeyImpose(\mathbf{g}, \mathbf{e}, \mathbf{d}, Q_s^i)$. Let QPub_i have any pk s.t. $\mathbf{w}(pk) = \top$ and $(\langle \mathfrak{g}, * \rangle, pk) \in Q_s^i$.
3. If $(\mathsf{QPub}_0 \cap \mathsf{QPub}_1) \setminus \mathsf{FreqPub} \neq \emptyset$ then halt and return $\bot$.
4. Compute $S_{out}^1 = D^{\widetilde{d}_0}(SK_0', C_0)$ and $S_{out}^0 = D^{\widetilde{d}_1}(SK_1', C_1)$. Return S_{out}^0 if the following condition holds for both $i = 0, 1$, and return $\bot$, otherwise: For all $pk \in \mathsf{QPub}_i \setminus \mathsf{FreqPub}$ the query/response $(\langle \mathbf{g}, * \rangle, pk)$ is made during the execution of $G^{\mathbf{g}}(S_{out}^i)$.

$\mathbf{T_2}$ does not break 1-seed-circular security of $\mathbf{O}$. We show any adversary $\mathcal{A}^{\mathbf{O}, \mathbf{T_2}}(1^n, pk, c_1, \ldots, c_n, c)$, against 1-seed-circular-security of $\mathbf{O} = (\mathbf{g}, \mathbf{e}, \mathbf{d})$ can be simulated by a CCA-valid adversary $\mathcal{B}^{\mathbf{O},\mathbf{u},\mathbf{w}}(1^n, pk, c_1, \ldots, c_n, c)$ that makes a poly-related number of queries. By Lemma 1 we then obtain our desired result.

The main challenge for $\mathcal{B}$ is to handle $\mathcal{A}$'s $\mathbf{T_2}$ queries. Fix a $\mathbf{T_2}$ query $Tqu = \langle \mathbf{T_2}, (1^{n_1}, PK, C_1, C_2) \rangle$ of $\mathcal{A}$. To reply to Tqu, $\mathcal{B}$ forms FreqPub, $\mathsf{Q_s^0}$, $\mathsf{Q_s^1}$, SK_0' and SK_1' as in $\mathbf{T_2}$'s computation, which $\mathbf{B}$ can perfectly do. Without loss of generality assume $pk \notin$ FreqPub, since otherwise $\mathcal{B}$ has found its challenge secret key. Also, assume for some $i \in \{0,1\}$ $pk \notin$ QPub$_i$ because otherwise by Line 3 the answer to Tqu is $\bot$. In what follows assume $pk \notin$ QPub$_1$. (The same argument goes through if $pk \notin$ QPub$_0$.)

$\mathcal{B}$ forms $S_{out}^0 = D^{\widetilde{\mathbf{d_1}}}(SK_1', C_1)$, where $(\widetilde{\mathbf{g_1}}, \widetilde{\mathbf{d_1}}) = KeyImpose(\mathbf{g}, \mathbf{e}, \mathbf{d}, \mathsf{Q_s^1})$. By Fact 1, $\mathcal{B}$ is perfectly able to run this decryption. Now consider two cases:

1. $pk \notin$ QPub$_0$: in this case again $\mathcal{B}$ can compute $S_{out}^1 = D^{\widetilde{\mathbf{d_0}}}(SK_0', C_0)$, where $(\widetilde{\mathbf{g_0}}, \widetilde{\mathbf{d_0}}) = KeyImpose(\mathbf{g}, \mathbf{e}, \mathbf{d}, \mathsf{Q_s^0})$. Having both S_{out}^0 and S_{out}^1 $\mathcal{B}$ can easily perform the rest of $\mathbf{T_2}$'s computation (which only involves $\mathbf{g}$ queries).

2. $pk \in$ QPub$_0$: in this case by Line 4 of $\mathbf{T_2}$'s computation, either the answer to Tqu is $\bot$ or pk's corresponding secret key turns up during the execution of $G^{\mathbf{O}}(S_{out}^0)$. (Recall that $pk \notin$ FreqPub.) Thus, $\mathbf{B}$ either finds its challenge secret key or finds out that the answer to Tqu is $\bot$.

$\mathbf{T_2}$ **breaks 2-seed-circular security:** For $\mathcal{O} = (\mathbf{g}, \mathbf{e}, \mathbf{d}, \mathbf{u}, \mathbf{w}) \leftarrow \Psi$, $S_i \leftarrow \{0,1\}^n$, $(SK_i, PK_i) = G^{\mathbf{g}}(S_i)$ for $i = 0,1$, and $C_1 \leftarrow E^{\mathbf{e}}(PK_1, S_0)$ and $C_0 \leftarrow E^{\mathbf{e}}(PK_0, S_1)$ we show the probability that $\mathbf{T_2}(1^n, PK_0, PK_1, C_0, C_1)$ does not return S_0 is exponentially small. First, as in the corresponding proof in Subsect. 4.1 we can easily show it is always the case that $S_i = S_{out}^i$ for $i = 0,1$. Thus, the probability that $\mathbf{T_2}(1^n, PK_0, PK_1, C_0, C_1)$ does not return S_0 is the probability that one of the bad events in Lines 3 and 4 of $\mathbf{T_2}$'s computation holds. Let Ev be the event that $\mathbf{T_2}(1^n, PK_0, PK_1, C_0, C_1)$ does not return S_0.

The bad events in Lines 3 and 4 correspond to events $Ev1$ and $Ev2$, defined as follows: $Ev1 = ($QPub$_0 \cap$ QPub$_1) \setminus$ FreqPub $\neq \emptyset$ and

$$Ev2 = (($QPub$_0 \not\subseteq$ RealPub$_0 \cup$ FreqPub$) \vee ($QPub$_1 \not\subseteq$ RealPub$_1 \cup$ FreqPub$)), \quad (2)$$

where RealPub$_i = \{pk \mid$ the query/response $(\langle \mathbf{g}, * \rangle, pk)$ occurs during $G^{\mathbf{g}}(S_i)\}$. Note that for $Ev2$ we use the fact that $S_i = S_{out}^i$. We have $\Pr[Ev] \leq \Pr[Ev2] + \Pr[Ev1 \wedge \overline{Ev2}]$.

First, using the same technique as in Subsect. 4.1.2 we can show $\Pr[Ev2]$ is exponentially small. To bound $\Pr[Ev1 \wedge \overline{Ev2}]$, note whenever $Ev1 \wedge \overline{Ev2}$ happens, the event $Ev3$, defined below, happens:

$$Ev3 = ($RealPub$_0 \cap$ RealPub$_1) \setminus$ FreqPub $\neq \emptyset.$$

Thus, we show how to bound the probability of $Ev3$. That is, the probability that there exists pk such that $pk \in$ RealPub$_0 \cap$ RealPub$_1$, but $pk \notin$ FreqPub. Intuitively, this probability should be small because if $pk \in$ RealPub$_0 \cap$ RealPub$_1$— namely, the query/response $(\langle \mathbf{g}, * \rangle, pk)$ occurs during both $G^{\mathbf{g}}(S_0)$ and $G^{\mathbf{g}}(S_1)$— then $(\langle \mathbf{g}, * \rangle, pk)$ should also occur at least once during the many random executions of $G^{\mathbf{g}}(1^n)$ performed in Step 1 of $\mathbf{T_2}$'s computation, and thus it should be that $pk \in$ FreqPub. Using this line of reasoning we can use Chernoff Bounds to upperbound the probability of $Ev3$ by any arbitrary inverse-polynomial, by instantiating the value of ϱ (Step 1 of $\mathbf{T_2}$'s computation) accordingly.

5   CPA Security $\not\Rightarrow$ Bit 1-seed-circular Security: General

In this section we describe the oracle $\mathbf{T}$ for general bit-encryption constructions of 1-seed circular security, and in the following two sections we prove that this oracle provides a separation.

Intuition. As in the previous section the main idea is to have $\mathbf{T}$, on input $(PK, C_1, \ldots, C_n)$, decrypt $C_1, \ldots, C_n$ relative to some $\widetilde{\mathbf{O}} = (\widetilde{\mathbf{g}}, \widetilde{\mathbf{e}}, \widetilde{\mathbf{d}})$, satisfying (a) $G^{\widetilde{\mathbf{O}}}$ produces $(*, PK)$ and (b) for any b with high probability $E^{\mathbf{O}}(PK, b, R) = E^{\widetilde{\mathbf{O}}}(PK, b, R)$. To obtain $\widetilde{\mathbf{O}}$ we may be tempted to proceed exactly as before, by sampling a set of query/response pairs Q and a seed S' such that $G^Q(S') = (*, PK)$ and then *superimposing* Q (which now has all types of queries) on $\mathbf{O}$. While the resulting $\widetilde{\mathbf{O}}$ satisfies Condition (a) it is not clear if Condition (b) is satisfied: The problem is there may be queries q asked quite frequently during random executions of $E^{\mathbf{O}}(PK, b)$ (call them *heavy*), and which may also occur in Q and receive a different response there. To overcome this problem we first run $E^{\mathbf{O}}(PK, b)$ for $b = 0, 1$ many times and collect all observed query/response pairs in a set Freq. (This is formalized in Definition 6.) We then force the sampled set Q to be *consistent* with Freq. Finally, we show how to superimpose Q on $\mathbf{O}$ to obtain $\widetilde{\mathbf{O}}$.

Setting things up. Fix the proposed construction (G, E, D). We now give an assumption to make our analysis easier and then give definitions formalizing the steps sketched above. We then use these definitions to define the oracle $\mathbf{T}$.

Assumption 3. We assume any oracle algorithm that has access to both $\mathbf{g}$ and $\mathbf{d}$ always queries $\langle \mathbf{g}, sk \rangle$ before querying $\langle \mathbf{d}, (sk, *) \rangle$. Also, we assume w.l.o.g. that G never calls the decryption algorithm of the base scheme, $\mathbf{O} = (\mathbf{g}, \mathbf{e}, \mathbf{d})$. (For $\langle \mathbf{d}, (sk, c) \rangle$: letting $pk = \mathbf{g}(sk)$, either the query/response $(\langle \mathbf{e}, (pk, *, *) \rangle, c)$ was already made in which case G knows the answer, or the answer w.h.p. is $\perp$.) For ease of notation we keep $\mathbf{d}$ as a superscript to G and write $G^{\mathbf{O}}$.

Definition 6. *We define the following probabilistic procedure, $FreqQue$.*

- *Oracles:* $\mathbf{O} = (\mathbf{g}, \mathbf{e}, \mathbf{d}, \mathbf{u})$
- *Input:* *A security parameter* 1^n *(left implicit), public key* PK *and* $p \in \mathbb{N}$.
- *Output:* *A set* Freq *formed as follows. For both* $b = 0, 1$ *run* $E^{\mathbf{O}}(1^n, PK, b)$ *independently p times and add the symbolic versions of all query/response pairs to* Freq. *Moreover, for any* $(\langle \mathfrak{d}, (sk, c) \rangle, *) \in$ Freq *if* $\mathbf{u}(\mathbf{g}(sk), c) = (b', r') \neq \perp$ *add* $(\langle \mathfrak{e}, (pk, b', r') \rangle, c)$ *to* Freq.
  Note that by Assumption 1 $|$Freq$| \leq 2pn^{\vartheta} + 2pn^{\vartheta} = 4pn^{\vartheta}$.

In the above definition apart from the actual observed query/response pairs we also enhanced Freq with some pairs obtained based on $(\langle\mathfrak{d}, (sk, c)\rangle, *)$ query/response pairs. This enhancement is only made to make some of the proofs simpler.

We say that oracle $\mathbf{O} = (\mathbf{g}, \mathbf{e}, \mathbf{d})$ is *consistent* with (or *agrees with*) a symbolic query/response pair $(\langle\mathfrak{g}, sk\rangle, pk)$ if $\mathbf{g}(sk) = pk$. The same definition can be given for other types of query/response pairs. We say $\mathbf{O}$ is consistent with a set of query/response pairs if $\mathbf{O}$ agrees with each element in the set.

Definition 7. *We define the following procedure we call* ConsOrc.

- **Input:** *a public key PK and a set* Freq *of symbolic query/response pairs.*
- **Output:** *a secret key SK' and query/response sets* $\mathsf{Q_s}$, $\mathsf{Q_c}$ *sampled as follows.*
  - *Sample $(\mathbf{g}', \mathbf{e}', \mathbf{d}', S')$ uniformly at random under the constraints that $\mathbf{O}' = (\mathbf{g}', \mathbf{e}', \mathbf{d}')$ is Ψ-valid and is consistent with* Freq *and that $G^{\mathbf{O}'}(S') = (*, PK)$. If no such a tuple exists, return $\perp$.*
  - *Let SK' be the secret-key outputted by $G^{\mathbf{O}'}(S')$ and let the sets $\mathsf{Q_s}$ and $\mathsf{Q_c}$ contain, respectively, the symbolic versions of all query/response pairs made to $\mathbf{g}'$ and $\mathbf{e}'$. (Recall by Assumption 3 no $\mathbf{d}'$-query is made.)*

In Definition 5 we defined the task of superimposing a set of $\mathfrak{g}$ type query/response pairs on an oracle $(\mathbf{g}, \mathbf{e}, \mathbf{d})$. We now define the task of superimposing a set $\mathsf{Q_c}$ of $\mathfrak{e}$ queries on $(\mathbf{g}, \mathbf{e}, \mathbf{d})$: the result will be $(\mathbf{e}_{imp}, \mathbf{d}_{imp})$, perturbed versions of $(\mathbf{e}, \mathbf{d})$. Intuitively, we want $(\mathbf{g}, \mathbf{e}_{imp}, \mathbf{d}_{imp})$ to form a PKE, $\mathbf{e}_{imp}$ to agree with $\mathsf{Q_c}$ and $(\mathbf{e}_{imp}, \mathbf{d}_{imp})$ to agree as much as possible with $(\mathbf{e}, \mathbf{d})$.

Definition 8. *We define the following procedure we call* EncImpose.

- **Input:** *a Ψ-valid $(\mathbf{g}, \mathbf{e}, \mathbf{d})$ and a set*

$$\mathsf{Q_c} = \{(\langle\mathfrak{e}, (pk_1, b_1, r_1)\rangle, c_1), \ldots, (\langle\mathfrak{e}, (pk_p, b_p, r_p)\rangle, c_p)\},$$

Note that pk_i's above are not necessarily distinct.
- **Output:** $(\mathbf{e}_{imp}, \mathbf{d}_{imp})$, *defined as follows. First, let* $\mathsf{W} = \{(pk_1, c_1), \ldots, (pk_p, c_p)\}$ *and* $\mathsf{W}' = \{(pk_1, \mathbf{e}(pk_1, b_1, r_1)), \ldots, (pk_p, \mathbf{e}(pk_p, b_p, r_p))\}$. *Define*

$$\mathbf{e}_{imp}(pk, b, r) = \begin{cases} c_i & \text{if } (pk, b, r) = (pk_i, b_i, r_i), \text{ for some } 1 \le i \le p \\ \hat{c} & \text{if } (pk, \mathbf{e}(pk, b, r)) \in \mathsf{W} \\ \mathbf{e}(pk, b, r) & \text{otherwise} \end{cases} \tag{3}$$

where $\hat{c}$ is defined as follows: Letting x be the smallest integer such that $(pk, \mathbf{e}(pk, b, r + x)) \notin \mathsf{W} \cup \mathsf{W}'$ we set $\hat{c} = \mathbf{e}(pk, b, r + x)$. Here, $r + x$ is done using a standard method.

$$\mathbf{d}_{imp}(sk, c) = \begin{cases} b_i & \text{if } \mathbf{g}(sk) = pk_i \text{ and } c = c_i \text{ for some } 1 \le i \le p \\ \mathbf{d}(sk, c) & \text{otherwise} \end{cases} \tag{4}$$

We justify the second case of $\mathbf{e}_{imp}$'s definition: if $(pk, \mathbf{e}(pk, b, r)) \in \mathsf{W}$, say $(pk, \mathbf{e}(pk, b, r)) = (pk_i, c_i)$, we cannot set $\mathbf{e}_{imp}(pk, b, r) = \mathbf{e}(pk, b, r)$ as we have already set $c_i = \mathbf{e}_{imp}(pk_i, b_i, r_i)$: in particular, $\mathbf{e}_{imp}$ will be rendered incorrect if $b_i \neq b$. Thus, we keep shifting $\mathbf{e}(pk, b, r)$ (by adding x to r) until we hit a ciphertext $\hat{c}$ s.t. $(pk, \hat{c}) \notin \mathsf{W} \cup \mathsf{W}'$. The requirement $(pk, \hat{c}) \notin \mathsf{W}'$ is stronger than necessary, but will simplify some proofs. Note $\mathbf{e}_{imp}$ is not necessarily injective.

Description of T. We define the oracle **T**. We first describe the output distribution of a random **T** on a single input-call, $(1^n, PK, C_1, \ldots, C_n)$, and then describe the underlying distribution from which **T** is chosen.
Oracles: $\mathcal{O} = (\mathbf{g}, \mathbf{e}, \mathbf{d}, \mathbf{u}, \mathbf{w})$. Denote $\mathbf{O} = (\mathbf{g}, \mathbf{e}, \mathbf{d})$.
Input: $(1^n, PK, C_1, \ldots, C_n)$ **Operations:**

1. **Learning frequent queries:** Let $\mathsf{Freq} \leftarrow FreqQue^{\mathbf{O}, \mathbf{u}}(PK, n^{23\vartheta})$. Define $\mathsf{FreqPub}$ to be the set of public keys pk such that $(\langle \mathbf{g}, * \rangle, pk) \in \mathsf{Freq}$.
2. **Sampling oracle/secret-key consistent with PK and Freq:** Sample

$$(SK', \mathsf{Q_s}, \mathsf{Q_c}) \leftarrow ConsOrc(PK, \mathsf{Freq}). \tag{5}$$

3. **Defining intermediate oracles:** Define

$$(\mathbf{e}_{imp}, \mathbf{d}_{imp}) = EncImpose(\mathbf{g}, \mathbf{e}, \mathbf{d}, \mathsf{Q_c})$$
$$(\widetilde{\mathbf{g}}, \widetilde{\mathbf{d}}) = KeyImpose(\mathbf{g}, \mathbf{e}_{imp}, \mathbf{d}_{imp}, \mathsf{Q_s}).$$

Let $\widetilde{\mathbf{e}} = \mathbf{e}_{imp}$, and $\widetilde{\mathbf{O}} = (\widetilde{\mathbf{g}}, \widetilde{\mathbf{e}}, \widetilde{\mathbf{d}})$. Let QPub contain any pk such that $\mathbf{w}(pk) = \top$ and $(\langle \mathbf{g}, * \rangle, pk) \in \mathsf{Q_s}$.
4. **Decrypting the encrypted input:** Compute $S_{out} = D^{\widetilde{\mathbf{O}}}(SK', C_1 \ldots C_n)$.
5. **Returning S_{out} subject to a check:** Run $G^{\mathbf{O}}(S_{out})$ and let $\mathsf{EmbedPub}$ contain any pk such that the query/response $(\langle \mathbf{g}, * \rangle, pk)$ is made during $G^{\mathbf{O}}(S_{out})$. If $\mathsf{QPub} \subseteq \mathsf{EmbedPub} \cup \mathsf{FreqPub}$ return S_{out}; else, return $\bot$.

Notation. $Tvars^{\mathcal{O}}(PK)$ denotes the random variable $(\mathsf{Freq}, SK', \mathsf{Q_s}, \mathsf{Q_c}, \widetilde{\mathbf{O}})$ obtained in the execution of **T** above w.r.t. $\mathcal{O}$ and PK. Note none of these random variables depend on $(C_1, \ldots, C_n)$. For the reader's convenience, we provide a table summary of how all these variables sampled in the last page of the paper.

Remark about T. Note that the only part of the oracle **T** that involves making random choices are Step 1 (sampling from $FreqQue^{\mathbf{O}, \mathbf{u}}(PK, n^{23\vartheta})$) and Step 2 (sampling from $ConsOrc(PK, \mathsf{Freq})$). The number of random coins required to do the sampling in Step 1 is obviously finite. For Step 2 recall that the output of $ConsOrc(PK, \mathsf{Freq})$ is formed based on sampling a Ψ-valid random oracle $\mathbf{O}'$ that is consistent with Freq and also that $G^{\mathbf{O}'}(1^n)$ generates PK (based on some seed). By default, $\mathbf{O}'$ should be defined for all security parameters. However, by Assumption 2 it suffices to sample $\mathbf{O}'$ only for security parameters n. Thus, for any fixed input $(1^n, PK, C_1, \ldots, C_n)$, the amount of randomness used by a random **T** to compute $\mathbf{T}(1^n, PK, C_1, \ldots, C_n)$ is finite.

Sampling space of T. We now explain how to choose a random **T**. In particular, we would like a randomly chosen **T**, if queried under a single input many times, to always return the same output. To this end, every possible **T** comes with a collection of random-coin strings, where for every possible query $qu = (1^n, PK, C_1, \ldots, C_n)$ to **T**, the collection has a corresponding random-coin string $Coin_{qu}$, used by **T** to make the random choices that appear during the computation of $\mathbf{T}(qu)$. When we write $\Pr_\mathbf{T}[]$ we mean the probability is computed over a **T** chosen uniformly at random from the above-mentioned space.

5.1   T Breaks 1-seed-circular Security of (G, E, D)

We show if **T** is called honestly (i.e., on a random public key and a random encryption of the underlying seed) it will return the seed with high probability. To formalize the statement we define the following *environment* that specifies a random choice of $(\mathbf{O}, \mathbf{u}, \mathbf{w})$ plus those underlying an honest random input to **T**.

Environment: $Env(n)$: Output $(\mathbf{O}, \mathbf{u}, \mathbf{w}, S, PK, C_1, \ldots, C_n)$, where:

1. $(\mathbf{g}, \mathbf{e}, \mathbf{d}, \mathbf{u}, \mathbf{w}) \leftarrow \Psi$ and $\mathbf{O} = (\mathbf{g}, \mathbf{e}, \mathbf{d})$;
2. $S \leftarrow \{0,1\}^n$, $(SK, PK) \leftarrow G^\mathbf{O}(S)$ and $(C_1, \ldots, C_n) \leftarrow E^\mathbf{O}(PK, S)$.

Convention. Sometimes that we are interested only in a specific part of the output of $Env(n)$ we may use notation such as $(\mathbf{O}, \mathbf{u}, \mathbf{w}, PK) \leftarrow Env(n)$.

The following theorem shows **T**'s usefulness in breaking seed-circular security.

Theorem 2. *It holds that*

$$\Pr_{\mathbf{Env},\mathbf{T}}[\mathbf{T}(PK, C_1, \ldots, C_n) = S] \geq 1 - \frac{1}{n^5}, \tag{6}$$

where

$$\mathbf{Env} = (\mathbf{O}, \mathbf{u}, \mathbf{w}, S, PK, C_1, \ldots, C_n) \leftarrow Env(n).$$

Proof layout. The proof consists of two parts. First, we show (Lemma 3) that with high probability $S_{out} = S$, where S_{out} is the string decoded in Step 5 of the execution of $\mathbf{T}(PK, C_1, \ldots, C_n)$. Next we show, conditioned on $S_{out} = S$, the probability that $\mathbf{T}(PK, C_1, \ldots, C_n)$ outputs $\perp$ is small (Lemma 4).

Lemma 3. *It holds that*

$$\alpha(n) = \Pr[D^{\tilde{\mathbf{O}}}(SK', C_1 \ldots C_n) \neq S] \leq \frac{1}{2n^5},$$

where the probability is taken over $(\mathbf{O}, \mathbf{u}, \mathbf{w}, S, PK, C_1, \ldots, C_n) \leftarrow Env(n)$ *and* $(\mathsf{Freq}, SK', \mathsf{Q_s}, \mathsf{Q_c}, \tilde{\mathbf{O}}) \leftarrow Tvars^{\mathbf{O},\mathbf{u},\mathbf{w}}(PK)$.

Lemma 4. *It holds that*

$$\alpha(n) \overset{def}{=} \Pr_{\mathbf{Env},\mathbf{T}}\left[\left(D^{\tilde{\mathbf{O}}}(SK', C_1 \ldots C_n) = S\right) \wedge (\mathbf{T}(PK, C_1, \ldots, C_n) = \perp)\right] \leq \frac{1}{2^{2n}},$$

where $\mathbf{Env} = (\mathbf{O}, \mathbf{u}, \mathbf{w}, S, PK, C_1, \ldots, C_n) \leftarrow Env(n)$, *and* SK' *and* $\tilde{\mathbf{O}}$ *are the random variables sampled inside* $\mathbf{T}(PK, C_1, \ldots, C_n)$.

The proof of Theorem 2 follows in a straightforward way by combining Lemmas 3 and 4. We prove Lemma 3 in Subsect. 5.2 and Lemma 4 in Subsect. 5.3.

5.2   Proof of Lemma 3

We start with a simple fact: Informally, it states, in particular, that the string SK' built during an execution of $\mathbf{T}(PK, C_1, \ldots, C_n)$ is a matching secret key of PK relative to $G^{\widetilde{\mathbf{O}}}$.

Fact 3. *For any* $(\mathbf{O}, \mathbf{u}, \mathbf{w}, PK) \in Env(n)$ *and any* $(\mathsf{Freq}, SK', \mathsf{Q_s}, \mathsf{Q_c}, \widetilde{\mathbf{O}}) \in Tvars^{\mathcal{O}}(PK)$, *(a)* $\widetilde{\mathbf{O}}$ *is a correct PKE, and (b)* $(SK', PK) \in G^{\widetilde{\mathbf{O}}}(1^n)$.

Equipped with Fact 3, toward proving Lemma 3 we bound the probability that $E^{\mathbf{O}}(PK, S; R) \neq E^{\widetilde{\mathbf{O}}}(PK, S; R)$, for a random R. If this probability is small then with high probability $D^{\widetilde{\mathbf{O}}}(SK', C_1 \ldots C_n)$ results in S, as desired. We will actually bound a related probability, where S above is replaced with $0^n 1^n$. (Recall that $|S| = n$.) To this end we need the following lemma.

Lemma 5. *Fix* $(\mathbf{O}, \mathbf{u}, \mathbf{w}, PK) \in Env(n)$ *and let* $M = 0^n 1^n$. *Let* $(qu_1, \ldots, qu_{2n^{\vartheta+1}})$ *denote the oracle queries asked during the execution of* $E^{\mathbf{O}}(PK, M; R)$, *for a random R. Then, for any query index $1 \leq i \leq 2n^{\vartheta+1}$*

(A) $\Pr\left[(qu_i \text{ is } \mathbf{g}\text{- or } \mathbf{e}\text{-type}) \wedge \left(\forall j < i, \mathbf{O}(qu_j) = \widetilde{\mathbf{O}}(qu_j)\right) \wedge \left(\mathbf{O}(qu_i) \neq \widetilde{\mathbf{O}}(qu_i)\right)\right] \leq \dfrac{1}{n^{8\vartheta}}$,

(B) $\Pr\left[(qu_i \text{ is } \mathbf{d}\text{-type}) \wedge \left(\forall j < i, \mathbf{O}(qu_j) = \widetilde{\mathbf{O}}(qu_j)\right) \wedge \left(\mathbf{d}(qu_i) \neq \widetilde{\mathbf{d}}(qu_i)\right)\right] \leq \dfrac{1}{n^{8\vartheta}}$,

where $(\mathsf{Freq}, SK', \mathsf{Q_s}, \mathsf{Q_c}, \widetilde{\mathbf{O}}) \leftarrow Tvars^{\mathcal{O}}(PK)$ *and R chosen at random.*

We slightly abused notation above by writing $\widetilde{\mathbf{O}}(qu_j)$, since qu_j is a query to $\mathbf{O}$ (e.g., $qu_j = \langle \mathbf{g}, sk' \rangle$); the meaning, however, should be clear.

We first show how to derive Lemma 3 from Lemma 5.

Proof of Lemma 3. All probabilities that appear below are taken over the choices $(\mathbf{O}, \mathbf{u}, \mathbf{w}, S, PK, C_1, \ldots, C_n) \leftarrow Env(n)$ and $(\mathsf{Freq}, SK', \mathsf{Q_s}, \mathsf{Q_c}, \widetilde{\mathbf{O}}) \leftarrow Tvars^{\mathbf{O}, \mathbf{u}, \mathbf{w}}(PK)$. Let QS be the set of all queries asked during the execution under which $(C_1, \ldots, C_n) \leftarrow E^{\mathbf{O}}(PK, S)$ was produced. We claim

$$\Pr[D^{\widetilde{\mathbf{O}}}(SK', C_1 \ldots C_n) \neq S] \leq \beta(n) \stackrel{\text{def}}{=} \Pr\left[\exists qu \in \mathsf{QS} \colon \mathbf{O}(qu) \neq \widetilde{\mathbf{O}}(qu)\right].$$

The reason is: if the event inside the right-hand side probability does not hold, then $(C_1, \ldots, C_n)$ is also a valid output of $E^{\widetilde{\mathbf{O}}}(PK, S)$. Also, by Fact 3 we know that $(SK', PK) \in G^{\widetilde{\mathbf{O}}}(1^n)$ and that $\widetilde{\mathbf{O}}$ is a correct PKE. Thus, by the correctness of the blackbox construction (G, E, D), we obtain $D^{\widetilde{\mathbf{O}}}(SK', C_1 \ldots C_n) = S$.

Let QS' denote the set of all queries asked during a random execution of $E^{\mathbf{O}}(PK, M)$, where $M = 0^n 1^n$. We claim

$$\beta(n) \leq \beta'(n) \stackrel{\text{def}}{=} \Pr[\exists qu \in \mathsf{QS}' \colon \mathbf{O}(qu) \neq \widetilde{\mathbf{O}}(qu)]. \tag{7}$$

Equation 7 holds because: if S has k 0's, then QS is identically distributed to the set of queries asked during a random execution of $E^{\mathbf{O}}(PK, 0^k 1^{n-k})$. Moreover, since $k \leq n$, the probability that during a random execution of $E^{\mathbf{O}}(PK, 0^k 1^{n-k})$

a query qu, with $\mathbf{O}(qu) \neq \widetilde{\mathbf{O}}(qu)$, is asked is less than the probability that during a random execution of $E^{\mathbf{O}}(PK, M)$ a query qu, with $\mathbf{O}(qu) \neq \widetilde{\mathbf{O}}(qu)$, is asked.

To conclude the proof of Lemma 3 we show $\beta'(n) \leq \frac{1}{2n^5}$. We have

$$\beta'(n) = \Pr\left[\exists qu \in \mathsf{QS}' : \mathbf{O}(qu) \neq \widetilde{\mathbf{O}}(qu)\right] \leq 2n^{\vartheta+1} \times \frac{1}{n^{8\vartheta}} \leq \frac{1}{2n^5};$$

the first inequality is obtained by applying Lemma 5 and a union bound.   □

We now describe the main lemma and tools we need to prove Lemma 5.

Lemma 6. *For any* $(\mathbf{O}, \mathbf{u}, \mathbf{w}, PK) \in Env(n)$ *and* $(\mathsf{Freq}, SK', \mathsf{Q_s}, \mathsf{Q_c}, \widetilde{\mathbf{O}}) \in Tvars^{\mathbf{O}}(PK)$ *all the following hold: (1) for any* $\mathbf{h} \in \{\mathbf{g}, \mathbf{e}\}$ *if* $((\mathfrak{h}, q), ans) \in \mathsf{Freq}$*, then* $\mathbf{h}(q) = \widetilde{\mathbf{h}}(q) = ans$*; (2) if* $\mathbf{g}(sk) \neq \widetilde{\mathbf{g}}(sk)$ *for some sk then* $((\mathfrak{g}, sk), *) \in \mathsf{Q_s}$*; (3) if* $\mathbf{e}(pk, b, r) \neq \widetilde{\mathbf{e}}(pk, b, r)$ *for some* pk, b *and* r *then either (a)* $((\mathfrak{e}, (pk, b, r)), *) \in \mathsf{Q_c}$ *or (b) for some* c*:* $((\mathfrak{e}, (pk, *, *)), c) \in \mathsf{Q_c}$ *and* $\mathbf{u}(pk, c) = (b, r)$*.*

We require the following standard result [32].

Theorem 4. *(A Chernoff-Hoeffding bound) Let* $x_1, \ldots, x_{n^t}$ *be independent boolean random variables all identically distributed to* x*, and suppose* $\Pr[x = 1] = p$*. Then for* $x_{av} = (x_1 + \cdots + x_{n^t})/n^t$

$$\Pr[|x_{av} - p| \geq \frac{1}{n^k}] \leq \frac{1}{2^{2n^{t-2k}}}. \tag{8}$$

We defer the proof of Lemma 5 to the full version.

5.3   Proof of Lemma 4

Proof. Let $\alpha(n)$ be as in the lemma. To bound $\alpha(n)$, suppose $\mathbf{T}(PK, C_1, \ldots, C_n) = \bot$ and $D^{\widetilde{\mathbf{O}}}(SK', C_1 \ldots C_n) = S$. Then by Step 5 of $\mathbf{T}$'s computation it must hold

$$\mathsf{QPub} \not\subseteq \mathsf{EmbedPub} \cup \mathsf{FreqPub}. \tag{9}$$

Thus,

$$\alpha(n) \leq \Pr_{\mathbf{Env}, \mathbf{T}}[\mathsf{QPub} \not\subseteq \mathsf{EmbedPub} \cup \mathsf{FreqPub})], \tag{10}$$

where $\mathbf{Env} = (\mathbf{O}, \mathbf{u}, \mathbf{w}, S, PK, C_1, \ldots, C_n) \leftarrow Env(n)$. We show whenever Eq. 9 holds we can forge a public key in the sense of Lemma 2. Specifically, our forger $\mathcal{B}$, provided with random oracles $(\mathbf{O}, \mathbf{u}, \mathbf{w})$, samples all the variables pertaining to $\mathbf{T}$ by itself and checks whether Eq. 9 holds. Details follow.

The adversary $\mathcal{B}^{\mathbf{O}, \mathbf{u}, \mathbf{w}}(1^n)$ works as follows:

1. $\mathcal{B}$ samples $S \leftarrow \{0, 1\}^n$ and runs $G^{\mathbf{O}}(S)$ to get (SK, PK), and for any query/response $((\mathfrak{g}, *), pk)$ made, it adds pk to $\mathsf{EmbedPub}$.
2. $\mathcal{B}$ samples $\mathsf{Freq} \leftarrow FreqQue^{\mathbf{O}, \mathbf{u}}(PK, n^{23\vartheta})$ and then samples $(SK', \mathsf{Q_s}, \mathsf{Q_c})$ by running $ConsOrc(PK, \mathsf{Freq})$.

3. $\mathcal{B}$ forms $\mathsf{FreqPub} = \{pk \mid (\langle \mathbf{g}, * \rangle, pk) \in \mathsf{Freq}\}$ and $\mathsf{QPub} = \{pk \mid (\langle \mathbf{g}, * \rangle, pk) \in$ $\mathsf{Q_s}$ and $\mathbf{w}(pk) = \top\}$. If there is $pk \in \mathsf{QPub} \setminus (\mathsf{EmbedPub} \cup \mathsf{FreqPub})$, $\mathcal{B}$ returns pk; else it returns $pk \leftarrow \{0,1\}^{5n}$.

Let Que be the set of all query/response pairs that $\mathcal{B}$ makes and note that $|\mathsf{Que}|$ is poly-bounded. To analyze $\mathcal{B}$'s success probability, note that for all pk: $pk \in \mathsf{EmbedPub} \cup \mathsf{FreqPub}$ iff $(\langle \mathbf{g}, * \rangle, pk) \in \mathsf{Que}$. Also, by definition if $pk \in \mathsf{QPub}$ then $\mathbf{w}(pk) = \top$. Thus, from Eq. 10, $\mathcal{B}$'s success probability is at least $\alpha(n)$. Applying Lemma 2 the desired bound for $\alpha(n)$ follows. $\qquad \square$

5.4 T Does Not Break the CPA Security of the Base Scheme

Theorem 5. *Suppose $\mathcal{A}$ is a CPA adversary with access to oracles $\mathbf{O} = (\mathbf{g}, \mathbf{e}, \mathbf{d})$ and $\mathbf{T}$ that makes at most $2^{n/8}$ queries. We have*

$$\Pr_{\mathcal{O}, \mathbf{T}, b, sk, c}[\mathcal{A}^{\mathbf{g},\mathbf{e},\mathbf{d},\mathbf{T}}(1^n, pk, c) = b] \leq \frac{1}{2} + \frac{1}{2^{n/4}}, \tag{11}$$

where $\mathcal{O} = (\mathbf{g}, \mathbf{e}, \mathbf{d}, \mathbf{u}, \mathbf{w}) \leftarrow \Psi$, $b \leftarrow \{0,1\}$, $sk \leftarrow \{0,1\}^n$, $pk = \mathbf{g}(sk)$ and $c \leftarrow \mathbf{e}(pk, b)$.

The following lemma is used in the proof of Theorem 5: it shows how to simulate responses to queries to $\widetilde{\mathbf{O}}$ via oracle access to $(\mathbf{O}, \mathbf{u}, \mathbf{w})$.

Lemma 7. *Fix $(\mathbf{g}, \mathbf{e}, \mathbf{d}, \mathbf{u}, \mathbf{w}, PK) \in Env(n)$ and $(\mathsf{Freq}, SK', \mathsf{Q_s}, \mathsf{Q_c}, \widetilde{\mathbf{g}}, \widetilde{\mathbf{e}}, \widetilde{\mathbf{d}}) \in Tvars^{\mathcal{O}}(PK)$. Assuming $\mathsf{Q_c} = \{(\langle \mathbf{e}, (pk_1, b_1, r_1)\rangle, c_1), \dots, (\langle \mathbf{e}, (pk_p, b_p, r_p)\rangle, c_p)\}$ let $W = \{(pk_i, c_i) \colon 1 \leq i \leq p\}$ and $W' = \{(pk_i, \mathbf{e}(pk_i, b_i, r_i)) \colon 1 \leq i \leq p\}$. Then*

(a) *Both $\widetilde{\mathbf{g}}$ and $\widetilde{\mathbf{e}}$ can be computed efficiently (on all points) given access to oracles $\mathbf{O} = (\mathbf{g}, \mathbf{e}, \mathbf{d})$ and having $\mathsf{Q_s}$ and $\mathsf{Q_c}$ as input.*
(b) *For any (sk, c), if $(\langle \mathbf{g}, sk \rangle, *) \notin \mathsf{Q_s}$, the value $\widetilde{\mathbf{d}}(sk, c)$ can be efficiently computed given access to oracle $\mathbf{O}$ and having $\mathsf{Q_c}$ as input.*
(c) *If for (sk, c) it holds that $(\langle \mathbf{g}, sk \rangle, pk) \in \mathsf{Q_s}$ for some pk and that $(pk, c) \notin W \cup W'$, then $\widetilde{\mathbf{d}}(sk, c)$ can be determined as follows: if $\mathbf{u}(pk, c) = (b, *) \neq \bot$ then $\widetilde{\mathbf{d}}(sk, c) = b$; otherwise, $\widetilde{\mathbf{d}}(sk, c) = \bot$.*
(d) *If for (sk, c) it holds that $(\langle \mathbf{g}, sk \rangle, pk) \in \mathsf{Q_s}$, for some pk, and that $(pk, c) \in W' \setminus W$, then $\widetilde{\mathbf{d}}(sk, c) = \bot$.*
(e) *If for (sk, c) it holds that $(\langle \mathbf{g}, sk \rangle, pk) \in \mathsf{Q_s}$, for some pk, and that $(pk, c) = (pk_i, c_i)$ for some $i \leq p$, then $\widetilde{\mathbf{d}}(sk, c) = b_i$.*

Proof sketch of Theorem 5. As in Sect. 4 the idea is to give an adversary $\mathcal{B}$, where $\mathcal{B}^{\mathcal{O}}(1^n, pk, c)$ can simulate responses to $\mathbf{T}$ queries of $\mathcal{A}^{\mathbf{O},\mathbf{T}}(1^n, pk, c)$, without calling $\langle \mathbf{u}, (pk, c) \rangle$. Let $Tqu = \langle \mathbf{T}, (1^n, PK, C_1, \dots, C_n) \rangle$ be an $\mathcal{A}$'s query. As per $\mathbf{T}$'s computation, $\mathcal{B}$ first samples $\mathsf{Freq} \leftarrow FreqQue^{\mathbf{O},\mathbf{u}}(PK, n^{23\vartheta})$. This may seem problematic since this step involves making $\mathbf{u}$ queries. By inspecting Definition 6, however, we can see for any query $\langle \mathbf{u}, (pk', *) \rangle$ that needs to be made, $\mathcal{B}$ already knows $\mathbf{g}^{-1}(pk')$. Finally, let $\mathsf{FreqPub}$ contain any pk' such that

$(\langle \mathfrak{g}, * \rangle, pk') \in \mathsf{Freq}$, and assume w.l.o.g. $pk \notin \mathsf{FreqPub}$ (because otherwise $\mathcal{B}$ has already found $\mathbf{g}^{-1}(pk)$.)

Next, $\mathcal{B}$ samples $(SK', \mathsf{Q_s}, \mathsf{Q_c}, \widetilde{\mathbf{O}})$ as in $\mathbf{T}$'s execution and starts simulating $D^{\widetilde{\mathbf{O}}}(SK', C_1 \ldots, C_n)$. Again the idea is to see if $(\langle \mathfrak{g}, * \rangle, pk) \in \mathsf{Q_s}$ or not. If $(\langle \mathfrak{g}, * \rangle, pk) \notin \mathsf{Q_s}$: by Lemma 7 we can see $\mathcal{B}$ can handle all $\widetilde{\mathbf{O}}$ queries. In particular, $\mathcal{B}$ will never need to call $\langle \mathbf{u}, (pk, *) \rangle$. After the decryption $\mathcal{B}$ performs Step 5 of $\mathbf{T}$'s computation, which $\mathcal{B}$ can efficiently do, since no $\mathbf{u}$ queries are involved.

If, however, $(\langle \mathfrak{g}, * \rangle, pk) \in \mathsf{Q_s}$, assuming $S_{out} = D^{\widetilde{\mathbf{O}}}(SK', C_1 \ldots, C_n)$, since $pk \notin \mathsf{FreqPub}$, the answer to Tqu is $\bot$ unless $(\langle \mathbf{g}, * \rangle, pk)$ occurs during $G^{\mathbf{O}}(S_{out})$. Now as before the idea is to show $\mathcal{B}$ can find S_0 and S_1 s.t. $S_{out} \in \{S_0, S_1\}$. To this end $\mathcal{B}$ starts simulating $D^{\widetilde{\mathbf{O}}}(SK', C_1 \ldots, C_n)$. By Lemma 7 all $\widetilde{\mathbf{g}}$ and $\widetilde{\mathbf{e}}$ queries can be handled. Let the sets W and W' be formed based on $\mathsf{Q_c}$ as in Lemma 7. For $\widetilde{\mathbf{d}}$ queries: $\mathcal{B}$ will be unable to simulate the answer to a query $qu = \widetilde{\mathbf{d}}(sk', c')$ only if $(\langle \mathfrak{g}, sk' \rangle, pk) \in \mathsf{Q_s}$, $c' = c$ and $(pk, c) \notin \mathsf{W} \cup \mathsf{W}'$ (Case (c) of Lemma 7): in this case, knowing that the answer is the challenge bit b, $\mathcal{B}$ starts two branches of simulation, where it replies to qu with 0 on one branch and with 1 on the other. As before, we need to make sure $\mathcal{B}$ provides a consistent reply on either branch if the same query shows up in the future. The two strings decoded on the two branches at the end satisfy the above claim.   □

5.5   Putting All Together

We may now use our two main established results to obtain our main result.

Theorem 6. *There exists no fully-blackbox construction of 1-seed-circular-secure bit-encryption schemes from CPA-secure encryption schemes.*

5.6   Extensions of the Separation Result

We briefly discuss why our separation holds even if E is a $(c \log n)$-bit encryption scheme and where our separation fails if E is allowed to be full length.

Remark 1. We first sketch how to adjust $\mathbf{T}$ to make our separation work for the case that (G, E, D) is an η-bit PKE, for $\eta = c \log n$. To this end, we need to change Definition 6 (i.e., the procedure $FreqQue$), so that instead of encrypting 0 and 1 many times (as in the bit encryption case), it encrypts all messages $m \in \{0, 1\}^{\eta}$, each many times. The total number of queries still remains polynomial. The description of the oracle $\mathbf{T}$ remains unchanged except that in the first step we call this new version of $FreqQue$. We can now prove the exact same version as Lemma 5, by changing M to have n copies of each string $z \in \{0, 1\}^{\eta}$ (instead of having n copies of 0 and n of 1 as in the bit-encryption case). Now the proofs of the rest of the lemmas that lead up to Theorem 2, as well as the proof of Theorem 5, remain unchanged.

Finally, note that the above extension heavily relies on the fact that the message size is $O(\log n)$, and so it does not apply if the constructed scheme is full-length, as expected.

6 Bit t-seed-circular Security $\not\Rightarrow$ Full-Length $(t+1)$-seed-circular Security

In this section we present our results for separating full-length $(t+1)$-seed-circular security from bit t-seed-circular security. To this end we define a weakening oracle $\mathbf{T_{t+1}}$, for a fixed candidate construction (G, E, D), generalizing a similar oracle given in Subsect. 4.2. Throughout this section note that (G, E, D) has the same plaintext and seed space. Mots of the tools underlying $\mathbf{T_{t+1}}$ have been presented before, but we need the following extension of Definition 6.

Definition 9. *We define the following probabilistic procedure, ExtFreqQue.*

- *Oracles:* $\mathbf{O} = (\mathbf{g}, \mathbf{e}, \mathbf{d}, \mathbf{u})$
- *Input: A security parameter 1^n (left implicit), public key PK and $p \in \mathbb{N}$.*
- *Output: A set Freq formed as follows.*
  - *Do the following independently p times and add the symbolic versions of all query/response pairs to Freq: Sample $S \leftarrow \{0,1\}^n$, and run $G^{\mathbf{O}}(S)$ and $E^{\mathbf{O}}(1^n, PK, S)$.*
  - *Finally, for any $(\langle \mathfrak{d}, (sk, c)\rangle, *) \in \mathsf{Freq}$ if $\mathbf{u}(\mathbf{g}(sk), c) = (b, r) \neq \perp$ add $(\langle \mathfrak{e}, (pk, b, r)\rangle, c)$ to Freq.*

Remark 2. Throughout the remaining sections we continue to use Assumption 1. In particular, since our focus right now is on schemes (G, E, D) with plaintext space $\{0,1\}^n$ (i.e., the same as the seed space) we assume that E on any plaintext $m \in \{0,1\}^n$ makes exactly n^ϑ queries.

Description of $\mathbf{T}_{t+1}$: We present the oracle $\mathbf{T}_{t+1}$. This new oracle shares many aspects with the oracle $\mathbf{T}$, and so we leave out details whenever appropriate.
Notation. Let t_0 be such that $t \leq n^{t_0}$.
Oracles: $\mathcal{O} = (\mathbf{g}, \mathbf{e}, \mathbf{d}, \mathbf{u}, \mathbf{w})$. Denote $\mathbf{O} = (\mathbf{g}, \mathbf{e}, \mathbf{d})$.
Input: $(1^n, PK_1, \ldots, PK_{t+1}, C_1, \ldots, C_{t+1})$

1. **Learning heavy queries:** For $i \leq t+1$ let $\mathsf{Freq}_i \leftarrow ExtFreqQue^{\mathbf{O}, \mathbf{u}}$ $(PK_i, n^{23\vartheta + 4t_0})$, and let $\mathsf{FreqPub}_i$ be the set of public keys pk s.t. $(\langle \mathfrak{g}, *\rangle, pk) \in \mathsf{Freq}_i$.
2. **Sampling consistent oracles/secret-keys:** For $i \leq t+1$ sample

$$(\widetilde{SK}_i, \mathsf{Q}_s^i, \mathsf{Q}_c^i) \leftarrow ConsOrc(PK_i, \mathsf{Freq}_i), \tag{12}$$

   and let QPub_i contain any pk such that $\mathbf{w}(pk) = \top$ and $(\langle \mathfrak{g}, *\rangle, pk) \in \mathsf{Q}_s^i$.
3. If for some distinct $i, j \in [t+1]$ $(\mathsf{QPub}_i \cap \mathsf{QPub}_j) \setminus (\mathsf{FreqPub}_i \cup \mathsf{FreqPub}_j) \neq \emptyset$ halt and return $\perp$.
4. **Defining intermediate oracles:** For $i \leq t+1$ define

$$(\mathbf{e}_{imp,i}, \mathbf{d}_{imp,i}) = EncImpose(\mathbf{g}, \mathbf{e}, \mathbf{d}, \mathsf{Q}_c^i)$$
$$(\widetilde{\mathbf{g}}_i, \widetilde{\mathbf{d}}_i) = KeyImpose(\mathbf{g}, \mathbf{e}_{imp,i}, \mathbf{d}_{imp,i}, \mathsf{Q}_s^i).$$

Let $\widetilde{\mathbf{e}}_i = \mathbf{e}_{imp,i}$, and $\widetilde{\mathbf{O}}_i = (\widetilde{\mathbf{g}}_i, \widetilde{\mathbf{e}}_i, \widetilde{\mathbf{d}}_i)$.

5. **Decrypting the ciphertexts:** Set $S_{out}^1 = D^{\widetilde{\mathbf{O}_{t+1}}}(\widetilde{SK_{t+1}}, C_{t+1})$ and for $2 \leq i \leq t+1$ set $S_{out}^i = D^{\widetilde{\mathbf{O}_{i-1}}}(\widetilde{SK_{i-1}}, C_{i-1})$.

6. **Forming the output.** For $i \leq t+1$ run $G^{\mathbf{O}}(S_{out}^i)$ and let $\mathsf{EmbedPub}_i$ contain any pk such that the query/response $(\langle \mathbf{g}, * \rangle, pk)$ is made during the execution. If for all $i \leq t+1$, $\mathsf{QPub}_i \subseteq \mathsf{EmbedPub}_i \cup \mathsf{FreqPub}_i$, then return S_{out}^1; otherwise, return $\bot$.

To state the main results we define the following environment, specifying a random choice of $(\mathbf{O}, \mathbf{u}, \mathbf{w})$ plus those underlying an honest input to $\mathbf{T}_{t+1}$.

Environment: $Env_t(n)$: Output

$$(\mathbf{O}, \mathbf{u}, \mathbf{w}, S_1, \ldots, S_t, PK_1, \ldots, PK_t, E^{\mathbf{O}}(PK_1, S_2), \ldots, E^{\mathbf{O}}(PK_t, S_1)),$$

where $(\mathbf{g}, \mathbf{e}, \mathbf{d}, \mathbf{u}, \mathbf{w}) \leftarrow \Psi$, $\mathbf{O} = (\mathbf{g}, \mathbf{e}, \mathbf{d})$, $S_i \leftarrow \{0,1\}^n$ and $(SK_i, PK_i) = G^{\mathbf{O}}(S_i)$, for $1 \leq i \leq t$.

We now state the two main results leading to our claimed separation.

Theorem 7. *It holds that*

$$\Pr_{\mathbf{Env}, \mathbf{T}_{t+1}} [\mathbf{T}_{t+1}(PK_1, \ldots, PK_{t+1}, C_1, \ldots, C_{t+1}) = S_1] \geq 1 - \frac{1}{n^5}, \qquad (13)$$

where

$$\mathbf{Env} = (\mathbf{O}, \mathbf{u}, \mathbf{w}, S_1, \ldots, S_{t+1}, PK_1, \ldots, PK_{t+1}, C_1, \ldots, C_{t+1}) \leftarrow Env_{t+1}(n).$$

Theorem 8. *Suppose $\mathcal{A}$ is a t-seed circular security adversary with access to oracles $\mathbf{O} = (\mathbf{g}, \mathbf{e}, \mathbf{d})$ and $\mathbf{T}$ that makes at most $2^{n/8}$ queries. We have*

$$\Pr_{\mathcal{O}, \mathbf{T}, b, sk_1, \ldots, sk_t} [\mathcal{A}^{\mathbf{O}, \mathbf{T}}(1^n, pk_1, \ldots, pk_t, \mathbf{e}(pk_1, sk_2), \ldots, \mathbf{e}(pk_t, sk_1), \mathbf{e}(pk_1, b)) = b] \leq \frac{1}{2} + \frac{1}{2^{n/4}},$$

where $\mathcal{O} = (\mathbf{O}, \mathbf{u}, \mathbf{w}) \leftarrow \Psi$, $b \leftarrow \{0,1\}$, $sk_1, \ldots, sk_t \leftarrow \{0,1\}^n$ and $pk_i = \mathbf{g}(sk_i)$ for $i \leq t$.

By combining the above two theorems we have the following.

Theorem 9. *There exists no fully-blackbox construction of full-length $(t+1)$-seed-circular secure encryption from t-seed-circular secure bit encryption.*

7   Constructions Based on Circular Security

We show how our results on seed-circular security extend to rule out a class of constructions for circular security that we call *key-isolating constructions*. To define this class we first define it in a related model we call the *canonical model*, and then we define it in the standard model. We start with some definitions.

Canonical-Form (CF) PKE. We call $\mathbf{O} = (\mathbf{gs}, \mathbf{gp}, \mathbf{e}, \mathbf{d})$ a CF PKE if the domain of $\mathbf{gp}$ (excluding 1^n) is the range of $\mathbf{gs}$ and $(\mathbf{g}, \mathbf{e}, \mathbf{d})$, where

$\mathbf{g}(s) = (\mathbf{gs}(s), \mathbf{gp}(\mathbf{gs}(s)))$, is a PKE. That is, the key-generation algorithm of a CF scheme first deterministically maps a seed to a secret key, and then *deterministically* maps the secret key to a public key.

CF-based blackbox model. A blackbox construction in the CF model is a tuple of oracle algorithms (GS, GP, E, D) s.t. for any CF PKE $\mathbf{O} = (\mathbf{gs}, \mathbf{gp}, \mathbf{e}, \mathbf{d})$, $(GS^{\mathbf{O}}, GP^{\mathbf{O}}, E^{\mathbf{O}}, D^{\mathbf{O}})$ is a CF PKE. Proving a syntactically-unrestricted impossibility result in the CF model implies one in the standard model, since any CPA-secure CF PKE can be turned into a CPA-secure standard PKE and that any circular-secure standard PKE can be put into a circular-secure CF PKE.

CF Key-isolating constructions. We call (GS, GP, E, D) *key-isolating* if GS never calls $\mathbf{gp}$ of the base scheme, i.e., GS only has access to $(\mathbf{gs}, \mathbf{e}, \mathbf{d})$.

Ruling-out key-isolating constructions. Our earlier results extend to rule out CF key-isolating constructions for circular security. To do this, we first need to change the distribution of Ψ, by replacing $\mathbf{g}$ with $(\mathbf{gs}, \mathbf{gp})$, for $\mathbf{gs}_n \colon \{0,1\}^n \to \{0,1\}^{3n}$ and $\mathbf{gp}_n \colon \{0,1\}^{3n} \to \{0,1\}^{5n}$. As for $\mathbf{T}$, which now takes as input a public key and an encryption of a PK's secret key, all we need to change is that in Step 5 of $\mathbf{T}$'s description the set EmbedPub should be formed by executing $GP^{\mathbf{O}}$ on the intermediate, decrypted string (which is now a secret key). All our proofs about $\mathbf{T}$ not breaking the semantic security of the base scheme go through with only making obvious modifications. The proofs about $\mathbf{T}$ being helpful in breaking the circular security of the constructed scheme follow by noting that all access to $\mathbf{gp}$ during key generation is only made by GP. This fact only becomes essential in the proof of Lemma 4, and is the reason behind the above way of defining EmbedPub. Other lemmas follow by making only obvious changes.

Interpretation w.r.t. standard constructions. Our above result also rules out *standard key-isolating* constructions. To define this notion for a standard construction (G, E, D) we first need to slightly change the standard model so that (G, E, D) takes as oracles a CF PKE. Again, as explained above this is w.l.o.g. Now we call $\mathcal{E} = (G, E, D)$ a *standard key-isolating construction* if $\mathcal{E}$ admits a key-isolating CF *counterpart* in the following sense: there exists algorithms GS and GP s.t. (GS, GP, E, D) is key-isolating and (GS, GP) induces the same distribution as G, i.e., for any $\mathbf{O} = (\mathbf{gs}, \mathbf{gp}, \mathbf{e}, \mathbf{d})$ it holds that (SK, PK) is identically distributed to (SK', PK'), where $(SK, PK) \leftarrow G^{\mathbf{O}}(1^n)$, $SK' \leftarrow GS^{\mathbf{gs}, \mathbf{e}, \mathbf{d}}(1^n)$ and $PK' = GP^{\mathbf{O}}(SK')$. Now the impossibility of CF key-isolating constructions extends to standard ones, by how the notion counterpart is defined.

Examples. Any standard construction $\mathcal{E} = (G, E, D)$ under which seeds and secret keys are the same is key-isolating: defining $GS(S) = S$ and $GP^{\mathbf{gs}, \mathbf{gp}, \mathbf{e}, \mathbf{d}}(S) = G_2^{\mathbf{gs}, \mathbf{gp}, \mathbf{e}, \mathbf{d}}(S)$, where G_2 is the algorithm corresponding to the public-key output of G, the construction (GS, GP, E, D) is the CF-counterpart of (G, E, D) and is key-isolating since GS makes no oracle calls at all. The class of key-isolating constructions is larger than this; we only wanted to give a concrete example.

We leave a more comprehensive and formal discussion to the full version.

8 Discussion Related to Impossibility for Circular Security

In this section we briefly explain why we were not able to fully extend our results to the circular-security case. For simplicity, we highlight the difficulties encountered w.r.t. the simple type of constructions discussed in Sect. 4. In what follows all mentions of the oracle $\mathbf{T}$ for the seed-circular-security case refers to the oracle $\mathbf{T}$ defined in Sect. 4.

As discussed previously, the main challenge in designing an appropriate Oracle $\mathbf{T}$ is to make sure that responses to queries to $\mathbf{T}$ do not leak information about the challenge secrets of a CPA adversary $\mathcal{A}$ against $\mathbf{O} = (\mathbf{g}, \mathbf{e}, \mathbf{d})$. We proved this for the seed-circular-security case by providing a CCA2 adversary $\mathcal{B}$ in such a way that $\mathcal{B}^{\mathbf{O},\mathbf{u},\mathbf{w}}(pk, c)$ is able to simulate $\mathcal{A}^{\mathbf{O},\mathbf{T}}(pk, c)$.

Roughly speaking, the only part of the execution of $\mathbf{T}(PK, C_1, \dots, C_n)$ that is not simulatable by a CCA2-adversary $\mathcal{B}^{\mathbf{O},\mathbf{u},\mathbf{w}}(pk, c)$ is when during the computation of $D^{\tilde{\mathbf{d}}}(SK', C_1 \dots C_n)$ a query $\tilde{\mathbf{d}}(sk', c)$ shows up and $(\langle \mathbf{g}, sk' \rangle, pk) \in \mathsf{Q_s}$. We fixed this non-simulatability problem by adding an extra check at the end of $\mathbf{T}$'s computation that ensures the following: either the value of $\mathbf{g}^{-1}(pk)$ is *embedded* in S_{out} (i.e., the query/response pair $(\langle \mathbf{g}, * \rangle, pk)$ shows up during $G^{\mathbf{g}}(S_{out})$) or the answer to the underlying $\mathbf{T}$ query is $\perp$.

To define a circular-security weakening oracle $\mathbf{T}$ we may be tempted to proceed as before: $\mathbf{T}$ accepts inputs of the form $(PK, C_1, \dots, C_n)$, where now $C_1, \dots, C_n$ are (supposedly) bit-wise encryption of a PK's secret key under PK itself. (For simplicity, assume that the length of the secret key is n.) Then, everything remain unchanged, as in $\mathbf{T}$ in Sect. 4, until $\mathbf{T}$ obtains $S_{out} = D^{\tilde{\mathbf{d}}}(SK', C_1 \dots C_n)$, which now is supposedly a PK's matching secret key. Now in order to make sure that the oracle $\mathbf{T}$ is simulatable (i.e., it does not leak non-simulatable information to a CPA adversary against $\mathbf{O}$) it seems that, as before, we need to make sure that $\mathbf{g}^{-1}(\mathsf{Q_{pub}})$ is "embedded" in S_{out}, before releasing S_{out}. (Recall the definition of $\mathsf{Q_{pub}}$ from $\mathbf{T}$'s definition in Sect. 4.) But this "embedding condition" seems hard to check. This check was easy for the seed-circular-security case since we can simply run $G^{\mathbf{g}}(S_{out})$ and monitor all sk' for which we observe a query/response $(\langle \mathbf{g}, sk' \rangle, *)$. For the circular-security case one idea is to run $D^{\mathbf{d}}(S_{out}, \cdot)$ on many random encryptions produced as $C \leftarrow E^{\mathbf{e}}(PK, b; R)$ for randomly chosen b and R, and record in a set $\mathsf{EmbedSec}$ all sk' for which we encounter a query $\langle \mathbf{d}, (sk', *) \rangle$. We then return S_{out} if $\mathsf{Q_{pub}} \subseteq \mathbf{g}(\mathsf{EmbedSec})$; otherwise, we return $\perp$. While this check makes the oracle $\mathbf{T}$ simulatable, it makes $\mathbf{T}$ unfortunately too weak in that we cannot anymore guarantee in general that $\mathbf{T}(PK, C_1 \dots C_n)$ will return SK with non-negligible probability, for $(SK, PK) \leftarrow G^{\mathbf{g}}(1^n)$ and $(C_1, \dots, C_n) \leftarrow E^{\mathbf{e}}(PK, SK)$, i.e., $\mathbf{T}$ is not useful in general for breaking circular security. (Contrived constructions (G, E, D) for which this is the case can be given.)

Acknowledgements. We would like to thank Mohammad Mahmoody for useful conversations in an early stage of this work.

References

1. Alamati, N., Peikert, C.: Three's compromised too: circular insecurity for any cycle length from (Ring-)LWE. In: Robshaw and Katz [37], pp. 659–680
2. Applebaum, B.: Key-dependent message security: generic amplification and completeness. J. Cryptol. **27**(3), 429–451 (2014)
3. Applebaum, B., Cash, D., Peikert, C., Sahai, A.: Fast cryptographic primitives and circular-secure encryption based on hard learning problems. In: Halevi, S. (ed.) CRYPTO 2009. LNCS, vol. 5677, pp. 595–618. Springer, Heidelberg (2009). doi:10.1007/978-3-642-03356-8_35
4. Asharov, G., Segev, G.: Limits on the power of indistinguishability obfuscation and functional encryption. In: Guruswami, V. (ed.) FOCS 2015, pp. 191–209. IEEE Computer Society (2015)
5. Baecher, P., Brzuska, C., Fischlin, M.: Notions of black-box reductions, revisited. In: Sako, K., Sarkar, P. (eds.) ASIACRYPT 2013. LNCS, vol. 8269, pp. 296–315. Springer, Heidelberg (2013). doi:10.1007/978-3-642-42033-7_16
6. Bishop, A., Hohenberger, S., Waters, B.: New circular security counterexamples from decision linear and learning with errors. In: Iwata, T., Cheon, J.H. (eds.) ASIACRYPT 2015. LNCS, vol. 9453, pp. 776–800. Springer, Heidelberg (2015). doi:10.1007/978-3-662-48800-3_32
7. Boneh, D., Halevi, S., Hamburg, M., Ostrovsky, R.: Circular-secure encryption from decision Diffie-Hellman. In: Wagner, D. (ed.) CRYPTO 2008. LNCS, vol. 5157, pp. 108–125. Springer, Heidelberg (2008). doi:10.1007/978-3-540-85174-5_7
8. Boneh, D., Papakonstantinou, P.A., Rackoff, C., Vahlis, Y., Waters, B.: On the impossibility of basing identity based encryption on trapdoor permutations. In: FOCS 2008, pp. 283–292. IEEE Computer Society (2008)
9. Brakerski, Z., Goldwasser, S.: Circular and leakage resilient public-key encryption under subgroup indistinguishability. In: Rabin, T. (ed.) CRYPTO 2010. LNCS, vol. 6223, pp. 1–20. Springer, Heidelberg (2010). doi:10.1007/978-3-642-14623-7_1
10. Brakerski, Z., Katz, J., Segev, G., Yerukhimovich, A.: Limits on the power of zero-knowledge proofs in cryptographic constructions. In: Ishai, Y. (ed.) TCC 2011. LNCS, vol. 6597, pp. 559–578. Springer, Heidelberg (2011). doi:10.1007/978-3-642-19571-6_34
11. Canetti, R., Lin, H., Tessaro, S., Vaikuntanathan, V.: Obfuscation of probabilistic circuits and applications. In: Dodis, Y., Nielsen, J.B. (eds.) TCC 2015. LNCS, vol. 9015, pp. 468–497. Springer, Heidelberg (2015). doi:10.1007/978-3-662-46497-7_19
12. Cash, D., Green, M., Hohenberger, S.: New definitions and separations for circular security. In: Fischlin, M., Buchmann, J., Manulis, M. (eds.) PKC 2012. LNCS, vol. 7293, pp. 540–557. Springer, Heidelberg (2012). doi:10.1007/978-3-642-30057-8_32
13. Cheng, C.-M., Chung, K.-M., Persiano, G., Yang, B.-Y. (eds.): PKC 2016, (II). LNCS, vol. 9615. Springer, Heidelberg (2016)
14. Choi, S.G., Dachman-Soled, D., Malkin, T., Wee, H.: Black-box construction of a non-malleable encryption scheme from any semantically secure one. In: Canetti, R. (ed.) TCC 2008. LNCS, vol. 4948, pp. 427–444. Springer, Heidelberg (2008). doi:10.1007/978-3-540-78524-8_24
15. Dolev, D., Dwork, C., Naor, M.: Non-malleable cryptography (extended abstract). In: STOC 1991, pp. 542–552 (1991)
16. Gentry, C.: Fully homomorphic encryption using ideal lattices. In: Mitzenmacher, M. (ed.) STOC 2009, pp. 169–178. ACM (2009)

17. Gertner, Y., Kannan, S., Malkin, T., Reingold, O., Viswanathan, M.: The relationship between public key encryption and oblivious transfer. In: FOCS 2000, pp. 325–335. IEEE Computer Society (2000)
18. Gertner, Y., Malkin, T., Myers, S.: Towards a separation of semantic and CCA security for public key encryption. In: Vadhan, S.P. (ed.) TCC 2007. LNCS, vol. 4392, pp. 434–455. Springer, Heidelberg (2007). doi:10.1007/978-3-540-70936-7_24
19. Gertner, Y., Malkin, T., Reingold, O.: On the impossibility of basing trapdoor functions on trapdoor predicates. In: FOCS 2001, pp. 126–135. IEEE Computer Society (2001)
20. Goldreich, O., Micali, S., Wigderson, A.: How to play any mental game. In: Proceedings of the Nineteenth Annual ACM Symposium on Theory of Computing, pp. 218–229. ACM (1987)
21. Haitner, I., Holenstein, T.: On the (Im)Possibility of key dependent encryption. In: Reingold, O. (ed.) TCC 2009. LNCS, vol. 5444, pp. 202–219. Springer, Heidelberg (2009). doi:10.1007/978-3-642-00457-5_13
22. Hajiabadi, M., Kapron, B.M.: Reproducible circularly-secure bit encryption: applications and realizations. In: Gennaro, R., Robshaw, M. (eds.) CRYPTO 2015. LNCS, vol. 9215, pp. 224–243. Springer, Heidelberg (2015). doi:10.1007/978-3-662-47989-6_11
23. Hajiabadi, M., Kapron, B.M., Srinivasan, V.: On generic constructions of circularly-secure, leakage-resilient public-key encryption schemes. In: Cheng et al.[13], pp. 129–158
24. Hsiao, C.-Y., Reyzin, L.: Finding collisions on a public road, or do secure hash functions need secret coins? In: Franklin, M. (ed.) CRYPTO 2004. LNCS, vol. 3152, pp. 92–105. Springer, Heidelberg (2004). doi:10.1007/978-3-540-28628-8_6
25. Impagliazzo, R., Rudich, S.: Limits on the provable consequences of one-way permutations. In: Johnson, D.S. (ed.) STOC 1989, pp. 44–61. ACM (1989)
26. Ishai, Y., Kushilevitz, E., Lindell, Y., Petrank, E.: Black-box constructions for secure computation. In: Proceedings of the Thirty-Eighth Annual ACM Symposium on Theory of Computing, pp. 99–108. ACM (2006)
27. Koppula, V., Waters, B.: Circular security separations for arbitrary length cycles from LWE. In: Robshaw and Katz [37], pp. 681–700
28. Mahmoody, M., Mohammed, A.: On the power of hierarchical identity-based encryption. In: Fischlin, M., Coron, J.-S. (eds.) EUROCRYPT 2016. LNCS, vol. 9666, pp. 243–272. Springer, Heidelberg (2016). doi:10.1007/978-3-662-49896-5_9
29. Mahmoody, M., Pass, R.: The curious case of non-interactive commitments – on the power of black-box vs. non-black-box use of primitives. In: Safavi-Naini, R., Canetti, R. (eds.) CRYPTO 2012. LNCS, vol. 7417, pp. 701–718. Springer, Heidelberg (2012). doi:10.1007/978-3-642-32009-5_41
30. Malkin, T., Teranishi, I., Yung, M.: Efficient circuit-size independent public key encryption with KDM security. In: Paterson, K.G. (ed.) EUROCRYPT 2011. LNCS, vol. 6632, pp. 507–526. Springer, Heidelberg (2011). doi:10.1007/978-3-642-20465-4_28
31. Marcedone, A., Pass, R., Shelat, A.: Bounded KDM security from iO and OWF. In: Zikas, V., Prisco, R. (eds.) SCN 2016. LNCS, vol. 9841, pp. 571–586. Springer, Cham (2016). doi:10.1007/978-3-319-44618-9_30
32. Motwani, R., Raghavan, P.: Randomized Algorithms. Cambridge University Press, New York (1995)
33. Myers, S., Shelat, A.: Bit encryption is complete. In: Foundations of Computer Science, 2009, FOCS 2009, pp. 607–616. IEEE (2009)

34. Naor, M., Yung, M.: Public-key cryptosystems provably secure against chosen ciphertext attacks. In: Proceedings of the Twenty-Second Annual ACM Symposium on Theory of Computing, pp. 427–437. ACM (1990)

35. Pass, R., Shelat, A., Vaikuntanathan, V.: Construction of a non-malleable encryption scheme from any semantically secure one. In: Dwork, C. (ed.) CRYPTO 2006. LNCS, vol. 4117, pp. 271–289. Springer, Heidelberg (2006). doi:10.1007/11818175_16

36. Reingold, O., Trevisan, L., Vadhan, S.: Notions of reducibility between cryptographic primitives. In: Naor, M. (ed.) TCC 2004. LNCS, vol. 2951, pp. 1–20. Springer, Heidelberg (2004). doi:10.1007/978-3-540-24638-1_1

37. Robshaw, M., Katz, J. (eds.): CRYPTO 2016 (II). LNCS, vol. 9815. Springer, Heidelberg (2016)

38. Rosen, A., Segev, G.: Chosen-ciphertext security via correlated products. SIAM J. Comput. **39**(7), 3058–3088 (2010)

39. Rothblum, R.D.: On the circular security of bit-encryption. In: Sahai, A. (ed.) TCC 2013. LNCS, vol. 7785, pp. 579–598. Springer, Heidelberg (2013). doi:10.1007/978-3-642-36594-2_32

40. Simon, D.R.: Finding collisions on a one-way street: can secure hash functions be based on general assumptions? In: Nyberg, K. (ed.) EUROCRYPT 1998. LNCS, vol. 1403, pp. 334–345. Springer, Heidelberg (1998). doi:10.1007/BFb0054137

41. Vahlis, Y.: Two is a crowd? a black-box separation of one-wayness and security under correlated inputs. In: Micciancio, D. (ed.) TCC 2010. LNCS, vol. 5978, pp. 165–182. Springer, Heidelberg (2010). doi:10.1007/978-3-642-11799-2_11

42. Dijk, M., Gentry, C., Halevi, S., Vaikuntanathan, V.: Fully homomorphic encryption over the integers. In: Gilbert, H. (ed.) EUROCRYPT 2010. LNCS, vol. 6110, pp. 24–43. Springer, Heidelberg (2010). doi:10.1007/978-3-642-13190-5_2

43. Wee, H.: KDM-security via homomorphic smooth projective hashing. In: Cheng et al. [13], pp. 159–179

A Note on Perfect Correctness
by Derandomization

Nir Bitansky$^{(\boxtimes)}$ and Vinod Vaikuntanathan

MIT, Cambridge, USA
nbitansky@gmail.com, vinodv@csail.mit.edu

Abstract. We show a general compiler that transforms a large class of
erroneous cryptographic schemes (such as public-key encryption, indistinguishability obfuscation, and secure multiparty computation schemes)
into perfectly correct ones. The transformation works for schemes that
are *correct on all inputs with probability noticeably larger than half*, and
are secure under parallel repetition. We assume the existence of one-way
functions and of functions with deterministic (uniform) time complexity
$2^{O(n)}$ and non-deterministic circuit complexity $2^{\Omega(n)}$.

Our transformation complements previous results that showed how
public-key encryption and indistinguishability obfuscation that err on a
noticeable fraction of inputs can be turned into ones that *for all inputs*
are often correct.

The technique relies on the idea of "reverse randomization" (Naor,
Crypto 1989) and on Nisan-Wigderson style derandomization, previously used in cryptography to remove interaction from witness-
indistinguishable proofs and commitment schemes (Barak, Ong and Vadhan, Crypto 2003).

1 Introduction

Randomized algorithms are often faster and simpler than their state-of-the-art
deterministic counterparts, yet, by their very nature, they are error-prone. This
gap has motivated a rich study of *derandomization*, where a central avenue has
been the design of *pseudo-random generators* [BM84, Yao82a, NW94] that could
offer one universal solution for the problem. This has led to surprising results,
intertwining cryptography and complexity theory, and culminating in a derandomization of **BPP** under worst-case complexity assumptions, namely, the exis-
tence of functions in $\mathbf{E} = \mathbf{Dtime}(2^{O(n)})$ with *worst-case* circuit complexity $2^{\Omega(n)}$
[NW94, IW97].

For cryptographic algorithms, the picture is somewhat more subtle. Indeed,
in cryptography, randomness is almost always *necessary* to guarantee any sense
of security. While many cryptographic schemes are *perfectly correct* even if randomized, some do make errors. For example, in some encryption algorithms,

Research supported in part by NSF Grants CNS-1350619 and CNS-1414119, Alfred
P. Sloan Research Fellowship, Microsoft Faculty Fellowship, the NEC Corporation,
and a Steven and Renee Finn Career Development Chair from MIT.

J.-S. Coron and J.B. Nielsen (Eds.): EUROCRYPT 2017, Part II, LNCS 10211, pp. 592–606, 2017.
DOI: 10.1007/978-3-319-56614-6_20

notably the lattice-based ones [AD97, Reg05], most but not all ciphertexts can be decrypted correctly. Here, however, we cannot resort to general derandomization, as a (completely) derandomized version will most likely be totally insecure.

It gets worse. While for general algorithms infrequent errors are tolerable in practice, for cryptographic algorithms, errors can be (and have been) exploited by adversaries (see [BDL01] and a long line of followup works). Thus, the question of eliminating errors is ever more important in the cryptographic context. This question was addressed in a handful of special contexts in cryptography. In the context of interactive proofs, [GMS87, FGM+89] show how to turn any interactive proof into one with perfect completeness. In the context of encryption schemes, Goldreich, Goldwasser, and Halevi [GGH97] showed how to *partially* eliminate errors from lattice-based encryption schemes [AD97, Reg05]. Subsequent works, starting from that of Dwork, Naor and Reingold [DNR04a], show how to *partially* eliminate errors from *any* encryption scheme [HR05, LT13]. Here, "partial" refers to the fact that they eliminate errors from the encryption and decryption algorithms, but not the key generation algorithm. That is, in their final *immunized* encryption scheme, it could still be the case that there are bad keys that always cause decryption errors. In the context of indistinguishability obfuscation (IO), Bitansky and Vaikuntanathan [BV16] recently showed how to partially eliminate errors from any IO scheme: namely, they show how to convert any IO scheme that might err on a fraction of the inputs into one that is correct on all inputs, with high probability over the coins of the obfuscator.

This Work. We show how to *completely immunize* a large class of cryptographic algorithms, turning them into algorithms that make no errors at all. Our most general result concerns cryptographic algorithms (or protocols) that are "secure under parallel repetition". We show:

Theorem 1.1 (Informal). *Assume that one-way functions exist and functions with deterministic (uniform) time complexity $2^{O(n)}$ and non-deterministic circuit complexity $2^{\Omega(n)}$ exist. Then, any encryption scheme, indistinguishability obfuscation scheme, and multiparty computation protocol that is secure under parallel repetition can be completely immunized against errors.*

More precisely, we show that perfect correctness is guaranteed when the transformed scheme or protocol are executed honestly. The security of the transformed scheme or protocol is inherited from the security of the original scheme under parallel repetition. In the default setting of encryption and obfuscation schemes, encryption and obfuscation are always done honestly, and security under parallel repetition is well known to be guaranteed automatically. Accordingly, we obtain the natural notion of perfectly-correct encryption and obfuscation. In contrast, in the setting of MPC, corrupted parties may in general affect any part of the computation. In particular, in the case of corrupted parties, the transformed protocol does not provide a better correctness guarantee, but only the same correctness guarantee as the original (repeated) protocol. We find that perfect correctness is a natural requirement and the ability to generically achieve it for a large class of cryptographic schemes is aesthetically appealing.

In addition, while in many applications almost perfect correctness may be sufficient, some applications do require *perfectly correct* cryptographic schemes. For example, using public-key encryption as a commitment scheme requires perfect correctness, the construction of non-interactive witness-indistinguishable proofs in [BP15] requires a perfectly correct indistinguishability obfuscation, and the construction of 3-message zero knowledge against uniform verifiers [BCPR14], requires perfectly correct delegation schemes.

Our tools, perhaps unsurprisingly given the above discussion, come from the area of derandomization, in particular we make heavy use of Nisan-Wigderson (NW) type pseudorandom generators. Such NW-generators were previously used by Barak, Ong and Vadhan [BOV07] to remove interaction from commitment schemes and ZAPs. We use it here for a different purpose, namely to immunize cryptographic algorithms from errors. Below, we elaborate on the similarities and differences.

1.1   The Basic Idea

We briefly explain the basic idea behind the transformation, focusing on the case of public-key encryption. Imagine that we have an encryption scheme given by randomized key-generation and encryption algorithms, and a deterministic decryption algorithm (Gen, Enc, Dec), where for any message $m \in \{0,1\}^n$, there is a tiny decryption error:

$$\Pr_{(r_g, r_e) \leftarrow \{0,1\}^{\mathrm{poly}(n)}} [\mathsf{Dec}_{sk}(\mathsf{Enc}_{pk}(m; r_e)) \neq m \mid (pk, sk) = \mathsf{Gen}(r_g)] \leq 2^{-n} .$$

Can we deterministically choose "good randomness" (r_g, r_e) that leads to correct decryption? This question indeed seems analogous to the question of derandomizing **BPP**. There, the problem can be solved using Nisan-Wigderson type pseudo-random generators [NW94]. Such generators can produce a poly(n)-long pseudo-random string using a short random seed of length $d(n) = O(\log n)$. They are designed to fool distinguishers of some prescribed polynomial size $t(n)$, and may run in time $2^{O(d)} \gg t$. Derandomization of the **BPP** algorithm is then simply done by enumerating over all $2^d = n^{O(1)}$ seeds and taking the majority.

We can try to use NW-type generators to solve our problem in a similar way. However, the resulting scheme wouldn't be secure – indeed, it will be *deterministic*, which means it cannot be semantically secure [GM84]. To get around this, we use the idea of reverse randomization from [Lau83, Nao91, DN07, DNR04a]. For each possible seed $i \in \{0,1\}^d$ for the NW-generator NWPRG, we derive corresponding randomness

$$(r_e^i, r_g^i) = \mathsf{NWPRG}(i) \oplus \left(\mathsf{BMYPRG}(s_e^i), \mathsf{BMYPRG}(s_g^i)\right) .$$

Here BMYPRG is a Blum-Micali-Yao (a.k.a cryptographic) pseudo-random generator [BM82, Yao82b], and the seeds $(s_g^i, s_e^i) \in \{0,1\}^\ell$ are chosen independently for every i, with the sole restriction that their image is sparse enough (say, they are of total length $\ell = n/2$). Encryption and decryption for any given message are now done in parallel with respect to all 2^d copies of the original scheme, where the final result of decryption is defined to be the majority of the 2^d decrypted messages.

Security is now guaranteed by the BMY-type generators and the fact that public-key encryption can be securely performed in parallel. Crucially, the pseudo-randomness of BMY strings is guaranteed despite the fact that their image forms a sparse set. The fact that the set of BMY string is sparse will be used to the perfect correctness of the scheme. In particular, when shifted at random, this set will evade the (tiny) set of "bad randomness" (that lead to decryption errors) with high probability $1 - 2^{\ell-n} \geq 1 - 2^{-n/2}$.

In the actual construction, the image is not shifted truly at random, but rather by an NW-pseudo-random string, and we would like to argue that this suffices to get the desired correctness. To argue that NW-pseudo-randomness is enough, we need to show that with high enough probability (say 0.51) over the choice of the NW string, the shifted image of the BMY generator still evades "bad randomness". This last property may not be efficiently testable deterministically, but can be tested non-deterministically in fixed polynomial time, by guessing the seeds for the BMY generator that would lead to bad randomness. We accordingly rely on NW generators that fool non-deterministic circuits. Such pseudo-random generators are known under the worst case assumption that there exist functions in $\mathbf{E}$ with non-deterministic circuit complexity $2^{\Omega(n)}$ [SU01].

Relation to [BOV07]. Barak, Ong, and Vadhan were the first to demonstrate how NW-type derandomization can be useful in cryptography. They showed how NW generators can be used to derandomize Naor's commitments [Nao91] and Dwork and Naor's ZAPs [DN07]. In the applications they examined, "reverse randomization" is already encapsulated in the constructions of ZAPs and commitments that they start from, and they show that "the random shift" can be derandomized, using the fact that ZAPs and commitments are secure under parallel repetition.

There, they were not interested in the correctness of a specific computation *per se*, but rather in the *existence* of an "incorrect object", namely an accepting proof for a false statement in ZAPs, or a commitment with inconsistent openings. Another difference is that in the applications they consider, it is in fact enough to use hitting set generators (against co-non-determinism) rather than pseudorandom generators. Intuitively, the reason is that in these applications there is one-sided error. For example, in a ZAP system, one already assumes that true statements are always accepted by the verifier, so when derandomizing they only need to recognize false statements. This is analogous to having an encryption system that is always correct on encryptions of zero, but may make mistakes on encryptions of one.

Organization. In Sect. 2, we give the required preliminaries. Section 3 presents the transformation itself. In Sect. 4, we discuss several examples of interest where the transformation can be applied.

2 Preliminaries

In this section, we give the required preliminaries, including standard computational concepts, cryptographic schemes and protocols, and the derandomization tools that we use.

2.1 Standard Computational Concepts

We recall standard computational concepts concerning Turing machines and Boolean circuits.

- By *algorithm* we mean a uniform Turing machine. We say that an algorithm is PPT if it is probabilistic and polynomial time.
- A polynomial-size circuit family $\mathcal{C}$ is a sequence of circuits $\mathcal{C} = \{C_\lambda\}_{\lambda \in \mathbb{N}}$, such that each circuit C_λ is of polynomial size $\lambda^{O(1)}$ and has $\lambda^{O(1)}$ input and output bits.
- We follow the standard habit of modeling any efficient adversary strategy $\mathcal{A}$ as a family of polynomial-size circuits. For an adversary $\mathcal{A}$ corresponding to a family of polynomial-size circuits $\{\mathcal{A}_\lambda\}_{\lambda \in \mathbb{N}}$, we often omit the subscript λ, when it is clear from the context. For simplicity, we shall simply call such an adversary a *polynomial-size adversary*.
- We say that a function $f : \mathbb{N} \to \mathbb{R}$ is negligible if it decays asymptotically faster than any polynomial.
- Two ensembles of random variables $\mathcal{X} = \{X_\lambda\}_{\lambda \in \mathbb{N}}$ and $\mathcal{Y} = \{Y_\lambda\}_{\lambda \in \mathbb{N}}$ are said to be computationally indistinguishable, denoted by $\mathcal{X} \approx_c \mathcal{Y}$, if for all polynomial-size distinguishers $\mathcal{D}$, there exists a negligible function ν such that for all λ,

$$|\Pr[\mathcal{D}(X_\lambda) = 1] - \Pr[\mathcal{D}(Y_\lambda) = 1]| \leq \nu(\lambda).$$

2.2 Cryptographic Schemes and Protocols

We consider a simple model of cryptographic schemes and protocols that will allow to describe the transformation generally. In Sect. 4, we give several examples of such schemes and protocols.

Executions: Let λ be a security parameter and let $m = m(\lambda), n = n(\lambda), \ell = \ell(\lambda)$ be polynomially-bounded functions. An (honest) execution of an m-party scheme (or protocol) Π involves interaction between m PPT parties with inputs $(x_1, \ldots, x_m) \in \{0,1\}^{n \times m}$ and randomness $(r_1, \ldots, r_m) \in \{0,1\}^{\ell \times m}$, at the end of which they each produce outputs $(y_1, \ldots, y_m) \in \{0,1\}^{n \times m}$. Abstracting out, we will think of Π as a single PPT process that runs in some fixed polynomial time and denote it by $y \leftarrow \Pi(1^\lambda, x, r)$, where $x = (x_1, \ldots, x_m), y = (y_1, \ldots, y_m)$, and $r = (r_1, \ldots, r_m)$.

Definition 2.1 ($(1 - \alpha)$-Correctness). *Let $f : \{0,1\}^{n \times m} \to \{0,1\}^{n \times m}$ be a polynomial-time computable function. Π computes f $(1 - \alpha)$-correctly if for any λ and any $x \in \{0,1\}^{n \times m}$,*

$$\Pr_{r \leftarrow \{0,1\}^{\ell \times m}} \left[y \neq f(x) \mid y \leftarrow \Pi(1^\lambda, x, r) \right] \leq \alpha(\lambda).$$

Repeated Executions: For a function $k = k(\lambda)$, inputs $x = (x_1, \ldots, x_m) \in \{0,1\}^{n \times m}$ and randomness $r = (r_{ij})_{i \in [m], j \in [k]}$, and $r_{i,j} \in \{0,1\}^\ell$, the repeated execution $y \leftarrow \Pi_{\otimes k}(1^\lambda, x, r)$ consists of executing $\Pi(1^\lambda, x, r_1), \ldots, \Pi(1^\lambda, x, r_k)$, where $r_j = (r_{1j}, \ldots, r_{mj})$, in parallel and obtaining the corresponding outputs, namely, $y = (y_{ij})_{i \in [m], j \in [k]}$.

2.3 NW and BMY PRGs

We now define the basic tools required for the main transformation — NW-type PRGs [NW94] and BMY-type PRGs [BM82, Yao82b]. The transformation itself is given in the next section.

Definition 2.2 (Nondeterministic Circuits). *A nondeterministic boolean circuit $C(x, w)$ takes x as a primary input and w as a witness. We define $C(x) := 1$ if and only if there exists w such that $C(x, w) = 1$.*

Definition 2.3 (NW-Type PRGs against Nondeterministic Circuits). *An algorithm $\mathsf{NWPRG} : \{0,1\}^{d(n)} \to \{0,1\}^n$ is an NW-generator against non-deterministic circuits of size $t(n)$ if it is computable in time $2^{O(d(n))}$ and any non-deterministic circuit C of size at most $t(n)$ distinguishes $U \leftarrow \{0,1\}^n$ from $\mathsf{NWPRG}(s)$, where $s \leftarrow \{0,1\}^{d(n)}$, with advantage at most $1/t(n)$.*

We shall rely on the following theorem by Shaltiel and Umans [SU01] regarding the existence NW-type PRGs as above assuming worst-case hardness for non-deterministic circuits.

Theorem 2.4 ([SU01]). *Assume there exists a function $f : \{0,1\}^n \to \{0,1\}$ in $\mathbf{E} = \mathbf{Dtime}(2^{O(n)})$ with nondeterministic circuit complexity $2^{\Omega(n)}$. Then, for any polynomial $t(\cdot)$, there exists an NW-generator against non-deterministic circuits of size $t(n)$ $\mathsf{NWPRG} : \{0,1\}^{d(n)} \to \{0,1\}^n$, where $d(n) = O(\log n)$.*

We remark that the above is a worst-case assumption in the sense that the function f needs to be hard in the worst-case (and not necessarily in the average-case). The assumption can be seen as a natural generalization of the assumption that $\mathbf{EXP} \not\subseteq \mathbf{NP}$. We also note that there is a universal candidate for the corresponding PRG, by instantiating the hard function with any $\mathbf{E}$-complete language under linear reductions. See further discussion in [BOV07].

We now define BMY-type (a.k.a cryptographic) PRGs.

Definition 2.5 (BMY-Type PRGs). *An algorithm $\mathsf{BMYPRG} : \{0,1\}^{d(n)} \to \{0,1\}^n$ is a BMY-generator if it is computable in time $\mathrm{poly}(d(n))$ and any polynomial-size adversary distinguishes $U \leftarrow \{0,1\}^n$ from $\mathsf{BMYPRG}(n)$, where $s \leftarrow \{0,1\}^{d(n)}$, with negligible advantage $n^{-\omega(1)}$.*

Theorem 2.6 ([HILL99]). *BMY-type pseudo-random generators can be constructed from any one-way function.*

3 The Error-Removing Transformation

We now describe a transformation from any $(1 - \alpha)$-correct scheme Π for a function f into a perfectly correct one. For a simpler exposition, we restrict attention to the case that the error α is tiny. We later explain how this restriction can be removed.

Ingredients. In the following, let λ be a security parameter, let $m = m(\lambda), n = n(\lambda), \ell = \ell(\lambda)$ be polynomials, and $\alpha = \alpha(\lambda) \leq 2^{-\lambda m - 2}$. We rely on the following:

- A $(1 - \alpha)$-correct scheme Π computing $f : \{0,1\}^{n \times m} \to \{0,1\}^{n \times m}$ where each party uses randomness of length ℓ.
- A BMY-type pseudo-random generator $\mathsf{BMYPRG} : \{0,1\}^\lambda \to \{0,1\}^\ell$.
- An NW-type pseudo-random generator $\mathsf{NWPRG} : \{0,1\}^d \to \{0,1\}^{\ell \times m}$ against nondeterministic circuits of size $t = t(\lambda)$, where t and d depend on the parameters $m, n, \ell, \Pi, f, \mathsf{BMYPRG}$, $t = \lambda^{O(1)}$, $d(\lambda) = O(\log \lambda)$, and will be specified later on. We shall denote $k = 2^d$.

The New Scheme:

Given security parameter 1^λ and input $x \in \{0,1\}^{n \times m}$:

1. **Randomness Generation:** Each party $i \in [m]$
    - samples k BMY strings $(r_{i1}^{\mathsf{BMY}}, \ldots, r_{ik}^{\mathsf{BMY}})$, where $r_{ij}^{\mathsf{BMY}} = \mathsf{BMYPRG}(s_{ij})$ and $s_{ij} \leftarrow \{0,1\}^\lambda$.
    - computes (all) k NW strings $(r_1^{\mathsf{NW}}, \ldots, r_k^{\mathsf{NW}})$, where $r_j^{\mathsf{NW}} = \mathsf{NWPRG}(j)$, and derives $(r_{i1}^{\mathsf{NW}}, \ldots, r_{ik}^{\mathsf{NW}})$, where r_{ij} is the i-th ℓ-bit block of r_j^{NW}.
    - compute $r_{i1}, \ldots, r_{ik}$ where $r_{ij} = r_{ij}^{\mathsf{BMY}} \oplus r_{ij}^{\mathsf{NW}}$.
2. **Emulating the Parallel Scheme:**
    - the parties emulate the repeated scheme $\Pi_{\otimes k}(1^\lambda, x, r)$, with randomness $r = (r_{ij})_{i \in [m], j \in [k]}$.
    - each party i obtains outputs $(y_{i1}, \ldots, y_{ik})$, and in turn computes and outputs $y_i = \mathsf{majority}(y_{i1}, \ldots, y_{ik})$.

Correctness. We now turn to show that the new scheme is perfectly correct.

Proposition 3.1. *The new scheme is perfectly-correct.*

Proof. We first note that had r^{NW} been chosen at truly random (instead of using NWPRG) then for any input, with high probability over the choice of r^{NW}, the corresponding scheme would have been perfectly correct.

Claim. For any $x \in \{0,1\}^{n \times m}$,

$$\Pr_{r^{\mathsf{NW}} \leftarrow \{0,1\}^{\ell \times m}} \left[\exists s_1, \ldots, s_m \in \{0,1\}^{\lambda} : \begin{matrix} f(x) \neq \Pi(1^{\lambda}, x, r) \\ r = r_s^{\mathsf{BMY}} \oplus r^{\mathsf{NW}} \end{matrix} \right] \leq \frac{1}{4},$$

where $r_s^{\mathsf{BMY}} = (\mathsf{BMYPRG}(s_1), \ldots, \mathsf{BMYPRG}(s_m))$.

Proof. Fixing any such x and $s = (s_1, \ldots, s_m)$, the string $r = r_s^{\mathsf{BMY}} \oplus r^{\mathsf{NW}}$ is distributed uniformly at random. In this case, the scheme is guaranteed to err with probability at most $\alpha \leq 2^{-\lambda m}/4$. The claim now follows by taking a union bound over all $2^{\lambda m}$ tuples $s_1, \ldots, s_m$. $\qquad\square$

We now claim that a similar property holds with roughly the same probability when r^{NW} is pseudorandom as in the actual transformation.

Claim. For any $x \in \{0,1\}^{n \times m}$,

$$\Pr_{j \leftarrow \{0,1\}^d} \left[\exists s_1, \ldots, s_m \in \{0,1\}^{\lambda} : \begin{matrix} f(x) \neq \Pi(1^{\lambda}, x, r) \\ r = r_s^{\mathsf{BMY}} \oplus r_j^{\mathsf{NW}} \end{matrix} \right] \leq \frac{1}{4} + \frac{1}{t},$$

where $r_s^{\mathsf{BMY}} = (\mathsf{BMYPRG}(s_1), \ldots, \mathsf{BMYPRG}(s_m))$ and $r_j^{\mathsf{NW}} = \mathsf{NWPRG}(j)$.

Proof. Assume towards contradiction that the claim does not hold for some $x \in \{0,1\}^{n \times m}$. We construct a non-deterministic distinguisher that breaks NWPRG. The distinguisher, given r^{NW}, non-deterministically guesses $s_1, \ldots, s_m$, computes $r^{\mathsf{BMY}} = (\mathsf{BMYPRG}(s_1), \ldots, \mathsf{BMYPRG}(s_m))$, $r = r^{\mathsf{NW}} \oplus r^{\mathsf{BMY}}$, and checks whether $f(x) \neq \Pi(1^{\lambda}, x, r)$. As we just proved in the previous claim, when r^{NW} is truly random, such a witness $s_1, \ldots, s_m$ exists with probability at most $1/4$, whereas, by our assumption towards contradiction, when r^{NW} is pseudo-random such a witness exists with probability larger than $\frac{1}{t} + \frac{1}{4}$.

The size of the above distinguisher is some fixed polynomial $t'(\lambda)$ that depends only on m, n, ℓ and the time required to compute Π, f, BMYPRG. Thus, in the construction we choose $t > \max(t', 8)$, meaning that the constructed distinguisher indeed breaks NWPRG. $\qquad\square$

With the last claim, we now conclude the proof of Proposition 3.1. Indeed, for any input x, when emulating the k-fold repetition $\Pi_{\otimes k}(1^{\lambda}, x, r)$, the randomness used for the j-th copy $\Pi(1^{\lambda}, x, r_j)$ is $r_j = r_j^{\mathsf{NW}} \oplus r_{s_j}^{\mathsf{BMY}}$ where $r_j^{\mathsf{NW}} = \mathsf{NWPRG}(j)$ and $r_{s_j}^{\mathsf{BMY}} = (\mathsf{BMYPRG}(s_{j1}), \ldots, \mathsf{BMYPRG}(s_{j1}))$. By the last claim, for all but a $\frac{1}{4} + \frac{1}{t} \leq \frac{3}{8}$ fraction of the NW-seeds j, any choice of BMY-seeds s_j yields the correct result $y_j = f(x)$ in the corresponding execution $\Pi(1^{\lambda}, x, r_j)$. In particular, it is always the case that the majority of executions results in $y = f(x)$, as required. $\qquad\square$

Security. We now observe that the randomness generated according to the transformation is indistinguishable from real randomness. Intuitively, this means that if the original scheme was secure under parallel-repetition, when the honest parties use real randomness, it will remain as secure when using randomness generated according to the transformation. Examples are given in the next section.

Concretely, we consider two distributions r^{tra} and r^{uni} on randomness for the parties in $\Pi_{\otimes k}$:

1. In $r^{\mathsf{tra}} = \left(r_{ij}^{\mathsf{tra}} : i \in [m], j \in [k]\right)$, each r_{ij}^{tra} is computed as in the above transformation; namely $r_{ij} = r_{ij}^{\mathsf{BMY}} \oplus r_{ij}^{\mathsf{NW}}$, where $r_{i,j}^{\mathsf{BMY}} = \mathsf{BMYPRG}(s_{ij})$ for a random seed $s_{ij} \leftarrow \{0,1\}^\lambda$ and r_{ij}^{NW} is the i-th ℓ-bit block of $\mathsf{NWPRG}(j)$.
2. In $r^{\mathsf{uni}} = \left(r_{ij}^{\mathsf{uni}} : i \in [m], j \in [k]\right)$, each r_{ij}^{uni} is sampled uniformly at random; namely $r_{ij}^{\mathsf{uni}} \leftarrow \{0,1\}^\ell$.

Proposition 3.2. *r^{tra} and r^{uni} are computationally indistinguishable.*

Proof. By the security of the BMY PRG, for any i, j:

$$r_{ij}^{\mathsf{tra}} = r_{ij}^{\mathsf{BMY}} \oplus r_{ij}^{\mathsf{NW}} = \mathsf{BMYPRG}(s_{ij}) \oplus r_{ij}^{\mathsf{NW}} \approx_c r_{ij}^{\mathsf{uni}} \oplus r_{ij}^{\mathsf{NW}} \equiv r_{ij}^{\mathsf{uni}}.$$

Since r_{ij}^{tra} (respectively r_{ij}^{uni}) is generated independently from all other $r_{i'j'}^{\mathsf{tra}}$ (respectively $r_{i'j'}^{\mathsf{uni}}$), the proposition follows by a standard hybrid argument. $\square$

Removing the Assumption Regarding Tiny Error. Above we assumed that $\alpha(\lambda) \leq 2^{-\lambda m - 2}$. We can start from any $\alpha \leq \frac{1}{2} - \eta$, for $\eta = \lambda^{-O(1)}$, perform $k' = O(\lambda m \eta^{-2})$ repetitions to reduce the error, and then apply the above transformation.

The amount of randomness $\ell(\lambda)$, and the execution time, grow proportionally, but are still polynomial in λ. Also, the same security guarantee as above holds, except that we should consider the $(k \times k')$-fold repetition of Π, rather than the k-fold one. This is sufficient as long as the original scheme was secure for *any* polynomial number of repetitions.

4   Examples of Interest

We now discuss three examples of interest.

Public-Key Encryption. Our first example concerns public-key encryption. We start by recalling the definition.

Definition 4.1 (Public-Key Encryption). *For a message space $\mathcal{M}$, and function $\alpha(\cdot) \leq 1$, a triple of algorithms (Gen, Enc, Dec), where the first two are PPT and third is deterministic polynomial-time, is said to be a public-key encryption scheme for $\mathcal{M}$ with $(1 - \alpha)$-correctness if it satisfies:*

1. $(1 - \alpha)$-**Correctness:** *for any $m \in \mathcal{M}$ and security parameter λ,*

$$\Pr_{\mathsf{Gen},\mathsf{Enc}} \left[\mathsf{Dec}_{sk}(\mathsf{Enc}_{pk}(m)) = m \mid (pk, sk) \leftarrow \mathsf{Gen}(1^\lambda) \right] \geq 1 - \alpha(\lambda).$$

2. **Semantic security:** *for any polynomial-size distinguisher $\mathcal{D}$ there exists a negligible function $\mu(\cdot)$, such that for any two messages $m, m' \in \mathcal{M}$ of the same size:*

$$|\Pr[\mathcal{D}(\mathsf{Enc}_{pk}(m)) = 1] - \Pr[\mathsf{Enc}_{pk}(m')) = 1]| \leq \mu(\lambda) \,,$$

where the probability is over the coins of Enc *and the choice of pk sampled by* $\mathsf{Gen}(1^\lambda)$.

Public-key encryption can be modeled as a three-party scheme Π consisting of a generator, an encryptor, and a decryptor. The generator has no input, and uses its randomness r_1 to generate pk and sk, which are sent to the encryptor and decryptor, respectively. The encryptor has as input a message m, and uses its randomness r_2 in order to generate an encryption $\mathsf{Enc}_{pk}(m; r_2)$, which is sent to the decryptor. The decryptor has no input nor randomness, it uses the secret key to decrypt and outputs the decrypted message. (In this case the function computed by Π is $f(\bot, m, \bot) = (\bot, \bot, m)$.)

In the repeated scheme $\Pi_{\otimes k}$, the generator $\mathsf{Gen}(1^\lambda; r_{1j})$ is applied k independent times, with fresh randomness r_{1j} for each $j \in [k]$, to generate corresponding keys $pk = \{pk_j\}, sk = \{sk_j\}$. Encryption involves k independent encryptions:

$$\mathsf{Enc}_{pk}^{\otimes k}(m; r_2) := \mathsf{Enc}_{pk_1}(m; r_{21}), \ldots, \mathsf{Enc}_{pk_k}(m; r_{2k}) \,.$$

As defined in Sect. 3, when applying the error-removal transformation, the randomness $r = (r_{ij} : i \in [2], j \in [k])$ is sampled according to r^{tra} instead of truly at random according to r^{uni}. Decryption is done by decrypting each encryption with the corresponding sk_j and outputting the majority.

The correctness of the new scheme given by the transformation, follows as in Proposition 3.1. We next observe that the new scheme is also secure. Concretely, for any (infinite sequence of) two messages $m, m' \in \mathcal{M}$,

$$\mathsf{Enc}_{pk}^{\otimes k}(m; r_2^{\mathsf{tra}}) \approx_c \mathsf{Enc}_{pk}^{\otimes k}(m; r_2^{\mathsf{uni}}) \approx_c \mathsf{Enc}_{pk}^{\otimes k}(m'; r_2^{\mathsf{uni}}) \approx_c \mathsf{Enc}_{pk}^{\otimes k}(m'; r_2^{\mathsf{tra}}) \,.$$

The fact that $\mathsf{Enc}_{pk}^{\otimes k}(m; r_2^{\mathsf{uni}}) \approx_c \mathsf{Enc}_{pk}^{\otimes k}(m'; r_2^{\mathsf{uni}})$ follows from the semantic security of the underlying encryption scheme and a standard hybrid argument. The first and last indistinguishability relations follow from the fact that $r_2^{\mathsf{tra}} \approx_c r_2^{\mathsf{uni}}$ (by Proposition 3.2).

In [DNR04a], Dwork, Naor, and Reingold show how public-key encryption where decryption errors may even occur for a large fraction of messages, can be transformed into ones that only have a tiny decryption error over the randomness of the scheme. Applying our transformation, we can further turn such schemes into perfectly correct ones.

Indistinguishability Obfuscation. Our second example concerns indistinguishability obfuscation (IO) [BGI+12]. We start by recalling the definition.

Definition 4.2 (Indistinguishability Obfuscation). *For a class of circuits $\mathcal{C}$, and function $\alpha(\cdot) \leq 1$, a PPT algorithm $\mathcal{O}$ is said to be an* indistinguishability obfuscator *for $\mathcal{C}$ with $(1 - \alpha)$-correctness if it satisfies:*

1. $(1 - \alpha)$-**Correctness:** *for any $C \in \mathcal{C}$ and security parameter λ,*

$$\Pr_{\mathcal{O}} \left[\forall x : \mathcal{O}(C, 1^\lambda)(x) = C(x) \right] \geq 1 - \alpha(\lambda).$$

2. **Indistinguishability:** *for any polynomial-size distinguisher $\mathcal{D}$ there exists a negligible function $\mu(\cdot)$, such that for any two circuits $C, C' \in \mathcal{C}$ that compute the same function and are of the same size:*

$$\left| \Pr[\mathcal{D}(\mathcal{O}(C, 1^\lambda)) = 1] - \Pr[\mathcal{D}(\mathcal{O}(C', 1^\lambda)) = 1] \right| \leq \mu(\lambda),$$

where the probability is over the coins of $\mathcal{D}$ and $\mathcal{O}$.

IO can be modeled as a two-party scheme Π consisting of an obfuscator and an evaluator. The obfuscator has as input a circuit C, and uses its randomness r_1 in order to create an obfuscated circuit $\widetilde{C} = \mathcal{O}(C, 1^\lambda; r_1)$, which is sent to the evaluator. The evaluator has an input x for the circuit, and no randomness, it computes $\widetilde{C}(x)$ and outputs the result. (In this case the function computed by Π is $f(C, x) = (\bot, C(x))$.)

In the repeated scheme $\Pi_{\otimes k}$, obfuscation involves k independent obfuscations:

$$\mathcal{O}^{\otimes k}(C, 1^\lambda; r_1) := \mathcal{O}(C, 1^\lambda; r_{11}), \ldots, \mathcal{O}(C, 1^\lambda; r_{1k}).$$

As defined in Sect. 3, when applying the error-removal transformation, the randomness $r = (r_{1j} : j \in [k])$ is sampled according to r^{tra} instead of truly at random according to r^{uni}. Evaluation for input x is done by running each obfuscated circuit on the input x and outputting the majority of outputs.

The correctness of the new scheme given by the transformation, follows as in Proposition 3.1. We now observe that the new scheme is also secure, which follows similarly to the case of public-key encryption considered above. Concretely, for any (infinite sequence of) two equal-size circuits $C, C' \in \mathcal{C}$,

$$\mathcal{O}^{\otimes k}(C, 1^\lambda; r_1^{\mathsf{tra}}) \approx_c \mathcal{O}^{\otimes k}(C, 1^\lambda; r_1^{\mathsf{uni}}) \approx_c \mathcal{O}^{\otimes k}(C', 1^\lambda; r_1^{\mathsf{uni}}) \approx_c \mathcal{O}^{\otimes k}(C', 1^\lambda; r_1^{\mathsf{tra}}).$$

The fact that $\mathcal{O}^{\otimes k}(C, 1^\lambda; r_1^{\mathsf{uni}}) \approx_c \mathcal{O}^{\otimes k}(C', 1^\lambda; r_1^{\mathsf{uni}})$ follows from the security of the underlying obfuscation scheme and a standard hybrid argument. The first and last indistinguishability relations follow from the fact that $r_1^{\mathsf{tra}} \approx_c r_1^{\mathsf{uni}}$ (by Proposition 3.2).

In [BV16], Bitansky and Vaikuntanathan show how indistinguishability obfuscation [BGI+12] where the obfuscated circuit may err also on a large fraction of *inputs* can be transformed into one that only has a tiny error over the randomness of the obfuscator as required here. Applying our transformation, we can further turn such schemes into perfectly correct ones.

MPC. Our third and last example concerns multi-party computation (MPC) protocols. There are several models for capturing the adversarial capabilities in an MPC protocol. Roughly speaking, our transformation can be applied whenever the protocol is secure against parallel repetition. In the new protocol, perfect correctness will be guaranteed when all the parties behave honestly. The security guarantee given by the new protocol will be inherited from the original repeated protocol. We stress that, in the case of corrupted parties, the transformed protocol does not provide any correctness guarantees beyond those given by the original (repeated) protocol. In particular, if the adversary can inflict a certain correctness error in the original (repeated) protocol, it may also be able to do so in the transformed protocol.

We now give more details. Since we rely on standard MPC conventions, we shall keep our description relatively light (for further reading, see for instance [Can01, Gol04]). We consider protocols with security against *static corruptions* according to the *real-ideal paradigm*. For simplicity of exposition, we restrict attention to the single-execution setting. (Later, we explain how the transformation can also be applied in the setting of multiple executions, for example, in the UC model [Can01].) In this setting, the adversary $\mathcal{A}$ corrupts some set of parties $C \subseteq [m]$, which it fully controls throughout the protocol, and can also choose the inputs for honest parties at the onset of the computation. The *adversarial view* in the protocol consists of all the communication generated by the honest parties and their respective outputs. We denote by $\mathsf{Real}_{\Pi}^{\mathcal{A}}(1^{\lambda}, z; r)$ the polynomial-time process that generates the adversarial view and the outputs of the honest parties in $[m] \setminus C$ when these parties execute protocol Π for functionality f with randomness $r = (r_{i_1}, \ldots, r_{i_{m-|C|}})$, and a PPT adversary $\mathcal{A}$ with auxiliary input z controlling the parties in C.

The requirement is that the output of this process can be simulated by a PPT process $\mathsf{Ideal}_{f}^{\mathcal{S}}(1^{\lambda}, z)$ called the ideal process where $\mathcal{A}$ is replaced by an efficient *simulator* $\mathcal{S}$. The simulator can only submit inputs $x_1, \ldots, x_m$ to f, learn the outputs of the corrupted parties in C, and has to generate the adversarial view. The ideal process outputs the view generated by the simulator as well as the output generated by f for the honest parties.

As before, we denote by $\Pi_{\otimes k}$ the k-fold parallel repetition of a protocol Π for computing $f_{\otimes k}(x) = (f(x))^k$, where each honest party $i \in [m] \setminus C$, given input x_i, runs k parallel copies of Π, all with the same input x_i and obtains outputs $y_{i1}, \ldots, y_{ik}$. We consider protocols that are secure under parallel repetition in the following sense.

Definition 4.3. *We say that an MPC protocol Π (for some functionality f) is secure under parallel repetition with respect to an ideal process Ideal if for any PPT adversary $\mathcal{A}$ and polynomial $k(\lambda)$ there exists a PPT simulator $\mathcal{S}$ such that for any (infinite sequence of) security parameter $\lambda \in \mathbb{N}$ and auxiliary input in $z \in \{0,1\}^*$,*

$$\mathsf{Real}_{\Pi_{\otimes k}}^{\mathcal{A}}(1^{\lambda}, z) \approx_c \mathsf{Ideal}_{f_{\otimes k}}^{\mathcal{S}}(1^{\lambda}, z).$$

We denote by Π^{tra} the protocol Π for computing f after applying the transformation from Sect. 3 where Π is repeated in k times in parallel, the randomness of parties is derived as defined in the transformation, and the final output of party i is set to $\mathsf{majority}(y_{i1}, \ldots, y_{ik})$. When all the parties act honestly, the correctness of the new protocol Π^{tra} given by the transformation, follows as in Proposition 3.1.

We show that if the original protocol is secure under parallel repetition then the transformed protocol is as secure.

Claim. Assume that Π is a protocol for f that is secure under parallel repetition (in the sense of Definition 4.3). For any PPT adversary $\mathcal{A}$ against Π^{tra}, viewing $\mathcal{A}$ as an adversary against $\Pi_{\otimes k}$, let $\mathcal{S}$ be its simulator given by Definition 4.3. Then for any (infinite sequence of) security parameter λ, and auxiliary input z,

$$\mathsf{Real}^{\mathcal{A}}_{\Pi^{\mathsf{tra}}}(1^\lambda, z) \approx_c \mathsf{Ideal}^{\mathcal{S}}_f(1^\lambda, z).$$

Proof. Let $\Pi^{\mathsf{maj}}_{\otimes k}$ be the protocol where the parties first execute the k-fold repetition of $\Pi_{\otimes k}$ and then each party sets its final output to be the majority of the outputs obtained in that execution. Then we first note that

$$\mathsf{Real}^{\mathcal{A}}_{\Pi^{\mathsf{tra}}}(1^\lambda, z) \equiv \mathsf{Real}^{\mathcal{A}}_{\Pi^{\mathsf{maj}}_{\otimes k}}(1^\lambda, z; r^{\mathsf{tra}}),$$

where r^{tra} is the randomness of the honest parties, generated according to our transformation. By Proposition 3.2, it holds that:

$$\mathsf{Real}^{\mathcal{A}}_{\Pi^{\mathsf{tra}}}(1^\lambda, z) \equiv \mathsf{Real}^{\mathcal{A}}_{\Pi^{\mathsf{maj}}_{\otimes k}}(1^\lambda, z; r^{\mathsf{tra}}) \approx_c \mathsf{Real}^{\mathcal{A}}_{\Pi^{\mathsf{maj}}_{\otimes k}}(1^\lambda, z; r^{\mathsf{uni}}),$$

where r^{tra} is randomness generated according to our transformation and r^{uni} is truly random. It is left to note that

$$\mathsf{Real}^{\mathcal{A}}_{\Pi^{\mathsf{maj}}_{\otimes k}}(1^\lambda, z; r^{\mathsf{uni}}) \approx_c \mathsf{Ideal}^{\mathcal{S}}_f(1^\lambda, z).$$

Indeed, recall that by Definition 4.3,

$$\mathsf{Real}_{\Pi_{\otimes k}}{}^{\mathcal{A}}(1^\lambda, z; r^{\mathsf{uni}}) \approx_c \mathsf{Ideal}^{\mathcal{S}}_{f_{\otimes k}}(1^\lambda, z),$$

and each of the first two distributions can be efficiently computed from the respective distribution in the second two, by fixing the (single) output of each honest party to be the majority of its outputs.

Applying the Transformation in More General Models. Above, we have considered a model with a single execution. The analysis naturally extends to more general models such as the model of universally composable (UC) protocols [Can01], where multiple executions controlled by an adversarial *environment* can be performed. Indeed, the only feature of the model we have relied on is that the real world view can be generated using the randomness of honest parties as external input (regardless of how the randomness was generated), which is the case as long corruptions are static, and the adversary is never exposed to the randomness of honest parties, but only to the communication between parties. This is also the case in the UC model.

Acknowledgements. We thank Stefano Tessaro for pointing out [HR05,LT13]. We also thank the reviewers of EUROCRYPT 2017 for their valuable comments.

References

[AD97] Ajtai, M., Dwork, C.: A public-key cryptosystem with worst-case/average-case equivalence. In: Leighton, F.T., Shor, P.W. (eds.) Proceedings of the Twenty-Ninth Annual ACM Symposium on the Theory of Computing, El Paso, Texas, USA, 4–6 May 1997, pp. 284–293. ACM (1997)

[BCPR14] Bitansky, N., Canetti, R., Paneth, O., Rosen, A.: On the existence of extractable one-way functions. In: Shmoys, D.B. (ed.) Symposium on Theory of Computing, STOC 2014, New York, NY, USA, 31 May–03 June, pp. 505–514. ACM (2014)

[BDL01] Boneh, D., DeMillo, R.A., Lipton, R.J.: On the importance of eliminating errors in cryptographic computations. J. Cryptology **14**(2), 101–119 (2001)

[BGI+12] Barak, B., Goldreich, O., Impagliazzo, R., Rudich, S., Sahai, A., Vadhan, S.P., Yang, K.: On the (im)possibility of obfuscating programs. J. ACM **59**(2), 6 (2012)

[BM82] Blum, M., Micali, S.: How to generate cryptographically strong sequences of pseudo random bits. In: 23rd Annual Symposium on Foundations of Computer Science, Chicago, Illinois, USA, 3–5 November, pp. 112–117 (1982)

[BM84] Blum, M., Micali, S.: How to generate cryptographically strong sequences of pseudo-random bits. SIAM J. Comput. **13**(4), 850–864 (1984)

[BOV07] Barak, B., Ong, S.J., Vadhan, S.P.: Derandomization in cryptography. SIAM J. Comput. **37**(2), 380–400 (2007)

[BP15] Bitansky, N., Paneth, O.: ZAPs and non-interactive witness indistinguishability from indistinguishability obfuscation. In: Dodis, Y., Nielsen, J.B. (eds.) TCC 2015. LNCS, vol. 9015, pp. 401–427. Springer, Heidelberg (2015). doi:10.1007/978-3-662-46497-7_16

[BV16] Bitansky, N., Vaikuntanathan, V.: Indistinguishability obfuscation: from approximate to exact. In: Kushilevitz, E., Malkin, T. (eds.) TCC 2016. LNCS, vol. 9562, pp. 67–95. Springer, Heidelberg (2016). doi:10.1007/978-3-662-49096-9_4

[Can01] Canetti, R.: Universally composable security: a new paradigm for cryptographic protocols. In: 42nd Annual Symposium on Foundations of Computer Science, FOCS 2001, Las Vegas, Nevada, USA, 14–17 October 2001, pp. 136–145. IEEE Computer Society (2001)

[CC04] Cachin, C., Camenisch, J.L. (eds.): EUROCRYPT 2004. LNCS, vol. 3027. Springer, Heidelberg (2004)

[DN07] Dwork, C., Naor, M.: Zaps and their applications. SIAM J. Comput. **36**(6), 1513–1543 (2007)

[DNR04a] Dwork, C., Naor, M., Reingold, O.: Immunizing encryption schemes from decryption errors. In: Cachin, Camenisch, pp. 342–360

[FGM+89] Furer, M., Goldreich, O., Mansour, Y., Sipser, M., Zachos, S.: On completeness and soundness in interactive proof systems. Adv. Comput. Res. Res. Ann. **5**, 429–442 (1989). (Randomness and Computation, S. Micali, ed.)

[GGH97] Goldreich, O., Goldwasser, S., Halevi, S.: Eliminating decryption errors in the Ajtai-Dwork Cryptosystem. In: Kaliski, B.S. (ed.) CRYPTO 1997. LNCS, vol. 1294, pp. 105–111. Springer, Heidelberg (1997). doi:10.1007/BFb0052230

[GM84] Goldwasser, S., Micali, S.: Probabilistic encryption. J. Comput. Syst. Sci. **28**(2), 270–299 (1984)

[GMS87] Goldreich, O., Mansour, Y., Sipser, M.: Interactive proof systems: provers that never fail and random selection (extended abstract). In: 28th Annual Symposium on Foundations of Computer Science, Los Angeles, California, USA, 27–29 October 1987, pp. 449–461. IEEE Computer Society (1987)

[Gol04] Goldreich, O.: The Foundations of Cryptography. Basic Applications, vol. 2. Cambridge University Press, New York (2004)

[HILL99] Håstad, J., Impagliazzo, R., Levin, L.A., Luby, M.: A pseudorandom generator from any one-way function. SIAM J. Comput. **28**(4), 1364–1396 (1999)

[HR05] Holenstein, T., Renner, R.: One-way secret-key agreement and applications to circuit polarization and immunization of public-key encryption. In: Shoup, V. (ed.) CRYPTO 2005. LNCS, vol. 3621, pp. 478–493. Springer, Heidelberg (2005). doi:10.1007/11535218_29

[IW97] Impagliazzo, R., Wigderson, A.: P = BPP if E requires exponential circuits: derandomizing the XOR lemma. In: Proceedings of the Twenty-Ninth Annual ACM Symposium on the Theory of Computing, El Paso, Texas, USA, 4–6 May 1997, pp. 220–229 (1997)

[Lau83] Lautemann, C.: BPP and the polynomial hierarchy. Inf. Process. Lett. **17**(4), 215–217 (1983)

[LT13] Lin, H., Tessaro, S.: Amplification of chosen-ciphertext security. In: Johansson, T., Nguyen, P.Q. (eds.) EUROCRYPT 2013. LNCS, vol. 7881, pp. 503–519. Springer, Heidelberg (2013). doi:10.1007/978-3-642-38348-9_30

[Nao91] Naor, M.: Bit commitment using pseudorandomness. J. Cryptology **4**(2), 151–158 (1991)

[NW94] Nisan, N., Wigderson, A.: Hardness vs randomness. J. Comput. Syst. Sci. **49**(2), 149–167 (1994)

[Reg05] Regev, O.: On lattices, learning with errors, random linear codes, and cryptography. In: Gabow, H.N., Fagin, R. (eds.) Proceedings of the 37th Annual ACM Symposium on Theory of Computing, Baltimore, MD, USA, 22–24 May, pp. 84–93. ACM (2005)

[SU01] Shaltiel, R., Umans, C.: Simple extractors for all min-entropies and a new pseudo-random generator. In: 42nd Annual Symposium on Foundations of Computer Science, FOCS 2001, Las Vegas, Nevada, USA, 14–17 October 2001, pp. 648–657 (2001)

[Yao82a] Yao, A.C.-C.: Theory and applications of trapdoor functions (extended abstract). In: 23rd Annual Symposium on Foundations of Computer Science, Chicago, Illinois, USA, 3–5 November 1982, pp. 80–91. IEEE Computer Society (1982)

[Yao82b] Yao, A.C.-C.: Theory and applications of trapdoor functions (extended abstract). In: 23rd Annual Symposium on Foundations of Computer Science, Chicago, Illinois, USA, 3–5 November 1982, pp. 80–91 (1982)

Blockchain

Decentralized Anonymous Micropayments

Alessandro Chiesa[1]([✉]), Matthew Green[2], Jingcheng Liu[1], Peihan Miao[1],
Ian Miers[2], and Pratyush Mishra[1]

[1] UC Berkeley, Berkeley, CA, USA
{alexch,liuexp,peihan,pratyush}@berkeley.edu
[2] Johns Hopkins University, Baltimore, MD, USA
mgreen@cs.jhu.edu,imiers@cs.jhu.edu

Abstract. Micropayments (payments worth a few pennies) have numerous potential applications. A challenge in achieving them is that payment networks charge fees that are high compared to "micro" sums of money.

Wheeler (1996) and Rivest (1997) proposed probabilistic payments as a technique to achieve micropayments: a merchant receives a macro-value payment with a given probability so that, in expectation, he receives a micro-value payment. Despite much research and trial deployment, micropayment schemes have not seen adoption, partly because a trusted party is required to process payments and resolve disputes.

The widespread adoption of decentralized currencies such as Bitcoin (2009) suggests that decentralized micropayment schemes are easier to deploy. Pass and Shelat (2015) proposed several micropayment schemes for Bitcoin, but their schemes provide no more privacy guarantees than Bitcoin itself, whose transactions are recorded in plaintext in a public ledger.

We formulate and construct *decentralized anonymous micropayment* (DAM) schemes, which enable parties with access to a ledger to conduct offline probabilistic payments with one another, directly and privately. Our techniques extend those of Zerocash (2014) with a new privacy-preserving probabilistic payment protocol. One of the key ingredients of our construction is *fractional message transfer* (FMT), a primitive that enables probabilistic message transmission between two parties, and for which we give an efficient instantiation.

Double spending in our setting cannot be prevented. Our second contribution is an economic analysis that bounds the additional utility gain of any cheating strategy, and applies to virtually any probabilistic payment scheme with offline validation. In our construction, this bound allows us to deter double spending by way of advance deposits that are revoked when cheating is detected.

1 Introduction

We formulate and construct *decentralized anonymous micropayments*, by way of probabilistic payments.

This work was supported in part by the Center for Long-Term Cybersecurity at UC Berkeley.

© International Association for Cryptologic Research 2017
J.-S. Coron and J.B. Nielsen (Eds.): EUROCRYPT 2017, Part II, LNCS 10211, pp. 609–642, 2017.
DOI: 10.1007/978-3-319-56614-6_21

Micropayments. A *micropayment* is a payment of a small amount, e.g., a fraction of a penny [Whe96, Riv97]. Micropayments have many potential applications, including advertisement-free content delivery, spam protection, rewarding nodes of P2P networks, and others. Achieving micropayments involves at least two main challenges. First, payment processing fees dwarf "micro" payment values. Second, micropayment applications often require *fast merchant responses*, which, in many settings, are achieved via *offline payments*, which are vulnerable to double spending.

Probabilistic Payments. A technique to reduce processing fees is to amortize them over multiple payments by way of *probabilistic payments* [Whe96, Riv97].[1] These are protocols that enable a customer to pay V units of currency to a merchant with probability p: with probability $1 - p$ the merchant receives a *nullpayment* that is not processed, and with probability p the merchant receives a *macropayment* that is processed. In expectation, the merchant receives pV units per micropayment, but the overhead and processing fees of these "lottery tickets" is p times smaller as only the infrequently generated macropayments are actually handled by the payment network. Constructing probabilistic payments is an area of ongoing interest in cryptography.

Centralized vs. decentralized systems. Despite extensive research and trial deployments [Whe96, Riv97, LO98, MR02, Riv04, Mic14], micropayment schemes have not seen widespread usage. This is perhaps due to them being *centralized systems*: a trusted third party is tasked with processing payments and punishing cheaters. Appointing such a party raises deployment costs, requires establishing complex business relationships between all involved (the trusted party, merchants, and customers), and makes participation conditional on certain requirements being met [vOR+03].

Recent work in digital currencies has focused on *decentralized systems*, as the cost of entry and deployment appears to be lower. The most notable such currency is Bitcoin [Nak09], a widely adopted peer-to-peer payment system. Unlike traditional banking and e-cash schemes [Cha82, CHL05, ST99] where transactions are processed by a trusted party, Bitcoin utilizes a distributed public ledger known as the *blockchain* to store all transactions; these transactions are verified by network nodes in a peer-to-peer fashion.

Decentralized systems are thus potentially attractive for micropayments, because the overhead involving trusted parties is no longer a factor. However, Bitcoin processing fees are still relatively high (as of May 2016 the fee for a 1kB-transaction is $\approx$ \$0.20), with present fees believed to be well below the cost of performing a transaction on the Bitcoin network [MB15]. Thus, fee amortization is still necessary. Caldwell [Cal12] first sketched probabilistic payments for Bitcoin. Recently, Pass and Shelat [PS15, PS16] also proposed three probabilistic payment schemes for Bitcoin, where, informally, the customer first puts V bitcoins in escrow, and then the customer and merchant engage in a coin-flipping protocol that allows the merchant to retrieve the escrow with probability p.

---

[1] Another technique is micropayment channels, which we discuss in Sect. 1.2.

Their three schemes differ in how payments are processed and how disputes are resolved.

Our privacy goal and limitations of prior work. We study the question of how to construct *decentralized anonymous micropayments* via the technique of (offline) probabilistic payments. The aforementioned prior work [PS15, PS16] provides *no more privacy than the underlying Bitcoin protocol*. And Bitcoin itself provides little to no privacy because every transaction is publicly broadcast and contains a payment's origin, destination, and amount; a user's payment history is thus readily available to any passive observer who can can link pseudonyms together or to real world identities.[2] This lack of privacy is particularly dangerous for micropayment applications because they typically involve high-volume pattern-rich payments (e.g., per-click payments while surfing the web), and sometimes necessitate user anonymity (e.g., bandwidth payments for Tor relays [BP15]).

Privacy is not merely an issue of individual users: if each coin's history is public, a customer may not be able to spend a coin at its 'declared' value due to its past. For example, a merchant may not accept coins whose past owners include certain political organizations. Privacy thus ensures a fundamental property of the currency: *fungibility*, which means that any two sets of coins with the same 'declared' total value are interchangeable, regardless of their provenance.

Prior work on privacy-preserving analogues of Bitcoin [MGGR13, DFKP13, BCG+14] does not achieve probabilistic payments, and merely "plugging" these schemes into [PS15, PS16]'s approach results in subtle problems. Consider the following natural modification to Pass and Shelat's coin-flipping protocol: instead of a Bitcoin transaction, the sender probabilistically transmits to the merchant a Zerocash transaction [BCG+14]. Despite the strong anonymity guarantees provided by Zerocash, merchants *still* learn information about their customers' spending habits, because each Zerocash transaction includes a unique serial number corresponding to the spent "coin". Since the customer sends to the merchant information about the escrow, this serial number is revealed in each micropayment. Since the same escrow is used across multiple probabilistic payments (to amortize fees), privacy of the customer is compromised because the merchant learns (1) which (macro or null) payments to him were made with the same escrow; and (2) which macropayments to other merchants were made with an escrow used for payments to him. This breach of privacy worsens if merchants share information with one another. In sum, while the above natural approach achieves "macropayment unlinkability", *micropayments are still linkable*, and thus customers have little privacy.

Double spending in offline probabilistic payments. Micropayment applications often require fast responses. In many settings, these in turn require *offline* validation: a merchant responds to a payment after only a local "offline" check,

---

[2] This is not merely a theoretical concern: extracting information from Bitcoin transactions is the subject of applied research [RH11, BBSU12, RS13, MPJ+13] and commercial ventures [Ell13, Blo14, Cha15].

because he cannot wait for the payment network to validate the payment (this validation instead completes after the merchant's response). For example, valida-tion takes a few minutes in Bitcoin, while responding to unconfirmed `zero-conf` transactions takes only a few seconds. We thus focus on *offline probabilistic payments*.

However, such payments are *vulnerable to double spending*, as we now explain. First, double spending *cannot be prevented* for offline payments, because, to prevent it, a merchant would have to refrain from responding to any payment before all payments up to, and including, this payment have been validated. One fallback is to detect and punish all double-spending customers. However, for offline *probabilistic* payments, not all double spending can even be detected.

Indeed, there are two types of double spending when using the same lottery ticket in two probabilistic payments: (1) both payments result in macropay-ments; or (2) the first payment results in a macropayment (thereby 'consuming' the ticket) while the second payment results in a nullpayment. While detecting the first type is easy, detecting the second type requires the payment network to 'know' the temporal order of all payments, because whether the nullpayment or the macropayment occurred first determines whether the two payments corre-spond to honest behavior (nullpayment first) or not (macropayment first). But knowing the global order of all payments (with high precision) is a strong syn-chronization property that is unrealistic in many decentralized settings, includ-ing that of Bitcoin, because information does not instantly reach everyone in the network.

Given that not all double spending can be detected, the "detect-and-punish" approach is effective only if the disadvantages of being punished (upon detection) outweigh the advantages of double spending. This may be plausible in the cen-tralized setting, where customers have registered with a trusted party that can permanently ban and legally prosecute them. In the decentralized setting, how-ever, banning has few consequences, if any: anyone can abandon old identities and use fresh new identities in their place.

Ruffing, Kate, and Schröder [RKS15] introduce "accountable assertions", which enable timelocked deposits in Bitcoin that are revoked upon evidence of double spending. Pass and Shelat [PS16] also suggest a Bitcoin-specific penalty mechanism to deter rational customers and merchants from cheating.[3] Unfortu-nately, both of these works do not provide an economic analysis to indicate how large a penalty should be to deter double spending. Such an analysis is crucial: how could detect-and-punish be a deterrent if double spending were to yield *unbounded* additional utility?

1.1 Our Contributions

We overcome the aforementioned limitations via a combination of cryptographic and economic techniques. We adopt a "detect-and-punish" approach in which

---

[3] We also note that two of the three schemes in [PS15] do not support offline payments, and the remaining one only provides "fast online payments" where an online (publicly verifiable) trusted party assists the ledger by processing macropayments faster.

cryptography is used to retroactively detect and economically punish double spending and, separately, an economic analysis clarifies how much to punish so as to deter double spending in the first place. More precisely, we present the following three contributions.

1.1.1 Economic Analysis of Double Spending for Offline Probabilistic Payments

We characterize the additional utility that can be gained by double spending via offline probabilistic payments. We suppose that: (i) every probabilistic payment is backed by an advance deposit;[4] (ii) all macropayment double spends can be detected; and (iii) if a merchant detects a double spend then he reports it, and doing so results in the revocation of the cheating customer's deposit. (Our cryptographic constructions will provide suitable mechanisms for these tasks.) We then ask: *how large must the deposit be in order to deter double spending?*

We provide a simple yet powerful analysis that answers this question under reasonable network behavior. Namely, let T denote the time it takes to catch a macropayment double spend (e.g., in Bitcoin one could take T to be the network's broadcast time). Within any period of time T, let A denote the maximum cumulative value of probabilistic payments and W the maximum cumulative value of macropayments; our analysis will show that imposing bounds on these quantities is *necessary*. To simplify discussions, we make the assumption that *only* macropayment (and not nullpayment) double spends are detectable; our analysis extends to the case where nullpayment double spends may also be detected eventually (see Remark 1).

Below we informally state our theorem, for simplicity in the special case where the macropayment value V and the payment probability p are fixed across all probabilistic payments, and all merchants share the same detection time T. The formal statement that we prove is in fact more general, because it applies even when these quantities are chosen dynamically and arbitrarily across different payments.

Theorem 1 (informal statement of Theorem 4)

(a) If the deposit is at least W, then there is no <u>worst-case utility</u> gain in double spending.

(b) If the deposit is at least $(1-p)V + A$, then there is no <u>average-case utility</u> gain in double spending.

(c) Both bounds above are tight.

Our theorem has a simple interpretation: the required deposit amount equals the maximum financial activity that can happen within any time period of T. Namely, if macropayments have maximum total worth W within time T, the deposit must be at least W (w.r.t. worst-case utility); and if probabilistic payments have maximum total worth A within time T, the deposit must be at

---

[4] One deposit may back multiple payments; in particular, an honest customer may use a single deposit to back all of his payments.

least $\approx A$ (w.r.t. average-case utility). Note that it is unsurprising that the two statements in the theorem depend on the two different quantities W and A, because they target different notions of utility; also note that, while one can take $pW \leq A$ without loss of generality, a bound on W does *not* always imply a bound on A (there could still be arbitrarily many probabilistic payments, though with extremely small probability).

But which of the two bounds should one use in practice? Naturally, the worst-case bound is safer than the average-case bound; however, an appropriate setting of W will be $\Omega(1/p)$ larger than A, which implies a substantial increase in the required deposit. The choice between the two depends on whether one cares about malicious customers that are lucky with even very small probability (as opposed to focusing on their average gains possibly across many deposits).

As already mentioned, bounding the value of probabilistic payments (via A) or macropayments (via W) within time T is *necessary* because our bounds are tight (i.e., there exist double-spending strategies that achieve them). In the "real world" these bounds may be imposed by the environment (e.g., limited network throughput), or the merchants (e.g., they accept up to a given number of payments within time T).

In terms of analysis, our proof shows that any additional utility gained via double spending must come from *macropayment* double spending. This may be surprising because, superficially, one may think that *nullpayment* double spending also contributes to additional utility; e.g., one may think that a malicious customer gains pV for every nullpayment double spend. This proposition is alarming: in the worst case there could be infinitely-many nullpayment double spends (which imply infinite additional utility); and in the average case there could be clever strategies that leverage double spends across multiple merchants to lower the probability of detection. We prove that this is not the case: we use a simulation argument to show that *the naive strategy of double spending as much as possible is the best strategy* (i.e., maximizes additional utility), both in the worst case and in the average case. In particular, we learn that the best strategy always leads to detection (after a time period of T) and that additional utility *is finite even in the worst case* (if W is finite). Details of our analysis are in Sect. 3.

We believe our theorem to be of independent interest because it applies to virtually any (centralized or decentralized) setting that enforces a deposit mechanism for offline payments. One such setting could be probabilistic *smart contracts* (an application suggested by [PS15, PS16]). A thorough understanding of the economic benefits of double spending is necessary to ensure that such smart contracts, as well as other applications, function as intended.

Example. As a demonstration, we invoke our theorem on parameters that could fit the application of *advertisement-free content delivery*, to see what conclusions our economic analysis gives us. Suppose that we consider a Bitcoin-like setting, where (i) transaction fees are typically a few cents; and (ii) we could take the detection time T to be, e.g., 20 min, which is typically two blocks (ideal block generation follows an exponential distribution with a mean of 10 min).

Suppose further that we fix the deposit to be $D := \$200$ and the expected value of the probabilistic payment to be $\$0.1$ (similar size as a transaction fee); concentration bounds then suggest that, subject to the condition $pV = \$0.1$, good choices are $V := \$10$ and $p := 1\%$. Note that these settings imply that we can take W up to $D = \$200$ and A up to $D - (1 - p)V = \$190.1$. Then our theorem implies that: (1) Even the *luckiest* double spending user has no extra utility gain if the cumulative value of macropayments every 20 min is less than $\$200$ (that is, the number of macropayments every 20 min is less than 20), regardless of how much nullpayment double spending occurred. (2) A double spending user has no extra utility gain on average if the cumulative value of probabilistic payments every 20 min is less than $\$190.1$ (that is, the number of probabilistic payments every 20 min is less than 1901).

1.1.2   Decentralized Anonymous Micropayments

We formulate the notion of a *decentralized anonymous micropayment* (DAM) scheme. This notion formalizes the functionality and security properties of an offline probabilistic payment scheme that enables parties with access to a ledger to conduct transactions with one another, directly and privately. To realize the requirements of our economic analysis, a DAM scheme enables parties to set up deposits, which are revoked when macropayments reveal that double spending has occurred. Crucially, the security guarantees of a DAM scheme guarantee anonymity not only across macropayments but also across nullpayments, so that even the "offline stream of payments" remains unlinkable.

We construct a DAM scheme and prove its security under specific cryptographic assumptions. Our two main building blocks are decentralized anonymous payment (DAP) schemes [BCG+14] and fractional message transfer schemes (see below).

Theorem 2 (informal). *Given a decentralized anonymous payment scheme and a fractional message transfer scheme (and other standard cryptographic primitives) there exists a DAM scheme.*

Formally capturing the notion of a DAM scheme and proving security of our construction was quite challenging due to the combination of rich functionality and strong anonymity guarantees. Parties can mint standard coins, deposits, or lottery tickets; they can withdraw deposits; they can pay each other with deterministic payments, switch coin types; they can also pay each other with probabilistic payments; they can revoke deposits of cheating parties — all of this while essentially revealing no information about origins, destinations, and amounts of money transfers. In particular, two features of our construction required particular attention: (1) revocation of an unknown cheating party's deposit when two macropayments with the same ticket are detected; and (2) monitoring of payment value rates (as required by our economic analysis) despite deposits being anonymous. Deterministic payments in our construction are non-interactive, while probabilistic payments consist of a 3-message protocol between a sender and a receiver; it is an interesting open question whether these can be made non-interactive as well.

We express the security of a DAM scheme via the ideal-world/real-world paradigm, specifying a suitable ideal functionality, and we prove our construction's security via a simulator against non-adaptive corruptions of parties. We consider security in the standalone setting, and leave security under composition to future work (that perhaps can build upon the work of [KMS+16]).

1.1.3   Fractional Message Transfer

A key ingredient in our construction of DAM schemes is *fractional message transfer* (FMT): a primitive that enables probabilistic message transmission between two parties, called the 'sender' and the 'receiver'. Informally, FMT works as follows: (i) the receiver samples a one-time key pair based on a transfer probability p; (ii) the sender uses the receiver's public key to encrypt a message m into a ciphertext c; (iii) the receiver uses the secret key to decrypt c, thereby learning m, but only with the pre-defined probability p (and otherwise learns no information about m).

We thus (1) formulate the notion of an *FMT scheme*, which formally captures the functionality and security of probabilistic message transmission, and (2) present an efficient construction that works for probabilities that are inverses of positive integers.

Theorem 3 (informal). *In the random oracle model and assuming the hardness of* DDH *in prime-order groups, there exists an FMT scheme that works for transfer probabilities* $p = 1/n$ *with* $n \in \mathbb{N}$. *Moreover, the number of group elements and scalars in the public key and ciphertext is constant (independent of* n); *see Table 1.*

Our definition of FMT is closely related to *non-interactive fractional oblivious transfer* (NFOT), which was studied in the context of 'translucent cryptography' as an alternative to key escrow [BM89,BR99]. Namely, prior definitions target *one-way security*, which protects *random* messages. While one-way security suffices to encapsulate random secret keys (the setting of translucent cryptography), it does not suffice for probabilistically transmitting non-random messages (as needed in our construction). Therefore, our definition of an FMT scheme targets a fractional variant of *semantic security*, which we express via two properties: *fractional hiding* and *fractional binding*. Furthermore, since in our system any party can act as both sender and receiver, we require the FMT scheme to be *composable*. Our construction achieves this via simulation-extractability.

Our construction of FMT is loosely related to the constructions in [BM89, BR99], which (like our construction) build on the Elgamal encryption scheme [Elg85]. In fact, such constructions, if analyzed under the hardness of DDH rather than CDH, are likely to yield FMT according to our stronger definition. We did not carry out such an analysis, but instead chose to construct a scheme that is more efficient than prior work for the case of $p = 1/n$ (these probabilities suffice for our application); we assume hardness of DDH and work in the random

Table 1. Comparison of prior NFOT schemes vs. our FMT scheme. All constructions assume a common random string.

scheme	security	assumption	transfer probability	size of public key group elts.	scalars	size of ciphertext group elts.	scalars	# exponentiations to encrypt	to decrypt
[BM89] one-way	CDH		$1/2$	2	—	2	—	2	1
[BR99, § 5.1] one-way	CDH		$1/n$	n	—	2	—	2	1
[BR99, § 5.1] one-way	CDH		$(n-1)/n$	n	—	2	—	2	1
[BR99, § 5.2] one-way	CDH		a/n^*	$2\log_2 n$	—	$2\log_2 n$	—	$4\log_2 n$	$2\log_2 n$
[BR99, § 5.3] one-way	CDH		a/n	$a+n$	—	2	—	2	1
our FMT	semantic	DDH + RO	$1/n$	2	3	2	2	4	4

* n is restricted to be a power of 2.

oracle model in order to take advantage of certain Σ-protocols. See Table 1 for a comparison of our construction with prior work.

1.2 Prior Work on Micropayment Channels

Micropayment channels were introduced by Hearn and Spilman [HS12, Bit13], and further studied by Poon and Dryja [PD16] and Decker and Wattenhofer [DW15]. Roughly, a micropayment channel enables a sender and a receiver to set up a contract by way of an online (slow) transaction that escrows funds, after which the sender and receiver can update the contract, and thus the relative split of the escrowed funds, without recording the new contract on the blockchain. Thus payments can be made instantaneously. These can be dynamically combined to obtain multi-hop "payment channel networks" that go through several intermediaries, by using hashed timelock contracts; this technique amortizes the cost of setting up a new channel for new receivers. From the perspective of our work, micropayment channels have several limitations in terms of economics, functionality, and privacy.

Economic limitations of payment channels. First, payment channels in general require a channel to be established in advance with a party: payments are only instantaneous with advanced preparation. To alleviate this constraint, payment channel networks allow transactions with arbitrary new parties provided there exists a path of existing channels between the payer and payee.

Such networks have limitations. First, considerable capital is escrowed in the many pairwise channels forming the network. The capital requirements may exceed those required for deposits in probabilistic micropayments. Both settings require escrowed funds proportional to a user's economic activity (either for the double spend deposit or the "last mile" channel between the user and the payment network), but payment channel networks escrow similar amounts in each edge of the network. Second, a variety of pressures, including minimizing the capital escrowed, may centralize such networks into a hub-and-spoke model.

Privacy limitations of payment channels. Payment channels reveal to the world that a given pair of parties have a channel between them, the opening value of that channel, and the final closing value. More importantly, especially for applications like advertisement-free content delivery, payment channels provide no privacy between the parties on the channel: if Alice pays say Wikipedia every time she views a page, then each of those views is linked to the channel she established just as effectively as if she had a tracking cookie in her browser.

Attempts to add privacy, either from intermediate nodes in the network [HAB+16] or from recipients and intermediaries [GM16], to payment channels hit some seemingly fundamental limitations of the payment channel setting. First, the anonymity set when paying a given receiver is composed only of those users who have opened channels with the receiver. This is likely far smaller than the global anonymity set provided by probabilistic payments. Moreover, the receiving party can arbitrarily reduce the anonymity set further by closing channels. This leaves open a range of attacks that are not present in a system with a global anonymity set.

Finally it is unclear if non-hub-and-spoke private payment networks are scalable or can provide privacy for payment values from intermediary nodes in the network. When a payment is made via two intermediaries (i.e. $A \rightarrow I_1 \rightarrow I_2 \rightarrow B$), some combination of I_1 and I_2 must know the balance of their pairwise channel at any given time or they could not close the channel. Thus the value of any payment relayed through multiple parties cannot be completely private. Moreover, discovering a multi-hop route between two parties in a diverse and large network without leaking any identifying information seems costly at scale. While [GM16] extend their point-to-point channel protocol to a hub-and-spoke model that alleviates both these concerns, such a network is inherently centralized.

2 Techniques

We discuss the intuition and techniques behind our results, first for our cryptographic construction (Sect. 2.1) and then for our economic analysis of double spending (Sect. 2.2).

2.1 Constructing Decentralized Anonymous Payments

We discuss our design of a decentralized anonymous micropayment (DAM) scheme via a sequence of candidate constructions, each fixing problems of the previous one; the last one is a sketch of our construction.

2.1.1 Attempt 1: Non-anonymous Probabilistic Payments + DAP

We begin with a natural candidate construction for a DAM scheme. The idea is to combine two primitives, one providing probabilistic payments and the other anonymity. For example, consider: (1) the scheme MICROPAY1 of [PS15], which provides probabilistic payments for Bitcoin; and (2) a *decentralized anonymous payment* (DAP) scheme [BCG+14], which provides privacy-preserving payments for Bitcoin-like currencies.

To make MICROPAY1 privacy-preserving, we could try to replace its Bitcoin payments with DAP payments, which hide the payment's origin, destination, and amount. Thus, when a probabilistic payment goes through, and the corresponding DAP (macro-)payment is broadcast, others cannot learn this information about the payment. However, this idea does *not* provide the strong anonymity guarantees that we seek, as we now explain.

Problem: not fully anonymous. Despite the anonymity guarantees provided by the DAP scheme, merchants still learn information about their customers' spending habits. Each DAP payment includes a unique serial number corresponding to the underlying "coin" that was spent by that payment; this is used to prevent double spending of DAP coins. In the above proposal, the customer sends the merchant this serial number regardless of whether the payment becomes a nullpayment or a macropayment. Since the same underlying DAP payment and serial number are used across multiple probabilistic payments (to amortize fees), this compromises customer anonymity because a merchant learns (1) which (macro or null) payments to him were made with the same escrow; and (2) which macropayments to other merchants were made with an escrow used for payments to him. This compromise in anonymity gets even worse if merchants share such information with one another.

Moreover, recall (from Sect. 1.1) that it is not possible to prevent double spending in the setting of offline probabilistic payments. Pass and Shelat note this in the full version of their paper [PS16], and propose adding a 'penalty escrow' to the scheme MICROPAY1; the escrow is burned upon evidence of double spending. But observe that anonymity for penalty escrows poses a similar challenge: to prove that a penalty escrow is unspent, a merchant reveals its serial number, once again enabling merchants to link probabilistic payments by learning about their escrows.

Overall, while the above ideas do achieve unlinkability of macropayments, customers have little meaningful privacy until nullpayments and escrows are also unlinkable.

2.1.2   Attempt 2: Commit to DAP Payment + Probabilistic Opening + Private Deposit Coins

One way to address the anonymity problems of the previous attempt is to ensure that the merchant learns the serial number only when the payment turns into a macropayment (and, conversely, learns nothing otherwise). Then, to enable the aforementioned penalty escrow mechanism, a customer creates a special 'deposit' coin.

Then, the modified protocol works as follows: (1) the customer sends to the merchant a commitment to a DAP payment and to a 2-out-of-n share of the deposit serial number; (2) the customer and merchant engage in a protocol that opens the commitment with probability p (opening thus corresponds to a macropayment, and not opening corresponds to a nullpayment); (3) when publishing a macropayment to the ledger, the merchant also publishes the secret share.

The probabilistic opening hides the serial number of the coin in the DAP payment until a macropayment occurs, and the secret share hides the deposit serial number until a macropayment double spend occurs. To punish a double spending customer, the merchant obtains (from the network or from the ledger) two secret shares of the deposit serial number from two macropayments and reconstructs the serial number. He then publishes this to the ledger, thereby blacklisting the deposit.

One issue that must be addressed is ensuring that the secret shared deposit serial number corresponds to a valid deposit. To do this, first notice that there are two kinds of blacklisted deposits: those whose serial number appears on the ledger (in previous 'punish' transactions), and those that have been revoked in the current epoch. The serial numbers of the latter kind are broadcast across the network, but have not yet appeared on the ledger.

To prevent users from using blacklisted deposits of the first kind, a customer must prove to the merchant that his deposit's serial number does not appear on the ledger (this can be done efficiently [MRK03]). To prevent use of deposits of the second kind, customers must also send to the merchant a tag derived from the deposit's serial number. Since anyone with access to this serial number can compute this tag, merchants can deduce if a deposit has been revoked by checking if this tag has been computed with a blacklisted deposit's serial number. The customer accompanies the tag with a zero-knowledge proof that the deposit used for this tag is consistent with the share inside the commitment.

The aforementioned proposal, however, is still vulnerable to attacks.

Problem: front-running deposit revocation. While deposits are intended to deter double spending, customers may try to withdraw a deposit before it is blacklisted, thereby rendering punishment ineffective.

Problem: merchant aborts. At the end of the commitment opening protocol, the merchant can refuse to inform the customer of whether or not the commitment was opened. This poses a problem for the customer because if the commitment was in fact opened, the merchant has learned the serial number and a share of the deposit, enabling him to: (i) track the customer and learn when they spend the coin with another merchant, and (ii) revoke the customer's deposit after the (honest) customer next spends the coin, with another merchant or the same one.

2.1.3   Outline of Our Construction

The deposit mechanism described so far is insufficient to deter double spending. The problem is that there is no restriction on how and when coins used for probabilistic payments and for deposits can be transferred; in particular, a cheating customer can double spend these back to himself while at the same time engaging in a probabilistic payment with a merchant. We address this problem by (i) partitioning coins into different types depending on their different uses, and (ii) restricting transfers between coins depending on their types. We now outline how we carry out this plan.

First, we extend the notion of a DAP scheme to allow users to associate public and private information strings when minting a coin. Users can now store a coin's type in its public information string, and we allow three types of coins: in addition to the "standard" coin type, we introduce deposits and tickets. A ticket is bound to a deposit by storing the deposit inside the ticket's private information string. We thus have the following semantics:

- *Coins* are used for deterministic DAP payments (whose processing fees are not amortized).
- *Deposits* are used to back tickets and are revoked when two macropayments using the same ticket are detected.
- *Tickets* are used for probabilistic payments; every ticket is bound to a single deposit at minting time, and can be spent provided that the associated deposit is valid (i.e., has not be transferred to a coin, or revoked).

We also restrict the set of possible transactions depending on the types of coins involved, as follows.

- *Transactions with coins:* Coins can be used to create other coins, deposits, or tickets. In particular, coin-to-coin transactions preserve the deterministic payment functionality of the underlying DAP scheme.
- *Transactions with deposits:* Deposit-to-coin transactions let customers withdraw deposits, though not immediately, since these transactions become active only after an *activation delay* Δ_w that is a parameter of the system.
- *Transactions with tickets:* Ticket-to-coin transactions enable probabilistic payments; they are associated with a secret share of the ticket's deposit and with a deposit-derived tag that allows merchants to detect the validity of the ticket's deposit. Ticket-to-ticket transactions omit the secret share and tag and (like deposit-to-coin transactions) become active only after an *activation delay* Δ_r that is a parameter of the system.

Restrictions on inter-type transactions are achieved via a *pour predicate* that checks that input and output coin types satisfy the above restrictions. Having made these modifications, we can now resolve the issues of the previous proposal.

Preventing deposit theft. Deposit-to-coin transactions now have a delayed activation, so customers can no longer withdraw deposits before they are blacklisted, as merchants have enough time to post deposit revocations to the ledger.

Recovering from merchant aborts. Since we cannot know what is the utility gain of a merchant for learning about the spending patterns of a customer, we cannot effectively deter merchant aborts by economic means. Instead, at the end of our commitment opening protocol, we require the merchant to prove to the customer whether or not he could open the commitment. If the merchant fails to do so, we allow customers to "refresh" their tickets by creating a ticket-to-ticket payment to themselves. Since the new ticket has a different serial number that merchants have not yet seen, they cannot track the new ticket's transaction history. Finally, since ticket-to-ticket transactions become active only after a delay, the new tickets cannot be spent immediately, thus allowing merchants to post macropayments over the old ticket.

The above sketch omits many technical details, including how a DAM scheme interacts with the economic analysis. See the full version.

2.2   Intuition for Our Economic Analysis of Double Spending

Our economic analysis characterizes the additional utility that customers can gain by double spending in offline probabilistic payments. We discuss the intuition for the analysis via an example; details of the analysis are in Sect. 3 (the formal statement is Theorem 4). Recall that we assume that: (i) every probabilistic payment is backed by an advance deposit, and (ii) macropayment double spends are detected within time T, and result in deposit revocation.

At a high level, the deposit must be at least as large as the additional utility that a malicious customer gains by double spending until that deposit is revoked; additional utility occurs when the customer double spends, and accumulates until cheating is detected and every merchant has blacklisted the customer. If we can bound the value of payments in this period of time, then we can derive a corresponding bound on the additional utility gained, and thus bound the deposit.

A naive analysis, however, yields an impractically large bound, because the natural definition of "additional utility" is too coarse. We illustrate this issue via an example: a malicious customer $\tilde{\mathsf{C}}$ selects two merchants $\mathsf{M}_1, \mathsf{M}_2$, and uses the same "lottery ticket" to conduct parallel probabilistic payments $\widetilde{\mathsf{pay}}_1, \widetilde{\mathsf{pay}}_2$ to $\mathsf{M}_1, \mathsf{M}_2$ respectively. The merchants cannot immediately detect that $\tilde{\mathsf{C}}$ is cheating because $\tilde{\mathsf{C}}$ is indistinguishable from an honest user so far. If both $\widetilde{\mathsf{pay}}_1$ and $\widetilde{\mathsf{pay}}_2$ become macropayments, which happens with probability p^2, then the merchants (eventually) catch $\tilde{\mathsf{C}}$ cheating, and revoke $\tilde{\mathsf{C}}$'s deposit of value D. Consider the following two analyses.

(i) A naive analysis. The malicious customer $\tilde{\mathsf{C}}$ earns an additional utility of pV compared to an honest customer, and is caught and punished by D with probability p^2. Hence, to deter $\tilde{\mathsf{C}}$ from cheating, the deposit amount should be such that $p^2 D > pV$, which is equivalent to $D > V/p$.

(ii) A better analysis. The average-case utility $\mathbb{E}[\mathcal{U}(\mathsf{C})]$ of an honest customer C for any probabilistic payment is zero: C gains pV with probability $1 - p$, and $pV - V$ with probability p. Instead, the utility $\mathcal{U}(\tilde{\mathsf{C}})$ of the malicious customer $\tilde{\mathsf{C}}$ has four cases, as given in Table 2; also, $\tilde{\mathsf{C}}$ is caught and punished by D with probability p^2. Thus, the deposit amount should be such that $p^2 D > \mathbb{E}[\mathcal{U}(\tilde{\mathsf{C}})] = 2pV - (1 - (1 - p)^2)V$, which is equivalent to $D > V$.

How do the two analyses differ? The first analysis states that the deposit amount D must be greater than V/p while the second states that it must be greater than V, which is a much smaller lower bound. This is because the first analysis adopted an intuitive, but coarse, definition of additional utility, which

Table 2. Utility $\mathcal{U}(\tilde{\mathsf{C}})$ of the malicious customer $\tilde{\mathsf{C}}$.

$\widetilde{pay}_1$ \ $\widetilde{pay}_2$	null	macro
null	$2pV$	$2pV - V$
macro	$2pV - V$	$2pV - V$

Table 3. Utility $\mathcal{U}(\mathsf{C})$ of the honest customer C.

pay_1 \ pay_2	null	macro
null	$2pV$	$2pV - V$
macro	$2pV - V$	$2pV - 2V$

did not consider the fact that a malicious customer does not gain additional utility unless two macropayments with the same ticket occur. Indeed, the utility $\mathcal{U}(\mathsf{C})$ of an honest user C that uses two *different* tickets to make two parallel probabilistic payments pay_1, pay_2 is in Table 3. By comparing $\mathcal{U}(\tilde{\mathsf{C}})$ and $\mathcal{U}(\mathsf{C})$, one can see that the utility function differs *only when two macropayments occur*, where, if there is no deposit/punishment, $\tilde{\mathsf{C}}$ gains extra utility of V by paying only one macropayment instead of paying two as C does. In sum, any additional utility gained via double spending *must come from macropayment double spends*.

Towards a general analysis. The above discussion suggests that the additional utility of $\tilde{\mathsf{C}}$, which we denote by $\mathcal{U}'(\tilde{\mathsf{C}})$, should be defined as follows:

$$\mathcal{U}'(\tilde{\mathsf{C}}) := \begin{cases} V \text{ if } \widetilde{pay}_1, \widetilde{pay}_2 \text{ are macropayments} \\ 0 \text{ otherwise} \end{cases}.$$

More generally, the additional utility of any malicious customer $\tilde{\mathsf{C}}$ is the extra gain compared to an honest customer achieving the same outcome. This can be computed by considering an honest customer C that simulates the behavior of $\tilde{\mathsf{C}}$ while only using unspent tickets; the extra gain arises from the fact that C has "paid" for these other unspent tickets while $\tilde{\mathsf{C}}$ has not. By understanding the maximum of this refined notion of additional utility we can derive the minimum amount of deposit needed such that, for any double spending attack, there is a non-double-spending strategy that achieves better utility, in the worst-case and in the average-case respectively. See Sect. 3 for a formal argument of this intuition, as well as a discussion of the implications of our economic analysis.

3   Economic Analysis of Double Spending for Offline Probabilistic Payments

We provide the economic analysis that characterizes the additional utility that can be gained by double spending via offline probabilistic payments. This section is organized as follows. First, we informally describe dynamics that model offline probabilistic payments (Sect. 3.1). Then, we define a formal game that captures these dynamics and analyze this game (Sect. 3.2). Finally, we discuss the interpretation and consequences of our economic analysis (Sect. 3.3).

3.1   Informal Description of Payment Dynamics

We informally describe the dynamics of arbitrary probabilistic payments from customers to merchants. A concrete example is the setting of *advertisement-free Internet*: a customer is a user surfing the Internet; a merchant is a web server; every HTTP request by a user to a web server is accompanied by a probabilistic payment from that user to the web server (to buy an ad-free HTTP response).

Abstraction of probabilistic payments. A probabilistic payment is an interactive protocol between a customer and a merchant. The customer's input is a *ticket* $\mathbf{t} = (\mathsf{t}, p, V, \mathbf{d})$ where $\mathsf{t} \in \{0,1\}^*$ is the unique *ticket identifier*, $p \in [0,1]$ is the *payment probability*, $V \in \mathbb{R}_{\geq 0}$ is the *macropayment value*, and $\mathbf{d} = (\mathsf{d}, D)$ is the *deposit*, which consists of a unique *deposit identifier* $\mathsf{d} \in \{0,1\}^*$ and a *deposit value* $D \in \mathbb{R}_{\geq 0}$. Informally, the customer first convinces the merchant that the deposit is not "invalid", and then the customer pays V to the merchant with probability p. The two outcomes are called a *nullpayment* and a *macropayment*, and involve different protocol outputs.

Detectable double spends. At any moment in time, a deposit is in one of two states: *valid* or *invalid*. Each deposit is initially valid. When two macropayments occur on the same ticket $\mathbf{t}$, the associated deposit $\mathbf{d}$ becomes invalid, once and for all. We call this event a *macropayment double spend*, and we assume that, in this case, the underlying probabilistic payment protocol enables merchants to eventually learn that $\mathbf{d}$ (more precisely, its identifier) has become invalid;[5] we denote by T_{M} the time for merchant M to learn this from the moment the macropayment double spend occurred. The fact that $\max_{\mathsf{M}} T_{\mathsf{M}} > 0$ is the fundamental reason that allows a malicious customer to gain any additional utility.

Finally, we make the simplifying assumption that, while macropayment double spends are detectable, nullpayment double spends are undetectable. Our analysis does extend to the case where (not necessarily all) nullpayment double spends are also detectable; see Remark 1.

Honesty of merchants. We assume that *merchants behave honestly*. Thus, every merchant (a) rejects aborted payments (e.g., due to invalid deposits); (b) honors successful payments (e.g., replies with an ad-free HTTP response) regardless of whether the payment resulted in a nullpayment or macropayment; (c) reports detected double spends; more generally, (d) follows the probabilistic payment protocol (e.g., uses fresh randomness in each instance of the protocol, broadcasts any messages to all other merchants as instructed, and so on).

In principle, merchants may deviate from the aforementioned honest behavior in a variety of ways. For instance, a merchant may "honor" an aborted payment (e.g., regardless of the validity of the customer's deposit); or the merchant may not honor a successful payment (e.g., does not reply to the HTTP request); or the merchant may abort and prevent the customer from learning the payment's outcome; or the merchant may not report a detected double spend.

---

[5] Exactly *how* merchants learn $\mathbf{d}$'s identifier depends on the details of a construction, and is orthogonal to our economic analysis; ditto for exactly how the monetary funds escrowed in $\mathbf{d}$ are revoked after $\mathbf{d}$ becomes invalid.

However, we assume that all merchants behave honestly because the only incentive for a merchant to deviate comes from colluding with malicious customers, and we cannot prevent such collusions. Indeed, if a merchant does not collude with any malicious customer, then for the merchant it is individually rational to behave honestly, because: (i) some malicious merchant behavior (e.g., "honoring" an aborted payment, or using correlated randomness across payments) does not increase the merchant's utility; (ii) other malicious merchant behavior (e.g., not honoring a successful payment) decreases the customer's utility, but taking into account this possibility does not affect a customer's maximum additional utility (the quantity we study) and ruling it out significantly simplifies the analysis. However, a malicious customer could convince a merchant to *not* report a double spend by offering side payments as compensation; if the merchant has already replied to the customer's payment then this collusion may indeed be economically attractive, but we cannot systematically prevent such side payments in all applications. (In the setting of micropayments, V is small so a merchant may prefer to see the malicious customer punished, after losing V, rather than receiving compensation.)

Honest vs. malicious customers. Our goal is to characterize the additional utility obtained by any malicious customer, when compared to what is possible by honest customers. We now discuss both kinds of customers.

Honest customers. For an honest customer, a ticket **t** is in one of three states: it is *spent* if a probabilistic payment on it has resulted in a macropayment; otherwise, it is *occupied* if it is being used in a probabilistic payment; otherwise, it is *unspent* (i.e., it never resulted in a macropayment, nor is it being used in a probabilistic payment).

At any moment in time, an honest customer may select any number of merchants, and initiate any number of probabilistic payments in parallel to every one of them. Each probabilistic payment uses a distinct unspent ticket, which immediately becomes occupied, and at the end of the payment protocol becomes either unspent or spent. The selected tickets may or may not have different deposits that back them; deposits are never invalidated for honest customers. In sum, an honest customer maintains the invariant that an occupied ticket does not participate in more than one payment at a time, and a spent ticket does not participate in future payments.

Malicious customers. A malicious customer may deviate from the aforementioned honest behavior in a variety of ways, as we now describe. Like an honest customer, a malicious customer owns an arbitrary number of tickets and deposits; unlike an honest customer, a malicious customer may use an occupied ticket in multiple payments, or may use a spent ticket in future payments (hence, a ticket of a malicious customer could be in *both spent and occupied states at the same time*). We give some examples of malicious behavior.

- **One-ticket-one-merchant attack.** A malicious customer $\tilde{\mathsf{C}}$ has a ticket **t** and selects a merchant M; then $\tilde{\mathsf{C}}$ initiates multiple probabilistic payments to M in parallel, and continues using the same ticket **t** even after it is spent.

The merchant M cannot detect that $\tilde{\mathsf{C}}$ is cheating until M receives two macro-payments relative to the same ticket $\mathbf{t}$.

- **One-ticket-multiple-merchant attack.** A malicious customer $\tilde{\mathsf{C}}$ has a ticket $\mathbf{t}$ and selects two merchants $\mathsf{M}_1, \mathsf{M}_2$; then $\tilde{\mathsf{C}}$ conducts a sequence of probabilistic payments to M_1, using $\mathbf{t}$ until it is spent to M_1. In parallel, $\tilde{\mathsf{C}}$ adopts the same strategy with M_2, until $\mathbf{t}$ is spent to M_2. Observe that $\tilde{\mathsf{C}}$ acts like an honest customer to M_1 and M_2 individually; hence, the two merchants cannot detect that $\tilde{\mathsf{C}}$ is cheating until they communicate.

- **Multiple-ticket-multiple-merchant attack.** More generally, a malicious customer $\tilde{\mathsf{C}}$ has multiple tickets $\mathbf{t}_1, \mathbf{t}_2, \ldots$ and selects multiple merchants $\mathsf{M}_1, \mathsf{M}_2, \ldots$; then $\tilde{\mathsf{C}}$ conducts a sequence of probabilistic payments to M_1, using $\mathbf{t}_1$ until it is spent to M_1. Then $\tilde{\mathsf{C}}$ switches to $\mathbf{t}_2$ and continues making probabilistic payments to M_1 until $\mathbf{t}_2$ is spent. The customer $\tilde{\mathsf{C}}$ continues in this way until all the tickets are spent to M_1. In parallel, $\tilde{\mathsf{C}}$ adopts the same strategy with every other merchant. Observe again that $\tilde{\mathsf{C}}$ acts like an honest customer to each merchant individually; hence, the merchants cannot detect that $\tilde{\mathsf{C}}$ is cheating until they communicate.

Recall that, no matter what a malicious customer does, whenever two macro-payments relative to the same ticket $\mathbf{t}$ occur, the deposit of $\mathbf{t}$ becomes invalid, and eventually (after at most time $\max_{\mathsf{M}} T_{\mathsf{M}}$) all merchants learn about this.

Towards a formal game. The above discussion leads us to the following informal description of arbitrary dynamics of probabilistic payments from a potentially-malicious customer to honest merchants; this description is only an intermediate step that we provide for intuition, because we formally define an abstract game in Sect. 3.2 below.

For each time t, let $\mathcal{I}(t)$ denote the set of deposit identifiers of invalid deposits at time t. This set is not maintained by anyone: by definition it contains the correct identifiers at any time. It is public and, hence, known to the customer.

Suppose that a customer initiates a probabilistic payment with merchant M at time t, using a ticket $\mathbf{t} = (\mathbf{t}, V, p, (\mathsf{d}, D))$. If $\mathsf{d} \in \mathcal{I}(t - T_{\mathsf{M}})$ (the deposit identifier belongs to an invalid deposit) then the payment aborts. Otherwise, (i) with probability $1 - p$, both parties receive the output `null`; (ii) with probability p, both parties receive the output `macro`.

Crucially, the decision of whether a payment aborts depends only on the global information from T_{M} units of time "into the past", because, in the worst case, there is a delay of T_{M} for merchant M to learn that a deposit has been invalidated. Of course, the merchant M may happen to learn this information faster than that; though modeling this fact does not ultimately change the maximum additional utility, so we ignore this for simplicity. This means that all merchants "behave the same" and thus we replace them with a single abstract player, 'Nature', in the next section.

Note that a construction of a probabilistic payment should also involve a check of whether the deposit value D is "large enough" to back the payment (as informed by our economic analysis). We ignore this check (and how it can be performed) because it is irrelevant to the economic analysis.

3.2   The Game and Its Analysis

We define a single-player game against Nature that captures the dynamics described in Sect. 3.1, namely, the dynamics of a customer $\tilde{C}$ conducting arbitrary probabilistic payments with all merchants. We prove tight bounds on $\tilde{C}$'s additional utility, in the worst case and in the average case. Note that, due to the additive nature of utility, we only need to analyze $\tilde{C}$'s additional utility *per deposit*; hence, we restrict $\tilde{C}$ to backing all his probabilistic payments with a single deposit.

As mentioned in Sect. 1.1.1, our analysis involves two parameters A and W, which denote the (per-deposit) maximum value of probabilistic payments and of macropayments, within any "detection time period". More precisely, let T_M denote the time for a merchant M to detect a detectable double spend, and let $\mathfrak{a}_M(t)$ be the (cumulative) value of probabilistic payments accepted by M within the time period $[t, t + T_M]$; similarly, let $\mathfrak{w}_M(t)$ be the (cumulative) value of macropayments accepted by M within the time period $[t, t+T_M]$. The parameters A and W are defined as $\max_t \sum_M \mathfrak{a}_M(t)$ and $\max_t \sum_M \mathfrak{w}_M(t)$ respectively. We defer to Sect. 3.3 a discussion of the interpretation of these parameters, and for now we focus on analyzing the additional utility in terms of these.

We argue that it suffices to study $\tilde{C}$'s additional utility across merchants within a certain time period, and to consider only probabilistic payments that use spent tickets.

- *Starting point.* It suffices to analyze $\tilde{C}$'s additional utility from the first time when two macropayments occur relative to the same ticket; denote by $\widetilde{\text{pay}}$ the payment among these that terminates later (if they terminate simultaneously then break ties arbitrarily). Indeed, recall that $\tilde{C}$'s additional utility is the extra gain compared to any honest customer achieving the same outcome. So consider the honest customer C that uses unspent tickets for every probabilistic payment that terminates before $\widetilde{\text{pay}}$ does: the utilities up to then for $\tilde{C}$ and C are the same. Thus, we only need to consider $\tilde{C}$'s additional utility from when $\widetilde{\text{pay}}$ terminates.

- *Ending point.* It suffices to analyze $\tilde{C}$'s additional utility from when $\widetilde{\text{pay}}$ terminates until when every merchant M has detected $\tilde{C}$'s cheating. Indeed, $\widetilde{\text{pay}}$ is a detectable double spend, so within time T_M merchant M detects $\tilde{C}$'s cheating (i.e., has learned that $\tilde{C}$'s deposit is invalid) and will not accept $\tilde{C}$'s probabilistic payment anymore. Moreover, $\tilde{C}$'s deposit is eventually revoked.

- *Which payments.* It suffices to consider every probabilistic payment that terminates within the aforementioned time period and uses a ticket that is spent before the termination of that payment (if multiple payments terminate simultaneously then pick an arbitrary termination order for them). Throughout this section we say that these probabilistic payments *use spent tickets*, and say that the other probabilistic payments *use unspent tickets*. Indeed, consider again the honest customer C that uses unspent tickets for every probabilistic payment: the utilities for $\tilde{C}$ and C are the same on probabilistic payments that use unspent tickets.

In conclusion, we only need to worry about $\tilde{C}$'s additional utility from when $\widetilde{\text{pay}}$ terminates until when every merchant has detected $\tilde{C}$'s cheating, and it suffices to consider only probabilistic payments that use spent tickets.

Suppose that during this time period $\tilde{C}$ has finished $C + 1$ probabilistic payments, including $\widetilde{\text{pay}}$, using spent tickets: $\widetilde{\text{pay}}$ is fixed to be a macropayment, while the remaining C payments are probabilistic (i.e., turn into nullpayments or macropayments with the appropriate probability). Perhaps $\tilde{C}$ only made $C+1$ payments, or perhaps the merchants accepted only the first $C + 1$ and rejected the rest due to invalid or insufficient deposit. (We assume $C < \infty$ for ease of exposition, but we could replace C with ∞ and our analysis would still hold.) Either way, note that $\tilde{C}$ may select the payment probability and macropayment value of a probabilistic payment based on the outcomes of prior probabilistic payments. Below we define a game that captures these payments.

Definition 1. *Consider the following single-player game against Nature.*

- *The set of randomness choices is $[0,1]^C$; Nature samples $\boldsymbol{\lambda}$ uniformly at random from $[0,1]^C$. We denote by $\boldsymbol{\lambda}_{<i}$ the first $(i-1)$ coordinates of $\boldsymbol{\lambda}$ (and define $\boldsymbol{\lambda}_{<0}$ and $\boldsymbol{\lambda}_{<1}$ to be the empty string).*
- *The player strategies Σ consist of tuples $\boldsymbol{\sigma} = (p_i, V_i)_{i=0}^C$ consisting of computable functions that, based on Nature's randomness choice, output parameters for all the probabilistic payments. More precisely, for each i, $p_i(\boldsymbol{\lambda}_{<i}) \in [0,1]$ is the payment probability of the i-th probabilistic payment, and $V_i(\boldsymbol{\lambda}_{<i}) \in \mathbb{R}_{\geq 0}$ is its macropayment value.*

The game proceeds as follows. The player selects a strategy $\boldsymbol{\sigma} \in \Sigma$; afterwards, Nature samples $\boldsymbol{\lambda}$, whose coordinates are revealed to the player round by round. More precisely, the game is played in rounds, as follows: in round i, the player learns $\boldsymbol{\lambda}_{<i}$, and conducts a probabilistic payment (using a spent ticket) with payment probability $p_i(\boldsymbol{\lambda}_{<i})$ and macropayment value $V_i(\boldsymbol{\lambda}_{<i})$. The outcome of the i-th round is given by the indicator $\mathbb{I}[\lambda_i \leq p_i(\boldsymbol{\lambda}_{<i})]$, stating whether the payment resulting in a macropayment (the indicator equals 1) or nullpayment (the indicator equals 0).

Observe that *all* strategies in the above game are double-spending strategies: as discussed, it suffices to consider only probabilistic payments that use spent tickets. We now turn to define additional utility. Comparing an honest customer with a malicious one, we observe that any additional utility comes only from macropayments that involve spent tickets. More precisely, the first such macropayment (which is $\widetilde{\text{pay}}$) contributes additional utility V_0 and, after that, if the i-th probabilistic payment results in a macropayment then additional utility increases by $V_i(\boldsymbol{\lambda}_{<i})$. As for nullpayments, neither an honest nor a malicious customer loses tickets, hence additional utility does not increase. Therefore, we define additional utility as follows.

Definition 2. *The* **additional utility** *of a strategy* $\sigma \in \Sigma$ *on randomness* $\boldsymbol{\lambda} \in [0,1]^C$ *is*

$$\mathcal{U}'_{\boldsymbol{\lambda}}(\sigma) := V_0 + \sum_{i=1}^{C} \mathbb{I}[\lambda_i \leq p_i(\boldsymbol{\lambda}_{<i})]V_i(\boldsymbol{\lambda}_{<i}).$$

(Additional utility is a random variable, as it depends on Nature's randomness $\boldsymbol{\lambda}$, which is a random variable.)

We analyze the *maximum* additional utility achievable by any strategy, in the worst case and in the average case, for the game from Definition 1; these maximum values bound from below the required deposit value D (for the goal of deterring double spending). Below we define two subsets of strategies in which the bounds A or W are respected. (Note that if $C < \infty$, then $(\min\{p_i\}_{i=0}^{C}) \cdot W \leq A$ so that if A is bounded then so is W.)

Definition 3. *We define the following two sets of strategies, which respectively capture the condition that the total worth of probabilistic payments is at most A and the total worth of macropayments is most W:*

$$\Sigma_A^{\mathsf{pp}} := \left\{ \sigma \in \Sigma : \forall \boldsymbol{\lambda}, \ p_0 V_0 + \sum_{i=1}^{C} p_i(\boldsymbol{\lambda}_{<i}) V_i(\boldsymbol{\lambda}_{<i}) \leq A \right\},$$

$$\Sigma_W^{\mathsf{mp}} := \left\{ \sigma \in \Sigma : \forall \boldsymbol{\lambda}, \ V_0 + \sum_{i=1}^{C} \mathbb{I}[\lambda_i \leq p_i(\boldsymbol{\lambda}_{<i})] V_i(\boldsymbol{\lambda}_{<i}) \leq W \right\}.$$

We now state and prove our worst-case and average-case bounds on additional utility. (Recall that, by Yao's minimax principle, it suffices to consider only deterministic strategies [Yao77], and thus we ignore randomized ones.)

Theorem 4 (formal statement of Theorem 1). *For the game described above, the following holds.*

(a) WORST CASE: *for every randomness choice $\boldsymbol{\lambda} \in [0,1]^C$ and strategy $\sigma \in \Sigma_W^{\mathsf{mp}}$, it holds that $\mathcal{U}'_{\boldsymbol{\lambda}}(\sigma) \leq W$.*
(b) AVERAGE CASE: *for every strategy $\sigma \in \Sigma_A^{\mathsf{pp}}$, it holds that $\mathbb{E}_{\boldsymbol{\lambda}}[\mathcal{U}'_{\boldsymbol{\lambda}}(\sigma)] \leq (1 - p_0)V_0 + A$.*
(c) *Both bounds are tight.*

Proof. We prove the three statements in order.

Part (a). By definition of Σ_A^{pp} (see Definition 3), for every randomness choice $\boldsymbol{\lambda} \in [0,1]^C$ and strategy $\sigma \in \Sigma_W^{\mathsf{mp}}$, it holds that $V_0 + \sum_{i=1}^{C} \mathbb{I}[\lambda_i \leq p_i(\boldsymbol{\lambda}_{<i})]V_i(\boldsymbol{\lambda}_{<i}) \leq W$; but the quantity on the left-hand side of the inequality is $\mathcal{U}'_{\boldsymbol{\lambda}}(\sigma)$ (see Definition 2), and the claimed statement follows.

Part (b). Recall that Nature samples $\boldsymbol{\lambda}$ uniformly at random from $[0,1]^C$, so the coordinates of $\boldsymbol{\lambda}$ are independent from one another. Therefore, for every strategy $\sigma \in \Sigma_A^{\mathsf{pp}}$,

$$\mathbb{E}_{\boldsymbol{\lambda}}\left[\mathcal{U}_{\boldsymbol{\lambda}}'(\sigma)\right] = V_0 + \mathbb{E}_{\boldsymbol{\lambda}}\left[\sum_{i=1}^{C} \mathbb{I}\left[\lambda_i \le p_i(\boldsymbol{\lambda}_{<i})\right] V_i(\boldsymbol{\lambda}_{<i})\right]$$

$$= V_0 + \mathbb{E}_{\lambda_1}\cdots\mathbb{E}_{\lambda_C}\left[\sum_{i=1}^{C} \mathbb{I}\left[\lambda_i \le p_i(\boldsymbol{\lambda}_{<i})\right] V_i(\boldsymbol{\lambda}_{<i})\right] \qquad \text{(by independence)}$$

$$= V_0 + \sum_{i=1}^{C} \mathbb{E}_{\boldsymbol{\lambda}_{<i}}\left[p_i(\boldsymbol{\lambda}_{<i}) V_i(\boldsymbol{\lambda}_{<i})\right]$$

$$\le (1 - p_0) V_0 + A. \qquad \text{(by definition of } \Sigma_A^{\mathrm{pp}})$$

as claimed.

Part (c). Consider the following two strategies consisting of a single probabilistic payment after $\widetilde{\text{pay}}$ (of value V_0):

- Choose σ such that $C := 1$, $p_1 := 1$, and $V_1 := W - V_0$. Note that $\sigma \in \Sigma_W^{\mathrm{mp}}$ and, for every randomness choice $\boldsymbol{\lambda} \in [0,1]^C$, it holds that $\mathcal{U}_{\boldsymbol{\lambda}}'(\sigma) = W$.
- Choose σ such that $C := 1$, $p_1 := 1$, and $V_1 := A - p_0 V_0$. Note that $\sigma \in \Sigma_A^{\mathrm{pp}}$ and $\mathbb{E}_{\boldsymbol{\lambda}}\left[\mathcal{U}_{\boldsymbol{\lambda}}'(\sigma)\right] = (1 - p_0) V_0 + A$.

In sum, the first strategy shows that our worst-case bound is tight, while the second strategy shows that our average-case bound is tight.

Remark 1 (detectable nullpayment double spends). So far our analysis assumes that macropayment double spends are detectable, but nullpayment double spends are not. What if some nullpayment double spends *are* detectable? For example, merchants could maintain a partial order of all payments via a synchronous clock that ticks every second, even if the broadcast time is 10 s; this partial order would give chronological information on some nullpayment vs. macropayment pairs. But does such a stronger detection guarantee improve the economic bounds?

Our analysis *does* extend to this setting, and the answer is yes, but not by much. First, if some nullpayment double spends are also detectable, the additional utility of a malicious customer can only go down, so the upper bounds of our theorem continue to hold. However, the upper bounds are not tight; nevertheless, below we sketch modifications to our analysis that do recover a tight result.

- *Starting point:* the first time a detectable double spend occurs, i.e., a macropayment *or detectable nullpayment* occurs after another macropayment on the same ticket.
- *Ending point:* every merchant has detected that double spend.
- *Additional utility:* if the starting point is a macropayment double spend, the additional utility is the same, but if the starting point is a detectable nullpayment double spend, the additional utility goes down by V_0.

The rest of the analysis follows, for parameters A and W that are now defined for this new time interval. The only difference is in the initial cost of detection, due to different detection guarantees. Afterwards, only macropayment double spends provide additional utility, which are detectable in both settings. Overall, even if we had the stronger guarantee of detecting *all* nullpayment double spends, it would only save V_0 in the average-case bound.

3.3   Interpreting the Payment Value Rates

Our analysis in Sect. 3.2 can be viewed as a reduction from the required deposit amount to certain per-deposit payment value rates: A (for the average case analysis), which is the maximum cumulative value of probabilistic payments across merchants within any detection time period; or W (for the worst case analysis), which is the maximum cumulative value of macropayments across merchants within the same period. Our analysis is *tight*, so leaving these parameters unbounded enables a malicious customer to gain unbounded additional utility (and rules out the possibility of deterring malicious behavior via economic means such as advance deposits). The purpose of this section is to discuss the meaning of bounding payment value rates, and what are the implications of such bounds. Throughout, recall that our analysis is *per deposit*, so we fix a single deposit $\mathbf{d}$ that backs all the probabilistic payments discussed below.

Interpretation of the parameters. We first discuss the detection time (used to define the rate), and then discuss how W and A may arise as a sum, across all merchants, of corresponding payment value rates.

– *Detection time.* We denote by T_M the *time for a merchant* M *to detect a detectable double spend*. For example, T_M can be the network's broadcast time, that is, the time for a message sent by a merchant to reach all other merchants (this is true, e.g., if the network contains enough honest nodes to provide reliable and timely broadcast, or if merchants have the same view of the ledger). In a Bitcoin-like system the broadcast time is much smaller than the validation time (the time for a broadcast transaction to appear in the ledger): a few seconds as opposed to a few minutes.
– *Merchants (per deposit).* We denote by N the *number of merchants* that accept probabilistic payments (backed by the deposit $\mathbf{d}$). For example, N could be the number of all merchants. (Though this need not be the case, see below.)
– *Payment value rates (per deposit).* For every merchant M, $\mathfrak{a}_M := \max_t \mathfrak{a}_M(t)$ is the maximum (cumulative) value of probabilistic payments (backed by the deposit $\mathbf{d}$) accepted by M within any time period of T_M; similarly, $\mathfrak{w}_M := \max_t \mathfrak{w}_M(t)$ is the maximum (cumulative) value of macropayments accepted by M within any time period of T_M. Then one sets A equal to $\sum_M \mathfrak{a}_M$, and W equal to $\sum_M \mathfrak{w}_M$ (or consider these as upper bounds to A and W).

Necessity of bounds. We now explain why simultaneous bounds on the aforementioned parameters are necessary. First, if there is no bound on the number

N of merchants that accept probabilistic payments backed by $\mathbf{d}$, a malicious customer can use $\mathbf{d}$ to gain unbounded additional utility via a one-ticket-multiple-merchant attack (see Sect. 3.1), even in the average case. Second, even if N is bounded (and greater than 1) but $\max_{\mathsf{M}} T_{\mathsf{M}}$ is unbounded (e.g., a large-scale eclipse attack is underway [HKZG15]), a malicious customer can gain unbounded additional utility via a multiple-ticket-multiple-merchant attack (see Sect. 3.1), even in the average case. Third, even if N and $\max_{\mathsf{M}} T_{\mathsf{M}}$ are bounded but some $\mathfrak{a}_{\mathsf{M}}$ is unbounded, our analysis implies that a malicious customer can again gain unbounded additional utility in the average case; similarly, if some $\mathfrak{w}_{\mathsf{M}}$ is unbounded, our analysis implies that a malicious customer can again gain unbounded additional utility in the worst case. In sum, if either $\max_{\mathsf{M}} T_{\mathsf{M}}$ or N are unbounded, then $A = \sum_{\mathsf{M}} \mathfrak{a}_{\mathsf{M}}$ and $W = \sum_{\mathsf{M}} \mathfrak{w}_{\mathsf{M}}$ are also unbounded; but even if $\max_{\mathsf{M}} T_{\mathsf{M}}$ and N are bounded, either A or W could still be unbounded, and so we must explicitly bound A or W (depending if we target average or worst case, or both).

Finally, observe that the above discussion assumes that there is no a-priori bound on how many tickets a single deposit can back; see Remark 2 below for a discussion of what happens if a deposit is restricted to only back macropayments up to a certain maximum total value.

Respecting the bounds. Whose responsibility is it to ensure that the bounds A or W are respected? One answer to this question could be that there are exogenous reasons (e.g., spending patterns, network behavior, and so on) that justify this statement. Another answer to this question is to say that every merchant M is responsible "for his own share": he needs to monitor that $\mathfrak{a}_{\mathsf{M}}$ and $\mathfrak{w}_{\mathsf{M}}$ are locally respected for him (and if they are about to be exceeded, he defers further payments to the next period of time T_{M}). This second answer raises an interesting technical problem: how does M know which payments are backed by the same deposit? If a payment's deposit is not private (as in [PS16]) this is not a problem. But if a payment's deposit is private, this could be tricky. In our DAM scheme construction, when engaging in a probabilistic payment, a merchant does not learn any information about the deposit that backs it, beyond the bit of whether the deposit is valid or not. Nevertheless, we still enable a merchant to get around this problem, by leveraging the notion of a *rate limit tag* within a probabilistic payment.

Implications: good news and bad news. The good news about our economic analysis is that it gives a tight characterization of the additional utility that can be gained via double spending. The bad news is that bounding A or W may impact usability. (Perhaps this is not surprising because offline probabilistic payments are a "tough" setting since double spending cannot be fully prevented.) Namely, if all $\mathfrak{a}_{\mathsf{M}}$ (resp., $\mathfrak{w}_{\mathsf{M}}$) are large, then A (resp., W) is even larger; but this impacts usability because the required deposit is large. Conversely, if many $\mathfrak{a}_{\mathsf{M}}$ (resp., $\mathfrak{w}_{\mathsf{M}}$) are small then A (resp., W), and thus the required deposit, is not as large; but the amount of value transacted with many merchants is limited, and this impacts usability because a user may not be able to transact large amounts with his "favorite" merchants.

Mitigations. A way to mitigate the above problem is to associate to each deposit a subset $\mathcal{R}$ of allowed "receiver merchants" so that the sum is taken only over this subset: $A = \sum_{M \in \mathcal{R}} \mathfrak{a}_M$ and $W = \sum_{M \in \mathcal{R}} \mathfrak{w}_M$. Then, any particular user would only have to cover his spending habits with one (or more) deposits that cover one (or more) not-too-large subsets of merchants. The subset $\mathcal{R}$ can even be private and chosen by the user; in fact, we take this approach both when defining and constructing a DAM scheme.

Another way to mitigate the above problem is for merchants to group together into *micropayment agencies*. Such an agency acts as a proxy to the subset of merchants it serves, and its only task is to "monitor" the cumulative values of $\mathfrak{a}_M$ and $\mathfrak{w}_M$ for merchants in the agency. This approach does not affect any privacy guarantees from the perspective of the customer (since every probabilistic payment is anonymous from the perspective of a single merchant or any coalition of merchants). In the extreme, one could even think of a *single* micropayment agency, and the only obstacle would be coordinating and keeping track of A and W across the network.

Remark 2 (bounded macropayments per deposit). So far we have assumed that there is no a-priori bound on how many tickets a single deposit can back. Suppose instead that a deposit $\mathbf{d}$ can only back tickets with total macropayment value up to V_{tot}. To analyze this other setting, we can reuse ideas from our economic analysis: again, one can define additional utility by comparing the utilities of a malicious merchant and a corresponding honest merchant. We omit the analysis and simply state that the additional utility is bounded by $(2N - 1)V_{tot}$, where N is the number of merchants that accept probabilistic payments backed by $\mathbf{d}$ (note that in this case $\max_M T_M$, A, W could all be unbounded). Moreover, the bound is tight; intuitively, the maximum additional utility is achieved via a multiple-ticket-multiple-merchant attack until two macropayments with the same ticket occur for each of the N merchants.

4 Efficient Fractional Message Transfer

A key ingredient in our construction of a DAM scheme is *fractional message transfer* (FMT): a primitive that enables probabilistic message transmission between two parties, called the 'sender' and the 'receiver'. Informally, the receiver samples a one-time key pair based on a transfer probability p; then, the sender uses the receiver's public key to encrypt a message m into a ciphertext c; finally, the receiver uses the secret key to decrypt c, thereby learning m, but only with the pre-defined probability p (and learns no information about m with probability $1 - p$). Our definition and construction of FMT are closely related to *non-interactive fractional oblivious transfer* (NFOT), which was studied in the context of 'translucent cryptography' as an alternative to key escrow [BM89, BR99]; see Sect. 1.1.3 for a discussion.

In this work we formulate the notion of an *FMT scheme*, which formally captures the functionality and security of probabilistic message transmission; we rely on this tool (and others) in our construction of a DAM scheme. Moreover, we give an efficient construction of an FMT scheme that works for transfer probabilities $p = 1/n$ with $n \in \mathbb{N}$; this construction is in the random oracle model and assumes the hardness of the DDH problem in prime-order groups. Finally, since probabilistic message transmission is of independent interest, we also define the notion of an *FMT protocol* via an ideal functionality, and show that the security definition of FMT schemes does imply security relative to that ideal functionality. (Our DAM scheme relies on an FMT scheme, rather than an FMT protocol, because we interleave the FMT scheme with other building blocks.)

We defer the definitions, constructions, and proofs about FMT to the full version. In the rest of this section, we informally describe the syntax, correctness, and security of FMT schemes, and then sketch our FMT construction.

Syntax. An FMT scheme is a quintuple of algorithms (FMT.Setup, FMT.Keygen, FMT.Encrypt, FMT.Decrypt) with the following syntax.

- *Parameter setup (executed by a trusted party):* FMT.Setup(1^λ) $\rightarrow$ pp$_{\mathsf{FMT}}$. On input a security parameter λ, FMT.Setup outputs the public parameters pp$_{\mathsf{FMT}}$ for the scheme.
- *Key generation (executed by the receiver):* FMT.Keygen(pp$_{\mathsf{FMT}}, p$) $\rightarrow$ (pk$_{\mathsf{FMT}}$, sk$_{\mathsf{FMT}}$). On input public parameters pp$_{\mathsf{FMT}}$ and a transfer probability p, FMT.Keygen outputs a one-time key pair (pk$_{\mathsf{FMT}}$, sk$_{\mathsf{FMT}}$).
- *Message encryption (executed by the sender):* FMT.Encrypt(pp$_{\mathsf{FMT}}$, pk$_{\mathsf{FMT}}, m$) $\rightarrow c$. On input public parameters pp$_{\mathsf{FMT}}$, a public key pk$_{\mathsf{FMT}}$ and a message m, FMT.Encrypt outputs a ciphertext c.
- *Message decryption (executed by the receiver):* FMT.Decrypt(pp$_{\mathsf{FMT}}$, sk$_{\mathsf{FMT}}, c$) $\rightarrow m'$. On input public parameters pp$_{\mathsf{FMT}}$, a secret key sk$_{\mathsf{FMT}}$ and a ciphertext c, FMT.Decrypt outputs a message m' that equals m or $\varnothing$. (The special symbol $\varnothing$ denotes that decryption resulted in no message.)

An FMT scheme satisfies the correctness and security properties defined below.

Correctness. An FMT scheme is *correct* if for every security parameter λ, public parameters pp$_{\mathsf{FMT}} \in$ FMT.Setup(1^λ), transfer probability $p \in \mathcal{P} \subseteq [0, 1]$, key pair (pk$_{\mathsf{FMT}}$, sk$_{\mathsf{FMT}}$) $\in$ FMT.Keygen(pp$_{\mathsf{FMT}}, p$), and message $m \in \mathcal{M}$,

$$\mathsf{FMT.Decrypt}(\mathsf{pp}_{\mathsf{FMT}}, \mathsf{sk}_{\mathsf{FMT}}, \mathsf{FMT.Encrypt}(\mathsf{pp}_{\mathsf{FMT}}, \mathsf{pk}_{\mathsf{FMT}}, m)) = \begin{cases} m & \text{w.p.} \quad p \\ \varnothing & \text{w.p.} \quad 1 - p \end{cases}$$

where the probability is taken over the randomness of FMT.Encrypt (and FMT.Decrypt is deterministic).

Security. An FMT scheme is *secure* if it has the properties of *fractional hiding* and *fractional binding*. Informally, fractional hiding says that an honest encryptor transferring a message m can be sure that the decryptor, who knows the secret key, learns m with probability exactly p (and $\varnothing$ with probability $1 - p$), even if the public key was generated maliciously. Fractional binding says that, for every $p' \neq p$, a malicious encryptor cannot produce a valid ciphertext that decrypts with probability p' to a valid message (i.e., not $\varnothing$).

An efficient FMT scheme. Our construction of an FMT scheme targets the case where p equals $1/n$ for some positive integer n; this case suffices within our construction of a DAP scheme. As in prior work [BM89,BR99], our starting point is the Elgamal encryption scheme [Elg85], whose semantic security relies on the hardness of DDH in prime-order groups. We now give an informal sketch of our construction.

- FMT.Setup(1^λ): sample a group $\mathbb{G}$ of prime order q (depending on λ), along with two generators $g, g_0 \in \mathbb{G}$.
- FMT.Keygen(pp_{FMT}, p): the public key contains a Pedersen commitment [Ped91] to a random s in $\{1, \ldots, n\}$ and the secret key contains the commitment's randomness; that is, the commitment is $h = g_0^{-s} g^\alpha$ for random $\alpha \in \mathbb{Z}_q$.
- FMT.Encrypt(pp_{FMT}, pk_{FMT}, m): sample random $r \in \mathbb{Z}_q$ and random $t \in \{1, \ldots, n\}$, and use h as an Elgamal public key to encrypt the message $m' := m \cdot g_0^{rt}$; the resulting ciphertext is $c = (t, c_1, c_2) = (t, g^r, m'h^r)$.
- FMT.Decrypt(pp_{FMT}, sk_{FMT}, c): use the secret key α to decrypt the ciphertext by setting $m'' := c_2/c_1^\alpha = mg_0^{r(t-s)}$.

The above sketch omits several important details. In particular, our construction also includes NIZKs (obtained via the Fiat–Shamir transform applied to simple Σ-protocols) to prove correctness of key generation and encryption. Informally, our FMT's correctness and security follow from the fact that $m'' = m$ only when $t = s$, which occurs with probability $p = 1/n$. The full construction and proof of security (based on hardness of DDH) are the full version.

5 Informal Construction Description

Recall that a DAM scheme is a tuple of algorithms:

$$
DAM = \begin{pmatrix}
\text{Setup} & \text{MintCoin}^L & \text{PourCoinToCoin}^L & & \\
\text{CreateAddr} & \text{MintDeposit}^L & \text{PourCoinToDeposit}^L & \text{WithdrawDeposit}^L & \text{Punish} \\
\text{Receive}^L & \text{MintTicket}^L & \text{PourCoinToTicket}^L & \text{RefreshTicket}^L & \text{VerifyTransaction}^L \\
& & \text{PourTicket}^L & &
\end{pmatrix}.
$$

We sketch the construction of these in Sects. 5.1 and 5.2, and then separately discuss security intuition in Sect. 5.3 and 'pour regulation' in Sect. 5.4.

5.1 Informal Algorithm Descriptions

Setup. The algorithm DAM.Setup samples public parameters for the various building blocks that we use, which includes DAP schemes and FMT schemes, as well as one-time signature schemes and NIZKs

Creating addresses. The algorithm DAM.CreateAddr samples a new address key pair by running DAP.CreateAddr with the address information set to the probabilistic payment specification and outputting its result; in other words, DAM addresses are simply addresses of the underlying DAP scheme. Receivers must bind their intended payment rates, probability, and value to the address by passing it as input to DAM.CreateAddr (other users can set it to $\perp$ if they do not intend to receive probabilistic payments).

Receiving coins. The algorithm DAM.Receive, given an address key pair, retrieves all the unspent coins sent to this address by simply running DAP.Receive. Indeed, DAP pour-coin, pour-ticket, withdraw, and refresh transactions can be viewed as DAP pour transactions, and so DAP.Receive may retrieve from these any relevant coins.

Minting notes. Each of the minting algorithms DAM.MintCoin, DAM.MintDeposit, and DAM.MintTicket first sets the public information string pub to the type of the note being minted (respectively, cn, dp, or tk), and sets the secret information string sec accordingly: for coins, sec equals $\perp$; for deposits, sec is a commitment to the deposit's receiver address set $\mathcal{R}$; for tickets, sec equals the (already-minted) deposit that backs it. Then, the algorithm mints the note by running DAP.Mint.

Pouring coins. Each of the algorithms DAM.PourCoinToCoin, DAM.PourCoinToDeposit, DAM.PourCoinToTicket first sets the public and secret information strings similarly to above, and then runs DAP.Pour to generate the new notes. A DAM scheme also includes a protocol for pouring tickets into coins, which we discuss separately in Sect. 5.2 because it is the most complex part of the construction.

Withdrawing deposits. The algorithm DAM.WithdrawDeposit, given a deposit **d** (and its address secret key) and address public key apk, pours the deposit into a new coin **c** with address apk by running DAP.Pour. The output consists of the new coin **c**, as well as a withdraw transaction tx_{wd} that is just a DAP pour transaction having activation delay Δ_w. Since pour transactions reveal the serial numbers of input notes, it is easy to blacklist the withdrawn deposit (see Sect. 5.2).

5.2   A 3-message Protocol for Probabilistic Payments

We outline the construction of DAM.PourTicket, a 3-message protocol that realizes an offline probabilistic payment between a sender (customer) and receiver (merchant). For simplicity, we only discuss enforcement of the worst-case payment rate bounds; enforcement of the average-case bound is achieved via essentially the same ideas. Recall that the worst-case bound limits the number of macropayments that occur in a particular time window tw.

1st message (sender $\leftarrow$ receiver). The first message of the protocol is from the receiver to the sender and consists of the receiver's *session identifier* sid, *session public key* spk, the list of deposits $\mathcal{D}$ that have been blacklisted in this epoch, and the desired public value v_{pub}. These are constructed as follows.

Suppose that the receiver has an address key pair $(\mathsf{apk_c}, \mathsf{ask_c})$ and wishes to receive payments at this address with payment probability p_r and macropayment value V_r; moreover, suppose that the receiver's per-deposit maximum cumulative average-case payment value rate is $\mathfrak{a}_r$. Then the receiver constructs his session identifier as $\mathsf{sid} := (\mathsf{apk_c}, \mathsf{tw_r})$. To construct the session public key, the receiver samples a new key pair $(\mathsf{pk_{SIG}}, \mathsf{sk_{SIG}})$ for the one-time signature scheme, and a new key pair $(\mathsf{pk_{FMT}}, \mathsf{sk_{FMT}})$ for the fractional message transfer scheme and sets $\mathsf{spk} := (\mathsf{pk_{FMT}}, \mathsf{pk_{SIG}})$. Finally, the deposit blacklist $\mathcal{D}$ consists of the identifiers of deposits seen in punish transactions within the *current* epoch.

2nd message (sender → receiver). The sender now pours his ticket $\mathbf{t}$ into a new coin $\mathbf{c}$ using DAP.Pour, and then uses fractional message transfer to probabilistically transmit the new coin $\mathbf{c}$ to the receiver, while also proving, in zero knowledge, that he did so correctly. We now expand on this description, which hides subtle aspects of our construction.

After pouring his ticket $\mathbf{t}$ into a new coin $\mathbf{c}$ (which results in a DAP pour transaction $\mathsf{tx_p}$), the sender uses the deposit $\mathbf{d}$ backing $\mathbf{t}$ to generate two crucial quantities: the worst-case *rate limit tag* wrlt and the *double spend tag* dst. The rate limit tag allows the receiver to enforce the payment value rate bounds required by the economic analysis. The double spend tag allows the receiver to extract deposit revocation information if and only if $\mathbf{t}$ is spent in two macropayments.

A natural strategy would be for the sender to send to the receiver, in the clear, the rate limit tag wrlt, and a FMT ciphertext c_{FMT} containing $\mathsf{tx_p}$ and dst, along with a non-interactive zero knowledge proof that both were generated correctly. However, doing so does not preserve privacy. Indeed, to ensure that the sender cannot double spend the ticket to herself and escape punishment, the ledger needs to check that the double spend tag was generated correctly. This can be done by verifying the NIZK proof, but to do this would require including the FMT ciphertext, blacklist detection tag, and rate limit tag as part of the NP instance being verified. This is problematic, since *publishing these leaks information about the transfer probability p_r and the deposit, both of which are private information.*

To fix this problem, the sender hides wrlt and c_{FMT} inside two commitments ω_0 and ω_1, and then computes a proof of correctness relative to these commitments. More precisely, the first commitment ω_0 hides $m_0 := (\mathsf{sid}, \mathsf{spk}, v_{\mathsf{pub}}, c_{\mathsf{FMT}})$, where sid, spk, and v_{pub} are the receiver's session identifier, session public key, and public value respectively, and c_{FMT} is a FMT ciphertext. The FMT ciphertext c_{FMT}, as before, contains wrlt, $\mathsf{tx_p}$ and dst, but now also contains randomness r_1 that opens the second commitment ω_1, which in turn hides $m_1 := (\mathsf{tx_p}, \mathsf{dst}, \mathsf{wrlt})$. Thus opening the FMT ciphertext allows the receiver to open ω_1 and obtain the correct $\mathsf{tx_p}, \mathsf{dst}, \mathsf{wrlt}$. Next, the sender generates a non-interactive zero knowledge proof of knowledge π_{pt} asserting that he performed all these steps correctly. The NIZK also asserts (a) that the deposit $\mathbf{d}$'s receiver address set $\mathcal{R}$ contains the receiver's address public key apk, (b) that $\mathbf{d}$'s identifier has not appeared in punish transactions in the current epoch, and (c) that $\mathbf{d}$'s serial number has not appeared on the ledger prior to the current epoch (that is, $\mathbf{d}$ was not revoked or withdrawn in prior epochs).

Finally, he sends $(\omega_0, m_0, \omega_1, \pi_{\mathsf{pt}})$ and randomness r_0 for opening ω_0 to the receiver. Since the proof is now computed relative to ω_0 and ω_1, and not c_{FMT} and wrlt, it can safely be published to the ledger.

3rd message (sender $\leftarrow$ receiver). The receiver uses r_0 and m_0 to open ω_0 and checks that the committed sid, spk and v_{pub} are indeed the correct ones (which were sent in the first message). Next, he checks the correctness of π_{pt}, and finally, using the rate limit tag, he checks that the payment value rate a_r has not been exceeded. If these checks pass, he tries to open the FMT ciphertext c_{FMT} inside ω_0. If he is able to successfully open it, he can open ω_1 to obtain tx_p and dst. If the ticket $\mathbf{t}$ has already been spent (i.e., the deposit $\mathbf{d}$ has been blacklisted), the receiver recovers the deposit and creates a punish transaction $\mathsf{tx}_{\mathsf{pun}}$. If not, he posts tx_p to obtain his payment. Finally, he sends to the sender the secret key $\mathsf{sk}_{\mathsf{FMT}}$ used for decryption, and m', which is the outcome of decryption, to communicate whether the outcome was 'macropayment' or 'nullpayment'.

Outcome verification. Upon receiving the FMT secret key, the sender checks that the FMT ciphertext c_{FMT} decrypts to claimed message m' under the key $\mathsf{sk}_{\mathsf{FMT}}$; this reveals whether the claimed outcome was the correct one. If the receiver sends an incorrect secret key, or does not send anything at all, the sender refreshes his ticket, thereby generating a new ticket $\mathbf{t}'$ and a refresh transaction $\mathsf{tx}_{\mathsf{ref}}$.

5.3   Security Considerations

We give an intuitive justification of why the probabilistic payment protocol is secure.

Sender security. The fractional hiding property of the FMT scheme ensures that the receiver can only open c_{FMT} with probability p_r. Since the commitment ω_1 is hiding and the proof π_{pt} is zero knowledge, the rest of the sender's message is indistinguishable from random. Finally, the security of the "outcome verification" step is guaranteed by the fractional hiding property of the FMT scheme; if the receiver could generate two different secret keys that can decrypt the same FMT ciphertext to different messages, then he could bias the probability of opening the ciphertext in his favor, thus breaking fractional hiding.

The above ensures "intra-protocol" sender security. Post-protocol security requires that the receiver cannot compromise the honest sender's anonymity or cause monetary loss by aborting. This is achieved by allowing the sender to refresh tickets by pouring them into new ones. This breaks the link between the ticket that the receiver has seen and the ticket that the sender can now spend, enabling the sender to freely spend his new ticket.

Receiver security. Opening ω_0 allows the receiver to check that the sender generated the rate limit tags relative to the true session identifier and public key. The fractional binding property of the FMT scheme ensures that the sender cannot alter the probability of opening c_{FMT}. The NIZK proof ensures the correctness of each step.

The above ensures "intra-protocol" receiver security. Achieving post-protocol security is trickier, since we need to ensure that the sender can only create double spend tags that are consistent across independent pour-ticket transactions. In our construction, the sender can attempt to bypass this requirement by manipulating the three inputs that create a double spend tag: the randomness x used for generating the tag, and the deposit $\mathbf{d}$ that is hidden in the tag, and the ticket $\mathbf{t}$ that $\mathbf{d}$ backs.

Preventing reuse of randomness. To prevent recovery of the deposit serial number from multiple double spend tags, the sender could attempt to reuse randomness across each tag. This would prevent recovery, since each receiver would possess the same tag. To prevent this, our construction of a double spend tag dst uses a special one-time signature public key $\mathsf{pk}_{\mathsf{SIG}}$ as randomness. Later, upon receiving the tag, the receiver signs the tag (among other things) with the secret key $\mathsf{sk}_{\mathsf{SIG}}$ corresponding to $\mathsf{pk}_{\mathsf{SIG}}$. To create two different pour-ticket transactions with the same double spend tag (one to an honest receiver and one back to himself), the sender would thus have to forge a signature, which is computationally infeasible by the security of the signature scheme.

Ensuring $\mathbf{d}$ backs $\mathbf{t}$. The NIZK proof created by the sender ensures that the deposit $\mathbf{d}$ hidden in the double spend tag is the one backing $\mathbf{t}$.

'Identical' tickets backed by different deposits. In principle, one could construct two tickets $\mathbf{t}, \mathbf{t}'$ that have the same serial number (and are thus indistinguishable from the point of view of double spending), but are backed by different deposits. Since $\mathbf{t}$ and $\mathbf{t}'$ would share serial numbers, only one of the two could be successfully spent. This could lead to the following attack: the sender generates two such tickets, and pays himself with one, and pays a receiver with the other. When a macropayment occurs, he front runs the receiver to get his self-payment onto the ledger first. The receiver is then robbed of his payment, but also cannot punish the sender, since the double spend tags hide different deposits, making revocation impossible.

However, our construction prevents such an attack by ensuring that the serial number of a note is derived (in part) from its secret information string sec. This property is guaranteed by the DAP scheme.

5.4   Regulating Type Transitions When Pouring

The definition of a DAM scheme restricts fund transfers between different note types: coins can be poured into coins, deposits, or tickets; a deposit can be poured into a coin; a ticket can be poured into a coin or ticket. Moreover, some type transitions are handled differently from others: for example, pouring from a set of coins yields a pour-coin transaction that is immediately valid, while pouring from a ticket to a ticket yields a refresh transaction that only becomes valid after a waiting period (the activation delay). We realize most of these fund transfers via DAP pours, but we must also somehow meet the aforementioned restrictions.

The first obstacle is that a note's type is not necessarily known, because we store the type of note in its public information string pub, which is not revealed by a DAP pour transaction. But remember that a DAP scheme allows us to choose, at parameter setup time, a pour predicate that regulates all pour transactions. We thus engineer a pour predicate Π_{p}^*, tailored for our application, that (i) allows only the aforementioned type transitions, and (ii) ensures that the information string info in a DAP pour transaction correctly exposes the type of the note from which we are pouring.

References

[BBSU12] Barber, S., Boyen, X., Shi, E., Uzun, E.: Bitter to better — how to make bitcoin a better currency. In: Keromytis, A.D. (ed.) FC 2012. LNCS, vol. 7397, pp. 399–414. Springer, Heidelberg (2012). doi:10.1007/978-3-642-32946-3_29

[BCG+14] Ben-Sasson, E., Chiesa, A., Garman, C., Green, M., Miers, I., Tromer, E., Virza. M.: Zerocash: decentralized anonymous payments from Bitcoin. In: SP 2014 (2014)

[Bit13] Bitcoinj: Working with micropayment channels (2013). https://bitcoinj.github.io/working-with-micropayments

[Blo14] Block Chain Analysis: Block chain analysis (2014). http://www.block-chain-analysis.com/

[BM89] Bellare, M., Micali, S.: Non-interactive oblivious transfer and applications. In: Brassard, G. (ed.) CRYPTO 1989. LNCS, vol. 435, pp. 547–557. Springer, New York (1990). doi:10.1007/0-387-34805-0_48

[BP15] Biryukov, A., Pustogarov, I.: Proof-of-work as anonymous micropayment: rewarding a tor relay. In: Böhme, R., Okamoto, T. (eds.) FC 2015. LNCS, vol. 8975, pp. 445–455. Springer, Heidelberg (2015). doi:10.1007/978-3-662-47854-7_27

[BR99] Bellare, M., Rivest, R.L.: Translucent cryptography - an alternative to key escrow, and its implementation via fractional oblivious transfer. J. Cryptology **12**(2), 117–139 (1999)

[Cal12] Caldwell, M.: Sustainable nanopayment idea: probabilistic payments (2012). https://bitcointalk.org/index.php?topic=62558.0

[Cha82] Chaum, D.: Blind signatures for untraceable payments. In: Chaum, D., Rivest, R.L., Sherman, A.T., (eds.) CRYPTO 1982. Springer, New York (1982)

[Cha15] Chainalysis: Chainalysis inc. (2015). https://chainalysis.com/

[CHL05] Camenisch, J., Hohenberger, S., Lysyanskaya, A.: Compact e-cash. In: Cramer, R. (ed.) EUROCRYPT 2005. LNCS, vol. 3494, pp. 302–321. Springer, Heidelberg (2005). doi:10.1007/11426639_18

[DFKP13] Danezis, G., Fournet, C., Kohlweiss, M., Parno, B.: Pinocchio Coin: building Zerocoin from a succinct pairing-based proof system. In: PETShop 2013 (2013)

[DW15] Decker, C., Wattenhofer, R.: A fast and scalable payment network with Bitcoin duplex micropayment channels. In: Pelc, A., Schwarzmann, A.A. (eds.) SSS 2015. LNCS, vol. 9212, pp. 3–18. Springer, Cham (2015). doi:10.1007/978-3-319-21741-3_1

[Elg85] Elgamal, T.: A public key cryptosystem and a signature scheme based on discrete logarithms. IEEE Trans. Inf. Theor. **31**(4), 469–472 (1985)

[Ell13] Elliptic: Elliptic enterprises limited (2013). https://www.elliptic.co/

[GM16] Green, M., Miers, I.: Bolt: anonymous payment channels for decentralized currencies. ePrint 2016/701 (2016)

[HAB+16] Heilman, E., Alshenibr, L., Baldimtsi, F., Scafuro, A., Goldberg, S.: TumbleBit: an untrusted Bitcoin-compatible anonymous payment hub. ePrint 2016/575 (2016)

[HKZG15] Heilman, E., Kendler, A., Zohar, A., Goldberg, S.: Eclipse attacks on Bitcoin's peer-to-peer network. In: Security 2015 (2015)

[HS12] Hearn, M., Spilman, J.: Bitcoin contracts (2012). https://en.bitcoin.it/wiki/Contract

[KMS+16] Kosba, A.E., Miller, A., Shi, E., Wen, Z., Papamanthou, C.: Hawk: the blockchain model of cryptography and privacy-preserving smart contracts. In: SP 2016 (2016)

[LO98] Lipton, R.J., Ostrovsky, R.: Micro-payments via efficient coin-flipping. In: Hirchfeld, R. (ed.) FC 1998. LNCS, vol. 1465, pp. 1–15. Springer, Heidelberg (1998). doi:10.1007/BFb0055469

[MB15] Möser, M., Böhme, R.: Trends, tips, tolls: a longitudinal study of bitcoin transaction fees. In: Brenner, M., Christin, N., Johnson, B., Rohloff, K. (eds.) FC 2015. LNCS, vol. 8976, pp. 19–33. Springer, Heidelberg (2015). doi:10.1007/978-3-662-48051-9_2

[MGGR13] Miers, I., Garman, C., Green, M., Rubin, A.D.: Zerocoin: anonymous distributed e-cash from Bitcoin. In: SP 2013 (2013)

[Mic14] Micali, S.: Universal payment systems (2014). https://www.youtube.com/watch?v=xgA6TO7drok

[MPJ+13] Meiklejohn, S., Pomarole, M., Jordan, G., Levchenko, K., McCoy, D., Voelker, G.M., Savage, S.: A fistful of Bitcoins: characterizing payments among men with no names. In: IMC 2013 (2013)

[MR02] Micali, S., Rivest, R.L.: Micropayments revisited. In: Preneel, B. (ed.) CT-RSA 2002. LNCS, vol. 2271, pp. 149–163. Springer, Heidelberg (2002). doi:10.1007/3-540-45760-7_11

[MRK03] Micali, S., Rabin, M.O., Kilian, J.: Zero-knowledge sets. In: FOCS 2003 (2003)

[Nak09] Nakamoto, S.: Bitcoin: a peer-to-peer electronic cash system (2009). http://www.bitcoin.org/bitcoin.pdf

[PD16] Poon, J., Dryja, T.: The Bitcoin lightning network: scalable off-chain instant payments (2016). https://lightning.network/lightning-network-paper.pdf

[Ped91] Pedersen, T.P.: Non-interactive and information-theoretic secure verifiable secret sharing. In: Feigenbaum, J. (ed.) CRYPTO 1991. LNCS, vol. 576, pp. 129–140. Springer, Heidelberg (1992). doi:10.1007/3-540-46766-1_9

[PS15] Pass, R., Shelat, A.: Micropayments for decentralized currencies. In: CCS 2015 (2015)

[PS16] Pass, R., Shelat, A.: Micropayments for decentralized currencies. ePrint 2016/332 (2016)

[RH11] Reid, F., Harrigan, M.: An analysis of anonymity in the Bitcoin system. In: SocialCom/PASSAT 2011 (2011)

[Riv97] Rivest, R.L.: Electronic lottery tickets as micropayments. In: Hirschfeld, R. (ed.) FC 1997. LNCS, vol. 1318, pp. 307–314. Springer, Heidelberg (1997). doi:10.1007/3-540-63594-7_87

[Riv04] Rivest, R.L.: Peppercoin micropayments. In: Juels, A. (ed.) FC 2004. LNCS, vol. 3110, pp. 2–8. Springer, Heidelberg (2004). doi:10.1007/978-3-540-27809-2_2

[RKS15] Ruffing, T., Kate, A., Schröder, D.: Liar, liar, coins on fire!: penalizing equivocation by loss of Bitcoins. In: CCS 2015 (2015)

[RS13] Ron, D., Shamir, A.: Quantitative analysis of the full bitcoin transaction graph. In: Sadeghi, A.-R. (ed.) FC 2013. LNCS, vol. 7859, pp. 6–24. Springer, Heidelberg (2013). doi:10.1007/978-3-642-39884-1_2

[ST99] Sander, T., Ta-Shma, A.: Auditable, anonymous electronic cash. In: Wiener, M. (ed.) CRYPTO 1999. LNCS, vol. 1666, pp. 555–572. Springer, Heidelberg (1999). doi:10.1007/3-540-48405-1_35

[vOR+03] Someren, N., Odlyzko, A., Rivest, R., Jones, T., Goldie-Scot, D.: Does anyone really need micropayments? In: Wright, R.N. (ed.) FC 2003. LNCS, vol. 2742, pp. 69–76. Springer, Heidelberg (2003). doi:10.1007/978-3-540-45126-6_5

[Whe96] Wheeler, D.: Transactions using bets. In: Lomas, M. (ed.) Security Protocols 1996. LNCS, vol. 1189, pp. 89–92. Springer, Heidelberg (1997). doi:10.1007/3-540-62494-5_7

[Yao77] Chi-Chih Yao, A.: Probabilistic computations: toward a unified measure of complexity. In: FOCS 1977 (1977)

Analysis of the Blockchain Protocol in Asynchronous Networks

Rafael Pass[1]([✉]), Lior Seeman[2], and Abhi Shelat[3]

[1] Cornell Tech, New York City, USA
rafael@cornell.edu
[2] Uber, San Francisco, USA
lior.seeman@gmail.com
[3] Northeastern, Boston, USA
abhi@neu.edu

Abstract. Nakamoto's famous *blockchain* protocol enables achieving consensus in a so-called *permissionless setting*—anyone can join (or leave) the protocol execution, and the protocol instructions do not depend on the identities of the players. His ingenious protocol prevents "sybil attacks" (where an adversary spawns any number of new players) by relying on *computational puzzles* (a.k.a. "moderately hard functions") introduced by Dwork and Naor (Crypto'92).

The analysis of the blockchain consensus protocol (a.k.a. Nakamoto consensus) has been a notoriously difficult task. Prior works that analyze it either make the simplifying assumption that network channels are *fully synchronous* (i.e. messages are instantly delivered *without delays*) (Garay et al. Eurocrypt'15) or only consider specific attacks (Nakamoto'08; Sampolinsky and Zohar, FinancialCrypt'15); additionally, as far as we know, none of them deal with players joining or leaving the protocol.

In this work we prove that the blockchain consensus mechanism satisfies a strong forms of *consistency* and *liveness* in an *asynchronous network* with adversarial delays that are *a-priori* bounded, within a formal model allowing for adaptive corruption and spawning of new players, assuming that the computational puzzle is modeled as a random oracle. (We complement this result by showing a simple attack against the blockchain protocol in a fully asynchronous setting, showing that the "puzzle-hardness" needs to be appropriately set as a function of the maximum network delay; this attack applies even for static corruption.)

As an independent contribution, we define an abstract blockchain protocol and identify appropriate security properties of such protocols;

R. Pass—Supported in part by NSF Award CNS-1561209, NSF Award CNS-1217821, AFOSR Award FA9550-15-1-0262, a Microsoft Faculty Fellowship, and a Google Faculty Research Award.

L. Seeman—This work was done while Lior was at Cornell Tech and a postdoctoral fellow at the Harvard University Center for Research on Computation and Society, supported by Simons Foundation grant 315783.

A. Shelat—Supported in part by NSF grants 0845811, 0939718, 1565412, a Microsoft Faculty Fellowship, an SAIC Faculty Award, and a Google Faculty Research Award.

© International Association for Cryptologic Research 2017
J.-S. Coron and J.B. Nielsen (Eds.): EUROCRYPT 2017, Part II, LNCS 10211, pp. 643–673, 2017.
DOI: 10.1007/978-3-319-56614-6_22

we prove that Nakamoto's blockchain protocol satisfies them and that these properties are sufficient for typical applications; we hope that this abstraction may simplify further applications of blockchains.

1   Introduction

Distributed systems have been historically analyzed in a *closed* setting in which both the number of participants in the system, as well as their identities, are common knowledge. A departure from this model started with the design of *peer-to-peer* systems, e.g. with systems such as *Napster* and *Gnutella* for file sharing. The success of those systems led to academically designed systems such as Freenet [CSWH00], CAN [RFH+00], Chord [SMK+01], and Pastry [DR01] which offered redundant file storage, distributed hashing, selection of nearby servers, and hierarchical naming.

A novel aspect of these peer systems is that they are *permissionless*—anyone can join (or leave) the protocol execution (without getting permission from a centralized or distributed authority), and the protocol instructions do not depend on the identities of the players. As participants may continuously join and leave the system, successful permissionless systems require a fault-tolerant design. Unfortunately, the mentioned systems, while "robust" with respect to measures such as connectivity [DLN02], were not designed to tolerate against adversarial behavior. For example, there were no guarantee that one participant's experience with the system was *consistent* with another's: Two participants requesting the same file may end up receiving different versions *and never know that they did.* At first, one may think that using standard consensus/Byzantine agreement methods (e.g., [CL99, MA05, Lam10, Lam11]) could help overcome this issue. The problem is that such protocols require that a large fraction of the participating players are honest, but in the permissionless setting an attacker can trivially mount a "sybil attack"—it simply spawns players (that it controls) and can thus ensure that it controls a majority of all players. Indeed, Barak *et al.* [BCL+05] prove that this is a fundamental problem with the permissionless model.

Nakamoto's Blockchain. In 2008, Nakamoto [Nak08] proposed his celebrated "blockchain protocol" which overcomes the above-mentioned problems by relying on the idea of computational puzzles—a.k.a. *moderately hard functions* or *proofs of work*—put forth by Dwork and Naor [DN92]. Rather than attempting to provide robustness whenever the majority of the participants are honest (since participants can be easily spawned in the permissionless setting), it attempts to provide robustness as long as a *majority of the computing power* is held by honest participants. It explicitly claims *consistency* properties that are strong enough to support a financial transaction system; indeed, the first application of a blockchain is the Bitcoin digital currency which needs strong properties to prevent fraud and double-spending attacks. A number of follow-up digital currencies [Lit], micro-payment schemes [PS15, PD15], time-stamping [BTP], naming [Nam], fair secure computation [BK14] and secure messaging and PKI applications [FVY14] are based on the blockchain idea. Additionally, financial firms have announced intentions of using the blockchain to lower transaction costs, remove geopolitical barriers to transferring assets, and reconcile differences between systems.

The core blockchain protocol (a.k.a. "Nakamoto consensus", or the "Barebones blockchain protocol"), roughly speaking, is a method for maintaining a *public*, *immutable* and *ordered* ledger of records (for instance, in the bitcoin application, these records are simply transactions); that is, records can be added to the *end* of the ledger at any time (but only to the end of it); additionally, we are guaranteed that records previously added cannot be removed or reordered and that all honest users have a *consistent view* of the ledger. While standard consensus/Byzantine agreement mechanisms could be used to achieve such an immutable ordered sequence of records, the amazing aspect of Nakamoto's consensus mechanism is that it functions in a fully permissionless setting.

Roughly speaking, in his protocol each participant maintains its own local "chain" of "blocks" of records/messages—called the *blockchain*. Each block consist of a triple (h_{-1}, η, m) where h_{-1} is a pointer to the previous block in chain, m is the record component of the block, and η is a "proof-of-work"—a solution to a computational puzzle that is derived from the pair (h_{-1}, m). The proof of work can be thought of as a "key-less digital signature" on the whole blockchain up until this point.

Concretely, Nakamoto's protocol is parametrized by a parameter p—which we refer to as the *mining hardness parameter*, and a proof-of-work is deemed valid if η is a string such that $H(h_{-1}, \eta, m) < D_p$, where H is a hash function (modeled as a random oracle) and D_p is set so that the probability that an input satisfies the relation is less than p. In practice, the hardness parameter p is adaptively modified through some external process to incorporate an estimate of the number of participants in the system and the network delays; we shall return to the choice of p later. At any point of the protocol execution, each participant attempts to increase the length of its own chain by "mining" for a new block: upon receiving some record m, it picks a random η and checks whether η is a valid proof of work w.r.t. m and h_{-1}, where h_{-1} is a pointer to the last block of its current chain; if so, it extends is own local chain and broadcast it to the all the other participants (the broadcast takes places through some gossip protocol, which we do not discuss here). Whenever a participant receives a chain that is longer than its own local chain, it replaces its own chain with the longer one.

The fundamental question with such an approach is whether honest participants eventually end up with the same longest chain of blocks, and thus, the same ordered list of records, or whether the system devolves into a state where participants have *inconsistent* local chains.

1.1 Does Nakamoto's Protocol Achieve Consistency?

Requiring that all participants agree on the *whole* chain is a too strong consistency requirement if the protocol is executed on a network with message delays (as Nakamoto's protocol is intended to be)—for instance, some players may have received the last block whereas other have not. Rather, as discussed by Nakamoto [Nak08], the appropriate notion of consistency for the blockchain—which we refer to as *T-consistency*—should require that honest players agree on the current chain, *except* for potentially a small number, T, of *unconfirmed* blocks at the end of the chain. If we can show this property holds except with exponentially small probability in T, honest parties are guaranteed that for a sufficiently large choice of T (except with tiny probability), *confirmed* blocks will never be lost from the chain (which is the property needed for all the above-mentioned applications; for instance, in bitcoin, it ensures that players cannot double-spend money).

Nakamoto provides an initial analysis of consistency assuming that the adversary only mounts a particular attack strategy (namely, an attacker tries to generate a chain faster than the honest players); for instance, his analysis does not consider more sophisticated attack strategies where the adversary may attempt to "split the players" and have them work on different chains.

A beautiful recent work by Garay, Kiayas and Leonardos [GKL15] provides a more formal model for studying Nakamoto's blockchain protocol; their analysis, however, only considers a *synchronous network* with a rushing adversary—that is, messages sent in a particular round arrive in the next round *without any delays*, but the adversary sees all messages sent by honest parties before having to send its own message. In this model, they demonstrate that the blockchain protocol satisfies consistency (under appropriate assumptions on the mining hardness and the relative computational power held by the attacker), in a setting with a fixed number of players (but the protocol is not aware of the exact number of players).

Assuming a synchronous network, however, is a very strong, possibly unrealistic assumption; indeed, Nakamoto's protocol is explicitly designed to work in a network *with message delays*, and indeed is executed on such a network (i.e., the Internet).

The Power of Network Delays. Consequently, we are interested in analyzing to what extent the blockchain protocol satisfies consistency in the more realistic setting of an *asynchronous network* in which an adversary controls the scheduling/delivery of messages between honest parties. As we observe (and formally prove in Theorem 10), in a *fully* asynchronous setting, where an adversary can arbitrarily delay messages, consistency cannot be satisfied: an adversary controlling a small percentage of the computational power can simply delay messages from honest parties for sufficiently long to ensure that the adversary can find its own chain (containing *any* set of records it desires) which is longer than the chain held by all honest players, and consequently it can make the honest players switch to the adversarial chain at any point. In fact, our attack works even in the setting of *partial synchrony* (see e.g. [DLS88]) where there is an *a-priori* bound Δ on the network latency (that is, the adversary may arbitrary delay messages as long as it delivers them within time Δ), as long as the mining hardness parameter p exceeds[1] $\frac{1}{\rho n \Delta}$, where ρ is the fraction of the computational power held by the adversary and n is the number of players. Indeed, Decker and Wattenhofer [DW13] already experimentally observed that increasing the networks delays in Nakamoto's protocol leads to increased forks, and they noted (through heuristic calculations) that an attacker could use these delays to violate consistency with an attack that requires less than 50% of the mining power.

Motivated by the work by Decker and Wattenhofer, an elegant work by Sompolinsky and Zohar [SZ15] provides some initial analysis of the blockchain protocol even in a network with (bounded) delays. They show how to extend Nakamoto's analysis to deal with (bounded) delays, but again (just like Nakamoto) they only consider particular attack strategies—e.g., they do not consider "block-withholding or pre-mining attacks" where the attacker withholds blocks for later use [mtg10, ES14]; furthermore, their analysis only shows that consistency holds in the limit (when T goes to infinity), and consequently their bounds (even for the restricted attacker setting) are not useful for applications.

This leaves open the question of analyzing Nakamoto's blockchain protocol—or in fact *any* consensus protocol in the permissionless setting—with respect to *arbitrary* attack strategies in networks with Δ-bounded delays.

Does Nakamoto's blockchain protocol satisfy consistency when executed in asynchronous networks with Δ-bounded delays?

As mentioned above, Garay *et al.* provide a positive answer for the special case when $\Delta = 1$ (i.e., messages are delivered in the next time step[2]), and Sompolinsky and Zohar show that certain (natural, but restricted) strategies cannot be employed to break consistency of Nakamoto's protocol (in the limit) in Δ-bounded delay networks.

Let us highlight why dealing with network delays in the "proof-of-work" setting (where we assume that a majority of the computing power is honest) is significantly more challenging than in the standard permissioned setting. In the standard model, any synchronous protocol can be turned into a protocol that is secure also in Δ-delay networks by simply requiring that all honest players always *wait* (without doing anything) for Δ time steps before responding to any message, effectively emulating synchronous rounds. This approach completely fails in the proof-of-work setting—the adversary can now increase its computational resources by a factor Δ (since it can try to solve puzzles when the honest players are waiting).

1.2   Main Results

In this paper, we resolve the above-mentioned problem and demonstrate that (assuming puzzles are modeled as random oracles) Nakamoto's protocol satisfies

---

[1] Recall that a *larger* hardness parameter means that it is easier to find a block.

[2] Alternatively, one way to interpret the result of Garay *et al.* is that it shows consistency of Nakamoto's protocol also with Δ delays, but with a *particular delay structure* where time is divided into intervals of length Δ, and any message sent within an interval is delayed to the end of it.

consistency (under appropriate assumptions on the mining hardness and the relative computational power held by the attacker) also in networks with message delays. We emphasize that our analysis is not just a combination of the techniques/ideas from [GKL15] and [SZ15]—in fact, the bulk of our proof consists of dealing with the attack strategies which are omitted from the analysis in [SZ15], and dealing with them requires us to consider an altogether different proof technique. Additionally, our analysis considers adaptive corruption and spawning of new players (i.e., new players joining); as far as we know, it is the first analysis to formally deal with spawning of new players (which is a crucial desiderata of the blockchain protocol).

A Consistency Theorem with Delays. We provide a rough overview of our model and consistency theorem. Consider Nakamoto's protocol with mining-hardness p (that is, a single random oracle query is successful "in mining" with probability p), and consider an execution with n players, each of them with identical computing power—we assume the protocol proceeds in rounds (timesteps), and in each round each player gets a single random oracle query and the adversary controlling a ρ fraction of the players gets ρn random oracles queries (as in [GKL15], the honest players need to make their queries in parallel, but we allow the adversary to makes the queries sequentially). Let $\alpha = 1 - (1-p)^{(1-\rho)n}$ be the probability that some honest player succeeds in solving a puzzle in one round, and let $\beta = \rho n p$ be the expected number of blocks that an attacker can mine in a round. When $p \ll 1/n$ (which is the case considered in practice), we have that $\alpha \approx p(1-\rho)n$ and thus $\frac{\alpha}{\beta} \approx \frac{1-\rho}{\rho}$.

Theorem 1. *Assume there exists some $\delta > 0$ such that*

$$\alpha(1 - (2\Delta + 2)\alpha) \geq (1 + \delta)\beta.$$

Then, except with exponentially small probability (in T), Nakamoto's protocol satisfies T-consistency in the random oracle model, assuming the network's latency is bounded by Δ.

As a consequence we have that as long as $\rho < \frac{1}{2}$ (i.e., the adversary controls less than half of the computational power), for every Δ there exists some (sufficiently small) p, such that Nakamoto's protocol satisfies consistency. (Note that as mentioned above, if $p > \frac{1}{\rho n \Delta}$, Nakamoto's protocol fails to satisfy consistency.)

1.3 What Is a Blockchain?

As an independent contribution, we formally define an *abstract* notion of a blockchain (as opposed to *the* blockchain protocol proposed by Nakamoto) and put forward desired security properties of such a blockchain. We believe that having such a notion will (a) simplify applications of blockchains (as we can ignore the implementation details of the blockchain protocol) and (b) enable formally studying to what extent the protocol can be improved. (As we explain below, both of these points have been illustrated in subsequent works

[PS16a, PS16b].) We mention that while abstract models for *higher-level applications* of the blockchain (e.g., a "smart contract" abstraction) were provided in the UC framework—see [KMS+15, BK14]—it is not clear to what extent those abstractions can be satisfied by Nakamoto's protocol; rather, we are here interested in having a simple notion of the blockchain itself that we can prove is satisfied by Nakamoto's protocol and yet is useful for applications.

Roughly speaking, a blockchain is an interactive protocol where each participant has a local variable state which contains a list of messages $\vec{m}$, called the "chain". Players receive inputs, called records/batches/messages, that they attempt to include in the chain of themselves and of others. We require the following properties from a secure blockchain:

- *consistency*: with overwhelming probability (in T), at any point, the chains of two honest players can differ only in the last T blocks;
- *future self-consistence*: with overwhelming probability (in T), at any two points r, s the chains of any honest player at r and s differ only within the last T blocks;
- *g-chain-growth*: with overwhelming probability (in T), at any point in the execution, the chain of honest players grows by at least T messages in the last $\frac{T}{g}$ rounds; g is called the chain-growth of the protocol.
- the *μ-chain quality* with overwhelming probability (in T), for any T consecutive messages in any chain held by some honest player, the fraction of messages that were "contributed by honest players" is at least μ.

The consistency property is just the plain one considered by Nakamoto [Nak08] (and formalized by Garay *et al.* [GKL15]). As we note, however, this consistency property is typically not sufficient for applications. In particular, it does not rule out a protocol that oscillates between two different chains $\vec{m}_1, \vec{m}_2$; on even rounds all players have $\vec{m}_1$ as their chain, and on odd rounds $\vec{m}_2$. Clearly such a protocol does not suffice for typical applications (e.g., bitcoin, or achieving a public ledger). Thus, to prevent it, we introduce the *future self-consistency* property.

The lower bound on chain-growth was explicitly considered by Sampolinsky and Zohar [SZ15] (but they only consider growth in expectation); Garay *et al.* [GKL15] implicitly show a lower bound on chain growth within one of their proofs, and [KP15] explicitly introduce it as a desideratum. In this paper, we additionally introduce an *upper-bound* on chain growth as a desirable property; as shown in subsequent work [PS16a, PS16b], this property is useful in applications.

Finally, the chain quality property was first discussed on the Bitcoin forum [mtg10] and made explicit in the selfish mining attacks by Eyal and Sirer [ES14] with respect to the bitcoin application of the blockchain.[3] The property was first formalized, and given the name "chain quality" by Garay *et al.* who also show new applications of it (as we discuss shortly).

---

[3] In the bitcoin application of the blockchain, each player receives a reward whenever if mines a block; the chain quality thus dictates a bound on how much more reward an adversary can get by deviating from the protocol.

We show the usefulness of these properties by demonstrating that *any* blockchain protocol satisfying them can be used to achieve a *public ledger (i.e., consensus)* satisfying (a) *persistence* (namely, if a message gets added to the public ledger, it never gets removed) and (b) *liveness* (that is, if all honest players want to add a some message to the ledger, the message should eventually appear on it). We mention that Garay *et al.* already noted that, *intuitively*, the chain quality property implies liveness (since, by chain quality the adversary cannot monopolize the chain), and consistency implies persistence. However, although they show how to use Nakamoto's protocol to obtain a public ledger (in the synchronous model), they use those two properties *and* additional properties of the concrete protocol to establish it. Kiayias and Panagiotakos [KP15] demonstrate that additionally requiring chain growth suffices to prove liveness in a black-box way, but proving persistence still requires an analysis of the concrete protocol. We highlight that it is our notion of future-self consistency that allows us to obtain also persistence in a black-box way. Subsequent works by Pass and Shi [PS16a, PS16b] give further evidence to the usefulness of our abstract notion of a blockchain (and its security properties).

Main theorem. Our main result demonstrates that Nakamoto's protocol achieves consistency as well as all of our other desiderata. Let $\gamma = \frac{\alpha}{1+\Delta\alpha}$; think of γ as a "discounted" version of α due to delays on the network. Intuitively, by delaying messages the adversary gets additional computation time.

Theorem 2. *Assume there exists some* $\delta > 0$ *such that*

$$\alpha(1 - 2(\Delta + 1)\alpha) \geq (1 + \delta)\beta.$$

Let $g = \frac{\gamma}{1+\delta}$ *and* $\mu = 1 - (1+\delta)\frac{\beta}{\gamma}$. *Then Nakamoto's protocol satisfies consistency, future self consistency,* μ-*chain quality and* g-*chain growth.*

Note that when $p \ll 1/n\Delta$ (which is the case considered in practice), we have that $\gamma \approx \alpha \approx (1 - \rho)np$ and thus $\frac{\gamma}{\beta} \approx \frac{1-\rho}{\rho}$. As a consequence, we have the following corollary:

Corollary 3. *Assume* $\rho < \frac{1}{2}$. *Then for every* n, Δ, *there exists some sufficiently small* $p_0 = \Theta(\frac{1}{\Delta n})$ *such that Nakamoto's protocol with mining parameter* $p \leq p_0$ *satisfies consistency, future self consistency,* $1 - \frac{\rho}{1-\rho}$-*chain quality and* $\frac{pn}{2}$-*growth.*

Thus, as long as $\rho < \frac{1}{2}$, Nakamoto's protocol guarantees that messages contributed by honest players will eventually end up on the chain, and as long as $\rho < \frac{1}{3}$, we have that half of the messages on the chain will be contributed by honest players. We mention that our chain quality bound matches that established by Garay *et al.* assuming *no delays* (i.e., $\Delta = 1$), and is tight due to the selfish mining (a.k.a. "mining-cartel") attacks of [mtg10, ES14].

A natural question left open by our main theorem is whether there exist protocols satisfying our abstract notion of a blockchain that improve upon the parameters achieved by Nakamoto's protocol (i.e., is Nakamoto's protocol "optimal"?). A subsequent result by Pass and Shi [PS16a] shows how we can "amplify" the chain quality in Nakamoto's protocol to achieve a "close-to-optimal" chain quality of $1 - (1 - \delta)\rho$, where δ is an arbitrary small constant.[4] We highlight that the results in [PS16a] relies on the analysis from this paper in a blackbox way.

1.4 Is Nakamoto's Protocol Really Permissionless?

Our theorem only shows that for every n, Δ, there exists some mining-hardness parameter p that makes the protocol secure, so it might seem like the protocol needs to know n and therefore cannot be "permissionless"; see Sect. 1.5 for an experimental evaluation of how the level of security depends on the choice of p. As we pointed out above, this is not an anomaly of our analysis; when $p > \frac{1}{n\rho\Delta}$ the protocol is insecure. The point, however, is that the protocol only needs to know a *very rough upper-bound* on the number of players n (but the worse the upper-bound gets, the worse the efficiency of the protocol becomes.)

We additionally remark that our theorem regarding the lower bound on the chain growth actually does not make any assumption about p; this means that the honest players can use an initial set-up phase to estimate the chain growth and from this deduce a weak upper-bound on the number of players n, and then use this new upperbound to run the protocol. Indeed, as we hinted to before, the bitcoin protocol recalibrates the mining hardness parameter p every 2016 blocks (roughly 2 weeks) based on the time it took to find 2016 blocks. We leave a formal analysis of this update procedure for future work.

1.5 An Experimental Interpretation

In this section, we provide an experimental interpretation of our theorems by using estimates of parameters in a real world setting. Using estimates of hardware hashing rates (10^{12} h/s), we consider $n = 10^5$ participants and $\Delta = 10^{13}$, which corresponds to roughly $10s$ delay for the network at the given hashing rates. These numbers roughly coincide with estimates of the number of hash operations per second occurring in the Bitcoin network (7×10^{17}) at the beginning of 2016 [Blo16]. The 10 s estimation, under *network assumptions*, roughly aligns with the empirical measurements made by Decker and Wattenhofer [DW13] and their bitcoinstats.com website.[5]

The hardness parameter p in Nakamoto's protocol reflects the expected time between the discovery of blocks among all participants. Here, we explore how consistency is related to this parameter $p = \frac{1}{n\Delta \cdot c}$ by changing c. One can interpret c as the scale-free *expected block-time* in terms of the number of network delays.

---

[4] An "optimal" chain quality of $1 - \rho$ means a ρ fraction attacker gets a ρ fraction of the blocks.

[5] However, in both cases, they measure connectivity *by number of nodes* instead of by computational resources; thus their "95th percentile" estimations are biased larger because they include many hobby nodes which are connected by slow network connections and do not contribute any noticeable computation to the system.

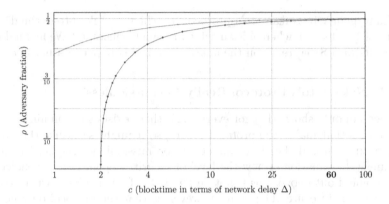

Fig. 1. For $n = 10^5$ and $\Delta = 10^{13}$ (i.e., 10 s delays at 1TH/s for commercially available mining hardware—these parameters roughly coincide with estimates of hashrate at the start of 2016), we set hardness parameter $p = \frac{1}{c \cdot n\Delta}$ where c varies along the x-axis. We can interpret c as the expected blocktime in terms of the network delay Δ. The blue graph depicts a numerically-computed maximum value of ρ for which $\alpha(1 - (2\Delta + 2)\alpha) > \beta$, i.e. parameters under which Theorem 6 shows consistency of the Nakamoto protocol. The red plot shows when our best attack succeeds in violating consistency. When $c = 60$, the hardness roughly corresponds to an expected 10-minute blocktime, and our theorem shows that Nakamoto tolerates a $\rho < 49.57\%$ attack, and our best attack succeeds when $\rho > 49.79\%$. (Color figure online)

For these choices, Fig. 1 depicts when our consistency theorem holds in Nakamoto's protocol by graphing c against the fraction (ρ) of computation controlled by the adversary. The blue graph depicts a numerically-computed maximum value of ρ for which $\alpha(1 - (2\Delta + 2)\alpha) > \beta$, i.e. parameters under which our Theorem 6 shows consistency of the Nakamoto protocol. The red plot shows when our best attack succeeds in violating consistency.

Nakamoto's protocol attempts to maintain a 10-minute blocktime by varying hardness p. For a delay $\Delta \sim 10$ s, this corresponds to a setting of $c = 60$. In this range, the Nakamoto protocol, as well as our attack give essentially the same result: Nakamoto tolerates an adversary with $\rho < 49.57\%$ and our best attack succeeds when $\rho > 49.79\%$. If we make a very conservative estimate of network delays being 1m, then $c = 10$, and Nakamoto remains consistent with respect to a 47.2% coalition.

Finally, our analysis is not tight when c is small because our attack only analyzes the probability that the adversary is able to completely control the chain. When c is small, there is also a large probability that honest players do not converge on a chain even without any adversarial messages.

1.6 Proof Highlights

Although our high-level approach follows similar intuitions as the analyses from Garay *et al.* [GKL15] and Sompolinsky and Zohar [SZ15], our actual proof uses a quite different strategy. The bulk of our proof consists of dealing with the attack strategies which are omitted from the analysis in [SZ15], and dealing with them requires us to consider an altogether different proof technique: instead of *directly* analyzing the whole blockchain process, we consider a sequence of simplified processes which are "dominated" by the original one but are simpler to analyze. For instance, we aim to show that in the optimal attack, the adversary should always delay messages for as long as possible (so that messages are always delivered after Δ steps). An obstacle in performing such a stochastic domination analysis is that once we start delaying messages, honest parties start to "mine" different blocks and the executions of our two processes diverge and become hard to compare: Ideally, to perform the domination argument we would like to consider a *fixed* execution (where the randomness of all parties are fixed) and to show by induction that delaying messages *less than* Δ never helps the attacker in that *particular* execution. The problem is that such a domination claim is not true: in some *lucky* scenarios (where the randomness is fixed), delaying messages in fact *improves* the situation for the honest parties (they now start mining blocks that magically lead to more successes). Of course, the probability of this happening should be small, but formally showing this would require us to somehow couple the experiments with and without maximum delays which is non-trivial (due to dependencies created by the random oracle).

The $\mathcal{F}_{tree}$ model. To overcome this issue, we rely on "simulation techniques" from the cryptographic literature on secure computation [GMW87, Can00]: we first consider an idealized scenario where the players do not mine blocks but instead have access to an idealized "mining" functionality, which we call $\mathcal{F}_{tree}$. This functionality determines whether honest parties succeed in mining (at random) and the success probability is independent of the current chain an honest party is trying to extend. In this model, we can now perform a domination argument for every *fixed* randomness for the experiment. One of our main technical lemmas, which turns out to be quite subtle to prove, shows that any attack that succeeds in the "real-life" protocol in the random oracle model can be turned into (i.e., simulated by) an attack in the idealized $\mathcal{F}_{tree}$ model. The key technical issue here is to deal with the dependencies created by the random oracle. (As an independent contribution, we believe that our $\mathcal{F}_{tree}$ simulation lemma can be helpful in formalizing some steps left informal in e.g., [GKL15, KP15, SZ15].)

The chain growth lowerbound. Armed with the above-mentioned techniques, the next crucial step is demonstrating a lowerbound on the chain growth. Roughly speaking, we prove by induction that (in the $\mathcal{F}_{tree}$ model) the chain grows at least as fast in the real execution of the protocol, as in a "hybrid" experiment where (a) all messages are maximally delayed, (b) honest parties "freeze" and stop mining for Δ steps whenever some honest player mines a block and (c) all messages sent by the adversary are removed. The advantage of this hybrid experiment is that the chain growth process can now be described as a simple Markov chain—there are no longer any "adversarial transitions" and due to the "freezing", honest players never have any chain conflict. This process can next be analyzed using standard Chernoff bounds. We emphasize that for the induction proof to work, we crucially rely on the fact that our analysis is in the $\mathcal{F}_{tree}$-model.

No "long" block withholding. We next use the chain growth lowerbound to demonstrate a central property of the blockchain protocol, which we refer to as the "no long block withholding" property: an adversary cannot withhold a block that it has mined for too long. Unless the adversary broadcasts the block to the honest players within some short amount of time, the block becomes "irrelevant" and will never be accepted by the honest players. Roughly speaking, we prove this by showing that, assuming that the adversary controls less than half of the computational power in the network, the chain of honest players will grow at a faster rate than any private chain the adversary can create, and thus unless it releases any block it finds quickly, the honest players' chain will be too long for the block to ever be relevant.

Proving consistency. Finally, proving consistency is the most challenging part of our proof. We start by first considering an execution *without adversarial messages*, and with deterministic delays, and identify a "pattern" which ensures that the chain of honest players converges: roughly, the pattern—which we refer to as a "convergence opportunity"—is that (1) there is a period of "silence" for Δ rounds where no honest player mines a block, (2) this is followed by a round where a *single* honest player mines a block, (3) which is followed by another Δ rounds of silence. (This notion of a convergence opportunity is closely related to a notion considered in [SZ15], and can be thought of a generalization of the notion of a "uniquely successful round" considered in the synchronous setting in [GKL15].) Whenever such a pattern occurs, all honest players converge on the chain (thus we call it a convergence opportunity): after the first period of silence, they all agree on the length of the chain (but may still have different chains), and thus the lone miner who finds a new block extends this longest chain by 1, and finally after the second period of silence this chain has propagated to all honest players (and since it is longer than all their current chains, they will switch to it). We are now interested in understanding how many times this patterns occurs within some specific period of time t. The crucial point here is that the process we now analyze is memoryless, and thus can be described by a (somewhat simple) Markov chain. On the negative side, the Markov chain that arises from this problem is too complicated to be analyzed with standard concentration bounds for Markov chains (see e.g., [CLLM12]); we instead, provide a direct analysis of a simplified experiment (which, roughly speaking, instead analyzes the times between successful mining of honest players.) and we then use this to provide a lower bound on the number of convergence opportunities.

Finally, once we have established a strong concentration bound on the number of convergence opportunities, we argue that the only way that an attacker can *ruin* such a pattern is by itself mining a block that is accepted by the honest players during it. We here rely on the block-withholding lemma to argue that any block that the attacker can use to ruin a convergence opportunity must have been mined by the adversary not long before the beginning of the period of time we are analyzing; we then show that the number of adversarial block mined during this (slightly extended) period of time is smaller than the number of convergence opportunities, and thus conclude that at least one convergence opportunity will remain even in the presence of the adversary, and thus honest parties still converge on their chain.

1.7   Related Work

The problem of reaching agreement in the presence of faulty participants, described first by Pease, Shostak, and Lamport [PSL80], and also known as distributed consensus has been very well studied over the past 40 years. The basic problem considers a set of n parties connected by reliable and authenticated pairwise network channels who wish to agree on a common output in the presence of an adversary who controls a fraction of the participants. Many aspects of the problem have been studied, with relaxations concerning the fraction of corrupted parties, the channels available to the participants, whether the protocols are deterministic or randomized and whether the participants are computationally bounded. Some protocols only consider fail-stop adversaries, while others consider a Byzantine setting in which some of the participants are malicious adversaries who attempt to disrupt the agreement. In the Byzantine agreement (BA) version of the problem, Castro and Liskov [CL99] implemented a replication library that was practical enough to use for a file system; subsequently, other works have considered *fast* or *simpler* versions of the Paxos protocol [MA05, Lam10, Lam11]. All of these works assume common knowledge of the number of participants n, as well as identities for the participants.

Okun [Oku05a, Oku05b, OB08] considers BA in an "anonymous [synchronous] model without port awareness" in which processors do not have identifiers and cannot correlate messages to their sources; Okun shows both an impossibility result for deterministic protocols, and a feasibility result for probabilistic ones. Aspnes et al. [AJK05] shows how to use a proof-of-work in a pre-processing step for this model to assign interim identities to parties so that the number of identities assigned is proportional to computational power. After the pre-processing, a standard authenticated BA protocol is used. Neither results, however, are in the peer-to-peer setting in which new users can join and leave during the execution.

Miller and LaViola [ML14] show that a variant of Nakamoto's protocol can be used to solve the single-shot Byzantine agreement problem in the presence of a minority of faults in an asynchronous setting. The single-shot setting is substantially easier, since the adversary is limited, and for example, cannot mount block-withholding attacks. Garay, Kiayias, and Leonardas [GKL15] provide a better analysis of Nakamoto's protocol, and also propose two protocols based on Nakamoto's protocol that satisfy all the properties of BA in the multiple-instance setting. They only consider synchronous networks (and no spawning of new honest players). As mentioned above, in synchronous networks, simpler solutions are possible.

2   Blockchain Protocols and Executions

In this section, we present an abstract model for blockchain protocols which aims to cover many variants of blockchain protocols.

2.1   Blockchain Protocols

A blockchain protocol is a pair of algorithms $(\Pi, \mathcal{C})$ where Π is a stateful algorithm that receives a security parameter κ as inputs and maintains a local state state. The algorithm $\mathcal{C}(\kappa, \text{state})$ outputs an *ordered* sequence of "records", or "batches", $\vec{m}$ (e.g., in the bitcoin protocol, each such record is an ordered sequence of transactions). We call $\mathcal{C}(\kappa, \text{state})$ the "record chain" of a player with security parameter κ and local variable state; to simplify notation, whenever κ is clear from context we often write $\mathcal{C}(\text{state})$ to denote $\mathcal{C}(\kappa, \text{state})$.

Algorithm Π is parameterized by a *validity* predicate V (denoted by Π^V) that encapsulates the semantic properties (e.g., "no double spending") that a blockchain application aims to achieve. $V(\vec{m})$ returns 1 if and only if the chain $\vec{m}$ is *valid* for some notion of validity.

A Blockchain Execution. Following the framework for Universal Composability [Can00], we consider the execution of a blockchain protocol $(\Pi^V, \mathcal{C})$ that is directed by an environment $Z(1^\kappa)$ (where κ is a security parameter), which activates a number of parties $1, 2, \ldots, n$ as either "honest" or corrupted parties. Honest parties execute Π on input 1^κ with an empty local state state; corrupt parties are controlled by an attacker A which reads all their inputs/message and sets their outputs/messages to be sent.

- The execution proceeds in *rounds* that model time steps. In round r, each honest player i receives a message (a "record") m from Z (that it attempts to "add" to its chain) and potentially receives incoming network messages (delivered by A). It may then perform any computation, *broadcast* a message to all other players (which will be delivered by the adversary; see below) and update its local state state_i.
- A is responsible for delivering all messages sent by parties (honest or corrupted) to *all* other parties. A cannot modify the content of messages broadcast by honest players, *but it may delay or reorder the delivery of a message* as long as it eventually delivers all messages. (Later, we shall consider restrictions on the delivery time.) The identity of the sender is not known to the recipient.[6]
- At any point, Z can communicate with adversary A or access $\mathcal{C}(\text{state}_i)$ (i.e., the current record chain of the player) where state_i is the local state of player i.

---

[6] We could also consider a seemingly weaker model where messages sent by corrupted parties need not be delivered to all honest players. We can easily convert the weaker model to the stronger model by having honest parties "gossip" all messages they receive.

– At any point, Z can *corrupt* an honest party j which means that A gets access
  to its local state and subsequently, A controls party j. (In particular, this
  means we consider a model with "erasures"; random coin tosses that are no
  longer stored in the local state of j are not visible to A.)[7]
– At any point, Z can *uncorrupt* a corrupted player j, which means that A no
  longer controls j and instead player j starts executing $\Pi(1^\kappa)$ with a fresh
  state state$_j$. (This is also how we model Z spawning a "new" honest player.)
  A gets informed of all such uncorrupt messages and is required to deliver all
  messages previously sent by (currently alive) honest players.[8]

Let $\mathsf{EXEC}^{(\Pi^V, \mathcal{C})}(A, Z, \kappa)$ be a random variable denoting the joint view of all
parties (i.e., all their inputs, random coins and messages received, including
those from the random oracle) in the above execution; note that this joint view
fully determines the execution.

Admissible Environments. We consider executions with restricted adversaries
and environments; these restrictions will be specified by a predicate $\Gamma(\cdot, \cdot, \cdot)$.

Definition 1 (Admissible Environments). *We say that the tuple of para-
meters* $(n(\cdot), \rho, \Delta(\cdot), A, Z)$ *is* Γ-*admissible w.r.t.* $(\Pi^V, \mathcal{C})$ *if* A *and* Z *are non-
uniform probabilistic polynomial-time algorithms,* $\Gamma(n(\cdot), \rho, \Delta) = 1$ *and for every*
$\kappa \in N$, *every view* view *in the support of* $\mathsf{EXEC}^{(\Pi^V, \mathcal{C})}(A, Z, \kappa)$, *the following
holds:*

1. Z *activates* $n = n(\kappa)$ *parties in* view;
2. A *delays messages by at most* $\Delta = \Delta(\kappa)$ *rounds (and in the case of newly
   spawned players, instantly delivers messages that were sent more than* Δ
   rounds ago);
3. *at any round* r *in* view, A *controls at most* $\rho \cdot n(\kappa)$ *parties; and*
4. *in every round* r *in* view, Z *only sends local inputs* m *to an honest player* i, *if*
   $V(\mathcal{C}(\mathsf{state}_i) \| m) = 1$, *where* state$_i$ *is player* i's *local state at round* r *in* view.

Whenever the protocol $(\Pi^V, \mathcal{C})$ *is clear from context, we simply call*
(n, ρ, Δ, A, Z) Γ-*admissible.*

2.2   A Remark About the Communication Model

Our model assumed that any player can send a message to all other players in
the network, and that those messages arrive within Δ rounds, no matter how
long they are. This is clearly not a realistic model. In real-life, players communicate their messages through a gossip network, and thus we need to assume

---

[7] Our proof actually extends also to the model "without erasures".

[8] This models the fact that a player is not considered "honest" before it has joined the
   network and gotten "initialized". In the real-life execution of bitcoin, new players
   joining send out a message to the network, request to be initialized and download
   the longest chain known to the network. We only consider them honest once this
   process is over.

that this network is sufficiently connected and has sufficiently many honest players to ensure Δ delivery time. This model remains infeasible if messages can be arbitrary long. However, in the applications we consider—assuming that records m provided by the environment are of length $O(\kappa)$ (i.e., there is a "block-size limit"[9])—honest players only communicate messages that differ in the last $O(\kappa)$ bits from messages that they have previously received. For such cases it seems reasonable to assume that a sufficiently connected routing network has the desired property of ensuring delivery of all messages within Δ rounds.

2.3   Blockchain Protocols in the ROM

To study Nakamoto's blockchain protocol, we need to extend the model with a random oracle. In an execution with security parameter κ, we assume all parties have access to a random function $H : \{0,1\}^* \rightarrow \{0,1\}^\kappa$ which they can access through two oracles: H(x) simply outputs $H(x)$ and H.ver(x,y) output 1 iff $H(x) = y$ and 0 otherwise. In any round r, the players (as well as A) may make *any* number of queries to H.ver. On the other hand, in each round r, honest players can make only a *single* query to H, and an adversary A controlling q parties, can make q *sequential* queries to H. (This modeling is meant to capture the assumption that we only charge for the effort of finding a solution to a proof of work [DN92], but checking the validity of a solution is cheap. We discuss this further after introducing Nakamoto's protocol.) We emphasize that the environment Z does not get direct access to the random oracle (but can instruct A to make queries).

2.4   Nakamoto's Protocol

We turn to describing Nakamoto's protocol [Nak08], which we refer to as $(\Pi^p_{Nak}, \mathcal{C}^p_{Nak})$. The local state state maintained by Π^p_{Nak} is a sequence of *(mined) blocks* $\vec{b}$, where each mined block is a tuple $(h_{-1}, \eta, \mathsf{m}, h)$ that consists of a hash h_{-1} (a pointer to the previous record), a nonce η, a record m, and a hash h (a pointer to the current record[10]) and is initialized to a special "genesis" block: $(0, 0, \perp), \mathsf{H}(0, 0, \perp)$. Let $\mathcal{C}(\mathsf{state})$ be the sequence of records $\vec{\mathsf{m}}$ contained in the sequence of blocks state. The protocol is parameterized by a hardness function $p(\cdot)$ which defines a constant $D_p = p(\kappa) \cdot 2^\kappa$ such that for all (h, b), $\Pr_\eta[\mathsf{H}(h, \eta, b) < D_p] = p(\kappa)$. Whenever p is clear for context, we simply denote the protocol $(\Pi_{Nak}, \mathcal{C}_{Nak})$ (without the p superscript); additionally, whenever κ is clear from context, we let $p = p(\kappa)$.

---

[9] In Bitcoin's instantiation of the blockchain protocol, there is currently a severe restriction on the block-size. There is currently an active debate whether to raise the block-size limit or to leave it small.

[10] In reality (as well as in the description in the introduction), h is not included in the block (as it can be easily determined from the remaining elements); we include it to ensure that we can verify validity of a block using only H.ver.

We say a block $b = (h_{-1}, \eta, m, h)$ is *valid with respect to (a predecessor block)* $b_{-1} = (h'_{-1}, \eta', m', h')$ if three conditions hold:

1. $h_{-1} = h'$,
2. $h = H(h_{-1}, \eta, m)$,
3. and $h < D_p$.

A sequence of blocks state $= (b_0, \ldots, b_\ell)$ is *valid* if (a) $b_0 = (0, 0, \perp, H(0, 0, \perp))$ is the genesis block, (b) for all $i \in [\ell]$, b_i is valid with respect to b_{i-1}, and (c) $V(\mathcal{C}(\text{state})) = 1$.

Each round of Π_{Nak}^V proceeds as follows:

- Read all incoming messages (delivered by A). If any incoming message state′ is a valid sequence of blocks that is longer than its local state state, replace state by state′. (Note that checking the validity of state′ can be done using only H.ver queries)
- Read local message m (from Z). If m is such that $V(\mathcal{C}(\text{state})\|m) \neq 1$, proceed to the next round. Otherwise, pick a random nonce $\eta \in \{0,1\}^\kappa$ and issue query $h = H(h_{-1}, \eta, m)$ where h_{-1} is the 4'th element in the last block in state. If $h < D_p$, then Π adds the *newly mined* block (h_{-1}, η, b, h) to state and broadcasts the updated state.

Depending on the definition of V, one can instantiate either Bitcoin, e.g., by having V enforce that m can be parsed into a sequence of well-formed *transactions* each of which is *authorized* and spends money from a source account to a destination account at most once without deficit, etc., as well as other cryptocurrencies with different semantics such as Namecoin. We may also consider a simpler predicate $V_{\mathcal{L}}$ that simply accepts all messages; that is $V_{\mathcal{L}}(\vec{m}) = 1$; such a predicate is useful, for instance, to use a blockchain to provide a public ledger.

A Remark on our use of the Random Oracle. Recall that in our model, we restrict players to a single evaluation query H per round, but allow them any number of verification queries H.ver in the same round. We do this to model the fact that checking the validity of mined blocks is "cheap" whereas the mining process is expensive. (To enable this, we have included a pointer h to the current record in every mined block in the description of Nakamoto; thus a player need not spend an H query to compute the pointer to the previous record.)

In practice, the cost of evaluating a hash function (which is used to instantiate the random oracle) is the same as verifying its outputs, but our modeling attempts to capture the phenomena that a miner typically use various heuristics (such as black lists of IP addresses that have sent invalid blocks) and different hardware to check the validity of a mined block versus to mine a new block.

3   Formal Definitions of the Desiderata

In this section, we provide formal definitions of the desiderata mentioned in the introduction. We start with some notation and preliminaries.

Notation. For some A, Z, consider some view in the support of $\mathsf{EXEC}^{(\Pi^V, \mathcal{C})}$ (A, Z, κ). We use the notation $|\mathsf{view}|$ to denote the number of rounds in the execution, view^r to denote the prefix of view up until round r, $\mathsf{state}_i(\mathsf{view})$ denotes the local state of player i in view, $\mathcal{C}_i(\mathsf{view}) = \mathcal{C}(\mathsf{state}_i(\mathsf{view}))$ and $\mathcal{C}_i^r(\mathsf{view}) = \mathcal{C}_i(\mathsf{view}^r)$.

(Strongly) Negligible Functions. A function $\epsilon(\cdot)$ is said to be *negligible* if for every polynomial $p(\cdot)$, there exists some κ_0 such that $\epsilon(\kappa) \leq \frac{1}{p(\kappa)}$ for all $\kappa \geq \kappa_0$. Our bounds will actually also apply to an *exponentially-strong* interpretation of what it means for a function to be negligible. A function $\epsilon(\cdot)$ is said to be *(strongly) negligible* if there exists constants $c_0 > 0, c_1$ such that for all κ, $\epsilon(\kappa) \leq e^{-c_0 \kappa + c_1}$. In the rest of the paper, we simply use the term "negligible", but all uses of it can be replaced by strongly negligible. We often use the shorthand $\mathsf{neg}(\kappa)$ to denote a function that is negligible as a function of κ.

3.1   Chain Growth

Our first desiderata is that the chain grows proportionally with the number of rounds of the protocol. This intuitive property was explicitly considered by Sompolinsky and Zohar [SZ15] but only *in expectation*; it was also implicitly considered in Garay *et al.* within one of their proofs (but was not highlighted as a desideratum), and it was explicitly highlighted as a desideratum by Kiayias and Panagiotakos [KP15]. We here generalize these definitions to abstract blockchain protocols, and add a useful length-consistency property.[11] (Looking forward, in Sect. 3.4, we also consider an *upper-bound* on chain growth.) Let,

$$\mathsf{min\text{-}chain\text{-}increase}_{r,t}(\mathsf{view}) = \min_{i,j} |\mathcal{C}_j^{r+t}(\mathsf{view})| - |\mathcal{C}_i^r(\mathsf{view})|$$

where we quantify over players i, j such that i is honest at view^r and j is honest at view^{r+t}.

Let $\mathsf{growth}^t(\mathsf{view}, \Delta, T) = 1$ iff the following two properties hold:

- **(consistent length)** for all rounds r, r' such that $r \leq |\mathsf{view}| - \Delta$ and $r + \Delta \leq r' \leq |\mathsf{view}|$, for every two players i, j such that in view, i is honest at r and j is honest at r', we have that

$$|\mathcal{C}_j^{r'}(\mathsf{view})| \geq |\mathcal{C}_i^r(\mathsf{view})|$$

---

[11] The length-consistency requirement is actually not needed for any of our applications, but having it enables achieving sharper bounds, and this property is trivially satisfied by Nakamoto's protocol.

– **(chain growth)** for every round $r \leq |\text{view}| - t$, we have

$$\text{min-chain-increase}_{r,t}(\text{view}) \geq T.$$

In other words, growth^t is a predicate which tests that (a) honest parties have chains of roughly the same length, and (b) during any t rounds in the execution, all honest parties' chains increase by at least T.

Definition 2. *A blockchain protocol $(\Pi, \mathcal{C})$ has chain growth rate $g(\cdot, \cdot, \cdot, \cdot)$ in Γ-environments if for all Γ-admissible $(n(\cdot), \rho, \Delta(\cdot), A, Z)$, there exists some constant c and negligible functions ϵ_1, ϵ_2 such that for every $\kappa \in \mathbb{N}, T \geq c\log(\kappa)$, and $t \geq \frac{T}{g(n(\kappa),\rho,\Delta(\kappa))}$, the following holds:*

$$\Pr\left[\text{view} \leftarrow \text{EXEC}^{(\Pi^V, \mathcal{C})}(A, Z, \kappa) : \text{growth}^t(\text{view}, \Delta(\kappa), T) = 1\right] \geq 1 - \epsilon_1(\kappa) - \epsilon_2(T)$$

If $\epsilon_1 = 0$, we say that $(\Pi, \mathcal{C})$ has error-less chain growth rate g in Γ-environments.

3.2 Chain Quality

Our second desideratum is that the number of records contributed by the adversary is proportional to its relative power. This property was first discussed on the Bitcoin forum [mtg10] and made explicit in the selfish mining attacks by Eyal and Sirer [ES14] w.r.t. the bitcoin application of the blockchain.[12] The property was first formalized, and given the name "chain quality" by Garay et al. [GKL15]. We generalize their definition to abstract blockchain protocols. Doing so is somewhat non-trivial in that it is not directly clear what it means for a record to be adversarial (Garay et al. only provide a definition of an adversarial block for the particular protocol of Nakamoto, and their definition only applies in the random oracle model).

We say that a *record* m *is non-adversarial (or honest) w.r.t. view and prefix* m̄ if there exists a player j and some round r' such that in $\text{view}^{r'}$, j is honest, the environment provided m as input to j, and m̄ is a prefix of $\mathcal{C}_j(\text{view}^{r'})$. (That is, there exists some honest player that received m as an input when their chain contained m̄).

Let $\text{quality}^T(\text{view}, \mu) = 1$ iff for every round r and every player i such that i is honest in view^r, among any consecutive sequence of T records M in $\mathcal{C}_i^r(\text{view})$, the fraction of records m that are honest w.r.t. view^r and m̄, where m̄ is the prefix of $\mathcal{C}_i^r(\text{view})$ preceeding M, is at least μ.

---

[12] In the bitcoin application of the blockchain, each player receives a reward whenever if mines a block; the chain quality thus dictates a bound on how much more reward an adversary can get by deviating from the protocol.

Definition 3. *A blockchain protocol* $(\Pi, \mathcal{C})$ *has* chain quality $\mu(\cdot, \cdot, \cdot, \cdot)$ *in* Γ *environments, if for all* Γ*-admissible* $(n(\cdot), \rho, \Delta(\cdot), A, Z)$, *there exists some constant* c *and negligible functions* ϵ_1, ϵ_2 *such that for every* $\kappa \in \mathbb{N}, T > c\log(\kappa)$ *the following holds:*

$$\Pr\left[\text{view} \leftarrow \text{EXEC}^{(\Pi^V, \mathcal{C})}(A, Z, \kappa) : \text{quality}^T(\text{view}, \mu(\kappa, n(\kappa), \rho, \Delta(\kappa))) = 1 \right]$$
$$\geq 1 - \epsilon_1(\kappa) - \epsilon_2(T)$$

If $\epsilon_1 = 0$, *we say that* $(\Pi, \mathcal{C})$ *has* errorless chain quality μ *in* Γ*-environments.*

3.3   Consistency

The *common-prefix* property by Garay *et al.* [GKL15], which was already considered and studied by Nakamoto [Nak08], requires that in any round r, the record chains of any two honest players i, j agree on all, but potentially the last T, records. We note that this property (even in combination with the other two desiderata) provides quite weak guarantees: even if any two honest parties perfectly agree on the chains, the chain could be completely different on, say, even rounds and odd rounds. We here consider a stronger notion of consistency which additionally stipulates players should be consistent with their "future selves".[13]

Let $\text{consistent}^T(\text{view}) = 1$ iff for all rounds $r \leq r'$, and all players i, j (potentially the same) such that i is honest at view^r and j is honest at $\text{view}^{r'}$, we have that the prefixes of $\mathcal{C}_i^r(\text{view})$ and $\mathcal{C}_j^{r'}(\text{view})$ consisting of the first $\ell = |\mathcal{C}_i^r(\text{view})| - T$ records are identical.[14]

Definition 4. *A blockchain protocol* $(\Pi, \mathcal{C})$ *satisfies* consistency *in* Γ *environments, if for all* Γ*-admissible* $(n(\cdot), \rho, \Delta(\cdot), A, Z)$, *there exists some constant* c *and negligible functions* ϵ_1, ϵ_2 *such that for every* $\kappa \in \mathbb{N}, T > c\log(\kappa)$ *the following holds:*

$$\Pr\left[\text{view} \leftarrow \text{EXEC}^{(\Pi^V, \mathcal{C})}(A, Z, \kappa) : \text{consistent}^T(\text{view}) = 1 \right] \geq 1 - \epsilon_1(\kappa) - \epsilon_2(T)$$

If $\epsilon_1 = 0$, *we say that* $(\Pi, \mathcal{C})$ *has* errorless consistency *in* Γ*-environments.*

Note that a direct consequence of consistency is that the chain *length* of any two honest players can differ by at most T (except with negligible probability in T).

---

[13] This stronger notion of consistency combines what we called "plain" consistency and "future-self" consistency in the introduction.

[14] Pedantically, the "first ℓ records of $\mathcal{C}_j^{r'}(\text{view})$ is not defined if $\mathcal{C}_j^{r'}(\text{view}) < \ell$; to formalize it, we may represent the chains as infinite sequences of records, where all records after the end of the chain is a special "nil" symbol. In particular, this ensures that $\text{consistent}^T(\text{view}) = 0$ if $\mathcal{C}_j^{r'}(\text{view}) < \ell$.

3.4   Chain Growth Upperbound

Our final desiderata is the existence of an *upperbound on the chain growth*. While we do not present any applications of this property in the current paper, it is an intuitively useful property—for instance, combined with the chain growth lower bound, it implies we can use a blockchain as a "partially-synchronized clock". (Additionally, subsequent work by Pass and Shi [PS16a, PS16b] demonstrate the usefulness of this property.)

Let,

$$\text{max-chain-increase}_{r,t}(\text{view}) = \max_{i,j} |\mathcal{C}_j^{r+t}(\text{view})| - |\mathcal{C}_i^r(\text{view})|$$

where we quantify over players i, j such that i is honest at view^r and j is honest at view^{r+t}. Let $\text{upper-growth}^t(\text{view}, \Delta, T) = 1$ iff for every round $r \leq |\text{view}| - t$, we have

$$\text{max-chain-increase}_{r,t}(\text{view}) \leq T.$$

Definition 5. *A blockchain protocol* $(\Pi, \mathcal{C})$ *has* upper-bound on chain growth rate $g'(\cdot, \cdot, \cdot, \cdot)$ *in* Γ-environments *if for all* Γ-admissible $(n(\cdot), \rho, \Delta(\cdot), A, Z)$, *there exists some constant* c *and negligible functions* ϵ_1, ϵ_2 *such that for every* $\kappa \in \mathbb{N}, T \geq c \log(\kappa)$, *and* $t = \frac{T}{g'(n(\kappa), \rho, \Delta(\kappa))}$, *the following holds:*

$$\Pr\left[\text{view} \leftarrow \text{EXEC}^{(\Pi^V, \mathcal{C})}(A, Z, \kappa) : \text{upper-growth}^t(\text{view}, \Delta(\kappa), T) = 1 \right]$$
$$\geq 1 - \epsilon_1(\kappa) - \epsilon_2(T)$$

If $\epsilon_1 = 0$, *we say that* $(\Pi, \mathcal{C})$ *has* error-less upper-bound on chain growth rate g' *in* Γ-environments.

4   Main Theorem Statements

Our main results will be most convenient to parameterize in the following two quantities (which are defined for some fixed mining hardness function $p(\cdot)$; recall that Nakamoto's protocol is parametrized by p):

- let $\alpha(\kappa, n, \rho, \Delta) = 1 - (1 - p(\kappa))^{(1-\rho)n}$. That is, α is the probability that *some* honest player succeeds in mining a block in a round;
- let $\beta(\kappa, n, \rho, \Delta) = \rho n p(\kappa)$. That is, β is the expected number of blocks that an attacker can mine in a round.

Whenever κ, n, ρ, Δ are clear from the context, we simply write α, β. In essence, the quantities capture the per-round *expected increase* in chain length by the honest parties and the adversary; the reason the quantities are defined differently is that we assume that the adversary can sequentialize its queries in a round, whereas honest players make a single parallel query (they each act independently), and thus even if they manage to mine several blocks, the longest chain held by honest players can increase by at most 1. Note, however, that when p is small (in comparison to $1/n$), which is case for the Bitcoin protocol, α is well-approximated by $(1-\rho)np$ and thus $\frac{\alpha}{\beta} \approx \frac{1-\rho}{\rho}$, so this difference is minor.

We also consider the following quantity:

– let $\gamma(\kappa, n, \rho, \Delta) = \frac{\alpha}{1+\Delta\alpha}$ (When clear from context, we simply write γ.)

Roughly speaking, γ should be thought of as a *discounted* version of α due to the fact that messages sent by honest parties can be delayed by Δ rounds and this may lead to honest players redoing work; γ corresponds to their *effective* mining power. Note that if p is sufficiently small then $\gamma \approx \alpha$ and thus $\frac{\gamma}{\beta} \approx \frac{1-\rho}{\rho}$.

We are now ready to state our main theorems. The proof of these theorems are all given in the Appendix (see Sect. 1.6 for a high-level overview of key ideas). We will consider two environments:

– In the least restrictive environment, Γ_0, we make *no restrictions* on the parameters (more than them being "valid"). Namely, let $\Gamma_0(n(\cdot), \rho, \Delta(\cdot)) = 1$ iff $n(\cdot), \Delta(\cdot)$ are functions $\mathcal{N} \to \mathcal{N}^+$ and $0 \leq \rho \leq 1$.
– In the more restrictive environment, we additionally assume that the adversary controls a sufficiently small fraction of the computational power. Let $\Gamma_\lambda^p(n(\cdot), \rho, \Delta(\cdot)) = 1$ iff $\Gamma_0(n(\cdot), \rho, \Delta(\cdot))) = 1$ and for all $\kappa, n = n(\kappa), \Delta = \Delta(k)$,

$$\alpha(1 - 2(\Delta + 1)\alpha) \geq \lambda\beta$$

The following three theorems formalize Theorem 2 from the introduction (which in turn implies Theorem 1). We first prove a lower bound on the chain growth.

Theorem 4 (Chain growth). *For any $\delta > 0$, any $p(\cdot)$, $(\Pi_{Nak}^p, \mathcal{C}_{nak}^p)$ has chain growth rate*

$$g_\delta^p(\kappa, n, \rho, \Delta) = (1-\delta)\gamma$$

in Γ_0 environments.

We next prove a lower bound on the chain quality.

Theorem 5 (Chain quality). *For all $\delta > 0$, any $p(\cdot)$, $(\Pi_{Nak}^p, \mathcal{C}_{nak}^p)$ has chain quality*

$$\mu_\delta^p(\kappa, n, \rho, \Delta) = 1 - (1+\delta)\frac{\beta}{\gamma}$$

in Γ_0 environments.

We finally show consistency.

Theorem 6 (Consistency). *For any $\lambda > 1$, any $p(\cdot)$, $(\Pi_{nak}^p, \mathcal{C}_{nak}^p)$ satisfies consistency in Γ_λ^p environments.*

Chain growth upperbound. We additionally present an upperbound on the chain growth. (As mentioned before, this property is not needed for any of the applications that we present in the current paper, nor for the statement of the main result in the introduction, but may be useful in other contexts such as [PS16a, PS16b].)

Theorem 7 (Upper-bound on Chain growth). *For any $\delta > 0$, any $p(\cdot)$, $(\Pi^p_{Nak}, \mathcal{C}^p_{nak})$ has the upper-bound on chain growth rate*

$$\hat{g}^p_\delta(\kappa, n, \rho, \Delta) = (1 + \delta)np$$

in Γ^p_λ environments.

5   Application: Public Ledger

In this section, we demonstrate how to use *any* blockchain satisfying the growth, quality, and consistency properties defined in Sect. 3 to construct a secure *public ledger* system. Garay *et al.* [GKL15] show a similar theorem, in the synchronous setting, for the *specific* blockchain of Nakamoto.

Informally, a *public ledger* serves as an immutable "bulletin board" to which anyone can post a message, and everyone can read all messages posted. As described by Garay *et al.* [GKL15], such a bulletin board ought to satisfy two properties, *liveness* and *persistence*:[15]

- *Liveness:* The liveness property stipulates that from any given round r, if a sufficiently long period of time t elapses—we refer to this time as the *wait-time* of the ledger—*every* honest player will output a message m as part of their (local) ledger, where m was provided as an input to some honest player between rounds r and $r + t$. (In particular, this implies the liveness condition of [GKL15] which requires that if the same message was provided to all honest players between rounds r and $r + t$, this messages will be output in the ledger.)
- *Persistence:* The persistence property stipulates that if some honest player i outputs a message m at position i in its local ledger, then (1) m is the only message that can ever be output at position i of any other honest player's ledger and (2) every honest player will eventually output m at position i.

Let us turn to a formal definition.

5.1   Definition of a Public Ledger

Just like the blockchain protocol, a *public ledger* is pair of algorithms $(\Pi, \mathcal{L})$ where Π is a stateful algorithm that maintains a local state state. The algorithm $\mathcal{L}(\kappa, \text{state})$ outputs *ordered* sequence of messages $\vec{m}$. We call $\mathcal{L}(\kappa, \text{state})$

---

[15] The notion of Garay *et al.* [GKL15] is actually somewhat different and weaker: for instance, (1) they only require these properties to hold for records that are sufficiently "deep" in the ledger (we feel it is more natural/simpler to require it for *all* records in the ledger), and (2) they only require the liveness property to hold if all players received the *same* message.

the (local) ledger of a player with security parameter κ and local variable state. We define the execution of a public ledger protocol in exactly the same way as the execution of a blockchain protocol (see Sect. 2.1), and define the random variable $\mathsf{EXEC}^{(\Pi,\mathcal{L})}(A, Z, \kappa)$ in exactly the same way. Let $\mathcal{L}_i(\mathsf{view})$ denote the ledger of player i in the view view and let $\mathcal{L}_i^r(\mathsf{view}) = \mathcal{L}_i(\mathsf{view}^r)$.

Liveness. Let $\mathsf{live}(\mathsf{view}, t) = 1$ iff for any t consecutive rounds $r, \ldots, r + t$ in view there exists some round $r' \in [r, r + t]$ and players i such that in view, (1) i is honest at r', (2) i received a message m as input at round r', and (3) for every player j that is honest at $r + t$ in view, $\mathsf{m} \in \mathcal{L}_j^{r+t}(\mathsf{view})$.

Definition 6 (Liveness). *We say that public ledger* $(\Pi, \mathcal{L})$ *is* live *with wait-time* $w(\cdot, \cdot, \cdot, \cdot)$ *in* Γ *environments if for all* Γ*-admissible* $(n(\cdot), \rho, \Delta(\cdot), A, Z)$, *there exists a negligible function* ϵ *in the security parameter* $\kappa \in \mathbb{N}$, *such that*

$$\Pr\left[\mathsf{view} \leftarrow \mathsf{EXEC}^{(\Pi,\mathcal{L})}(A, Z, \kappa) : \mathsf{live}(\mathsf{view}, w(\kappa, n(\kappa), \rho, \Delta(\kappa))) = 1\right] \geq 1 - \epsilon(\kappa)$$

Persistence. Let $\mathsf{persist}_\Delta(\mathsf{view}) = 1$ iff for every round $r \leq |\mathsf{view}| - \Delta$, every player i that is honest at view^r and every position $\mathsf{pos} \leq |\mathcal{L}_i^r(\mathsf{view})|$, if $\mathcal{L}_i^r(\mathsf{view})$ contains the message m at position pos, then for every round r' such that $r + \Delta \leq r'$ and every honest player j (possibly the same as i) we have that m is also at position pos in $\mathcal{L}_j^{r'}(\mathsf{view})$.

Definition 7 (Persistence). *We say that* $(\Pi, \mathcal{L})$ *is* persistent *in* Γ *environments if for all* Γ*-admissible* $(n(\cdot), \rho, \Delta(\cdot), A, Z)$, *there exists a negligible function* ϵ *such that for every security parameter* $\kappa \in \mathbb{N}$,

$$\Pr\left[\mathsf{view} \leftarrow \mathsf{EXEC}^{(\Pi,\mathcal{L})}(A, Z, \kappa) : \mathsf{persist}_{\Delta(\kappa)}(\mathsf{view}) = 1\right] \geq 1 - \epsilon(\kappa)$$

5.2   Constructing a Public Ledger from a Blockchain

We turn to constructing a public ledger from any blockchain protocol. Let TRUE be the predicate that always outputs 1 (on any input).

Definition 8. *Given a blockchain protocol* $(\Pi, \mathcal{C})$, *we call* $(\Pi', \mathcal{L})$ *the public ledger* $T(\kappa)$*-induced by* $(\Pi, \mathcal{C})$, *where* $\Pi' = \Pi^{TRUE}$ *and* $\mathcal{L}(\kappa, \mathsf{state})$ *computes* $\mathcal{C}(\kappa, \mathsf{state})$, *truncates the last* $T(\kappa)$ *records of it, and outputs the results.*

Theorem 8. *Let* $T(\cdot)$ *be a strictly positive, super-constant, polynomial,* $(\Pi, \mathcal{C})$ *a blockchain protocol satisfying chain growth* g, *chain quality* μ *and chain consistency in* Γ*-environments, where* μ *and* g *are strictly positive. Then, for every* $\delta > 0$, *the public ledger* $(\Pi', \mathcal{L})$ $T(\cdot)$*-induced by* $(\Pi, \mathcal{C})$ *is persistent and live with wait-time* $w(\kappa, n, \rho, \Delta) = (1 + \delta)\frac{T(\kappa)}{g(\kappa, n, \rho, \Delta)}$ *in* Γ*-environments.*[16]

Proof. Consider Γ-admissible $n(\cdot), \rho, \Delta(\cdot), A, Z$, some $\delta > 0$, some κ, and some view $\mathsf{view} \leftarrow \mathsf{EXEC}^{(\Pi,\mathcal{L})}(A, Z, \kappa)$. Let $n = n(\kappa), \Delta = \Delta(\kappa), g = g(\kappa, n, \rho, \Delta)$ and $\mu = \mu(\kappa, n, \rho, \Delta), T = T(\kappa)$. We now separately show liveness and persistence.

---

[16] We are grateful to Elaine Shi for pointing out that a variant of our proof for the liveness property works with a sharper wait-time bound. Our original theorem and

Liveness. Let $T' = (1 + \delta)T$ and let $t = \frac{T'}{g}$. Pick δ' such that $0 < \delta' < \delta$. Condition on the events that $\mathsf{growth}^t(\mathsf{view}, \Delta(\kappa), T') = 1$, $\mathsf{consistent}^{\delta'T-1}(\mathsf{view}) = 1$, and $\mathsf{quality}^{T'-T}(\mathsf{view}) = 1$; by our assumptions and the union bound, these events occur with probability $1 - \mathsf{neg}(T)$; since T is polynomial in κ, these events occur except with probability $\mathsf{neg}(\kappa)$. Let j, j' be players such that in view, j' is honest at r and j is honest at $r + t$ such that $r + t \leq |\mathsf{view}|$.

By the conditioning, we have that:

- By chain growth, $|\mathcal{C}_j^{r+t}(\mathsf{view})| - |\mathcal{C}_{j'}^r(\mathsf{view})| \geq T + \delta T$; thus $|\mathcal{C}_j^{r+t}(\mathsf{view})| \geq T + \delta T$
- By "truncation", at least δT records that were not part of the chain of j' at r are thus output as part of j's ledger.
- By consistency, before round r, no honest player has *ever* had a chain whose length exceeds $|\mathcal{C}_{j'}^r(\mathsf{view})| + \delta'T - 1$.
- Thus, we have a segment of length at least $(\delta - \delta')T$ of records in $\mathcal{C}_j^{r+t}(\mathsf{view})$ which is output as part of j's ledger such that each record appears at a position which exceeds $|\mathcal{C}_{j'}^r(\mathsf{view})| + \delta'T$. By (strictly positive) chain quality, at least $(\delta - \delta')T\mu > 0$ records at a position exceeding $|\mathcal{C}_{j'}^r(\mathsf{view})| + \delta'T$ are "non-adversarial"; since no honest player ever had a chain of length $|\mathcal{C}_{j'}^r(\mathsf{view})| + \delta'T$ before round r, these non-adversarial records must have been provided by the environment at or after round r.

Persistence. Let $t = \frac{T}{g}$. Condition on the events that $\mathsf{growth}^t(\mathsf{view}, \Delta(\kappa), T) = 1$ and $\mathsf{consistent}^T(\mathsf{view}) = 1$; by our assumptions and the union bound, these events occur with probability $1 - \mathsf{neg}(T)$; since T is polynomial in κ, these events occur except with probability $\mathsf{neg}(\kappa)$.

Consider players i, j such that in view, i is honest at round r, and j is honest at round r' such that $r' \geq r + \Delta$. By the conditioning, we have that:

- Because $\mathsf{consistent}^T(\mathsf{view}) = 1$, prefixes of $\mathcal{C}_i^r(\mathsf{view})$ and $\mathcal{C}_j^{r'}(\mathsf{view})$ consisting of the first $|\mathcal{C}_i^r(\mathsf{view})| - T$ records are identical.
- By the consistent-length property of the chain-growth property, it also follows that $|\mathcal{C}_j^{r'}(\mathsf{view})| \geq |\mathcal{C}_i^r(\mathsf{view})|$.

By the above two statements, and the fact that $\mathcal{L}$ simply truncates the last T records of the chain, it follows that $\mathcal{L}_i^r(\mathsf{view})$ is a *prefix* of $\mathcal{L}_j^{r'}(\mathsf{view})$. Therefore, if $\mathcal{L}_i^r(\mathsf{view})$ contains a message m at position p, then so does $\mathcal{L}_j^{r'}(\mathsf{view})$. Because this holds for all such $r, r' > r + \Delta, i, j$, it follows that $\mathsf{persist}_{\Delta(\kappa)}(\mathsf{view}) = 1$.

---

proof (which set parameters in a non-optimal way) only claimed $w(\kappa, n, \rho, \Delta) = \frac{(1+\delta)T(\kappa)}{\mu(\kappa,n,\rho,\Delta)\cdot g(\kappa,n,\rho,\Delta)}$. The reason we do not need a dependency on μ is that by our definition of chain quality, it suffices for the fraction of non-adversarial blocks to be *positive* (as opposed to greater than $\frac{1}{T}$) to conclude the existence of at least one non-adversarial block.

Corollary 9. *For any $\lambda > 1$, any $\delta > 0$, any $p(\cdot)$, and any strictly positive, super-constant, polynomial $T(\cdot)$, the public ledger $(\Pi_{\mathsf{Nak}}, \mathcal{L}_{\mathsf{Nak}})$ that is $T(\cdot)$-induced by the blockchain protocol $(\Pi_{\mathsf{Nak}}^p, \mathcal{C}_{\mathsf{Nak}}^p)$ is persistent and live with waittime*

$$w(n, \kappa, \rho, \Delta) = (1 + \delta) \frac{T(\kappa)}{\gamma}$$

in Γ_λ^p environments.

Proof. From Theorems 4, 5 and 6, for every δ', δ'', $(\Pi_{\mathsf{Nak}}^p, \mathcal{C}_{\mathsf{Nak}}^p)$ has growth $(1 - \delta')\gamma$, chain quality $1 - (1 + \delta'')\frac{\beta}{\gamma}$, and satisfies consistency. It can be shown that the chain quality is thus strictly positive. From Theorem 8, for every δ'', $(\Pi_{\mathsf{Nak}}, \mathcal{L}_{\mathsf{Nak}})$ thus has rate

$$w(n, \kappa, \rho, \Delta) = (1 + \delta''') \frac{T(\kappa)}{(1 - \delta')\gamma} < (1 + \delta) \frac{T(\kappa)}{\gamma}$$

where the last inequality follows by picking sufficiently small δ', δ'''.

6   An Attack on Nakamoto with Long Delays

In this section, we formally demonstrate that Nakamoto's protocol satisfies neither consistency nor positive chain quality in a fully asynchronous network without an upperbound Δ on the network delay, even if the adversary controls just a tiny fraction of computational power. More specifically, we show that for every hardness parameter p, $\Pi_{Nak}^p, \mathcal{C}_{Nak}^p$, satisfies neither consistency nor chain quality when $\Delta = \frac{1+\delta}{\rho np}$ for some $\delta > 0$. This demonstrates why our consistency theorem needs to rely on the assumption that $p \leq \frac{\Theta(1)}{\Delta n}$, and why the chain quality is $1 - \frac{\beta}{\gamma}$ as opposed to just $1 - \frac{\beta}{\alpha}$ (recall that $\gamma = \frac{\alpha}{1+\Delta\alpha}$ is a discounted version of α that takes delays into account.) In particular, we present a "51%" attack a la Nakamoto in which the attacker at some point in the future replaces the *whole chain* with a chain of its choice, even if it only controls a small fraction of the computational power.

Intuitively, in every segment of Δ rounds, if we delay all messages between honest players until the end of the segment, honest players are effectively "mining on their own" and thus are unlikely to extend their chain by more than 1. The adversary, on the other hand, coordinates its mining and thus in expectation extends its chain by $\Delta \cdot \rho np$; if we set $\Delta > \rho np$ the adversary can mine its own *longer* chain (without sending it to the honest player).

Theorem 10 (Inconsistency of Nakamoto with Unbounded Delays).
Let $\widehat{\Gamma}_{\rho', \delta}^p(\kappa, n, \rho, \Delta) = 1$ iff (1) $n = \frac{2}{\rho^2} \cdot \kappa$, (2) $\rho = \rho'$ and (3) $\Delta = \frac{1+\delta}{\rho np}$. For every $0 < \delta < \frac{1}{2}, 0 < \rho' < 1$, and every inverse polynomial $p(\cdot)$, $(\Pi_{Nak}^p, \mathcal{C}_{Nak}^p)$ does not satisfy neither consistency nor chain quality q in $\widehat{\Gamma}_{\rho', \delta}^p$-valid environments, where $q > 0$.

Proof. Consider an environment Z that invokes $n = \frac{2}{\rho^2} \cdot \kappa$ players, a fraction ρ of them being adversarial, and sends messages $m_1, m_2 \ldots$ to the honest players; for simplicity, assume $m_i = 0$ for all i. The environment runs for $\kappa\Delta + 1$ steps.

The attacker A proceeds as follows:

- A divides the rounds into κ segments of Δ rounds and delays all messages sent by honest players within such a segment to the end of it (note that this means no messages are delayed more than Δ);
- A ignores the content of the messages sent by honest players and tries to independently build its own chain $\hat{C}$ with messages $m'_1, m'_2, \ldots$ such that $m_i \neq m'_i$ for $\kappa\Delta$ rounds (for simplicity, assume $m'_i = 1$ for all i);
- In the next to last round $r = \kappa\Delta$, it sends $\hat{C}$ to any (strict) subset of the honest players (and delivers it instantly).

Note that in any view $\text{view} \in \text{EXEC}^{(\Pi^V, \mathcal{C})}(A, Z, \kappa)$ where (1) $|\hat{C}| > \kappa$ and (2) $\hat{C}$ is longer than the longest chain known to the honest players, we have that $\text{consistent}^\kappa(\text{view}) = 0$ and $\text{quality}^\kappa(\text{view}, 1) = 0$. We show that the probability that both events happen is constant, which proves the theorem.

The following two claims bound the probability that either event does not happen; by a union bound we can then conclude that the probability that both happen is constant.

Claim. Let $\hat{C}(\text{view})$ denote the length of the adversary's chain in the next to last round (i.e., round $\kappa\Delta$) of view. Then,

$$\Pr[\text{view} \leftarrow \text{EXEC}^{(\Pi^V, \mathcal{C})}(A, Z, \kappa) : |\hat{C}(\text{view})| < (1 + \frac{\delta}{2})\kappa] \leq e^{-\Omega(\kappa)}.$$

Proof. In the $\kappa\Delta$ rounds, the adversary has $\rho np \cdot \kappa\Delta$ chances to mine a block; each chance succeeds with probability p; since $\Delta = \frac{(1+\delta)}{\rho np}$, the expected number of mined blocks is thus $(1 + \delta)\kappa$. The desired bound thus follows directly from the Chernoff bound.

Claim. Let $\ell(\text{view})$ denote the length of the longest chain known to the honest players in the last round of view. Then,

$$\Pr[\text{view} \leftarrow \text{EXEC}^{(\Pi^V, \mathcal{C})}(A, Z, \kappa) : \ell(\text{view}) \geq \kappa] \leq \frac{3}{4}.$$

Proof. In every fixed segment of Δ rounds, the number of blocks mined by a *single* honest player is distributed as a binomial distribution with parameters Δ (trials) and p (success probability). Let X be such a random variable. The probability that some *fixed* single honest player mines more than 1 block in any *fixed* segment is

$$\Pr[X > 1] = 1 - \Pr[X \leq 1] = 1 - \Pr[X = 0] - \Pr[X = 1]$$
$$= 1 - (1-p)^\Delta - \Delta p(1-p)^{\Delta-1}$$
$$= 1 - (1-p)^{\Delta-1}(1 + (\Delta - 1)p)$$

$$\leq 1 - (1 - (\Delta - 1)p)(1 + (\Delta - 1)p)$$
$$= (\Delta - 1)^2 p^2 \leq \frac{(1 + \delta)}{\rho^2 n^2} \leq \frac{(1 + \delta)}{2n\kappa}$$

By a union bound over the number of players n and the number of segments κ, we have that except with probability $\frac{1+\delta}{2} \leq \frac{3}{4}$, no honest player mines more than one block in any segment, and whenever that happens, the length of the longest chain grows by at most 1 for each segment and thus becomes of length at most κ after κ segments.

Remark 11. *We note that our proof applies even in the setting of static corruptions, and already to a weaker notion of consistency which ignores "future-self consistency". In addition, the attacker never looks at the messages sent by honest players.*

Remark 12. *We additionally point out that at the cost of complicating the proof (and increasing the number of players), we can obtain an even stronger attack—which works also when $\Delta > \frac{1}{c \cdot np}$ where $\frac{1}{c} > \frac{1}{\rho} - \frac{1}{1-\rho}$ (as opposed to just $\frac{1}{\rho}$ as in our previous proof)—as follows: instead or partitioning the rounds into segments, simply always delay messages between honest players by Δ. Intuitively (but significantly over-simplifying), when we delay the messages between honest parties by Δ, the expected time they need to wait until finding and propagating a block is roughly $\frac{1}{(1-\rho)np} + \Delta$, whereas the adversary only needs to wait $\frac{1}{\rho np}$ in expectation; thus, the attacker succeeds whenever it mines faster (i.e., when $\frac{1}{\rho} < \frac{1}{1-\rho} + \Delta np$), and since $\Delta np = \frac{1}{c}$, the attack succeeds when $\frac{1}{c} > \frac{1}{\rho} - \frac{1}{1-\rho}$.*

We turn to describe how to formalize this attack (following the proof of the second claim above). We, in fact, show an attack that works as long as $\beta > \gamma$ (i.e., the adversary mining rate is higher than the "discounted" honest player mining rate), and then use this to deduce that the attack applies when $\frac{1}{c} > \frac{1}{\rho} - \frac{1}{1-\rho}$.

It follows using exactly the same proof as the *lowerbound* on chain growth in the "hybrid" model (see the full version of this paper) that we can get $(1 + \delta)\gamma$ as an *upperbound* on the chain growth of the honest players in a modified game where all honest players "freeze" for Δ rounds whenever some honest player mines a block. Since successes in each round are independent, it follows that *conditioned* on no single player ever mining two blocks within Δ rounds, the chain growth of honest players is upperbounded by $(1 + \delta)\gamma$, whereas the chain growth of the adversary is lowerbounded by $(1-\delta)\beta$. Thus when $\beta > (1+\delta')\gamma$, if we run the experiment for t steps (and condition on no single player ever mining two blocks within Δ rounds), we get an attack except with probability $e^{-\Omega(\gamma t)}$. Since γ is monotonically increasing in α and $\alpha \leq (1 - \rho)np$, it follows that the above also holds when[17]

$$\beta = \rho np > \frac{(1 - \rho)np}{1 + \Delta(1 - \rho)np}$$

---

[17] For readability, we ignore the $(1 + \delta')$ term.

and thus when

$$\Delta np > \frac{1}{\rho} - \frac{1}{1-\rho}$$

So if we set $\Delta = \frac{1}{cnp}$, we get an attack (conditioned on no single player ever mining two blocks within Δ rounds) when $\frac{1}{c} > \frac{1}{\rho} - \frac{1}{1-\rho}$.

Finally, as in the proof of the second Claim above we have that at any given round r, for any fixed player j, the probability of j mining more that 1 block within the next Δ rounds is upperbounded by $(\Delta - 1)^2 p^2 \leq \frac{1}{c^2 n^2}$. Thus, if we set $n > 2t$, it follows that no player every mines more than 1 block within Δ rounds, except with probability $1/2$ (by the union bound).

Acknowledgements. We are extremely grateful to Elaine Shi for many helpful comments on an earlier draft of this paper, and in particular for suggestion of how to sharpen the parameters in the construction of a public ledger from a blockchain.

References

[AJK05] Aspnes, J., Jackson, C., Krishnamurthy, A.: Exposing computationally-challenged byzantine impostors (2005)

[BCL+05] Barak, B., Canetti, R., Lindell, Y., Pass, R., Rabin, T.: Secure computation without authentication. In: Shoup, V. (ed.) CRYPTO 2005. LNCS, vol. 3621, pp. 361–377. Springer, Heidelberg (2005). doi:10.1007/11535218_22

[BK14] Bentov, I., Kumaresan, R.: How to use bitcoin to design fair protocols. In: Garay, J.A., Gennaro, R. (eds.) CRYPTO 2014. LNCS, vol. 8617, pp. 421–439. Springer, Heidelberg (2014). doi:10.1007/978-3-662-44381-1_24

[Blo16] Blockchain.info. Hash rate for blockchain, February 2016. https://blockchain.info/charts/hash-rate

[BTP] BTProof. https://www.btproof.com

[Can00] Canetti, R.: Universally composable security: a new paradigm for cryptographic protocols. Cryptology ePrint Archive, Report 2000/067 (2000). http://eprint.iacr.org/2000/067

[CL99] Castro, M., Liskov, B.: Practical byzantine fault tolerance. In: OSDI 1999 (1999)

[CLLM12] Chung, K.-M., Lam, H., Liu, Z., Mitzenmacher, M.: Chernoff-hoeffding bounds for markov chains: generalized and simplified. In: 29th International Symposium on Theoretical Aspects of Computer Science, STACS 2012, 29th February – 3rd March 2012, Paris, France, pp. 124–135 (2012)

[CSWH00] Clarke, I., Sandberg, O., Wiley, B., Hong, T.W.: Freenet: a distributed anonymous information storage and retrieval system. In: Proceedings of the ICSI Workshop on Design Issues in Anonymity and Unobservability (2000)

[DLN02] Karger, D., Liben-Nowell, D., Balakrishnan, H.: Analysis of the evolution of peer-to-peer systems. In: PODC 2002 (2002)

[DW13] Decker, C., Wattenhofer, R.: Information propagation in the bitcoin network. In: IEEE International Conference on Peer-to-Peer Computing, pp. 1–10 (2013)

[DR01] Druschel, P., Rowstron, A.: Past: persistent and anonymous storage in a peer-to-peer networking environment. In: HotOS 2001, pp. 65–70 (2001)

[DLS88] Dwork, C., Lynch, N., Stockmeyer, L.: Consensus in the presence of partial synchrony. J. ACM (JACM) **35**(2), 288–323 (1988)

[DN92] Dwork, C., Naor, M.: Pricing via processing or combatting junk mail. In: Brickell, E.F. (ed.) CRYPTO 1992. LNCS, vol. 740, pp. 139–147. Springer, Heidelberg (1993). doi:10.1007/3-540-48071-4_10

[ES14] Eyal, I., Sirer, E.G.: Majority is not enough: bitcoin mining is vulnerable. In: Christin, N., Safavi-Naini, R. (eds.) FC 2014. LNCS, vol. 8437, pp. 436–454. Springer, Heidelberg (2014). doi:10.1007/978-3-662-45472-5_28

[FVY14] Fromknecht, C., Velicanu, D., Yakoubov, S.: A decentralized public key infrastructure with identity retention. IACR Cryptology ePrint Archive 2014, 803 (2014)

[GKL15] Garay, J., Kiayias, A., Leonardos, N.: The bitcoin backbone protocol: analysis and applications. In: Oswald, E., Fischlin, M. (eds.) EUROCRYPT 2015. LNCS, vol. 9057, pp. 281–310. Springer, Heidelberg (2015). doi:10.1007/978-3-662-46803-6_10

[GMW87] Goldreich, O., Micali, S., Wigderson, A.: How to play any mental game or a completeness theorem for protocols with honest majority. In: STOC, pp. 218–229 (1987)

[KP15] Kiayias, A., Panagiotakos, G.: Speed-security tradeoffs in blockchain protocols (2015)

[KMS+15] Kosba, A., Miller, A., Shi, E., Wen, Z., Papamanthou, C.: Hawk: the blockchain model of cryptography and privacy-preserving smart contracts. Technical report, Cryptology ePrint Archive, Report 2015/675 (2015). http://eprint.iacr.org

[Lam10] Lamport, L.: Byzantizing paxos by refinement (2010)

[Lam11] Lamport, L.: Leaderless Byzantine Paxos. In: DISC 2011 (2011)

[Lit] Litecoin. https://litecoin.org

[MA05] Martin, J.-P., Alvisi, L.: Fast Byzantine consensus. In: DSN 2005 (2005)

[ML14] Miller, A., LaViola, J.J.: Anonymous Byzantine consensus from moderately-hard puzzles: a model for bitcoin (2014)

[mtg10] mtgox (2010). https://bitcointalk.org/index.php?topic=2227.msg29606#msg29606

[Nak08] Nakamoto, S.: Bitcoin: a peer-to-peer electronic cash system (2008)

[Nam] Namecoin. https://www.namecoin.org

[Oku05a] Okun, M.: Agreement among unacquainted Byzantine generals. In: Fraigniaud, P. (ed.) DISC 2005. LNCS, vol. 3724, pp. 499–500. Springer, Heidelberg (2005). doi:10.1007/11561927_40

[Oku05b] Okun, M.: Distributed computing among unacquainted processors in the presence of byzantine failures (2005)

[OB08] Okun, M., Barak, A.: Efficient algorithms for anonymous Byzantine agreement. Theor. Comp. Sys. **42**, 222–238 (2008)

[PS15] Pass, R., Shelat, A.: Micropayments for decentralized currencies. In: CCS 2015 (2015)

[PS16a] Pass, R., Shi, E.: Fruitchains: an (almost) optimally fair blockchain (2016)

[PS16b] Pass, R., Shi, E.: Hybrid consensus (2016)

[PSL80] Pease, M.C., Shostak, R.E., Lamport, L.: Reaching agreement in the presence of faults. J. ACM **27**, 228–234 (1980)

[PD15] Poon, J., Dryja, T.: The bitcoin lightning network: scalable off-chain instant payments draft 0.5.9.1 (2015). https://lightning.network/lightning-network-paper.pdf

[RFH+00] Ratanasamy, S., Francis, P., Handley, M., Karp, R., Shenker, S.: A scalable content-addressable network. In: SIGCOMM 2000 (2000)

[SZ15] Sompolinsky, Y., Zohar, A.: Secure high-rate transaction processing in bitcoin. In: Böhme, R., Okamoto, T. (eds.) FC 2015. LNCS, vol. 8975, pp. 507–527. Springer, Heidelberg (2015). doi:10.1007/978-3-662-47854-7_32

[SMK+01] Stoica, I., Morris, R., Karger, D., Kaashoek, M.F., Balakrishnan, H.: Chord: a scalable peer-to-peer lookup service for internet applications. In: SIGCOMM 2001 (2001)

Author Index

Printed in the United States
By Bookmasters